Allen-Bradley PLCs:

An Emphasis on Design and Application

Second Edition

Kelvin T. Erickson
Missouri University of Science and Technology

Dogwood Valley Press, LLC

This book was set in Times New Roman and printed on acid-free paper.

Printed in the United States of America

ISBN 978-0-9766259-5-7

Dogwood Valley Press, LLC
1604 Lincoln Lane
Rolla, MO 65401
1-573-426-3507

http://www.DogwoodValleyPress.com

10 9 8 7 6 5 4 3 2 1

Dedicated to Fran, Esther, David, Amanda,
Ezra, and Elijah

CONTENTS

PREFACE

Programmable logic controllers (PLCs) are the workhorses of modern manufacturing automation. Automatic control allows the production of a consistent product at a reasonable cost and the PLC is the most prevalent control technology in manufacturing..

This book presents the subject of programming Allen-Bradley PLCs with an emphasis on the design of the programs. Many texts teach one how to program the PLC in its languages, but little, if any, attention is paid to how does one attack the problem: "Given a set of operational specifications, how does one develop the PLC program?" This book develops the design process: the tasks involved, breaking the program into manageable pieces, standard code for the various parts, and handling the sequential parts of the problem. The emphasis is toward those who will be programming PLCs.

The text emphasizes the following Allen-Bradley controllers: ControlLogix, CompactLogix, MicroLogix, and SLC-500. Furthermore, because of its popularity (now and in the future), ladder logic is the language that is used for the text. The industry trend is toward using the IEC 61131-3 (formerly IEC 1131-3) standard, which also defines four other languages: function block diagram, structured text, instruction list, and sequential function chart. One interested in the other languages and in programming PLCs from Modicon, Siemens, and Emerson/GE, should see Erickson (2023).

Since a typical manufacturing plant may contain discrete, continuous, and batch processes, all of these applications are treated in this text, although the emphasis is on discrete and continuous processes. The emphasis is on a methodology that can be applied to any automation project, regardless of the size.

Throughout, the book contains example problems demonstrating good design practice. In addition, these problems are solved with each PLC covered in the book.

This book takes a practical approach to the design of PLC control systems. Some mathematical theory is used to backup the presentation on PID controllers. However, the theory is not detailed and can be omitted.

Except for Chapter 1, every chapter begins with a scenario that reflects the experience of the author and his colleagues in the challenging world of factory automation. These scenarios present a small problem and the solution and are intended to illustrate troubleshooting techniques.

Objectives

The main objectives of this text are to teach:

- PLC ladder logic programming for Allen-Bradley PLCs
- Approach to sequential problems
- Good program design practice
- Simple PID control tuning

- Introduction to sensors and actuators
- Human-machine interface (HMI) concepts

Content Overview

The book starts by introducing programmable logic controllers (PLCs) and their distinguishing characteristics. Chapters 2 – 5 cover basic ladder logic programming: contact, timer, and counter instructions. As part of the basics, the memory structure of the three particular PLC families and installation topics are treated. Chapter 6 covers ladder logic program design for sequential applications, probably the most significant contribution of the text. Chapters 7 and 8 treat computation, comparison, and advanced ladder logic instructions. PLC troubleshooting is covered in Chapter 9 and PID controller tuning is covered in Chapter 10. Sensors and actuators appear in Chapter 11. Chapter 12 introduces factory communication networks. Operator interface, often called human-machine interface (HMI), issues are treated in Chapter 13. PLC selection is introduced in Chapter 14 and covers the PLCs from Allen-Bradley/Rockwell Automation. Chapter 15 presents the perspective of an entire automation project, bringing together the various pieces of PLC control design and then outlines a full-length project case study. Details about number systems and drawing symbols are included as appendices, rather than interrupt the flow of the text material.

Throughout the text, any reference to ControlLogix also applies to the CompactLogix, which is a smaller version of the ControlLogix. Also, much of the MicroLogix and SLC-500 ladder logic programming also apply to the older PLC-5 processors.

The Audience

This book primarily serves the academic market, at the two-year technical school level, though the material on PID controllers will be more challenging than the typical PLC textbook for this level of student. This text is also suitable for junior or senior undergraduate electrical, mechanical, or industrial engineering or engineering technology level.

In addition, this text serves the professional market. Economic and regulatory pressures in the manufacturing, chemical, petrochemical, pharmaceutical, and food industries have forced control engineers to design new systems or retrofit existing control systems. Hence, there are many control engineers (primarily chemical and electrical) who need to rapidly educate themselves in an area of technology in which they are probably only somewhat familiar. This book is valuable to this audience.

Online Content

Supplementary online content for the text is posted at **www.dogwoodvalleypress.com** and includes:

1. PLC project files and print-outs for the text examples.
2. Additional problems with solutions.
3. PID tuning demonstrator program.
4. Design documentation for the coal handler project in Chapter 22.

Second Edition

The second edition primarily adds the Micro800 controller, but there are other changes throughout. Though obsolete, the MicroLogix and SLC-500 processors have been retained. Information on input and output buffering has been added to Chapter 2 and addressing for 5069- and 5094-series I/O modules has been added to Chapter 3. The Chapter 11 problems have been deleted, as they are irrelevant since most currently available analog input modules perform the conversion. The thermocouple and RTD conversion examples are retained in case the reader encounters an older system that performs the conversion in the PLC program. Chapter 13 has extensive updates to reflect recent developments in operator interfaces, called "High Performance HMI." The section on ControlLogix simulation in Chapter 15 has been revised to utilize I/O module inhibits to disable the I/O scan. The appearance of the ControlLogix blocks has been updated to match the changes effective with Studio 5000 version 31. Lastly, all of the chapter problems have been replaced with new problems.

Acknowledgements

The author wishes to acknowledge the beneficial suggestions and comments of many colleagues. Steve Ingracia provided the sample panel specification in Chapter 4. Guy Mace provided information about the control of a low-pressure high-volume aeration air turbine, forming the basis of problem P7-27. Ken Ball provided more information on the history of the PLC. I especially thank Esther and Fran Erickson for correcting the manuscript for grammatical errors and Fran for doing the initial typesetting.

Portions of this material were taught in industrial short courses and university courses and the students are acknowledged for their help in pointing out errors in the text and where the presentation was unclear.

Disclaimer

Information furnished herein is believed to be accurate and reliable; however no responsibility is assumed for any errors. The user assumes full responsibility for the accuracy and appropriateness of this information.

Reference

Erickson, Kelvin T., 2023. *Programmable Logic Controllers: An Emphasis on Design and Application,* 4th Edition, Dogwood Valley Press, Rolla, MO.

1 Introduction to PLCs

Chapter Topics:

- Control system categories
- History of PLCs
- PLC versus other technologies
- PLC architecture

OBJECTIVES

Upon completion of this chapter, you will be able to understand:

- Two ways to categorize a control system
- PLC history and developments
- The differences between PLC, relay and PC-based control
- Basic PLC architecture

1.1 INTRODUCTION

Programmable logic controllers (PLCs) are definitely the workhorses of modern manufacturing automation. Automatic control allows the manufacture of a consistent product at reasonable cost and the PLC is the most prevalent control technology in manufacturing. In order to establish a context for PLCs, the need for automatic control in manufacturing is justified along with the major ways control systems are classified. The history of the PLC is outlined and placed in the context of the parallel development of computer control in the chemical industry. The use of the PLC in automation is compared with other automation technologies, with special attention to a comparison of PLC versus personal computer-based control. Finally, the architecture of the PLC is described and its place in the context of an automation system is shown.

1.2 AUTOMATIC CONTROL IN MANUFACTURING

There are many reasons automatic control is indispensable to manufacturing. The most important reasons are:

- To improve the quality and lower the cost of production.
- To attain optimal performance.
- To relieve the drudgery of many routine, repetitive manual operations.

Over the last four decades, manufacturing has been receiving more emphasis from the engineering community. To an enterprise, profitability usually depends on productivity and automation is a means towards greater productivity. Usually, the most important question is not, "Should we automate production?" but "What should be automated to increase productivity?"

As an example, throughout the 1950's and 1960's the American automakers were placing their emphasis on the power and style of their automobiles and not on producing a high quality product at the lowest possible cost. The gasoline shortages of the early 1970's forced many people to buy smaller automobiles and they discovered that the small automobiles produced by the Japanese automakers were of better quality and at a lower cost than the equivalent American automobiles. To remain competitive, the American automakers started placing more emphasis on manufacturing quality automobiles at the lowest possible cost and automation was the means toward their goal. This renewed emphasis on manufacturing extended to other industries and will not soon diminish.

The basic elements of an automatic control system are shown in Figure 1.1. A control system has three basic elements:

Input sensors:	Convert physical phenomena (for example, position) to electrical signal.
Programmable Logic Controller:	Using the measurements, calculate control actions.
Output actuators:	Convert electrical signal to a physical action (for example, motor rotation).

In this text, the PLC is the controller, which was initially developed out of the needs of the automotive industry. In the chemical industry, a distributed control system (DCS) is typically used in place of the PLC. As will be shown later, DCS and PLC systems are merging into one device often called a programmable automation controller (PAC).

1.3 CONTROL SYSTEM CLASSIFICATIONS

Control systems are commonly categorized two ways: (1) type of sensors/actuators, and (2) type of process. Either way is valid, but in certain contexts one is often preferred over the other. For example, many people choose whether to use a PLC versus a DCS based on the type of sensors/actuators.

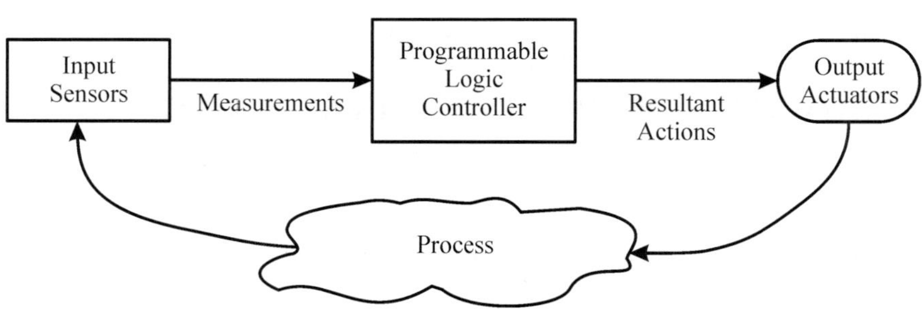

Figure 1.1. Basic elements of an automatic control system.

1.3.1 Type of Sensors/Actuators

When considering the sensor or actuator, there are two major types: (1) discrete, and (2) analog. A discrete control system has mainly discrete sensors and actuators. Similarly, an analog system has mainly analog sensors and actuators. A discrete sensor or actuator has one of two values, for example: on/off, open/closed, present/not present, running/stopped, or extended/retracted. An analog sensor or actuator theoretically has an infinite number of values. For example, an analog sensor value could represent position, velocity, acceleration, temperature, pressure, flow, or weight. An analog actuator value could represent valve position, motor speed (for motor controller), damper position, or strip-chart recorder value. Because an analog sensor or actuator value is converted to an integer or real value within the input/output (I/O) module, an analog value practically assumes a finite number of values. This point becomes more apparent when analog I/O modules are discussed in Chapter 4.

Bear in mind that most real programmable logic control systems are a combination of analog and discrete elements. For example, Figure 1.2 depicts some of the elements of a heating, ventilation and air conditioning (HVAC) system. The proximity switches that detect the presence of a person in a room are discrete sensors and the control valve for an air solenoid is a discrete actuator. As an example of the PLC operation, the damper is opened whenever anyone is present in the room and closed when no one is present for 15 minutes. The airflow sensor is an analog sensor and the fan motor speed command is an analog actuator signal. The PLC calculates the desired fan speed in order to maintain the airflow at a desired value.

1.3.2 Type of Process

A process is a sequence of chemical, physical, or biological activities for the conversion, transport, or storage of material or energy (ISA, 2010). Industrial manufacturing processes are generally classified as continuous, batch, or discrete-parts manufacturing. This classification stems from the way the output appears: as a continuous flow (continuous); in finite quantities of material (batch); or in finite quantities of parts (discrete-parts manufacturing).

Continuous Process. In a continuous process, material passes in a continuous stream through the processing equipment. Once the process has established a steady operating state, the nature of the process does not depend on the length of time the process is operating

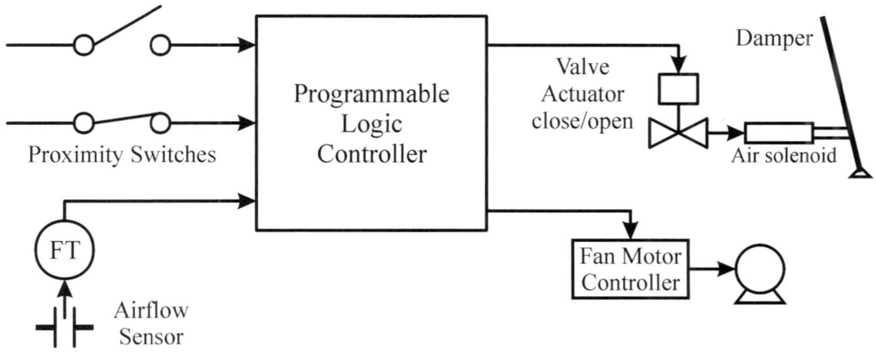

Figure 1.2. Control elements of HVAC system.

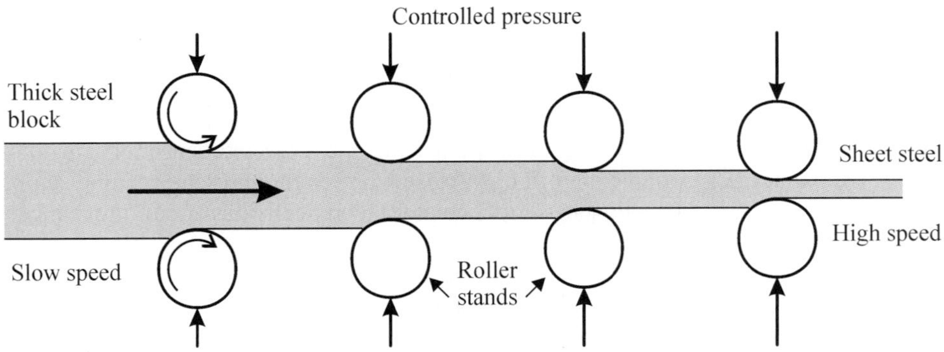

Figure 1.3. Continuous steel rolling mill.

(ISA, 2010). Commodity chemical manufacturing typically falls into this category. A continuous steel rolling mill, simplified in Figure 1.3, is an example continuous process. The control system manipulates the force at each stand and the speed of each roller in order to control the thickness and speed of the steel sheet at the exit. As the thickness of the steel sheet is reduced by each roller stand, the speed of the sheet must be increased proportionally to maintain a constant steel mass flow rate. If the last roller stand rotates too slow, material will accumulate between the third and fourth stands and eventually cause a jam. If the last roller stand rotates a little too fast, the sheet will be torn. Either scenario is not desirable.

Batch Process. In a batch process, finite quantities (batches) of material are produced by subjecting quantities of input materials to a defined order of processing actions using one or more pieces of equipment. Batch processes are discontinuous processes from a material flow standpoint. Batch processes are neither discrete nor continuous, though they have characteristics of both (ISA, 2010). Food, beverage, pharmaceutical, and specialty chemical processes are usually encompassed by this category. An example of a small batch process is illustrated in Figure 1.4. In this process, ingredients A and B are measured into tanks. Ingredient A is heated to a desired temperature and then the two ingredients are pumped to a third vessel where a chemical reaction occurs. After a certain length of time, the resultant product is pumped through a filter and then on to storage.

Discrete-Parts Manufacturing Process. In a discrete-parts manufacturing process, a specified quantity of material moves as a unit (part or group of parts) between workstations, and each unit maintains its unique identity (ISA, 2010). At a workstation, a unit may be modified (drilled, machined, painted, etc.) or may be combined with one or more other parts (assembly). Packaging operations also fall into this category. An example of a discrete-parts workstation is shown in Figure 1.5. Incoming parts are received on the input conveyor, moved to the drilling station where a few holes are drilled, moved to the computer numerical control (CNC) station where the part is shaped, and then moved to the output conveyor. A simple packaging operation is shown in simplified form in Figure 1.6. An empty carton arrives on the incoming conveyor. When the loading area is clear, the carton is moved into position. The parts are marshaled to form a row and the row is pushed into the carton. The carton is indexed down and succeeding rows of parts are placed in the carton. When fully loaded, the carton is moved out to the next station where the flaps are sealed.

Processes are often primarily one of the three main types, though they often contain at least one of the other types. For example, the output of many batch processes is packaged in

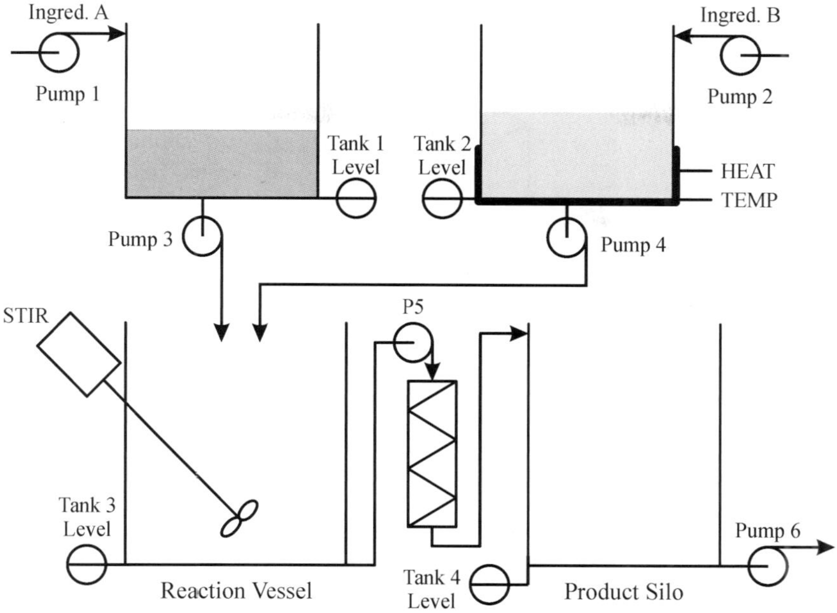

Figure 1.4. Small batch process.

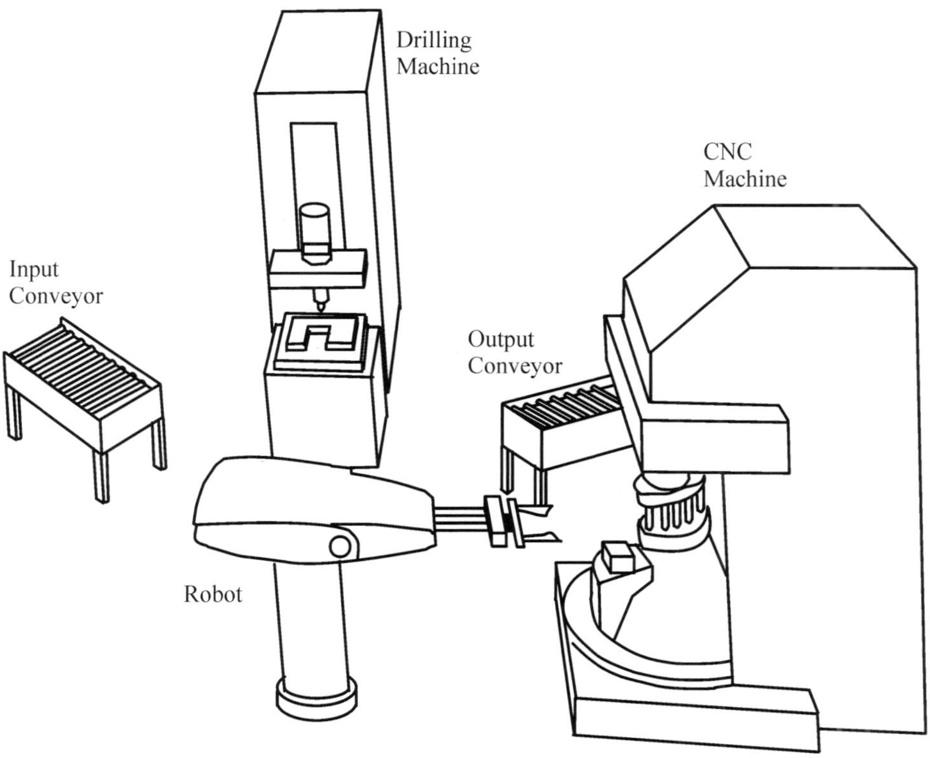

Figure 1.5. Discrete parts workcell.

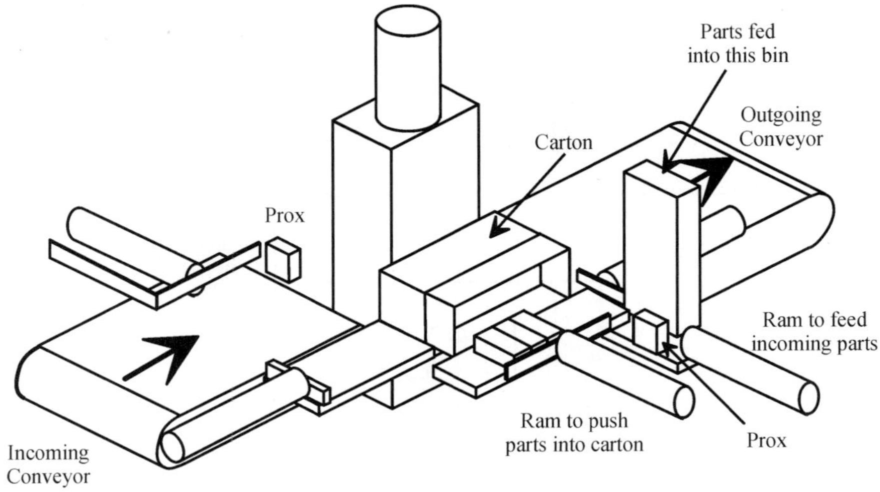

Figure 1.6. Packaging station.

containers smaller than a batch. The packaging operation is a discrete-parts operation. Even a discrete-parts manufacturing plant, for example, a small gasoline engine manufacturing facility, has waste recycling or treatment facilities that often contain continuous or batch processes.

1.4 HISTORY OF THE PLC

The development of automatic control technology has two very definite parallel tracks. Distributed control system (DCS) technology evolved from the needs of continuous processes. Programmable logic controllers evolved from the needs of batch and discrete manufacturing processes.

Controllers for continuous processes (e.g., flow, level, pressure) can be traced to float regulator mechanisms in Greece in the period of 300 to 1 B.C. (Mayr, 1971). These primitive controllers eventually evolved to pneumatic control hardware of the 1900's where air pressure was the method of actuation and signal transmission. Electronic controllers started to supplant these pneumatic controllers in the 1960's (Merritt, 1999). Electronics also made the development of computer systems possible and computers were applied to continuous control problems in the late 1950's. These devices were business computers retrofitted for process control service and mainly used for supervisory control and data acquisition. There were some attempts to implement direct digital control of a process but the reliability and performance of early computer technology was inadequate. In the mid-1970's the microprocessor was introduced in direct control for PID loops and eventually led to the development of the DCS in the late 1970's where the control was distributed among several devices rather than being concentrated in one central computer. Current DCS systems are basically smaller and more powerful versions of the early DCS control systems.

Early technology for discrete and batch control employed the relays developed for the telegraph industry of the 1800's. In 1836 Samuel F. B. Morse invented the electromechanical relay as a means to increase the station-to-station distance of his newly

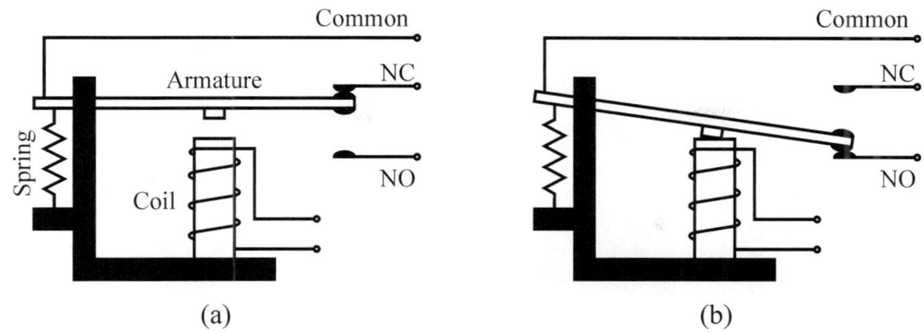

Figure 1.7. Electromechanical relay: *(a)* unenergized; *(b)* energized.

developed telegraph system, then limited to about 20 miles. This electromechanical device sensed the small telegraphic signal current and "relayed" or amplified this signal to carry the information to the next station or relay. An electromechanical relay consists of a coil of wire magnetically coupled to a moveable piece of iron, called an armature (Figure 1.7). A spring holds the armature in its resting position in which there is continuity between the "Common" and "NC" connections (Figure 1.7*a*). When a sufficient current passes through the coil, the armature is forced downward causing continuity to be established between the "Common" and the "NO" connections (Figure 1.7*b*). Eventually, interconnections of electromechanical relays were used to implement logic functions, as explained in Chapter 2.

In concert with electromechanical relays, drum sequencers were developed to control sequential processes, where the operation is expressed as a sequence of steps. An electromechanical drum sequencer (Figure 1.8) is a cylinder with pegs strategically placed to make or break contacts. When the contact is closed, power is applied to relay logic or a final control element. The drum advances one row at a time triggered by an external event. As the drum advances, it makes or breaks control circuits in a pattern prescribed by the position of each peg in a row.

Relay and sequencer technology remained relatively stagnant until the late 1960's. The only real change was the replacement of electromechanical relays with solid-state (electronic) relays in the 1960's. Computers were tried but were even less reliable than relay systems because of the harsh environment of the typical factory. A state-of-the-art manufacturing control system consisted of racks of relays (Figure 1.9*a*) whose wiring was documented by drawings called relay ladder logic diagrams. Relay systems were

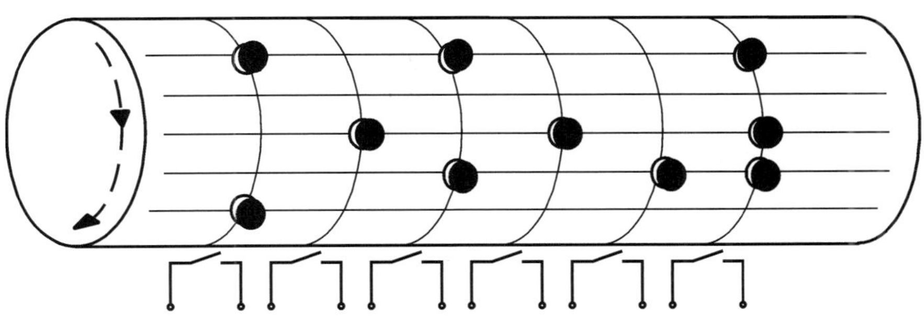

Figure 1.8. Electromechanical drum sequencer.

(a) (b)

Figure 1.9. Small control system: *(a)* relay rack; *(b)* PLC replacement.

maintenance headaches: prone to failure and hard to troubleshoot. In addition, a change to the function of the control system meant a wiring change at a minimum and often meant additional relays.

The machine tool and automotive industries were large users of relay control systems. A simple machine tool would require six months to a year to completely debug (Morley, 2001). Every year, automotive manufacturing facilities would be shut down for two to three months in order to implement the changes due to the new automobile models. The lost production due to these changes was significant.

Richard Morley is credited with inventing the initial concept for the PLC in January, 1968 while working for Bedford Associates (Morley, 2001). Faced with the design of a machine tool control for a high-performance lathe, he decided to try something different from the solid-state relays or minicomputers which were state-of-the-art at that time. Initially, he conceived a machine or box that was rugged, that contained its own direct interface to the machine tool and that had its own application-specific language. Eventually, the language was changed to ladder logic since that is what the control engineers already knew.

Bedford Associates chose to "spin off" the part of the company dealing with this new controller as a separate company called Modicon. The Modicon (MOdular DIgital CONtroller) 084 was introduced in 1969 and is generally credited as the first PLC and proved the concept of the PLC.

In parallel with the developments at Bedford Associates in 1968, a group of engineers at the Hydramatic Division of General Motors specified a *Standard Machine Controller* with similar features (Stone et al, 1968). A few companies were interested in developing such a device. In addition to Modicon, Information Instruments, Inc. (3I) and Digital Equipment Corp (DEC) responded to GM's specification. DEC developed the PDP-14 programmable controller based on the PDP-8 series computer and 3I introduced the PDQ-II (Program Data Quantizer) logic controller. Controllers from all three vendors were installed to control various. machines in 1969 (Ball, 2008a). Though all three met GM's requirements, the Modicon 084 became the controller of choice because it was programmed in ladder logic and extremely rugged.

A third parallel development of the PLC happened at Struthers-Dunn (Ball, 2008b). Though this effort started in 1967, development was delayed for a year to resolve trade secret litigation. The effort was not publically known until Struthers-Dunn introduced the VIP programmable controller in 1970.

Initially, this new controller was named the programmable controller, abbreviated as the PC. When the personal computer was introduced, it was also called a PC. To avoid confusion, the programmable controller was renamed to the programmable *logic* controller (PLC).

Two aspects of the initial PLCs are generally credited with their success. First, the electronic circuits were designed to be highly reliable and to work in the harsh industrial environment. Dick Morley tells the following tale about the Modicon 084 (Morley, 2001):

> Landis [a machine tool company] decided to purchase the MODICON units and not use the PDP-14. When DIGITAL [*sic*] tried to get back into Landis, Landis wrapped a welder cable (operating) around the 084 and poured Coke over the unit. The 084 kept right on trucking. Digital retreated with grace.

Second, the programming language was based on standard electrical relay ladder logic drawings. Some earlier computer applications in manufacturing had failed because the plant technicians and engineers had difficulty learning the computer language. In contrast, most of them already knew relay ladder logic design and so could quickly learn PLC ladder logic.

Though adoption by industry was initially slow, other well-known electrical controls manufacturers began either developing their own PLCs, or reselling someone else's PLC. Of the original three PLC developers, only Digital Equipment Corp. no longer produces PLCs. Modicon is now a part of Schneider Electric. Information Instruments, Inc. was later purchased by Allen-Bradley, which is now part of Rockwell Automation. Struthers-Dunn later became Uticor. One web site, http://www.plcs.net/chapters/links.htm, lists over 90 PLC vendors.

The original PLC language was ladder logic. However, each PLC manufacturer did not implement ladder logic in the same fashion; therefore, programs could not be easily copied between PLC manufacturers. In addition, ladder logic is not very well suited to arithmetic calculations, PID control, or motion control. This need to incorporate other control languages and to copy programs among PLC vendors led to the development of the IEC 61131-3 (originally numbered 1131-3) standard (IEC, 2013). The industry trend is toward using the IEC 61131-3 standard. However, 61131-3 is only a voluntary standard and individual manufacturers have some freedom in the implementation.

Table 1.1. Comparison of Control System Hardware

Characteristic	Relay Systems	Computers	PLC Systems
Price per function	Moderate	Low	Low
Physical size	Bulky	Fairly compact	Compact
Operating speed	Slow	Fairly fast	Fast
Industrial environment	Excellent	Fair to good	Good
Design	Time-consuming	Usually simple	Simple
Complicated operations	No	Yes	Yes
Installation	Time-consuming	Simple to complex	Simple
Easy to change function	Difficult	Usually simple	Very simple
Ease of maintenance	Poor - many contacts	Good if std. I/O modules	Good - few std. cards
Power failure recovery	0 sec.	10 - 100 sec.	1 - 3 sec.

The PLC evolved from its initial role as a "relay replacer" and the DCS evolved from its initial role as a "PID loop replacer". The distinction between PLC and DCS systems persisted throughout the 1990's. While both control technology types claimed to do some of the functions of the other type, it was not until the 61131-3 standard blended them that the distinction became obsolete. This merging of the PLC and DCS technology suggests a new term to describe control technology: *programmable automation controller (PAC)*. This term is now used by some vendors to describe their factory automation controllers.

1.5 PLC VERSUS OTHER TECHNOLOGIES

As demonstrated by the history of the PLC, it is not the only technology available for implementing factory automation. Table 1.1 shows a comparison of available control system technologies. This table clearly shows why PLCs quickly supplanted relay systems.

Except for recovery from power failure, the PLC has a clear advantage over relay systems. Figure 1.9*b* shows the PLC replacement of the relay rack of Figure 1.9*a* and illustrates the smaller space required for the PLC system. Less clear are the advantages of PLCs over computers.

In many respects, the architecture of the PLC resembles a general-purpose computer. In fact, some of the early PLCs (e.g., PDP-14) were computer-based. However, there are some important characteristics of a PLC that distinguish it from a general-purpose or personal computer.

A PLC can be placed in an **industrial environment**. In contrast to the typical computer, a PLC can be placed in areas with substantial amount of:

Electrical noise

Electromagnetic interference

Mechanical vibration

Extreme temperatures (e.g., 140° F)

Non-condensing humidity (95%)

Most PLCs are placed in cabinets in order to protect them and the electrical wiring from dust and other airborne contaminants. However, some manufacturers offer PLC modules with a conformal coating that protects the electronics from dust and certain other airborne contaminants.

A PLC is **more reliable**. Manufacturing is often an unforgiving environment. General Motors estimates that production line downtime costs $15,000 a minute (Garber, 2001). In some processes, a millisecond mistiming can mean a multimillion-dollar loss or personnel casualties. The mean time between failure of most PLCs is measured in years whereas the mean time between failures of a Windows-based PC is on the order of a few months (in the author's experience). The PLC also does not have any inherently less-reliable mechanical components (e.g., disk drives).

A PLC is **easily maintained by plant technicians**. The hardware interfaces are standardized and are easily connected. The interface circuitry is modular and self-diagnosing to pinpoint malfunctions and allow easy replacement. The PLC is programmed using ladder logic, which was already established as the standard documentation tool for relay systems. In addition, ladder logic is easily learned.

The PLC **executes a single program in an orderly and sequential fashion**. However, most medium to large PLCs have instructions that allow subroutine calling, interrupt routines, and bypass of certain instructions. Also, many PLCs incorporate the IEC 61131-3 languages or have modules that implement higher-level languages (e.g., C).

The PLC **recovers quickly from a power failure**. There are no boot-up procedures or a need to load programs on power-up. If the PLC is running when the power fails, when power is restored, it quickly runs some diagnostic self-tests and resumes running the program without operator intervention.

Currently, PC-based control (sometimes called "soft" PLC) is working to address these weaknesses. Typically, the I/O modules are the same as used by PLCs. So, the environmental, reliability, and maintenance aspects of these modules are retained. The reliability of the PC is partly addressed by using a proprietary operating system that is more reliable than Microsoft Windows and does not access the disk drives after boot-up (Murray, 1998). According to Murray (1998) PC-based control systems must:

- Provide deterministic operation.
- Survive a operating system crash and continue to operate in a safe manner.
- Be isolated from poorly behaved applications and drivers.
- Survive a hard disk crash.
- Be based on a proven, real-time engine for mission-critical applications.

The main advantages of PC-based control are the ability to use other programming languages when implementing sophisticated control and to include the operator interface

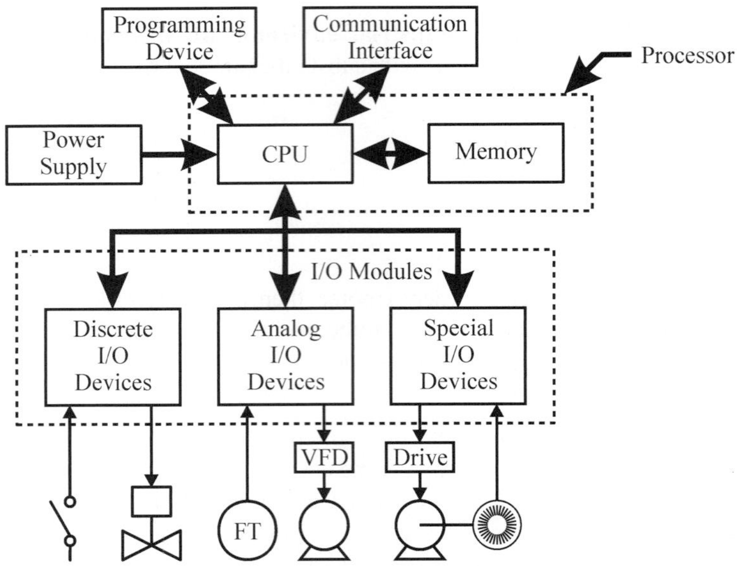

Figure 1.10. Basic PLC architecture.

with the control program. When PC-based control was first introduced in the mid-1980's, the demise of the PLC was predicted. This demise has not happened and probably will not happen. Both PC-based control and PLCs will coexist and find their application niches.

1.6 BASIC PLC ARCHITECTURE

The basic architecture of a PLC is shown in Figure 1.10. The main components are the processor module, the power supply and the input/output (I/O) modules. The processor module consists of the central processing unit (CPU) and memory. In addition to a microprocessor, the CPU also contains at least an interface to a programming device and may contain interfaces to remote I/O and other communication networks. The power supply is usually a separate module and the I/O modules are separate from the processor. The types of I/O modules include discrete (on/off), analog (continuous variable), and special modules

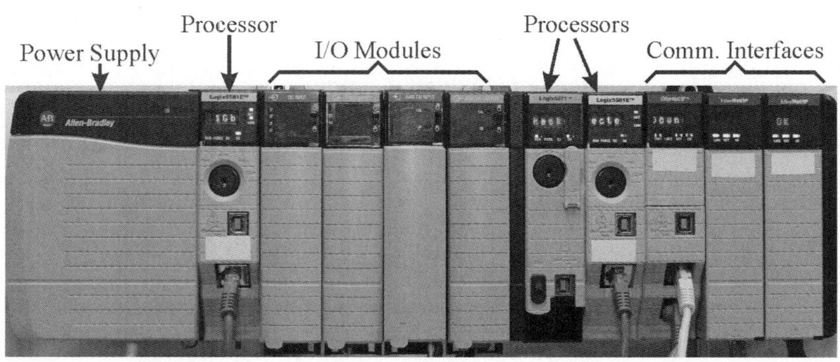

Figure 1.11. Allen-Bradley Control Logix chassis.

like motion control or high-speed counters. The field devices are connected to the I/O modules.

Depending on the amount of I/O and the particular PLC processor, the I/O modules may be in the same chassis as the processor and/or in one or more other chassis. Up until the late 1980's, the I/O modules in a typical PLC system were in chassis separate from the PLC processor. In the more typical present-day PLC, some of the I/O modules may be present in the chassis that contains the processor (Figure 1.11). Note that the Allen-Bradley Control Logix system (Figure 1.11) allows more than one processor in the same chassis. Smaller PLCs (Figures 1.12, 1.13, and 1.14) are often mounted on a DIN rail.

Within the context of an automation system, the PLC appears as shown in Figure 1.15. In this system, the I/O modules are in the same chassis as the processor. A typical medium-size PLC installation is shown in Figure 1.16. This figure shows the placement of the PLC processor, I/O modules, motor drives, power supplies, and wiring terminal blocks within a cabinet. There are also fuses or circuit breakers, one or more per input or output

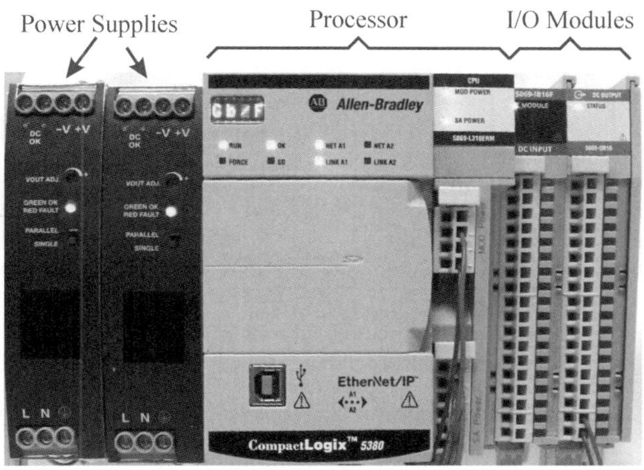

Figure 1.12. Allen-Bradley CompactLogix 5380.

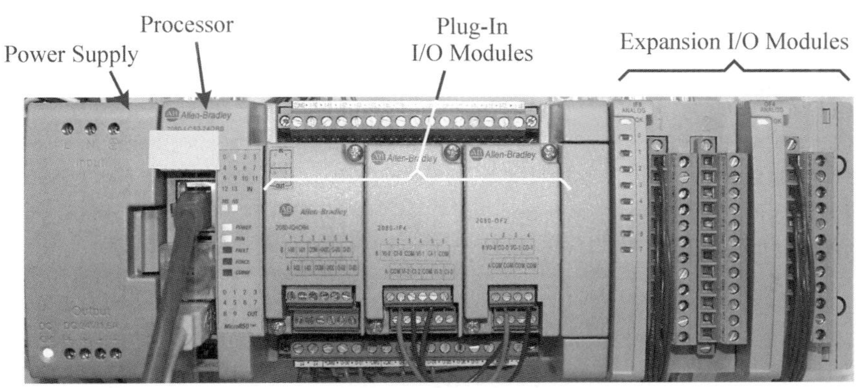

Figure 1.13. Allen-Bradley Micro850.

module for protection of the I/O module circuitry. The I/O modules are wired to one side of the terminal blocks instead of being wired directly to the field devices. A cabinet is typically

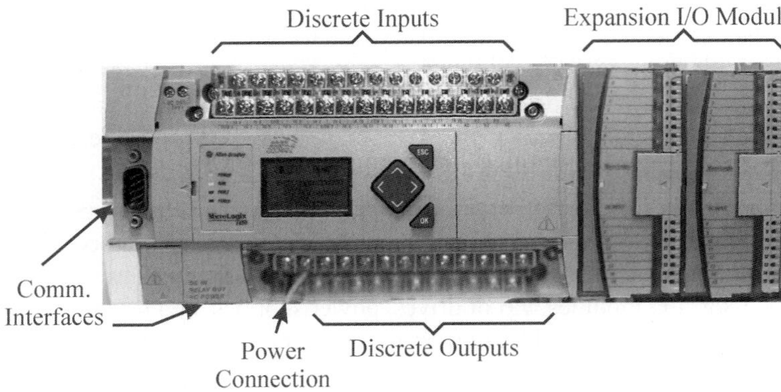

Figure 1.14. Allen-Bradley MicroLogix1400.

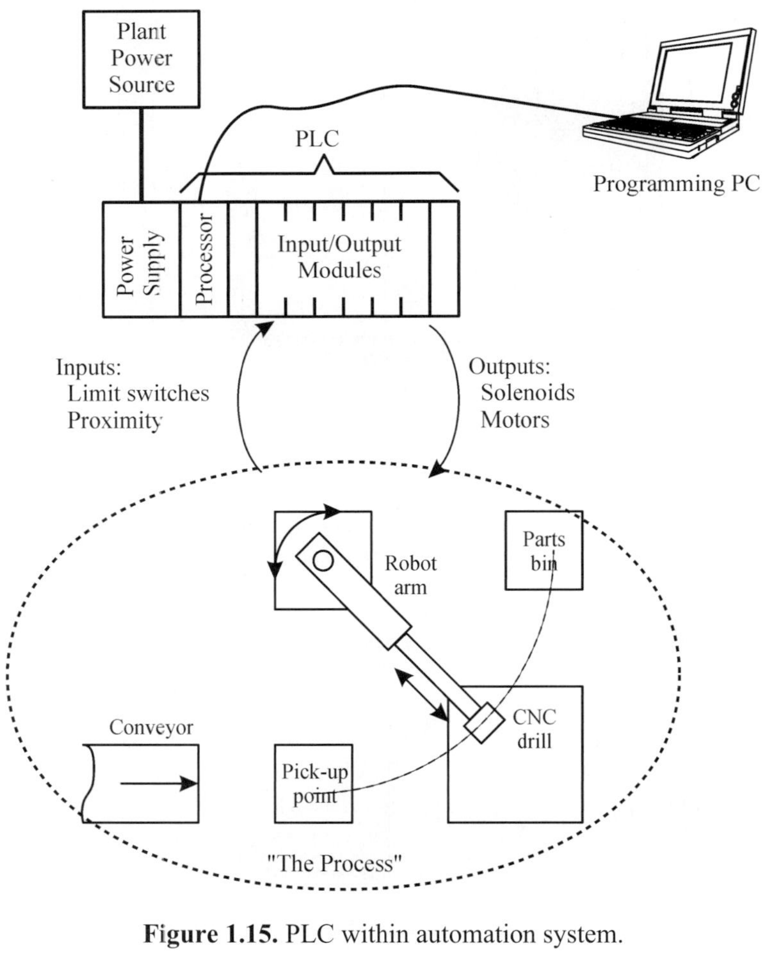

Figure 1.15. PLC within automation system.

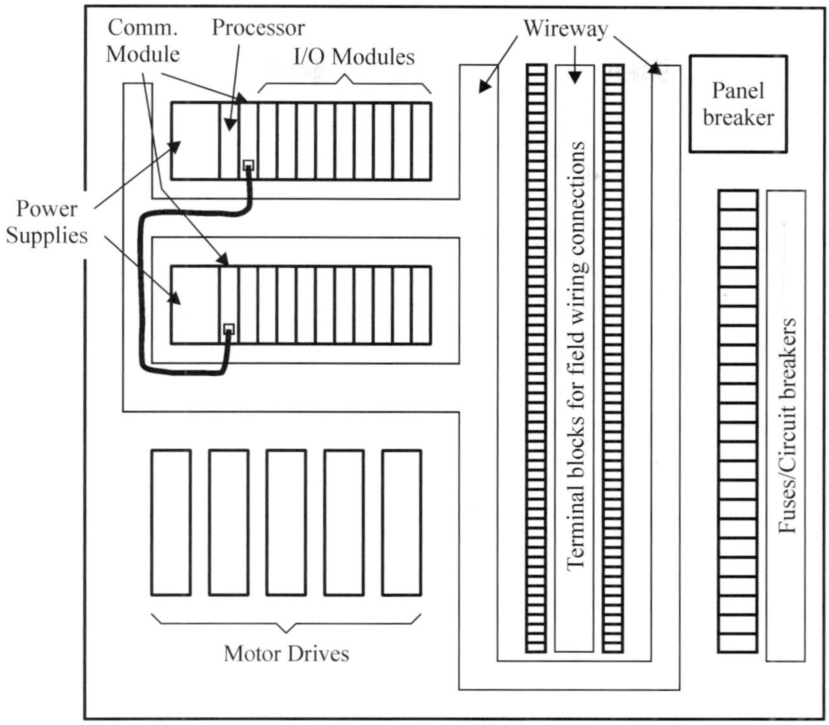

Figure 1.16. Typical PLC system cabinet installation.

assembled and wired at a location away from the process. When ready to install, the cabinet is brought to the plant location and only the plant power and field devices need to be connected to the other side of the terminal blocks.

1.7 CHAPTER SUMMARY

This chapter introduced the programmable logic controller and placed it in the context of manufacturing automation. The evolution of the PLC was outlined and compared to other automation technologies, especially personal computers. With this basic PLC knowledge, one is ready to learn its unique language, ladder logic.

REFERENCES

Ball, Ken, 2008a. "How programmable logic controllers emerged from industry needs," *Control Engineering*, pp. 66-70, September.

Ball, Ken, 2008b. "PLCs: Their time had arrived," paper draft, July 27.

Dummermuth, Ernst, 2002. "A Bit Twiddler's History of the PLC, " *A-B Journal*, pp. 26-30, November.

Garber, Joseph R., 2001. "The PLC versus the PC," http://www.barn.org/FILES/PLCvsPC.htm.

IEC, 2013. *IEC 1131-3: Programmable Logic Controllers - Part 3: Programming Languages*, Edition 3.0, International Electrotechnical Commission, Geneva, Switzerland.

ISA, 2010. *ANSI/ISA-S88.00.01-2010, Batch Control, Part 1: Models and Terminology*, International Socciey of Automation, Research Triangle Park, NC, Dec.

Mayr, Otto, 1971. *Feedback Mechanisms in the Historical Collections of the National Museum of History and Technology*, Smithsonian Institution Press, Washington, D.C.

Merritt, Rich, 1999. "The electronic age," *Control Magazine*, Dec.

Morley, Richard, 2001. "History of the PLC," R. Morley Incorporated.

Murray, Charles J., 1998. "PC vs. PLC: The lines blur", Special Supplement: Motion Control, *Design News*, March 23.

Stone, William S., David C. Emmett, Edward J. O'Connell, Leonard Radionoff, William Wegryn, and Clifford H. Wilford, "Standard Machine Controller," General Motors, June 6, 1968.

2 Basic Ladder Logic Programming

Chapter Topics:

- Basic ladder logic symbols
- Ladder logic diagram
- Ladder logic evaluation
- Start/stop logic

OBJECTIVES

Upon completion of this chapter, you will be able to:

- Understand basic ladder logic symbols
- Write ladder logic for simple applications

Scenario: A program with a long scan time may not detect short-duration events.

A manufacturer of small gasoline engines had an intermittent problem on the final assembly line. Sometimes, a defective engine would not be automatically removed from the line for repair at a "kick-out" station. If an operator noticed a problem with an engine, he/she inserted a bolt into a certain hole in the engine carrier. A proximity sensor before the kick-out station sensed the presence of the bolt, and the PLC activated a pneumatic cylinder to push the carrier (and engine) off the main conveyor and into the repair area. A view of this station is shown in Figure 2.1. Further investigation revealed that the duration of the **on** pulse of the proximity sensor was approximately 3/4 seconds. One PLC controlled all of the stations on the assembly line and its ladder logic program was quite large. As indicated in the PLC status, the time to scan the ladder logic program was slightly less than 1 second. Hence, it was very likely that a pulse from the proximity sensor could be undetected by the PLC processor. The proximity sensor could be **off** at the start of the ladder scan, generate an **on** pulse from a passing bolt in the carrier, and be **off** at the start of the next ladder scan.

Solution: Logic to examine the proximity sensor is placed in a ladder logic routine that is executed every ½ second. If the proximity sensor is detected to be on, an internal coil is turned on for at least 1.5 seconds. The main PLC program is changed to examine this internal coil to determine when to activate the pneumatic cylinder and push a carrier off the main conveyor.

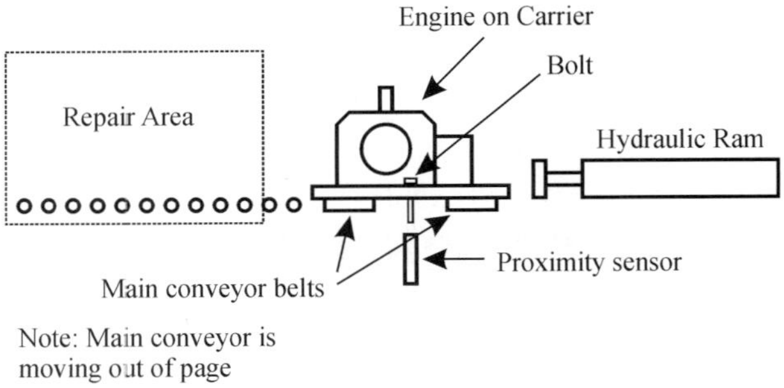

Figure 2.1. Kick-out station.

2.1 INTRODUCTION

Now that the PLC has been introduced, let us move on to programming the PLC. The first, and still most popular programming language, is ladder logic. Using examples, the language is developed from the electromechanical relay system-wiring diagram. After describing the basic symbols, they are combined into a ladder diagram. The subsequent section details the process of scanning a program and accessing the physical inputs and outputs. Programming with the normally closed contact is given particular attention because it is often misapplied by novice programmers. To solidify these concepts, the start/stop of a physical device is considered. Start/stop is a very common PLC application and occurs in many other contexts. A section on transitional contacts and coils concludes the chapter.

2.2 SIMPLE LADDER LOGIC

Ladder logic is the primary programming language of programmable logic controllers. Since the PLC was developed to replace relay logic control systems, it was only natural that the initial language closely resembles the diagrams used to document the relay logic. By using this approach, the engineers and technicians using the early PLCs did not need retraining to understand the program. To introduce ladder logic programming simple switch circuits are converted to relay logic and then to PLC ladder logic.

In all of the ladder logic examples used in this chapter, tags (also called variables or symbols) are used for all inputs, outputs, and internal memory in the examples to avoid having to deal with input/output addressing. This addressing, treated in Chapter 3, is generally different for each PLC processor family.

Example 2.1. OR Circuit. Two switches labeled A and B are wired in parallel controlling a lamp as shown in Figure 2.2*a*. Implement this function as PLC ladder logic where the two switches are separate inputs.

Solution. The switch circuit action is described as, "The lamp is **on** when switch A is **on** (closed) or switch B is **on** (closed)." All possible combinations of the two switches and the consequent lamp action is shown as a truth table in Figure 2.2*b*.

To implement this function using relays, the switches A and B are not connected to the lamp directly, but are connected to relay coils labeled AR and BR whose normally-open

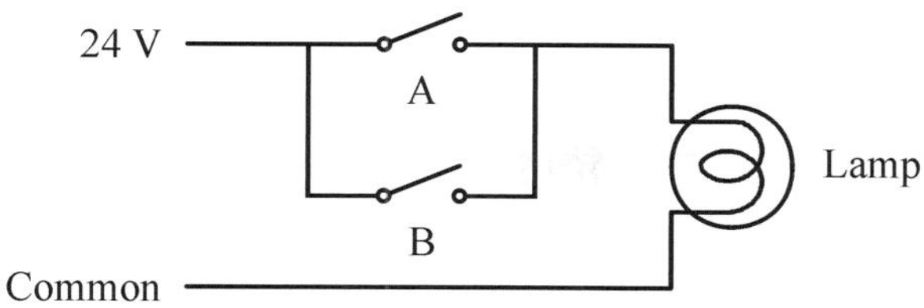

(a)

A	B	Lamp
off	off	off
off	on	on
on	off	on
on	on	on

(b)

Figure 2.2. Parallel switch circuit: *(a)* switch circuit; *(b)* truth table.

(NO) contacts control a relay coil, LR, whose contacts control the lamp, Figure 2.3*a*. The switches, A and B, are the inputs to the circuit. When either switch A or B is closed, the corresponding relay coil AR or BR is energized, closing a contact and supplying power to the LR relay coil. The LR coil is energized, closing its contact and powering the lamp.

The output (lamp in this case) is driven by the LR relay to provide voltage isolation from the relays implementing the logic. The switches, A and B, control relay coils (AR and BR) to isolate the inputs from the logic. Also, with this arrangement, the one switch connection to an input relay can be used multiple times in the logic. A typical industrial control relay can have up to 12 poles, or sets of contacts, per coil. For example, if the AR relay has six poles (only one shown in Figure 2.3*a*), then the other five poles are available for use in the relay logic without requiring five other connections to switch A.

Before the PLC was developed, engineers had already developed a graphical electrical circuit shorthand notation for the relay circuit of Figure 2.3*a*. This notation was called a *relay ladder logic diagram*, shown in Figure 2.3*b*. The switches are shown as their usual symbol, the circles indicate the relay coils, and the NO relay contacts are shown as the vertical parallel bars.

The *PLC ladder logic* notation (Figure 2.3*c*)[1] is shortened from the relay wiring diagram to show only the third line, the relay contacts and the coil of the output relay. The PLC ladder logic notation assumes that the inputs (switches in this example) are connected to discrete input channels (equivalent to the relay coils AR and BR in Figure 2.3*b*). Also, the actual output (lamp) is connected to a discrete output channel (equivalent to the normally open contacts of LR in Figure 2.3*b*) controlled by the coil. The label shown above a contact symbol is not the contact label, but the control for the coil that drives the contact. Also, the output for the rung occurs on the extreme right side of the rung and power is assumed to flow from left to right. The PLC ladder logic rung is interpreted as: "When input (switch) A is **on** OR input (switch) B is **on** then the lamp is **on**," which is the same as the statement describing the switch circuit in Figure 2.2*a*.

Notice that the original description of the switch circuit in Figure 2.2*a*,

The lamp is **on** when switch A is **on** <u>or</u> switch B is **on**.

translates into a relay circuit described as

1 The contact and coil symbols shown in Figure 2.3*c* are for the ControlLogix PLC. Other vendors use contact and coil symbols like those in Figure 2.3*b*.

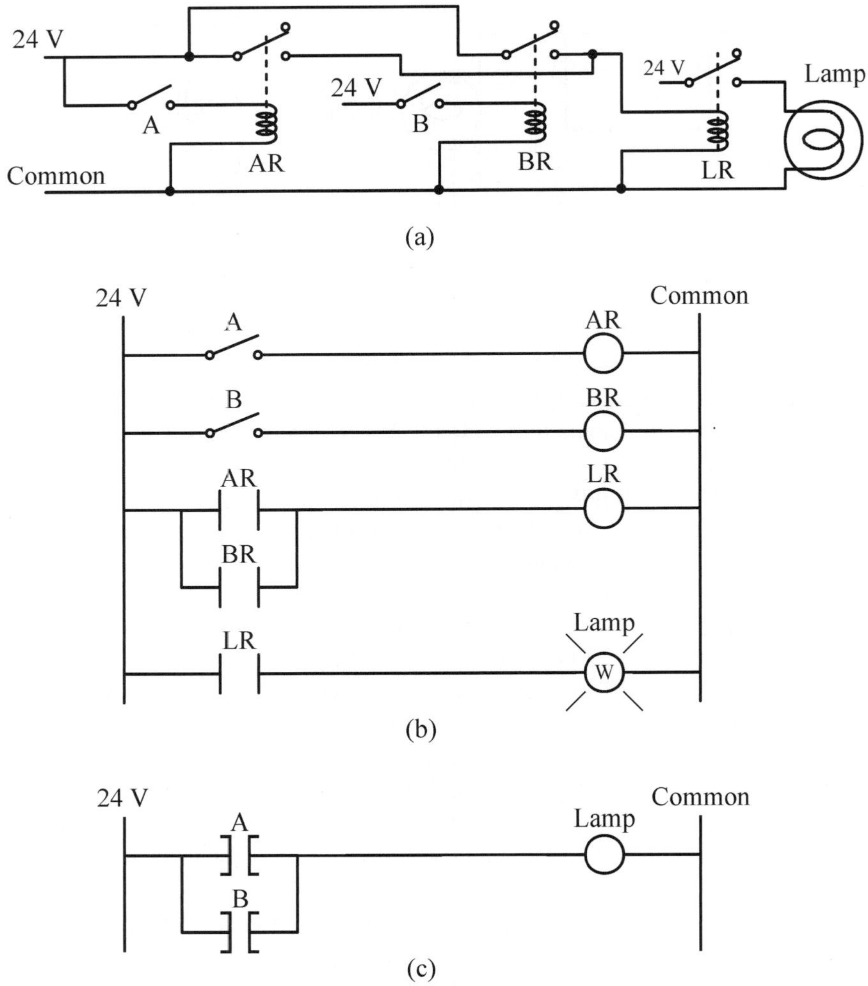

Figure 2.3. Parallel switch relay and ladder logic circuits: *(a)* equivalent relay circuit; *(b)* equivalent relay ladder logic circuit; *(c)* equivalent PLC ladder logic.

A parallel connection of **normally-open contacts**,

which describes the PLC ladder logic in Figure 2.3*c*.

Example 2.2. AND Circuit. Two switches labeled A and B are wired in series controlling a lamp as shown in Figure 2.4*a*. Implement this function as PLC ladder logic where the two switches are separate inputs.

Solution. The switch circuit action is described as, "The lamp is **on** when switch A is **on** (closed) and switch B is **on** (closed)." All possible combinations of the two switches and the consequent lamp action is shown as a truth table in Figure 2.4*b*. To implement this function using relays, the only change from Example 2.1 is to wire the normally-open contacts of control relays AR and BR in series to control the light, Figure 2.5*a*. The wiring of switches A and B and the wiring of the lamp do not change. The relay circuit diagram, shown in Figure

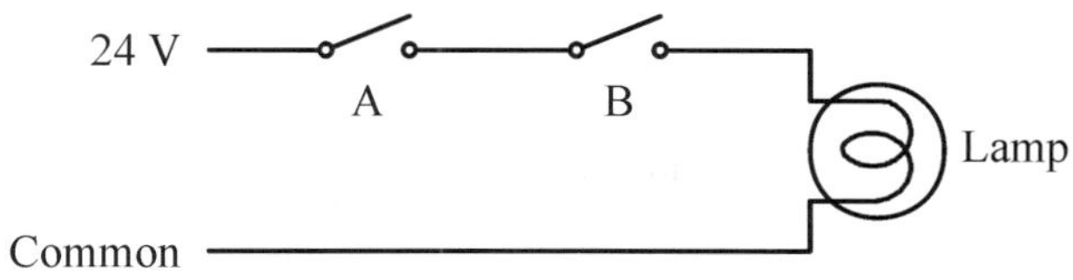

(a)

A	B	Lamp
off	off	off
off	on	off
on	off	off
on	on	on

(b)

Figure 2.4. Series switch circuit: *(a)* switch circuit; *(b)* truth table.

2.5*b* is different from Figure 2.3*b* only in the third line. As for example 2.1, the PLC ladder logic notation (Figure 2.5*c*) is shortened from the relay wiring diagram to show only the third line, the relay contacts and the coil of the output relay. The PLC ladder logic rung is interpreted as: "When input (switch) A is **on** AND input (switch) B is **on** then the lamp is **on**."

Notice that the original description of the switch circuit in Figure 2.4*a*,

The lamp is **on** when switch A is **on** <u>and</u> switch B is **on**.

translates into a relay circuit described as

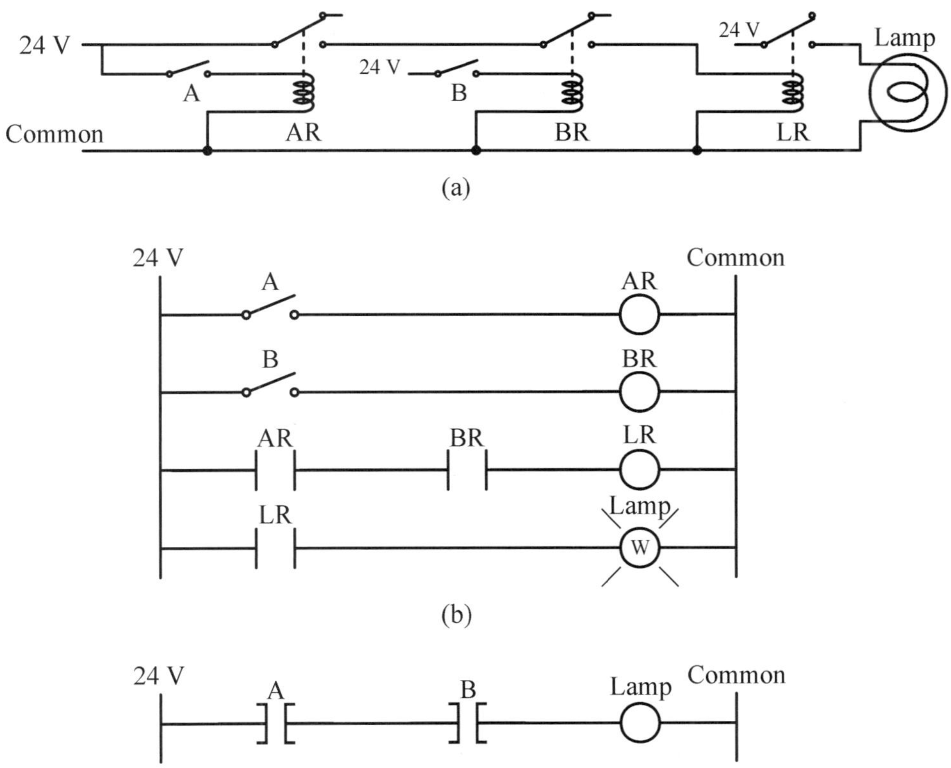

Figure 2.5. Series switch relay and ladder logic circuits: *(a)* equivalent relay circuit; *(b)* equivalent relay ladder logic circuit; *(c)* equivalent PLC ladder logic.

A series connection of **normally-open contacts**,

which describes the PLC ladder logic in Figure 2.5*c*.

Example 2.3. As a third example, consider the implementation of a logical NOT function. Suppose a lamp needs to be turned **on** when switch A is **on** (closed) and switch B is **off** (open). Implement this function as PLC ladder logic where the two switches are separate inputs.

Solution. Figure 2.6 shows the truth table, relay implementation and ladder logic for this example. The only difference between the relay implementation in Figure 2.6*b* and Figure 2.5*a* is the wiring of the relay BR contacts. The logical NOT for switch B is accomplished with the normally closed (NC) contact of relay BR. The PLC ladder logic rung in Figure 2.6*c* is different from Figure 2.5*c* only in the second contact symbol. The PLC ladder logic is interpreted as: "When input (switch) A is **on** (closed) and input (switch) B is **off** (open) then the lamp is **on**." This particular example is impossible to implement with a combination of only two normally open switches and no relays.

Notice that the original description of the Example 2.3,

The lamp is **on** when switch A is **on** and switch B is **off**.

translates into a relay circuit described as

A series connection of a **normally-open contact** and a **normally-closed contact**,

A	B	Lamp
off	off	off
off	on	off
on	off	on
on	on	off

(a)

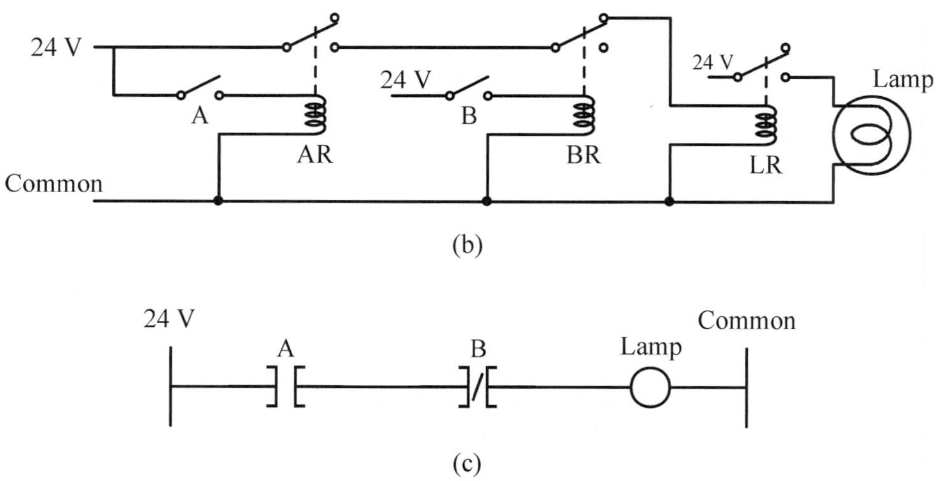

Figure 2.6. NOT function ladder logic circuits; *(a)* truth table; *(b)* equivalent relay circuit; *(c)* equivalent PLC ladder logic.

which describes the PLC ladder logic in Figure 2.6*c*.

Summarizing these three examples, one should notice that key words in the description of the operation translate into certain aspects of the solution:

and	→	series connection of contacts
or	→	parallel connection of contacts
on	→	**normally-open** contact
off	→	**normally-closed** contact

These concepts are key to being able to understand and write ladder logic. To many people these concepts appear strange and foreign at first. However, they will become more natural as one works problems. Ladder logic is a very visual and graphical language. It is very different from textual languages like C++, Fortran, Basic, and Python. In contrast, one can become proficient at ladder logic much quicker than with textual languages.

2.3 BASIC LADDER LOGIC SYMBOLS

At this point, one should start interpreting ladder logic directly and not think of its implementation with relays. As introduced by the examples in the previous section, the basic ladder logic symbols are

*** —] [—	*Normally open (NO) contact.* Passes power (**on**) if *** is **on** (closed).
*** —]/[—	*Normally closed (NC) contact.* Passes power (**on**) if *** is **off** (open).
*** —( )—	Output or *coil.* If any left-to-right path of contacts passes power, the *** output is energized. If there is no continuous left-to-right path of contacts passing power, *** is de-energized.

These symbols are ladder logic instructions that are scanned (executed) by the PLC. In order to avoid confusion, the contact symbols should be equated with certain concepts as follows:

—] [— = **on** = Closed = True = 1

—]/[— = **off** = Open = False = 0

This crucial point will be repeated later when the use of the NC contact is clarified. Figure 2.7 is an example ladder logic diagram with the basic instructions. The first line (also called a *rung*) that determines output labeled Out1 is interpreted as follows: Out1 is **on** if inputs A, B, and C are all **on**, or if inputs A and C are **on** and input D is **off**. For Out1 to be **on** there must be a continuous electrical path through the contacts.

The ControlLogix, CompactLogix, MicroLogix and SLC-500 procesors use the contact and coil symbols shown in the previous paragraph, though the Micro800 procesor

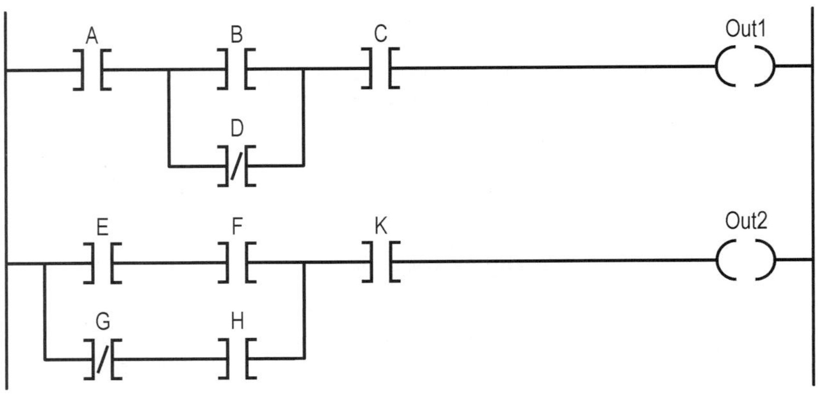

Figure 2.7. Ladder logic diagram with basic instructions.

shows the coil as a circle. There are other contact and coil symbols, but there is no universal graphic representation for these other symbols among PLC vendors. The IEC 61131-3 standard (IEC, 2013) has the most contact and coil symbols and many manufacturers do not implement the full set of symbols.

The industry trend is toward using the IEC 61131-3 (formerly IEC 1131-3) standard. However, this text emphasizes the Allen-Bradley ControlLogix and Micro800 implementations of IEC 61131-3. Since IEC 61131-3 is only a voluntary standard, individual manufacturers have some freedom in the implementation. Because of their widespread use, the Allen-Bradley MicroLogix and SLC-500 ladder logic languages are also covered.

2.3.1 ControlLogix and MicroLogix/SLC-500

The ControlLogix and MicroLogix/SLC-500 basic ladder logic contact symbols are

Normally open (NO) contact. Passes power (**on**) if *** is **on** (closed). Also called XIC (eXamine If Closed).

Normally closed (NC) contact. Passes power (**on**) if *** is **off** (open). Also called XIO (eXamine If Open).

One-shot contact. If conditions before this contact change from **off** to **on**, this contact passes power for only one scan (ControlLogix, PLC-5, and certain MicroLogix only). It is analogous to the IEC positive transition sensing contact except that this contact <u>follows</u> the contact(s) whose transition is being sensed. The *** is a storage Boolean that retains the previous state of the contact input (left side).

One-shot rising contact. If conditions before this contact change from **off** to **on**, this contact passes power for only one scan (SLC-500 and certain MicroLogix only). It must immediately precede an output coil. It is analogous to the IEC positive transition sensing contact except that this contact <u>follows</u> the contact(s) whose transition is being sensed. The *** is a storage Boolean that retains the previous state of the contact input (left side).

The basic ladder logic coil (output) symbols are

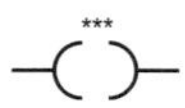

Output or *coil.* If any left-to-right rung path passes power, *** is energized (**on**). If there is no continuous left-to-right rung path passing power, the output is de-energized (**off**). Also called OTE (OuTput Energize).

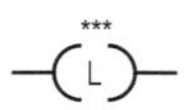

Latch coil. If any rung path passes power, *** is energized and remains energized, even when no rung path passes power. Also called OTL (OuTput Latch).

Unlatch coil. If any rung path passes power, *** is de-energized and remains de-energized, even when no rung path passes power. Also called OTU (OuTput Unlatch).

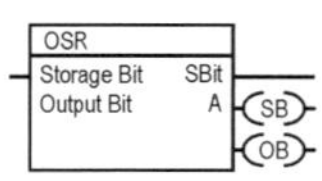

One shot rising output. If conditions before this block change from **off** to **on**, the specified output bit is turned **on** for one scan (ControlLogix and CompactLogix only). This is more appropriately a function block because of its appearance. The storage bit retains the previous state of the block input .

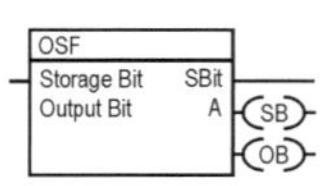

One shot falling output. If conditions before this block change from **on** to **off**, the specified output bit is turned **on** for one scan (ControlLogix and CompactLogix only). This is more appropriately a function block because of its appearance. The storage bit retains the previous state of the block input.

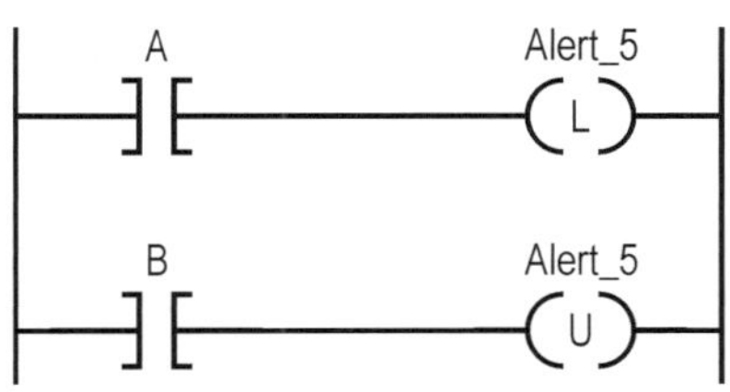

A latches (turns on) Alert_5

B unlatches (turns off) Alert_5

Figure 2.8. Latch and unlatch coil example.

Comments about the basic instructions

1. The transition sensing contacts and one-shot blocks are useful for initialization and detecting input transitions, for example, a push button press (Section 2.8).
2. The latch and unlatch coils are used in conjunction with each other. Figure 2.8 is a short example using these two coils in conjunction to control an alert.

2.3.2 Micro800

The Micro800 basic ladder logic contact symbols are

Normally open (NO) contact. Passes power (**on**) if *** is **on** (closed).

Normally closed (NC) contact. Passes power (**on**) if *** is **off** (open).

Positive transition sensing contact. If the state of *** changes from **off** to **on**, this contact passes power for only one scan (until rung is scanned again).

Negative transition sensing contact. If the state of *** changes from **on** to **off**, this contact passes power for only one scan (until rung is scanned again).

The basic ladder logic coil (output) symbols are

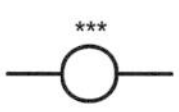

Output or *coil.* If any left-to-right rung path passes power, *** is energized (**on**). If there is no continuous left-to-right rung path passing power, the output is de-energized (**off**).

Negated coil. If any left-to-right rung path passes power, *** is de-energized (**off**). If there is no continuous left-to-right rung path passing power, *** is energized (**on**).

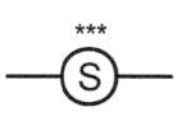

Set coil. If any rung path passes power, *** is energized and remains energized, even when no rung path passes power.

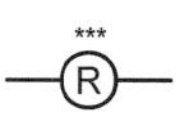

Reset coil. If any rung path passes power, *** is de-energized and remains de-energized, even when no rung path passes power.

Positive transition sensing coil. If conditions before this coil change from **off** to **on**, *** is turned **on** for one scan (until rung is scanned again).

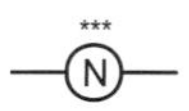

Negative transition sensing coil. If conditions before this instruction change from **on** to **off**, coil is turned **on** for one scan (until rung is scanned again).

Comments about the basic instructions

1. The transition sensing contacts and coils are useful for initialization and detecting input transitions, for example, a push button press (Section 2.8).
2. The set and reset are analogous to the ControlLogix latch and unlatch coils and are used in conjunction with each other.
3. The author strongly discourages use of the negated coil for the following reason. In most systems the safe state is one in which the output from the PLC is **off**. Generally, contacts (often called permissives) are placed in series with the coil, indicating multiple conditions must be satisfied before the output is allowed to be energized. With the negated coil the rung conditions must be satisfied to turn **off** the output which is contrary to most safety concepts.

2.4 LADDER LOGIC DIAGRAM

An example PLC ladder logic diagram appears in Figure 2.9. The vertical lines on the left and right are called the power rails. The contacts are arranged horizontally between the power rails, hence the term *rung*. The ladder diagram in Figure 2.9 has three rungs. The arrangement is similar to a ladder one uses to climb onto a roof. In addition, Figure 2.9 shows an example diagram like one would see if monitoring the running program in the PLC. The thick lines indicate continuity and the state (**on**/**off**) of the inputs and outputs is shown next to the tag. Regardless of the contact symbol, if the contact is closed (continuity through it), it is shown as thick lines. If the contact is open, it is shown as thin lines. In a relay ladder diagram, power flows from left to right. In PLC ladder logic, there is no real power flow, but there still must be a continuous path through closed contacts in order to energize an output. In Figure 2.9 the output on the first rung is **off** because the contact for C is open, blocking continuity through the D and E contacts. Also notice that the E input is **off**, which means the NC contact in the first rung is closed and the NO contact in the second rung is open.

Figure 2.9 also introduces the concept of *function block instructions*. Any instruction that is not a contact or a coil is called a function block instruction because of its appearance in the ladder diagram. The most common function block instructions are timer, counter, comparison, and computation operations. More advanced function block instructions include sequencer, shift register, and first-in first-out operations.

Allen-Bradley groups the instructions into two classes: input instructions and output instructions. This distinction is made because in relay ladder logic, outputs were never

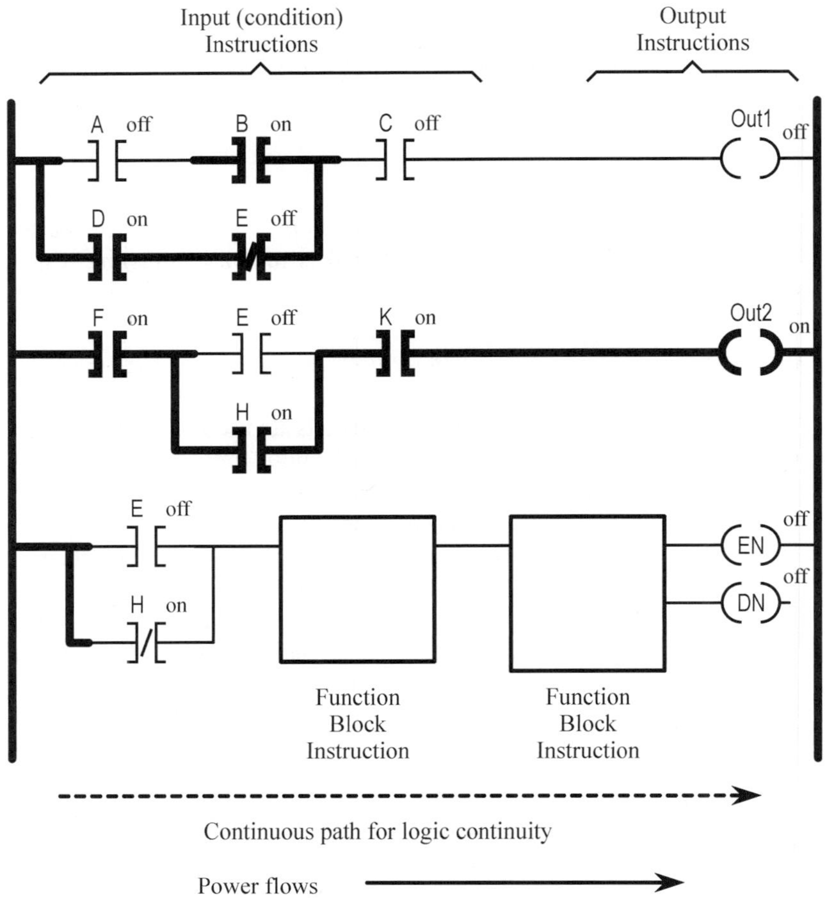

Figure 2.9. Sample ladder logic diagram.

connected in series and always occurred on the extreme right hand side of the rung. Contacts always appeared on the left side of coils and never on the right side. To turn on multiple outputs simultaneously, coils are connected in parallel. This restriction was relaxed in ControLogix and CompactLogix processors and outputs for these processors may be connected in series. Also, contacts can occur on the right side of a coil as long as a coil is the last element in the rung.

This text uses a series connection of coils for the ControlLogix and CompactLogix processors because it is allowed and is becoming common in practice. Series connections of coils are not allowed on Micro800 and MicroLogix processors and so for these processors, coils are connected in parallel.

Also, in ControlLogix and CompactLogix PLCs, all function block instructions are input instructions because the only output instructions are the coils. The MicroLogix and SLC-500 have function block output instructions (e.g., timer, counter, and computation) which must be remembered when constructing ladder logic programs for these PLCs.

Figure 2.10. Solution to Example 2.4.

Example 2.4. Draw a ladder diagram that will cause the output, pilot light PL2, to be **on** when selector switch SS2 is **closed**, push-button PB4 is **closed** and limit switch LS3 is **open**. (Note: no I/O addresses yet.)

Solution. The first question to answer is "What is the output?" The output is PL2, so the coil labeled as PL2 is put on the right side of the rung. Secondly, consider the type of connection of contacts to use. Since **all** three switches must be in a certain position to turn on the pilot light, a series connection is needed. Thirdly, the type of contact is determined by the switch position to turn on the pilot light:

SS2 closed → –] [–

PB4 closed → –] [–

LS3 open → –]/[–

Putting all the pieces together, only one rung of ladder logic is needed, as shown in Figure 2.10.

Design Tip

The concept of placing the output on the rung first and then "looking back" to determine the input conditions is very important. Because of the way the diagram is configured, one has a tendency to consider the input conditions first and then position the output coil as the last step. As will be shown later, the ordinary coil referring to a particular output must only occur **once** in a ladder program. Considering the output coil first and the conditions for which it is active (on) will avoid repeating output coils.

Example 2.5. Draw a ladder diagram that is equivalent to the digital logic diagram in Figure 2.11, which is the same as the following descriptions.

In words:

Y is **on** when (A is **on** and B is **on** and C is **off**) or D is **on** or E is **off**.

Boolean logic equation:

$$Y = AB\overline{C} + D + \overline{E}$$

Solution. First, answer, "What is the output?" The output is Y, so the coil labeled as Y is put on the right side of the rung. Secondly, consider the type of connection of contacts to use. For this problem, there is more than one type of connection. The three inputs within the

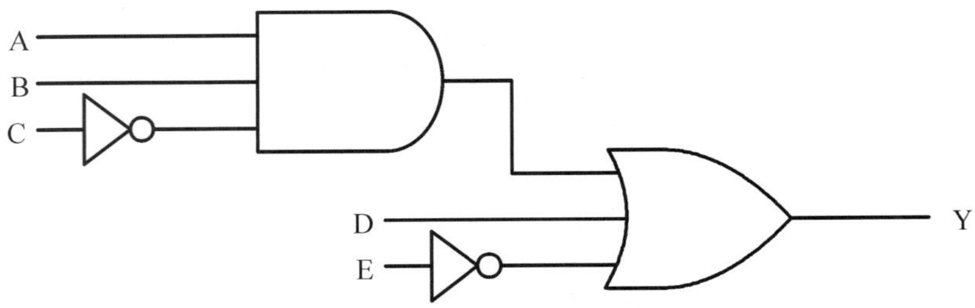

Figure 2.11. Digital logic for Example 2.5.

parentheses (the AND gate in Figure 2.11) are connected with "and," so a <u>series</u> connection is required for these three contacts. The other two inputs (D and E) are connected with the three series contacts by "or" (the OR gate inputs), so a <u>parallel</u> connection is required. Thirdly, the type of contact is determined by the input state that turns **on** the output, Y:

A **on** → ┤ ├   D **on** → ┤ ├

B **on** → ┤ ├   E **off** → ┤/├

C **off** → ┤/├

Putting all the parts together, only one rung of ladder logic is needed, as shown in Figure 2.12.

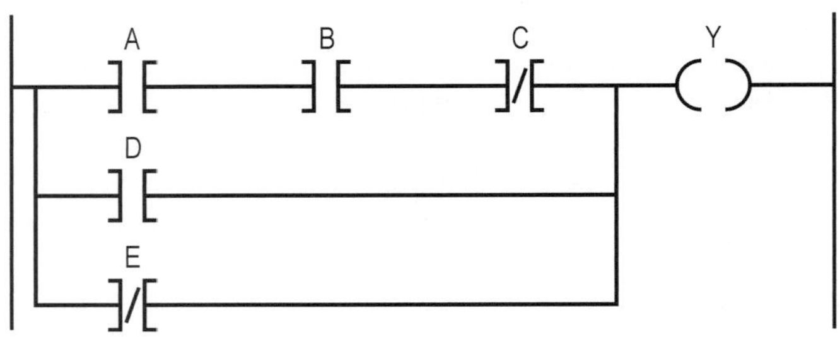

Figure 2.12. Solution to Example 2.5.

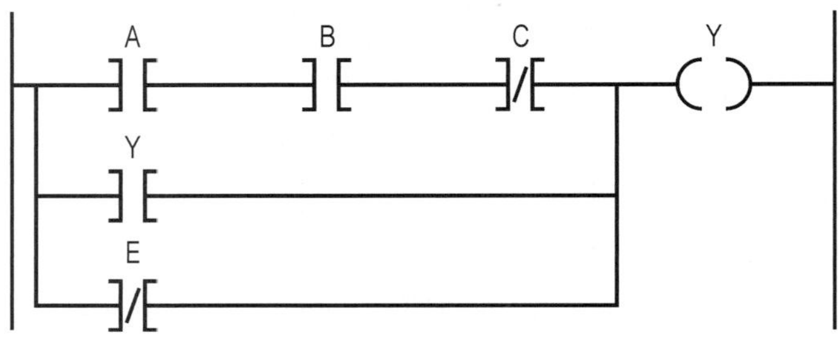

Figure 2.13. Output coil also used as an input contact.

Suppose one changes the D contact in Figure 2.12 to refer to Y, the output (shown as Figure 2.13). Is this legitimate? Yes, it is legitimate, though probably not something one would want to do for this example. Even in relay ladder logic, it is legal and there is no wiring short because the coil for relay Y and its NO contact are not electrically connected. This concept is called *sealing* or latching an output without using the set (or latch) coil instruction. In this example, it is not a good idea because once Y is sealed **on**, there is no provision to turn it off. Why?

There are some precautions to observe when programming in ladder logic:

1. **DO NOT** repeat normal output coils or latch/unlatch coils that refer to the same tag. To illustrate what happens when this is done, consider the ladder logic diagram in Figure 2.14. This is the ladder of Figure 2.9, modified for this illustration. Note that the coils for both the first and second rung refer to Out1. When the first rung of the ladder is scanned, Out1 is turned **on**. However, when the second rung is scanned, Out1 is turned **off**, overriding the logic in the first rung. If all of these conditions are needed to turn on Out1, then they all should be placed in parallel, as in Figure 2.15. In this illustration, it was obvious there is a problem. Normally, when this problem occurs, the rungs are not adjacent, and it is not so obvious. Fortunately, all Allen-Bradley PLC programming software warns about this situation. Still, the best way to prevent this problem is to consider the output coil **first** and then consider all of the conditions that drive that output.
2. Use the latch/set coil and unlatch/reset coils together. If a latch/set coil refers to an output, there should also be an unlatch/reset coil for that output. Also, for the same reason that ordinary output coils should not be repeated, do not mix the latch/unlatch coils with an ordinary output coil that refers to the same output.
3. Be careful when using the latch/unlatch coils to reference PLC physical outputs. If the system involves safety and a latch/set coil is used for a PLC physical output,

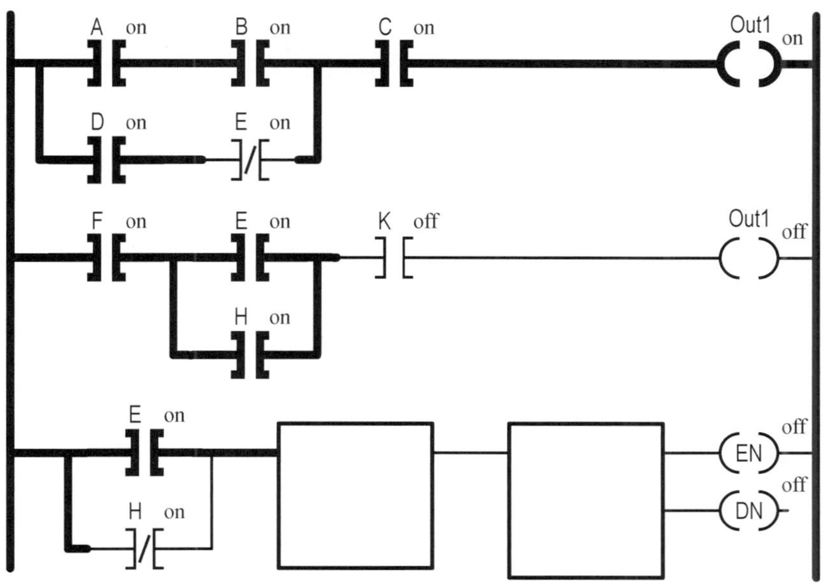

Figure 2.14. Ladder logic with repeated output.

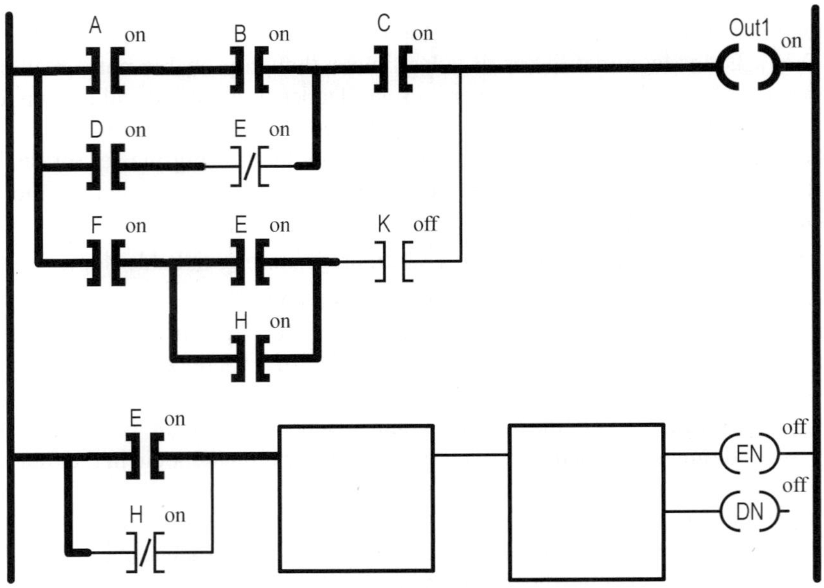

Figure 2.15. Repeated output corrected.

simply interrupting the condition on the latch/set coil rung **will not** turn off the physical output. All of the conditions that prevent the device from being turned on must also appear on a rung with an unlatch/reset coil output. For this reason, some companies forbid the use of the latch/unlatch coils controlling an actual physical output.

4. Reverse power flow in the contact matrix is **not** allowed. When electromechanical relays implement ladder logic, power can flow either way through the contacts. For example, consider the ladder logic in Figure 2.16. If implemented with electromechanical relays, power may flow right-to-left through the SS2 contact. When solid state relays replaced electromechanical relays for ladder logic, power can flow only one way (left-to-right) through the contacts. This restriction was carried to PLC ladder logic. If the reverse power flow path is truly needed, then insert it as a separate path, where the power flows from left to right. The reverse

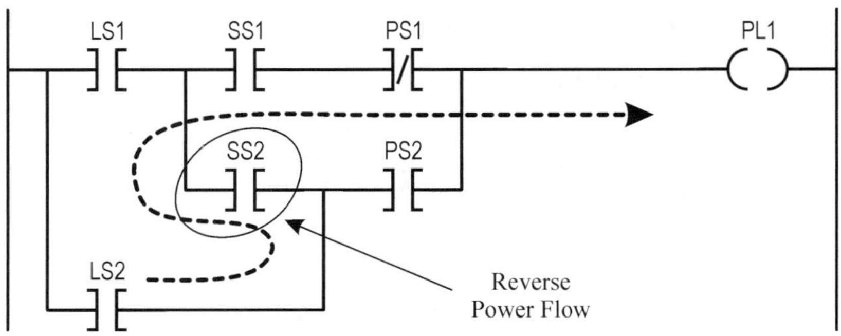

Figure 2.16. Reverse power flow in ladder logic.

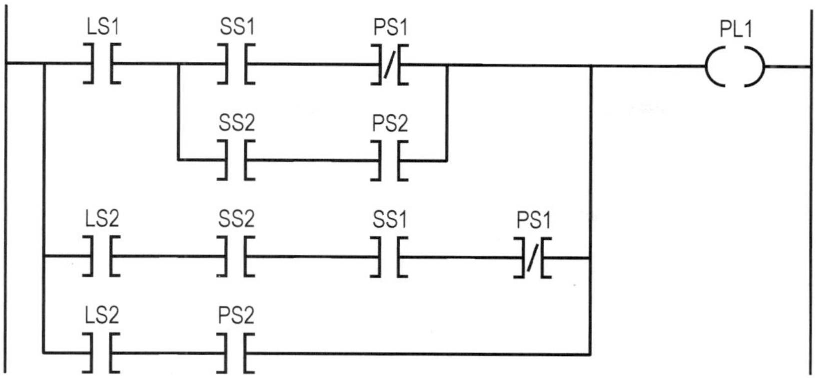

Figure 2.17. Reverse power flow in ladder logic corrected.

power flow path in Figure 2.16 is added as a separate path in Figure 2.17. Actually, one cannot program Allen-Bradley PLCs with ladder logic shown in Figure 2.16. One is not permitted to start a branch on one row of contacts and finish it on another row of contacts.

2.5 PLC PROCCESSOR SCAN

Previously, the process that the PLC uses to scan the ladder logic has only been implied. Now it will be discussed in detail. In addition to scanning the ladder logic, the PLC processor must also read the state of its physical inputs and set the state of the physical outputs. Historically, these three major tasks in a PLC processor scan are executed in the following order:

Read the physical inputs
Scan the ladder logic program
Write the physical outputs

The processor repeats these tasks as long as it is running, as shown pictorially in Figure 2.18. The time required to complete these three tasks is defined as the *scan time* and is typically 1 - 200 milliseconds, depending on the length of the ladder logic program. For very large ladder logic programs, the scan time can be more than one second. When this happens, the PLC program may miss transient events, especially if they are shorter than one second. In this situation, the possible solutions are:

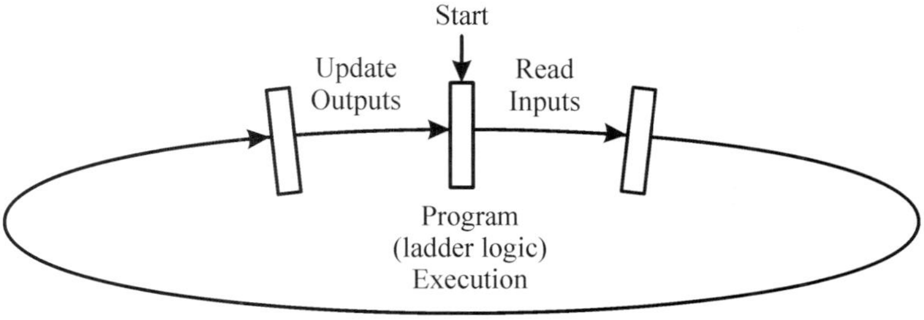

Figure 2.18. PLC processor scan.

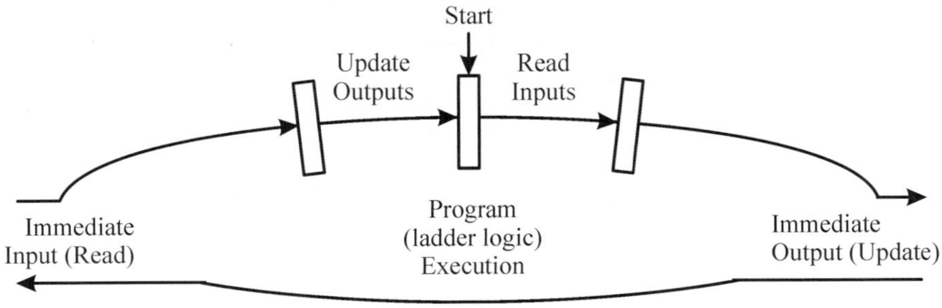

Figure 2.19. PLC processor scan with immediate input/output.

1. Break ladder logic into programs (POUs for Micro800) that are executed at a slower rate and execute the logic to detect the transient event on every scan.
2. Lengthen the time of the transient event so that it is longer than the maximum scan time. If the event is counted, both the on time and off time of the event must be longer than the scan time. A counter must sense both values to work correctly.
3. Place the logic examining the transient in a program or POU that is executed at a fixed time interval, smaller than the length of the transient event.
4. Partition long calculations. For example, if calculating the solution to an optimization, do one iteration per scan cycle rather than execute the entire algorithm every scan.

Depending on the PLC processor, one or more of these solutions may be unavailable.

Note that in the ControlLogix processor, the reading of the physical inputs and writing of the physical outputs is not coordinated with the scanning of the ladder logic. The Micro800, MicroLogix and SLC-500 processors follow the traditional scan method and that is described first. Examples 2.7 and 2.8 illustrate the differences between these two scan methods.

Traditionally, during the ladder logic program scan, changes in physical inputs cannot be sensed, nor can physical outputs be changed at the output module terminals. However, most Allen-Bradley processors have an instruction that can read the current state of a physical input and another instruction that can immediately set the current state of a physical output, as shown in Figure 2.19. However, using the immediate input/output instruction incurs a severe time penalty on the program scan. For example, to scan one contact in the ladder logic typically requires less than one microsecond. The time to execute an immediate input/output instruction typically requires 200 to 300 microseconds. Consequently, these instructions are used sparingly.

Another way to view the processor scan is shown in Figure 2.20. In this figure the state of the actual physical inputs is copied to a portion of the PLC memory, commonly called the *input image table*. When the ladder logic is scanned, it examines the input image table to read the state of a physical input. When the ladder logic determines the state of a physical output, it writes to a portion of the PLC memory commonly called the *output image table*. The output image may also be examined during the ladder logic scan. To update the physical outputs, the output image table contents are copied to the physical outputs **after** the ladder logic is scanned.

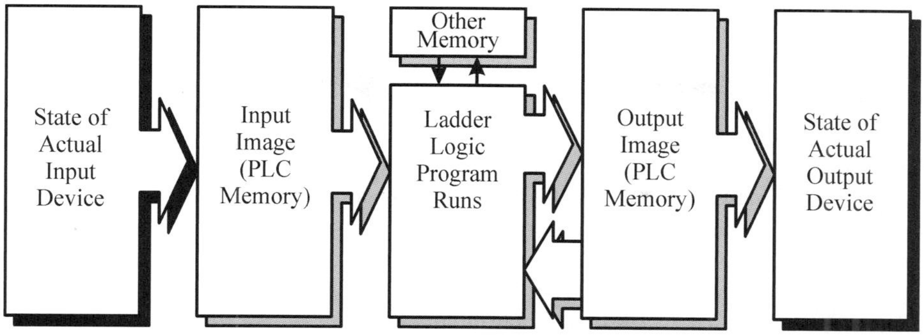

Figure 2.20. Alternate view of PLC processor scan.

An actual PLC processor executes other tasks than the three listed above. At a minimum, it communicates with other devices and checks its own hardware for faults. These other tasks occur in the background and have some influence on the scan time but are generally not considered part of the scan time.

Most PLC processors have a *watchdog timer* that monitors the scan time. If the processor scan time exceeds the watchdog timer time-out value, the processor halts program execution and signals a fault. This type of fault usually indicates the presence of an infinite loop in the ladder program or too many interrupts to the ladder scan.

For the Rockwell ControlLogix processors, the scan is not synchronized to the reading of the inputs and the update of the outputs. In this situation, an input can change its value at any time during the scan of the ladder. Therefore, logic that depends on an input value remaining constant during the entire scan of the ladder may not function properly. To correct this potential problem, the inputs are *buffered* to ensure that no inputs can change during a scan. Similarly, to apply any output changes simultaneously, the outputs are also buffered and applied at the end of a scan. Buffering is shown in Figure 2.21. At the start of the ladder logic scan, the ladder logic executes a routine or function block that copies the input image to the *input buffer*. The ladder logic then examines the inputs as necessary from the input buffer. Similarly, the ladder logic writes to the output buffer and then the very last

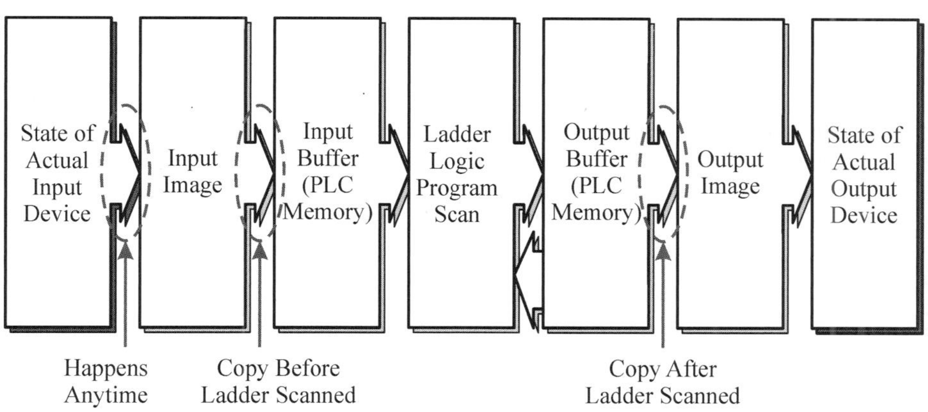

Figure 2.21. PLC processor scan with input and output buffering.

rung of the ladder executes a routine or function block that copies the output buffer to the output image. Example 2.8 shows the potential effect of not buffering inputs. More details about the ControlLogix scan are contained in section 3.2.2.

The overall execution of the PLC processor scan is controlled by the processor *mode*. When the PLC processor is in the *run mode*, the physical inputs, physical outputs, and ladder logic are scanned as described previously. When the processor is in *program mode* (sometimes called *stopped*), the ladder logic is **not** scanned. Depending on the particular PLC processor the physical inputs may be copied into the input image, but the physical outputs are disabled. Some processors have a *test mode*, where the physical inputs and ladder logic are scanned. The output image table is updated, but the physical outputs remain disabled.

The ladder logic program can be evaluated in one of two ways. Most PLC processors scan in rung order, also called top-to-bottom scan or "scan from the top". Some older PLC processors scan the ladder logic in column order, also called left-to-right scan or "scan from the left." Because all of the Allen-Bradley processors use the top-to-bottom scan, that is the only method described and illustrated here. Left-to-right scan is described in Erickson (2023).

In top-to-bottom scan, the ladder scan starts at the top of the ladder and proceeds to the bottom of the ladder, examining each rung from left to right. Once a rung is examined, it is not examined again until the next ladder scan. The rungs are not examined in reverse order. However, most processors have a jump instruction that one could use to jump back up the ladder and execute previous rungs. However, that use of the instruction is **not** recommended, because the PLC could be caught in an infinite loop. Even if the processor is caught in an infinite loop, the watchdog timer will cause a processor halt so the problem can be corrected.

Examples 2.6 and 2.7 illustrate the traditional method of scanning ladder logic used in the Micro800, MicroLogix and SLC-500 processors, where the ladder logic scan is synchronized to the reading of the inputs and the updated of the outputs. Example 2.8 illustrates the ControlLogix method of ladder logic scanning, which is not synchronized to the transfer to the physical inputs and outputs.

Example 2.6. Show the scan timing for the ladder diagram in Figure 2.22 when it is scanned top-to-bottom as in the Micro800, MicroLogix and SLC-500 processors. The only physical input is PB, a push button. The physical outputs are PL1, PL2, PL3, and PL4.

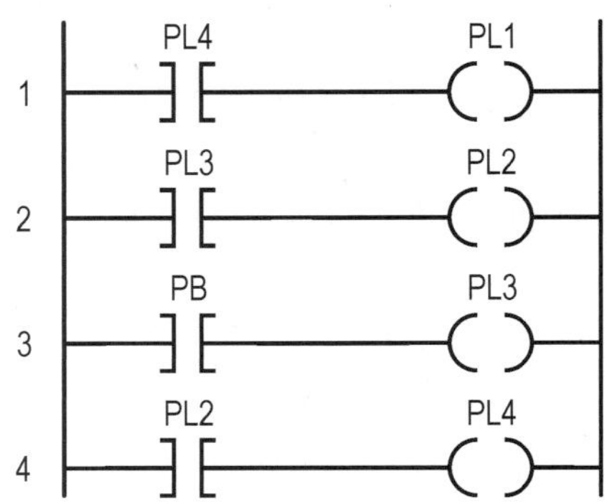

Figure 2.22. Ladder logic for Example 2.6.

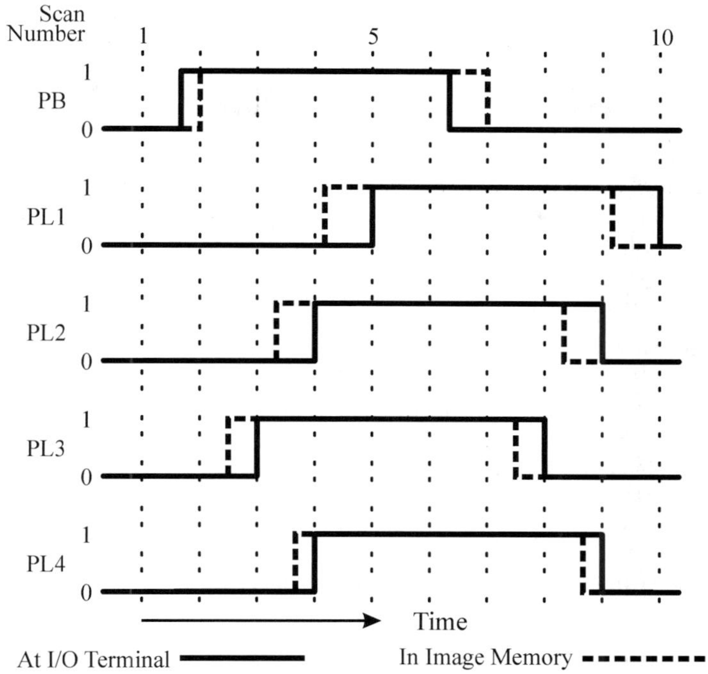

Figure 2.23. Timing diagram for Micro800 and MicroLogix scan of Figure 2.22.

Solution. When constructing a diagram showing scan timing, remember that the ladder scan uses the input and output image tables, the input image is updated before the ladder is scanned and the physical outputs are updated after the ladder is scanned. The timing diagram is shown in Figure 2.23. The widely spaced dashed vertical lines indicate the start of the processor scans. The solid lines indicate the state of the physical input or output at the module terminal. The dashed lines indicate the state of the PLC image memory corresponding to the physical input or output. Assume everything starts **off** (a value of 0). The **on** state is indicated by a value of 1. The ladder logic rungs are scanned in this order: 1, 2, 3, and 4. The actions happening during each scan are as follows:

Scan 1: Nothing happens even though physical input PB changes during the ladder scan. The new state of PB is not copied into the PLC input image until the start of the next scan.

Scan 2: The new state of PB is copied into the PLC input image before the ladder logic is scanned. When rungs 1 and 2 are scanned, the values of PL1 and PL2 in the output image remain at 0 (**off**) because the values of PL4 and PL3 are 0 in the output image. When rung 3 is scanned, the value of PB in the input image is now 1 (**on**), so there is continuity through the contact and the value of PL3 in the output image changes to 1 (**on**). When rung 4 is scanned, the value of PL4 in the output image remains at 0. At the end of scan 2, the output image is copied to the physical outputs, and the physical output PL3 turns **on**.

Scan 3: When rung 1 is scanned, the value of PL4 in the output image is 0, so the value of PL1 in the output image remains at 0. When rung 2 is scanned, the value of PL3 in the output image is now 1 so there is continuity through the contact and the value of PL2 in the output image changes to 1. When rung 3 is scanned, the value of PB in the input image is still 1, so the value of PL3 in the output image remains at 1. When rung 4 is scanned, the value of PL2 in the output image is now 1, so there is continuity through the contact and the value of PL4 in the output image changes to 1. At the end of scan 3, the output image is copied to the physical outputs and physical outputs PL2 and PL4 turn **on** simultaneously. So, even though the output image copy of PL2 and PL4 were not turned on simultaneously, they are turned on simultaneously at the physical terminal.

Scan 4: When rung 1 is scanned, the value of PL4 in the output image is now 1, so there is continuity through the contact and the value of PL1 in the output image changes to 1. When rung 2 is scanned, the value of PL3 in the output image is still 1 so the value of PL2 in the output image remains at 1. When rung 3 is scanned, the value of PB in the input image is still 1 so the value of PL3 in the output image remains at 1. When rung 4 is scanned, the value of PL2 in the output image is still 1 so the value of PL4 in the output image remains at 1. At the end of scan 3, the output image is copied to the physical outputs and physical output PL1 turns **on**.

Scans 5 and 6: Nothing changes.

Scans 7 - 9: Similar to scans 2 - 4, except that states change from 1 (**on**) to 0 (**off**).

Notice that the arrangement of the rungs has forced it to operate in the manner shown in Figure 2.23. How would one change the program so that the physical outputs turned **on** and **off** simultaneously? Answer: arrange the rungs so the output coils appear in the following order: PL3, PL2, PL4, PL1.

This example, though trivial, serves to illustrate two concepts: the way ladder logic is scanned and the difference between image memory and physical input/output. The next example is a little more complicated and introduces the concept of internal coils.

Example 2.7. Show the scan timing for the ladder diagram in Figure 2.24 when it is scanned top-to-bottom in the Micro800, MicroLogix and SLC-500 processors. The only physical input is PB1, a push button. The only physical output is PL1, a pilot lamp. Int1, Int2, and Int3 are internal one-bit memory locations, often called *internal coils*. They are not attached to any physical outputs. What does this ladder logic program do?

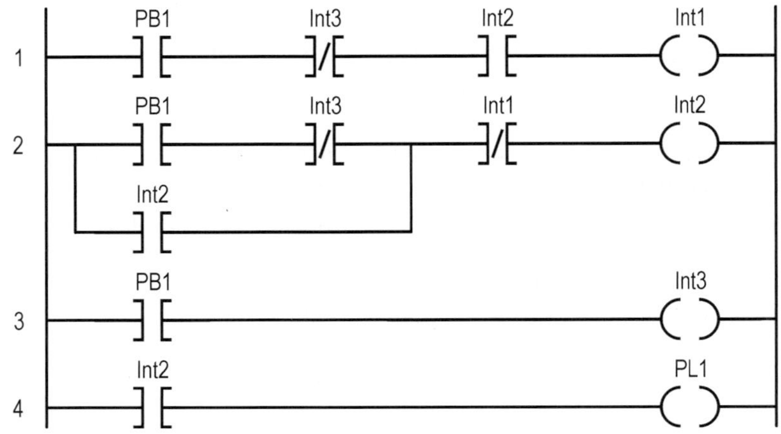

Figure 2.24. Ladder logic for Examples 2.7 and 2.8.

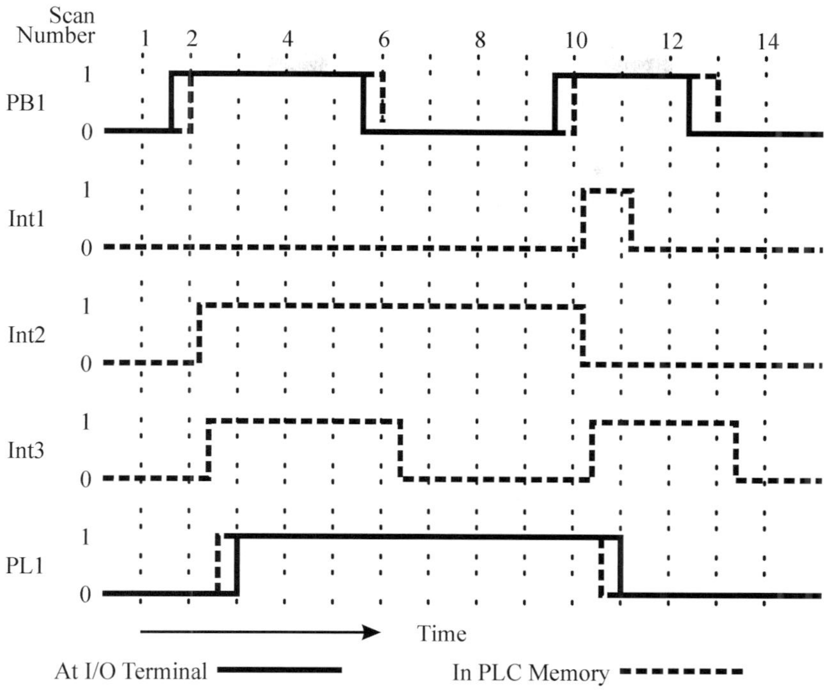

Figure 2.25. Scan timing diagram for Micro800 and MicroLogix scan of Figure 2.24

Solution. The scan timing diagram is shown in Figure 2.25 and has the same representation as in the previous example. Assume everything starts **off** (a value of 0). The ladder logic rungs are scanned in this order: 1, 2, 3, and 4. The actions happening during each scan are as follows:

Scan 1: Nothing happens even though physical input PB1 changes during the ladder scan. The new state of PB1 is not copied into the PLC input image until the start of the next scan.

Scan 2: The new state of PB1 is copied into the PLC input image before the ladder logic is scanned. When rung 1 is scanned, PB1 is **on**, and Int3 is **off**, so the first two contacts pass power, but Int2 is off. So, Int1 remains **off**. When rung 2 is scanned, there is continuity along the top path, so Int2 is turned **on**. When rung 3 is scanned, the value of PB1 in the input image is now **on**, so there is continuity through the contact and the Int3 changes to **on**. Note that this change in Int3 will not be sensed by the first two rungs until the next scan. When rung 4 is scanned, the value of Int2 in the PLC memory is **on**, so PL1 is turned **on** in the output image. At the end of scan 2, the output image is copied to the physical outputs and the physical output PL1 turns **on**.

Scan 3: There is no change in the rung output coils, though some of the contacts in the first two rungs have changed since last scanned. When rung 1 is scanned, there is now continuity through the first and third contact, but not through the second (Int3 is **off**). When rung 2 is scanned, there is no continuity through the top branch, but there is continuity through the bottom branch, and so Int2 remains **on**.

Scans 4-5: There is no change in the rung output coils because there is no change in any of the contacts.

Scan 6: The new state of PB1 is copied into the PLC input image before the ladder logic is scanned. When rungs 1 and 2 are scanned, there is no change in the output coils even though PB1 has changed. When rung 3 is scanned, the value of PB1 in the input image is now **off** so there is no continuity through the contact and the Int3 changes to **off**. Since the value of Int2 has not changed, PL1 remains **on**.

Scans 7-9: There is no change in the rung output coils because there is no change in any of the contacts.

Scan 10: The new state of PB1 is copied into the PLC input image before the ladder logic is scanned. When rung 1 is scanned, PB1 is **on**, Int3 is **off**, and Int2 is **on** so Int1 is turned **on**. Because Int1 is **on**, when rung 2 is scanned, there is no continuity through the last contact and so Int2 is turned **off**. When rung 3 is scanned, the value of PB1 in the input image is now **on** so there is continuity through the contact and the Int3 changes to **on**. When rung 4 is scanned, the value of Int2 in the PLC memory is **off**, so PL1 is turned **off** in the output image. At the end of scan 10, the output image is copied to the physical outputs and the physical output PL1 turns **off**.

Scan 11: When rung 1 is scanned, the second contact is open since Int3 is now **on**, and so Int1 is turned **off**. Because Int3 is **on** and Int2 is **off** when rung 2 is scanned, there is no continuous path to the output coil and Int2 remains **off**. When rungs 3 and 4 are scanned, the output coils do not change since the contacts have not changed since the last scan.

Scan 12: There is no change in the rung output coils because there is no change in any of the contacts.

Scan 13: The new state of PB1 is copied into the PLC input image before the ladder logic is scanned. When rungs 1 and 2 are scanned, there is no change in the output coils even though PB1 has changed. When rung 3 is scanned, the value of PB1 in the input image is now **off** so there is no continuity through the contact and the Int3 changes to **off**. Since the value of Int2 has not changed, PL1 remains **off**.

What does this ladder program do? This is called a "toggle" or "push-to-start, push-to-stop" circuit. Every time the push button PB1 is pushed (turned **on**), PL1 changes its state.

Example 2.8. Show the scan timing for the ladder diagram in Figure 2.24 when it is scanned top-to-bottom in the ControlLogix processor. Assuming the only physical input, PB1, is in the same chassis as the processor, any change in the value is sent immediately to the processor input image. Furthermore, assuming the only physical output, PL1, is also in the same chassis as the processor, any change to the output because of a rung scan is sent immediately to the channel on the output module.

In order to show the scan timing, the relationship of changes in PB1 to the rung scan must be known. Assume that the initial press of PB1 in scan 1 occurs *after* the first 3 rungs are scanned and the release of PB1 in scan 5 also occurs *after* the first three rungs are scanned. However, assume that the second press of PB1 in scan 9 occurs *between* the scan of the second and third rungs and the release of PB1 in scan 12 also occurs *between* the scan of the second and third rungs.

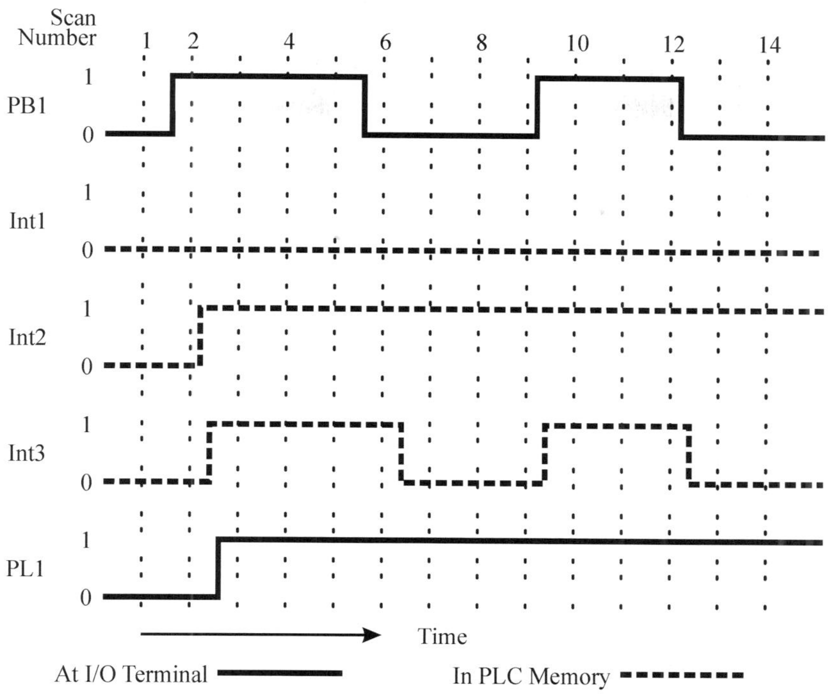

Figure 2.26. Scan timing diagram for ControlLogix scan of Figure 2.23.

Solution. The scan timing diagram is shown in Figure 2.26 and has the same representation as in the previous example. Assume everything starts **off** (a value of 0). The actions happening during each scan are as follows:

Scans 1 - 8: Same as in Example 2.7, except that there are no dashed lines for the PB1 and PL1 timing. A change in PB1 is immediately transferred to the input image and a change in PL1 in the output image is immediately transferred to the output terminal. However, since PB1 changes after the first 3 rungs are scanned, the effect of the change is seen during the next scan of the ladder logic rungs.

Scan 9: When rung 1 is scanned, PB1 is **off**, Int3 is **off**, and Int2 is **on** so Int1 remains **off**. When rung 2 is scanned, there is still no continuity through the top branch, but there is continuity through the bottom branch and Int1 is **off**, so Int2 remains **on**. When rung 3 is scanned, the value of PB1 in the input image is now **on** so there is continuity through the contact and the Int3 changes to **on**. When rung 4 is scanned, the value of Int2 in the PLC memory is **on**, so PL1 remains **on** in the output image and at the output terminal.

Scan 10: When rung 1 is scanned, PB1 is **on**, Int3 is **on**, and Int2 is **on** so Int1 remains **off**. When rung 2 is scanned, there is continuity through the both branches and Int1 is **off**, so Int2 remains **on**. When rungs 3 and 4 are scanned, the output coils do not change since the contacts have not changed since the last scan.

Scan 11: There is no change in the rung output coils because there is no change in any of the contacts.

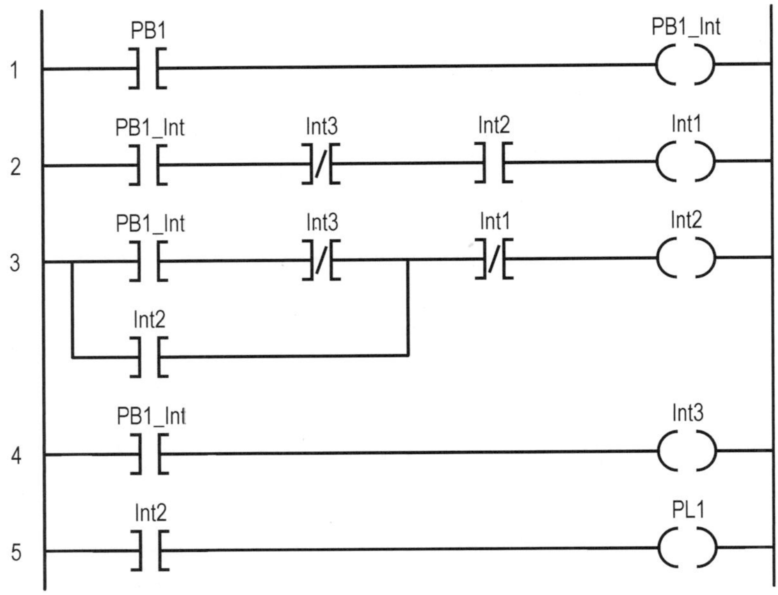

Figure 2.27. Corrected ladder logic for Example 2.8.

Scan 12: When rungs 1 and 2 are scanned, there is no change in the output coils because the PB1 change has not been sensed. When rung 3 is scanned, the value of PB1 in the input image is now **off** so there is no continuity through the contact and the Int3 changes to **off**. Since the value of Int2 has not changed, PL1 remains **on**.

Scan 13: When rungs 1 and 2 are scanned, there is no change in the output coils even though PB1 is now sensed **off**.

What happened? The timing diagram is not the same as in Figure 2.25 and in fact does not operate as expected. The key problem is that the state of PB1 changed in scan 9 between the scan of rungs 2 and 3. Actually, the timing diagram would have been the same if PB1 had changed between the scan of rungs 1 and 2. In this particular ladder logic, it is essential that the state of PB1 remain constant during the scan of rungs 1, 2, and 3. The historic scan method enforced that consistency because of the way input image updates were handled. But, the ControlLogix scan method does not guarantee that the state of physical inputs remains constant during the ladder logic scan. So, how does one fix this problem? Copy the state of all physical inputs to internal coils at the start of the ladder logic scan. In order to fix the problem for this particular ladder logic, the state of PB1 must be copied to an internal coil, PB1_Int, before using it in the logic. The changed ladder logic is shown in Figure 2.27. The timing diagram for the ladder logic in Figure 2.27 is very similar to the timing diagram in Figure 2.25 with the following differences:

The dashed line showing the input image of PB1 is now the value of PB1_Int.

The solid line for PL1 is shifted to be the same as the dashed line for PL1.

Note that simply moving the third rung in Figure 2.24 to be before the first two rungs will not solve the problem. That change will ensure that Int1 is never pulsed **on** to "unlatch" Int2, which is what happens in Figure 2.26. It turns out that the logic for Int1, Int2 and Int3 must occur in the order shown in Figure 2.24 or Figure 2.27.

2.6 PROGRAMMING WITH NC CONTACT

Many novice PLC ladder logic programmers tend to use the same type of contact (NO or NC) in the ladder that corresponds to the type of field switch (or sensor) wired to the discrete input channel. While this is true in many cases, this is not the best way to think of the concept.

A NO switch does not require a —] [— **in the ladder logic.**

A NC switch does not require a —]/[— **in the ladder logic.**

The PLC only knows one thing about the discrete input: whether it is **on** (closed) or **off** (open). The PLC knows nothing about how the switch is wired in the field; whether it is NO held open, NO held closed, NC held open, or any other combination.

For example, safety switches are generally NC switches, but the contact in the ladder logic that refers to the safety switch is generally a NO contact. The reason for this type of safety switch will be explained in the next section.

Regardless of the type of switch in the field, the principle to remember is

If one wants "action" (turn ON) when the switch is closed, use —] [—

If one wants "action" (turn ON) when the switch is open, use —]/[—

As explained earlier in this chapter, think of the contact as a symbol

—] [— = **on** = Closed = True = 1

—]/[— = **off** = Open = False = 0

An alternative view of this concept is to consider the contact symbol in the ladder as the unenergized state of the discrete input:

—] [— has no continuity (**off**, or false) when the discrete input is **off**. Consequently, when the discrete input is **on**, the contact has continuity (**on**, or true).

—]/[— has continuity (**on**, or true) when the discrete input is **off**. Consequently, when the discrete input is **on**, the contact has no continuity (**off**, or false).

To illustrate this concept, consider the system in Figure 2.28*a* that is a simple water level control system. S1 is a NC safety switch and FS1 is a NO float switch. The water inlet valve V1 should be **on** (fill tank) when S1 is **closed** and FS1 is **open**. The ladder logic rung to accomplish this operation is shown in Figure 2.28*b*. Notice that for both switches, the contact used in the ladder is the opposite of the type of field switch. The type of field switch is irrelevant to the PLC ladder logic. The type of contact in the PLC ladder logic is determined by the operation description.

V1 is **on** when S1 is **closed** and FS1 is **open**.

S1 closed → —] [— FS1 open → —]/[—

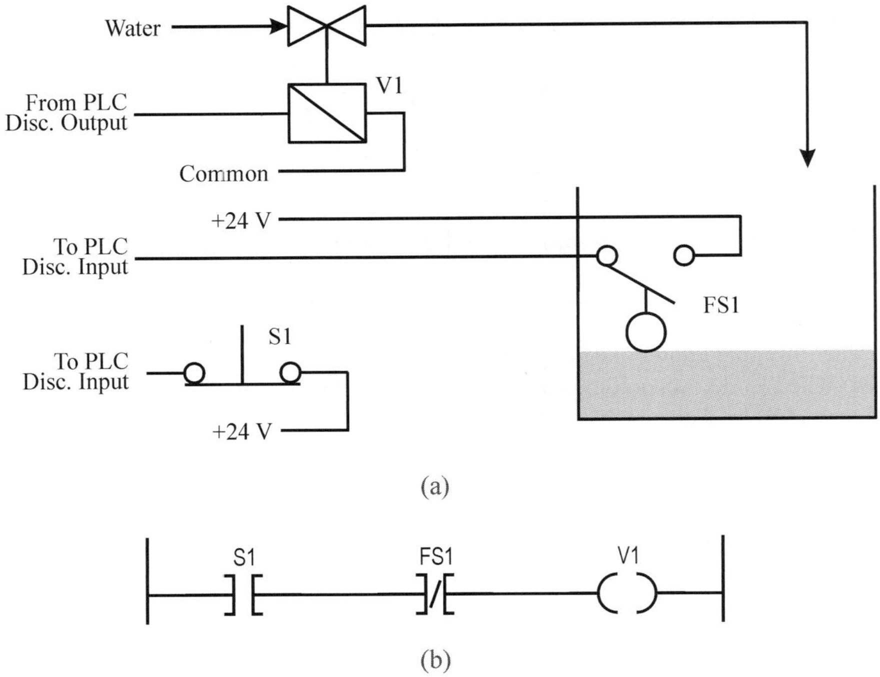

Figure 2.28. Example showing differences in field switch and ladder logic contact: *(a)* simple level control system; *(b)* PLC ladder logic.

> **Design Tip**
>
> The type of contact (NO, NC) in the field is determined by safety or fail-safe factors, but **these factors are not relevant to the PLC ladder logic**. The PLC is concerned only about the current state of the discrete input (open or closed).

2.7 START/STOP

The concept described in the previous section will be reinforced when we consider the problem of starting and stopping a motor with momentary switches. The motor is representative of any device that must run continuously but started and stopped with momentary switches.

This problem is handled in two parts: starting and stopping. How do I start a motor and keep it running with a momentary normally open push button switch? The switch closes to start the motor and when it is released (opened) the motor should continue to run. Obviously, one needs to add a contact in parallel with the start switch contact that maintains continuity when the start switch is opened. One way is to use the PLC output that drives the motor to seal around the start switch contact, as shown in Figure 2.29. Note that the

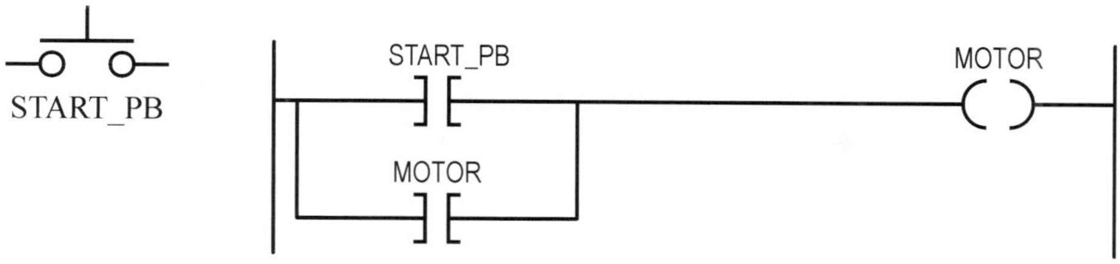

Figure 2.29. Ladder logic to start motor with momentary switch.

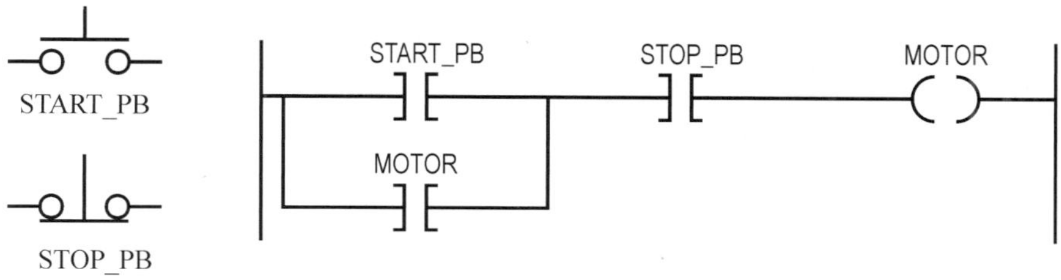

Figure 2.30. Ladder logic to start and stop motor.

MOTOR output does not directly control the motor (the PLC output cannot handle high currents) but represents the input to a motor contactor or motor controller.

The second part of the problem is to add a stop switch. The stop switch is a normally closed push button switch, (push) open to stop, closed to run. A contact for this stop switch needs to be added in series with the motor coil, as shown in Figure 2.30. Notice that the contact in the ladder logic is a NO contact, which is not the same as the NC push button stop switch. This is a common situation where the novice PLC programmer makes a mistake. Many people do not resolve the issue for themselves until they actually program it the wrong way (use a NC contact for the stop switch) and then understand the correct concept.

> **Design Tip**
>
> The type of contact (NO, NC) in the field is determined by safety or fail-safe factors, but **these factors are not relevant to the PLC ladder logic**. The PLC is concerned only about the current state of the discrete input (open or closed).

> **Design Tip**
>
> In PLC ladder logic, use the NO contact instruction for physical stop switches. This is not universal, but any physical stop switch that is not closed to run probably should be replaced for safety reasons.

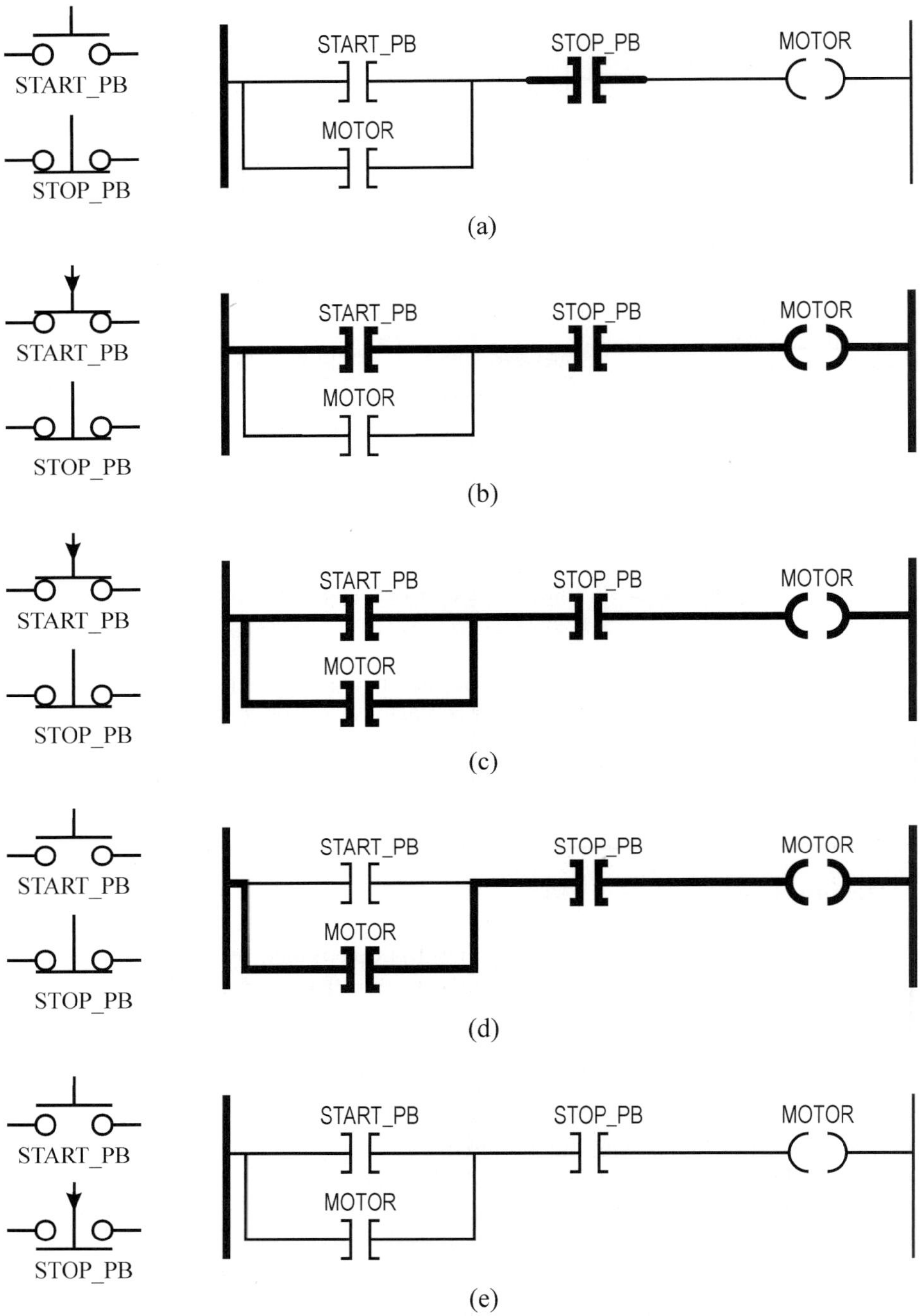

Figure 2.31. Operation of motor start/stop: *(a)* initial conditions; *(b)* start switch pressed; *(c)* scan after start switch pressed; *(d)* start switch released; *(e)* stop switch

The operation of the start/stop ladder logic is detailed in Figure 2.31. Assume one starts with the motor stopped and both push buttons not pressed (Figure 2.31*a*). In this situation, the contact for STOP_PB has continuity because the STOP_PB is **closed**. Now when the START_PB is pressed, the switch is **closed**, and the START_PB contact in the ladder has continuity. Therefore, a continuous path exists through the START_PB and STOP_PB contacts, and the MOTOR output (coil) is energized (Figure 2.31*b*). When this rung is next scanned, the path through the MOTOR contact has continuity (Figure 2.31*c*). This situation exists as long as START_PB is held down. When START_PB is released, the switch is **opened**, and the START_PB contact in the ladder does not have continuity (Figure 2.31*d*). However, a continuous path exists through the MOTOR contact and the STOP_PB contact. Hence, the MOTOR output remains energized. This situation exists until STOP_PB is pressed. When STOP_PB is pressed, the switch is **opened**, and the STOP_PB contact in the ladder does not have continuity (Figure 2.31*e*). Thus, there is no continuous path to the MOTOR output, and it is de-energized, which stops the motor. Now the system is in the same state as when the description of its operation commenced.

The questions arise: "Why are the switches wired this way?"; "Why not use normally open push button switches for the start switch and the stop switch?"; "Why not use normally closed switches for both?" The switches are wired this way for safety reasons. If any part of the system fails (switch or wiring), the motor will go to a safe state. If the start switch wiring is faulty (open wire), then the motor cannot be started because the PLC will not sense a closed start switch. If the stop switch wiring is faulty (open wire), then the motor will immediately stop if it is running. Also, the motor cannot be started with an open wire to the stop switch. Since the operator has lost control over stopping the motor, the safe state is one in which the motor is stopped.

However, these types of switches do not automatically handle all failures. If the start switch fails so that its contact is always closed, the stop push button must be held in to prevent the motor from starting. In this case, the circuit breaker at the motor controller must be opened to stop the motor. If the stop push button fails so that the contact is always closed, the ladder logic cannot de-energize the motor. To handle this situation, many systems have another stop switch contact in series with the stop switch on the PLC rung. Also, most systems have an emergency-stop wired external to the PLC that will cut power to all control devices. It should be noted that a switch failing closed is extremely rare. An open wire or switch failing open is much more common.

The ladder logic circuit in Figure 2.30 is also called a *seal circuit*, and it will appear in other contexts. In many real systems, the start and stop of a device, like a motor, have more conditions that must be satisfied in order for the motor to run. These conditions are referred to as *permissives*, *permits*, *lockouts*, *inhibits*, or *restrictions*. These conditions appear on the start/stop rung as shown in Figure 2.32. Permissives allow the motor to start, and lockouts will stop the motor, as well as prevent it from being started.

An alternate way of starting/stopping the motor is shown in Figure 2.33 with the latch and unlatch coils. However, this way of controlling the motor is **not recommended**. Any lockout logic will need to be in series with START_PB contact and the **logical inverse** of the lockout logic will need to be placed in parallel with the STOP_PB contact. This task is not impossible, but error-prone, especially if someone other than the original programmer must make a change to the lockout logic.

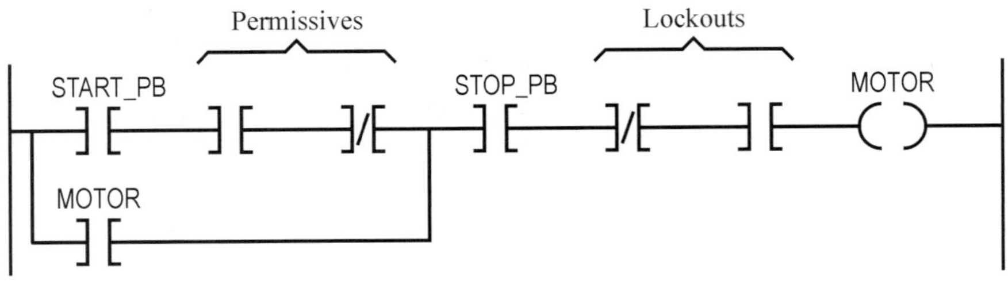

Figure 2.32. Start/stop with permissives and lockouts.

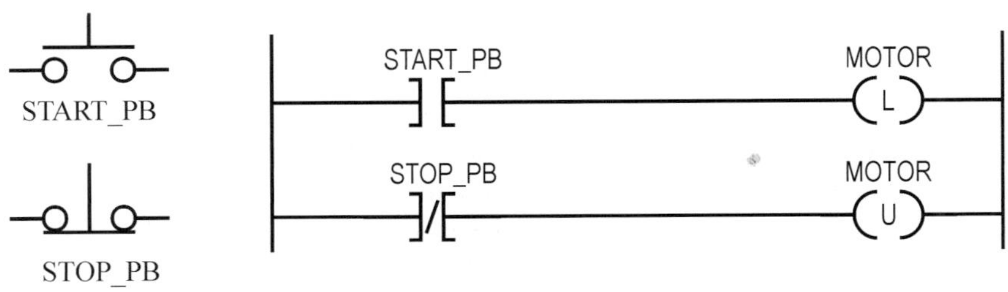

Figure 2.33. Start/stop with latch and unlatch coils (**not recommended**).

> **Design Tip**
>
> Design the logic so that a change to lockouts, permissives, and so on, affects only one rung. If a change to a lockout forces one to change multiple rungs, it is almost guaranteed that the changes will be wrong because (1) someone other than the original programmer will be making the change and (2) he/she will be under time pressure to make the change.

There is a slight problem with the ladder logic of Figure 2.30. Motor overload contacts are always located on the wires leading directly to the motor and external to the PLC. If the motor stops because of an overload, the seal circuit of Figure 2.30 will continue to energize the rung output (input to the motor controller). When the overload contacts cool and reclose, the motor will start automatically, with possible injury to personnel. To solve this problem, the sealing contact (in parallel with START_PB) is provided by a motor auxiliary contact that closes when the motor has power, or by a motor speed sensor that is **on** when the motor or the equipment it is driving is running. If the motor stops because of an overload, the seal contact in the ladder loses continuity and the MOTOR output from the PLC is de-energized. Thus, the start push button must be pressed after the overload has been cleared to restart the motor. If the speed sensor on the driven equipment is used as the seal contact, then a broken shaft will automatically stop the motor.

Note that the stop switch has priority in the ladder logic of Figure 2.30. If the stop switch is pressed, no amount of pushing the start switch will start the motor. What if one wanted the start condition to have priority over the stop condition? Obviously, this is not

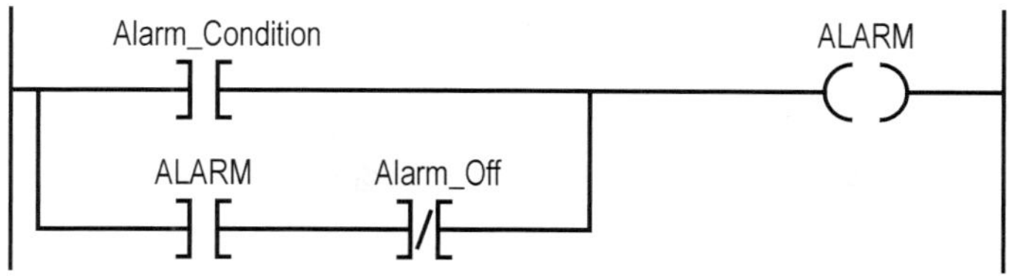

Figure 2.34. Start/stop with start priority.

wanted for a device like a motor, but is extremely useful for alarms. For alarms, the alarm condition (start condition) should have priority over the alarm turn-off condition (stop condition). As long as the alarm condition is still happening, the alarm cannot be turned off. This type of start/stop ladder is shown in Figure 2.34. Note that Alarm_Off is **on** to turn off the alarm (as long as Alarm_Condition is **off**).

Example 2.9. Add a jog switch to the motor start/stop ladder logic of Figure 2.30. The jog switch is a normally open push button switch that when pressed (closed) runs the motor as long as it is held in. If the stop push button switch is pressed, the motor should immediately stop if the jog or start switches are closed. The jog switch should have no effect if the motor is already running.

Solution. One may be tempted to just place the JOG_PB contact in parallel with the START_PB contact, but then the jog switch functions as another start switch. When it is released, the motor will still be running. So the seal circuit will need to drive an internal coil INT_MOTOR instead of the motor coil directly. The INT_MOTOR internal coil, in parallel with the jog and stop switches, drives the motor coil. This ladder logic is shown in Figure 2.35.

Figure 2.36 shows one incorrect solution to this problem. At first glance, it seems correct because the motor is not sealed when the JOG_PB contact is pressed. However, when the JOG_PB is released, there is continuity through the middle branch because at that instant MOTOR is **on** and JOG_PB is **off**, and so MOTOR remains **on**.

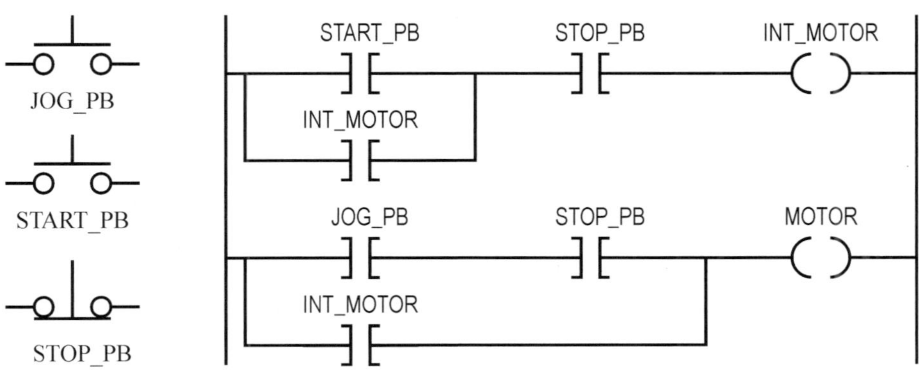

Figure 2.35. Start/stop/jog motor control.

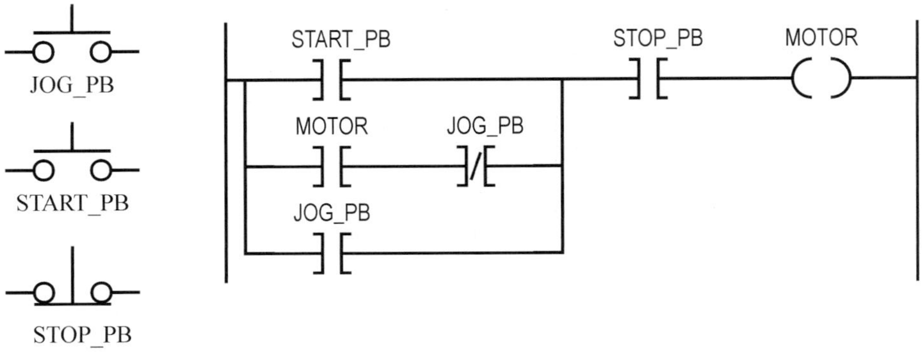

Figure 2.36. Incorrect start/stop/jog motor control.

2.8 TRANSITIONAL CONTACTS AND COILS

Transitional contacts and coils are generally not frequently used in ladder logic, but there are certain situations where their use is essential. Transitional contacts are used to detect the change in a discrete signal whether it is a discrete input (such as a push button switch) or the result of some logic operation (such as a parallel/series connection of contacts). For example, consider a system where an operator pushes an "increase" button (PB_INC) to increase a desired tank level (LT100_DES) on a display and pushes a "decrease" button (PB_DEC) to decrease the desired tank level (Figure 2.37). Whether the display is implemented by two pushbuttons and a 3-digit LED display, or by a touch screen with a graphical representation of the two buttons and the 3-digit number, the ladder logic is the same. One might use the ladder logic as shown in Figure 2.38*a*, where the detailed implementation of the blocks "Increment LT100_DES by 0.1" and "Decrement LT100_DES by 0.1" are not shown. However, this logic is incorrect. Suppose the PB_INC button is depressed. Every time the rung is scanned when PB_INC is pressed (PB_INC is **on**) LT100_DES is incremented by one-tenth. In practice, typical PLC scan times are on the order of a few hundred milliseconds, which means that LT100_DES may increase by more than 0.1, even for very short button presses. If one wants to only change LT100 by 0.1 for each button press, then transitional contacts must be used for PB_INC and PB_DEC, as in Figure 2.38*b*. The ONS transitional contact passes power for only one scan when the button (PB_INC or PB_DEC) is pressed.

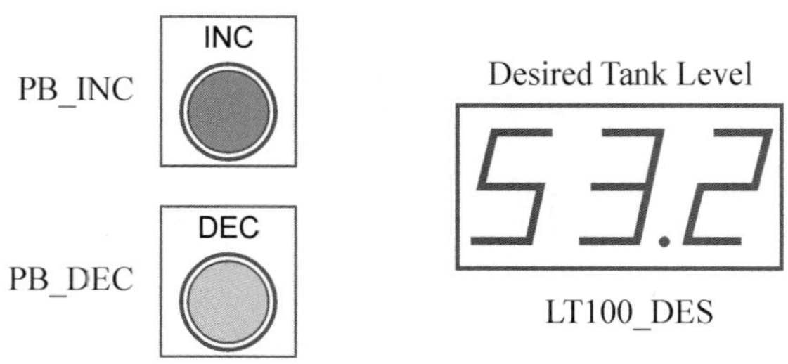

Figure 2.37. Simple operator interface to change desired tank level.

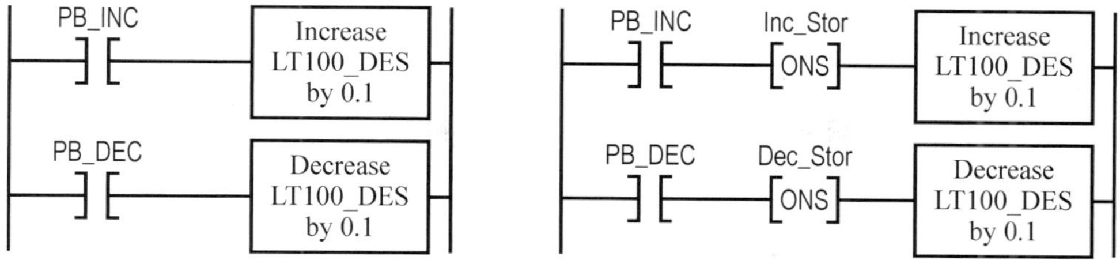

(a) (b)

Figure 2.38. Ladder logic to increase/decrease desired tank level: *(a)* incorrect; *(b)*

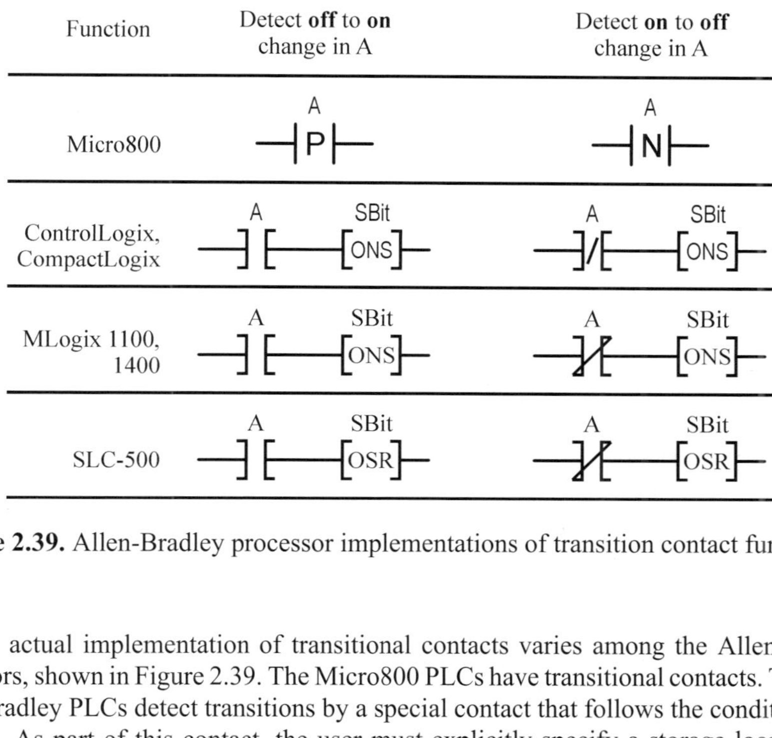

Figure 2.39. Allen-Bradley processor implementations of transition contact functions.

The actual implementation of transitional contacts varies among the Allen-Bradley processors, shown in Figure 2.39. The Micro800 PLCs have transitional contacts. The other Allen-Bradley PLCs detect transitions by a special contact that follows the condition to be detected. As part of this contact, the user must explicitly specify a storage location that holds the last value of the detected condition. Note that tag/symbol above the transitional contact is the location of the Boolean that stores the value of the left side of the transitional contact at the last scan. When the transitional contact is scanned on the rung, it is evaluated in the following manner:

If the left side of the contact is **on**, and
the previous value of the contact (value in the tag above contact) is **off**, then
  The right side of the contact is set to **on**
Otherwise, reset the right side of the contact to **off**.
Store the value of the right side of the contact in the tag/symbol above the contact.

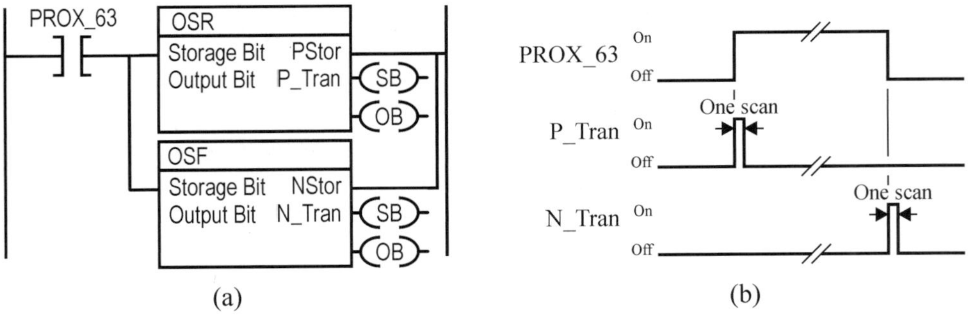

Figure 2.40. Transitional output blocks: *(a)* ladder logic; *(b)* timing diagram.

The result is that the contact passes power for only one scan when the condition before the contact (left of the contact) changes from **off** to **on**.

> **Design Tip**
>
> Transitional contacts are useful for detecting push button presses.

Transitional output blocks provide a way to activate a discrete signal (internal Boolean variable or discrete output) for one scan of the ladder logic when a change is sensed on the input (left) side of the block. An example use of ControlLogix transitional blocks is shown in Figure 2.40*a*. The two output coils change their value in response to changes in the PROX_63 contact as shown in Figure 2.40*b*. Notice that the positive transition-sensing OSR block output P_Tran is **on** for only one scan in response to an **off**-to-**on** change in PROX_63. The negative transition-sensing OSF block output N_Tran is **on** for only one scan in response to an **on**-to-**off** change in PROX_63.

As for the transitional contacts, the actual implementation of transitional output blocks varies among the Alled-Bradley processors. The Micro800 PLCs have transitional coils. For the other processors, the location of a storage bit (SBit tag/symbol) must be specified. The MicroLogix 1400 also has OSR and OSF blocks similar to the ControlLogix and CompactLogix blocks except that the output bit is specified as a bit within a word address. The other MicroLogix and SLC-500 processors do not have a direct equivalent to the transitional output blocks, but their function can be realized with appropriate ladder logic as shown in Figure 2.41.

> **Design Tip**
>
> Be careful when implementing transitional contacts/blocks for internal tags/variables since these internal tags are **on** (activated) for only one scan of the ladder logic. In these cases, one must ensure that the ladder logic is resolved on one scan. If the ladder logic needs more than one scan to resolve the logic, the internal tage may not be **on** long enough for consistent operation. In addition, problems with transitional internal tags are notoriously difficult to troubleshoot.

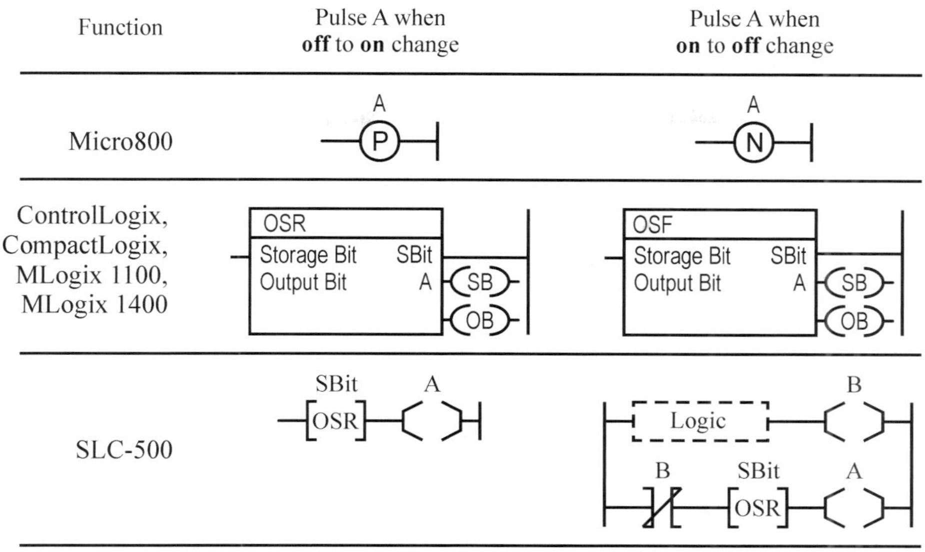

Figure 2.41. Allen-Bradley processor implementations of transitional coil functions.

Scenario. Improper use of one-shot contacts. You are called to solve a problem with a shared sampling system.

Figure 2.42 shows a simplified diagram of a QA (quality assurance) sampling system. The product tanks (Tank #1, Tank #2, Tank #3) are each filled with the product of a chemical reactor and emptied to a storage tank. The sampling system pumps the product in a particular tank out of the tank through a chemical analyzer and then back into the tank. The chemical analyzer checks the product for a key component and reports the results to the operator or supervisory controller. Instead of continuously sampling the product tank, the product is sampled at certain points in the process. Therefore, a sampling system is usually shared among a number of tanks. When a particular tank needs to be sampled, the sampling system control must open and close valves in the proper sequence, as well as control the pump.

In normal operation, a tank is filled with product as it is being produced. When a tank reaches a level of 30%, the reactor control system directs the sampling system to sample product from the tank and report the results. If the sampling system is not in the process of obtaining an analyzer reading for another tank and is not locked to another tank, then the sampling system switches to that tank. Otherwise, the request is queued until the sampling system is released from the other tank. After the analyzer has switched to a tank and reports the analyzer results, the reactor directs the sampling system to release the tank. However, unless another reactor request the sampling system, the sampling system continues to sample the product tank. If the tank level > 95% then the reactor control system directs the sampling system to lock to its product tank. A lock request is handled similarly to a regular request to sample, except that it overrides any regular sampling operation in progress and the sampling system remains connected to the tank until that reactor control system sends a release command to the sampling system. The operator can also manually direct the sampling system to sample a particular tank.

The normal operation of the sampling system works fine. However, when a device fails (for example, a valve fails open/close, or the pump fails to run) with the sampling system

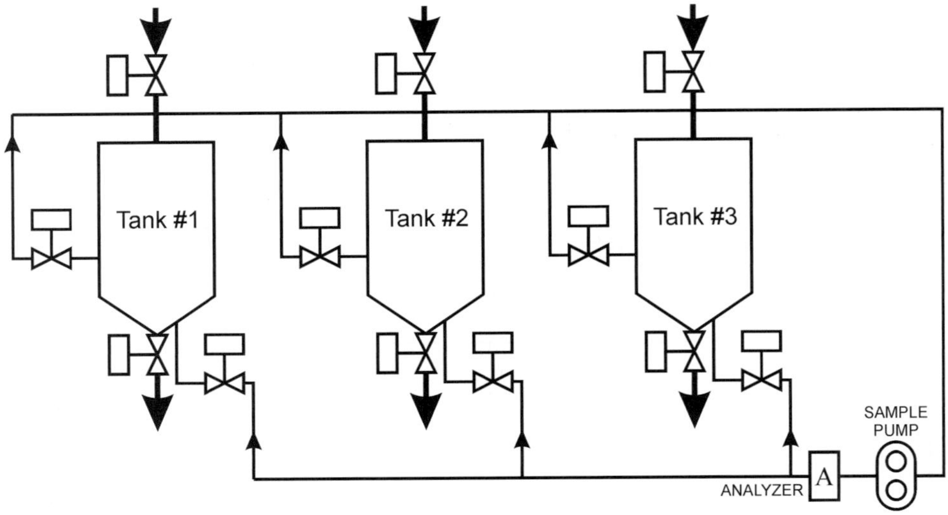

Figure 2.42. QA sampling system.

locked to a tank, the sampling system switches to another tank, but then the fault is cleared and it switches back to the first one, which drops out because of the failure and the cycle repeats.

Separating the reset of the device failures into 3 separate resets, one for each tank, helps, but the problem still occurs for a pump failure. Therefore, attention is focused on the logic to clear failures. After a couple of hours of checking the ladder logic and observing its operation, you discover why it was not recovering gracefully from a pump failure. The command to clear a failure is handled as a one-shot, that is, an internal coil is **on** for only one scan; and this internal coil is used to reset the fault and to release the tank from sampling. You discover that it takes two scans to completely clear a fault condition after the operator manually selects a tank (operator manual selection clears any faults). Therefore, the ladder logic is changed to release the tank after the fault condition has been reset, regardless of how many scans are required to clear the fault. As part of the process, the one-shot contacts are removed. With this program change, the system correctly initiates sampling of a tank after the fault is cleared, rather than switching to another tank and then switching back to the correct tank.

2.9 CHAPTER SUMMARY

This chapter introduced the basics of PLC ladder logic: the contacts, coils, and their assemblage into a ladder logic program. The process of scanning a program, the use of the NC contact, and transitional contacts/coils received special attention. For the most part, this chapter has taken a generic approach to ladder logic and avoided treating the memory addressing details specific to each PLC model. This topic cannot be ignored any further and is the subject of the next chapter, before timer and counter instructions are described.

REFERENCES

Erickson, Kelvin T., 2023. *Programmable Logic Controllers: An Emphasis on Design and Applications*, 4th Ed., Dogwood Valley Press, Rolla, MO.

IEC, 2013. *IEC 1131-3: Programmable Logic Controllers - Part 3: Programming Languages*, Edition 3.0, International Electrotechnical Commission, Geneva, Switzerland.

Rockwell Automation, 2008. *SLC Instruction Set Reference Manual*, pub. 1747-RM001G-EN-P, Rockwell Automation Inc., Milwaukee, WI, Nov.

Rockwell Automation, 2014. *MicroLogix™ 1200 and MicroLogix 1500 Programmable Controllers Instruction Set Reference Manual*, pub. 1762-RM001H-EN-P, Rockwell Automation Inc., Milwaukee, WI, July.

Rockwell Automation, 2019. *MicroLogix 1400 Programmable Controllers Instruction Set Reference Manual*, pub. 1766-RM001I-EN-P, Rockwell Automation Inc., Milwaukee, WI, June.

Rockwell Automation, 2021. *Micro800 Programmable Controllers General Instructions*, pub. 2080-RM001J-EN-E, Rockwell Automation Inc., Milwaukee, WI, Mar.

Rockwell Automation, 2022. *Logix5000 Controllers General Instructions Reference Manual*, pub. 1756-RM003W-EN-P, Rockwell Automation Inc., Milwaukee, WI, Mar.

PROBLEMS

P2-1. Implement the digital logic gate circuit in Figure P2.1 in PLC ladder logic. EX221 and EX222 are the only physical outputs. Do not use any internal coils.

In words, EX221 is **on** when LS234 is **off** and either:

LS233 is **on**

Or

LS232 is **off** and (EX221 is **on** or LS231 is **off**)

EX222 is **on** when EX221 is **off** and LS241 is **on**

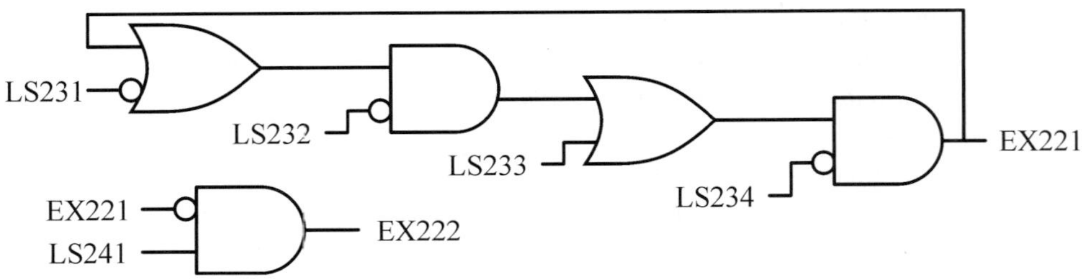

Figure P2.1. Digital logic for problem P2-1.

P2-2. Implement the digital logic gate circuit in Figure P2.2 in PLC ladder logic. CNV2105 is the only physical output. Do not use any internal coils.

In words, CNV2105 is **on** when PB2102 is **on** and AS2104 is **on** and ALM2100 is **off** and either:

SS2105 is **on** and CNV2105 is **on**

Or

PB2101 is **on** and TS2103 is **on** and (LS2110 is **off** or LS2111 is **on**)

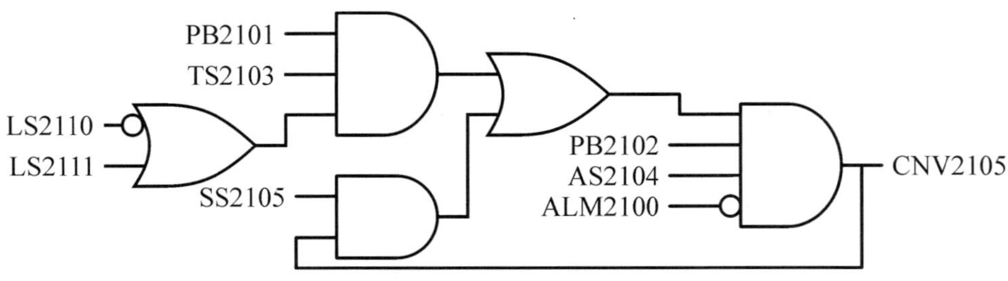

Figure P2.2. Digital logic for problem P2-2.

P2-3. Implement the digital logic gate circuit in Figure P2.3 in PLC ladder logic. XV201 is the only physical output. Do not use any internal coils.

In words, XV201 is **on** when OL101 is **on** and (FS201 is **off** or XV201is **on)** and:

LS203 is **on** and LS232 is **off** and either

PS341 is **on** or PS342 is **off**

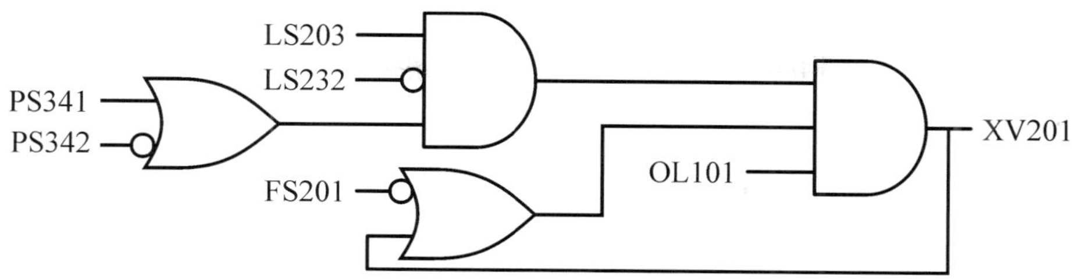

Figure P2.3. Digital logic for problem P2-3.

P2-4. Implement the digital logic gate circuit in Figure P2.4 in PLC ladder logic. LA321 and HN325 are the only physical outputs. Do not use any internal coils.

In words, LA321 is **on** when PS315 is **on** or PB327 is **on** or LS328 is **off**.
HN325 is **on** when LA321 is **off** and SW352 is **on** and PS354 is **off**

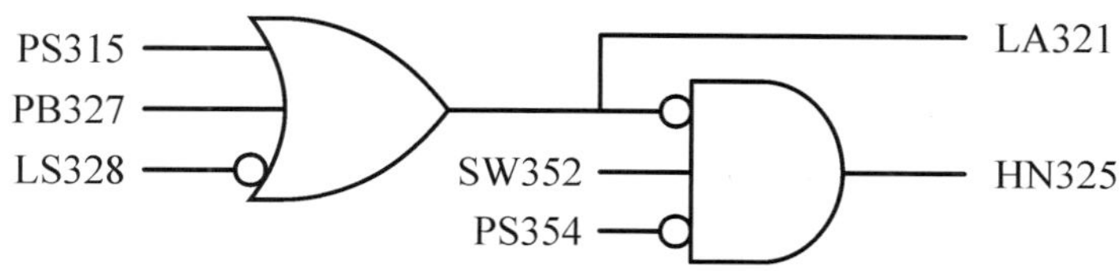

Figure P2.4. Digital logic for problem P2-4.

P2-5. Implement the digital logic gate circuit in Figure P2.5 in PLC ladder logic. CNV466 is the only physical output. Do not use any internal coils.

In words, CNV466 is **on** when either:
PB465 is **on**
Or
LS464 is **on** and either:
LS462 is **off**
Or
LS463 is **off**
Or
SS461 is **on** and CNV466 is **on**

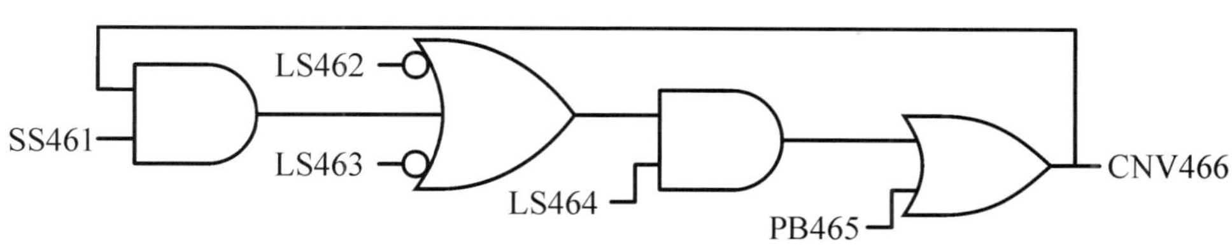

Figure P2.5. Digital logic for problem P2-5.

P2-6. Implement the digital logic diagram in Figure P2.6 in PLC ladder logic. XV425 is the only physical output. Do not use any internal coils.

In words, XV425 is **on** when LS442 is **on** or LS445 is **off** or:
LS432 is **off** and LS413 is **on** and LS421 is **on** and XV425 is **on** and LS424 is **off**

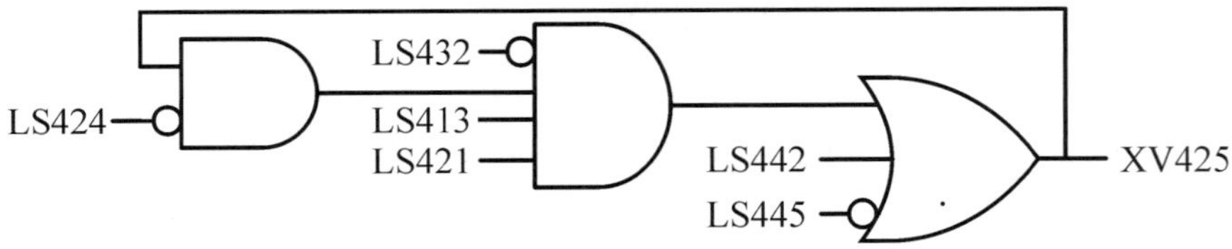

Figure P2.6. Digital logic for problem P2-6.

P2-7. Implement the digital logic gate circuit in Figure P2.7 in PLC ladder logic. MTR1401 is the only physical output. Do not use any internal coils.

In words, MTR1401 is **on** when LS1407 is **off** and <u>either</u>:
SS1401 is **on** and MTR1401 is **on**
Or
LS1408 is **on** and LS1403 is **off** and (SW1405 is **on** or PS1404 is **off**)

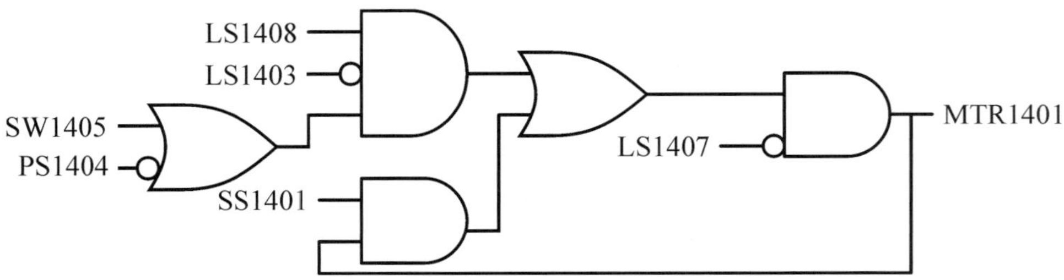

Figure P2.7. Digital logic for problem P2-7.

P2-8. Implement the digital logic gate circuit in Figure P2.8 in PLC ladder logic. EX631 and EX632 are the only physical outputs. Do not use any internal coils.

In words, EX632 is **on** when LS624 is **on** and either:
PS233 is **off**
Or
HS623 is **on** and EX632 is **on**.
EX631 is **on** when PS642 is **off** and EX632 is **on**

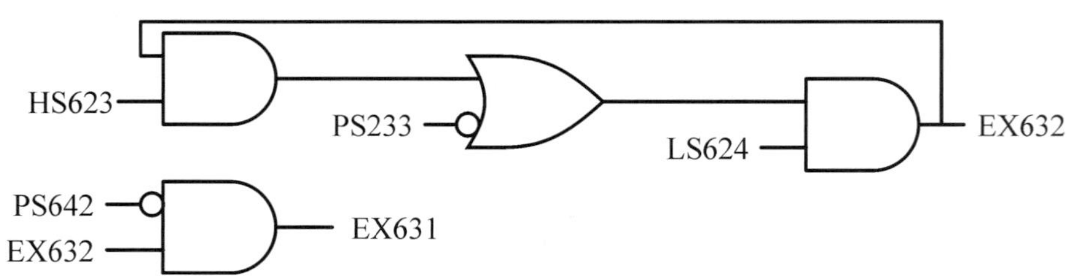

Figure P2.8. Digital logic for problem P2-8.

P2-9. Implement the digital logic gate circuit in Figure P2.9 in PLC ladder logic. XV1311 is the only physical output. Do not use any internal coils.

In words, XV1311 is **on** when LS1305 is **off** and (LS1307 is **on** or XV1311 is **on)** and either:
PS1302 is **on** and PB1310 is **off**
Or
LS1304 is **on**
Or
LS1303 is **off**

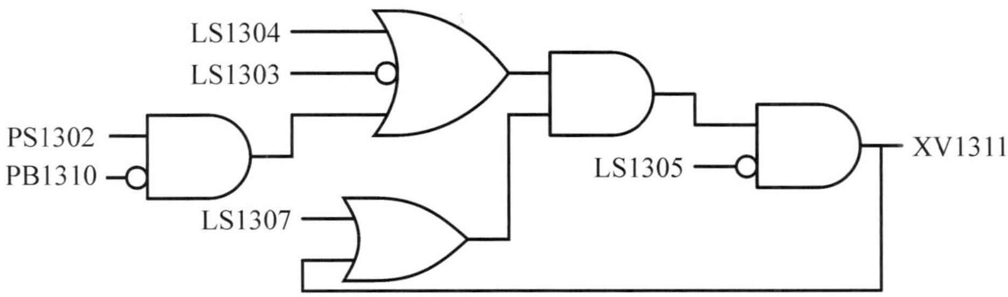

Figure P2.9. Digital logic for problem P2-9.

P2-10. Implement the digital logic gate circuit in Figure P2.10 in PLC ladder logic. XV270 is the only physical output. Do not use any internal coils.

In words, XV270 is **on** when LS271 is **on** or PS274 is **on**
OR
LS272 is **off** and ALM273 is **off** and (LS275 is **off** or XV270 is **on**)

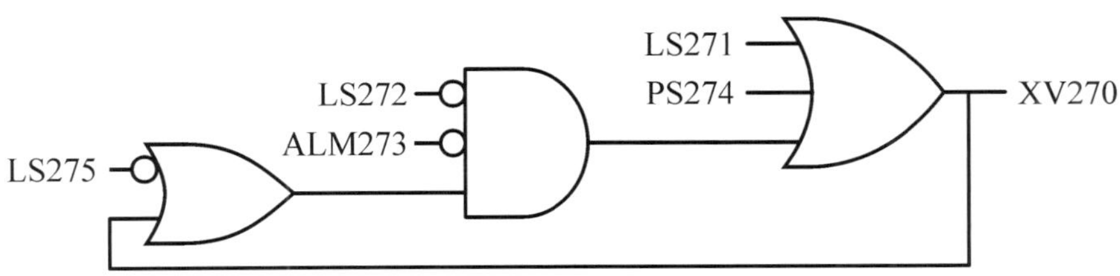

Figure P2.10. Digital logic for problem P2-10.

P2-11. Implement the digital logic gate circuit in Figure P2.11 in PLC ladder logic. P1001 is the only physical output. Do not use any internal coils.

In words, P1001 is **on** when OL1001 is **off** and (TD1003 is **off** or FS1002 is **on)** and either:

P1001 is **on**

Or

LS2510 is **on**

Or

PS1612 is **on** and PS1613 is **off**

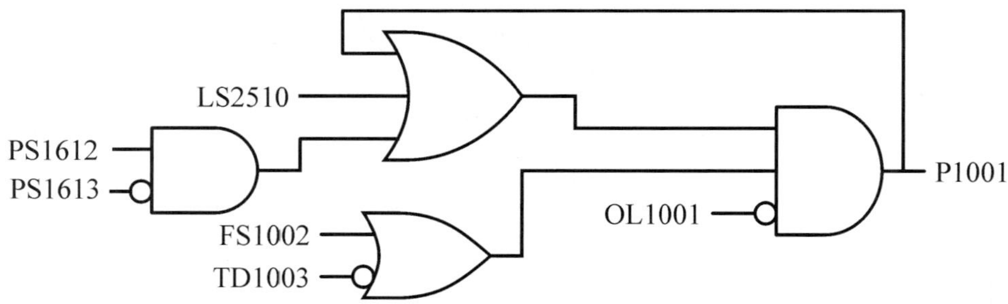

Figure P2.11. Digital logic for problem P2-11.

P2-12. Implement the digital logic gate circuit in Figure P2.12 in PLC ladder logic. LA480 and EX490 are the only physical outputs. Do not use any internal coils.

In words, LA480 is **on** when LS481 is **off** and PB482 is **on** and PS483 is **on**. EX490 is **on** when LA480 is **on** or LS491 is **off** or LS493 is **off**.

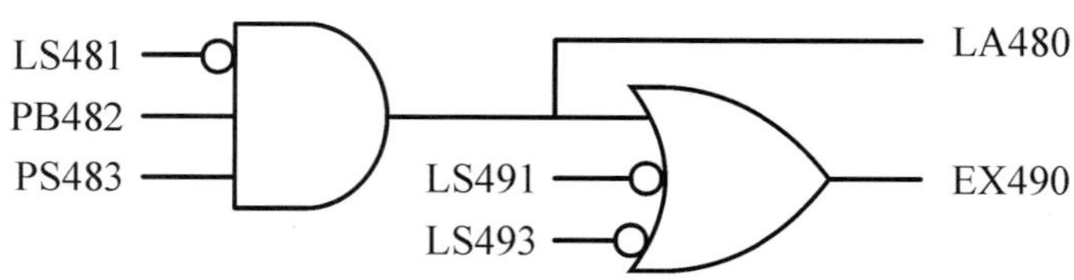

Figure P2.12. Digital logic for problem P2-12.

P2-13. Use ladder logic to implement a simple *low flow indicator and alarm* for a pipe.

A panel implements a simple interface showing the pipe flow. The panel has three lamps (green, yellow, and red), an alarm horn, and an alarm acknowledge button that turns off the alarm horn. No more than one light is **on** at any time:

Green lamp **on** when flow >= 50 gpm
Yellow lamp **on** when 40 gpm <= flow < 50 gpm
Red lamp **on** when flow <= 40 gpm

In addition, when the flow is less than 30 gpm, the alarm horn is turned **on**, and remains **on** until the flow is greater than or equal to 30 gpm and the alarm acknowledge (ALM_ACK) button is pressed. If the flow is less than 30 gpm, pressing the alarm acknowledge button should not silence the alarm horn, even momentarily.

There are 3 flow sensors, each corresponding to one of the "boundary" flows:

FS_1 is **on** when the pipe flow >= 30 gpm
FS_2 is **on** when the pipe flow >= 40 gpm
FS_3 is **on** when the pipe flow >= 50 gpm

Assume the following physical inputs and outputs. Only tags are used to avoid any PLC-specific addressing. **DO NOT assign any more inputs!!**

Physical Inputs:

Tag	Description
ALM_ACK	Alarm acknowledge pushbutton switch, N. O., **on** when acknowledging (resetting) alarm horn
FS_1	**On** when the tank flow >= 30 gpm
FS_2	**On** when the tank flow >= 40 gpm
FS_3	**On** when the tank flow >= 50 gpm

Physical Outputs:

Tag	Description
GRN_LA	**On** to light green indicator lamp
YEL_LA	**On** to light yellow indicator lamp
RED_LA	**On** to light red indicator lamp
ALARM	**On** to sound alarm horn

P2-14. Use ladder logic to implement a simple *high flow indicator and alarm* for a pipe.

A panel implements a simple interface showing the pipe flow. The panel has three lamps (blue, yellow, and red), an alarm horn, and an alarm acknowledge button that turns off the alarm horn. No more than one light is **on** at any time:

Blue lamp **on** when flow < 30 gpm
Yellow lamp **on** when 80 gpm <= flow < 100 gpm
Red lamp **on** when flow >= 100 gpm

In addition, when the flow is greater than or equal to 100 gpm, the alarm horn is turned **on**, and remains **on** until the flow becomes less than 100 gpm and the alarm acknowledge (ALM_ACK) button is pressed. If the flow is greater than or equal to 100 gpm, pressing the alarm acknowledge button should not silence the alarm horn, even momentarily.

There are 3 flow sensors, each corresponding to a "boundary" flow:

FS_1 is **on** when the pipe flow >= 30 gpm
FS_2 is **on** when the pipe flow >= 80 gpm
FS_3 is **on** when the pipe flow >= 100 gpm

Assume the following physical inputs and outputs. Only tags are used to avoid any PLC-specific addressing. **DO NOT assign any more inputs!!**

Physical Inputs:

Tag	Description
ALM_ACK	Alarm acknowledge pushbutton switch, N. O., **on** when acknowledging (resetting) alarm horn
FS_1	**On** when the flow >= 30 gpm
FS_2	**On** when the flow >= 80 gpm
FS_3	**On** when the flow >= 100 gpm

Physical Outputs:

Tag	Description
BLU_LA	**On** to light blue indicator lamp
YEL_LA	**On** to light yellow indicator lamp
RED_LA	**On** to light red indicator lamp
ALARM	**On** to sound alarm horn

P2-15. Use ladder logic to implement a simple *low temperature indicator and alarm* for a tank.

A panel implements a simple interface showing the tank temperature. The panel has three lamps (green, yellow, and red), an alarm horn, and an alarm acknowledge button that turns off the alarm horn. No more than one light is **on** at any time:

Green lamp **on** when 10 degrees C <= temperature < 50 degrees C
Yellow lamp **on** when 2 degrees C <= temperature < 10 degrees C
Red lamp **on** when temperature < 2 degrees C

In addition, when the temperature is less than 2 degrees C, the alarm horn is turned **on**, and remains **on** until the temperature becomes greater than or equal to 2 degrees C and the alarm acknowledge (ALM_ACK) button is pressed. If the level is less than 2 degrees C, pressing the alarm acknowledge button should not silence the alarm horn, even momentarily.

There are 3 temperature sensors, each corresponding to one of the "boundary" temperatures:

TS_1 is **on** when the tank temperature ≥ 2 degrees C
TS_2 is **on** when the tank temperature ≥ 10 degrees C
TS_3 is **on** when the tank temperature ≥ 50 degrees C

Assume the following physical inputs and outputs. Only symbols are used to avoid any PLC-specific addressing. **DO NOT assign any more inputs!!**

Physical Inputs:

Tag	Description
ALM_ACK	Alarm acknowledge pushbutton switch, N. O., **on** when acknowledging (resetting) alarm horn
TS_1	**On** when the tank temperature ≥ 2 degrees C
TS_2	**On** when the tank temperature ≥ 10 degrees C
TS_3	**On** when the tank temperature ≥ 50 degrees C

Physical Outputs:

Tag	Description
GRN_LA	**On** to light green indicator lamp
YEL_LA	**On** to light yellow indicator lamp
RED_LA	**On** to light red indicator lamp
ALARM	**On** to sound alarm horn

P2-16. Use ladder logic to implement a simple *high temperature indicator and alarm* for a tank.

A panel implements a simple interface showing the tank temperature. The panel has three lamps (blue, yellow, and red), an alarm horn, and an alarm acknowledge button that turns off the alarm horn. No more than one light is **on** at any time:

Blue lamp **on** when temperature < 70 degrees C

Yellow lamp **on** when 70 degrees C <= temperature < 100 degrees C

Red lamp **on** when temperature >= 100 degrees C

In addition, when the temperature is greater than or equal to 120 degrees C, the alarm horn is turned **on**, and remains **on** until the temperature becomes less than 120 degrees C and the alarm acknowledge (ALM_ACK) button is pressed. If the temperature is greater than or equal to 120 degrees C, pressing the alarm acknowledge button should not silence the alarm horn, even momentarily.

There are 3 temperature sensors, each corresponding to one of the "boundary" temperatures:

TS_1 is **on** when the temperature >= 70 degrees C

TS_2 is **on** when the temperature >= 100 degrees C

TS_3 is **on** when the temperature >= 120 degrees C

Assume the following physical inputs and outputs. Only tags are used to avoid any PLC-specific addressing. **DO NOT assign any more inputs!!**

Physical Inputs:

Tag	Description
ALM_ACK	Alarm acknowledge pushbutton switch, N. O., **on** when acknowledging (resetting) alarm horn
TS_1	**On** when the temperature >= 70 degrees C

TS_2	**On** when the temperature >= 100 degrees C
TS_3	**On** when the temperature >= 120 degrees C

Physical Outputs:

Tag	Description
BLU_LA	**On** to light blue indicator lamp
YEL_LA	**On** to light yellow indicator lamp
RED_LA	**On** to light red indicator lamp
ALARM	**On** to sound alarm horn

P2-17. Use ladder logic to implement a simple *high weight indicator and alarm* for a crane.

A panel implements a simple interface showing the weight on a crane. The panel has three lamps (blue, yellow, and red), an alarm horn, and an alarm acknowledge button that turns off the alarm horn. No more than one light is **on** at any time:

Blue lamp **on** when weight < 2000 kg

Yellow lamp **on** when 9000 kg <= weight < 10000 kg

Red lamp **on** when weight >= 10000 kg

In addition, when the weight is greater than or equal to 10000 kg, the alarm horn is turned **on**, and remains **on** until the weight becomes less than 10000 kg and the alarm acknowledge (ALM_ACK) button is pressed. If the weight is greater than or equal to 10000 kg, pressing the alarm acknowledge button should not silence the alarm horn, even momentarily.

There are 3 weight sensors, each corresponding to one of the "boundary" weights:

WT_1 is **on** when the weight >= 2000 kg

WT_2 is **on** when the weight >= 9000 kg

WT_3 is **on** when the weight >= 10000 kg

Assume the following physical inputs and outputs. Only tags are used to avoid any PLC-specific addressing. **DO NOT assign any more inputs!!**

Physical Inputs:

Tag	Description
ALM_ACK	Alarm acknowledge pushbutton switch, N. O., **on** when acknowledging (resetting) alarm horn
WT_1	**On** when the weight >= 2000 kg
WT_2	**On** when the weight >= 9000 kg
WT_3	**On** when the weight >= 10000 kg

Physical Outputs:

Tag	Description
BLU_LA	**On** to light blue indicator lamp
YEL_LA	**On** to light yellow indicator lamp
RED_LA	**On** to light red indicator lamp
ALARM	**On** to sound alarm horn

P2-18. Use ladder logic to implement a simple *voltage indicator and alarm*.

A panel implements a simple interface showing the status of the voltage bus feeding the motors in the system. The panel has three lamps: yellow, green, and red. The panel also has an alarm horn and an alarm acknowledge button that turns **off** the alarm horn. The green lamp indication:

Green lamp **on** when 430 volts < voltage < 530 volts (normal range)

The yellow and red lamps and the the ALARM output is handled in the following manner. ALARM and one of the yellow/red lamps is turned **on** for following conditions:

1. When voltage <= 430 volts, yellow lamp and ALARM turned **on**
2. The voltage >= 530 volts, red lamp and ALARM turned **on**

ALARM is turned **off** for the following condition:

The operator acknowledges the alarm by pushing ALM_ACK

Even if the voltage moves into the normal range while ALARM is **on**, the ALARM output remains **on** until acknowledged by the operator. If the voltage is outside of the normal range when ALM_ACK is pushed, ALARM must remain **off** when the button is released.

The yellow lamp is turned **off** for the following condition:

The operator acknowledges the alarm by pushing ALM_ACK and the voltage is > 430 volts. Note that if the voltage <= 430 volts when the ALM_ACK is pressed, the lamp remains **on**.

The red lamp is turned **off** for the following condition:

The operator acknowledges the alarm by pushing ALM_ACK and the voltage is < 530 volts. Note that if the voltage >= 530 volts when the ALM_ACK is pressed, the lamp remains **on**.

There are two voltage sensors, each corresponding to one of the "boundary" voltages:

LO_480 is **on** when the voltage <= 430 volts

HI_480 is **on** when the voltage >= 530 volts

Do not use latch/unlatch or set/reset coils for your solution.

Hint: this solution involves transitional contacts.

Assume the following physical inputs and outputs.

Physical Inputs:

Tag	Description
ALM_ACK	Alarm acknowledge pushbutton switch, N. O., **on** when acknowledging (resetting) alarm horn
LO_480	**On** when the voltage <= 430 volts
HI_480	**On** when the voltage >= 530 volts

Physical Outputs:

Tag	Description
GRN_LA	**On** to light green indicator lamp
YEL_LA	**On** to light yellow indicator lamp
RED_LA	**On** to light red indicator lamp
ALARM	**On** to sound alarm horn

P2-19. Use ladder logic to detect the direction of a conveyor belt.

The figure below shows the top view of a conveyor belt. Two retroreflective photoelectric sensors, PE1 and PE2, monitor objects on the conveyor. Use PE1 and PE2 to determine the direction of the conveyor. If an object is moving left, the LEFT internal coil is turned **on** (and the RIGHT internal coil is turned **off**). If an object is moving right, RIGHT is turned **on** (and LEFT is turned **off**). Assume the objects are placed far enough apart that both PE1 and PE2 are **off** between objects. You may also assume the conveyor will not reverse direction when either PE1 or PE2 are **on**. Your logic must control LEFT and RIGHT.

Once LEFT is turned **on**, it remains **on** until a part is sensed moving right. Once RIGHT is turned **on**, it remains **on** until a part is sensed moving left. You may assume the processor starts with both LEFT and RIGHT **off**.

Hint: this solution involves transitional contacts.

Assume the following physical inputs and outputs. Only tags are used to avoid any PLC-specific addressing.

Physical Inputs:

Tag	Description
PE1	Left sensor. **On** when object sensed
PE2	Right sensor. **On** when object sensed

Internal Coils:

Tag	Description
LEFT	**On** if objects moving in the left direction
RIGHT	**On** if objects moving in the right direction

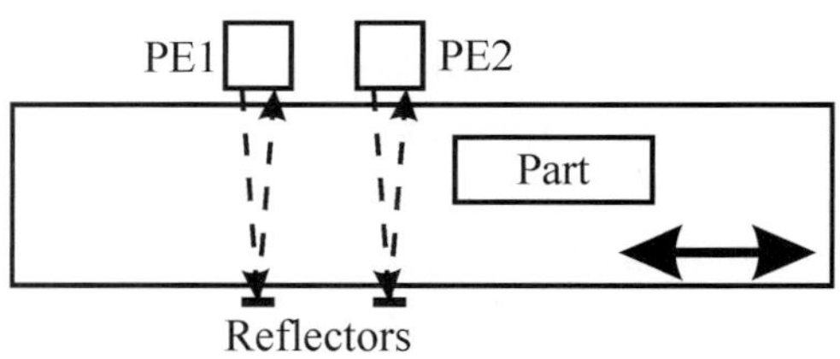

Figure P2.19. Photoelectric sensors for problem P2-19.

P2-20. Use ladder logic to implement to indicate how many of three pumps are currently selected to run.

A control system has three pumps, P-11, P-12, and P-13, to control the water flow. Depending on how many pumps are selected by the operator to run (zero, one, two, or three) a different set of PID tuning parameters must be selected. This ladder logic only determines if zero, one, two, or three pumps are selected to run and turns on the appropriate internal coil. The only inputs used by your ladder are the three pump selects, P11_SEL, P12_SEL, and P13_SEL, which have been set/cleared by the operator from an HMI (human-machine interface). Your logic needs to determine the proper value (**on** or **off**) of the following internal coils:

PUMP_0_SEL **on** when **no** pumps selected to run
PUMP_1_SEL **on** when **only one** pump selected to run
PUMP_2_SEL **on** when **two** pumps selected to run
PUMP_3_SEL **on** when **all three** pumps selected to run

Assume the following inputs and outputs. Even though P11_SEL, P12_SEL, and P13_SEL are internal PLC memory, there **must not** be any coils referring to these tags. Only the operator can change these tags. Only tags are used to avoid any PLC-specific addressing. **DO NOT assign any more inputs!!**

Inputs:

Tag	Description
P11_SEL	**On** when operator has selected P-11 to run. Otherwise, it is **off**.
P12_SEL	**On** when operator has selected P-12 to run. Otherwise, it is **off**.
P13_SEL	**On** when operator has selected P-13 to run. Otherwise, it is **off**.

Outputs:

Symbol (Tag)	Description
PUMP_0_SEL	**On** when no pumps selected to run
PUMP_1_SEL	**On** when any one pump selected to run
PUMP_2_SEL	**On** when any two pumps selected to run
PUMP_3_SEL	**On** when all three pumps selected to run

P2-21. Develop the ladder logic to control a simple plastic pellet transfer system shown in Figure P2.21 to transfer plastic pellets into a molding machine from a main bin. The control panel has 2 push buttons and 2 lamps. The START_PB is pushed to turn on the system, enabling automatic control. The STOP_PB is pushed to turn off the system, disabling automatic control. When T_BLWR is running, B_LAMP is **on**. When T_VALVE is open, V_LAMP is **on**.

When the system is on (automatic control enabled), and the machine bin low-level indicator, LO_LVL, turns **on**, the slide valve, T_VALVE, is opened and the blower motor, T_BLWR, is started to blow pellets from the main bin to the machine bin. When the machine bin high-level indicator, HI_LVL, turns **on**, the blower motor should stop and the slide valve closed.

When pressed, the STOP_PB should also stop the blower motor if it is running and close the valve if it is open.

Assume the following physical inputs and outputs. Only tags are used to avoid any PLC-specific addressing. **DO NOT assign any more physical inputs!!**

Physical Inputs:

Tag	Description
START_PB	Start pushbutton switch, N. O., **on** to enable transfer system
STOP_PB	Stop pushbutton switch, N. C., **off** to disable transfer system
LO_LVL	**On** when the machine bin level is low
HI_LVL	**On** when the machine bin level is high

Physical Outputs:

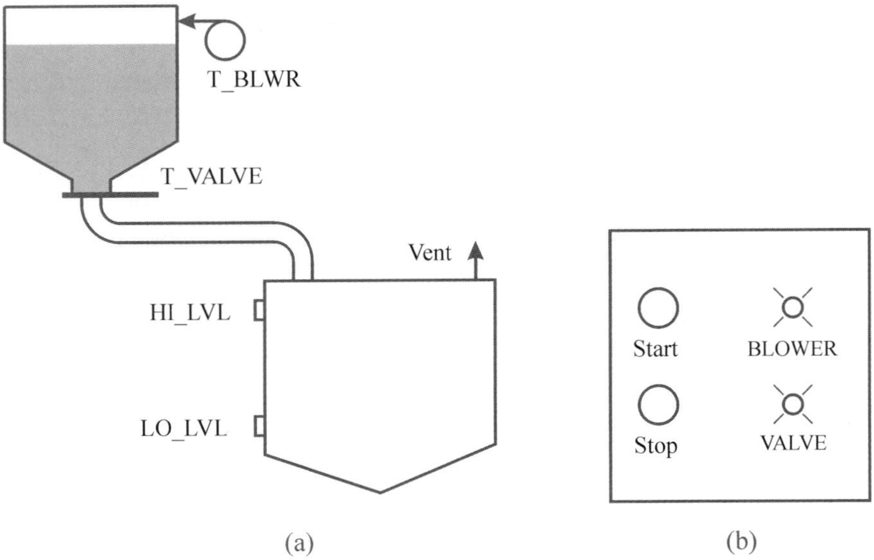

Figure P2.21. Simple plastic pellet transfer system: *(a)* system; *(b)* control panel.

Tag	Description
T_BLWR	**On** to run transfer blower motor. **Off** stops motor.
T_VALVE	**On** to open transfer slide valve. **Off** closes valve
B_LAMP	**On** to light blower indicator lamp. Should be lit when blower motor is running.
V_LAMP	**On** to light slide valve indicator lamp. Should be lit when slide valve is open.

3 Memory Organization and Addressing

Chapter Topics:

- PLC memory organization
- PLC memory addressing

OBJECTIVES

Upon completion of this chapter, you will be able to understand how the memory is organized and addressed for the following Allen-Bradley PLCs:

- ControlLogix and CompactLogix
- Micro800
- MicroLogix and SLC-500

Scenario: Every control engineer's nightmare.

The phone call came at 2 a.m. The shift supervisor is on the line. "We're having trouble with the A2438 agitator. It keeps faulting out. The maintenance electrician has traced it to a faulty vibration sensor. It is occasionally turning on, even when there isn't any noticeable vibration, causing the batch procedure to go to hold. This problem started occurring this afternoon, but now it seems to happen every 10 minutes. We do not have a spare sensor in stock. The earliest we can get one is in two days." You then remember the particular batch system was recently commissioned and the contractor left a week ago. The operator breaks into your thoughts. "Is there anything you can do in the PLC? I don't have a spare operator who can restart the batch procedure every 10 minutes." In your half-awake state you cannot recall exactly what the supervisor says next. Something about losing $10,000 if the batch procedure does not run to completion in 3 hours. You mumble something in reply and then hang up the phone to hurriedly dress and drive to the plant.

While driving to the plant, you think of some possible temporary solutions. You could force the vibration sensor discrete input off, but then if a vibration is truly occurring equipment could be damaged, forcing costly downtime and repairs. The best solution would be to add a 2-second time delay to the sensor signal. The vibration sensor signal would have to remain on for 2 seconds before it would cause a fault and put the batch procedure in hold. "At least it's a SLC-500," you think to yourself. "I can make online changes to the ladder logic while the PLC is in run mode." When you arrive at the control room, you discuss these solutions with the shift supervisor and he okays the additional delay on the sensor. "The contractor should have added this delay for the sensor anyway," you add. The supervisor agrees.

With the engineering workstation you connect to the PLC and find the discrete input address. You check the cross-reference and find that the discrete input is used in only two places in the program. "This will be easy," you think to yourself. "I need to find a spare timer, replace the references to the sensor with the timer done bit, drive the timer with the sensor, and I can go home."

When you check the data files, the only file of timers is the default, T4. You check the usage table and discover that every timer is being used. There are no spare timers. You must put the PLC in program mode in order to expand the timer data file. However, the plant rules dictate that you obtain the shift supervisor's permission since operations could be disrupted when the PLC is put in program mode. When you ask the supervisor for permission to take the PLC to program mode for 1 minute in order to add a few timers, he asks, "Will it turn anything off?" "I don't know," you reply. "It depends on the jumper position on the discrete output modules." The supervisor responds, "Then you may not. The PLC also controls your steam generator unit. If the PLC causes the steam generator to shut down, it takes a minimum of an hour to get it back on-line and so your batch will certainly be ruined." You sigh, tell him that you will force the vibration sensor off and request a radio so that you can inform the operator if the agitator starts vibrating. You get a radio and a chair and post yourself by the agitator, ready to radio the control room if a vibration occurs. It is a long night.

Solution: The programmer should have left spare timers in the data files to handle these situations. As a rule, if the processor does not allow run-mode tag creation, one should ALWAYS put spare timers, counters, internal coils, and so on, to allow code corrections without requiring the PLC to be taken out of run mode. In this situation, there was another mistake made by the programmer who originally programmed the PLC. Every sensor that can cause a fault should be delayed in the PLC. In other words, the sensor signal should persist for a small time (e.g., one second) before the PLC turns on an alarm and causes some other action. Frequently, fault sensors are plagued with intermittent false readings. The addition of the timer will help prevent false alarms.

3.1 INTRODUCTION

This chapter summarizes the organization and addressing of the memory for the PLCs covered by this text. Each of the sections is written as stand-alone. That is, only the section(s) pertaining to the PLC(s) of interest need to be studied.

3.2 CONTROLLOGIX AND COMPACTLOGIX MEMORY

3.2.1 Memory Organization

The Allen-Bradley ControlLogix and CompactLogix PLC memory model is presented in Figure 3.1. Each of the main elements is now described.

The **project** is the entire body of software (program and data) that corresponds to one PLC.

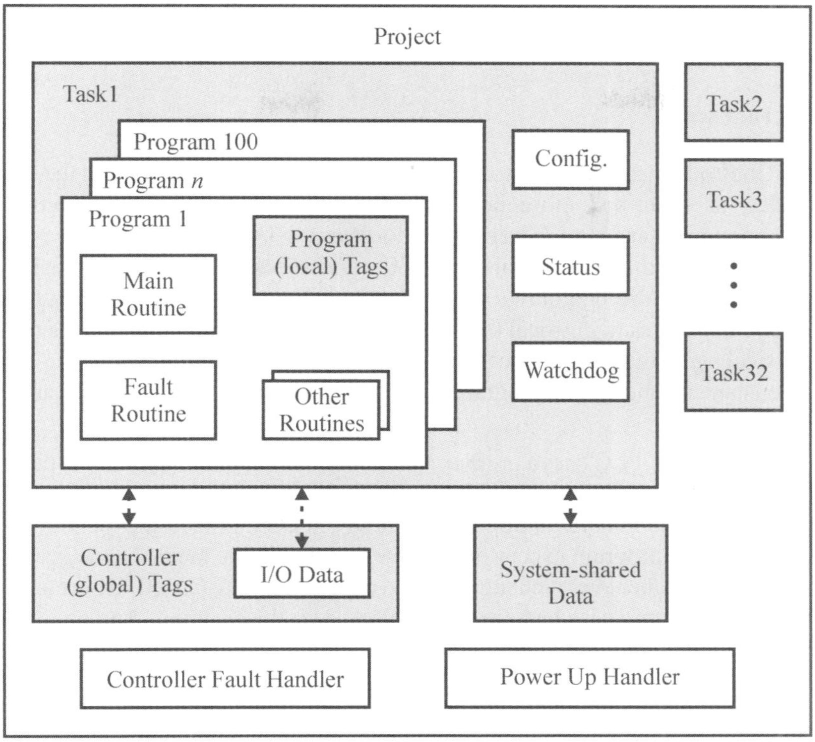

Figure 3.1. ControlLogix/CompactLogix memory model.

A **task** is configured to control the execution of one or more programs, similar to IEC 61131-3. There is a maximum of 32 tasks. No more than one task can be continuously executed. The other tasks are periodic or event-driven. A periodic task is executed at a fixed time interval. An event task is triggered by an event such as a change in a discrete input channel, new analog data, or certain motion operations. Priority may be assigned to the periodic and event tasks and so a task may be interrupted by a higher-priority task. A task can contain up to 100 programs, executed in the specified order. Each task also contains status information and a watchdog timer to detect if the task does not finish execution within a specified interval. ControlLogix processors prior to the L6*x*–series are limited to 32 programs per task and do not have event tasks. Earlier CompactLogix controllers were limited to a maximum number of 3 to 16 tasks with up to 16 programs per task.

A **program** consists of **routines**. The main routine, usually named "MainRoutine" is always executed first and must call the other routines (with a Jump to Subroutine block). The "FaultRoutine" is executed if a program fault is encountered when executing any of the routines in the program. The main routine, the fault routine, and the other routines may be written in either ladder logic, function block diagram, sequential function chart, or structured text. Each routine is coded in a single programming language. Each program has an area of local memory called **program tags**. Only the program for which they are defined can access program tags. Every routine in the program can access the program tags.

The **controller tags** are the global controller memory. The I/O data is considered part of the controller tags.

3.2.2 Program scan

In the ControlLogix PLC, the program execution is not coordinated with the scan of the I/O modules, as described in section 2.5. The programs in the continuous task are continuously scanned and data is transferred to/from the I/O modules any time during the logic scan. In order to duplicate the historic PLC scan process, the input data must be copied to an *input buffer* at the beginning of the first program and used throughout the scan. Additionally, any writes to physical outputs are made to an *output buffer* that is transferred to the physical outputs at the end of the scan. These two changes ensure the input data remains constant throughout a program scan and that the physical outputs are updated nearly simultaneously.

The ControlLogix PLC uses a producer/consumer model of I/O updates. Input modules produce data. Output modules, controllers, and intelligent modules produce and consume data. In this model, the transfer of information between I/O modules and the processor(s) is not coordinated with program execution. A discrete input in the local chassis (same chassis as processor) is sent whenever it changes, called change-of-state (COS). Discrete outputs in the local chassis are set/cleared whenever set/cleared by the program. Analog input/output modules in the local chassis send/receive their information at a fixed rate, called the real time sample (RTS) period. For I/O modules in a remote chassis, I/O data is sent/received at the requested packet interval (RPI).

For a 1769 CompactLogix processor, the I/O is scan is controlled by a priority 6 periodic task that executes at the RPI of the CompactBus, the connection to the I/O modules.

3.2.3 Tags and data types

In the ControlLogix PLC, the symbolic name of a tag can be any string of letters, digits, and underlines up to 40 characters long provided that:

1. The first character is not a digit.
2. There are not two or more consecutive underlines.
3. The last character is not an underline.

There is no distinction between upper and lower case characters. If a particular tag is defined as a program tag and a controller tag, the program uses the program tag, and not the controller tag. In order to avoid potential problems, there should not be a program tag with the same name as any controller tag.

When a tag is declared, it must be assigned a data type. The **elementary data types** are as follows:

Name	Description	Range of values
BOOL	1-bit Boolean	0=cleared, 1=set
SINT	1-byte integer	-128 to 127
INT	2-byte integer	-32,768 to 32,767
DINT	4-byte integer	-2^{31} to $2^{31}-1$ (-2,147,483,648 to 2,147,483,647)
LINT	8-byte integer	-2^{63} to $2^{63}-1$

REAL	4-byte float. pt.	$\pm 8.43\times10^{-37}$ to $\pm 3.36\times10^{+38}$

The LINT data type has limited use on the ControlLogix 5570 and CompactLogix 5370 processors and on the ControlLogix 5580 and CompactLogix 5380 processors prior to Rev 32.

The following data types are only available for the ControlLogix 5580 and CompactLogix 5380 processors beginning with Rev 32:

Name	**Description**	**Range of values**
LREAL	8-byte float. pt.	$\pm 2.23\times10^{308}$ to $\pm 1.80\times10^{+308}$
USINT	1-byte integer	0 to 255
UINT	2-byte integer	0 to 65,535
UDINT	4-byte integer	0 to 2^{32}-1
ULINT	8-byte integer	0 to 2^{64}-1

The following data types are only available for the ControlLogix 5580 and CompactLogix 5380 processors starting with Rev 34:

Name	**Description**	**Range of values**
TIME32	4-bytes	± about 35 minutes, in microseconds
TIME	8-bytes	± about 31 days, in microseconds
LTIME	8-bytes	± about 31 days, in nanoseconds
DT	8-bytes	Jan. 1, 1970 to Dec. 31, 2250, in microseconds
LDT	8-bytes	Jan. 1, 1970 to Dec. 31, 2250, in nanoseconds

To identify a bit in an integer tag use "." between the tag name and the bit number, like,

"MyInt.2" identifies bit 2 in "MyInt"(MyInt data type is any *INT)

Literals are used to directly provide values for inputs of function blocks, transition conditions, and so on. These values cannot be overwritten by the program logic. Literals are also used to assign a constant or an initial value to a tag.

The BOOL (Boolean) data type has the literals 0 and 1. Literals for the integer data types can be expressed as an integer literal, base 2 literal (binary), base 8 literal (octal), or base 16 (hexadecimal) literal. The formats are as follows:

Literal type	**Examples**
integer	-12, 0, +23, 14_639
base-2	2#111 (7 decimal)
	2#1011_0011 (179 decimal)
base-8	8#177 (127 decimal)
	8#2001 (1025 decimal)
base-16	16#FFFF (65,535 decimal)
	16#AA (170 decimal)

The underline character may be inserted into numeric literals to aid readability; otherwise it has no significance. Literals for the REAL or LREAL types may be expressed as normal decimal numbers,

5.341, +14_639.7, -0.0041

or with "E" or "e" indicating the tens' exponential value,

7.45e-12, 1.27E+4, 1.27E4.

Example literals for the time data types are as follows:

Data type	Examples
TIME32	T32#2m_15s_2ms_340us
TIME	T#4d_30m_15s_200ms_125us
LTIME	LT#6d_30m_15s_1ms_690ns
DT	DT#2022-06-01-10:09:00.000_000
LDT	LDT#2032-01-01-7:30:00.230_125_500

A **structured data type** is generally a collection of data elements. Some of the more common predefined structured data types are the following:

Data Type	Use
AXIS	axis in motion instructions
CONTROL	bit shift, sequencer, array (file) instructions
COUNTER	counter instructions
MESSAGE	MSG instruction
MOTION_GROUP	one per controller (if doing motion)
MOTION_INSTRUCTION	motion instructions
PID	PID instruction
TIMER	timer instructions

The CONTROL, COUNTER and TIMER data types are constructed from 3 DINT's and are shown in Figure 3.2.

A **user-defined data type** (UDT) allows one to define a new structured data type from the elementary and complex data types. After creating the user-defined data type, one can use it multiple times. Figure 3.3 illustrates the structure of a UDT named Machine_Info. An individual element is accessed by using a "." between the tag name and the name of the element. For example, if the tag "Mach423" is of type Machine_Info, the serial number is accessed with "Mach423.Serial_Num."

Arrays are defined by following the data type with the maximum subscript enclosed in square brackets. Up to three dimensions can be defined. For example,

Dimensions	Example Data Type	Example Use
One	DINT[20]	DintTag[3]
Two	REAL[10, 10]	RealTag[3,7]
Three	BOOL[5, 10, 5]	BoolTag[idx1,idx2,idx3]

When used in instructions, the subscript can be a constant, tag, or expression (the latter 2 are similar to MicroLogix indirect addressing). The array indices are zero-based, that is, the first element has an index of zero. One can make arrays of structures, except for the AXIS, MESSAGE, and MOTION_GROUP data types.

Some useful predefined controller tags that represent control status information are:

S:FS	First scan of program
S:N	Negative sign
S:V	Overflow
S:Z	Zero

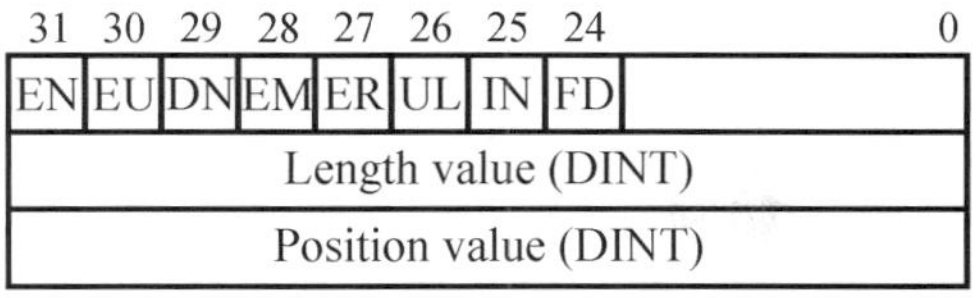

Individual CONTROL fields:

Name.LEN	Length value
Name.POS	Position value
Name.EN	Enable bit
Name.EU	Enable-unload bit
Name.DN	Done bit
Name.EM	Empty bit
Name.ER	Error bit
Name.UL	Unload bit
Name.IN	Inhibit bit
Name.FD	Found bit

(a)

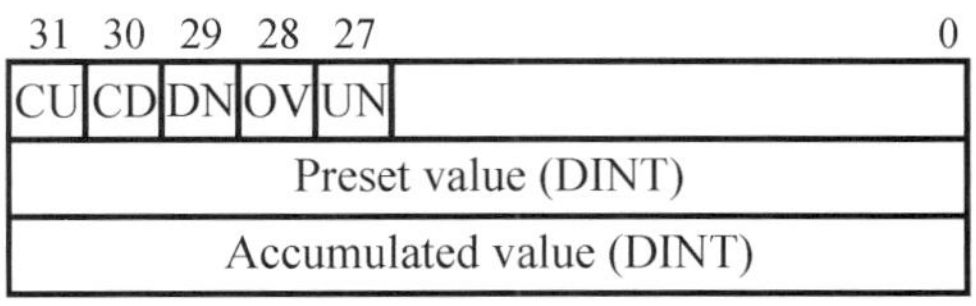

Individual COUNTER fields:

Name.PRE	Preset value
Name.ACC	Accumulated value
Name.CU	Count-up bit
Name.CD	Count-down bit
Name.DN	Counter-done bit
Name.OV	Counter overflow bit
Name.UN	Counter underflow bit

(b)

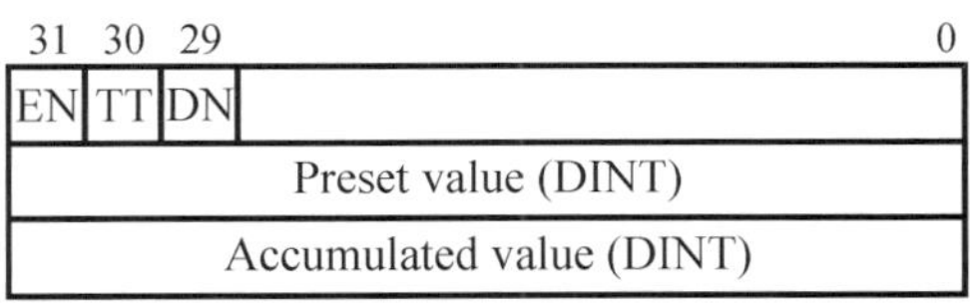

Individual TIMER fields:

Name.PRE	Preset value
Name.ACC	Accumulated value
Name.EN	Timer-enabled bit
Name.TT	Timer-timing bit
Name.DN	Timer-done bit

(c)

Figure 3.2. ControlLogix data type structures: *(a)* CONTROL; *(b)* COUNTER; *(c)* TIMER.

Data type Name: MACHINE_INFO
Description:
Machine information

Size 100 bytes

Name	Data Type	Style	Description
Serial_Num	DINT	Decimal	Serial number
Nam	STRING		Name of machine
Oper_Time	DINT	Decimal	Operating time in ms
Shift_Faults	INT	Decimal	Number of faults during shift
Last_Fault	INT	Decimal	Last recorded fault #

Figure 3.3. Example "Machine_Info" user-defined data type.

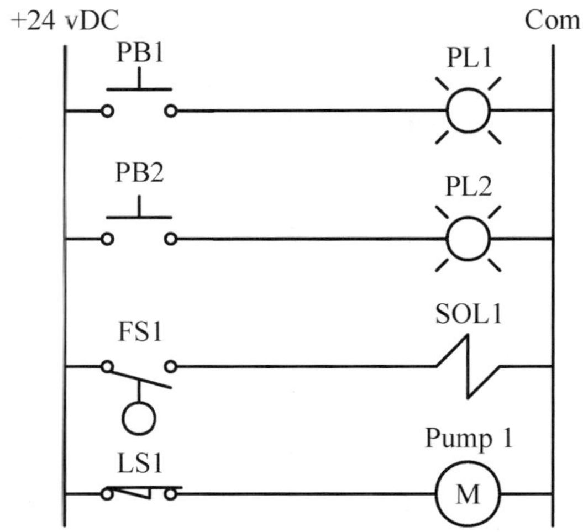

Figure 3.4. Sample hard-wired circuit.

3.2.4 Input/Output Tags

For the ControlLogix PLC the physical I/O channels are addressed directly and have pre-defined tags. The user does need to specify the location and type of each I/O module that will be used by the processor.

To illustrate addressing for discrete I/O, a sample hardwired circuit is shown in Figure 3.4. A ControlLogix PLC implementation of the hardwired circuit of Figure 3.4 might appear as in Figure 3.5. The switches are connected to channels 0 through 3 of a discrete input module and the lights, solenoid, and motor are connected to channels 0 through 3 of a discrete output module. The tag of the input channel is shown with the contact in the ladder. Normally, one would define a symbolic tag that refers to the I/O tag (called an alias tag) and the I/O address tag appears with the alias tag. However, for the purpose of this example only the address tags are shown. The tag "Local:4:I.Data.2" means the contact is controlled by the discrete input point (channel) 2, in slot 4 of the local chassis. In general, the tag for a discrete input channel is

AD_Name:*Slot*:I.Data.*x*

where *AD_Name* is the name of the communication adapter for a remote chassis, which is "Local" for modules in the same chassis as the processor, *Slot* is the chassis slot in which the module resides, and *x* is replaced by the channel number. Similarly, the tag "Local:9:O.Data.3" means the output (coil) is connected to output channel 3, in slot 9 of the local chassis. The most common ControlLogix I/O tags are:

Type of Channel	General Tag Name
Discrete Input	*AD_Name*:*Slot*:I.Data.*x*
Discrete Output	*AD_Name*:*Slot*:O.Data.*x*
Analog Input	*AD_Name*:*Slot*:I.Ch*x*Data
Analog Output	*AD_Name*:Slot:O.Ch*x*Data

where

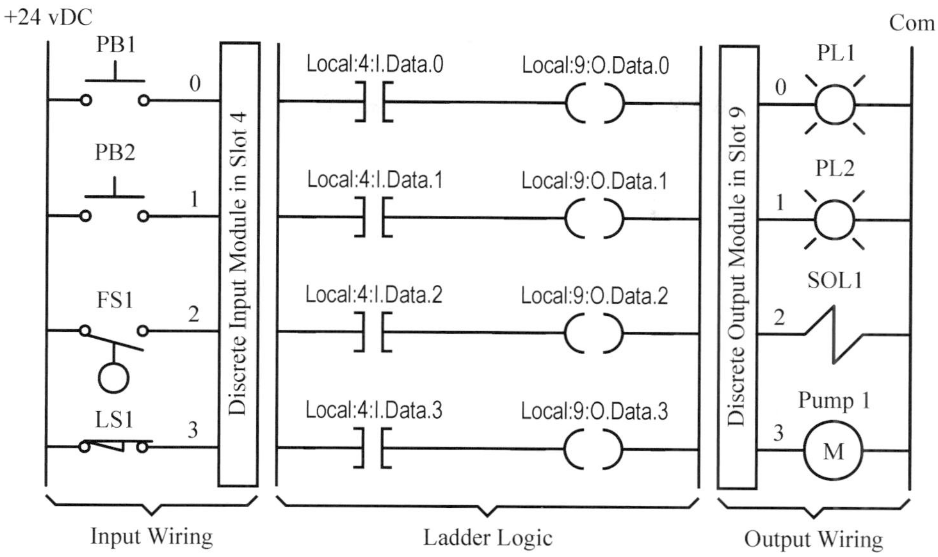

Figure 3.5. ControlLogix PLC implementation for Figure 3.4 showing I/O tags.

AD_Name is the name of the communication adapter for the chassis, which is "Local" for inputs in the same chassis as the processor,
Slot is the chassis slot in which the module resides, and
x is replaced by the channel number.

The chassis slots and module channels are numbered in decimal. An example ControlLogix system with two remote chassis and 1756-series I/O modules is shown in Figure 3.6. Each I/O module has information other than just the channel data (e.g., channel errors). Consult the module documentation for information about the other pre-defined tags for that module.

Example 3.1. Draw a ControlLogix ladder diagram that will cause the output, pilot light PL2, to be **on** when selector switch SS1 is closed, push button switch PB5 is open, and limit switch LS7 is open. Show only the I/O tags in the logic. The input/output devices are wired to the following locations:

PL2:	Output module, local I/O, slot 7, channel 13
SS1:	Input module, remote I/O adapter 'REM_1', slot 2, channel 4
PB5:	Input module, remote I/O adapter 'REM_2', slot 1, channel 8
LS7:	Input module, remote I/O adapter 'REM_2', slot 5, channel 7

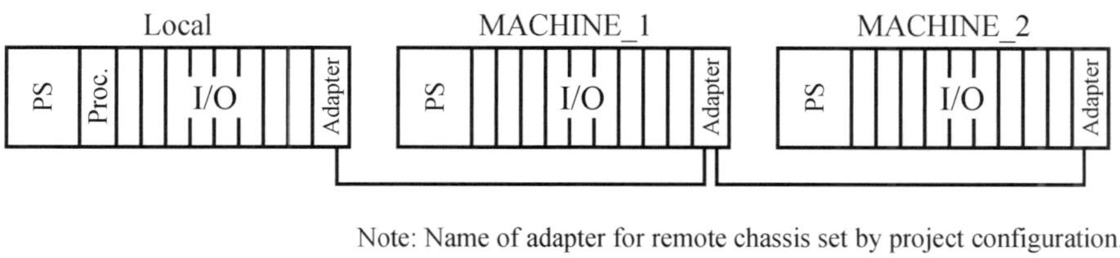

Figure 3.6. Example ControlLogix system with remote chassis.

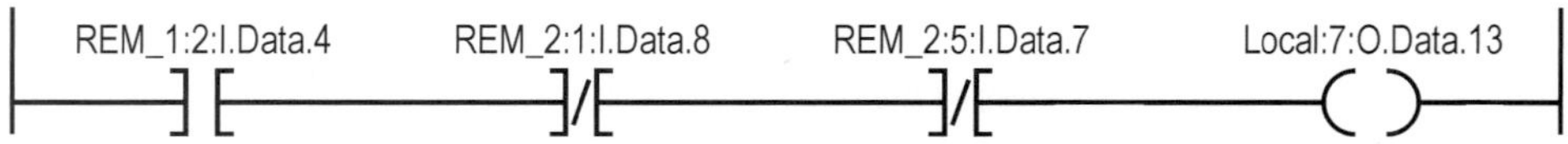

Figure 3.7. Solution to Example 3.1.

The remote I/O modules are 1756-series modules.

Solution. The tags are predefined and are,

PL2:	Local:7:O.Data.13
SS1:	REM_1:2:I.Data.4
PB5:	REM_2:1:I.Data.8
LS7:	REM_2:5:I.Data.7

The ladder logic is shown in Figure 3.7. The tags are shown above the contact and coil instructions.

In contrast to the ControlLogix system, that places the processor and I/O modules in chassis, the modules in a CompactLogix system mount on a DIN rail. There are currently four versions of CompactLogix processors: 5370 L1, 5370 L2, 5370 L3, and 5380. A CompactLogix 5370 L1 processor has some built-in input/output channels and uses series 1734 Point I/O™ modules for any local expansion I/O. A CompactLogix 5370 L2 processor has some built-in input/output channels and uses series 1769 Compact I/O™ modules for any local expansion I/O. A CompactLogix 5370 L3 processor has no built-in input/output channels and uses series 1769 Compact I/O™ modules for any local expansion I/O. A CompactLogix 5380 processor has no built-in input/output channels and uses series Compact 5000™ I/O modules (also called 5069-series modules) for any local expansion I/O. All of the CompactLogix processors have two built-in Ethernet ports for remote I/O modules.

An example CompactLogix 5370 L3 system is shown in Figure 3.8. The allowable number of modules depends on the particular processor. There are a few constraints to the placement of the modules. The processor mounts on the left side of the power supply and there can be no more than 3 modules between the processor and the power supply and no more than 8 modules on the right side of the power supply, subject to the power supply capacity. For the other banks, there can be no more than 8 modules on either side of the power supply, subject to the capacity of the power supply. The tag names for the I/O local to the processor are like those of a ControlLogix system. Embedded I/O (5370 L1 and L2 processors) are in slot 1. Tags for remote I/O, accessed through communication adapter modules, are the same as for the ControlLogix.

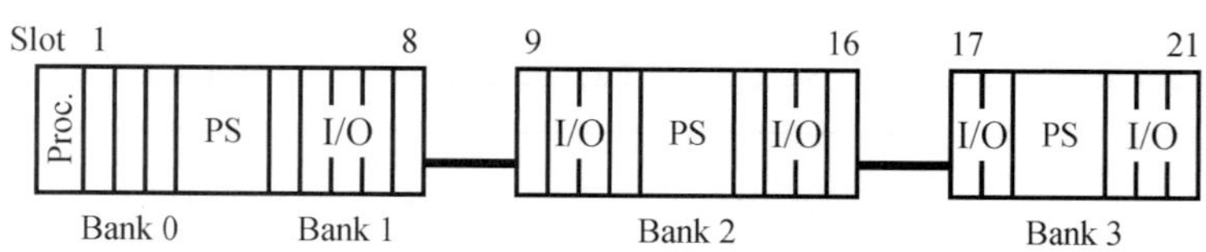

Figure 3.8. Example CompactLogix system with I/O banks.

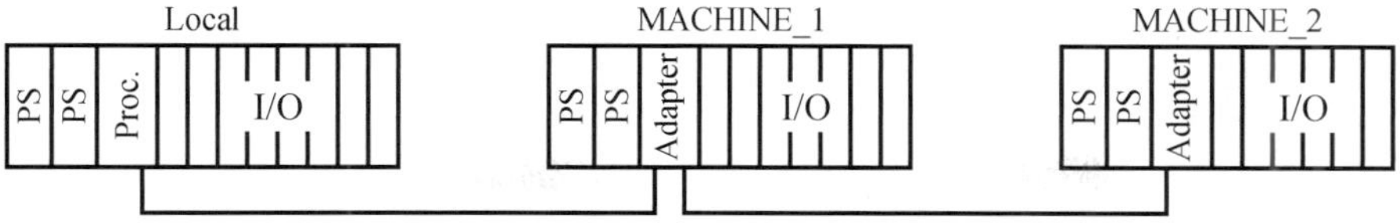

Note: Name of adapter for remote chassis set by project configuration.

Figure 3.9. Example CompactLogix 5380 system with remote I/O.

An example CompactLogix 5380 system is shown in Figure 3.9. The allowable number of modules depends on the particular processor. The tag names for the 5069-series I/O module channels are different than those of a ControlLogix system:

Type of Channel	5069-Series General Tag Name
Discrete Input	*AD_Name*:*Slot*:I.Pt*xx*.Data
Discrete Output	*AD_Name*:*Slot*:O.Pt*xx*.Data
Analog Input	*AD_Name*:*Slot*:I.Ch*xx*.Data
Analog Output	*AD_Name*:*Slot*:O.Ch*xx*.Data

where

AD_Name is the name of the remote adapter, which is "Local" for those modules local to the processor,
Slot is the slot number, and
xx is replaced by the two-digit channel number

Each I/O module has information other than just the channel data (e.g., channel errors). Consult the module documentation for information about the other pre-defined tags for that module.

Example 3.2. Draw a CompactLogix 5380 ladder diagram that will cause the output, pilot light PL2, to be **on** when selector switch SS1 is closed, push button switch PB5 is open, and limit switch LS7 is open. Show only the I/O tags in the logic. The input/output devices are wired to the following locations:

PL2: Output module, local I/O, slot 7, channel 13
SS1: Input module, remote I/O adapter 'REM_1', slot 2, channel 4
PB5: Input module, remote I/O adapter 'REM_2', slot 1, channel 8
LS7: Input module, remote I/O adapter 'REM_2', slot 5, channel 7

The remote I/O modules are 5069-series modules.

Solution. The tags are predefined and are,

PL2: Local:7:O.Pt13.Data
SS1: REM_1:2:I.Pt04.Data
PB5: REM_2:1:I.Pt08.Data
LS7: REM_1:5:I.Pt07.Data

The ladder logic is shown in Figure 3.10.

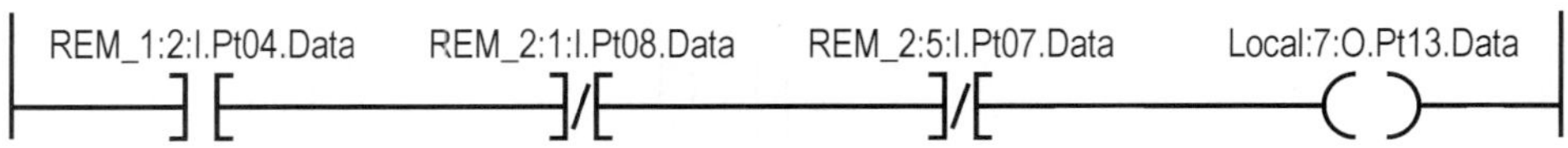

Figure 3.10. Solution to Example 3.2.

In many systems, the input/output modules are remote, that is, connected via a communication network to the processor (Figures 3.6 and 3.9) . Rather than describe all of the available families of remote I/O modules, the ones that are chassis-based or modular are as follows:

1734-series Point I/O
1756-series ControlLogix I/O
1769-series Compact I/O
1794-series Flex I/O
5069-series Compact 5000™ I/O
5094-series Flex 5000 ™ I/O

The 5580-series ControlLogix and the 5380-series CompactLogix processors can communicate with all of these I/O module families. The 5570-series ControlLogix processors and the 5370-series CompactLogix processors cannot communicate with the 5069-series and 5094-series modules.

The 5094-series I/O modules share the same format for the I/O channel tag names with the 5069-series I/O modules. The I/O channel tag names for the 1734-series, the 1769-series, and the 1794-series I/O modules are the same as for the 1756-series modules.

3.3 MICRO800 MEMORY

3.3.1 Memory Organization

The Allen-Bradley Micro800 PLC memory model is presented in Figure 3.11. Each of the main elements is now described.

The **project** is the entire body of software (program and data) that corresponds to one PLC processor.

The **program organization units (POUs)** are the programs for the project. A project may contain up to 256 programs, subject to memory limitations, executed in the specified order. Each program may contain **local variables**, that is, variables that can only be accessed by the program/POU. A POU can call either a **user defined function** (UDF) or a **user defined function block** (UDFB). A UDF is similar to a traditional subroutine and is limited to have only one output connection. UDFBs can call other UDFBs, but the maximum nesting depth is five. UDFs and UDFBs can also have local variables, accessible only to the UDF or UDFB.

By default, POUs are cyclic, executed once per program scan, but an **interrupt** can control the execution of a POU. The available interrupts are:

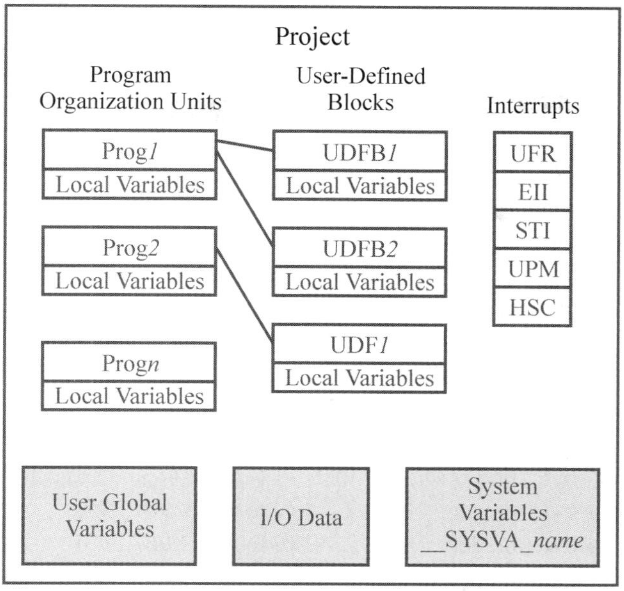

Figure 3.11. Micro800 memory model.

Name	Description
EII	Event input interrupt: execute a POU on the rising or falling edge of a discrete input. Maximum of 8 events
HSC	High-speed counter. Maximum of 6 counters
STI	Selectable timed interrupt: execute a POU on a fixed time period. Maximum of 4
UFR	User Fault Routine: execute a POU when a fault is detected. No more than one
UPM	Plug-in Module

3.3.2 Program scan

The program execution is basically the same as introduced in section 2.5. In addition to reading the inputs, scanning the programs, and updating the outputs, additional housekeeping tasks are performed at the end of scan: communications, data logging, and recipe handling.

3.3.3 Variables and data types

In the Micro800 PLC, the symbolic name of a variable can be any string of letters, digits, and underlines up to 40 characters long provided that:

1. The first character is not a digit.
2. There are not two or more consecutive underlines.

There is no distinction between upper and lower case characters. If a particular variable is defined as a program local variable and a global variable, the program uses the local variable, and not the global variable.

When a tag is declared, it must be assigned a data type. The **simple data types** are as follows:

Name	Description	Range of values
BOOL	1-bit Boolean	0=cleared, 1=set
BYTE	8-bit	0 to 255
SINT	1-byte integer	-128 to 127
USINT	1-byte integer	0 to 255
INT	2-byte integer	-32,768 to 32,767
UINT	2-byte integer	0 to 65,535
WORD	2-byte integer	0 to 65,535
DINT	4-byte integer	-2^{31} to $2^{31}-1$ (-2,147,483,648 to 2,147,483,647)
UDINT	4-byte integer	0 to $2^{32}-1$ (0 to 4,294,967,295)
DWORD	4-byte integer	0 to $2^{32}-1$ (0 to 4,294,967,295)
LINT	8-byte integer	-2^{63} to $2^{63}-1$
ULINT	8-byte integer	0 to $2^{64}-1$
LWORD	8-byte integer	0 to $2^{64}-1$
REAL	4-byte float. pt.	$\pm 3.403 \times 10^{37}$ to $\pm 3.403 \times 10^{+38}$
LREAL	8-byte float. pt.	$\pm 1.798 \times 10^{308}$ to $\pm 1.798 \times 10^{+308}$
TIME	4-byte integer	0 to $2^{31}-1$ milliseconds (>24 days)
DATE	4-byte integer	Seconds since Jan. 1, 1970 at 0:00 GMT
STRING	variable length	1 byte per character

To identify a bit in an integer tag use "." between the tag name and the bit number, for example,

"MyInt.2" identifies bit 2 in "MyInt" (MyInt data type is *INT, any integer type)

Literals are used to directly provide values for inputs of function blocks, transition conditions, and so on. These values cannot be overwritten by the program logic. Literals are also used to assign a constant or an initial value to a variable.

The BOOL (Boolean) data type has the literals "false" and "true." Literals for the integer data types can be expressed as an integer literal, base 2 literal (binary), base 8 literal (octal), or base 16 (hexadecimal) literal. The formats are as follows:

Literal type	Examples
integer	-12, 0, +23, 14_639
base-2	2#111 (7 decimal)
	2#1011_0011 (179 decimal)
base-8	8#177 (127 decimal)
	8#2001 (1025 decimal)
base-16	16#FFFF (65,535 decimal)
	16#AA (170 decimal)

The underline character may be inserted into numeric literals to aid readability; otherwise it has no significance. Literals for the REAL or LREAL types may be expressed as normal decimal numbers,

5.341, +14_639.7, -0.0041

or with "E" or "e" indicating the tens' exponential value,

7.45e-12, 1.27E+4, 1.27E4.

Constants for the TIME type (called duration constants) are specified in days (D), hours (H), minutes (M), seconds (S), and milliseconds (MS), or combinations thereof. The duration must be identified by the prefix T# or TIME#. Some examples are:

T#10MS	10 milliseconds
T#3s500ms	3 seconds, 500 milliseconds
TIME#18M	18 minutes
TIME#5D_14H_12M	5 days, 14 hours, 12 minutes

Arrays are defined by specifying a dimension range for the variable. Dimensions can be defined for a variable of one of the simple data types. For example,

Dimensions	**Dim. Range**	**Example Use**
One	[1..10]	DintTag[3]
Two	[1..10,1..10]	RealTag[3,7]
Three	[1..5,1..10,1..5]	BoolTag[2,5,Idx]

When used in instructions, the subscript can be a constant, DINT tag, or DINT expression. The array index range is specified. The number of dimensions is not limited, though the size of the array is limited to the amount of available controller memory.

A **structured data type** is generally a collection of data elements. Some of the more common predefined structured data types are the following:

Data Type	**Use**
CTU/CTD/CTUD	counter blocks
IPIDCONTROLLER	PID block
TON/TOF/TP	timer blocks

A user-defined structured data type allows one to define a new data type from the elementary and complex data types. After creating the user-defined data type, one can use it multiple times. Figure 3.12 illustrates the structure of a structured data type named MACHINE_INFO. An individual element is accessed by using a "." between the tag name and the name of the element. For example, if the tag "Mach423" is of type MACHINE_INFO, the serial number is accessed with "Mach423.Serial_Num." Structured data types may only be defined when using the Developer version of the Connected Components Workbench (CCW), the programming software for the Micro800 processors.

Name	Data Type	String Size	Comment
MACHINE_INFO	MACHINE_INFO		
Serial_Num	DINT		Serial number
Nam	STRING	80	Name of machine
Oper_Time	DINT		Operating time in ms
Shift_Faults	INT		Number of faults during shift
Last_Fault	INT		Last recorded fault #

Figure 3.12. Example "MACHINE_INFO" user-defined data type.

Some useful predefined controller tags that represent control status information are:

_SYSVA_FIRST_SCAN	First scan of program. True only on first scan.
_SYSVA_POWER_UP_BIT	Power-up. True only on first scan after a power-up, or running a new program for the first time

3.3.4 Input/Output Variables

For the Micro800 PLC the physical I/O channels are addressed directly and have pre-defined variables. The user does need to specify the location and type of each I/O plug-in and expansion module that will be used by the processor.

A Micro800 PLC implementation of the hardwired circuit of Figure 3.4 might appear as in Figure 3.13. The switches are connected to channels 0 through 3 of the built-in discrete input channels. The lights, solenoid, and motor are connected to channels 0 through 3 of the built-in discrete output channels. The variable of the input channel is shown with the contact in the ladder. Normally, one would define a symbolic variable that refers to the I/O variable (called an alias variable) and the I/O address variable appears with the alias variable. However, for the purpose of this example only the address variables are shown. The variable "_IO_EM_DI_02" means the contact is controlled by the embedded discrete input point (channel) 2 of the processor. In general, the variable for a discrete input channel is

_IO_*Loc*_DI_*xx*

where *Loc* designates the channel location and *xx* is replaced by the two-digit channel number. Possible Loc designations are:

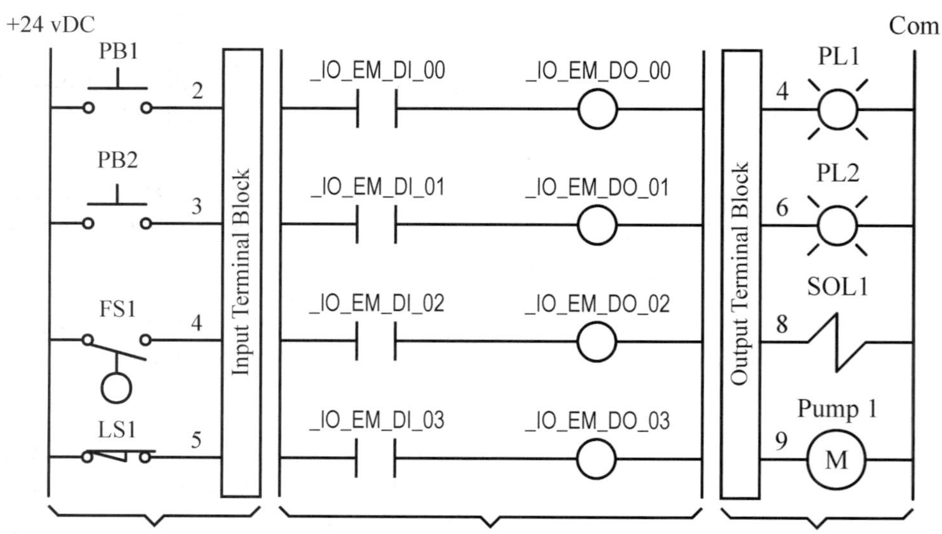

Figure 3.13. Micro800 PLC implementation for Figure 3.4 showing I/O variables.

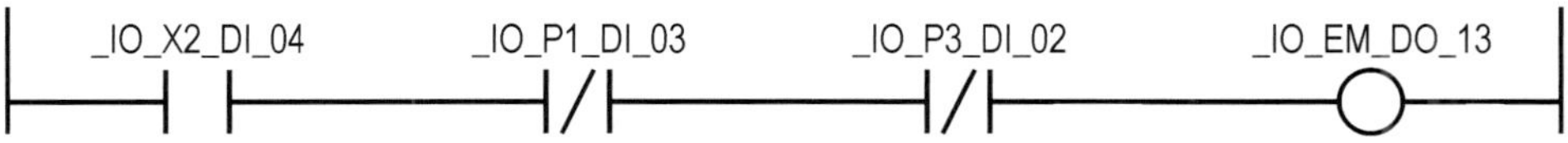

Figure 3.14. Solution to Example 3.3.

Loc	Channel Location
EM	Embedded (built-in) channel
P*y*	Plug-in module *y*
X*y*	Expansion module *y*

Similarly, the variable "_IO_EM_DO_03" means the output (coil) is connected to embedded output channel 3. The most common I/O variables are:

Type of Channel	General Variable Name
Discrete Input	_IO_*Loc*_DI_*xx*
Discrete Output	_IO_*Loc*_DO_*xx*
Analog Input	_IO_*Loc*_AI_*xx*
Analog Output	_IO_*Loc*_AO_*xx*

Each I/O module has information other than just the channel data (e.g., channel status). Consult the module documentation for information about the other pre-defined variables for that module.

Example 3.3. Draw a Micro800 ladder diagram that will cause the output, pilot light PL2, to be **on** when selector switch SS1 is closed, push button switch PB5 is open, and limit switch LS7 is open. Show the I/O variables in the logic. The input/output devices are wired to the following locations:

PL2:	Output module, embedded, channel 13
SS1:	Input module, expansion 2, channel 4
PB5:	Input module, plug-in 1, channel 3
LS7:	Input module, plug-in 3, channel 2

Solution. The variables are predefined and are,

PL2:	_IO_EM_DO_13
SS1:	_IO_X2_DI_04
PB5:	_IO_P1_DI_03
LS7:	_IO_P3_DI_02

The ladder logic is shown in Figure 3.14. The variables are shown above the contact and coil instructions.

3.4 MICROLOGIX AND SLC-500 MEMORY

The memory organization of the MicroLogix and the SLC-500 are very similar. The only major differences are the addition of function files and new file types in more recent MicroLogix pocessors.

3.4.1 Memory Organization

The memory model of the Allen-Bradley MicroLogix and SLC-500 PLC processors is presented in Figure 3.15. The model represents the most structured form of PLCs prior to IEC 61131-3 and the introduction of the ControlLogix processors. When the memory model was developed for the PLC-3 and PLC-5 and later adopted by the SLC-500, it was the most sophisticated memory model for its time and is suited to projects with large programs and large amounts of data. Each of the main elements of the memory is now described.

The **project** memory is divided into 6 major parts: program files, data files, function files, speciality files, communication channel configuration, and unused memory. Technically, the project is more than the program, data files, and configuration that reside in the PLC. The project also includes program documentation that resides in the programming device only.

The **program files** contain controller information, the main program(s), and any subroutine programs. The purpose of the program files is as follows:

File Number	Use
0	Password and program identification
1	Reserved
2	Main ladder logic program
3-255	Subroutine file

For most of the MicroLogix processors and the SLC-500, the maximum file number is 255. The MicroLogix 1000 has 14 available program files, of which files 3 through 5 may be

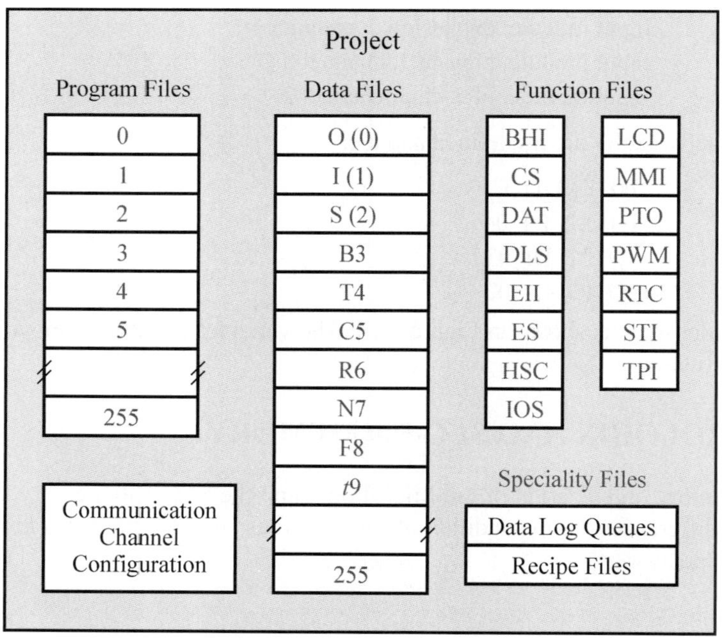

Figure 3.15. MicroLogix/SLC-500 memory model.

used for pre-defined functions (high-speed counter, fault handling, timed interrupt) or as subroutines..

All of the MicroLogix processors and the SLC-5/02 and later processors support the definition of a selectable timed interrupt (STI) program file and a fault program file. An STI program file is executed on a fixed periodic time interval, which should be longer than the main program scan time. A fault program file is executed when the processor detects a major fault. The fault file allows one to record the fault and restart the processor without operator intervention. The method of specifying the STI and fault file differs between the MicroLogix and SLC-500 processors. For the MicroLogix 1000 processor, the STI file is program file 5 and the fault file is program file 3. For the other MicroLogix processors, the STI file is specified as part of the STI function file and the fault file is specified as part of the EII function file. For the SLC-500, the STI and fault files are specified as part of the status data file.

Most MicroLogix/SLC-500 processors can have up to 256 **data files** (numbered 0 - 255). However, the MicroLogix 1000 has only 8 data files. The address format of these files is explained in section 3.4.3. The data files can be protected from inadvertently being overwritten by a program download or protected from being overwritten by the ladder logic.

The **function files** available on the MicroLogix processors provide access to certain processor functions. In general these functions are available on the MicroLogix 1100, 1200, 1400, and 1500 processors, though exceptions are noted below. The list of available function files and their description follows.

Names	Description
BHI	Base hardware information, such as catalog number and revision.
CS	Communication status of a channel, such as the node table and diagnostic counters.
DAT	Data access tool configuration information (1500 only)
DLS	Data log status (not 1200).
EII	Event input interrupt configuration, such as the discrete input to monitor and the program file to execute
ES	Ethernet communication channel status(1100 and 1400 only)
HSC	High-speed counter
IOS	Input/output status of embedded channels and expansion modules.
LCD	LCD display status and information (1100 and 1400 only)
MMI	Memory module information, such as catalog number and status.
PTO	Pulse-train output
PWM	Pulse-width modulated output
RTC	Real-time clock
STI	Selectable timed interrupt configuration, such as the period and the program file number.
TPI	Trim pot information; value and error code.

The function files are described in the user manual for the appropriate processor (Rockwell Automation, 2011, 2014, 2015, 2019).

The **specialty files** are available for MicroLogix 1500 processors to load/store recipe records for batch processing. Certain MicroLogix 1500 processors have data log queues to allow one to store application data in records for later retrieval. For processors that support data logs, the recipe records can be stored in the data logs rather than use regular data files. For more information on the specialty files, see Rockwell Automation (2014).

The **communication channel configuration** contains information specific to the processor communication channels: serial channel baud rates, remote I/O addresses, and channel protocols.

3.4.2 Program scan

The program execution is basically the same as introduced in section 2.5. For all of the MicroLogix and SLC-500 processors, immediate input and output instructions are supported.

3.4.3 Data files

The organization of the data files is shown in Figure 3.16. Except for the first three files, every data file is designated by a letter and a number. Within a project, the letter/number combination is unique. For example, once data file 12 is designated as a timer file (T12), then it cannot be referenced as an integer file (N12). The first three data files are

Number of Elements	Data File	File Type	File Number	Words per Element	Addressed As
varies	Output Image	O	0	1	O:*ss*/*cc* Discrete O:*ss*.*w*/*b* Analog
varies	Input Image	I	1	1	I:*ss*/*cc* Discrete I:*ss*.*w*/*b* Analog
33 to 164	Status	S	2	1	S:*n* or S:*n*/*m*
1-256	Bit	B	3	1	B3/*n* or B3:*n*/*m*
1-256	Timer	T	4	3	T4:*n*
1-256	Counter	C	5	3	C5:*n*
1-256	(File) Control	R	6	3	R6:*n*
1-256	Integer	N	7	1	N7:*n*
1-256	Floating Point	F	8	2	F8:*n*
1-256	String Message	ST	9-255	42	ST*x*:*n*
1-256	ASCII Message	A	9-255	1	A*x*:*n*
1-256	Any Type (except O, I, S)	any (not O, I, S)	9-255	varies	

Figure 3.16. MicroLogix/SLC-500 data file organization.

only designated by their letter (O, I, or S) since these types cannot be assigned to any other data files.

Data files 0 - 8 are the default data files that are defined for a new project. The type for data files 0 - 2 cannot be changed and their length depends on the processor model. The type for data files 3 - 8 also cannot be changed, but the length can be changed. Data files 9 - 255 can be defined as any of the possible types and can be created as needed.

Except for the O, I, and S files, the length of a data file can be from 1 to 256 elements (addressed as 0 - 255), subject to the available processor memory. When programming offline, the lengths of the data files are increased as needed. For example, if the counter file C5 contains 24 elements (last address is C5:23) and a counter instruction is programmed to refer to C5:24, the C5 file length is increased by one. Typically, one does not size the data files in this manner. Normally, the data files are manually created and sized before the program is entered. After the programming is complete, the data file sizes should be increased to accommodate online editing changes.

Data files 9-255 may be assigned any file type (except O, I, or S), as needed by the program. These data files are often used to organize the data files by process area or function. For example,

Separate bit/timer/counter/integer files for each process section or machine.

PID controller parameter storage.

File for messages.

The format of each data file type is now explained. The format of the output and input image files are explained in section 3.4.7.

The **status** file contains detailed information regarding the processor status and the specific information varies by the processor. Some of the status file addresses that are common to many of the MicroLogix and SLC-500 processors:

Address	Description
S:0	Arithmetic flags, such as carry, sign, and zero
S:1	Processor mode and force flags
S:1/15	"First pass" flag, set if processor is executing ladder for first time
S:5	Minor error bits
S:6	Major error code

A **bit** data file is internal one-bit memory used for internal coils and bit-shift instructions. For the default bit file, individual bits (coils) are addressed as

B3/*n*

where *n* is the bit number, 0-4095. A bit file can be addressed as words

B3:*n*

where *n* is the word number, 0-255. One can address individual bits in a word as

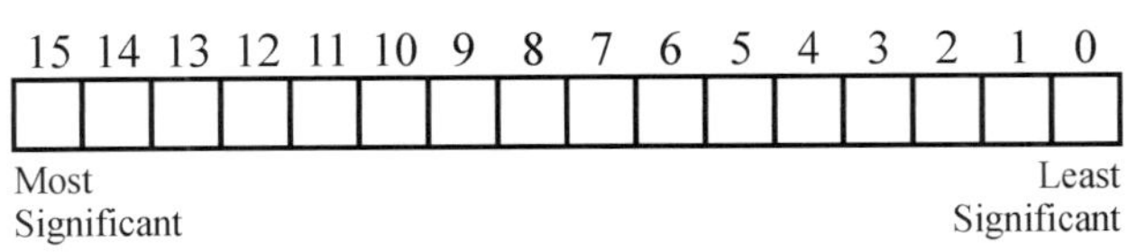

Figure 3.17. Bit numbering within bit file word or integer.

Bit number in word

15	14	13	12	11	10	9	8	7	6	5	4	3	2	1	0	
15	14	13	12	11	10	9	8	7	6	5	4	3	2	1	0	B3:0
31	30	29	28	27	26	25	24	23	22	21	20	19	18	17	16	B3:1
47	46	45	44	43	42	41	40	39	38	37	36	35	34	33	32	B3:2
⋮															⋮	⋮
159	158	157	156	155	154	153	152	151	150	149	148	147	146	145	144	B3:9
175	174	173	172	171	170	169	168	167	166	165	164	163	162	161	160	B3:10

Figure 3.18. Relationship between bit and word addressing in bit file type.

B3:*n*/*m*

where *n* is the word number and *m* is the bit number, 0-15. The bits within a word are numbered as shown in Figure 3.17. Other data files (>8) can also be bit files. For example, if data file 9 is a bit file, each coil in that file is addressed as B9/*n*.

The relationship between the two ways to address bits in a bit file is shown in Figure 3.18. For example, B3:2/5 addresses the same bit as B3/37 and B3:10/11 and B3/171 address the same bit. The programming software can be configured to display bit addresses in either format.

A **timer** file consists of timer elements whose structure appears in Figure 3.19*a*. For the default timer file, each individual timer is addressed as

T4:*n*

where *n* is the timer number, 0-255. The individual parts of each timer element can be addressed as shown in Figure 3.19*a*. The "." can be used in place of the "/" For example, the addresses T4:14.DN and T4:14/DN are equivalent. Other data files (>8) can also be timer files. For example, if data file 10 is a timer file, each timer in that file is addressed as T10:*n*.

A **counter** file consists of counter elements whose structure is shown in Figure 3.19*b*. For the default counter file, each individual counter is addressed as

C5:*n*

where *n* is the counter number, 0-255. The individual parts of each counter element can be addressed as shown in Figure 3.19*b*. The "." can be used in place of the "/". Other data files (>8) can also be counter files. For example, if data file 14 is a counter file, each counter in that file is addressed as C14:*n*.

A **file control** data file element (Figure 3.19*c*) is used to store information for shift registers, queues, and various other instructions. For the default control file, each individual element is addressed as

R6:*n*

where *n* is the element number, 0-255. The individual parts of each file control element can be addressed as shown in Figure 3.19*c*. Other data files (>8) can also be file control data files.

An **integer** data file stores integer data values, with a range from -32,768 through 32,767. Individual integers are addressed as

15	14	13	... 0
EN	TT	DN	
Preset value (0 to 32,767)			
Accumulated value (0 to 32,767)			

Individual timer parts addressed as:

T4:*n*/PRE	Preset value (integer)
T4:*n*/ACC	Accumulated value (integer)
T4:*n*/EN	Timer-enabled bit
T4:*n*/TT	Timer-timing bit
T4:*n*/DN	Timer-done bit

(a)

15	14	13	12	11	... 0
CU	CD	DN	OV	UN	
Preset value (-32,768 to 32,767)					
Accumulated value (-32,768 to 32,767)					

Individual counter parts addressed as:

C5:*n*/PRE	Preset value (integer)
C5:*n*/ACC	Accumulated value (integer)
C5:*n*/CU	Count-up bit
C5:*n*/CD	Count-down bit
C5:*n*/DN	Counter-done bit
C5:*n*/OV	Counter overflow bit
C5:*n*/UN	Counter underflow bit

(b)

15	14	13	12	11	10	9	8	... 0
EN	EU	DN	EM	ER	UL	IN	FD	
Length value								
Position value								

Individual parts addressed as:

R6:*n*/LEN	Length value (integer)
R6:*n*/POS	Position value (integer)
R6:*n*/EN	Enable bit
R6:*n*/EU	Enable-unload bit
R6:*n*/DN	Done bit
R6:*n*/EM	Empty bit
R6:*n*/ER	Error bit
R6:*n*/UL	Unload bit
R6:*n*/IN	Inhibit bit
R6:*n*/FD	Found bit

(c)

Word	Upper Byte (15 ... 9)	Lower Byte (8 ... 0)
0	String Length (number of characters)	
1	Character 0	Character 1
2	Character 2	Character 3
41	Character 80	Character 81

Individual parts addressed as:

ST*x*:*n*.LEN	Length value (integer)
ST*x*:*n*.DATA[*m*]	Chars in word *m*-1
ST*x*:*n*.*m*	Chars in word *m*

(d)

Figure 3.19. MicroLogix/SLC-500 element structures: *(a)* timer; *(b)* counter; *(c)* file control; *(d)* string.

N7:*n*

where *n* is the element number, 0-255. One can address individual bits within the integer as

N7:*n*/*m*

where n is the word number and m is the bit number, 0-15. The bits within a word are numbered in the same manner as for the words in a bit file (Figure 3.17). Other data files (>8) can also be integer files. For example, if data file 21 is an integer file, each integer in that file is addressed as N21:*n*.

A **floating-point** data file stores floating-point values, with a range from $\pm 1.1755\times10^{-38}$ through $\pm 3.4028\times10^{+38}$. Individual floating-point numbers are addressed as

F8:*n*

where n is the element number, 0-255. Each floating-point number occupies 2 words, but individual words or bits cannot be addressed within the floating-point file. To access the individual words or bits, the CPW block instruction must be used to copy the floating-point number to a bit or integer file. Floating-point data files are not supported on earlier MicroLogix processors.

A **string** data file element (Figure 3.19*d*) is used to store ASCII messages. Individual elements are accessed as words. The length of the message string is addressed as

ST*x*:*n*.LEN or ST*x*:*n*.0

where x is the file number, 9-255 and n is the string number, 0-255. Characters 2 and 3 are addressed as

ST*x*:*n*.2

where x is the file number, 9-255 and n is the string number, 0-255. The string file type can be used to store messages for communication with certain display modules. Message-type data can be received from certain devices (for example, bar-code readers) and stored in this file type.

An **ASCII** data file is used mainly for the storage of messages. Each word in an ASCII data file contains two ASCII characters, but only words can be addressed as

A*x*:*n*

where x is the file number (9-255) and n is the word number, 0-255. Figure 3.20 shows how the message "PLC" is stored starting at address A15:10. This data file type can be used to store messages for communication with certain display modulesand to store embedded web pages for the SLC-5/05. This file type is only supported for the SLC-5/03, -5/04, and -5/05 processors.

The MicroLogix 1100, 1200, 1400, and 1500 processors have additional file types, which can be used for data files 9-255:

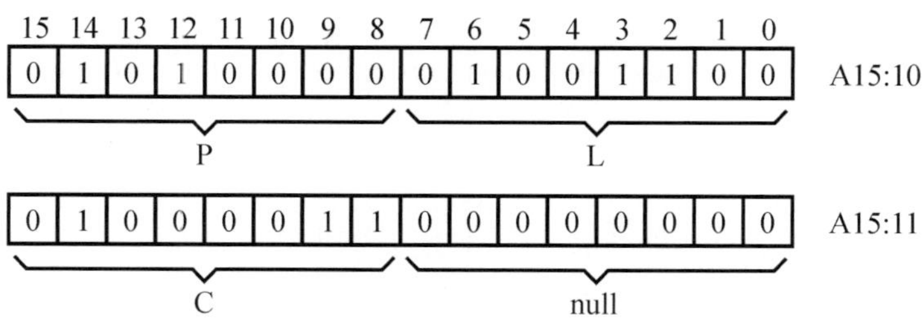

Figure 3.20. Representation of "PLC" in ASCII data file.

Letter	Meaning	Use	Max. elements
L	Long words	32-bit data	256
MG	Message	MSG instruction	256
PD	PID loop	PID block parameters	256
PLS	Prog. Limit Sw.	Programmable limit switch	256
RI	Routing Info	For MSG block	256
RIX	Ext. Route Info	For MSG block	256

Not all file types are available on all MicroLogix processors.

Design Tip

Use the data files to organize your program data. For example, a processor that controls multiple assembly machines could have data files organized in blocks of 10 files as:

B10	Machine 1 bits
T11	Machine 1 timers
C12	Machine 1 counters
R13	Machine 1 file control
N14	Machine 1 integers
F15	Machine 1 floating-point
ST16	Machine 1 messages
B20	Machine 2 bits
T21	Machine 2 timers
C22	Machine 2 counters

3.4.4 Indirect and Indexed Addressing

In order to aid in structuring and accessing large amounts of data, the SLC-500, and most MicroLogix processors support indirect and indexed addressing. Indirect addressing is not supported on the earlier SLC processors. Indexed addressing is not supported on the MicroLogix 1100, 1200, 1400, and 1500 processors.

In **indirect addressing**, an address may substitute for the numerical part of an address. Indirect addressing is identified by square brackets around the address supplying the number. Indirect addressing is useful for accessing arrays of elements (bits, integers, etc.). For example, in the address B3/[N7:3], the N7:3 address determines the bit number in B3 that will be accessed. The programmer must ensure that the value of N7:3 does not cause a reference beyond the number of words in B3, or the processor will fault.

As another application of indirect addressing, the data file number can be supplied by an integer address. For example, in the address N[N7:4]:10, the value in N7:4 determines which integer file is accessed to get the word 10 value. This type of indirect addressing could be used in a batch control application where there are a series of integer files, each holding a separate batch recipe parameter set. An integer location specifies the particular integer file, and thus the current recipe.

In **indexed addressing**, the value of the index register (S:24) is added to the address to obtain a new address. Indexed addressing is identified by a "#" character in front of an address reference. For example, an address of #N30:4 means the actual address is obtained by the index register as an offset to N30:4. If the value of S:24 is 9, then the actual address is N30:13 (N30:4+9).

3.4.5 Constants

Constants are used to directly provide values for instructions. These values cannot be overwritten by the program logic.

Integer constants may be expressed in decimal, binary, octal or hexadecimal notation by using the appropriate suffix. The formats are as follows:

Type	Suffix	Examples	
decimal	none, D	-12, 0, 23D, 14639	
binary	B	101B (5 decimal)	1110B (14 decimal)
octal	O	177O (127 decimal)	2001O (1025 decimal)
hexadecimal	H	FFFFH (65,535 decimal)	AAH (170 decimal)

Floating-point constants may be expressed as normal decimal numbers,

5.341, +1439.7, -0.0041

or with "E" indicating the tens' exponential value,

7.45E-12, 1.27E+4, 1.27E4.

3.4.6 Symbols

The variables in a MicroLogix and SLC-500 program are basically address-based. In other words, only addresses are needed to construct a program, though a program constructed in such a manner would be difficult to understand. Therefore, symbols are associated with each address used in a program. The database of symbols and descriptions for each address is maintained separately on the programming device and it does not reside in the controller. Therefore, when connecting to a MicroLogix/SLC-500 in order to examine the online operation of the program, the connecting computer must also have the project file in order to display the symbols with the program addresses.

The symbol associated with an address can be any string of letters, digits, and underlines up to 20 characters long. The symbol must have at least one letter and must not end in one or more digits followed by the single letter B, D, O, H, or E, because it is interpreted as a value.

3.4.7 MicroLogix/SLC-500 Input/Output Addressing

In the MicroLogix and SLC-500 PLCs, the discrete I/O and the analog I/O are part of the I/O image. The O data file contains the image for all output channels, which is transferred to the physical output modules at the end of a program scan. The I data file contains the image of all input channels, which is read from the physical input modules at the beginning of a scan.

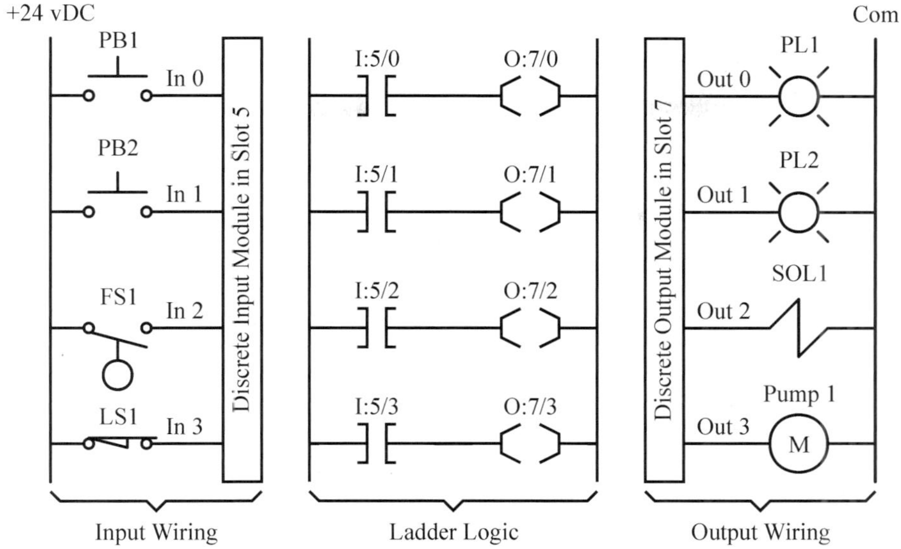

Figure 3.21. MicroLogix or SLC-500 implementation for Figure 3.4 showing I/O addressing.

A MicroLogix or SLC-500 implementation of the hardwired circuit of Figure 3.4 might appear as in Figure 3.21. The switches are connected to channels 0 through 3 of a discrete input module and the lights, solenoid, and motor are connected to channels 0 through 3 of a discrete output module. The address of the input channel is shown with the contact in the ladder. Normally, one defines a symbol that refers to the I/O address and both the address and its symbol appears on the listing. However, for the purpose of this example only the addresses are shown. The address format shown in Figure 3.21 is called the "single line" format. The address "I:5/2" means the contact is controlled by the discrete input channel 2, in slot 5. In general, the address for a discrete input channel is

I:*s*/*c*

where *s* is the slot number (1-30) in which the module resides and *c* is the channel number. Similarly, the address "O:7/1" means the output (coil) is connected to output channel 1, in slot 7.

An alternate format of the discrete I/O address, shows the address as a bit within a word, for example, I:*s*.*w*/*b*, where *w* is the word number and *b* is the bit within the word. For both forms of the address, an alternate way of showing the discrete I/O address on the display is called the "split line" format. In this format, the bit part of the address is shown below the contact or coil.

Summarizing, the SLC-500 and MicroLogix I/O addresses are:

Type of Channel	General Address
Discrete Input	I:*s*/*c* or I:*s*.*w*/*b*
Discrete Output	O:*s*/*c* or O:*s*.*w*/*b*
Analog Input	I:*s*.*w*[/*b*]
Analog Output	O:*s*.*w*[/*b*]

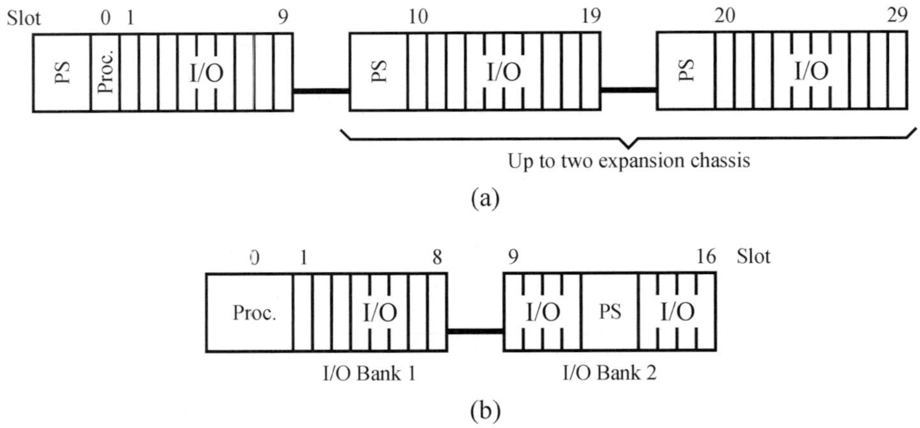

Figure 3.22. Expansion I/O: *(a)* SLC-500 processors; *(b)* Micrologix processors.

where

s is the slot number (0 - 30)
c is the channel number (0 - 31)
w is the word number
b is the bit number (0-15), optional for analog.

Slot 0 is reserved for input/output channels resident in a MicroLogix processor module (base unit). In a SLC-500 system, slots are numbered starting with 1. For analog modules in a SLC-500 system, the word number is generally the channel number. For analog modules in a MicroLogix system, the word number does not correspond to the channel number. The allowable maximum number of slots depends on the particular processor. The numbering of slots for an example SLC-500 system with expansion chassis is shown in Figure 3.22*a*. In contrast to the SLC-500 system, that places the processor and I/O modules in a chassis, the modules in a Micrologix mount on a DIN rail. An example MicroLogix system is shown in Figure 3.22*b*. The allowable number of modules depends on the particular processor.

Example 3.4. Draw a MicroLogix/SLC-500 ladder diagram that will cause the output, pilot light PL2, to be **on** when selector switch SS1 is closed, push button switch PB5 is open, and limit switch LS7 is open. Show only the I/O addresses in the logic. The input/output devices are wired to the following locations:

PL2: Output module, slot 7, channel 15
SS1: Input module, slot 5, channel 4
PB5: Input module, slot 5, channel 10
LS7: Input module, slot 4, channel 7

Solution. The addresses are predefined, so there is no need to map channels to addresses. The addresses are,

PL2: O:7/15
SS1: I:5/4
PB5: I:5/10
LS7: I:4/7

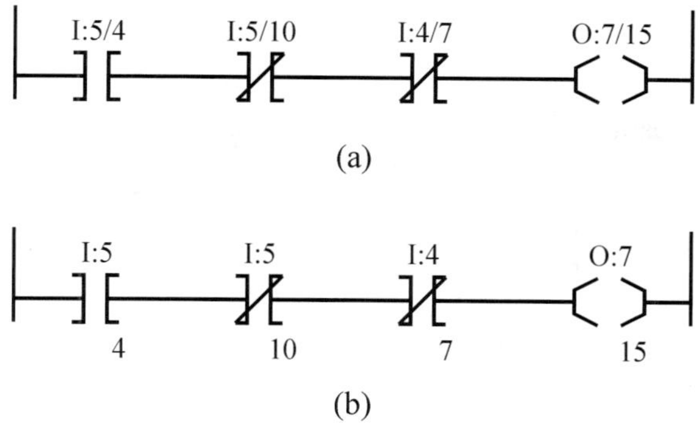

Figure 3.23. Solution to Example 3.4: *(a)* single line address format; *(b)* split line address format.

The ladder logic is shown in Figure 3.23. The addresses are shown above the contact and coil instructions. The addresses are shown in the single line format in Figure 3.23*a* and in the split line format in Figure 3.23*b*.

3.5 CHAPTER SUMMARY

This chapter summarized the organization and addressing of the memory and I/O for the ControlLogix, CompactLogix, Micro800, MicroLogix, and SLC-500 PLCs covered by this text. With the basic PLC knowledge from Chapter 2 and the knowledge about the internal organization of the PLC, one is ready to tackle timers and counters, the most common block instructions.

REFERENCES

Erickson, Kelvin T., 2023. *Programmable Logic Controllers: An Emphasis on Design and Application,* 4th Edition, Dogwood Valley Press, Rolla, MO.

IEC, 2013. *IEC 1131-3: Programmable Logic Controllers - Part 3: Programming Languages*, Edition 3.0, International Electrotechnical Commission, Geneva, Switzerland.

Rockwell Automation, 2008. *SLC 500 Modular Hardware Style User Manual*, pub. 1747-UM011G-EN-P, Rockwell Automation, Inc., Milwaukee, WI, June.

Rockwell Automation, 2011. *MicroLogix 1100 Programmable Controllers Instruction Set Reference Manual*, pub. 1763-RM001D-EN-P, Rockwell Automation Inc., Milwaukee, WI, Sep.

Rockwell Automation, 2014. *MicroLogix™ 1200 and MicroLogix 1500 Programmable Controllers Instruction Set Reference Manual*, pub. 1762-RM001H-EN-P, Rockwell Automation Inc., Milwaukee, WI, July.

Rockwell Automation, 2015. *MicroLogix 1500™ Programmable Controllers User Manual*, pub. 1764-UM001C-EN-P, Rockwell Automation, Inc., Milwaukee, WI, June.

Rockwell Automation, 2017. *ControlLogix Controllers User Manual*, pub. 1756-UM001P-EN-P, Rockwell Automation, Inc., Milwaukee, WI, May.

Rockwell Automation, 2018. *CompactLogix 5370 Controllers User Manual*, pub. 1769-UM021I-EN-P, Rockwell Automation, Inc., Milwaukee, WI, May.

Rockwell Automation, 2018. *Micro800 Expansion I/O Modules User Manual*, pub. 2080-UM003B-EN-P, Rockwell Automation, Inc., Milwaukee, WI, Mar.

Rockwell Automation, 2019. *MicroLogix 1400 Programmable Controllers Instruction Set Reference Manual*, pub. 1766-RM001I-EN-P, Rockwell Automation, Inc., Milwaukee, WI, June.

Rockwell Automation, 2022a. *Compact 5000 I/O Analog Modules User Manual*, pub. 5069-UM005C-EN-P, Rockwell Automation, Inc., Milwaukee, WI, Jan.

Rockwell Automation, 2022b. *Compact 5000 I/O Digital Modules User Manual*, pub. 5069-UM004E-EN-P, Rockwell Automation, Inc., Milwaukee, WI, Jan.

Rockwell Automation, 2022c. *CompactLogix 5380 and Compact GuardLogix 5380 Controllers User Manual*, pub. 5069-UM001I-EN-P, Rockwell Automation, Inc., Milwaukee, WI, Mar.

Rockwell Automation, 2022d. *Logix5000 Controllers Tasks, Programs, and Routines Programming Manual*, pub. 1756-PM005J-EN-P, Rockwell Automation, Inc., Milwaukee, WI, Mar.

Rockwell Automation, 2022e. *Micro830,Micro850, and Micro870 Programmable Controllers User Manual*, pub. 2080-UM002M-EN-P, Rockwell Automation, Inc., Milwaukee, WI, Apr.

PROBLEMS

P3-1. Draw a ladder diagram that will cause the output, pneumatic solenoid PS4250, to be ON when push button switch PB4251 is open and pushbutton PB4252 is closed, and either limit switch LS4112 is open or limit switch LS4479 is closed. Do this problem for the ControlLogix, CompactLogix, Micro800, MicroLogix, and/or SLC-500 PLCs. Show only the I/O address with the ladder contacts/coils.

a) For a ControlLogix processor, the input/output devices are wired to:

PS4250: Output module, chassis 'REM_1', slot 5, channel 12
PB4251: Input module, local chassis, slot 6, channel 2
PB4252: Input module, local chassis, slot 6, channel 3
LS4112: Input module, chassis 'REM_4', slot 8, channel 2
LS4479: Input module, chassis 'REM_2', slot 1, channel 8

b) For a CompactLogix 5370, the input/output devices are wired to:

PS4250: Output module, bank 3, slot 18, channel 12
PB4251: Input module, bank 3, slot 18, channel 2
PB4252: Input module, bank 3, slot 18, channel 3
LS4112: Input module, remote chassis 'REM_4', slot 8, channel 2
LS4479: Input module, remote chassis 'REM_2', slot 1, channel 8

c) For a CompactLogix 5380 processor, the input/output devices are wired to:

PS4250: Output module, remote I/O adapter 'REM_1', slot 5, channel 12
PB4251: Input module, local I/O, slot 6, channel 2
PB4252: Input module, local I/O, slot 6, channel 3
LS4112: Input module, remote I/O adapter 'REM_4', slot 8, channel 2
LS4479: Input module, remote I/O adapter 'REM_2', slot 1, channel 8

d) For a Micro800, the input/output devices are wired to:

PS4250: Output module, plug-in 4, channel 3
PB4251: Input module, plug-in 3, channel 2
PB4252: Input module, plug-in 3, channel 3
LS4112: Input module, expansion 3, channel 2
LS4479: Input module, embedded, channel 8

e) For a MicroLogix, the I/O devices are wired to (use the single-line format):

PS4250: Output module, slot 4, channel 12
PB4251: Input module, slot 1, channel 2
PB4252: Input module, slot 1, channel 3
LS4112: Input module, slot 2, channel 2
LS4479: Input module, base (slot 0), channel 8

f) For a SLC-500, the I/O devices are wired to (use the single-line format):

PS4250: Output module, slot 14, channel 12
PB4251: Input module, slot 7, channel 2
PB4252: Input module, slot 7, channel 3
LS4112: Input module, slot 28, channel 2
LS4479: Input module, slot 17, channel 8

P3-2. Draw a ladder diagram that will cause the output, pneumatic solenoid PSOL2, to be ON when both push button switches PB1 and PB2 are closed (pushed), and either limit switch LS20 is closed or limit switch LS101 is open. Do this problem for the ControlLogix, CompactLogix, Micro800, MicroLogix, and/or SLC-500 PLCs. Show only the I/O address with the ladder contacts/coils.

a) For a ControlLogix processor, the input/output devices are wired to:

PSOL2:	Output module, chassis 'REM_1', slot 5, channel 10
PB1:	Input module, local chassis, slot 4, channel 6
PB2:	Input module, local chassis, slot 4, channel 7
LS20:	Input module, chassis 'REM_1', slot 2, channel 2
LS101:	Input module, chassis 'REM_2', slot 5, channel 15

b) For a CompactLogix 5370 processor, the input/output devices are wired to:

PSOL2:	Output module, bank 1, slot 5, channel 10
PB1:	Input module, bank 2, slot 9, channel 6
PB2:	Input module, bank 2, slot 9, channel 7
LS20:	Input module, remote chassis 'REM_1', slot 2, channel 2
LS101:	Input module, remote chassis 'REM_2', slot 5, channel 15

c) For a CompactLogix 5380 processor, the input/output devices are wired to:

PSOL2:	Output module, remote I/O adapter 'REM_1', slot 5, channel 10
PB1:	Input module, local I/O, slot 4, channel 6
PB2:	Input module, local I/O, slot 4, channel 7
LS20:	Input module, remote I/O adapter 'REM_1', slot 2, channel 2
LS101:	Input module, remote I/O adapter 'REM_2', slot 5, channel 15

d) For a Micro800, the input/output devices are wired to:

PSOL2:	Output module, embedded, channel 10
PB1:	Input module, plug-in 2, channel 2
PB2:	Input module, plug-in 2, channel 3
LS20:	Input module, plug-in 1, channel 1
LS101:	Input module, expansion 3, channel 7

e) For a MicroLogix, the I/O devices are wired to (use the single-line format):

PSOL2:	Output module, base (slot 0), channel 10
PB1:	Input module, slot 2, channel 7
PB2:	Input module, slot 2, channel 8
LS20:	Input module, slot 1, channel 1
LS101:	Input module, slot 1, channel 15

f) For a SLC-500, the I/O devices are wired to (use the single-line format):

PSOL2:	Output module, slot 7, channel 10
PB1:	Input module, slot 21, channel 7
PB2:	Input module, slot 21, channel 8
LS20:	Input module, slot 14, channel 1
LS101:	Input module, slot 17, channel 15

P3-3. Draw a ladder diagram that will cause the output, pneumatic solenoid PS6124, to be ON when both push button switches PB6121 and PB6122 are closed (pushed), and either limit switch LS6123 is open or limit switch LS6125 is closed. Do this problem for the ControlLogix, CompactLogix, Micro800, MicroLogix, and/or SLC-500 PLCs. Show only the I/O address with the ladder contacts/coils.

a) For a ControlLogix processor, the input/output devices are wired to:

PS6124: Output module, chassis 'REM_2', slot 5, channel 12
PB6121: Input module, local chassis, slot 0, channel 7
PB6122: Input module, local chassis, slot 0, channel 8
LS6123: Input module, chassis 'REM_2', slot 4, channel 5
LS6125: Input module, chassis 'REM_3', slot 7, channel 0

b) For a CompactLogix 5370 processor, the input/output devices are wired to:

PS6124: Output module, remote chassis 'REM_2', slot 5, channel 12
PB6121: Input module, bank 0, slot 0, channel 7
PB6122: Input module, bank 0, slot 0, channel 8
LS6123: Input module, bank 1, slot 4, channel 5
LS6125: Input module, bank 1, slot 7, channel 0

c) For a CompactLogix 5380 processor, the input/output devices are wired to:

PS6124: Output module, remote I/O adapter 'REM_2', slot 5, channel 4
PB6121: Input module, local I/O, slot 0, channel 7
PB6122: Input module, local I/O, slot 0, channel 8
LS6123: Input module, remote I/O adapter 'REM_2', slot 4, channel 5
LS6125: Input module, remote I/O adapter 'REM_3', slot 7, channel 0

d) For a Micro800, the input/output devices are wired to:

PS6124: Output module, expansion 3, channel 7
PB6121: Input module, plug-in 2, channel 7
PB6122: Input module, plug-in 2, channel 8
LS6123: Input module, plug-in 3, channel 1
LS6125: Input module, plug-in 3, channel 2

e) For a MicroLogix, the I/O devices are wired to (use the single-line format)

PS6124: Output module, slot 5, channel 15
PB6121: Input module, slot 1, channel 7
PB6122: Input module, slot 1, channel 8
LS6123: Input module, slot 2, channel 12
LS6125: Input module, slot 2, channel 0

f) For a SLC-500, the I/O devices are wired to (use the single-line format):

PS6124: Output module, slot 26, channel 15
PB6121: Input module, slot 11, channel 7
PB6122: Input module, slot 11, channel 8
LS6123: Input module, slot 23, channel 12
LS6125: Input module, slot 30, channel 0

P3-4. Draw a ladder diagram that will cause the output, solenoid XV302, to be ON when toggle switch SW321 is closed, and either limit switch LS306 is open or limit switch LS308 is open. Do this problem for the ControlLogix, CompactLogix, Micro800, MicroLogix, and/or SLC-500 PLCs. Show only the I/O address with the ladder contacts/coils.

a) For a ControlLogix processor, the input/output devices are wired to:

XV302: Output module, chassis 'REM_2', slot 7, channel 2
SW321: Input module, local chassis, slot 6, channel 1
LS306: Input module, chassis 'REM_2', slot 1, channel 10
LS308: Input module, chassis 'REM_2', slot 1, channel 11

b) For a CompactLogix 5370 processor, the input/output devices are wired to:

XV302: Output module, bank 1, slot 7, channel 2
SW321: Input module, remote chassis 'REM_2', slot 6, channel 1
LS306: Input module, bank 0, slot 1, channel 10
LS308: Input module, bank 0, slot 1, channel 11

c) For a CompactLogix 5380 processor, the input/output devices are wired to:

XV302: Output module, remote I/O adapter 'REM_2', slot 7, channel 2
SW321: Input module, local I/O, slot 6, channel 1
LS306: Input module, remote I/O adapter 'REM_2', slot 1, channel 10
LS308: Input module, remote I/O adapter 'REM_2', slot 1, channel 11

d) For a Micro800, the input/output devices are wired to:

XV302: Output module, embedded, channel 2
SW321: Input module, expansion 2, channel 1
LS306: Input module, expansion 2, channel 12
LS308: Input module, expansion 2, channel 13

e) For a MicroLogix, the I/O devices are wired to (use the single-line format):

XV302: Output module, base (slot 0), channel 2
SW321: Input module, slot 6, channel 1
LS306: Input module, slot 2, channel 12
LS308: Input module, slot 2, channel 13

f) For a SLC-500, the I/O devices are wired to (use the single-line format):

XV302: Output module, slot 23, channel 2
SW321: Input module, slot 6, channel 1
LS306: Input module, slot 17, channel 10
LS308: Input module, slot 17, channel 11

P3-5. Draw a ladder diagram that will cause the output, pump P604, to be ON when toggle switch SW601 is closed, and either limit switch PS605 is open or limit switch PS607 is open. Do this problem for the ControlLogix, CompactLogix, Micro800, MicroLogix, and/or SLC-500 PLCs. Show only the I/O address with the ladder contacts/coils.

a) For a ControlLogix processor, the input/output devices are wired to:

P604: Output module, chassis 'REM_1', slot 6, channel 3
SW601: Input module, local chassis, slot 5, channel 2
PS605: Input module, chassis 'REM_1', slot 2, channel 11

PS607: Input module, chassis 'REM_1', slot 2, channel 13

b) For a CompactLogix 5370 processor, the input/output devices are wired to:

P604: Output module, bank 1, slot 6, channel 3
SW601: Input module, bank 1, slot 5, channel 2
PS605: Input module, remote chassis 'REM_1', slot 2, channel 11
PS607: Input module, remote chassis 'REM_1', slot 2, channel 13

c) For a CompactLogix 5380 processor, the input/output devices are wired to:

P604: Output module, remote I/O adapter 'REM_1', slot 6, channel 3
SW601: Input module, local I/O, slot 5, channel 2
PS605: Input module, remote I/O adapter 'REM_1', slot 2, channel 11
PS607: Input module, remote I/O adapter 'REM_1', slot 2, channel 13

d) For a Micro800, the input/output devices are wired to:

P604: Output module, expansion 6, channel 3
SW601: Input module, embedded, channel 3
PS605: Input module, plug-in 1, channel 1
PS607: Input module, plug-in 2, channel 3

e) For a MicroLogix, the I/O devices are wired to (use the single-line format).

P604: Output module, slot 6, channel 3
SW601: Input module, base (slot 0), channel 3
PS605: Input module, slot 2, channel 11
PS607: Input module, slot 2, channel 13

f) For a SLC-500, the I/O devices are wired to (use the single-line format):

P604: Output module, slot 16, channel 3
SW601: Input module, slot 5, channel 3
PS605: Input module, slot 12, channel 11
PS607: Input module, slot 12, channel 13

P3-6. Draw a ladder diagram that will cause the output, alarm lamp ALRM1534, to be ON when (1) either the infeed conveyor CONV1541 is OFF or the outfeed conveyor CONV1542 is OFF and (2) the limit switch LS1535 is open and (3) the proximity switch PS1536 is closed. Do this problem for the ControlLogix, CompactLogix, Micro800, MicroLogix, and/or SLC-500 PLCs. Show only the I/O address with the ladder contacts/coils.

a) For a ControlLogix processor, the input/output devices are wired to:

LS1535: Input module, chassis 'REM_1', slot 2, channel 8
PS1536: Input module, chassis 'REM_1', slot 2, channel 14
CONV1541: Output module, chassis 'REM_2', slot 6, channel 12
CONV1542: Output module, chassis 'REM_2', slot 6, channel 13
ALRM1534: Output module, local chassis, slot 1, channel 3

b) For a CompactLogix 5370 processor, the input/output devices are wired to:

LS1535: Input module, bank 2, slot 10, channel 8
PS1536: Input module, bank 2, slot 11, channel 14
CONV1541: Output module, remote chassis 'REM_2', slot 6, channel 12

CONV1542: Output module, remote chassis 'REM_2', slot 6, channel 13
ALRM1534: Output module, bank 0, slot 1, channel 3

c) For a CompactLogix 5380 processor, the input/output devices are wired to:

LS1535: Input module, remote I/O adapter 'REM_1', slot 2, channel 8
PS1536: Input module, remote I/O adapter 'REM_1', slot 2, channel 14
CONV1541: Output module, remote I/O adapter 'REM_2', slot 6, ch. 12
CONV1542: Output module, remote I/O adapter 'REM_2', slot 6, ch. 13
ALRM1534: Output module, local I/O, slot 1, channel 3

d) For a Micro800, the input/output devices are wired to:

LS1535: Input module, embedded, channel 8
PS1536: Input module, embedded, channel 14
CONV1541: Output module, plug-in 2, channel 0
CONV1542: Output module, plug-in 2, channel 1
ALRM1534: Output module, embedded, channel 3

e) For a MicroLogix, the I/O devices are wired to (use the single-line format):

LS1535: Input module, slot 4, channel 8
PS1536: Input module, slot 4, channel 14
CONV1541: Output module, slot 6, channel 14
CONV1542: Output module, slot 6, channel 15
ALRM1534: Output module, base (slot 0), channel 3

f) For a SLC-500, the I/O devices are wired to (use the single-line format):

LS1535: Input module, slot 12, channel 8
PS1536: Input module, slot 12, channel 14
CONV1541: Output module, slot 26, channel 14
CONV1542: Output module, slot 26, channel 15
ALRM1534: Output module, slot 2, channel 3

P3-7. Draw a ladder diagram that will cause the output, alarm lamp ALM492, to be ON when (1) the outfeed conveyor CNV465 is OFF and (2) either limit switch LS423 is open or limit switch LS425 is open and (3) the toggle switch SW436 is closed. Do this problem for the ControlLogix, CompactLogix, Micro800, MicroLogix, and/or SLC-500 PLCs. Show only the I/O address with the ladder contacts/coils.

a) For a ControlLogix processor, the input/output devices are wired to:

LS423: Input module, chassis 'REM_1', slot 1, channel 2
LS425: Input module, chassis 'REM_1', slot 1, channel 3
SW436: Input module, chassis 'REM_1', slot 5, channel 9
CNV465: Output module, chassis 'REM_3', slot 2, channel 13
ALM492: Output module, chassis 'REM_3', slot 3, channel 3

b) For a CompactLogix 5370 processor, the input/output devices are wired to:

LS423: Input module, bank 2, slot 15, channel 2
LS425: Input module, bank 2, slot 15, channel 3
SW436: Input module, remote chassis 'REM_1', slot 5, channel 9
CNV465: Output module, remote chassis 'REM_3', slot 2, channel 13

ALM492: Output module, remote chassis 'REM_3', slot 3, channel 3

c) For a CompactLogix 5380 processor, the input/output devices are wired to:

LS423: Input module, remote I/O adapter 'REM_1', slot 1, channel 2
LS425: Input module, remote I/O adapter 'REM_1', slot 1, channel 3
SW436: Input module, remote I/O adapter 'REM_1', slot 5, channel 9
CNV465: Output module, remote I/O adapter 'REM_3', slot 2, channel 13
ALM492: Output module, remote I/O adapter 'REM_3', slot 3, channel 3

d) For a Micro800, the input/output devices are wired to:

LS423: Input module, plug-in 1, channel 2
LS425: Input module, plug-in 1, channel 3
SW436: Input module, plug-in 3, channel 1
CNV465: Output module, embedded, channel 3
ALM492: Output module, embedded, channel 5

e) For a MicroLogix, the I/O devices are wired to (use the single-line format):

LS423: Input module, slot 1, channel 2
LS425: Input module, slot 1, channel 3
SW436: Input module, slot 5, channel 9
CNV465: Output module, base (slot 0), channel 3
ALM492: Output module, base (slot 0), channel 5

f) For a SLC-500, the I/O devices are wired to (use the single-line format):

LS423: Input module, slot 11, channel 2
LS425: Input module, slot 11, channel 3
SW436: Input module, slot 15, channel 9
CNV465: Output module, slot 26, channel 13
ALM492: Output module, slot 22, channel 3

P3-8. Draw a ladder diagram that will cause the output, solenoid CYL201, to be ON when limit switch LS204 is closed and LS205 is open and either push button switch PB210 is closed (pushed) or push button switch PB211 is closed. Do this problem for the ControlLogix, CompactLogix, Micro800, MicroLogix, and/or SLC-500 PLCs. Show only the I/O address with the ladder contacts/coils.

a) For a ControlLogix processor, the input/output devices are wired to:

CYL201: Output module, local chassis, slot 6, channel 7
LS204: Input module, chassis 'REM_1', slot 0, channel 21
LS205: Input module, chassis 'REM_1', slot 0, channel 22
PB210: Input module, chassis 'REM_2', slot 3, channel 9
PB211: Input module, chassis 'REM_2', slot 3, channel 10

b) For a CompactLogix 5370 processor, the input/output devices are wired to:

CYL201: Output module, bank 1, slot 6, channel 7
LS204: Input module, bank 2, slot 9, channel 21
LS205: Input module, bank 2, slot 9, channel 22
PB210: Input module, remote chassis 'REM_2', slot 3, channel 9
PB211: Input module, remote chassis 'REM_2', slot 3, channel 10

c) For a CompactLogix 5380 processor, the input/output devices are wired to:

CYL201: Output module, local I/O, slot 6, channel 7
LS204: Input module, remote I/O adapter 'REM_1', slot 0, channel 5
LS205: Input module, remote I/O adapter 'REM_1', slot 0, channel 6
PB210: Input module, remote I/O adapter 'REM_2', slot 3, channel 9
PB211: Input module, remote I/O adapter 'REM_2', slot 3, channel 10

d) For a Micro800, the input/output devices are wired to:

CYL201: Output module, expansion 2, channel 7
LS204: Input module, plug-in 3, channel 1
LS205: Input module, plug-in 3, channel 2
PB210: Input module, embedded, channel 9
PB211: Input module, embedded, channel 10

e) For a MicroLogix, the I/O devices are wired to (use the single-line format):

CYL201: Output module, slot 6, channel 7
LS204: Input module, slot 3, channel 5
LS205: Input module, slot 3, channel 6
PB210: Input module, slot 5, channel 9
PB211: Input module, slot 5, channel 10

f) For a SLC-500, the I/O devices are wired to (use the single-line format):

CYL201: Output module, slot 6, channel 7
LS204: Input module, slot 13, channel 21
LS205: Input module, slot 13, channel 22
PB210: Input module, slot 25, channel 9
PB211: Input module, slot 25, channel 10

4 Input/Output Modules and Installation

Chapter Topics:

- Discrete input and output modules
- Analog input and output modules
- Specialized modules
- Installation wiring

OBJECTIVES

Upon completion of this chapter, you will be able to understand:

- Wiring of typical sensors and actuators to PLC I/O modules
- Surge protection for certain types of devices
- Typical control cabinet layout
- Typical installation wiring

Scenario: Plug-chute switch problem.

In a power plant coal-handling system, a plug-chute switch is placed at the end of conveyor belt and detects when coal is "piling up" and plugging the system. Obviously, when a plug is detected, the conveyor should be immediately stopped.

A simplified diagram of the system is shown in Figure 4.1. The plug-chute switch is a normally-closed contact that opens as the coal presses against a diaphragm. The switch wire eventually reaches a terminal block in the control cabinet where it is wired to a PLC AC discrete input module and to a time-delay relay. The PLC replaces a relay control system, and the relay system is being dismantled but is still connected to many of the field sensors. In the PLC program, the plug-chute switch is the input to a timer. The timer protects against intermittent pieces of coal hitting the diaphragm and shutting down the system. The switch must be open for 2 seconds before the conveyor is shut off.

Yesterday, one of the workers discovered a pile of coal around the end of the conveyor and the chamber at the end of the conveyor was completely full of coal. Apparently, the plug-chute switch had not stopped the conveyor.

***Solution*:** After cleaning out the coal, the conveyors are run without coal. The PLC properly senses the closed plug-chute switch (or else the conveyor would not run). However, pushing the diaphragm on the plug-chute switch does not stop the conveyor. The plug-chute switch is replaced with a known good switch, and it still does not stop the conveyor when

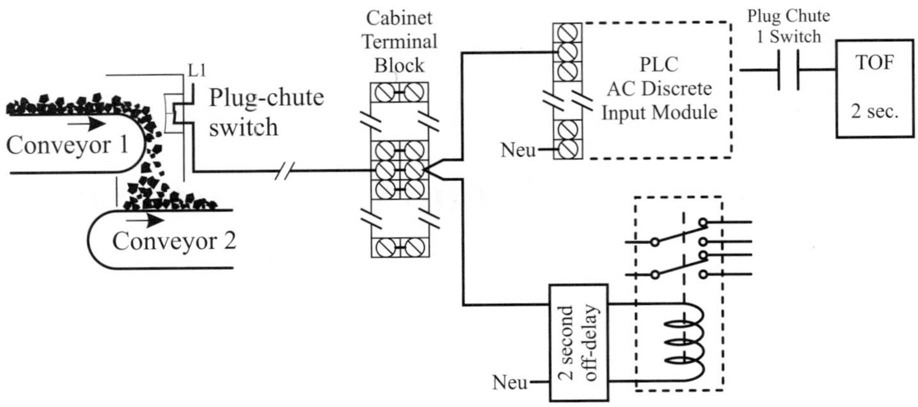

Figure 4.1. Plug-chute switch wiring.

pressed. A broomstick is wedged in to constantly press the diaphragm. The PLC program operation is then observed. The plug-chute input intermittently flashes off but does not stay off longer than 0.01 second. The intensity of the channel indicator lamp on the discrete input module is about half of other discrete inputs that are on. At this point the problem is in one of three places: the PLC module, the time-delay relay, or the wiring between the PLC and the switch. Disconnecting the time-delay relay is the easiest thing to try, and so it is tried first. When the time-delay relay is disconnected, the PLC correctly senses the open plug-chute switch. Since the time-delay relay is no longer used, it is left disconnected.

4.1 INTRODUCTION

In order for a PLC to be functional, it must connect to the "real world," involving wiring to discrete, analog, and specialized field devices. This chapter describes these connections and covers other installation issues such as power distribution and control panel layout.

The input and output (I/O) modules provide the physical interface between the PLC processor and the field devices, such as switches, lamps, and valves. There are various types of modules to handle the different types of sensors and actuators. Regardless of the type of module, there are some common features: removable terminal blocks, isolation, and diagnostic indicators.

All wiring to the I/O module is connected to a removable terminal block, and the terminal block is plugged onto the module. If the module must be replaced, the terminal block is removed, the old module is removed, the new module is inserted, and the terminal block is plugged onto the new module. This process of changing the module requires less than a minute and does not involve rewiring the module, an important maintenance consideration.

The I/O modules provide isolation between the field wiring and PLC internal circuitry. Figure 4.2 shows this isolation barrier in both the input and output module. Optical isolators provide the isolation barrier, shown in the circuit diagrams in the succeeding sections. This isolation is important since the field wiring is generally powered from a different circuit and/or voltage level than the PLC processor. The PLC power supply provides low-voltage power to the right side of the input module and the left side of the output module. In addition

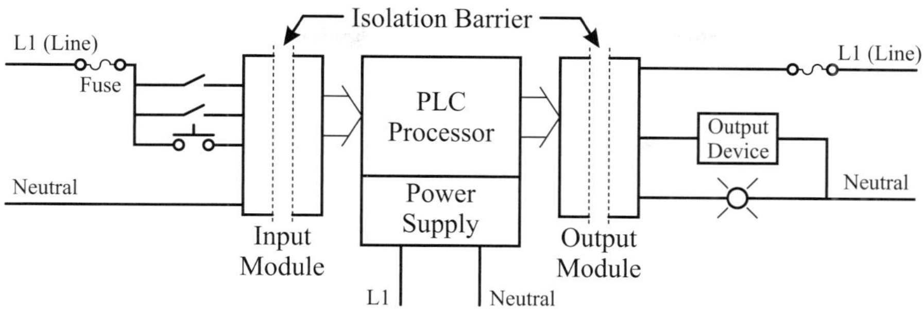

Figure 4.2. Isolation between I/O devices and PLC.

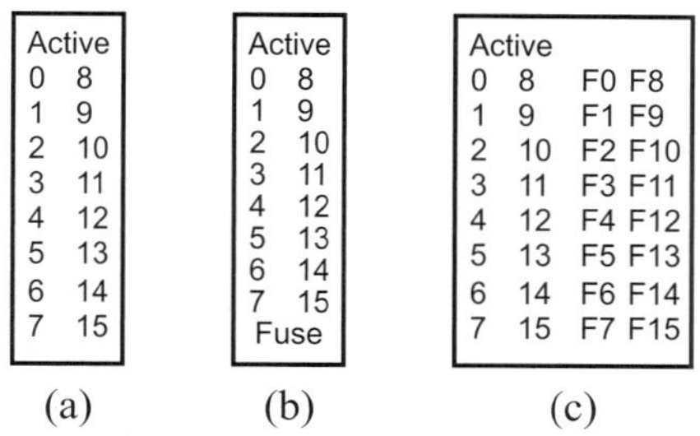

Figure 4.3. Example module indicators: *(a)* discrete input; *(b)* discrete output; *(c)* discrete output with individual fuse indicators.

to providing the interface between the field and PLC voltage levels, the isolation prevents ground loops and noisy electrical signals from interfering with the PLC operation. Typically, there are also differences in the earth (ground) potential between the field wiring circuits and the PLC power wiring. Also, during an abnormal event, such as a near lightning strike, circuit commons that are normally at the same potential may have a short-term (< 1 second) potential difference. The isolation prevents damage to the PLC even if a surge damages the module. For example, a wiring fault that causes 480-volt AC to be imposed on a 24-volt DC input channel would only damage the input module. The PLC itself and the other modules in the system will not be affected.

I/O modules have indicators useful for troubleshooting. For example, discrete input modules have one indicator for every channel that is **on** when its corresponding input channel is **on** (Figure 4.3*a*). Similarly, discrete output modules have one indicator for each channel that is **on** when the corresponding output channel is being commanded to turn **on** (Figure 4.3*b*). A blown (open) fuse indicator on an output module may be turned on if any output fuse is open ("Fuse" in Figure 4.3*b*) or they may be an indication for each fuse ("Fx" in Figure 4.3*c*). Modules may have indicators that turn **on** when the module is functioning properly. For example, in Figure 4.3*a*, the "Active" indicator on a discrete input module lights when the module is properly communicating with the processor. For a discrete output module, the "Active" indicator may also indicate that the processor is in the run mode.

One aspect of PLC I/O modules that is often confusing is the designation of a device connection as being *sinking* or *sourcing*. Sinking and sourcing refer to the flow of current (power) when the device is active. An output is called sourcing if current flows **out** when the output is active. An output is called sinking if current flows **in** when the output is active.

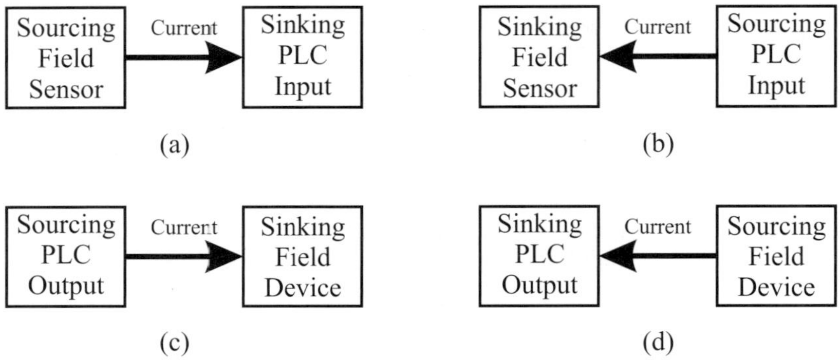

Figure 4.4. Sourcing and sinking connections: *(a)* sinking PLC input; *(b)* sourcing PLC input; *(c)* sourcing PLC output; *(d)* sinking PLC output.

An input is called sourcing if current flows **out** when the input is active. An input is called sinking if current flows **in** when the input is active. Normally, a sourcing field sensor is connected to a sinking PLC input (Figure 4.4*a*) and a sinking field sensor is connected to a sourcing PLC input (Figure 4.4*b*). Similarly, a sourcing PLC output is connected to a sinking field output device (Figure 4.4*c*) and a sinking PLC output is connected to a sourcing field output device (Figure 4.4*d*). For DC devices, the arrows in Figure 4.4 indicate the direction of current flow when the field sensor or PLC output is active (**on**). For AC devices, the arrows indicate the direction of power flow.

Only the basic operation of the modules is described. Many older I/O modules have switches and/or jumpers that set up, or configure, the module to operate as desired. Most newer I/O modules have an included microprocessor and so the configuration parameters are set up with the PLC programming software. For discrete modules, the configuration information consists of such items as channel filter time constants, on-to-off transition delay times, and off-to-on transition delay times. For analog modules, the configuration information consists of such items as number of channels, type of channel (single-ended or differential, current or voltage), data format (integer or real), voltage/current range, and integer range.

Also, the process of configuring the PLC processor for the actual I/O modules is not presented. Some processors require the user to set switches that determine the modules in each chassis position. For other processors, the user configures the types of modules that are expected in each chassis slot as part of the PLC software set up. In this case, when the PLC is placed in run mode, the processor verifies that the expected modules are present and functioning before commencing the program scan. Other processors automatically identify the I/O modules that are present and adjust the I/O scan accordingly.

Finally, the connection diagrams in this chapter are typical and are not specific to a particular manufacturer. One should always refer to the manufacturer's documentation for details.

4.2 DISCRETE MODULES

Since PLCs were originally developed to solve discrete sequential control problems, discrete modules remain the largest number of modules in a PLC system. Even though

PLCs are capable of performing continuous control functions, their mainstay is still discrete control.

There is a limited set of standard discrete voltage levels handled by discrete input and output modules:

120 volt AC or DC	10 - 30 volt DC
220 volt AC or DC	20 - 60 volt DC
10 - 30 volt AC or DC	5 volt digital logic

Historically, most discrete input/output signals in non-chemical control applications were nominally 120 volts. Due to safety considerations, most chemical control applications used 24-volt signals. However, recent safe electrical work practices (NFPA, 2021a) require protective clothing for electricians working on exposed wiring having voltages higher than 50 volts. Because of these regulations, newer systems use 24-volt control signals, and many older systems are being converted to 24-volt signals.

4.2.1 Discrete Input Modules

A discrete input module senses the status of a device that has only two states, on/off, open/closed, running/stopped, and so on. The most common discrete devices sensed by a PLC module are some kind of a switch, for example,

Pushbutton	Level	Starter auxiliary
Selector	Pressure	Relay contact
Limit	Position	Zero-speed
Proximity	Thumbwheel	

The block diagram of a typical discrete input is shown in Figure 4.5. The optical isolator provides electrical isolation between the field wiring and the PLC internal circuitry. The electronic circuitry to the left of the optical isolator is powered from the field, and the circuitry to the right of the optical isolator is powered from the internal PLC voltage source.

Each discrete input module has a threshold detector that senses when the input device is on. Threshold sensing is different for sinking and sourcing modules. For sinking input modules, the device is sensed **on** when the voltage produced by the input device is above a minimum voltage, commonly called the "minimum on-state voltage." If the input device produces a voltage below the "maximum off-state voltage," then the device is sensed to be **off**. The maximum off-state voltage is less than the minimum on-state voltage. For example, a 120-volt AC/DC sinking input module may have a minimum on-state voltage of 80 volts AC/DC and a maximum off-state voltage of 45 volts AC/DC. A 24-volt DC sinking module may have a minimum on-state voltage of 10 volts and a maximum off-state voltage of 6 volts. Obviously, module operation is not guaranteed if the device produces a voltage between the maximum off-state and minimum on-state voltages. For sourcing input

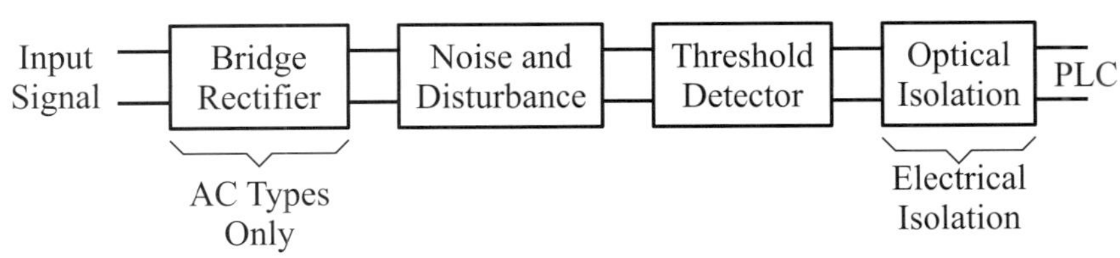

Figure 4.5. Discrete input module block diagram.

modules, the device is sensed **on** when the voltage produced by the input device is below a maximum voltage, commonly called the "maximum on-state voltage." If the input device produces a voltage above the "minimum off-state voltage", then the device is sensed to be **off**. For example, a 24-volt DC sourcing module may have a maximum on-state voltage of 14 volts and a minimum off-state voltage of 20 volts.

The circuitry of a typical AC discrete input module, along with connections to typical input devices, is shown in Figure 4.6. The capacitor and resistors R_1 and R_2 form a simple noise filter. The bridge rectifier converts the AC to full-wave rectified DC. The resistor R_3 and the zener diode (D_Z) form the threshold detection. An optoisolator, consisting of a light-emitting diode and a photoelectric transistor, provides electrical isolation between the field wiring and the PLC internal circuitry. When a switch is closed, producing an input voltage of sufficient magnitude, current flows through the light-emitting diode in the optoisolator, generating light and triggering current flow through the photoelectric transistor which is sensed by the PLC digital logic. For fault diagnosis, an input state indication is **on** when the input switch is **closed**. This indicator may be on the field wiring side of the optoisolator or may be on the PLC side of the optoisolator. Input modules that handle AC signals are typically *sinking* modules, that is, power flows **into** the module when the field device is **on**. This module can also handle DC voltages.

The module shown in Figure 4.6 has one common connection for all channels. This common is the return path for all of the input devices. Some modules provide multiple commons, for example, one common for every four input channels. Other modules provide isolated channels, that is, the common of each channel is separated from the other channel commons. Isolated channels or multiple commons are needed when

1. Mixing voltages on the same module.
2. Mixing different AC phases on the same module. This is especially needed when the 3-phase plant wiring is in a delta configuration.
3. Mixing AC and DC circuits on the same module.

There may be a problem when connecting a solid-state sensor with a triac-type output to an AC discrete input module. This sensor produces a small leakage current even when the device is off. Unfortunately, this leakage current may be large enough to trigger the input

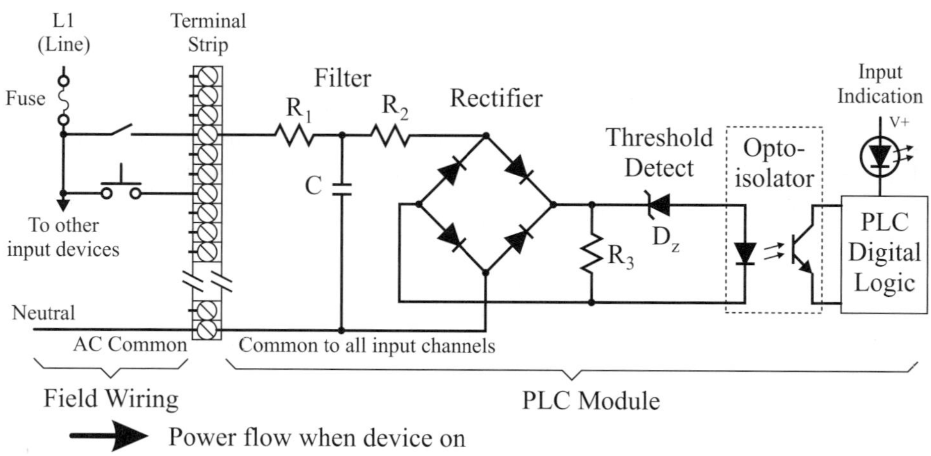

Figure 4.6. Typical AC discrete input channel.

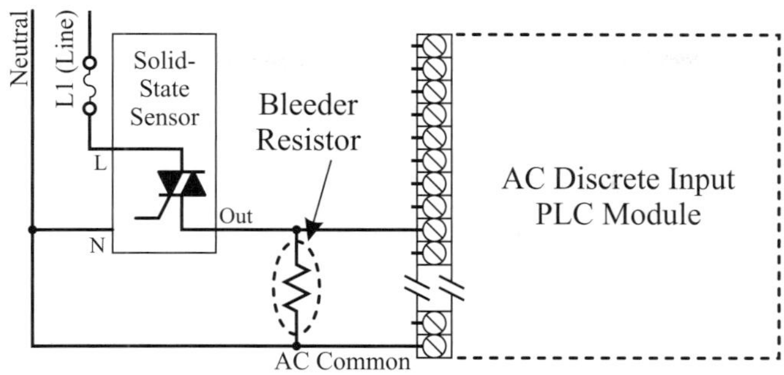

Figure 4.7. Using bleeder resistor with solid-state sensor.

module channel. In this case, a bleeder resistor is added as shown in Figure 4.7 to "bleed" the leakage current to the AC common. The bleeder resistor assures that the PLC input channel senses the **on** state of the device only when the device is truly **on**.

Figure 4.8 shows circuitry of a typical *sinking* DC discrete input module channel. Connections to typical switches and a 3-wire sensor are also illustrated. The two resistors form a simple threshold detector, and the capacitor is a simple noise filter. The diode D_1 provides protection for the optoisolator in case the DC power supply polarity is accidentally reversed. When a switch is closed, current flows through the optoisolator LED and is sensed by the PLC digital logic. The 3-wire sensor (having a PNP transistor output) is a sourcing device. When active, the PNP output transistor conducts, causing current to flow into the discrete input channel and then be sensed by the PLC logic. The module shown in Figure 4.8 has one DC common connection for all channels. Some modules may provide one DC common connection for a group of channels, for example, four or eight channels. Isolated channels may also be used and are discussed below.

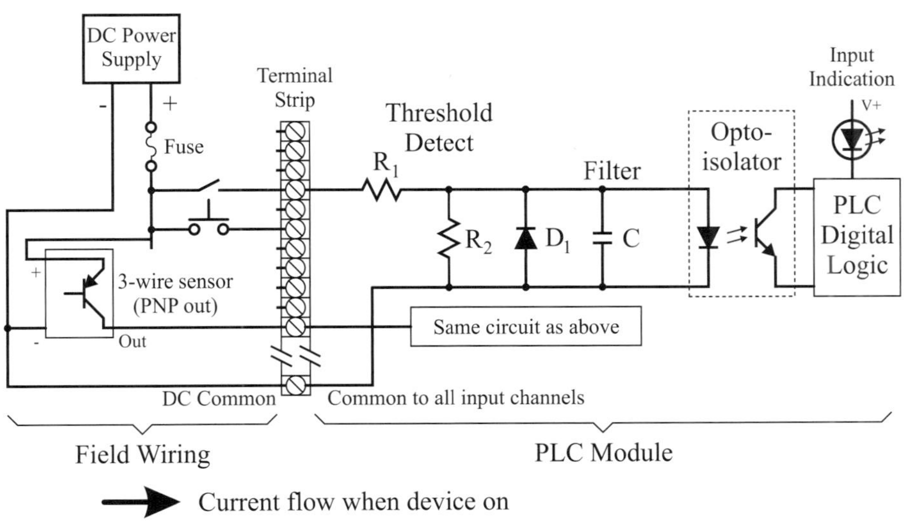

Figure 4.8. Typical sinking DC discrete input channel.

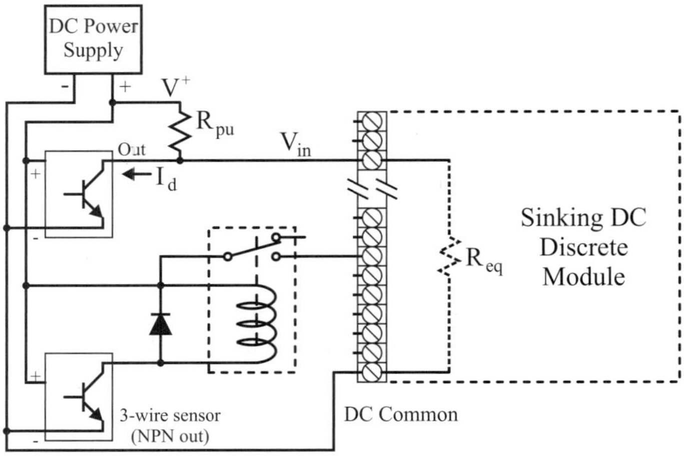

Figure 4.9. Connect a sinking sensor to a sinking DC discrete input channel.

It is possible to connect a sinking field input to a sinking DC discrete input channel with a pull-up resistor or interposing relay, as shown in Figure 4.9. The interposing relay is less reliable and introduces an additional delay that may be unacceptable for a rapidly switching discrete device. A pull-up resistor is less expensive, but may not be possible if there is no resistor that satifies the criteria below. The pull-up resistor (R_{pu}) in Figure 4.9 ensures the voltage sensed by the input channel is sufficiently high when the device is **off**. When using the pull-up resistor, the PLC state of the input is opposite to the input sensor state. When the sensor is **on**, the PLC detects it as **off** since the voltage at the discrete input channel is below the threshold for detection of the **on** state. When the sensor is **off**, the PLC detects it as **on** since the pull-up resistor ensures the voltage at the input channel is above the threshold voltage. When specifying the value of R_{pu}, it must be selected so

1. The voltage at the PLC discrete input is above the minimum **on** voltage when the device is **off** (inactive), and
2. The sinking current into the device is always below a maximum value.

These criteria specify the minimum and maximum value of R_{pu}. In equation form, the first criterion means

$$V_{in} = V^{+}\frac{R_{eq}}{R_{pu}+R_{eq}} > V_{th} \text{ or } R_{pu} < R_{eq}\frac{V^{+}-V_{th}}{V_{th}} \tag{4.1}$$

where V^{+} is the power supply voltage, R_{eq} is the equivalent input resistance of the input channel, and V_{th} is the minimum "on" state voltage of the input channel. In order to accommodate situations when the power supply voltage is degraded, the minimum power supply voltage should be used in (4.1). In equation form, the second criterion means

$$I_{d} = \frac{V^{+}-V_{out,\min}}{R_{pu}} < I_{d,\max} \text{ or } R_{pu} > \frac{V^{+}-V_{out,\min}}{I_{d,\max}} \tag{4.2}$$

where $I_{d,\max}$ is the maximum device sink current and $V_{out,\min}$ is the minimum output voltage of the sensor when it is active. In order to accommodate a worst-case scenario, the maximum power supply voltage should be used in (4.2). Combining equations 4.1 and 4.2 to specify the acceptable range of the pull-up resistor,

$$\frac{V_{\max}^{+} - V_{out,\min}}{I_{d,\max}} < R_{pu} < R_{eq} \frac{V_{\min}^{+} - V_{th}}{V_{th}} \tag{4.3}$$

where $V_{\min}^{+}$ and $V_{\max}^{+}$ are the minimum and maximum power supply voltages, respectively. In general, the largest acceptable resistance will be the best since it will require the smallest power rating.

Example 4.1. Specify an acceptable value and power rating of a pull-up resistor to be used to connect a sinking sensor device to a PLC sinking discrete input channel. The power supply is nominally 24 volts, but can range between 23 and 25 volts, depending on the current draw. The minimum "on" state voltage of the discrete input channel is 10 volts, and the equivalent input resistance is 2500 ohms. The maximum device sink current is 20 mA, and the minimum device output voltage is 0.3 volts.

Solution. The values to be used in equation 4.3 are:

$$V_{\max}^{+} = 25 \qquad R_{eq} = 2500 \qquad I_{d,\max} = 0.02$$
$$V_{\min}^{+} = 23 \qquad V_{th} = 10 \qquad V_{out,\min} = 0.3$$

Substituting into (4.3),

$$\frac{25-0.3}{0.02} < R_{pu} < 2500\frac{23-10}{10}$$
$$1235 < R_{pu} < 3250$$

If the standard 5% tolerance resistor of 3000 ohms is selected, its range of 2850 to 3150 ohms is acceptable. The required power rating is determined by squaring the maximum voltage across the resistor and dividing it by the minimum possible resistance,

$$P_{pu} = \frac{(25-0.3)^2}{2850} = 0.214 \text{ watts}$$

Therefore, a 3000 ohm resistor with a 1/4-watt power rating must be used.

The circuit for a typical *sourcing* DC discrete input channel is shown in Figure 4.10. Note that when a switch is closed, current flows **out** of the discrete input module. The two resistors form a simple threshold detector, and the capacitor is a simple noise filter. The diode D_1 provides protection for the optoisolator in case the DC power supply polarity is accidentally reversed. When a switch is closed, current flows through the optoisolator LED and is sensed by the PLC digital logic. The 3-wire sensor (having a NPN transistor output) is a sinking device. When active, the NPN output transistor conducts, causing current to flow into NPN collector (and out of the discrete input channel). The fuse shown internal to the module may or may not be present. If an internal fuse is not provided, one should be provided external to the module. The module shown in Figure 4.10 has one DC power supply connection for all channels. Some modules may provide one power supply connection for a group of channels, for example, four or eight channels. Isolated channels may also be used and are discussed below.

It is possible to connect a sourcing field sensor to a sourcing DC discrete input channel. In this case, an interposing NPN transistor or relay is added as shown in Figure 4.11. When using a transistor, the NPN transistor Q_1 conducts when the device is **on**, causing current to

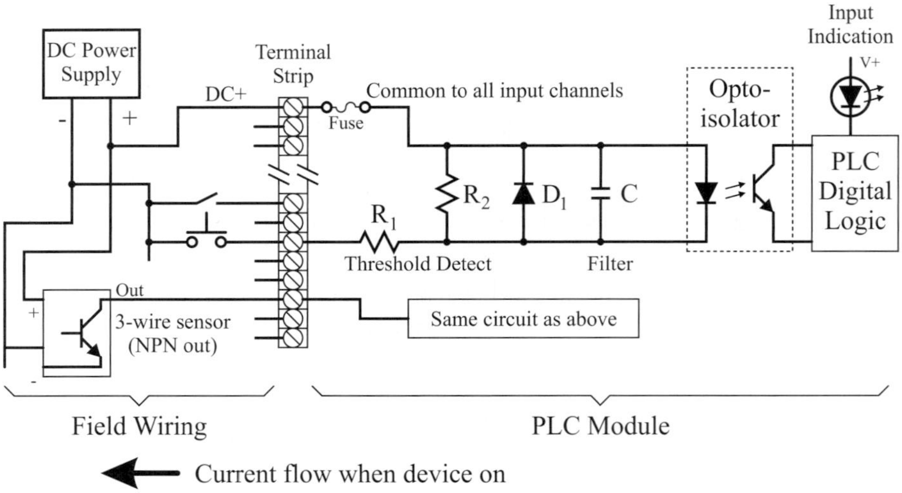

Figure 4.10. Typical sourcing DC discrete input channel.

flow **out** of the discrete input channel. This approach is not recommended because (1) many maintenance electricians will have trouble diagnosing device and/or PLC channel problems and (2) a replacement transistor may not be readily available. An interposing relay is a better approach from the maintenance standpoint. However, the interposing relay introduces an additional delay that may be unacceptable if the output changes rapidly. An even better approach is to use isolated discrete input modules if both sourcing and sinking input devices must be connected to the same discrete input module.

Isolated discrete input modules are most useful when sourcing and sinking input devices must be connected to the same discrete input module. As shown in Figure 4.12, isolated DC discrete input channels may be wired for either sinking or sourcing operation. The channels do not share either DC power or DC common.

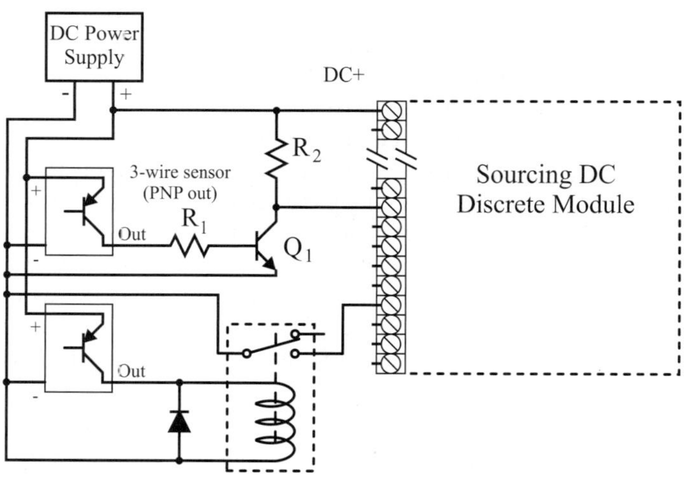

Figure 4.11. Connecting a sourcing sensor to sourcing DC discrete input channel.

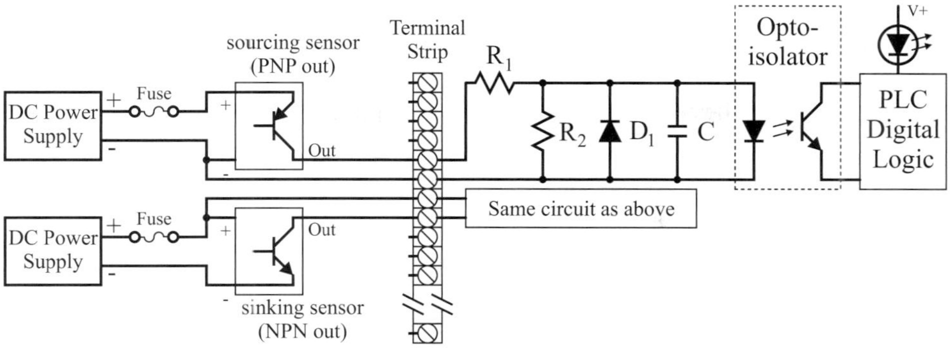

Figure 4.12. Isolated DC discrete input module for mixture of sinking and sourcing input devices.

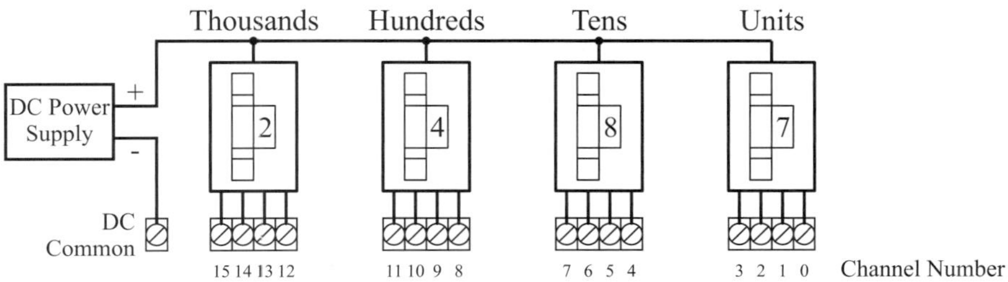

Figure 4.13. Thumbwheel switches connected to DC discrete input module.

Thumbwheel switches can be connected to multiple DC discrete input channels to allow inexpensive operator input of numerical values. Figure 4.13 shows four thumbwheel switches, each with four connections to a discrete input module. A sinking DC discrete input module is shown, but a sourcing module can be used by connecting the DC- power supply connection as the common to the thumbwheel switches and connecting the DC+ power supply to the DC+ terminal of the module. In either case, the 16-bit discrete input is read as a 4-digit binary-coded decimal (BCD) value. Some manufacturers have special modules that have multiplexing and handle three or four groups of four thumbwheel switches. In this case, each group of four digits is handled in a manner similar to analog input cards. That is, each group of digits is a BCD number.

Most PLC vendors also provide a discrete input module that handles 5-volt TTL (transistor-transistor logic) devices. This type of module is generally used for discrete sensors that produce TTL-level signals. This module may be required to receive numerical output from specialized equipment. For example, the thickness reading from a high-precision thickness-measuring device could be presented as 32 TTL signals, representing 8 BCD digits.

4.2.2 Discrete Output Modules

A discrete output module switches a device that has only two states, on/off, open/closed, and so on. Common discrete devices controlled by a PLC are,

Indicating lamps	Motor starters	Heater relay
Alarm lamps	Electric valves	Control relays
Alarm horns	Solenoid valves	
Annunciators	Electric fans	

The voltage ranges are basically the same as for the discrete input modules, except that the AC discrete output modules can only handle AC signals and DC output modules can only handle DC signals. Only relay output modules can handle either AC or DC signals.

The block diagram of a typical discrete output is shown in Figure 4.14. The optical isolator provides electrical isolation between the PLC internal circuitry and the field wiring. The electronic circuitry to the left of the optical isolator is powered from the internal PLC voltage source and the circuitry to the right of the optical isolator is powered from the field. In the case of a relay output module, the optical isolation is not needed since the isolation is provided by the relay. Except for relay output modules, the switch on all other discrete output modules is not perfect and introduces a voltage drop. Depending on the actual circuitry, this voltage drop will range from 0.3 to 2 volts. This voltage drop reduces the voltage available to drive the actuator connected to the discrete module.

The circuitry of a typical AC discrete output module, along with connections to typical output devices, is shown in Figure 4.15. The capacitor and resistor R_1 form a filtered DC power source for the field side of the optoisolator and triac trigger. Resistors R_2 and R_3 are the trigger for the triac, which is a bilateral AC switch. The triac trigger often has a

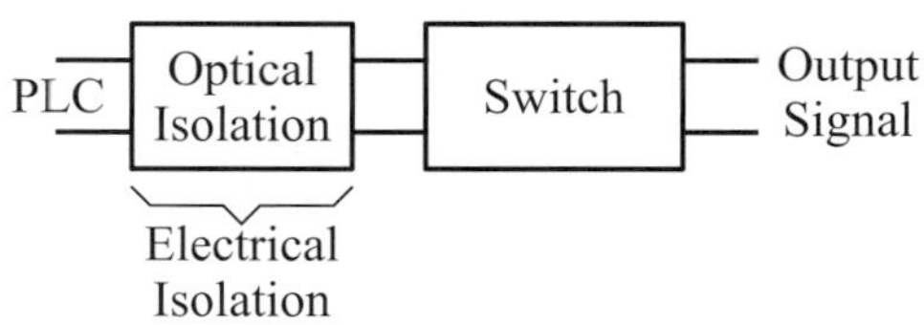

Figure 4.14. Discrete output module block diagram.

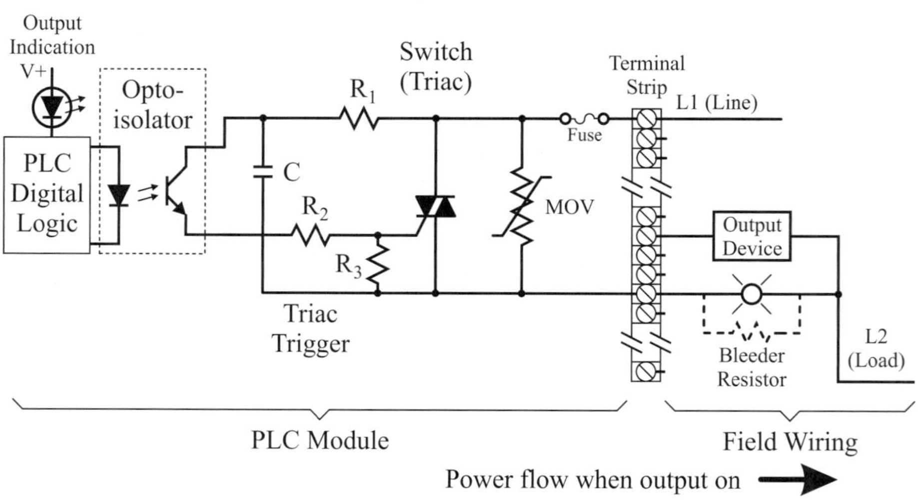

Figure 4.15. Typical AC discrete output channel.

zero-crossing detector so that the triac is switched on when the AC waveform is close to zero. This feature prolongs the life of the triac. The MOV (metal-oxide varistor) suppresses high-voltage transients from the AC source. When the PLC logic turns on the light-emitting diode in the optoisolator, current flows through the photoelectric transistor and triggers the triac. When the triac is triggered, it conducts AC current in either direction. After it is triggered, the triac will continue to conduct current until the next zero crossing of the AC source. A triac cannot be used to switch a DC signal since the triac continues to conduct current until the DC power source is turned off. For fault diagnosis, an output state indication is **on** when the PLC is commanding the output switch (triac) to conduct power to the output channel. This indicator may be on the field wiring side of the optoisolator or may be on the PLC side of the optoisolator. Output modules that handle AC signals are *sourcing* modules, that is, power flows **out of** the module when the triac is **on**.

Figure 4.15 shows an optional bleeder resistor in parallel with the indicator lamp. A triac-type output produces a small leakage current even when the triac is off. Unfortunately, this leakage current may be large enough to trigger the output device. For example, this leakage current causes neon indicator lamps to glow. The bleeder resistor "bleeds" the leakage current to the AC common and assures that the output device senses the on state of the output channel only when the PLC logic is turning it on.

If the load has a rectifying diode in series with a resistor or inductor (Figure 4.16*a*), a shunt resistor should be added to allow proper operation of the PLC output channel. If the shunt resistor is absent from Figure 4.16*a*, the circuit operation is described as follows. When L1 is positive with respect to L2, the capacitor is charged. When L2 is positive with respect to L1, the diode prevents the capacitor from discharging. Therefore, the zero-crossing detector will not detect a zero voltage due to the charged capacitor and the triac will not be triggered when the PLC logic attempts to turn on the output. Reversing the polarity of the load diode only changes the polarity of the charged capacitor voltage. The shunt resistor allows the capacitor voltage to bleed off, ensuring proper operation of the zero-crossing detector.

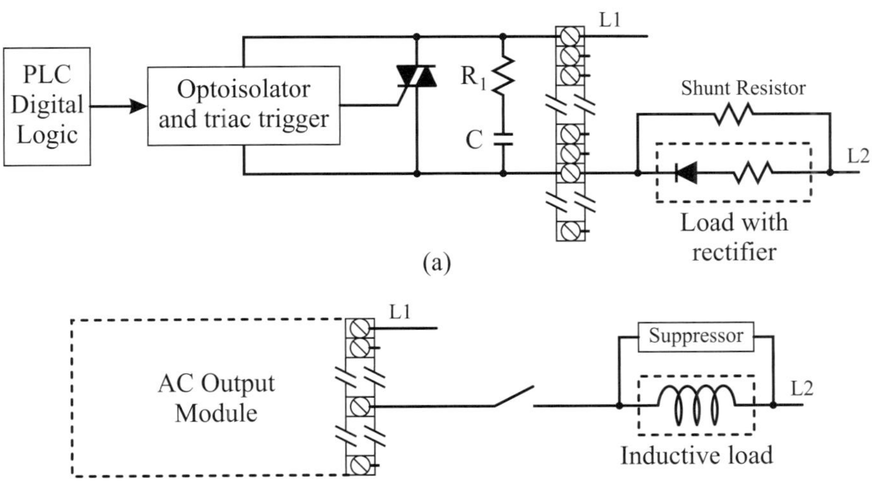

Figure 4.16. AC discrete output channel suppression devices: *(a)* rectifier in load; *(b)* driving contact in series with inductive load.

If the output device is primarily inductive (for example, a relay coil or a pilot light with step-down transformer) and is in series with or parallel to a "hard contact" such as a push button or selector switch, a suppression device (or snubber) should be added in parallel with the coil (Figure 4.16*b*). The suppressor prevents arcing across the switch contact, prolonging contact life.

The module may have a fuse for every output channel, or one fuse for a group of channels. Even if a fuse is provided for every channel, the maximum current that the module can handle is generally less than the sum of the individual channel maximum currents. For example, each channel of a certain 16-output module can handle a maximum of 2 amperes, but the module can handle a maximum of 8 amperes. It is generally up to the user to ensure the module current limits are met. The module may not have any built-in current limit protection. In this case, violating the module current limit often causes the module to overheat. Normally, the outputs will not each draw the maximum channel current, not will all outputs be on simultaneously. Nevertheless, the user should consider the worst case when choosing the output module.

There is a very definite reason why discrete outputs are typically sourcing: safety. When the PLC is not in run mode or commanding the output to be **off**, there is no power at the field device. This is the same reason that in residential or commercial wiring the switch to control an overhead lamp is placed in the line (L1) wire, not the neutral wire. When the switch is in the off position, the lamp may be safely changed because the socket is completely unpowered.

The circuit for a typical *sourcing* DC discrete output channel, including devices, is shown in Figure 4.17. The optoisolator transistor switches a PNP power transistor to control the output device. The diode D_1 provides protection for the circuit in case the DC power supply polarity is accidentally reversed. When the PLC logic turns on the light-emitting diode in the optoisolator, current flows through the photoelectric transistor and triggers a PNP transistor. A field-effect transistor (FET) may be used in place of the PNP transistor. When the PNP output transistor conducts, current flows out of the discrete output channel. Diode D_2 suppresses the negative voltage spike that inductive loads produce when they are turned off. The fuse shown internal to the module may be common to all channels, or there

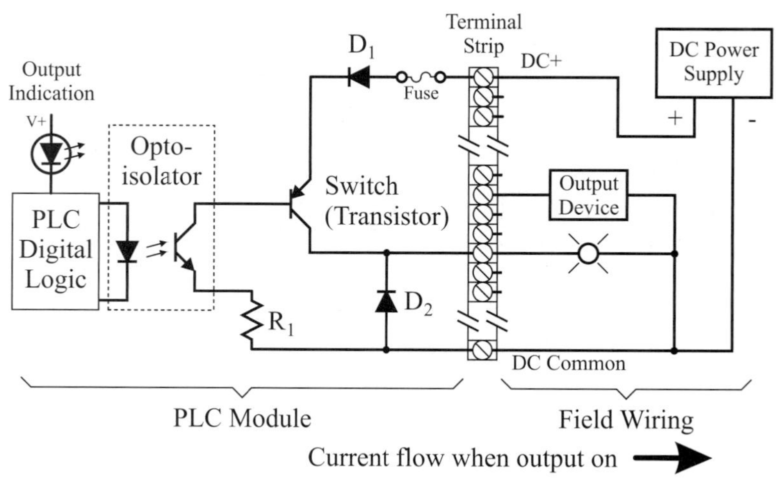

Figure 4.17. Typical sourcing DC discrete output channel.

may be one fuse per channel. The module shown in Figure 4.17 has one DC power supply connection for all channels. Some modules may provide one power supply connection for a group of channels, for example, four or eight channels. Isolated channels may also be used and are discussed below.

A sourcing DC discrete output module is normally used with sinking output devices. However, a sourcing discrete output module can drive a sourcing output device by interposing a transistor circuit or a relay, similar to that shown in Figure 4.11.

The circuit for a typical *sinking* DC discrete output channel and associated devices, is shown in Figure 4.18. The only basic difference with the sourcing output module is that the output switch is a NPN power transistor. A FET may be used in place of the NPN transistor. Diode D_1 protects the circuit in case the DC power supply polarity is accidentally reversed. When the NPN output transistor conducts, current flows into the discrete output channel. Diode D_2 suppresses the positive voltage spike that inductive loads produce when they are turned off. As for the DC sourcing discrete output module, the fuse shown internal to the module may be common to all channels, or there may be one fuse per channel.

A sinking DC discrete output module is normally used with sourcing output devices. However, a sinking discrete output module can drive a sinking output device by using a pull-up resistor or interposing relay, similar to that shown in Figure 4.9. The pull-up resistor value is determined in the same manner as when connecting a sinking sensor device with a sinking discrete input channel.

Isolated discrete output modules are most useful when sourcing and sinking output devices must be connected to the same discrete output module. As shown in Figure 4.19, isolated DC discrete output channels may be wired for either sinking or sourcing operation. A transformer provides the isolation so that power can be transferred across the barrier to provide a positive gate voltage to the FET and thus allow current to flow from the drain to the source. The channels do not share either DC power or DC common. An alternate way to provide an isolated output with either a sourcing or sinking discrete module is to use an output channel to drive a relay. The relay contacts provide an isolated output.

The circuitry for a relay discrete output module is shown in Figure 4.20. In contrast to the other types of discrete output modules, there is no optoisolator. The lack of an electrical

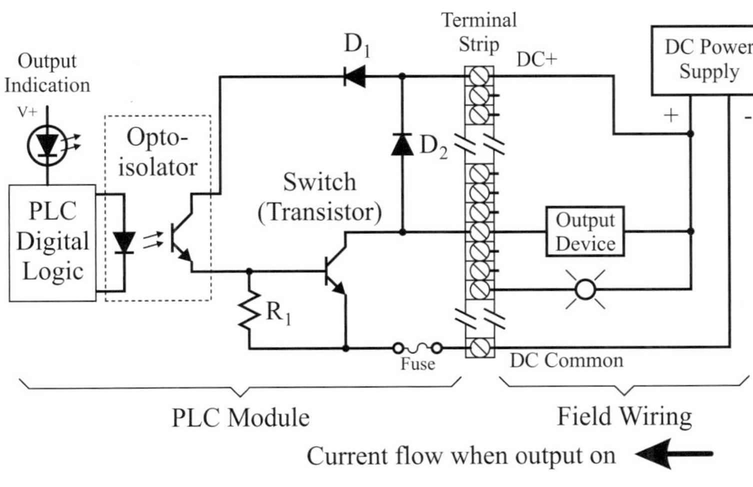

Figure 4.18. Typical sinking DC discrete output channel.

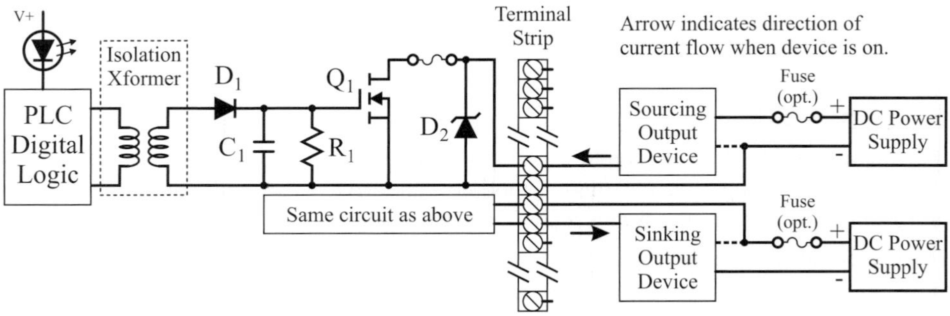

Figure 4.19. Isolated DC discrete output module for mixture of sinking and sourcing output devices.

connection between the coil and contacts provides the electrical isolation. Relay outputs are used for output devices that have a power source isolated from the PLC. A typical example is a motor control center (MCC) where each motor starter has a separate 120-volt supply derived from the 480-volt bus, shown in Figure 4.21. Note that according to the National Electrical Code® section 430.102 (NFPA, 2020), the disconnect must be in sight of the motor.If the output device is an inductive load, then a suppressor device should be placed in parallel with the inductive load to protect the relay contacts. Figure 4.22 shows typical suppression devices for an inductive load powered by AC and DC sources (Emerson, 2022).

Most systems also provide some capability to manually operate a device in conjunction with the PLC control. A typical device is called a Hand-Off-Auto (HOA) switch, illustrated in Figure 4.23. The "HAND" and "OFF" positions override the state of the PLC output. When the switch is in the "HAND" position, the P-101 pump runs, regardless of the state of the PLC output. In the "OFF" position, the pump stops, regardless of the state of the PLC output. When the HOA switch is in the "AUTO" position, the PLC controls the pump. An indication is also provided to the PLC that is **on** when the HOA switch is in the "AUTO" position. The PLC logic turns off the pump control whenever the switch is moved out of the "AUTO" position.

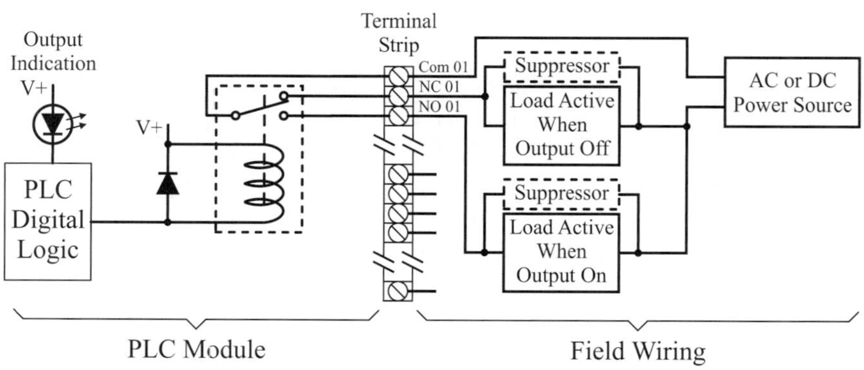

Figure 4.20. Typical relay discrete output channel.

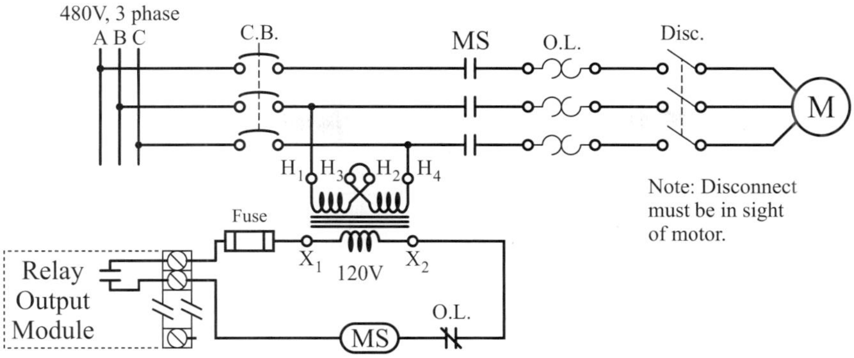

Figure 4.21. Relay discrete output channel controlling a MCC.

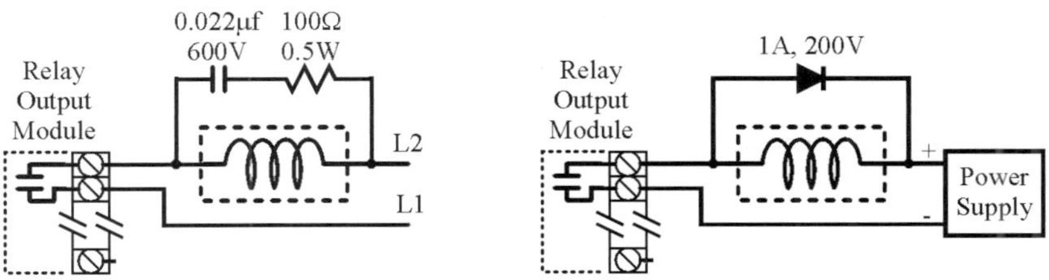

(a) (b)

Figure 4.22. Relay output suppression devices: *(a)* AC powered load; *(b)* DC powered load.

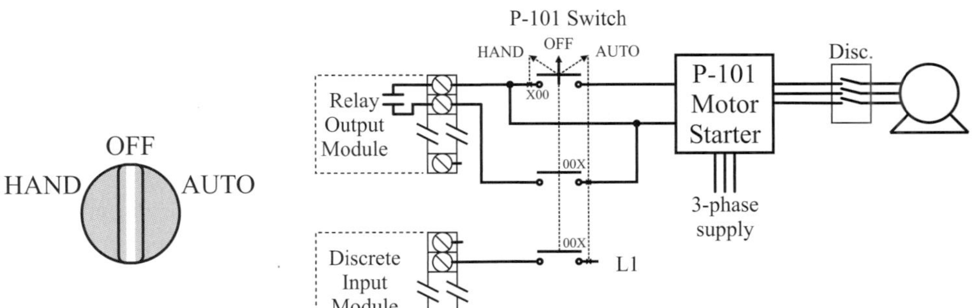

(a) (b)

Figure 4.23. Hand-Off-Auto switch between PLC output and MCC: *(a)* HOA switch; *(b)* switch wiring.

Multiple DC discrete output channels can be connected to LED 7-segment displays to provide inexpensive operator display of numerical values. Figure 4.24 shows four LED 7-segment displays, each with four connections to a discrete output module. A sinking DC discrete output module is shown since most of these modules have a TTL (sourcing) input. In either case, the 16-bit discrete output must be formatted as a 4-digit BCD value.

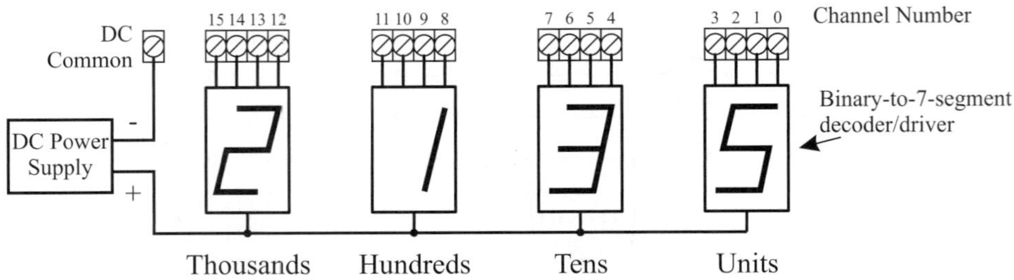

Figure 4.24. LED displays connected to DC discrete output module.

Most PLC vendors provide a discrete output module that interfaces to 5-volt TTL devices. This module may be required to send numerical input to specialized equipment. For example, the desired linear position for a hydraulic positioning controller could be sent as 16 TTL signals, representing 4 BCD digits.

4.3 ANALOG MODULES

Even though PLCs were originally developed to handle discrete signals, there is often the need to handle continuous, or analog, signals. An analog signal represents a physical quantity that can have an infinite number of values. The typical analog signal ranges are:

1 to 5 V	4 to 20 mA
0 to 5 V	0 to 20 mA
0 to 10 V	−20 to 20 mA
−5 to +5 V	
−10 to +10 V	

For reasons detailed later in this section, the 4 to 20 mA range is the most popular signal range.

4.3.1 Analog Input Modules

An analog input module receives the measured quantity (voltage or current) from process transmitters and other devices that produce analog signals. Common physical quantities measured by a PLC analog input module are:

Position	Flow	Distance
Speed	Level	Analytic instruments
Acceleration	Pressure	Opacity
Weight	Temperature	

The block diagram of a typical analog input is shown in Figure 4.25. The optical isolator provides electrical isolation between the field wiring and the PLC internal circuitry. The electronic circuitry to the left of the optical isolator is typically powered from the internal PLC, but through an isolated power supply to maintain isolation between the field wiring and the PLC voltage source. The sensor may be powered by an external power supply or by the PLC isolated power supply. The analog signal is filtered to remove noise. The actual field signal may be a single-ended voltage, single-ended current, differential voltage, or differential current. In all cases, the analog signal is converted into a

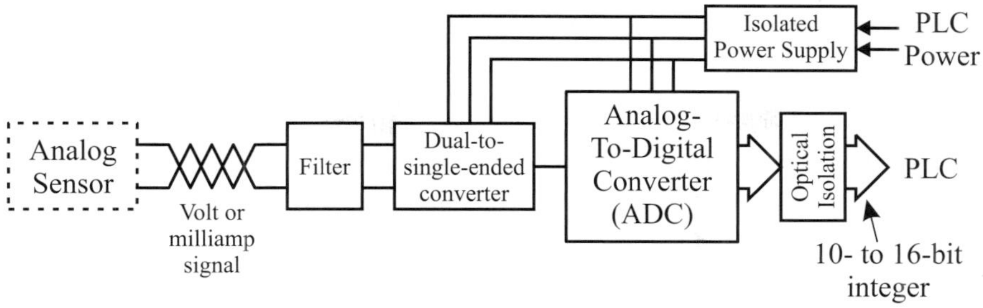

Figure 4.25. Analog input module block diagram.

single-ended voltage for the analog-to-digital converter (ADC). There is generally only one ADC per module, regardless of the number of input channels. The individual channel voltages are multiplexed into the ADC input.

The ADC converts the analog voltage into an integer number, which has a finite resolution. A common ADC outputs a 16-bit integer, which means the output integer ranges from 0 to 65,535. The relationship between the analog input voltage and the ADC output integer is illustrated in Figure 4.26. In this example, the output integer is assumed to be 4 bits, which means the output integer value ranges from 0 to 15. If the conversion were perfect for this example, the ADC output would be three times the input voltage. The actual relationship between the input voltage and the output integer resembles "stair steps" (Figure 4.26*b*). For example, an input voltage between 0.833 volts and 1.166 volts is read as the integer 3. The *resolution* of an analog input channel is the smallest change that can be sensed and corresponds to the width of each step in Figure 4.26*b*. The resolution is calculated as the maximum span of the signal divided by the integer range of the ADC output. For example, a 16-bit ADC with a signal range of 5 volts has a resolution of 5/65,535 = 0.076 mV.

The ADC requires a finite time to perform the analog-to-digital conversion, so the analog signal is sampled repeatedly and the conversion is performed on each sample. Any

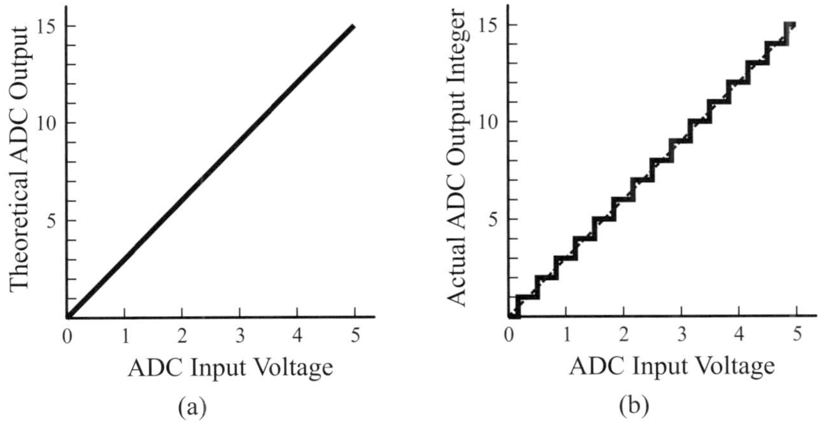

Figure 4.26. ADC input/output relationship: *(a)* theoretical output; *(b)* actual output integer.

change in the analog signal between samples is ignored. Since multiple signals are multiplexed into one ADC, the time between successive samples of one channel is the ADC conversion time multiplied by the number of channels. The maximum frequency of the measured signal must be considered when selecting the analog input module. The sampling rate of the input module should be at least twice the highest frequency of the measured signal.

The channels on an analog input module are sinking. Figure 4.27 shows the connections of voltage sensors to analog input modules. Analog input channels are generally configured as single-ended (Figure 4.27*a*) or differential (Figure 4.27*b*). Single-ended channels share a common connection that may also be the common of the sensor power supply. In contrast, differential channels do not share a common connection and are isolated from each other and from the sensor power supply. In either case, the signals are shielded to attenuate electromagnetic noise. The shield is connected to earth ground on only one end of the cable, preferably close to the point where the measured signal is connected to earth ground (Morrison, 1986) usually at the source (sensor) end. If the shield cannot be grounded at the sensor, then it should be connected to an earth ground at the PLC module.

There are two disadvantages to using voltage to represent the analog signal:

1. Even with careful shielding, the signals easily pick up electromagnetic noise.
2. The signal cannot be transmitted long distances. The wire resistance causes a significant voltage drop for long cables. For example, the resistance of 18-gauge wire is 0.0064 ohms per foot. If the cable length is 2000 feet, the combined resistance of the signal and common is 12.8 ohms. If the current through the differential signal loop is 0.1 mA, the resulting voltage drop is 1.28 mV, much larger than the resolution of a 16-bit ADC when the maximum signal is 5 volts.

The solution to the disadvantages of voltage signals is to use current to transmit the signal, as shown in Figure 4.28. As for the voltage sensors, the input channels are generally connected as single-ended (Figure 4.28*a*) or differential (Figure 4.28*b*). Single-ended channels share a common connection that is also the common of the loop (and sensor) power supply. In contrast, differential channels do not share a common connection in the analog input module. However, the power supply is still part of the loop, and if the same power supply is used for all sensors, the channels are not isolated from each other. The loop

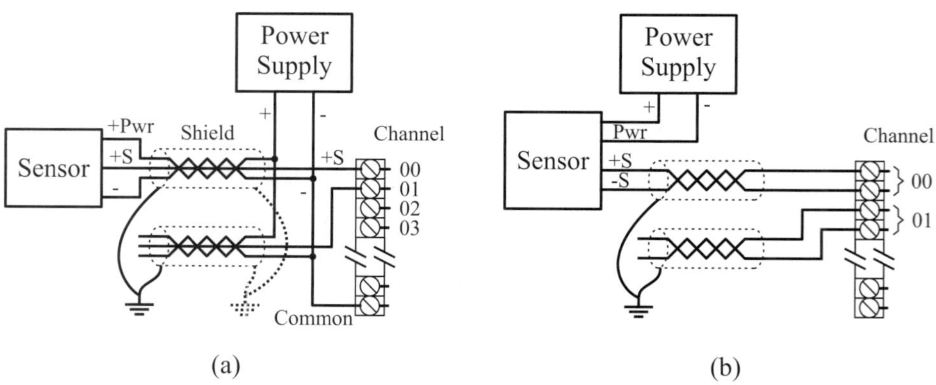

Figure 4.27. Connection of sensors to voltage-input analog input module: *(a)* single-ended; *(b)* differential.

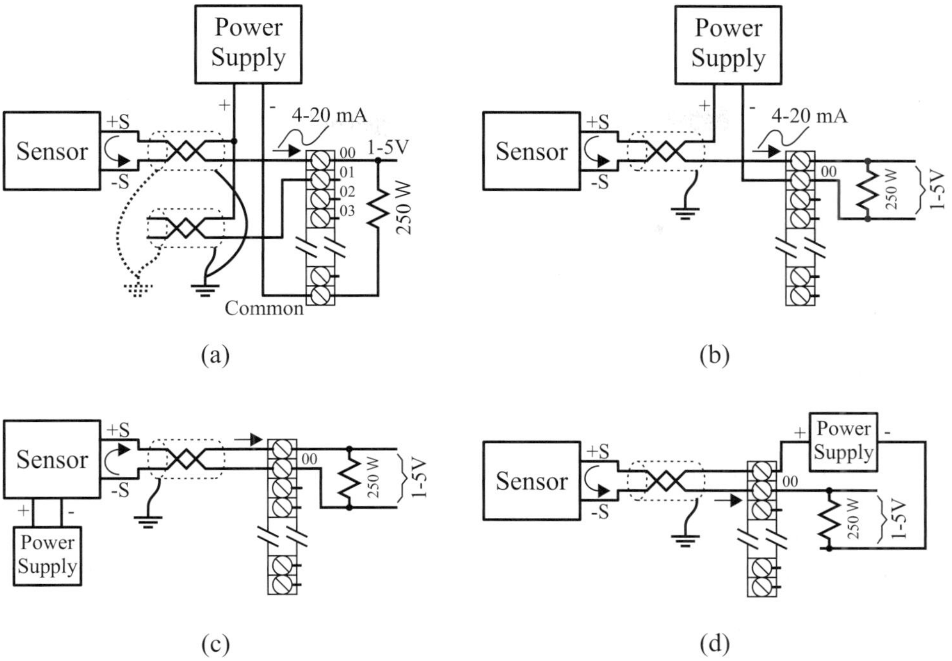

Figure 4.28. Connection of sensors to current-input analog input module: *(a)* single-ended; *(b)* differential; *(c)* separate sensor power supply; *(d)* module-supplied power.

power may be supplied by the sensor (Figure 4.28*c*) or may be provided by the analog output module (Figure 4.28*d*). In all cases, the signals are shielded, and the shield is grounded to attenuate electromagnetic noise.

The advantages to using current to transmit the signal are:

1. Current signals are less susceptible to electromagnetic interference.
2. The signal can be transmitted long distances (thousands of feet) limited only by power supply voltage and wire resistance. For long wire runs, the power supply voltage is increased to be at least 2 volts higher than the maximum voltage drop (wire resistance times 20 mA) plus the minimum sensor voltage drop plus the 5-volt drop at the analog input channel. For the circuit in Figure 4.28*a*, if the minimum sensor voltage drop is 5 volts and the sensor wiring run is 5000 feet of 18-gauge wire, the minimum power supply voltage is

   $$V_{min} = 2 + (10000 \text{ ft.}) \times (0.0064 \text{ ohms/ft.}) \times (0.020 \text{ amps}) + 5 + 5 = 13.28 \text{ volts}$$

3. Power and signal are carried by two wires.
4. A broken loop (wire) is indicated by zero mA, assuming the normal range of the signal is 4 to 20 mA.

A current signal can easily be converted to a voltage signal by installing a resistor across the voltage input terminals, as shown in Figure 4.29. This method is often used for sensors that produce non-standard current signal ranges. For example, if a sensor outputs a 0 to 1 mA signal, a resistor value of 5000 ohms in Figure 4.29 converts the current signal into a 0 to 5 volt signal to be measured by the analog input module.

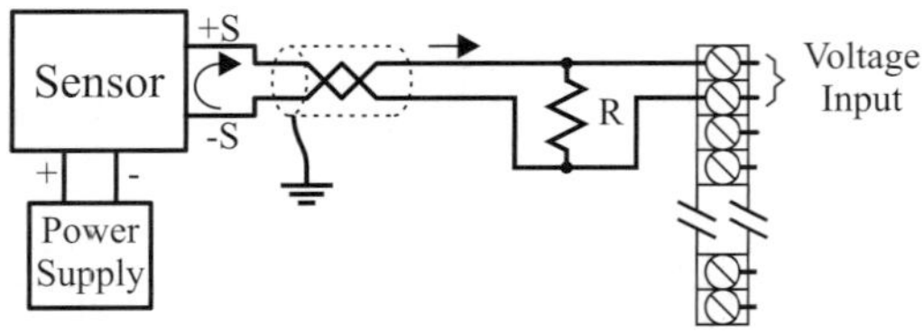

Figure 4.29. Conversion of current signal into voltage measurement.

The disadvantage of using a current loop to transmit the signal is that it may be difficult to diagnose problems with the current loop. The loop may need to be broken in order to check the signal value and special equipment may be required to find the location of wire breaks. For this reason, some installations place a 100 to 250 ohm resistor in the loop so measuring the voltage drop across the resistor indirectly checks the current. Of course, the power supply voltage must be large enough to accommodate the increased loop voltage drop due to the additional resistor.

Current loops will not function properly if more than one point in the loop is connected to the power supply common. For example, this problem occurs if the sensor in Figure 4.28*a* receives its power from the loop power supply and the "S-" output is connected internally to the power supply common. In this case, no current will flow into the analog input module. In order to prevent this problem, sensors are either powered totally from the 5-volt drop across the output terminals or from a power supply completely isolated from the loop power supply.

A separate visual indicator may be part of the wiring of the sensor to the PLC (Figure 4.30). This indicator provides inexpensive local display of the sensor value and is often useful for diagnostics. For a sensor that provides a voltage signal, this indicator is connected in parallel to the PLC channel wiring, as shown in Figure 4.30*a*. For a current signal, the indicator is placed in series with the current loop wiring, as in Figure 4.30*b*. Indicators that measure current may be completely powered by the current loop and not require a separate power supply. If a current indicator uses a separate power supply, one should verify that there is no internal connection between the indicator power supply and the sensing inputs. This internal connection may cause more than one point in the loop to be connected to the common of the analog input module.

Thermocouple and resistance temperature device (RTD) sensors are generally connected to specialized analog input modules. Figure 4.31 shows typical connections. A

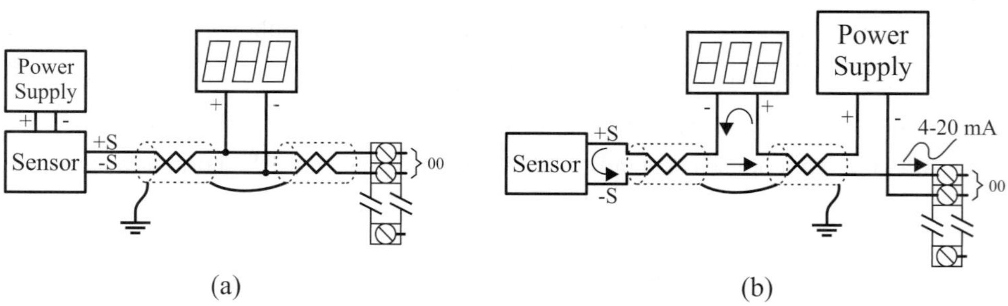

Figure 4.30. Connection of indicator to analog input signal: *(a)* voltage sensor; *(b)* current sensor.

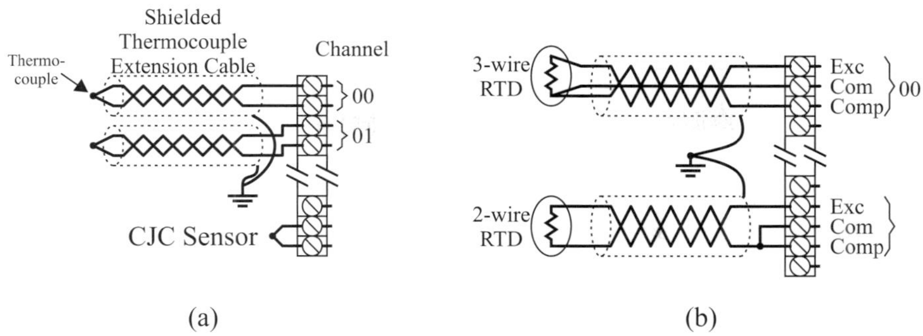

Figure 4.31. Connection of temperature sensors: *(a)* thermocouples; *(b)* RTD sensors.

thermocouple generates a small voltage (-10 to 80 mV) and so requires a high-gain amplifier. Proper voltage-to-temperature conversion requires cold-junction compensation (CJC), provided by a separate sensor measuring the terminal block temperature. A RTD is basically a temperature-sensitive resistor. The most common conversion of RTD resistance to temperature places the RTD in a Wheatstone bridge circuit before amplification. A three-wire RTD provides compensation for the resistance of the cable wires. Chapter 11 contains more information about these temperature sensors.

4.3.2 Analog Output Modules

An analog output module generates a voltage or current signal that drives another device to manipulate a physical quantity. Common devices controlled by a PLC analog output module are:

Variable-speed motor drive (speed)
Current-to-pneumatic converter (pneumatic regulating valve position)
Electronic valve (valve position)
Analog meter
Chart recorder

The block diagram of a typical analog output module is shown in Figure 4.32. A digital-to-analog converter (DAC) converts the integer number from the PLC into a single-ended voltage. If the output signal is a current, then the DAC output voltage must be converted into a current signal. Like the analog input modules, the actual field signal may be a single-ended voltage, single-ended current, differential voltage, or differential current. There is generally a separate DAC for each channel. The optical isolator provides electrical isolation between the field wiring and the PLC internal circuitry. The electronic circuitry to the right of the optical isolator is typically powered from the internal PLC, but through an isolated power supply to maintain isolation between the field wiring and the PLC voltage source. The field device may be powered by an external power supply or by the PLC isolated power supply.

The channels on an analog output module are sourcing. Figure 4.33 shows the connections of voltage and current devices to analog output modules. Analog output channels are generally configured as single-ended (Figure 4.33*a*,*c*) or differential (Figure 4.33*b*,*d*). Single-ended channels share a common connection. In contrast, differential channels do not share a common connection and are isolated from each other. As for the

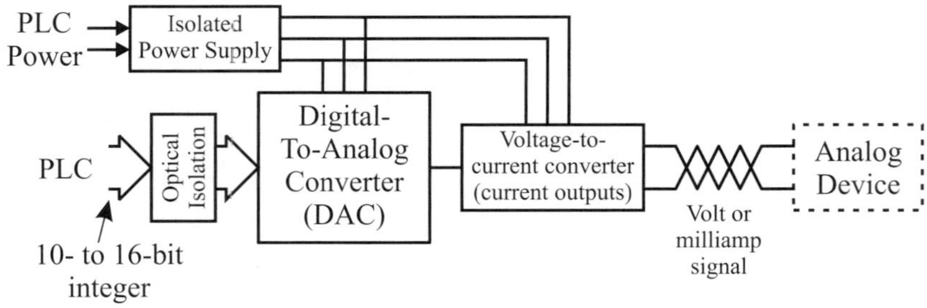

Figure 4.32. Analog output module block diagram.

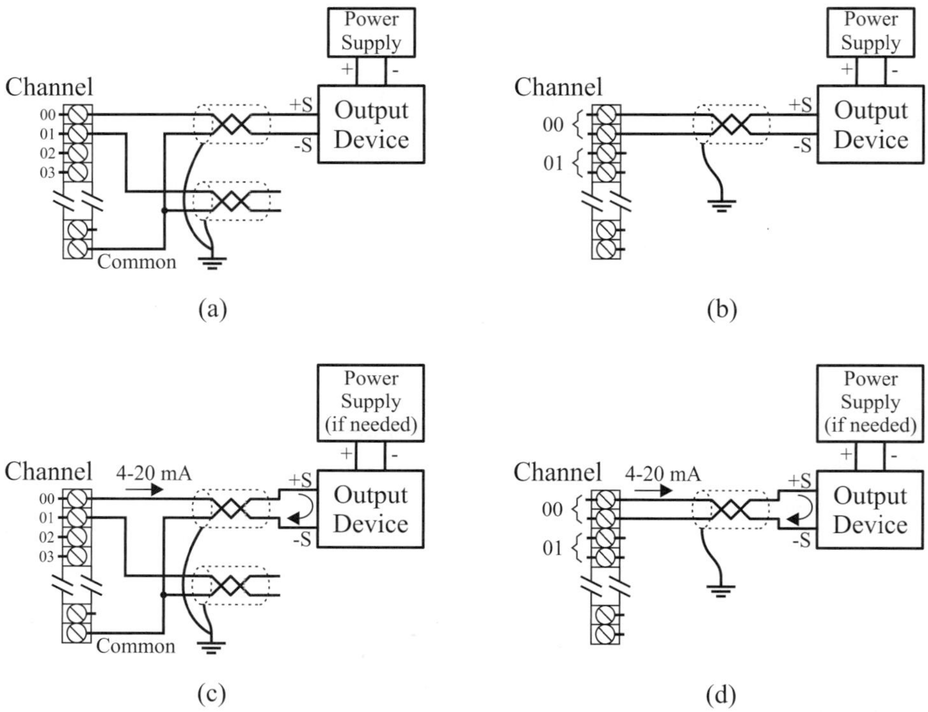

Figure 4.33. Connection of devices to analog output module: *(a)* single-ended voltage; *(b)* differential voltage; *(c)* single-ended current; *(d)* differential current.

analog inputs, the signals are shielded, and the shield is connected to earth ground at the source (module) end.

Some systems also provide the capability to manually operate an analog output device in conjunction with the PLC control. A typical device is called an auto/manual station, illustrated in Figure 4.34. When the switch is in the "MAN" position, the P-101 pump speed is set by the potentiometer, regardless of the value of the PLC output. When the switch is in the "AUTO" position, the PLC analog output controls the pump speed setting. The panel also shows the pump speed. The actual pump speed setting and an indication of the switch position is also provided to the PLC. When the selector switch is not in the "AUTO"

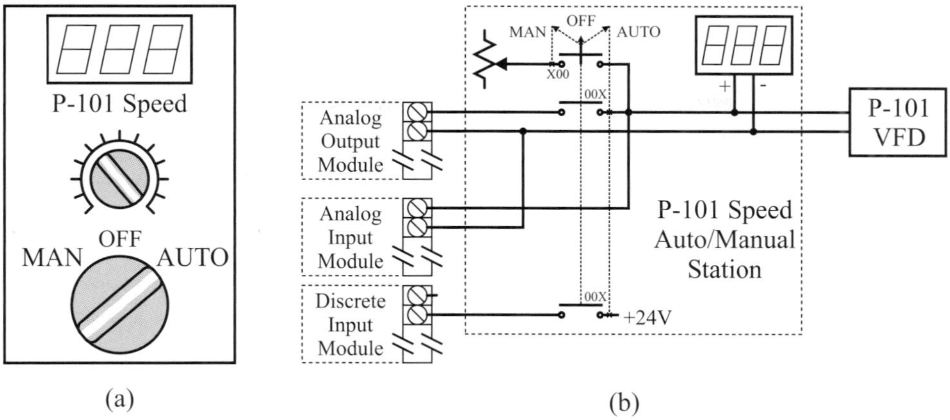

Figure 4.34. Analog output bypass station: *(a)* panel; *(b)* wiring.

position, the PLC logic disables any automatic control of the pump and adjusts the PLC analog output to match the operator setting.

4.4 SPECIALIZED MODULES

Many control systems need special-purpose I/O modules. For many systems, these modules are a minority of the total number of modules in the system but are essential for certain applications. Among specialized I/O modules, the high-speed counter and motion control modules are probably the most popular.

4.4.1 High-speed counter

A high-speed counter module provides a means to count pulses and time events independent of the program or I/O scan. A high-speed counter module is generally used as an interface for position encoders and positive-displacement flow meters. If a normal discrete input channel is used to count pulses, the highest pulse frequency is limited by the program scan time. For example, if the maximum program scan time is 1 ms, the minimum pulse width that can be reliably sensed is 2 ms (twice the maximum program scan time). Thus the maximum frequency that can be reliably sensed is 250 Hz (the reciprocal of twice the minimum pulse width). Many position encoder signals operate at 20 kHz and higher. Therefore, a normal discrete input channel is inadequate to count position encoder pulses. In addition, a high-speed counter module provides a differential input interface required by many encoders.

Figure 4.35 shows the connections of typical devices to a high-speed counter module used in simple counting applications. The connection to a single-ended device is similar to a sourcing discrete input device (Figure 4.8). Whenever possible, differential devices are preferred since the differential signal provides the best immunity to electrical noise. Many high-speed counter modules provide options to set the signal voltage level and signal filtering.

In most motion applications, a position encoder generates two-phase quadrature signals, shown in Figure 4.36*a* connected to a high-speed counter module. The relative

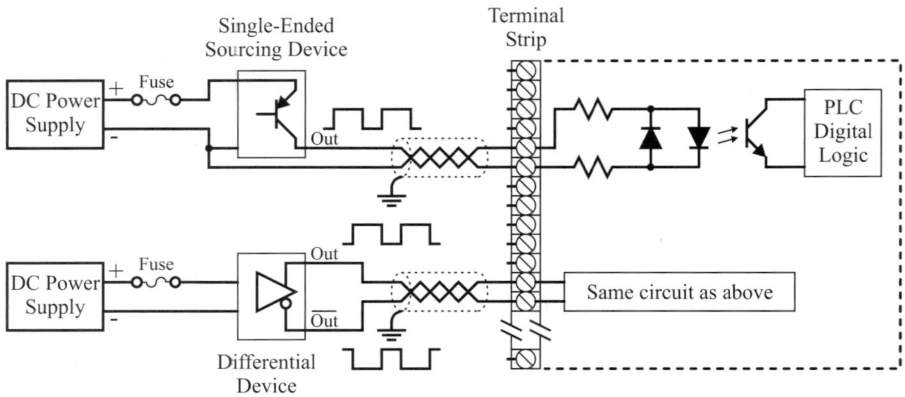

Figure 4.35. Single-ended and differential connections to high-speed counter module.

phasing of the A and B signals indicate the direction of encoder rotation. If the positive edge of A leads the positive edge of B, the encoder is rotating clockwise (Figure 4.36*b*). For counterclockwise rotation, the positive edge of A lags the positive edge of B (Figure 4.36*c*). The Z signal is the index pulse, which pulses once for each encoder revolution.

Many high-speed counter modules can also be programmed for operations more complex than simple counting. For example, many modules can calculate the frequency of

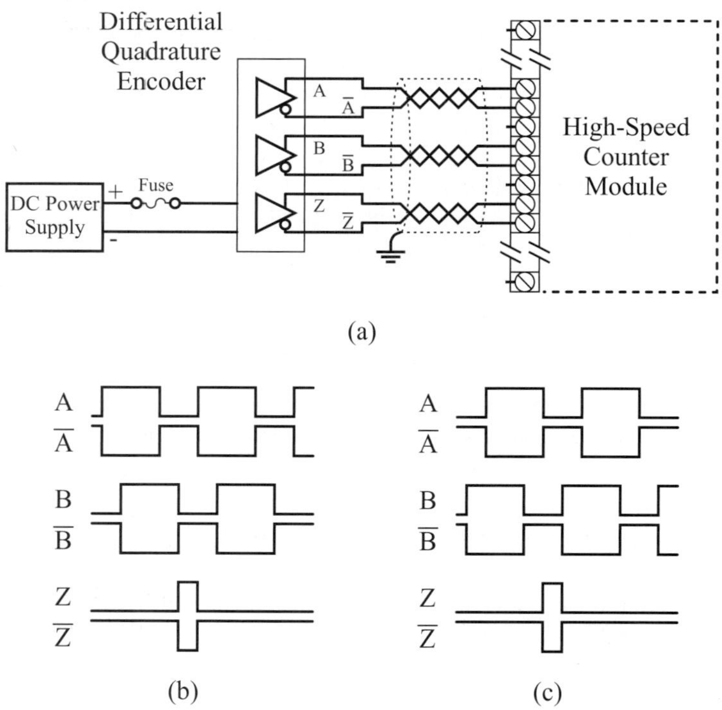

Figure 4.36. Quadrature encoder connections: *(a)* wiring to high-speed counter module; *(b)* timing for clockwise rotation; *(c)* timing for counterclockwise rotation.

the pulses by measuring the pulse period (averaged over a specified number of pulses) or by counting the number of pulses during a fixed interval.

4.4.2 Motion control

Motion control modules meet the needs of accurate, high-speed machining, packaging, and other types of production machinery. There are two types of motors used in motion control systems: servomotors and stepper motors. Generally, servomotors are used in applications that require high torque or power and stepper motors are used in low-torque, low-power applications. Both types of motors require a ***drive***, which contains the power electronics that translate the signals from the PLC motion module into the signals required by the motor. The drive for a stepper motor is usually less expensive than the drive for a servomotor.

A servomotor is a specially-designed brushed DC motor or brushless AC motor. The traditional servomotor is a brushed DC motor, where the coil is wound on the armature and the stator is a permanent magnet. In this motor, accurate speed control is obtained by manipulating the armature current while keeping the field current constant. By manipulating the field current with a constant armature current, the torque of the servomotor can be controlled. In a brushless AC motor, the rotor is a permanent magnet, and the coils are wound on the stator. A brushless servomotor requires complicated stator drive signals, and hence its drive is more complicated. Chapter 11 contains more details about the differences between brushed and brushless servomotors. Regardless of the type of servomotor, the speed and/or position is typically measured and fed back to the motion module, and the motor speed is adjusted accordingly. Figure 4.37 shows typical connections from the PLC servo motion module to the servo drive and to the servomotor when the servo is used in positioning control. In the example shown, an encoder is used for feedback, but a tachometer or a resolver could be used in place of the encoder. Also shown is a brake, which prevents motion when the motor is not powered. A brake is especially needed in certain vertical motion applications to hold a load at a certain position when motor power is removed. The over-travel limit switches cut off the motor power when the device moves too far in either direction.

The electrical connections in Figure 4.37 are sensitive to electrical noise and cannot be located very far away (Rockwell Automation, 2001). In order to alleviate these electrical problems in motion control systems, many recent systems use the Ethernet connections

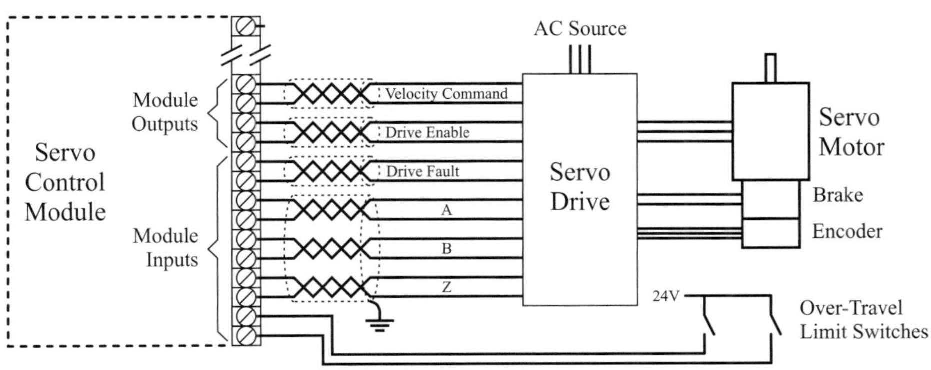

Figure 4.37. Typical servo-motor control module wiring.

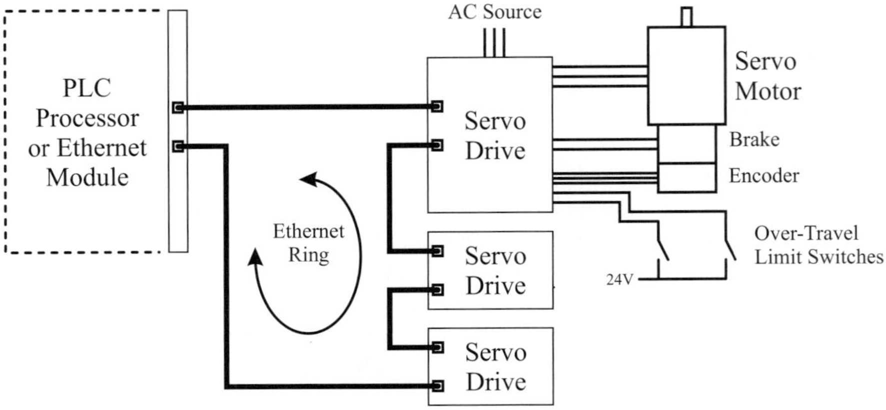

Figure 4.38. Typical Ethernet multi-axis motion module connections.

between the PLC and the drives (Figure 4.38). This figure shows an Ethernet ring connection, though the standard tree/star connection is also valid. The SERCOS (SErial Real-time COmmunication System) fiber optic communication network (IEC, 1995) is another option. A fiber optic medium assures reliable high-speed data transmission with excellent noise immunity.

A stepper motor is a brushless permanent-magnet type motor that translates incoming pulses into mechanical motion. Because the stepper motor makes a fixed angular motion in response to the pulses, stepper motors are often used in an open-loop configuration. In this case, the motion is controlled by generating the proper pulses to the motor, and encoders are not used. Figure 4.39 shows typical connections from the PLC stepper control module to the stepper drive and to the stepper motor. This figure also shows the optional encoder connections. Note that a brake may not be needed, since the signals from the drive to the stepper will hold the stepper motor shaft in a fixed position. However, when the power to the drive is removed, the motor is free to spin. In vertical applications, a brake will hold a load at a certain position when power to the stepper drive is lost.

Stepper motors have two disadvantages. First, if the stepper motor encounters a torque load higher than its rating, the motor "cogs" and does not move. For this reason, encoders

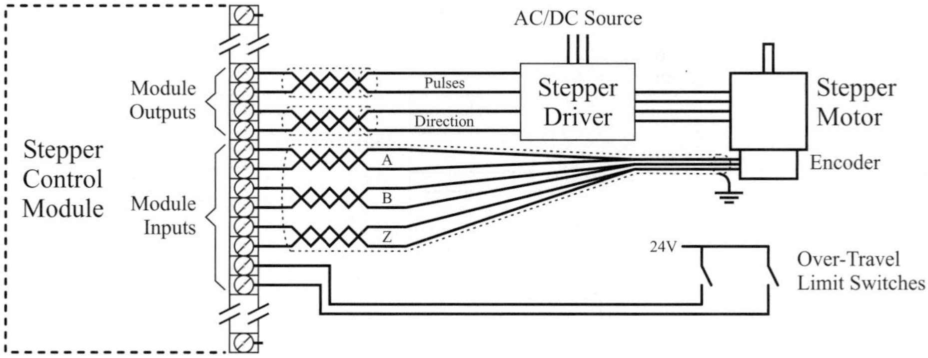

Figure 4.39. Typical stepper motor control module wiring.

are utilized in high-performance motion applications. Second, one must be careful when using encoders in a stepper motor application. Because the motor is driven with pulses, the actual shaft rotational motion is also "pulsed" and not smooth as for a servomotor. If the stepper motor is driving a flexible belt, the pulsed shaft motion is magnified and will cause false encoder pulses to be generated. For this reason, servomotors are a better choice when driving a flexible belt.

4.4.3 Other specialized devices

Other types of specialized I/O devices include bar code readers, vision systems, message displays, and operator terminals. These types of devices generally do not have a separate PLC I/O module, and the easiest way to connect them to the PLC is through the Ethernet communications network. Many of these devices provide a direct connection to Ethernet, like the camera in Figure 14.40*a*. Others provide only a RS-232 serial port. For the latter case, one must have a serial server to provide the serial-to-Ethernet conversion, as shown for the barcode reader and message display in Figure 4.40*a*. As an alternative to serial port connections to each device, a multi-drop communication network may be used to connect the devices to the serial server, as shown in Figure 4.40*b*. Modbus and RS-485 are two popular simple multi-drop communications networks. Communication networks are discussed in Chapter 12.

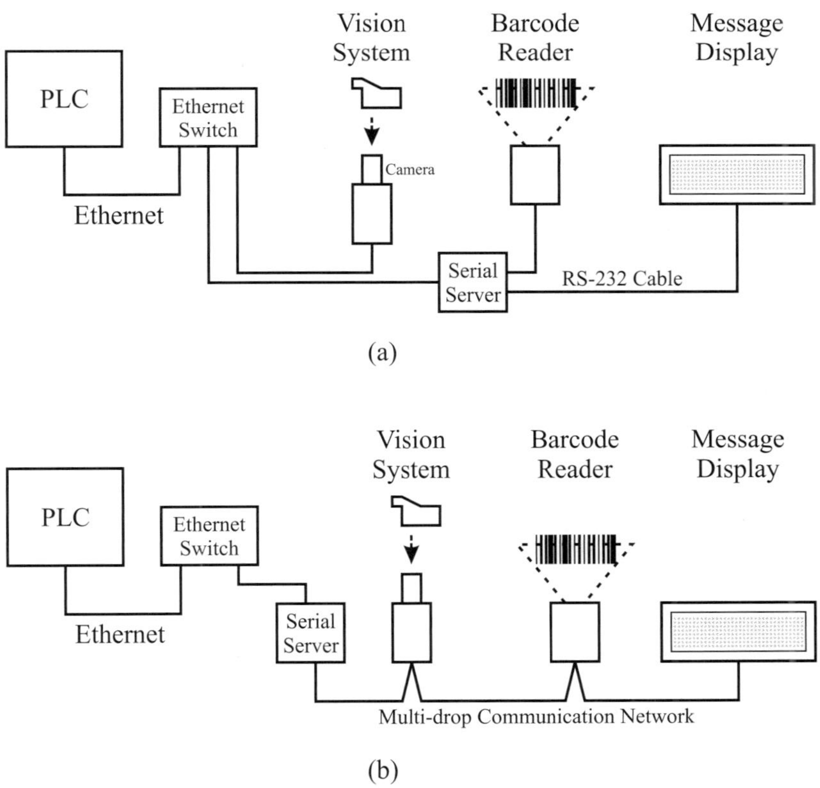

Figure 4.40. Connections to other intelligent devices: *(a)* serial communications; *(b)* communication network.

4.5 INSTALLATION WIRING

This section outlines general guidelines for the wiring required to install a PLC-based control system. It covers power distribution, control cabinet and control panel layout. Consult any vendor documentation and the appropriate standards (e.g., NFPA 70E and NFPA 79) for specific information.

4.5.1 Power distribution

If the control system power supply source is 120 AC volts, a typical distribution of power is shown in Figure 4.41. A step-down transformer decreases the higher voltage for the motor starters (typically 480 volts) to the 120 volts for the controller power supplies and DC power supplies. The power supplies for the PLC processor and all chassis that contain I/O modules are powered directly from the 120-volt source, protected by a fuse (or circuit breaker). The DC power supply in this configuration is used only to supply power to the I/O circuitry. For a system where DC powers the PLC processor and I/O modules, a typical power distribution is shown in Figure 4.42.

For safety and reliability in both AC and DC power distribution systems, all equipment must be properly grounded. There are typically three types of circuit grounds:

1. Safety grounds for chassis, panels, and junction boxes
2. Dirty grounds for EMI-generating devices such as inductive relay coils
3. Clean grounds for low current signals.

All of these grounds should meet at a common ground bus bar mounted in the panel to prevent ground loops. The panel ground bus bar is connected to an earth ground.

The master-control relay provides some of the personnel and equipment protection for the system since it de-energizes the I/O devices when any of the emergency-stop switches are pressed. The Start push-button is pressed to energize the master-control relay (MCR). Pressing any one of the emergency-stop switches de-energizes the master-control relay and thus de-energizes the I/O devices. In general, the master-control relay does not control power to the PLC processor. Depending on the system, a separate master-control relay may be provided for each section of the process. For the AC-powered master-control relay (Figure 4.41), note that a suppressor is needed across the line at the load side of the MCR contacts. The suppressor is needed since the AC I/O circuits and the DC power supply form an inductive load.

Protection for input and output devices is illustrated in Figure 4.43. For input devices, a fuse is used to protect each group of input devices. The devices may be organized so that a fuse protects all devices connected to one discrete input module. Alternatively, all of the discrete devices in one area of the process may be protected by one fuse, even if the devices are connected to different discrete input modules. For the discrete output devices, a separate fuse protects each output module, though in some applications, each discrete output device is fused separately. Fusing for the analog modules is treated in a manner similar to the DC discrete devices. The fuses also allow one to de-energize certain inputs or outputs during commissioning or troubleshooting.

Whether fuses or circuit breakers are used for circuit protection is often subject to debate. Fuses offer low-cost protection, but that protection is compromised if it is replaced with a higher-rated fuse when maintenance personnel tire of frequent blown fuses. Circuit breakers are often more convenient since a trip does not necessitate a trip to the storeroom

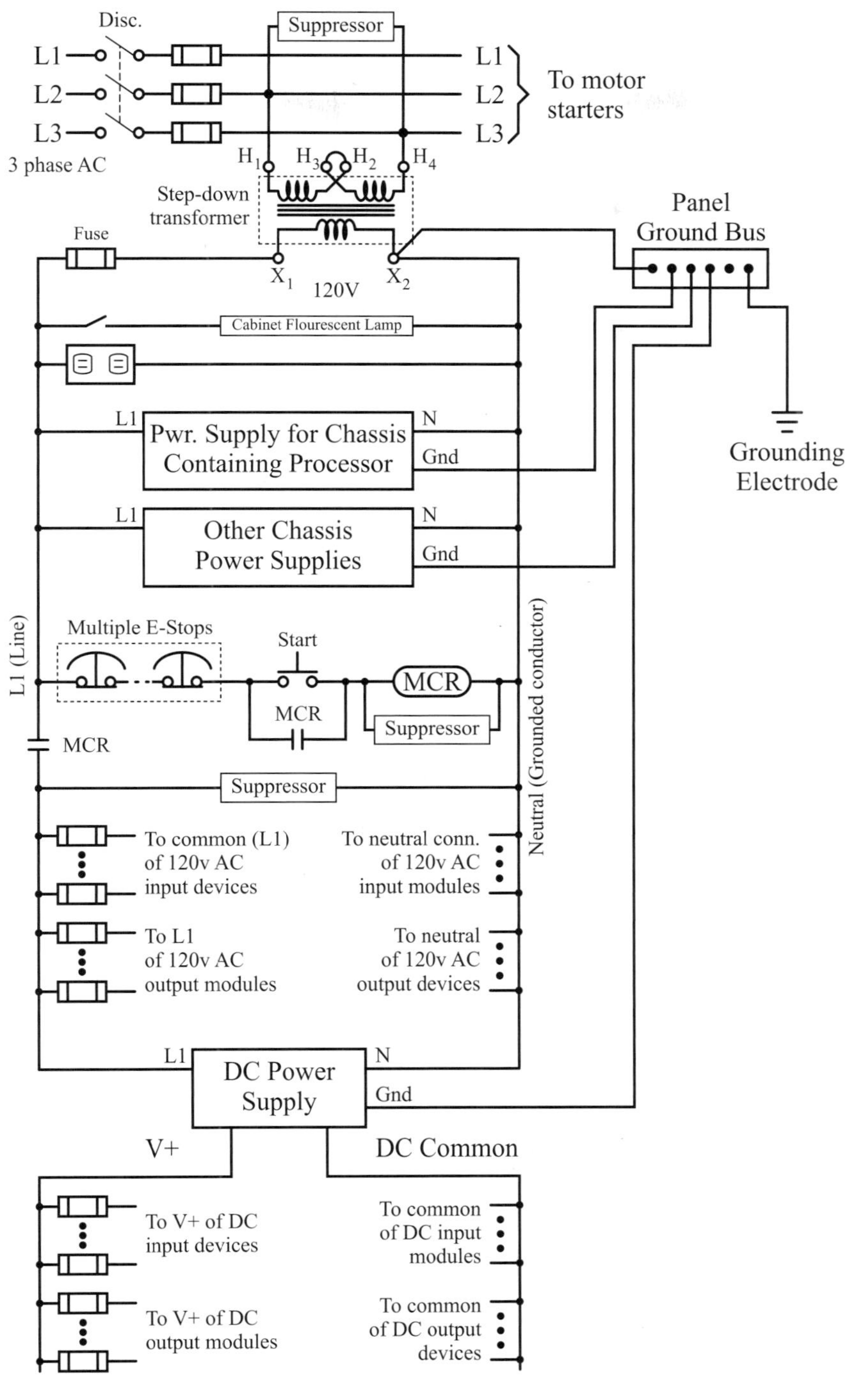

Figure 4.41. AC power distribution system with master-control relay.

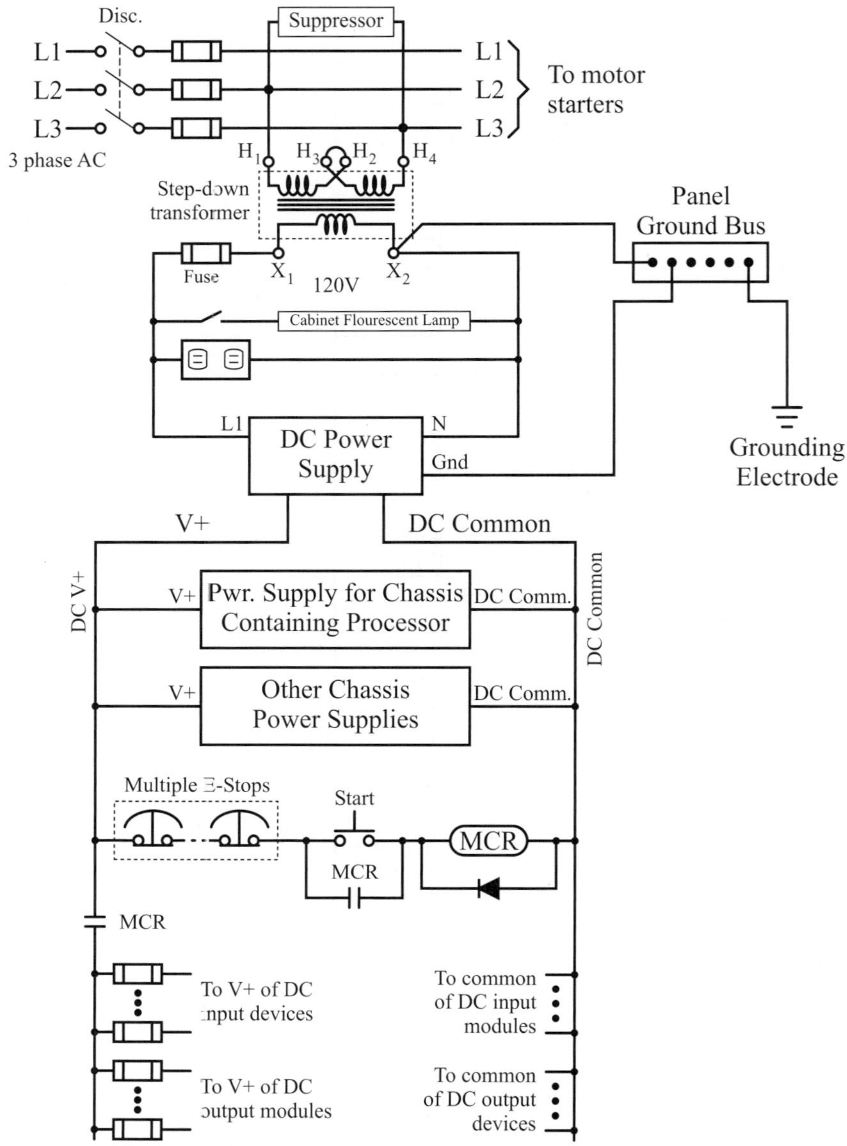

Figure 4.42. DC power distribution system with master-control relay.

for a replacement. However, circuit breakers act somewhat slower that a fuse. If fuses are used, indicating fuse holders allow one to quickly locate a blown fuse. Also, when possible all fuses in a particular cabinet protecting PLC I/O should be of the same physical size and current rating. A panel with fuses of varying sizes or ratings in identical holders may be confusing to maintenance personnel.

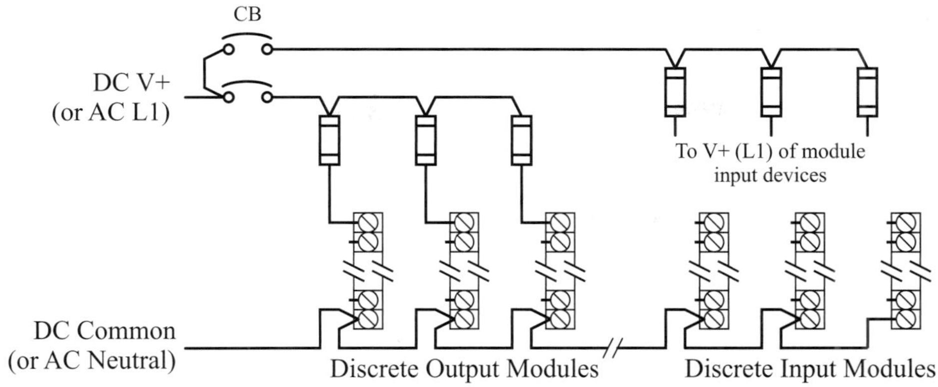

Figure 4.43. Protection of input and output devices.

4.5.2 Enclosure types

Even though PLCs are designed to operate in harsh industrial environments, most are installed in a metal enclosure, also called a control panel. The panel protects the components from atmospheric contaminants such as dust, moisture, oil and corrosive vapors. The panel also deters tampering and protects operating personnel from dangerous voltages. Finally, a panel reduces the effect of electromagnetic radiation generated by other electrical equipment, especially welders.

All enclosures installed in industrial applications must meet certain standards. In the United States, the types of enclosures are often described by the NEMA type number (NEMA, 2014). Table 4.1 summarizes the definitions of the various NEMA type numbers and their common applications. The most common types for industrial applications are types 12 and 4X. The IEC enclosure designation (IEC, 2001) consists of the letters "IP" followed by two digits and is summarized in Table 4.2. The first digit indicates the degree of protection against persons and solid objects entering the enclosure. The second digit indicates the degree of protection against water ingress. Common ratings are IP11, IP21, IP22, IP44, IP54, and IP55.

The IEC enclosure classifications (IEC, 2001) specify the degrees of protection against solid bodies and liquids. The IEC classifications do not specify protection against mechanical damage, risk of explosion, condensation, or corrosive vapors. The NEMA enclosure type numbers (NEMA, 2014) consider environmental conditions such as rust, corrosion, icing, and oil. Because of these differences, the IEC and NEMA enclosure classifications cannot be exactly equated. However, the replacement NEMA enclosure for certain IEC IP classifications can be designated as in Table 4.3. In general, the NEMA type numbers meet or exceed the associated IEC IP classification. For example, a NEMA type 12 enclosure can be used where an IEC IP52 classification is specified. However, an IEC IP52 enclosure cannot be used where a NEMA 12 enclosure is specified, since IP52 does not specify enclosure rust resistance.

4.5.3 Control Panel Design

A well-designed control panel interior and exterior is simple to understand and easily installed and maintained. Example layouts are shown in Figure 4.44 – 4.46. A panel where

Table 4.1. NEMA Enclosure Types

NEMA Enclosure Type no.	Abbreviated NEMA Definition	Application
1	Indoor use. Some protection to incidental personnel contact.	General purpose indoors
2	Type 1 plus drip-tight	Cooling rooms
3	Outdoor use. Protection against weather hazards such as rain and sleet.	General purpose outdoors Ship docks, tunnels
3R	Type 3 minus dust protection	
3S	Type 3 plus external mechanisms remain operable when ice laden	
4	Watertight. Protection against wash-down	Food processing
4X	Type 4 plus corrosion protection	Food processing
5	Indoor use. Dust-tight	Cement, powder blending
6	Submersible	Mines, quarries
6P	Prolonged submersible	Mines, quarries
7	Hazardous. Indoor use in NEC Class I, Groups A, B, C, D (certain gases and vapors which may be explosive)	Petrochemical plants
8	Hazardous. Type 7 plus outdoors	Petrochemical plants
9	Hazardous. Indoor/outdoor use in NEC Class II, Groups E, F, G (combustible dust	Petrochemical plants
10	Meets requirements of the Mine Safety and Health Administration, 10 CFR, Part 18 (1978)	Mines
12	Type 1 plus dust-tight, drip-tight, protection against lint and fibers	Industrial indoors
12X	Type 12 with knockouts	
13	Type 12 plus protection against spraying water, oil, noncorrosive coolant	Metal machining

the field I/O connections and high-volt AC enter from the top is shown in Figure 4.44. This panel also includes a transformer and a disconnect. Figure 4.45 shows the layout of a panel where the field I/O wiring enters from the bottom. A vertical mounting of a PLC chassis is shown in Figure 4.46. The more important recommendations are outlined below and then followed by a typical control panel specification.

Table 4.2. IEC Ingress Protection (IP) Ratings

First No.	Protection against:
0	No protection
1	Solid objects more than 50 mm (e.g., hands)
2	Solid objects more than 12 mm (e.g., fingers)
3	Solid objects more than 2.5 mm (e.g., tools)
4	Solid objects more than 1 mm
5	Totally enclosed. Dust may enter but not in harmful quantities
6	Totally dust-tight
Second No.	**Protection against:**
0	No protection
1	Vertically falling drops of water
2	Falling liquid up to 15° from vertical
3	Falling liquid up to 60° from vertical
4	Splashing from any direction. Limited ingress permitted
5	Strong jets of water. Limited ingress permitted
6	Occasional immersion. Water must not enter
7	Permanent submersion up to 1 meter
8	Permanent submersion up to specified depth and pressure

Table 4.3. Conversion of NEMA Enclosure Type Numbers to IEC Classification Designations (NEMA, 2014).

NEMA Enclosure Type Number	Meets or Exceeds IEC Enclosure Classification
1	IP10
2	IP11
3	IP54
3R	IP14
3S	IP54
4 and 4X	IP56
5	IP52
6 and 6P	IP67
12 and 12K	IP52
13	IP54

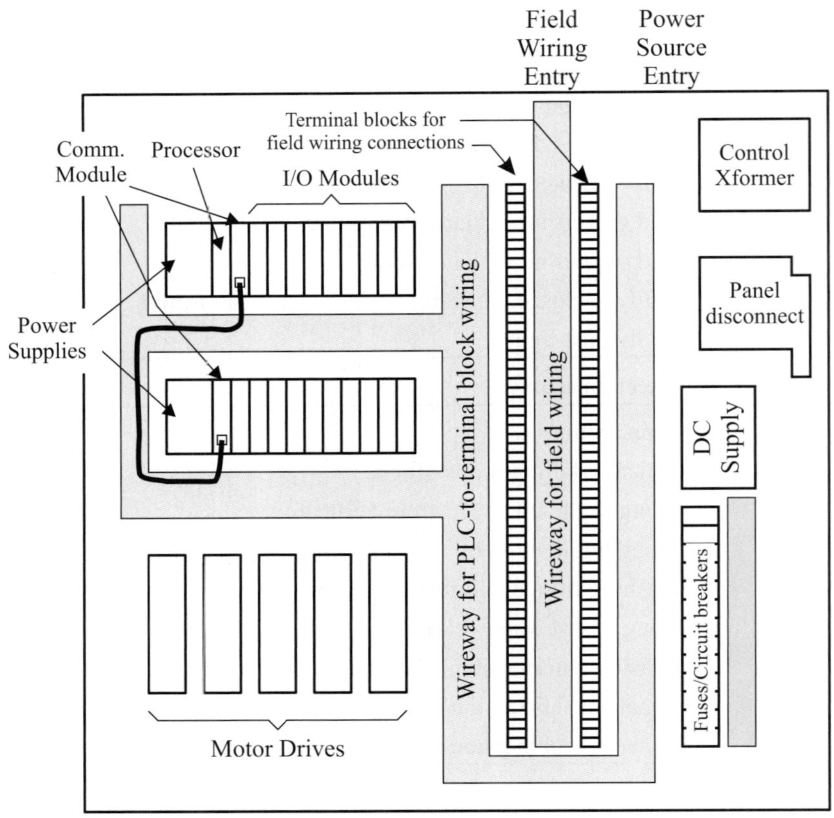

Figure 4.44. Typical panel layout that includes transformer and disconnect.

General. Some general aspects of the physical particulars of a PLC enclosure are:

- The enclosure should be located so that the doors can be fully opened.
- The enclosure depth should provide adequate clearance between the enclosed equipment, including cables and the closed door of the enclosure. If switches and/or indicators are mounted on the door or the door contains a drawing pocket, then these items should also factor into the depth requirements.
- The enclosure should contain lighting and convenience outlets to be used for troubleshooting and maintenance.
- All components should be mounted on a removable back panel.
- An emergency disconnect switch should be provided. If the enclosure contains a voltage greater than 120 volts, then the panel should have a through-the-door disconnect switch.

Environmental. The effects of temperature, humidity, electrical noise, and vibration are important when designing the panel layout. In particular:

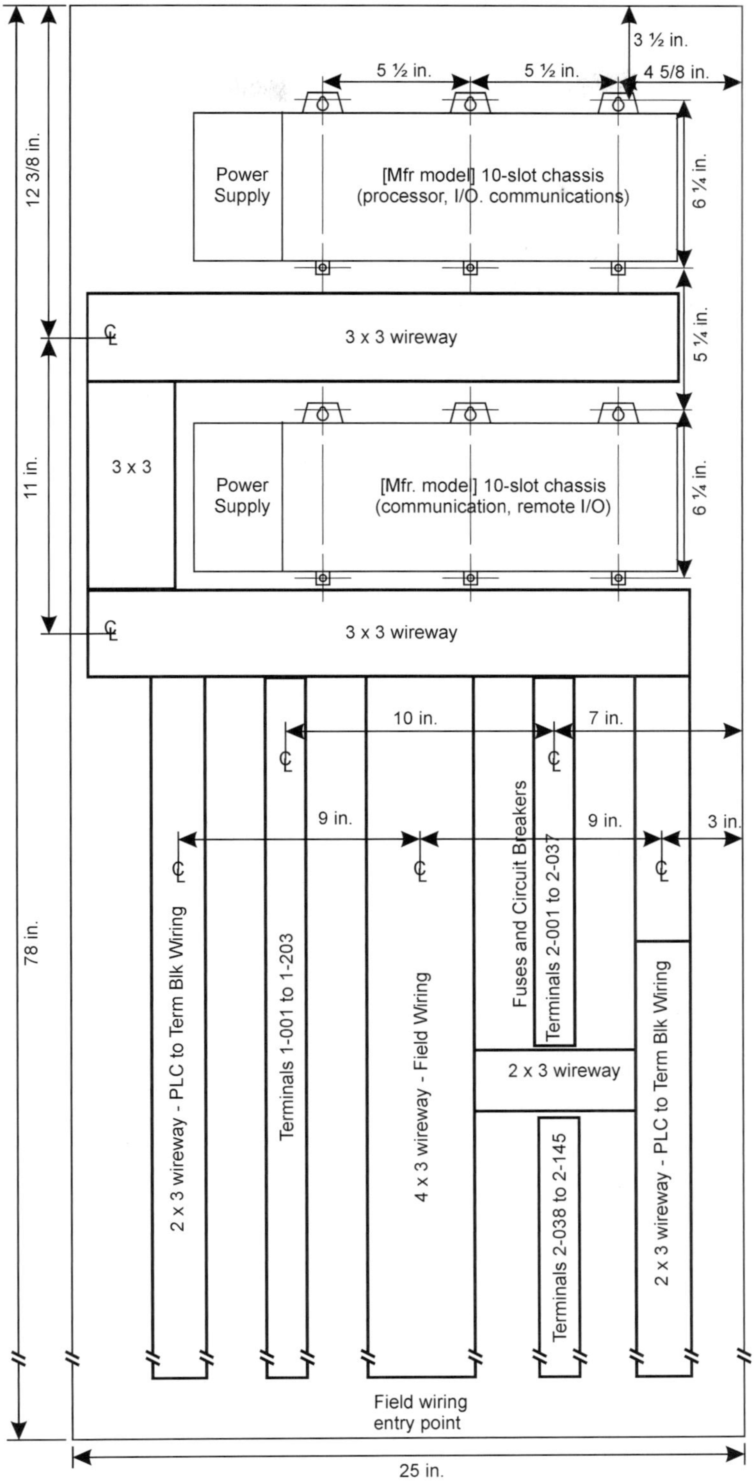

Figure 4.45. Typical panel where field wiring enters from bottom.

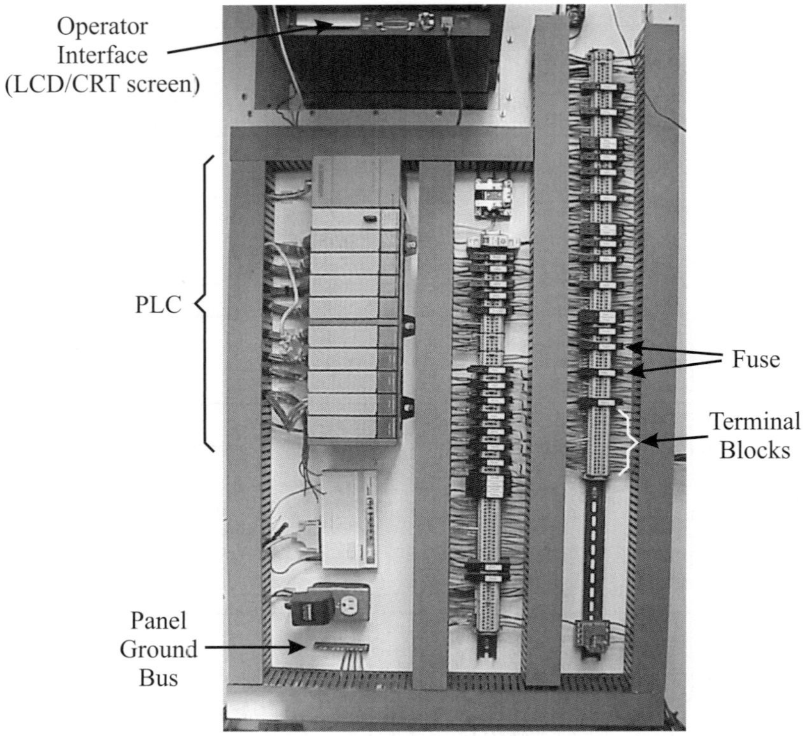

Figure 4.46. Panel with vertically-oriented PLC chassis.

- The temperature inside the enclosure must not exceed the maximum temperature allowed by the electronic equipment (typically 55 – 60 °C or 130 – 140 °F).
- If convection cooling is inadequate to maintain that temperature, then a cooling fan may be installed to help dissipate the heat. If the enclosure is dust-tight or watertight, then a vortex cooler or refrigeration unit may be required. Many control enclosure manufacturers provide information so that one can calculate the temperature rise inside the enclosure, given the amount of electrical power consumed by the devices in the enclosure.
- If condensation is likely, then the enclosure should contain a thermostatically-controlled heater.
- The panel should be located away from large motor starters, welders, and inductive heaters. These devices generate excessive electromagnetic interference or radio frequency interference.

Placement of Components. The size of the panel depends on the size and amount of equipment to be installed in the enclosure. The layout of the equipment should conform to the manufacturer's recommendations.

- The manufacturer generally specifies the minimum spacing allowed between a PLC chassis and the enclosure walls, between multiple PLC chassis, between a PLC chassis and a motor drive module, and so on. Most importantly, the minimum spacing allows adequate cooling but also reduces electrical noise coupling between the high-power signals and the low-voltage DC signals.

- System components should be mounted in a position that allows effective convective cooling. For example, any cooling fins should be oriented vertically. Even though the layout in Figure 4.46 is not optimal in this regard, the chassis is oriented so the power supply is at the top.
- Incoming line devices, such as constant voltage transformers and local power disconnects should be mounted close to the top of the enclosure, which is a common entry point for power.
- Power supplies generate heat and so are mounted toward the top of the panel and close to the incoming line devices.
- The PLC processor and I/O modules should be mounted close to eye level, either adjacent to or below the power supply.
- Motor contactors, relays and other electromechanical components should be mounted in an area well away from the PLC components, typically toward the top of the cabinet.
- Motor drives should also be located well away from the PLC components. Large drives should be in a separate cabinet.
- The I/O modules should be grouped according to the voltage level. Whenever possible, the 120-volt AC discrete input or output modules should be segregated from the DC discrete input or output modules. The analog modules should be separated from the AC or DC discrete modules.
- If possible, all I/O modules in a chassis should be the same general type (AC discrete, DC discrete, or analog). If general types must be mixed, then they should be grouped together in the chassis, and any unused slots should be allocated between the groups.

Wireway and Wiring Layout. Wires within a cabinet are routed in wireways that organize the wiring and give a neat appearance to the panel. All wires should be organized into the three categories in Table 4.4 (IEEE, 1982). In order to guard against coupling noise from

Table 4.4. Conductor categories (IEEE, 1982).

Category	Description	Examples
1	Control and AC power	AC power lines for I/O circuits and power supplies High-power digital AC I/O lines High-power digital DC I/O lines (typically connection to relays)
2	Signal and communication	Analog I/O lines and DC power for analog circuits Low-power digital DC I/O lines Communication cables (Ethernet, ControlNet, remote I/O, etc.)
3	Intra-enclosure	Low-voltage DC power wires (power modules in cabinet) Communication cables

Table 4.5. Wire routing guidelines to reduce noise coupling (IEEE, 1982).

Category	Routing Guidelines
1	Can be routed in the same cable tray or raceway with machine power conductors of up to 600V AC (feeding up to 100HP devices).
2	If it must cross power feed lines, it should do so at a right angle. Route at least 5 ft. from high-voltage enclosures or sources of RF/microwave radiation. Properly shield (where applicable) and route in a wireway separate from category-1 conductors. If **not in a contiguous metallic wireway or conduit**, route at least 0.15m (6 in.) from category-1 conductors of less than 20A; 0.3m (1 ft.) from AC power lines of 20A or more, but only up to 100 kVA; 0.6m (2 ft.) from AC power lines of greater than 100 kVA. If **in a contiguous metallic wireway or conduit**, route at least 0.08m (3 in.) from category-1 conductors of less than 20A; 0.15m (6 in.) from AC power lines of 20A or more, but only up to 100 kVA; 0.3m (1 ft.) from AC power lines of greater than 100 kVA.
3	Route conductors external to all wireways in the enclosure or in a wireway separate from any category-1 conductors with the same spacing listed from category-2 conductors, where possible.

one conductor to another, follow the routing guidelines in Table 4.5. Some other considerations are:

- Analog I/O signals should be shielded and routed separate from the discrete signals. A foil shield is preferable to a braided shield. When possible, the shield on shielded cable should be grounded at the signal source. When a shielded cable passes through intermediate junction boxes, shield continuity must be maintained.
- I/O rack interconnection cables (not the same as remote I/O cables) should be in a wireway separate from other wiring. If this is not practical, then they should be routed external to all wireways and fastened directly to the panel.
- The field devices are generally not wired directly to the I/O modules. Instead, the field devices are wired to interposing terminal blocks. The connections between the I/O module to the interposing terminal blocks can be done with individual wires to terminal blocks (Figure 4.47*a*) or with a prewired cable between the I/O module and a prefabricated termination module (Figure 4.47*b*).
- There should be at least 1 inch between wireways and terminals to allow for wiring room and to show the wire labels.
- All wires should be clearly labeled with the same designation on both ends. Wire labels should be easily matched to the control panel drawings. All labels in a PLC-based system should be labeled in such a way that relates to the PLC addressing, that is, the label should indicate the chassis, module, and channel.
- Color-code wires according to their function in order to aid troubleshooting. NFPA 79 (NFPA, 2021b) defines the following colors:

Green/yellow stripe	Equipment grounding conductor.
Black	Ungrounded line, load, control conductors at line voltage.
Red	Ungrounded AC control conductors at less than line voltage.
Blue	Ungrounded DC control conductors.
Yellow	Ungrounded control circuit conductors that remain energized when the main disconnect switch is in the OFF position.
White or gray	Grounded circuit conductor.
White/blue stripe	Grounded (current-carrying) DC circuit conductor.
White/yellow stripe	Grounded (current-carrying) AC control circuit conductor that remains energized when the main disconnect switch is in the OFF position.

Others define colors differently and add colors for analog signals.

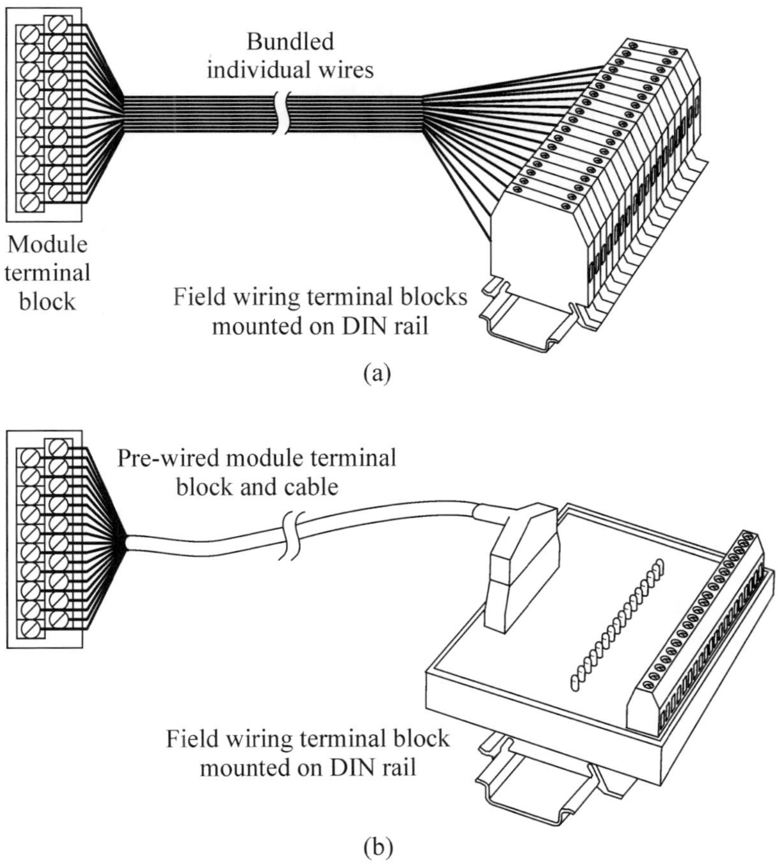

Figure 4.47. Interconnection between I/O module and field-wiring terminal blocks: *(a)* individual wires; *(b)* prefabricated termination module.

Operator Control Station Layout. If the panel door has operator controls or indications, then it should be organized for ease of understanding and use. Considerations (Lawrence, 1998):

- Locate controls at a comfortable height for viewing and operating. Indicators, alarm pilot lights, and meters should be at eye level, generally 62" to 67" from the floor. Push-button and selector switches should be about 48" to 60" from the floor.
- Group operator control devices into master, manual, and automatic functions. Within these groups, arrange the controls into a logical order to initiate machine functions in the proper sequence.
- Remember to check for the presence of a door stiffener or rib when placing the indicators and controls, since they cannot be placed there.
- The legend plates should clearly describe the action to occur and the mechanism's direction of movement.
- Do not abbreviate a word on the legend plate unless necessary.

Sample Panel Specification. A sample specification follows. Any reference to a manufacturer and model number has been removed.

SECTION 1 – PRODUCTS

1.1 CONTROL ENCLOSURES

1.1.1 Control enclosures shall comply with the following requirements:

a. Control enclosures located indoors in a protected environment shall be NEMA 12 steel with a gray powder-coated finish.

b. Control enclosures located in harsh or wet indoor locations shall be NEMA 4X fiberglass or stainless steel.

1.1.2 Control enclosures shall be [Mfr] or equal.

1.2 PLC PROCESSORS

1.2.1 PLC processors shall be furnished and installed per the model numbers and quantities on the drawings.

1.2.2 Provide a 13-slot chassis minimum. Use slot fillers in unused slots.

1.3 I/O MODULES

1.3.1 The following I/O modules shall be provided for [Mfr model] processors as indicated on the drawings:

a. Digital Input Modules: [Mfr model], without exception.

b. Digital Output Modules: [Mfr model], without exception.

c. Analog Input Modules: [Mfr model], without exception.

d. Analog Output Modules: [Mfr model], without exception.

1.4 TERMINAL BLOCKS

1.4.1 Terminal blocks for power distribution and digital signals shall comply with the following requirements:

a. Terminal blocks shall be UL rated for 600V, 30A minimum.

b. Terminal blocks shall have a compression-style screw clamp connection.

c. Terminal blocks shall be capable of accepting two #12 AWG wires.

1.4.2 Terminal blocks for analog signals shall comply with the following requirements:

a. Terminal blocks shall be UL rated for 300V, 20A minimum.

b. Terminal blocks shall have a compression-style screw clamp connection.

c. Terminal blocks shall be capable of accepting two #16 AWG wires.

d. Terminal blocks shall be three level sensor blocks for termination of signal positive, negative, and shield.

1.4.3 Fuse blocks shall comply with the following requirements:

a. Fuse blocks shall be UL rated for 600V, 10A minimum.

b. Fuse blocks shall incorporate a hinged lever that accepts 5x20 mm fuses.

c. Fuse blocks shall have a compression-style screw clamp connection.

d. Fuse blocks shall be capable of accepting two #12 AWG wires.

e. Fuse blocks shall contain blown-fuse indication through the use of a neon lamp or an LED.

1.4.4 All terminal blocks and fuse blocks shall be designed for DIN rail mounting. Extra deep 25 mm DIN rail shall be used.

1.4.5 Contractor shall provide terminal block end sections and end stops as necessary for a complete installation.

1.4.6 Terminal blocks and fuse blocks shall be provided with snap-on label strips. Label shall be consistent with Contractor's control panel drawings.

1.4.7 Terminal blocks and fuse blocks shall be [Mfr] or equal.

1.5 CIRCUIT BREAKERS

1.5.1 Circuit Breakers shall comply with the following requirements:

a. Single pole circuit breakers shall be [Mfr model] series or approved equal.

b. Double and triple pole circuit breakers shall be [Mfr model] series or approved equal.

1.6 RELAYS

1.6.1 Relays shall comply with the following requirements:

a. Relays shall be plug-in style with a DIN-rail mountable base.

b. Relays shall have on/off indication.

c. Relays shall have a manual operator.

1.6.2. Relays shall be [Mfr model] or equal.

1.7 PUSHBUTTONS AND SELECTOR SWITCHES

1.7.1 Pushbuttons and selector switches shall comply with the following requirements:

a. Pushbuttons and selector switches shall be suitable for a 30.5 mm mounting hole.
b. Pushbuttons and selector switches shall be NEMA type 4/13.
c. Pushbuttons used for non-emergency service shall be black flush-head.
d. Pushbuttons used for emergency stop service shall be red mushroom-head.
e. Pushbuttons and selector switches shall be provided with engraved legend plates indicating their function.
f. Pushbuttons and selector switches shall be provided with normally-open and normally-closed contact blocks as necessary. For pushbuttons, a minimum of 1 normally-open contact block and 1 normally-closed contact block shall be provided. For selector switches, a minimum of 1 normally-open contact block and 1 normally-closed contact block shall be provided in each position other than the "off" position.

1.7.2 Pushbuttons and selector switches shall be [Mfr model] or equal.

1.8 INDICATORS

1.8.1 Indicators shall comply with the following requirements:
a. Indicators shall be full-voltage incandescent.
b. Indicators shall be suitable for a 30.5 mm mounting hole.
c. Indicators shall be NEMA type 4/13.
d. Indicators shall be provided with engraved legend plates indicating their meaning.

1.8.2. Indicators shall be [Mfr model] or equal.

1.9 CONTROL PANEL FUSED DISCONNECT SWITCHES

1.9.1 Fused disconnect switches for use in control panels shall comply with the following requirements:
a. Disconnect switches shall be rated for 600V service.
b. Disconnect switches shall incorporate a blown fuse indication.
c. Disconnect switches shall have an option for handles that can be through-the-door mounted or direct mounted.

1.9.1 Control panels containing voltages in excess of 120V shall have a through-the-door disconnect switch. Control panels containing voltages of 120V or less shall have a disconnect switch with a direct-mounted handle that does not penetrate the enclosure.

1.9.2 Control panel disconnect switches shall be [Mfr model] or equal.

1.10 INSTRUMENT POWER SUPPLIES

1.10.1 Instrument power supplies shall comply with the following requirements:

a. Power supplies that feed voltage to DC devices or PLC inputs at DC voltage levels shall be either linear or switching.
b. Power supplies that feed analog voltage inputs or analog current loops shall be linear.

1.10.2 Instrument power supplies shall be [Mfr] or equal.

SECTION 2 – EXECUTION

2.1 CONTROL PANEL CONSTRUCTION

2.1.1 The Contractor shall size the control panels to allow sufficient space for all interior mounted equipment and wiring. Control panels shall be sized for ease of maintenance.

2.1.2 All control panels shall have 10% spare terminal blocks and fuse blocks, with a minimum of 10 spare terminal blocks and 2 spare fuse blocks.

2.1.3 All I/O including spare shall be wired out to terminal blocks.

2.1.4 All major panel equipment including PLC power supply, I/O chassis power supply, 24VDC power supply and power to each I/O module shall be sub-fused.

2.1.5 A minimum of two 120-VAC 60-Hz utility outlets should be installed in the panel.

2.1.6 An internal fluorescent light with a conveniently located on/off switch shall be installed in the panel.

2.2 I/O MODULE WIRING

2.2.1 Connections from the control panel to field devices shall only be made from interposing terminal blocks, not directly from the PLC I/O modules. Provide the following interposing methods:

a. For digital input or digital output modules provide prefabricated termination module, [Mfr model].
b. For analog input modules wire each channel to terminal blocks per requirements of section 1.4.2. Provide minimum of 4 terminal block connections per channel for (1) power, (2) signal +, (3) signal -, and (4) shield. Distribute 24VDC power for each channel. Tie all shield blocks together and wire to enclosure ground.
c. For analog output modules, wire each channel to terminal blocks per requirements of section 1.4.2. Provide a minimum of 3 terminal block connections per channel for (1) signal +, (2) signal -, and (3) shield. Configure analog output module to provide 24VDC power from the module. Tie all shield blocks together and wire to enclosure ground.

2.3 CONTROL PANEL WIRING PRACTICES

2.3.1 All internal control panel wiring shall be a minimum of #16 AWG. Power distribution wiring shall be stranded copper with MTW insulation, rated 600V. Digital input/output wiring shall be stranded copper UL 1061, rated 300V.

2.3.2 As an exception to the previous paragraph, analog input/output wiring shall be stranded copper, twisted, shielded pair, #24 AWG minimum, rated 300V, [Mfr model] or equal.

2.3.3 Every effort shall be made to segregate analog signals from power distribution and digital signal wiring. Where analog wiring must cross other wiring, it shall cross at right (90 degree) angles.

2.3.4 Use the following wire colors:

Green	Equipment grounding conductor.
White	Neutral (grounded) 120 VAC circuit conductor.
Black	Line (120 VAC) control conductor.
Red	AC Discrete output control conductor.
Orange	AC Discrete input conductor.
Brown	DC positive voltage control conductor
White/blue stripe	DC negative (common) voltage control conductor.
Blue	DC Discrete output control conductor.
Violet	DC Discrete input

2.3.5 All wiring shall be firmly supported inside the control panel. A wireway system such as [Mfr model] shall be utilized in each control panel. Wire ties shall be used as needed to support cabling and to provide a clean installation. The Contractor shall size the wireways to allow ample space for both internal control panel wiring and field wiring.

2.3.6 All wiring in each control panel shall be clearly labeled with the same designation on both ends. Wire labels shall be easily matched to the control panel drawings. Wire labels shall be [Mfr model] heat-shrink or equivalent.

2.3.7 All wiring shall be in compliance with the National Electrical Code and other applicable state and local building codes.

4.6 CHAPTER SUMMARY

This chapter describes the wiring of PLC I/O modules to discrete, analog, and specialized field devices. In addition, other installation issues such as power distribution and control panel layout is covered.

REFERENCES

Allen-Bradley, 1986. *DC (5V) Multiplexer Input Module: User Manual*, pub. 1771-6.5.10, Allen-Bradley Company, Inc, Milwaukee, WI.

Allen-Bradley, 1992. *1771 Discrete I/O DC Input and Output Modules*, pub. 1771-2.180, Allen-Bradley Company, Inc, Milwaukee, WI.

Allen-Bradley, 1993. *Very High Speed Counter Module: User Manual*, pub. 1771-6.5.74, Allen-Bradley Company, Inc, Milwaukee, WI.

Allen-Bradley, 1996a. *1771 Discrete I/O AC Input and Output Modules*, pub. 1771-2.182, Allen-Bradley Company, Inc, Milwaukee, WI.

Allen-Bradley, 1996b. *High Speed Counter Module: User Manual*, pub. 1746-6.5, Allen-Bradley Company, Inc, Milwaukee, WI.

Allen-Bradley, 1996c. *SLC 500 TM Analog I/O Modules: User Manual*, pub. 1746-6.4, Allen-Bradley Company, Inc, Milwaukee, WI.

Allen-Bradley, 1996d. *SLC 500 TM Thermocouple/mV Input Module: User Manual*, pub. 1746-6.6.1, Allen-Bradley Company, Inc, Milwaukee, WI.

Allen-Bradley, 1996e. *Stepper Controller Module: User's Manual*, pub. 1746-999-121, Allen-Bradley Company, Inc., Milwaukee, WI.

Allen-Bradley, 1998a. *Analog Output Module: User Manual*, pub. 1771-6.5.30, Allen-Bradley Company, Inc, Milwaukee, WI.

Allen-Bradley, 1998b. *Discrete Input and Output Modules: Product Data*, pub. 1746-2.35, Allen-Bradley Company, Inc, Milwaukee, WI.

Allen-Bradley, 1998c. *SLC 500 TM RTD/Resistance Input Module: User Manual*, pub. 1746-6.7, Allen-Bradley Company, Inc, Milwaukee, WI.

Allen-Bradley, 1999a. *Analog Input Module: User Manual*, pub. 1771-6.5.115, Allen-Bradley Company, Inc, Milwaukee, WI.

Allen-Bradley, 1999b. *High Resolution Isolated Analog Modules: User Manual*, pub. 1771-6.5.127, Allen-Bradley Company, Inc, Milwaukee, WI.

Emerson, 2022. *PACSystems RX3i System Manual*, pub. GFK-2314AD, Emerson Electric Co., St. Louis, MO, Nov.

IEEE, 1982. *IEEE Guide for the Installation of Electrical Equipment to Minimize Electrical Noise Inputs to Controllers from External Sources,* (IEEE Std. 518-1982), Institute of Electrical and Electronic Engineers, New York, NY.

IEEE, 2001. *IEEE Recommended Practice for Protection and Coordination of Industrial and Commercial Power Systems – IEEE Buff Book TM,* (IEEE Std. 242-2001), Institute of Electrical and Electronic Engineers, New York, NY.

IEC, 2001. *IEC 60529: Classification of Degrees of Protection Provided by Enclosures (IP Code)*, International Electrotechnical Commission, Geneva, Switzerland.

Lawrence, Ken, 1998. "Successful Control Panel Design", *The Dynamic Engineer*, Dynamic Engineering, Inc.

Morrison, Ralph, 1986. *Grounding and Shielding Techniques in Instrumentation*, 3rd Ed., John Wiley & Sons, 1986.

NEMA, 2014. *Enclosures for Electrical Equipment (1000 Volts Maximum)*, pub. 250-2014, National Electrical Manufacturers Association, Rosslyn, VA.

NFPA, 2020. *NFPA 70: National Electrical Code®*, National Fire Protection Association, Quincy, MA.

NFPA, 2021a. *NFPA 70E: Standard for Electrical Safety in the Workplace®*, National Fire Protection Association, Quincy, MA.

NFPA, 2021b. *NFPA 79: Electrical Standard for Industrial Machinery*, 2015 Edition, National Fire Protection Association, Quincy, MA.

Rockwell Automation, 1998. *Industrial Automation Wiring and Grounding Guidelines*, pub. 1770-4.1, Rockwell Automation, Inc., Milwaukee, WI.

Rockwell Automation, 2001. *System Design for Control of Electrical Noise*, pub. GMC-RM001A-EN-P, Rockwell Automation, Inc., Milwaukee, WI.

Rockwell Automation, 2017a. *ControlLogix Controllers User Manual*, pub. 1756-UM001P-EN-P, Rockwell Automation, Inc., Milwaukee, WI, May.

Rockwell Automation, 2017b. *ControlLogix Digital I/O Modules: User Manual*, pub. 1756-UM058I-EN-P, Rockwell Automation, Inc., Milwaukee, WI, Dec.

Rockwell Automation, 2021. *ControlLogix Configurable Flowmeter Module: User Manual*, pub. 1756-UM010D-EN-P, Rockwell Automation, Inc., Milwaukee, WI, Sep.

Rockwell Automation, 2021. *ControlLogix High Speed Counter Module: User Manual*, pub. 1756-UM007E-EN-P, Rockwell Automation, Inc., Milwaukee, WI, Aug.

Rockwell Automation, 2022. *ControlLogix Analog I/O Modules: User Manual*, pub. 1756-UM009E-EN-P, Rockwell Automation, Inc., Milwaukee, WI, Sep.

Siemens, 1998. *S7-300 and M7-300 Programmable Controllers Module Specifications: Reference manual,* pub. EWA 4NEB 710 6067-02 01, Siemens AG, Nürnberg, Germany.

Siemens, 2000. *S7-400 and M7-400 Programmable Controllers Module Specifications: Reference manual,* July, 2000 edition, pub. A5E00069467-03, Siemens AG, Nürnberg, Germany.

PROBLEMS

P4-1. Specify an acceptable value and power rating of a pull-up resistor, R_{pu}, to be used to connect a sinking sensor device to a PLC sinking discrete input channel (Figure P4.1). The power supply is nominally 24 volts but can range between 23 and 25 volts, depending on the current draw. The minimum "on" state voltage of the discrete input channel is 10 volts, and the equivalent input resistance is 1000 ohms. The maximum device sink current is 100 mA, and the minimum device output voltage is 2.0 volts.

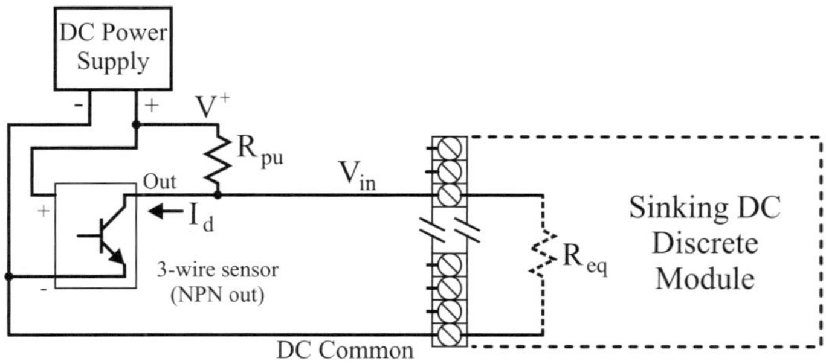

Figure P4.1. Sinking sensor connected to sinking DC discrete input channel.

P4-2. Specify an acceptable value and power rating of a pull-up resistor, R_{pu}, to be used to connect a sinking sensor device to a PLC sinking discrete input channel (Figure P4.1). The power supply is nominally 24 volts but can range between 23 and 25 volts, depending on the current draw. The minimum "on" state voltage of the discrete input channel is 13 volts, and the equivalent input resistance is 2200 ohms. The maximum device sink current is 200 mA, and the minimum device output voltage is 2.0 volts.

P4-3. Specify an acceptable value and power rating of a pull-up resistor, R_{pu}, to be used to connect a sinking sensor device to a PLC sinking discrete input channel (Figure P4.1). The power supply is nominally 24 volts but can range between 23 and 25 volts, depending on the current draw. The minimum "on" state voltage of the discrete input channel is 20 volts, and the equivalent input resistance is 10,000 ohms. The maximum device sink current is 10 mA, and the minimum device output voltage is 2.0 volts.

P4-4. Specify an acceptable value and power rating of a pull-up resistor, R_{pu}, to be used to connect a sinking sensor device to a PLC sinking discrete input channel (Figure P4.1). The power supply is nominally 12 volts but can range between 11 and 13 volts, depending on the current draw. The minimum "on" state voltage of the discrete input channel is 10 volts, and the equivalent input resistance is 2500 ohms. The maximum device sink current is 50 mA, and the minimum device output voltage is 2.0 volts.

P4-5. Specify an acceptable value and power rating of a pull-up resistor, R_{pu}, to be used to connect a sinking sensor device to a PLC sinking discrete input channel (Figure P4.1). The power supply is nominally 12 volts but can range between 11 and 13 volts, depending on the current draw. The minimum "on" state voltage of the discrete input channel is 10 volts, and the equivalent input resistance is 2200 ohms. The maximum device sink current is 50 mA, and the minimum device output voltage is 2.0 volts.

P4-6. Specify an acceptable value and power rating of a pull-up resistor, , to be used to connect a sinking sensor device to a PLC sinking discrete input channel (Figure P4.1). The power supply is nominally 12 volts but can range between 11 and 13 volts, depending on the current draw. The minimum "on" state voltage of the discrete input channel is 10 volts, and the equivalent input resistance is 2200 ohms. The maximum device sink current is 100 mA, and the minimum device output voltage is 2.0 volts.

P4-7. Specify an acceptable value and power rating of a pull-up resistor, R_{pu}, to be used to connect a sinking sensor device to a PLC sinking discrete input channel (Figure P4.1). The power supply is nominally 24 volts but can range between 23 and 25 volts, depending on the current draw. The minimum "on" state voltage of the discrete input channel is 10 volts, and the equivalent input resistance is 2200 ohms. The maximum device sink current is 50 mA, and the minimum device output voltage is 2.0 volts.

P4-8. Specify an acceptable value and power rating of a pull-up resistor, R_{pu}, to be used to connect a sinking sensor device to a PLC sinking discrete input channel (Figure P4.1). The power supply is nominally 24 volts but can range between 22 and 25 volts, depending on the current draw. The minimum "on" state voltage of the discrete input channel is 20 volts, and the equivalent input resistance is 10,000 ohms. The maximum device sink current is 50 mA, and the minimum device output voltage is 0.7 volts.

5 Timers and Counters

Chapter Topics:

- Timer blockss
- Counter blocks
- Timing and counting applications

OBJECTIVES

Upon completion of this chapter, you will be able to:

- Describe the function of the timers and counters of the PLCs covered by this text
- Describe the major types of industrial timing and counting applications
- Apply PLC timers and counters to industrial situations

Scenario: Sum total of pieces on multiple conveyors not totalizing correctly.

You are called in to troubleshoot a problem with a part of the program that is tracking the sum of bottles on two conveyors (Figure 5.1). The CompactLogix program is shown in Figure 5.2. The programming was done quite awhile ago. Someone checking production data notices a discrepancy between the number of bottles being processed by the packaging machines on a shift versus the tracking sum that this part of the program is calculating. Each time a bottle passes one of the proximity sensors, the counter increments its accumulator.

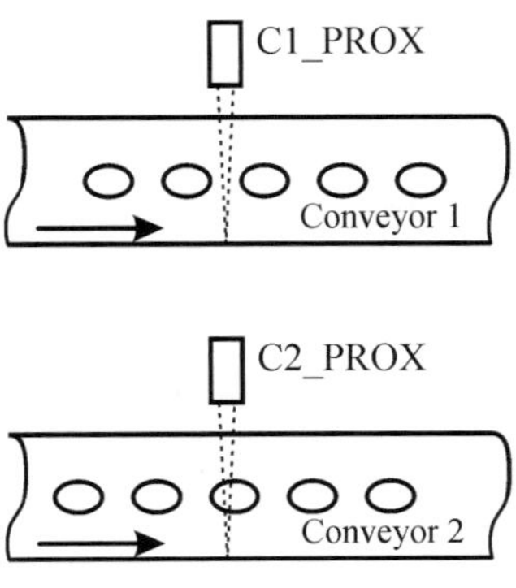

Figure 5.1. Two conveyors for counter scenario.

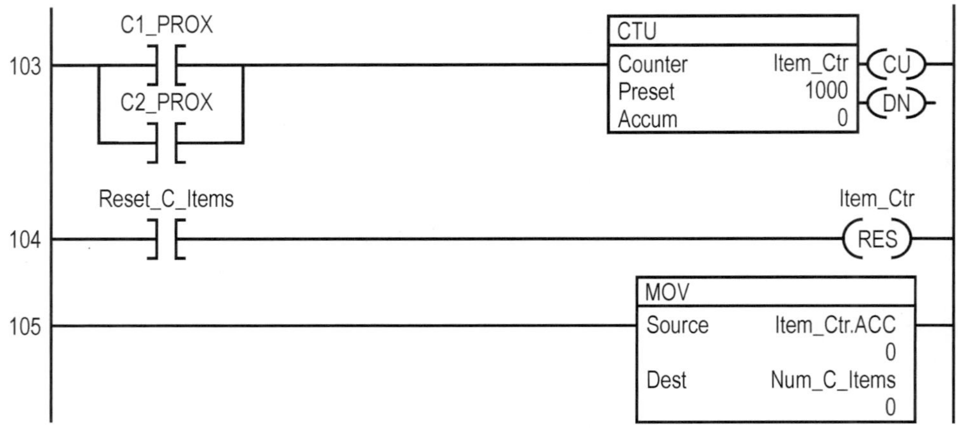

Figure 5.2. Incorrect CompactLogix ladder logic.

The accumulator value is copied to Num_C_Items. At the end of a shift, Num_C_Items is logged on a data server and the Reset_C_Items internal coil is pulsed on briefly to zero the counter accumulator.

Solution: You expect some discrepancy due to the location of the sensors for the tracking sum being some distance from the packaging machines. However, you would expect the discrepancy to be random and you notice that the tracking sum is usually less than the total of the bottles being processed on the two packaging machines. Checking the data further, you notice that if only one packing machine is running, the discrepancy can be either positive or negative. Therefore, you rule out a sensor problem. Monitoring the operation of the ladder logic, you notice that if a bottle on each conveyor passes its sensor simultaneously, or nearly simultaneously, the counter increments only once, counting only one bottle. Considering the operation of a CTU block, that makes sense. A CTU increments the counter only when its input transitions from **off** to **on** and in the situation where each sensor turns on simultaneously, or when their on-times overlap, the CTU senses only one transition. You briefly consider placing a ONS transitional contact after each of the proximity sensors, but that solution only corrects the situation where the on-time of the two sensors overlap. If both sensors transition **on** at the same time, the CTU increments just once. You delve into the operation of the CTU block and discover that the .CU bit stores the value of the CTU input at the last scan and thus is used by the CTU to sense a transition at its input. So, if each proximity sensor, followed by a transitional contact, drives a separate CTU block (both blocks referring to the same counter) and the CU bit is always unlatched after the CTU is scanned, every transition of both proximities will be counted. The implemented solution is shown in Figure 5.3.

Over the next few days, you check the discrepancy between the number of bottles being processed by the packaging machines on a shift versus the tracking sum from this part of program and it appears to be random, with both positive and negative discrepancies.

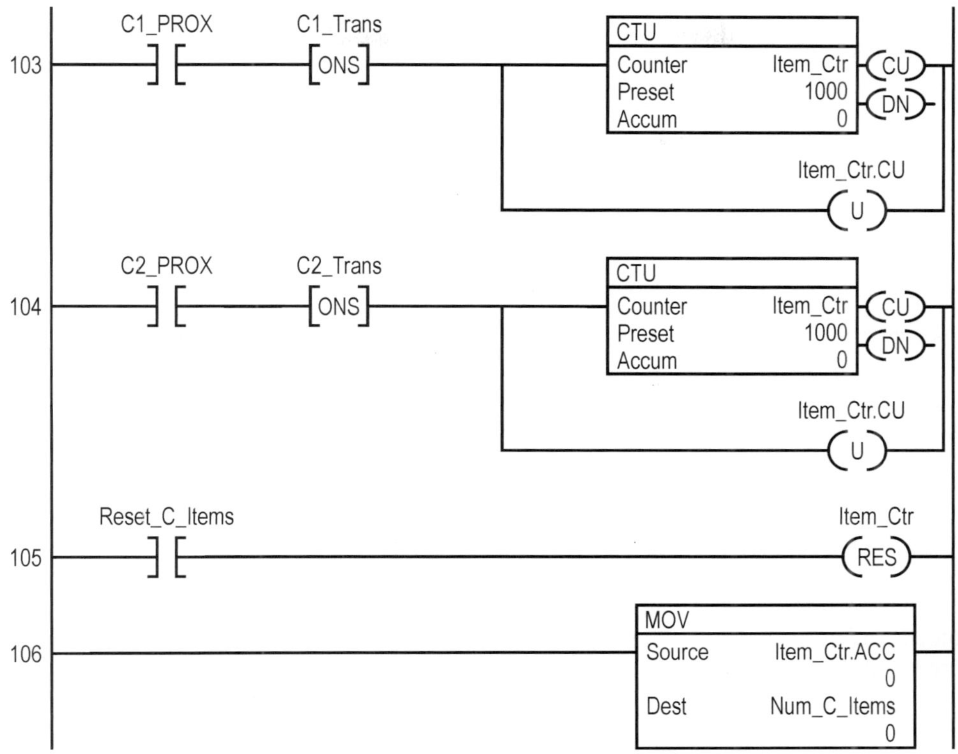

Figure 5.3. Corrected CompactLogix ladder logic.

5.1 INTRODUCTION

After coils and contacts, timers and counters are the next most commonly used ladder logic instruction. This chapter describes the timer and counter function blocks for the ControlLogix/CompactLogix, Micro800, and MicroLogix/SLC-500 PLCs covered by this text. As for Chapter 3, the timer and counter function blocks for a particular PLC are described in separate sections. That is, only the section(s) pertaining to the PLC(s) of interest need to be studied. Following these timer and counter descriptions, common timer and counter situations are described and illustrated with industrial examples.

There are two basic types of timers: on-delay and off-delay. The off-delay timer is often confusing to PLC programmers. The on-delay timer basically delays the turn-on of a discrete signal and the off-delay timer basically delays the turn-off of a discrete signal. This difference is illustrated in Figure 5.4. The PROX_1 discrete input is the input to an on-delay timer and an off-delay timer. The output of the on-delay timer delays the off-to-on transition of PROX_1, but does not delay the on-to-off transition. The output of the off-delay timer delays the on-to-off transition of PROX_1, but does not delay the off-to-on transition. Which type is appropriate? It depends on the application. For example, if an action needs to be initiated x seconds after PROX_1 turns on, then an on-delay timer is appropriate. If an action needs to be stopped x seconds after PROX_1 turns off, then an off-delay timer is appropriate.

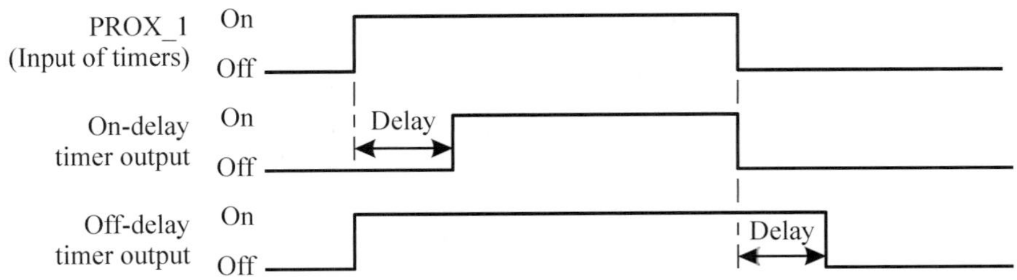

Figure 5.4. Comparison of on-delay and off-delay timers.

There are three basic types of counters: up-counter, down-counter, and up/down counter. The three types are largely self-explanatory and their operation is explained in the following sections.

5.2 CONTROLLOGIX TIMERS AND COUNTERS

The Allen-Bradley ControlLogix/CompactLogix processor has three timer function blocks and two counter function blocks. The names and description of these function blocks are as follows:

ControlLogix Function Block	Description
TON	On-delay timer
TOF	Off-delay timer
RTO	Retentive on-delay timer
CTU	Up counter
CTD	Down counter

The CTU and CTD function blocks can be combined to construct an up/down counter. The operation of each of these function blocks is described and illustrated.

5.2.1 TON On-Delay Timer

The TON on-delay timer is shown in Figure 5.5. The TON timer basically delays the turn-on of a signal and does not delay the turn-off. The tag specified in the "Timer" field must be a TIMER data type (Figure 3.2*c*). For the tag defined in Figure 5.5, the individual parts (fields) of the TIMER structure are addressed as follows:

Reference	Description
Sampl_Ton.DN	Done
Sampl_Ton.TT	Timer timing bit
Sampl_Ton.EN	Enable
Sampl_Ton.ACC	Accumulator
Sampl_Ton.PRE	Preset value

When the timer input is **on**, the accumulator is counted up once each millisecond. When the accumulator equals the preset value, the timer is "timed out" and the .DN bit is set to **on**. If the timer input turns **off** during the timing interval, the accumulator is set to zero

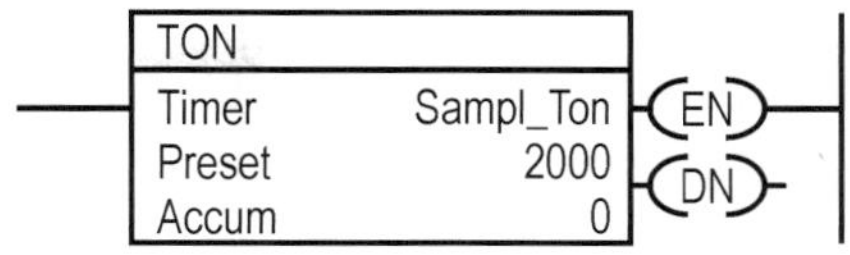

Figure 5.5. ControlLogix TON timer.

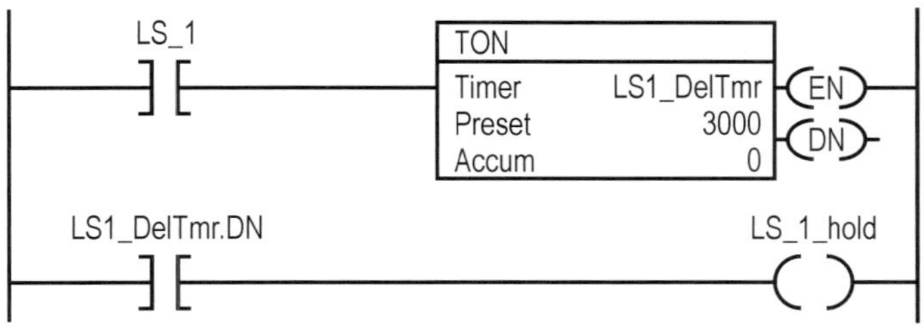

(a)

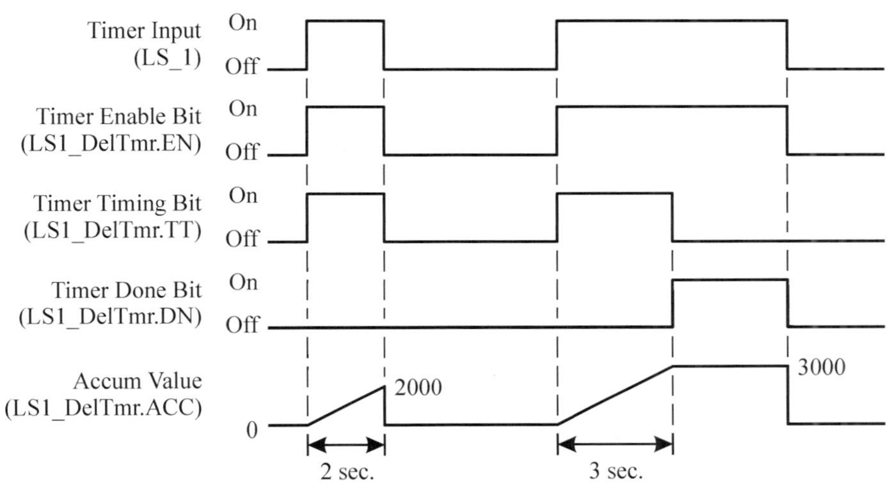

(b)

Figure 5.6. Example ControlLogix TON timer: *(a)* ladder logic; *(b)* timing diagram.

and timing recommences when the input turns **on**. The .TT bit is on while the accumulator is counting, but not yet at the preset value. The .EN bit is identical to the timer input condition. The maximum interval is 2,147,483 seconds (≈24.8 days). The value of the accumulator (.ACC) or the preset value (.PRE) can be accessed as DINT.

An example application of the TON block is shown in Figure 5.6*a* and its associated timing diagram is shown in Figure 5.6*b*. The preset time is 3 seconds (3000 ms). The input, LS_1 must remain **on** for at least 3 seconds before the LS1_DelTmr.DN bit and the LS_1_hold coil are turned **on**. When LS_1 is turned **off** after 2 seconds, the accumulator is set to zero and the LS1_DelTmr.DN bit remains **off**.

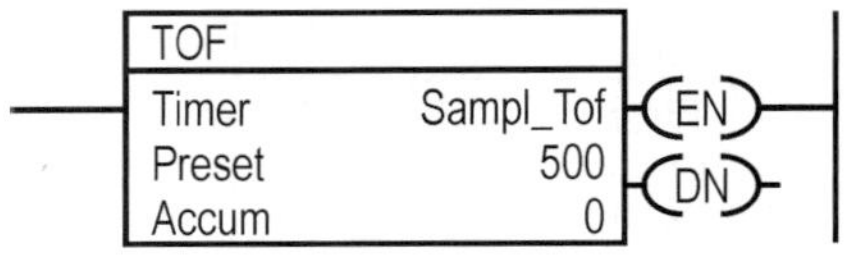

Figure 5.7. ControlLogix TOF timer.

5.2.2 TOF Off-Delay Timer

The TOF off-delay timer is shown in Figure 5.7. Looking at the timer input and the .DN bit, the TOF timer operates exactly opposite of the TON timer. The TOF timer basically delays the turn-off of a signal and does not delay the turn-on. The descriptions of the function block fields and the tag references are the same as for the TON function block.

When the timer input is **off**, the accumulator is counted up once each millisecond. When the accumulator equals the preset value, the timer is "timed out" and the .DN bit is turned **off**. If the timer input turns **on** anytime, the accumulator is set to zero and timing recommences when the input turns **off**. The operation of the .EN and .TT bits is the same as for the TON timer instruction.

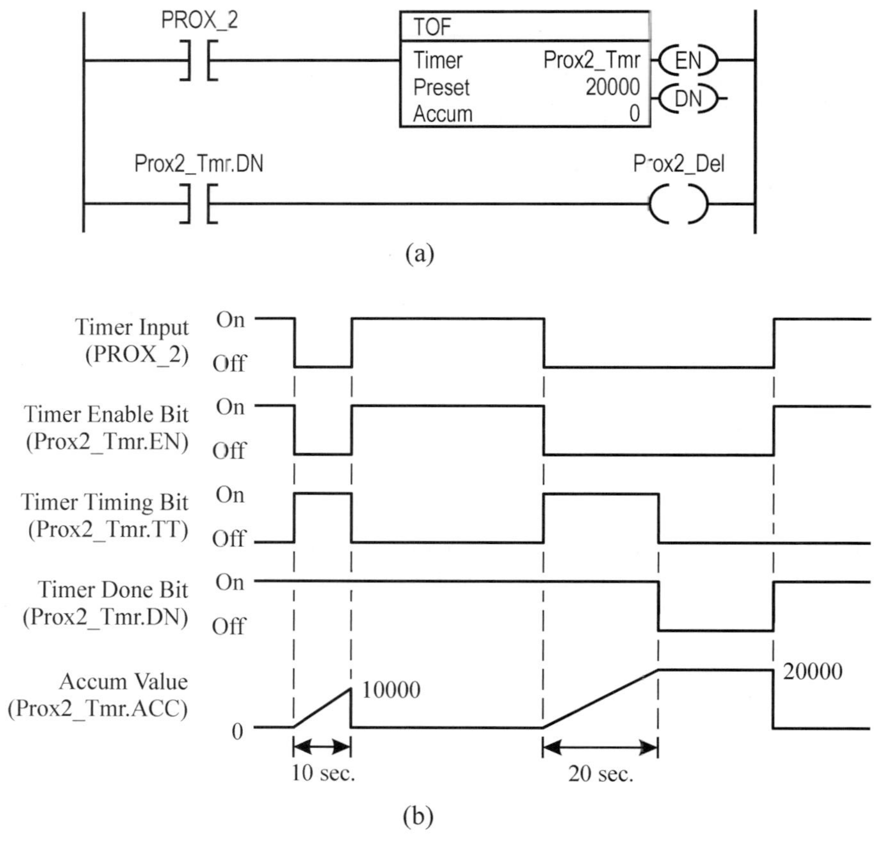

Figure 5.8. Example ControlLogix TOF timer: *(a)* ladder logic; *(b)* timing diagram.

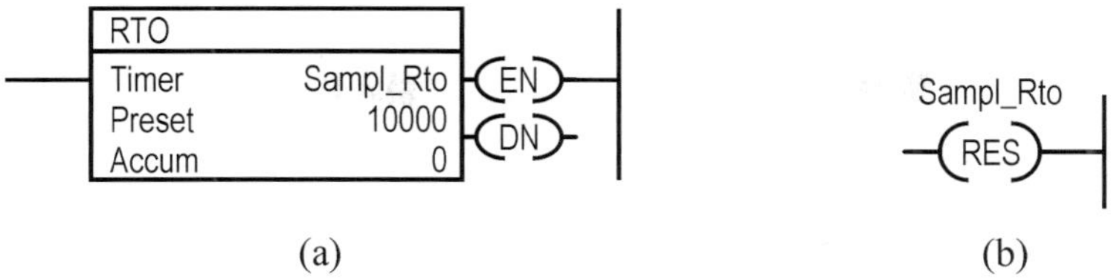

Figure 5.9. ControlLogix RTO timer: *(a)* RTO instruction; *(b)* RES coil to reset RTO.

An example application of the TOF block is shown in Figure 5.8*a* and its associated timing diagram is shown in Figure 5.8*b*. The preset time is 20 seconds (20,000 ms). The input, PROX_2 must remain **off** for at least 20 seconds before the Prox2_Tmr.DN bit and the Prox2_Del coil are turned **off**. When PROX_2 is turned **on** after 10 seconds, the accumulator is set to zero and the Prox2_Tmr.DN bit remains **on**.

5.2.3 RTO Retentive On-Delay Timer

The RTO retentive on-delay timer is shown in Figure 5.9*a*. The RTO timer is like the TON timer except that the accumulator is not reset when the timer input turns **off**. When the timer input is **on**, the accumulator is counted up once each millisecond. When the accumulator equals the preset value, the .DN bit is set to **on**. If the timer input turns **off** during the timing interval, the accumulator remains at its current value and timing resumes when the input turns **on**. A reset coil that refers to the timer (Figure 5.9*b*) may be used to set the accumulator to zero and thus turn **off** the .DN bit. Alternatively, one can use a MOV block (Chapter 7) to copy a zero to the timer accumulator. The operation of the .EN and .TT bits is the same as for the TON and TOF timer instructions.

An example application of the RTO block is shown in Figure 5.10 and its associated timing diagram is shown in Figure 5.11. The preset time is 200 seconds (200,000 ms). The LS3_OnTmr.DN bit is turned **on** when the cumulative time that LS_3 has been **on** is at least 200 seconds. After the accumulator has reached 200 seconds, the LS3_OnTmr.DN bit remains **on** even after LS_3 turns **off**. The All_Reset must be turned **on** to reset the timer.

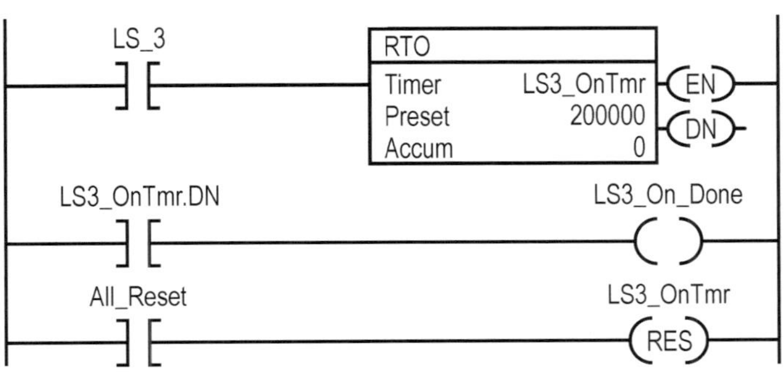

Figure 5.10. Example ControlLogix RTO timer ladder logic.

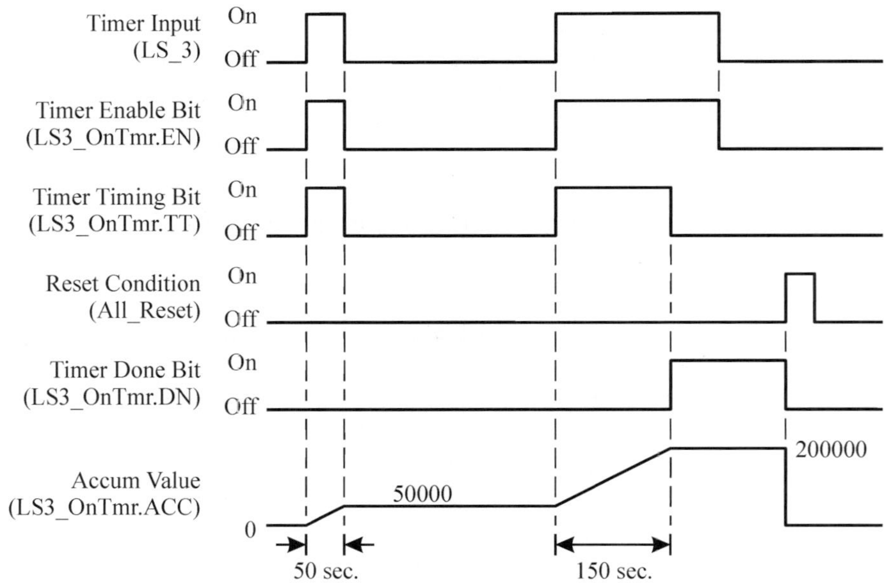

Figure 5.11. Example ControlLogix RTO timing diagram.

5.2.4 CTU Up Counter

The CTU up counter function block is shown in Figure 5.12*a*. The tag specified in the "Counter" field must be a COUNTER data type (Figure 3.2*b*). For the tag defined in Figure 5.12*a*, the individual parts (fields) of the COUNTER structure are addressed as follows:

Reference	Description
Sampl_Ctu.DN	Done
Sampl_Ctu.CU	Count-up bit
Sampl_Ctu.CD	Count-down bit
Sampl_Ctu.OV	Accumulator overflow
Sampl_Ctu.UN	Accumulator underflow
Sampl_Ctu.ACC	Accumulator
Sampl_Ctu.PRE	Preset value

The .CU and .OV bits are modified only by the CTU function block. The .CD and .UN bits are modified only by the CTD function block.

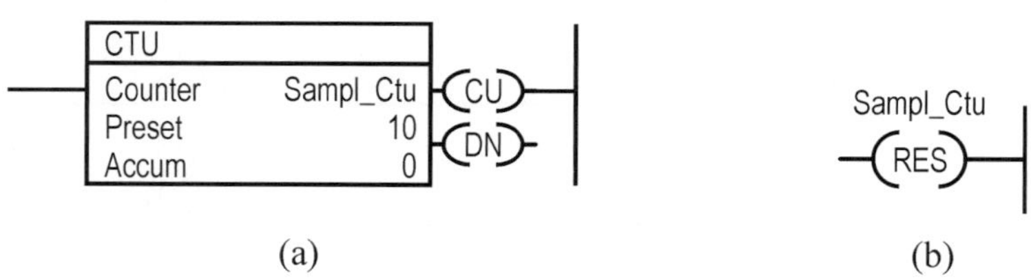

Figure 5.12. ControlLogix CTU counter: *(a)* CTU block; *(b)* RES coil to reset counter.

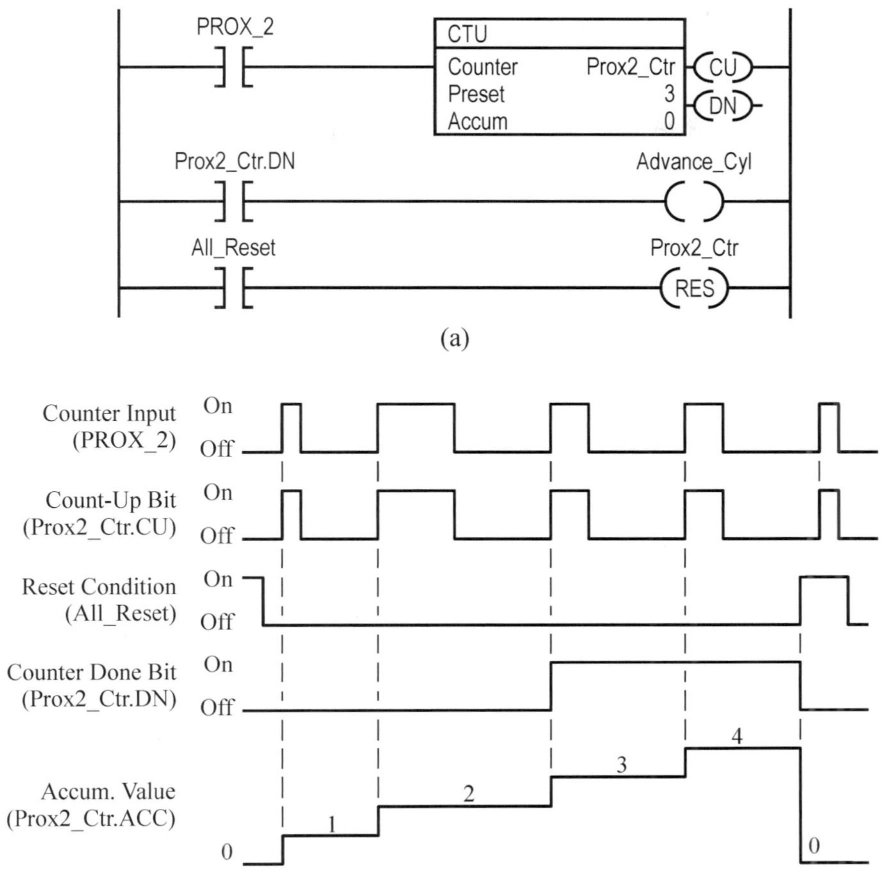

Figure 5.13. Example ControlLogix CTU counter: *(a)* ladder logic; *(b)* timing diagram.

For the CTU, the counter accumulator increments by one for every **off**-to-**on** transition of the counter input. When the accumulator ≥ preset value, then the .DN bit is turned **on**. A reset coil that refers to the counter (Figure 5.12*b*) may be used to set the accumulator to zero and thus turn **off** the .DN bit (if the preset value is greater than zero). The .CU bit holds the value of the CTU input at the last scan. The preset value and accumulator are both DINT data types. The counter can count beyond the Preset value. The overflow bit (.OV) is turned **on** if the accumulator exceeded the upper limit of DINT (approximately 2.15×10^9) and wrapped around to the lower limit of DINT.

An example application of the CTU function block is shown in Figure 5.13*a* and its associated timing diagram is shown in Figure 5.13*b*. As long as All_Reset is **on**, the accumulator is held at zero. When All_Reset is **off**, at every **off**-to-**on** transition of PROX_2, the accumulator, Prox2_Ctr.ACC, increases by one. When the accumulator is equal to or greater than 3 (the preset value), the Prox2_Ctr.DN bit is turned **on**. When Prox2_Ctr.DN is already **on**, any more positive transitions of PROX_2 cause the value of Prox2_Ctr.ACC to change, and Prox2_Ctr.DN remains **on**.

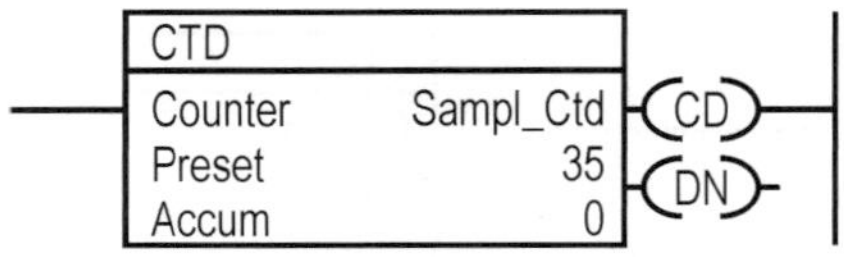

Figure 5.14. ControlLogix CTD counter.

5.2.5 CTD Down Counter

The CTD down counter function block is shown in Figure 5.14. The descriptions of the function block fields and the tag references are the same as described with the CTU function block.

For the CTD function block, the counter accumulator decrements by one for every **off**-to-**on** transition of the counter input. When the accumulator ≥ preset value, then the .DN bit is turned **on**, which is the same behavior as for CTU. A reset coil that refers to the counter (Figure 5.12*b*) may be used to set the accumulator to zero and thus turn **off** the .DN bit (if the preset value is greater than zero). A MOV instruction (Chapter 7) must be used to set the accumulator to a nonzero value. The .CD bit holds the value of the CTD input at the last scan. The preset value and accumulator are both DINT data types. The counter can count below zero. The underflow bit (.UN) is turned **on** if the accumulator went below the lower limit of DINT (approximately -2.15×10^9) and wrapped around to the upper limit of DINT.

An example application of the CTD function block is shown in Figure 5.15*a* and its associated timing diagram is shown in Figure 5.15*b*. When Prox3_CtLd is **on**, Prox_3_Ctr.ACC is loaded with the value of 4. When All_Reset is **off**, at every **off**-to-**on** transition of PROX_3, the accumulator, Prox3_Ctr.ACC, decreases by one. When the accumulator is less than 2 (the preset value), the Prox3_Ctr.DN bit is turned **off**. Regardless of the value of Prox3_Ctr.DN, any positive transitions of PROX_3 cause the value of the accumulator to change. To reset the accumulator to zero, All_Reset is turned **on**.

5.2.6 Up/Down Counter

For the ControlLogix, an up/down counter is constructed by combining a CTU and a CTD function block that operate on the same counter. Both the CTU and CTD function blocks basically act on a counter structure, incrementing or decrementing the accumulator and setting or clearing the bits in the structure. There is no conflict if the CTU function block and the CTD function block refer to the same counter structure. The CTU block provides the condition that increases the accumulator and the CTD block provides the condition that decreases the accumulator. Note that the condition that determines the value of the .DN bit (accumulator ≥ preset value) is the same, for both counter blocks.

An example application of the combination the CTU and CTD function blocks is shown in Figure 5.16*a* and its associated timing diagram is shown in Figure 5.16*b*. When All_Reset is **off**, at every **off**-to-**on** transition of Box_In, the accumulator increases by one. At every **off**-to-**on** transition of Box_Out, the accumulator decreases by one. When the accumulator is equal to or greater than 3 (the preset value), the Box_Ctr.DN bit and the Remove_Box coil are turned **on**. When All_Reset is **on**, the accumulator is cleared. When Boxes_Ld is **on**, the accumulator is set to 3.

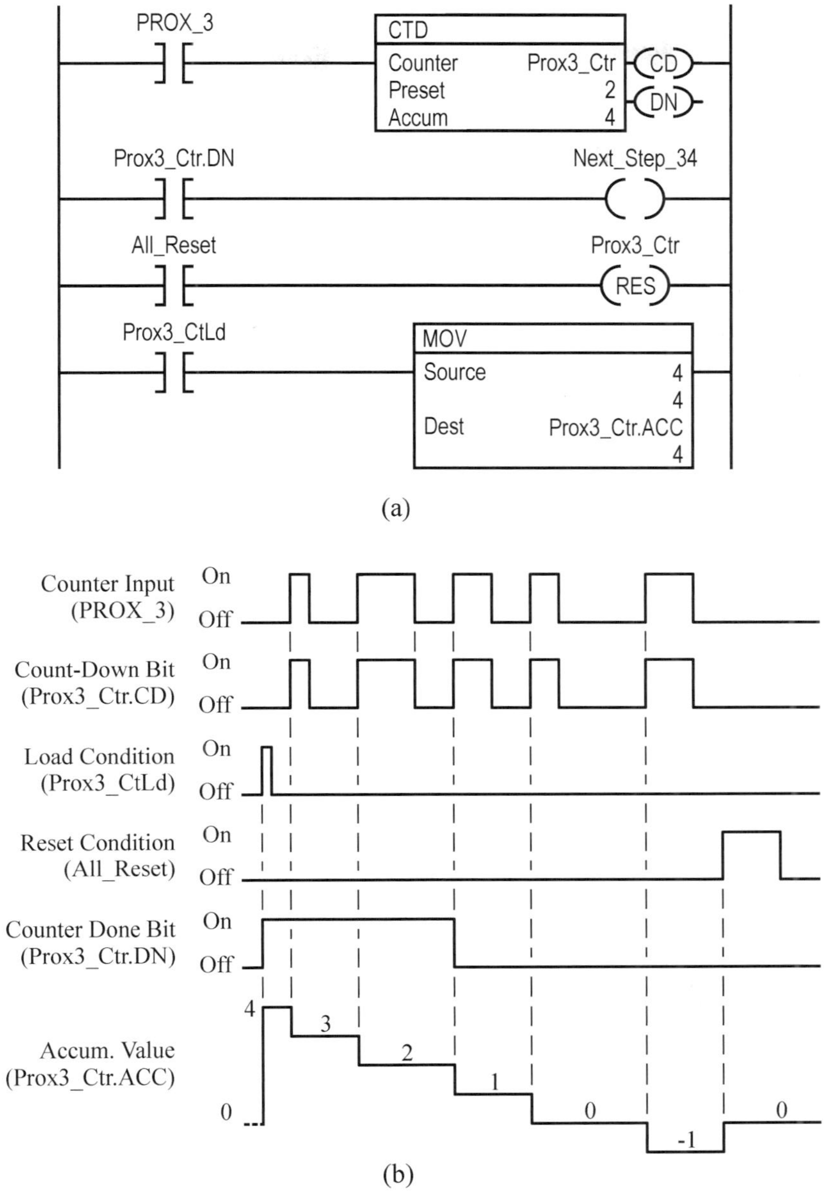

Figure 5.15. Example ControlLogix CTD counter: *(a)* ladder logic; *(b)* timing diagram.

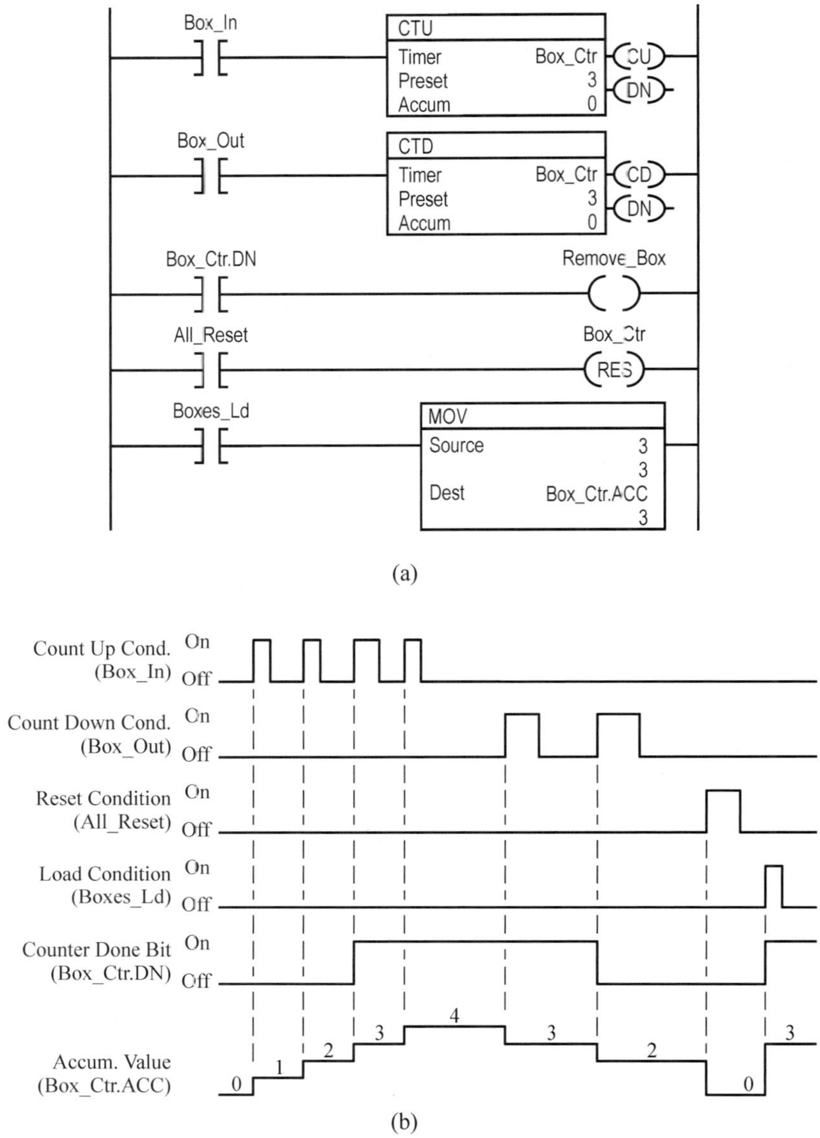

Figure 5.16. Example ControlLogix up/down counter: *(a)* ladder logic; *(b)* timing diagram.

5.1 MICRO800 TIMERS AND COUNTERS

The Allen-Bradley Micro800 processors have five timer function blocks and three counter function blocks. The names and description of these function blocks are as follows:

Micro800 Function Block	**Description**
TON	On-delay timer
TOF	Off-delay timer
RTO	Retentive on-delay timer
TONOFF	On-off-delay timer
TP	Pulse timer
CTU	Up counter
CTD	Down counter
CTUD	Up/down counter

5.3.1 TON On-Delay Timer

The TON on-delay timer is shown in Figure 5.17. The TON timer basically delays the turn-on of a signal and does not delay the turn-off. The variable specified above the "TON" in the block field must be a TON data type. For the variable defined in Figure 5.17, the individual parts (fields) of the TON data type that can be addressed by other logic are as follows:

Reference	**Description**
Sampl_Ton.Redge	Input
Sampl_Ton.Q	Done
Sampl_Ton.ET	Elapsed time (type TIME)

All of these fields are read-only. The ET cannot be the destination of a MOV block or calculation and the Booleans cannot be used by a coil. The .Redge field is the same value as the IN input.

When the IN timer input is **on**, the elapsed time (ET) increases. When ET equals the preset time (PT), the timer is "timed out" and the Q output turns **on**. If IN turns **off** at any time, the timer resets, that is, ET is set to zero and Q is turned **off**. The preset time can be a variable, or a constant of type TIME. The accuracy is 1 millisecond. The maximum interval is 2,147,483 seconds (≈24.8 days). The ET output can be connected to a variable (of type TIME) in order to monitor the elapsed time, or simply referenced as <timer name>.ET in a calculation or MOV block.

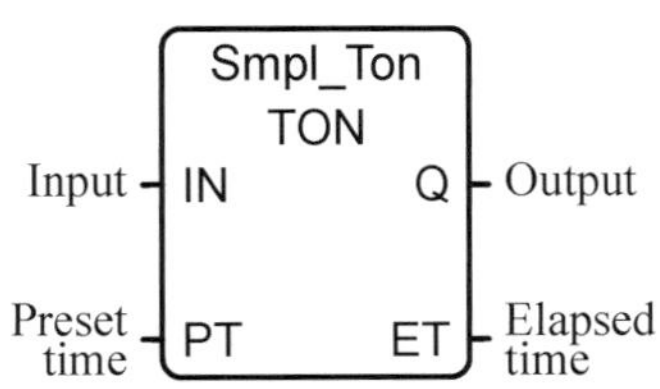

Figure 5.17. Micro800 TON timer.

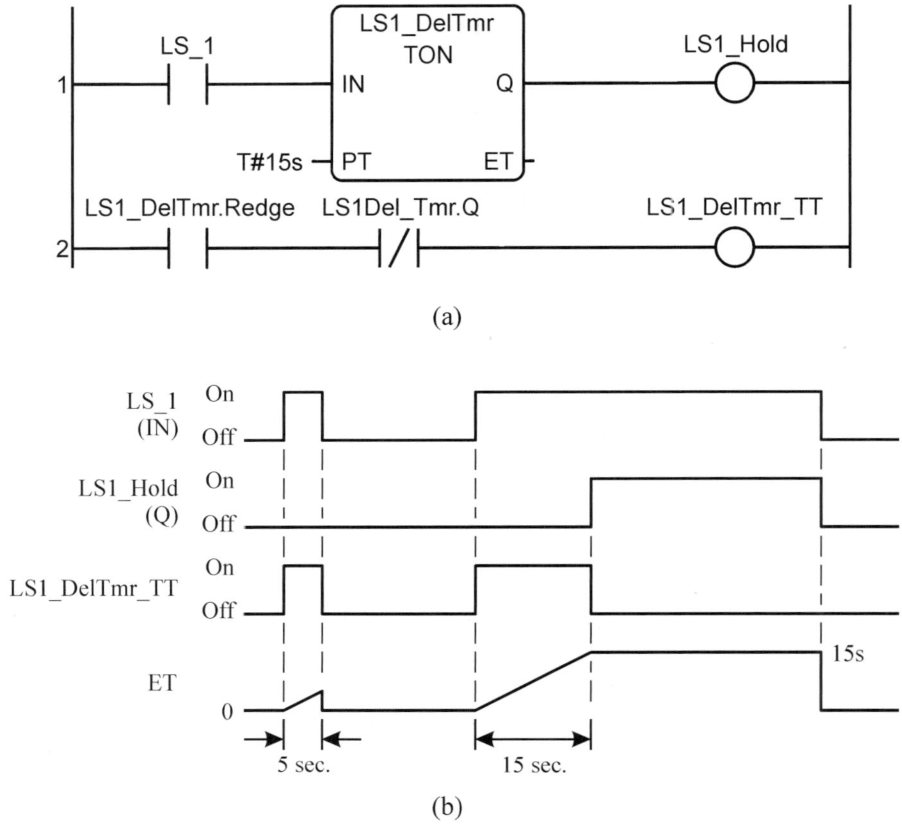

Figure 5.18. Example Micro800 TON timer: *(a)* ladder logic; *(b)* timing diagram..

An example application of the TON block is shown in Figure 5.18*a* and its associated timing diagram is shown in Figure 5.18*b*. The preset time is 15 seconds. The input, LS_1 must remain **on** for at least 15 seconds before the LS1_Hold coil is turned **on**. When LS_1 is turned **off** after 5 seconds, the elapsed time is set to zero and the LS1_Hold remains **off**. The second rung illustrates the logic to generate a "timer timing" Boolean that is automatically provided by the ControlLogix and MicroLogix timers.

5.3.2 TOF Off-Delay Timer

The TOF off-delay timer is shown in Figure 5.19. The TOF timer basically delays the turn-off of a signal and does not delay the turn-on. The variable specified above the "TOF" in the block field must be a TOF data type. The variable references to the individual fields of the TOF data type are the same as for the TON data type, except the .Fedge field is the program-addressable tag that indicates the value of the IN input.

When the IN timer input turns **off**, the internal time (ET) increases. When ET equals the preset time (PT), the timer is "timed out" and the Q output turns **off**. If IN turns **on** at any time, the timer resets, that is, ET is set to zero and Q is tuned **on**.

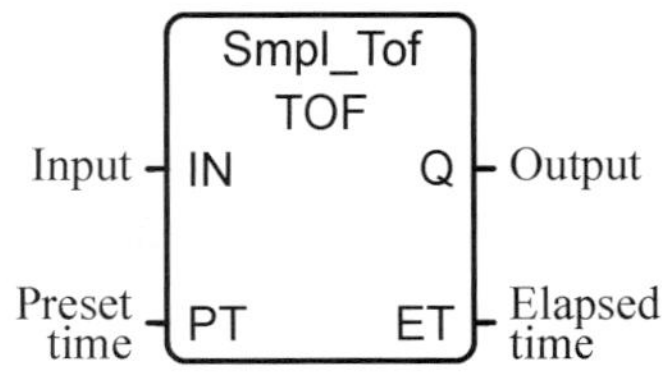

Figure 5.19. Micro800 TOF timer.

An example application of the TOF block is shown in Figure 5.20*a* and its associated timing diagram is shown in Figure 5.20*b*. The preset time is 15 seconds. The PROX_2 discrete input must remain **off** for at least 20 seconds before the Prox2_Del coil is turned **off**. When PROX_2 is turned **on** after 5 seconds, the elapsed time is set to zero and the Prox2_Del coil remains **on**. The LS1_DelTmr_TT coil on rung 2 is a "timer timing" Boolean that is automatically provided by the ControlLogix and MicroLogix timers.

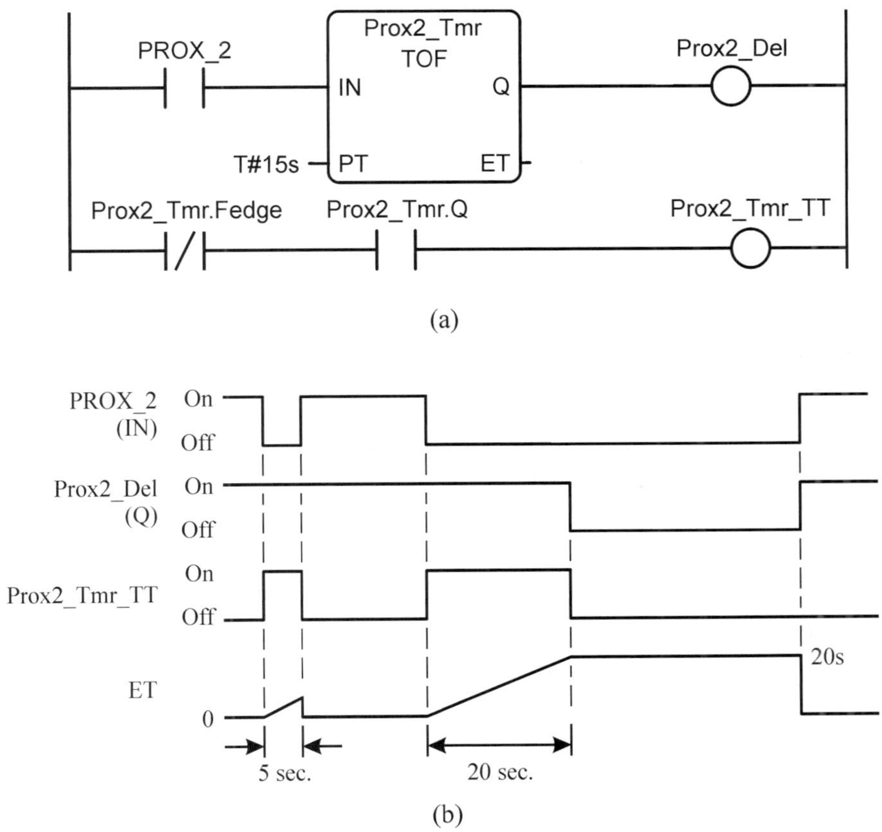

Figure 5.20. Example Micro800 TOF timer: *(a)* ladder logic; *(b)* timing diagram.

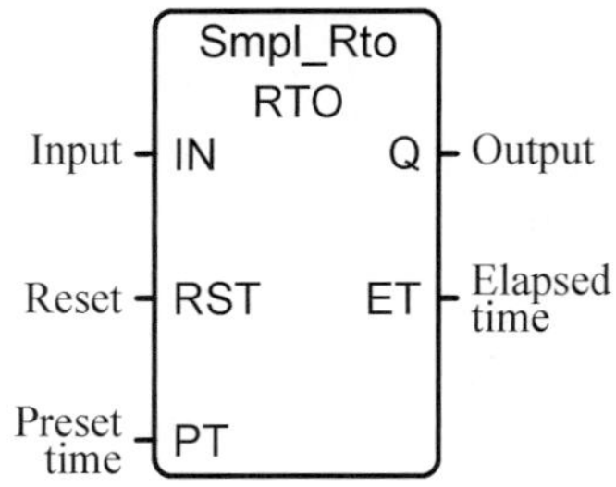

Figure 5.21. Micro800 RTO timer.

5.3.3 RTO Retentive On-Delay Timer

The RTO retentive on-delay timer is shown in Figure 5.21. The variable specified above the "RTO" in the block field must be an RTO data type. The variable references to the individual fields of the RTO data type are the same as for the TON data type.

The RTO timer is like the TON timer except that the ET is not reset when the timer input turns **off**. When the timer input, IN, is **on**, the ET increases. When ET equals the preset time (PT), the timer is "timed out" and the Q output turns **on**. If IN turns **off** during the timing interval, ET remains at its current value and timing resumes when the input turns **on**.

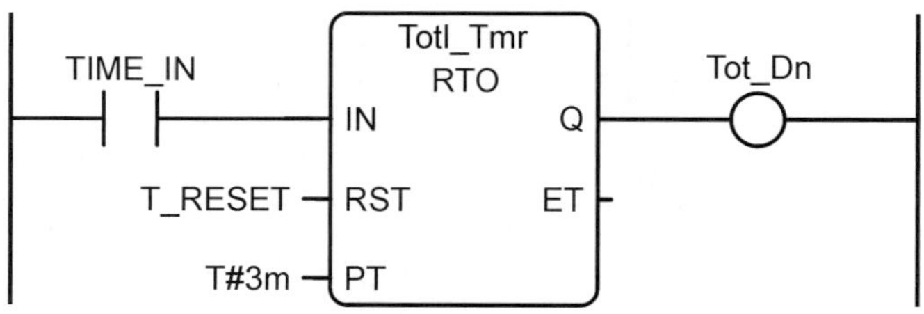

(a)

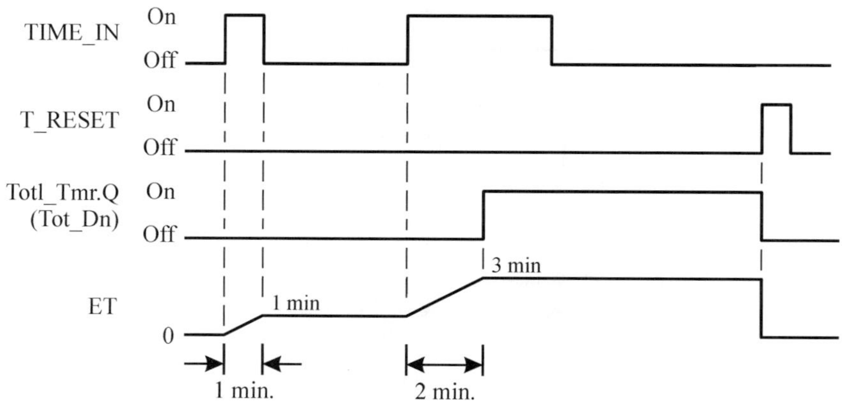

(b)

Figure 5.22. Example Micro800 RTO timer: *(a)* ladder logic; *(b)* timing diagram.

The RTO has a separate reset input, RST, which sets the ET to zero when **on**. An example application of the RTO block is shown in Figure 5.22*a* and its associated timing diagram is shown in Figure 5.22*b*. The preset time is 3 minutes. The Totl_Tmr.Q output is turned **on** when the cumulative time that TIME_IN has been **on** is at least 3 minutes. After the elapsed time has reached 3 minutes, Tot_Dn remains **on** even after TIME_IN turns off. The RST input of the timer must be turned **on** to reset the timer.

5.3.4 TONOFF On-Off-Delay Timer

The TONOFF timer is shown in Figure 5.23. The variable specified above the "TONOFF" in the block field must be a TONOFF data type. The variable references to the individual fields of the TONOFF data type are the same as for the TON data type, except there is an additional .Fedge field. The .Redge field is the same value as the IN input. The .Fedge field is normally **on** and is **off** only during the off-delay interval.

The TONOFF timer is a combination of a TON and a TOF timer. The on-delay time and the off-delay time are set separately. In general, when the IN input transitions **on** (and stays on), the Q output turns **on** at the end of the PT time. When the IN input transitions **off** (and remains off), the Q output turns **off** at the end of the PTOF time. If IN does not remain **on** during the on-delay time or IN does not remain **off** during the off-delay, the timer operation is as follows.

- If IN transitions **off** during the on-delay time (PT), the off-delay (PTOF) starts and if IN subsequently transitions **on** before the end of the PTOF time, the on-delay time restarts and the Q output remains **off**.
- If IN transitions **on** during the off-delay time (PTOF), the Q output turns **off** and the on-delay interval restarts.

An example application of the TONOFF block is shown in Figure 5.24*a* and its associated timing diagram is shown in Figure 5.24*b*. The on-delay preset time is 3 seconds and the off-delay preset time is 4 seconds. Part (1) of Figure 5.24*b* shows the on-delay operation and part (2) shows the normal off-delay timing. Part (3) illustrates the operation of the timer if IN turns off during the on-delay and turns back on before the off-delay has elapsed. Part (4) shows the operation if IN turns on during the off-delay time and thus restarts the on-delay. Note that for parts (3) and (4), the ET restarts at every transition of PROX_347.

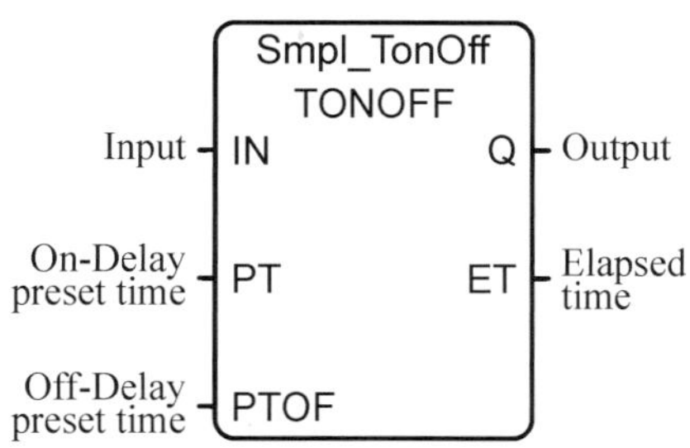

Figure 5.23. Micro800 TONOFF timer.

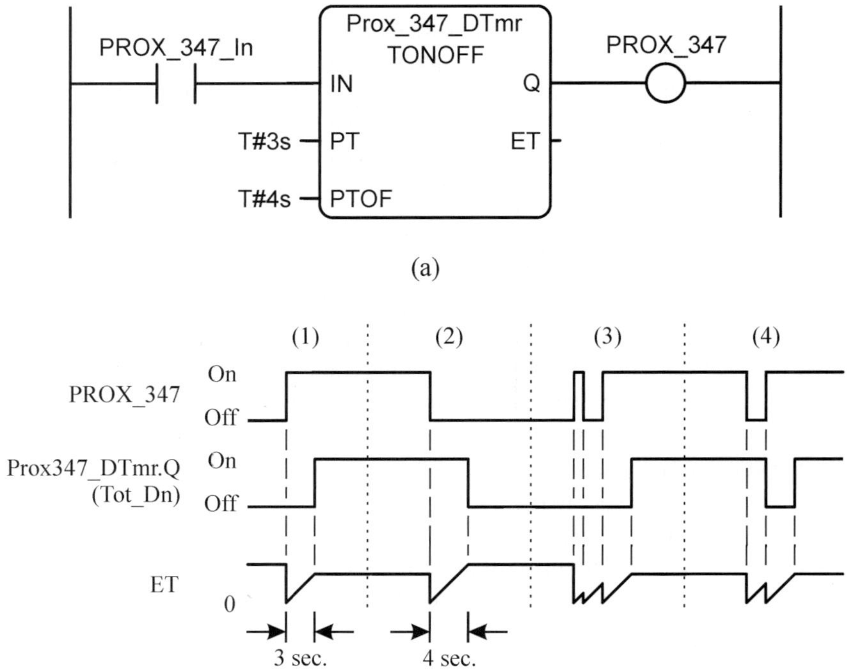

Figure 5.24. Example Micro800 TONOFF timer: *(a)* ladder logic; *(b)* timing diagram.

5.3.5 TP Pulse Timer

The pulse timer is shown in Figure 5.25. The variable specified above the "TP" in the block field must be a TP data type. The variable references to the individual fields of the TP data type are the same as for the TON data type. However, the .Redge field operates a little differently. It is **on** while the .Q output is **on** and then is the value of the IN input after .Q turns **off**.

The TP timer generates a pulse at the Q output with duration of the preset time (PT) when the IN input transitions from **off**-to-**on**. When IN changes from **off**-to-**on**, Q is turned **on** and ET starts increasing. When ET equals PT, Q turns **off**. The internal time is set to zero only if ET equals PT and IN is **off**. While the ET is increasing toward PT, any transitions on

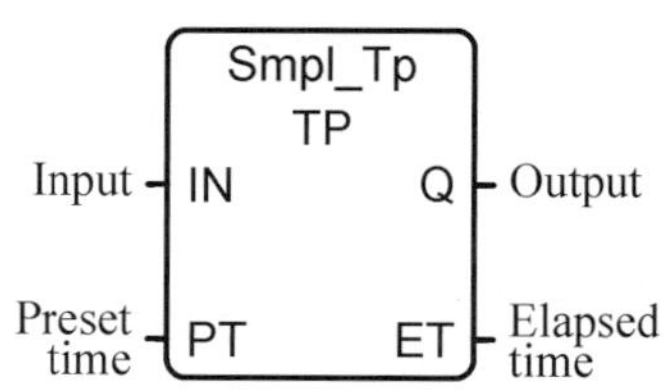

Figure 5.25. Micro800 TP timer.

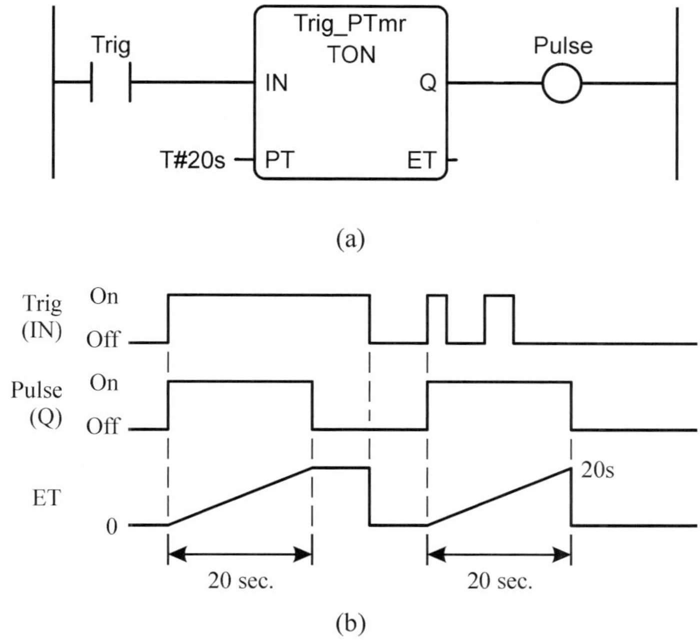

Figure 5.26. Example Micro800 TP timer: *(a)* ladder logic; *(b)* timing diagram.

IN are ignored. The preset time is specified in the same manner as for the TON and other timer blocks.

An example application of the TP block is shown in Figure 5.26*a* and the associated timing diagram is shown in Figure 5.26*b*. An **off**-to-**on** transition of TRIG causes the PULSE output to be turned **on** for 20 seconds. Keeping TRIG on longer than 20 seconds does not alter the duration of PULSE, it only prevents ET from being set to zero at the end of the 20-second duration. Changes in TRIG during the 20-second pulse duration are ignored.

5.3.6 CTU Up Counter

The CTU up counter function block is shown in Figure 5.27. The variable specified above the "CTU" in the block must be a CTU data type. For the variable defined in Figure 5.27, the individual parts (fields) of the CTU data type are addressed as follows:

Reference	Description
Sampl_Ctu.P_CU	Previous CU value
Sampl_Ctu.Q	CV = PV
Sampl_Ctu.CV	Count value, accumulator (DINT)

All of these fields are read-only. The CV cannot be the destination of a MOV or calculation and the Booleans cannot be used by a coil.

For the CTU block, the counter accumulator (CV) increments by one for every **off**-to-**on** transition of the CU counter input. When the CV = PV, then the Q output is turned **on**. When the RESET input is **on**, the CV is zeroed. The counter does not increment beyond the PV.

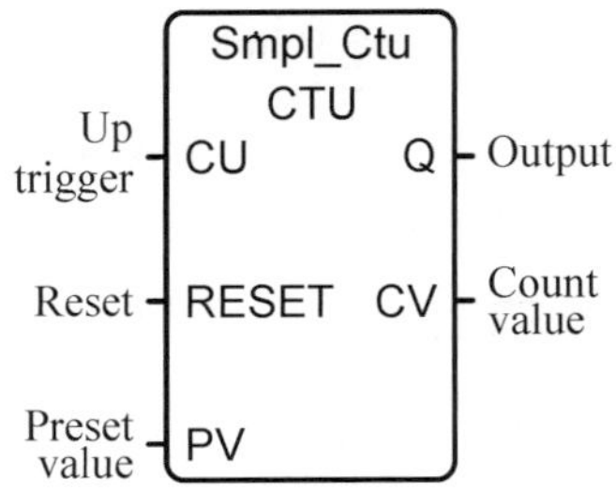

Figure 5.27. Micro800 CTU counter.

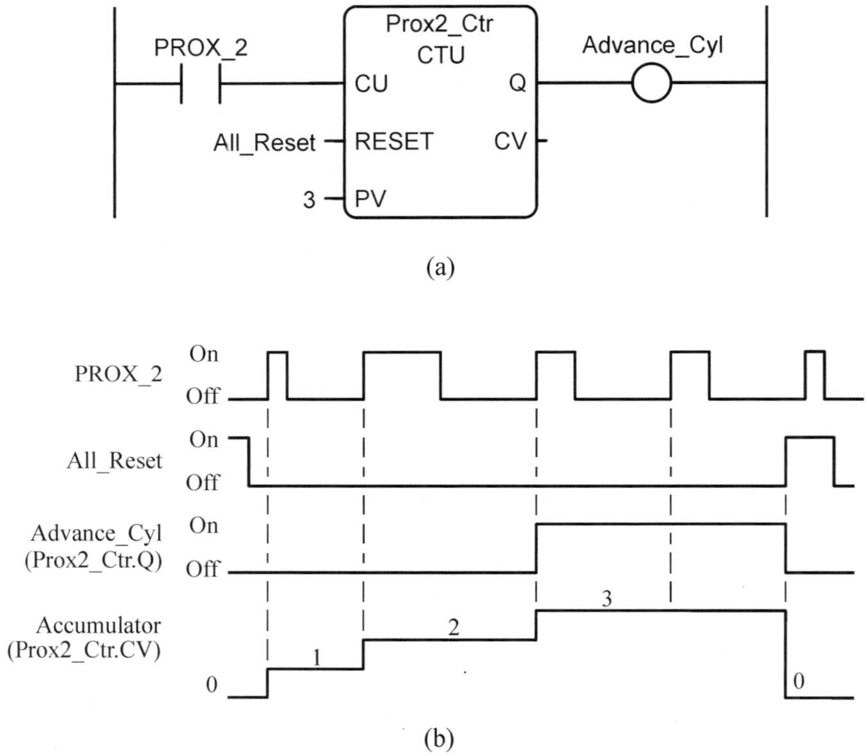

Figure 5.28. Example Micro800 CTU counter: *(a)* ladder logic; *(b)* timing diagram.

An example application of the CTU function block is shown in Figure 5.28*a* and its associated timing diagram is shown in Figure 5.28*b*. As long as All_Reset is **on**, the accumulator is held at zero. When All_Reset is **off**, at every **off**-to-**on** transition of PROX_2, the accumulator, Prox2_Ctr.CV, increases by one. When the accumulator is equal to 3 (the preset value), the Prox2_Ctr.Q output is turned **on**. When Prox2_Ctr.Q is already **on**, any more positive transitions of PROX_2 do not change the value of Prox2_Ctr.CV and Prox2_Ctr.Q remains **on**.

5.3.7 CTD Down Counter

The CTD up counter function block is shown in Figure 5.29. The variable specified above the "CTD" in the block must be a CTD data type. For the variable defined in Figure 5.29, the individual parts (fields) of the CTD data type are addressed as follows:

Reference	Description
Sampl_Ctu.P_CD	Previous CD value
Sampl_Ctd.Q	CV = 0
Sampl_Ctd.CV	Count value, accumulator (DINT)

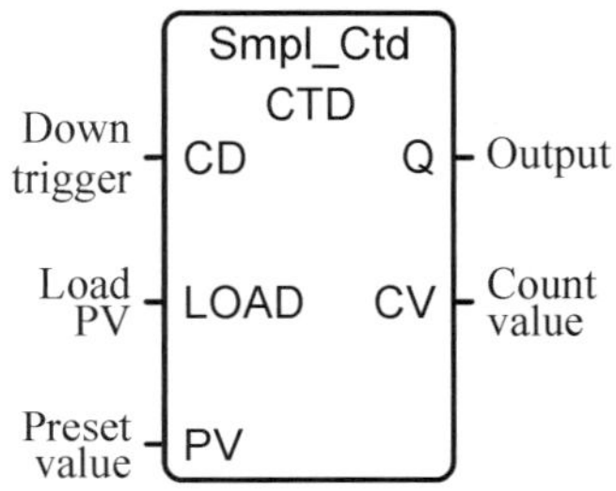

Figure 5.29. Micro800 CTD counter.

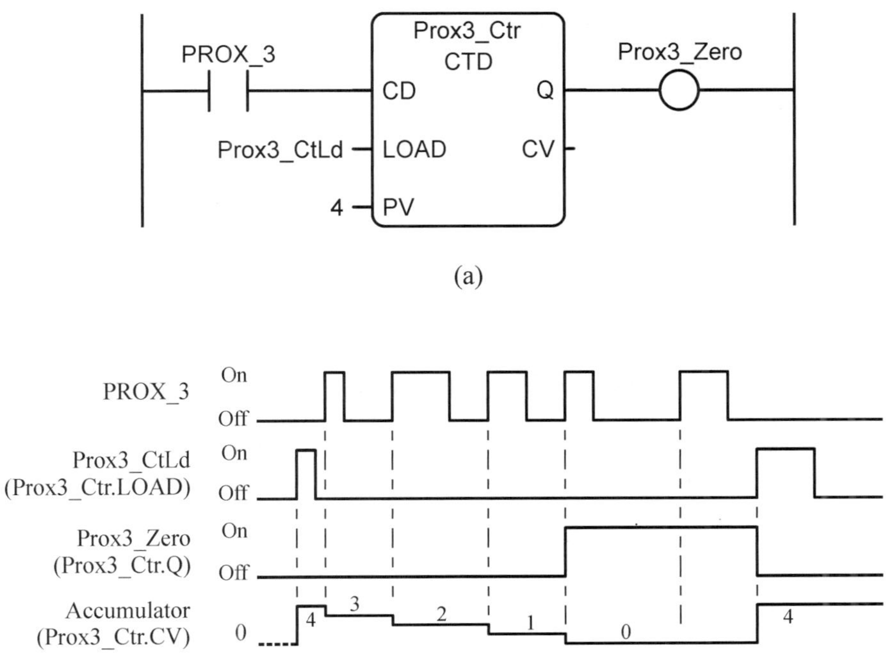

Figure 5.30. Example Micro800 CTD counter: *(a)* ladder logic; *(b)* timing diagram.

As for the CTU block, all of these fields are read-only. Note that the .CD field may be viewed online, but cannot be used in program logic.

For the CTD block, the counter accumulator (CV) decrements by one for every **off**-to-**on** transition of the CD counter input. When the CV = zero, then the Q output is turned **on**. When the LOAD input is **on**, the PV value is transferred to the CV. The counter does not decrement below zero.

An example application of the CTD function block is shown in Figure 5.30*a* and its associated timing diagram is shown in Figure 5.30*b*. As long as Prox3_CtLd is **on**, the CV is set to the PV value. When Prox3_CtLd is **off**, at every **off**-to-**on** transition of PROX_3, the accumulator, Prox3_Ctr.CV, decreases by one. When the accumulator is equal to zero, the Prox3_Zero.Q output is turned **on**. When Prox3_Ctr.Q is already **on**, any more positive transitions of PROX_3 do not change the value of Prox3_Ctr.CV and Prox3_Ctr.Q remains **on**.

5.3.8 Up/Down Counter

The CTUD up/down counter is shown in Figure 5.31. The variable specified above the "CTUD" in the block must be a CTUD data type. For the variable defined in Figure 5.31, the individual parts (fields) of the CTUD data type are addressed as follows:

Reference	Description
Sampl_Ctud.P_CU	Previous CU value
Sampl_Ctud.P_CD	Previous CD value
Sampl_Ctud.QU	CV = PV
Sampl_Ctud.QD	CV = 0
Sampl_Ctud.CV	Count value, accumulator (DINT)

As for the other counter blocks, these fields are read-only.

The CTUD function block is basically a combination of the up counter and down counter. The counter accumulator (CV) increments by one for every **off**-to-**on** transition of the count up (CU) input and CV decrements by one for every **off**-to-**on** transition of the count down (CD) input. When the CV = PV, then the QU output is turned **on**. When the CV = 0, then the QD output is turned **on**. When the RESET input is **on**, the CV is zeroed. When

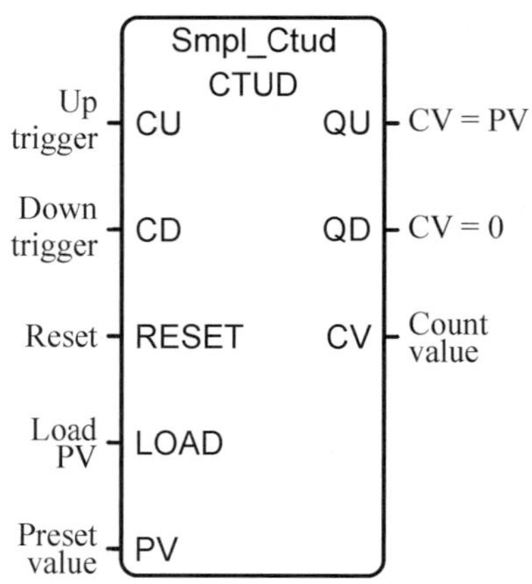

Figure 5.31. Micro800 CTUD counter.

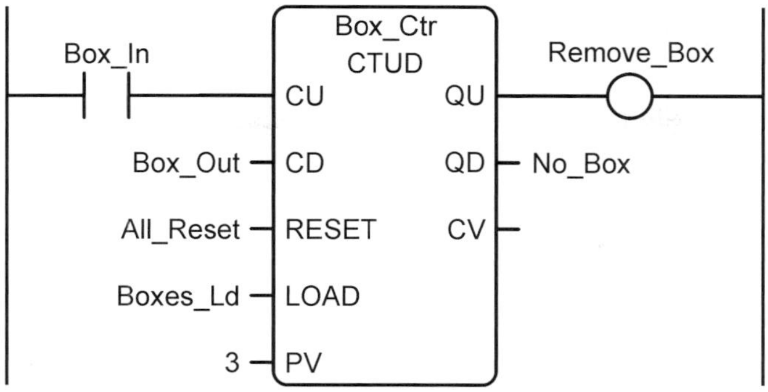

Figure 5.32. Example Micro800 CTUD counter ladder logic.

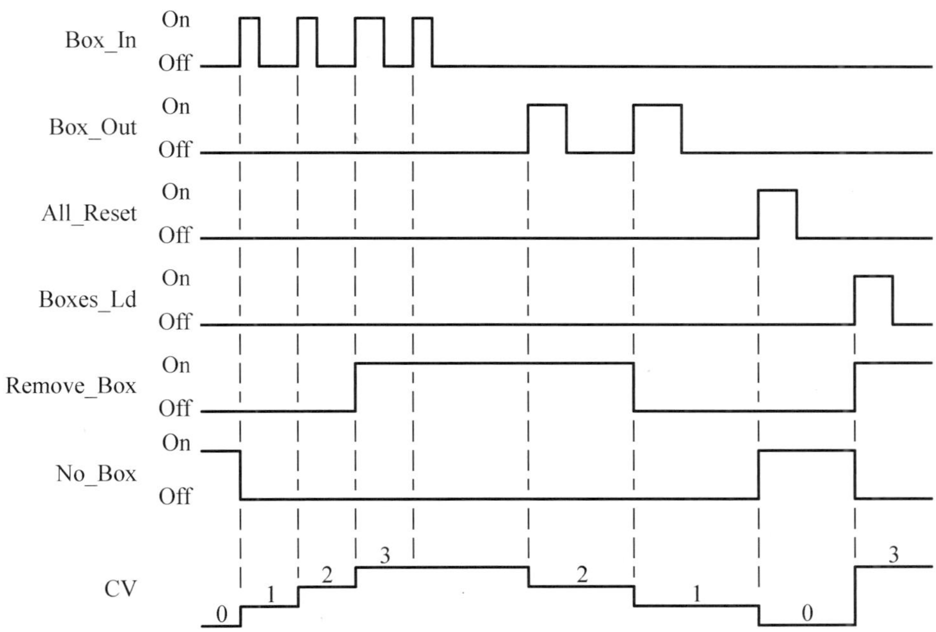

Figure 5.33. Example Micro800 CTUD counter timing diagram.

the LOAD input is **on**, the CV is set to the PV value. If both RESET and LOAD are **on** simultaneously, the RESET input has precedence and sets the CV to zero. The counter cannot count above the preset value, nor count below zero.

The timing diagram for an example application of the CTUD block in Figure 5.32 appears in Figure 5.33. When both All_Reset and Boxes_Ld are **off**, at every **off**-to-**on** transition of Box_In, CV increments by one. When both All_Reset and Boxes_Ld are **off**, at every **off**-to-**on** transition of Box_Out, CV decrements by one. When the CV is equal to 3 (the PV input value), the Remove_Box coil is turned **on**. When the CV is equal to zero, the No_Box coil is turned **on**. When All_Reset is **on**, CV is loaded with zero. When Boxes_Ld is **on**, the CV is loaded with the PV (3).

5.4 MICROLOGIX/SLC-500 TIMERS AND COUNTERS

The Allen-Bradley MicroLogix and SLC-500 processors have three timer function blocks and two counter function blocks. The names and description of these function blocks are as follows:

MLogix/SLC-500 Function Block	Description
TON	On-delay timer
TOF	Off-delay timer
RTO	Retentive on-delay timer
CTU	Up counter
CTD	Down counter

As for the ControlLogix PLC, the CTU and CTD function blocks can be combined to construct an up/down counter.

The MicroLogix and SLC-500 timers and counters function blocks are nearly identical to the ControlLogix function blocks of the same name. The major difference is that the MicroLogix/SLC-500 timers and counters are output instructions, meaning they must always occur at the right side of the rung. Other differences are as follows:

	ControlLogix	MLogix/SLC-500
Timer accumulator	DINT	INT
Timer time base	1 ms	.01 or 1 sec.
Counter preset	DINT	INT
Counter accumulator	DINT	INT

Therefore, the approach in this section will be to show the function block and describe the operation of each one, but only the operation of the TON and CTU function blocks is illustrated.

Most of the MicroLogix PLCs have special high-speed counters for motion control applications. These counters are not covered by the text.

5.4.1 TON On-Delay Timer

The TON on-delay timer is shown in Figure 5.34. The TON timer basically delays the turn-on of a signal and does not delay the turn-off. The address specified in the "Timer" field must be an element of a timer (T*x*) data file (Figure 3.19*a*). For the symbol and address

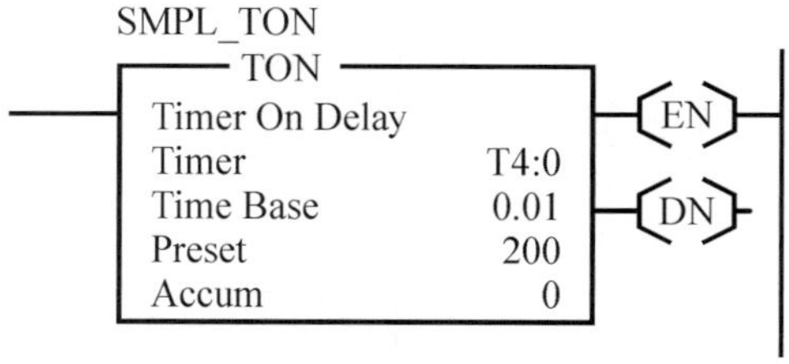

Figure 5.34. MicroLogix/SLC-500 TON timer.

defined in Figure 5.34, the individual parts (fields) of the timer data type are referenced as follows:

Symbol Reference	Address Reference	Description
SMPL_TON/DN	T4:0/DN	Done
SMPL_TON/TT	T4:0/TT	Timer timing bit
SMPL_TON/EN	T4:0/EN	Enable
SMPL_TON.ACC	T4:0.ACC	Accumulator
SMPL_TON.PRE	T4:0.PRE	Preset value

When entering the reference into a block, a "/" or "." can be used to separate the symbol or address from the field name, but the above table shows how they are displayed.

When the timer input is **on**, the accumulator is counted up once for each time base interval (1 or 0.01 seconds). When the accumulator equals the preset value, the timer is "timed out" and the /DN bit is set to **on**. If the timer input turns **off** during the timing

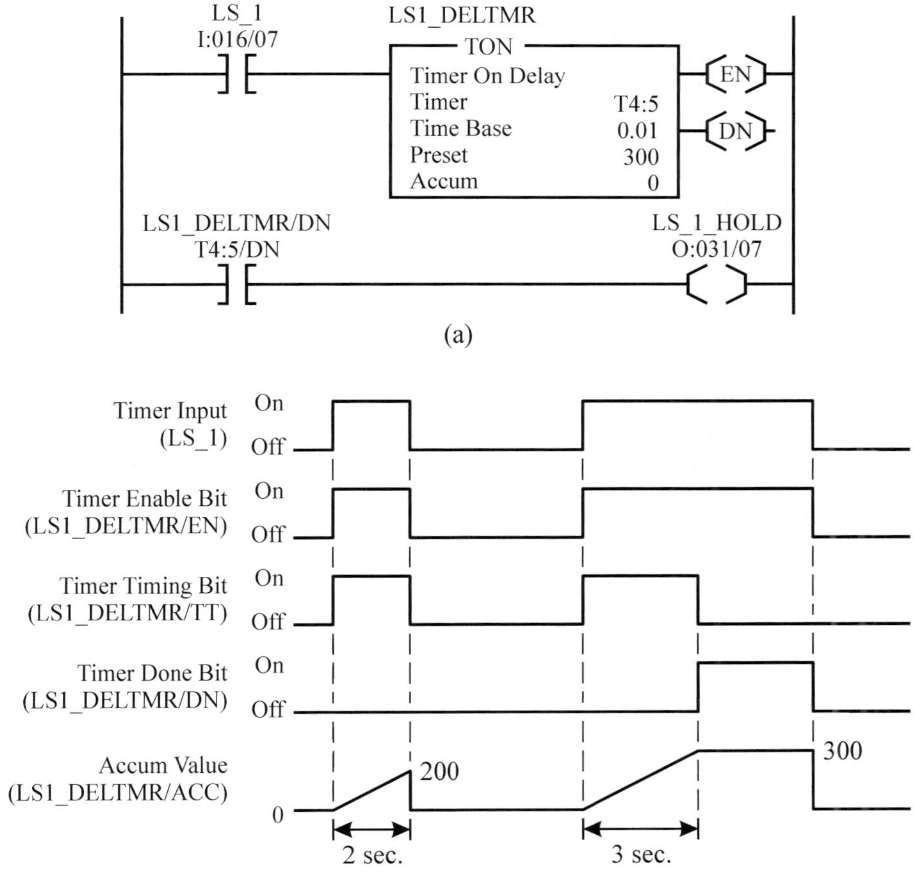

Figure 5.35. Example MicroLogix/SLC-500 TON timer: *(a)* ladder logic; *(b)* timing diagram.

interval, the accumulator is set to zero and timing restarts when the input turns **on**. The /TT bit is on while the accumulator is counting, but not yet at the preset value. The /EN bit is the same as the timer input condition. The maximum interval is 32,767 seconds (with a 1 second time base). The value of the accumulator (.ACC) or the preset value (.PRE) can be accessed as integers.

The accuracy of the timer is basically ±1 time base interval. If the time base is 0.01 second, the timer could finish as much as 0.01 second before desired interval or as much as 0.01 second after the desired interval. For a one second time base, the timer could finish as much as one second earlier or one second later than the desired interval. This behavior also means that a timer with a preset value of one may have **no** delay.

An example application of the TON block is shown in Figure 5.35*a* and its associated timing diagram is shown in Figure 5.35*b*. The preset time is 3 seconds (300×0.01 sec.). The input, LS_1 must remain **on** for at least 3 seconds before the LS1_DELTMR/DN bit and the LS_1_HOLD coil are turned **on**. When LS_1 is turned **off** after 2 seconds, the accumulator is set to zero and the LS1_DELTMR/DN bit remains **off**. Note that the symbol, LS1_DELTMR, associated with the timer address, T4:0, is shown above the TON function block.

5.4.2 TOF Off-Delay Timer

The TOF off-delay timer is shown in Figure 5.36. Looking at the timer input and the /DN bit, the TOF timer operates exactly opposite of the TON timer. The TOF timer basically delays the turn-off of a signal and does not delay the turn-on. The descriptions of the function block fields and the symbol/address references are the same as for the TON function block.

When the timer input is **off**, the accumulator is counted up once for each time base interval (1 or 0.01 seconds). When the accumulator equals the preset value, the timer is "timed out" and the /DN bit is turned **off**. If the timer input turns **on** anytime, the accumulator is set to zero and timing recommences when the input turns **off**. The operation of the /EN and /TT bits is the same as for the TON timer instruction.

5.3.3 RTO Retentive On-Delay Timer

The RTO retentive on-delay timer is shown in Figure 5.37*a*. The RTO timer is like the TON timer except that the accumulator is not reset when the timer input turns **off**. When the timer input is **on**, the accumulator is counted up once for each time base interval.When the accumulator equals the preset value, the /DN bit is set to **on**. If the timer input turns **off** during the timing interval, the accumulator remains at its current value and timing resumes

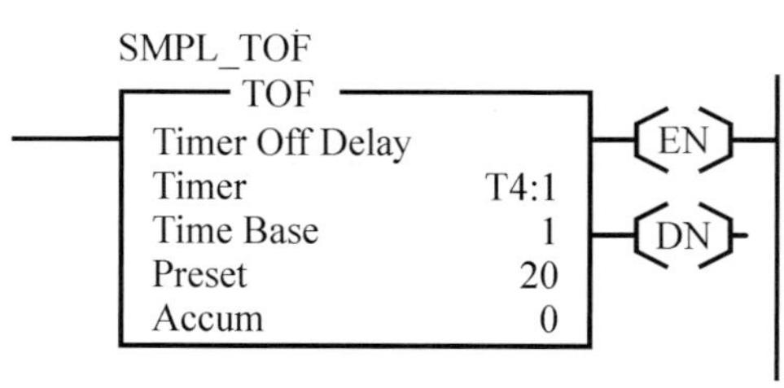

Figure 5.36. MicroLogix/SLC-500 TOF timer.

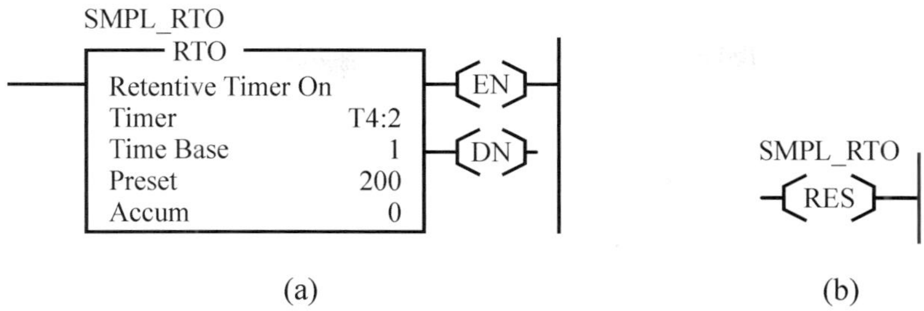

Figure 5.37. MicroLogix/SLC-500 RTO timer: *(a)* RTO block; *(b)* RES coil to reset RTO.

when the input turns **on**. A reset coil that refers to the timer (Figure 5.37*b*) may be used to set the accumulator to zero and thus turn **off** the /DN bit. Alternatively, one can use a MOV block (Chapter 7) to copy a zero to the timer accumulator. The operation of the /EN and /TT bits is the same as for the TON and TOF timer instructions.

5.4.4 CTU Up Counter

The CTU up counter function block is shown in Figure 5.38*a*. The address specified in the "Counter" field must be an element of a counter (C*x*) data file (Figure 3.19*b*). For the symbol defined in Figure 5.38, the individual parts (fields) of the counter structure are addressed as follows:

Symbol Reference	**Address Reference**	**Description**
SMPL_CTU/DN	C5:5/DN	Done
SMPL_CTU/CU	C5:5/CU	Count-up bit
SMPL_CTU/CD	C5:5/CD	Count-down bit
SMPL_CTU/OV	C5:5/OV	Accumulator overflow
SMPL_CTU/UN	C5:5/UN	Accumulator underflow
SMPL_CTU.ACC	C5:5.ACC	Accumulator
SMPL_CTU.PRE	C5:5.PRE	Preset value

The /CU and /OV bits are modified only by the CTU function block. The /CD and /UN bits are modified only by the CTD function block. When entering the reference into a block, a "/" or "." can be used to separate the symbol or address from the field name, but the above table shows how they are displayed.

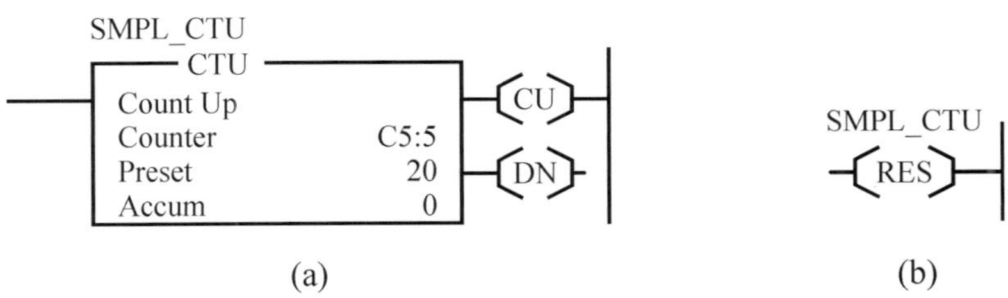

Figure 5.38. MicroLogix/SLC-500 CTU counter: *(a)* CTU block; *(b)* RES coil to reset counter.

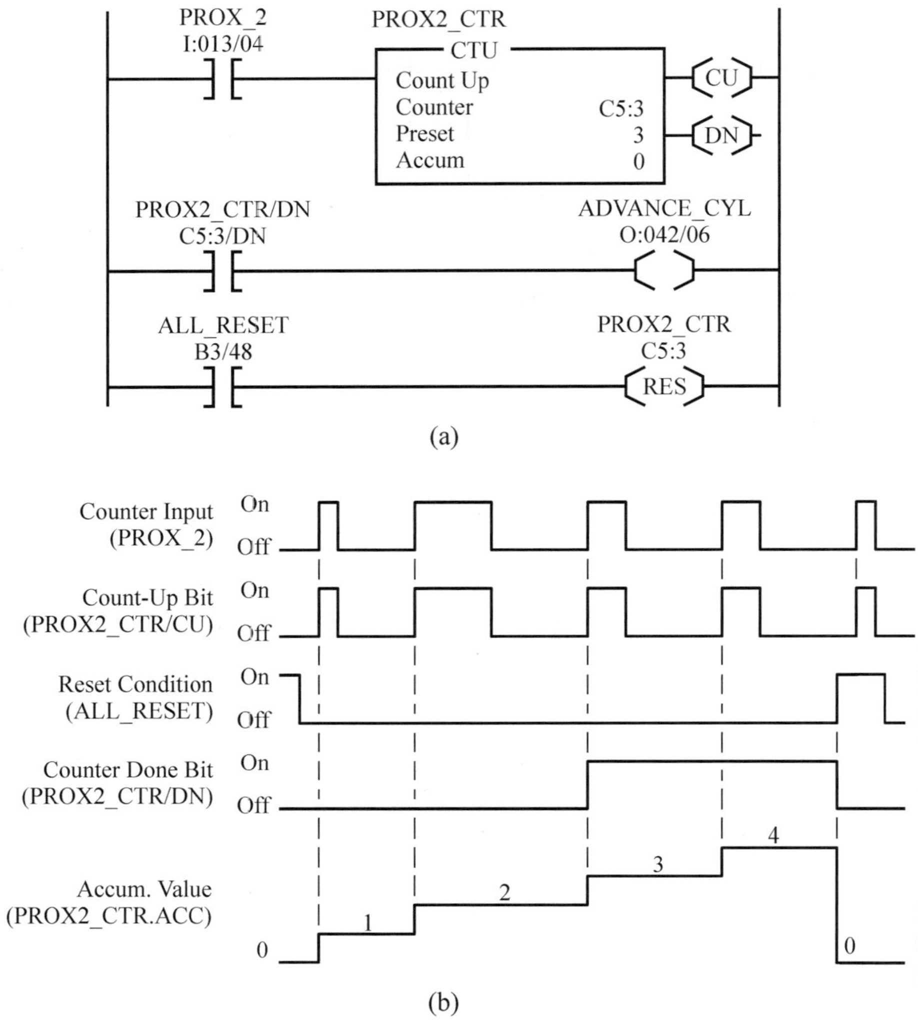

Figure 5.39. Example MicroLogix/SLC-500 CTU counter: *(a)* ladder logic; *(b)* timing diagram.

For the CTU function block, the counter accumulator increments by one for every **off**-to-**on** transition of the counter input. When the accumulator ≥ preset value, then the /DN bit is turned **on**. A reset coil that refers to the counter (Figure 5.38*b*) may be used to set the accumulator to zero and thus turn **off** the /DN bit. The /CU bit holds the last value of the CTU input at the last scan. The preset value and accumulator are both integer data types. The counter can count beyond the preset value. The overflow bit (/OV) is turned **on** if the accumulator exceeded the upper limit of 32,767 and wrapped around to –32,768.

An example application of the CTU function block is shown in Figure 5.39*a* and its associated timing diagram is shown in Figure 5.39*b*. As long as All_Reset is **on**, the accumulator is held at zero. When All_Reset is **off**, at every **off**-to-**on** transition of

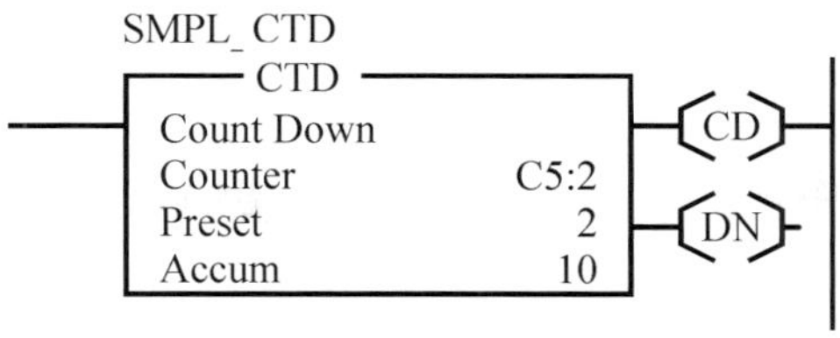

Figure 5.40. MicroLogix/SLC-500 CTD counter.

PROX_2, the accumulator, PROX2_CTR.ACC, increases by one. When the accumulator is equal to or greater than 3 (the preset value), the PROX2_CTR/DN bit is turned **on**. When PROX2_CTR/DN is already **on**, any more positive transitions of PROX_2 cause the value of PROX2_CTR.ACC to change, and PROX2_CTR/DN remains **on**.

5.4.5 CTD Down Counter

The CTD down counter function block is shown in Figure 5.40. The function block fields and the symbol/address references are described with the CTU function block.

The counter accumulator decrements by one for every **off**-to-**on** transition of the counter input. When the accumulator ≥ preset value, then the /DN bit is turned **on**, which is the same behavior as for CTU. A reset coil that refers to the counter (Figure 5.38*b*) may be used to set the accumulator to zero and thus turn **off** the /DN bit. The /CD bit holds the value of the CTD input at the last scan. The preset value and accumulator are both integer data types. The counter can count below zero. The underflow bit (/UN) is turned **on** if the accumulator went below –32,768 and wrapped around to 32,767.

5.4.6 Up/Down Counter

For the MicroLogix and SLC-500, an up/down counter is constructed by combining a CTU and a CTD function block that operate on the same counter. Both the CTU and CTD function blocks basically affect a counter structure, incrementing or decrementing the accumulator and setting or clearing the bits in the structure. There is no conflict if the CTU function block and the CTD function block refer to the same counter structure. The CTU block provides the condition that increases the accumulator and the CTD block provides the condition that decreases the accumulator. Note that the condition that determines the value of the .DN bit (accumulator ≥ preset value) is the same, regardless of the type of counter block.

An example application of the combination the CTU and CTD function blocks is shown in Figure 5.16 for the ControlLogix PLC. The only difference is that for the MicroLogix/SLC-500 the counter address is shown in the function block and the symbol appears above the function block.

5.5 GENERAL TIMER AND COUNTER SITUATIONS

When formulating the ladder logic for an application involving timers, some general questions are:

What is being delayed?

Is it an on-delay or an off-delay?

The appropriate timer most often depends on whether an action is started or stopped after the delay. An on-delay timer is appropriate if an action is started after a delay. If an action needs to be stopped after a delay, then an off-delay timer is usually appropriate. However, since many find the off-delay timer confusing, many novice programmers avoid the off-delay timer entirely.

When formulating the ladder logic for an application involving counters, some general questions are:

What is being counted?

Is the count up, down, or both?

What are the reset conditions for each counter?

The counter reset conditions are important, but often overlooked by the neophyte programmer.

Common timer situations are the following:

1. *On delay*. B turns **on** at a specified time after A turns **on**. Both turn **off** simultaneously.
2. *Cascaded on delay*. A is turned **on**. After a specified time, B turns **on**. After a specified time, C turns **on**. The process is repeated as necessary. All turn **off** simultaneously. An example application: turning on multiple devices (e.g., motors) that would cause voltage sag if turned on simultaneously.
3. *Off delay*. Both A and B are turned **on** simultaneously. When A is turned **off**, B remains **on** for a specified time and then turns **off**.
4. *Cascaded off delay*. All devices turn on simultaneously. A is turned **off** and all others remain **on**. After a specified time, B is turned **off**. After a specified time, C is turned **off**. The process is repeated as necessary. An example application: turning off the transfer devices in a conveyor (or other material handling) system. The feeding device is turned off immediately and successive conveyors are turned off as material clears them.
5. *Timed clock "ticks."* A is **on** for one scan at a periodic interval. An example application: generate an enable signal to execute a PID controller or trigger a message to another PLC.
6. *One-shot*. A is **on** for a specified time after a trigger goes **on**. A stays **on** for the specified interval even if the trigger turns **off** during the interval. An example application: control a pneumatic cylinder to eject a bad part from a conveyor.
7. *Alternate cycling*. Also know as "flashing lights." A and B alternate being **on**. The "on" time of A and B can be the same or different. An example application: alarm lamps or alarm horn.
8. *Delayed one-shot*. A specified time after a trigger turns **on**, A stays **on** for a specified interval. The initial delay and duration that A stays **on** do not change

even if the trigger turns **off** before the end of the duration that A stays **on**. Example application: Marking a part on a moving assembly line where the part-sensing device is some distance before the marking device.

9. *Monitor cumulative time*. Determine the total time that A has been **on**. Often used with a counter for long durations. Example application: determine the "run" time for a pump motor and generate a message for maintenance when it has a total run time of 2000 hours.

Common counter situations are the following:

1. *Straight counting*. Counter output turns **on** after a specified number of pulses at the counter input.
2. *Cascaded counters*. The output of one counter serves as the input to another counter. An example application: determine the number of items in a lot when the total number will overflow the accumulator.
3. *Difference between two counts*. Example application: determine the number of parts on a conveyor system by finding the difference between the count of parts going in and coming out of a conveyor system.
4. *Sum of multiple counts*. Count multiple events with the same counter. Example application: determine the sum of parts on multiple conveyors where each conveyor has its own part sensor.
5. *Determine rate*. Count pulses within a certain time interval. This type of application uses a timer and a counter. Example application: determine the number of parts per hour being produced by a machine.

5.6 EXAMPLES

Of the timer situations listed in the previous section, #1 and #3 are handled with the examples given with the timer function block descriptions. Situations #6 and #7 are relegated to the problems. The others will be demonstrated by the following examples:

Example 5.1 Turn on three pumps (timer situation #2)

Example 5.2 Turn off two conveyors (timer situation #4)

Example 5.3 Message send tick (timer situation #5)

Example 5.4 Marking defective parts (timer situation #8)

Example 5.5 Monitor pump run time (timer situation #9)

Of the counter situations listed in the previous section, #1 is handled with the description of the up counter function block and #4 is shown in the scenario at the beginning of this chapter. The others will be demonstrated with the following examples:

Example 5.6 Clock (counter situation #2)

Example 5.7 Determine amount of work in progress (counter situation #3)

Example 5.8 Determine number of parts per hour produced (counter situation #5)

All of the examples are worked for the ControlLogix and Micro800 PLCs. The MicroLogix solutions are nearly identical to the ControlLogix solutions and are shown in most of the examples. In the solutions, ControlLogix and CompactLogix solutions are identical, though stated as ControlLogix solutions. Similarly, the MicroLogix and SLC-500 solutions are identical, though stated as MicroLogix solutions..

Example 5.1. Control three pumps (timer situation #2). Design a ladder logic program to sequentially start three pump motors. When the start button is pressed, Pump 1 should be started by turning **on** (closing) its run contact. After a delay of at least 5 seconds, Pump 2 should be started. After a delay of at least another 5 seconds, Pump 3 should be started. When the stop button is pressed, all three pumps should be stopped by turning **off** (opening) all three pump run contacts.

Assume the following physical input and physical output assignments:

Tag/Var,/Symbol	Description
PSTRT_PB	Start push button, N. O., **on** when starting
PSTOP_PB	Stop push button, N. C., **off** when stopping
PUMP_1	Pump 1 run contact, **on** to close and start pump
PUMP_2	Pump 2 run contact, **on** to close and start pump
PUMP_3	Pump 3 run contact, **on** to close and start pump

The addresses associated with the physical inputs and outputs are:

Tag/Var./Symbol	ControlLogix	Micro800	MLogix	SLC-500
PSTRT_PB	Local:1:I.Data.0	_IO_EM_DI_00	I:0/0	I:1/0
PSTOP_PB	Local:1:I.Data.1	_IO_EM_DI_01	I:0/1	I:1/0
PUMP_1	Local:2:O.Data.0	_IO_EM_DO_00	O:0/0	O:2/0
PUMP_2	Local:2:O.Data.1	_IO_EM_DO_01	O:0/1	O:2/1
PUMP_3	Local:2:O.Data.2	_IO_EM_DO_02	O:0/2	O:2/2

Solution. Basically, the solution needs two on-delay timers: one to delay between Pump 1 started and Pump 2 started, and one to delay between Pump 2 started and Pump 3 started. The start/stop buttons only directly control Pump 1. The ControlLogix code is shown in Figure 5.41. Note that the output connection of a timer block **does not** indicate that the timer is "done." The .DN bit must be referenced with a contact in order to control a pump. The Micro800 implementation is shown in Figure 5.42. For a MicroLogix implementation

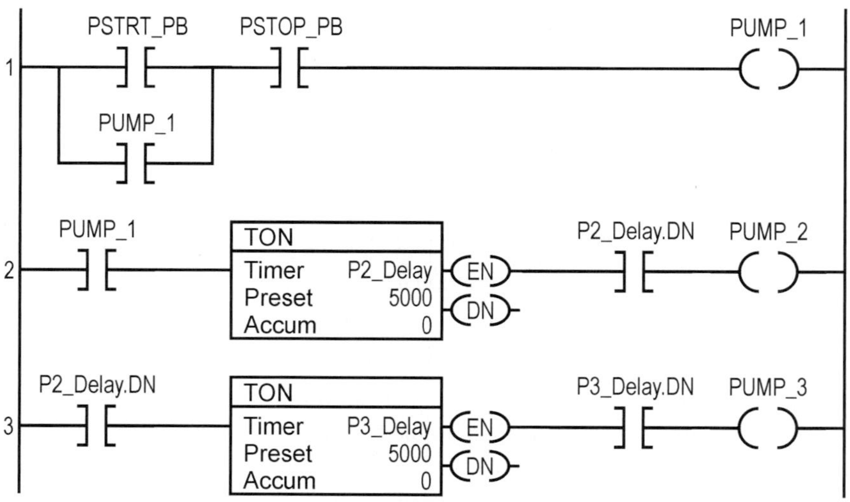

Figure 5.41. ControlLogix solution to Example 5.1.

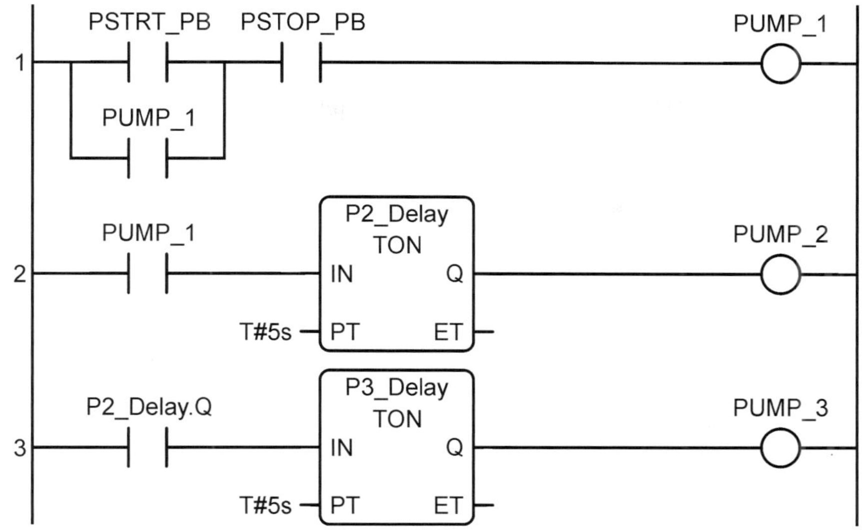

Figure 5.42. Micro800 solution to Example 5.1.

(Figure 5.43), there are two differences with the ControlLogix implementation. First, the timers have a time base of 0.01 and therefore the preset value is 1/10 the preset value of the ControlLogix timers. Second, since the MicroLogix/SLC-500 timers are output instructions, the timers cannot be placed in series with a coil.

The addresses or data types associated with the internal tags/variables/symbols:

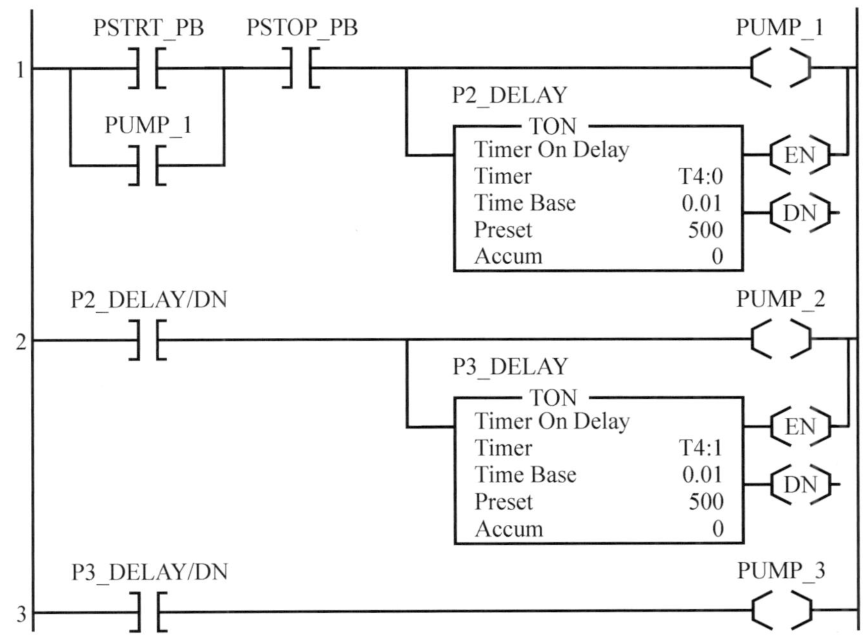

Figure 5.43. MicroLogix solution to Example 5.1.

Tag/Var./Symbol	CLogix Data Type	Micro800 Data Type	MLogix/SLC Address
P2_Delay	TIMER	TON	T4:0
P3_Delay	TIMER	TON	T4:1

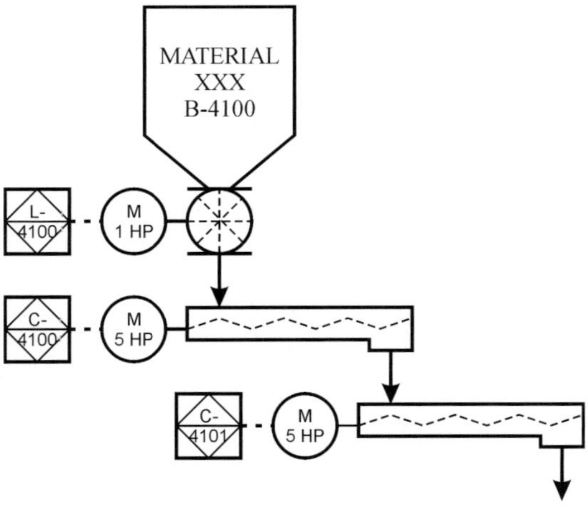

Figure 5.44. Bin and conveyor system.

Example 5.2. Control conveyors (timer situation #4). Design a ladder logic program to sequentially stop an airlock motor and two conveyor motors, shown in Figure 5.44. When the start button is pressed, the airlock motor and the two conveyor motors are started. When the stop button is pressed, the airlock motor, L-4100, should stop immediately. After a delay of at least 45 seconds (to allow material to be removed from the conveyor), conveyor motor C-4100 should be stopped. After a delay of at least another 60 seconds (to allow material to be completely removed from the conveyor), conveyor motor C-4101 should be stopped.

Assume the following physical input and physical output assignments:

Tag/Var./Symbol	Description
CSTRT_PB	Start push button, N. O., **on** when starting
CSTOP_PB	Stop push button, N. C., **off** when stopping
L_4100	L-4100 airlock motor run contact, **on** to close and start motor
C_4100	C-4100 conveyor motor run contact, **on** to close and start motor
C_4101	C-4101 conveyor motor run contact, **on** to close and start motor

The addresses associated with the physical inputs and outputs are:

Tag/Var./Symbol	ControlLogix	Micro800	MLogix	SLC-500
CSTRT_PB	Local:1:I.Data.2	_IO_EM_DI_00	I:0/2	I:1/2
CSTOP_PB	Local:1:I.Data.3	_IO_EM_DI_01	I:0/3	I:1/3
L_4100	Local:2:O.Data.3	_IO_EM_DO_00	O:0/3	O:2/3
C_4100	Local:2:O.Data.4	_IO_EM_DO_01	O:0/4	O:2/4
C_4101	Local:2:O.Data.5	_IO_EM_DO_02	O:0/5	O:2/5

Solution. Basically, the solution needs two off-delay timers: one to delay between the airlock stopped and conveyor C-4100 stopped, and one to delay between C-4100 stopped and C-4101 stopped. The start/stop buttons only directly control the airlock motor. The ControlLogix code is shown in Figure 5.45, the Micro800 ladder logic is shown in Figure 5.46, and the MicroLogix ladder logic is shown in Figure 5.47. Note that the delay timers are placed in parallel with the motor controls. For a MicroLogix/SLC-500 implementation, the other difference is that the timers have a 1.0 time base and the preset values are 46 and 61. The timer preset times are 1 second longer than needed since the accuracy of the timers is ±1 second and since the specification specifies a minimum time, then 1 second is added to the specified delay time.

The addresses or data types associated with the internal tags/variables/symbols:

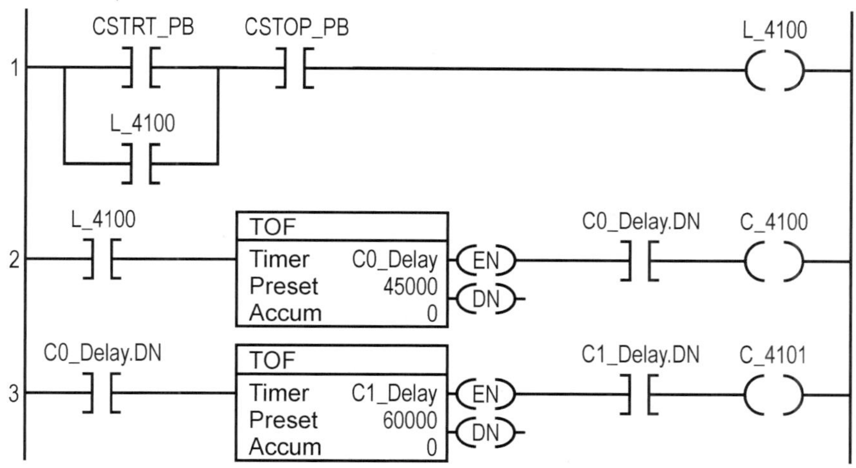

Figure 5.45. ControlLogix solution to Example 5.2.

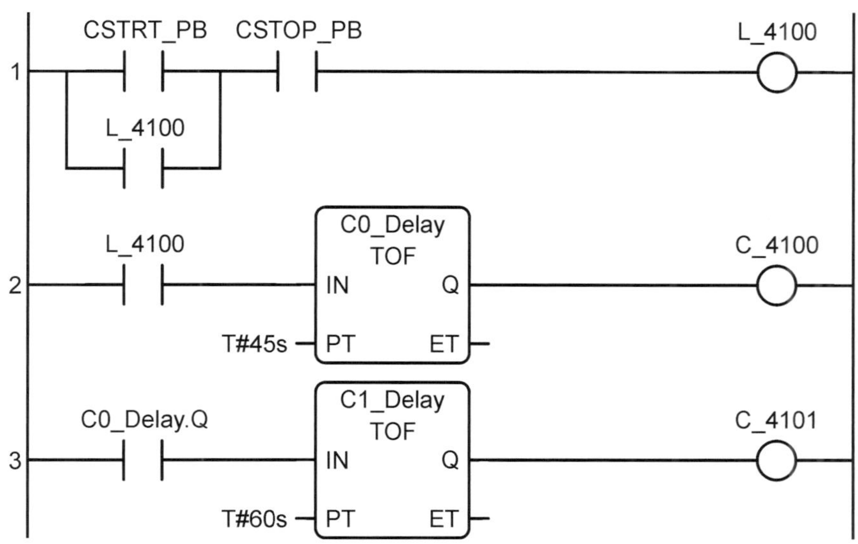

Figure 5.46. Micro800 solution to Example 5.2.

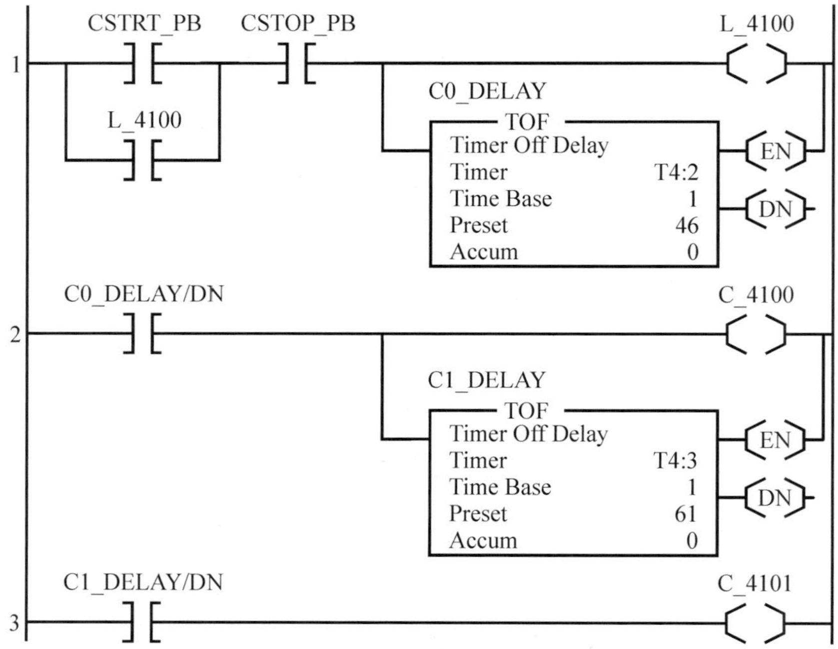

Figure 5.47. MicroLogix solution to Example 5.2.

Tag/Var./Symbol	CLogix Data Type	Micro800 Data Type	MLogix/SLC Address
C0_Delay	TIMER	TOF	T4:2
C1_Delay	TIMER	TOF	T4:3

Example 5.3. Message send tick (timer situation #5). Design a ladder logic program to generate a 1-second "tick", called Msg_Tic, that will be used to trigger a message to another PLC. The tick should only be one scan in duration.

Solution. The solution needs one on-delay timer and Msg_Tic is its output. When Msg_Tic turns **on**, it is used to reset the timer on the next scan, which also turns Msg_Tic off. The ControlLogix code is shown in Figure 5.48, the Micro800 code is shown in Figure 5.49, and the MicroLogix code is shown in Figure 5.50. The addresses or data types associated with the internal tags/variables/symbols:

Tag/Var./Symbol	CLogix Data Type	Micro800 Data Type	MLogix/SLC Address
Msg_Tic	BOOL	BOOL	B3/163
MTic_Tmr	TIMER	TON	T4:4

Example 5.4. Marking defective parts (timer situation #8). Design a ladder logic program to mark defective parts for removal from the assembly line. A simplified diagram of the station is shown in Figure 5.51. If the PDEFECT internal coil is **on** when the part passes the PROX5 proximity sensor, then the part must be marked with a 1-second pulse of spray

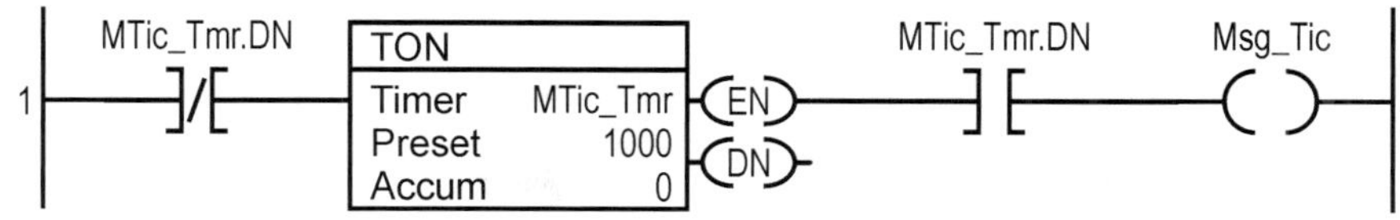

Figure 5.48. ControlLogix solution to Example 5.3.

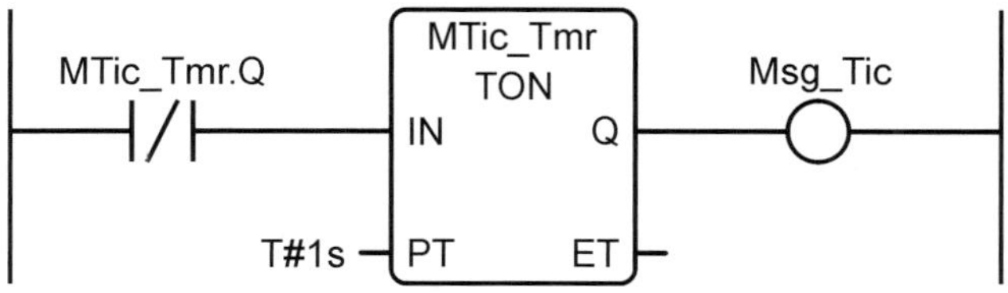

Figure 5.49. Micro800 solution to Example 5.3.

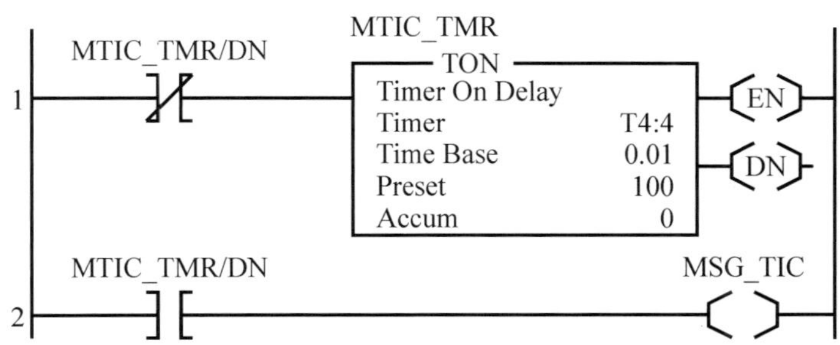

Figure 5.50. MicroLogix solution to Example 5.3.

paint. The spray head is 5.5 seconds after the PROX5 sensor, so there must be a 5.5-second delay before the sprayer is activated. The PROX5 input turns **off** when a part passes. Make no assumption about the off-time duration of PROX5 except that it is at least one program scan. Also, assume that the PDEFECT internal coil is **on** at least as long as PROX5 senses the part.

Assume the following physical input, physical output, and internal tag/variable/symbol assignments:

Tag/Var./Symbol	Description
PROX5	Senses part on conveyor, **off** when part is present
SPRAY	Paint sprayer control valve, **on** to spray
PDEFECT	Defective part internal coil, if **on** when part is sensed, then the part needs to be marked

The addresses and/or data types associated with the physical I/O and internal tags/variables/symbols are:

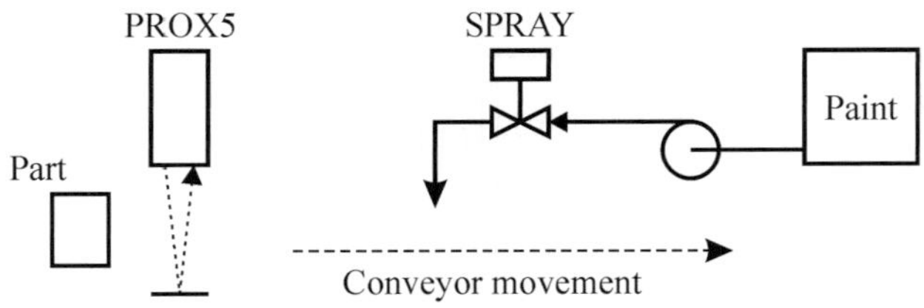

Figure 5.51. Defective part marking system.

Tag/Var./Symbol	CLogix Addr/Type	Micro800 Addr./Type	MLogix Addr.	SLC-500 Addr,
PROX5	Local:1:I.Data.4	_IO_EM_DI_04	I:0/4	I:1/4
SPRAY	Local:2:O.Data.6	_IO_EM_DO_06	O:0/6	O:2/6
PDEFECT	BOOL	BOOL	B3/102	B3/106

Solution. The ControlLogix code is shown in Figure 5.52 and the Micro800 code is shown in Figure 5.53. The MicroLogix code is very similar to the ControlLogix code and is not shown. Since the trigger is only momentary, the first part of the code is a start/stop rung. The timing cycle is triggered (start condition) when a defect is indicated and a part is detected at the proximity sensor. For the ControlLogix (Figure 5.52), the EN bit of the St_Delay timer is turned **on** when the defect is detected and is also used to seal the input to the St_Delay timer. When the St_Delay timer finishes the 5.5 second interval, the 1-second interval is started and the spray is started. At the end of the 1-second interval, the .DN bit of the Pls_Tmr timer is turned **on**, interrupting the input to the St_Delay timer, and thus resetting both timers. The cycle does not start again until the next defective part is detected. The addresses or data types associated with the internal tags/variables/symbols:

Tag/Var./Symbol	CLogix Data Type	Micro800 Data Type	MLogix/SLC Address
St_Delay	TIMER	TON	T4:5
Pls_Tmr	TIMER	TP	T4:6

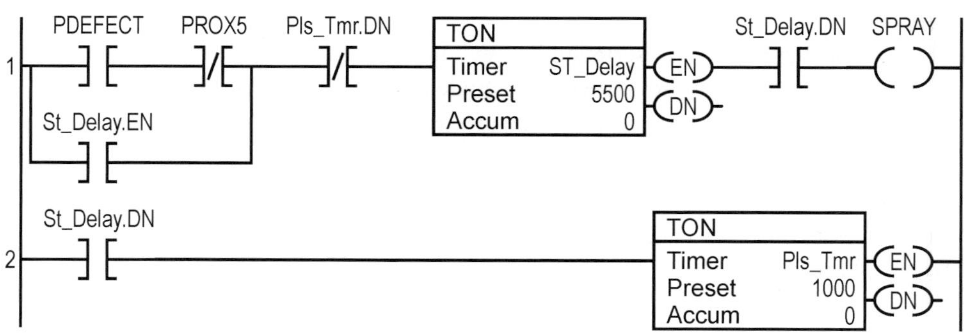

Figure 5.52. ControlLogix solution to Example 5.4.

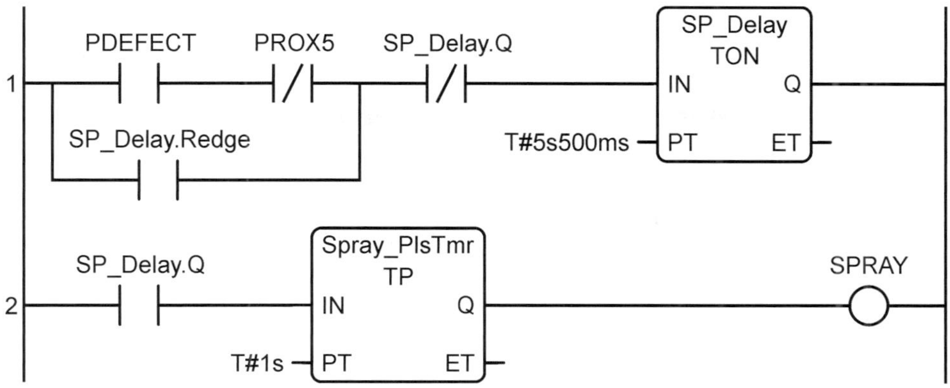

Figure 5.53. Micro800 solution to Example 5.4.

Example 5.5. Monitor pump run time (timer situation #9). Design a ladder logic program to track the total time Pump 1 is running and turn **on** the P1_MAINT internal coil when it has been running for 1000 hours. When the P1_MT_RES internal coil is **on**, reset the total time to zero and turn **off** P1_MAINT. The total time still needs to be monitored even after it has been running for 1000 hours.

Assume the following physical input, physical output, and internal tag/variable/symbol assignments:

Tag/Var./Symbol	Description
P1_STAT	Pump 1 run indication, **on** when pump running
P1_MAINT	Turn **on** when cumulative run time of pump 1 is 1000 hours
P1_MT_RES	Pump 1 maintenance indication reset, **on** to reset timer

The addresses and/or data types associated with the physical I/O and internal tags/variables/symbols are:

Tag/Var./Symbol	CLogix Addr/Type	Micro800 Addr./Type	MLogix Addr.	SLC-500 Addr,
P1_STAT	Local:1:I.Data.10	_IO_EM_DI_02	I:0/7	I:1/7
P1_MAINT	BOOL	BOOL	B3/103	B3/102
P1_MT_RES	BOOL	BOOL	B3/104	B3/104

Solution. The ControlLogix and Micro800 code is shown in Figures 5.54 and 5.55, respectively. The MicroLogix code is similar to the ControlLogix code and not shown. A retentive timer generates a “tick” every hour, which increments a counter and resets itself. When the P1_TH_CTR accumulator reaches 1000, the P1_MAINT internal coil is turned **on**. The P1_MT_RES internal coil resets the counting and the retentive timing functions. The addresses or data types associated with the internal tags/variables/symbols:

Tag/Var./Symbol	CLogix Data Type	Micro800 Data Type	MLogix/SLC Address
P1_Hr_Tic	TIMER	RTO	T4:7
P1_Th_Ctr	COUNTER	CTU	C5:0

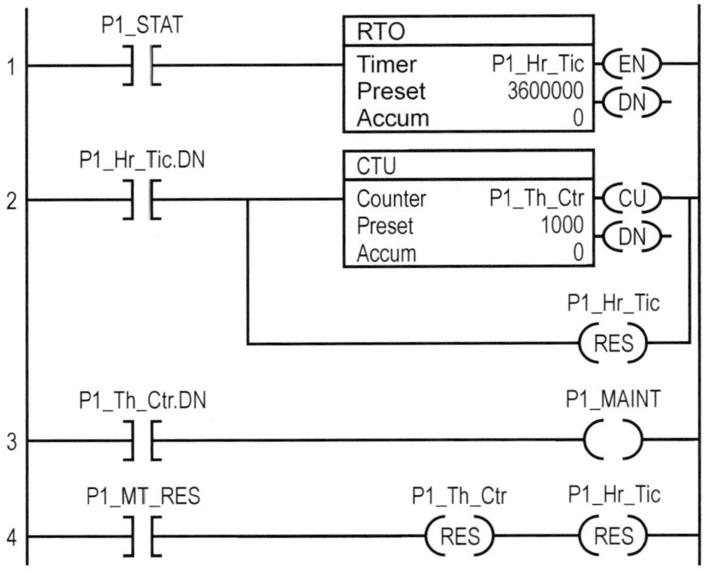

Figure 5.54. ControlLogix solution to Example 5.5.

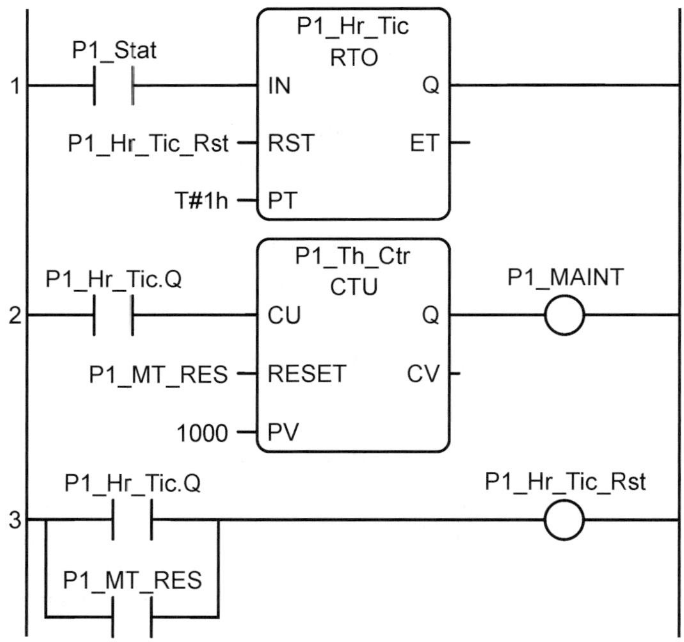

Figure 5.55. Micro800 solution to Example 5.5.

Example 5.6. Clock (counter situation #2). Design a ladder logic program to keep track of the age of the material in tank T-28 in days, hours, and minutes. There are only two controls for this clock: T28_CLK_STT starts the clock by resetting it and then allowing it to run; and

T28_CLK_STP stops the clock and holds the current time (does not change). The days, hours, and minutes are retained in T28_DAY, T28_HR, and T28_MIN, respectively.

Assume the following internal tag/variable/symbol assignments:

Tag/Var./Symbol	Description
T28_CLK_STT	T-28 age clock start command. Transition to **on** resets the current hours, minutes, and seconds and starts the clock running.
T28_CLK_STP	T-28 age clock stop command. Transition to **on** stops the clock at its current time, but does not reset the time. The time remains constant until the start command is turned **on**.
T28_MIN	Age of tank T-28 in minutes
T28_HR	Age of tank T-28 in hours
T28_DAY	Age of tank T-28 in days

The addresses and/or data types associated with the internal tags/variables/symbols are:

Tag/Var./Symbol	CLogix Data Type	Micro800 Data Type	MLogix/SLC Address
T28_CLK_STT	BOOL	BOOL	B3/105
T28_CLK_STP	BOOL	BOOL	B3/106
T28_MIN	DINT	DINT	N7:45
T28_HR	DINT	DINT	N7:46
T28_DAY	DINT	DINT	N7:47

Solution. The ControlLogix code is shown in Figure 5.56. Rung 1 has the clock start/stop and generates a one-minute tick. Rung 2 has the counter that maintains the clock minute value. Rungs 3 and 4 are the hour and day counters, respectively. The minute and hour counters are reset when they have reached their preset value or when the clock is started. The day counter is reset only when the clock is started. Note that the preset for the day counter is irrelevant since the counter .DN bit is unused. Rung 5 resets the counters when the clock is started. Rung 6 transfers the accumulator values to the correct clock value locations. The Micro800 code is shown in Figure 5.57. The MicroLogix code is very similar to the ControlLogix and thus not shown. The addresses or data types associated with the internal tags/variables/symbols:

T28_Dy_Ctr COUNTER C5:3

Tag/Var./Symbol	CLogix Data Type	Micro800 Data Type	MLogix/SLC Address
T28_Clk_Run	BOOL	BOOL	B3/107
T28_Clk_Ons	BOOL	BOOL	B3/109
Tic_Tmr	TIMER	TON	T4:8
T28_Mn_Ctr	COUNTER	CTU	C5:1
T28_Hr_Ctr	COUNTER	CTU	C5:2
T28_Dy_Ctr	COUNTER	CTU	C5:3

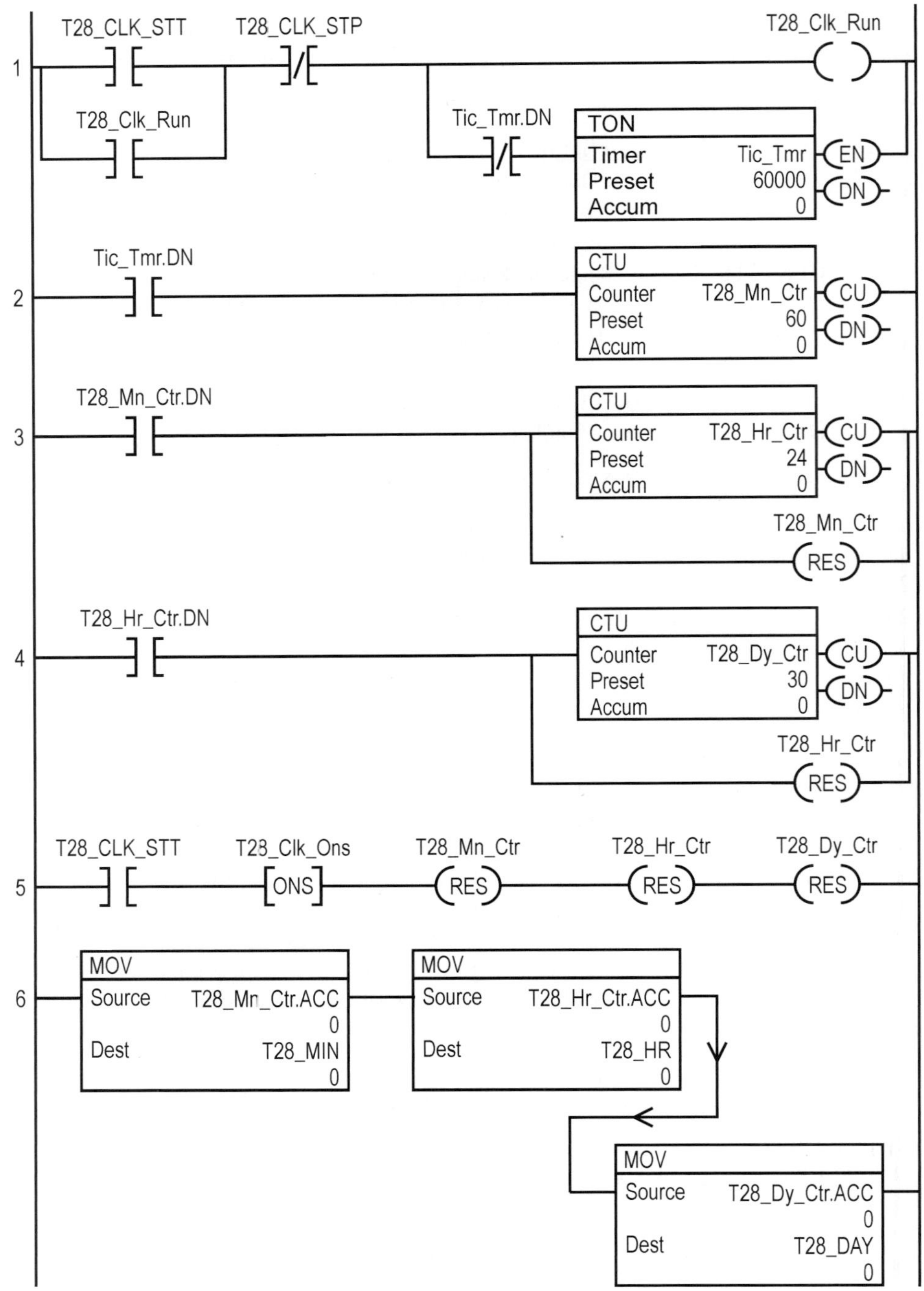

Figure 5.56. ControlLogix solution to Example 5.6

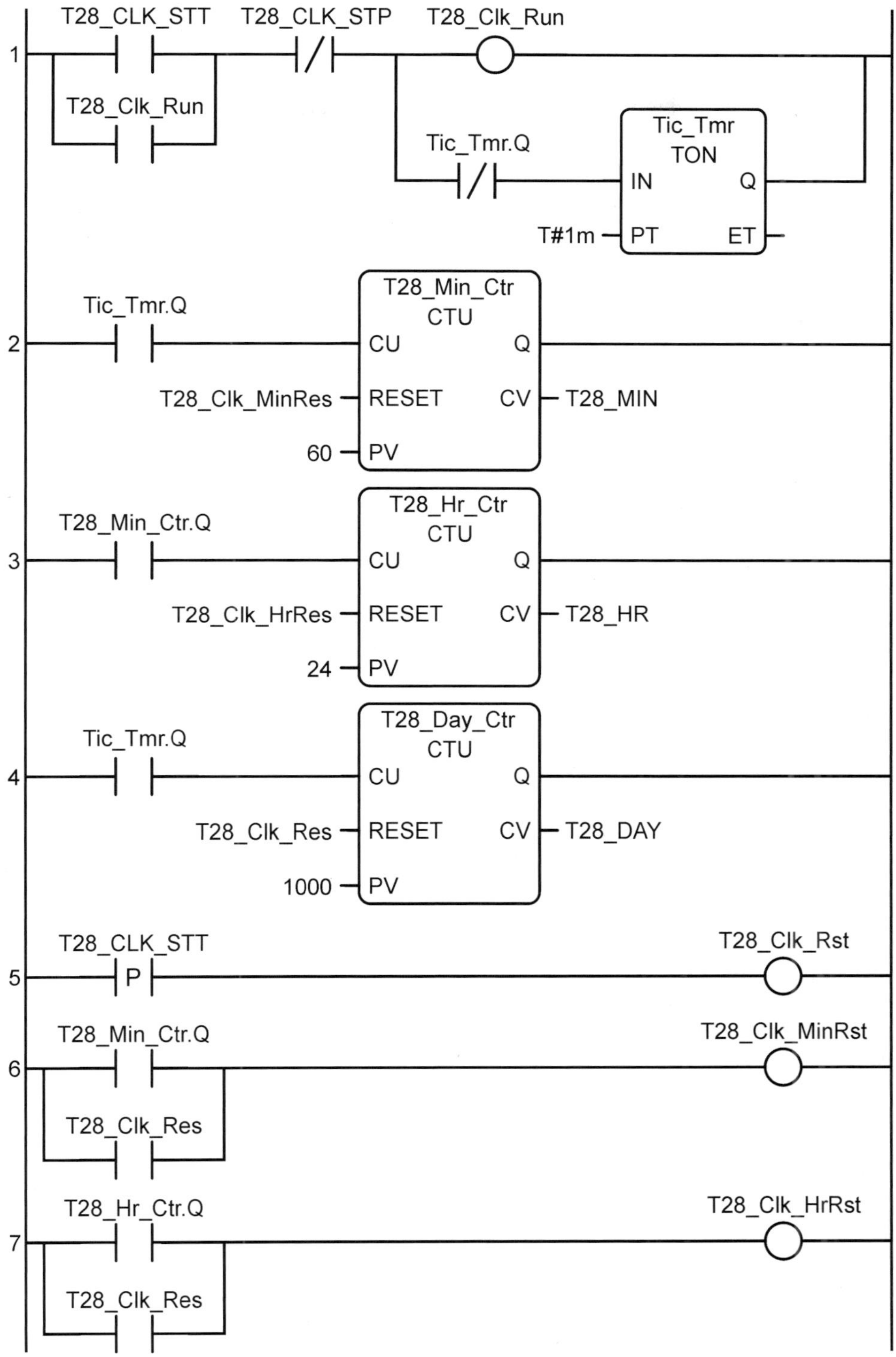

Figure 5.57. Micro800 solution to Example 5.6.

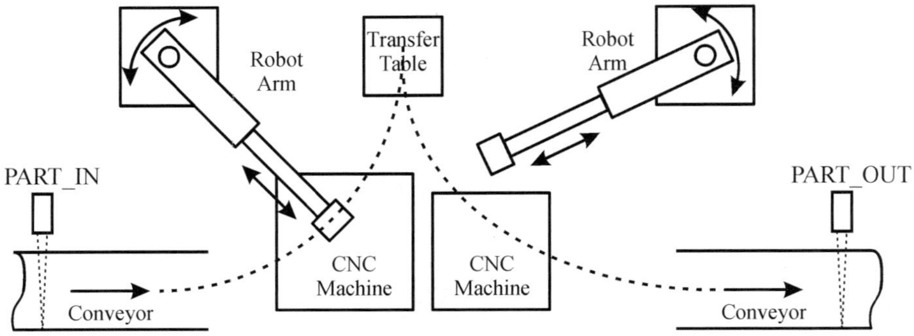

Figure 5.58. Work cell for Example 5.7.

Example 5.7. Determine amount of work in progress (counter situation #3). Parts move into and out of a work cell on conveyors (Figure 5.58). Design a ladder logic program to keep track of the number of parts that are in the work cell and provide status/alarm indications. The PART_IN discrete input is momentarily energized when a part enters the work cell and PART_OUT is energized when a part leaves the work cell. When the work cell contains 25 parts, the CELL_25P indicator is turned on. When the work cell has no parts, the CELL_ZRP indicator is turned on. When the Reset_Ctr internal coil is on, the part count should be zeroed.

Assume the following physical input, physical output, and internal tag/variable/symbol assignments:

Tag/Var./Symbol	Description
PART_IN	Senses part entering work cell, energized (**on**) and then de-energized as part enters work cell.
PART_OUT	Senses part exiting work cell, energized (**on**) and then de-energized as part exits work cell.
Reset_Ctr	**On** to clear part count.
CELL_25P	**On** when at least 25 parts are in work cell.
CELL_ZRP	**On** when no parts are in work cell.

The addresses and/or data types associated with the physical I/O and internal tags/variables/symbols are:

Tag/Var./Symbol	CLogix Addr/Type	Micro800 Addr./Type	MLogix Addr.	SLC Addr,
PART_IN	Local:1:I.Data.11	_IO_EM_DI_02	I:1/1	I:1/11
PART_OUT	Local:1:I.Data.12	_IO_EM_DI_03	I:1/11	I:1/12
Reset_Ctr	BOOL	BOOL	B3/111	B3/111
CELL_25P	Local:2:O.Data.8	_IO_EM_DO_03	O:1/3	O:2/4
CELL_ZRP	Local:2:O.Data.9	_IO_EM_DO_04	O:1/4	O:2/5

Solution. The ControlLogix code is shown in Figure 5.59. The first four rungs implement a basic up/down counter. The last rung generates the indication that no parts are present by checking if the counter accumulator is zero. The CMP block is explained in Chapter 7. The Micro800 code is shown in Figure 5.60. The MicroLogix solution is nearly identical to the

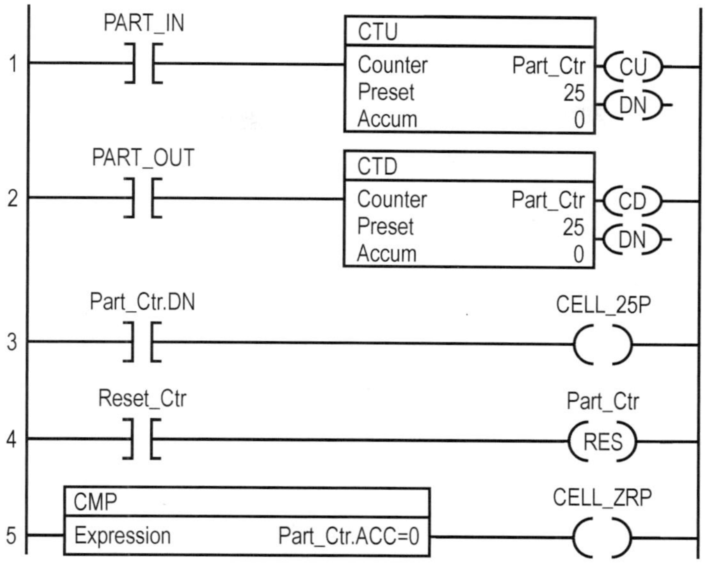

Figure 5.59. ControlLogix solution to Example 5.7.

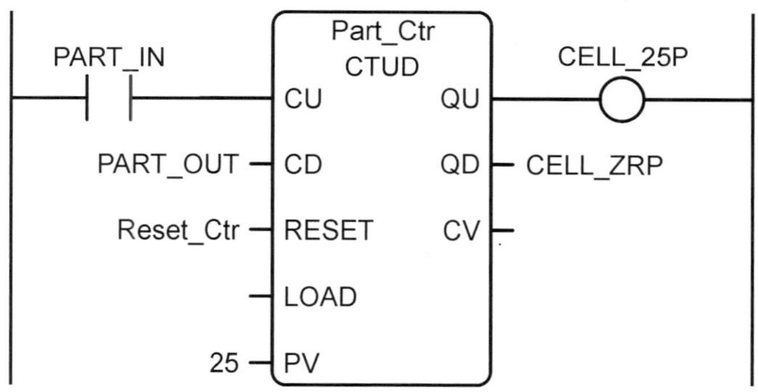

Figure 5.60. Micro800 solution to Example 5.7.

ControlLogix solution and therefore is not shown. The addresses or data types associated with the internal tag/variable/symbol:

Tag/Var./Symbol	CLogix Data Type	Micro800 Data Type	MLogix/SLC Address
Part_Ctr	COUNTER	CTUD	C5:4

Example 5.8. Determine number of parts per hour produced (counter situation #5). For the work cell depicted in Figure 5.58, determine the production rate in parts per hour.

Assume the following physical input, physical output, and internal tag/variable/symbol assignments:

Tag/Var./Symbol	Description
PART_OUT	Senses part exiting work cell, energized (**on**) and then de-energized as part exits work cell.
Cell_PRate	Cell production rate, in parts per hour.

The addresses and/or data types associated with the physical I/O and internal tag/variables/symbols are:

Tag/Var./Symbol	CLogix Addr./Type	Micro800 Addr. Type	MLogix Addr.	SLC Addr.
PART_OUT	Local:1:I.Data.12	_IO_EM_DI_02	I:1/14	I:1/14
Cell_PRate	DINT	DINT	N7:50	N7:50

Solution. Basically, in order to determine the production rate, one must count the number of parts produced in a certain time interval and then translate that number into parts per hour. To keep the solution simple, the time interval is one hour and at the end of each hour, the counter accumulator is transferred to the appropriate location and then reset. The ControlLogix solution is shown in Figure 5.61. The MOV block is explained in Chapter 7. The Micro800 code is shown in Figure 5.62. Since the MicroLogix code is similar to the ControlLogix solution, it is not shown.

The addresses or data types associated with the internal tag/variables/symbols:

Tag/Var./Symbol	CLogix Data Type	Micro800 Data Type	MLogix/SLC Address
CHr_Tmr	TIMER	TON	T4:10
Cell_PRate_Ctr	COUNTER	CTU	C5:6

5.7 CHAPTER SUMMARY

For the ControlLogix/CompactLogic and the MicroLogix/SLC-500 PLCs covered by this text, this chapter described timers and counters, the most common function block

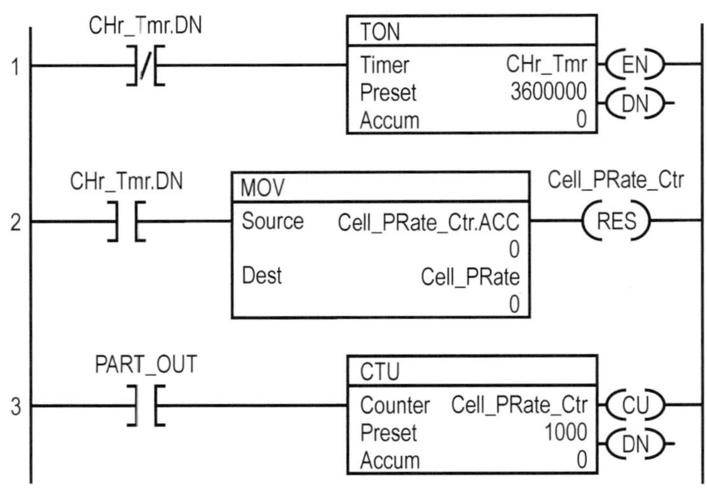

Figure 5.61. ControlLogix solution to Example 5.8.

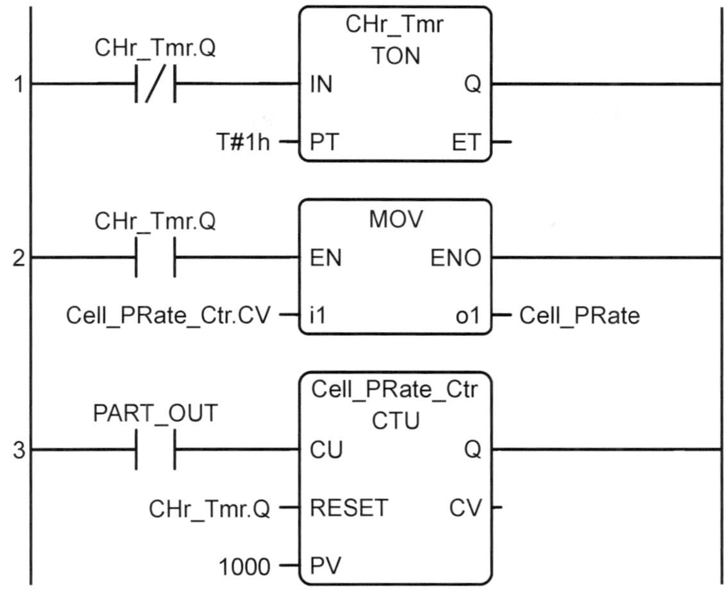

Figure 5.62. Micro800 solution to Example 5.8.

instructions. The major types of industrial timing and counting applications are also described and illustrated with industrial examples. With this background, the reader has the knowledge to move on to sequential operations.

REFERENCES

Erickson, Kelvin T., 2016. *Programmable Logic Controllers: An Emphasis on Design and Application,* 3rd Edition, Dogwood Valley Press, Rolla, MO.

Rockwell Automation, 2008. *SLC Instruction Set Reference Manual*, pub. 1747-RM001G-EN-P, Rockwell Automation Inc., Milwaukee, WI, Nov.

Rockwell Automation, 2011. *MicroLogix 1100 Programmable Controllers Instruction Set Reference Manual*, pub. 1763-RM001D-EN-P, Rockwell Automation Inc., Milwaukee, WI, Sep.

Rockwell Automation, 2014. *MicroLogix™ 1200 and MicroLogix 1500 Programmable Controllers Instruction Set Reference Manual*, pub. 1762-RM001H-EN-P, Rockwell Automation Inc., Milwaukee, WI, Jul.

Rockwell Automation, 2019. *MicroLogix 1400 Programmable Controllers Instruction Set Reference Manual*, pub. 1766-RM001I-EN-P, Rockwell Automation, Inc., Milwaukee, WI, June.

Rockwell Automation, 2022a. *Logix5000 Controllers General Instructions Reference Manual*, pub. 1756-RM003W-EN-P, Rockwell Automation Inc., Milwaukee, WI, Mar.

Rockwell Automation, 2022b. *Micro800 Programmable Controllers General Instructions Reference Manual*, pub. 2080-RM001K-EN-P, Rockwell Automation Inc., Milwaukee, WI, Mar.

PROBLEMS

General instructions for the problems:

Write a ladder logic program for the application and implement it for one or more of the following PLC ladder logic languages:

ControlLogix/CompactLogix, **or**

Micro800, **or**

MicroLogix/SLC-500

If any part of the operation is ambiguous, write down your additional assumptions.

The physical inputs, physical outputs, and internal tags/variables/symbols for each problem are given in the problem. Note that for the problems, the ControlLogix addresses assume 1756-, 1734-, 1769-, or 1794-series I/O modules. If using 5069- or 5094-series modules, the address tags will need to be modified. **DO NOT** assign any more physical inputs!

Your solution should include the following:

1. Specify the PLC processor used.
2. Ladder logic diagram (with comments). For consistency among the different PLCs, use only tags/variables/symbols/tags in the ladder logic. Use instructions consistent with the PLC processor.
3. Table listing additional internal memory (tags/variables/symbols) used and a brief description of their use. For the ControlLogix, CompactLogiz, and Micro800 processors, list the tags/variables and the data type. For MicroLogix/SLC-500 processors, list the symbols and the memory address.

P5-1. Develop a ladder logic program that will **open** (turn **on**) valve XV235 3.7 seconds after limit switch LS233 is **closed**. Pump motor P237 should turn **on** 1.5 seconds after XV235 is **opened**. Pilot light PL241 should turn **on** 5.0 seconds after P237 is turned **on**. Activating (turning **on**) limit switch LS231 should reset all timers, **close** the valve, turn **off** the pump and turn **off** the pilot light. The program must ensure that turning **on** LS231 is ignored if PL241 is **off**. Assume that LS233 is **on** only momentarily (for only one scan). Timers must be accurate to the nearest 0.01 second.

Assume the following physical input and output tags/variables/symbols:

Tag/Var./Symbol	Description
LS231	Limit switch, **on** when closed
LS233	Limit switch, **on** when closed
XV235	Valve control, **on** to open valve
P237	Pump motor contacter, **on** to run motor
PL241	Pilot light

The addresses associated with the physical I/O are:

Tag/Var./Symbol	ControlLogix	Micro800	MLogix	SLC-500
LS231	Local:1:I.Data.0	_IO_EM_DI_00	I:0/0	I:1/0
LS233	Local:1:I.Data.1	_IO_EM_DI_01	I:0/1	I:1/1

XV235	Local:2:O.Data.0	_IO_EM_DO_00	O:0/0	O:2/0
P237	Local:2:O.Data.1	_IO_EM_DO_01	O:0/1	O:2/1
PL241	Local:2:O.Data.2	_IO_EM_DO_02	O:0/2	O:2/2

P5-2. Develop a ladder logic program that will turn **on** pump motor P1441 1.2 seconds after limit switch LS1444 is closed. Pump motor P1442 should turn **on** 4.5 seconds after P1441 is turned **on**. Pilot light PL1443 should turn **on** 8.0 seconds after P1442 is turned **on**. Activating (turning **on**) limit switch LS1445 should reset all timers and turn off all motors and the pilot light. The program must ensure that turning **on** LS1445 is ignored if PL1443 is **off**. Assume that LS1444 is **on** only momentarily (for only one scan). Timers must be accurate to the nearest 0.01 second.

Assume the following physical input and output tags/variables/symbols:

Tag/Var./Symbol	Description
LS1444	Limit switch, **on** when closed
LS1445	Limit switch, **on** when closed
P1441	Pump motor contacter, **on** to run motor
P1442	Pump motor contacter, **on** to run motor
PL1443	Pilot light

The addresses associated with the physical I/O are:

Tag/Var./Symbol	ControlLogix	Micro800	MLogix	SLC-500
LS1444	Local:1:I.Data.0	_IO_EM_DI_00	I:0/0	I:1/0
LS1445	Local:1:I.Data.1	_IO_EM_DI_01	I:0/1	I:1/1
P1441	Local:2:O.Data.0	_IO_EM_DO_00	O:0/0	O:2/0
P1442	Local:2:O.Data.1	_IO_EM_DO_01	O:0/1	O:2/1
PL1443	Local:2:O.Data.2	_IO_EM_DO_02	O:0/2	O:2/2

P5-3. Develop a ladder logic program that will turn **on** pump motor P832 4.5 seconds after limit switch LS832 is closed. Output XV835 should turn **on** (to open valve) 2.4 seconds after P832 is turned **on**. Pilot light PL831 should turn **on** 2.0 seconds after XV835 is turned **on**. Activating (turning **on**) pressure switch PS837 should reset all timers and turn **off** the motor, valve, and pilot light. The program must ensure that the state of PS837 is ignored if PL831 is **off**. Assume that LS832 is **on** only momentarily (for only one scan). Timers must be accurate to the nearest 0.01 second.

Assume the following physical input and output tags/variables/symbols:

Tag/Var./Symbol	Description
LS832	Limit switch, **on** when closed
PS837	Pressure switch, **on** when closed
P832	Pump motor contacter, **on** to run motor
XV835	Valve control, **on** to open valve
PL831	Pilot light

The addresses associated with the physical I/O are:

Tag/Var./Symbol	ControlLogix	Micro800	MLogix	SLC-500
LS832	Local:1:I.Data.2	_IO_EM_DI02	I:0/2	I:1/2
PS837	Local:1:I.Data.3	_IO_EM_DI_03	I:0/3	I:1/3
P832	Local:2:O.Data.3	_IO_EM_DO_03	O:0/3	O:2/3
XV835	Local:2:O.Data.4	_IO_EM_DO_04	O:0/4	O:2/4
PL831	Local:2:O.Data.5	_IO_EM_DO_05	O:0/5	O:2/5

P5-4. Develop a ladder logic program that will turn **on** cylinder CYL214 1.5 seconds after proximity switch PROX211 is activated (turned **on**). Pilot light PL216 should turn **on** 2.0 seconds after CYL214 is turned **on**. Cylinder CYL215 should turn **on** 4.5 seconds after PL216 is turned **on**. Activating (turning **on**) limit switch LS212 will reset all timers and turn off all cylinders and the pilot light. The program must ensure that turning **on** LS212 is ignored if CYL215 is **off**. Assume that PROX211 is **on** for at least one scan, but no longer than 1 second. Timers must be accurate to the nearest 0.01 second. Assume the following input and output tags/variables/symbols:

Tag/Var./Symbol	Description
PROX211	Proximity switch, **on** when closed
LS212	Limit switch, **on** when closed
CYL214	Air cylinder control, **on** to extend cylinder
CYL215	Air cylinder control, **on** to extend cylinder
PL216	Pilot light

The addresses associated with the physical I/O are:

Tag/Var./Symbol	ControlLogix	Micro800	MLogix	SLC-500
PROX211	Local:1:I.Data.2	_IO_EM_DI_02	I:0/2	I:1/2
LS212	Local:1:I.Data.3	_IO_EM_DI_03	I:0/3	I:1/3
CYL214	Local:2:O.Data.3	_IO_EM_DO_03	O:0/3	O:2/3
CYL215	Local:2:O.Data.4	_IO_EM_DO_04	O:0/4	O:2/4
PL216	Local:2:O.Data.5	_IO_EM_DO_05	O:0/5	O:2/5

P5-5. Develop a ladder logic program that will turn **on** pilot lamp PL712 1.5 seconds after limit switch LS711 is activated (turned **on**). Cylinder CYL721 should turn **on** 4.0 seconds after PL712 is turned **on**. Cylinder CYL725 should turn **on** 2.5 seconds after CYL721 is turned **on**. When PROX727 is activated (turns **on**) while CYL725 is **on**, all timers are reset and all outputs (cylinders and pilot light) are turned **off**. The program must ensure that turning **on** PROX727 is ignored if CYL725 is **off**. Assume that LS711 must remain **on** for 1.5 seconds for the operation to commence. If LS711 turns **off** during the 1.5 second interval, the operation is cancelled (reset). However, after the initial 1.5 seconds, the state of LS711 is not guaranteed to be **on**. Timers must be accurate to the nearest 0.01 second. Assume the following input and output tags/variables/symbols:

Tag/Var./Symbol	Description
LS711	Limit switch, **on** when closed
PROX727	Proximity switch, **on** when closed
PL712	Pilot light
CYL721	Air cylinder control, **on** to extend cylinder

CYL725	Air cylinder control, **on** to retract cylinder

The addresses associated with the physical I/O are:

Tag/Var./Symbol	ControlLogix	Micro800	MLogix	SLC-500
LS711	Local:1:I.Data.0	_IO_EM_DI_00	I:0/0	I:1/0
PROX727	Local:1:I.Data.1	_IO_EM_DI_01	I:0/1	I:1/1
PL712	Local:2:O.Data.0	_IO_EM_DO_00	O:0/0	O:2/0
CYL721	Local:2:O.Data.1	_IO_EM_DO_01	O:0/1	O:2/1
CYL725	Local:2:O.Data.2	_IO_EM_DO_02	O:0/2	O:2/2

P5-6. Develop a ladder logic program that will turn **on** pilot light PL202 5 seconds after pressure switch PS201 is turned **on**. Valve XV203 should be opened (turned **on**) 6 seconds after PL202 turns **on**. Pilot light PL204 should turn **on** 8 seconds after XV203 is opened. Pressing PB205 will reset all timers and turn **off** the valve and all pilot lights. The operation of PB205 is allowed only when PL204 is **on**. Assume that PS201 is **on** for at least 6 seconds. After 6 seconds, PS201 may be **on** or **off**. In either case, the operation is not triggered until PS201 turns **on** again after PL204 has been turned **off**. Timers must be accurate to the nearest 0.01 second.

Assume the following input and output tags/variables/symbols:

Tag/Var./Symbol	Description
PS201	Pressure switch, **on** when closed.
PB205	Push button switch, **on** when pressed.
PL202	Pilot light
PL204	Pilot light
XV203	Air valve, **on** to open valve.

The addresses associated with the physical I/O are:

Tag/Var./Symbol	ControlLogix	Micro800	MLogix	SLC-500
PS201	Local:1:I.Data.0	_IO_EM_DI_00	I:0/0	I:1/0
PB205	Local:1:I.Data.4	_IO_EM_DI_04	I:0/4	I:1/4
PL202	Local:2:O.Data.2	_IO_EM_DO_02	O:0/2	O:2/2
XV203	Local:2:O.Data.3	_IO_EM_DO_03	O:0/3	O:2/3
PL204	Local:2:O.Data.5	_IO_EM_DO_05	O:0/5	O:2/5

P5-7. Develop a ladder logic program that will turn **on** cylinder CYL122 0.7 seconds after limit switch LS123 is turned **on**. Cylinder CYL125 should turn **on** 2.5 seconds after CYL122 is turned **on**. Pilot light PL132 should turn **on** 3.0 seconds after CYL125 is turned **on**. Activating (turning **on**) proximity switch PX131 will reset all timers and turn off all cylinders and the pilot light. The program must ensure that the turn **on** of PX131 is ignored if PL132 is **off**. Assume that LS123 is **on** for at least one scan, but no longer than 1 second. Timers must be accurate to the nearest 0.01 second. Assume the following input and output tags/variables/symbols:

Tag/Var./Symbol	Description
LS123	Limit switch, **on** when closed
PX131	Proximity switch, **on** when closed

CYL122	Air cylinder control, **on** to extend cylinder
CYL125	Air cylinder control, **on** to extend cylinder
PL132	Pilot light

The addresses associated with the physical I/O are:

Tag/Var./Symbol	ControlLogix	Micro800	MLogix	SLC-500
LS123	Local:1:I.Data.0	_IO_EM_DI_00	I:0/0	I:1/0
PX131	Local:1:I.Data.1	_IO_EM_DI_01	I:0/1	I:1/1
CYL122	Local:2:O.Data.0	_IO_EM_DO_00	O:0/0	O:2/0
CYL125	Local:2:O.Data.1	_IO_EM_DO_01	O:0/1	O:2/1
PL132	Local:2:O.Data.2	_IO_EM_DO_02	O:0/2	O:2/2

P5-8. Develop a ladder logic program that will turn **on** pneumatic cylinder CYL521 0.5 seconds after limit switch LS513 is activated (turned **on**). Motor MTR522 should turn **on** 3.0 seconds after CYL521 is turned **on**. Pilot light PL523 should turn **on** 6.0 seconds after MTR522 is turned **on**. When PROX514 is activated (turns **on**) while PL523 is **on**, all timers are reset and all outputs (cylinder, motor, and pilot light) are turned **off**. The program must ensure that turning **on** PROX514 is ignored if PL523 is **off**. Assume that LS513 is **on** for at least one scan, but no longer than 1 second. Timers must be accurate to the nearest 0.01 second. Assume the following input and output tags/variables/symbols:

Tag/Var./Symbol	Description
LS513	Limit switch, **on** when closed
PROX514	Proximity switch, **on** when closed
CYL521	Air cylinder control, **on** to extend cylinder
MTR522	Motor control, **on** to run motor
PL523	Pilot light

The addresses associated with the physical I/O are:

Tag/Var./Symbol	ControlLogix	Micro800	MLogix	SLC-500
LS513	Local:1:I.Data.7	_IO_EM_DI_07	I:0/0	I:1/0
PROX514	Local:1:I.Data.8	_IO_EM_DI_08	I:0/1	I:1/1
CYL521	Local:2:O.Data.0	_IO_EM_DO_00	O:0/0	O:2/0
MTR522	Local:2:O.Data.1	_IO_EM_DO_01	O:0/1	O:2/1
PL523	Local:2:O.Data.2	_IO_EM_DO_02	O:0/2	O:2/2

P5-9. Develop a ladder logic program that will turn **on** pilot light PL903 7.0 seconds after pressure switch PS901 is turned **on**. Relief valve XV905 should be opened (turned **on**) 2.0 seconds after PL903 turns **on**. Pilot light PL904 should turn **on** 10 seconds after XV905 is opened. Pressing PB907 should reset all timers, turn **off** all pilot lights and close the relief valve. The operation of PB907 is allowed only when PL904 is **on**. PS901 must remain on for 7 seconds before turning on PL903. Though, after the initial 7-second period, PS901 may be **on** or **off**. In either case, the operation is not triggered again until PS901 turns **on** again after PL904 has been turned **off**. Timers must be accurate to the nearest 0.01 second.

Assume the following input and output tags/variables/symbols:

Tag/Var./Symbol	Description
PS901	Pressure switch, **on** when closed
PB907	Push button switch, **on** when pressed
PL903	Pilot light
PL904	Pilot light
XV905	Relief valve, **on** to open valve

The addresses associated with the physical I/O are:

Tag/Var./Symbol	ControlLogix	Micro800	MLogix	SLC-500
PS901	REM_2:1:I.Data.0	_IO_X2_DI_00	I:4/0	I:4/0
PB907	REM_2:1:I.Data.5	_IO_X2_DI_05	I:4/5	I:4/5
PL903	REM_2:2:O.Data.3	_IO_X3_DO_03	O:5/3	O:5/3
PL904	REM_2:2:O.Data.4	_IO_X3_DO_04	O:5/4	O:5/4
XV905	REM_2:2:O.Data.5	_IO_X3_DO_05	O:5/5	O:5/5

P5-10. Develop a ladder logic program that will extend (turn **on**) air cylinder CYL312 10 seconds after limit switch LS311 is turned **on**. Valve XV313 should be opened (turned **on**) 5 seconds after CYL312 turns **on**. Pilot light PL314 should turn **on** 8 seconds after XV313 is opened. Pressing PB315 will reset all timers and turn **off** the cylinder, valve and pilot light. The operation of PB315 is allowed only when PL314 is **on**. Assume that LS311 is **on** for no longer than one second. Timers must be accurate to the nearest 0.01 second.

Assume the following input and output tags/variables/symbols:

Tag/Var./Symbol	Description
LS311	Limit switch, **on** when closed.
PB315	Push button switch, **on** when pressed.
CYL312	Air cylinder control, **on** to extend cylinder
XV313	Air valve, **on** to open valve.
PL314	Pilot light

The addresses associated with the physical I/O are:

Tag/Var./Symbol	ControlLogix	Micro800	MLogix	SLC-500
LS311	Local:1:I.Data.5	_IO_EM_DI_05	I:0/5	I:1/5
PB315	Local:1:I.Data.6	_IO_EM_DI_06	I:0/6	I:1/6
CYL312	Local:2:O.Data.6	_IO_EM_DO_06	O:0/6	O:2/6
XV313	Local:2:O.Data.7	_IO_EM_DO_07	O:0/7	O:2/7
PL314	Local:2:O.Data.8	_IO_EM_DO_08	O:0/8	O:2/8

P5-11. Develop a ladder logic program to produce a 10-second one-shot pulse. When limit switch LS431 turns **on**, the CY458 output should be **on** for 10±0.01 seconds and then turn **off**. Activating of LS431 should have no effect when CY458 is **on** (the output should not turn **off**, nor should it remain **on** any longer). When CY458 turns **off**, another activation (turn **on**) of LS431 should start the pulse over. If LS431 remains **on**, CY458 should be **on** for twenty seconds and turn **off**. CY458 should not turn **on** until the LS431 start switch is released and pressed again.. Use only one timer.

Assume the following input and output tags/variables/symbols:

Tag/Var./Symbol	Description
LS431	Limit switch, **on** when activated.
CY458	Cylinder control, **on** to move piston

The addresses associated with the physical I/O are:

Tag/Var./Symbol	ControlLogix	Micro800	MLogix	SLC-500
LS431	Local:1:I.Data.7	_IO_EM_DI_07	I:0/7	I:1/7
CY458	Local:2:O.Data.9	_IO_EM_DO_09	O:0/6	O:2/6

P5-12. Develop a ladder logic program to make lamp LA103 to flash. Lamp LA103 should be **on** for one-half second and **off** for one second, as shown in Figure P5.12. The ALARM internal coil controls the flashing lamp. When ALARM is **on**, the lamp flashes. When ALARM is **off**, the lamp should be **off**. When ALARM turns on, LA103 should immediately start its **on** period, as shown in Figure P5.12. Timers must be accurate to the nearest 0.01 second.

Assume the following input and output tags/variables/symbols:

Tag/Var./Symbol	Description
ALARM	Flashing lamp enable; **on** to cause lamp to flash, **off** turns lamp **off**.
LA103	Pilot light

The addresses associated with the physical I/O are:

Tag/Var./Symbol	ControlLogix	Micro800	MLogix	SLC-500
ALARM	Local:1:I.Data.9	_IO_EM_DI_09	I:0/11	I:1/11
LA103	Local:2:O.Data.12	_IO_EM_DO_12	O:0/4	O:2/4

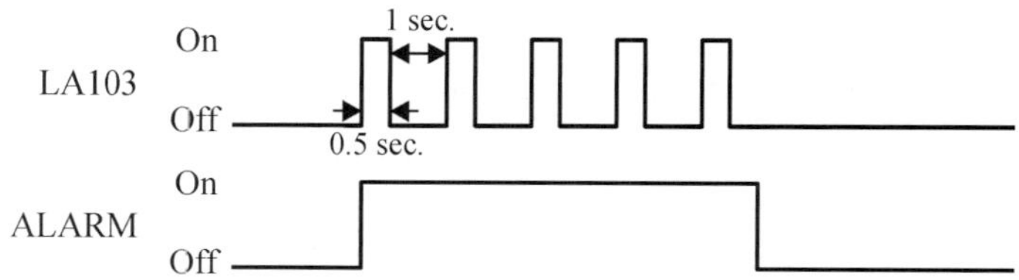

Figure P5.12. Timing diagram for flashing lamp.

P5-13. Use ladder logic to implement a simple *level indicator and alarm* for a tank.

A panel implements a simple interface showing the tank level. The panel has four lamps: blue, green, yellow, and red. The panel also has an alarm horn and a reset button that turns **off** the alarm horn. No more than one light is **on** at any time:

Blue lamp **on** when tank level < 5 inches

Green lamp **on** when 5 inches <= tank level < 50 inches

Yellow lamp **on** when 50 inches <= tank level < 60 inches

Red lamp **on** when tank level >= 60 inches

In addition, when the level is greater than or equal to 60 inches for 10 seconds, the alarm horn is turned **on**, and remains **on** until the RESET_PB button is pressed. In addition, the red lamp flashes at 1 Hz (0.5 sec **on**, 0.5 sec **off**) when the alarm is **on**. Pressing the RESET_PB button should silence the alarm horn, even if the level is still greater than or equal to 60 inches. The level must drop below 60 inches and then increase to 60 inches before the alarm is re-activated.

There are 3 level sensors, each corresponding to one of the "boundary" levels listed above:

LS_1 is **on** when the tank level >= 5 inches

LS_2 is **on** when the tank level >= 50 inches

LS_3 is **on** when the tank level >= 60 inches

Assume the following physical input and output tags/variables/symbols:

Tag/Var./Symbol	Description
RESET_PB	Push button switch, N. O., **on** when resetting alarm
LS_1	Level switch, **on** when the tank level >= 5 inches
LS_2	Level switch, **on** when the tank level >= 50 inches
LS_3	Level switch, **on** when the tank level >= 60 inches
BLUE	**On** to light blue indicator lamp
GREEN	**On** to light green indicator lamp
YELLOW	**On** to light yellow indicator lamp
RED	**On** to light red indicator lamp
ALARM_HORN	**On** to sound alarm horn

The addresses associated with the physical I/O are:

Tag/Var./Symbol	ControlLogix	Micro800	MLogix	SLC-500
RESET_PB	Local:1:I.Data.0	_IO_EM_DI_00	I:0/0	I:1/0
LS_1	Local:1:I.Data.1	_IO_EM_DI_01	I:0/1	I:1/1
LS_2	Local:1:I.Data.2	_IO_EM_DI_02	I:0/2	I:1/2
LS_3	Local:1:I.Data.3	_IO_EM_DI_03	I:0/3	I:1/3
BLUE	Local:2:O.Data.0	_IO_EM_DO_00	O:0/0	O:2/0
GREEN	Local:2:O.Data.1	_IO_EM_DO_01	O:0/1	O:2/1
YELLOW	Local:2:O.Data.2	_IO_EM_DO_02	O:0/2	O:2/2
RED	Local:2:O.Data.3	_IO_EM_DO_03	O:0/3	O:2/3
ALARM_HORN	Local:2:O.Data.4	_IO_EM_DO_04	O:0/4	O:2/4

P5-14. Develop a ladder logic program to handle a low bunker alarm for a coal-fired power plant coal bunker. The devices are shown in Figure P5.14. Normal sequence of events:

Coal bunker 2 becomes low for 3 seconds (BUNK2_LLEV turns off for 3 sec.) which causes

- BUNK2_LLMP starts flashing (1 sec. on, 1 sec. off)
- ALARM_HORN sounds (turns on).

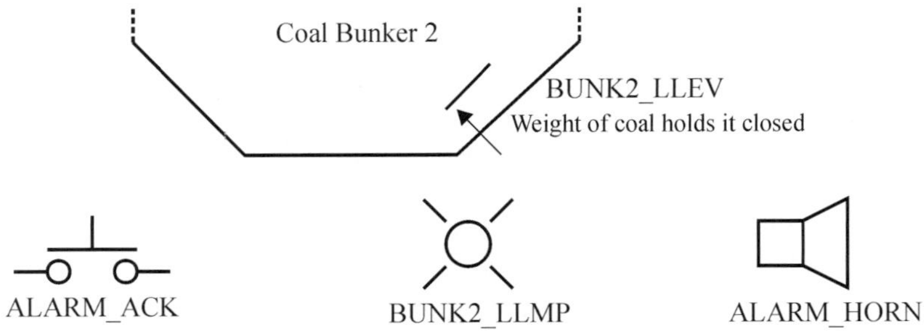

Figure P5.14. Coal bunker devices associated with low level alarm.

Operator hears horn and/or sees lamp and presses the ALARM_ACK button which causes

- ALARM_HORN is silenced
- BUNK2_LLMP constantly on

Operator activates the conveyor(s) to cause bunker 2 to be filled with coal. When coal level is high enough to make the BUNK2_LLEV sensor to turn on for 5 seconds, then the BUNK2_LLMP turns off.

Other specifications:

- The BUNK2_LLMP is extinguished only when BUNK2_LLEV turns on and the alarm has been acknowledged (operator has pressed ALARM_ACK). If the level becomes high enough to turn on BUNK2_LLEV, but the operator has not pressed ALARM_ACK, the lamp remains flashing.
- The alarm horn must stay on until ALARM_ACK is pressed. If the level becomes high enough to turn **on** BUNK2_LLEV but the operator has not pressed ALARM_ACK, the horn must remain **on**.
- If BUNK2_LLEV has been **on** for 5 seconds while the alarm is unacknowledged, the BUNK2_LLMP should turn **off** immediately when the ALARM_ACK button is pressed.
- All timing intervals must be accurate to ± 0.1 seconds.

Assume the following physical input and output tags/variables/symbolssignments:

Tag/Var./Symbol	Description
ALARM_ACK	Alarm acknowledge; N.O. push button, **on** to acknowledge alarm
BUNK2_LLEV	Bunker 2 low level indication; **off** indicates low level
ALARM_HORN	Alarm horn; **on** to sound alarm
BUNK2_LLMP	Bunker 2 low level indicator; **on** when bunker 2 level is low

The addresses associated with the physical I/O are:

Tag/Var./Symbol	ControlLogix	Micro800	MLogix	SLC-500
ALARM_ACK	Local:1:I.Data.1	_IO_EM_DI_01	I:0/1	I:1/1
BUNK2_LLEV	Local:1:I.Data.9	_IO_EM_DI_09	I:0/9	I:1/9

ALARM_HORN	Local:2:O.Data.7	_IO_EM_DO_07	O:0/7	O:2/7
BUNK2_LLMP	Local:2:O.Data.11	_IO_EM_DO_11	O:0/8	O:2/8

P5-15. Implement an alarm for the generator room cooling fans.

The university power plant basically generates steam for the campus. But, it has a 1 MW backup diesel-powered generator to supply electric power to run the motors and other steam boiler equipment when the normal electric feed into the plant fails. In the summer, the backup generator is also used to "shave" the campus peak electric demand. If the campus demand goes above a certain value, the utility company assesses a hefty "demand charge." Therefore, when the electric demand of the campus approaches this peak, the backup generator is started and supplies power to the campus, avoiding the demand charge. Earlier in the summer, the generator was being used for peak shaving and failed. The failure was traced to one of the generator room cooling fans "tripping out" and causing some of the generator control equipment to overheat. The fan indicators are lamps on a control cabinet in the generator room. When the diesel-powered generator is running, the room is closed and one must put on ear protectors to enter the room. Consequently, the status of the cooling fans is not normally checked, which led to the generator failure.

You need to program ladder logic that monitors the four fans and turns **on** an alarm horn if any of the four fans stops running while the diesel motor is running. There are four cooling fans and two of these are two-speed fans. The fans and their run indications are:

Supply (two-speed)	SUP_HI and SUP_LO
Exhaust (two-speed)	EXH_HI and EXH_LO
Cooling #1	COOL_1
Cooling #2	COOL_2

The program needs to monitor these six indications. The GEN3_RUN input is **on** when the diesel is running. The ALARM physical output controls the alarm horn.

(a) If the diesel is running and any of the four fans is **off** for 2 seconds, the ALARM output is 2 seconds **on** and then 2 seconds **off**, in a repeating cycle. There is no separate reset button. The alarm continues to sound as long as at least one of the four fans is not running. When the diesel is not running or the diesel is running and all four fans are on, the ALARM output should be **off**.

(b) The alarm is to be fail-safe, that is the alarm sounds if the PLC fails or one of the fans stops running. To sound the alarm when the PLC fails, the ALARM physical output connection to the alarm horn is modified from part (a). Now, the ALARM output drives a relay whose NC contact drives the alarm horn. When ALARM is **on**, the relay coil is energized, which causes the NC contact to stay open and the alarm does not sound. If the PLC fails, then the PLC output turns **off** and the NC relay contact closes, causing the alarm horn to sound steadily. Modify your program from (a) to invert the ALARM output. If the diesel is running and any of the four fans is **off** for 2 seconds, the ALARM output is 2 seconds **off** and then 2 seconds **on**, in a repeating cycle. When the diesel is not running or the diesel is running and all four fans are on, the ALARM output should be **on**.

Assume the following physical input and output tags/variables/symbols:

Tag/Var./Sym.	Description
SUP_HI	Supply fan high-speed indication, **on** when fan running at high speed.
SUP_LO	Supply fan low-speed indication, **on** when fan is running at low speed.
EXH_HI	Exhaust fan high-speed indication, **on** when fan running at high speed.
EXH_LO	Exhaust fan low-speed indication, **on** when fan running at low speed.
COOL_1	Cooling fan #1 run indication, **on** when cooling fan #1 is running.
COOL_2	Cooling fan #2 run indication, **on** when cooling fan #2 is running.
GEN3_RUN	Diesel running indication; **on** indicates diesel is running, **off** indicates diesel is not running.
ALARM	Alarm horn; operation described above.

The addresses associated with the physical I/O are:

Tag/Var./Symbol	ControlLogix	Micro800	MLogix	SLC-500
SUP_HI	Local:1:I.Data.0	_IO_EM_DI_00	I:0/0	I:1/0
SUP_LO	Local:1:I.Data.1	_IO_EM_DI_01	I:0/1	I:1/1
EXH_HI	Local:1:I.Data.2	_IO_EM_DI_02	I:0/2	I:1/2
EXH_LO	Local:1:I.Data.3	_IO_EM_DI_03	I:0/3	I:1/3
COOL_1	Local:1:I.Data.4	_IO_EM_DI_04	I:0/4	I:1/4
COOL_2	Local:1:I.Data.5	_IO_EM_DI_05	I:0/5	I:1/5
GEN3_RUN	Local:1:I.Data.6	_IO_EM_DI_06	I:0/6	I:1/6
ALARM	Local:2:O.Data.4	_IO_EM_DO_04	O:0/0	O:2/0

P5-16. The Devil's Well, south of Salem, MO, is a sinkhole opening into an underground lake (Figure P5.16*a*). Visitors go down a set of stairs to enter the very top of the cave. In order to allow visitors to see the lake below, the cave is lit with a floodlight. However, the floodlight is not **on** constantly, but a visitor presses (turns **on**) a button to activate (turn on) the floodlight. The control for the floodlight turns off the floodlight 10 minutes after the button has been pressed. Design a PLC-based control for this system.

The timing diagram is shown in Figure P5.16*b*. The 10 minute interval can be started when either the button is pressed or when it is released. If the visitor presses the button during the 10-minute interval, the 10-minute interval must be restarted. In other words, the floodlight is **on** for 10 minutes after any time the button is pressed (and then released).

Assume the following physical input/output tags/variables/symbols:

Tag/Var./Symbol	Description
START_PB	When pressed, starts 10-minute timing interval. When button is pressed, this input is **on**; when released, this input is **off**.
FLAMP	Floodlight control. When **on**, the lamp is lit; when **off**, the lamp is not lit.

The addresses associated with the physical I/O are:

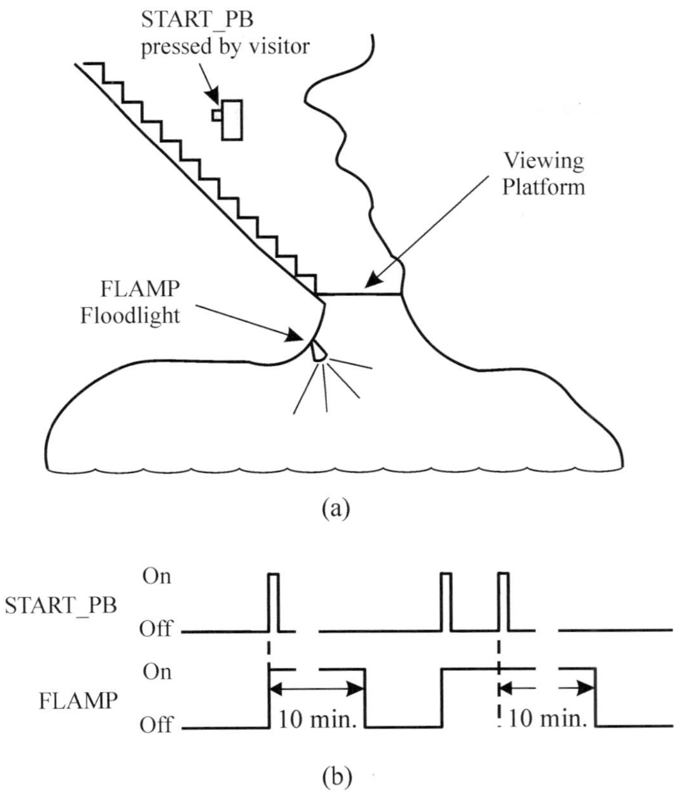

Figure P5.16. Devil's Well light control: (a) view of system; (b) timing diagram.

Tag/Var./Symbol	ControlLogix	Micro800	MLogix	SLC-500
START_PB	Local:1:I.Data.0	_IO_EM_DI_00	I:0/0	I:1/0
FLAMP	Local:2:O.Data.0	_IO_EM_DO_00	O:0/0	O:2/0

P5-17. Use ladder logic to implement a *tank level indicator* described below.

The T101 tank level indicator is a bar graph on an HMI, like a cell phone battery charge indicator. It consists of 2 bars stacked vertically, as shown in Figure P5.17. The two bars basically indicate if the tank level is low, in the normal range, and/or being filled. You are programming the logic that controls the display of the 2 bars. The bars are controlled by two internal coils, T101_LO and T101_NM.

When the pump to fill T101 is not running (P101_AUX is **off**), the bars indicate the relative level of the tank (Figure P5.17*a*). When the pump to fill T101 is running (P101_AUX is **on**), the bars indicate the relative level of the tank and the top bar flashes to indicate that the tank is being filled (Figure 5.17*b*).

The indicator bars operate as follows when P101_AUX is **off**:

Indicator	Indicator **steady on** if
T101_LO	Always

T101_NM T101 level >= minimum

The indicator bars operate as follows when P101_AUX is **on**:

Indicator	Indicator **steady on** if	Indicator **flashing** if
T101_LO	T101 level >= minimum	T101 level < minimum
T101_NM	T101 level >= maximum	T101 level >= minimum**and** T101 level < maximum

When flashing, the indicator bar should be **on** for 0.60±0.01 second, and **off** for 0.40±0.01 second. The start of the **on** part of a cycle does not need to be coordinated with any level change. For example, when P101_AUX is **on** and the level changes from less than the minimum to be equal to or greater than the minimum, do not synchronize the start of the **on** part of the flash cycle with this change.

There are two level sensors:

T101LS_LO is **on** when T101 level is >= minimum

T101LS_HI is **on** when T101 level is >= maximum

Assume the following physical input, physical output, and internal tags/variables/symbols:

Tag/Var./Symbol	Description
T101LS_LO	Tank T101 low level switch, **on** when level >= minimum.
T101LS_HI	Tank T101 high level switch, **on** when level >= maximum.
P101_AUX	Pump running indication: **on** when pump running, **off** when pump stopped.
T101_LO	Tank level below minimum level indication. Operation described above. Your logic must control this tag/variable.
T101_NM	Tank level at least minimum level indication. Operation described above. Your logic must control this tag/variable.

The addresses associated with the physical inputs are:

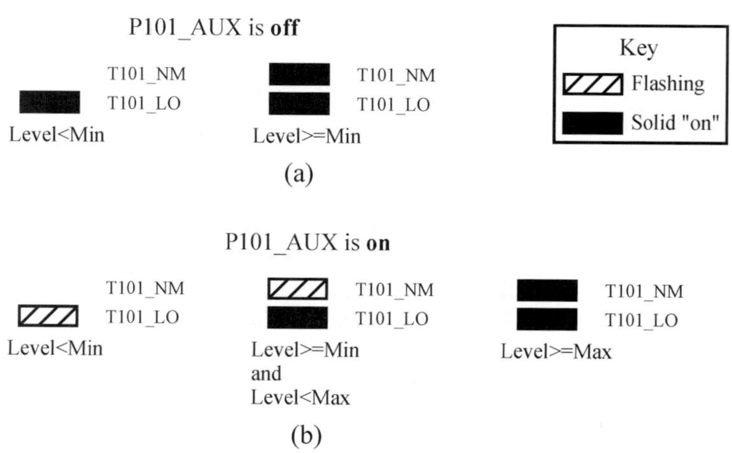

Figure P5.17. T101 level indicator operation: *(a)* When P101_AUX is **off**; *(b)* When P101_AUX is **on**.

Tag/Var./Symbol	ControlLogix	Micro800	MLogix	SLC-500
T101LS_LO	Local:1:I.Data.0	_IO_EM_DI_00	I:0/0	I:1/0
T101LS_HI	Local:1:I.Data.1	_IO_EM_DI_01	I:0/1	I:1/1
P101_AUX	Local:1:I.Data.9	_IO_EM_DI_09	I:0/9	I:1/9

The addresses and/or data types associated with the internal tags/variables/symbols are:

Tag/Var./Symbol	CLogix Data Type	Micro800 Data Type	MLogix/SLC Address
T101_LO	BOOL	BOOL	B3/123
T101_NM	BOOL	BOOL	B3/124

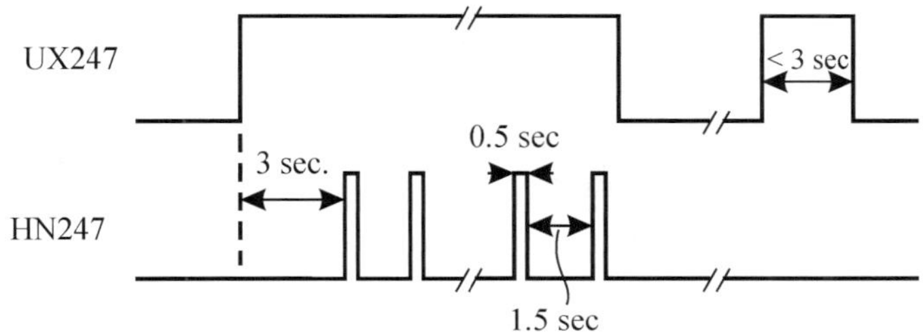

Figure P5.18. Timing diagram.

P5-18. Use ladder logic to implement a simple *hazard horn* system described below.

A certain area of the plant contains hazardous materials and so an auditory warning should be installed to indicate the danger when a person is approaching the area. Your job is to add that function to the PLC. When ultrasonic proximity UX247 is sensed **on** for 3 seconds, indicating a person is approaching this hazardous area, the HN247 horn must be beeped. The period of the horn beep should be 0.5 second **on** and 1.5 seconds **off**. The horn should continue to beep until the UX247 sensor is **off** (indicating the person has left). The first horn beep should commence at the end of the 3 second delay interval and not 1.5 seconds later. The timing diagram is shown in Figure P5.18.

Assume the following physical I/O tags/variables/symbols:

Tag/Var./Symbol	Description
UX247	Ultrasonic proximity sensor that detects a person close to hazardous area, N. O., **on** to indicate presence.
HN247	Hazardous area approach horn indication. Beeps 0.5 sec. **on** and 1.5 sec **off** when a person is close for more than 3 seconds.

The addresses associated with the physical input and output are:

Tag/Var./Symbol	ControlLogix	Micro800	MLogix	SLC-500
UX247	Local:1:I.Data.0	_IO_EM_DI_00	I:0/0	I:1/0
HN247	Local:2:O.Data.0	_IO_EM_DO_00	O:0/0	O:2/0

P5-19 Use ladder logic to *track the bin number* for an overhead crane system.

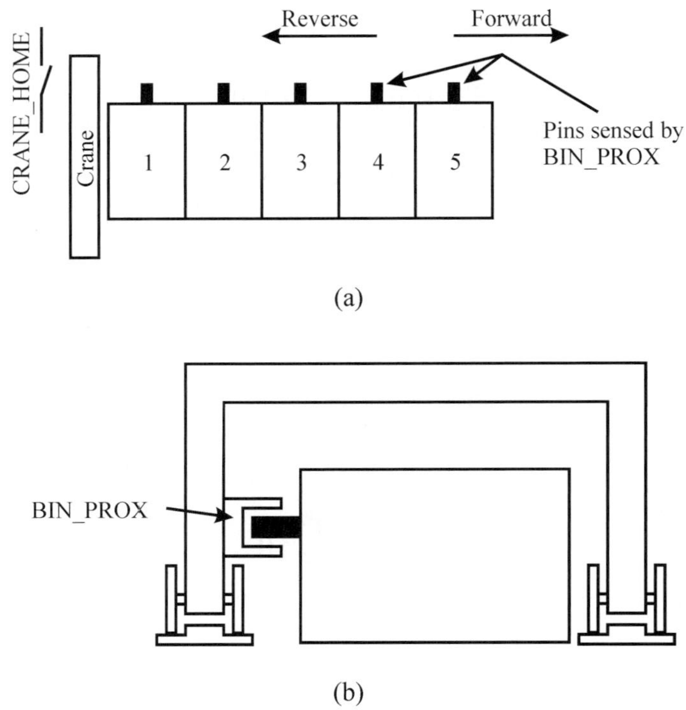

Figure P5.19. Crane system: (*a*) top view of system; (*b*) end view.

Figure P5.19 shows the layout of an overhead crane system that moves material from the home position to one of a series of bins. Your task is to provide a simple means of determining the bin number beneath the crane when it moves in the forward direction only. Each bin has a pin that interrupts an optical through-bean proximity sensor whose output is connected to the BIN_PROX input to the PLC. The BIN_PROX input is normally **on**. As the crane travels along its track, the BIN_PROX input to the PLC turns **off** as the BIN_PROX sensor's beam is interrupted by the pin at a particular bin and then back **on** after it has passed the pin. The BIN_PROX senses the center of each bin.

The home position is on the left side of Figure P5.19*a* and the bins are numbered as shown. As the crane moves to the right, the forward direction, the bin number increases. As the crane moves right, the bin number should increase on the transition of BIN_PROX from **on** to **off**.

The bin number must be saved in the BIN_NUMBER internal variable. Hint: Example 5.6 saves counter accumulators in tags/variables/symbols.

The CRANE_FWD tag/variable is controlled by another part of the ladder. Your logic **must not** have a coil controlling CRANE_FWD. CRANE_FWD indicates that the crane is traveling in the forward direction and remains **on** if the crane movement is paused while moving in the forward direction. There is a separate PLC output that actually turns on the motor. Your ladder is not concerned about the crane moving in the reverse direction. Problem P5-23 adds the reverse direction to this situation.

The CRANE_HOME PLC input is **on** when the crane is at the home position and thus certainly at bin number zero. You will need to determine how to reset your counter.

Assume the following physical I/O and internal tags/variables/symbols:

Tag/Var./Symbol	Description
BIN_PROX	Plate proximity sensor; **off** as pin passes, **on** otherwise.
CRANE_HOME	Limit switch that is on when the crane is at the home position and thus the bin number should be zero.
CRANE_FWD	Internal coil that indicates the crane is traveling in the forward direction. **On** indicates forward direction.
BIN_NUMBER	Holds bin number to be used by another part of the program.

The addresses associated with the physical inputs are:

Tag/Var./Symbol	ControlLogix	Micro800	MLogix	SLC-500
BIN_PROX	Local:1:I.Data.3	_IO_EM_DI_03	I:0/3	I:1/3
CRANE_HOME	Local:1:I.Data.7	_IO_EM_DI_07	I:0/7	I:1/7

The addresses or data types associated with the internal tags/variables/symbols are:

Tag/Var./Symbol	CLogix Data Type	Micro800 Data Type	MLogix/SLC Address
CRANE_FWD	BOOL	BOOL	B3/523
BIN_NUMBER	DINT	DINT	N7:5

P5-20 Use ladder logic to *measure steel beam lengths* by counting encoder pulses.

Figure P5.20*a* shows the layout of a station that uses a proximity detector (PROX_264) along with the pulses from a shaft encoder (ENC_264) to determine the length of the steel beam that passes the proximity detector. When PROX_264 is **on**, indicating the beam is passing, your ladder should be counting the encoder pulses to determine the beam length. Each pulse represents 0.1" of beam length, but your ladder does not actually calculate the length in inches, it only determines the length in pulses. When PROX_264 **transitions off**, your ladder needs to generate a 2-second **on** pulse on the BLEN_RDY internal coil. BLEN_RDY is used by another part of the ladder to use the counter accumulator contents to calculate the length. **Your ladder portion does not do this calculation.**

You will need to determine how to reset your counter. However, it cannot be reset while counting the steel beam length, nor during the time BLEN_RDY is **on**.

Assume the following physical I/O and internal tags/variables/symbols:

Tag/Var./Symbol	Description
PROX_264	Steel beam proximity sensor; **on** as steel beam passes, **off** otherwise.
ENC_264	Encoder output, pulses **on** and **off** as conveyor moves. Each encoder pulse is 0.1 inches of conveyor travel.

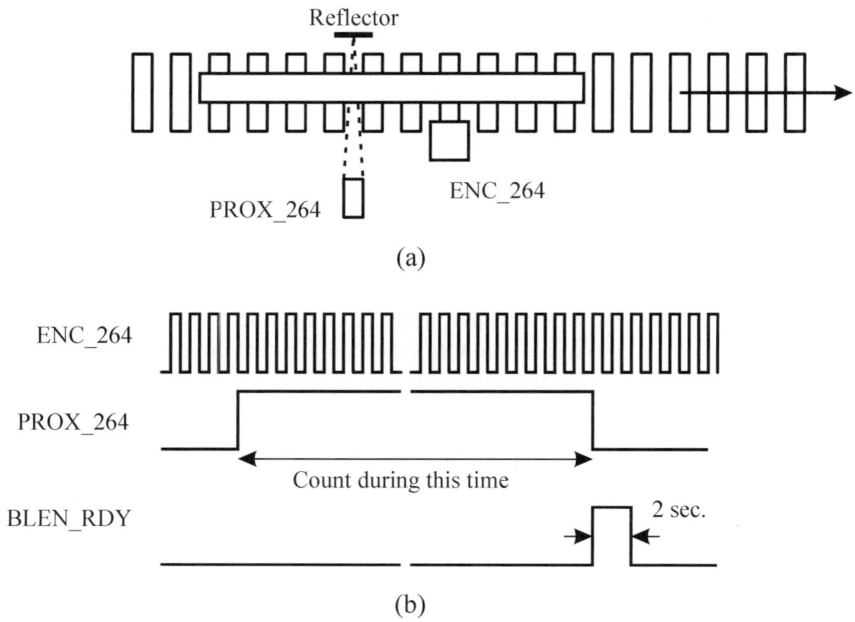

Figure P5.20. Steel beam measuring system: *(a)* top view of system; *(b)* relationship between signals.

BLEN_RDY — "Length ready" internal coil used by another part of the ladder to know when the steel beam length in encoder pulses can be extracted from the counter accumulator. Must be pulsed **on** for 2 seconds after steel beam has passed the proximity sensor. **Your ladder must generate this pulse**.

Addresses and/or data types associated with the physical I/O and internal tags/variables/symbols are:

Tag/Var./Symbol	ControlLogix	Micro800	MLogix	SLC-500
PROX_264	Local:1:I.Data.3	_IO_EM_DI_03	I:0/3	I:1/3
ENC_264	Local:1:I.Data.7	_IO_EM_DI_07	I:0/7	I:1/7
BLEN_RDY	BOOL	BOOL	B3/523	B3/523

P5-21. A parking lot has a capacity of 2000 cars. A FULL sign is **on** when the parking lot is full, and a VACANCY sign is **on** when parking space is available. Also, a CLOSED sign is **on** (and the other two signs are **off**) when no attendant is on duty. When the attendant leaves, there may be cars in the parking lot. To handle imperfect sensing of the automobiles, the attendant presses the RST_CNT push button **only if** the lot is empty when s/he leaves. Draw the ladder logic program, showing the proper values that must be configured for each instruction.

Assume the following physical I/O tags/variables/symbols:

Tag/Var./Symbol	Description
AT_PRESNT	Attendant present, **on** when attendant present
RST_CNT	Reset car count, **on** to zero the number of cars in the lot
CAR_IN	Car enter detector, energized (**on**) and then de-energized as car drives over platform at entrance
CAR_OUT	Car exit detector, energized (**on**) and then de-energized as car drives over platform at exit
FULL_SIGN	Illuminates "full" sign when **on**.
VACY_SIGN	Illuminates "vacancy" sign when **on**
CLOS_SIGN	Illuminates "closed" sign when **on**

The addresses associated with the physical I/O are:

Tag/Var./Symbol	ControlLogix	Micro800	MLogix	SLC-500
AT_PRESNT	Local:1:I.Data.0	_IO_EM_DI_00	I:0/0	I:1/0
CAR_IN	Local:1:I.Data.1	_IO_EM_DI_01	I:0/1	I:1/1
CAR_OUT	Local:1:I.Data.2	_IO_EM_DI_02	I:0/2	I:1/2
RST_CNT	Local:1:I.Data.3	_IO_EM_DI_03	I:0/3	I:1/3
FULL_SIGN	Local:2:O.Data.0	_IO_EM_DO_00	O:0/0	O:2/0
VACY_SIGN	Local:2:O.Data.1	_IO_EM_DO_01	O:0/1	O:2/1
CLOS_SIGN	Local:2:O.Data.2	_IO_EM_DO_02	O:0/2	O:2/2

P5-22. Building security needs a means of determining if there is anyone present in a particular office building. Each person entering the building passes though a turnstile that trips the PERSON_IN detector. Also, each person exiting the building passes through another turnstile that trips the PERSON_OUT detector. Develop a ladder logic program that will track the number of persons in the building and provides two indicators: OCCUPIED is **on** when anyone is in the building and VACANT is **on** when there are no people in the building. These indicators are visible from outside the building and so one does not need to enter the building to view these indicators. A guard checks the building late at night and makes sure the building is unoccupied. To handle the situation where there is a miscount, the guard always resets the occupancy count by pressing the RESET button when s/he leaves. The guard than has 30 seconds to actually leave the building before the occupancy count is zeroed. The guard does check that the VACANT indicator is **on** at the end of the 30 seconds (after exiting the building). Assume that there can be no more than 450 people in the building. Develop a ladder logic program to handle this occupancy system.

Assume the following physical I/O tags/variables/symbols:

Tag/Var./Symbol	Description
PERSON_IN	Person enter detector, **on** for about 0.5 second when person passes through entrance turnstile
PERSON_OUT	Person exit detector, **on** for about 0.5 second when person passes through exit turnstile
RESET	Reset button switch, **on** to reset occupancy count in 30 seconds
OCCUPIED	Illuminates when there is anyone in the building
VACANT	Illuminates when there is no one in the building

The addresses associated with the physical I/O are:

Tag/Var./Symbol	ControlLogix	Micro800	MLogix	SLC-500
PERSON_IN	Local:1:I.Data.0	_IO_EM_DI_00	I:0/0	I:1/0
PERSON_OUT	Local:1:I.Data.1	_IO_EM_DI_01	I:0/1	I:1/1
RESET	Local:1:I.Data.2	_IO_EM_DI_02	I:0/2	I:1/2
OCCUPIED	Local:2:O.Data.4	_IO_EM_DO_04	O:0/4	O:2/4
VACANT	Local:2:O.Data.5	_IO_EM_DO_05	O:0/5	O:2/5

P5-23 Use ladder logic to *track the bin number* for an overhead crane system moving in both directions.

Figure P5.19 shows the layout of an overhead crane system that moves material from the home position to one of a series of bins. Your task is to provide a simple means of determining the bin number beneath the crane. Each bin has a pin that interrupts an optical through-bean proximity sensor whose output is connected to the BIN_PROX input to the PLC. The BIN_PROX input is normally **on**. As the crane travels along its track, the BIN_PROX input to the PLC turns **off** as the BIN_PROX sensor's beam is interrupted by the pin at a particular bin and then back **on** after it has passed the pin. BIN_PROX senses the center of each bin.

The home position is on the left side of Figure P5.19*a* and the bins are numbered as shown. As the crane moves to the right, the forward direction, the bin number increases. As the crane moves right, the bin number should increase on the transition of BIN_PROX from **on** to **off**. As the crane moves to the left, the reverse direction, the bin number decreases. As the crane moves left, the bin number should decrease on the transition of BIN_PROX from **off** to **on**. The case when the crane reverses direction while the BIN_PROX input is **off** will be handled properly if the bin number is changed according to the previous description.

Two internal variables, CRANE_FWD and CRANE_REV, indicate that direction of crane movement. CRANE_FWD indicates that the crane is traveling in the forward direction though it may be temporarily paused. In a similar manner, CRANE_REV indicates that the crane is traveling in the reverse direction. Your logic **must not** have coils referencing CRANE_FWD or CRANE_REV, as these tags are controlled by another part of the ladder logic. There are separate PLC outputs that actually turn on the motor for the proper direction. CRANE_FWD and CRANE_REV only indicate the direction of travel and do not change value if the crane movement is paused while moving.

The CRANE_HOME PLC input is **on** when the crane is at the home position and thus certainly at bin number zero. You will need to determine how to reset your counter.

Assume the following physical I/O and internal variable tags/variables/symbols:

Tag/Var./Symbol	Description
BIN_PROX	Pin proximity sensor; **off** as pin passes, **on** otherwise.
CRANE_HOME	Limit switch that is on when the crane is at the home position and thus the bin number should be zero.
CRANE_FWD	Internal coil that indicates the crane is traveling in the forward direction. **On** indicates forward direction.

CRANE_REV	Internal coil that indicates the crane is traveling in the reverse direction. **On** indicates reverse direction.
BIN_NUMBER	Holds bin number to be used by another part of the program.

The addresses associated with the physical inputs are:

Tag/Var./Symbol	ControlLogix	Micro800	MLogix	SLC-500
BIN_PROX	Local:1:I.Data.3	_IO_EM_DI_03	I:0/3	I:1/3
CRANE_HOME	Local:1:I.Data.7	_IO_EM_DI_07	I:0/7	I:1/7

The addresses or data types associated with the internal tags/variables/symbols are:

Tag/Var./Symbol	CLogix Data Type	Micro800 Data Type	MLogix/SLC Address
CRANE_FWD	BOOL	BOOL	B3/523
CRANE_REV	BOOL	BOOL	B3/524
BIN_NUMBER	DINT	DINT	N7:5

P5-24 Use ladder logic to *measure plate lengths* by counting encoder pulses.

Figure P5.24*a* shows the layout of a station that uses a proximity detector (PROX_325) along with the pulses from a shaft encoder (ENC_325) to determine the length of the plate that passes the proximity detector. When PROX_325 is **off**, indicating the plate is passing, your ladder should be counting the encoder pulses to determine the plate length. Each pulse represents 0.1" of plate length, but your ladder does not actually calculate the length in inches, it only determines the length

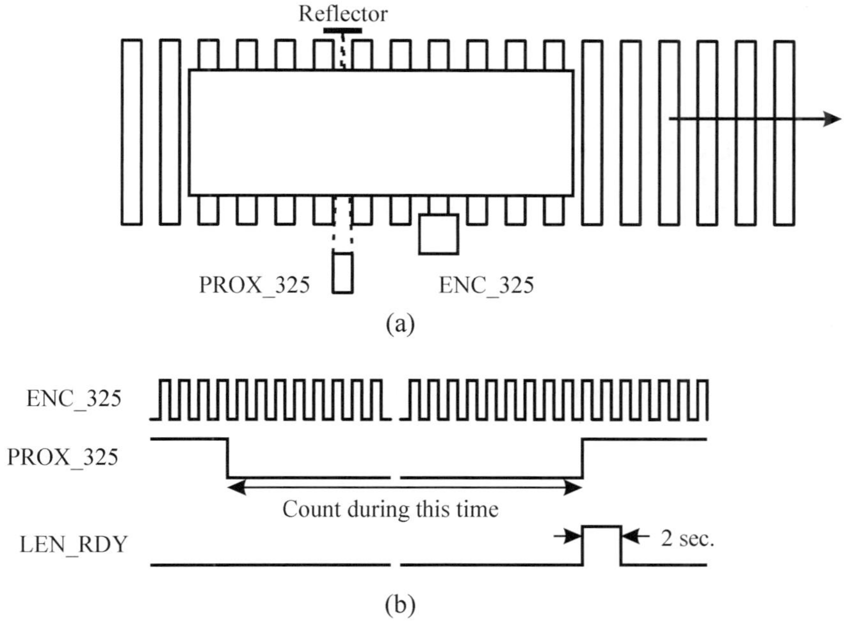

Figure P5.24. Plate measuring system: *(a)* top view of system; *(b)* relationship between signals.

in pulses. When PROX_325 **transitions on**, your ladder needs to generate a 2-second **on** pulse on the LEN_RDY internal variable. LEN_RDY is used by another part of the ladder to use the counter accumulator contents to calculate the length. **Your ladder portion does not do this calculation.**

You will need to determine how to reset your counter. However, it cannot be reset while counting the plate length, nor during the time LEN_RDY is **on**.

Assume the following physical I/O and internal variable tags/variables/symbols:

Tag/Var./Symbol	Description
PROX_325	Plate proximity sensor; **off** as plate passes, **on** otherwise.
ENC_325	Encoder output, pulses on and off as conveyor moves. Each encoder pulse is 0.1 inches of conveyor travel.
LEN_RDY	"Length ready" internal coil used by another part of the ladder to know when the plate length in encoder pulses can be extracted from the counter accumulator. Must be pulsed **on** for 2 seconds after plate has passed the proximity sensor.

Addresses and/or data types associated with the physical I/O and internal tags/variables/symbols are:

Tag/Var./Symbol	ControlLogix	Micro800	MLogix	SLC-500
PROX_325	Local:1:I.Data.3	_IO_EM_DI_03	I:0/3	I:1/3
ENC_325	Local:1:I.Data.7	_IO_EM_DI_07	I:0/7	I:1/7
LEN_RDY	BOOL	BOOL	B3/105	B3/105

P5-25. Use ladder logic to implement a *warning indicator* described below.

Generate a flashing light indicator that signifies that a labeling machine has processed 10,000 labels and that it is time to change the reel of labels (Figure P5.25).

When 10,000 labels have passed the PE245 photoelectric eye, the CHANGE_LAMP should flash **on** for 0.6 seconds and **off** for 0.4 seconds until it is acknowledged by pressing the CHANGE_ACK pushbutton. When CHANGE_ACK is pressed, the CHANGE_LAMP should immediately be turned **off** and the checking for the next 10,000 labels begins.

When 10,000 labels have been processed, CHANGE_LAMP should start the **on** flash immediately. There should not be an additional delay after 10,000 labels have passed.

Pressing the CHANGE_ACK button should not change the stored count if the alarm lamp is not flashing.

You do not have any control over the mechanism that moves the labels. The labels are spaced so that the photoelectric eye can sense the notch. **Do not add any start/stop switches.**

Assume the following physical I/O tags/variables/symbols:

Tag/Var./Symbol	Description
PE245	Photoelectric eye, **off** when label notch passes.
CHANGE_ACK	Alarm acknowledge push-button. **On** when pressed.

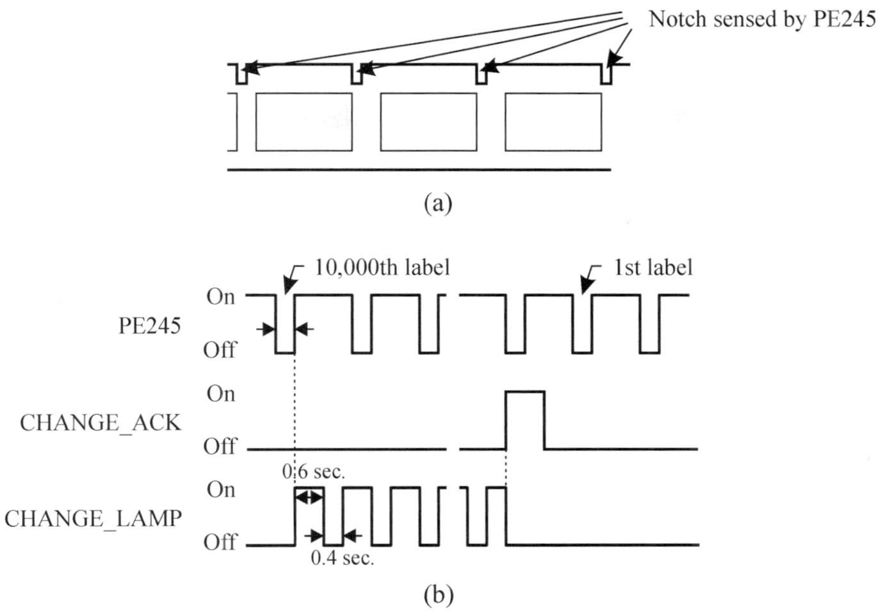

Figure P5.25. Labels low warning: *(a)* sensed labels; *(b)* timing.

CHANGE_LAMP Alarm lamp, **on** to light lamp.

The addresses associated with the physical I/O are:

Tag/Var./Symbol	ControlLogix	Micro800	MLogix	SLC-500
PE245	Local:1:I.Data.0	_IO_EM_DI_00	I:0/0	I:1/0
CHANGE_ACK	Local:1:I.Data.1	_IO_EM_DI_01	I:0/1	I:1/1
CHANGE_LAMP	Local:2:O.Data.0	_IO_EM_DO_00	O:0/0	O:2/0

P5-26. Use ladder logic to implement a simple chemical injection system that injects a pulse of chemical for every 1000 liters of flow in the main water line.

For every 1000 liters that flow through a water line, a small amount of chemical must be injected. To inject the chemical, a valve must be opened for 0.8 seconds. After injecting the chemical, your logic must reset the count and produce another injection pulse after the next 1000 liters.

The amount of water that has passed through the flow sensor is indicated by a pulse, FPLS_501. For every liter, the FPLS_501 input pulses **on** for 0.25 seconds. The valve control, CINJ_502, is pulsed **on** for 0.8 seconds in order to inject the chemical. The general relationship between the FPLS_501 input and the CINJ_502 output is shown in Figure P5.26.

You may **not** assume the water line flow is no faster than 1 liter per second, even though it is shown that way in Figure P5.26. In other words, the next pulse after the 1000th FPLS_501 pulse could occur during the CINJ_502 pulse and your logic must count **all** FPLS_501 pulses.

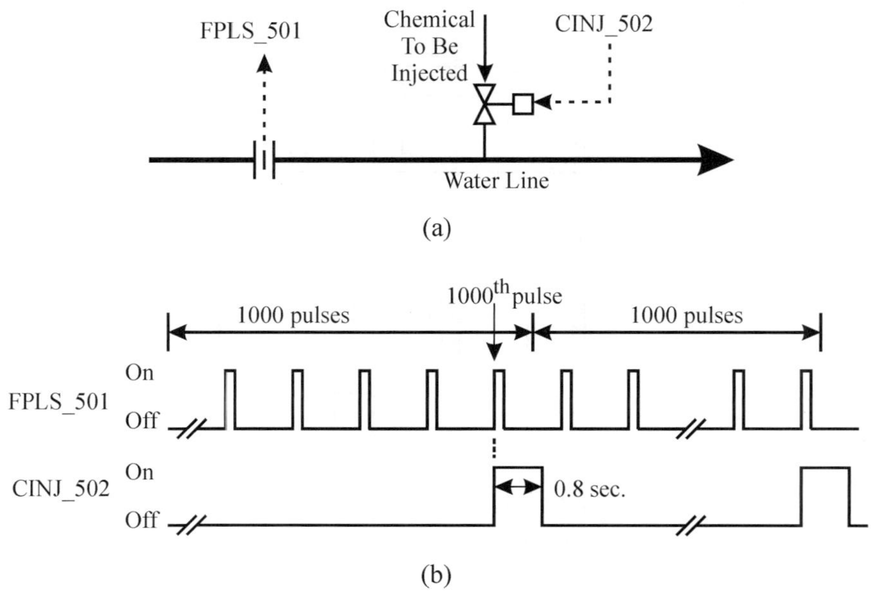

Figure P5.26. Chemical injection system: (a) system layout; (b) timing diagram.

A RSTCNT_502 internal coil is provided to your logic in order to reset the count. When RSTCNT_502 is **on**, the number of liters counted is set to zero and the chemical is injected 1000 liters after RSTCNT_502 is pulsed **on**. RSTCNT_502 is on for only one scan.

Checking for the 1000 pulses starts when the PLC is switched to Run mode for the first time. No special initialization must be needed. RSTCNT_502 can be used to reset the count, but your logic must not wait for RSTCNT_502 to start checking.

Assume the following physical input, physical output, and internal variable tags/variables/symbols:

Tag/Var./Symbol	Description
FPLS_501	Water flow pulse signal. Pulses **on** for each liter of water that flows through the main water line.
CINJ_502	Chemical additive valve control. **On** for 0.8 seconds to inject chemical.
RSTCNT_502	Reset amount count. **On** for only one scan. **Your logic does not control this internal coil**. Treat it as a physical input.

The addresses and/or data types associated with the physical I/O and internal tags/variables/symbols are:

Tag/Var./Symbol	ControlLogix	Micro800	MLogix	SLC-500
FPLS_501	Local:1:I.Data.0	_IO_EM_DI_00	I:0/0	I:1/0
CINJ_502	Local:2:O.Data.0	_IO_EM_DO_00	O:0/0	O:2/0
RSTCNT_502	BOOL	BOOL	B3/123	B3/123

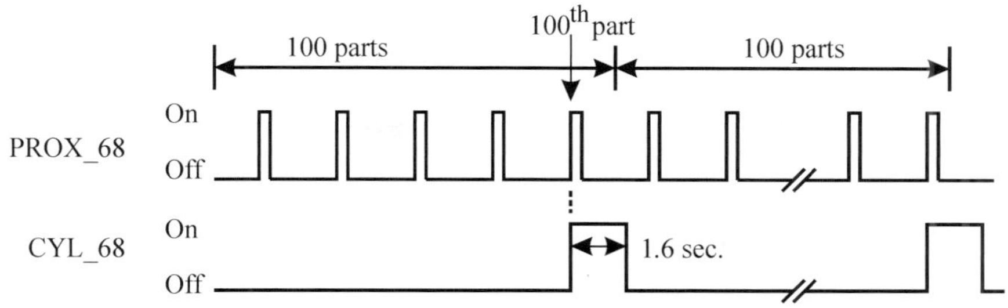

Figure P5.27. Timing relationship between PROX_68 (input) and CYL_68 (output).

P5-27. Use ladder logic to implement a simple quality check station that pushes parts off a conveyor into a check station.

Every 100th part needs to be pushed off a conveyor into an inspection machine where it is checked for manufacturing defects. To push off the part, a pneumatic cylinder must be activated for 1.6 seconds to remove the part from the conveyor. After removing a part, your logic resets the count and removes the next 100th part.

The presence of a part is sensed by the PROX_68 input, which pulses **on** for 0.25 seconds as the part passes. The parts are spaced a minimum of 2 seconds apart on the conveyor. The pneumatic cylinder control, CYL_68, is pulsed **on** for 1.6 seconds in order to remove the part from the conveyor. The relationship between the PROX_68 input and the CYL_68 output is shown in Figure P5.27.

A RSTCNT_68 internal coil is provided to reset the count. When RSTCNT_68 is pulsed **on**, the number of parts counted is set to zero and the 100th part after RSTCNT_68 is pulsed **on** is removed from the conveyor.

Assume the following physical input, physical output, and internal tags/variables/symbols:

Tag/Var./Symbol	Description
PROX_68	Part presence signal. Pulses **on** as part passes.
CYL_68	Removal cylinder control. **On** for 1.6 seconds to remove part from conveyor.
RSTCNT_68	Reset part count. **On** for at least one scan, but no longer than 0.1 sec.

Addresses and/or data types associated with the physical I/O and internal tags/variables/symbols are:

Tag/Var./Symbol	ControlLogix	Micro800	MLogix	SLC-500
PROX_68	Local:1:I.Data.0	_IO_EM_DI_00	I:0/0	I:1/0
CYL_68	Local:2:O.Data.0	_IO_EM_DO_00	O:0/0	O:2/0
RSTCNT_68	BOOL	BOOL	B3/123	B3/123

P5-28. Use ladder logic to implement a *bearing oil injection system* described below.

Every 8 hours of pump operation, generate 2 pulses to inject oil into the two pump bearings. Figure P5.28 shows the timing diagram.

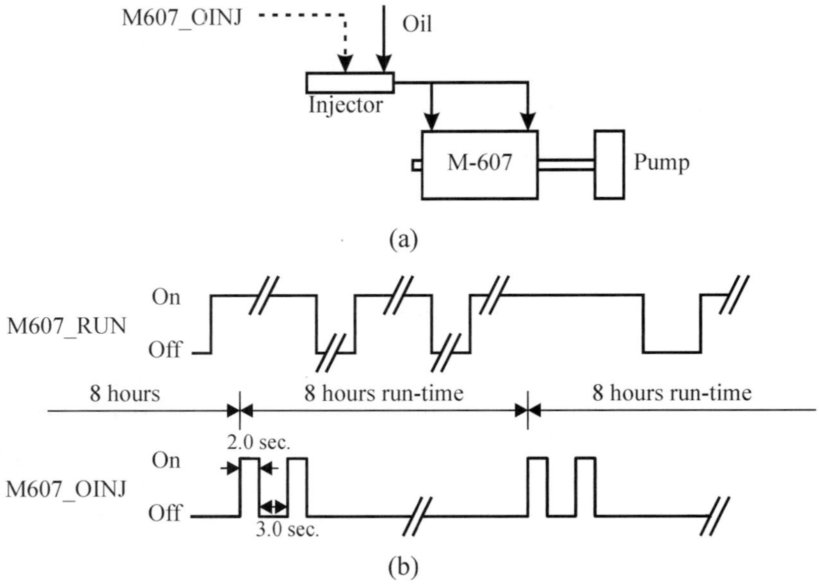

Figure P5.28. Oil injection system: (a) system layout; (b) timing diagram.

Every 8 hours of pump operation (indicated by M607_RUN), generate 2 pulses of the M607_OINJ signal to inject oil into the bearings. The M607_OINJ pulses should be **on** for 2.0 seconds and **off** for 3.0 seconds for 2 times. For each on-off period, the M607_OINJ output should be **on** for 2.0±0.01 second, and **off** for 3.0±0.01 second. When the oil is not being injected, M607_OINJ should remain **off**.

The logic used to track the time the pump is running should accumulate time only when the pump is running. The pump operation time needs to be no less accurate than to the nearest second. Also, the next 8-hour interval should be started at the end of the previous interval. Do not wait for the injection pulses to be complete before starting the 8-hour interval. One cannot assume the motor remains running during the pulses.

The ladder logic does not require a separate reset switch. Anything that should be reset at the end of the 2 pulses should be handled automatically. Also, assume the first 8-hour interval starts when the PLC starts running for the first time. No special initialization must be needed.

Assume the following physical I/O tags/variables/symbols:

Tag/Var./Symbol	Description
M607_RUN	Pump motor run status: **on** when pump running, **off** otherwise.
M607_OINJ	Oil injection control, must be pulsed as described above.

The addresses associated with the physical I/O are:

Tag/Var./Symbol	ControlLogix	Micro800	MLogix	SLC-500
M607_RUN	Local:1:I.Data.0	_IO_EM_DI_00	I:0/0	I:1/0
M607_OINJ	Local:2:O.Data.0	_IO_EM_DO_00	O:0/0	O:2/0

P5-29. Use ladder logic to implement a *warning indicator* described below.

Generate a flashing light indicator that signifies that 5 defective parts are in a reject station (Figure P5.29).

Rejected parts are pushed off the conveyor onto a small roller conveyor (reject station), which can only hold 7 parts. When there are at least 5 parts in the reject station, the ALM_LMP_25R should flash **on** for 0.3 seconds and **off** for 0.7 seconds until it is acknowledged by pressing an "alarm acknowledge" (ALM_ACK_25R) pushbutton.

When the number of parts in the reject bin changes from 4 to 5, the "on" interval of the ALM_LMP_25R lamp should start immediately.

When ALM_ACK_25R is **on**, the ALM_LMP_25R should immediately be turned **off** and the stored count of defective parts zeroed. ALM_LMP_25R should remain off until the number of parts changes from 4 to 5. ALM_ACK_25R should be ignored if there are less than 5 parts in the reject bin. Actually, the alarm acknowledge push button exists on an HMI touch screen, which turns **on** the ALM_ACK_25R internal coil when the screen icon is pressed.

PROX_25R senses a part as it is pushed into the reject station. It turns **off** and then back **on** (a pulse) as a part is pushed into the defect station.

Assume the following physical input, physical output and internal tags/variables/symbols:

Tag/Var./Symbol	Description
PROX_25R	Photoelectric eye. Pulses **off** as part is pushed into the defect station.
ALM_LMP_25R	Alarm lamp, **on** to light lamp for reject station.
ALM_ACK_25R	Alarm acknowledge internal coil. **On** to stop flashing light and zero count of items in reject bin. **Your ladder must have no coil referring to this tag/symbol.**

Addresses and/or data types associated with the physical I/O and internal tags/variables/symbols are:

Tag/Var./Symbol	ControlLogix	Micro800	MLogix	SLC-500
PROX_25R	Local:1:I.Data.0	_IO_EM_DI_00	I:0/0	I:1/0
ALM_LMP_25R	Local:2:O.Data.0	_IO_EM_DO_00	O:0/0	O:2/0
ALM_ACK_25R	BOOL	BOOL	B3/123	B3/123

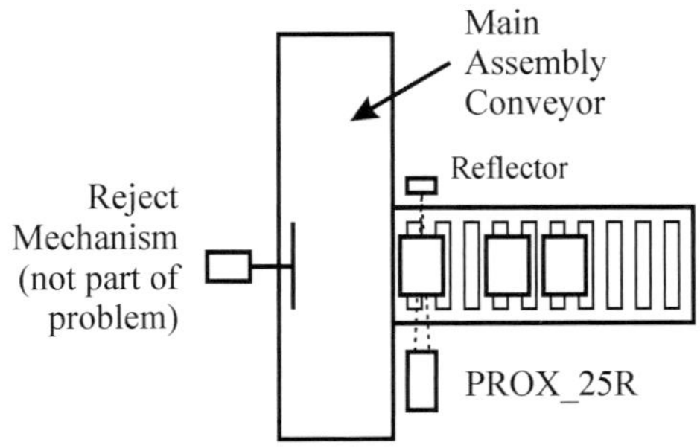

Figure P5.29. Part reject station.

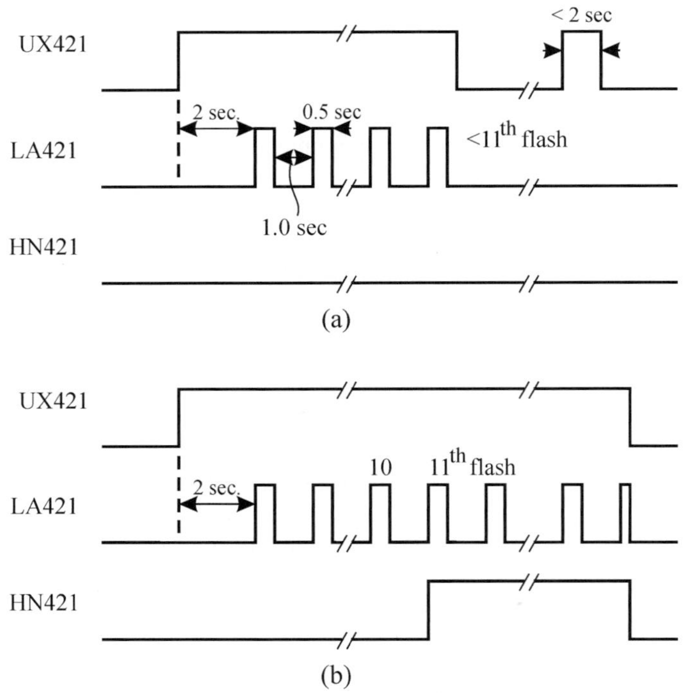

Figure P5.30. Timing diagram when system is armed: *(a)* person leaves before 11th flash; *(b)* horn sounding.

P5-30. Use ladder logic to implement a simple *security warning* system described below.

A certain area of the plant contains secure materials and so a visual and auditory warning should be installed to indicate the danger when the system is armed and a person is approaching the area. Your job is to add that function to the PLC.

The security warning system only sounds when it is armed. This feature allows authorized personnel to come into the secure area without triggering the warning. The system is armed when the AR421 is **on**. You are not concerned with how AR421 is generated.

With the system armed, when ultrasonic proximity UX421 is sensed **on** for 2 seconds, indicating a person is approaching this hazardous area, the LA421 lamp must be flashed. The period of the lamp flash should be 0.5 second **on** and 1.0 seconds **off**. The flashing should continue until the UX421 sensor is **off** (indicating the person has left). The first lamp flash should commence at the end of the 2 second delay interval and not 1.0 seconds later. The timing is shown in Figure P5.30

In addition, if the person is still present after the 10th flash, the HN421 horn must sound and remain **on** as long as the person is still present. The horn starts its sounding at the beginning of the 11th flash.

Assume the following physical I/O tags/variables/symbols:

Tag/Var./Symbol	Description
AR421	System armed indication, N. O., **on** enables the warning system. **Off** disables the warning.
UX421	Ultrasonic proximity sensor that detects a person close to hazardous area, N. O., **on** to indicate presence.
LA421	Hazardous area approach lamp indication. Flashes 0.5 sec. **on** and 1.0 sec **off** when a person is close for more than 2 seconds.
HN421	Hazardous area approach horn indication. **On** to sound horn. Starts sounding at start of 11th flash of LA421 if person is still present.

The addresses associated with the physical I/O are:

Tag/Var./Symbol	ControlLogix	Micro800	MLogix	SLC-500
AR421	Local:1:I.Data.0	_IO_EM_DI_00	I:0/0	I:1/0
UX421	Local:1:I.Data.1	_IO_EM_DI_01	I:0/1	I:1/1
LA421	Local:2:O.Data.0	_IO_EM_DO_00	O:0/0	O:2/0
HN421	Local:2:O.Data.1	_IO_EM_DO_01	O:0/1	O:2/1

P5-31. Use ladder logic to implement a *tornado siren warning indicator* described below.

Generate a siren warning indicator that sounds a pattern while the "tornado siren" switch is **on**. Figure P5.31 shows the timing.

When the XA_1000 tornado siren switch is **on**, generate an alarm signal that sounds a siren 3 times, is silent for 10 seconds, and then sounds 3 more times. The siren continues its pattern as long as XA_1000 is **on**. The siren control (XV_1001) should be **on** for 3.0 seconds and **off** for 4.0 seconds for 3 times. For each on-off period, the horn should be **on** for 3.0±0.01 seconds, and **off** for 4.0±0.01 seconds. If XA_1000 is still **on** after the 10 second silent interval after the siren has sounded 3 times, then the 3 "honks" should resume, followed by 10 more seconds of silence. Make no assumption about the length of time XA_1000 is **on** (except that it must be **on** for one scan to initially trigger the horn).

When the XA_1000 switch turns **on**, the siren (controlled by XV_1001) should start sounding immediately. There should not be an additional delay after the XA_1000 switch turns **on**.

If XA_1000 turns **off** during the siren sounding, the siren should continue until it sounds 3 times.

The ladder logic does not require a separate reset switch. Anything that should be reset at the end of the 3 siren soundings should be handled automatically.

Assume the following physical I/O tags/variables/symbols:

Tag/Var./Symbol	Description
XA_1000	Tornado alarm enable switch, if **on** the siren needs to be sounded
XV_1001	Siren horn control valve, **on** to sound siren.

The addresses associated with the physical I/O are:

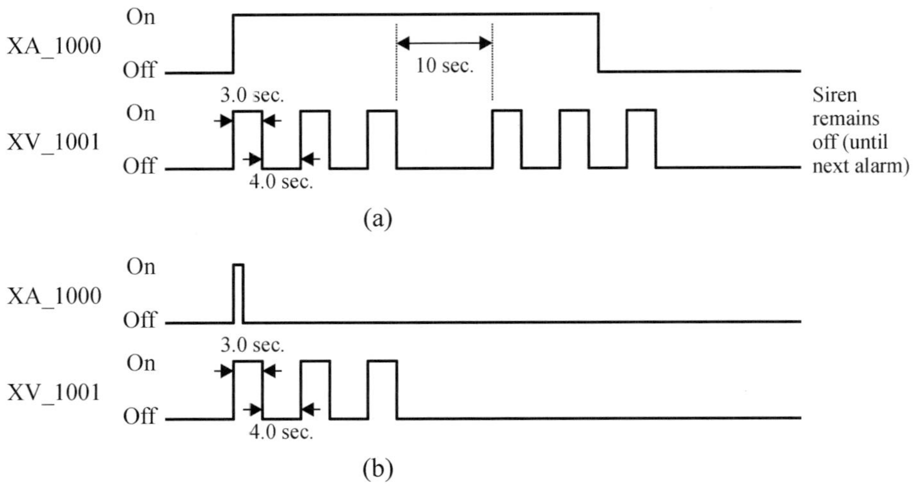

Figure P5.31. Tornado alarm siren (a) long XA_1000; (b) short XA_1000.

Tag/Var./Symbol	ControlLogix	Micro800	MLogix	SLC-500
XA_1000	Local:1:I.Data.0	_IO_EM_DI_00	I:0/0	I:1/0
XV_1001	Local:2:O.Data.0	_IO_EM_DO_00	O:0/0	O:2/0

P5-32. Use ladder logic to implement the following simple tank level control.

Figure P5.32 shows the schematic diagram of a tank with level sensor that is supplied by a pump, T100_PUMP. There is only one level sensor, T100_LEVEL. Currently, the pump is controlled directly by the level sensor. When the RUN switch is **on** and the level sensor is **off**, the pump is **on**. However, the pump motor wears out quickly due to excessive cycling (short on, short off periods). You need to do a fair job of level control, and yet not wear out the motor. Therefore, the pump motor control is to be changed to the following:

When the RUN switch is **on** and the level sensor is **off** for 1 minute, turn **on** the pump motor. When the level sensor is **on** for 2 minutes, or the RUN switch is turned **off**, turn **off** the pump motor.

In addition, as an aid for maintenance personnel, the number of pump cycles is counted, and P_CYC_ALM is turned **on** at the beginning of the 1001st cycle. A cycle is defined as pump **on** and pump **off**. A reset switch, PCYC_RESET, is provided to reset (turn **off** P_CYC_ALM and reset the counter). The counter should not be reset unless the P_CYC_ALM is already **on**.

Assume the following physical I/O tags/variables/symbols:

Tag/Variable	Description
RUN	Pump run toggle switch, **on** to run, **off** to stop.
T100_LEVEL	Level switch, N. O., **on** when fluid at or above proper level, **off** when fluid below proper level.
PCYC_RESET	Reset for pump cycle alarm, **on** to reset alarm and reset counter.
T100_PUMP	Pump motor, **on** to run, **off** to stop.

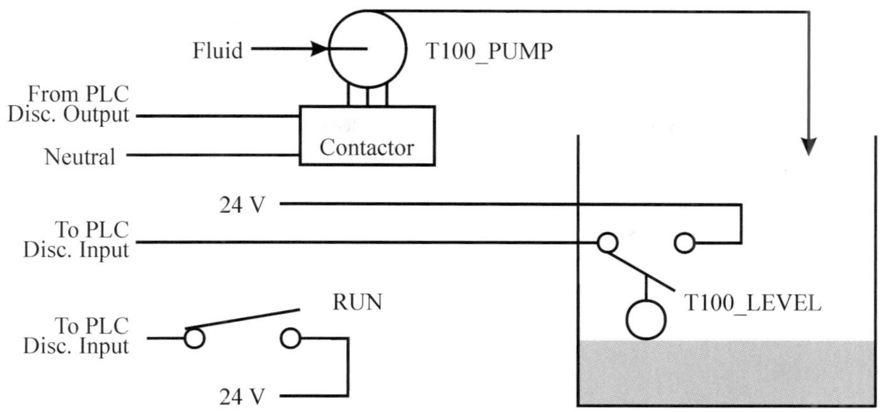

Figure P5.32. Simple tank level control.

P_CYC_ALM Pump cycle alarm, on to indicate 1000 cycles have run.

The addresses associated with the physical I/O are:

Tag/Var./Symbol	ControlLogix	Micro800	MLogix	SLC-500
RUN	Local:1:I.Data.0	_IO_EM_DI_00	I:0/0	I:1/0
T100_LEVEL	Local:1:I.Data.1	_IO_EM_DI_01	I:0/1	I:1/1
PCYC_RESET	Local:1:I.Data.3	_IO_EM_DI_03	I:0/3	I:1/3
T100_PUMP	Local:2:O.Data.0	_IO_EM_DO_00	O:0/0	O:2/0
P_CYC_ALM	Local:2:O.Data.2	_IO_EM_DO_02	O:0/2	O:2/2

P5-33. Use ladder logic to implement a *baggage carousel start warning* that warns that the baggage carousel is about to start and then starts the carousel.

Generate horn and light indicators that sound/flash a pattern when the baggage handler presses the C101_START button. At the end of the pattern, the baggage carousel motor control, M101, is turned **on** and remains **on** until the C101_STOP button is pressed. Figure P5.33 illustrates the operation.

When the C101_START button turns **on**, generate an alarm horn and lamp signal that honks/flashes 6 times and then is silent until the next time the carousel is started. The alarm horn (C101_AHRN) and lamp (C101_ALMP) should both be **on** for 1.0 seconds and **off** for 1.0 seconds for 6 times. For each on-off period, the horn should be **on** for 1.0±0.01 second, and **off** for 1.0±0.01 second.

If C101_START is still **on** after the 6th honk/flash, the horn and lamp should remain **off**. The horn and lamp are retriggered the next time C101_START turns **on** after the 6 honks are complete. Make no assumption about the length of time C101_START is **on** (except that it must be on for at least one scan to initially trigger the horn). If C101_START turns **off** during the honking/flashing, the honks/flashes should continue until 6 honks have sounded.

If C101_STOP is **off**, the carousel motor is stopped. If C101_STOP is **off** during the warning, the honks/flashes should stop immediately.

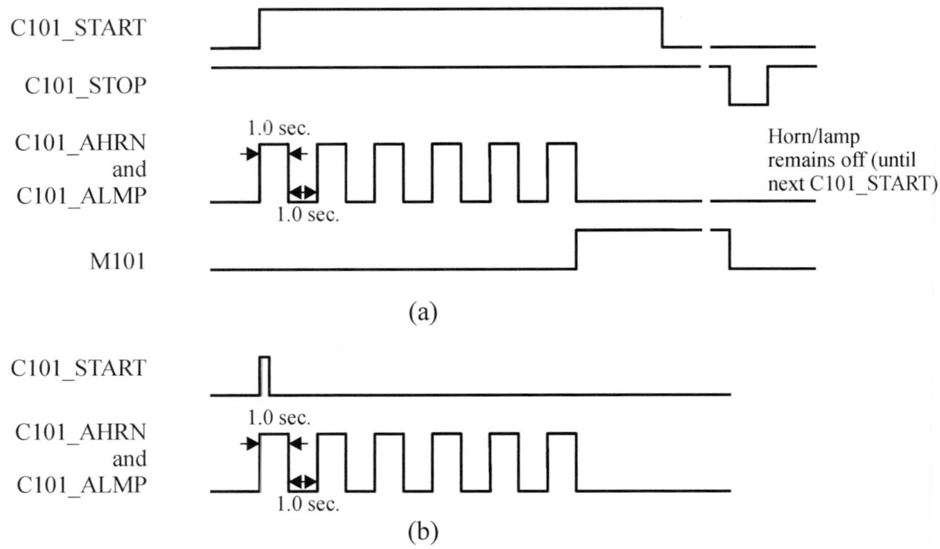

Figure P5.33. Baggage carousel start warning (a) long C101_START; (b) short C101_START.

When the C101_START switch turns **on**, C101_AHRN should start sounding immediately and C101_ALMP should start flashing immediately. There should not be an additional delay.

The ladder logic does not require a separate reset switch. Anything that should be reset at the end of the 6 honks/flashes should be handled automatically by your ladder logic.

Hint: You will find the transitional contact useful.

Assume the following physical input and physical output tags/variables/symbols:

Tag/Var./Symbol	Description
C101_START	Carousel start pushbutton, N.O. switch, **on** to start warning and then start motor after 6 flashes/beeps.
C101_STOP	Carousel stop pushbutton, N.C. switch, **off** to stop warning and motor immediately.
C101_AHRN	Warning horn, **on** to sound horn.
C101_ALMP	Warning lamp, **on** to light lamp.
M101	Carousel motor, **on** to run motor.

The addresses associated with the physical I/O are:

Tag/Var./Symbol	ControlLogix	Micro800	MLogix	SLC-500
C101_START	Local:1:I.Data.4	_IO_EM_DI_04	I:0/4	I:1/4
C101_STOP	Local:1:I.Data.5	_IO_EM_DI_05	I:0/5	I:1/5
C101_AHRN	Local:2:O.Data.13	_IO_EM_DO_13	O:0/5	O:2/0
C101_ALMP	Local:2:O.Data.14	_IO_EM_DO_14	O:0/6	O:2/6
M101	Local:2:O.Data.15	_IO_EM_DO_15	O:0/7	O:2/7

6 Sequential Applications

Chapter Topics:

- Function charts
- Simple ladder logic implementation of function charts
- Parallel operations

OBJECTIVES

Upon completion of this chapter, you will be able to:

- Draw a function chart, given the operational description of a sequential process
- Translate the function chart to ladder logic
- Handle pause and reset of the sequential operation

Scenario: Using a proximity sensor to detect material moving into and out of a station.

Commonly, only one sensor is used to detect the presence of material as it moves into a station, is processed, and moves out of the station. As an example, consider the problem of applying a label to each box as it travels on a conveyor, shown in Figure 6.1. A retro-reflective proximity sensor (PROX) is used to detect the presence of the box. Assume PROX turns **on** when the box is in the proper position to have the label applied. When the labeling station is started, the presence of a box must be detected, the label is applied (involving multiple steps), and then the operation is repeated. A chart of the steps of this process is shown in Figure 6.2. The process starts waiting for a box to be detected. When PROX turns **on**, then the machinery applies the label (multiple steps). When the labeling steps are finished, then the station waits for the next box. However, as depicted in Figure 6.2, the operation **does not work!** After applying the label, PROX remains **on**, (box still in

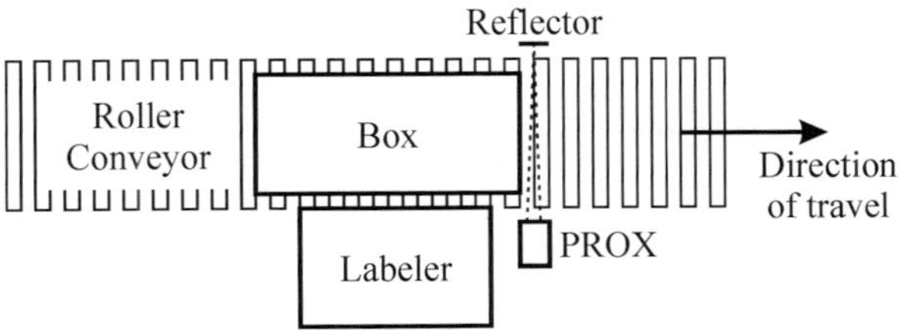

Figure 6.1. Labeling station.

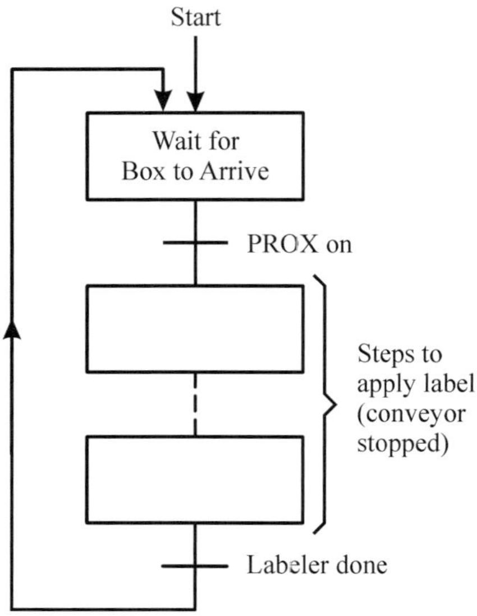

Figure 6.2. Chart of labeling station operation.

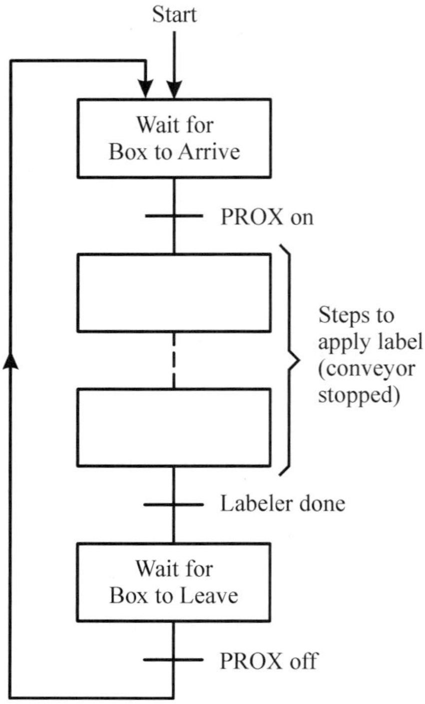

Figure 6.3. Corrected chart of labeling station operation.

station) and so the condition that indicates a new box (PROX **on**) is true. Therefore, a new label is applied to the box before it leaves the station. If left to run unattended, this station will continue to apply multiple labels to the first box that enters the station!

Solution: The station must detect that a labeled box has exited the station before detecting that a new box has entered. Therefore, a step must be added to wait for the box to leave the station, detecting PROX is **off**. The correct chart of this process is shown in Figure 6.3. The moral of this scenario is to remember that one must detect that a proximity sensor is first **off** before it can be detected to be **on**.

> **Design Tip**
>
> When one proximity sensor is used to sense material moving in/out , one must detect that the sensor is **off** before detecting the sensor is **on** (or vice versa).

6.1 INTRODUCTION

With the basic ladder logic contact, timer, and counter instructions, one is able to tackle more significant problems. This chapter introduces ladder logic program design for sequential applications, a significant contribution of the text. More advanced techniques for sequential control are treated in Chapters 9 and 21 of Erickson (2023).

The sequential design technique is based on describing the operation as a function chart and then translating the function chart to ladder logic code. The ladder logic primarily uses the basic contact and coil instructions. Timers and counters are used only when explicitly needed by the operation. The ability to pause and reset an operation is added to the basic sequential design. Operations with parallel steps and machine control involving manual and single-step modes are also considered. The chapter concludes by presenting an unstructured approach to a sequential control solution. Unfortunately, one will often encounter this approach, often when modifying existing programs. The most important disadvantage of this approach is that is requires more time to design, program, and troubleshoot the initial solution and any future modifications.

6.2 FUNCTION CHART

The basic tool used to design sequential control applications is the function chart. This method of describing sequential operations is described in the IEC 848 standard (IEC, 1988) and incorporated as one of the IEC 61131-3 languages (IEC, 2013). The form of the function chart described in this chapter is a simplified version of the IEC 61131-3 SFC (sequential function chart) language. The full IEC sequential function chart language is described in Chapter 14 of Erickson (2023).

The general form of the function chart is shown in Figure 6.4. The function chart has the following major parts:

Steps of the sequential operation,

Transition conditions to move to next step

Actions of each step

The **initial step** is indicated by the double-line rectangle. The initial step is the initial state of the ladder logic when the PLC is first powered up or when the operator resets the operation. The **steps** of the operation are shown as rectangles on the left side of the diagram. Unless shown by an arrow, the progression of the steps proceeds from top to bottom. Each step rectangle contains a short description of what is happening during the step. To the left of the step rectangle is the variable/symbol/tag name of the step-in-progress coil (or bit) that is **on** when that step is active. The **transition condition** is shown to the right of a horizontal bar between steps. If a step is active and the transition condition below that step becomes true, the step becomes inactive, and the next step becomes active. The stepwise flow continues until the bottom of the diagram. At the bottom, the sequencing may end, as indicated by a filled black circle within another circle, or it may repeat by going back to a previous step. The **actions** associated with a step are shown in the rectangle to the right of the step. The actions are output(s) that are **on** when a step is active. Any outputs not listed are assumed to be **off**. However, the set/reset of outputs may be indicated. Any timer or counter active during a particular step is also listed as an action.

The function chart is prepared from the operational description of the system. Often, the hardest part about formulating the function chart is making a distinction between the transitions and the steps. Also, one must remember that physical outputs are actions associated with a step. In order to help in the recognition of the steps and transitions within an operational description, use the following definitions:

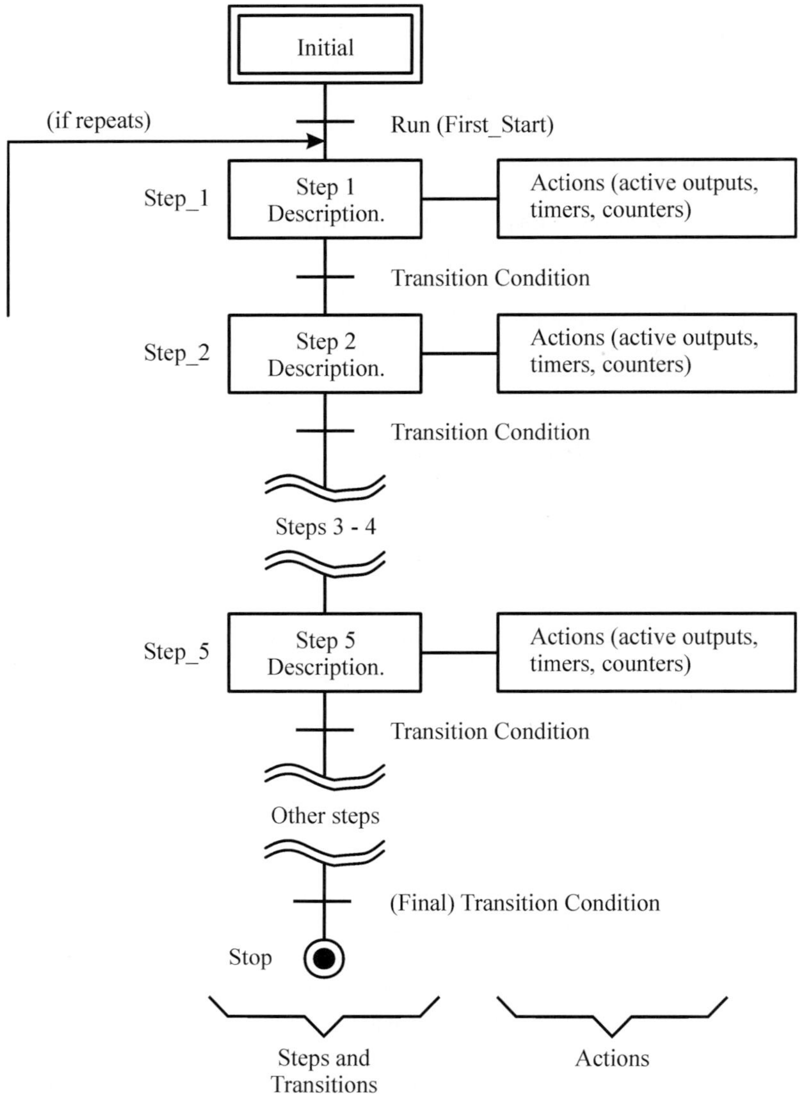

Figure 6.4. General function chart.

Step:

Operation spanning a length of time (however long or short). The time period may be defined or undefined.

Transition:

Physical input device or internal coil turning **on** or **off**

— or —

Timer or counter **done**

A transition condition is recognized when the narrative describes a physical input device or internal coil turning **on** or **off**. Alternatively, the end of a defined time period also signals a transition to the next step. If the narrative describes a physical output being turned on/off, that is not a transition. A physical output is considered a step action and the turning on/off of a physical output is handled by a change in the active step. For example, if "Output1" is being turned on as the active step is changed from "Step1" to "Step2," it is accomplished by **not** listing "Output1" as a "Step1" action and by listing it as a "Step2" action. Output1 is **not** a transition condition. The change in Output1 does not cause a change in the active step, but is a consequence of the change in the active step.

> **Design Tip**
>
> When constructing the function chart, remember that physical outputs **never** occur as part of the transition condition. Also, physical inputs are **never** an action.

These concepts are illustrated by the following example.

Example 6.1. Metal Shear Control. Design the function chart of the program to control the metal shear shown in Figure 6.5 and whose operation is described as:

> The shear cuts a continuous length of steel strip. Two conveyors (driven by CONV1_MTR and CONV2_MTR) move the strip into position. Inductive proximity sensor PROX turns **on** to indicate that the strip is in position to be sheared. When the strip is in position, both conveyors should stop. A hydraulic cylinder (controlled by SHEAR_CYL_RET) is then retracted to move the shear down to cut the material. Limit switch DOWN_LS closes (turns **on**) when the shear is fully down. The cylinder is then extended to move the shear blade up. Limit switch UP_LS closes when the shear blade is fully up. Conveyor 2 (controlled by CONV2_MTR) is now turned **on** to move the cut sheet out of the station. The proximity sensor PROX turns **off** when the sheet has been moved out of the station. Both conveyors are now operated to move the strip into position, and the operation repeats.
>
> Your program is not controlling conveyor 3, so assume it is always running.
>
> The shear is controlled by SHEAR_CYL_RET, a single action linear hydraulic cylinder. Once SHEAR_CYL_RET is energized, the shear blade moves down to cut the material until a mechanical stop is reached and remains in the "down" position as long as power is applied (turned **on**). The shear blade moves up when power is removed (turned **off**).
>
> Upon initial startup, no material is in the shear and the conveyors operate to bring the material into the shearing position (PROX turns **on**). The start switch should have no effect if the process is already running. If the stop switch is pressed at any time, the station operation should pause, except when the shear blade is moving. If the stop switch is pressed when the shear blade is moving, the blade movement must complete. When the start switch is pressed while the operation is paused, the station should resume the suspended step. When the station is paused, the conveyor drive motors should be shut off.

Assume the following physical input and physical outputs:

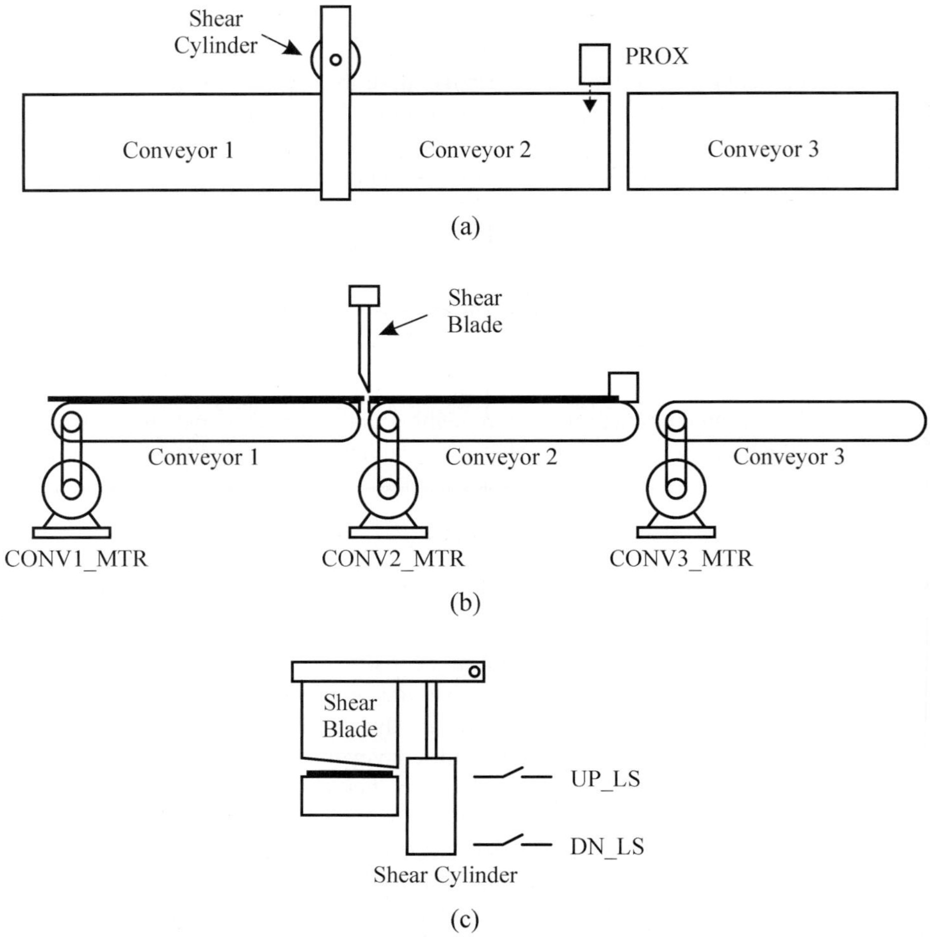

Figure 6.5. Metal shear: *(a)* top view; *(b)* front view; *(c)* side view.

Variable	Description
START_PB	Start push button, N. O., **on** when starting
STOP_PB	Stop push button, N. C., **off** when stopping
PROX	Proximity sensor, **on** when strip in shearing position
DOWN_LS	Limit switch, N. O., **on** (closed) when blade fully down
UP_LS	Limit switch, N. O., **on** (closed) when blade fully up
CONV1_MTR	Conveyor 1 control, **on** to move material on conveyor 1
CONV2_MTR	Conveyor 2 control, **on** to move material on conveyor 2
SHEAR_CYL_RET	Shear cylinder control, **on** to retract cylinder and move blade down

Solution. There are two main steps to develop the function chart:

1. Identify the steps and transition conditions.
2. Add step actions.

To identify the steps and transitions, the first paragraph of the process description is repeated, with the steps identified by the underlined phrases and the transition conditions identified by the *italicized phrases*. Often, it is easier to identify the first transition condition (signaled by an input sensor change) and then recognize the step before and the step after the transition condition. Also, many times the steps and transition conditions alternate during the narrative.

> The shear cuts a continuous length of steel strip. Two conveyors (driven by CONV1_MTR and CONV2_MTR) move the strip into position. Inductive proximity sensor *PROX turns **on*** to indicate that the strip is in position to be sheared. When the strip is in position, both conveyors should stop. A hydraulic cylinder (controlled by SHEAR_CYL_RET) is then retracted to move the shear down to cut the material. Limit switch *DOWN_LS closes (turns **on**)* when the shear is fully down. The cylinder is then extended to move the shear blade up. Limit switch *UP_LS closes* when the shear blade is fully up. Conveyor 2 (controlled by CONV2_MTR) is now turned **on** to move the cut sheet out of the station. The proximity sensor *PROX turns **off*** when the sheet has been moved out of the station. Both conveyors are now be operated to move the strip into position, and the operation repeats.

Notice that the phrase "... both conveyors should stop." is not marked as a transition condition. This phrase describes a physical output being turned on/off and that will be handled by a change in the active step.

So, the steps and the transition conditions that indicate the end of each step are:

Step	Transition Condition (out of step)
Move strip into position	PROX on
Move shear down	DOWN_LS on
Move shear up	UP_LS on
Move cut sheet out	PROX off

These steps and the transition conditions between them are shown in Figure 6.6. The "off" state of PROX that signals the end of the fourth step is shown with the "/" in front of the variable name. The variable name of the step-in-progress bit for each step is also shown beside the step box. This particular operation repeats, indicated by the line from the fourth step back to the first step.

The next part of the function chart development is to add the actions to each step. Reading back through the metal shear narrative, the process actions for each step are:

Step	Action
Move strip into position	CONV1_MTR and CONV2_MTR
Move shear down	SHEAR_CYL_RET
Move shear up	
Move cut sheet out	CONV2_MTR

These actions are added to the steps and the transition conditions to form the function chart shown in Figure 6.7.

The part of the narrative that describes the operation pause is handled in the ladder logic code and is considered in the following section.

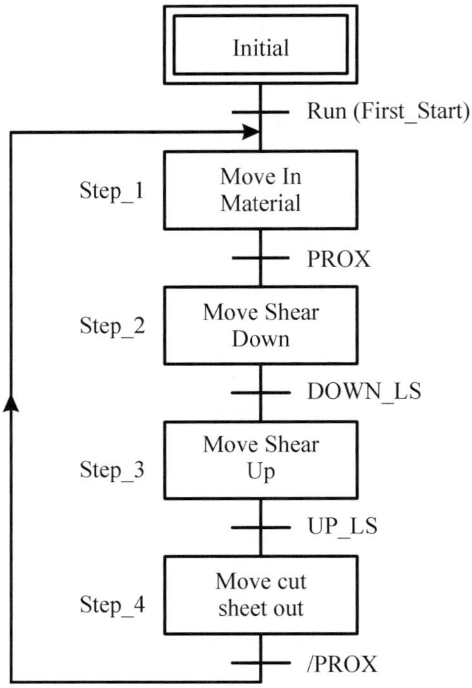

Figure 6.6. Steps and transitions for metal shear.

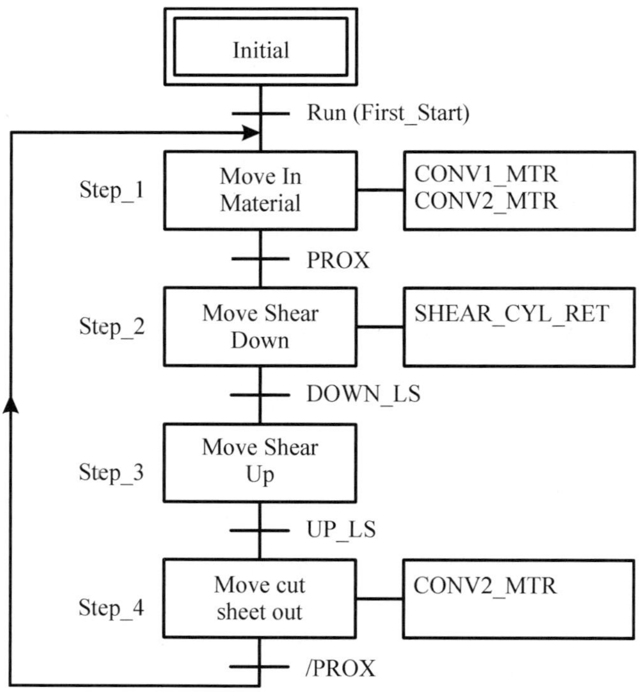

Figure 6.7. Function chart for metal shear.

6.3 IMPLEMENTING FUNCTION CHART IN LADDER LOGIC

Once a function chart has been developed, it needs to be implemented in ladder logic. There are multiple ways to accomplish this task. The design technique described in this chapter utilizes only the basic ladder logic instructions to implement the step and transition logic. Other methods are shown in Chapter 9 of Erickson (2023).

The author calls this method the "cookie cutter" or "template-based" approach because the form of the ladder logic code is the same, regardless of the application. Also, this approach aids in debugging because the logic that handles the transitions and the logic that handles the step actions are distinct. The latter advantage is apparent when comparing this approach to the ad-hoc approach of section 6.8.

The code is broken into the following sections:

Start/stop/pause of overall operation

First start

Transitions between steps

Step actions

Each of these code templates is covered in detail and then applied to the metal shear of Example 6.1.

The start/stop/pause of the overall operation is handled as the rung in Figure 6.8, which is the same general format as the start/stop rung shown in section 2.7. An internal coil (tag/variable/symbol) named Run controls the overall operation of the function chart. It will be used to turn **off** physical outputs that need to be **off** when pausing the operation. The Run may be part of a transition condition. The optional permissive conditions must be satisfied to allow the operation to be started or restarted after an abnormal condition. The optional lockout conditions cause the operation to pause or stop in addition to preventing a restart.

The "first start" transition condition causes the first step of the operation to be initiated when no steps are currently active. The ladder logic to latch (set) the first step (Step_1) is shown in Figure 6.9*a*. When the Run internal tag/variable/symbol is turned **on** (start push button pressed) and no steps are active (Step_N is the last step), the Step_1 internal coil is latched (set). START_PB could be used in place of Run in Figure 6.9, but if the run rung has permissive and/or lockout conditions, these conditions should also be repeated on the rung that starts the operation for the first time. As explained in section 2.7, a change to lockouts and permissives should affect only one rung.

Transitions between steps are handled as shown in Figure 6.10. The logic implements the transition condition below the step in the function chart, which is the transition condition **out** of a step. When the current step is active (Current_Step is **on**) and the transition condition is true, then the step-in-progress bit of the current step is

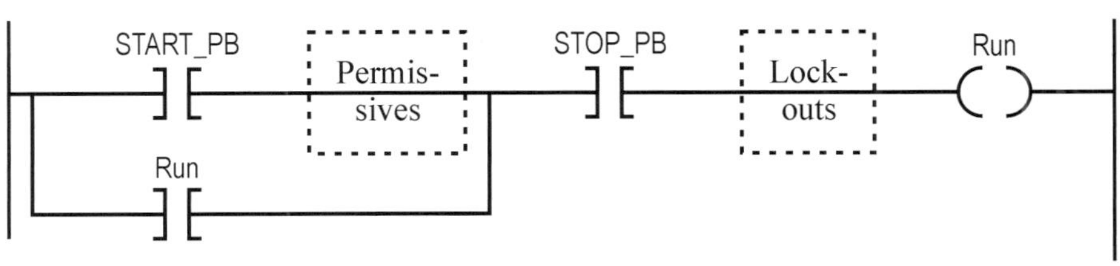

Figure 6.8. General start/stop/pause rung.

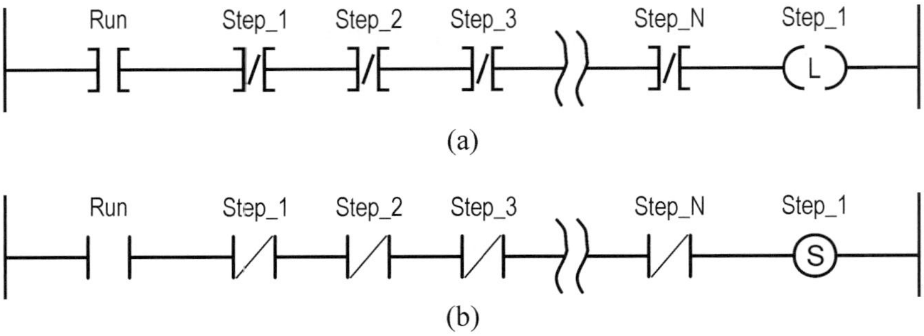

Figure 6.9. General first start rung: *(a)* ControlLogix and MicroLogix; *(b)* Micro800.

unlatched/reset and the step-in-progress-bit of the next step is latched/set. Thus, the next step becomes active and the current step becomes inactive. For the ControlLogix processors, the coils may be in series (as shown in Figure 6.10*a*) or in parallel. For the Micro800 and MicroLogix processors, the coils must be in parallel, as shown in Figures 6.10*b* and 6.10*c*.

The step-in-progress internal coils are used to control the step actions. The appropriate step-in-progress bits turn **on** the outputs and timers that are the step actions. The Run internal coil is also used as part of the condition for those actions that must be **off** when the operation is paused. For example, if the MOTOR_ON output should be **on** in steps 4 and 15 of the function chart (represented by Step_4 and Step_15), then the logic driving MOTOR_ON appears as shown in Figure 6.11. The Run internal coil turns **off**

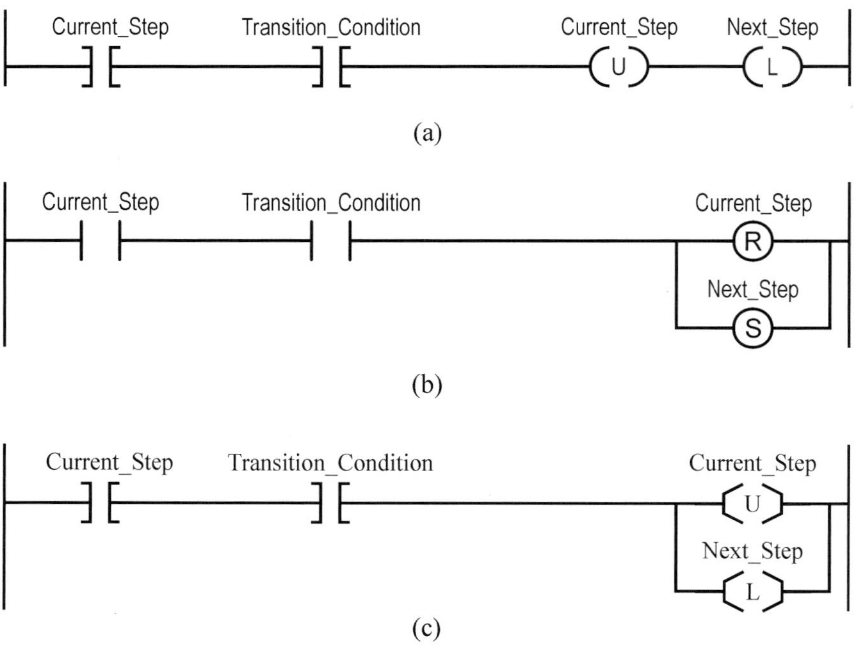

Figure 6.10. General transition between steps: *(a)* ControlLogix; *(b)* Micro800; *(c)* MicroLogix.

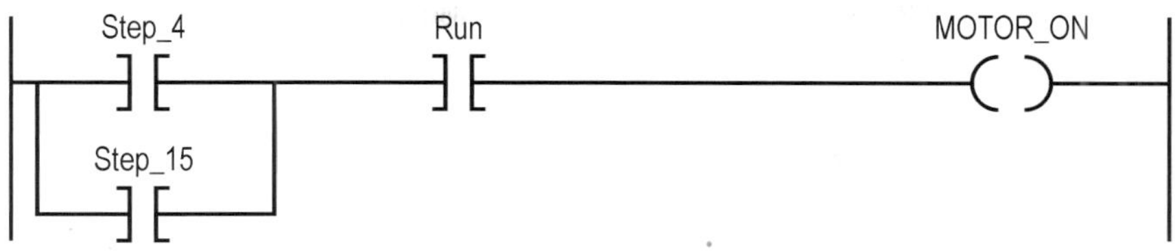

Figure 6.11. Example step action.

MOTOR_ON if step 4 or step 15 is active and the stop push button is pressed to pause the operation. When the operation is resumed (by pressing the start push button), then MOTOR_ON is reactivated. If the Run internal coil is omitted from the rung in Figure 6.11, then MOTOR_ON will remain **on** when the operation is paused when in step 4 or step 15.

Note that in Figure 6.11, the Step_4 and Step_15 step-in-progress bits are in parallel, meaning that MOTOR_ON is an action in steps 4 and 15. The MOTOR_ON output is **off** for any other steps.

If the action associated with a step is a latch/unlatch (set/reset) of an output, then the output coil of Figure 6.11 is replaced by a latch/set or unlatch/reset coil.

Design Tip

Repeating outputs is a common mistake when implementing a function chart where a particular output is the action for more than one step. Consider the output first and then the steps for which it is **on** to avoid repeating output instructions.

Example 6.2. Metal Shear Control. Use ladder logic to implement the metal shear operation described in Example 6.1.

The physical inputs and physical outputs are:

Tag/Var./Symbol	Description
START_PB	Start push button, N. O., **on** when starting
STOP_PB	Stop push button, N. C., **off** when stopping
PROX	Proximity sensor, **on** when strip in shearing position
DOWN_LS	Limit switch, N. O., **on** (closed) when blade fully down
UP_LS	Limit switch, N. O., **on** (closed) when blade fully up
CONV1_MTR	Conveyor 1 control, **on** to move material on conveyor 1
CONV2_MTR	Conveyor 2 control, **on** to move material on conveyor 2
SHEAR_CYL_RET	Shear cylinder control, **on** to retract cylinder and move blade down

The addresses associated with the inputs and outputs:

Tag/Var./Symbol	ControlLogix	Micro800	MLogix	SLC-500
START_PB	Local:1:I.Data.0	_IO_EM_DI_00	I:0/0	I:1/0
STOP_PB	Local:1:I.Data.1	_IO_EM_DI_01	I:0/1	I:1/1
PROX	Local:1:I.Data.2	_IO_EM_DI_02	I:0/2	I:1/2
DOWN_LS	Local:1:I.Data.3	_IO_EM_DI_03	I:0/3	I:1/3
UP_LS	Local:1:I.Data.4	_IO_EM_DI_04	I:0/4	I:1/4

CONV1_MTR	Local:2:O.Data.0	_IO_EM_DO_00	O:0/0	O:2/0
CONV2_MTR	Local:2:O.Data.1	_IO_EM_DO_01	O:0/1	O:2/1
SHEAR_CYL_RET	Local:2:O.Data.2	_IO_EM_DO_02	O:0/2	O:2/2

Solution. The function chart for the shear operation is shown in Figure 6.7. Before developing the ladder logic code, the internal tags/variables/symbols should be identified:

Tag/Var./Symbol	Description
Run	Indicates operation running
Step_1 to Step_4	Step-in-progress bits for steps

The addresses or data types associated with the internal tags/variables/symbols:

Tag/Var./Symbol	CLogix Data Type	Micro800 Data Type	MLogix/SLC Address
Run	BOOL	BOOL	B3/0
Step_1 to	BOOL	BOOL	B20/1 to
Step_4	BOOL	BOOL	B20/4

The ladder logic code is broken into the following sections:

Start/stop/pause of overall operation

First start

Transitions between steps

Step actions

The ControlLogix code for the metal shear, shown in Figure 6.12, is developed using the code templates shown in Figures 6.8 - 6.11. A rung comment is shown within a rectangle above the rung. The function of each rung is as follows:

1. Start/stop/pause of overall operation
2. First start (starting the operation for the very first time)
3. Transition from step 1 to step 2
4. Transition from step 2 to step 3
5. Transition from step 3 to step 4
6. Transition from step 4 to step 1
7. Control of conveyor 1 (an action for step 1)
8. Control of conveyor 2 (an action for steps 1 and 4)
9. Control of shear cylinder (an action for step 2)

The Micro800 ladder logic is similar with set/reset coils in parallel instead of larch/unlatch coils in series. The MicroLogix ladder logic is similar except the latch/unlatch coils are in parallel rather than series.

The initial start of the operation is handled like Figure 6.9*a*. Note the use of the Run in rungs 7 and 8 to turn **off** the conveyors when the station is paused. Since the shear cylinder operation should not stop if the stop push button is pressed while it is moving, Run is not used as a condition in rung 9.

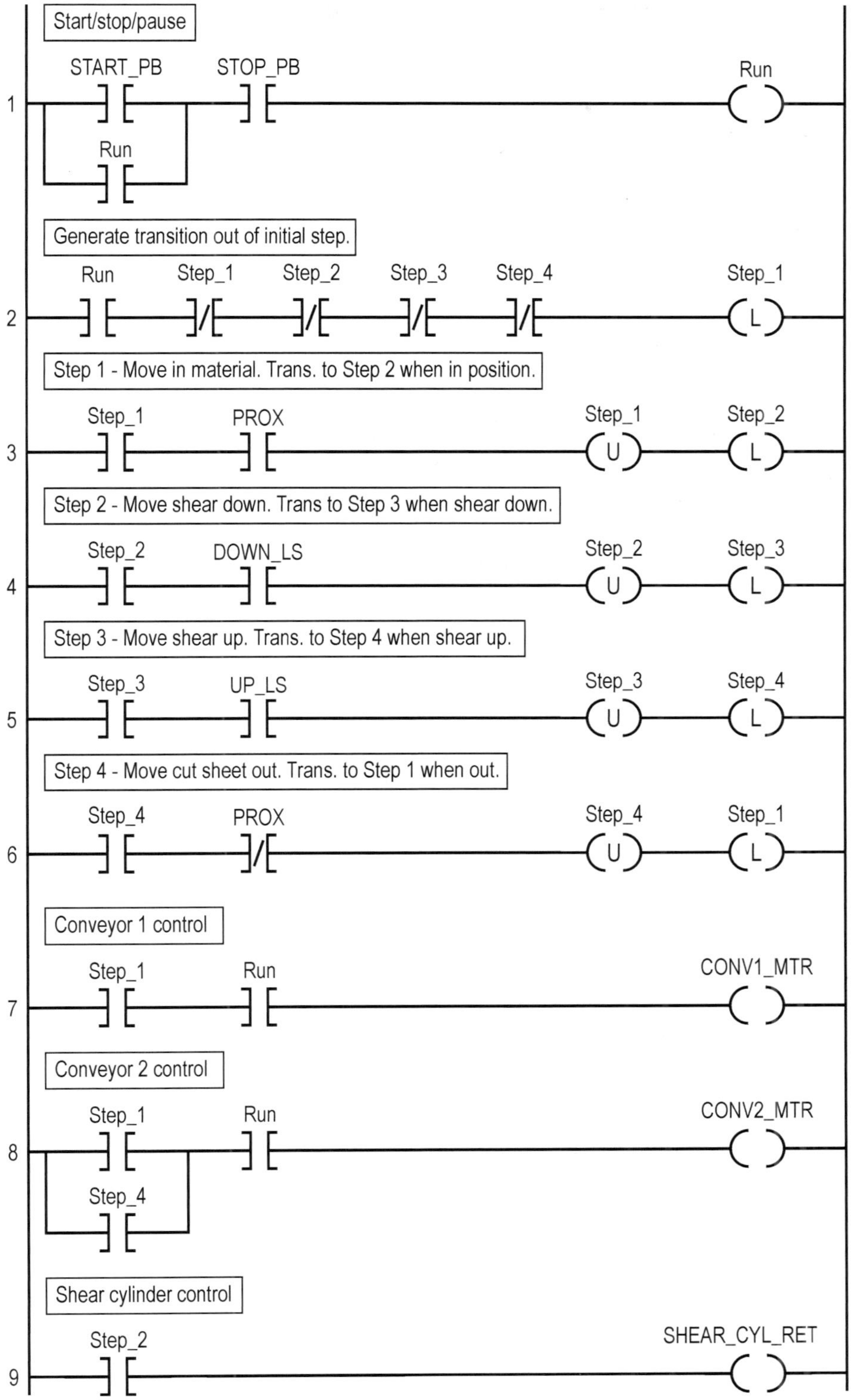

Figure 6.12. Ladder logic for metal shear.

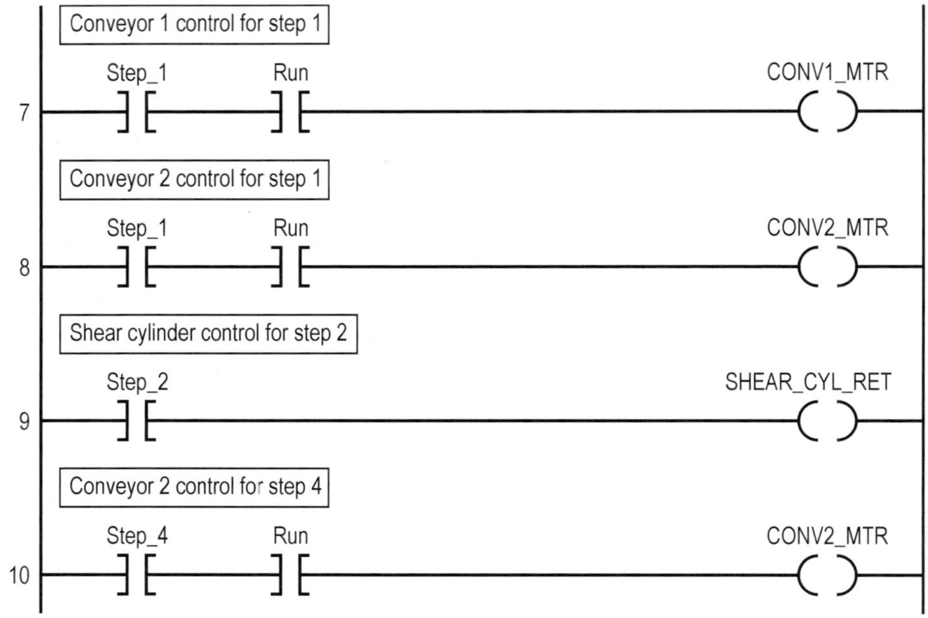

Figure 6.13. Incorrect output logic for metal shear.

The CONV2_MTR output is an action for 2 steps, as shown in rung 8 of Figure 6.12. Note that if a particular output is an action for multiple steps, then the step-in-progress bits of each step are placed in parallel. When a particular output is the action for more than one step, novice programmers often repeat the outputs. If one did not consider the output first and then steps for which it is **on** then the ladder logic driving the physical outputs for the shear may appear as in Figure 6.13. Rungs 8 and 10 both drive the CONV2_MTR output. What is the result? Since rung 10 is scanned after rung 8, the logic of rung 10 will override the logic of rung 8. Consequently, CONV2_MTR is never **on** in step 1, causing the material to jam as it is conveyed into position.

Example 6.2 does not have all of the features of a real application, but serves to illustrate the basic approach to implementing a function chart in ladder logic. The next example adds timers, counters, and reset to an application.

Example 6.3. Tub Loader Control. Design the function chart of the program to control the tub loader described below. Also, implement the control with ladder logic.

> Figure 6.14 shows the layout of a parts tub loader machine. Parts are placed on the belt conveyor by a milling machine. The parts move down the conveyor and drop into the parts tub. Parts on the belt conveyor are detected by a photoelectric sensor, PE272, which is **off** as a part interrupts the beam. Assume PE272 detects the part as it falls into the tub. After 100 parts are deposited in the tub, the tub is moved out and a new, empty tub moves into position. To change the tub, the following operation must take place:
>
> Open Gate 1 (GATE1_OPLS senses when open).

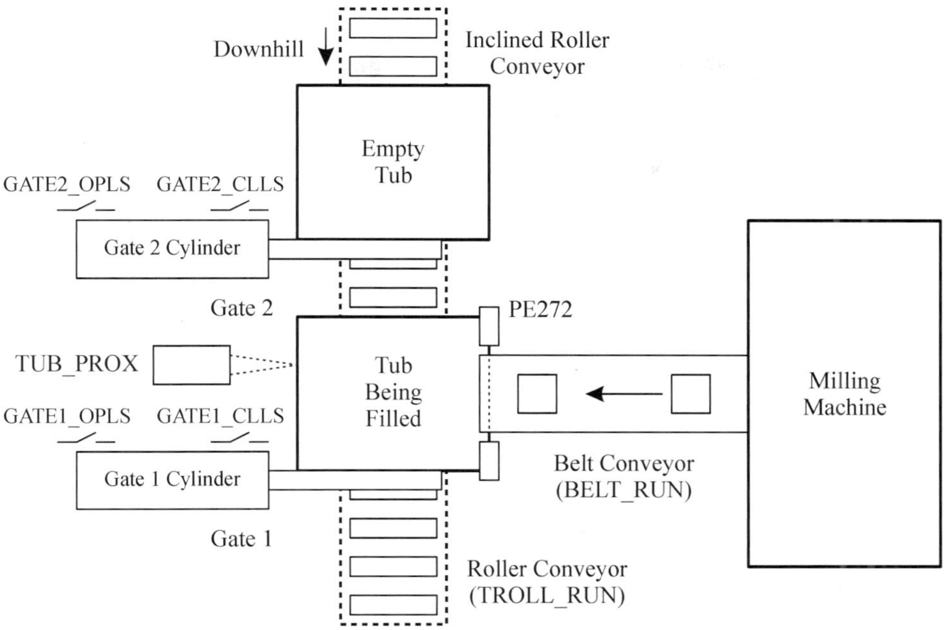

Figure 6.14. Parts tub loading station.

Hold Gate 1 open and wait for TUB_PROX to be **off** for 3 seconds to allow the full tub to be moved out of the loading station. Run the tub roller conveyor to move out the full tub.

Gate 1 is closed (GATE1_CLLS senses when closed).

Gate 2 is opened (GATE2_OPLS senses when open).

Hold Gate 2 open to allow an empty tub to move down a slight incline into the loading station. When the tub contacts the tub roller conveyor, the tub roller conveyor moves the tub into position. When TUB_PROX is **on** for 5 seconds, the tub is in position (front resting on Gate 1).

Gate 2 is closed (GATE2_CLLS senses when closed).

The TUB_PROX proximity sensor is **on** when the tub is present, though not necessarily in position. Hence, the delays ensure the empty tub has moved in and the full one has moved out.

While the tub is being changed, the belt conveyor motor must be stopped (BELT_RUN **off**) and an internal coil, Tub_Permissive, must be turned **off**. After a new tub is in position, BELT_RUN is turned **on**, the Tub_Permissive coil is turned **on**, and the counting of parts is resumed. The Tub_Permissive is used by the milling machine ladder logic. When Tub_Permissive is **on**, the machine produces parts.

The roller conveyor for the tubs has two sections. The section between the two gates and extending out of the station is powered and controlled by the TROLL_RUN output. The roller conveyor section before Gate 1 is unpowered and inclined to allow new tubs to move into the station. In order to completely move the empty tub into the station, the powered section must be running.

Single-action pneumatic cylinders control Gate 1 and Gate 2. Once GATE1_RET is energized, gate 1 opens and remains in the open position as long as power is applied (turned **on**). The gate closes when power is removed (turned **off**). Limit switches GATE1_OPLS and GATE1_CLLS sense the open and closed positions, respectively. Similarly, GATE2_RET controls Gate 2. The GATE2_OPLS and GATE2_CLLS limit switches sense the position of Gate 2.

Single-speed motors drive the two conveyors. When BELT_RUN is **on**, the conveyor moves parts from the milling machine to the tub. When BELT_RUN is **off**, the conveyor is stopped. When TROLL_RUN is **on**, the powered section of the roller conveyor moves. When TROLL_RUN is **off**, the powered section of the roller conveyor is stopped.

There is an internal coil, Run, that is **on** when the operation is enabled. The Run internal coil is set by another part of the ladder logic. When the Run coil is **off**, the tub loading operation should be paused at the current step. When paused, do not advance to the next step. When the Run coil turns **on** while the operation is paused, the tub loader should resume the suspended step. When paused, both conveyors must be stopped, all counter and timer accumulator values must be retained, and the ladder logic program must remain in the step in which the Run coil changed from **on** to **off**. If the Run coil turns **off** when changing tubs, the pneumatic cylinder controls must continue to be activated, holding the gate open (otherwise, a tub may be damaged).

There is another internal coil, Reset, that when **on**, restarts the operation. The Reset internal coil is set by another part of the ladder logic. When Reset is **on**, internal counters and timers are reset and the internal state is set so that the ladder logic program assumes an empty tub is in position. The Reset internal coil must be ignored while Run is **on**.

Assume the following physical input, physical output, and internal coil assignments:

Tag/Var./Symbol	Description
PE272	Photoelectric sensor, **off** when part passes.
TUB_PROX	Proximity sensor, **on** (closed) when tub is present, though not necessarily in position to receive parts.
GATE1_OPLS	Limit switch, **on** (closed) when Gate 1 is open.
GATE1_CLLS	Limit switch, **on** (closed) when Gate 1 is closed.
GATE2_OPLS	Limit switch, **on** (closed) when Gate 2 is open.
GATE2_CLLS	Limit switch, **on** (closed) when Gate 2 is closed.
BELT_RUN	Belt conveyor control, **on** to run conveyor to move parts from milling machine to the parts tub.
TROLL_RUN	Powered roller conveyor control, **on** to run conveyor to move parts tub.
GATE1_RET	Gate 1 cylinder control, **on** to retract cylinder and open gate; **off** closes gate.
GATE2_RET	Gate 2 cylinder control, **on** to retract cylinder and open gate; **off** closes gate.

Run	Internal coil, **on** when loading enabled to operate (controlled by another part of the ladder logic).
Reset	Internal coil, **on** to reset tub loader operation (controlled by another part of the ladder logic).
Tub_Permissive	Internal coil, **on** when milling machine is permitted to run (controlled by this part of the ladder logic).

The addresses associated with the physical inputs and outputs are:

Tag/Var./Symbol	ControlLogix	Micro800	MLogix	SLC-500
PE272	Local:1:I.Data.2	_IO_EM_DI_02	I:0/2	I:1/2
TUB_PROX	Local:1:I.Data.3	_IO_EM_DI_03	I:0/3	I:1/3
GATE1_OPLS	Local:1:I.Data.4	_IO_EM_DI_04	I:0/4	I:1/4
GATE1_CLLS	Local:1:I.Data.5	_IO_EM_DI_05	I:0/5	I:1/5
GATE2_OPLS	Local:1:I.Data.6	_IO_EM_DI_06	I:0/6	I:1/6
GATE2_CLLS	Local:1:I.Data.7	_IO_EM_DI_07	I:0/7	I:1/7
BELT_RUN	Local:2:O.Data.0	_IO_EM_DO_00	O:0/0	O:2/0
TROLL_RUN	Local:2:O.Data.1	_IO_EM_DO_01	O:0/1	O:2/1
GATE1_RET	Local:2:O.Data.2	_IO_EM_DO_02	O:0/2	O:2/2
GATE2_RET	Local:2:O.Data.3	_IO_EM_DO_03	O:0/3	O:2/3

The addresses/data types associated with the internal tags/variables/symbols are:

Tag/Var./Symbol	CLogix Data Type	Micro800 Data Type	MLogix/SLC Address
Run	BOOL	BOOL	B3/100
Reset	BOOL	BOOL	B3/101
Tub_Permissive	BOOL	BOOL	B3/102

Solution. This example introduces the following aspects of sequential problems:

- Using timers
- Using counters
- Using Run as part of the transition condition
- Reset of the operation

As illustrated in Example 6.1, there are two main steps to develop the function chart:

1. Identify the steps and transition conditions.
2. Add step actions.

To identify the steps and transitions, the first paragraph of the process description is repeated, with the steps identified by the underlined phrases and the transition conditions identified by the *italicized phrases*. As in example 6.1, many times it is easier to identify the first transition condition (signaled by an input sensor change) and then recognize the step before and the step after the transition condition. Often, the steps and transition conditions alternate during the narrative.

Figure 6.14 shows the layout of a parts tub loader machine. Parts are placed on the belt conveyor by a milling machine. The parts move down the conveyor and drop into the parts tub. Parts on the belt conveyor are detected by a photoelectric sensor, PE272, which is **off** as a part interrupts the beam. Assume PE272 detects the part as it falls into the tub. After *100 parts are deposited* in the tub, the tub is moved out and a new, empty tub moves into position. To change the tub, the following operation must take place:

Open Gate 1 (*GATE1_OPLS senses* when open).

Hold Gate 1 open and wait for *TUB_PROX to be **off** for 3 seconds* to allow the full tub to be moved out of the loading station. Run the tub roller conveyor to move out the full tub.

Gate 1 is closed (*GATE1_CLLS senses* when closed).

Gate 2 is opened (*GATE2_OPLS senses* when open).

Hold Gate 2 open to allow an empty tub to move down a slight incline into the loading station. When the tub contacts the tub roller conveyor, the tub roller conveyor moves the tub into position. When *TUB_PROX is **on** for 5 seconds*, the tub is in position (front resting on Gate 1).

Gate 2 is closed (*GATE2_CLLS senses* when closed).

Since the timer accumulator values must be retained when paused, retentive timers must be used for the time delays. Also, the Run coil must be one of the conditions that controls the timer.

The sentence, "When paused, do not advance to the next step." normally means that the internal Run coil is part of the transition condition. However, since retentive timers are used for the transition out of the steps holding the gates open, the Run coil is not needed for these steps. One could argue that the Run coil is not needed for the transitions out of the other steps since the conveyors are stopped when paused, but for the purposes of the example, the Run coil is used.

So, the steps and the transition conditions that indicate the end of each step are:

Step	Transition Condition (out of step)
Parts into tub	Part_Ctr done (100 parts detected) and Run
Open Gate 1	GATE1_OPLS on and Run
Hold Gate 1 open	G1_Hold_Tmr done (TUB_PROX off for 3 sec.)
Close Gate 1	GATE1_CLLS on and Run
Open Gate 2	GATE2_OPLS on and Run
Hold Gate 2 open	G2_Hold_Tmr done (TUB_PROX on for 5 sec.)
Close Gate 2	GATE2_CLLS on and Run

The next part of the function chart development is to add the actions to each step. Reading back through the tub loader narrative, the process actions for each step are:

Step	Actions
Parts into tub	BELT_RUN and Tub_Permissive and Part_Ctr (counts 100 parts with /PE272)
Open Gate 1	GATE1_RET
Hold Gate 1 open	GATE1_RET and TROLL_RUN and G1_Hold_Tmr (3 sec.)
Close Gate 1	
Open Gate 2	GATE2_RET
Hold Gate 2 open	GATE2_RET and TROLL_RUN and G2_Hold_Tmr (5 sec.)
Close Gate 2	

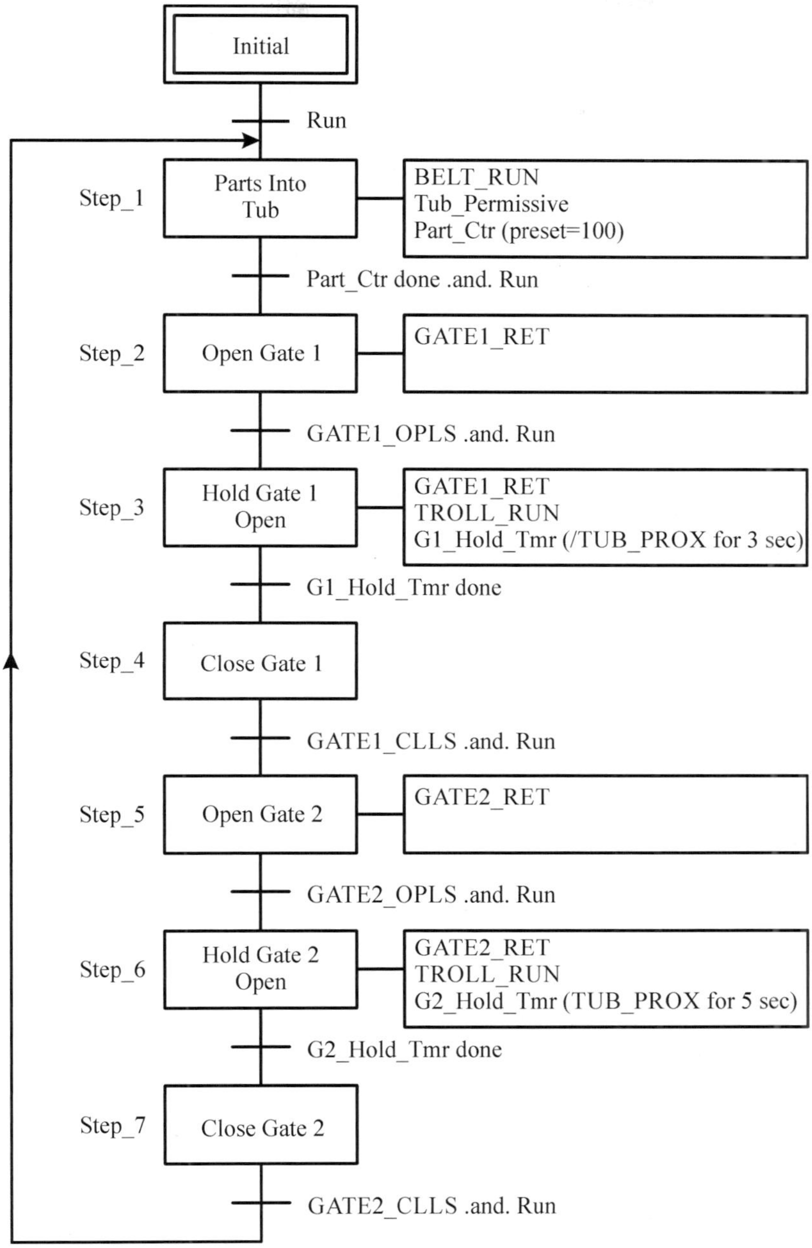

Figure 6.15. Function chart for parts tub loader.

The function chart for the tub loader is shown in Figure 6.15. This particular operation repeats, indicated by a line from the last step back to the first step. Before developing the ladder logic code, the internal tags/variables/symbols should be identified.

The addresses or data types associated with the tags/variables/symbols:

Tag/Var./Symbol	CLogix Data Type	Micro800 Data Type	MLogix/SLC Address
Step_1 to	BOOL	BOOL	B20/1 to
Step_7	BOOL	BOOL	B20/7
Part_Ctr	COUNTER	CTU	C5:1
G1_Hold_Tmr	TIMER	RTO	T4:1
G2_Hold_Tmr	TIMER	RTO	T4:2
Part_Ctr_Rst	n/a	BOOL	n/a
G1_Hold_Tmr_Rst	n/a	BOOL	n/a
G2_Hold_Tmr_Rst	n/a	BOOL	n/a

The ladder logic code is broken into the following sections:

Start/stop/pause of overall operation
First start
Transitions between steps
Step actions

Since the timers and counters are shown as actions, they may be placed with the rungs that drive the physical outputs. However, since they are also part of the transitions, they may also be placed with the rungs handling the transitions. The author favors the latter approach since the transition condition is more likely to be changed.

The ControlLogix ladder logic for the tub loader, shown in Figure 6.16, is developed using the code templates shown earlier in this chapter. A rung comment is shown within a rectangle above the rung. The function of each rung is as follows:

1. First start (starting the operation for the very first time)
2. Transition from step 1 to step 2 and counting parts
3. Transition from step 2 to step 3
4. Transition from step 3 to step 4 and delay tub prox. off
5. Transition from step 4 to step 5
6. Transition from step 5 to step 6
7. Transition from step 6 to step 7 and delay tub prox. on
8. Transition from step 7 to step 1
9. Reset of steps
10. Control of belt conveyor (an action for step 1)
11. Control of roller conveyor (an action for steps 3 and 6)
12. Control of gate 1 cylinder (an action for steps 2 and 3)
13. Control of gate 2 cylinder (an action for steps 5 and 6)
14. Control of Tub_Permissive (an action for step 1)

When the reset push button is pressed while the station is paused, all step-in-progress coils are reset. This action effectively places the station operation in the initial state.

Logic is connected to the right of the RTO timer blocks in rungs 4 and 7 to accomplish the transition to the next step, but a contact for the .DN Boolean must still be present, as the output of a timer block is the .EN Boolean, not the .DN Boolean. The Run internal coil is part of the input condition for each retentive timer, thus pausing the timer when the station

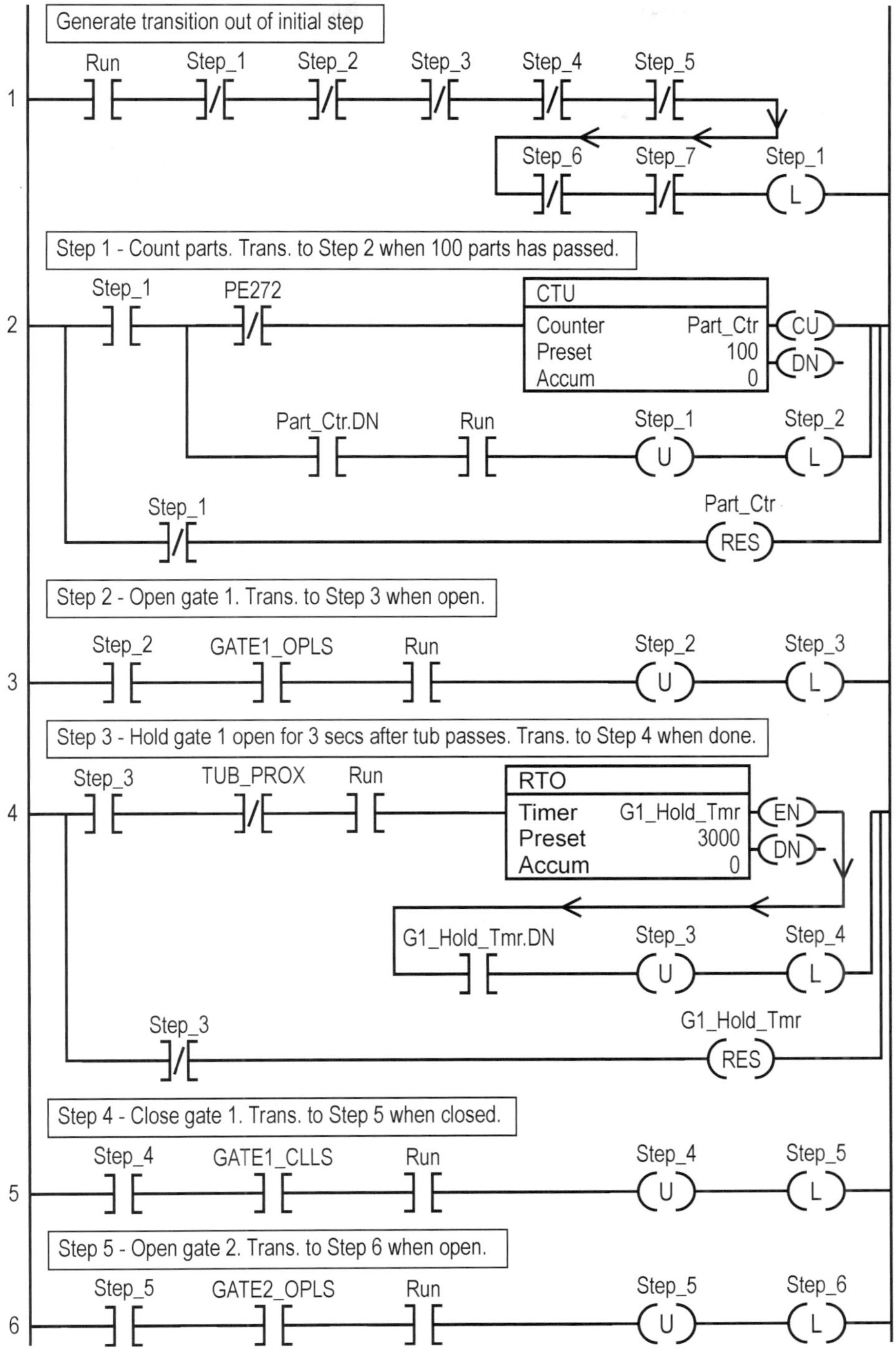

Figure 6.16. ControlLogix ladder logic for tub loader. *(continued)*

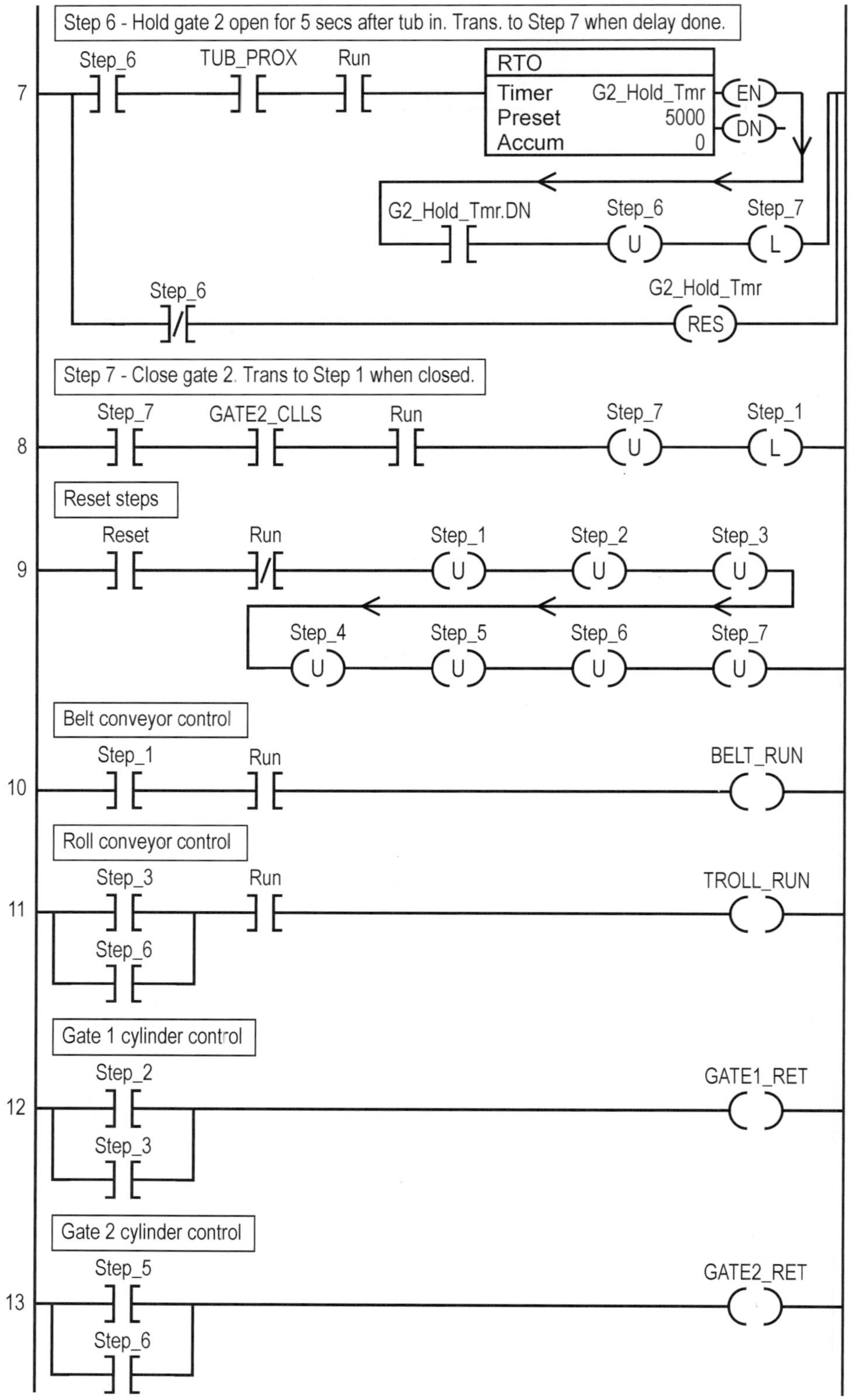

Figure 6.16. *(continued)*

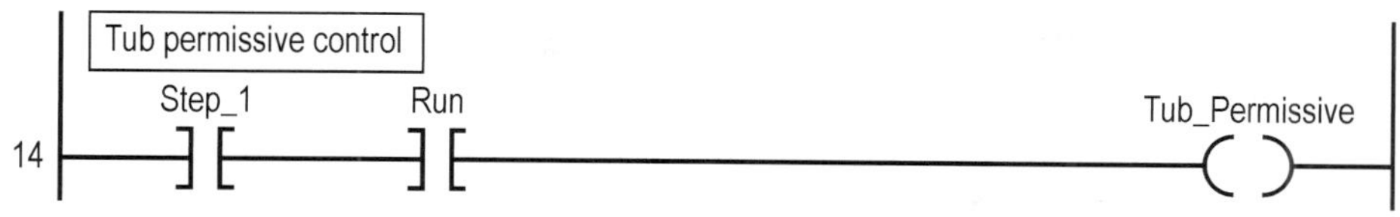

Figure 6.16. *(continued)*

operation is paused. The transition logic for a step can be connected to the right of a CTU block. However, this approach is not good practice and does not work for Step_1 in this application due to the Run contact that is also a part of the transition condition. If the logic that is parallel to the CTU in rung 2 is placed in series with the CU output of the CTU, and the operation is paused when Part_Ctr.DN is on, PE272 must remain off while the operation is paused. In general, this cannot be guaranteed, as the belt conveyor's inertia may carry the part past PE272 while it is stopping, and PE272 turn back on, blocking the transition when the operation is resumed.

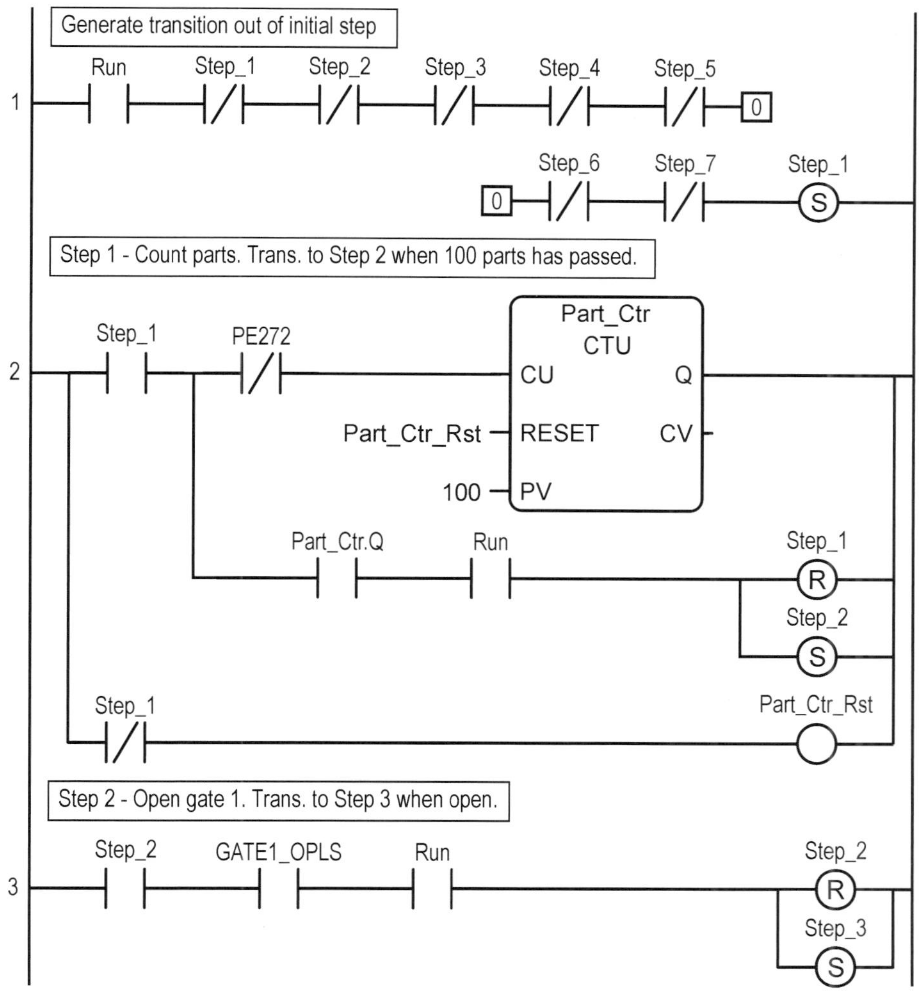

Figure 6.17. Micro800 ladder logic for tub loader. *(continued)*

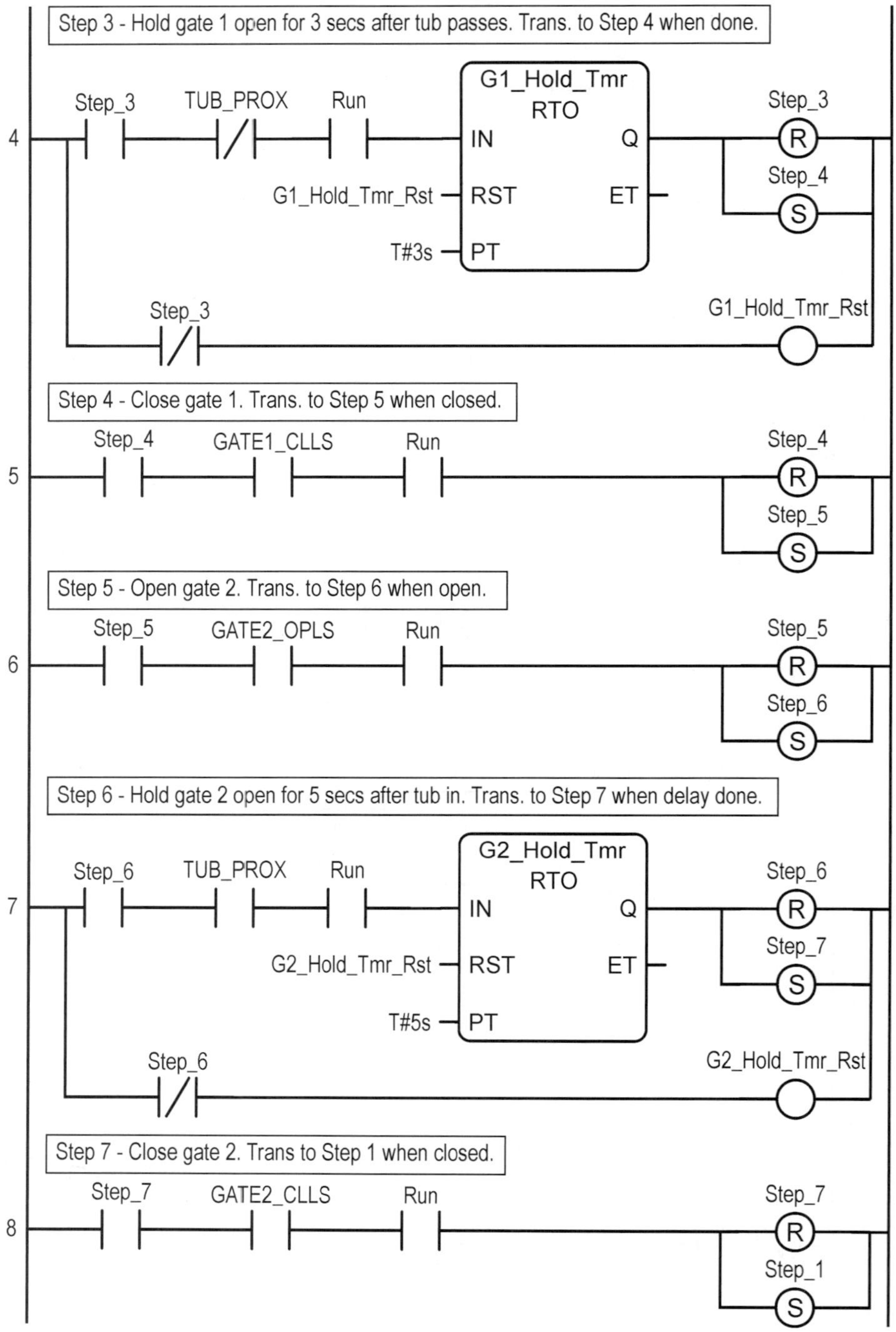

Figure 6.17. *(continued)*

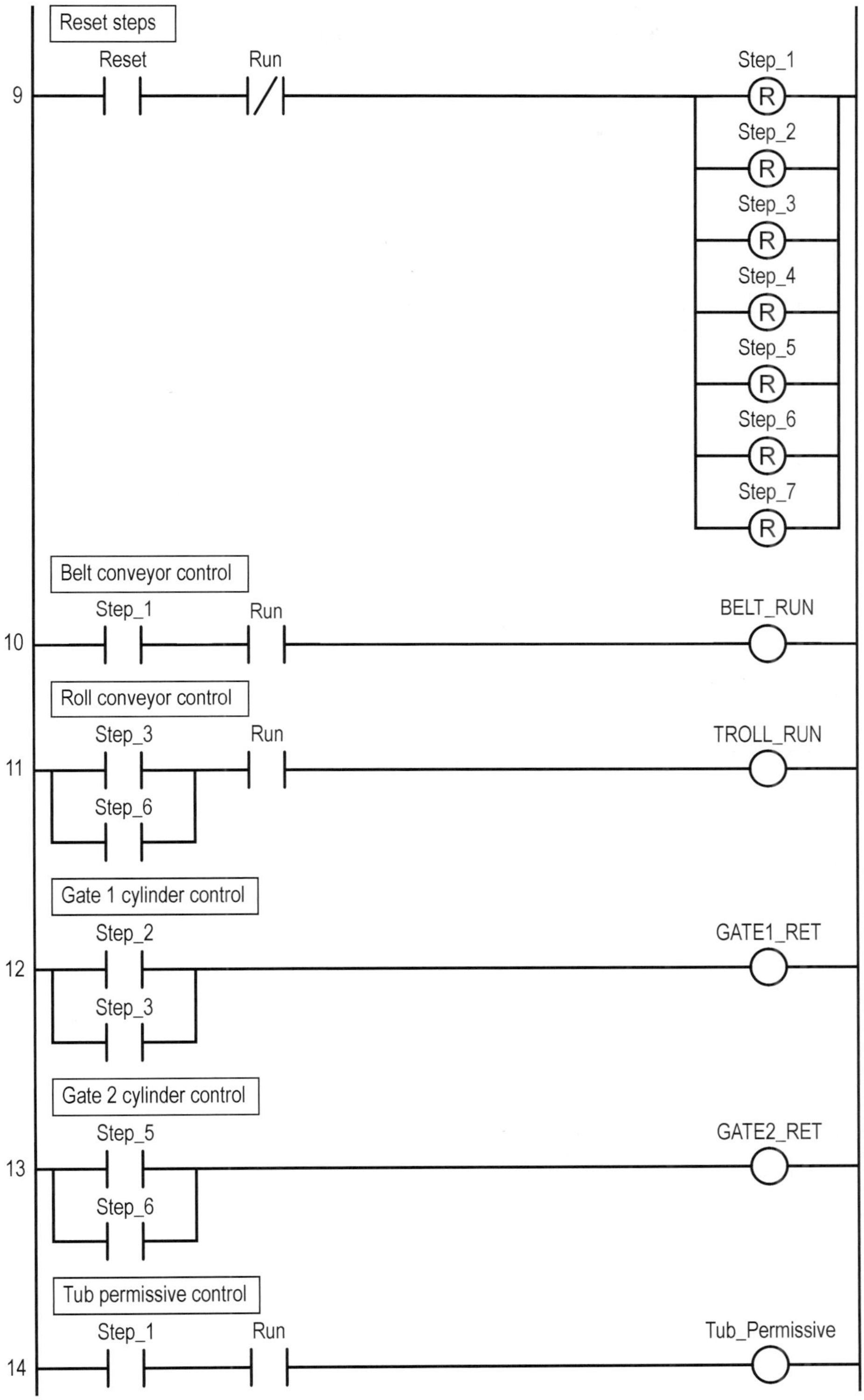

Figure 6.17. *(continued)*

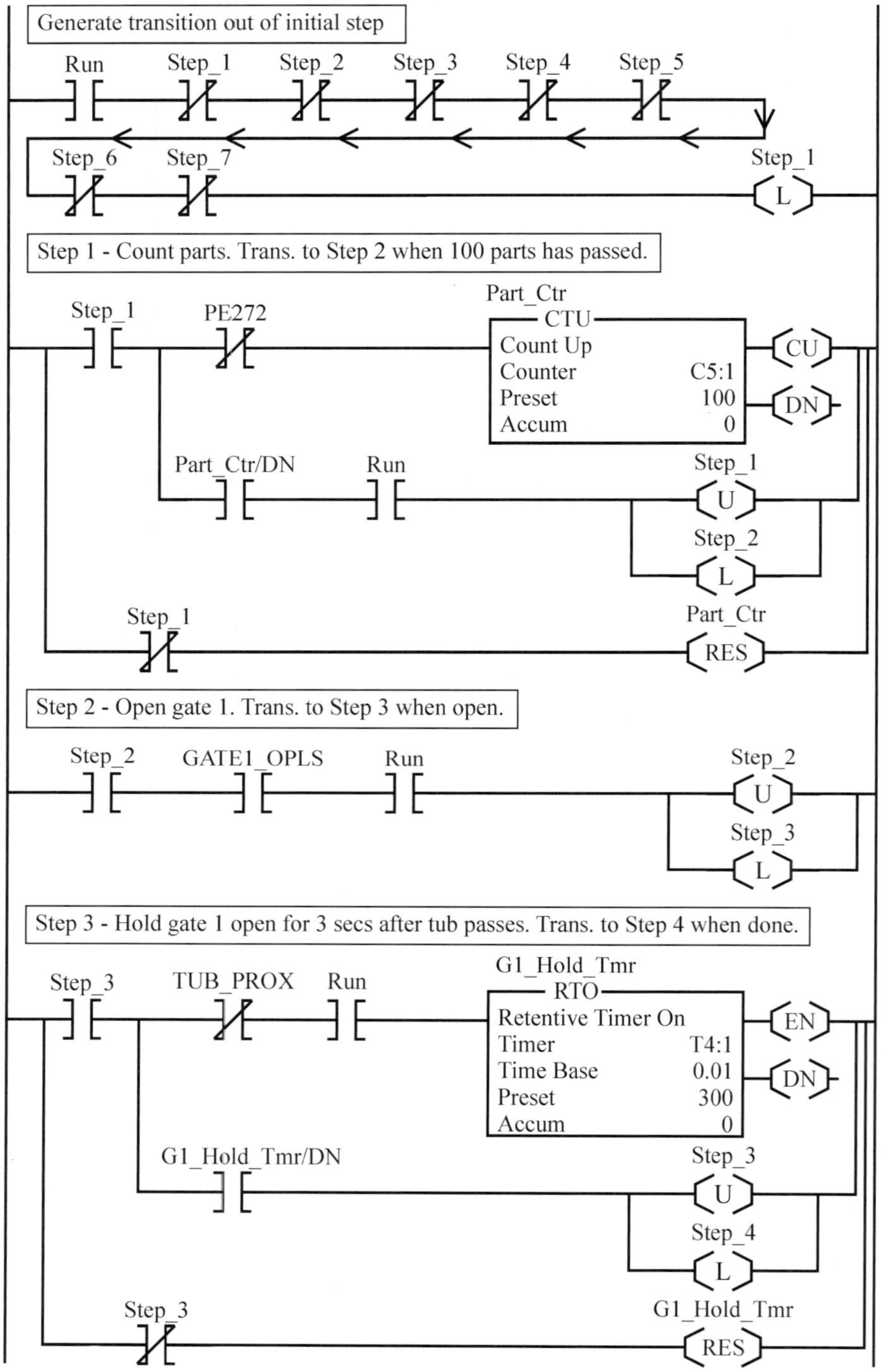

Figure 6.18. MicroLogix ladder logic for tub loader. *(continued)*

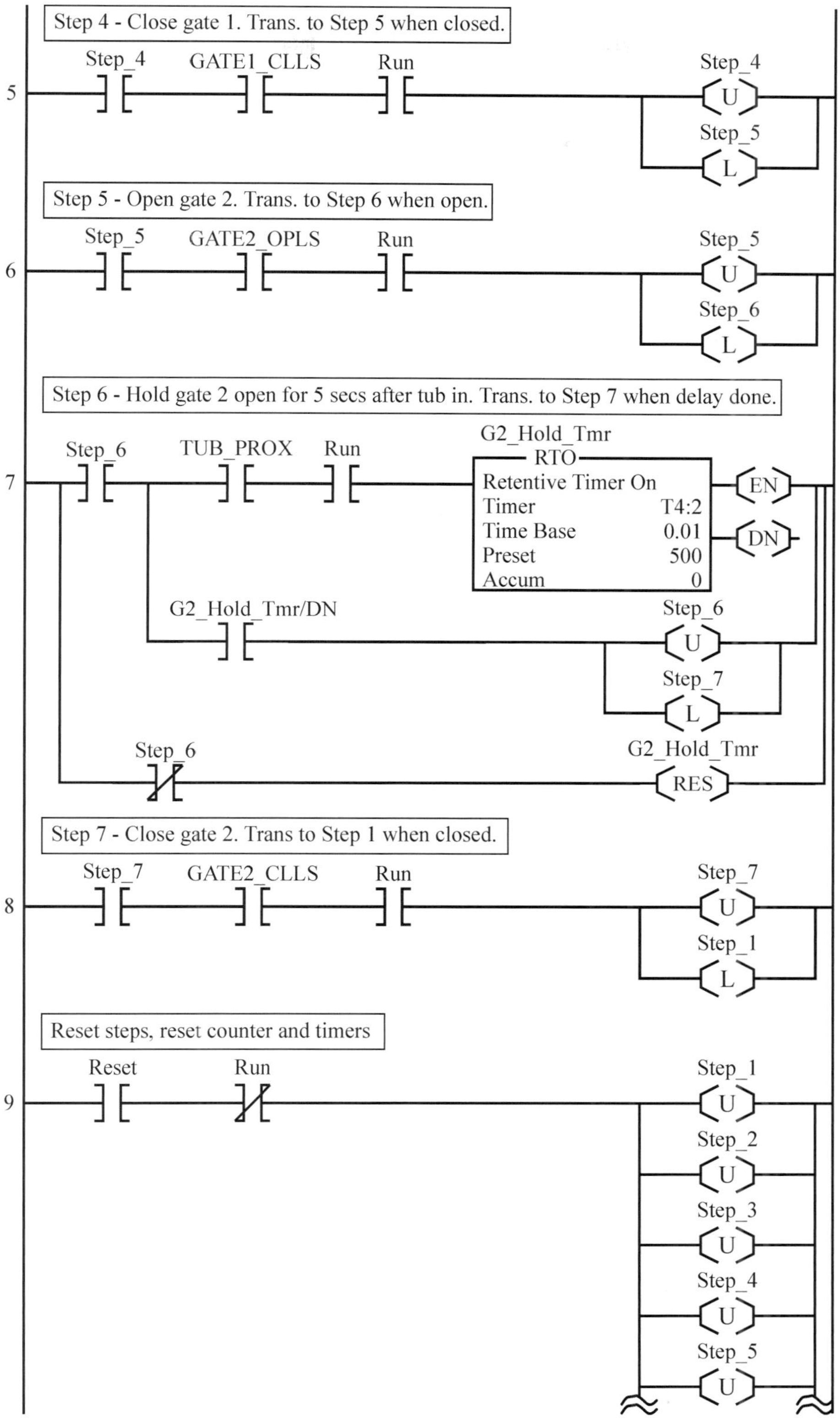

Figure 6.18. *(continued)*

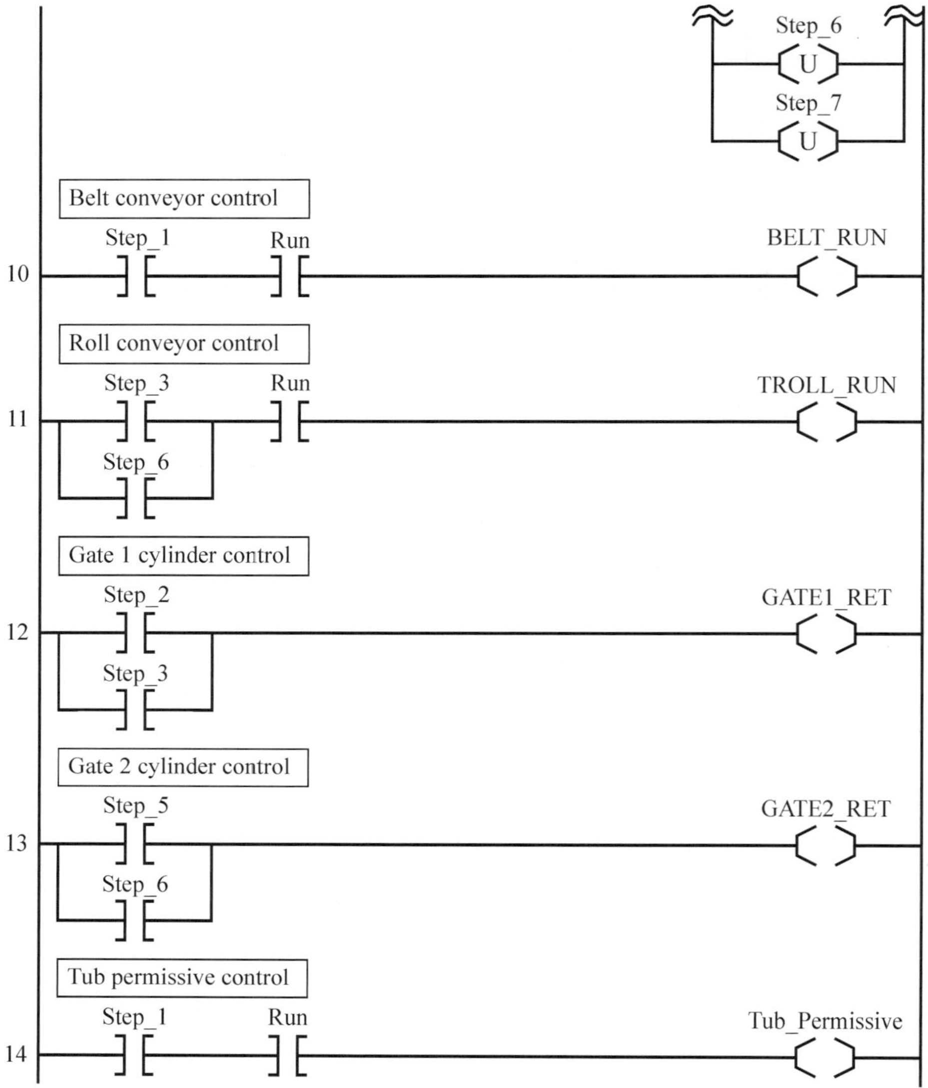

Figure 6.18. *(continued)*

The reset condition for the counter must also be defined in two situations: normal operation and operator-initiated reset. Both of these situations are handled if the counter is reset when the operation is not in the step where the counter is used. For example, the counter used to count parts in step 1 is reset when the operation is not in step 1, the step where parts are counted (Figure 6.16, rung 2). Also, the retentive timers are reset when the operation is not in the step using the timer (Figure 6.16, rungs 4 and 7).

The Micro800 ladder logic is shown in Figure 6.17 and differs from the ControlLogix implementation in the following ways:

1. Timers and counters are shown differently.
2. Set/reset coils used instead of latch/unlatch coils

3. An internal coil is used to reset a counter and a retentive timer since contacts cannot connect to the RESET input of these blocks.
4. Multiple coils on the end of a rung are placed in parallel rather than in series.

The MicroLogix/SLC-500 ladder logic is nearly identical to the ControlLogix ladder logic and is shown in Figure 6.18. The basic differences are:

1. Logic cannot be placed to the right of the counter and timer blocks and so the transition logic is placed in parallel with the counter or timer.
2. The "Part_Ctr" symbol appears above the CTU block in rung 2, and the address "C5:1" appears in the Counter field.
3. For the timers (rungs 4 and 7), the "Time Base" field is set to 0.01 and the Preset value is divided by 10. Also, the timer address appears in the Timer field and the timer symbol is above the block.
4. Multiple coils on a rung are placed in parallel instead of in series.

6.4 COMPLICATED RESET OPERATION

Example 6.3 has most of the features of a real application. The next example illustrates a problem in which the reset operation is more complicated than in the previous example.

Example 6.4. Engine Inverter Station Control. Design the function chart of the program to control the following station that inverts (turns over) gasoline engine assemblies and implement the control with ladder logic.

> Figure 6.19 shows the layout of a station that inverts gasoline engine assemblies riding on a pallet as they come down the conveyor. This station is only one in a series of stations along this conveyor. Implement ladder logic for this station only. The conveyor is controlled by another PLC, so assume it is always moving. This particular line is asynchronous, that is, each station processes assemblies at its own speed and does not coordinate its operation with any other station. Because this is an asynchronous line, the station contains two capturing mechanisms (engaging hooks) that control access to the station and allow pallets to queue up before the station.
>
> Upon initial startup, assume that there are no pallets waiting at engaging hook Engage 1. When a pallet is detected at Engage 1 (by PROX1), the following major steps are executed:
>
> > Lower the Engage 1 hook (by activating ENG1_RET) for 2 seconds to allow only one assembly to move into the station and be caught by the Engage 2 hook. When the Engage 1 hook is raised, it catches the next pallet.
> >
> > Raise the pallet off the conveyor.
> >
> > Lower the rotator mechanism to the correct position (ROTR_DNLS closed).
> >
> > Clamp the engine.
> >
> > Raise the rotator mechanism (until ROTR_UPLS closes).
> >
> > Rotate the engine one-half turn clockwise (until ROTR_CWLS closes).

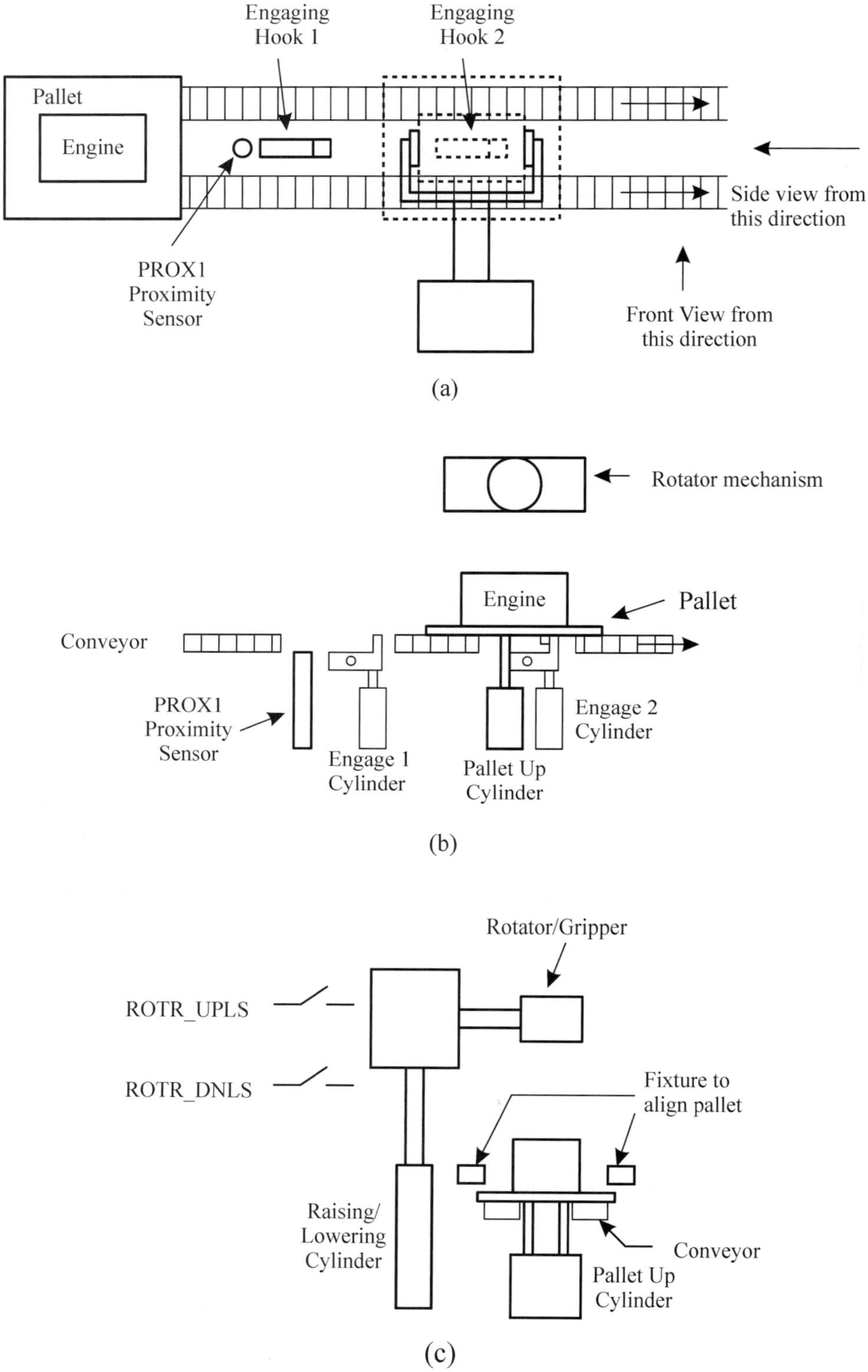

Figure 6.19. Engine inverter station: *(a)* top view; *(b)* front view; *(c)* side view.

Lower the rotating mechanism to the correct position (ROTR_DNLS closed).

Unclamp the engine.

Raise the rotator mechanism (until ROTR_UPLS closes).

Rotate the clamp one-half turn counterclockwise (until ROTR_CCWLS closes).

Activate the ENG2_RET for 3 seconds to allow pallet to move out.

The operation then repeats. Assume the conveyor is on at all times. The conveyor consists of two parallel tracks and slides beneath the pallets as they are held by the engaging hooks or raised off the conveyor.

The proximity sensor, PROX1, is inductive and senses the metal assembly pallet. PROX1 senses the pallet before the pallet reaches the engage position. You must assume that when Engage 1, the first engaging hook, captures the pallet, PROX1 remains **on**.

ENG1_RET and ENG2_RET are controls for single action pneumatic cylinders that move the engaging hooks. Once ENG1_RET is energized, the Engage 1 hook moves down and remains in the "down" position as long as power is applied (turned **on**). The hook moves up when power is removed (turned **off**). The engaging mechanism works in this manner to be fail-safe, that is, if electrical power or air pressure is interrupted because of a failure, no pallets proceed down the conveyor. ENG2_RET controls the second engaging hook, Engage 2, in a similar manner.

The pallet-raising mechanism is driven by a single-action pneumatic cylinder controlled by PALL_UPCTL. Once the PALL_UPCTL output is energized, the clamp moves the pallet (and engine) off the conveyor and into a fixture to properly align the engine to the gripper clamp. PALL_UPCTL must remain **on** to hold the assembly in the fixture. If PALL_UPCTL is turned **off**, the pallet falls back onto the conveyor. The PALL_UPLS is **on** when the pallet is in the proper position.

The mechanism used to lower and raise the rotating mechanism consists of a double-action linear hydraulic cylinder. When the ROTR_DOWN output is energized (turned **on**), the rotator moves down and continues to move down as long as it is energized and a mechanical stop is not reached. When the ROTR_UP output is energized, the rotator moves up and continues to move up as long as it is energized and a mechanical stop is not reached. The mechanism stops if neither output is **on**, or if they are energized simultaneously. ROTR_UPLS is **on** when the rotator is in the "up" position. ROTR_DNLS is **on** when the rotator is in the "down" position.

The gripper is powered by a single-action pneumatic cylinder. When the GRIP_CLOS output is energized, the gripper jaws close to clamp the engine and hold it in place as long as power is applied (turned **on**). GRIP_CLOS must remain **on** to hold the engine in the gripper. If GRIP_CLOS is turned **off**, the engine is released. There are no limit switches indicating the gripper is open or closed. Allow 1.5 seconds for the gripper to clamp (close) and 1.0 seconds for the gripper to unclamp (open).

A double-action pneumatic rotary cylinder controls the rotation action of the gripper. When the ROTAT_CW output is energized, the gripper rotates clockwise as long as power is applied (turned **on**) and the CW mechanical stop has not been reached. When the ROTAT_CCW output is energized, the gripper rotates counterclockwise as long as power is applied (turned **on**) and the CCW mechanical stop has not been reached. The rotation stops at its current position when power is removed (turned **off**). The rotation will not move if both opposing directions are energized simultaneously (e.g., CW and CCW). ROTR_CWLS is **on** when the gripper is fully clockwise. ROTR_CCWLS is **on** when the gripper is fully counterclockwise.

The start/stop switches are only for the station. They do not control any other stations or the conveyor. Upon initial startup, assume there are no pallets present in either of the engaging hooks. If the stop switch is pressed at any time, the station operation should pause, except when either engaging hook is activated. If the operation is paused when ENG1_RET is activated the station may contain two pallets with no space in between. When the start switch is pressed while the operation is paused, the station should resume the suspended step. When paused, do not advance to the next step. When the station is paused, the raise/lower and rotating cylinder controls should be turned off. The engine clamping gripper and the pallet raising cylinders must remain **on** when paused (or the engine may be dropped).

A separate reset switch is provided that when pressed, the clamp gripper is released, the rotating mechanism is raised, then rotated counterclockwise, and the process step is set as if the process is waiting for the next pallet. When the start switch is next pressed, no items are assumed present at the first engage position. The reset switch should have no effect unless the operation is already paused.

Assume the tolerance on all timer values is ±0.1 seconds.

Assume the following physical input and physical output assignments.

Tag/Var./Symbol	Description
START_PB	Start push button, N. O., **on** when starting.
STOP_PB	Stop push button, N. C., **off** when stopping.
RESET_PB	Reset push button, N. O., **on** when restoring station to initial state.
PROX1	Proximity sensor, **on** (closed) when pallet is either approaching or at the Engage 1 hook position.
PALL_UPLS	Limit switch, **on** when the pallet is lifted off conveyor and in the proper position.
ROTR_UPLS	Limit switch, **on** (closed) when rotating mechanism is up.
ROTR_DNLS	Limit switch, **on** (closed) when rotating mechanism is down (can clamp/unclamp engine).
ROTR_CWLS	Limit switch, **on** (closed) when rotary solenoid is clockwise.
ROTR_CCWLS	Limit switch, **on** (closed) when rotary solenoid is counterclockwise.
ENG1_RET	Engage hook 1 cylinder retract control, **on** to lower hook, **off** raises hook.
ENG2_RET	Engage hook 2 cylinder retract control, **on** to lower hook, **off** raises hook.

ROTR_UP	Rotating mechanism raise cylinder control, **on** to raise.
ROTR_DOWN	Rotating mechanism lower cylinder control, **on** to lower.
ROTAT_CW	Clockwise rotary cylinder control, **on** to rotate clockwise.
ROTAT_CCW	Counterclockwise rotary cylinder control, **on** to rotate counterclockwise.
GRIP_CLOS	Gripper cylinder control, **on** closes jaws, **off** opens jaws.
PALL_UPCTL	Pallet retainer cylinder control, **on** to move pallet up and off the conveyor and retain it there, **off** lowers pallet back onto conveyor.

The addresses associated with the physical inputs and outputs are:

Tag/Var./Symbol	ControlLogix	Micro800	MLogix	SLC-500
START_PB	Local:1:I.Data.0	_IO_EM_DI_00	I:0/0	I:1/0
STOP_PB	Local:1:I.Data.1	_IO_EM_DI_01	I:0/1	I:1/1
RESET_PB	Local:1:I.Data.2	_IO_EM_DI_02	I:0/2	I:1/2
PROX1	Local:1:I.Data.3	_IO_EM_DI_03	I:0/3	I:1/3
PALL_UPLS	Local:1:I.Data.4	_IO_EM_DI_04	I:0/4	I:1/4
ROTR_UPLS	Local:1:I.Data.5	_IO_EM_DI_05	I:0/5	I:1/5
ROTR_DNLS	Local:1:I.Data.6	_IO_EM_DI_06	I:0/6	I:1/6
ROTR_CWLS	Local:1:I.Data.7	_IO_EM_DI_07	I:0/7	I:1/7
ROTR_CCWLS	Local:1:I.Data.8	_IO_EM_DI_08	I:0/8	I:1/8
ENG1_RET	Local:2:O.Data.0	_IO_EM_DO_00	O:0/0	O:2/0
ENG2_RET	Local:2:O.Data.1	_IO_EM_DO_01	O:0/1	O:2/1
ROTR_UP	Local:2:O.Data.2	_IO_EM_DO_02	O:0/2	O:2/2
ROTR_DOWN	Local:2:O.Data.3	_IO_EM_DO_03	O:0/3	O:2/3
ROTAT_CW	Local:2:O.Data.4	_IO_EM_DO_04	O:0/4	O:2/4
ROTAT_CCW	Local:2:O.Data.5	_IO_EM_DO_05	O:0/5	O:2/5
GRIP_CLOS	Local:2:O.Data.6	_IO_EM_DO_06	O:0/6	O:2/6
PALL_UPCTL	Local:2:O.Data.7	_IO_EM_DO_07	O:0/7	O:2/7

Solution. The function chart for the station is shown in Figure 6.20. Run is not really needed as a transition condition out of steps 3, 4, 6, 7, 8, 10, and 11 since the motion ceases when paused. One could argue that Run is not needed as part of the transition condition out of steps 5 and 9 since advancing to the next step does not actually turn on any physical outputs (since they will remain off as long as the station is paused). Non-retentive timers are acceptable for steps 5 and 9 since the clamping/unclamping will still occur since the gripper continues to function when the operation is paused. Likewise, non-retentive timers are acceptable for steps 2 and 13 since the operation cannot be paused in these steps.

For this problem, the operator-initiated reset is not merely resetting all counters, retentive timers and step-in-progress bits. A sequential operation must restore the mechanical parts of the system to the initial state (clamp open, rotating mechanism in up and counterclockwise positions). The function chart of the reset operation is shown in Figure 6.21. Note that the last step exists only to unseal the Int_Reset internal coil, that indicates the reset operation is in progress and prevents the station from restarting.

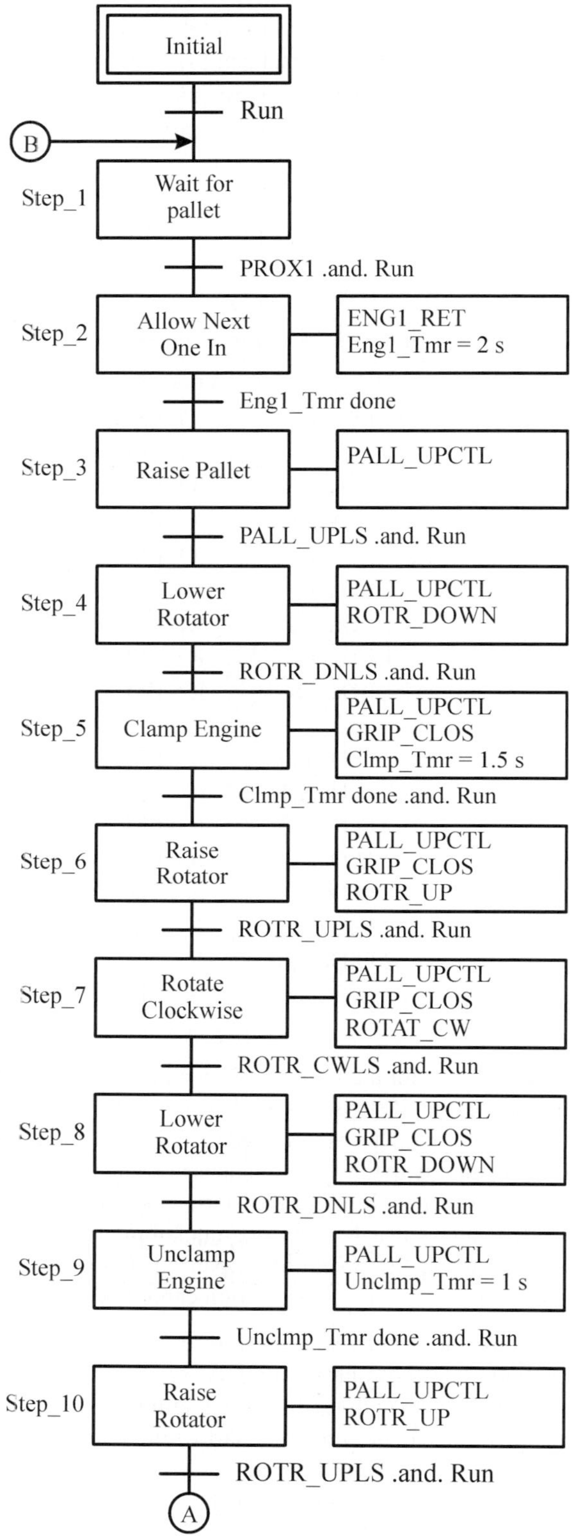

Figure 6.20. Function chart for engine inverter station. *(continued)*

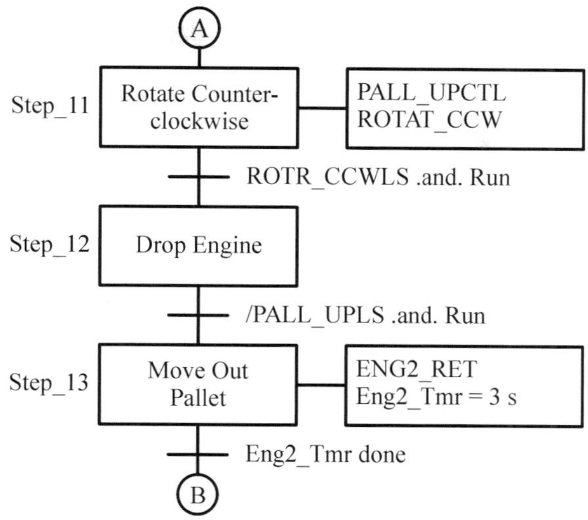

Figure 6.20. *(continued)*

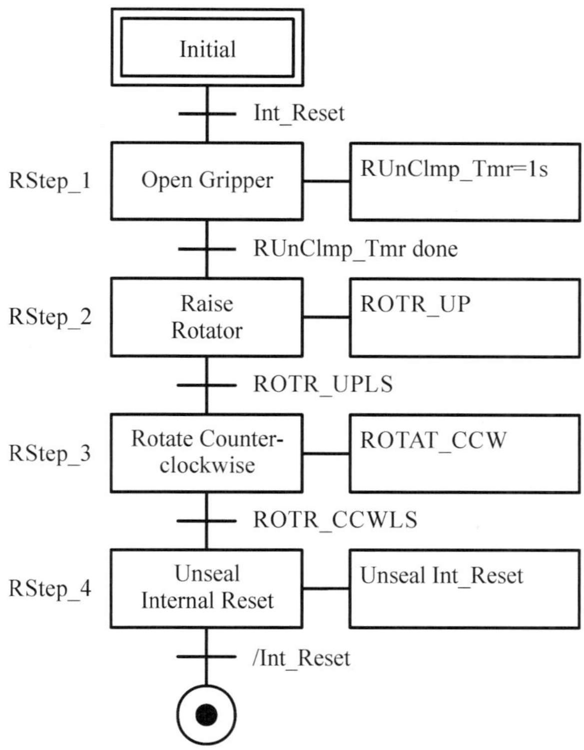

Figure 6.21. Function chart for reset.

Before developing the ladder logic code, the internal tag/variable/symbol addresses or data types should be identified:

Tag/Var./Symbol	CLogix Data Type	Micro800 Data Type	MLogix/SLC Address
Step_1 to Step_13	BOOL	BOOL	B20/1 to B20/13
RStep_1 to	BOOL	BOOL	B20/41 to B20/44
Run	BOOL	BOOL	B20/0
Int_Reset	BOOL	BOOL	B20/40
Eng1_Tmr	TIMER	TON	T4:1
Eng2_Tmr	TIMER	TON	T4:2
Clmp_Tmr	TIMER	TON	T4:3
UnClmp_Tmr	TIMER	TON	T4:4
RUnClmp_Tmr	TIMER	TON	T4:5

The ControlLogix ladder logic is shown in Figure 6.22. The ladder logic is broken into the following sections:

Start/stop/pause of normal operation
First start of normal operation
Normal operation transitions between steps
Reset operation transitions
Step and reset actions

The start and transitions for the reset operation (rungs 16 - 21) are handled similarly as for the normal operation steps. The last step (step 4) of the reset operation unlatches the Int_Reset coil which is **on** as long as the reset operation is in progress. The reset steps are also conditions to turn on the rotation mechanism up and rotate controls (rungs 25 and 28).

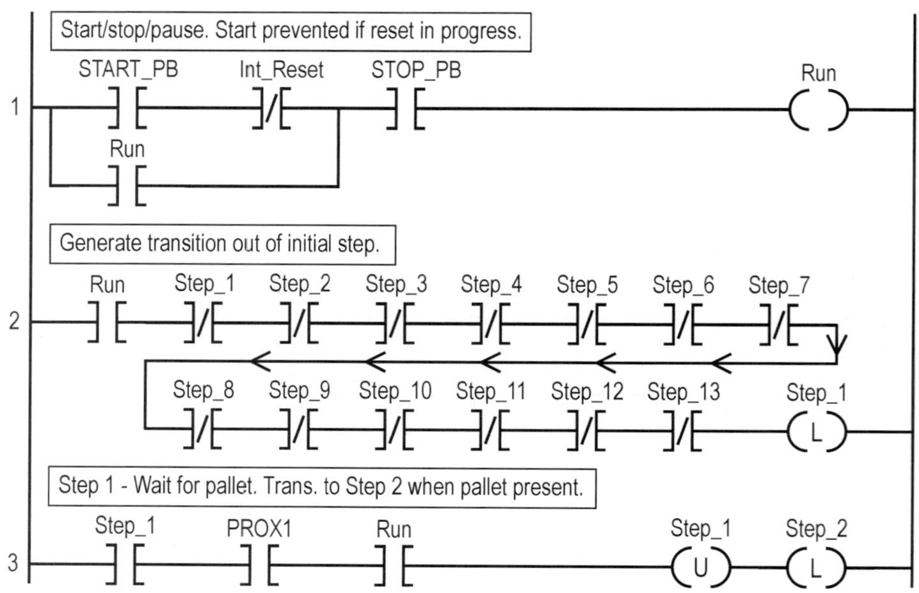

Figure 6.22. ControlLogix ladder logic for engine inverter station. *(continued)*

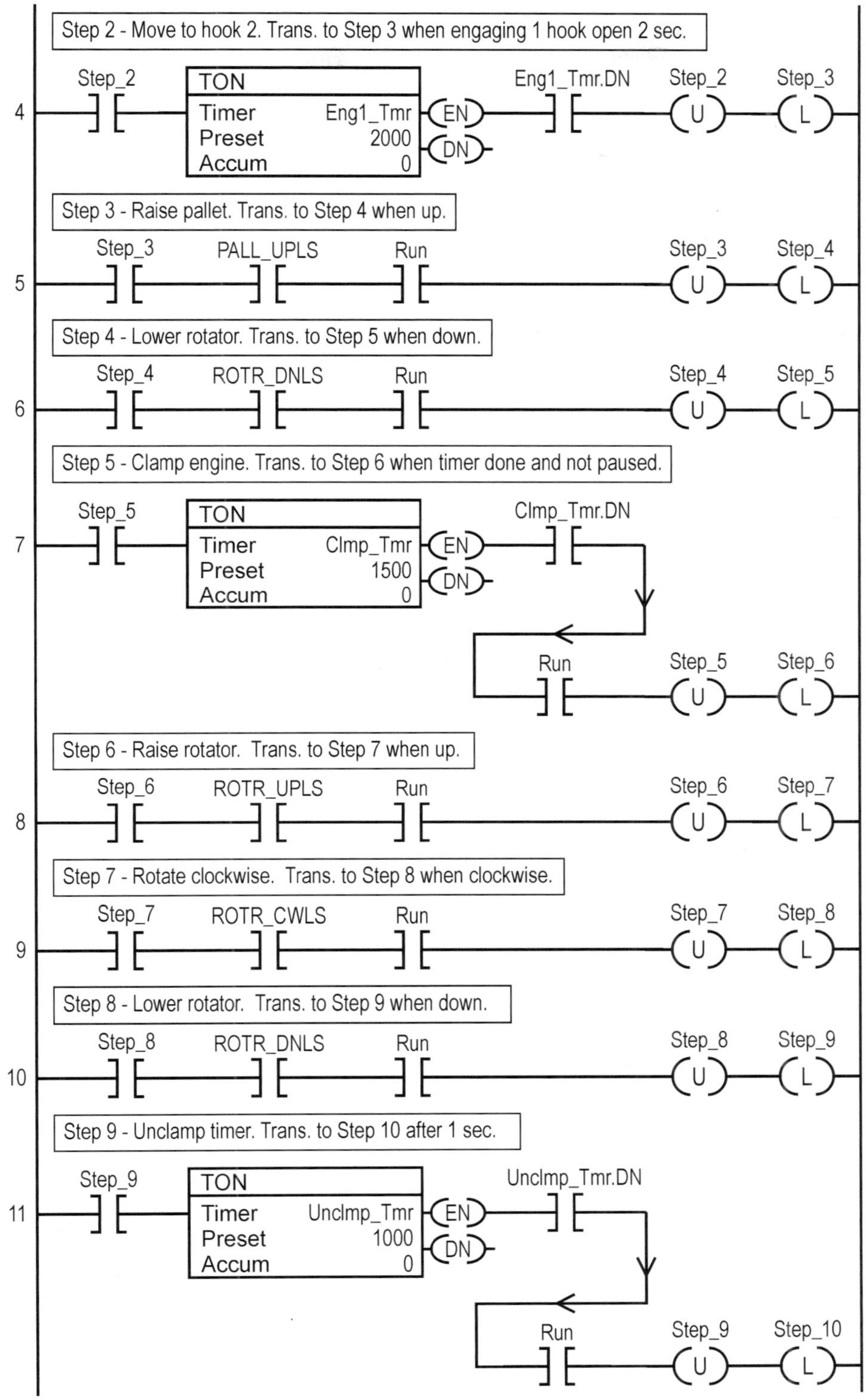

Figure 6.22. *(continued)*

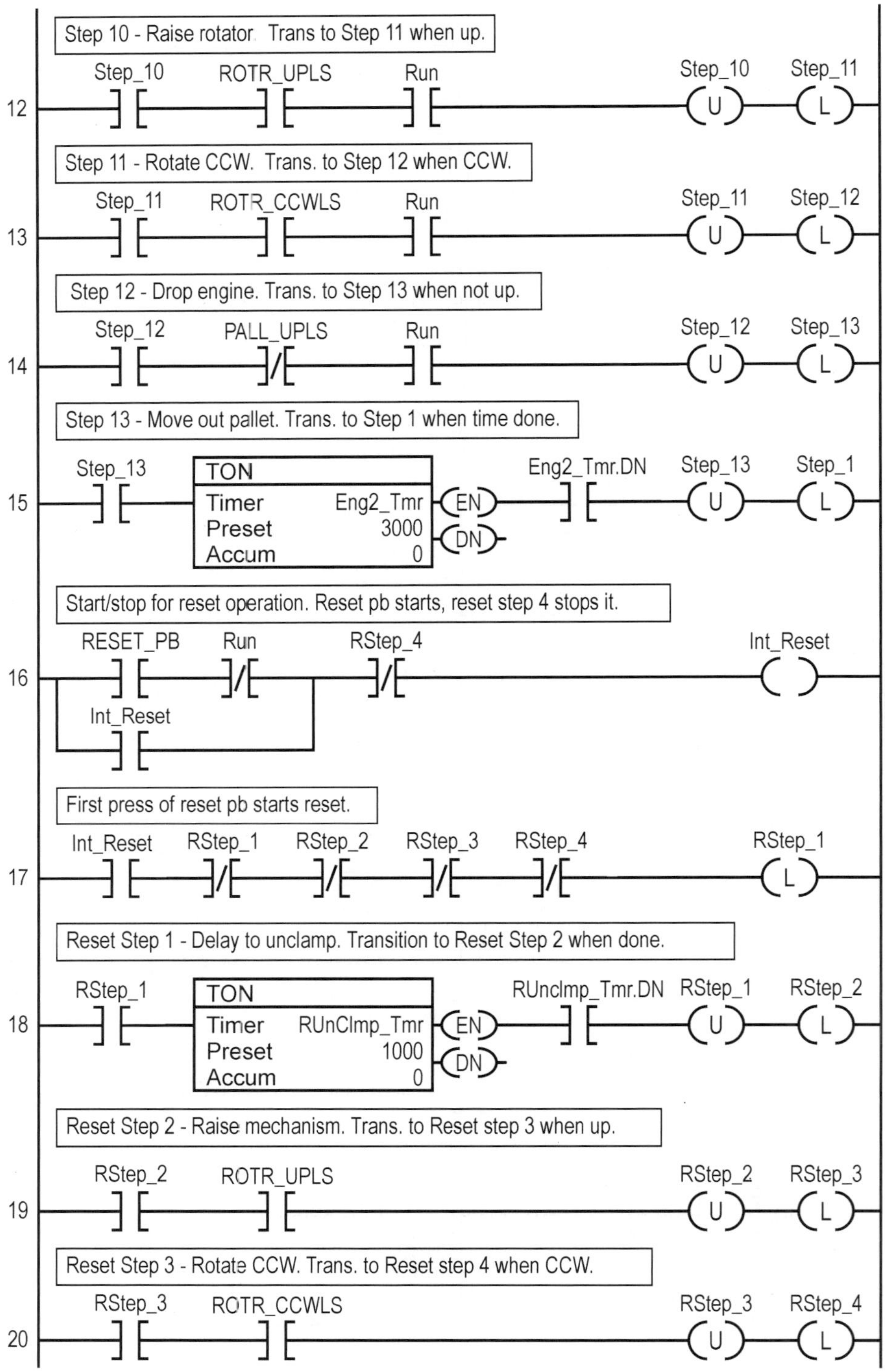

Figure 6.22. *(continued)*

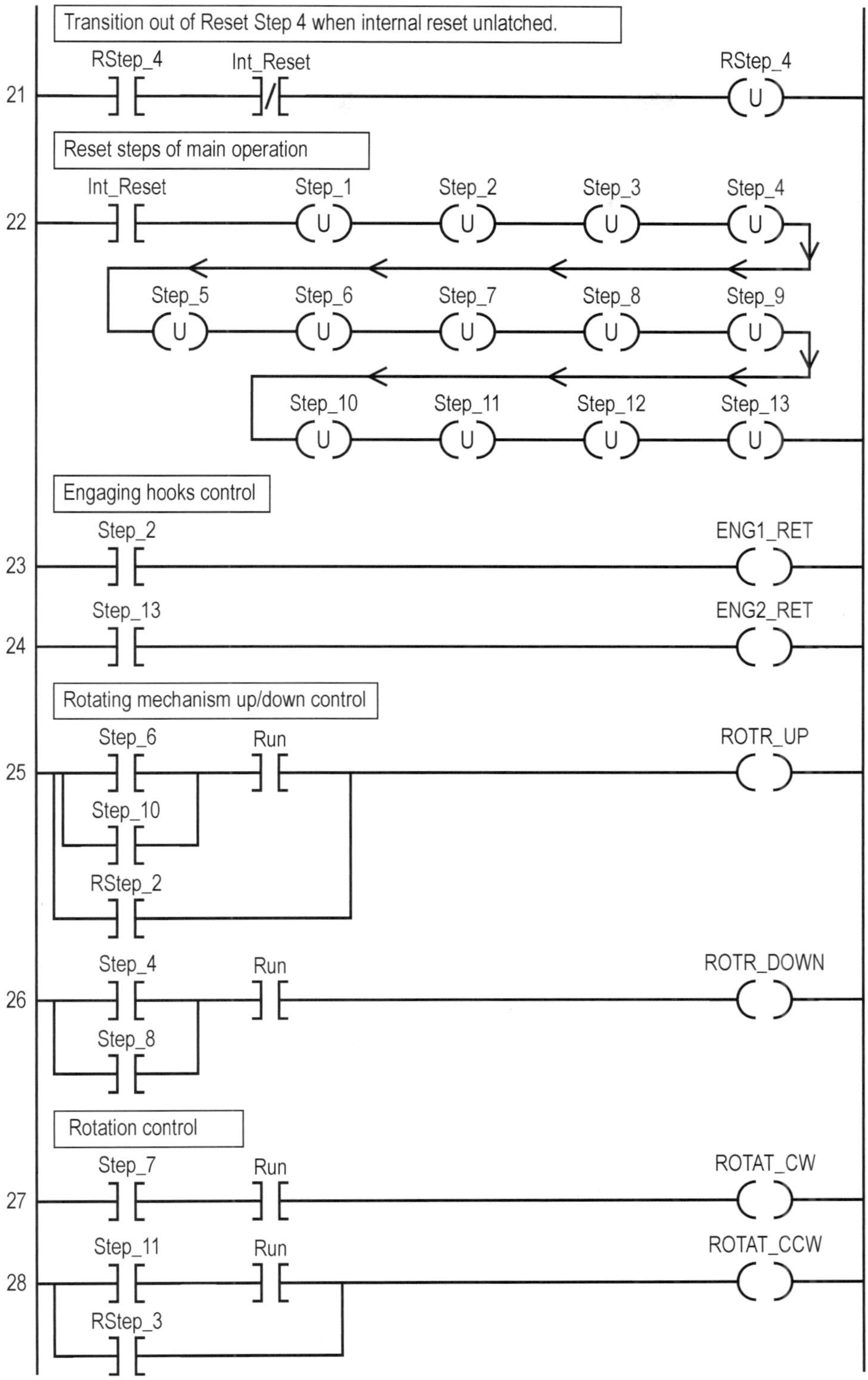

Figure 6.22. *(continued)*

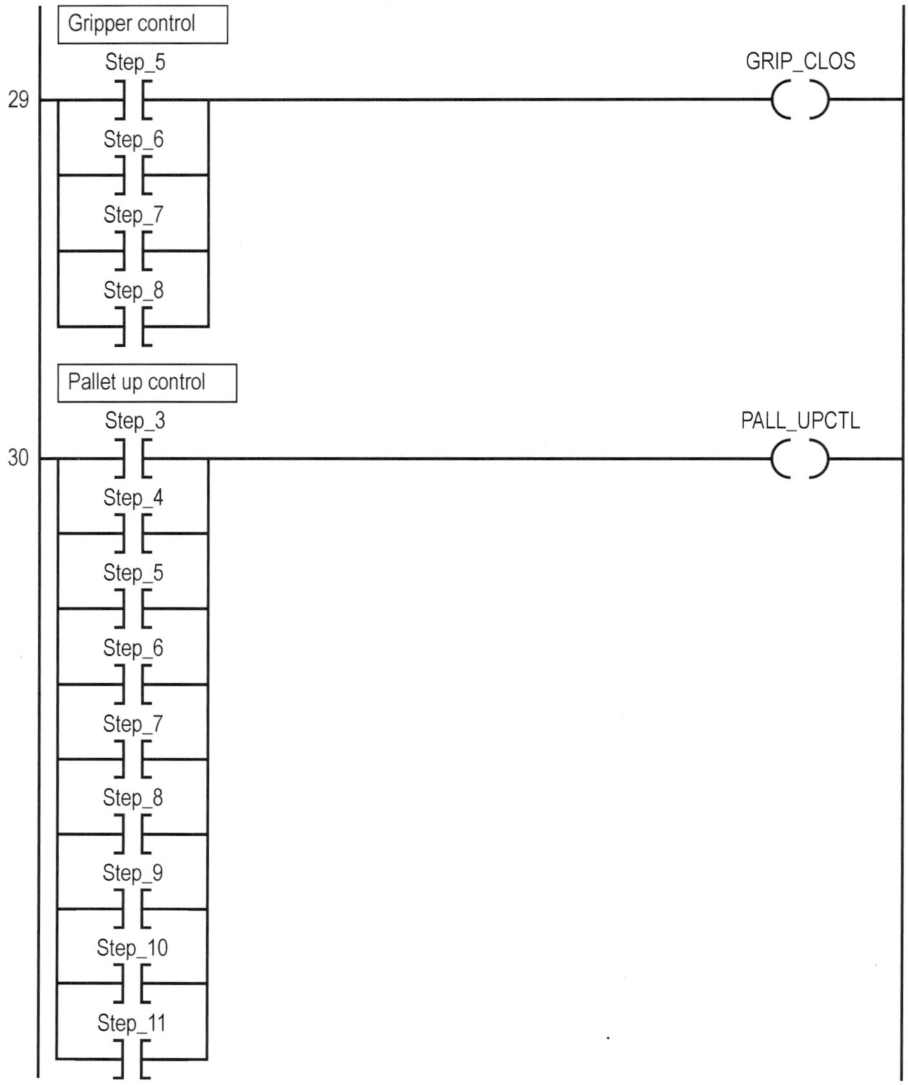

Figure 6.22. *(continued)*

The Micro800 ladder logic is shown in Figure 6.23 and differs from the ControlLogix implementation in the following ways:

1. Timers are shown differently.
2. Set/reset coils used instead of latch/unlatch coils
3. Multiple coils on a rung are placed in parallel instead of in series.

The MicroLogix/SLC-500 ladder logic is very similar to the ControlLogix ladder logic, with three exceptions. First, coils are always in parallel. Secondly, the logic on the right side of the ControlLogix timers is placed on a parallel branch. Thirdly, timers have a "Time Base" field (set to 0.01) and the Preset value is divided by 10. Also, the timer symbol appears above the TON block and the timer address is in the Timer field of the TON block.

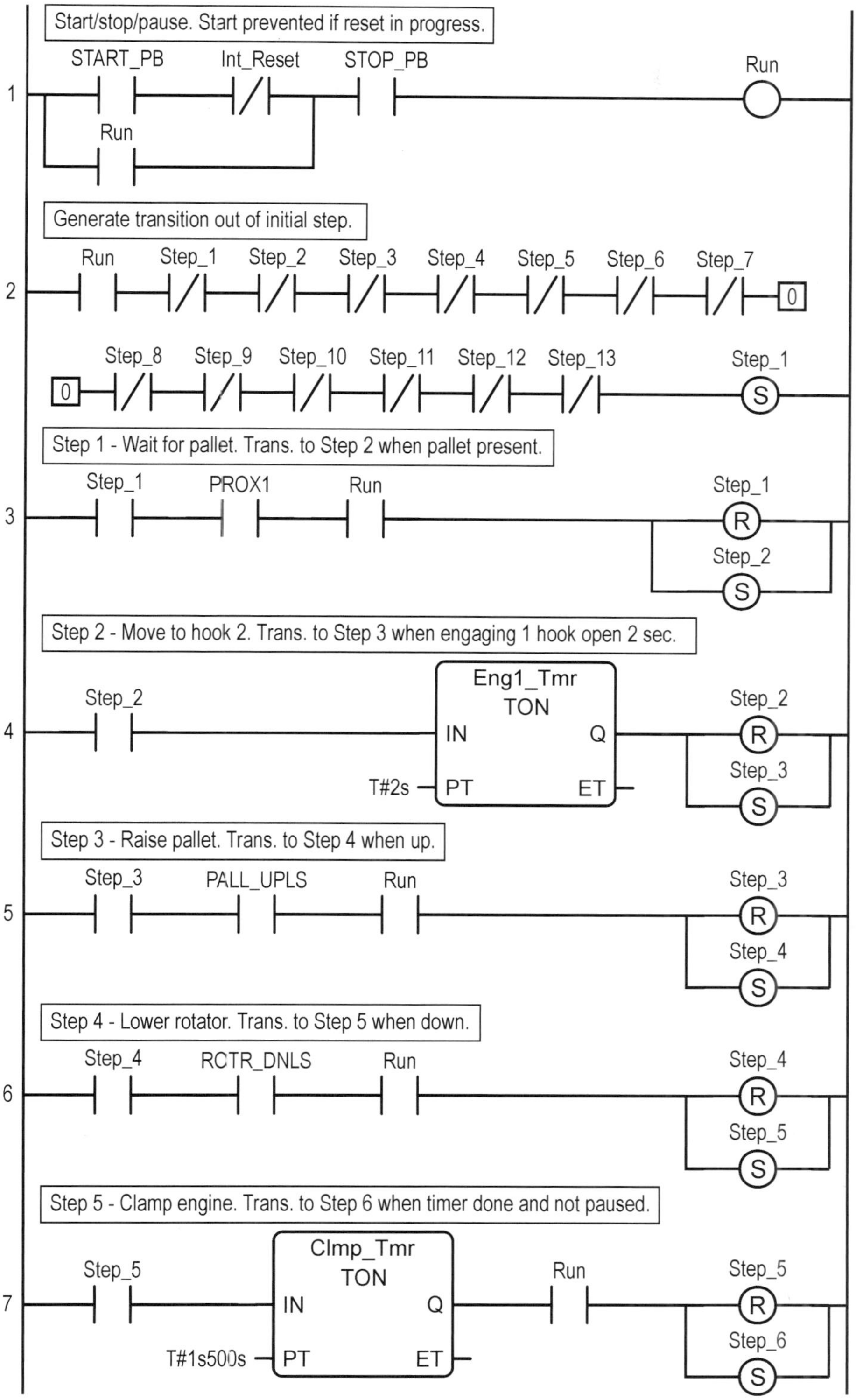

Figure 6.23. Micro800 ladder logic for engine inverter station. *(continued)*

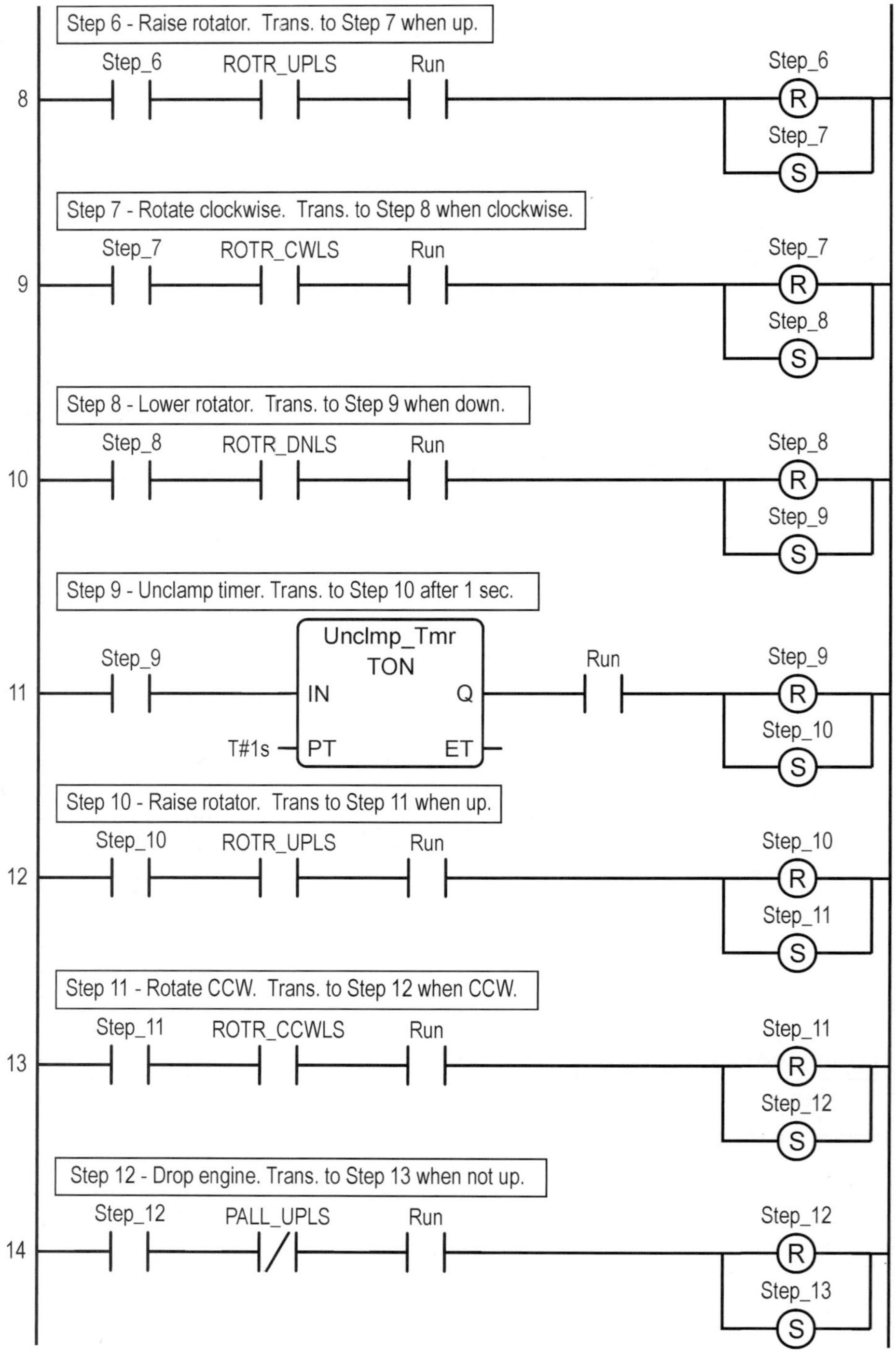

Figure 6.23. *(continued)*

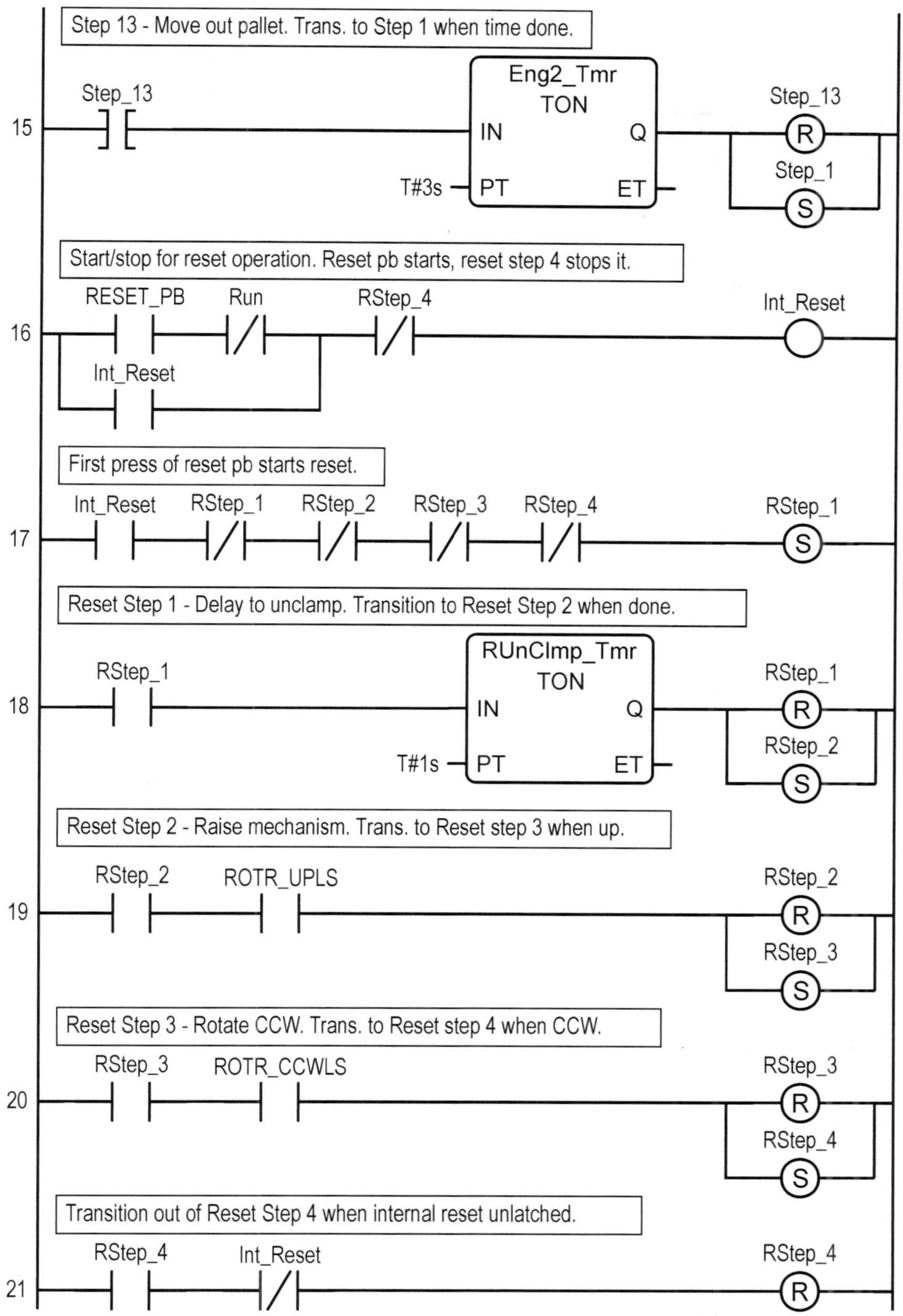

Figure 6.23. *(continued)*

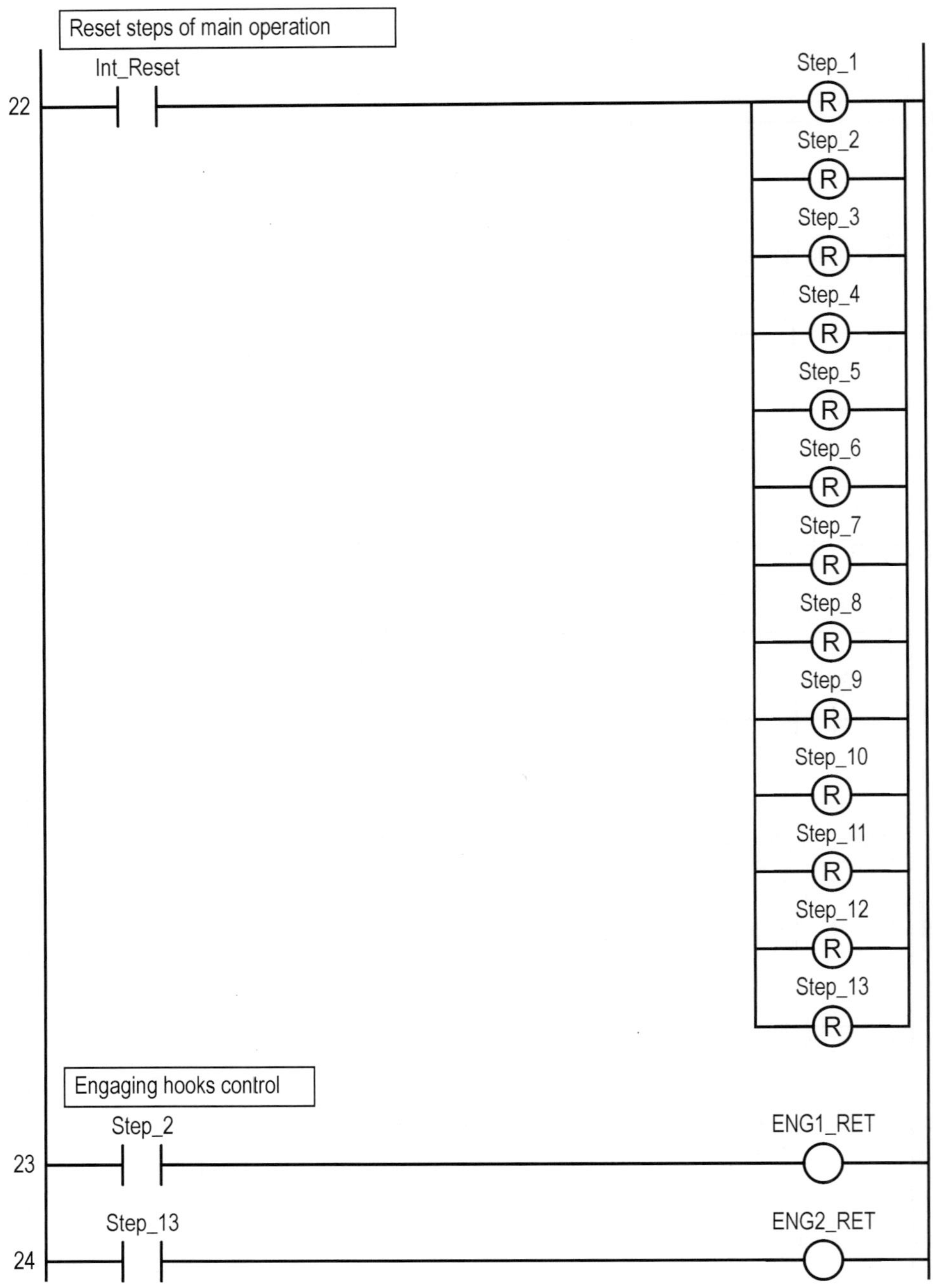

Figure 6.23. *(continued)*

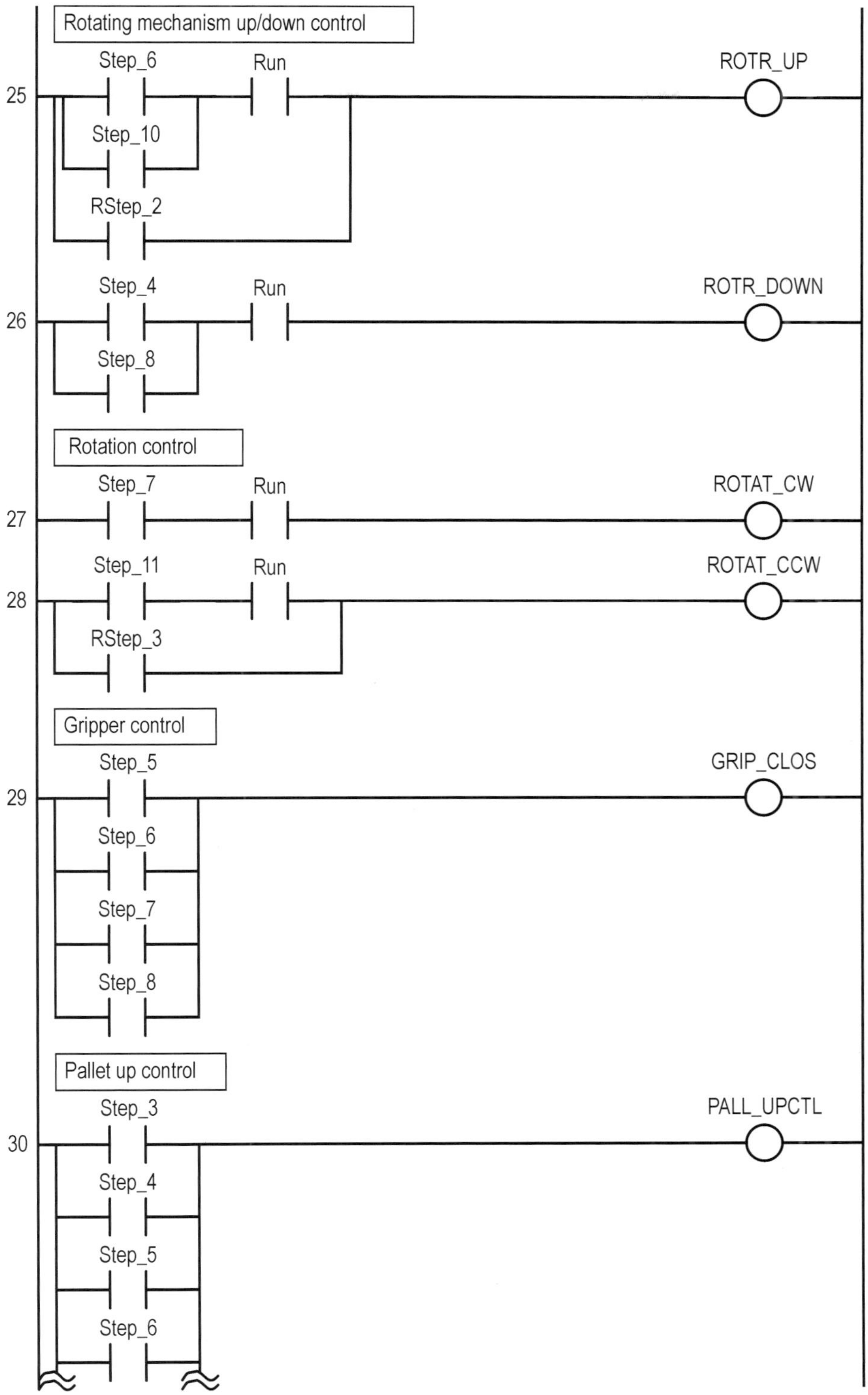

Figure 6.23. *(continued)*

Figure 6.23. *(continued)*

6.5 PARALLEL BRANCHING

Suppose the gasoline engine manufacturer wants to increase the throughput of the assembly line. Therefore, the cycle time of each station must be decreased. One way to meet this requirement for the engine inverter station is to allow certain steps to happen simultaneously. For example, the raising of the pallet (step 2) and the lowering of the rotator (step 3) steps can occur simultaneously. The first part of this revised function chart (without the actions) is shown in Figure 6.24. The double horizontal line indicates that both paths are executed simultaneously.

On a function chart, two kinds of branching are allowed. If the transition out of a step causes more than one step to be activated simultaneously, called ***simultaneous divergence***

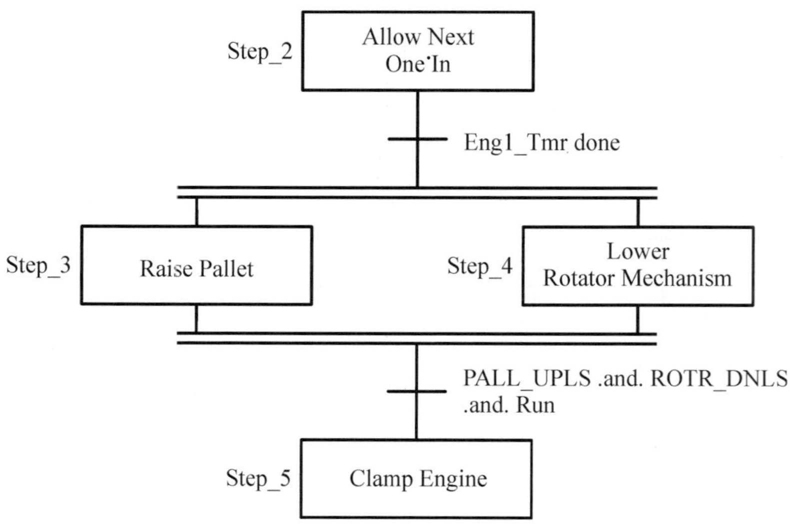

Figure 6.24. Example parallel steps.

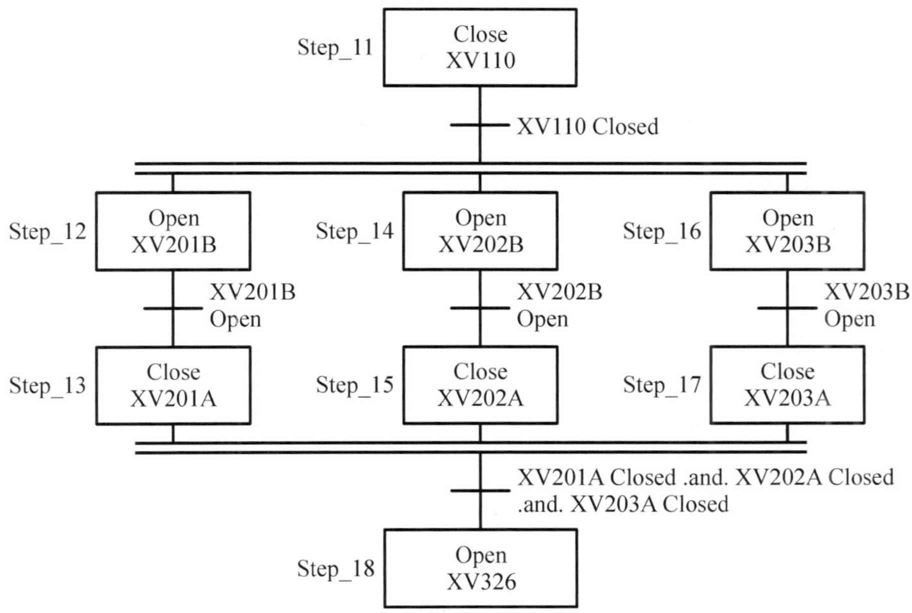

Figure 6.25. Example of simultaneous divergence (AND branching).

or ***AND branching***, these simultaneous sequences are represented as in Figure 6.25. The double horizontal lines distinguish this type of branching. Also, only one common transition condition is permitted above the top double horizontal line, and no transitions are permitted below the upper double horizontal line. When step 11 is active and the condition

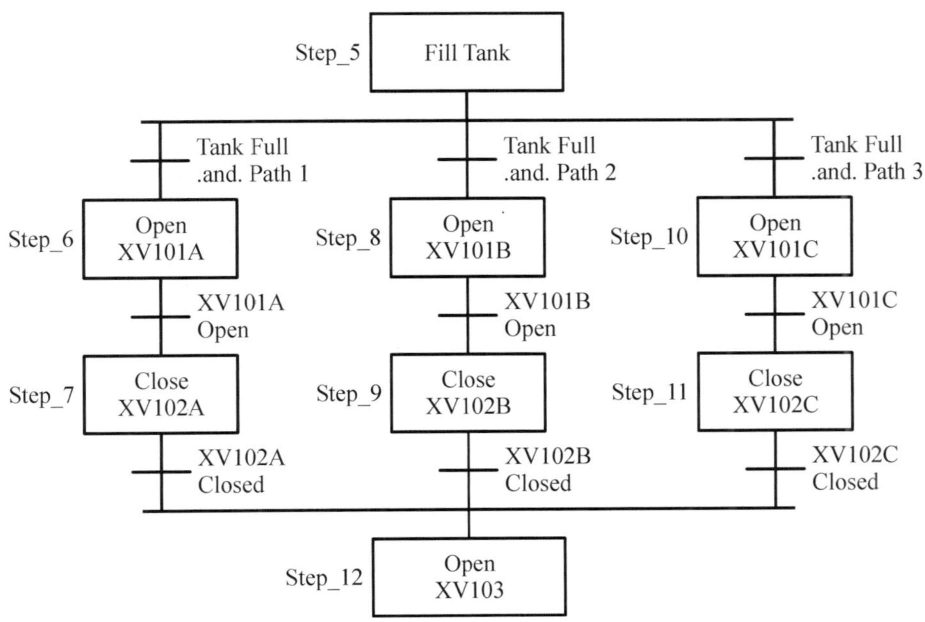

Figure 6.26. Example of exclusive divergence (OR branching).

"XV110 Closed" is true, then step 11 becomes inactive and steps 12,14, and 16 become active simultaneously. The sequences converge with a double horizontal line having a common transition symbol under the double horizontal line. Step 18 will become the active step only when all the steps above the double horizontal line are active and the transition condition "XV201A Closed .and. XV202A Closed .and. XV203A Closed" is true. For some systems, especially those that involve mechanical motion, the simultaneous steps do not finish at the same time. In this situation, an extra step must be added before the branch convergence (lower double horizontal line). This step serves to stop the motion and to wait for the other parallel steps to finish. This issue is considered in Example 6.5.

A selection of one sequence out of more than one sequence is called ***exclusive divergence*** or ***OR branching*** and is represented by multiple transitions below the single horizontal line, as shown in the upper part of Figure 6.26. Each possible sequence path

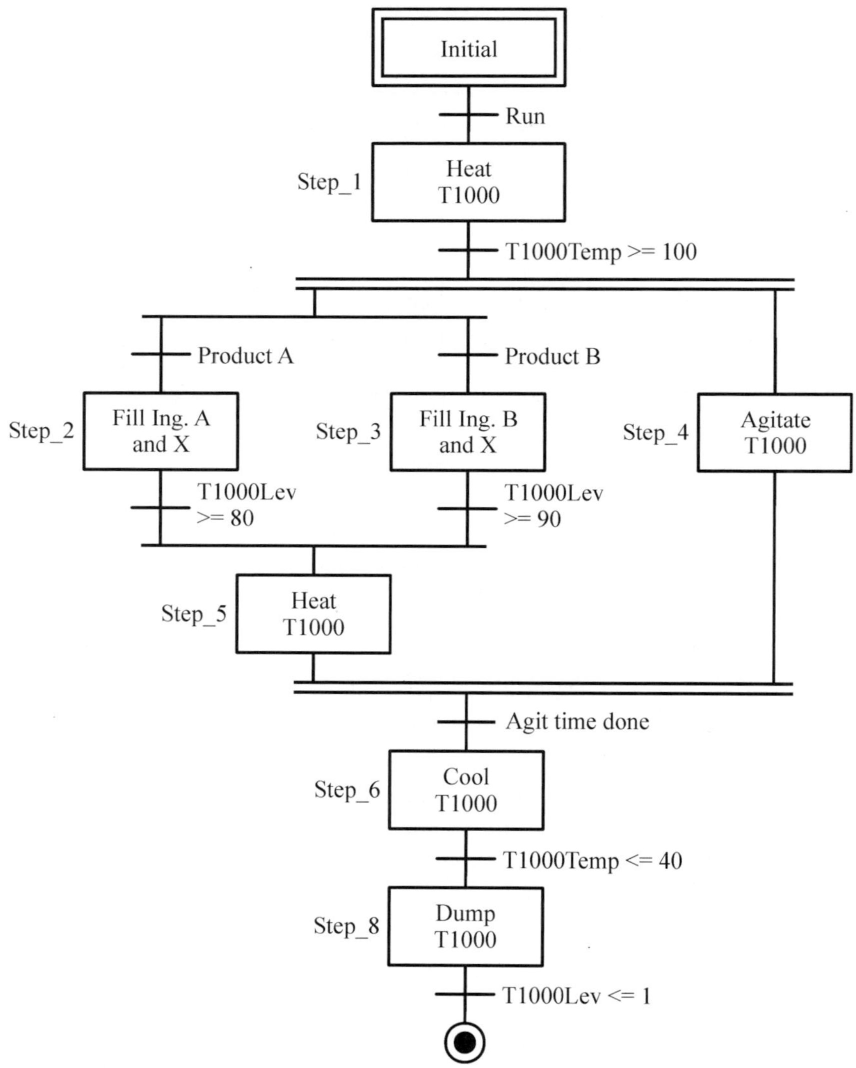

Figure 6.27. Example combined types of branching.

contains a transition condition. No common transition condition is permitted above the horizontal line. The exclusive divergence is differentiated from the simultaneous divergence by the single horizontal line. If step 5 is active and the tank is full, then there are three possible transition conditions. If the "Path 1" condition is true, then the logic advances to step 6. Otherwise, if "Path 2" is true, then the logic advances to step 8, or if "Path 3" is true, the logic advances to step 10. In order to select only one succeeding step, the transition conditions must be mutually exclusive. The several sequences must also converge to a common sequence, as in the lower part of Figure 6.26. There must be as many transitions above the horizontal line as sequences to be re-grouped. No common transition condition is permitted below the lower horizontal line. If step 7 is active and "XV102A Closed" is true, or if step 9 is active and "XV102B Closed" is true, or if step 11 is active and "XV102C Closed" is true, then step 12 becomes the active step.

Both types of branching can be combined on the same function chart as shown in Figure 6.27. Note that the beginning and ending of each branch must correspond. A simultaneous divergence (double horizontal line) cannot be finished with an exclusive convergence (single horizontal line).

The ControlLogix ladder logic to handle the transitions for the simultaneous divergence in Figure 6.25 is shown in Figure 6.28. The exclusive divergence of Figure 6.26 is handled in ladder logic as shown in Figure 6.29. For the Micro800 and MicroLogix ladder logic implementations of Figures 6.28 and 6.29, the coils are in parallel rather than in series.

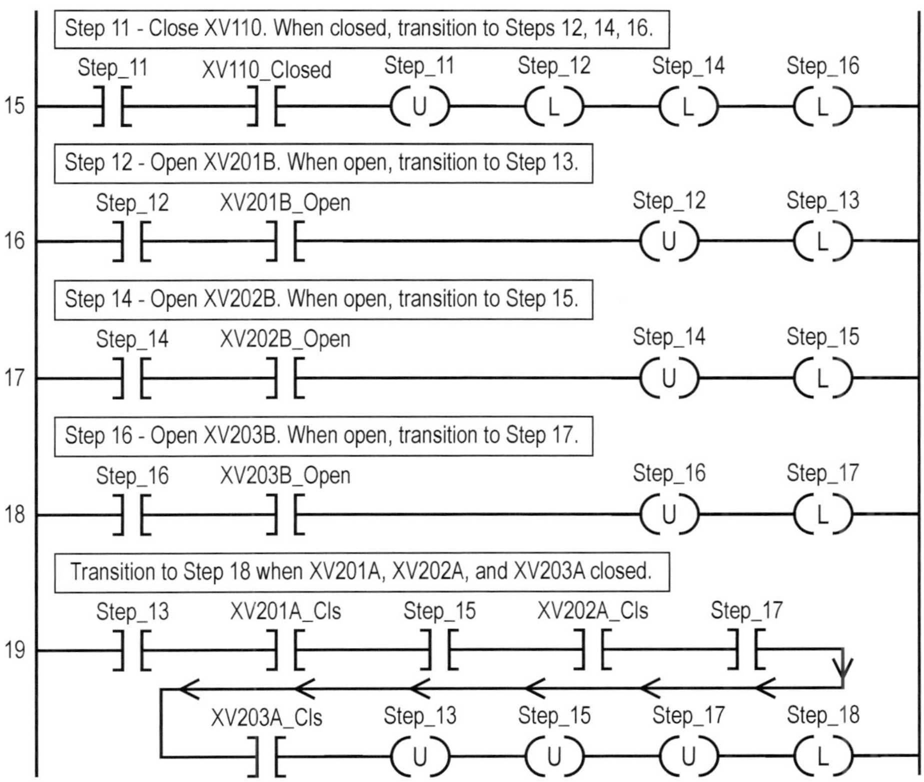

Figure 6.28. ControlLogix ladder logic for simultaneous divergence of Figure 6.23.

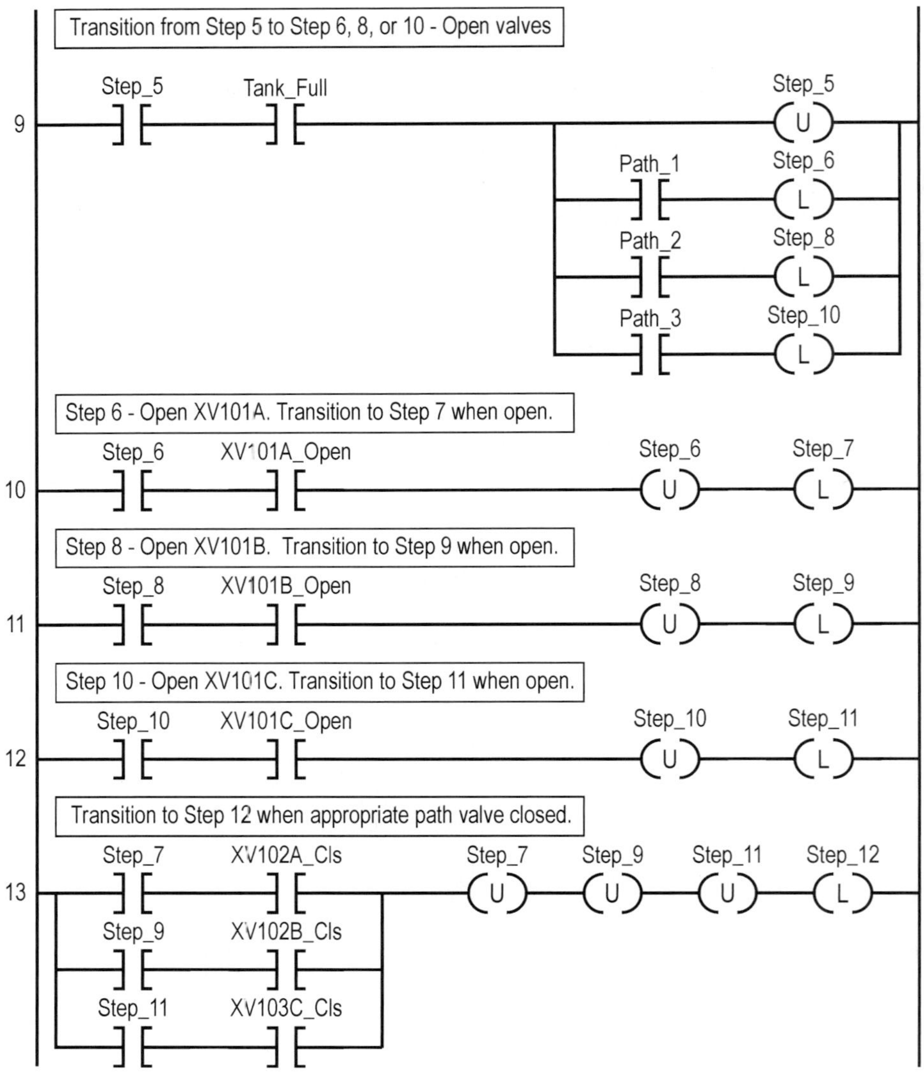

Figure 6.29. ControlLogix ladder logic for exclusive divergence of Figure 6.24.

Example 6.5. In order to decrease the cycle time of the engine inverter of Example 6.4, certain steps occur simultaneously:

1. The "raise pallet" (step 2) and the "lower rotator" (step 3) steps occur simultaneously.
2. After the engine is unclamped, the rotator is raised and rotated CCW (steps 10 and 11) while the engine is dropped and moved out (steps 12 and 13).

Solution. For both simultaneous divergence parts of this example, one must consider if an extra step must be added for each branch just before the convergence. For the first AND branch, no extra step is needed. If the "raise pallet" step finishes first, the PALL_UPCTL output continues to be activated while the rotator lowers, which is acceptable since the

PALL_UPCTL needs to be **on** until the engine is dropped back onto the conveyor. If the "lower rotator" step finishes first, the ROTR_DOWN output continues to be activated, holding it against the mechanical stop, which is also not a problem since it should be a short time. The "raise pallet" should finish first, since this motion travels a shorter distance. The second AND branch requires an extra step before the convergence. The "move out pallet" step must complete, even when the operation is paused. The "wait" step allows the ENG2_RET output to be turned **off** when the operation is paused while the rotator is moving. The extra step after the "rotate counterclockwise" step is not necessary, but it does make sure the ROTR_CCW output is **off** while the engine is moving out.

The revised function chart for the station is shown in Figure 6.30. Note the extra wait steps added to the second simultaneous divergence. Also, with the wait steps (Step_12 and Step_15), the transition condition out of the convergence of steps 12 and 15 uses the step-in-progress bits of the wait steps instead of the ROTR_CCWLS input and Eng2_Tmr done internal coil.

The ControlLogix ladder logic for the transitions is shown in Figure 6.31. The code for the reset operation and the physical outputs is the same as rungs 16 - 30 of Figure 6.22 with the exception that the contact labeled "Step_13" in rung 24 is replaced by "Step_14".The Micro800 ladder logic for the transitions is shown in Figure 6.32. The code for the reset operation and the physical outputs is the same as rungs 16 - 30 of Figure 6.23 with the exception that the contact labeled "Step_13" in rung 24 is replaced by "Step_14".

The MicroLogix/SLC-500 ladder logic is very similar to the ControlLogix ladder logic in Figure 6.31 with three exceptions. First, coils are always in parallel. Secondly, the logic

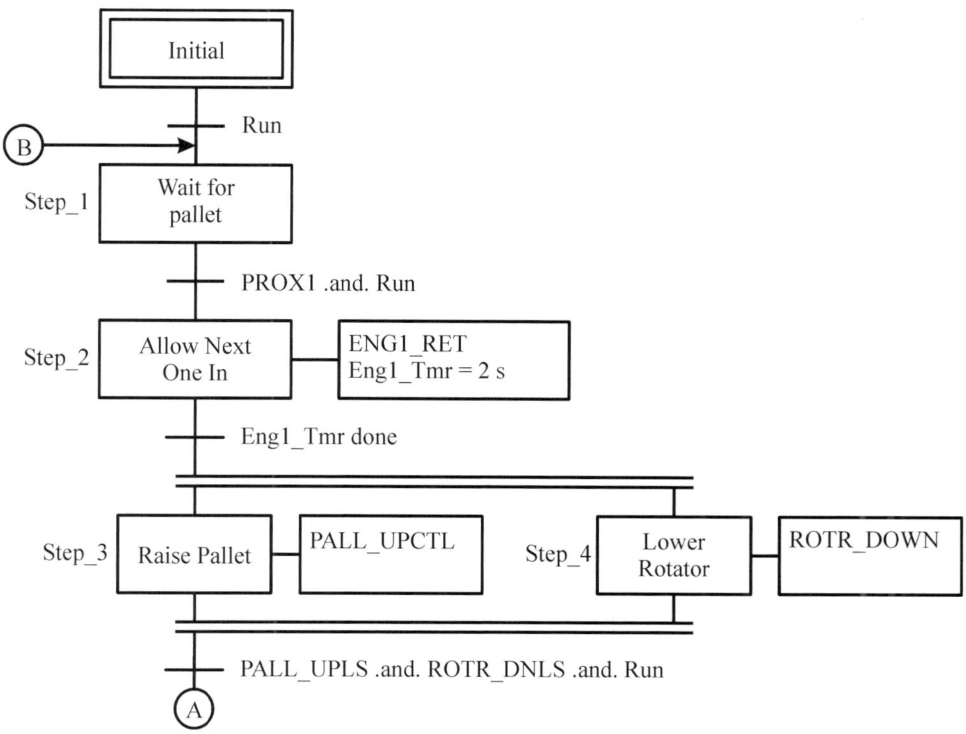

Figure 6.30. Function chart for revised engine inverter. *(continued)*

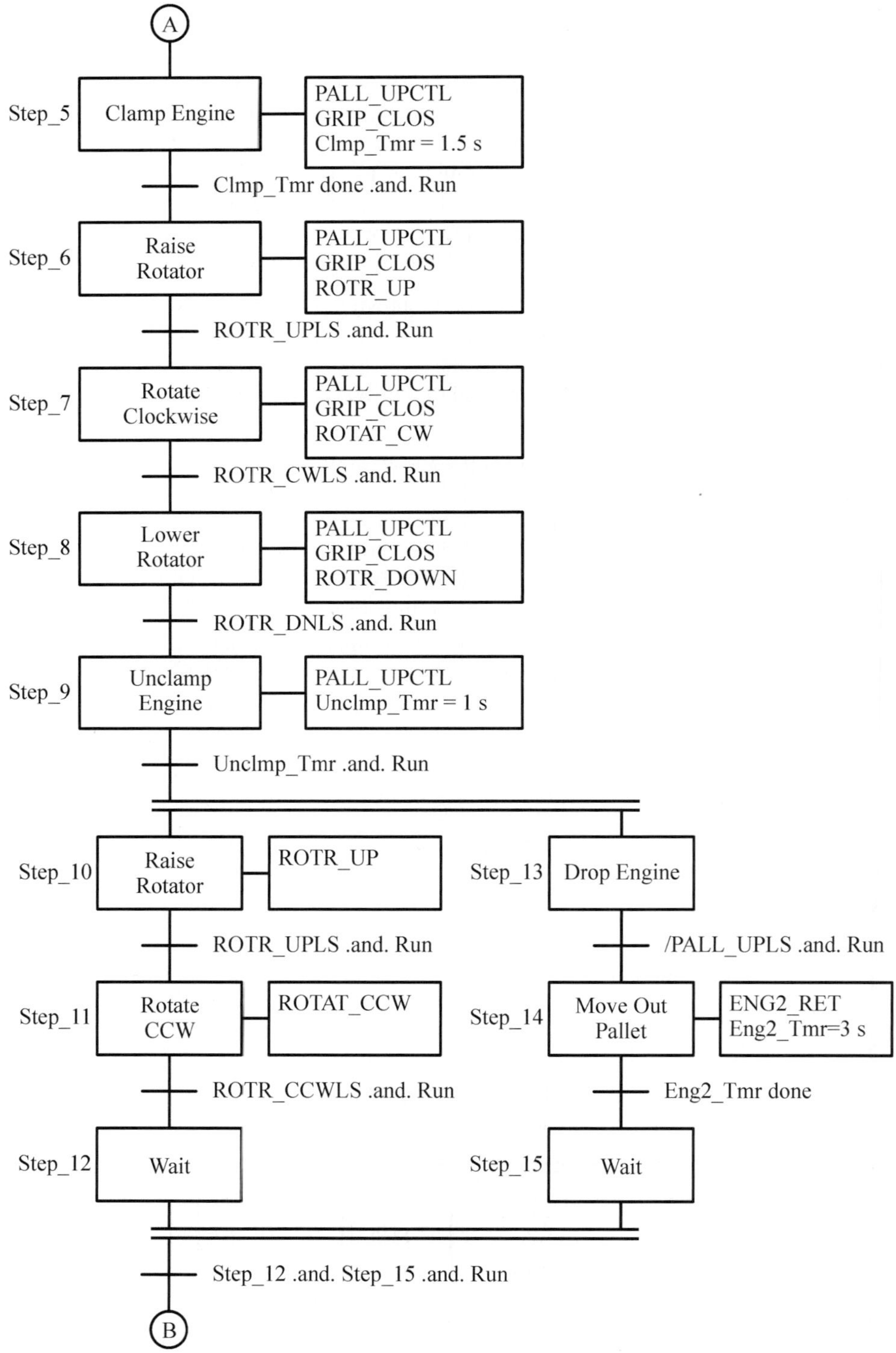

Figure 6.30. *(continued)*

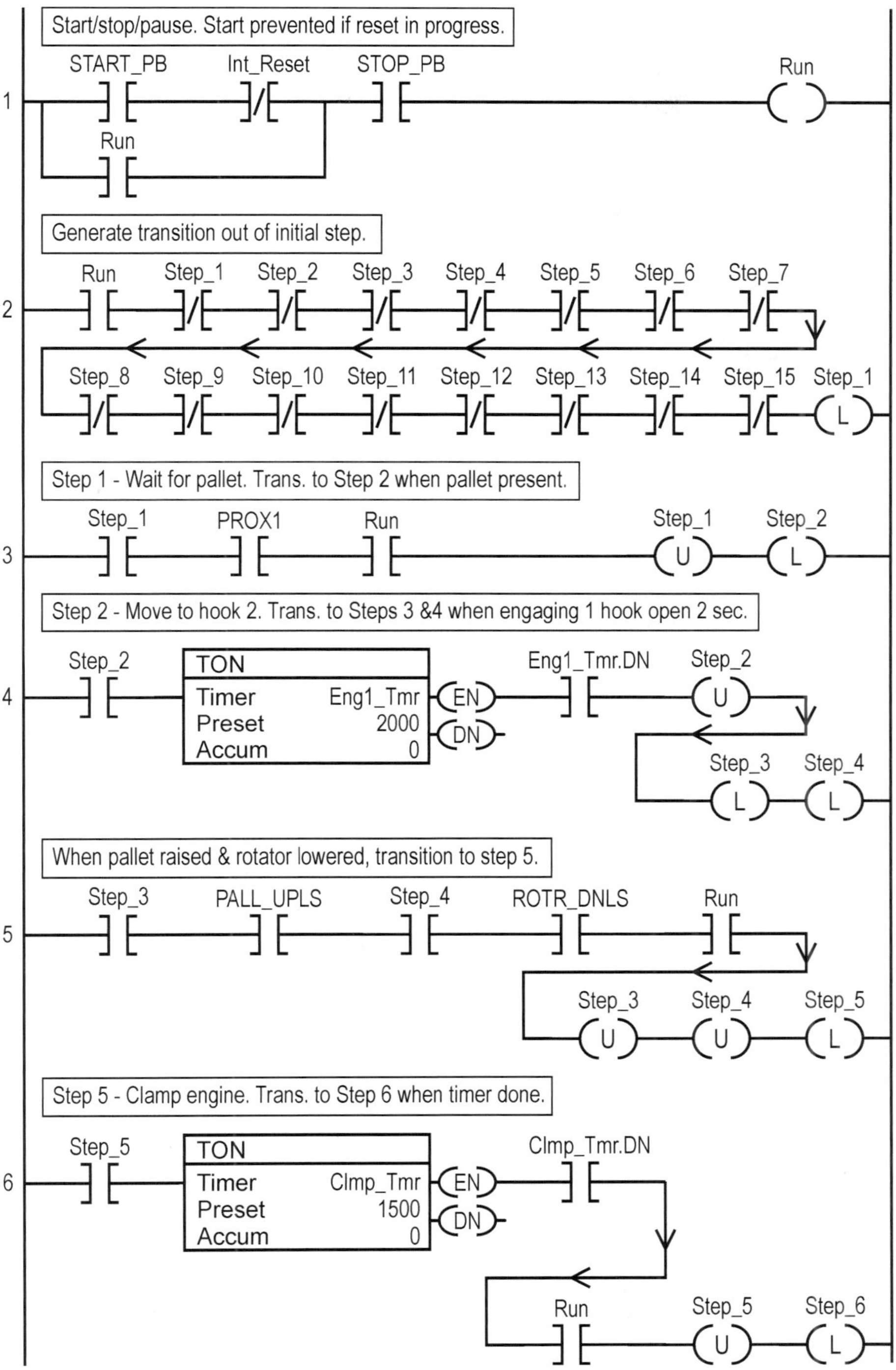

Figure 6.31. ControlLogix ladder logic for transitions of revised engine inverter station. *(continued)*

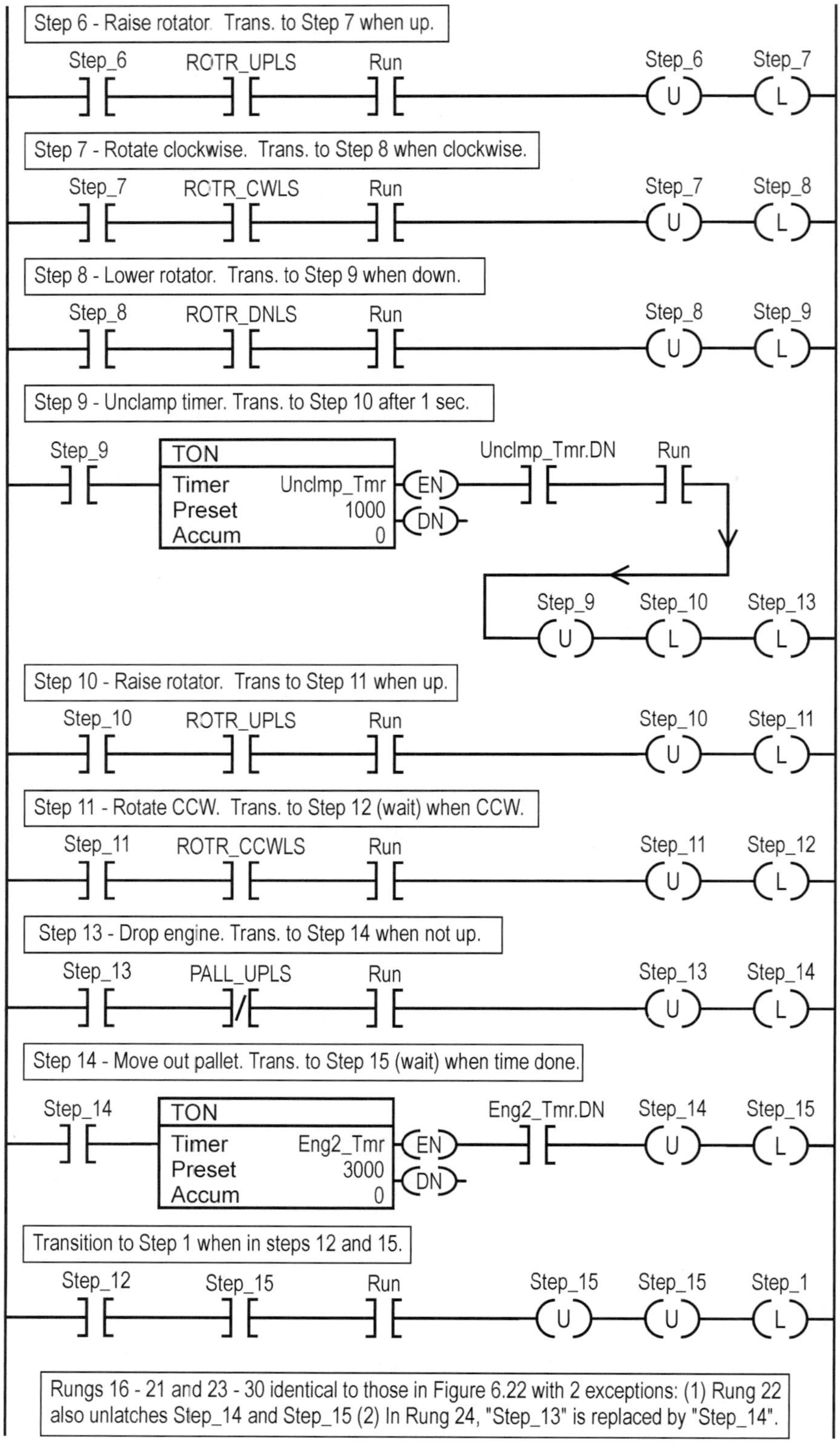

Figure 6.31. *(continued)*

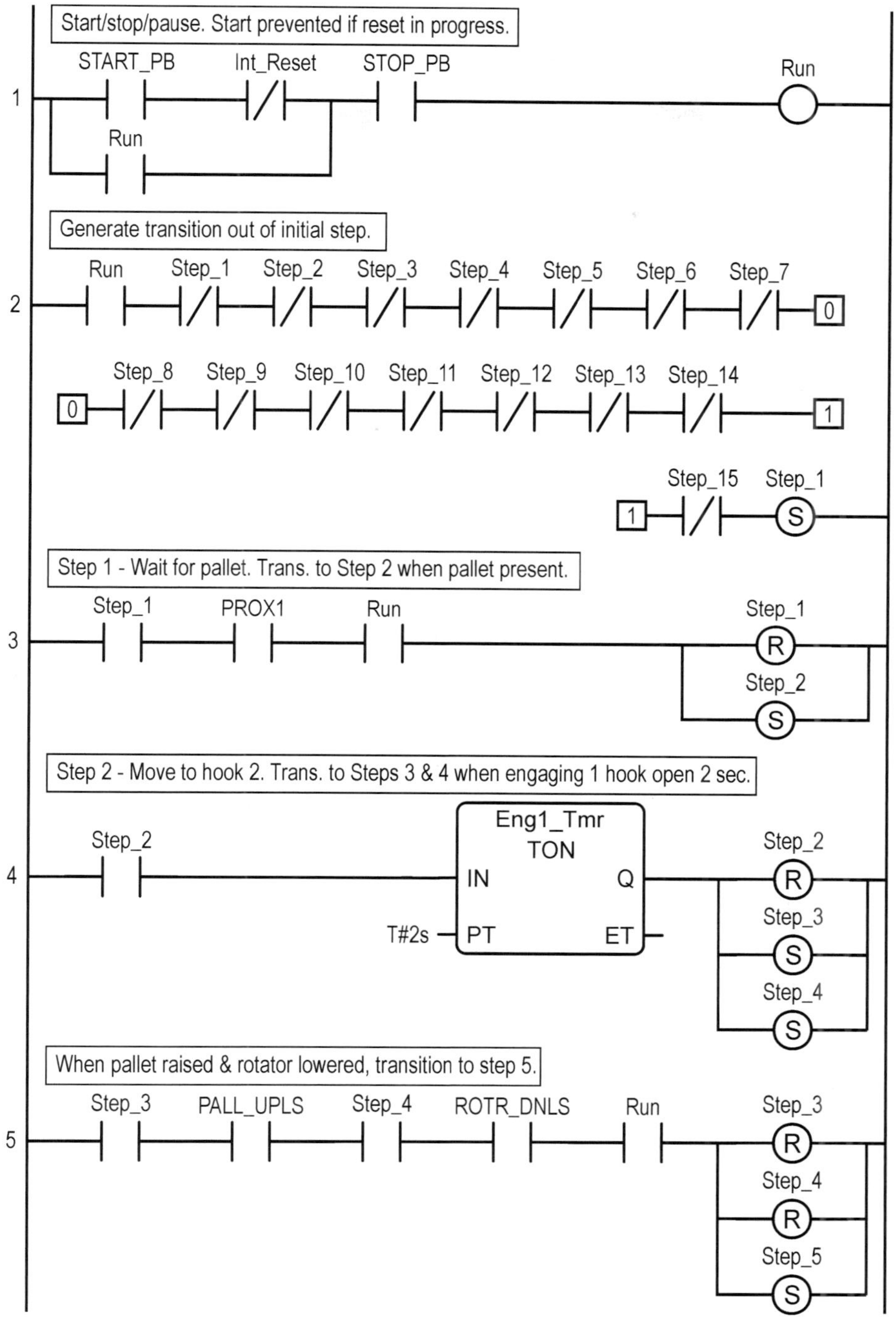

Figure 6.32. Micro800 ladder logic for transitions of revised engine inverter station.
(continued)

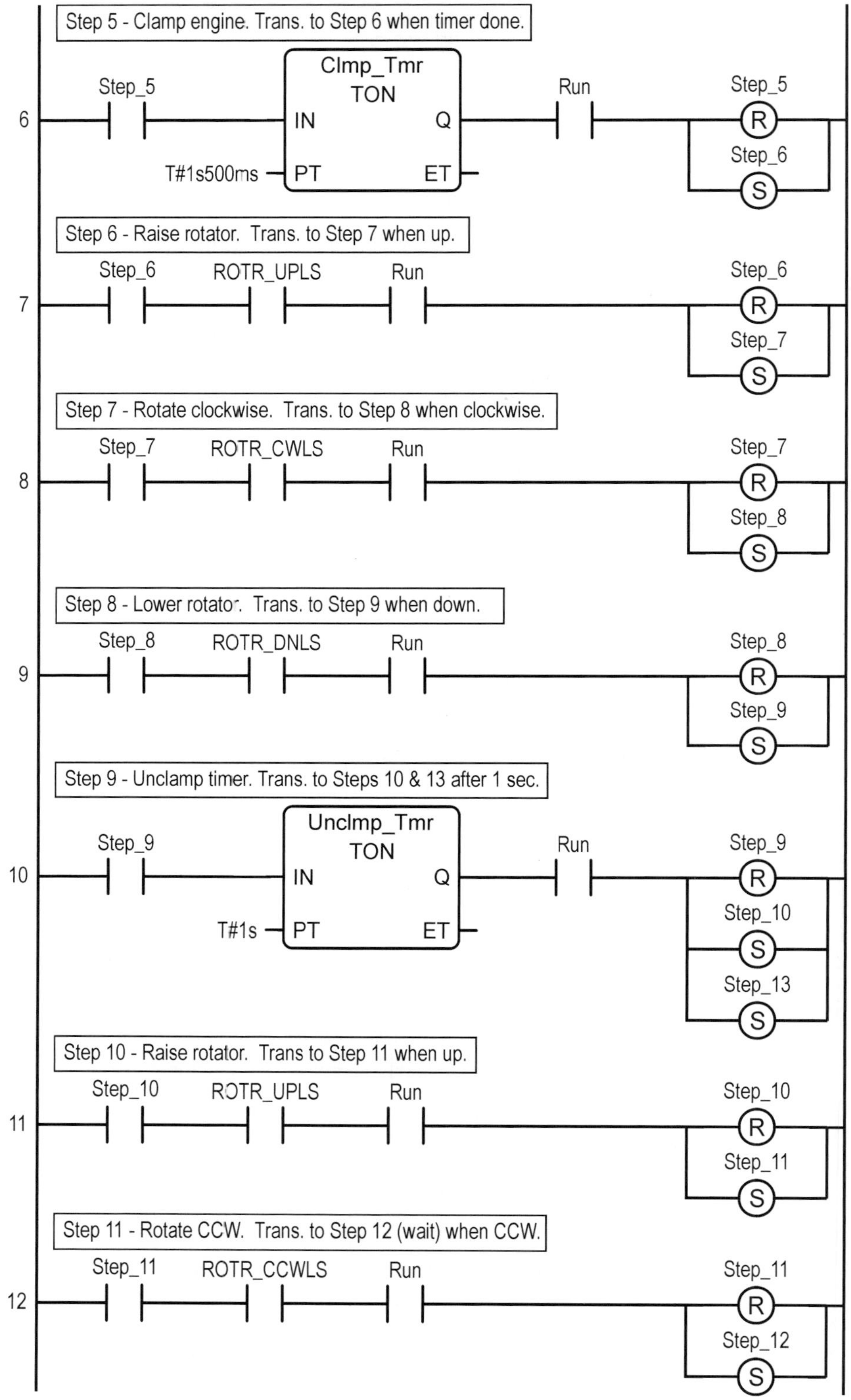

Figure 6.32. *(continued).*

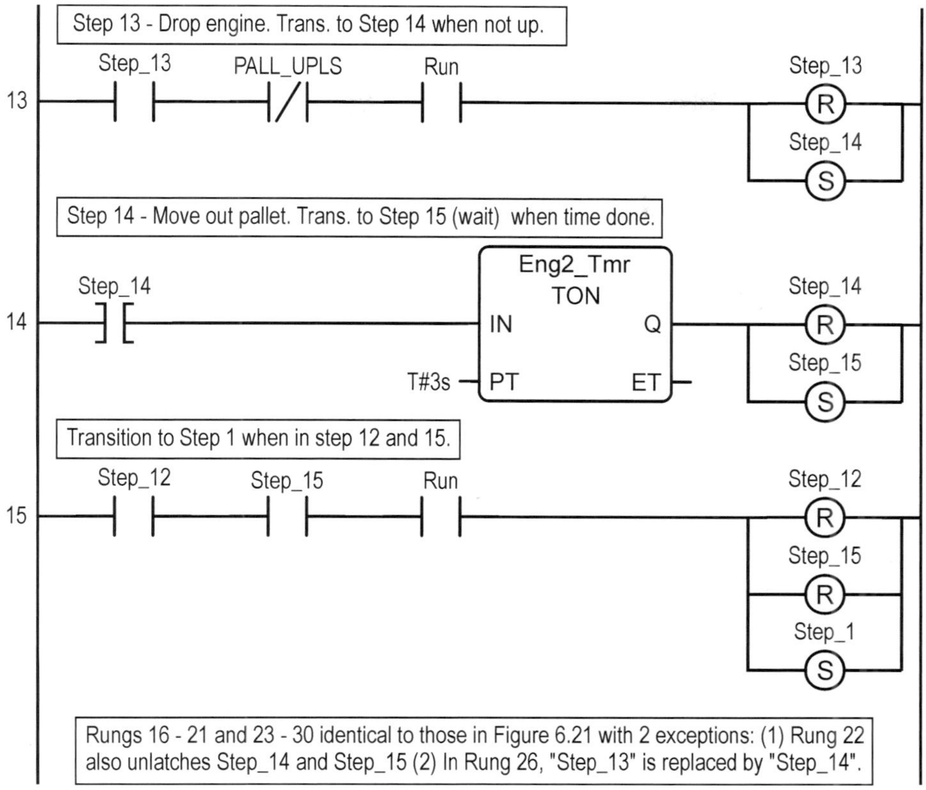

Figure 6.32. *(continued)*

on the right side of the ControlLogix timers is placed on a parallel branch. Thirdly, timers are represented differently. The "Time Base" field is present (set to 0.01) and the Preset value is divided by 10. Also, the timer symbol appears above the TON block and the timer address is in the Timer field of the TON block.

6.6 KEY QUESTIONS IN THE SEQUENTIAL DESIGN PROCESS

The examples in this chapter illustrate a simple method of programming sequential operations. The major part of the method is to develop a function chart of the operation:

1. Identify the steps and transition conditions.
2. Add step actions.

After the first draft of the function chart is developed, other key questions to ask:

Does the operation repeat?

If it does not repeat, then the last step may need to unseal the operation run status.

How is stop/pause handled?

Identify those physical outputs that must be off when paused.

Identify the timers that must retain their accumulator value.

Does pause prevent transitions?

Ignored in certain steps?

Does stop cause reset (or emergency stop)?

How is reset handled?

Reset only allowed when already paused?

Is there another sequential operation that must occur when the system is reset, to bring it back to an initial state?

6.7 MANUAL AND SINGLE-STEP SEQUENTIAL OPERATION

Up to this point, the control of a sequential machine has assumed mostly continuous operation. The operator can start and stop/pause the operation and reset it to the first step. This mode of operation is typically called the auto mode. However, when devices malfunction, there are two other modes of operation that are also useful. The first is the ability to single-step the operation. When in the single-step mode and the conditions to the next step are met, the program waits for the operator to press a "Continue" button in order for the operation to advance to the next step. The single-step mode allows personnel to monitor the operation of each step individually. The second mode that is useful is completely manual operation. In this mode, the operator may individually manipulate the physical outputs. Of course, a button or manual activation must be provided for each manipulated physical output.

The operator panel for this *three-mode* control of the machine operation is shown in Figure 6.33. The mode switch is a three-position selector switch. The other switches are all push button switches. The switches required for manual manipulation of the physical outputs are not shown.

In the single-step mode, the start/stop push buttons function the same as in the auto mode. When in the single-step mode, pressing the stop button pauses the operation. When paused, the start button must be pressed before pressing the continue button.

The transitions between modes are handled as follows:

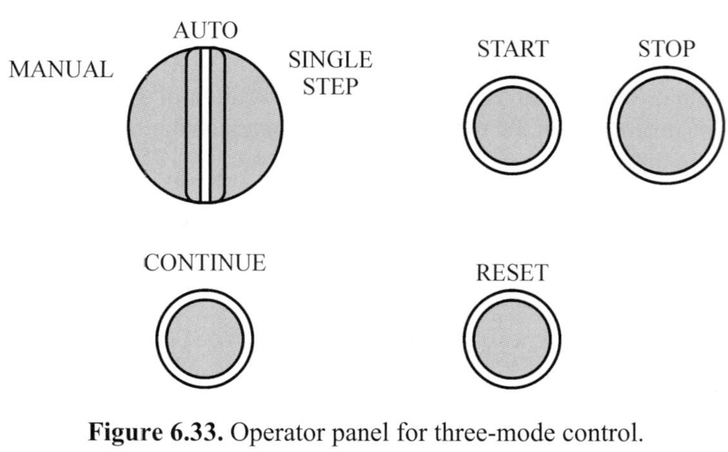

Figure 6.33. Operator panel for three-mode control.

Auto to Single-step: Operation completes the current step and then pauses, waiting for the "Continue" button to be pressed. The state of the Run coil does not change.

Single-step to Auto: Operation will resume from the current step. The state of the Run coil does not change.

Auto to Manual: Operation pauses at the current step (like the stop switch was pressed). The Run coil is turned off.

Manual to Auto: Operator must also press the Start button to resume automatic operation at the step from which it was switched to Manual. If the operation must resume from the first step, the Reset button must be pressed before the Start button is pressed.

The single-step and manual modes are easily added to the sequential operation. The ladder logic is modified as follows:

1. The start/stop rung is modified to disable the Run coil when in manual mode.
2. Auto and single-step mode logic is added in series with the step transition logic.
3. Manual mode logic is added in parallel to the conditions driving the physical outputs.

When adding the single-step and manual logic to the ladder logic, particular attention must be paid to those physical outputs that must be turned off when some physical limit is reached. For example, the command to extend a double-acting pneumatic cylinder may need to be turned off when the cylinder is fully extended, even while the operator is still pressing the "Extend" button in manual mode or before the "Continue" button is pressed when in single-step mode.

Example 6.6. Add three-mode control to the engine inverter of Example 6.4. Four additional physical inputs are required:

Variable	Description
AUTO	Mode selector switch in "Auto" position.
MANUAL	Mode selector switch in "Manual" position.
SSTEP	Mode selector switch in "Single Step" position.
CONTINU	Single-step mode continue push button, N. O., **on** transition causes transition to the next step when other transition conditions met and in single-step mode.

Solution. Rather than giving a complete solution to this problem, the following changes to the ladder logic of Figure 6.22 are shown in Figure 6.34:

Start/stop, rung 1 (Figure 6.34*a*)

Transition to step 2, rung 3 (Figure 6.34*b*)

Rotating mechanism down control, rung 26 (Figure 6.34*c*)

The dashed lines indicate the changes to the original ladder logic.

In rung 26, M_RTDN is an operator manual push button to move the mechanism. Also note that for both the single-step and manual modes, the control to move the mechanism down is turned off as soon as the ROTR_DNLS limit switch turns on.

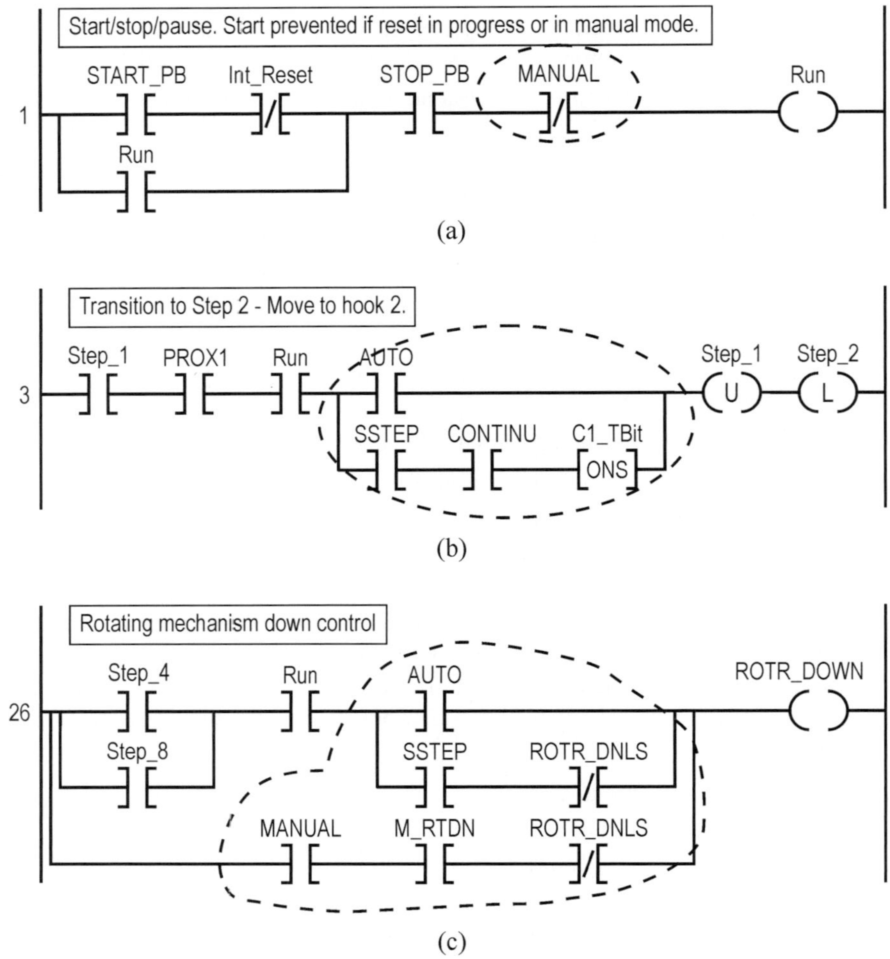

Figure 6.34. ControlLogix ladder logic changes for three-mode control (partial): *(a)* start/stop rung; *(b)* transition to step 2; *(c)* ROTR_DOWN control.

6.8 UNSTRUCTURED SEQUENCE

Unfortunately, one will often encounter existing ladder logic solutions that do not follow the structured approach advocated by this text. These solutions are typically unorganized and hard to follow. The most important disadvantage of this approach is that it requires more time to design, program, and troubleshoot the initial solution and any future modifications. These disadvantages are illustrated by the following example.

Example 6.7. Engine Inverter Station Control. Implement the engine inverter control of Example 6.4 without using the structured technique of this chapter.

Solution. Once one is exposed to the function chart design approach to design controls for sequential operations, it is hard to implement an unstructured approach. One approach to a solution that does not use the function chart as a design tool is described here. It is best described as "determine the start and stop conditions for each physical output and program the ladder logic accordingly." A ControlLogix solution for this problem is shown in Figure 6.35. The function of each rung is explained in the comments. Note that the normal operation start/stop/pause and the reset start/stop are handled as in previous examples. Internal coils are added as needed to define transitional conditions. Note the large number of transitional contacts. The logic of starting and stopping a particular physical output is largely determined by examining the state of the physical inputs. This solution does not exactly implement the operation as described in Example 6.4. On an initial start, there is a 2-second delay (engage 2 timer) before the next pallet is brought in, even if PROX1 is on when the operation is initially started.

With the function chart technique of this chapter, the step is explicitly defined and actions and transitions are associated with a step. In the unstructured approach of this example, the step is implicitly defined by the physical output actions. The sequence of steps in the unstructured approach is much harder to follow, particularly without detailed documentation. Also, it required about three times longer to design, program and troubleshoot this solution compared to the solution of Example 6.4.

Other examples of sequential control applications using unstructured approaches to the ladder logic appear in Bolton, 2015 (chapter 10 examples) and Petruzella, 2023 (section 6.11). These examples construct a flow chart to document the operation, but use an approach similar to this section to design the ladder logic.

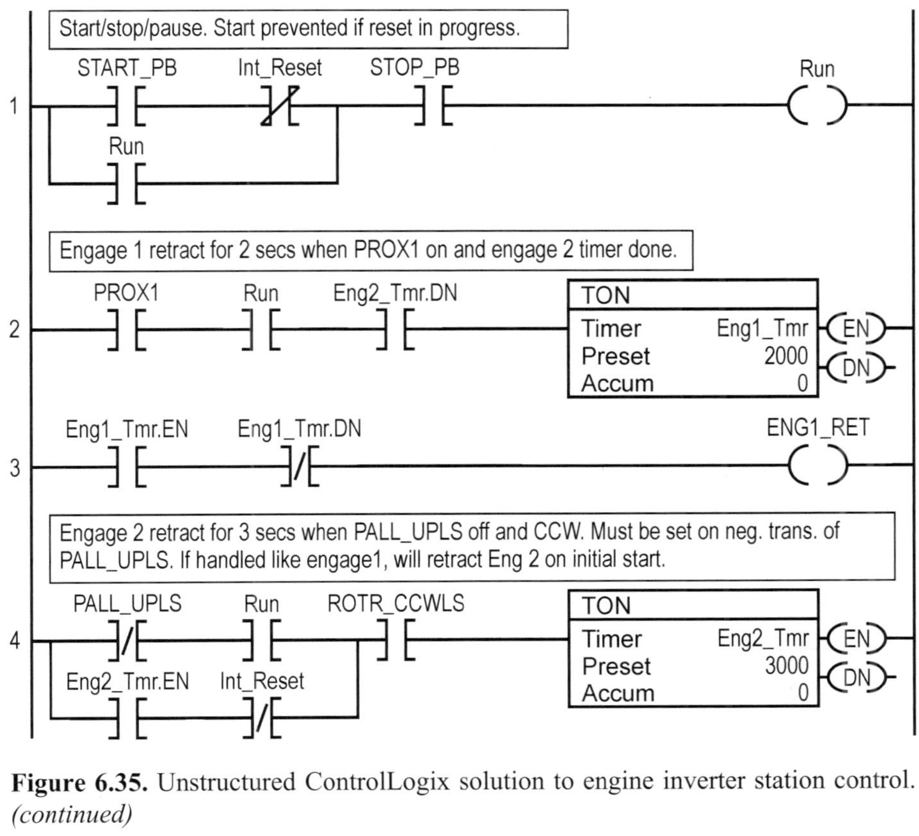

Figure 6.35. Unstructured ControlLogix solution to engine inverter station control. *(continued)*

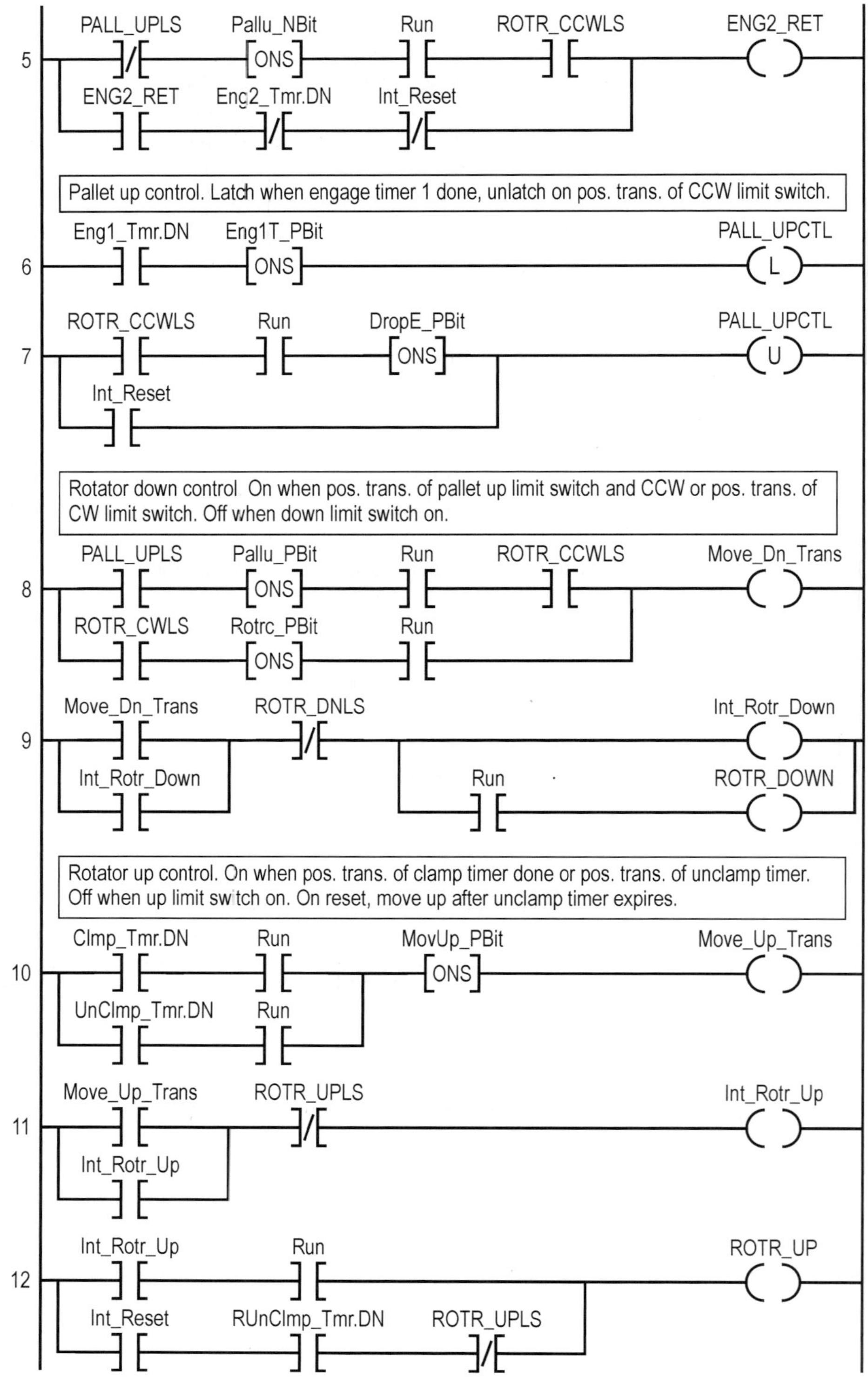

Figure 6.35. *(continued)*

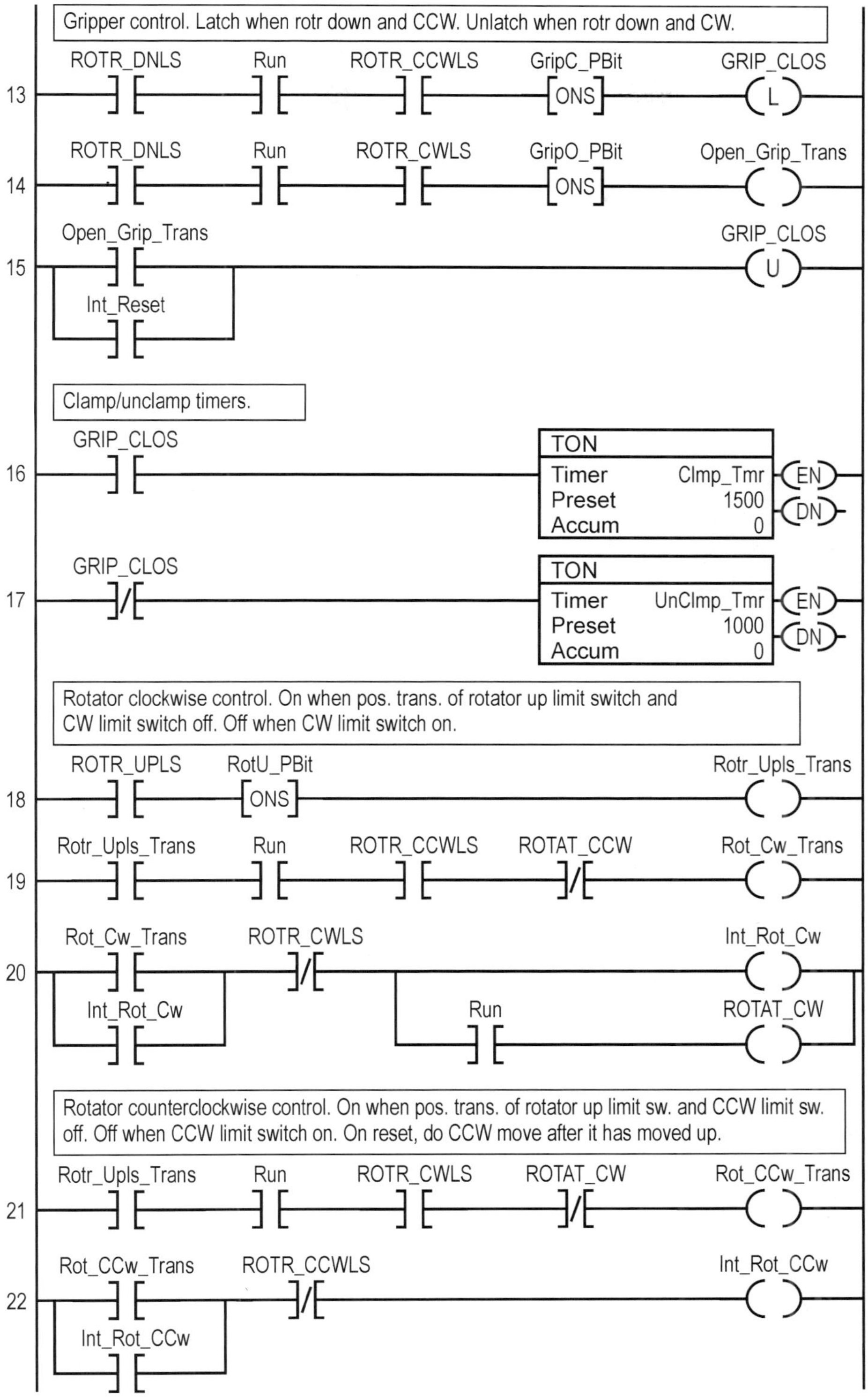

Figure 6.35. *(continued)*

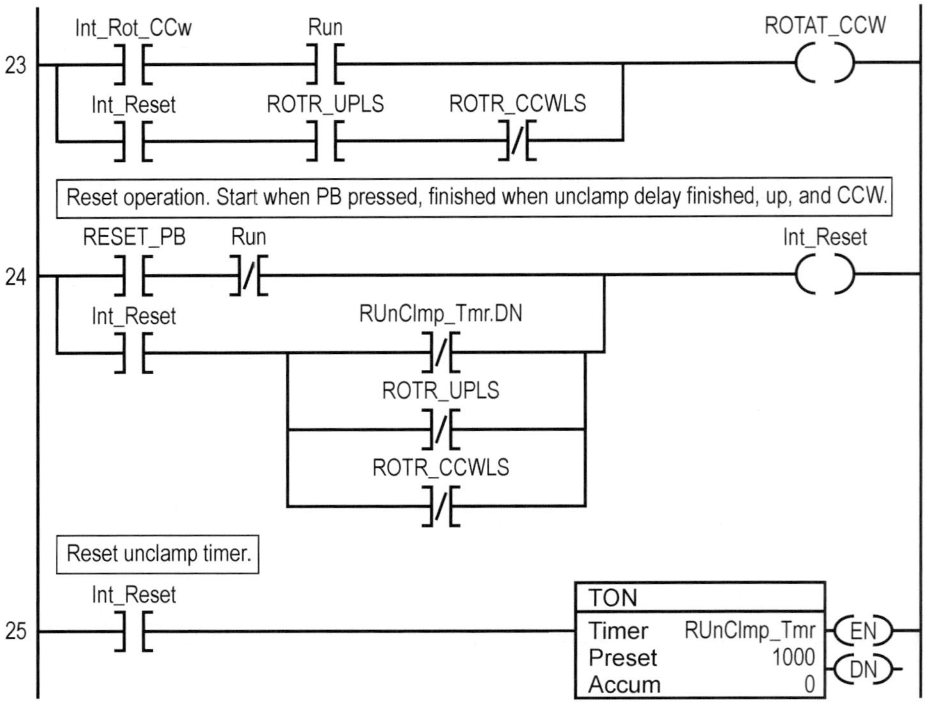

Figure 6.35. *(continued)*

6.9 CHAPTER SUMMARY

This chapter presents a technique for designing ladder logic programs to control sequential processes. The technique is based on describing the operation as a function chart and then translating the function chart to ladder logic code. The ladder logic uses the basic contact and coil instructions. Timers and counters are used only when explicitly needed by the operation. The ability to pause and reset an operation is also considered. The design of programs for operations with parallel steps and for operations that need single-step and manual control is examined. The chapter concludes by presenting an unstructured sequential control solution and emphasizing its drawbacks.

REFERENCES

Bolton, W., 2015. *Programmable Logic Controllers*, 6th Ed., Newnes.

Erickson, Kelvin T., 2023. *Programmable Logic Controllers: An Emphasis on Design and Application,* 4th Edition, Dogwood Valley Press, Rolla, MO.

IEC, 1988. *IEC 848: Preparation of Function Charts for Control Systems*, International Electrotechnical Commission.

IEC, 2013. *IEC 1131-3: Programmable Logic Controllers - Part 3: Programming Languages*, Edition 3.0, International Electrotechnical Commission.

Petruzella, Frank D., 2023. *Programmable Logic Controllers*, 6th Ed., McGraw-Hill.

Rockwell Automation, 2008. *SLC Instruction Set Reference Manual*, pub. 1747-RM001G-EN-P, Rockwell Automation Inc., Milwaukee, WI, Nov.

Rockwell Automation, 2011. *MicroLogix 1100 Programmable Controllers Instruction Set Reference Manual*, pub. 1763-RM001D-EN-P, Rockwell Automation Inc., Milwaukee, WI, Sep.

Rockwell Automation, 2014. *MicroLogix^{TM} 1200 and MicroLogix 1500 Programmable Controllers Instruction Set Reference Manual*, pub. 1762-RM001H-EN-P, Rockwell Automation Inc., Milwaukee, WI, Jul.

Rockwell Automation, 2019. *MicroLogix 1400 Programmable Controllers Instruction Set Reference Manual*, pub. 1766-RM001I-EN-P, Rockwell Automation, Inc., Milwaukee, WI, June.

Rockwell Automation, 2022a. *Logix5000 Controllers General Instructions Reference Manual*, pub. 1756-RM003W-EN-P, Rockwell Automation Inc., Milwaukee, WI, Mar.

Rockwell Automation, 2022b. *Micro800 Programmable Controllers General Instructions Reference Manual*, pub. 2080-RM001K-EN-P, Rockwell Automation Inc., Milwaukee, WI, Mar.

PROBLEMS

General instructions for the problems:

Using the function chart approach, write a ladder logic program for the application. Implement it for one of the following PLC ladder logic languages

ControlLogix/CompactLogix, **or**
Micro800, **or**
MicroLogix/SLC-500

If any part of the operation is ambiguous, write down your additional assumptions.

The physical inputs, physical outputs, and internal tags/variables/symbols for each problem are given in the problem. **DO NOT** assign any more physical inputs! Note that for the problems, the ControlLogix addresses assume 1756-, 1734-, 1769-, or 1794-series I/O modules. If using 5069- or 5094-series modules, the address tags will need to be modified.

Your solution should include the following:

1. Function chart of the process, showing the transitions between steps and the outputs active (or **on**) during each step.
2. Specify the PLC processor used.
3. Ladder logic diagram (with comments). For consistency among the different PLCs, use only tags/variables/symbols in the ladder logic. Use instructions consistent with the PLC processor.
4. Table listing additional internal memory (tags/variables/symbols) used and a brief description of their use. For the ControlLogix, CompactLogix, and Micro800, also indicate the data type for the internal memory. For MicroLogix and SLC-500, indicate the associated memory address.

Note to instructor: Break each problem into two assignments. For the first assignment, the students draw the function chart. The second assignment implements the ladder logic. For the second assignment, the students are allowed to use the correct function chart or their function chart if it is close to the correct solution. This approach will save the instructor from needing to grade many different ladder logic solutions.

P6-1. Using the function chart approach, implement the program for the following machine that injects ink into a highlighter marker.

Figure P6.1 shows the layout of a machine that injects ink into the felt "inside" of two highlighter markers. The markers are already mounted in a "belt" carrier and are moved in pairs through the stations that insert the felt, inject the ink, place the plug in the bottom, and then add the cap. You are only responsible for the station that injects the ink. The other aspects of the operation are controlled by other parts of the PLC program.

Also, assume that when the prox senses that the markers are indexed into position, both markers are present. Your solution will not be concerned with how to handle the case where no markers are present (for example, when starting a production run).

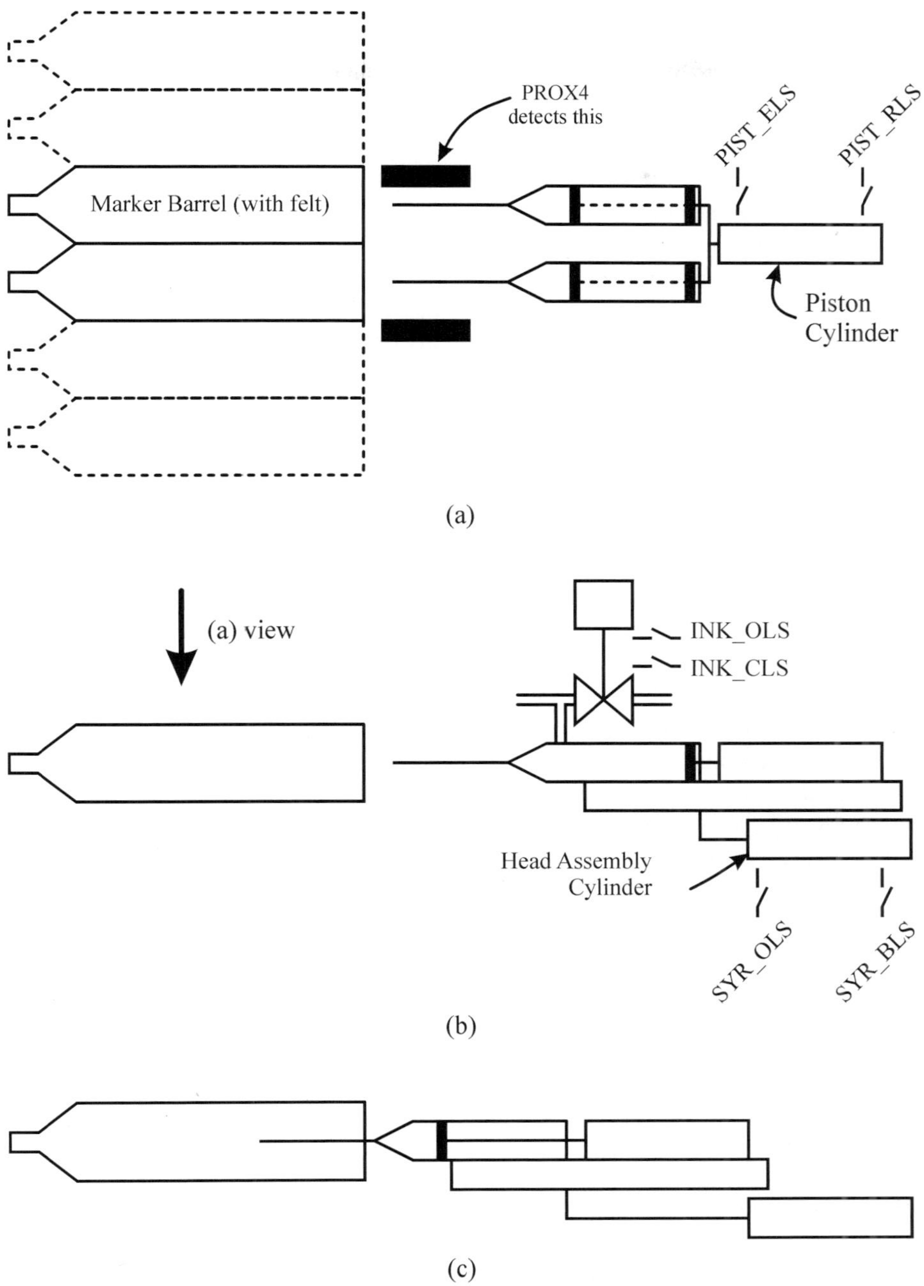

Figure P6.1. Marker ink injection station: *(a)* top view; *(b)* side view, head back; *(c)* side view, head out and piston extended.

Upon initial startup, wait for the prox to sense the pin on the belt carrier. In summary, after the prox senses the presence of the markers, the syringe head moves out, the syringe piston is extended, there is a small delay, the syringe head moves back, and the syringe piston is retracted. The initial states of the main mechanical controls are: syringe head back, syringe piston retracted, and ink valve closed.

The following steps are executed:

Wait for PROX4 to turn **on** to indicate that the belt has indexed the markers into position. The markers are now in position to be filled with ink.

Move syringe head out to insert the syringe needles into the felt inside the marker barrel. SYR_OLS senses when the syringe head is in position.

Extend syringe piston to inject ink into the marker felt. PIST_ELS senses when the piston is in its fully extended position. Note that the ink valve must be closed during this step (or the ink will be simply driven back to the reservoir).

Delay 0.5 seconds to allow the ink to finish flowing into the felt.

Move syringe head back. SYR_BLS senses when the syringe head assembly is fully retracted and the needles are out of the marker barrel.

Retract syringe piston (PIST_RLS senses). The ink valve must be open in this step to fill the syringe with ink for the next markers.

The operation then repeats.

PROX4: The PROX4 proximity switch senses a pin on the belt holding the markers and indicates that the markers are in position. When **on**, the markers are in position. When **off**, the markers are not in position (in the process of being moved).

Syringe Head Cylinder: The cylinder that controls the out/back movement of the syringe head assembly is a double-action pneumatic cylinder, controlled by the SYR_OUT and SYR_BACK outputs. When SYR_OUT is energized (turned **on**), the cylinder extends and moves the head assembly out, inserting the syringe needles into the marker barrel. When SYR_BACK is energized (turned **on**), the cylinder retracts, moving the head assembly back, retracting the needles. When SYR_OUT and SYR_BACK are both **off**, the cylinder stops at its present position. When SYR_OUT and SYR_BACK are both **on**, the cylinder operation is not consistent (it may stop or move in either direction slowly). The cylinder has two limit switches, SYR_OLS, which is **on** when the cylinder is extended and the needles are fully inserted into the marker barrels and SYR_BLS, which is **on** when the cylinder is retracted (fully back).

Syringe Piston Cylinder: The cylinder that controls the extend/retract movement of the syringe piston is a double-action pneumatic cylinder, controlled by the PIST_EXT and PIST_RET outputs. When PIST_EXT is energized (turned **on**), the syringe piston extends, injecting ink into the marker. When PIST_RET is energized (turned **on**), the syringe piston retracts. When PIST_EXT and PIST_RET are both **off**, the piston stops at its present position. When PIST_EXT and PIST_RET are both **on**, the piston operation is not consistent (it may stop or move in either direction slowly). The cylinder controlling the piston has two limit switches, PIST_ELS, which is **on** when the piston is extended and the ink has been injected into the marker and PIST_RLS, which is **on** when the piston is retracted (fully back).

The **ink valve** is basically a single-action solenoid controlled by the INK_VLV output. The INK_VLV physical output must remain **on** to open the valve. This output must be **off** to close the valve. The INK_OLS input is **on** when the valve is open, and the INK_CLS input is **on** when the valve is closed. The ink valve **must be open** when the syringe piston is **being retracted**, to bring ink into the syringe for the next marker. The ink valve must be closed all other times. If the piston is in its retracted position, the valve must be closed.

There is no start/stop switch for your part of the ladder. Instead, there is an internal coil, I_RUN that is **on** when the operation is to happen. When I_RUN is **off**, the operation should pause at its current step. **Do not advance** to the next step when paused. Also, when I_RUN is **off**, the **on** transition of PROX4 should be ignored (this is how the system disables ink injection at the start of a production run).

When paused, all outputs should be **off except** for the following:

1. The ink valve control, INK_VLV must not be turned **off** (if it is **on**).

When I_RUN turns back **on** while the operation of the station is paused, the station should resume its suspended step. Note that it is not necessary to retain the timer accumulator when paused.

The start and stop switches are absent from your part of the ladder because they control the overall operation of the marker production line. In order to reliably start/stop/pause the operation of the ink injection, the belt carrier and other production line stations must be controlled. These parts of the operation are outside the scope of this problem. The other part of the logic controls I_RUN. **You must not have any output coil referring to I_RUN.**

A separate reset internal coil, I_RESET, is provided which restores the production line to its initial state. The reset operation should cause the syringe head to be moved back, and the syringe piston retracted (valve open while retract). The reset operation is not complete until the head is back, the syringe piston is retracted and the ink valve is closed. There is no mechanical interference between the head and syringe piston. If the syringe piston is in the retracted position, the ink valve must be closed during the reset operation. Assume that an operator is responsible for actually removing any markers from the station after a reset. I_RESET should be ignored if I_RUN is **on**. The I_RUN internal coil used to restart the machine should be ignored when the reset operation is in progress. **You must not have any output coil referring to I_RESET.**

Do not add any more timed steps to those explicitly stated in the problem. In other words, do not put a timer in a step unless it is stated that the step duration is a specific time. Assume the tolerance on all timer values is at most ±0.01 seconds.

Assume the following physical inputs and output tags/variables/symbols.

Tag/Var./Symbol	Description
PROX4	Belt pin-sensing inductive proximity switch, **on** when markers in position, **off** otherwise.
SYR_OLS	Limit switch that closes (**on**) when the syringe head assembly is in the out position therefore the syringe needles are inserted into the marker barrel. **Off** otherwise.
SYR_BLS	Limit switch that closes (**on**) when the syringe head assembly is in the back position. **Off** otherwise.
PIST_ELS	Limit switch that closes (**on**) when the syringe piston is fully extended. **Off** otherwise.
PIST_RLS	Limit switch that closes (**on**) when the syringe piston is retracted. **Off** otherwise.
INK_OLS	Limit switch that closes (**on**) when ink valve is open. **Off** otherwise.
INK_CLS	Limit switch that closes (**on**) when ink valve is closed. **Off** otherwise.

SYR_OUT	Syringe head assembly cylinder out control, **on** to move assembly out. **Off** has no effect.
SYR_BACK	Syringe head assembly cylinder back control, **on** to move assembly back. **Off** has no effect.
PIST_EXT	Syringe piston cylinder extend control, **on** to extend piston and inject ink into marker. **Off** has no effect.
PIST_RET	Syringe piston cylinder retract control, **on** to retract piston. **Off** has no effect.
INK_VLV	Ink valve control. **On** to open valve and allow ink to be sucked into syringe. **Off** will close the valve. The control must be **on** while the syringe piston is being retracted. The control must be **off** while the syringe piston is being extended.

Assume the following internal tags/variables/symbols:

Tag/Var./Symbol	Description
I_RUN	Run/stop/pause control for the ink injection station; **on** to run machine, **off** to stop/pause. Controlled by another part of the ladder logic in the PLC. **You must not have any output coil referring to I_RUN.**
I_RESET	Reset control, **on** to restore station to initial state. Controlled by another part of the ladder logic in the PLC. **You must not have any output coil referring to I_RESET.**

The addresses associated with the physical input and output tags/variables/symbols are:

Tag/Var./Symbol	ControlLogix	Micro800	MLogix	SLC-500
PROX4	Local:1:I.Data.3	_IO_EM_DI_03	I:0/3	I:1/3
SYR_OLS	Local:1:I.Data.4	_IO_EM_DI_04	I:0/4	I:1/4
SYR_BLS	Local:1:I.Data.5	_IO_EM_DI_05	I:0/5	I:1/5
PIST_ELS	Local:1:I.Data.6	_IO_EM_DI_06	I:0/6	I:1/6
PIST_RLS	Local:1:I.Data.7	_IO_EM_DI_07	I:0/7	I:1/7
INK_OLS	Local:1:I.Data.8	_IO_EM_DI_08	I:0/8	I:1/8
INK_CLS	Local:1:I.Data.9	_IO_EM_DI_09	I:0/9	I:1/9
SYR_OUT	Local:2:O.Data.0	_IO_EM_DO_00	O:0/0	O:2/0
SYR_BACK	Local:2:O.Data.1	_IO_EM_DO_01	O:0/1	O:2/1
PIST_EXT	Local:2:O.Data.2	_IO_EM_DO_02	O:0/2	O:2/2
PIST_RET	Local:2:O.Data.3	_IO_EM_DO_03	O:0/3	O:2/3
INK_VLV	Local:2:O.Data.4	_IO_EM_DO_04	O:0/4	O:2/4

The addresses and/or data types associated with the internal tags/variables/symbols are:

Tag/Var./Synbol	CLogix Data Type	Micro800 Data Type	MLogix/SLC Address
I_RUN	BOOL	BOOL	B3/101
I_RESET	BOOL	BOOL	B3/102

P6-2. Using the function chart approach, implement the program for the following machine that unloads a part from a weld die.

Figure P6.2 shows the layout of a machine that removes a part from a weld die. The part is being removed from a rotary table. The part is placed onto a holder on the table, the table is indexed (turned one position), a piece is placed on the part, the table is indexed, the part is welded, the table is indexed, and then the part is removed. You are only programming the part of the operation that removes the welded part and places it in a bin. The other aspects of the operation are controlled by another part of the PLC program. Also, assume that when commanded to remove the part, the part is present. Your solution will not be concerned with how to handle the case where no part is present on the die.

Upon initial startup, assume a part is present at the "unload" position of the rotary table. In summary, the arm is swiveled to its CCW position, the arm is moved down, the piece is clamped, the arm is moved up, swiveled to its CW position, and the part is released. The initial states of the main mechanical controls are: arm CW (over bin), arm up, and the clamp released. The following steps are executed:

Wait for the INDEX_DN internal coil to turn **on**. This is the signal that the rotary table has indexed one position and a new part is ready to be unloaded.

Rotate the arm counter-clockwise (CCW) to place it over the part. ARM_CCWLS senses when the arm is in position over the part.

Move the arm down (ARM_DNLS senses).

Clamp the part (PART_PROX senses). Note that PART_PROX must be on for 0.5 seconds to reliably indicate the clamp has closed and grips the part.

Move the arm up (ARM_UPLS senses).

Rotate the arm clockwise (CW) to locate it over the part bin. ARM_CWLS senses when the arm is in position over the bin.

Release the part (CLAMP_OPLS and PART_PROX sense).

The operation then repeats.

The INDEX_DN internal coil transitions **on** when the rotary table has indexed one position and a new part is ready to be removed. This internal coil is controlled by another part of the ladder logic and thus your part of the program must not have any coil that refers to INDEX_DN. You may assume this internal coil is **off** by the time the part is released from the clamp.

Rotary Cylinder: The rotary cylinder that controls the swiveling of the arm is a double-action pneumatic cylinder, controlled by the ARM_CW and ARM_CCW outputs. When ARM_CW is energized (turned **on**), the arm rotates to the bin position, where the end of the arm is over the bin. When ARM_CCW is energized (turned **on**), the arm rotates to the part position, where the end of the arm is over the position to pick up the welded part. When ARM_CW and ARM_CCW are both **off**, the arm stops at its present position. When ARM_CW and ARM_CCW are both **on**, the rotary cylinder operation is not consistent (it may stop or move in either direction slowly). The cylinder has two limit switches, ARM_CWLS, which is **on** when the arm is over the bin position and ARM_CCWLS, which is **on** when the arm is over the part position.

Lift Cylinder: The cylinder that controls the up/down of the arm is a double-action pneumatic cylinder, controlled by the ARM_UP and ARM_DN outputs. When ARM_DN is energized (turned **on**), the cylinder retracts and moves the arm mechanism down. When ARM_UP is energized (turned **on**), the cylinder extends, moving the arm up. When

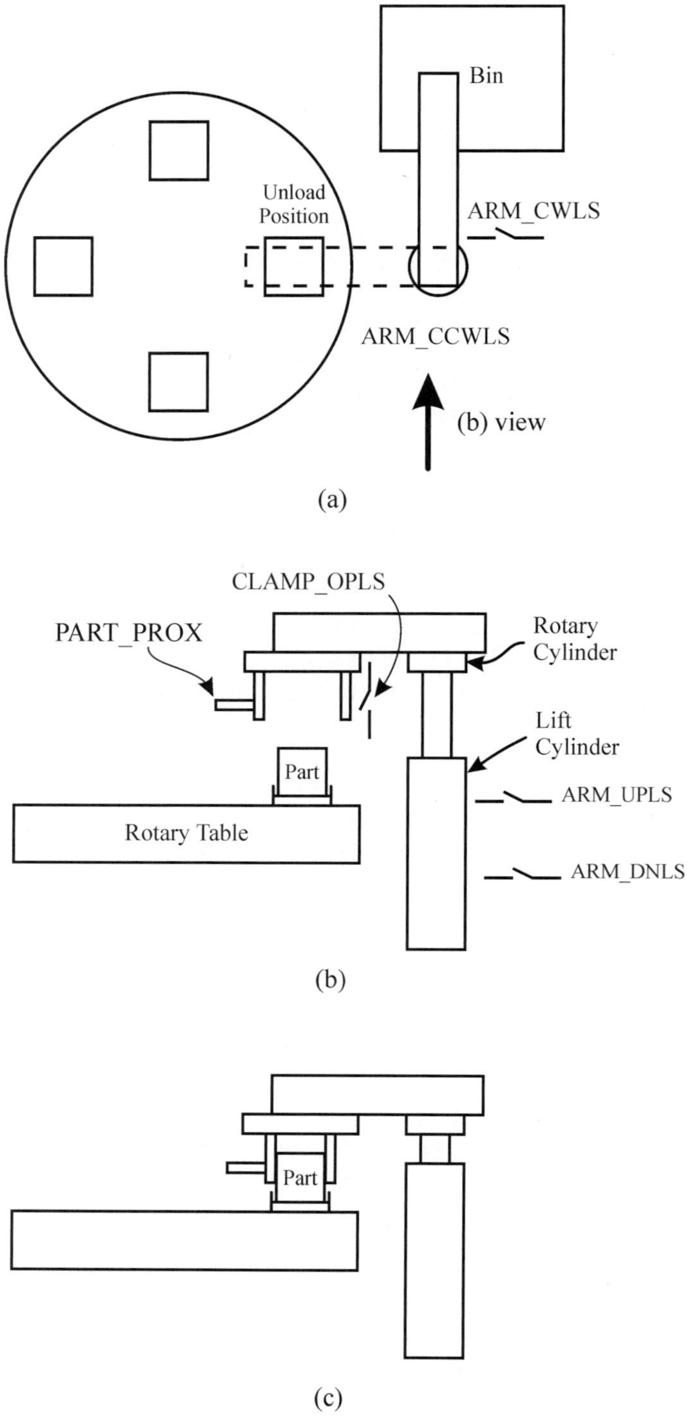

Figure P6.2. Weld die unload station: *(a)* top view; *(b)* side view, arm up; *(c)* side view, arm down and part clamped.

ARM_DN and ARM_UP are both **off**, the arm stops at its present position. When ARM_DN and ARM_UP are both **on**, the cylinder operation is not consistent (it may stop or move in either direction slowly). The cylinder has two limit switches, ARM_DNLS, which is **on** when the arm is in the position to pick up the part and ARM_UPLS, which is **on** when the cylinder is extended (fully up).

The **part clamp** is controlled by a single-action cylinder controlled by the CLAMP_CLOS output. The CLAMP_CLOS physical output must remain **on** to grip and hold the part. This output must be **off** to release the part. The CLAMP_OPLS input is **on** when the clamp is open, not holding the part and **off** otherwise.

The PART_PROX **capacitive sensor** senses the presence/absence of a part in the clamp. However, since the sense distance is a few millimeters, PART_PROX must be **on** for 0.5 seconds to reliably indicate the clamp has closed and grips the part. Also, PART_PROX is not a reliable indicator that the part is completely released and so it and the CLAMP_OPLS sensor must be used together to sense that the part has been released by the clamp and has fallen into the bin. Of course, PART_PROX remains **on** when the clamp is holding the part.

There is no start/stop switch for your part of the ladder. Instead, there is an internal coil, W_RUN that is **on** when the operation is to happen. When W_RUN is **off**, the operation should pause at its current step. **Do not advance** to the next step when paused. When paused, all outputs should be **off except** for the following exception:

1. The clamp control, CLAMP_CLOS must not be turned **off** (if it is **on**).

When W_RUN turns back **on** while the operation of the station is paused, the station should resume its suspended step. Note that it is not necessary to retain the timer accumulator when paused.

The start and stop switches are absent from your part of the ladder because they control the overall operation of the rotary table. In order to reliably start/stop/pause the operation of the machine, the rotary table and other stations around the perimeter must be controlled. These parts of the operation are outside the scope of this problem. The other part of the logic controls W_RUN. **You must not have any output coil referring to W_RUN.**

A separate reset internal coil, W_RESET, is provided which restores the rotary table machine to its initial state. The reset operation should cause the arm to be fully raised, the arm rotated so that it is over the bin, and the clamp released. The reset operation is not complete until the arm is up and over the bin (clockwise) and the clamp is released. There is some mechanical interference between the clamp and the part holder on the wheel. So, the arm should not be rotated nor the clamp released until the arm is in its fully up position. Assume that an operator is responsible for actually removing any parts from the station after a reset. W_RESET should be ignored if W_RUN is **on**. The W_RUN internal coil used to restart the machine should be ignored when the reset operation is in progress. **You must not have any output coil referring to W_RESET.**

Do not add any more timed steps to those explicitly stated in the problem. In other words, do not put a timer in a step unless it is stated that the step duration is a specific time. Assume the tolerance on all timer values is at most ±0.01 seconds.

Assume the following physical input and output tags/variables/symbols.

Tag/Var./Symbol	Description
PART_PROX	Part-sensing capacitive proximity switch, **on** when part in clamp, **off** otherwise. However, because of the sensing distance, there must be a delay before reliably detecting the part presence. Description above.
CLAMP_OPLS	Limit switch that closes (**on**) when part clamp is fully open and therefore cannot grip the part. **Off** otherwise.
ARM_CWLS	Limit switch that closes (**on**) when the arm is in the clockwise position therefore over the bin where the part is to be released. **Off** otherwise.
ARM_CCWLS	Limit switch that closes (**on**) when the arm is in the counter-clockwise position therefore in position to pick up the part. **Off** otherwise.
ARM_DNLS	Limit switch that closes (**on**) when the arm mechanism is in its down position. **Off** otherwise.
ARM_UPLS	Limit switch that closes (**on**) when the arm mechanism is in its fully up position. **Off** otherwise.
CLAMP_CLOS	Part clamp control. **On** to close the clamp and grip the part. **Off** will open the clamp, dropping the part from the clamp.
ARM_CW	Arm rotary cylinder clockwise control, **on** to move arm mechanism clockwise to bin position. **Off** has no effect.
ARM_CCW	Arm rotary cylinder counter-clockwise control, **on** to move arm mechanism counter-clockwise to part position. **Off** has no effect.
ARM_DN	Arm linear cylinder down control, **on** to move arm mechanism down. **Off** has no effect.
ARM_UP	Arm linear cylinder up control, **on** to move arm mechanism up. **Off** has no effect.

Assume the following internal tags/variables/symbols:

Variable	Description
W_RUN	Run/stop/pause control for the unload station (also for rotary table); **on** to run machine, **off** to stop/pause. Controlled by another part of the ladder logic in the PLC. **You must not have any output coil referring to W_RUN.**
W_RESET	Reset control, **on** to restore unload station (and rotary table) to initial state. Controlled by another part of the ladder logic in the PLC. **You must not have any output coil referring to W_RESET.**
INDEX_DN	Rotary index done. When **on**, table has indexed one position and a new part is ready to be removed. Controlled by another part of the ladder logic in the PLC. **You must not have any output coil referring to INDEX_DN.**

The addresses associated with the physical input and output tags/variables/symbols are:

Tag/Var./Symbol	ControlLogix	Micro800	MLogix	SLC-500
PART_PROX	Local:1:I.Data.3	_IO_EM_DI_03	I:0/3	I:1/3
CLAMP_OPLS	Local:1:I.Data.4	_IO_EM_DI_04	I:0/4	I:1/4
ARM_CWLS	Local:1:I.Data.5	_IO_EM_DI_05	I:0/5	I:1/5
ARM_CCWLS	Local:1:I.Data.6	_IO_EM_DI_06	I:0/6	I:1/6
ARM_DNLS	Local:1:I.Data.7	_IO_EM_DI_07	I:0/7	I:1/7
ARM_UPLS	Local:1:I.Data.8	_IO_EM_DI_08	I:0/8	I:1/8
CLAMP_CLOS	Local:2:O.Data.0	_IO_EM_DO_00	O:0/0	O:2/0
ARM_CW	Local:2:O.Data.1	_IO_EM_DO_01	O:0/1	O:2/1
ARM_CCW	Local:2:O.Data.2	_IO_EM_DO_02	O:0/2	O:2/2
ARM_DN	Local:2:O.Data.3	_IO_EM_DO_03	O:0/3	O:2/3
ARM_UP	Local:2:O.Data.4	_IO_EM_DO_04	O:0/4	O:2/4

The addresses and/or data types associated with the internal tags/variables/symbols are:

Tag/Var./Symbol	CLogix Data Type	Micro800 Data Type	MLogix/SLC Address
W_RUN	BOOL	BOOL	B3/101
W_RESET	BOOL	BOOL	B3/102
INDEX_DN	BOOL	BOOL	B3/151

P6-3. Using the function chart approach, implement the program for the following station that places a barcode label on a screwdriver.

Figure P6.3 shows the layout of a station that places a barcode label on a screwdriver. This station is only one in a series of stations along this conveyor. The stations upstream of this station manufacture the screwdrivers and each screwdriver comes to this station in a carrier. You are implementing ladder logic for this station only. You have no control over the conveyor so assume it is always moving. As a carrier is captured, the conveyor slides underneath. This particular line is asynchronous, that is, each station processes screwdrivers at its own speed and does not directly coordinate its operation with any other station.

Upon initial startup, assume that there are no carriers (with screwdrivers) in the station. The initial states of the main mechanical controls are: catch mechanism retracted, rotate head up and label ram retracted. The following steps are executed:

Rotate the label takeup reel until the label is in position (sensed by PX362). The catch cylinder must also be extended during the rotation.

Wait for next carrier (continue to extend the catch mechanism). When a carrier has been captured, PX361 senses.

Lower the rotate head into position (CY368_RLS senses). As the head is lowered, it engages the top of the screwdriver.

Extend the label cylinder to move the labels (on a backing tape) into contact with the screwdriver shaft (CY364_ELS senses).

Activate the motor on the rotate head to rotate the screwdriver so that the label is wrapped around the screwdriver shaft. The action also pulls the labels such that PX362 is turned **off**. Pulses from an encoder attached to the rotate motor are used to gauge the amount of rotation, which is 225 encoder pulses.

Retract the label cylinder to pull the label tape away from the screwdriver (CY364_RLS senses). Simultaneously, raise the rotate head (CY368_ELS senses). Do not advance to the next step until **both** of these movements are finished.

Allow the carrier to move out by retracting the catch mechanism for 2 seconds. This action allows the carrier to move down the conveyor enough so that the tab that is "caught" by the catch mechanism has passed the catch mechanism.

The operation then repeats.

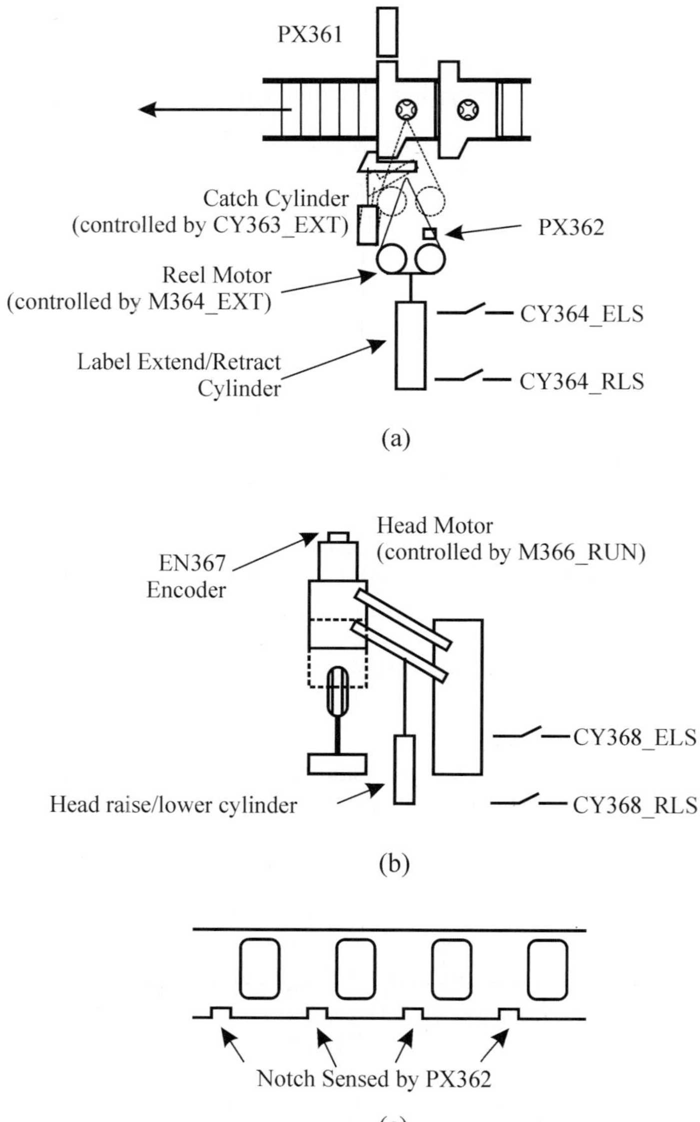

Figure P6.3. View of station: *(a)* from top; *(b)* from end; (c) barcode labels in backing tape.

The PX361 **proximity sensor** is an inductive proximity sensor that senses the presence/absence of a carrier in the station. PX361 is **on** when a carrier is present and **off** when there is no carrier in the station. PX361 actually senses the presence/absence of a tab extension on the carrier so that PX361 does turn **off** as the carrier moves out of the station. However, it turns **off** before the carrier has cleared the catching mechanism and so it is not a reliable indication that the carrier has moved out of the station.

The PX362 **proximity sensor** is a retro-reflective photoelectric proximity sensor that senses the presence/absence of a notch in the backing tape that holds the barcode label. PX362 is **on** when a label is in the proper position to be applied to the screwdriver shaft. You must assume that PX362 turns **off** while the label is being applied to the screwdriver, since the shaft rotation will pull the label tape.

The **rotate head motor** is controlled by the M366_RUN physical output. This output must be **on** in order to rotate the screwdriver to pull the label off the backing tape and wrap it around the shaft. The motor has an encoder on its shaft to measure the amount of rotation, 225 pulses of the EN367 physical input.

The **label reel motor** is controlled by the M365_RUN physical output. This output must be **on** in order to advance the label backing tape to the next label after the current one has been "peeled off" by the rotation of the screwdriver.

Catch Mechanism Cylinder: The catch cylinder is a single-action pneumatic cylinder, controlled by the CY363_EXT output. When CY363_EXT is energized, the cylinder extends and moves the catching "hook" to capture the next carrier. When CY363_EXT is de-energized, the cylinder and "hook" retracts. The cylinder has no limit switches.

The catch cylinder control must remain **on** while the head is being lowered/raised and while the label is being applied. Also, it must be **on** while the label tape is advanced in order to capture the next (or the first) carrier. Otherwise, if the carriers are spaced closely, the next one may be missed, which is not desirable.

Rotate Head Lower/Raise Cylinder: The cylinder that controls the lowering/raising of the rotation head is a double-action pneumatic cylinder, controlled by the CY368_RET and CY368_EXT outputs. When the CY368_RET output is energized (**on**), the cylinder retracts and lowers the rotating head down onto the screwdriver handle. When the CY368_EXT output is energized (**on**), the cylinder extends and raises the rotating head to clear the screwdriver handle. The mechanism stops if neither output is **on** or if they are energized simultaneously. The CY368_RLS limit switch is **on** when the cylinder is retracted (head down) and the CY368_ELS limit switch is **on** when the cylinder is extended (head up).

Label Extend/Retract Cylinder: The cylinder that controls the extend/retract of the label reel is a single-action pneumatic cylinder, controlled by the CY364_EXT output. When the CY364_EXT output is energized (**on**), the cylinder extends and moves the labels to contact the screwdriver shaft. When the CY364_EXT output is de-energized (**off**), the cylinder retracts and moves the labels away from the screwdriver. The CY364_ELS limit switch is **on** when the cylinder is extended (mechanism out) and the CY364_RLS limit switch is **on** when the cylinder is retracted (mechanism in). This cylinder must remain extended as the screwdriver shaft is rotated.

When the start switch is pressed (turned **on**) for the first time only, the station assumes there are no carriers in the station. When the stop switch is pressed (turned **off**) the operation should pause except when the screwdriver is being rotated. If the stop switch is pressed during the step that is rotating the screwdriver, this step must complete and advance to the next step. For all other steps **do not advance** to the next step when paused. When paused, all outputs **except** for the catch cylinder control (CY363_RET), label cylinder (CY364_EXT), and head motor must be turned **off**. Pressing the start switch while the operation of the station is paused causes the station to resume its suspended step.

A separate reset switch, RESET_PB, is provided which restores the system to its initial state. The reset operation should cause the rotate head to raise (cylinder extend) and the label cylinder to retract and the reset operation is not complete until the head is up and the labeling reel is retracted. Assume that an operator is responsible for actually removing any carrier from the station after a reset. The reset switch should be ignored if the station is running or during the screwdriver rotation step. The start switch should be ignored when the reset operation is in progress. If the station is paused and RESET_PB is **on**, the operator must release the RESET_PB before the start pushbutton switch can be used to restart the station.

Do not add any more timed steps to those explicitly stated in the problem. In other words, do not put a timer in a step unless it is stated that the step duration is a specific time. Assume the tolerance on all timer values is ±0.01 seconds.

Assume the following physical input and output tags/variables/symbols:

Tag/Var./Symbol	Description
START_PB	Start push button, N. O., **on** when starting.
STOP_PB	Stop push button, N. C., **off** when stopping.
RESET_PB	Reset push button, N. O., **on** when restoring station to initial state.
PX361	Carrier-sensing inductive proximity switch, **on** when carrier in station, **off** otherwise.
PX362	Photo-electric proximity switch that detects a notch in the backing tape, when **on** the barcode label is in position to be applied, **off** otherwise.
CY364_ELS	Limit switch that closes (**on**) when the labeling mechanism is extended and therefore in contact with the screwdriver shaft.
CY364_RLS	Limit switch that closes (**on**) when the labeling mechanism is retracted and therefore clear of the carrier.
CY368_ELS	Limit switch that closes (**on**) when the rotate head is extended and in the up position.
CY368_RLS	Limit switch that closes (**on**) when the rotate head is retracted and in the down position.
EN367	Head motor encoder pulses to count to measure amount of rotation.
CY363_EXT	Carrier catch cylinder control, **on** to extend and capture carrier, **off** to retract and release the carrier.
CY364_EXT	Labeling mechanism extend/retract control, **on** to extend, **off** to retract.
M365_RUN	Label reel motor, **on** to advance to next barcode label.

M366_RUN	Screwdriver rotation motor, **on** to turn screwdriver shaft.
CY368_RET	Rotate head down cylinder control, **on** to move head down, **off** has no effect.
CY368_EXT	Rotate head up cylinder control, **on** to move head up, **off** has no effect.

The addresses associated with the physical input and output tags/variables/symbols are:

Tag/Var./Symbol	ControlLogix	Micro800	MLogix	SLC-500
START_PB	Local:1:I.Data.0	_IO_EM_DI_00	I:0/0	I:1/0
STOP_PB	Local:1:I.Data.1	_IO_EM_DI_01	I:0/1	I:1/1
RESET_PB	Local:1:I.Data.2	_IO_EM_DI_02	I:0/2	I:1/2
PX361	Local:1:I.Data.3	_IO_EM_DI_03	I:0/3	I:1/3
PX362	Local:1:I.Data.4	_IO_EM_DI_04	I:0/4	I:1/4
CY364_ELS	Local:1:I.Data.5	_IO_EM_DI_05	I:0/5	I:1/5
CY364_RLS	Local:1:I.Data.6	_IO_EM_DI_06	I:0/6	I:1/6
CY368_ELS	Local:1:I.Data.7	_IO_EM_DI_07	I:0/7	I:1/7
CY368_RLS	Local:1:I.Data.8	_IO_EM_DI_08	I:0/8	I:1/8
EN367	Local:1:I.Data.9	_IO_EM_DI_09	I:0/9	I:1/9
CY363_EXT	Local:2:O.Data.0	_IO_EM_DO_00	O:0/0	O:2/0
CY364_EXT	Local:2:O.Data.1	_IO_EM_DO_01	O:0/1	O:2/1
M365_RUN	Local:2:O.Data.2	_IO_EM_DO_02	O:0/2	O:2/2
M366_RUN	Local:2:O.Data.3	_IO_EM_DO_03	O:0/3	O:2/3
CY368_RET	Local:2:O.Data.4	_IO_EM_DO_04	O:0/4	O:2/4
CY368_EXT	Local:2:O.Data.5	_IO_EM_DO_05	O:0/5	O:2/5

P6-4. Using the function chart approach, implement the program for the following carton erector that unfolds a six-pack carton and pushes it onto a conveyor.

Figures P6.4.1 and P6.4.2 show the layout of a station that erects (unfolds) a six-pack carton and pushes it onto a conveyor. The conveyor moves the cartons into a station that puts four cartons into a box and packs bottles of product into the boxes. You are implementing ladder logic for this station only. You have no control over the output conveyor, so assume it is always moving. Also assume that the carton erector runs as fast as it can and the box packer regulates its speed to match the production of cartons.

Upon initial startup, assume that there are no cartons in the station. The initial states of the main mechanical controls are: rotating hinge in the down position, unfolding cylinder in the up position, and the push ram in the retracted position. To erect a new carton, the following steps are executed:

Move the rotating hinge to the vertical position to pull off the next flat carton by activating ROTAT_EXT. At the same time, the ROT_VAC output must also be activated in order to hold the hinge to the flat carton. The ROTAT_ELS input senses when the hinge is vertical. The carton cannot be pulled off until **both** ROTAT_ELS and ROT_VAC_OK are **on**.

Move the rotating hinge to the horizontal position by activating ROTAT_RET, thus pulling down the flat carton. The ROTAT_RLS input senses when the hinge is horizontal. The ROT_VAC output must continue to be activated in order to hold the flat carton.

Release the vacuum holding the flat carton by turning **off** the ROT_VAC output. Delay 0.5 second to allow the feed vacuum to be released.

Push the flat carton into the position for unfolding by activating the PUSH_CYL output. The PUSH_ELS input senses when the cylinder has pushed the flat carton into position.

Lower the unfolding mechanism. RAISE_RLS senses when it is in the proper position. Simultaneously, the top and bottom vacuum (TOP_VAC and BOT_VAC) must be activated. Also, retract the push cylinder. The carton cannot be unfolded until RAISE_RLS, TOP_VAC_OK, BOT_VAC_OK, **and** PUSH_RLS are **all on**.

Erect the carton by raising the unfolding mechanism. RAISE_ELS senses when it is in the proper position.

The operation then repeats. The action of pushing in the next flat carton pushes out the previous unfolded carton.

Push Cylinder: The push cylinder is a single-action pneumatic cylinder, controlled by the PUSH_CYL output. When PUSH_CYL is energized, the cylinder extends and pushes the flat carton into the unfolding position (pushing out the previously unfolded carton). When PUSH_CYL is de-energized, the cylinder retracts. Two limit switches detect the position of the cylinder. PUSH_ELS is **on** when the cylinder is extended and the flat carton is in position to be unfolded. PUSH_RLS is **on** when the cylinder is retracted.

The **raising (unfolding)** and **rotation cylinders** are both double-acting pneumatic cylinders controlled by two outputs, one for each direction. Once a direction control is energized, the mechanism moves and keeps moving as long as power is applied (turned **on**) until the mechanical stop is reached. The mechanism stops at its current position when power is removed (turned **off**). The mechanism will not move if both opposing directions are energized simultaneously (e.g., extend and retract).

Rotation cylinder: When the ROTAT_EXT output is **on**, the hinge rotates to the vertical position and when the ROTAT_RET output is **on**, the hinge rotates to the horizontal

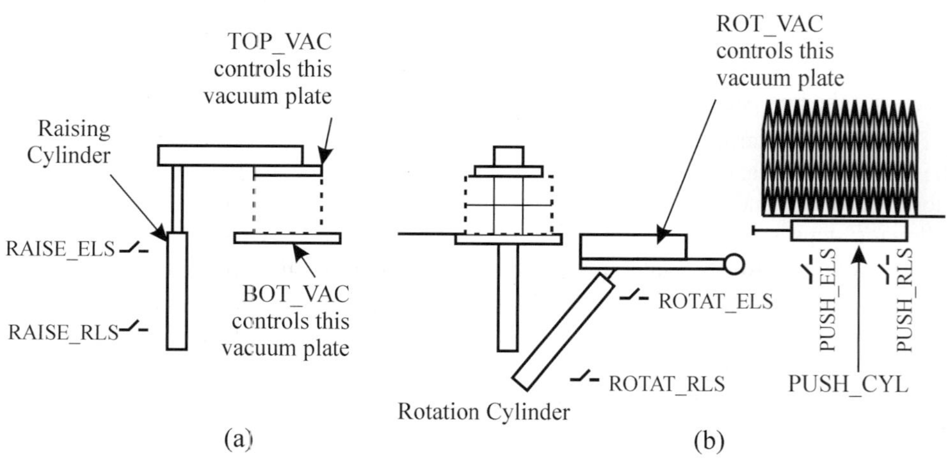

Figure P6.4.1. View of station at start of operation: *(a)* from front; *(b)* from side.

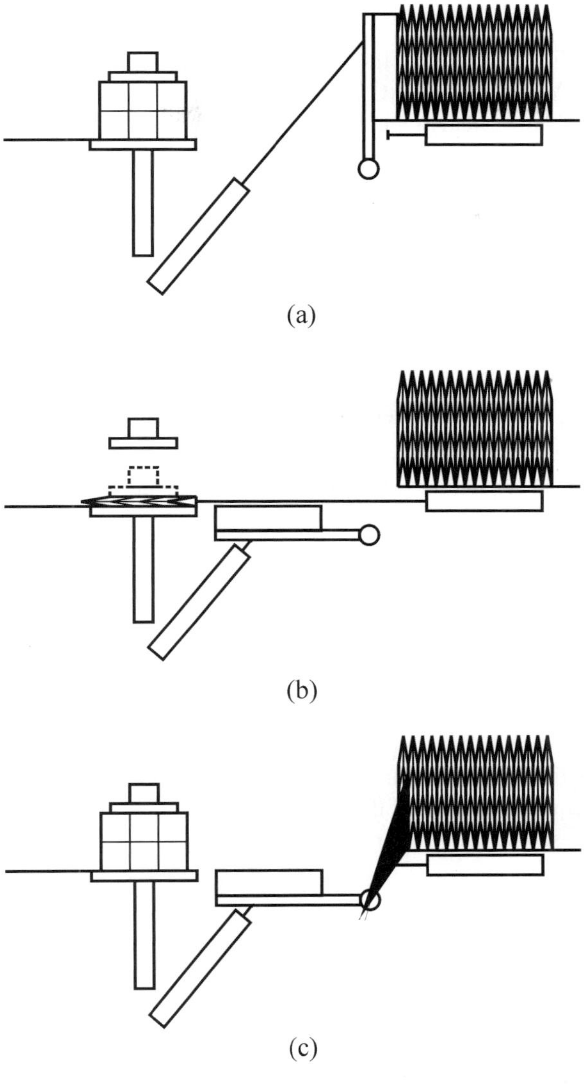

Figure P6.4.2. Station operation: *(a)* picking up flat carton; *(b)* carton pushed into position to be erected (unfolded); *(c)* erected carton.

position. The ROTAT_ELS limit switch is **on** when the hinge is vertical. The ROTAT_RLS limit switch is **on** when the hinge is horizontal.

Raising cylinder: When the RAISE_EXT output is **on**, the raising (unfolding) mechanism moves up and when the RAISE_RET output is **on**, the unfolding mechanism moves down. The RAISE_ELS limit switch is **on** when the mechanism is in the up position. The RAISE_RLS limit switch is **on** when the mechanism is down and contacting the flat carton.

Vacuums:

The flat carton is "gripped" by the **rotating hinge** using a vacuum, activated with the ROT_VAC output. When ROT_VAC is **on**, a valve is opened and air sucked through holes in a vacuum plate in the vertical hinge. When ROT_VAC is **off**, the valve is closed and there is no suction at the hinge. The ROT_VAC_OK input is **on** when the vacuum has been "made", that is the flat carton is now "stuck to" the hinge. You cannot assume that if ROT_VAC_OK is **off** the vacuum is totally released (hence the small time delay after releasing the vacuum).

The flat carton is "gripped" by the **top part** of the unfolding mechanism by a vacuum. The TOP_VAC output is used to activate the vacuum. When TOP_VAC is **on**, a valve is opened and air sucked through vacuum plate in the top part of the unfolding mechanism. When TOP_VAC is **off**, the valve is closed and there is no suction. The TOP_VAC_OK input is **on** when the vacuum has been "made", that is the top of the carton is "stuck to" the plate.

The flat carton is "gripped" at the **bottom part** of the unfolding position by a vacuum. The BOT_VAC output is used to activate the vacuum. When BOT_VAC is **on**, a valve is opened and air sucked through a vacuum plate below the carton. When BOT_VAC is **off**, the valve is closed and there is no suction. The BOT_VAC_OK input is **on** when the vacuum has been "made", that is the bottom of the carton is "stuck to" the vacuum plate.

The ROT_VAC output is **on** only when removing the next flat carton from the feed tray and when moving the hinge from the vertical to the horizontal position. The TOP_VAC and BOT_VAC outputs are **on** only when the unfolding mechanism is either moving down (to "grip" the flat carton) or moving up (to unfold the carton). All other times, these outputs must be **off**. The action of pushing in the next flat carton pushes out the previous unfolded carton. The time required to pull off the next flat carton allows time for the top and bottom vacuums to be released.

Start/Stop/Pause: When the start switch is pressed (turned **on**) for the first time only, the station assumes there is no carton (flat or unfolded) in the station. When the stop switch is pressed (turned **off**) the operation should pause. **Do not advance** to the next step when paused. When paused, all outputs **except** for the vacuum controls must be turned **off**. Pressing the start switch while the operation of the station is paused causes the station to resume its suspended step.

Reset: A separate reset switch, RESET_PB, is provided which restores the system to its initial state. The reset operation should cause all cylinders to retract and the reset operation is not complete until all cylinders are retracted. Mechanically, there is no interference between cylinders. Assume that an operator is responsible for actually removing any cartons (flat or folded) from the station after a reset. The reset switch should be ignored if the station is running. The start switch should be ignored when the reset operation is in progress. If the station is paused and RESET_PB is **on**, the operator must release the RESET_PB before the start pushbutton switch can be used to restart the station.

Do not add any more timed steps to those explicitly stated in the problem. In other words, do not put a timer in a step unless it is stated that the step duration is a specific time. Assume the tolerance on all timer values is ±0.01 seconds.

Assume the following physical input and output tags/variables/symbols.

Tag/Var./Symbol	Description
START_PB	Start push button, N. O., **on** when starting.
STOP_PB	Stop push button, N. C., **off** when stopping.
RESET_PB	Reset push button, N. O., **on** when restoring station to initial state.
ROTAT_ELS	Limit switch that closes (**on**) when the rotating hinge is in the vertical position.
ROTAT_RLS	Limit switch that closes (**on**) when the rotating hinge is horizontal.
RAISE_ELS	Limit switch that closes (**on**) when the raising (unfolding) cylinder is in the up position (carton unfolded).
RAISE_RLS	Limit switch that closes (**on**) when the raising (unfolding) cylinder is in the down position (ready to grab flattened carton).
PUSH_ELS	Limit switch that closes (**on**) when the push cylinder is extended.
PUSH_RLS	Limit switch that closes (**on**) when the push cylinder is retracted.
ROT_VAC_OK	Switch that closes (**on**) when the vacuum seal has been made between the rotating hinge and the next (flat) carton.
TOP_VAC_OK	Switch that closes (**on**) when the seal has been made between the top vacuum plate and the top of the (flat) carton.
BOT_VAC_OK	Switch that closes (**on**) when the seal has been made between the lower vacuum plate and the bottom of the (flat) carton.
PUSH_CYL	Push cylinder control, **on** extends, **off** retracts the cylinder.
ROTAT_EXT	Rotating hinge cylinder extend control, **on** to extend (pushing hinge to vertical position), **off** has no effect on cylinder position.
ROTAT_RET	Rotating hinge cylinder retract control, **on** to retract (pulling hinge to horizontal position), **off** has no effect on cylinder position.
RAISE_EXT	Raising linear cylinder extend control, **on** to extend (move up, opening held carton), **off** has no effect on cylinder position.
RAISE_RET	Raising linear cylinder retract control, **on** to retract (move down), **off** has no effect on cylinder position.
ROT_VAC	Control for vacuum plate on rotating hinge, **on** to activate vacuum, **off** releases vacuum.
TOP_VAC	Control for vacuum plate on top part of unfolding mechanism, **on** to activate vacuum, **off** releases vacuum.
BOT_VAC	Control for vacuum plate below unfolding position, **on** to activate vacuum, **off** releases vacuum.

The addresses associated with the physical input and output tags/variables/symbols are:

Tag/Var./Symbol	ControlLogix	Micro800	MLogix	SLC-500
START_PB	Local:1:I.Data.0	_IO_EM_DI_00	I:0/0	I:1/0
STOP_PB	Local:1:I.Data.1	_IO_EM_DI_01	I:0/1	I:1/1

RESET_PB	Local:1:I.Data.2	_IO_EM_DI_02	I:0/2	I:1/2
ROTAT_ELS	Local:1:I.Data.3	_IO_EM_DI_03	I:0/3	I:1/3
ROTAT_RLS	Local:1:I.Data.4	_IO_EM_DI_04	I:0/4	I:1/4
RAISE_ELS	Local:1:I.Data.5	_IO_EM_DI_05	I:0/5	I:1/5
RAISE_RLS	Local:1:I.Data.6	_IO_EM_DI_06	I:0/6	I:1/6
PUSH_ELS	Local:1:I.Data.7	_IO_EM_DI_07	I:0/7	I:1/7
PUSH_RLS	Local:1:I.Data.8	_IO_EM_DI_08	I:0/8	I:1/8
ROT_VAC_OK	Local:1:I.Data.9	_IO_EM_DI_09	I:0/9	I:1/9
TOP_VAC_OK	Local:1:I.Data.10	_IO_EM_DI_10	I:0/10	I:1/10
BOT_VAC_OK	Local:1:I.Data.11	_IO_EM_DI_11	I:0/11	I:1/11
PUSH_CYL	Local:2:O.Data.0	_IO_EM_DO_00	O:0/0	O:2/0
ROTAT_EXT	Local:2:O.Data.1	_IO_EM_DO_01	O:0/1	O:2/1
ROTAT_RET	Local:2:O.Data.2	_IO_EM_DO_02	O:0/2	O:2/2
RAISE_EXT	Local:2:O.Data.3	_IO_EM_DO_03	O:0/3	O:2/3
RAISE_RET	Local:2:O.Data.4	_IO_EM_DO_04	O:0/4	O:2/4
ROT_VAC	Local:2:O.Data.5	_IO_EM_DO_05	O:0/5	O:2/5
TOP_VAC	Local:2:O.Data.6	_IO_EM_DO_06	O:0/6	O:2/6
BOT_VAC	Local:2:O.Data.7	_IO_EM_DO_07	O:0/7	O:2/7

P6-5.Using the function chart approach, implement the program for the following machine that presses multiple items that have been stacked into a disk.

Figure P6.5 shows the layout of a machine that presses flat disk-shaped items that have been previously stacked into an assembly (Figure P6.5*a*) and then drops the assembly into a vibrating tray. In summary, the assembly is pressed, lifted, and then ejected into a vibrating tray that carries it to a bin. Upon initial startup, assume there is no assembly in position. The following steps are executed:

Wait for an assembly to move into position (sensed by ASSY_PROX). The assemblies are conveyed by a roller conveyor over which your part of the program has no control.

Move the eject rod up to be clear of the materials being pressed (EJCT_ULS senses).

Move the press down for 2 seconds to press the items together.

Move the press up (sensed by PRES_ULS)

Move the tray to place it beneath the press holding the pressed assembly (sensed by TRAY_ELS).

Move the eject rod down to push the pressed part off the press and drop into the tray (EJCT_DLS senses).

Move the tray back so that it is out of the way of the press (sensed by TRAY_RLS).

The operation then repeats.

Note that the assembly is "stuck" to the end of the press at the end of the pressing part of the operation and thus the need for the separate eject rod to push the pressed assembly off the press.

The ASSY_PROX **proximity sensor** is an inductive sensor placed between conveyor rollers that senses the presence/absence of an assembly in the position where it can be

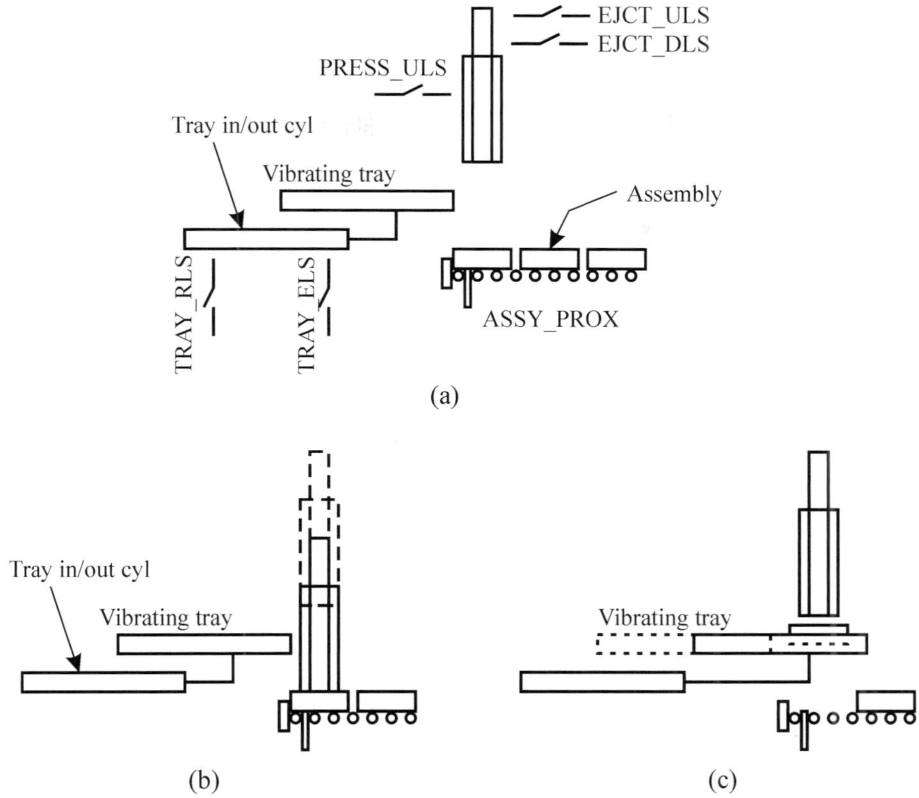

Figure P6.5. Assembly press machine: *(a)* view of station at start; *(b)* press down, tray in; *(c)* press up, tray out, eject down.

pressed. ASSY_PROX is **on** when an assembly is present and **off** when no assembly is present.

Press Cylinder: The cylinder controlling the press is a single-action pneumatic cylinder, controlled by the PRESS_EXT output. When PRESS_EXT is energized, the press cylinder extends, moving the press down, pushing the various pieces in the assembly together. When PRESS_EXT is de-energized, the press retracts. The PRESS_ULS limit switch is **on** when the press is in the "up" position (cylinder retracted. There is no limit switch that indicates that the press is extended (due to the possible small variations in the height of the assembly). Assume it requires 2 seconds for the press to move down and press the assembly.

Eject Rod Raise/Lower Cylinder: The cylinder that controls the lowering/raising of the eject rod is a double-action pneumatic cylinder, controlled by the EJCT_DOWN and EJCT_UP outputs. When the EJCT_DOWN output is energized (**on**), the cylinder extends and pushes the assembly off the press. When the EJCT_UP output is energized (**on**), the cylinder retracts and moves the eject rod up, allowing another assembly to be pressed. The mechanism stops if neither output is **on** or if they are energized simultaneously. The EJCT_DLS limit switch is **on** when the cylinder is extended (and assembly pushed off) and the EJCT_ULS limit switch is **on** when the cylinder is retracted.

Tray Extend/Retract: The cylinder that controls the in/out position of the vibrating tray that catches the pressed assembly is a double-action pneumatic cylinder, controlled by the TRAY_EXT and TRAY_RET outputs. When the TRAY_EXT output is energized (**on**), the cylinder extends and moves the tray out to catch the ejected assembly. When the TRAY_RET output is energized (**on**), the cylinder retracts and moves the tray in and thus not beneath the press. The mechanism stops if neither output is **on** or if they are energized simultaneously. The TRAY_ELS limit switch is **on** when the cylinder is extended (tray beneath press) and the TRAY_RLS limit switch is **on** when the cylinder is retracted (tray clear of press).

When the start switch is pressed (turned **on**) for the first time only, the station assumes there is no metal assembly in position in the station. When the stop switch is pressed (turned **off**) the operation should pause. **Do not advance** to the next step when paused **except** when the assembly is being pressed. When the assembly is being pressed, the step **must** complete and advance to the next step before pausing. When paused, all cylinder controls **except** for the press cylinder control (PRESS_EXT) and eject up control (EJCT_UP) must be turned **off**. Pressing the start switch while the operation of the station is paused causes the station to resume its suspended step.

A separate reset switch, RESET_PB, is provided which restores the system to its initial state. The reset operation should cause the press to raise, the eject rod to be retracted, and the tray moved in. The reset operation is not complete until the press is raised, the eject rod is retracted, and the tray is in. **Do not implement a function chart for the reset** operation. There is no mechanical interference since the press is not down with the tray out. RESET_PB should be ignored during the pressing step. Assume that an operator is responsible for actually removing any material from the machine after a reset. The reset switch should be ignored if the machine is running. The start switch should be ignored when the reset operation is in progress. If the station is paused and RESET_PB is **on**, the operator must release the RESET_PB before the start pushbutton switch can be used to restart the station.

Do not add any more timed steps to those explicitly stated in the problem. In other words, do not put a timer in a step unless it is stated that the step duration is a specific time. Assume the tolerance on all timer values is ±0.01 seconds.

Assume the following physical input and output tags/variables/symbols.

Tag/Var./Symbol	Description
START_PB	Start push button, N. O., **on** when starting.
STOP_PB	Stop push button, N. C., **off** when stopping.
RESET_PB	Reset push button, N. O., **on** when restoring station to initial state.
ASSY_PROX	Assembly-sensing proximity switch, **on** when assembly present in pressing position, **off** otherwise.
PRESS_ULS	Limit switch that closes (**on**) when the press is fully up, cylinder retracted, **off** otherwise.
EJCT_DLS	Limit switch that closes (**on**) when the eject rod is down (to push off assembly), **off** otherwise.
EJCT_ULS	Limit switch that closes (**on**) when the eject rod is up, **off** otherwise.

TRAY_ELS	Limit switch that closes (**on**) when the tray is in the extended position beneath the press, **off** otherwise.
TRAY_RLS	Limit switch that closes (**on**) when the tray is in the retracted position clear of the press, **off** otherwise.
PRESS_EXT	Press cylinder control, **on** to move press down, **off** to move press up to its mechanical stop.
EJCT_DOWN	Eject rod down (extend) cylinder control, **on** to move rod down, **off** has no effect.
EJCT_UP	Eject rod up (retract) cylinder control, **on** to move rod up, **off** has no effect.
TRAY_EXT	Catch tray mechanism extend cylinder control, **on** to move tray out, **off** has no effect.
TRAY_RET	Catch tray mechanism retract cylinder control, **on** to move tray in, **off** has no effect.

The addresses associated with the physical input and output tags/variables/symbols are:

Tag/Var./Symbol	ControlLogix	Micro800	MLogix	SLC-500
START_PB	Local:1:I.Data.0	_IO_EM_DI_00	I:0/0	I:1/0
STOP_PB	Local:1:I.Data.1	_IO_EM_DI_01	I:0/1	I:1/1
RESET_PB	Local:1:I.Data.2	_IO_EM_DI_02	I:0/2	I:1/2
ASSY_PROX	Local:1:I.Data.3	_IO_EM_DI_03	I:0/3	I:1/3
PRESS_ULS	Local:1:I.Data.4	_IO_EM_DI_04	I:0/4	I:1/4
EJCT_DLS	Local:1:I.Data.5	_IO_EM_DI_05	I:0/5	I:1/5
EJCT_ULS	Local:1:I.Data.6	_IO_EM_DI_06	I:0/6	I:1/6
TRAY_ELS	Local:1:I.Data.7	_IO_EM_DI_07	I:0/7	I:1/7
TRAY_RLS	Local:1:I.Data.8	_IO_EM_DI_08	I:0/8	I:1/8
PRESS_EXT	Local:2:O.Data.0	_IO_EM_DO_00	O:0/0	O:2/0
EJCT_DOWN	Local:2:O.Data.1	_IO_EM_DO_01	O:0/1	O:2/1
EJCT_UP	Local:2:O.Data.2	_IO_EM_DO_02	O:0/2	O:2/2
TRAY_EXT	Local:2:O.Data.3	_IO_EM_DO_03	O:0/3	O:2/3
TRAY_RET	Local:2:O.Data.4	_IO_EM_DO_04	O:0/4	O:2/4

P6-6. Using the function chart approach, implement the program for the following machine that turns a screw on an assembly a certain number of revolutions.

Figure P6.6 shows the layout of a station that turns a screw for 4 revolutions. The screw has been previously placed on every assembly that comes down the conveyor. The station is only one in a series of stations along this conveyor. You are implementing ladder logic for this station only. Your ladder program has control over the inbound conveyor, but does not control the outbound conveyor, so assume the outbound conveyor is always moving. Assume that the assemblies are spaced far enough apart that this operation can complete before the next assembly comes down the conveyor.

The inbound conveyor consists of rollers and the assembly rides upon the rollers.

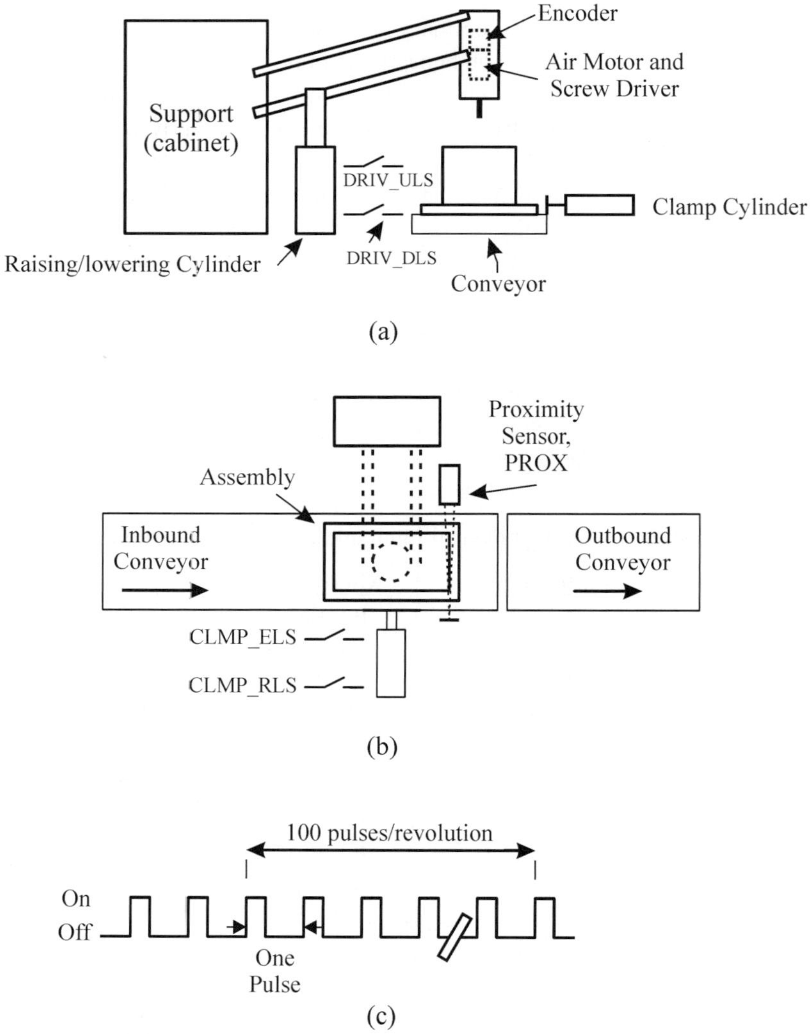

Figure P6.6. Station *(a)* end view; *(b)* top view; *(c)* encoder waveform.

Upon initial startup, assume there are no assemblies in the station. The following steps are executed:

Wait for a new assembly (PROX senses).

The assembly is clamped (CLMP_ELS senses).

The screw driver is lowered onto the screw. Assume that the mechanism on the end of the driver self-aligns to the screw head as it contacts the screw head.

The screw driver, powered by an air motor is turned for 4 revolutions. Pulses from an encoder are used to sense the number of revolutions.

Wait 0.5 seconds (to allow the driver to mechanically relax).

The screw driver is raised, and the assembly is unclamped (limit switches sense).

The next assembly is brought in.

The operation then repeats.

It is not necessary to use parallel steps. The screw driver mechanism can be moved up and the clamp retracted simultaneously without using parallel steps. Both mechanisms have a mechanical stop that prevents further movement past the limit switch. Do **not** assume that either movement finishes first or last. Both movements must be complete before moving the assembly out of the station.

Inbound Conveyor: The inbound conveyor transports assemblies into and out of the station. The inbound conveyor motor control, IN_MOTOR must be **on** to transport assemblies into and out of the station. IN_MOTOR must be **off** whenever the assembly is clamped. Assume that the assemblies are spaced far enough apart that this operation can complete before the next assembly comes down the conveyor.

Proximity sensor: PROX is a reflective type of proximity that at the instant it turns **off**, the assembly is at the correct position for screwing. Assume PROX remains **off** during the screwing operation and for at least 2 seconds after the assembly is unclamped. Assume the assemblies are spaced out far enough so that after one assembly is processed and moves out, the PROX sensor will detect the gap between assemblies.

Assembly Clamp: A single-action pneumatic cylinder drives the assembly clamp. Once the CLMP_EXT output is energized, the clamp extends and pushes the assembly about 5 mm into a "wall" to hold the assembly during the driving. The assembly is not pushed off the conveyor. The assembly must clamped while the driver is being lowered, the air motor is turning, and during the short wait after the air motor stops. Two limit switches detect the position of the clamp. CLMP_ELS is **on** when the clamp is extended (assembly clamped). CLMP_RLS is **on** when the clamp is retracted.

Raising/lowering Cylinder: A double-action pneumatic cylinder drives the vertical movements of the screw driver mechanism. It is controlled by two discrete outputs, DRIV_RET and DRIV_EXT. When DRIV_RET is energized (turned **on**), the screw driver mechanism moves down. When DRIV_EXT is energized (turned **on**),the screw driver mechanism moves up. When DRIV_RET and DRIV_EXT are both **off**, the mechanism stops at its position. When DRIV_RET and DRIV_EXT are both **on**, the cylinder operation is not consistent (it may stop or move slowly in either direction). Limit switches DRIV_ULS and DRIV_DLS indicate the upper and lower positions, respectively, of the driver mechanism.

Air Motor: The screw driver is turned by an air motor. The AIR_MOTOR discrete output controls the air motor. This output drives a valve whose opening has been set to generate the proper motor speed.

Encoder: The screw driver mechanism contains an encoder on the shaft that generates pulses (DRV_ENC) as the screw driver turns. The encoder generates 100 pulses for every revolution of the screw driver. Assume the PLC program scan time is sufficiently small to detect all pulses.

Start/Stop/Pause: When the start switch, START_PB, is pressed (turned **on**) for the first time only, assume there are no assemblies in the station. If the stop switch is pressed (turned **off**) at any time, the station operation should pause. When the start switch is pressed while the operation is paused, the station should resume the suspended step. When the station is paused, all outputs except for CLMP_EXT must be **off**. When paused, the clamp must continue to hold the assembly. When paused, **do not advance** to the next step.

Reset: A separate reset switch, RESET_PB, when **on**, must reset any internal states. After a press of RESET_PB, no assemblies are assumed present in the station. Assume that an operator has already removed any assembly before the start switch is pressed,. Also, the reset switch should cause the screw driver mechanism to raise and the clamp to retract. Do not implement a separate function chart for reset. Just do all the movements simultaneously until the driver and the clamp are in the correct positions. The reset switch should be ignored if the station is running. The start switch should be ignored when the reset operation is in progress. Also, the start switch should be ignored as long as RESET_PB is pressed.

Do not add any more timed steps to those explicitly stated in the problem. In other words, do not put a timer in a step unless it is stated that the step duration is a specific time. Assume the tolerance on all timer values is ±0.01 seconds.

Assume the following physical input and output tags/variables/symbols.

Tag/Var./Symbol	Description
START_PB	Start push button, N. O., **on** when starting.
STOP_PB	Stop push button, N. C., **off** when stopping.
RESET_PB	Reset push button, N. O., **on** when restoring station to initial state.
PROX	Proximity sensor, turns **off** (open) as assembly enters station (see above). Remains **off** as assembly is clamped and processed.
DRIV_DLS	Limit switch that closes (**on**) when the driver mechanism is in the lower (down) position.
DRIV_ULS	Limit switch that closes (**on**) when the driver mechanism is in the higher (up) position.
CLMP_ELS	Limit switch that closes (**on**) when clamp cylinder is extended (assembly clamped).
CLMP_RLS	Limit switch that closes (**on**) when clamp cylinder is retracted.
DRV_ENC	Screw driver encoder generates 100 pulses (Fig. 2*c*) for every revolution of the screw driver.
IN_MOTOR	Inbound roller conveyor motor control, **on** to move conveyor.
AIR_MOTOR	Screw mechanism air motor control, **on** to cause air motor to rotate, **off** stops rotation.
DRIV_RET	Screw driver mechanism down control, **on** to retract cylinder and move mechanism down, **off** has no effect.
DRIV_EXT	Screw driver mechanism up control, **on** to extend cylinder and move mechanism up, **off** has no effect.
CLMP_EXT	Clamp cylinder extend control, **on** to extend clamp, **off** causes clamp to retract.

The addresses associated with the physical input and output tags/variables/symbols are:

Tag/Var./Symbol	ControlLogix	Micro800	MLogix	SLC-500
START_PB	Local:1:I.Data.0	_IO_EM_DI_00	I:0/0	I:1/0
STOP_PB	Local:1:I.Data.1	_IO_EM_DI_01	I:0/1	I:1/1
RESET_PB	Local:1:I.Data.2	_IO_EM_DI_02	I:0/2	I:1/2
PROX	Local:1:I.Data.3	_IO_EM_DI_03	I:0/3	I:1/3

DRIV_DLS	Local:1:I.Data.4	_IO_EM_DI_04	I:0/4	I:1/4
DRIV_ULS	Local:1:I.Data.5	_IO_EM_DI_05	I:0/5	I:1/5
CLMP_ELS	Local:1:I.Data.6	_IO_EM_DI_06	I:0/6	I:1/6
CLMP_RLS	Local:1:I.Data.7	_IO_EM_DI_07	I:0/7	I:1/7
DRV_ENC	Local:1:I.Data.8	_IO_EM_DI_08	I:0/8	I:1/8
IN_MOTOR	Local:2:O.Data.0	_IO_EM_DO_00	O:0/0	O:2/0
AIR_MOTOR	Local:2:O.Data.1	_IO_EM_DO_01	O:0/1	O:2/1
DRIV_RET	Local:2:O.Data.2	_IO_EM_DO_02	O:0/2	O:2/2
DRIV_EXT	Local:2:O.Data.3	_IO_EM_DO_03	O:0/3	O:2/3
CLMP_EXT	Local:2:O.Data.4	_IO_EM_DO_04	O:0/4	O:2/4

P6-7. Using the function chart approach, implement the program for the following machine that polishes discs.

Figure P6.7 shows the layout of a station that polishes discs. The station is only one in a series of stations along this conveyor. You are implementing ladder logic for this station only. Your ladder program has control over the conveyor section in the station, and does not control the inbound or outbound conveyors. The inbound conveyor stops (**not controlled** by your logic) when your conveyor section stops. However, assume the outbound conveyor is always moving.

The conveyor consists of two parallel tracks that the discs ride on.

Upon initial startup, assume there are no discs in the station. The following steps are executed:

Wait for a new disc (PROX senses).

The spindle is moved up to place the disc in contact with polishing compound on a stationary wheel (SPNDL_ULS senses).

A rubber bladder on the end of the spindle is inflated to center the disc on the spindle and provide a grip to hold the disc as it is rotated. A pressure switch indicates that bladder has inflated sufficiently.

The spindle motor is turned for 10 seconds to polish the disc.

Stop the spindle motor and wait for it to completely stop (a zero-speed switch senses).

The spindle is moved down and the bladder is deflated (limit switches sense).

The disc is moved out.

The operation then repeats.

It is not necessary to use parallel steps. The spindle mechanism can be moved down and the bladder deflated simultaneously without using parallel steps. Do **not** assume that either action finishes first or last. The spindle must be down and the bladder deflated before moving the disc out of the station.

Station Conveyor: The conveyor transports discs into and out of the station. The conveyor motor control, CONV_MOTOR must be **on** only to transport discs into and out of the station. CONV_MOTOR must be **off** during all other parts of the operation. Another part of the ladder logic for the line controls the inbound conveyor and stops it when your conveyor is stopped. Therefore, this station can complete its operation before the next disc comes down the conveyor.

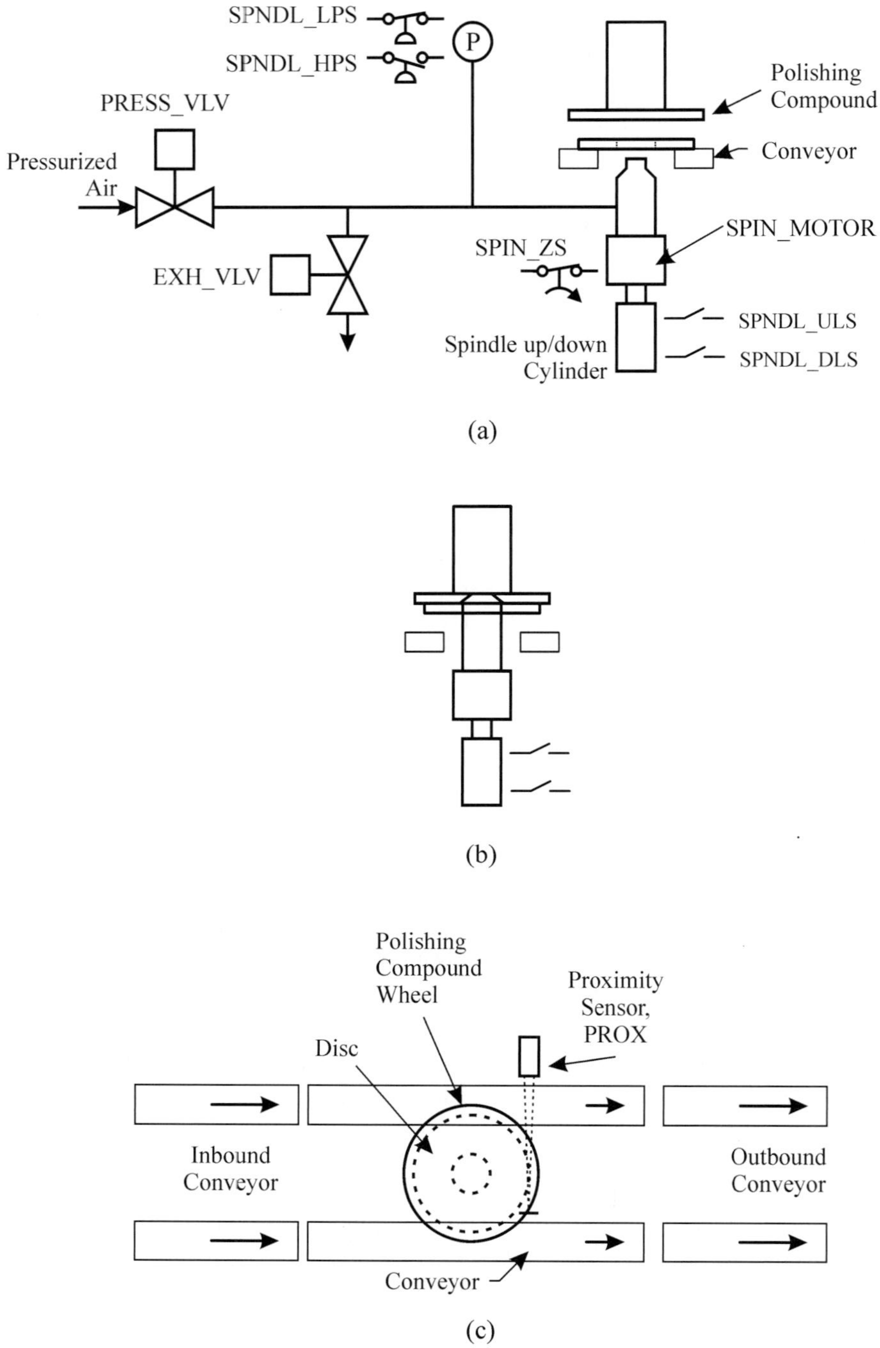

Figure P6.7. Station *(a)* end view; *(b)* spindle up and bladder inflated; *(c)* top view.

Proximity sensor: PROX is a reflective type of proximity that at the instant it turns **off**, the disc on the conveyor is at the correct position for polishing. PROX is **on** during the polishing operation and until the spindle is in the down position. PROX is **off** when the polished disk is set back on the conveyor. Assume the discs are spaced enough so that after one disc is processed and moves out, PROX will detect the gap between discs.

Bladder Inflate/deflate: Bladder inflation/deflation is controlled by two valves. When the PRESS_VLV output is energized, pressurized air enters the bladder and inflates it, holding the disc in its proper position. When the EXH_VLV output is energized, the air is exhausted from the bladder and it deflates and no longer holds the disc. The bladder must be pressurized while polishing the disc and therefore PRESS_VLV must continue to be energized while polishing. Two limit switches detect the pressure in the bladder. SPNDL_HPS is **on** when the bladder is at sufficient pressure to hold the disc in its proper position. SPNDL_LPS is **on** when the bladder pressure is low enough to have released the disc.

To prevent the problem where both PRESS_VLV and EXH_VLV are open (even briefly), PRESS_VLV must be **off** when the spindle motor is decelerating.

Spindle Up/down Cylinder: A double-action pneumatic cylinder drives the vertical movements of the spindle mechanism. It is controlled by two discrete outputs, SPNDL_UP and SPNDL_DOWN. When SPNDL_DOWN is energized (turned **on**), the spindle mechanism moves down. When SPNDL_UP is energized (turned **on**), the spindle mechanism moves up. When SPNDL_UP and SPNDL_DOWN are both **off**, the mechanism stops at its position. When SPNDL_UP and SPNDL_DOWN are both **on**, the cylinder operation is not consistent (it may stop or move slowly in either direction). Limit switches SPNDL_ULS and SPNDL_DLS indicate the upper and lower positions, respectively, of the spindle mechanism.

Spindle Motor: The spindle motor is a controlled by the SPIN_MOTOR discrete output. This output drives a variable-speed drive that has been configured to generate the proper motor speed. The SPIN_ZS switch indicates that the motor is stopped. SPIN_ZS is **on** when the motor is stopped and **off** otherwise.

Start/Stop/Pause: When the start switch, START_PB, is pressed (turned **on**) for the first time only, assume there are no discs in the station. If the stop switch is pressed (turned **off**) at any time, the station operation should pause, except when polishing. If the operation is paused when polishing, the polish step must complete. When the start switch is pressed while the operation is paused, the station should resume the suspended step. When the station is paused during the polishing part of the operation, the bladder must continue to be pressurized and the spindle motor must continue to run. All other outputs must be **off** when paused. If paused while the bladder is being inflated, the valve must be de-energized. When paused, **do not advance** to the next step, with the only exception noted above.

Reset: A separate reset switch, RESET_PB, when **on**, must reset any internal states. After a press of RESET_PB, no discs are assumed present in the station. Assume that an operator has already removed any disc before the start switch is pressed. Also, the reset switch should cause the spindle to move down, the bladder to deflate, and the motor to completely stop. **Do not implement a separate function chart for reset.** Just do all the actions simultaneously until the spindle is down, the bladder is deflated and the spindle motor is stopped. The reset switch should be ignored if the station is running. The reset switch should also be ignored in those steps in which pause is not allowed. The start switch should be ignored when the reset operation is in progress. Also, the start switch should be ignored as long as RESET_PB is pressed.

Do not add any more timed steps to those explicitly stated in the problem. In other words, do not put a timer in a step unless it is stated that the step duration is a specific time. Assume the tolerance on all timer values is ±0.01 seconds.

Assume the following physical input and output tags/variables/symbols.

Tag/Var./Symbol	Description
START_PB	Start push button, N. O., **on** when starting.
STOP_PB	Stop push button, N. C., **off** when stopping.
RESET_PB	Reset push button, N. O., **on** when restoring station to initial state.
PROX	Proximity sensor, turns **off** (open) as disc on conveyor enters station (see above). Remains **on** as disc is processed because disc is not on the conveyor. **Off** when disc is placed on conveyor after being polished.
SPNDL_DLS	Limit switch that closes (**on**) when the spindle mechanism is in the lower (down) position.
SPNDL_ULS	Limit switch that closes (**on**) when the spindle mechanism is in the higher (up) position.
SPNDL_HPS	Limit switch that closes (**on**) when bladder pressure is sufficiently high to hold the disc in position.
SPNDL_LPS	Limit switch that closes (**on**) when bladder pressure is sufficiently low to release the disc.
SPIN_ZS	Limit switch that closes (**on**) when spindle motor is at zero speed.
CONV_MOTOR	Station conveyor motor control, **on** to run conveyor, moving discs in and out of the station.
SPIN_MOTOR	Spindle motor control, **on** to cause spindle motor to rotate, **off** stops rotation, but not immediately.
SPNDL_DOWN	Spindle mechanism down control, **on** to retract cylinder and move mechanism down, **off** has no effect.
SPNDL_UP	Spindle mechanism up control, **on** to extend cylinder and move mechanism up, **off** has no effect.
PRESS_VLV	Bladder pressurized air valve control, **on** to admit pressurized air into spindle bladder, **off** closes valve.
EXH_VLV	Bladder exhaust air valve control, **on** to exhaust air out of spindle bladder, **off** closes valve.

The addresses associated with the physical input and output tags/variables/symbols are:

Tag/Var./Symbol	ControlLogix	Micro800	MLogix	SLC-500
START_PB	Local:1:I.Data.0	_IO_EM_DI_00	I:0/0	I:1/0
STOP_PB	Local:1:I.Data.1	_IO_EM_DI_01	I:0/1	I:1/1
RESET_PB	Local:1:I.Data.2	_IO_EM_DI_02	I:0/2	I:1/2
PROX	Local:1:I.Data.3	_IO_EM_DI_03	I:0/3	I:1/3
SPNDL_DLS	Local:1:I.Data.4	_IO_EM_DI_04	I:0/4	I:1/4

SPNDL_ULS	Local:1:I.Data.5	_IO_EM_DI_05	I:0/5	I:1/5
SPNDL_HPS	Local:1:I.Data.6	_IO_EM_DI_06	I:0/6	I:1/6
SPNDL_LPS	Local:1:I.Data.7	_IO_EM_DI_07	I:0/7	I:1/7
SPIN_ZS	Local:1:I.Data.8	_IO_EM_DI_08	I:0/8	I:1/8
CONV_MOTOR	Local:2:O.Data.0	_IO_EM_DO_00	O:0/0	O:2/0
SPIN_MOTOR	Local:2:O.Data.1	_IO_EM_DO_01	O:0/1	O:2/1
SPNDL_DOWN	Local:2:O.Data.2	_IO_EM_DO_02	O:0/2	O:2/2
SPNDL_UP	Local:2:O.Data.3	_IO_EM_DO_03	O:0/3	O:2/3
PRESS_VLV	Local:2:O.Data.4	_IO_EM_DO_04	O:0/4	O:2/4
EXH_VLV	Local:2:O.Data.5	_IO_EM_DO_05	O:0/5	O:2/5

P6-8. Using the function chart approach, implement the program for the following plastic injection molding machine.

Figure P6.8 shows the layout of a station that injection-molds plastic parts. The station clamps the two halves of the mold, injects melted plastic into the mold, allows it to cool and then removes the plastic part from the mold. Note that this problem is a simplified version of an actual machine.

In injection molding, plastic beads fall from a hopper into an auger-type screw. The screw melts (plasticizes) the polymer by virtue of the shearing action on the plastic beads as the screw turns. The heater on the outside of the barrel maintains the temperature, but the screw does the work to melt the plastic. To push the melt into the mold, the entire screw is moved forward.

Upon initial startup, assume that the two halves of the mold are apart, the eject cylinder is retracted, and the extrusion screw is retracted. The overall operation is first described, followed by the details. When E_RUN initially turns **on**, the following steps should be executed:

Clamp the two mold halves together by extending the clamp cylinder (CLAMP_ELS senses).

Feed the melted plastic by moving the extrusion screw forward a desired distance (SCREW_ELS senses).

Pull back the extrusion screw (SCREW_RLS senses)

Hold the mold clamped for 4.5 seconds. This delay allows the plastic to cool and solidify.

Retract the clamp cylinder (CLAMP_RLS senses). The part remains attached to the moveable side of the mold.

Extend the small eject cylinder (EJECT_ELS senses) to push the plastic piece out of the mold and into a chute below.

Retract the eject cylinder (EJECT_RLS senses)

The process repeats.

Assume that the extrusion screw is always running whenever the machine is running.

Mold Clamp Cylinder: The cylinder that controls the extension/retraction of the mold clamp is a double-action hydraulic cylinder, controlled by the CLAMP_EXT and CLAMP_RET outputs. When CLAMP_EXT is energized (turned **on**), the cylinder extends

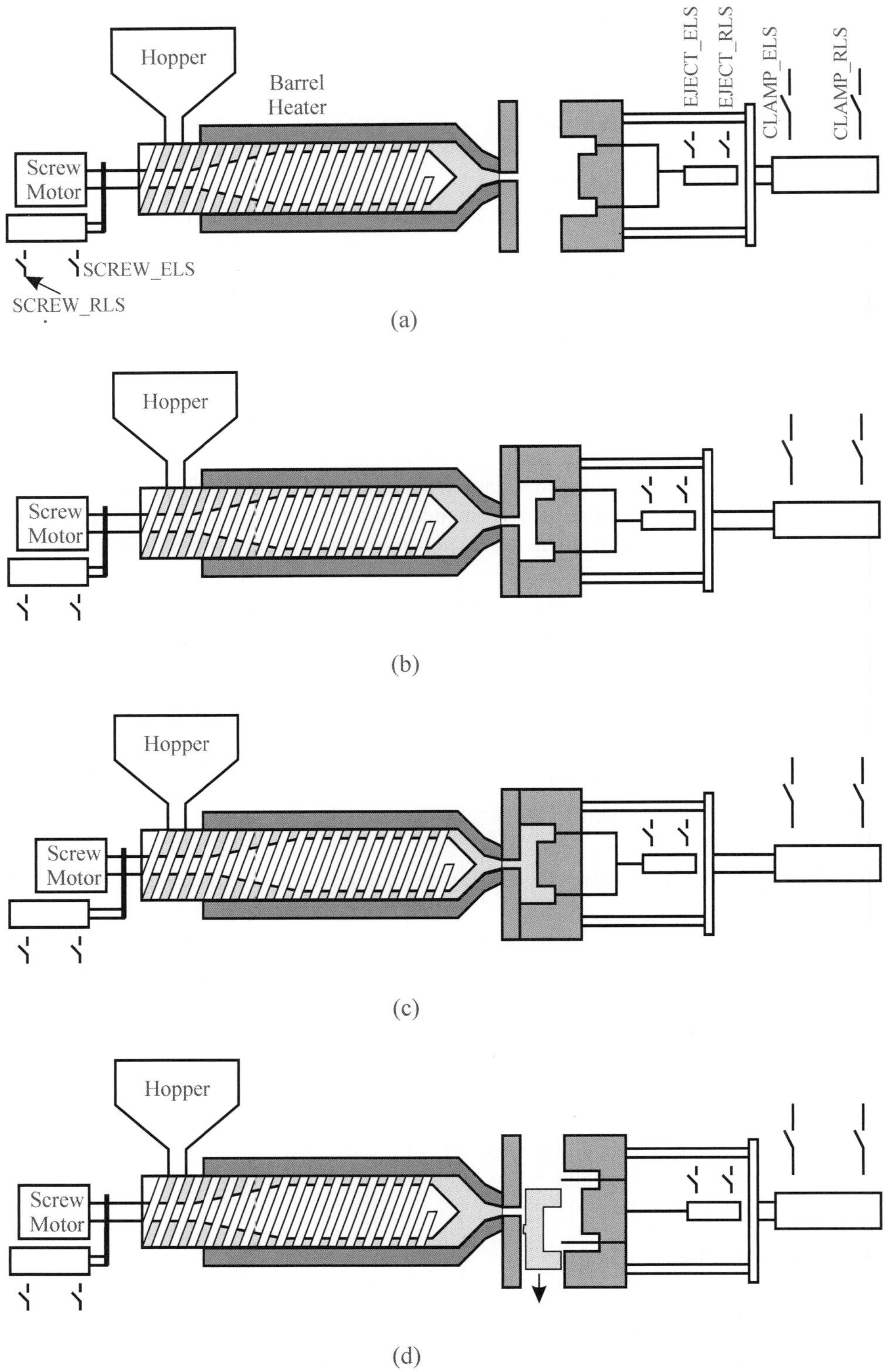

Figure P6.8. Injection molding machine: *(a)* start (mold open); *(b)* mold clamped; *(c)* push plastic into mold; *(d)* eject part.

and moves the clamp to push the two mold halves together to form the mold. When CLAMP_RET is energized (turned **on**), the cylinder retracts, pulling the two mold halves apart. When CLAMP_EXT and CLAMP_RET are both **off**, the clamp stops at its present position. When CLAMP_RET and CLAMP_EXT are both **on**, the cylinder operation is not consistent (it may stop or move in either direction slowly). The cylinder has two limit switches, CLAMP_ELS, which is **on** when the cylinder is extended (mold clamped) and CLAMP_RLS, which is **on** when the cylinder is retracted (mold apart).

Note: the CLAMP_EXT output must remain energized while feeding plastic into the mold and while cooling.

Part Eject Cylinder: The cylinder that controls the extension/retraction of the part eject mechanism is a single-action hydraulic cylinder, controlled by the EJECT_EXT output. When EJECT_EXT is energized (turned **on**), the cylinder extends and moves the pins of the eject mechanism to push the molded part out of the moveable part of the mold (and drop into a chute below). When EJECT_EXT is **off**, the cylinder retracts. The cylinder has two limit switches, EJECT_ELS, which is **on** when the cylinder is extended (part pushed out) and EJECT_RLS, which is **on** when the cylinder is retracted.

Extrusion Screw Cylinder: The cylinder that controls the extension/retraction of the extrusion screw is a double-action hydraulic cylinder, controlled by the SCREW_EXT and SCREW_RET outputs. When SCREW_EXT is energized (turned **on**), the cylinder extends and moves the screw head to push melted plastic into the mold. When SCREW_RET is energized (turned **on**), the cylinder retracts. When SCREW_EXT and SCREW_RET are both **off**, the extrusion screw stops at its present position. When SCREW_RET and SCREW_EXT are both **on**, the cylinder operation is not consistent (it may stop or move in either direction slowly). The cylinder has two limit switches. SCREW_RLS is **on** when the cylinder is retracted. SCREW_ELS is **on** when the cylinder is extended.

Extrusion Screw Rotation Motor: This motor is controlled by another part of the ladder logic and beyond the scope of this problem. **You must not have any output coil controlling the screw rotation**.

The extrusion screw barrel has electric heater bands surrounding it. The heater bands do not melt the plastic; the process of rotating the extruder screw melts the plastic. The heater bands maintain a uniform temperature profile along the extruder barrel. An actual extruder has at least three temperature control zones.

Extruder Barrel Temperature Control: The barrel temperature control is done by separate PID controllers. These PID controllers are not part of the PLC.

There is no start/stop switch for your part of the ladder. Instead, there is an internal coil, E_RUN that is **on** when the injection molding operation is making parts. When E_RUN is **off**, the operation should pause at its current step except for the following:

1. The screw is being advanced (ejecting plastic into mold)
2. The screw is being retracted.

Do not advance to the next step when paused. When paused, all outputs should be **off** **except** for the following exceptions:

1. The clamp cylinder extend control, CLAMP_EXT must not be turned **off**.
2. The barrel screw cylinder controls must not be turned **off**.

When E_RUN turns back **on** while the operation of the station is paused, the station should resume its suspended step. **DO NOT** retain the timer accumulator when paused.

The start and stop switches are absent from your part of the ladder because they control the overall operation of the extruder machine. In order to reliably start/stop/pause the operation of the machine, the initial fill of the plastic into the screw and barrel must be controlled. This part of the operation is outside the scope of this problem. The other part of the logic controls E_RUN. **You must not have any output coil referring to E_RUN.**

A separate reset internal coil, E_RESET, is provided which restores the extruder part of the machine to its initial state. The reset operation should cause the mold to be unclamped and the eject cylinder to be retracted. The reset operation is not complete until both cylinders are retracted. There is no mechanical interference between these two cylinders. Assume that an operator is responsible for actually removing anything from the station after a reset. E_RESET should be ignored if in **any step where the operation cannot be paused** or if E_RUN is **on**. The E_RUN internal coil used to restart the machine should be ignored when the reset operation is in progress. **You must not have any output coil referring to E_RESET.**

Do not add any more timed steps to those explicitly stated in the problem. In other words, do not put a timer in a step unless it is stated that the step duration is a specific time. Assume the tolerance on all timer values is at most ±0.01 second.

Assume the following physical input and output tags/variables/symbols.

Tag/Var./Symbol	Description
CLAMP_ELS	Limit switch that closes (**on**) when the mold clamp is extended and holding the two mold halves together.
CLAMP_RLS	Limit switch that closes (**on**) when the mold clamp is retracted and the two mold halves are apart.
EJECT_ELS	Limit switch that closes (**on**) when the part ejector is extended and pushed the part out of the mold.
EJECT_RLS	Limit switch that closes (**on**) when the part ejector is retracted.
SCREW_ELS	Limit switch that closes (**on**) when the extrusion screw is fully extended.
SCREW_RLS	Limit switch that closes (**on**) when the extrusion screw is fully retracted.
CLAMP_EXT	Mold clamp extend control, **on** to extend clamp, closing mold, **off** stops clamp.
CLAMP_RET	Mold clamp retract control, **on** to retract clamp, opening mold, **off** stops clamp.
EJECT_EXT	Mold eject extend control, **on** to extend pins, forcing part out of mold, **off** retracts pins.
SCREW_EXT	Extrusion screw extend control, **on** to extend screw, forcing plastic into mold, **off** stops movement.
SCREW_RET	Extruction screw retract control, **on** to retract screw, **off** stops movement.

Assume the following internal tags/variables/symbols:

Tag/Var./Symbol	Description
E_RUN	Run/stop/pause control for extrusion part of machine operation; **on** to run extrusion, **off** to stop/pause. Controlled by

another part of the ladder logic in the PLC. **You must not have any output coil referring to E_RUN.**

E_RESET Reset control, **on** to restore extrusion section to initial state. Controlled by another part of the ladder logic in the PLC. **You must not have any output coil referring to E_RESET.**

The addresses associated with the physical input and output tags/variables/symbols are:

Tag/Var./Symbol	ControlLogix	Micro800	MLogix	SLC-500
CLAMP_ELS	Local:1:I.Data.3	_IO_EM_DI_03	I:0/3	I:1/3
CLAMP_RLS	Local:1:I.Data.4	_IO_EM_DI_04	I:0/4	I:1/4
EJECT_ELS	Local:1:I.Data.5	_IO_EM_DI_05	I:0/5	I:1/5
EJECT_RLS	Local:1:I.Data.6	_IO_EM_DI_06	I:0/6	I:1/6
SCREW_ELS	Local:1:I.Data.7	_IO_EM_DI_07	I:0/7	I:1/7
SCREW_RLS	Local:1:I.Data.8	_IO_EM_DI_08	I:0/8	I:1/8
CLAMP_EXT	Local:2:O.Data.0	_IO_EM_DO_00	O:0/0	O:2/0
CLAMP_RET	Local:2:O.Data.1	_IO_EM_DO_01	O:0/1	O:2/1
EJECT_EXT	Local:2:O.Data.2	_IO_EM_DO_02	O:0/2	O:2/2
SCREW_EXT	Local:2:O.Data.3	_IO_EM_DO_03	O:0/3	O:2/3
SCREW_RET	Local:2:O.Data.4	_IO_EM_DO_04	O:0/4	O:2/4

The addresses and/or data types associated with the internal tags/variables/symbols are:

Tag/Var./Symbol	CLogix Data Type	Micro800 Data Type	MLogix/SLC Address
E_RUN	BOOL	BOOL	B3/101
E_RESET	BOOL	BOOL	B3/102

P6-9. BBQ Grill Handle Welder Control. Using the function chart approach, implement the program for the following machine that welds a handle onto a barbeque grill lid.

Figure P6.9 shows the layout of a machine that uses spot-welds the handle onto the top of a barbeque grill lid. This operation is semi-automatic. An operator loads the handle and grill lid into the machine and removes the completed assembly. You are implementing ladder logic for the automatic part of the welding operation only.

In summary, the operator loads the grill lid and a handle into the proper places in the machine, starts the operation, the handle is moved down, welded, and then the assembly is lifted for the operator to remove. Upon initial startup, assume the handle holder and welder are in the up position, the handle is unclamped, the welder is off, and there is no lid or handle in the station. The control should be in a state waiting for the operator to push the start buttons. The operation of the start buttons is detailed below. The following steps are executed:

Sense that the operator has pressed the start buttons (OPER_START senses) and that the operator has loaded a handle and the lid (sensed by HAN_PROX and LID_PROX). Do not continue until all three conditions are satisfied.

Clamp handle into holder (CLMP_LS senses).

Lower the handle to contact the lid (HAND_DLS senses).

Lower the welder contacts to contact the handle (WELD_DLS senses).
Activate the welder for 0.5 seconds.
Raise the welder contacts (WELD_ULS senses).
Raise the handle, which raises the assembly (HAND_ULS senses).
Unclamp handle (CLMP_LS senses).
Wait for the operator to release the start buttons and remove the assembly (sensed by HAN_PROX).

The operation repeats on the next press of the start buttons (and the operator has loaded the lid and handle).

The operator is required to press two buttons within a second of each other and hold them down until the welding operation is complete (keeps his/her hands out of the machine). The logic to accomplish this function is not a part of this problem. The result of this logic is the OPER_START internal discrete variable. OPER_START is **on** when the operator has pressed both buttons. When the OPER_START discrete input is **off**, treat it like a stop/pause. Basically, the OPER_START functions like a toggle switch. Your logic **must not** control OPER_START (it must not appear as a coil in your solution). Treat it like a discrete input.

Proximity sensors: The HAN_PROX discrete input is an optical proximity sensor that detects the presence of a handle in the handle holder. HAN_PROX is **on** when a handle is in

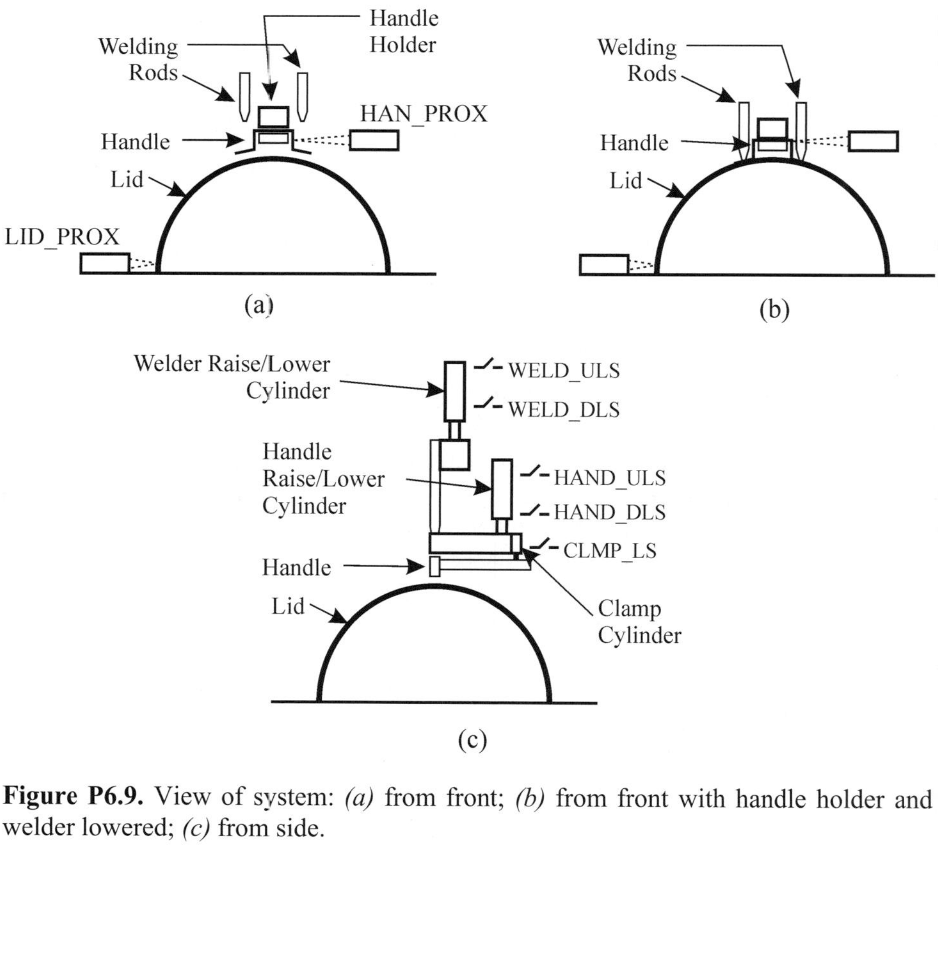

Figure P6.9. View of system: *(a)* from front; *(b)* from front with handle holder and welder lowered; *(c)* from side.

the holder and **off** otherwise. The LID_PROX discrete input is an optical proximity sensor that detects the presence of a lid in the station in the position to weld a handle. LID_PROX is **on** when a lid is in the station and **off** otherwise. Note that when the completed assembly is lifted, LID_PROX is **off**.

Handle Clamp Cylinder: The cylinder controlling the handle clamp is a single-action pneumatic cylinder, controlled by the CLMP_CYL output. When CLMP_CYL is energized (**on**), the mechanism clamps the handle into the handle holder. When CLMP_CYL is de-energized, the handle is unclamped. The CLMP_LS limit switch is **on** when the handle is clamped and the CLMP_LS limit switch is **off** when the handle is unclamped. The handle must remain clamped during the welding operation.

Handle Lower/Raise Cylinder: The cylinder that controls the lowering/raising of the handle holder is a double-action pneumatic cylinder, controlled by the HAND_DN and HAND_UP outputs. When the HAND_DN output is energized (**on**), the cylinder extends and moves the handle down to contact the lid. When the HAND_UP output is energized (**on**), the cylinder retracts and raises the handle (hopefully with lid attached). The mechanism stops if neither output is **on** or if they are energized simultaneously. The HAND_DLS limit switch is **on** when the cylinder is extended (down position) and the HAND_ULS limit switch is **on** when the cylinder is retracted (up position).

Welder Lower/Raise Cylinder: The cylinder controlling the welder contacts is a single-action pneumatic cylinder, controlled by the WELD_CYL output. When WELD_CYL is energized (**on**), the mechanism moves the welder contact assembly down to contact the handle. When WELD_CYL is de-energized, the mechanism moves up and out of the way. The WELD_DLS limit switch is **on** when the welder is in the "down" position and the WELD_ULS limit switch is **on** when the pickup is "up".

Welder control. The WELD_ON output controls the welder operation. When WELD_ON is **on**, a high current is sent through the welding contacts, to spot-weld the handle to the lid. When WELD_ON is **off**, the welder is not energized. The welder must only be energized when wanting to spot-weld. Otherwise, it must remain **off**.

The OPER_START internal variable functions as both a start and a stop switch, like a toggle switch. When OPER_START is **on**, station operates as above during the welding operation. Your logic must also sense when OPER_START is **off** (operator has released the buttons) at the end of the operation. When OPER_START is **off** during the steps that control any cylinder movement or welding, the operation should pause **except** when welding. If OPER_START is **off** during the welding step, the step should complete and the operation should advance to the next step. When paused **do not advance** to the next step (exception noted earlier). If paused during the timed delay step, the timer should not be paused and the handle must remain clamped. When paused, all other cylinder controls must be **off**. The welder control must not be affected by a pause. When OPER_START turns **on** while the operation of the station is paused, the station resumes its suspended step.

A separate reset pushbutton switch, RESET_PB, is provided to restore the system to its initial state. When RESET_PB is **on**, all internal steps should be cleared, the welder must be raised, the handle holder must be raised, and the clamp released. The reset operation is not complete until the welder and handle holder are up and the clamp is released. **Do not implement a function chart for the reset** operation. There is no mechanical interference between the movements. Assume that an operator is responsible for actually removing any material from the station after a reset. The reset switch should be ignored if the station is running. If the reset switch is pressed while paused and welding is happening, the weld step should complete before the system is reset. The OPER_START internal variable should be ignored when the reset operation is in progress. If the station is paused and RESET_PB is **on**, the operator must release the RESET_PB before the OPER_START internal variable can be used to restart the operation.

Do not add any more timed steps to those explicitly stated in the problem. In other words, do not put a timer in a step unless it is stated that the step duration is a specific time. Assume the tolerance on all timer values is 0.1 seconds.

Hint: The operator start (plus other conditions) is the transition into the first step.

Assume the following physical inputs/outputs and internal tags/variables/symbols.

Tag/Var./Symbol	Description
OPER_START	Start/pause internal coil, **on** to start/run; **off** to pause/unload handle.
RESET_PB	Reset push button, N. O., **on** when restoring station to initial state.
HAN_PROX	Proximity sensor detecting presence of handle in holder. **On** when handle present in holder (holder in the up and unclamped position); **off** otherwise.
LID_PROX	Proximity sensor detecting presence of lid in station. **On** when lid in position on table for start of operation; **off** otherwise.
HAND_DLS	Limit switch that closes (**on**) when the handle holder is in the down position to hold the handle against the lid, **off** otherwise.
HAND_ULS	Limit switch that closes (**on**) when the handle holder is in the up position, **off** otherwise.
WELD_DLS	Limit switch that closes (**on**) when the welding rods are in the down position for spot-welding the handle to the lid, **off** otherwise.
WELD_ULS	Limit switch that closes (**on**) when the welding rods are in the up position, **off** otherwise.
CLMP_LS	Limit switch that closes (**on**) when the handle clamp has clamped the handle, **off** otherwise.
CLMP_CYL	Handle clamp cylinder control, **on** to move mechanism up, **off** to move mechanism down.
HAND_DN	Handle holder assembly cylinder control, **on** to move handle holder down to place handle in contact with lid, **off** has no effect.
HAND_UP	Handle holder assembly cylinder control, **on** to move handle holder up, **off** has no effect.
WELD_CYL	Welder rod assembly cylinder control, **on** to extend cylinder, moving rods down into the weld position, **off** moves rods up to mechanical stop.
WELD_ON	Welder control, **on** to spot-weld, **off** stops welding.

The addresses associated with the physical input and output tags/variables/symbols are:

Tag/Var./Symbol	ControlLogix	Micro800	MLogix	SLC-500
RESET_PB	Local:1:I.Data.2	_IO_EM_DI_02	I:0/2	I:1/2
HAN_PROX	Local:1:I.Data.3	_IO_EM_DI_03	I:0/3	I:1/3
LID_PROX	Local:1:I.Data.4	_IO_EM_DI_04	I:0/4	I:1/4
HAND_DLS	Local:1:I.Data.5	_IO_EM_DI_05	I:0/5	I:1/5
HAND_ULS	Local:1:I.Data.6	_IO_EM_DI_06	I:0/6	I:1/6
WELD_DLS	Local:1:I.Data.7	_IO_EM_DI_07	I:0/7	I:1/7
WELD_ULS	Local:1:I.Data.8	_IO_EM_DI_08	I:0/8	I:1/8
CLMP_LS	Local:1:I.Data.9	_IO_EM_DI_09	I:0/9	I:1/9
CLMP_CYL	Local:2:O.Data.0	_IO_EM_DO_00	O:0/0	
HAND_DN	Local:2:O.Data.1	_IO_EM_DO_01	O:0/1	
HAND_UP	Local:2:O.Data.2	_IO_EM_DO_02	O:0/2	
WELD_CYL	Local:2:O.Data.3	_IO_EM_DO_03	O:0/3	
WELD_ON	Local:2:O.Data.4	_IO_EM_DO_04	O:0/4	

The data type/address associated with the internal tag/variable/symbol is:

Tag/Var./Symbol	CLogix Data Type	Micro800 Data Type	MLogix/SLC Address
OPER_START	BOOL	BOOL	B3/102

P6-10. Disk Install Machine Control. Using the function chart approach, implement the program for the following machine that inserts a disk into a component.

Figure P6.10 shows the layout of a machine that picks up the top disk from a stack of steel disks and places it into a component. The component moves to other stations where a spring is installed, a clip is installed, and then the assembly is crimped. You are implementing ladder logic for this part of the operation only.

In summary, the steel disk is picked up by vacuum and then placed in the component. Upon initial startup, assume the vacuum gripper is in the up position and over the components, the vacuum is off, and the magnets are in the lower position. There is a conveyor to move the components into and out of the station and is controlled separately. Your solution does not control this conveyor. When a component is detected by the proximity sensor, the following steps are executed:

Extend the grip horizontal cylinder to position the gripper over the stack of disks (GRIP_ELS senses). Simultaneously, raise the magnets to separate the disks at the top of the stack (MAGN_ULS senses). Both movements must be finished before advancing to the next step.

Lower the gripper to contact the top disk (GRIP_DLS senses).

Activate the vacuum to grip the top disk. The DVAC_OK discrete input senses when the vacuum pad has gripped the disk.

Raise the gripper (GRIP_ULS senses).

Retract the grip horizontal cylinder to position the gripper (with disk) over the component (GRIP_RLS senses). Simultaneously, lower the magnets (MAGN_DLS senses). Both movements must be finished before advancing to the next step.

Lower the gripper to place the disk in the component (GRIP_DLS senses).

Release the disk by deactivating (turning off) the vacuum. Use VAC_OK to determine when the disk has been released.

Raise the gripper (GRIP_ULS senses).

The operation then repeats.

The conveyor that moves the components has the components spaced out. Also, the conveyor is stopped while the component is being processed. The component is stopped long enough for the station to install the disk. Your logic does not control the conveyor. You only need to sense when the component has entered the station and then subsequently leaves after the disk is installed.

A **proximity sensor**, DISK_PROX, is used to determine when a component has entered/exited the station. You must assume that while the component is in the station, DISK_PROX remains **on**. The components are spaced far enough apart so that DISK_PROX is **off** after one component is moved out and before the next one is conveyed in.

Grip Horizontal Cylinder: The cylinder that controls the horizontal movement of the vacuum gripper is a double-action pneumatic cylinder, controlled by the GRIP_EXT and

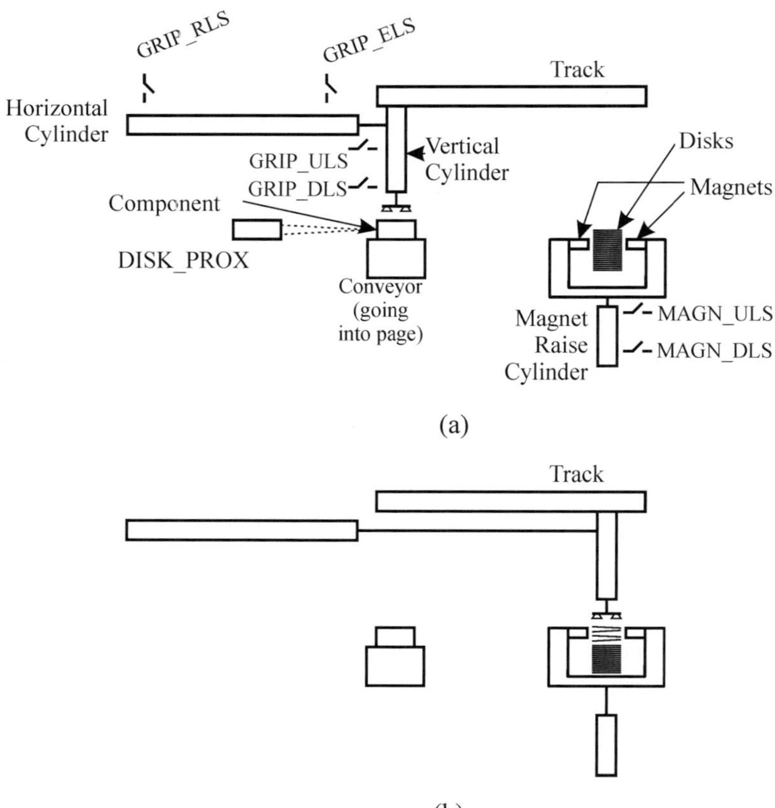

Figure P6.10. View of system: *(a)* from side; *(b)* horizontal cylinder extended and magnets up.

GRIP_RET outputs. When the GRIP_EXT output is energized (**on**), the cylinder extends and positions the gripper over the stack of disks. When the GRIP_RET output is energized (**on**), the cylinder retracts and positions the gripper over the component. The mechanism stops if neither output is **on** or if they are energized simultaneously. The GRIP_ELS limit switch is **on** when the cylinder is extended (gripper over stack of disks) and the GRIP_RLS limit switch is **on** when the cylinder is retracted (gripper over component).

Grip Vertical Cylinder: The cylinder that controls the vertical movement of the gripper is a double-action pneumatic cylinder, controlled by the GRIP_UP and GRIP_DN outputs. When the GRIP_UP output is energized (**on**), the cylinder retracts and raises the gripper to clear the disk stack and the component. When the GRIP_DN output is energized (**on**), the cylinder extends and lowers the gripper to contact either the stack of disks or the to place the disk inside the component. The mechanism stops if neither output is **on** or if they are energized simultaneously. The GRIP_ULS limit switch is **on** when the cylinder is retracted (grip up) and the GRIP_DLS limit switch is **on** when the cylinder is extended (grip down).

Magnets Raise/Lower Cylinder: Two magnets, one on either side of the disk stack, are used to separate the disks to ensure only one disk is gripped. The cylinder that controls the vertical movement of the magnets is a double-action pneumatic cylinder, controlled by the MAGN_UP and MAGN_DN outputs. When the MAGN_UP output is energized (**on**), the cylinder extends and raises the magnets. When the MAGN_DN output is energized (**on**), the cylinder retracts and lowers the magnets. The mechanism stops if neither output is **on** or if they are energized simultaneously. The MAGN_ULS limit switch is **on** when the cylinder is retracted (magnets up) and the MAGN_DLS limit switch is **on** when the cylinder is extended (magnets down).

Gripper vacuum. The gripper pickup mechanism has cups on the end through which a vacuum is drawn to grip the disk. The DVAC_ON discrete is **on** to pull a vacuum at the gripper. When it is **off**, there is no vacuum. DVAC_ON must be turned **on** as the top disk is picked up and is not turned **off** until the disk has been placed in the component. The DVAC_OK discrete indicates the state of the disk being gripped. If DVAC_OK is **on**, the disk is gripped. If DVAC_OK is **off**, the disk has been released from the gripper.

When the start switch (START_PB) is pressed (turned **on**) for the first time, assume the gripper is up and over the component, the vacuum is inactive and the magnets are down. When the stop switch (STOP_PB) is pressed (turned **off**) the operation should pause. When paused **do not advance** to the next step. When paused, all cylinder controls must be **off**. The vacuum control must not be affected by a pause. Pressing the start switch while the operation of the station is paused causes the station to resume its suspended step.

A separate reset switch, RESET_PB, is provided to restore the system to its initial state. When RESET_PB is **on**, all internal steps should be cleared, the gripper must be raised, placed over the component, the magnets must be lowered, and the vacuum off. Note that the gripper must be in the up position before any horizontal movements are attempted. The reset operation is not complete until the gripper is up and over the component, the magnets are lowered, and the vacuum is off. **Do not implement a function chart for the reset** operation. Assume that an operator is responsible for actually removing any material from the station after a reset. The reset switch should be ignored if the station is running. The start switch should be ignored when the reset operation is in progress. If the station is paused and RESET_PB is **on**, the operator must release the RESET_PB before the start pushbutton switch can be used to restart the operation.

Do not add any more timed steps to those explicitly stated in the problem. In other words, do not put a timer in a step unless it is stated that the step duration is a specific time. Assume the tolerance on all timer values is 0.1 seconds.

Assume the following physical input and output tags/variables/symbols.

Tag/Var./Symbol	Description
START_PB	Start push button, N. O., **on** when starting.
STOP_PB	Stop push button, N. C., **off** when stopping.
RESET_PB	Reset push button, N. O., **on** when restoring station to initial state.
DISK_PROX	Proximity switch that closes (**on**) when component is in station, open (**off**) when component is not in station.
GRIP_ELS	Limit switch that closes (**on**) when the gripper horizontal cylinder is extended, placing the gripper over the stack of disks, **off** otherwise.
GRIP_RLS	Limit switch that closes (**on**) when the gripper horizontal cylinder is retracted, placing the gripper over the component, **off** otherwise.
GRIP_DLS	Limit switch that closes (**on**) when the gripper is in the down position to either pick up a disk from the stack or install the disk into the component, **off** otherwise.
GRIP_ULS	Limit switch that closes (**on**) when the gripper is in the up position, **off** otherwise.
MAGN_DLS	Limit switch that closes (**on**) when the magnets are in the down position, **off** otherwise.
MAGN_ULS	Limit switch that closes (**on**) when the magnets are in the up position, **off** otherwise.
DVAC_OK	Vacuum switch that closed (**on**) when the vacuum pressure is of sufficient magnitude that indicates the disk is gripped, **off** indicates the disk is released.
GRIP_EXT	Gripper horizontal cylinder extend control, **on** to extend cylinder, moving it to the stack of disks, **off** has no effect.
GRIP_RET	Gripper horizontal cylinder retract control, **on** to retract cylinder, moving it to the component, **off** has no effect.
GRIP_DN	Gripper vertical cylinder down control, **on** to move gripper down, **off** has no effect.
GRIP_UP	Gripper vertical cylinder up control, **on** to move gripper up, **off** has no effect.
MAGN_DN	Magnet cylinder down control, **on** to move cylinder down, **off** has no effect.
MAGN_UP	Magnet cylinder up control, **on** to move cylinder up, **off** has no effect.
DVAC_ON	Pickup vacuum control, **on** to pull vacuum through vacuum cup, **off** stops vacuum.

The addresses associated with the physical input and output tags/variables/symbols are:

Tag/Var./Symbol	ControlLogix	Micro800	MLogix	SLC-500
START_PB	Local:1:I.Data.0	_IO_EM_DI_00	I:0/0	I:1/0
STOP_PB	Local:1:I.Data.1	_IO_EM_DI_01	I:0/1	I:1/1
RESET_PB	Local:1:I.Data.2	_IO_EM_DI_02	I:0/2	I:1/2
DISK_PROX	Local:1:I.Data.3	_IO_EM_DI_03	I:0/3	I:1/3
GRIP_ELS	Local:1:I.Data.4	_IO_EM_DI_04	I:0/4	I:1/4
GRIP_RLS	Local:1:I.Data.5	_IO_EM_DI_05	I:0/5	I:1/5
GRIP_DLS	Local:1:I.Data.6	_IO_EM_DI_06	I:0/6	I:1/6
GRIP_ULS	Local:1:I.Data.7	_IO_EM_DI_07	I:0/7	I:1/7
MAGN_DLS	Local:1:I.Data.8	_IO_EM_DI_08	I:0/8	I:1/8
MAGN_ULS	Local:1:I.Data.9	_IO_EM_DI_09	I:0/9	I:1/9
DVAC_OK	Local:1:I.Data.10	_IO_EM_DI_10	I:0/10	I:1/10
GRIP_EXT	Local:2:O.Data.0	_IO_EM_DO_00	O:0/0	O:2/0
GRIP_RET	Local:2:O.Data.1	_IO_EM_DO_01	O:0/1	O:2/1
GRIP_DN	Local:2:O.Data.2	_IO_EM_DO_02	O:0/2	O:2/2
GRIP_UP	Local:2:O.Data.3	_IO_EM_DO_03	O:0/3	O:2/3
MAGN_DN	Local:2:O.Data.4	_IO_EM_DO_04	O:0/4	O:2/4
MAGN_UP	Local:2:O.Data.5	_IO_EM_DO_05	O:0/5	O:2/5
DVAC_ON	Local:2:O.Data.6	_IO_EM_DO_06	O:0/6	O:2/6

P6-11. Soap Packaging Machine Control. Using the function chart approach, implement the program for the following machine that inserts a bar of soap into a folded sheet of plastic, forms the bag, and then pushes it out.

Figure P6.11 shows the layout of a machine that puts a bar of soap into a folded sheet of plastic, forms the bag, and then pushes the bag onto another conveyor where it is moved through a heat-shrink tunnel. You are implementing ladder logic for this part of the operation only.

In summary, after a soap bar is sensed at the infeed area, the plastic sheet is moved into position to accept the bar, the bar is pushed in, a heated knife is used to cut and seal the plastic sheet around the bar, and then the bagged bar is pushed out. There is also a rod mechanism that holds the top and bottom separated. The movement of this "hold open" rod must be coordinated with the other parts of the operation. Upon initial startup, assume the soap push cylinder (SOAP_CYL) is retracted, the "hold open" rod is in the right position and extended, and the heated knife is up. Upon initial startup, the following steps are executed:

Move soap bar into the infeed area (run the infeed conveyor) until detected by PE1.

Feed the plastic sheet into position (run the sheet vacuum conveyor) until 105 encoder pulses have been counted.

The "hold open" bar is moved left to keep the top and bottom separated (HOPEN_LLS senses).

Push the soap bar into the partially formed bag by extending the soap push cylinder (SOAP_MLS senses).

Simultaneously, retract both the hold-open bar and the soap push cylinder (sensed by HOPEN_OLS and SOAP_RLS). Note that both cylinders must be retracted before starting the next step. You cannot assume that one of them retracts before or after the other.

Move the hold-open rod right (HOPEN_RLS senses).

Extend the hold-open rod to move it between the top and bottom (HOPEN_ILS senses). The hold-open bar is now in position ready for the next soap bar.

Lower heated knife (KNIFE_DLS senses)

Hold knife down for 1 second to seal the edges of the bag.

Raise heated knife (KNIFE_ULS senses).

Push the bagged soap bar onto the outfeed conveyor (SOAP_ELS senses)

Retract the soap push cylinder (SOAP_RLS senses)

The operation then repeats.

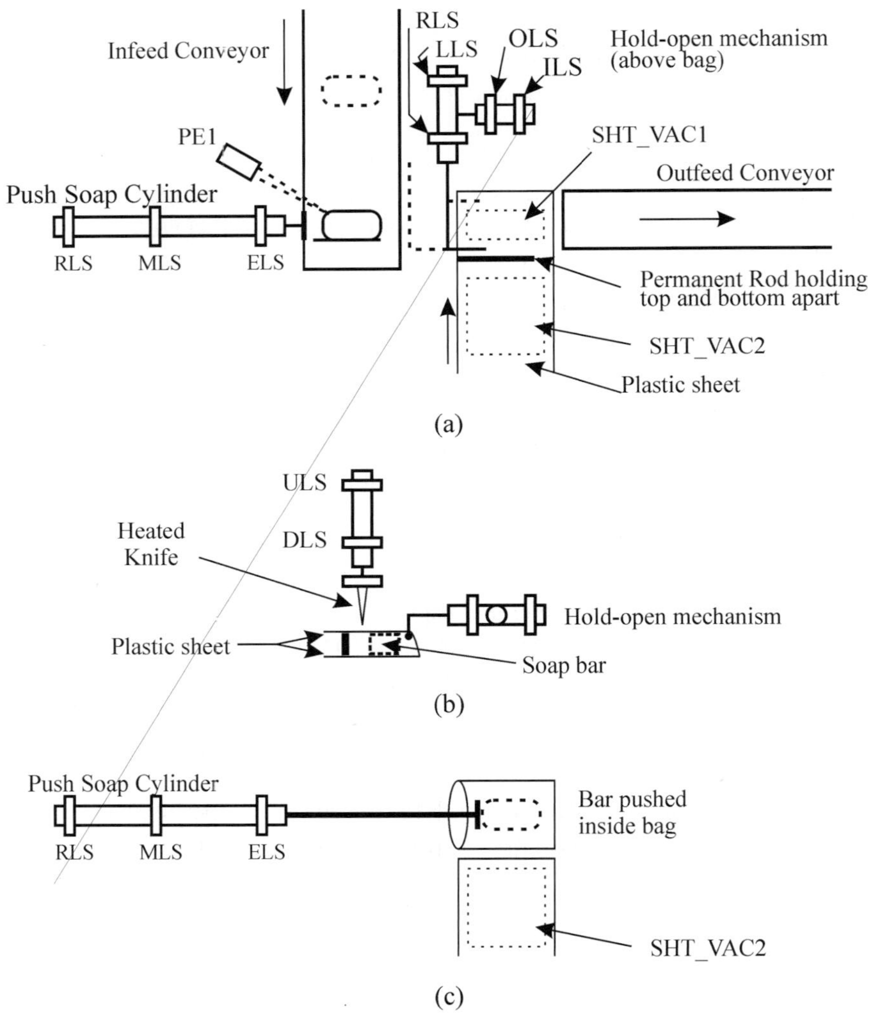

Figure P6.11. Soap packing machine: *(a)* overall view from top; *(b)* view from side showing hold-open mechanism and heated knife; *(c)* bar pushed inside bag.

The plastic sheet material has already been folded over so that there is a top and bottom. The back edge is the fold. There is a permanent rod that holds the top and bottom apart. The hold-open rod moves between this permanent rod and the end of the sheet to keep the top and bottom separated so the soap bar can be slipped between the top and bottom.

The **infeed conveyor** that moves the bars into the station is under your control. The INFEED_RUN output must be **on** for the conveyor to run. When INFEED_RUN is **off**, no bars are moved into position. The infeed conveyor must be stopped while packaging a soap bar.

The outfeed conveyor is not controlled by your logic and so assume it is always running.

A **proximity sensor**, PE1, is used to determine when a soap bar has entered the station. When PE1 turns **off**, a soap bar is in position to be pushed into the bag. The bars are spaced far enough apart so that PE1 is **on** after the soap push cylinder has been retracted. After the bar has been initially sensed, do not assume any particular state for PE1 as it senses the soap bar pusher and so will turn **on** and **off** as the operation progresses.

Sheet Conveyor. To move the plastic sheet for loading, the SHEET_RUN output must be **on**. The driving wheel of the conveyor has an encoder that generates pulses, ENC_PULSE, as it rotates. When ENC_PULSE has transitioned from **on** to **off** 105 times, the conveyor has move the proper distance. The pulses are slow enough so they can be counted by a counter block.

Sheet Conveyor Vacuum. The sheet conveyor contains a vacuum system to hold the bottom of the sheet to the conveyor. There are two vacuums that must be operated to hold down the sheet. One vacuum, controlled by SHT_VAC1, is directly under the place where the bag is loaded. SHT_VAC1 is turned **on** after the sheet is moved in to position and will be turned **off** when the bag is pushed off. The other vacuum, controlled by SHT_VAC2, is for the sheet conveyor and is **on** all of the time the station is running.

Soap Push Cylinder: The cylinder that controls the horizontal movement of the soap pusher mechanism is a single-action pneumatic cylinder, controlled by the SOAP_EXT output. When the SOAP_EXT output is energized (**on**), the cylinder extends and pushes the soap into the bag, or onto the outfeed conveyor. When the SOAP_EXT output is deenergized (**off**), the cylinder retracts. The SOAP_RLS limit switch is **on** when the cylinder is retracted (can accept a bar of soap), the SOAP_MLS limit switch is **on** then the soap bar is positioned in the bag, and the SOAP_ELS limit switch is **on** when the cylinder is extended (soap on outfeed).

Hold Open Right/Left Cylinder: The cylinder that controls the right/left horizontal movement of the hold open rod is a double-action pneumatic cylinder, controlled by the HOPEN_LEFT and HOPEN_RGHT outputs. When the HOPEN_LEFT output is energized (**on**), the cylinder extends and moves the rod left. When the HOPEN_RGHT output is energized (**on**), the cylinder retracts and moves the rod right. The mechanism stops if neither output is **on** or if they are energized simultaneously. The HOPEN_LLS limit switch is **on** when the cylinder is extended (rod left) and the HOPEN_RLS limit switch is **on** when the cylinder is retracted (rod right).

Hold Open In/Out Cylinder: The cylinder that controls the in/out horizontal movement of the hold open rod is a double-action pneumatic cylinder, controlled by the HOPEN_IN and HOPEN_OUT outputs. When the HOPEN_IN output is energized (**on**), the cylinder retracts and pushes the rod into the opened sheet. When the HOPEN_OUT output is energized (**on**), the cylinder extends and moves the rod out of the sheet. The mechanism stops if neither output is **on** or if they are energized simultaneously. The HOPEN_ILS limit switch is **on** when the cylinder is retracted (rod in) and the HOPEN_OLS limit switch is **on** when the cylinder is extended (rod out).

Heated Knife Cylinder: The cylinder that controls the vertical movement of the heated knife is a single-action pneumatic cylinder, controlled by the KNIFE_DWN output. When the KNIFE_DWN output is energized (**on**), the cylinder extends and moves the knife down to contact the plastic sheet. KNIFE_DWN must continue to be energized to hold the knife down. There is a mechanical stop and so nothing is damaged. When the KNIFE_DWN output is deenergized (**off**), the cylinder retracts, moving the knife up. The KNIFE_DLS limit switch is **on** when the cylinder is extended (knife down) and the KNIFE_ULS limit switch is **on** when the cylinder is retracted (knife up).

When the start switch (START_PB) is pressed (turned **on**) for the first time, assume the soap push cylinder is retracted, the "hold open" rod is in the right position and extended, and the heated knife is up. When the stop switch (STOP_PB) is pressed (turned **off**) the operation should pause at the current step except when holding the knife down or when the knife is being lowered/raised. The operation must not pause during these steps (otherwise there will be melted plastic mess). When paused **do not advance** to the next step. When paused, the infeed and vacuum conveyors must be **off** and all air cylinder controls must be **off**. The SHT_VAC1 vacuum control and the knife controls must not be affected by a pause. Pressing the start switch while the operation of the station is paused causes the station to resume its suspended step.

A separate reset switch, RESET_PB, is provided to restore the system to its initial state. When RESET_PB is **on**, all internal steps should be cleared, the soap push cylinder must be retracted, the hold-open rod in the right position and retracted, and the heated knife is up. Note that the soap push cylinder must be retracted and the knife must be in the up position before any hold-open rod movements are attempted. The reset operation is not complete until the soap push cylinder is retracted, the hold-open rod is retracted (in the sheet) in the right position, and the knife is up. **Do not implement a function chart for the reset operation**. It is fine for the hold-open rod to simultaneously move right and in. Assume that an operator is responsible for actually removing any material from the station after a reset. The reset switch should be ignored if the station is running and if the heated knife is moving or being held down. The start switch should be ignored when the reset operation is in progress. If the station is paused and RESET_PB is **on**, the operator must release the RESET_PB before the start pushbutton switch can be used to restart the operation.

Do not add any more timed steps to those explicitly stated in the problem. In other words, do not put a timer in a step unless it is stated that the step duration is a specific time. Assume the tolerance on all timer values is at most 0.1 seconds.

Assume the following physical input and output tags/variables/symbols.

Tag/Var./Symbol	Description
START_PB	Start push button, N. O., **on** when starting.
STOP_PB	Stop push button, N. C., **off** when stopping.
RESET_PB	Reset push button, N. O., **on** when restoring station to initial state.

PE1	Soap bar proximity sensor, **off** (open) when soap bar in position to be bagged. Also detects movement of bar that pushes soap into bag.
ENC_PULSE	Pulses to count for proper movement of sheet conveyor to feed correct amount of plastic sheet. The pulses are slow enough so they can be counted by a counter block.
SOAP_RLS	Limit switch that closes (**on**) when the soap pusher cylinder is retracted, ready to accept a new bar of soap, **off** otherwise.
SOAP_MLS	Limit switch that closes (**on**) when the soap pusher cylinder is extended to the position that places the soap bar into the sheet for bagging, **off** otherwise.
SOAP_ELS	Limit switch that closes (**on**) when the soap pusher cylinder is extended to the position that places the bagged soap on the outfeed conveyor, **off** otherwise.
KNIFE_DLS	Limit switch that closes (**on**) when the heated knife is in the down position, **off** otherwise.
KNIFE_ULS	Limit switch that closes (**on**) when the heated knife is in the up position, **off** otherwise.
HOPEN_LLS	Limit switch that closes (**on**) when the hold-open rod is in the left position, **off** otherwise.
HOPEN_RLS	Limit switch that closes (**on**) when the hold-open rod is in the right position, **off** otherwise.
HOPEN_ILS	Limit switch that closes (**on**) when the hold-open rod is in the in (extended) position, **off** otherwise.
HOPEN_OLS	Limit switch that closes (**on**) when the hold-open rod is in the out (retracted) position, **off** otherwise.

INFEED_RUN	Infeed conveyor control, **on** to run conveyor, moving soap bars into position, **off** stops conveyor.
SHEET_RUN	Sheet feed conveyor control, **on** to run conveyor, feeding plastic sheet material, **off** stops movement.
SHT_VAC1	Control for vacuum that holds down sheet when placing bar into it, **on** to turn on vacuum.
SHT_VAC2	Control for vacuum that holds plastic sheet onto conveyor, **on** to turn on vacuum.
SOAP_EXT	Soap pusher cylinder control, **on** to extend cylinder, **off** retracts cylinder.
HOPEN_LEFT	Hold-open rod mechanism cylinder "left" control, **on** to move rod left, **off** has no effect.
HOPEN_RGHT	Hold-open rod mechanism cylinder "right" control, **on** to move rod right, **off** has no effect.
HOPEN_IN	Hold-open rod mechanism cylinder "in" control, **on** to move rod between top and bottom sheet, **off** has no effect.
HOPEN_OUT	Hold-open rod mechanism cylinder "out" control, **on** to move rod out of sheet, **off** has no effect.
KNIFE_DWN	Heated knife lower/raise cylinder control, **on** to lower knife or hold it at its mechanical stop, **off** raises knife.

The addresses associated with the physical input and output tags/variables/symbols are:

Tag/Var./Symbol	ControlLogix	Micro800	MLogix	SLC-500
START_PB	Local:1:I.Data.0	_IO_EM_DI_00	I:0/0	I:1/0
STOP_PB	Local:1:I.Data.1	_IO_EM_DI_01	I:0/1	I:1/1
RESET_PB	Local:1:I.Data.2	_IO_EM_DI_02	I:0/2	I:1/2
PE1	Local:1:I.Data.3	_IO_EM_DI_03	I:0/3	I:1/3
ENC_PULSE	Local:1:I.Data.4	_IO_EM_DI_04	I:0/4	I:1/4
SOAP_RLS	Local:1:I.Data.5	_IO_EM_DI_05	I:0/5	I:1/5
SOAP_MLS	Local:1:I.Data.6	_IO_EM_DI_06	I:0/6	I:1/6
SOAP_ELS	Local:1:I.Data.7	_IO_EM_DI_07	I:0/7	I:1/7
KNIFE_DLS	Local:1:I.Data.8	_IO_EM_DI_08	I:0/8	I:1/8
KNIFE_ULS	Local:1:I.Data.9	_IO_EM_DI_09	I:0/9	I:1/9
HOPEN_LLS	Local:1:I.Data.10	_IO_EM_DI_10	I:0/10	I:1/10
HOPEN_RLS	Local:1:I.Data.11	_IO_EM_DI_11	I:0/11	I:1/11
HOPEN_ILS	Local:1:I.Data.12	_IO_EM_DI_12	I:1/0	I:1/12
HOPEN_OLS	Local:1:I.Data.13	_IO_EM_DI_13	I:1/1	I:1/13
INFEED_RUN	Local:2:O.Data.0	_IO_EM_DO_00	O:0/0	O:2/0
SHEET_RUN	Local:2:O.Data.1	_IO_EM_DO_01	O:0/1	O:2/1
SHT_VAC1	Local:2:O.Data.2	_IO_EM_DO_02	O:0/2	O:2/2
SHT_VAC2	Local:2:O.Data.3	_IO_EM_DO_03	O:0/3	O:2/3
SOAP_EXT	Local:2:O.Data.4	_IO_EM_DO_04	O:0/4	O:2/4
HOPEN_LEFT	Local:2:O.Data.5	_IO_EM_DO_05	O:0/5	O:2/5
HOPEN_RGHT	Local:2:O.Data.6	_IO_EM_DO_06	O:0/6	O:2/6
HOPEN_IN	Local:2:O.Data.7	_IO_EM_DO_07	O:0/7	O:2/7
HOPEN_OUT	Local:2:O.Data.8	_IO_EM_DO_08	O:0/8	O:2/8
KNIFE_DWN	Local:2:O.Data.9	_IO_EM_DO_09	O:0/9	O:2/9

P6-12. Using the function chart approach, implement the program for the following machine that puts a stripe on a cap.

Figures P6.12.1 and P16.12.2 shows the layout of a machine that places a stripe on a health/beauty aid cap. The caps are already mounted in a "belt" carrier and are moved through the station on the way to the station that places the cap on the spray bottle. You are only responsible for the station that adds the stripe. The other aspects of the operation are controlled by other parts of the PLC program.

Upon initial startup, wait for the prox to sense the pin on the belt carrier. In summary, after the prox senses the presence of the cap, the cap is picked up by vacuum, moved up and over to a mandrel, a wheel containing the stripe material contacts the cap, the cap rotates to transfer the stripe to the cap, the wheel is retracted and the cap is placed back in the carrier. The initial states of the main mechanical controls are: linear cylinder up (retracted), rotary cylinder clockwise, wheel cylinder retracted, bladder deflated, and vacuum deactivated.

The following steps are executed:

Move the belt until CPROX turns **on** to indicate that the belt has indexed the cap into position.

Extend linear cylinder so vacuum cup contacts top of cap (CAPL_ELS senses).
Activate the vacuum (VAC_OK senses).
Retract linear cylinder (CAPL_RLS senses).
Rotate cap 90 degrees counterclockwise (CAPR_CCWLS senses).
Extend linear cylinder to push cap into mandrel (CAPL_ELS senses).
Inflate mandrel bladder to grip the cap from the inside (BLD_OK senses).
Wait 1 second.
Extend the wheel cylinder to move the tape feed wheel to contact the cap. WHEEL_ELS senses when the wheel contacts the cap.
Rotate the mandrel (WSERVO_ON controls) to transfer the tape on the wheel to the cap. As the mandrel rotates, the tape wheel rotates, transferring tape to the cap. The servo controller executes slightly more than one revolution to get a complete stripe. The WSERVO_FIN internal tag/variable turns **on** when the correct rotation has been achieved and the servo has stopped.
Deflate mandrel bladder.
Retract the wheel cylinder. WHEEL_RLS senses when the wheel assembly is fully retracted.
Retract linear cylinder.
Rotate cap 90 degrees clockwise (CAPR_CWLS senses).
Extend linear cylinder.
Release (deactivate) the vacuum.
Retract linear cylinder.

The operation then repeats.

Your part of the program has indirect control over the belt carrier through an internal tag/variable. When this station wants to move the belt carrier, activate (turn **on**) the STR_BELT_MV tag/variable. This tag is one of multiple tags that all must be **on** to move the belt. Your part of the program just needs to maintain this tag/variable on as long as the

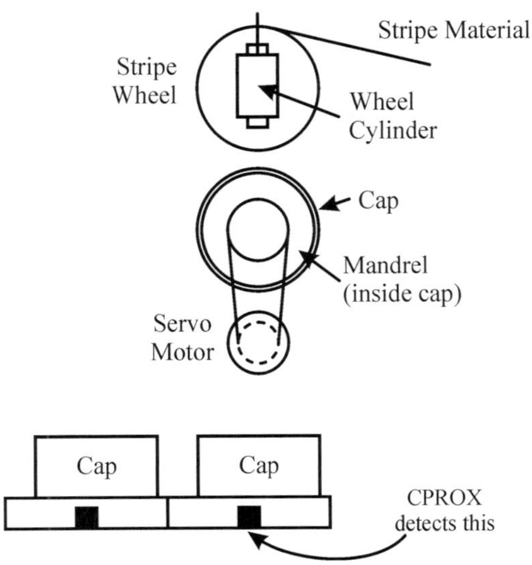

Figure P6.12.1. Cap striping station side view.

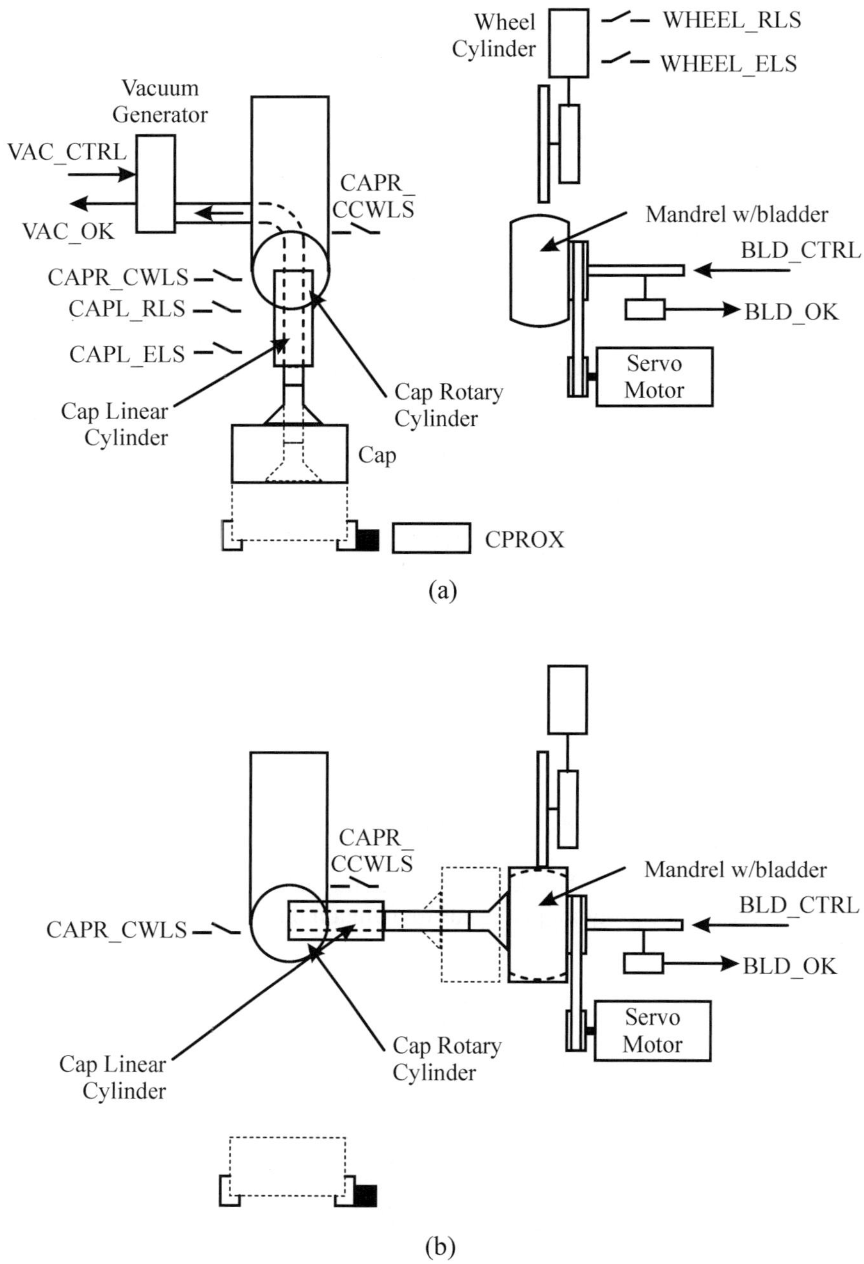

Figure P6.12.2. Cap striping station end view: *(a)* picked up cap; *(b)* cap on mandrel.

cap is being moved into and out of the station. This tag must be **off** during the striping operation. Your logic must control this tag.

CPROX: The CPROX proximity switch senses a pin on the belt holding the caps and indicates that a cap is in position. When **on**, the cap is in position. When **off**, the cap is not in position (in the process of being moved).

Cap Linear Cylinder: The cylinder that controls the linear extend/retract movement of the cap is a double-action pneumatic cylinder, controlled by the CAPL_EXT and CAPL_RET outputs. When CAPL_EXT is energized (turned **on**), the cylinder extends. When CAPL_RET is energized (turned **on**), the cylinder retracts. When CAPL_EXT and CAPL_RET are both **off**, the cylinder stops at its present position. When CAPL_EXT and CAPL_RET are both **on**, the cylinder operation is not consistent (it may stop or move in either direction slowly). The cylinder has two limit switches, CAPL_ELS, which is **on** when the cylinder is extended and CAPL_RLS, which is **on** when the cylinder is retracted (fully back). Even though the cylinder has mechanical stops at the end of its travel, it wastes energy to continue to activate the control once the limit is reached. For example, do not continue to activate the CAPL_RET output once the CAPL_RLS is **on**.

Cap Rotary Cylinder: The cylinder that controls the rotating movement of the cap is a double-action pneumatic cylinder, controlled by the CAPR_CW and CAPR_CCW outputs. When CAPR_CW is energized (turned **on**), the cylinder moves in the clockwise direction. When CAPR_CCW is energized (turned **on**), the cylinder moves in the counterclockwise direction. When CAPR_CW and CAPR_CCW are both **off**, the cylinder stops at its present position. When CAPR_CW and CAPR_CCW are both **on**, the cylinder operation is not consistent (it may stop or rotate in either direction slowly). The cylinder has two limit switches, CAPR_CWLS, which is **on** when the cylinder is clockwise and the linear cylinder is pointed downward and CAPR_CCWLS, which is **on** when the linear cylinder is pointed horizontal. Even though the cylinder has mechanical stops at the end of its travel, it wastes energy to continue to activate the control once the limit is reached.

Wheel Cylinder: The cylinder that controls the extend/retract movement of the stripe wheel is a single-action pneumatic cylinder, controlled by the WHEEL_EXT output. When WHEEL_EXT is energized (turned **on**), the cylinder extends, moving the stripe wheel assembly to contact the cap. When WHEEL_EXT is de-energized (turned **off**), the cylinder retracts. The cylinder has two limit switches, WHEEL_ELS, which is **on** when the cylinder is fully extended and WHEEL_RLS, which is **on** when the cylinder is retracted.

The **vacuum generator** is controlled by the VAC_CTRL output. The VAC_CTRL physical output must remain **on** to generate the vacuum. This output must be **off** to release the vacuum. While VAC_CTRL is **on**, compressed air is forced through a Venturi tube, creating a vacuum at the throat and thus at the cup holding the cap onto the end of the linear cylinder. The VAC_OK input is **on** when sufficient vacuum has been established and the vacuum cup is holding the cap. When the VAC_OK input is **off**, the vacuum has been released and the cap is not being held. A rotary coupling allows vacuum to be held while the cap is being rotated.

The **mandrel bladder** inflates to hold the cap on the mandrel, which rotates to put the strip on the cap. The bladder is controlled by the BLD_CTRL output. The BLD_CTRL physical output must remain **on** to inflate the bladder. This output must be **off** to deflate the bladder. The BLD_OK input is **on** when sufficient pressure is in the bladder and the mandrel is holding the cap. When the BLD_OK input is **off**, the bladder pressure has been released and the cap is not being held.

The **mandrel** is driven by a **servo motor**. When the WSERVO_ON internal tag/variable is **on**, another part of the ladder runs the servo motor to rotate the mandrel, transferring the stripe material to the cap. The mechanism that feeds the stripe material onto the wheel is not part of this problem. WSERVO_ON must be **on** only when transferring the stripe material to the cap. It must be **off** otherwise. The WSERVO_ON output needs to remain on during the entire rotation (unless paused). When the WSERVO_FIN internal tag/variable turns **on**, the mandrel has made a complete revolution.

There is no start/stop switch for your part of the ladder. Instead, there is an internal coil, S_RUN that is **on** when the operation is to happen. When S_RUN is **off**, the operation should pause at its current step. **Do not advance** to the next step when paused. Also, when S_RUN is **off**, the **on** transition of CPROX should be ignored (this is how the system disables putting the stripe on the cap).

When paused, all outputs should be **off except** for the following:

1. The vacuum control, VAC_CTRL, and bladder control, BLD_CTRL, must continue to be energized (remain **on**) while holding the cap.
2. The WHEEL_EXT control for the wheel cylinder must continue to be energized.

When S_RUN turns back **on** while the operation of the station is paused, the station should resume its suspended step.

The start and stop switches are absent from your part of the ladder because they control whether the stripe is being added and to reliably start/stop/pause the operation of the stripe addition as it fits into the overall packaging operation. These parts of the operation are outside the scope of this problem. The other part of the logic controls S_RUN. **You must not have any output coil referring to S_RUN.**

A separate reset internal coil, S_RESET, is provided which restores the striping station to its initial position. The reset operation should have the following steps:

Deflate bladder
Retract wheel cylinder
Retract linear cylinder
Rotate CW
Release vacuum

The steps must be done in the above order, as this system has mechanical interferences. Note that the bladder must be deflated and the wheel cylinder retracted before doing the remaining steps. Also, the vacuum must be maintained for all of the reset steps, except for the last one. Assume that an operator is responsible for actually removing any spoiled caps from the station after a reset. S_RESET should be ignored if S_RUN is **on**. The S_RUN internal coil used to restart the machine should be ignored when the reset operation is in progress. **You must not have any output coil referring to S_RESET.**

Do not add any more timed steps to those explicitly stated in the problem. In other words, do not put a timer in a step unless it is stated that the step duration is a specific time. Assume the tolerance on all timer values is at most 0.01 seconds.

Assume the following physical input and output tags/variables/symbols.

Tag/Var./Symbol	Description
CPROX	Belt pin-sensing inductive proximity switch, **on** when cap in position, **off** otherwise.
CAPL_ELS	Limit switch that closes (**on**) when the cap linear cylinder is extended. **Off** otherwise.
CAPL_RLS	Limit switch that closes (**on**) when the cap linear cylinder is retracted. **Off** otherwise.

CAPR_CCWLS	Limit switch that closes (**on**) when the cap rotary cylinder is in the counterclockwise (linear cylinder horizontal). **Off** otherwise.
CAPR_CWLS	Limit switch that closes (**on**) when the cap rotary cylinder is in the clockwise (linear cylinder vertical). **Off** otherwise.
WHEEL_ELS	Limit switch that closes (**on**) when the wheel cylinder is extended and thus the stripe wheel is contacting the cap. **Off** otherwise.
WHEEL_RLS	Limit switch that closes (**on**) when the wheel cylinder is retracted and thus the stripe wheel is pulled away from the cap. **Off** otherwise.
VAC_OK	**On** when vacuum is sensed and therefore the cap is being held onto the end of the cap linear cylinder.
BLD_OK	**On** when bladder pressure is sensed and therefore the bladder is holding the cap onto the mandrel.

CAPL_EXT	Cap linear cylinder extend control, **on** to extend cylinder. **Off** has no effect.
CAPL_RET	Cap linear cylinder retract control, **on** to retract cylinder. **Off** has no effect.
CAPR_CCW	Cap rotary cylinder counterclockwise control, **on** to move mechanism counterclockwise. **Off** has no effect.
CAPR_CW	Cap rotary cylinder clockwise control, **on** to move mechanism clockwise. **Off** has no effect.
WHEEL_EXT	Wheel cylinder extend control, **on** to extend cylinder and place strip wheel in contact with the cap. **Off** retracts cylinder.
VAC_CTRL	Vacuum control. **On** to generate vacuum. **Off** will deactivate the vacuum.
BLD_CTRL	Bladder control. **On** to pressurize mandrel bladder. **Off** will deflate bladder.

Assume the following internal tags/variables/symbols:

Tag/Var./Symbol	Description
S_RUN	Run/stop/pause control for the striping station; **on** to add stripe, **off** to stop/pause. Controlled by another part of the ladder logic in the PLC. **You must not have any output coil referring to S_RUN.**
S_RESET	Reset control, **on** to restore station to initial state. Controlled by another part of the ladder logic in the PLC. **You must not have any output coil referring to S_RESET.**
STR_BELT_MV	Striping operation belt move control. **Your logic must control this tag/variable**.
WSERVO_ON	Stripe wheel servo motor enable. **On** to signal to other ladder logic to rotate wheel, transferring stripe material to cap. **Off** will stop wheel rotation. **Your logic must control this tag/variable**.

WSERVO_FIN When **on**, the servo has rotated enough to transfer the proper stripe length to the cap. Controlled by another part of the ladder logic in the PLC. **You must not have any output coil referring to WSWERV_FIN.**

The addresses associated with the physical input and output tags/variables/symbols are:

Tag/Var./Symbol	ControlLogix	Micro800	MLogix	SLC-500
CPROX	Local:1:I.Data.3	_IO_EM_DI_03	I:0/3	I:1/3
CAPL_ELS	Local:1:I.Data.4	_IO_EM_DI_04	I:0/4	I:1/4
CAPL_RLS	Local:1:I.Data.5	_IO_EM_DI_05	I:0/5	I:1/5
CAPR_CCWLS	Local:1:I.Data.6	_IO_EM_DI_06	I:0/6	I:1/6
CAPR_CWLS	Local:1:I.Data.7	_IO_EM_DI_07	I:0/7	I:1/7
WHEEL_ELS	Local:1:I.Data.8	_IO_EM_DI_08	I:0/8	I:1/8
WHEEL_RLS	Local:1:I.Data.9	_IO_EM_DI_09	I:0/9	I:1/9
VAC_OK	Local:1:I.Data.10	_IO_EM_DI_10	I:0/10	I:1/10
BLD_OK	Local:1:I.Data.11	_IO_EM_DI_11	I:0/11	I:1/11
CAPL_EXT	Local:2:O.Data.0	_IO_EM_DO_00	O:0/0	O:2/0
CAPL_RET	Local:2:O.Data.1	_IO_EM_DO_01	O:0/1	O:2/1
CAPR_CCW	Local:2:O.Data.2	_IO_EM_DO_02	O:0/2	O:2/2
CAPR_CW	Local:2:O.Data.3	_IO_EM_DO_03	O:0/3	O:2/3
WHEEL_EXT	Local:2:O.Data.4	_IO_EM_DO_04	O:0/4	O:2/4
VAC_CTRL	Local:2:O.Data.5	_IO_EM_DO_05	O:0/5	O:2/5
BLD_CTRL	Local:2:O.Data.6	_IO_EM_DO_06	O:0/6	O:2/6

The addresses and/or data types associated with the internal tags/variables/symbols are:

Tag/Var./Symbol	CLogix Data Type	Micro800 Data Type	MLogix/SLC Address
S_RUN	BOOL	BOOL	B3/101
S_RESET	BOOL	BOOL	B3/102
WSERVO_FIN	BOOL	BOOL	B3/103
STR_BELT_MV	BOOL	BOOL	B3/151
WSERVO_ON	BOOL	BOOL	B3/152

P6-13. Thermoforming Machine Control. Using the function chart approach, implement the program for the following machine that forms a plastic tray from a plastic sheet.

Figure P6.13 shows the layout of a machine that uses heat and vacuum to form a plastic tray out of plastic sheet material. The formed trays are moved to another station where they are trimmed and stacked. You are implementing ladder logic for this part of the packaging operation only.

In summary, the sheet is pulled over the forming area, it is heated, a vacuum pulls the sheet over the mold, the formed tray is cooled, and then picked up and placed on an outfeed conveyor. Upon initial startup, assume the heater is in the upright position, the mold is down and the pickup is down. Also, assume someone has manually pulled some of the sheeting through the rollers, so there is something for the rollers to grip. Upon startup, the following steps are executed:

Activate the pinch rollers to move in a certain length of sheet.

Rotate the heater assembly so that it is on top of the sheet (sensed by HEAT_DLS). This action also cuts the sheet.

Activate the heater for 10 seconds to heat up the plastic in preparation for forming.

Move the forming table up into position (sensed by FORM_ULS). At the same time, FBLOW_ON is activated to blow air through the mold so that it does not yet contact the sheet.

Now pull a vacuum by activating the FVAC_ON output for 15 seconds. This step forms the tray.

Retract the heater assembly to remove it from the sheet (sensed by HEAT_ULS).

Activate the blower to cool the formed tray for 15 seconds.

Move the forming table back down (sensed by FORM_DLS).

Rotate the pickup up to contact the tray (sensed by PKUP_ULS). At the same time, activate PVAC_ON to so the tray is gripped immediately when the pickup is in position.

Rotate the pickup back so the tray is flipped over and rests on the outfeed conveyor (sensed by PKUP_DLS). At the end of this step, PVAC is turned off to let the conveyor move the tray to the trim/stack station. Assume the outfeed conveyor is always running.

The operation then repeats.

The **pinch rollers** are controlled with the PROLL_ON output. When **on**, the plastic sheet is pulled off the stock roller. When **off**, the plastic sheet is held in position. The pinch rollers contact the outside edges of the plastic sheet. When the heater assembly rotates into position, it punches out the middle part of the sheet, separating it from the outside edges. Hence, the pinch rollers have material with which to pull the sheet into position for the next tray.

An **encoder** is attached to one of the pinch rollers and is used to determine the appropriate length of sheet material to pull into position. The PINCH_ENC discrete input is the pulses from the encoder. The correct length of material is indicated by 150 pulses of PINCH_ENC. The pulses are slow enough so that an ordinary counter in the PLC can count the pulses.

Heater Up/Down Cylinder: The cylinder that controls the lowering/raising of the heater assembly is a double-action pneumatic cylinder, controlled by the HEAT_DN and HEAT_UP outputs. When the HEAT_DN output is energized (**on**), the cylinder extends and rotates the assembly so that it lies on top of the plastic sheet. When the HEAT_UP output is energized (**on**), the cylinder retracts and rotates the heater assembly so that it is vertical. The mechanism stops if neither output is **on** or if they are energized simultaneously. The HEAT_DLS limit switch is **on** when the cylinder is extended (heater down) and the HEAT_ULS limit switch is **on** when the cylinder is retracted (heater up).

Tray Form Raise/Lower Cylinder: The cylinder that controls the raising/lowering of the tray form is a double-action pneumatic cylinder, controlled by the FORM_UP and FORM_DN outputs. When the FORM_UP output is energized (**on**), the cylinder extends and raises the form assembly into position to form the tray. When the FORM_DN output is energized (**on**), the cylinder retracts and lowers the form to its down position. The mechanism stops if neither output is **on** or if they are energized simultaneously. The FORM_ULS limit switch is **on** when the cylinder is extended (form up) and the FORM_DLS limit switch is **on** when the cylinder is retracted (form down).

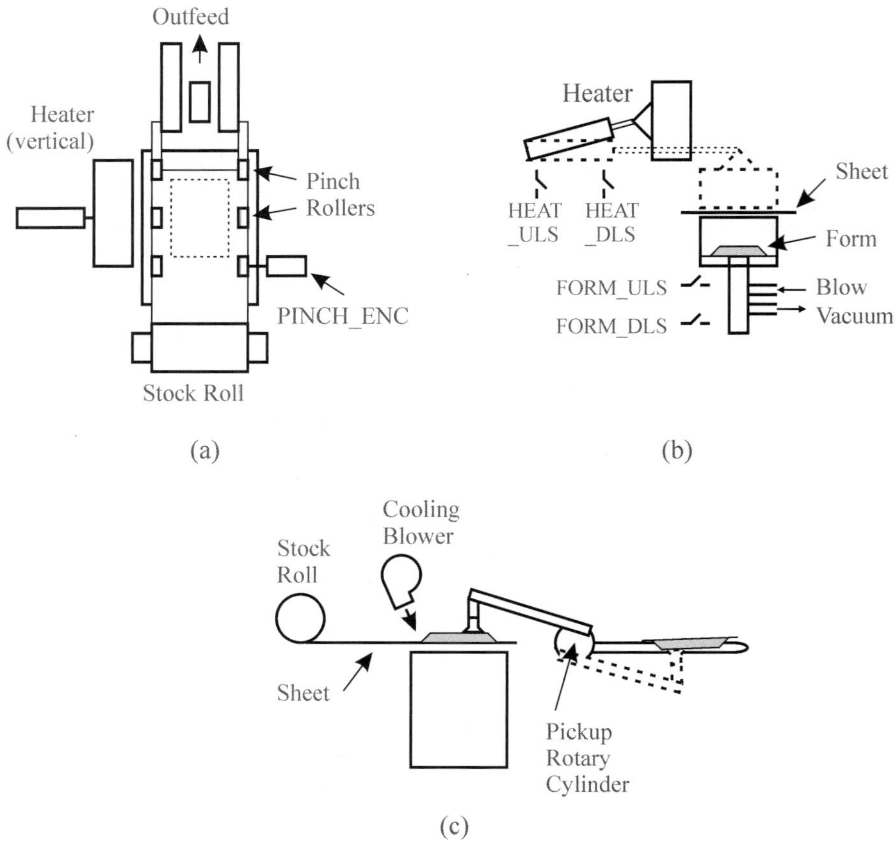

Figure P6.13. View of thermoforming machine: *(a)* from top; *(b)* from end; *(c)* from side.

Pickup Cylinder: The cylinder controlling the tray pickup is a single-action rotary pneumatic cylinder, controlled by the PKUP_CYL output. When PKUP_CYL is energized (**on**), the mechanism rotates up so that the vacuum cup contacts the tray on top of the form. When PKUP_CYL is de-energized, the mechanism rotates back to set the tray onto the outfeed conveyor. The PKUP_ULS limit switch is **on** when the pickup is in the "up" position (tray on form) and the PKUP_DLS limit switch is **on** when the pickup is "down" (tray on outfeed conveyor).

Heater control. The HEAT_ON output controls the heater operation. When HEAT_ON is **on**, the plastic sheet is heated in preparation for the vacuum molding step. When HEAT_ON is **off**, the heater is not energized.

Cooler control. The COOL_ON output controls the cooling blower operation. When COOL_ON is **on**, the formed tray is cooled before removing. When COOL_ON is **off**, the blower is not energized.

Form blow/vacuum. Air is initially blown through the form while it is being raised to achieve more uniform tray wall thickness. A vacuum is pulled through the form to pull the heated plastic to the form and thus make the tray shape. The FBLOW_ON discrete output is

on to blow air. When FBLOW_ON is **off**, no air is blown. The FVAC_ON discrete output is **on** to pull a vacuum. When FVAC_ON is **off**, no vacuum is pulled.

The FBLOW_ON is **on** only when the form is being raised into position. The FVAC_ON output is **on** when the form is in the "up" position, from tray formation through cooling.

Pickup vacuum. The pickup mechanism has a cup on the end through which a vacuum is drawn to grip the tray. The PVAC_ON discrete is **on** to pull a vacuum at the gripper. When it is **off**, there is no vacuum. PVAC_ON must be turned **on** as the pickup mechanism is rotated to grip the tray and is not turned **off** until the tray has been placed on the outfeed conveyor.

When the start switch (START_PB) is pressed (turned **on**) for the first time, assume the heater is upright, the mold is down and the pickup is down. Also, assume the rollers are gripping the plastic sheet. When the stop switch (STOP_PB) is pressed (turned **off**) the operation should pause **except** when heating. If the stop switch is pressed during the heating step, the step should complete and the operation should advance to the next step. When paused **do not advance** to the next step (exception noted earlier). If paused during the timed vacuum step, the timer should not be paused. When paused, all cylinder controls, the cooling blower, and the pinch roller control must be **off**. The air/vacuum and heater controls must not be affected by a pause. Pressing the start switch while the operation of the station is paused causes the station to resume its suspended step.

A separate reset switch, RESET_PB, is provided to restore the system to its initial state. When RESET_PB is **on**, all internal steps should be cleared, the heater assembly must be raised, the mold must be lowered, and the pickup mechanism should be moved "down." The reset operation is not complete until the heater assembly is up, the mold is down, and the pickup is down. **Do not implement a function chart for the reset** operation. There is no mechanical interference between the movements. Assume that an operator is responsible for actually removing any material from the station after a reset. The reset switch should be ignored if the station is running. The start switch should be ignored when the reset operation is in progress. If the station is paused and RESET_PB is **on**, the operator must release the RESET_PB before the start pushbutton switch can be used to restart the operation.

Do not add any more timed steps to those explicitly stated in the problem. In other words, do not put a timer in a step unless it is stated that the step duration is a specific time. Assume the tolerance on all timer values is 0.1 seconds.

Assume the following physical input and output tags/variables/symbols.

Tag/Var./Symbol	Description
START_PB	Start push button, N. O., **on** when starting.
STOP_PB	Stop push button, N. C., **off** when stopping.
RESET_PB	Reset push button, N. O., **on** when restoring station to initial state.
PINCH_ENC	Pinch roller encoder, pulses **on/off** as pinch rollers move. Used to determine proper amount of plastic sheet to feed.
HEAT_DLS	Limit switch that closes (**on**) when the heater assembly is in the down position to heat the plastic sheet, **off** otherwise.
HEAT_ULS	Limit switch that closes (**on**) when the heater assembly is in the vertical position, **off** otherwise.

FORM_DLS	Limit switch that closes (**on**) when the tray form is in the down position, **off** otherwise.
FORM_ULS	Limit switch that closes (**on**) when the tray form is in the up position, **off** otherwise.
PKUP_DLS	Limit switch that closes (**on**) when the pickup mechanism is in the down position to place the tray on the outfeed conveyor, **off** otherwise.
PKUP_ULS	Limit switch that closes (**on**) when the pickup mechanism is in the up position to pick up a tray, **off** otherwise.
PROLL_ON	Pinch rollers control, **on** to pull plastic sheet off stock roller, **off** holds the plastic sheet in position.
HEAT_DN	Heater assembly cylinder control, **on** to extend cylinder, rotating heater down, **off** has no effect.
HEAT_UP	Heater assembly cylinder control, **on** to retract cylinder, rotating heater up, **off** has no effect.
FORM_DN	Tray form position cylinder control, **on** to move form down, **off** has no effect.
FORM_UP	Tray form position cylinder control, **on** to move form up, **off** has no effect.
PKUP_CYL	Pickup mechanism up cylinder control, **on** to move mechanism up, **off** to move mechanism down.
HEAT_ON	Heater control, **on** to heat plastic sheet, **off** stops heater.
COOL_ON	Cooling blower control, **on** to cool plastic tray, **off** stops blower.
FBLOW_ON	Form air blow control, **on** to blow air through form, **off** stops blowing.
FVAC_ON	Form vacuum control, **on** to pull vacuum through form, **off** stops vacuum.
PVAC_ON	Pickup vacuum control, **on** to pull vacuum through vacuum cup, **off** stops vacuum.

The addresses associated with the physical input and output tags/variables/symbols are:

Tag/Var./Symbol	ControlLogix	Micro800	MLogix	SLC-500
START_PB	Local:1:I.Data.0	_IO_EM_DI_00	I:0/0	I:1/0
STOP_PB	Local:1:I.Data.1	_IO_EM_DI_01	I:0/1	I:1/1
RESET_PB	Local:1:I.Data.2	_IO_EM_DI_02	I:0/2	I:1/2
PINCH_ENC	Local:1:I.Data.3	_IO_EM_DI_03	I:0/3	I:1/3
HEAT_DLS	Local:1:I.Data.4	_IO_EM_DI_04	I:0/4	I:1/4
HEAT_ULS	Local:1:I.Data.5	_IO_EM_DI_05	I:0/5	I:1/5
FORM_DLS	Local:1:I.Data.6	_IO_EM_DI_06	I:0/6	I:1/6
FORM_ULS	Local:1:I.Data.7	_IO_EM_DI_07	I:0/7	I:1/7
PKUP_DLS	Local:1:I.Data.8	_IO_EM_DI_08	I:0/8	I:1/8
PKUP_ULS	Local:1:I.Data.9	_IO_EM_DI_09	I:0/9	I:1/9
PROLL_ON	Local:2:O.Data.0	_IO_EM_DO_00	O:0/0	O:2/0
HEAT_DN	Local:2:O.Data.1	_IO_EM_DO_01	O:0/1	O:2/1

HEAT_UP	Local:2:O.Data.2	_IO_EM_DO_02	O:0/2	O:2/2
FORM_DN	Local:2:O.Data.3	_IO_EM_DO_03	O:0/3	O:2/3
FORM_UP	Local:2:O.Data.4	_IO_EM_DO_04	O:0/4	O:2/4
PKUP_CYL	Local:2:O.Data.5	_IO_EM_DO_05	O:0/5	O:2/5
HEAT_ON	Local:2:O.Data.6	_IO_EM_DO_06	O:0/6	O:2/6
COOL_ON	Local:2:O.Data.7	_IO_EM_DO_07	O:0/7	O:2/7
FBLOW_ON	Local:2:O.Data.8	_IO_EM_DO_08	O:0/8	O:2/8
FVAC_ON	Local:2:O.Data.9	_IO_EM_DO_09	O:0/9	O:2/9
PVAC_ON	Local:2:O.Data.10	_IO_EM_DO_10	O:0/10	O:2/10

P6-14. Using the function chart approach, implement the program for the following machine that forms a part from a metal strip.

Figure P6.14 shows the layout of a station that forms a part from a strip of thin metal. The station moves the strip into position, punches out the part, bends the sides, moves the punch/bend head up, and removes the formed part. Note that this problem is a simplified version of an actual machine.

Upon initial startup, assume that the forming head is in the up position, all other cylinders are retracted, and the station has no metal strip in position. The operation steps are first described, followed by the details. When the machine is initially turned **on**, the following steps should be executed:

Feed metal strip into position (S_PROX senses).

Activate the punch cylinder to punch out the part from the metal strip (including punching some holes into the part). PUNCH_DLS senses.

Activate the bending cylinder to bend the sides of the part (the part is U-shaped). The BEND_DLS switch senses that the bending cylinder is in the down position.

Hold the bending and punch cylinders in the down position for an additional 1.5 seconds.

Retract the bending cylinder (BEND_ULS senses).

Retract the punch cylinder (PUNCH_ULS senses).

Extend the eject cylinder (EJECT_ELS senses). In addition, S_PROX must also be **off**, to make sure part has been completely ejected.

Retract the eject cylinder (EJECT_RLS senses)

The process repeats.

The metal strip is fed in by activating (turning **on**) the FEED_MOTOR output. This output activates a series of pinch rollers that move the metal strip into the forming area.

The S_PROX **proximity sensor** is an inductive sensor placed in the die where the part is punched out that senses the presence/absence of metal in the die. As the metal strip is being fed into the die, S_PROX turns **on** as the strip is in position to be punched. S_PROX remains on as the part is being punched/bent. S_PROX will turn **off** when the formed part is ejected.

Punch Cylinder: The cylinder that controls the extension/retraction of the punch is a double-action hydraulic cylinder, controlled by the PUNCH_DN and PUNCH_UP outputs. When PUNCH_DN is energized (turned **on**), the cylinder extends and moves the punch down to punch out the part, which separates the piece from the metal strip and punches out

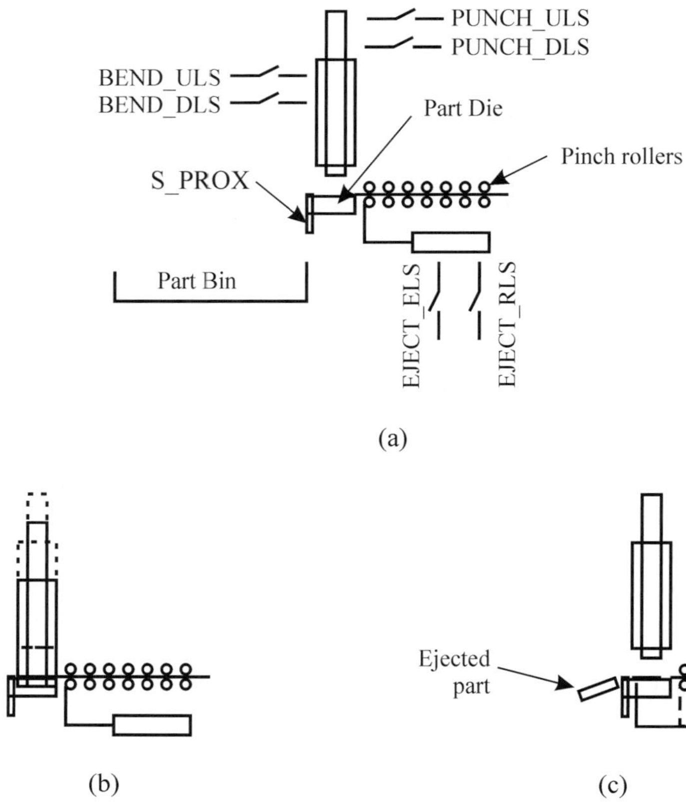

Figure 6.14. Part forming machine: *(a)* view of station at start; *(b)* punch down, bend down; *(c)* eject part cylinder extended.

some holes. When PUNCH_UP is energized (turned **on**), the cylinder retracts and moves up and out of the punching position. When PUNCH_DN and PUNCH_UP are both **off**, the punch stops at its present position. When PUNCH_DN and PUNCH_UP are both **on**, the cylinder operation is not consistent (it may stop or move in either direction slowly). The cylinder has two limit switches, PUNCH_DLS, which is **on** when the cylinder is extended (punch fully down) and PUNCH_ULS, which is **on** when the cylinder is retracted (punch fully up).

Bend Cylinder: The cylinder that controls the extension/retraction of the part bending mechanism is a single-action hydraulic cylinder, controlled by the BEND_DN output. When BEND_DN is energized (turned **on**), the cylinder extends and moves the mechanism down to bend the outside part of the part downward, forming a U-shaped part. When BEND_DN is **off**, the cylinder retracts. The cylinder has two limit switches, BEND_DLS, which is **on** when the cylinder is extended (part bent) and BEND_ULS, which is **on** when the cylinder is retracted. Note that the BEND_DN output must remain energized for an additional 1.5 seconds after the part is initially bent.

Part Eject Cylinder: The cylinder that controls the extension/retraction of the part eject mechanism is a single-action pneumatic cylinder, controlled by the EJECT_EXT output. When EJECT_EXT is energized (turned **on**), the cylinder extends and moves the formed part out of the die (and drop into a bin below). When EJECT_EXT is **off**, the cylinder retracts. The cylinder has two limit switches, EJECT_ELS, which is **on** when the cylinder is extended (part pushed out) and EJECT_RLS, which is **on** when the cylinder is retracted.

When the start switch is pressed (turned **on**) for the first time only, the station assumes there is no metal strip in position in the station. When the stop switch is pressed (turned **off**) the operation should pause. **Do not advance** to the next step when paused **except** when the part is being bent. When the assembly is being bent, the step **must** complete and advance to the next step before pausing. When paused, all cylinder controls **except** for the bend cylinder control (BEND_DN) must be turned **off**. Pressing the start switch while the operation of the station is paused causes the station to resume its suspended step. **DO NOT** retain the timer accumulator when paused.

A separate reset switch, RESET_PB, is provided which restores the system to its initial state. The reset operation should cause the punch cylinder to move up, the bend cylinder to move up, and the eject cylinder to be retracted. The reset operation is not complete until the punch cylinder is up, the bend cylinder is up and the eject cylinder is retracted. **Do not implement a function chart for the reset** operation. There is no mechanical interference. RESET_PB should be ignored when the bending cylinder is moving down. Assume that an operator is responsible for actually removing any material from the machine after a reset. The reset switch should be ignored if the machine is running. The start switch should be ignored when the reset operation is in progress. If the station is paused and RESET_PB is **on**, the operator must release the RESET_PB before the start pushbutton switch can be used to restart the station.

Do not add any more timed steps to those explicitly stated in the problem. In other words, do not put a timer in a step unless it is stated that the step duration is a specific time. Assume the tolerance on all timer values is at most ±0.01 second.

Assume the following physical input and output tags/variables/symbols.

Tag/Var./Symbol	Description
START_PB	Start push button, N. O., **on** when starting.
STOP_PB	Stop push button, N. C., **off** when stopping.
RESET_PB	Reset push button, N. O., **on** when restoring station to initial state.
S_PROX	Strip-sensing proximity switch**, turns on** when strip in position to be punched and while part is being punched/bent, **off** otherwise.
PUNCH_DLS	Limit switch that closes (**on**) when the punch cylinder is extended and in the down position, part is punched out.
PUNCH_ULS	Limit switch that closes (**on**) when the punch cylinder is retracted and in the up position.
BEND_DLS	Limit switch that closes (**on**) when the bend cylinder is extended and in the down position, part is bent into a U-shape.
BEND_ULS	Limit switch that closes (**on**) when the bend cylinder is retracted and in the up position.
EJECT_ELS	Limit switch that closes (**on**) when the part ejector is extended and pushed the part out of the die.
EJECT_RLS	Limit switch that closes (**on**) when the part ejector is retracted.

FEED_MOTOR Feed pinch roller control, **on** to activates a series of pinch rollers to move the metal strip into the forming area, **off** stops the feed.

PUNCH_DN Punch cylinder extend control, **on** to extend cylinder, moving it down to punch out part, **off** stops movement.

PUNCH_UP Punch cylinder retract control, **on** to retract cylinder, moving it up, **off** stops movement.

BEND_DN Bending cylinder extend control, **on** to extend mechanism, bending part into U-shape, **off** retracts cylinder.

EJECT_EXT Part eject extend control, **on** to force part out of die, **off** retracts.

The addresses associated with the physical input and output tags/variables/symbols are:

Tag/Var./Symbol	ControlLogix	Micro800	MLogix	SLC-500
START_PB	Local:1:I.Data.0	_IO_EM_DI_00	I:0/0	I:1/0
STOP_PB	Local:1:I.Data.1	_IO_EM_DI_01	I:0/1	I:1/1
RESET_PB	Local:1:I.Data.2	_IO_EM_DI_02	I:0/2	I:1/2
S_PROX	Local:1:I.Data.3	_IO_EM_DI_03	I:0/3	I:1/3
PUNCH_DLS	Local:1:I.Data.4	_IO_EM_DI_04	I:0/4	I:1/4
PUNCH_ULS	Local:1:I.Data.5	_IO_EM_DI_05	I:0/5	I:1/5
BEND_DLS	Local:1:I.Data.6	_IO_EM_DI_06	I:0/6	I:1/6
BEND_ULS	Local:1:I.Data.7	_IO_EM_DI_07	I:0/7	I:1/7
EJECT_ELS	Local:1:I.Data.8	_IO_EM_DI_08	I:0/8	I:1/8
EJECT_RLS	Local:1:I.Data.9	_IO_EM_DI_09	I:0/9	I:1/9
FEED_MOTOR	Local:2:O.Data.0	_IO_EM_DO_00	O:0/0	O:2/0
PUNCH_DN	Local:2:O.Data.1	_IO_EM_DO_01	O:0/1	O:2/1
PUNCH_UP	Local:2:O.Data.2	_IO_EM_DO_02	O:0/2	O:2/2
BEND_DN	Local:2:O.Data.3	_IO_EM_DO_03	O:0/3	O:2/3
EJECT_EXT	Local:2:O.Data.4	_IO_EM_DO_04	O:0/4	O:2/4

7 Comparison and Computation

Chapter Topics:

- Comparison blocks
- Computation blocks
- Comparison and computation applications

OBJECTIVES

Upon completion of this chapter, you will be able to:

- Describe the operation of the comparison and computation blocks of the PLCs covered by this text
- Apply comparison and computation to industrial situations

Scenario: Calibrating a differential pressure transmitter causes a major processor fault.

You are called by the power plant because they are having problems with the baghouse PLC. The baghouse (Figure 7.1) is essentially a filter for the boiler flue gas. It removes small particulates from the gas stream before it is released to the atmosphere. A technician has just recalibrated the differential pressure transmitter (PDT-016) and now the display on the operator panel shows "****" in place of the pressure reading and there is a message at

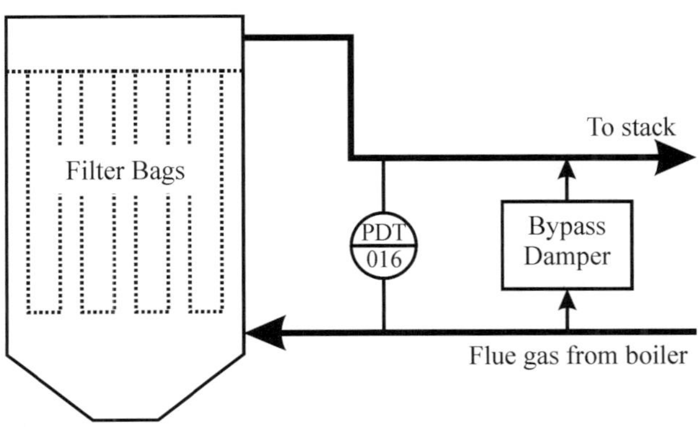

Figure 7.1. Power plant baghouse.

the top of the display that reads "PLC in Program Mode." They have opened the bypass damper so that the flue gases bypass the baghouse, but the incident will need to be reported to the regulatory agency if the baghouse is not placed in service within two hours.

***Solution*:** When you arrive, you notice from the indicators on the SLC 5/03 processor module that the processor is faulted and not running. You connect your notebook computer to the processor and examine the processor status. The fault code is 20 and is described as "A minor error bit is set at end of the scan." The fault is located in rung 223 of program file 2, the last rung of the program, which is to be expected from the description of the fault. The math overflow trap bit is also set, indicating that a math overflow occurred as a result of the calibration process. Using the cross-reference, you find a SCL function block that scales the transmitter reading. Consulting the instruction set reference manual, you discover that an overflow in one part of this calculation causes the math overflow trap bit to be set. However, the code checks for the transmitter value within range before using the SCL function block. So the SCL is not the problem. The last reading of the analog input channel is 17,352. With the cross-reference, you find another place where this transmitter value is used. It is multiplied by 1.9 and copied to an analog output integer that is connected to a chart recorder. The constant 1.9 is the ratio of the maximum value of the analog output integer divided by the maximum value of the transmitter value (31208/16384). However, the last transmitter value (17,352) multiplied by 1.9 is 32,969, which is an overflow! To prevent the problem from happening, you limit the analog input integer to 16,384 before multiplying it by 1.9.

7.1 INTRODUCTION

Comparison and computation function blocks are used less frequently than timers and counters, but are still important. This chapter describes these types of function blocks for the PLCs covered by this text. As done in previous chapters, the comparison and computation function blocks for a particular PLC are described in separate sections. That is, only the section(s) pertaining to the PLC(s) of interest need to be studied. Following these descriptions, examples are used to illustrate their application. To illustrate the need for computation, the next section describes the conversion of a physical quantity, as measured by an analog input module, to a value in the PLC.

The comparison function blocks provide the ability to compare two values and initiate some action based on that comparison. Computation function blocks are classified in the following categories: arithmetic, logical, conversion, and move. The arithmetic function blocks perform mathematical operations such as add, subtract, multiply, divide, square root, trigonometry, and exponentiation. The move function blocks transfer data from one location to another. The conversion function blocks provide conversion between various data formats.

7.2 CONVERSION OF PHYSICAL QUANTITY

To provide a motivation for the need of arithmetic operations in a PLC, consider the problem of converting the output signal of an analog transducer connected to an analog input module into a value in the appropriate units. The analog input module converts the transducer output signal into an integer or real value. However, in order to be useful this

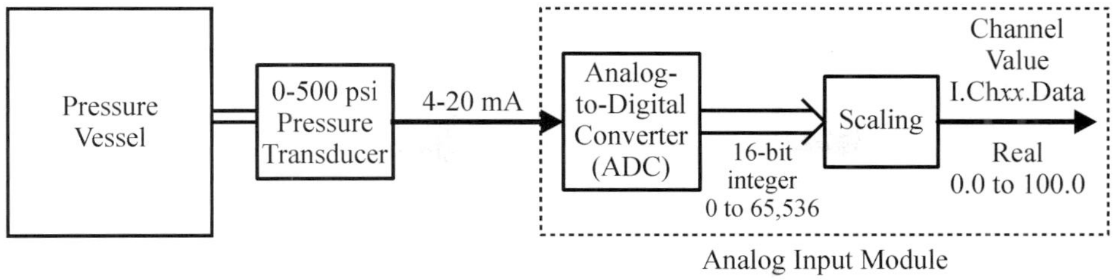

Figure 7.2. Pressure measurement example.

integer or real value should be converted into a value whose units are appropriate to the physical quantity being measured, for example, flow, level, or temperature. The following example illustrates this problem.

Example 7.1. A transducer measures pressure in the range of 0 to 500 pounds/square inch (psi) and transmits the pressure as a 4 to 20 mA signal to an analog input module having a 16-bit ADC whose output integer is scaled into a real number in the range of 0.0 to 100.0 (Figure 7.2). This particular example is for a 1769-IF8 analog input module whose channel scaling is configured for the default engineering unit range. Find the transmitted current and channel output value for pressures of 0, 125, 250, and 390 psi.

Solution. The table of signal currents and channel values for the various pressures is shown below. Unless known otherwise, one assumes a linear relationship between the physical quantity being measured and the output signal of the transducer. So, when the pressure is zero (the low end of the range), the transducer current and therefore the channel value are also at the low end of their respective ranges. A 250 psi pressure is in the middle of the measurement range and so the transducer current is also at the middle of its range (12 mA). The ADC output corresponding to 250 psi is also in the middle of its range, which is 50.0. The values for 125 psi are basically halfway between the values corresponding to zero and 250 psi. A pressure of 390 psi is at 78% (390/500=0.78) of the 0 to 500 psi range, so the transducer output is at 78% of its range, 4+0.78(20-4)=16.48. In a similar fashion, the channel value is at 78% of its range, 0.78(100)=78.0. The results are summarized as

Pressure	Transducer Current	Channel Value
0 psi	4.0 mA	0
125	8.0	25.0
250	12.0	50.0
390	16.48	78.0

Example 7.1 also illustrates that a given physical quantity may be represented in a control system by more than one type of signal. The pressure was not only represented by a real value, but also by a current signal. In addition, the ranges of these two representations are not identical. As another example, Figure 7.3 depicts the analog output of a PLC used to control the position of a regulatory control valve. The PLC analog output is a signal in the range of 4 to 20 mA which is converted into a 3 to 15 psi pressure signal. The pressure signal drives the valve which converts the 3 to 15 psi into 0 to 100% open (0% is closed). The valve position of 0 to 100% is represented three other ways (channel value, current, and pressure).

The need to represent a physical quantity with signals having different ranges is generalized as "percent of span." Even though a physical quantity takes on different signal

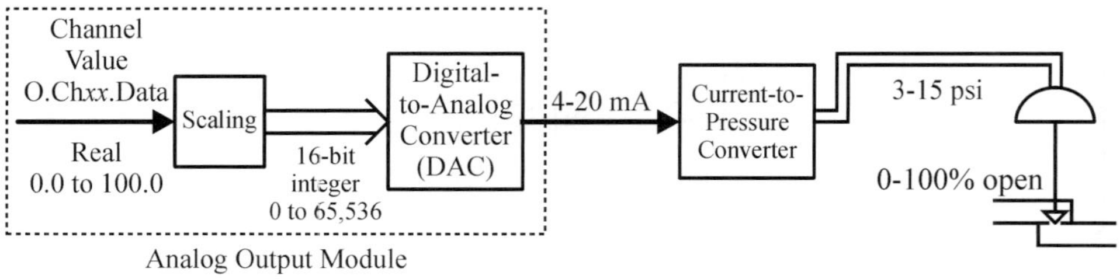

Figure 7.3. Valve position represented by multiple signals.

representations, the percent of span remains the same. Expressing physical quantities as percent of span is commonly used in chemical process industries to avoid confusion and mistakes. Older, pneumatic control systems use a pressure signal with a range of 3 to 15 psi to transmit quantities between instruments. In electronic control systems, the common signal between instruments is a 4 to 20 mA current. Figure 7.4 illustrates the relationship between the values of a physical quantity, expressed in any units to percent of span. As an equation, the conversion of a value to percent of span is

$$\text{Percent of span} = \frac{\text{Value} - \text{Low}}{\text{High} - \text{Low}} \times 100$$

The conversion of percent of span to a value is expressed as

$$\text{Value} = \frac{\text{Percent of span}}{100} \times (\text{High} - \text{Low}) + \text{Low}$$

The high and low values are often called Engineering Units (E.U.):

E.U. Low = Lowest physical quantity measured by transducer
E.U. High = Highest physical quantity measured by transducer
E.U. Span = E.U. High - E.U. Low

For Example 7.1, the E.U. Low is 0 psi and the E.U. High is 500 psi.

In the PLC, one is usually interested in obtaining a relationship between the analog module channel value and the value of the physical quantity. In this case, the value in % of span is not needed and the formula for the equation of a straight line between two points is used to find this relationship (Figure 7.5). Let

X = value of quantity in first units
X_1 = lowest value of first units (0%)
X_2 = highest value of first units (100%)

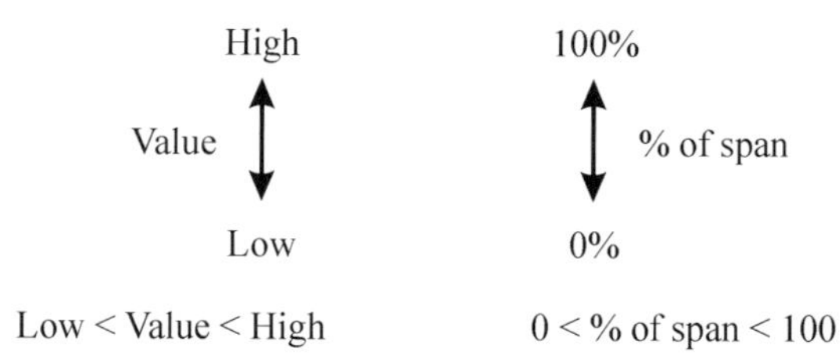

Figure 7.4. Relationship between value and % of span.

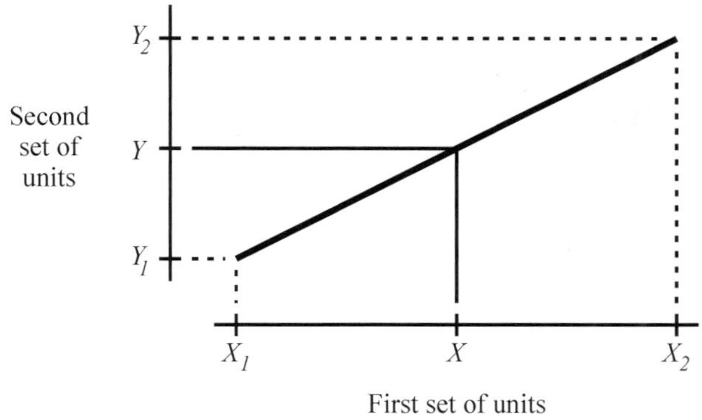

Figure 7.5. Linear relationship between two sets of units.

Y = value of quantity in second units
Y_1 = lowest value of second units (0%)
Y_2 = highest value of second units (100%)

To find the value of the quantity (Y) in the terms of the second units, given the value of the quantity (X) in terms of the first units,

$$Y = \frac{X - X_1}{X_2 - X_1} \times (Y_2 - Y_1) + Y_1 \tag{7.1}$$

The use of this formula is demonstrated by the following examples.

Example 7.2. A transducer measures pressure in the range of 0 to 500 pounds/square inch (psi) and transmits the pressure to an analog input module whose channel value is an integer in the range of 13,107 to 65,535 (Micro800 plug-in module). Find (a) the formula for the ADC output integer in terms of the pressure; and (b) the pressure (in psi), given the channel value integer. The ranges of the values are:

Pressure transducer: 0 to 500 psi
Channel value integer: 13,107 to 65,535

Solution. The formula given in (7.1) is applied for both of these situations. For (a) the variables for equation 7.1 are

X = Pressure (psi) Y = Channel value integer
X_1 = 0 (psi) Y_1 = 13,107
X_2 = 500 (psi) Y_2 = 65,535

The ADC output in terms of the pressure is then

$$\text{ADC output} = \frac{\text{Pressure} - 0}{500 - 0} \times (65{,}535 - 13{,}107) + 13{,}107 \text{ rounded to nearest integer}$$

which simplifies to

ADC output = (Pressure × 524.38) + 13,107 rounded to nearest integer

For (b) the variables for equation 7.1 are

X = Channel value integer $\quad Y$ = Pressure (psi)
X_1 = 13,107 $\quad Y_1$ = 0 (psi)
X_2 = 65,535 $\quad Y_2$ = 100 (psi)

The pressure in terms of the ADC output is then

$$\text{Pressure} = \frac{\text{Channel value} - 13{,}107}{65{,}535 - 13{,}107} \times (500 - 0) + 0$$

which simplifies to

$$\text{Pressure} = (\text{ADC output} - 13{,}107) \times 0.001907 \ \text{psi}$$

Part (b) of Example 7.2 is the usual situation that needs to be handled in the PLC. The channel value (integer or real) is converted to the proper units in order to be displayed on an operator display or to be used with a comparison function block to provide an alarm indication.

Example 7.3. A transducer measures temperature in the range of 90 to 700 °C and is connected to an analog input module whose channel value is a real tag in the range of 0.0 to 100.0 (ControlLogix range). Find the formula to convert the channel value to the temperature, in °C.

***Solution*.** Equation (7.1) is used to find the conversion equation. The variables for equation 7.1 are

X = Channel value $\quad Y$ = Temperature (°C)
X_1 = 0 $\quad Y_1$ = 90 (°C)
X_2 = 100 $\quad Y_2$ = 700 (°C)

The temperature in terms of the ADC output is then

$$\text{Temperature} = \frac{\text{Channel value} - 0}{100 - 0} \times (700 - 90) + 90$$

which simplifies to

$$\text{Temperature} = (\text{Channel value}) \times 6.1 + 90 \ \text{°C}$$

A note about the ranges for the ControlLogix analog modules (1756-series modules). The range of the integer output of the ADC and the range of integer input of the DAC is -20,030 to 30,920 for a 4 to 20 mA input signal. These modules can be configured to scale the analog value to a user-desired range. So, one could configure the analog input module in Example 7.3 to calculate the channel value (Ch*x*Data) in the range of 90 to 700, obviating the need to implement the formula of Example 7.3. However, if the temperature sensor is replaced with a sensor having a different range, say, 100 to 800, then the analog input module must be reconfigured, which means one must connect to the processor with the programming software, make the configuration changes, and download the changes to the module. A better approach is to define a tag for the lower end of the desired range (e.g., Temp_Low) and another tag for the upper end of the desired range (e.g., Temp_High). The value of these tags are initially set to a constant value and can be changed from an HMI, without needing the programming software. In addition, the analog input configuration is set to a range of 0 to 100, which provides an easier number to verify the analog value since it represents a percent of the range. The formula implemented in the processor converts the 0 to 100 Ch*x*Data range to the range defined by the two tags defining the desired range (e.g., Temp_Low and Temp_High). Then, if the sensor is changed or calibrated to a different range, the needed change in the scaling computation is accommodated without a change in the program, or in the module configuration.

7.3 CONTROLLOGIX COMPARISION AND COMPUTATION

The ControlLogix/CompactLogix comparison and computation function blocks are organized according to their function:

Category	Description
Comparison	Compare one or more values
Basic math	Basic arithmetic (+, -, ×, ÷)
Logical	Bit-wise logical operators
Advanced math	Trigonometry and exponentiation
Conversion	Integer/BCD and radian/degree
Move	Copy data

The comparison function blocks are input instructions, that is, they can be placed anywhere on a rung, except at the rightmost position. All of the computation and move function blocks are output instructions, that is, they are normally placed on the rightmost position of the rung. As explained in section 2.4, this distinction is made because in relay ladder logic, outputs are never connected in series and always occur on the extreme right side of the rung. To activate multiple output function blocks simultaneously, the blocks are connected in parallel. This restriction is relaxed in the ControlLogix PLC and output function blocks may be placed in series.

All of the comparison and computation function blocks have one input connection and one output connection. Multiple operands are handled as block parameters.

7.3.1 Comparison Function Blocks

For all of the comparison function blocks, integer and real data types may be mixed in the operands in the block. Any conversions between data types are handled automatically. Also, literals (constants) may be used in place of tags.

The most general comparison function block is CMP; an example is shown in Figure 7.6. The CMP block can evaluate any comparison operation, depending on the expression in the block. The output of the block is true (**on**) if the expression is true. The expression can contain only one comparison operation. The expression on either side of the comparison operation can include arithmetic operations. A table of valid expressions is in the instruction set reference (Rockwell Automation, 2022a).

As an alternative to the CMP function block, there are individual comparison function blocks; examples are shown in Figure 7.7. The current value of a tag is shown below the tag

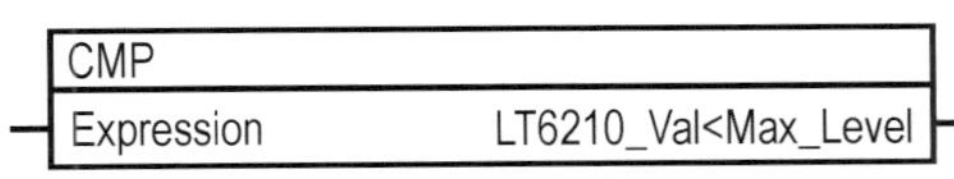

Figure 7.6. ControlLogix CMP comparison function block.

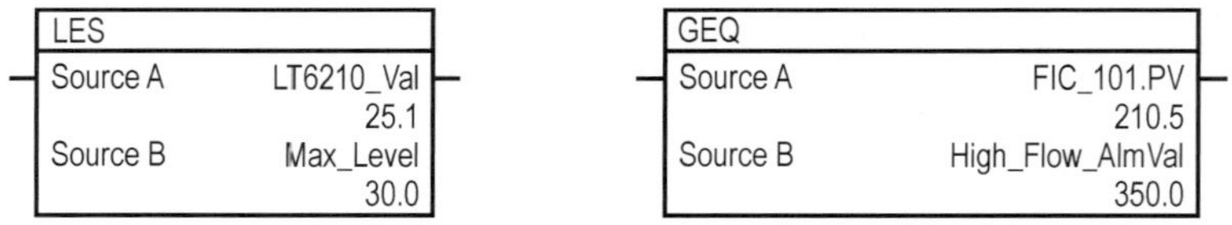

(a) (b)

Figure 7.7. Example ControlLogix comparison blocks: *(a)* less than; *(b)* greater than or equal.

name. The possible comparison function blocks and the condition that turns **on** the output are:

Block Name	**Output is On if:**
EQU	Source A = Source B
GEQ	Source A ≥ Source B
GRT	Source A > Source B
LEQ	Source A ≤ Source B
LES	Source A < Source B
LIM	Test value within Low Limit and High Limit
MEQ	Mask(Source A) = Mask(Source B)
NEQ	Source A ≠ Source B

The LIM and MEQ function blocks operate somewhat differently from the other comparison function blocks and are explained separately below.

At first glance, it would seem that the CMP block should always be used instead of the individual comparison function blocks. However, there are two important reasons that many PLC programmers prefer the individual blocks for simple comparisons. Most importantly, the values of the two tags being compared are displayed when viewing the individual function block on-line. The CMP function block displays only the expression without showing any of the values. Often, when troubleshooting a program, one needs to easily see the two values and thus why a certain comparison is true or false. For example, one can easily see why the comparison is true in Figure 7.7*a* and false for Figure 7.7*b*. In Figure 7.6 there is no indication why the comparison is true or false. Secondly, the CMP block requires a longer execution time and occupies more memory than the corresponding individual comparison blocks.

The MEQ is a masked equal comparison, an example is shown in Figure 7.8*a*. The MEQ block logically ANDs the Source and Compare values with the Mask value before comparing for equality. If the masked Source and Compare values are equal, the block output is turned on. The MEQ Source, Mask, and Compare operands must be integer data types (DINT, INT, SINT). A "1" in the mask means the particular bit is passed and a "0" in the mask means the particular bit is blocked. The example in Figure 7.8*a* compares the third digit of two thumbwheel switches. The mask isolates the third digit. The result of the MEQ with the values shown is true, so the output is **on**.

The LIM function block tests whether a value is either inside or outside of a specified range. An example LIM block is shown in Figure 7.8*b*. If Low Limit ≤ High Limit, then the block output functions as follows

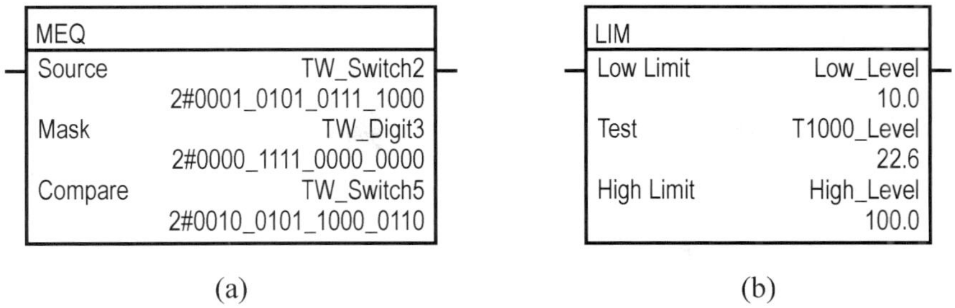

Figure 7.8. Advanced ControlLogix comparison blocks: *(a)* mask equal; *(b)* limit test.

Condition	**Output**
Low Limit ≤ Test ≤ High Limit	On
Test < Low Limit or Test > High Limit	Off

which means the block tests that a value is within the limits. For the example shown in Figure 7.8*b* the output is **on**. If the values of the limits are reversed such that Low Limit > High Limit, then the block output functions as follows

Condition	**Output**
Test > Low Limit or Test < High Limit	On
High Limit ≤ Test ≤ Low Limit	Off

which means the block tests that a value is outside the limits. If the low and high limit tags are reversed for the example in Figure 7.8*b* the output will be **off**.

7.3.2 Basic Math Function Blocks

For all of the computation and move function blocks, integer and real data types may be mixed in the operands. Any conversions between data types are handled automatically. However, when the result is an integer, fractions less than 0.5 are truncated and values outside the valid range are limited. The S:V status bit is set if the result is outside the valid range for the particular integer type. Also, literals (constants) may be used in place of source operand tags. If there are no contacts or other conditions attached to the block input, the function block will be executed on every scan.

The most general computation function block is CPT. An example is shown in Figure 7.9. The CPT block can evaluate any arithmetic, logical, conversion, and copy operation,

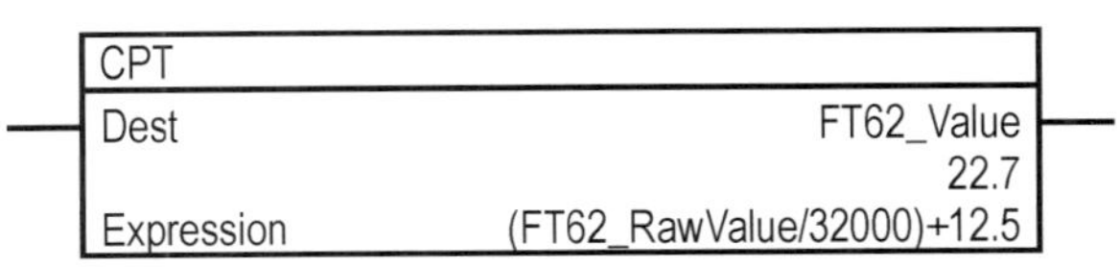

Figure 7.9. ControlLogix CPT compute function block.

depending on the expression in the block. A table of valid expressions is in the instruction set reference (Rockwell Automation, 2022a).

Design Tip

If the destination of the CPT is an integer and the expression contains all integer tags and/or constants, a decimal point in a constant forces the calculations to be done with real arithmetic, avoiding a possible overflow/rounding problem in an intermediate result. If the destination of the CPT is a real tag, then the calculations are done with real arithmetic.

As an alternative to the CPT function block, there are individual arithmetic function blocks. The possible basic arithmetic blocks are:

Block Name	Calculation:
ABS	Dest = absolute value of Source (|Source|)
ADD	Dest = Source A + Source B
DIV	Dest = Source A ÷ Source B
MOD	Dest = Source A modulo Source B
MUL	Dest = Source A × Source B
NEG	Dest = – Source
SQR	Dest = square root of Source
SUB	Dest = Source A – Source B

Examples of two-operand blocks are shown in Figure 7.10*a* and 7.10*b*. Example one-operand blocks are shown in Figure 7.10*c* and 7.10*d*. Note that when dividing two integers, the fractional part of the result is truncated and not rounded.

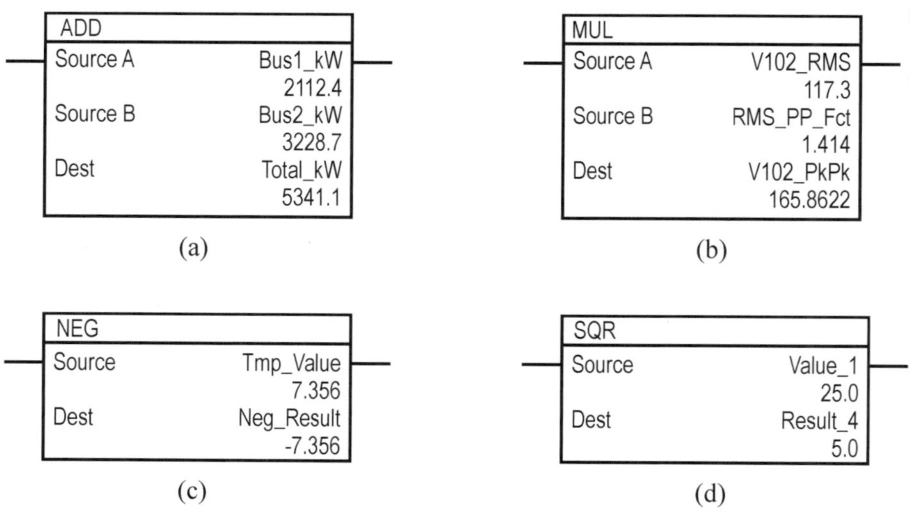

Figure 7.10. Example ControlLogix computation blocks: *(a)* addition; *(b)* multiply; *(c)* negation; *(d)* square root.

Initially, it would seem that the CPT function block should always be used instead of the individual computation function blocks. As for the comparison blocks, many PLC programmers prefer the individual function blocks for simple computations since the values of all tags in the block are displayed when viewing the individual function block on-line. The CPT function block displays only the value of the result and does not show any of the values that lead to the result. Often, when troubleshooting a program, one needs to easily see source operands to verify the reasons behind incorrect calculations (for example, overflow or integer truncation). Secondly, the CPT function block has a longer execution time and occupies more memory than the corresponding individual computation function blocks.

7.3.3 Advanced Math Function Blocks

The names of the advanced math function blocks and the calculation that occurs for each one are:

Block Name	Calculation Result
ACS	Dest = Arc cosine of Source (Dest in radians)
ASN	Dest = Arc sine of Source (Dest in radians)
ATN	Dest = Arctangent of Source (Dest in radians)
COS	Dest = Cosine of Source (Source in radians)
LN	Dest = Natural logarithm of Source
LOG	Dest = Common logarithm of Source
SIN	Dest = Sine of Source (Source in radians)
TAN	Dest = Tangent of Source (Source in radians)
XPY	Dest = $(\text{Source A})^{\text{Source B}}$

The XPY function block is a two-operand block, similar to the ones in Figure 7.10*a*. The other advanced math function blocks are one-operand blocks, similar in format to the block in Figure 7.10*c*. All of these function blocks are intended for use with REAL data types. Integer types may be used, but loss of accuracy and overflow may occur.

7.3.4 Logical Function Blocks

The logical function blocks provide bit-wise logical operations for the integer data types (SINT, INT, and DINT) and are shown in Figure 7.11. The logical function blocks and the calculation that occurs for each one are:

Block Name	Calculation Result
AND	Bitwise logical AND of Source A and Source B integers
NOT	Invert all bits of Source
OR	Bitwise logical OR of Source A and Source B integers
XOR	Bitwise logical exclusive-OR of Source A and Source A integers

If integer data types are mixed, the function block fills the upper bits of the smaller integer data types with zeros so that they are the same size as the largest data type.

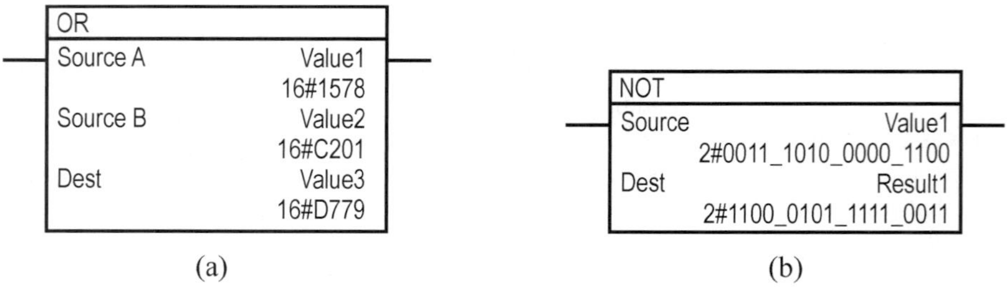

Figure 7.11. Example ControlLogix logical function blocks: *(a)* bit-wise OR; *(b)* bit-wise NOT.

7.3.5 Conversion Function Blocks

The names of the conversion function blocks and the conversion that occurs for each one are:

Block Name	Description
DEG	Convert source in radians to degrees.
RAD	Convert source in degrees to radians.
FRD	Convert source BCD (binary coded decimal) value to integer.
TOD	Convert source integer value to BCD.
TRN	Truncates (removes) fractional part of source.

The DEG and RAD function blocks are similar to the one-operand function blocks shown in Figure 7.10. An example FRD function block is shown in Figure 7.12. This example assumes that the bit pattern from two thumbwheel switches is transferred to the TW62 tag with a move type of function block. The FRD block converts a 2, 4, or 8-digit BCD value depending on the source type (SINT, INT, DINT) to a destination integer. An

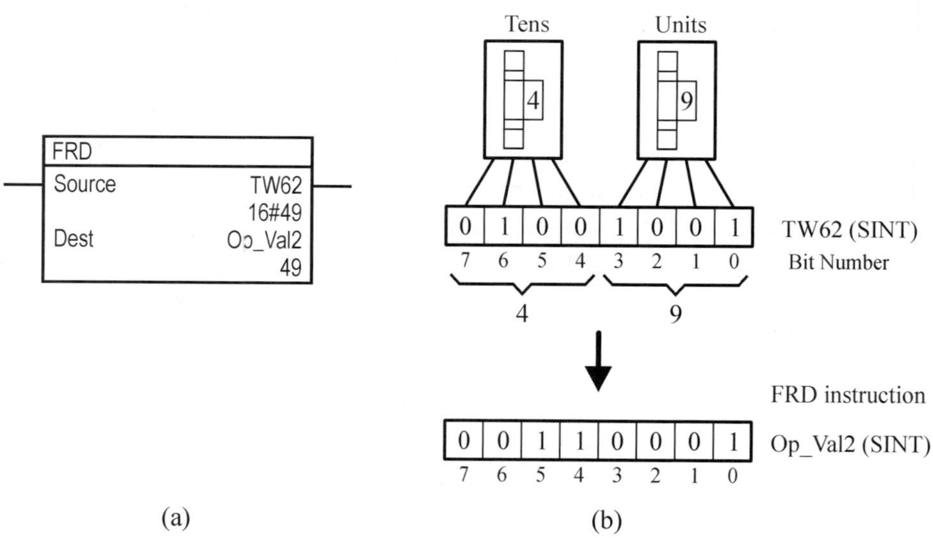

Figure 7.12. Example ControlLogix BCD-to-integer conversion: *(a)* function block; *(b)* operation.

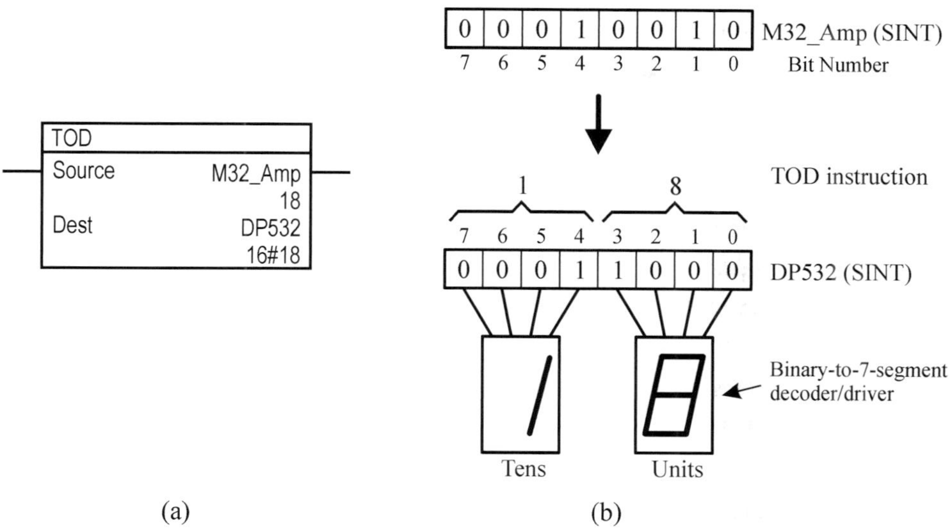

Figure 7.13. Example ControlLogix integer-to-BCD conversion: *(a)* function block; *(b)* operation.

example TOD function block is shown in Figure 7.13. The integer in M32_Amp is converted into the BCD format and placed in DP532. A separate function block is assumed to transfer the BCD number contained in DP532 to discrete outputs that are attached to binary-to-7-segment decoder drivers.

7.3.6 Move Function Blocks

The move function blocks transfer data between tags. The names and descriptions of the move function blocks are:

Block Name	Description
BTD	Move bits within an integer or between integers
CLR	Clear tag
MOV	Copy source tag to destination tag
MVM	Copy source integer to destination integer, allowing portions of the integer to be masked

The MOV and MVM function blocks are shown in Figure 7.14. For the MOV function block, the data types of the source and destination do not need to be identical. Conversion between the integer and real data types are handled automatically. However, when moving a REAL tag to an integer tag, one should consider overflow. The MVM Source, Mask, and Dest operands must be integer data types (DINT, INT, SINT). A "1" in the mask means the particular bit is passed from the Source integer to the Dest integer and a "0" in the mask means the particular bit is blocked. A "0" in the mask also means that those bits in the destination integer are not modified. The example in Figure 7.14*b* transfers the third digit of a thumbwheel switch. The mask isolates the third digit. If integer data types are mixed, the function block fills the upper bits of the smaller integer data types with zeros so that they are the same size as the largest data type.

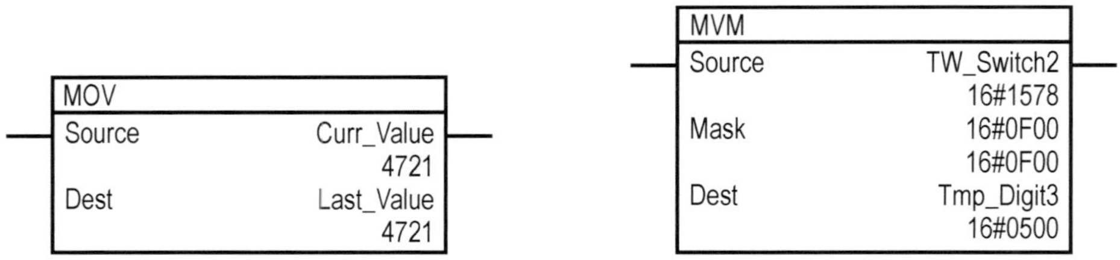

(a) (b)

Figure 7.14. ControlLogix move function blocks: *(a)* move; *(b)* masked move.

The BTD function block is useful for moving a group of bits. An example BTD function block is shown in Figure 7.15*a*. The Source Bit is the number of the least significant bit in the group of bits to be moved and the Length is the number of bits in the group. The group of bits are copied to the part of the Dest integer defined by the Dest Bit (the number of the least significant bit) and the Length. The other bits in the destination integer are not modified. The Source Bit and Dest Bit numbers must be in the valid range for the data types. If the length of the bit group extends beyond the destination most-significant bit, the extra bits are lost. The actual transfer for the BTD block in Figure 7.15*a* is illustrated in Figure 7.15*b*. The lower 4 bits of (BCD) BCD_Value are copied into bits 12 to 15 of output word Local:4:O.Data. The four outputs, 4:O.Data.12 to 4:O.Data.15, are connected to a 7 segment display that displays the (BCD) number represented by the 4 bits.

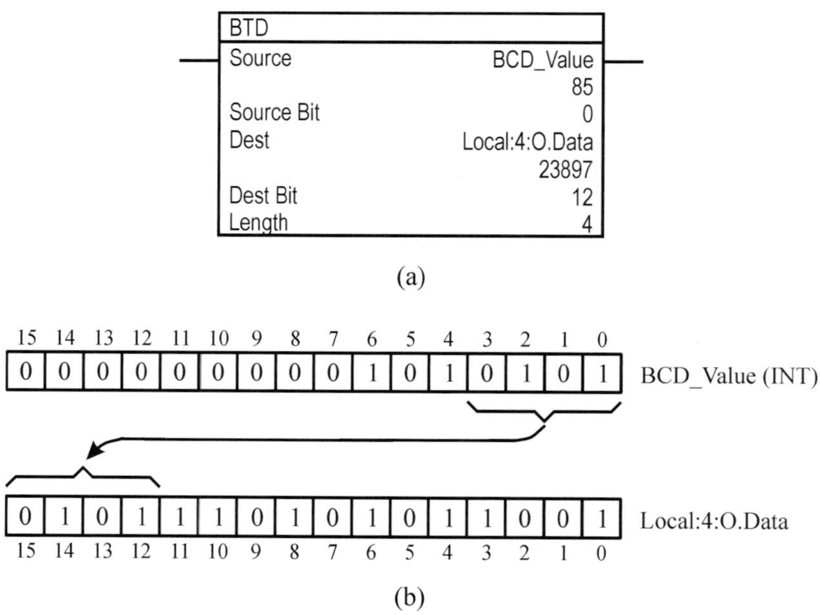

Figure 7.15. Example ControlLogix bit field distribute: *(a)* function block; *(b)* operation.

7.4 MICRO800 COMPARISION AND COMPUTATION

The Micro800 comparison and computation function blocks are organized according to their function:

Category	Description
Comparison	Compare one or more values
Basic math	Basic arithmetic (+, -, ×, ÷)
Advanced math	Trigonometry and exponentiation
Logical	Bit-wise logical operators
Data type conversion	Convert value to another data type
Move	Copy data
Selection	Select one of a set of values

The computation function blocks have EN and ENO connections as the topmost input and output, respectively (Figure 7.16). The EN input must be **on** ("1" or logical true) for the block to execute. The ENO output echoes the EN input and is **true (on)** if EN is **true**. If the EN input is **false (off)** the function block is **not** executed and all outputs remain with their old values, that is the value when the block was last executed. In this case, the value of ENO is automatically set to **false**.

Design Tip

The behavior of the function block outputs when EN is false must be considered in the ladder logic code. If the EN input is false, the block outputs remain at their last value. This behavior may produce unexpected results.

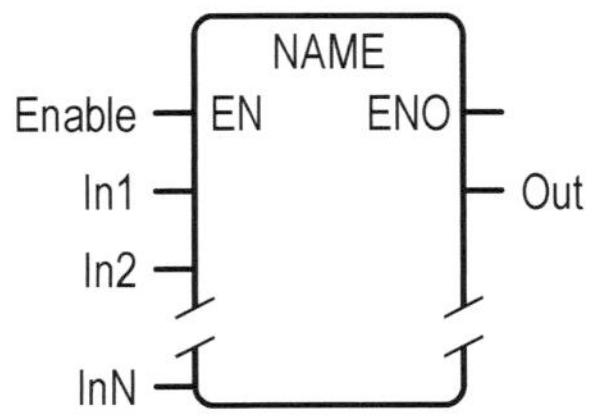

Figure 7.16. EN and ENO function block connections.

7.4.1 Comparison Function Blocks

Example comparison blocks are shown in Figure 7.17. Each comparison block has an enable (EN) input and two other inputs that are the variables to be compared. The result of the comparison, the "o1" output, is a Boolean and is **true** if the EN input is true and the comparison is true. The current value of a variable is shown below the variable name. The possible comparison function blocks and the condition that turns **on** the o1 output are:

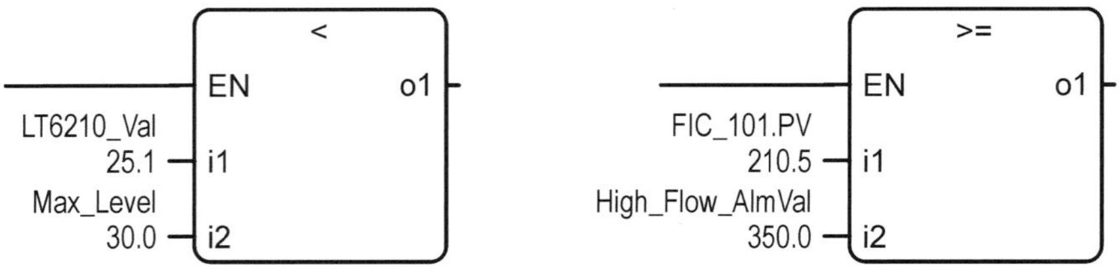

(a) (b)

Figure 7.17. Example Micro800 comparison blocks: *(a)* less than; *(b)* greater than or equal.

Block Name	Output is True if:
=	i1 = i2
>=	i1 ≥ i2
>	i1 > i2
<=	i1 ≤ i2
<	i1 < i2
<>	i1 ≠ i2

For all of the comparison function blocks, any of the simple data types may used for the i1 and i2 inputs, but the data types of the i1 and i2 inputs must be identical. Also, literals (constants) may be used in place of variables.

7.4.2 Basic Math Function Blocks

Example arithmetic function blocks are shown in Figure 7.18. The possible basic arithmetic blocks are:

Block Name	Calculation:
+	o1 = i1 + i2 + … (if more inputs selected)
-	o1 = i1 – i2
*	o1 = i1 × i2 × … (if more inputs selected)
/	o1 = i1 ÷ i2
ABS	ABS = absolute value of IN (|IN|)
MOD	MOD = IN modulo Base
Neg	o1 = – i1

For all of the computation function blocks, the data types of the inputs and outputs must be identical. Also, literals (constants) may be used in place of input variables. If there are no contacts or other conditions attached to the block EN input, the function block will be executed on every scan.

Examples of two-operand blocks are shown in Figure 7.18*a* and 7.18*b*. Example one-operand blocks are shown in Figure 7.18*c* and 7.18*d*. Note that when dividing two integers, the fractional part of the result is truncated and not rounded.

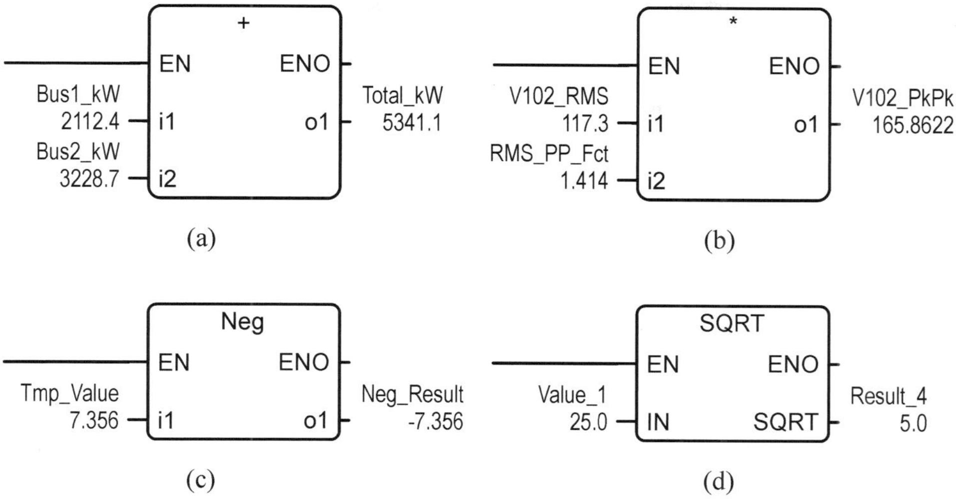

Figure 7.18. Example Micro800 computation blocks: *(a)* addition; *(b)* multiply; *(c)* negation; *(d)* square root.

7.4.3 Advanced Math Function Blocks

The names of the advanced math function blocks and the calculation that occurs for each one are:

Block Name	Calculation Result
ACOS	ACOS = Arc cosine of IN (ACOS in radians)
ASIN	ASIN = Arc sine of IN (ASIN in radians)
ATAN	ATAN = Arctangent of IN (ATAN in radians)
COS	COS = Cosine of IN (IN in radians)
EXPT	EXPT = $(\text{IN})^{\text{EXP}}$
LOG	LOG = Common logarithm of IN
POW	POW = $(\text{IN})^{\text{EXP}}$ (EXP must be whole number)
RAND	RAND = Random value in set [0..base-1]
SIN	SIN = Sine of IN (IN in radians)
SQRT	SQRT = square root of IN
TAN	TAN = Tangent of IN (IN in radians)
TRUNC	TRUNC = IN truncated, leaving only integer part

For the RAND block, the base input and the RAND output is DINT. For all of the other blocks, the inputs and outputs are REAL data types.

The EXPT and POW blocks are two-operand blocks, similar to the one in Figure 7.18*a*., except that the inputs are IN and EXP rather than i1 and i2. The other advanced math function blocks are one-operand blocks, similar in format to the SQRT block in Figure 7.18*d*.

The ACOS_LREAL, ATAN_LREAL, COS_LREAL, SIN_LREAL, and TAN_LREAL perform the same function as the blocks without the "_LREAL" suffix , but the data types of the input and output are LREAL.

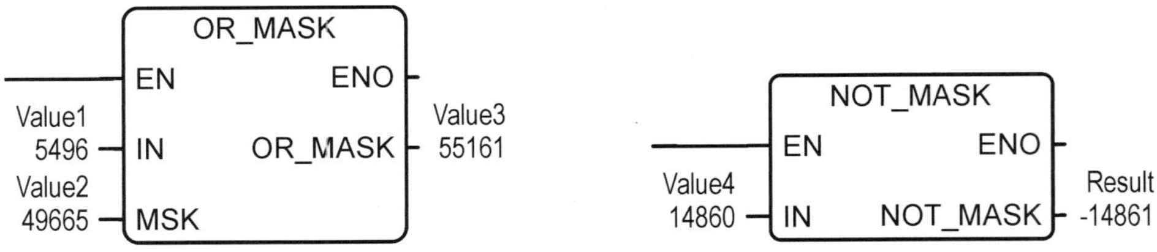

(a) (b)

Figure 7.19. Example Micro800 logical function blocks: *(a)* bit-wise OR_MASK; *(b)* bit-wise NOT_MASK.

7.4.4 Logical Function Blocks

The logical function blocks provide bit-wise logical operations for the DINT data type and are shown in Figure 7.19. The logical function blocks and the calculation that occurs for each one are:

Block Name	Calculation Result
AND_MASK	Bitwise logical AND of IN and MSK DINTs
NOT_MASK	Invert all bits of IN DINTs
OR_MASK	Bitwise logical OR of IN and MSK DINTs
XOR_MASK	Bitwise logical exclusive-OR of IN and MSK DINTs

If integer data types are mixed, the function block fills the upper bits of the smaller integer data types with zeros so that they are the same size as the largest data type.

7.4.5 Data Type Conversion Function Blocks

Function blocks are provided to convert from one data type to another. The conversion blocks are named ANY_TO_* with "*" replaced by the output data type. Examples are shown in Figure 7.20.

The data type of the i1 input must not be the same as the o1 output. For example, for the ANY_TO_INT block, the i1 input can be any data type, except that i1 cannot be an INT data type. If the value of the i1 input is larger than the maximum possible value of the o1 output data type, the result is truncated. For example, if the i1 for an ANY_TO_INT is a DINT with a value of 116,385 (16#0001C6A1), the o1 output is -14,687 (16#C6A1).

Specific information about some of the conversion function blocks:

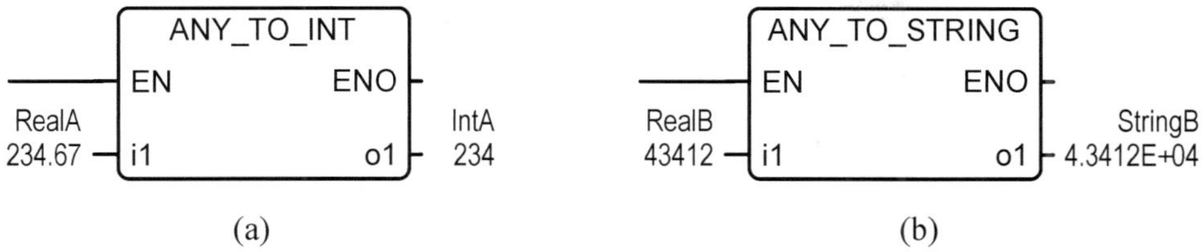

Figure 7.20. Example Micro800 conversion function blocks: *(a)* ANY_TO_INT; *(b)* ANY_TO_STRING.

Block Name	Description
ANY_TO_BOOL	Any non-zero input yields a true o1 There is no ENO output.
ANY_TO_DATE	If i1 is a STRING, it must be format D#YYY-MM-DD or DATE#YYYY-MM-DD. When i1 is an integer or real type, it must be positive and is interpreted as seconds and added to 1970-01-01 When i1 is TIME, it is added to 1970-01-01
ANY_TO_STRING	When i1 is REAL or LREAL, the string is in the exponential format. When i1 is TIME, o1 is of the format [seconds]s[milliseconds] If the length of the string is too short, the output is truncated at the string length.
ANY_TO_TIME	in1 is interpreted as the number of milliseconds and must be positive. The fractional part of any real types is ignored.

The SCALER block is not technically a conversion block, but is useful for converting and scaling the integer from a Micro800 plug-in or expansion module into a real number in useful units. An example SCALER function block is shown in Figure 7.21 that performs the conversion in Example 7.3 for a plug-in module analog input. All of the inputs and outputs are REAL data types. Since the analog input channel value is a UINT, an ANY_TO_REAL block converts the channel value into a REAL variable that the SCALER block can use. The

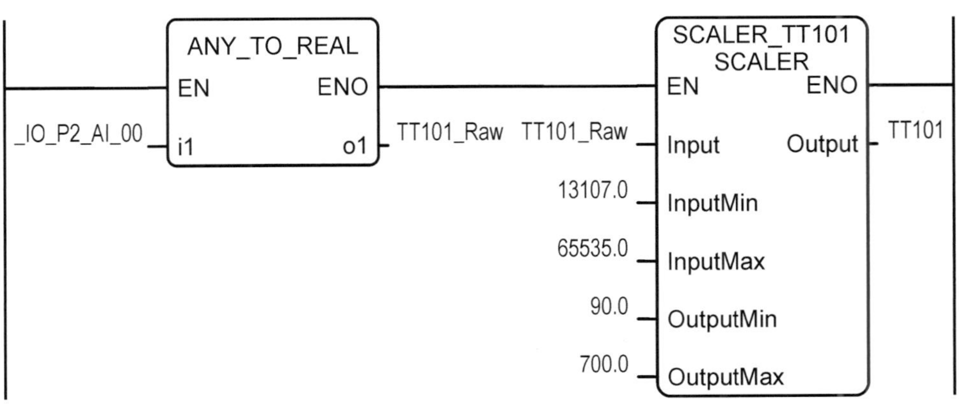

Figure 7.21. Example Micro800 SCALER block for plug-in analog input channel.

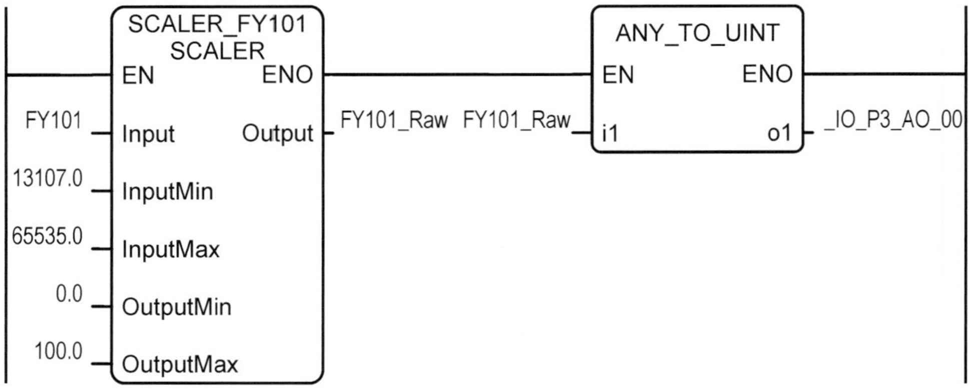

Figure 7.22. Example Micro800 SCALER block for plug-in analog output channel.

SCALER block implements the scaling equation (7.1). In terms of the block inputs and outputs, the equation is

$$\text{Output} = \frac{\text{Input} - \text{InputMin}}{\text{InputMax} - \text{InputMin}} \times (\text{OutputMax} - \text{OutputMin}) + \text{OutputMin}$$

The example SCALER block shown in Figure 7.22 performs the conversion for an analog output channel. The FY100 tag, in the range of 0.0 to 100.0 is converted to the integer for a 4 to 20 mA output channel for a plug-in analog output module.

7.4.6 Move Function Blocks

The move function blocks transfer data between variables. The operation that occurs for each one are:

Block Name	Operation Result
COP	Copy binary data
MOV	Copy source tag to destination tag

The MOV function block transfers data between two variables. An example is shown in Figure 7.23. The i1 input and the o1 output must be the same data type. The i1 input could be a literal instead of a variable.

The COP block copies the source data to the destination as a binary bit-for-bit copy. No conversion between data types is performed. Figure 7.24 shows an example COP. The block inputs are described as follows:

Enable – Execution control. Must be true to perform operation.

Src – Initial element to be copied. Can be an array.

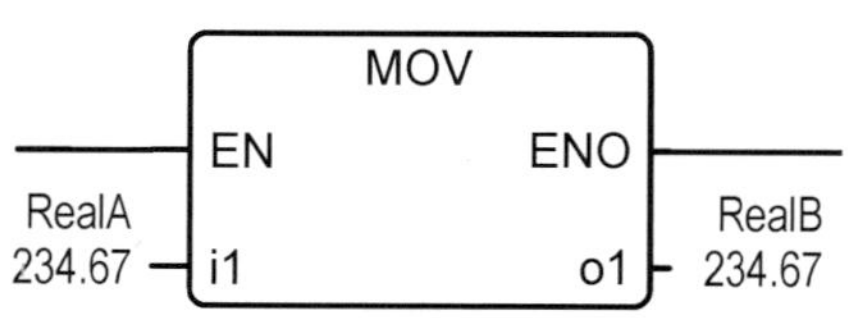

Figure 7.23. Example Micro800 move function block.

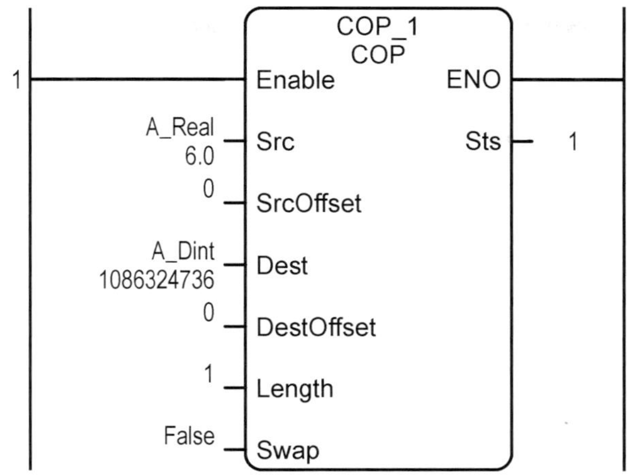

Figure 7.24. Micro800 sample COP block.

SrcOffset – Index of Src array element. If Src is not an array, set to zero.
Dest – Src is copied to this variable. Can be an array.
DestOffset – Index of Dest array element. If Dest is not an array, set to zero.
Length – Number of elements to copy. If the destination is a string, this input is the number of strings to copy.
Swap – If true, the byte order of the source data is swapped when writing to the destination. However, a swap does not occur if the source or destination is a STRING, or a 1-byte data type.

The output is described as follows:

Sts – Status of copy, 1 = copy success.

Other possible Sts values are listed in Rockwell Automation (2022b).

The Swap input needs more explanation and becomes important when reading/writing variables with other processors using a MSG_CIPSYMBOLIC block. In this case, the byte order of the data in the message packet is opposite of how it is stored internally in the processor. For example, reading a DINT with value 1 would appear as the following USINT values: 1, 0, 0, 0 (in order of index 0 to 3). If this USINT array is simply copied (with a COP, no swap) to a DINT variable, the value would be 16#1_00_00_00 = 16777216. When the Swap input is true, the result of the COP is 1 (16#00_00_00_01).

When dealing with copying of strings, the Src and Dest data types must be consistent. If the Src input is a STRING, then the Dest input must also be a STRING or a USINT data type. If the Dest input is a STRING, then the Src input must also be a STRING or a USINT data type.

The COP operation does not convert between data types. For the example in Figure 7.24 that shows a copy from a REAL variable (A_Real) to a DINT variable (A_Dint), the value of A_Dint is 16#40C0_0000, the binary representation of a floating-point 6.0. If the destination data type has fewer bits (for example, copying from a 32-bit data type to a 16-bit data type), the result is truncated. For example, if the A_Dint variable in Figure 8.30 is an

INT data type, the value of A_Dint is 16#40C0, the most-significant 16-bits of the binary representation of a floating-point 6.0.

7.4.7 Selection Function Blocks

The following function blocks allow one value to be selected from a given number of values:

Block Name	Description
LIMIT	Restrict IN between MIN and MAX (DINT only)
MAX	Maximum of IN1 and IN2 (both DINT)
MIN	Minimum of 2 IN1 and IN2 (both DINT)
MUX4B	Select n^{th} value from a list of 4 BOOLs
MUX8B	Select n^{th} value from a list of 8 BOOLs
TTABLE	Select n^{th} bit in UINT variable

The LIMIT block is shown in Figure 7.25*a* and ensures that a value is inside a specified range. When EN is true, the output of the block is determined as follows:

$$\text{LIMIT} = \text{IN if MIN} \leq \text{IN} \leq \text{MAX}$$

$$\text{LIMIT} = \text{MIN if IN} < \text{MIN}$$

$$\text{LIMIT} = \text{MAX if IN} > \text{MAX}$$

The MIN, IN, MAX, and LIMIT inputs are all DINTs.

The MAX block is shown in Figure 7.25*b* and when the EN input is true, the maximum of IN1 and IN2 is transferred to the MAX output. The MIN block is similar except the output is labelled as "MIN" and the minimum of IN1 and IN2 is transferred to the MIN output.

The MUX4B block is shown in Figure 7.25*c* and when the EN input is true, the Selector input determines which one of the inputs is transferred to the MUX4B output:

Selector	MUX4B
0	IN0
1	IN1
2	IN2
3	IN3

The value of the Selector input must be between 0 and 3. The MUX8B block is similar except the output is labelled "MUX8B" and there are 4 additional inputs named, IN4, IN5, IN6, and IN7. The Selector input must be between 0 and 7.

The TTABLE block is shown in Figure 7.25*d*. When the EN input is true the number of the bit in the Table input (UINT data type) transferred to the TTABLE output is determined by the IN0, IN1, IN2, and IN3 BOOL inputs as follows:

$$\text{Bit number} = \text{IN0} + 2\times\text{IN1} + 4\times\text{IN2} + 8\times\text{IN3}$$

For example, if Table=16#34D2 and IN0=1, IN1=0, IN2=0, and IN3=1, the 9th bit is selected, which is 0 (false).

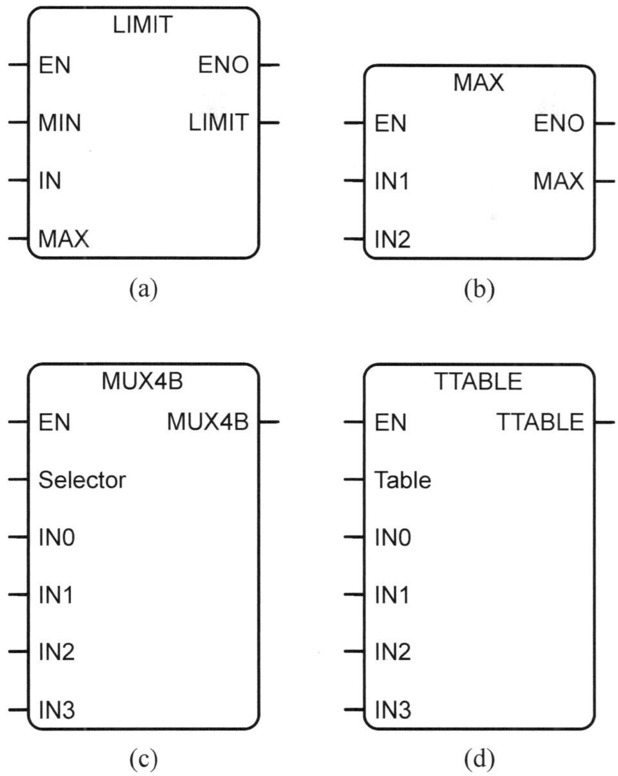

Figure 7.25. Example Micro800 selection function blocks: *(a)* LIMIT; *(b)* MAX; *(c)* MUX4B; *(d)* TTABLE.

7.5 MICROLOGIX/SLC COMPARISION AND COMPUTATION

The MicroLogix and SLC-500 comparison and computation function blocks are nearly identical to the ControlLogix function blocks of the same name. The major difference is that the operands are shown as addresses for the MicroLogix/SLC-500. The symbols can be used when specifying the operands in the function block, but the programming software displays the operand address in the function block. The comparison and computation function blocks are organized according to their function:

Category	Description
Comparison	Compare one or more values
Basic math	Basic arithmetic (+, -, ×, ÷)
Advanced math	Trigonometry and exponentiation
Logical	Bit-wise logical operators
Conversion	Integer/BCD and radian/degree
Move	Copy data

The comparison function blocks are input instructions, that is, they can be placed anywhere on a rung, except at the rightmost position. All of the computation and move function blocks are output instructions, that is, they are never connected in series and always occur on the extreme right side of the rung. To execute multiple output function blocks on one rung, the blocks are connected in parallel.

All of the comparison and computation function blocks have one input connection and one output connection. Multiple operands are handled as block parameters.

The specific function blocks that are supported vary among the different MicroLogix and SLC-500 processors. All processors support the basic two-operand comparison, basic arithmetic (+, -, ×, ÷), integer/BCD conversion, logical, and basic move function blocks.

7.5.1 Comparison Function Blocks

For all of the comparison function blocks, integer and floating-point data types may be mixed in the operands in the block. Any conversions between data types are handled automatically. Also, literals (constants) may be used in place of tags. There is one caution concerning SLC-500 data types. If words from ASCII file types are used as operands in comparison function blocks, they will be treated as integer numbers. Therefore, comparison of ASCII and integer values is not recommended..

The CMP general comparison block is not available on the MicroLogix and SLC-500. There are only individual comparison function blocks; examples are shown in Figure 7.26. The possible comparison function blocks and the condition that turns on the output are:

Block Name	Output is On if:
EQU	Source A = Source B
GEQ	Source A ≥ Source B
GRT	Source A > Source B
LEQ	Source A ≤ Source B
LES	Source A < Source B
LIM	Test value within Low Limit and High Limit
MEQ	Mask(Source A) = Mask(Source B)
NEQ	Source A ≠ Source B

The current value of an address is shown below the address and followed by a "<" symbol. Except for the LIM and MEQ blocks, the symbol of the Source A address is displayed above the function block. The LIM and MEQ function blocks operate somewhat differently from the other comparison function blocks and are explained separately below.

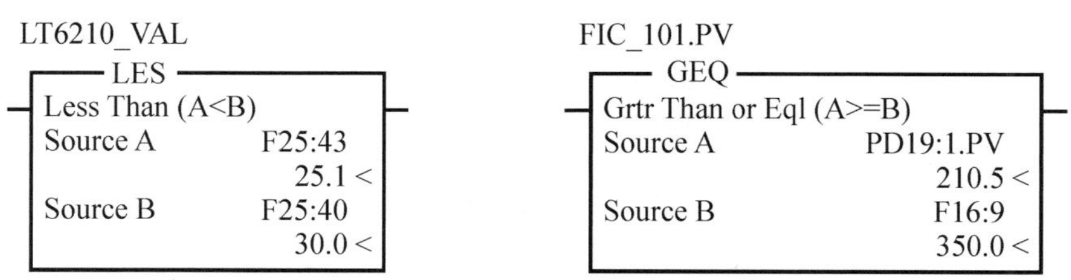

(a) (b)

Figure 7.26. Example MicroLogix/SLC-500 comparison blocks: *(a)* less than; *(b)* greater than or equal.

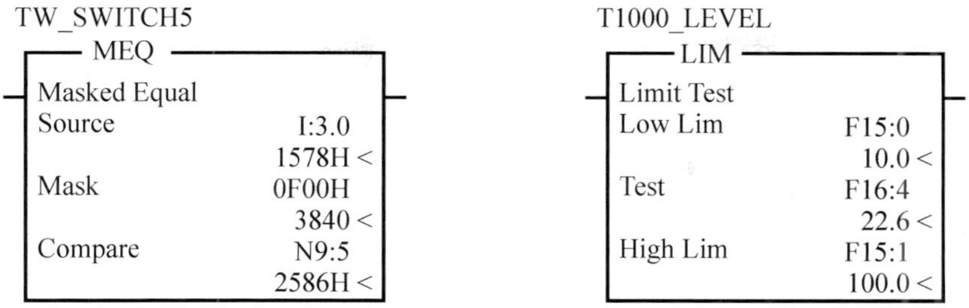

(a) (b)

Figure 7.27. Advanced MicroLogix/SLC-500 comparison blocks: *(a)* masked equal; *(b)* limit test.

The MEQ is a masked equal comparison, an example is shown in Figure 7.27*a*. The MEQ block logically ANDs the Source and Compare values with the Mask value before comparing for equality. If the masked Source and Compare values are equal, the block output is turned on. The MEQ Source, Mask, and Compare operands must be integers. If an ASCII word is used, it is interpreted as an integer. A "1" in the mask means the particular bit is passed and a "0" in the mask means the particular bit is blocked. The mask can be entered as a binary, octal, hexadecimal, or decimal constant, but is always displayed as hexadecimal. The symbol of the Compare address is displayed above the function block. The example in Figure 7.27*a* compares the third digit of two thumbwheel switches. The mask isolates the third digit. The result of the MEQ with the values shown is true, so the output is on.

The LIM function block tests whether a value is either inside or outside of a specified range. An example LIM block is shown in Figure 7.27*b*. If Low Lim ≤ High Lim, then the block output functions as follows

Condition	**Output**
Low Lim ≤ Test ≤ High Lim	On
Test < Low Lim or Test > High Lim	Off

which means the block tests that a value is within the limits. For the example shown in Figure 7.27*b* the output is on. If the values of the limits are reversed such that Low Limit > High Limit, then the block output functions as follows

Condition	**Output**
Test > Low Lim or Test < High Lim	On
High Lim ≤ Test ≤ Low Lim	Off

which means the block tests that a value is outside the limits. If the low and high limit tags are reversed for the example in Figure 7.27*b* the output will be off. The symbol of the Test address is displayed above the function block.

7.5.2 Basic Math Function Blocks

For all of the computation and move function blocks, integer and floating-point data types may be mixed in the operands. Any conversions between data types are handled

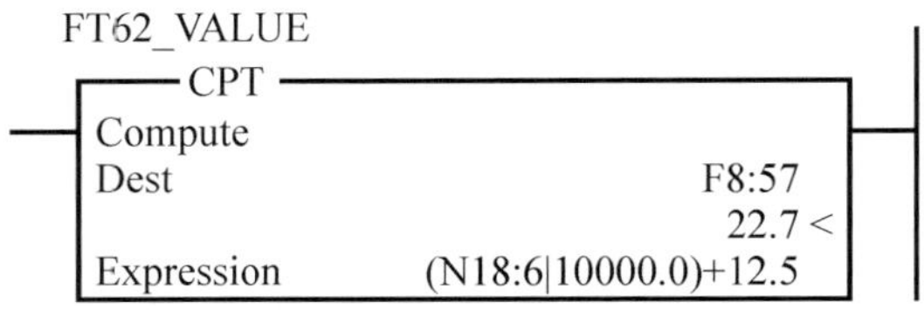

Figure 7.28. SLC-500 CPT computation function block.

automatically. However, when the result is an integer, fractions less than 0.5 are rounded to zero and values outside the valid range are limited. The S:0/01 (overflow) status bit is set if the result is outside the valid range for an integer. Also, literals (constants) may be used in place of source operand tags. As for the comparison function blocks, BCD or ASCII addresses are interpreted as integer numbers.

If there are no contacts or other conditions attached to the input of a computation or move function block, the function block will be executed on every scan. To execute the block only once, use an ONS contact (OSR for SLC-500) after any other conditions.

The most general computation function block is CPT; an example is shown in Figure 7.28. The CPT block can evaluate any arithmetic, logical, conversion, and copy operation, depending on the expression in the function block. A table of valid expressions is in the instruction set reference (Rockwell Automation, 2008). The CPT block is only valid on the SLC-5/03 or later processors and the expression is limited to 255 characters. The symbol of the Dest address is displayed above the function block. The CPT block is not available for MicroLogix processors. The operator for division is a vertical bar. For example, to divide N7:0 by F8:3, the expression is: N7:0|F8:3.

Design Tip

If the destination of the CPT is an integer (address N*x*:*n*) and the expression contains constants, a decimal point in a constant forces the calculations to be done with floating point numbers, avoiding a possible overflow/rounding problem in an intermediate result.

As alternative to the CPT function block on the SLC-500 and as the only option on MicroLogix processors, there are individual arithmetic function blocks; examples of two-operand blocks are shown in Figure 7.29. and one-operand blocks are shown in Figure 7.30. The possible basic arithmetic function blocks are:

Block Name	Calculation:
ABS	Dest = absolute value of Source (|Source|)
ADD	Dest = Source A + Source B
DDV	Dest = (S:14 and S:13) ÷ Source (MLogix 1000 and SLC-500 only)
DIV	Dest = Source A ÷ Source B
MUL	Dest = Source A × Source B
NEG	Dest = – Source

SQR	Dest = square root of Source
SUB	Dest = Source A – Source B

The current value of an address is shown below the address and followed by a "<" symbol. The symbol of the Dest address is displayed above the function block. The SQR function block is not valid for the SLC-5/01 processor. For the DIV block, if the Dest is an integer, the result is rounded. The DDV function block is only valid for MicroLogix and SLC-500 processors. For the DDV block, the 32-bit integer math register, consisting of words S:14 and S:13 is divided by the 16-bit source integer. The rounded result is placed in the Dest. The unrounded quotient is placed in S:14 and the remainder is placed in S:13.

For the SLC-500, there are two reasons that one may prefer the individual function blocks for simple computations. Most importantly, the values of all two or three addresses in the block are displayed when viewing the individual function block on-line. One can then view the source operands to verify the reasons behind incorrect calculations (for example, overflow). The CPT function block displays only the value of the result and does not show any of the values that lead to the result. Secondly, the CPT function block requires longer execution time and occupies more memory than the corresponding individual computation function blocks.

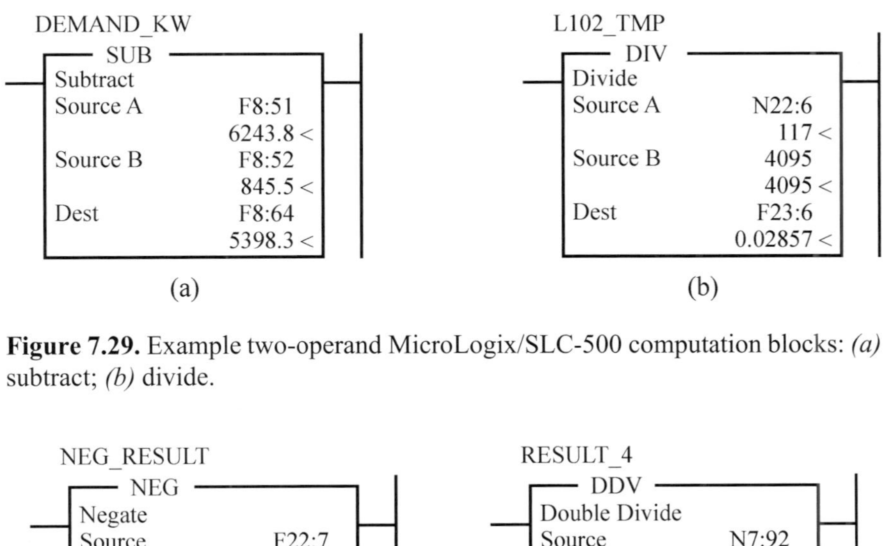

Figure 7.29. Example two-operand MicroLogix/SLC-500 computation blocks: *(a)* subtract; *(b)* divide.

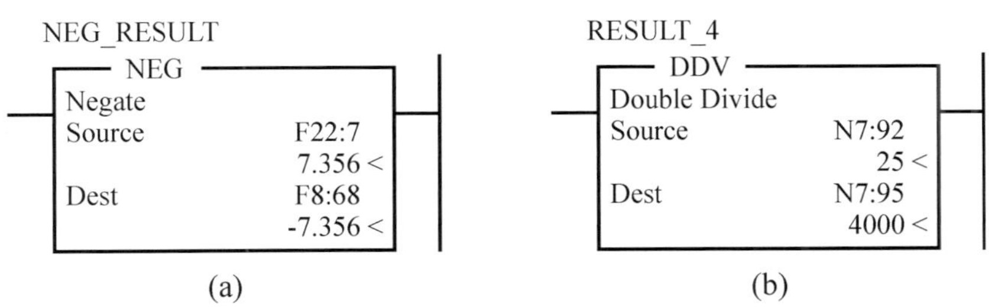

Figure 7.30. Example one-operand MicroLogix/SLC-500 computation blocks: *(a)* negation; *(b)* double divide (SLC-500 only).

7.5.3 Advanced Math Function Blocks

The advanced math function blocks are available for the MicroLogix 1400 and the SLC-5/03 and later processors only. Their names and the calculation that occurs for each one are:

Block Name	Calculation Result
ACS	Dest = Arc cosine of Source (Dest in radians)
ASN	Dest = Arc sine of Source (Dest in radians)
ATN	Dest = Arctangent of Source (Dest in radians)
COS	Dest = Cosine of Source (Source in radians)
LN	Dest = Natural logarithm of Source
LOG	Dest = Common logarithm of Source
SIN	Dest = Sine of Source (Source in radians)
TAN	Dest = Tangent of Source (Source in radians)
XPY	Dest = $(\text{Source A})^{\text{Source B}}$

The XPY function block is a two-operand block, similar to the ones in Figure 7.29. The other advanced math function blocks are one-operand blocks, similar in format to the blocks in Figure 7.30. All of these function blocks are intended for use with floating-point data. Integers may be used, but loss of accuracy and overflow may occur.

7.5.4 Logical Function Blocks

The logical function blocks provide bit-wise logical operations for integers. The logical function blocks and the calculation that occurs for each one are:

Block Name	Calculation Result
AND	Bitwise logical AND of Source A and Source B integers
NOT	Invert all bits of Source
OR	Bitwise logical OR of Source A and Source B integers
XOR	Bitwise logical exclusive-OR of Source A and Source A integers

Two example function blocks are shown in Figure 7.31.These function blocks may be used with ASCII data types, but the words are interpreted as if they are binary (integer) values.

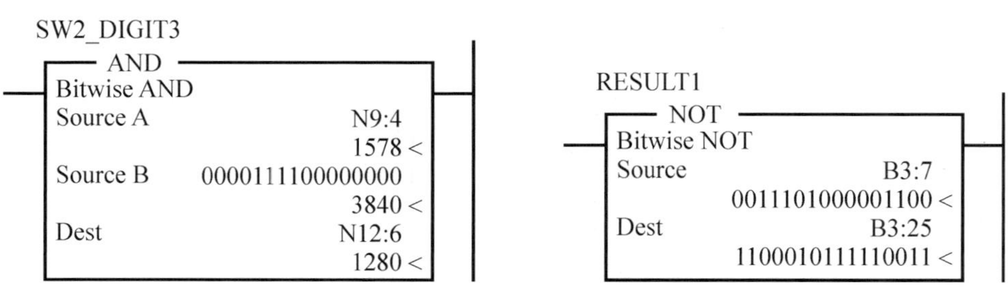

(a) (b)

Figure 7.31. Example MicroLogix/SLC-500 logical function blocks: *(a)* bit-wise AND; *(b)* bit-wise NOT.

7.5.5 Conversion Function Blocks

The names of the conversion function blocks and the conversion that occurs for each one are:

Block Name	Description
DCD	Decode to 1-of-16.
DEG	Convert source in radians to degrees.
ENC	Encode from 1-of-16
GCD	Convert source Gray code value to integer.
RAD	Convert source in degrees to radians.
FRD	Convert source BCD (binary coded decimal) value to integer.
TOD	Convert source integer value to BCD.
SCL	Scale source integer value.
SCP	Scale source integer value, with parameters.

The DEG and RAD function blocks are similar to the one-operand function blocks shown in Figure 7.30. These two function blocks are valid for the SLC-5/03 and later processors. The GCD block is also similar to the one-operand arithmetic blocks, but not available for SLC-500 processors. The SCL function block is valid for all MicroLogix processors and for the SLC-5/02 and higher processors. The SCP function block is valid for all MicroLogix processors and for the SLC-5/03 and higher processors.

An example FRD function block is shown in Figure 7.32. This example assumes that the bit pattern from two thumbwheel switches is transferred to the lower 8 bits of B3:25 with

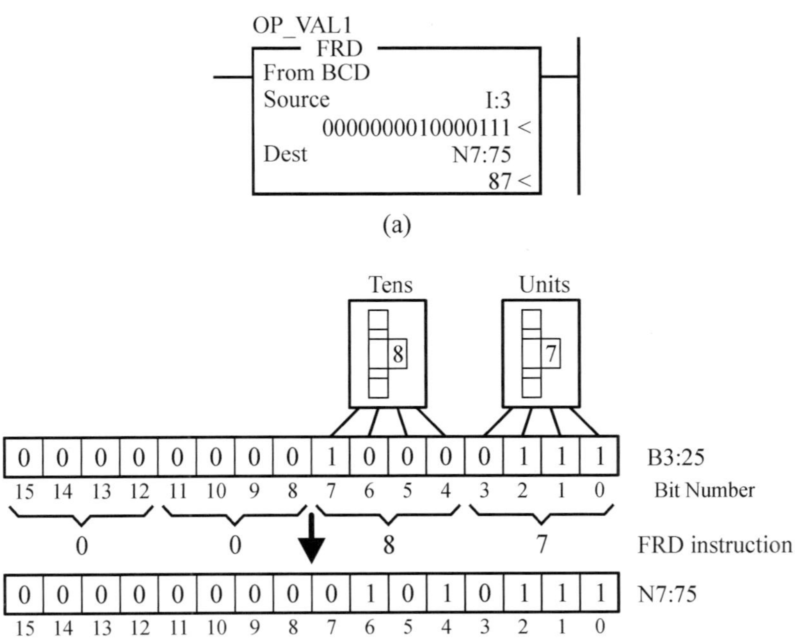

Figure 7.32. Example MicroLogix/SLC-500 BCD-to-integer conversion: *(a)* function block; *(b)* operation.

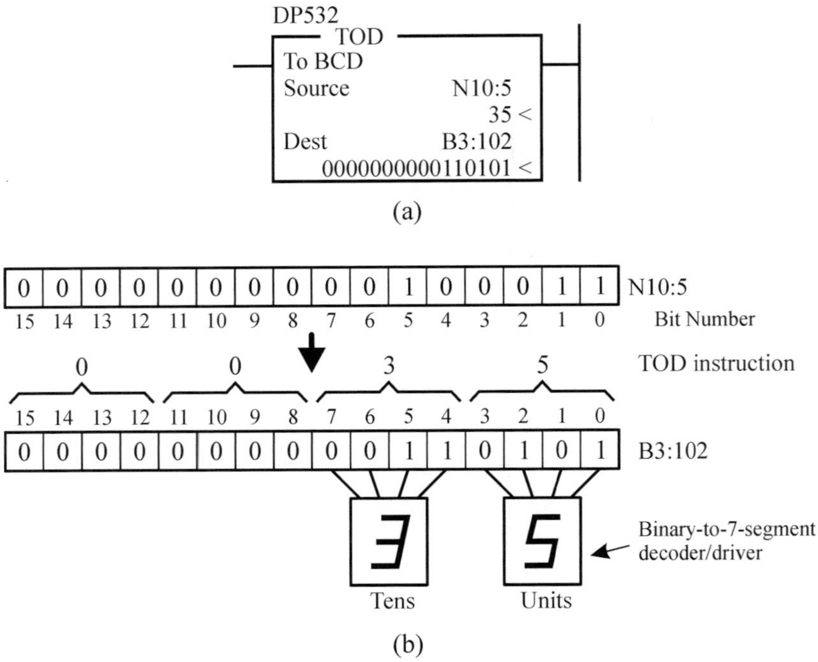

Figure 7.33. Example MicroLogix/SLC-500 integer-to-BCD conversion: *(a)* function block; *(b)* operation.

a move type of function block. The FRD block converts the 4-digit BCD value to an integer and places the result in N7:75. An example TOD function block is shown in Figure 7.33. The integer in N10:5 is converted into the BCD format and placed in B3:102. Additional ladder logic is assumed to transfer the lower 8 bits of the BCD number contained in B3:102 to discrete outputs that are attached to binary-to-7-segment decoder drivers.

An example SCP function block is shown in Figure 7.24*a* and implements the scaling equation (7.1). In terms of the block parameters, the equation is

$$\text{Output} = \frac{\text{Input} - \text{Input Min}}{\text{Input Max - Input Min}} \times (\text{Scaled Max} - \text{Scaled Min}) + \text{Scaled Min}$$

The parameters can be input image words (I), output image words (O), integers (N), floating-point (F), words (B), long words (L), or constants, however floating-point is not supported on the MicroLogix. The example SCP shown in Figure 7.34*a* performs the conversion in Example 7.3 for a MicroLogix analog input module.

An example SCL function block is shown in Figure 7.34*b*. The SCL function block scales an integer from an analog input module to desired limits. The function block multiplies the Source by the specified Rate, rounds the result, adds the Offset value, and places the final result in Dest. The Source, Rate, Offset, and Dest are integers. The equation for the Dest integer is:

$$\text{Dest} = (\text{Source} \times (\text{Rate}/10{,}000)) + \text{Offset}$$

The Rate and Offset can be calculated from the Source and Dest ranges. Let

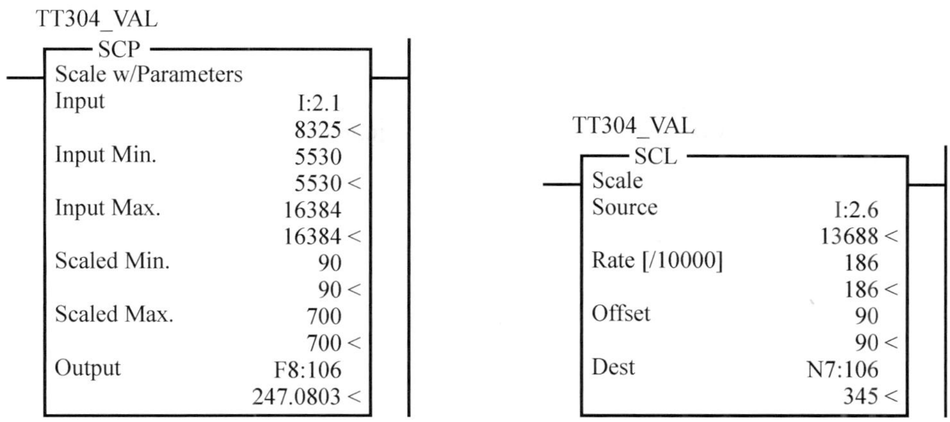

(a) (b)

Figure 7.34. MicroLogix/SLC-500 scale function blocks: *(a)* SCP; *(b)* SCL.

X_1 = minimum value of Source $\quad$ Y_1 = minimum value of Dest
X_2 = maximum value of Source $\quad$ Y_2 = maximum value of Dest

then

$$\text{Rate} = ((Y_2 - Y_1) / (X_2 - X_1)) \times 10{,}000$$
$$\text{Offset} = Y_1 - (X_1 \times \text{Rate}/10{,}000)$$

The example SCL shown in Figure 7.34*b* performs the conversion in Example 7.3 for a MicroLogix analog input module..

> **Design Tip**
>
> If the result of (Source × (Rate/10,000)) is greater than 32,767, the SCL function block overflows, setting the Overflow Trap Bit (S:5/0). The Overflow Trap Bit must be reset before the end of the scan, or the processor **will halt at the end of the program scan**.

7.5.6 Move Function Blocks

The move function blocks transfer data between addresses:

Block Name	Description
CLR	Clear address
MOV	Copy source address to destination address
MVM	Copy source integer to destination integer, allowing portions of the integer to be masked

The MOV and MVM function blocks are shown in Figure 7.35. For the MOV function block, the data types of the source and destination do not need to be identical. Conversion between the integer and floating-point data types are handled automatically. However, when moving a floating-point value to an integer, one should consider overflow. The MVM

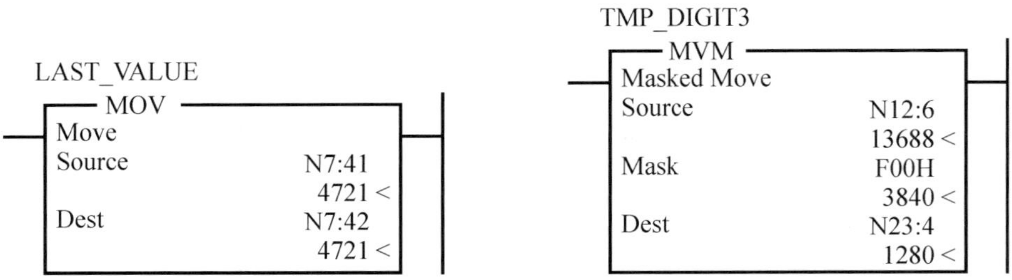

(a) (b)

Figure 7.35. MicroLogix/SLC-500 move function blocks: *(a)* move; *(b)* masked move.

Source, Mask, and Dest operands must be integers. If an ASCII word is used, it is interpreted as an integer. A "1" in the mask means the particular bit is passed from the Source integer to the Dest integer and a "0" in the mask means the particular bit is blocked. A "0" in the mask also means that those bits in the destination integer are not modified. The mask can be entered as a binary, octal, hexadecimal, or decimal constant, but is always displayed as hexadecimal. The symbol of the Compare address is displayed above the function block. The example in Figure 7.35*b* transfers the third digit of a thumbwheel switch. The mask isolates the third digit.

7.6 APPLICATION CAVEATS

Regardless of the particular PLC processor, there are some details one should heed when using comparison and/or computation function blocks:

1. Do not use equality (=) with real or floating-point numbers. The comparison rarely, if ever, is true. A very small change in either number will make the comparison false. The ≥, >, ≤, or < comparisons are much better choices.
2. Do not use equality with any number that represents the value from a sensor. Just as for the previous rule, the comparison rarely, if ever, is true. Some types of sensor readings (for example, pressure and level) are rarely stable and will vary somewhat even when the system is apparently at rest. Even for other sensors, when the system is in a dynamic state a sensor reading may be below the desired threshold during one program scan and above the threshold during the next scan. Thus, the threshold will not be detected if equality is used to test for the threshold.
3. When dividing two integers to yield another integer, remember that any fractional part of the result is lost.
4. In any complex calculation that has a mixture of multiplication and division, do the division operation last, especially when the calculation involves integers. If the integer division is done first, the fractional part is lost, affecting the correctness of the final result.
5. When the result of a computation is an integer, be mindful of overflow. On some MicroLogix/SLC-500 processors, an overflow will cause the processor to halt at the end of a program scan.

7.7 EXAMPLES

The comparison and computation blocks are illustrated by the following examples:

Example 7.4	Simple level control
Example 7.5	Simple operator interface and alarm
Example 7.6	Batch control

The first example uses the computation and comparison function blocks in a simple application. The second example adds alarming features and the conversion of numbers for simple operator display. The third example is a sequential application using comparison and computation as part of the solution.

The examples are worked for the following PLCs:

ControlLogix/CompactLogix
Micro800
MicroLogix/SLC-500

In addition, Example 7.4 is worked for the different ways of doing calculations with the MicroLogix/SLC-500 processors. Also, only a portion of the solution to Example 7.6 is given. The complete solution is in the supplementary online content.

Example 7.4. Simple tank level control. The PLC implements a simple level control for a tank (Figure 7.36). A level sensor (LT428) connected to an analog input channel measures the tank level. The PLC opens an inlet valve (XV427) to allow fluid to enter when the level is **below** a desired minimum and turns off the valve when the level is **above** a desired maximum. There is an internal coil, T428_Cntrl, which enables the level control operation. The desired minimum level, LT428_Min, and the desired maximum level, LT428_Max, are internal variables set by the operator.

The level transducer is calibrated to measure from 1.0 to 15.0 feet and is connected to an analog input module with a channel value (LT428_MEAS) corresponding to the lowest and highest levels as:

Tank Level (ft.)	ControlLogix/CompactLogix	Micro800	MicroLogix
1.0	0.0	13107	6240
15.0	100.0	65545	31200

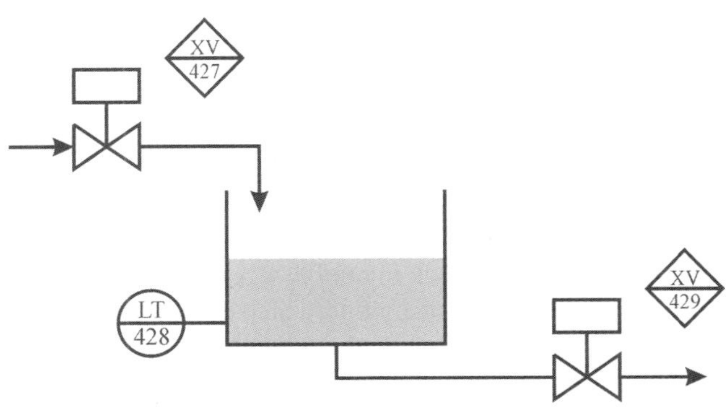

Figure 7.36. Tank level control.

The LT428_MEAS value should be converted into a tank level, in feet (a real number) and placed in the LT428_Val tag/variable/symbol. For the ControlLogix/CompactLogix, 1756- or 5069-series modules are used and the analog input channel is configured for a range of 0.0 to 100.0, a real number.

Assume the following physical input, physical output, and internal memory assignments:

Tag/Var./Symbol	Description
LT428_MEAS	Level measurement channel value, see range above
XV427_OPEN	Inlet valve control; **on** to open valve, **off** to close valve
LT428_Val	Level of tank, in feet
T428_Min	Minimum tank level for control, in feet
T428_Max	Maximum tank level for control, in feet
T428_Cntrl	Enables level control operation; **on** to run control, **off** keeps valve closed.

The addresses associated with the physical inputs and outputs are:

Tag/Var./Symbol	ControlLogix	CompactLogix	Micro800
LT428_MEAS	Local:3:I.Ch0Data	Local:3:I.Ch00.Data	_IO_P2_AI_00
XV427_OPEN	Local:2:O.Data.0	Local:2:O.Pt00.Data	_IO_EM_DO_00

Tag/Var./Symbol	MicroLogix	SLC-500
LT428_MEAS	I:1.0	I:3.0
XV427_OPEN	O:0/0	O:2/0

The addresses or data types associated with the internal tags/variables/symbols:

Tag/Var./Symbol	CLogix	Micro800	MLogix/SLC
LT428_Val	REAL	REAL	F8:20
T428_Min	REAL	REAL	F8:21
T428_Max	REAL	REAL	F8:22
T428_Cntrl	BOOL	BOOL	B3/121

Solution. There are two parts to this problem: (1) conversion of analog input module reading into the level in feet; (2) control of XV428 valve. Because of the different ranges of the ADC output, the conversion equation for the first part will be different for each PLC. The second part will be very similar for all of the PLCs.

The ladder logic for the ControlLogix/CompactLogix solution for this example is shown in Figure 7.37. The formula for converting the ADC output (LT428_MEAS) into the level in feet (LT428_Val) is

$$\text{LT428\_Val} = \frac{\text{LT428\_MEAS} - 0}{100 - 0} \times (15 - 1) + 1$$

which is implemented with the CPT block in rung 1. The formula is not simplified so that if the transducer is recalibrated, one can easily find which constants need to be changed. The conversion is done with the CPT block, but could also be done with individual arithmetic function blocks. Note that the constants do not have a decimal point, since LT428_MEAS and the destination are real tags. The second rung is the control for the valve and is implemented as a start/stop rung.

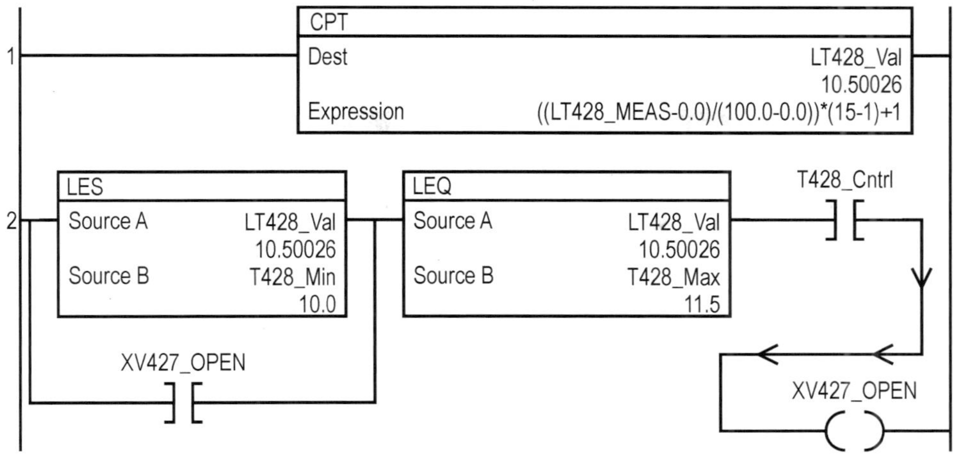

Figure 7.37. ControlLogix solution to Example 7.4, tank level control.

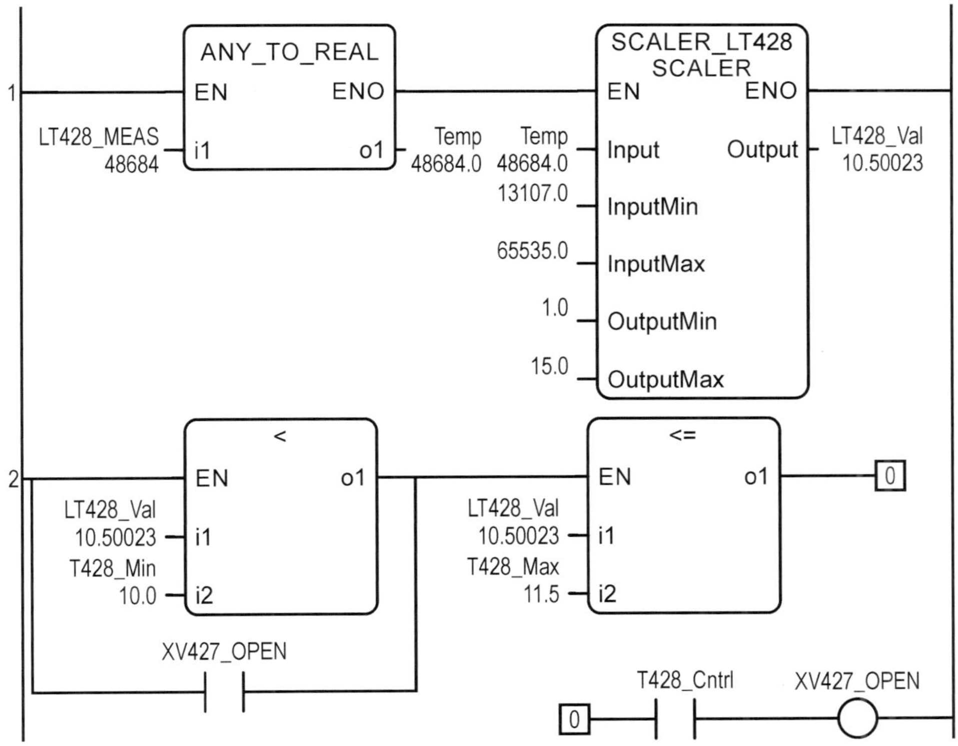

Figure 7.38. Micro800 solution to Example 7.4, tank level control.

Figure 7.38 contains the ladder logic for the conversion and valve control for a Micro800 processor. The SCALER block is used to convert the analog input integer (LT428_MEAS) into the level in feet (LT428_Val).

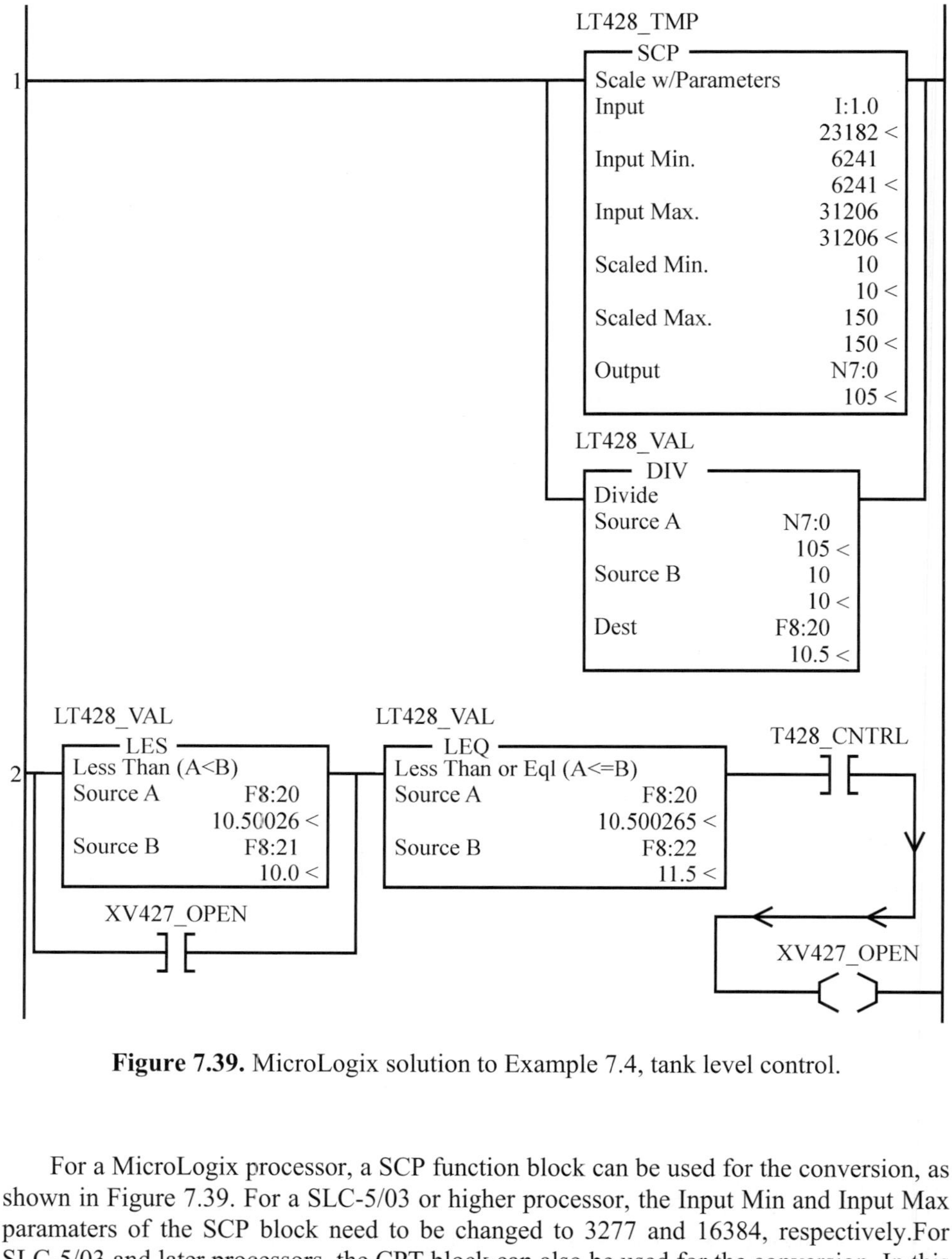

Figure 7.39. MicroLogix solution to Example 7.4, tank level control.

For a MicroLogix processor, a SCP function block can be used for the conversion, as shown in Figure 7.39. For a SLC-5/03 or higher processor, the Input Min and Input Max paramaters of the SCP block need to be changed to 3277 and 16384, respectively.For SLC-5/03 and later processors, the CPT block can also be used for the conversion. In this case, the CPT block would look similar to that of rung 1 of Figure 7.37, with the expression replaced by

((LT428_MEAS-3277.0)*(15.0-1.0))|(13107.0)+1.0

Note the "|" division operator and that all constants have decimal points, forcing the multiply and divide operations to be done as floating point operations.

Example 7.5. Simple operator interface/alarm system for tank. A small operator interface with the following features is added to the tank level control of Example 7.4:

1. Display of current level
2. Entry of desired minimum level. The maximum level is 1.5 feet higher than the minimum level.
3. Low level alarm indicator and horn

The current level should be shown on a small operator panel as a decimal number with one fractional digit. This feature requires no additions to the logic, as T428_Val is already a real value. The desired minimum level, LT428_Min is entered on the operator display as a real number. Logic in the processor must ensure this value is not so low that the low level indicator is triggered in normal operation and not so high that the maximum level is larger than the maximum tank level.

The low level indicator, T428_LOLA is on when the level is below 4.0 feet. The T428_HORN is handled in the following manner. T428_HORN is turned **on** for any of the following conditions:

1. The level becomes lower than 2.0 feet
2. The alarm has been acknowledged by the operator, but the level has remained below 2.0 feet for five minutes.

The T428_HORN output is turned **off** when the operator acknowledges the alarm by pushing the ALM_ACK push button. Even if the level goes above 2.0 feet while T428_HORN is **on**, the horn should remain **on** until acknowledged by the operator. If the level rises to 2.0 feet or above during the five minutes after alarm acknowledgment, the timing interval is completely canceled and the next time the level goes below 2.0 is treated as if it is the first time the level has gone below 2.0.

The ALM_ACK input is a pushbutton icon on the operator panel, **on** when pushed, **off** when released. Do not assume it is released shortly after it is pushed to acknowledge the alarm. The ALM_ACK switch must be released before it can be pushed to acknowledge the alarm again.

Assume the following additional input, output, and internal tags/variables/symbols:

Tag/Var./Symbol	Description
T428_LOLA	Low level indication; **on** when level < 4.0 feet
T428_HORN	Low level alarm horn, operation described above
ALM_ACK	Alarm acknowledge (panel pushbutton, **on** when pushed, **off** when released)

The addresses and data types associated with the additional physical output is:

The addresses associated with the additional physical output is:

Tag/Var./Symbol	ControlLogix	CompactLogix	Micro800
T428_HORN	Local:2:O.Data.2	Local:2:O.Pt02.Data	_IO_EM_DO_02

The addresses or data types associated with the internal tags/variables/symbols:

Tag/Var./Symbol	CLogix	Micro800	MLogix/SLC
T428_LOLA	BOOL	BOOL	B3/10
ALM_ACK	BOOL	BOOL	B3/11

Solution. The ladder logic is added to the ladder logic of Example 7.4. The ControlLogix ladder logic is shown in Figure 7.40. Rungs 3 and 4 implement the check of the LT428_Min value. If below 2.1, the value 2.1 is moved into the LT428_Min tag. If above 13.4, the value 13.4 is moved into LT428_Min. In Rung 5, the maximum level is calculated by adding 1.5 to the minimum level. The logic in rungs 6 – 10 handles the indicator and alarm. The comparison block in rung 6 drives the low level indicator lamp and the particular level that triggers the horn is determined by the CMP block in rung 7. Rung 8 detects the acknowledge button push transition. The horn is activated by the logic in rung 9. This logic is basically the start/stop rung of chapter 2, but with two starting conditions: (1) level falls below 2.0 feet, and (2) level has remained below 2.0 feet for 5 minutes after the acknowledge button has been pressed. The logic and timer to handle the second condition to turn on the horn are in rung 10. The ALM_ACK transition (Ack_Trans) starts the timer, which continues to run as long as the level stays below 2.0 feet. As soon as the timer finishes, it resets itself and generates a pulse (Ack_Tmr.DN) that turns **on** the horn.

The Micro800 ladder logic is shown in Figure 7.41 and closely follows the ControlLogix solution. Note the individual comparison blocks in rungs 3, 4, and 7. In rung 9, the positive transition contact senses the transition when the level falls below 2.0. In rung 10, the Ack_Tmr.REdge variable is used to seal the timer start with Ack_Trans.

The MicroLogix/SLC-500 ladder logic is shown in Figure 7.42 and closely follows the ControlLogix solution. Note that rungs 3, 5, 6 and 7 contain individual comparison blocks.

The addresses or data types associated with the additional internal tags/variables/symbols:

Tag/Var./Symbol	CLogix	Micro800	MLogix/SLC
T428_Hrn_Act	BOOL	BOOL	B3/122
Hrn_Ons	BOOL	n/a	B3/123
Ack_Trans	BOOL	BOOL	B3/124
Ack_Ons	BOOL	n/a	B3/125
Ack_Tmr	TIMER	TON	T4:0

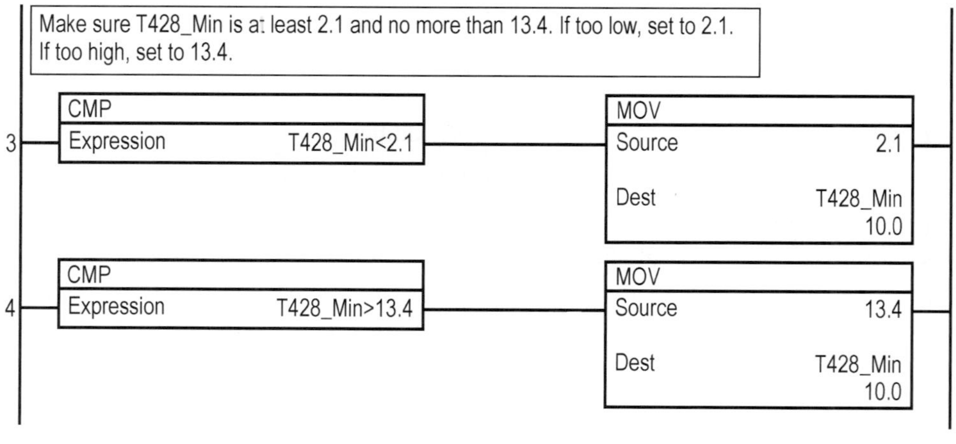

Figure 7.40. ControlLogix solution to Example 7.5. *(continued)*

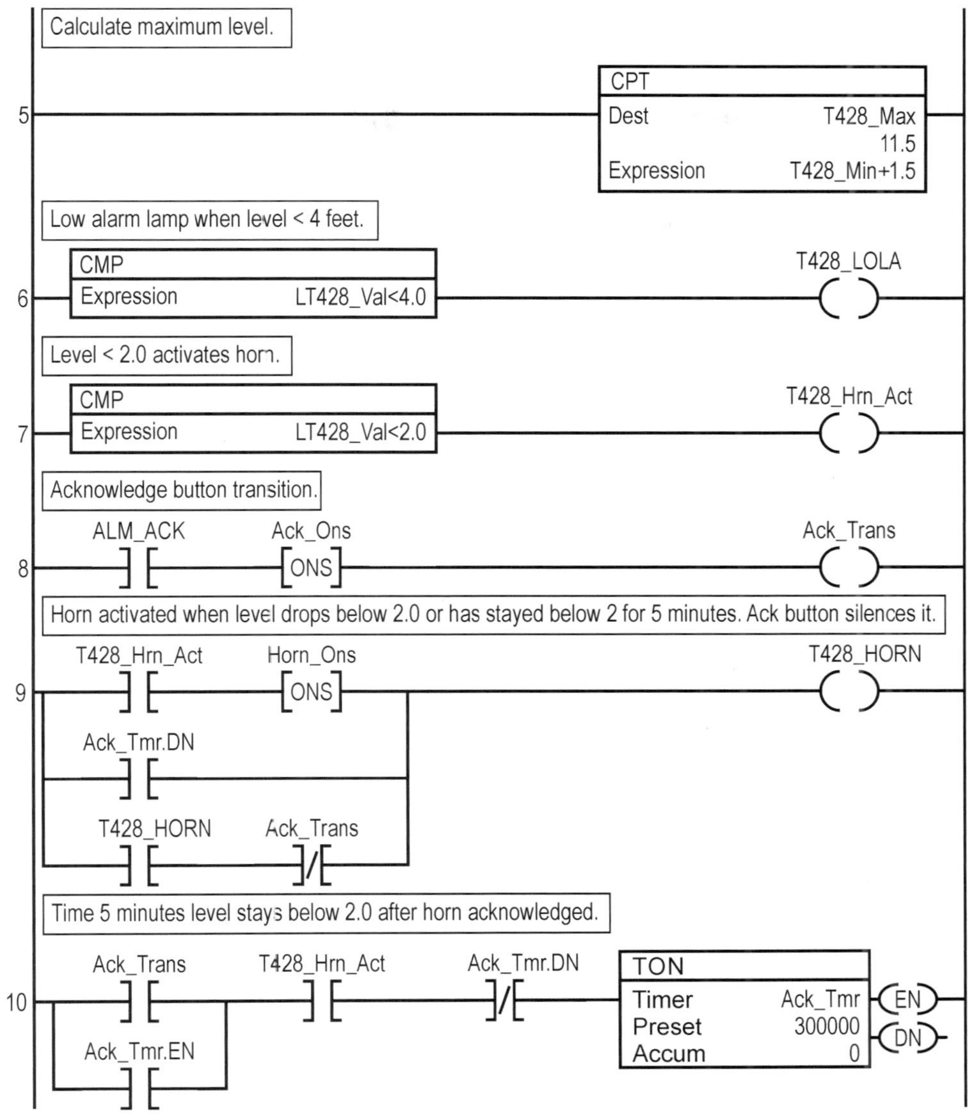

Figure 7.40. *(continued)*

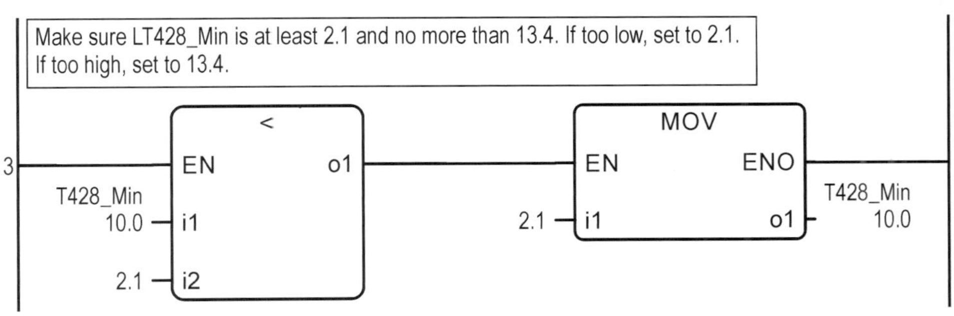

Figure 7.41. Micro800 solution to Example 7.5. *(continued)*

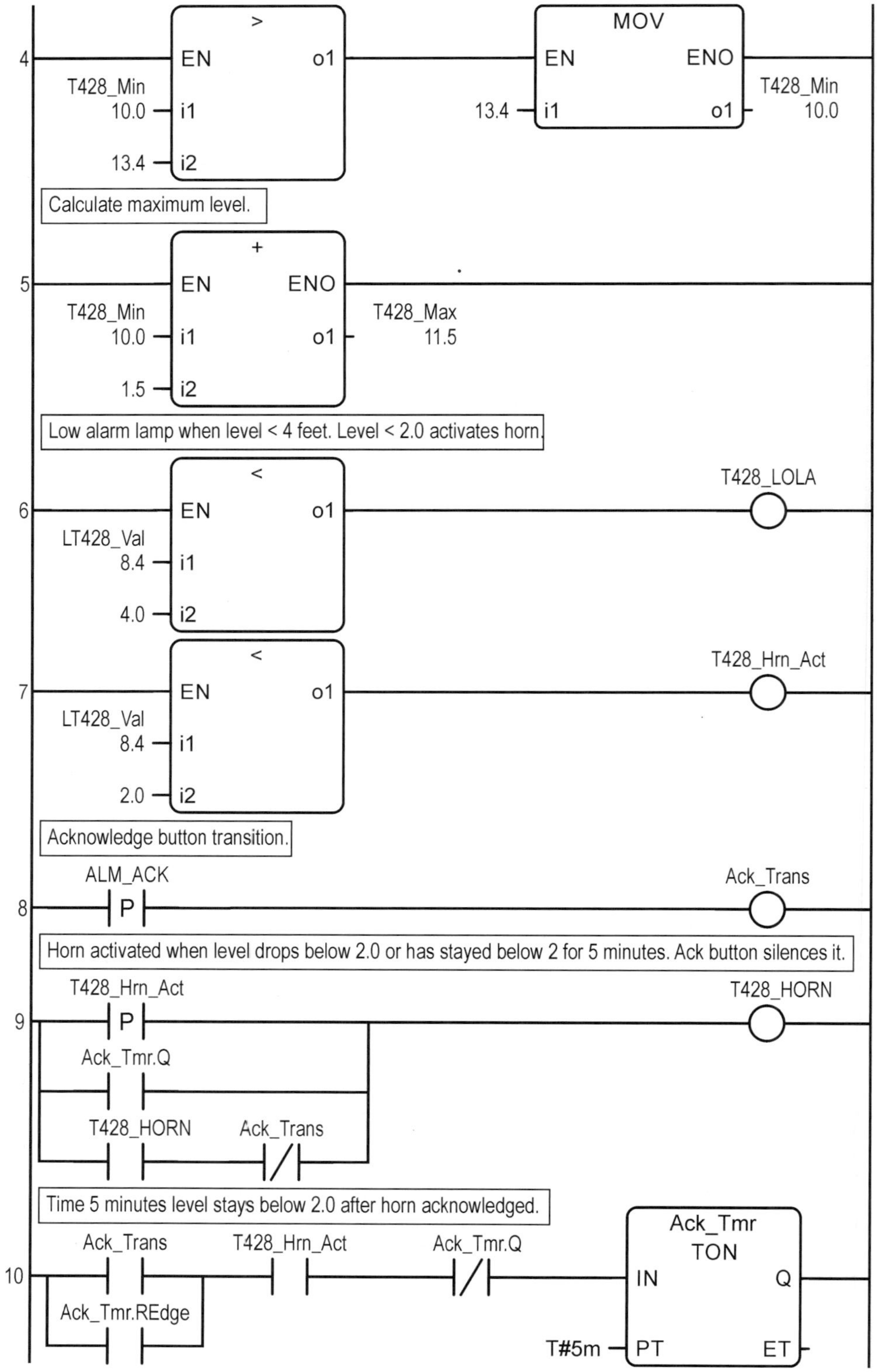

Figure 7.41. *(continued)*

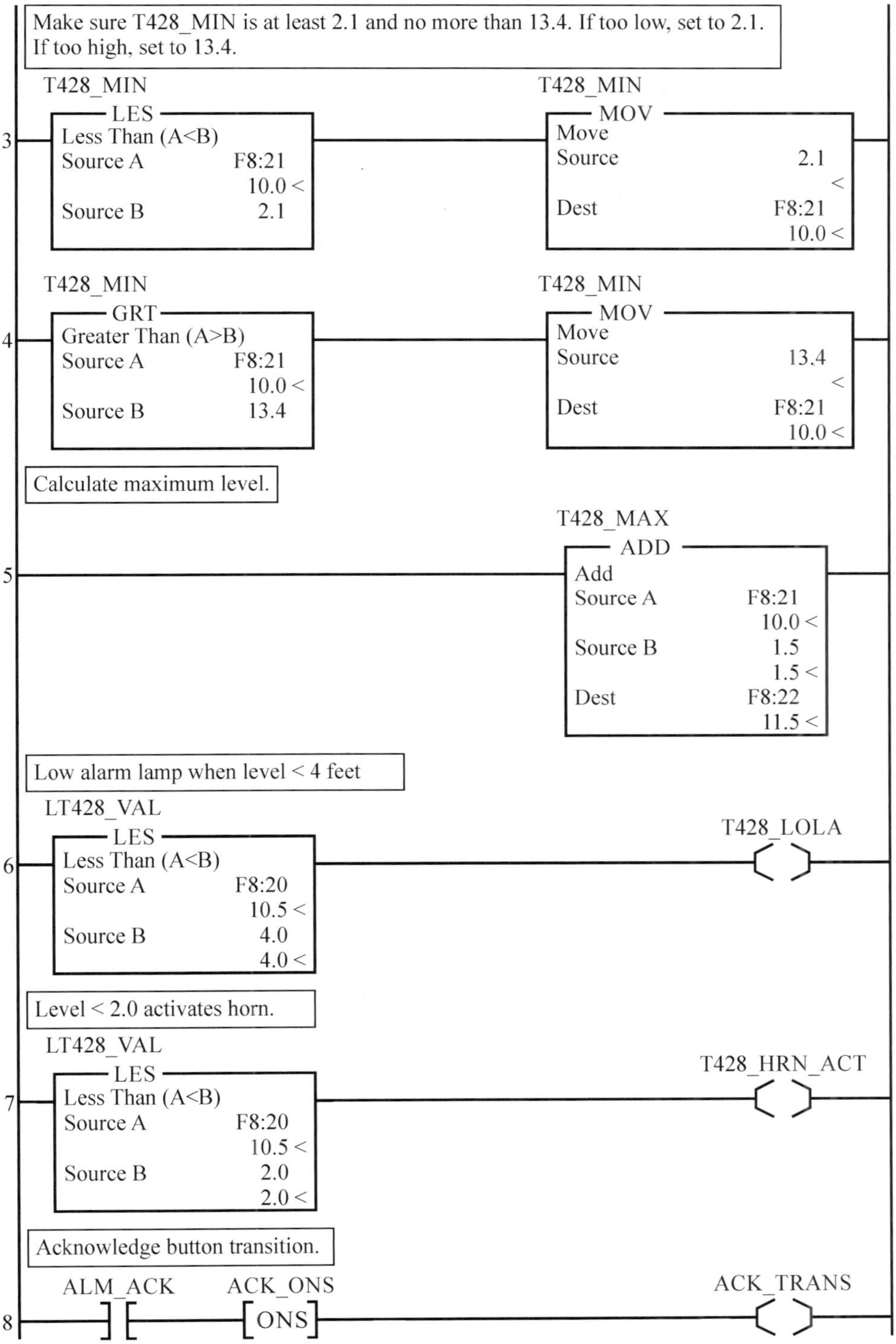

Figure 7.42. MicroLogix solution to Example 7.5. *(continued)*

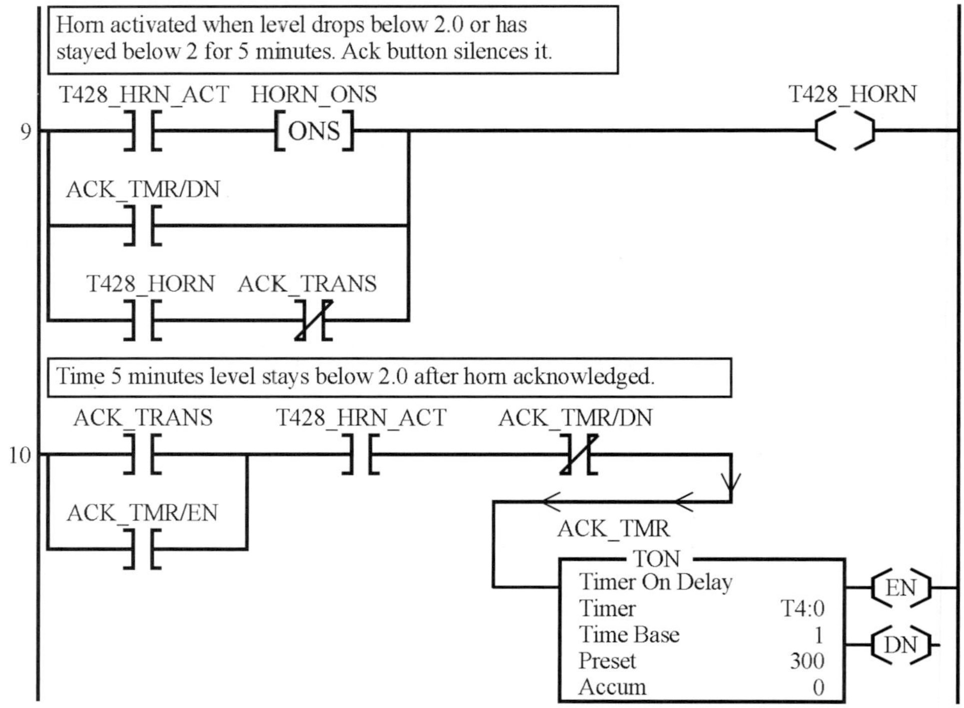

Figure 7.42. *(continued)*

Example 7.6. Batch process operation. Design the function chart for the program to control the batch process described below. Implement the control with ladder logic.

Figure 7.43 shows the batch process equipment. There are two ingredients, A and B that need to be mixed. PID controllers control the flow of the ingredients. You will not program the PID controller function blocks, only set the setpoint, and read the current flow and accumulated flow (amount of ingredient).

The overall operation of the process is first described, followed by the details. When the operator presses the start push button, the sequence of operation for the station follows these steps:

Prestart checks (empty tank, no flow of either ingredient, outlet closed, accumulations zero).
Add desired amount of ingredient A.
Add desired amount of ingredient B (agitate slowly).
Agitate mixture fast for 20 minutes.
Dump tank.

The operation **does not** repeat. The operator must press the start pushbutton to start the batch. At the end of a batch, **no** steps must be active and any internal run variable must be **off**.

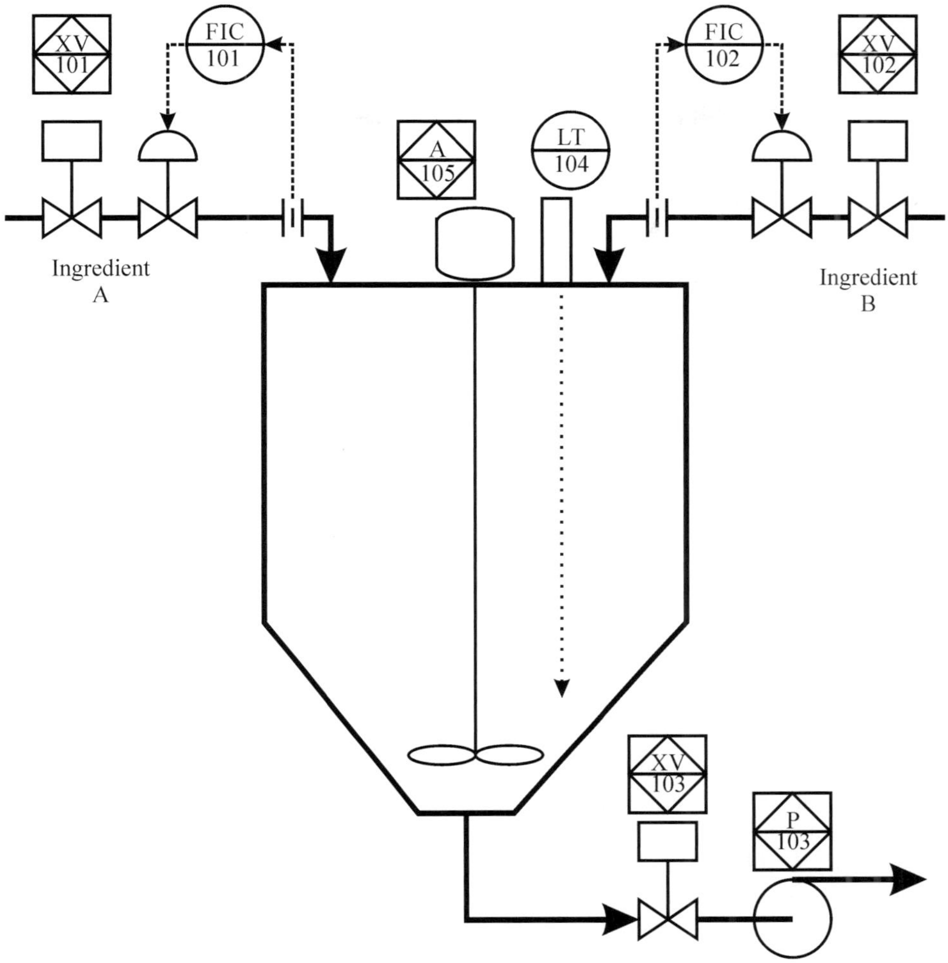

Figure 7.43. Batch process.

In order to fill the tank with an ingredient, a PID flow controller is used. Each controller has 3 variables that are used to control/monitor the flow:

FICxxx_SP	Controller setpoint, the desired flow of the ingredient, in gpm.
FICxxx_PV	Process variable, the flow in gpm.
FICxxx_ACC	Accumulation (integration over time) of the flow, which is the amount of the ingredient that has passed through the flow sensor, in gallons.

In order to fill the tank, the shutoff valve (XV102 or XV102) must be opened and the flow setpoint must be set to the desired value (INGA_Des_Flw or INGB_Des_Flw). Then, the accumulation (FICxxx_ACC) must be monitored to determine when the desired amount of the ingredient (INGA_Des_Amt or INGB_Des_Amt) has been put in the tank. In order to turn off the flow of an ingredient, the shutoff valve must be closed, the flow setpoint must set to -2.0 gpm (ensures flow valve positively shut off) and then the flow must be monitored until it is < 0.1 gpm before going to the next step in the process.

Both accumulations (FIC101_ACC and FIC102_ACC) must be zeroed (write a zero into location) as part of the prestart. The batch operation must not proceed until both of the accumulations are zero.

The agitator is driven by a two-speed motor, whose controls are A105_SLOW and A105_FAST. The agitator must be run slowly when adding ingredient B or dumping the tank. The agitator is run fast for the specified time to mix after both ingredients have been added.

In order to dump the tank, the outlet valve, XV103, must be opened (with XV103_OPEN) and the outlet pump, controlled by P103_RUN, must be run. The tank is empty when the tank level is less than 0.1 feet. The level transducer is calibrated to measure from 0.0 to 20.0 feet and is connected to an analog input module with a channle value (LT104_MEAS) corresponding to the lowest and highest levels as:

Tank Level (ft.)	ControlLogix/CompactLogix	Micro800	MLogix
0.0	0.0	13107	6240
20.0	100.0	65545	31200

The LT104_MEAS value should be converted into a tank level, in feet (a real number) and placed in the LT104_Val tag/variable'symbol. For the ControlLogix and CompactLogix, 1756- or 5069-series modules are used and the analog input channel is configured for a range of 0.0 to 100.0, a real data type.

When the start switch is pressed (turned **on**) for the first time only, the equipment must be checked for proper starting conditions. When the stop switch is pressed (turned **off**) the operation should pause and all outputs (**except** the agitator) must be turned **off**. An operation pause does not pause any timed steps. Pressing the start switch while the operation of the station is paused causes the station to resume its suspended operation.

There is a RESET_PB switch that when **on**, restarts the operation. When RESET_PB is **on**, all internal steps are turned **off** and all flow setpoints must be set to -2.0 gpm. START_PB must be pressed to actually restart the station. The RESET_PB switch must have no effect unless the operation is already paused. In other words, after RESET_PB is pressed and then released, the next press of the start switch is treated as the first time the start switch is pressed.

Assume the following physical input and physical output assignments:

Tag/Var./Symbol	Description
START_PB	Start push button, N. O., **on** (closed) to start
STOP_PB	Stop push button, N. C., **off** (open) to stop
RESET_PB	Reset push button, N. O., **on** (closed) when resetting
A105_SLOW	Agitator motor control, **on** to run agitator at slow speed.
A105_FAST	Agitator motor control, **on** to run agitator at fast speed.
XV101_OPEN	**On** to open XV101 valve, **off** closes valve.
XV102_OPEN	**On** to open XV102 valve, **off** closes valve.
XV103_OPEN	**On** to open XV103 valve, **off** closes valve.
P103_RUN	Outlet pump motor control, **on** to run pump.

The internal tags/variables/symbols are:

LT104_MEAS	Raw tank level measurement, channel value, represents 0-15 ft.; ranges listed above
INGA_Des_Flw	Desired flow rate setpoint (in gpm) for ingredient A flow (FIC101)
INGB_Des_Flw	Desired flow rate setpoint (in gpm) for ingredient B flow (FIC102)
INGA_Des_Amt	Desired amount of ingredient A (gallons) to put in tank.
INGB_Des_Amt	Desired amount of ingredient B (gallons) to put in tank.
FIC101_PV	Current value of ingredient A flow (gpm). Reading from FT101 flow transducer, converted to proper units.
FIC101_ACC	Accumulation for FIC101 PID loop. The amount of ingredient A that has passed through flow transducer FT101. Write a zero to this variable to reset it.
FIC102_PV	Current value of ingredient B flow (gpm). Reading from FT102 flow transducer, converted to proper units.
FIC102_ACC	Accumulation for FIC102 PID loop. The amount of ingredient A that has passed through flow transducer FT102. Write a zero to this variable to reset it.
FIC101_SP	Setpoint of FIC101, desired ingredient A flow (gpm).
FIC102_SP	Setpoint of FIC102, desired ingredient B flow (gpm).
LT104_VAL	Actual tank level, in feet
AGIT_TIME	Agitation time of step 6, in minutes.

The addresses associated with the physical inputs and outputs are:

Tag/Var./Symbol	ControlLogix	CompactLogix	Micro800
START_PB	Local:1:I.Data.0	Local:1:I.Pt00.Data	_IO_EM_DI_00
STOP_PB	Local:1:I.Data.1	Local:1:I.Pt01.Data	_IO_EM_DI_01
RESET_PB	Local:1:I.Data.2	Local:1:I.Pt02.Data	_IO_EM_DI_02
A105_SLOW	Local:2:O.Data.0	Local:2:O.Pt00.Data	_IO_EM_DO_00
A105_FAST	Local:2:O.Data.1	Local:2:O.Pt01.Data	_IO_EM_DO_01
XV101_OPEN	Local:2:O.Data.2	Local:2:O.Pt02.Data	_IO_EM_DO_02
XV102_OPEN	Local:2:O.Data.3	Local:2:O.Pt03.Data	_IO_EM_DO_03
XV103_OPEN	Local:2:O.Data.4	Local:2:O.Pt04.Data	_IO_EM_DO_04
P103_RUN	Local:2:O.Data.5	Local:2:O.Pt05.Data	_IO_EM_DO_05
LT104_MEAS	Local:3:I.Ch0Data	Local:3:I.Ch00.Data	_IO_P2_AI_00

Tag/Var./Symbol	MLogix	SLC-500
START_PB	I:0/0	I:1/0
STOP_PB	I:0/1	I:1/1
RESET_PB	I:0/2	I:1/2
A105_SLOW	O:0/0	O:2/0
A105_FAST	O:0/1	O:2/1
XV101_OPEN	O:0/2	O:2/2
XV102_OPEN	O:0/3	O:2/3
XV103_OPEN	O:0/4	O:2/4

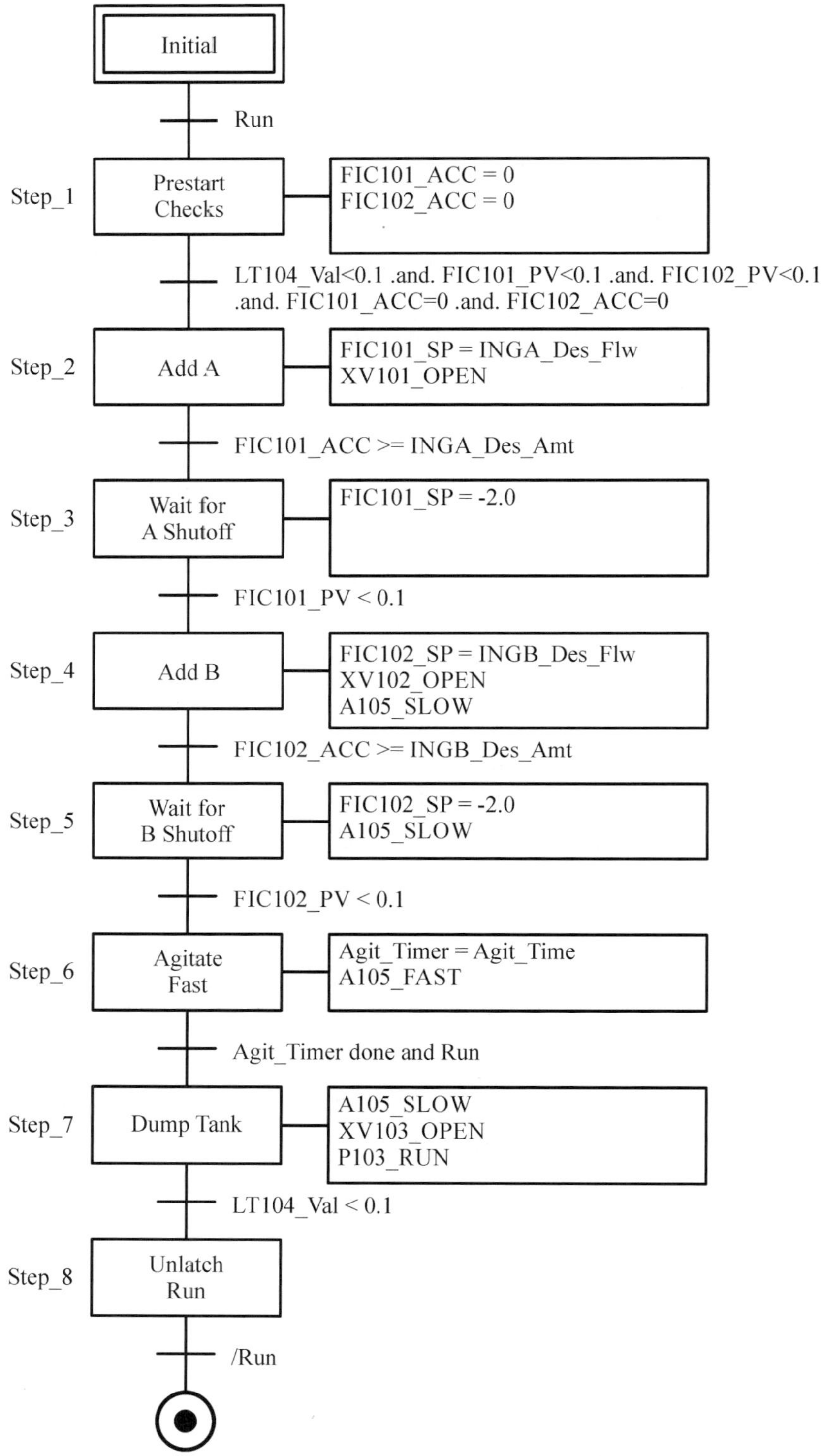

Figure 7.44. Function chart for batch process.

P103_RUN	O:0/5	O:2/5
LT428_MEAS	I:1.0	I:3.0

The addresses or data types associated with the internal tags/symbols:

Tag/Var./Symbol	CLogix	Micro800	MLogix/SLC
INGA_Des_Flw	REAL	REAL	F12:0
INGA_Des_Amt	REAL	REAL	F12:1
INGB_Des_Flw	REAL	REAL	F12:2
INGB_Des_Amt	REAL	REAL	F12:3
AGIT_TIME	REAL	REAL	F12:4
FIC101_PV	REAL	REAL	F12:5
FIC101_ACC	REAL	REAL	F12:6
FIC101_SP	REAL	REAL	F12:7
FIC102_PV	REAL	REAL	F12:8
FIC102_ACC	REAL	REAL	F12:9
FIC102_SP	REAL	REAL	F12:10
LT104_VAL	REAL	REAL	F13:5

Solution. The function chart for this batch process is shown in Figure 7.44. Rather than show a complete solution, only portions of the solution are shown. The complete solution is in the supplementary online content.

The ControlLogix ladder logic for the start and the transitions into steps 2 and 3 are shown in Figure 7.45. The handling of the FIC101 setpoint is shown in Figure 7.46. A CLR block could be used in place of the MOV function blocks in rung 12. The default value for the FIC101 setpoint is –2.0 and the setpoint is only set to a flow above zero when the ingredient is being added in step 2.

Figure 7.47 shows the Micro800 ladder logic for the start and the transitions into steps 2 and 3. The conversion of the level transducer reading, the initialization of the accumulators in step 1, and the handling of the FIC101 setpoint are shown in Figure 7.48.

Figure 7.49 shows the MicroLogix ladder logic for the start and the transitions into steps 2 and 3. The conversion of the level transducer reading, the initialization of the accumulators in step 1, and the handling of the FIC101 setpoint are shown in Figure 7.50.

Additional internal tag/variable/symbol addresses or data types in the solutions:

Tag/Var./Symbol	CLogix Data Type	Micro800 Data Type	MLogix/SLC Address
Step_1 to Step_8	BOOL	BOOL	B20/1 to B20/8
Run	BOOL	BOOL	B20/0
Int_Reset	BOOL	BOOL	B20/40
Agit_Timer	TIMER	TON	T4:1
Temp	n/a	REAL	n/a

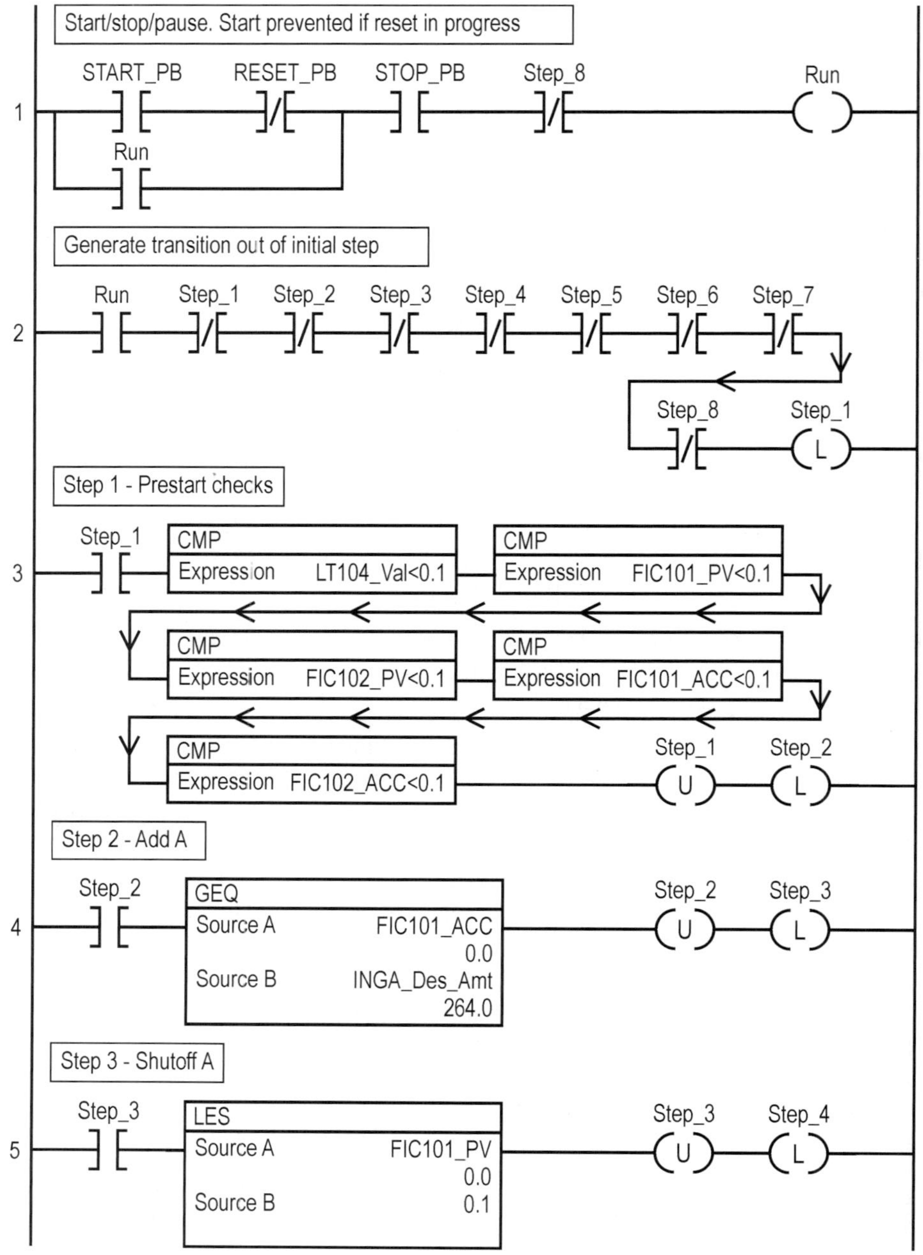

Figure 7.45. ControlLogix ladder logic for transitions of batch process (partial).

7.8 CHAPTER SUMMARY

For the PLCs covered by this text, this chapter described comparison and computation function blocks. Conversion, alarming, and batch control applications illustrated the use of these function blocks.

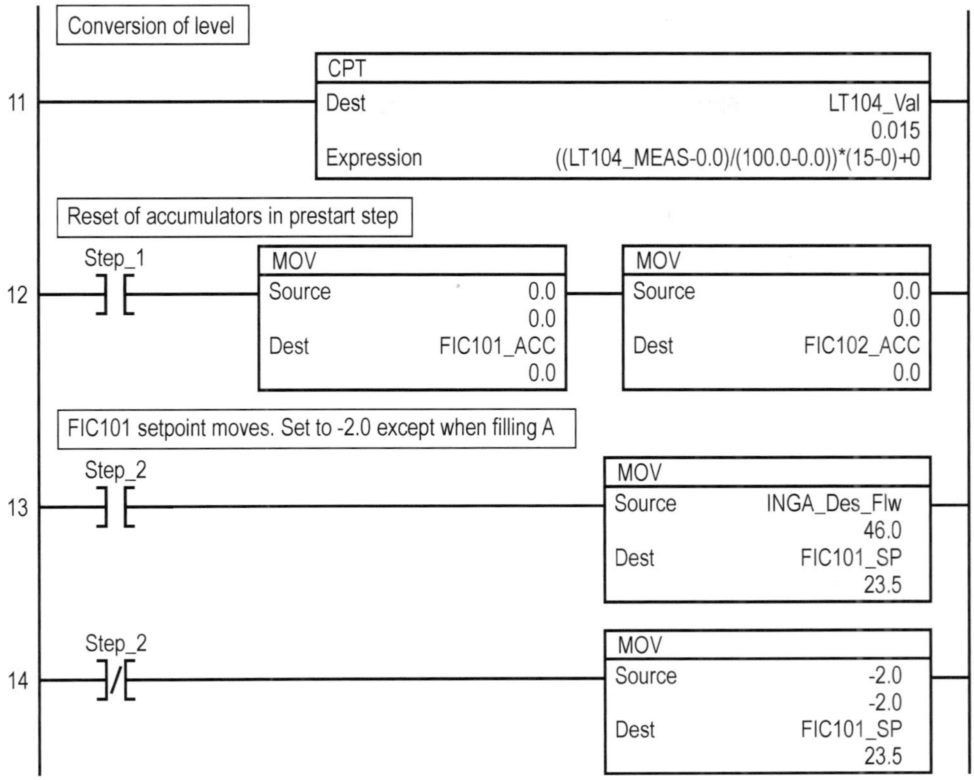

Figure 7.46. ControlLogix ladder logic for calculations of batch process (partial).

REFERENCES

Rockwell Automation, 2008. *SLC Instruction Set Reference Manual*, pub. 1747-RM001G-EN-P, Rockwell Automation Inc., Milwaukee, WI, Nov.

Rockwell Automation, 2011. *MicroLogix 1100 Programmable Controllers Instruction Set Reference Manual*, pub. 1763-RM001D-EN-P, Rockwell Automation Inc., Milwaukee, WI, Sep.

Rockwell Automation, 2014. *MicroLogix™ 1200 and MicroLogix 1500 Programmable Controllers Instruction Set Reference Manual*, pub. 1762-RM001H-EN-P, Rockwell Automation Inc., Milwaukee, WI, Jul.

Rockwell Automation, 2019. *MicroLogix 1400 Programmable Controllers Instruction Set Reference Manual*, pub. 1766-RM001I-EN-P, Rockwell Automation, Inc., Milwaukee, WI, June.

Rockwell Automation, 2022a. *Logix5000 Controllers General Instructions Reference Manual*, pub. 1756-RM003W-EN-P, Rockwell Automation Inc., Milwaukee, WI, Mar.

Rockwell Automation, 2022b. *Micro800 Programmable Controllers General Instructions Reference Manual*, pub. 2080-RM001K-EN-P, Rockwell Automation Inc., Milwaukee, WI, Mar.

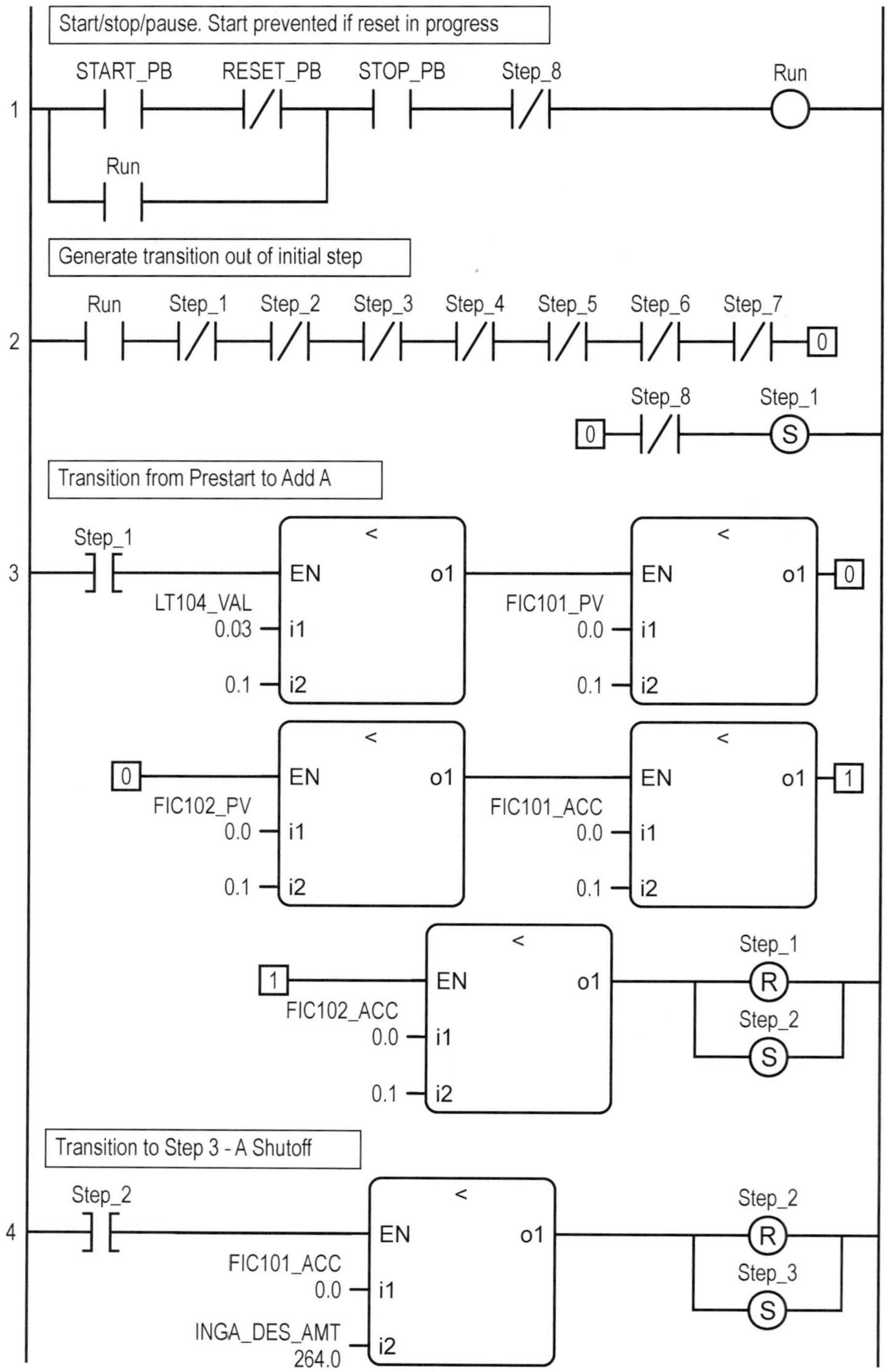

Figure 7.47. Micro800 ladder logic for transitions of batch process (partial).

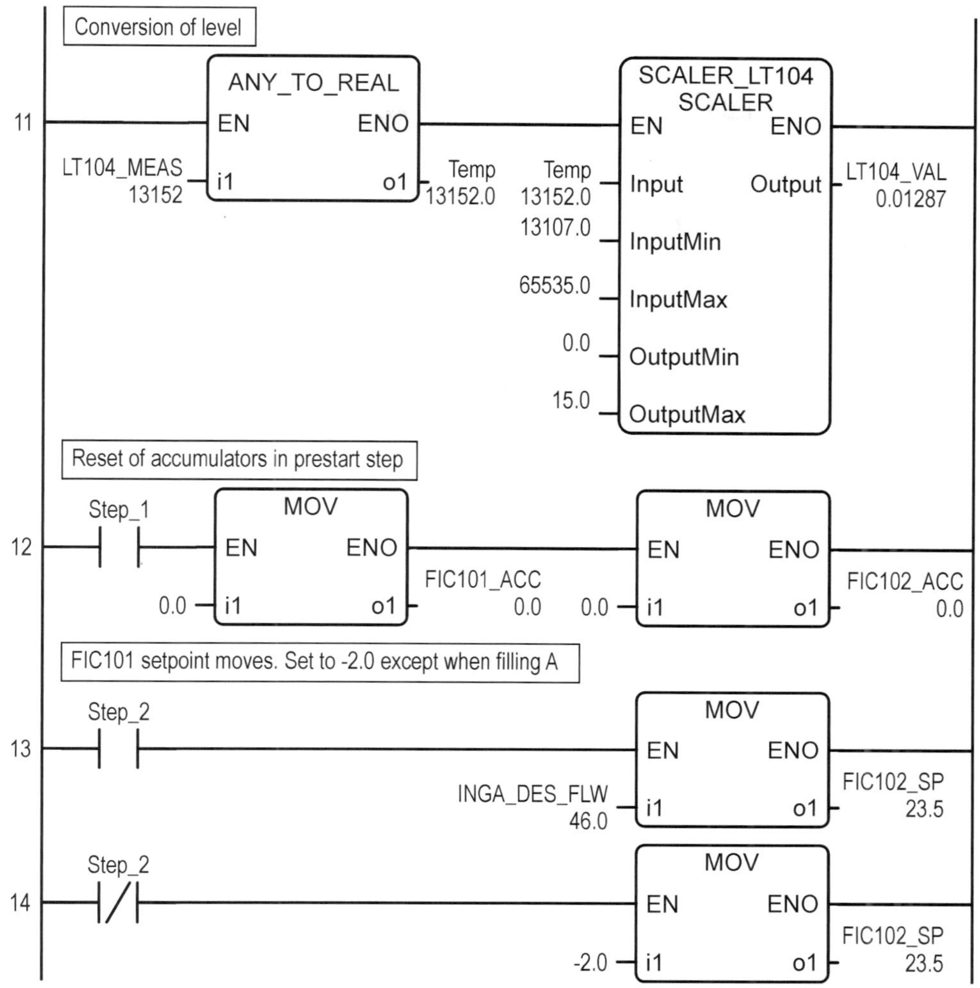

Figure 7.48. Micro800 ladder logic for calculations of batch process (partial).

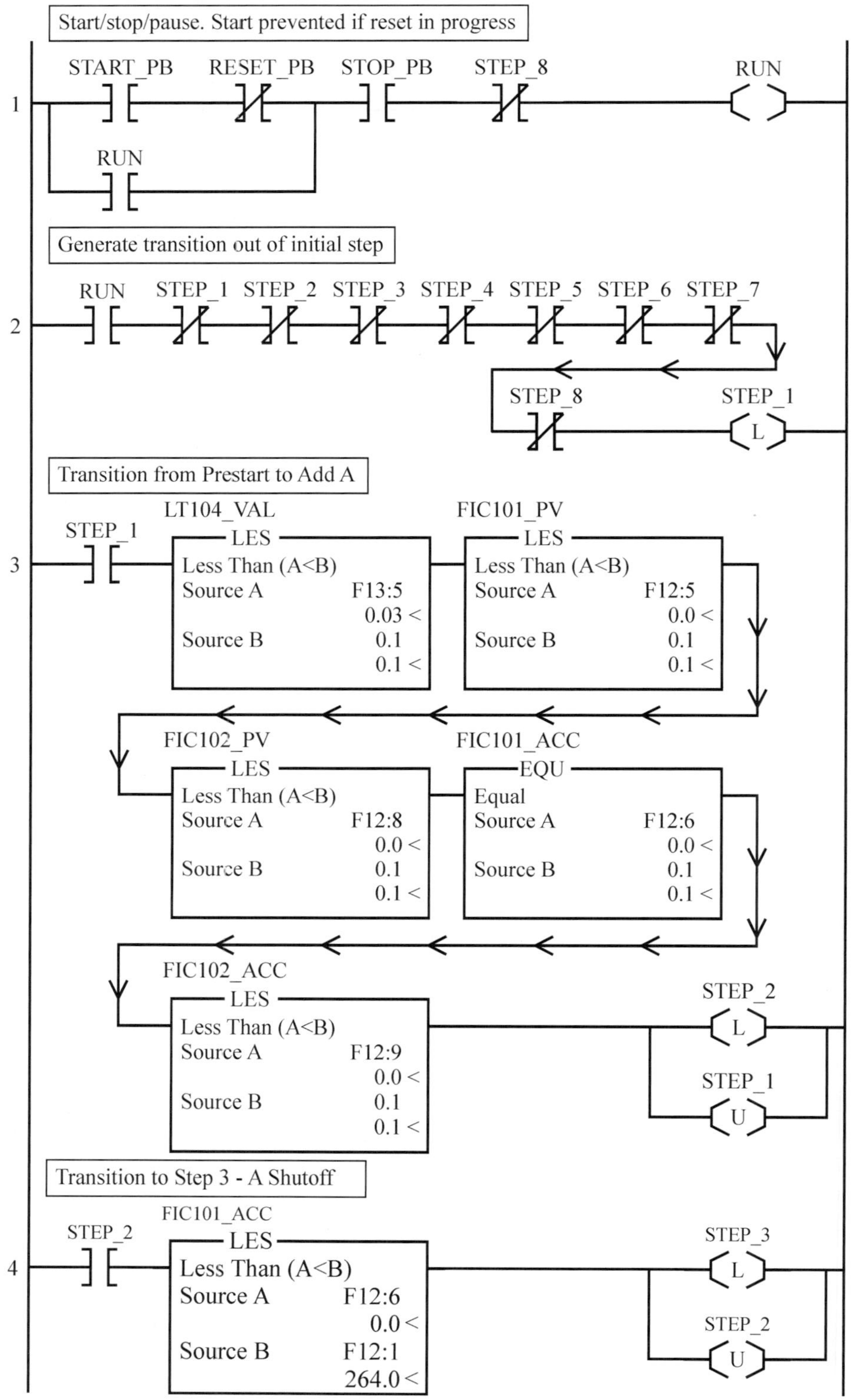

Figure 7.49. MicroLogix ladder logic for transitions of batch process (partial).

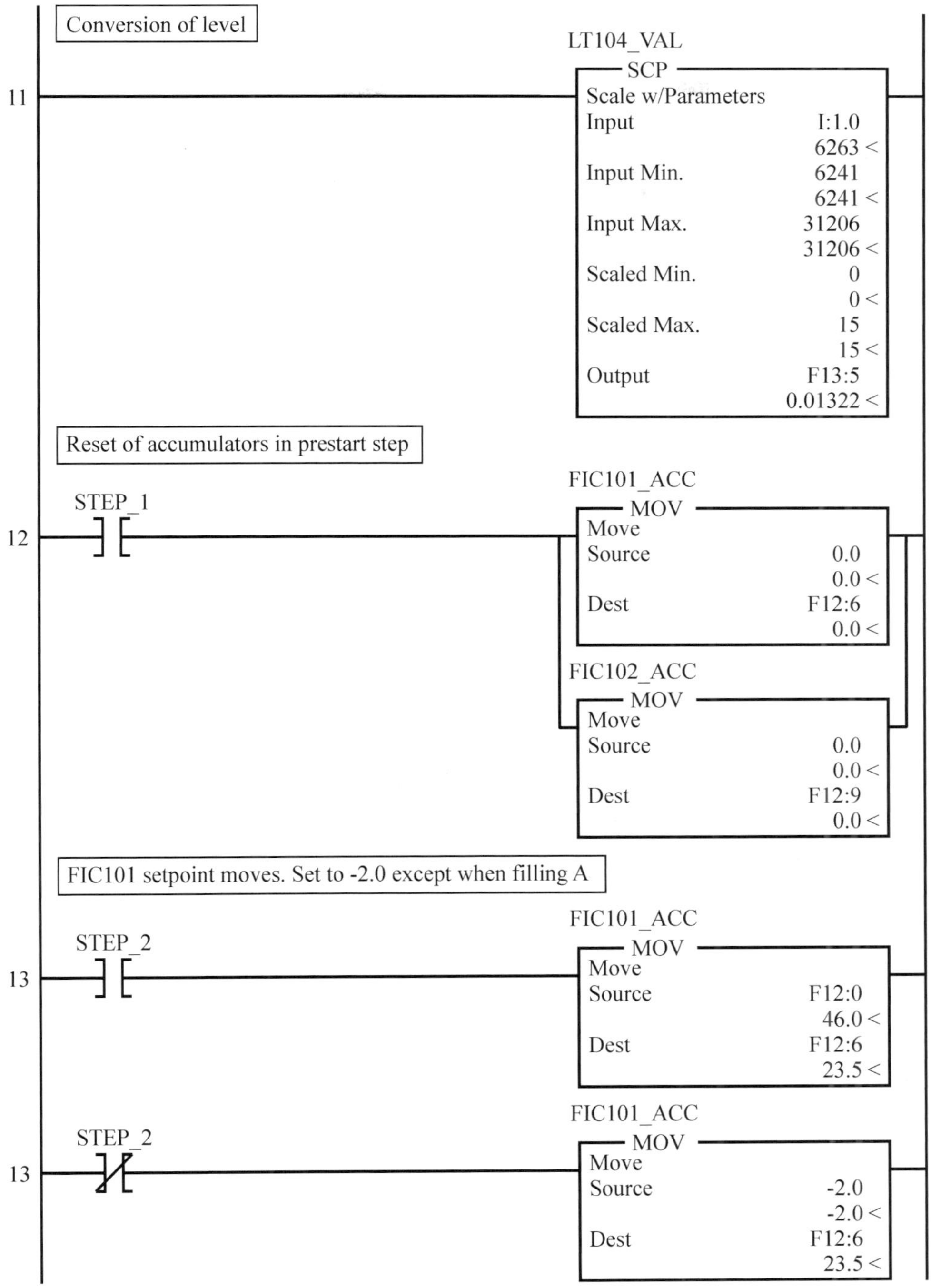

Figure 7.50. MicroLogix ladder logic for calculations of batch process (partial).

PROBLEMS

General instructions for problems:

Write a ladder logic program for the application and implement it for one of the following PLC ladder logic languages:

ControlLogix/CompactLogix, **or**

Micro800, **or**

MicroLogix/SLC-500

If any part of the operation is ambiguous, write down your additional assumptions.

The physical inputs, physical outputs, and internal tags/variables/symbols for each problem are given in the problem. **DO NOT** assign any more physical inputs! Note that for the problems, the ControlLogix addresses assume 1756-, 1734-, 1769-, or 1794-series I/O modules. If using 5069- or 5094-series modules, the address tags will need to be modified appropriately.

The ranges of the analog I/O channel values are given below.

Your solution should include the following:

1. Function chart (if appropriate)
2. Specify the PLC processor used.
3. Ladder logic diagram (with comments). For consistency among the different PLCs, use only tags/variables/symbols in the ladder logic. Use instructions and function blocks consistent with the PLC processor.
4. Table listing additional internal memory (tags/variables/symbols) used and a brief description of their use. For the ControlLogix, CompactLogix, and Micro800, also indicate the data type for the internal memory. For MicroLogix and SLC-500, indicate the associated memory address.

Analog Inputs

Unless otherwise specified, assume the analog input channel values corresponding to the lowest and highest sensor values for the different I/O module series are below. The data type of the channel value is also shown.

	1756/5069/ 5094	1734 (Default)	1747 (SLC)	1762 (MLogix)	1769 (EU)
Lowest Value	0.0	3277	3277	6240	4000
Highest Value	100.0	16,383	16,384	31,200	20,000
Data type	REAL	INT	INT	INT	INT

	2080	2085 (EU)
Lowest Value	13,107	4000
Highest Value	65,535	20,000
Data type	UINT	INT

The analog input is configured for a 4 to 20 mA signal from the sensor. In the case of the ControlLogix/CompactLogix PLC, the analog channel has been configured for the range above. The instructor may choose to specify that the channel has been configured to scale the channel value to be in the sensor units, for example, level in feet or temperature in °C.

Analog Outputs

Unless otherwise specified, assume the analog output channel values corresponding to the lowest and highest sensor values of the output device control for the different I/O module series are below. The data type of the channel value is also shown.

	1769/5069/ 5094	1734 (Default)	1747 (SLC)	1762 (MLogix)	1769 (EU)
Lowest Value	0.0	1638	6242	6240	4000
Highest Value	100.0	8191	31,208	31,200	20,000
Data type	REAL	INT	INT	INT	INT

	2080	2085 (EU)
Lowest Value	13,107	4000
Highest Value	65,535	20,000
Data type	UINT	INT

The analog output is configured for a 4 to 20 mA signal to the actuator. In the case of the ControlLogix/CompactLogix PLC, the analog channel has been configured for the range above. The instructor may choose to specify that the channel has been configured to scale the channel value to be in the actual units of the output device, for example, speed in rpm or valve position in percent of span.

P7-1. A pH probe measures the pH of the solution in a tank (Figure P7.1). The pH range is 0 to 14. The output of the pH transmitter, AT211, is a 4 to 20 mA signal that is connected to an analog input module channel whose channel value corresponds to the lowest and highest pH values as specified before P7-1. The tag/variable/symbol for the channel value is AT211_Raw. The tag for the conversion result is AT211_PH, a real.

Program ladder logic to do the conversion for

(a) ControlLogix/CompactLogix with 1756/5069/5094 input modules

(b) ControlLogix/CompactLogix with 1734 input modules

(c) CompactLogix with 1769 input modules

(d) MicroLogix with 1762 input modules

(e) Micro800 with 2080 input modules

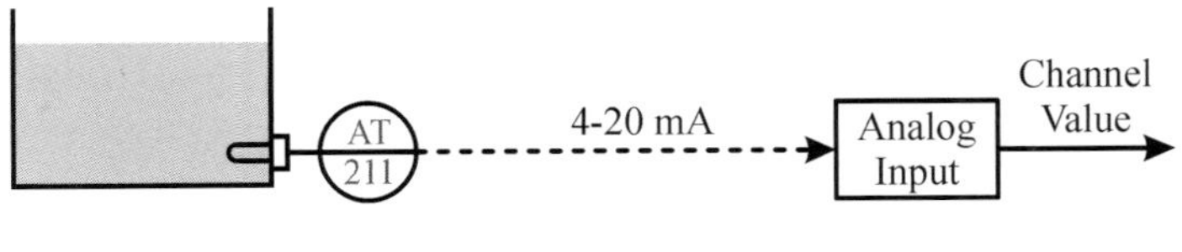

Figure P7.1. PH measurement.

P7-2. A pressure sensor measures vessel pressures between 5 and 400 pounds per square inch (psi). The output of the pressure transmitter is a 1 to 5 volt signal that is connected to an analog input module channel whose channel value corresponds to the lowest and highest pressure values as specified before P7-1. The tag/variable/symbol for the channel value is PT101_Raw. The tag for the conversion result is PT101_PSI, a real.

Program ladder logic to do the conversion for

(a) ControlLogix/CompactLogix with 1756/5069/5094 input modules

(b) ControlLogix/CompactLogix with 1734 input modules

(c) CompactLogix with 1769 input modules

(d) MicroLogix with 1762 input modules

(e) Micro800 with 2080 input modules

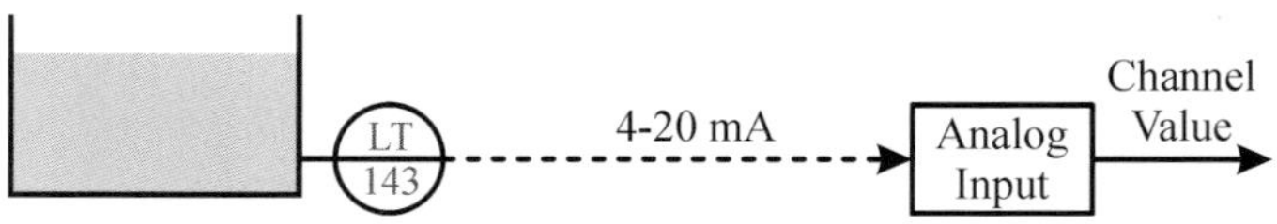

Figure P7.3. Level measurement.

P7-3. The LT143 level transmitter (Figure P7.3) measures water levels between 2 and 20 feet of water and converts the level into a 4 to 20 mA signal that is connected to an analog input module. The analog input channel value corresponds to the lowest and highest level values as specified before P7-1. The tag/variable/symbol for the channel value is LT143_Raw. The tag for the conversion result is LT143_FT, a real.

Program ladder logic to do the conversion for

(a) ControlLogix/CompactLogix with 1756/5069/5094 input modules

(b) ControlLogix/CompactLogix with 1734 input modules

(c) CompactLogix with 1769 input modules

(d) MicroLogix with 1762 input modules

(e) Micro800 with 2080 input modules

P7-4. The transmitter of the FT104 orifice flow meter (Figure P7.4) is adjusted so that the electrical output signal span of 4 to 20 mA represents 10 to 300 pounds/hour of steam flow. The 4 to 20 mA signal is connected to an analog input module channel whose channel value corresponds to the lowest and highest flow values as specified before P7-1. The

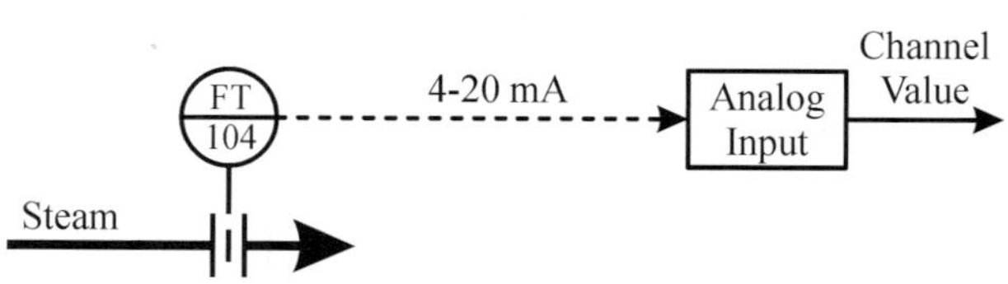

Figure P7.4. Orifice flow meter measurement.

tag/variable/symbol for the channel value is FT104_Raw. The tag for the conversion result is FT104_PPH, a real.

Program ladder logic to do the conversion for

(a) ControlLogix/CompactLogix with 1756/5069/5094 input modules

(b) ControlLogix/CompactLogix with 1734 input modules

(c) CompactLogix with 1769 input modules

(d) MicroLogix with 1762 input modules

(e) Micro800 with 2080 input modules

P7-5. A type K thermocouple measures the temperature of a vessel. The thermocouple is connected to a transmitter whose output of 4 to 20 mA is calibrated to represent 40 to 800 °F. The transmitter output is connected to an analog input module channel whose channel value corresponds to the lowest and highest temperature values as specified before P7-1. The tag/variable/symbol for the channel value is TT105_Raw. The tag for the conversion result is TT105_DEGC, a real.

Program ladder logic to do the conversion for

(a) ControlLogix/CompactLogix with 1756/5069/5094 input modules

(b) ControlLogix/CompactLogix with 1734 input modules

(c) CompactLogix with 1769 input modules

(d) MicroLogix with 1762 input modules

(e) Micro800 with 2080 input modules

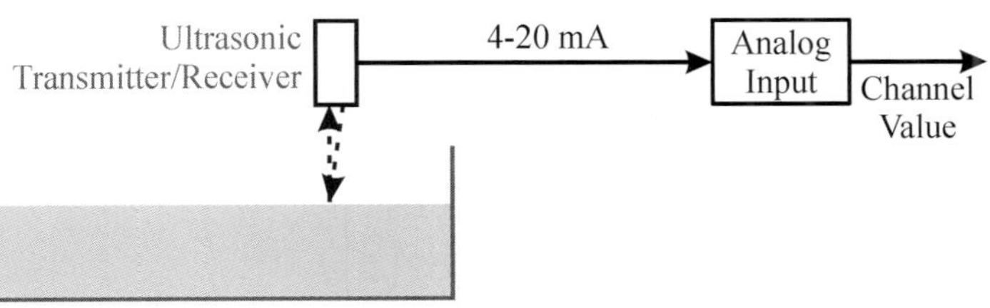

Figure P7.6. Ultrasonic level measurement.

P7-6. The ultrasonic level transmitter in Figure P7.6 measures the level of powder in a tank, in feet (range 0 to 44 ft.), and converts the level into a 4 to 20 mA signal that is connected to an analog input module channel whose channel value range is specified before P7-1. However, the relationship between the level and the channel value is reversed. A level of 44 ft. corresponds to the lowest channel value and a level of 0 feet corresponds to the highest channel value. The tag/variable/symbol for the channel value is LT206_Raw. The tag for the conversion result is LT206_FT, a real.

Program ladder logic to do the conversion for

(a) ControlLogix/CompactLogix with 1756/5069/5094 input modules

(b) ControlLogix/CompactLogix with 1734 input modules

(c) CompactLogix with 1769 input modules
(d) MicroLogix with 1762 input modules
(e) Micro800 with 2080 input modules

P7-7. The ultrasonic level transmitter in Figure P7.6 measures the level of liquid in a tank, in meters (range 2 to 21 meters), and converts the level into a 4 to 20 mA signal that is connected to an analog input module channel whose channel value range is specified before P7-1. However, the relationship between the level and the channel value is reversed. A level of 21 meters corresponds to the lowest channel value and a level of 2 meters corresponds to the highest channel value. The tag/variable/symbol for the channel value is LT307_Raw. The tag for the conversion result is LT307_M, a real.

Program ladder logic to do the conversion for
(a) ControlLogix/CompactLogix with 1756/5069/5094 input modules
(b) ControlLogix/CompactLogix with 1734 input modules
(c) CompactLogix with 1769 input modules
(d) MicroLogix with 1762 input modules
(e) Micro800 with 2080 input modules

P7-8. A load cell measures the force (10 – 2000 kgf) a hydraulic ram is exerting on an ject. The output of the load cell is a 4 to 20 mA signal that is connected to an analog input module channel whose channel value corresponds to the lowest and highest force values as specified before P7-1. The tag/variable/symbol for the channel value is WT408_Raw. The tag for the conversion result is WT408_KGF, a real.

Program ladder logic to do the conversion for
(a) ControlLogix/CompactLogix with 1756/5069/5094 input modules
(b) ControlLogix/CompactLogix with 1734 input modules
(c) CompactLogix with 1769 input modules
(d) MicroLogix with 1762 input modules
(e) Micro800 with 2080 input modules

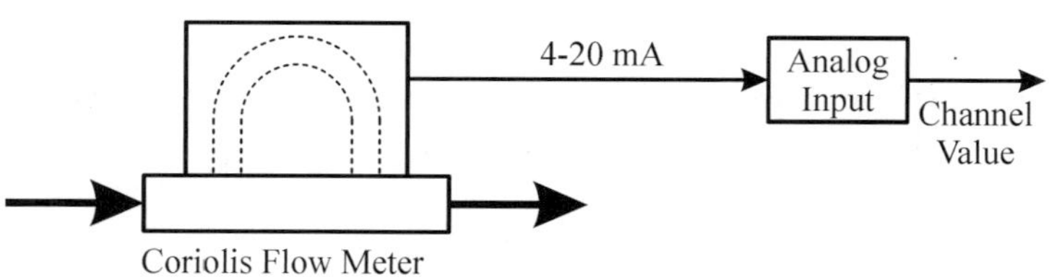

Figure P7.9. Coriolis flow measurement.

P7-9. The Coriolis flow meter in Figure P7.9 is adjusted so that the electrical output signal span of 4 to 20 mA represents 1 to 15 liters/minute of flow. The 4 to 20 mA signal is connected to an analog input module channel whose channel value corresponds to the lowest and highest flow values as specified before P7-1. The tag/variable/symbol for the channel value is FT409_Raw. The tag for the conversion result is FT409_LPM, a real.

Program ladder logic to do the conversion for

(a) ControlLogix/CompactLogix with 1756/5069/5094 input modules

(b) ControlLogix/CompactLogix with 1734 input modules

(c) CompactLogix with 1769 input modules

(d) MicroLogix with 1762 input modules

(e) Micro800 with 2080 input modules

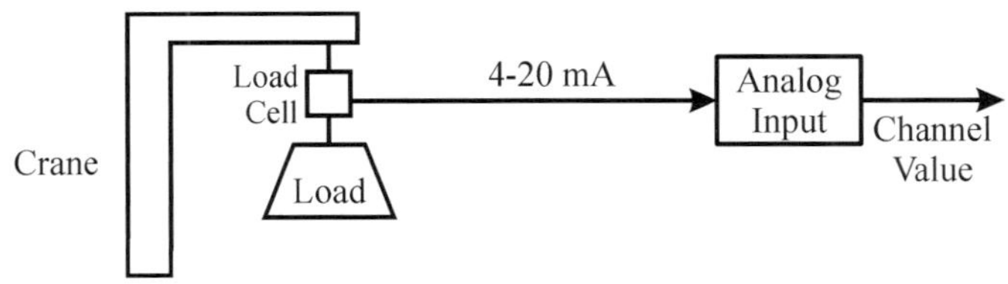

Figure P7.10. Weight measurement.

P7-10. A load cell on a crane (Figure P7.10) measures the weight (20 – 500 pounds) of whatever is hanging from the crane. The output of the load cell is a 4 to 20 mA signal that is connected to an analog input module channel whose channel value corresponds to the lowest and highest force values as specified before P7-1. The tag/variable/symbol for the channel value is WT410_Raw. The tag for the conversion result is WT410_LB, a real.

Program ladder logic to do the conversion for

(a) ControlLogix/CompactLogix with 1756/5069/5094 input modules

(b) ControlLogix/CompactLogix with 1734 input modules

(c) CompactLogix with 1769 input modules

(d) MicroLogix with 1762 input modules

(e) Micro800 with 2080 input modules

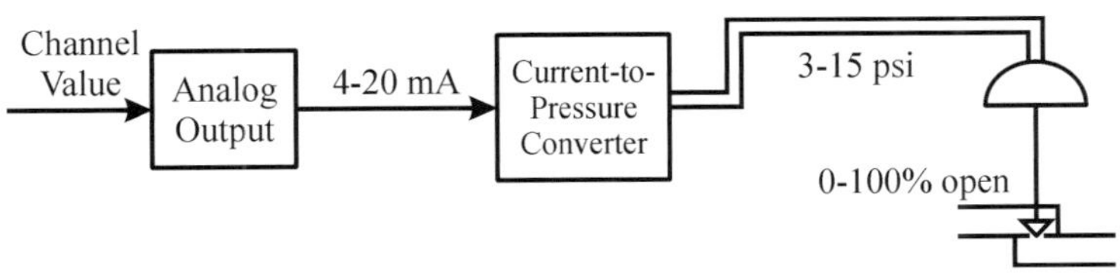

Figure P7.11. Valve position control by PLC analog output.

P7-11. Figure P7.11 depicts the analog output of a PLC used to control the position of a regulatory control valve. The PLC analog output is a signal in the range of 4 to 20 mA that is converted into a 3 to 15 psi pressure signal. The pressure signal drives the valve, which converts the 3 to 15 psi into 0 to 100% open (0% is closed). The analog output channel value has the range as specified before P7-1. However, the relationship between the position and the channel value is reversed. This is a "fail-open" valve that opens when the air supply fails. Therefore, the lowest valve position (zero) corresponds to the highest output channel value and the highest valve position (100.0) corresponds to the lowest output channel value. The tag/variable/symbol for the desired valve position is FY101, a real. The tag for the conversion result, the channel value, is FY101_Raw.

Program ladder logic to do the conversion for

(a) ControlLogix/CompactLogix with 1756/5069/5094 input modules

(b) ControlLogix/CompactLogix with 1734 input modules

(c) CompactLogix with 1769 input modules

(d) MicroLogix with 1762 input modules

(e) Micro800 with 2080 input modules

P7-12. Figure P7.11 depicts the analog output of a PLC used to control the position of a regulatory control valve. The PLC analog output is a signal in the range of 4 to 20 mA that is converted into a 3 to 60 psi signal (a different range than shown in Figure P7.11). The pressure signal drives the valve, which converts the 3 to 60 psi into 0 to 100% open (0% is closed). The analog output channel value corresponds to the lowest and highest valve position values as specified before P7-1. The tag/variable/symbol for the desired valve position is FY612, a real. The tag for the conversion result, the channel value, is FY612_Raw.

Program ladder logic to do the conversion for

(a) ControlLogix/CompactLogix with 1756/5069/5094 input modules

(b) ControlLogix/CompactLogix with 1734 input modules

(c) CompactLogix with 1769 input modules

(d) MicroLogix with 1762 input modules

(e) Micro800 with 2080 input modules

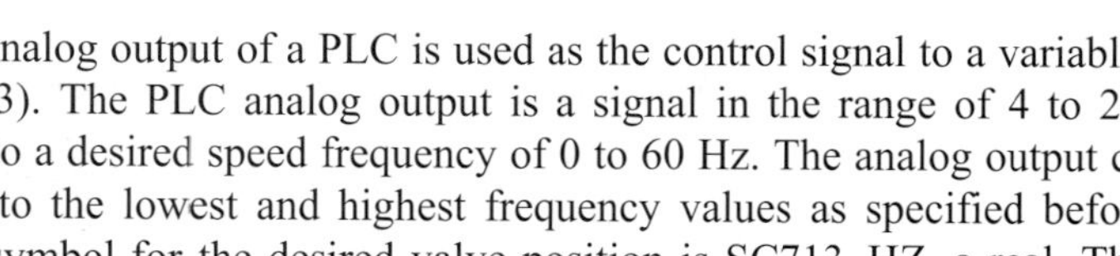

Figure P7.13. Motor controller.

P7-13. The analog output of a PLC is used as the control signal to a variable speed drive (Figure P7.13). The PLC analog output is a signal in the range of 4 to 20 mA that is converted into a desired speed frequency of 0 to 60 Hz. The analog output channel value corresponds to the lowest and highest frequency values as specified before P7-1. The tag/variable/symbol for the desired valve position is SC713_HZ, a real. The tag for the conversion result, the channel value, is SC713_Raw.

Program ladder logic to do the conversion for

(a) ControlLogix/CompactLogix with 1756/5069/5094 input modules
(b) ControlLogix/CompactLogix with 1734 input modules
(c) CompactLogix with 1769 input modules
(d) MicroLogix with 1762 input modules
(e) Micro800 with 2080 input modules

P7-14. The analog output of a PLC is used as the control signal to a variable speed drive (Figure P7.13) except that the frequency to the motor is different. The PLC analog output is a signal in the range of 4 to 20 mA that is converted into a desired speed frequency of 10 to 120 Hz. The analog output channel value corresponds to the lowest and highest frequency values as specified before P7-1. The tag/variable/symbol for the desired valve position is SC714_HZ, a real. The tag for the conversion result, the channel value, is SC714_Raw.

Program ladder logic to do the conversion for
(a) ControlLogix/CompactLogix with 1756/5069/5094 input modules
(b) ControlLogix/CompactLogix with 1734 input modules
(c) CompactLogix with 1769 input modules
(d) MicroLogix with 1762 input modules
(e) Micro800 with 2080 input modules

P7-15. The analog output of a PLC is used as the control signal to a variable speed drive (Figure P7.13) except that the frequency to the motor is different. The PLC analog output is a signal in the range of 4 to 20 mA that is converted into a desired speed frequency of 20 to 400 Hz. The analog output channel value corresponds to the lowest and highest frequency values as specified before P7-1. The tag/variable/symbol for the desired valve position is SC101_HZ, a real. The tag for the conversion result, the channel value, is SC101_Raw.

Program ladder logic to do the conversion for
(a) ControlLogix/CompactLogix with 1756/5069/5094 input modules
(b) ControlLogix/CompactLogix with 1734 input modules
(c) CompactLogix with 1769 input modules
(d) MicroLogix with 1762 input modules
(e) Micro800 with 2080 input modules

P7-16. The analog output of a PLC is used as the control signal to a vibratory feeder. The PLC analog output is a signal in the range of 4 to 20 mA that is converted into a desired flow rate of 100 to 2400 cubic meters per hour. The analog output channel value corresponds to the lowest and highest flow rate values as specified before P7-1. The tag/variable/symbol for the desired valve position is FC101_M3PH, a real. The tag for the conversion result, the channel value, is FC101_Raw.

Program ladder logic to do the conversion for
(a) ControlLogix/CompactLogix with 1756/5069/5094 input modules
(b) ControlLogix/CompactLogix with 1734 input modules

(c) CompactLogix with 1769 input modules
(d) MicroLogix with 1762 input modules
(e) Micro800 with 2080 input modules

P7-17. Tank Level Indicator and Alarm. Implement a level indicator and alarm system for a tank.

The tank level is measured at the top of the tank with an ultrasonic transducer (Figure P7.17). Hence, a longer distance measured by the transducer indicates a lower tank level. The level transmitter measures the tank level (0.0 to 15.0 ft.), converting the level into a 4 to 20 mA signal that is connected to an analog input module channel whose channel value range is specified before P7-1. However, the relationship between the level and the channel value is reversed. A level of 15 feet corresponds to the lowest channel value and a level of zero feet corresponds to the highest channel value. The tag/variable/symbol for the channel value is LT1632_Meas and should be converted into a level in feet (a real number) and placed in the LT1632_VAL location.

There are three indicator lights and an alarm on the operator panel that are operated in the following manner, depending on the level value:

Level (ft)	**LT1632_LO**	**LT1632_NM**	**LT1632_HI**	**LT1632_ALM**
< 2.0	On	Off	Off	Off
≥ 2.0 and ≤ 12.0	Off	On	Off	Off
> 12.0 and ≤ 14.0	Off	Off	On	Off
> 14.0	Off	Off	Flashing	See below

When a light is flashing, it should flash with an **on** period and **off** period of 0.5 second each.

The LT1632_ALM is handled in the following manner. LT1632_ALM is turned **on** for any of the following conditions:

1. The level becomes greater than 14.0 feet
2. The alarm has been acknowledged by the operator, but the level has remained above 14.0 feet for five minutes.

LT1632_ALM is turned **off** for the following condition:

The operator acknowledges the alarm by pushing ALM_ACK.

Even if the level drops below 14.0 feet while LT1632_ALM is **on**, the alarm should remain **on** until acknowledged by the operator. If the level falls to 14.0 ft or

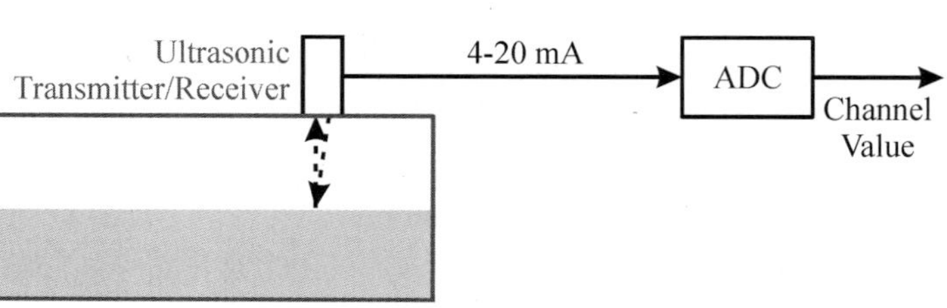

Figure P7.17. Tank level measurement..

below during the five minutes after alarm acknowledgment, the timing interval is completely canceled and the next time the level goes above 14.0 feet is treated as if it is the first time the level has gone above 14.0 feet.

The ALM_ACK input is a push button switch. Do not assume it remains **on** for the 5-minute interval that starts when the ALM_ACK switch is pushed. Also, do not assume ALM_ACK is **off** after 5 minutes (it could be taped down). It has to turn **off** in order for it to turn **on** to acknowledge the next alarm.

The solution should include the equation used to convert LT1632_MEAS into the level in feet.

Assume the following input, output, and internal memory tags/variables/symbols:

Tag/Var/Symbol	Description
ALM_ACK	Alarm acknowledge push button switch; **on** when pushed, **off** when released. Operation described above.
LT1632_HI	High level indication, high-high when flashing.
LT1632_NM	Normal level indication.
LT1632_LO	Low level indication
LT1632_ALM	Level alarm horn, **on** sounds horn.
LT1632_MEAS	Raw LT-1632 level measurement, represents 0 to 15.0 ft; ; see problem statement for conversion information.
LT1632_VAL	Tank level, in feet (a REAL). Your ladder must calculate this value.

The addresses associated with the physical input and output tags/variables/symbols:

Tag/Var./Symbol	ControlLogix	Micro800	MLogix	SLC
ALM_ACK	Local:1:I.Data.0	_IO_EM_DI_00	I:0/0	I:1/0
LT1632_HI	Local:2:O.Data.0	_IO_EM_DO_00	O:0/0	O:2/0
LT1632_NM	Local:2:O.Data.1	_IO_EM_DO_01	O:0/1	O:2/0
LT1632_LO	Local:2:O.Data.2	_IO_EM_DO_02	O:0/2	O:2/0
LT1632_ALM	Local:2:O.Data.3	_IO_EM_DO_03	O:0/3	O:2/0
LT1632_MEAS	Local:3:I.Ch0Data	_IO_P2_AI_00	I:1.0	I:3.0

The addresses or data types associated with the internal tag/variable/symbol:

Tag/Var./Symbol	CLogix Data Type	Micro800 Data Type	MLogix/SLC Address
LT1632_VAL	REAL	REAL	F12:0

P7-18. Level Indicator and Alarm. Implement a level indicator and alarm system for a tank.

The liquid tank level is measured with a level transducer at the bottom of the tank (Figure P7.18). The level transmitter is calibrated to measure from 0.0 to 8.0 meters and is connected to an analog input module having an ADC whose output value (LT2151_MEAS) corresponds to the lowest and highest values as specified in the general instructions prior to P7-1. The LT2151_MEAS value should be converted into a level in meters (a real number) and placed in the LT2151_VAL location.

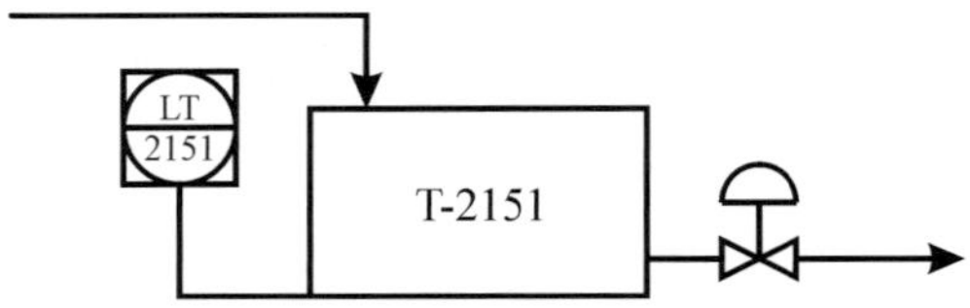

Figure P7.18. Tank level measurement.

There are three indicator lights and an alarm on the operator panel that are operated in the following manner, depending on the level value:

Level (m)	LT2151_LO	LT2151_NM	LT2151_HI	LT2151_ALM
< 1.0	Flashing	Off	Off	See below
≥ 1.0 and < 2.0	On	Off	Off	Off
≥ 2.0 and ≤ 7.0	Off	On	Off	Off
> 7.0	Off	Off	On	Off

When a light is flashing, it should flash with an **on** period and **off** period of 0.5 second each.

The LT2151_ALM is handled in the following manner. LT2151_ALM is turned **on** for any of the following conditions:

1. The level becomes less than 1.0 meter.
2. The alarm has been acknowledged by the operator, but the level has remained below 1.0 meter for four minutes.

LT2151_ALM is turned **off** for the following condition:

The operator acknowledges the alarm by pushing ALM_ACK.

Even if the level increases to 1.0 meter or above while LT2151_ALM is **on**, the alarm should remain **on** until acknowledged by the operator. If the level increases to 1.0 meter or above during the four minutes after alarm acknowledgment, the timing interval is completely canceled and the next time the level goes below 1.0 meter is treated as if it is the first time the level has gone below 1.0 meter.

The ALM_ACK input is a push button switch. Do not assume it remains **on** for the 4-minute interval that starts when the ALM_ACK switch is pushed. Also, do not assume ALM_ACK is **off** after 4 minutes (it could be taped down). It has to turn **off** in order for it to turn **on** to acknowledge the next alarm.

Assume the following input, output, and internal memory tags/variables/symbols:

Tag/Var/Symbol	Description
ALM_ACK	Alarm acknowledge push button switch; **on** when pushed, **off** when released. Operation described above.
LT2151_HI	High level indication.
LT2151_NM	Normal level indication.
LT2151_LO	Low level indication, low-low when flashing.
LT2151_ALM	Level alarm horn, **on** sounds horn.

LT2151_MEAS	Raw T-2151 level measurement, represents 0 to 8 meters; channel value range listed before P7-1.
LT2151_VAL	Tank level, in meters (a REAL). Your ladder must calculate this value.

The addresses associated with the physical input and output tags/variables/symbols:

Tag/Var./Symbol	ControlLogix	Micro800	MLogix	SLC
ALM_ACK	Local:1:I.Data.0	_IO_EM_DI_00	I:0/0	I:1/0
LT2151_HI	Local:2:O.Data.0	_IO_EM_DO_00	O:0/0	O:2/0
LT2151_NM	Local:2:O.Data.1	_IO_EM_DO_01	O:0/1	O:2/1
LT2151_LO	Local:2:O.Data.2	_IO_EM_DO_02	O:0/2	O:2/2
LT2151_ALM	Local:2:O.Data.3	_IO_EM_DO_03	O:0/3	O:2/3
LT2151_MEAS	Local:3:I.Ch0Data	_IO_P2_AI_00	I:1.0	I:3.0

The addresses or data types associated with the internal tag/variable/symbol:

Tag/Var./Symbol	CLogix Data Type	Micro800 Data Type	MLogix/SLC Address
LT2151_VAL	REAL	REAL	F12:0

P7-19. Tank Level Indicator and Alarm. The PLC implements a simple level indicator and alarm system for a tank.

The tank level is measured from the top of the tank with an ultrasonic transducer (Figure P7.17). Hence, a longer distance measured by the transducer indicates a lower tank level. The ultrasonic level transmitter measures the level in the tank and converts the level into a 4 to 20 mA signal that is connected to an analog input module. The ADC output integer value corresponds to the lowest and highest levels as:

Level (ft.)	CLogix	MLogix	SLC-500	Siemens	Modicon	GE
50.0 (full)	0	6241	3277	5530	0	0
0.0 (empty)	32,000	31,206	16,384	27,648	10,000	32,000

There are three indicator lights and an alarm on the operator panel that are operated in the following manner, depending on the tank level:

Level (ft.)	**LT1274_LO**	**LT1274_NM**	**LT1274_HI**	**LT1274_ALM**
< 2.0	Flashing	Off	Off	Off
≥ 2.0 and < 5.0	On	Off	Off	Off
≥ 5.0 and ≤ 40.0	Off	On	Off	Off
> 40.0 and ≤ 45.0	Off	Off	On	Off
> 45.0 and ≤ 50.0	Off	Off	Flashing	See below

When a light is flashing, it should flash with an **on** period and **off** period of 0.5 second each.

The LT1274_ALM is handled in the following manner. LT1274_ALM is turned **on** for any of the following conditions:

1. The level becomes greater than 45.0 feet
2. The alarm has been acknowledged by the operator, but the level has remained above 45.0 feet for five minutes.

LT1274_ALM is turned **off** for the following condition:

The operator acknowledges the alarm by pushing ALM_ACK.

Even if the level drops below 45.0 feet while LT1274_ALM is **on**, the alarm should remain **on** until acknowledged by the operator. If the level falls to 45.0 feet or below during the five minutes after alarm acknowledgment, the timing interval is completely canceled and the next time the level goes above 45.0 is treated as if it is the first time the level has gone above 45.0.

The ALM_ACK input is a push button switch. Do not assume it remains **on** for the 5-minute interval that starts when the ALM_ACK switch is pushed. Also, do not assume ALM_ACK is **off** after 5 minutes (it could be taped down). It has to turn **off** in order for it to turn **on** to acknowledge the next alarm.

Assume the following input, output, and internal memory tags/variables/symbols:

Tag/Var/Symbol	Description
ALM_ACK	Alarm acknowledge push button switch; **on** when pushed, **off** when released. Operation described above.
LT1274_HI	High level indication, high-high when flashing.
LT1274_NM	Normal level indication.
LT1274_LO	Low level indication, low-low when flashing.
LT1274_ALM	Level alarm horn, **on** sounds horn.
LT1274_MEAS	Raw T-1274 level measurement, represents 0 to 50 feet; channel value range listed before P7-1.
LT1274_VAL	Level of tank, in feet (a REAL).

The addresses associated with the physical input and output tags/variables/symbols are:

Tag/Var./Symbol	ControlLogix	Micro800	MLogix	SLC
ALM_ACK	Local:1:I.Data.0	_IO_EM_DI_00	I:0/0	I:1/0
LT1274_HI	Local:2:O.Data.0	_IO_EM_DO_00	O:0/0	O:2/0
LT1274_NM	Local:2:O.Data.1	_IO_EM_DO_01	O:0/1	O:2/1
LT1274_LO	Local:2:O.Data.2	_IO_EM_DO_02	O:0/2	O:2/2
LT1274_ALM	Local:2:O.Data.3	_IO_EM_DO_03	O:0/3	O:2/3
LT1274_MEAS	Local:3:I.Ch0Data	_IO_P2_AI_00	I:1.0	I:3.0

The addresses or data types associated with the internal tag/variable/symbol:

Tag/Var./Symbol	CLogix Data Type	Micro800 Data Type	MLogix/SLC Address
LT1274_VAL	REAL	REAL	F12:0

P7-20. Pressure Indicator and Alarm. Implement a pressure indicator and alarm system for a tank.

The gas tank pressure is measured at the top of the tank with a pressure transducer (Figure P7.20). The pressure transmitter is calibrated to measure from 0.0 to 1500.0 psi and is connected to an analog input module having an ADC whose output value (PT1632_MEAS) corresponds to the lowest and highest values as specified in the general instructions prior to P7-1. The PT1632_MEAS value should be converted into a pressure in psi (a real number) and placed in the PT1632_VAL location.

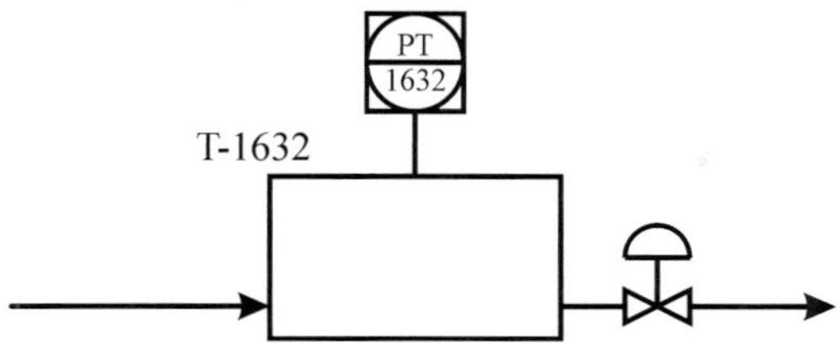

Figure P7.20. Tank pressure measurement.

There are three indicator lights and an alarm on the operator panel that are operated in the following manner, depending on the pressure value:

Pressure (psi)	**PT1632_LO**	**PT1632_NM**	**PT1632_HI**	**PT1632_ALM**
< 15.0	On	Off	Off	Off
≥ 15.0 and ≤ 1100	Off	On	Off	Off
> 1100 and ≤ 1200	Off	Off	On	Off
> 1200	Off	Off	Flashing	See below

When a light is flashing, it should flash with an **on** period and **off** period of 0.5 second each.

The PT1632_ALM is handled in the following manner. PT1632_ALM is turned **on** for any of the following conditions:

1. The pressure becomes greater than 1200 psi
2. The alarm has been acknowledged by the operator, but the pressure has remained above 1200 psi for five minutes.

PT1632_ALM is turned **off** for the following condition:

The operator acknowledges the alarm by pushing ALM_ACK.

Even if the pressure drops below 1200 psi while PT1632_ALM is **on**, the alarm should remain **on** until acknowledged by the operator. If the pressure falls to 1200 psi or below during the five minutes after alarm acknowledgment, the timing interval is completely canceled and the next time the pressure goes above 1200 psi is treated as if it is the first time the pressure has gone above 1200 psi.

The ALM_ACK input is a push button switch. Do not assume it remains **on** for the 5-minute interval that starts when the ALM_ACK switch is pushed. Also, do not assume ALM_ACK is **off** after 5 minutes (it could be taped down). It has to turn **off** in order for it to turn **on** to acknowledge the next alarm.

Assume the following input, output, and internal tags/variables/symbols:

Tag/Var/Symbol	Description
ALM_ACK	Alarm acknowledge push button switch; **on** when pushed, **off** when released. Operation described above.
PT1632_HI	High pressure indication, high-high when flashing.
PT1632_NM	Normal pressure indication.
PT1632_LO	Low pressure indication
PT1632_ALM	Pressure alarm horn, **on** sounds horn.
PT1632_MEAS	Raw PT-1632 pressure measurement, represents 0 to 1500 psi; channel value range listed before P7-1.

PT1632_VAL Tank pressure, in psi (a REAL). Your ladder must calculate this value.

The addresses associated with the physical input and output tags/variables/symbols:

Tag/Var./Symbol	ControlLogix	Micro800	MLogix	SLC
ALM_ACK	Local:1:I.Data.0	_IO_EM_DI_00	I:0/0	I:1/0
PT1632_HI	Local:2:O.Data.0	_IO_EM_DO_00	O:0/0	O:2/0
PT1632_NM	Local:2:O.Data.1	_IO_EM_DO_01	O:0/1	O:2/1
PT1632_LO	Local:2:O.Data.2	_IO_EM_DO_02	O:0/2	O:2/2
PT1632_ALM	Local:2:O.Data.3	_IO_EM_DO_03	O:0/3	O:2/3
PT1632_MEAS	Local:3:I.Ch0Data	_IO_P2_AI_00	I:1.0	I:3.0

The addresses or data types associated with the internal tag/variable'symbol:

Tag/Var./Symbol	CLogix Data Type	Micro800 Data Type	MLogix/SLC Address
PT1632_VAL	REAL	REAL	F12:0

P7-21. Pressure Indicator and Alarm. Implement a pressure indicator and alarm system for a reactor vessel.

The reactor vessel pressure is measured with a pressure transducer on the side of the vessel (Figure P7.21). The pressure transmitter is calibrated to measure from 14.7 to 250.0 psi and is connected to an analog input module having an ADC whose output value (PT2410_MEAS) corresponds to the lowest and highest values as specified in the general instructions prior to P7-1. The PT2410_MEAS value should be converted into a pressure in pounds per square inch (psi), a real number, and placed in the PT2410_VAL location.

There are three indicator lights and an alarm horn on the operator panel that are operated in the following manner, depending on the pressure value:

Pres. (psi)	**PT2410_LO**	**PT2410_NM**	**PT2410_HI**	**PT2410_ALM**
< 50	On	Off	Off	Off
≥ 50 and ≤ 200	Off	On	Off	Off
> 200 and ≤ 225	Off	Off	On	Off
> 225	Off	Off	Flashing	See below

When a light is flashing, it should flash with an **on** period and **off** period of 0.5 second each.

The PT2410_ALM is handled in the following manner. PT2410_ALM is turned **on** for any of the following conditions:

1. The pressure becomes greater than 225 psi.

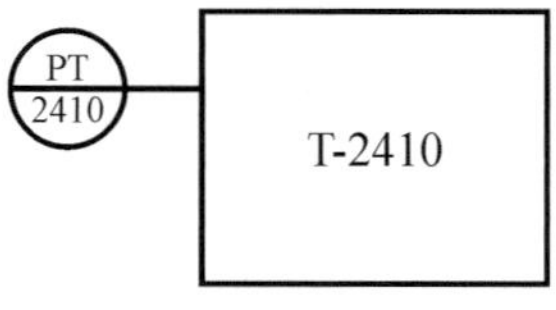

Figure P7.21. Reactor pressure measurement.

2. The alarm has been acknowledged by the operator, but the pressure has remained above 225 psi for two minutes.

PT2410_ALM is turned **off** for the following condition:

The operator acknowledges the alarm by pushing ALM_ACK.

Even if the pressure decreases to 225 psi or below while PT2410_ALM is **on**, the alarm should remain **on** until acknowledged by the operator. If the pressure decreases to 225 psi or below during the two minutes after alarm acknowledgment, the timing interval is completely canceled and the next time the pressure goes above 225 psi is treated as if it is the first time the pressure has gone above 225 psi.

The ALM_ACK input is a push button switch. Do not assume it remains **on** for the 2-minute interval that starts when the ALM_ACK switch is pushed. Also, do not assume ALM_ACK is **off** after 2 minutes (it could be taped down). It has to turn **off** in order for it to turn **on** to acknowledge the next alarm.

Assume the following input, output, and internal tags/variables/symbols:

Tag/Var/Symbol	Description
ALM_ACK	Alarm acknowledge push button switch; **on** when pushed, **off** when released. Operation described above.
PT2410_HI	High pressure indication, high-high when flashing.
PT2410_NM	Normal pressure indication.
PT2410_LO	Low pressure indication.
PT2410_ALM	Pressure alarm horn, **on** sounds horn.
PT2410_MEAS	Raw T-3250 pressure measurement, represents 14.7 to 250 psi; channel value range listed before P7-1.
PT2410_VAL	Tank pressure, in psi (a REAL). Your ladder must calculate this value.

The addresses associated with the physical input and output tags/variables/symbols:

Tag/Var./Symbol	ControlLogix	Micro800	MLogix	SLC
ALM_ACK	Local:1:I.Data.0	_IO_EM_DI_00	I:0/0	I:1/0
PT2410_HI	Local:2:O.Data.0	_IO_EM_DO_00	O:0/0	O:2/0
PT2410_NM	Local:2:O.Data.1	_IO_EM_DO_01	O:0/1	O:2/1
PT2410_LO	Local:2:O.Data.2	_IO_EM_DO_02	O:0/2	O:2/2
PT2410_ALM	Local:2:O.Data.3	_IO_EM_DO_03	O:0/3	O:2/3
PT2410_MEAS	Local:3:I.Ch0Data	_IO_P2_AI_00	I:1.0	I:3.0

The addresses or data types associated with the internal tag/variable/symbol:

Tag/Var./Symbol	CLogix Data Type	Micro800 Data Type	MLogix/SLC Address
PT2410_VAL	REAL	REAL	F12:0

P7-22. Flow Indicator and Alarm. Implement a flow indicator and alarm system.

The liquid chemical flow is measured with an orifice plate inserted into the pipe (Figure P7.22). The flow transmitter measures the chemical flow and is calibrated to measure from 0.0 to 1400.0 lbs/hr and is connected to an analog input

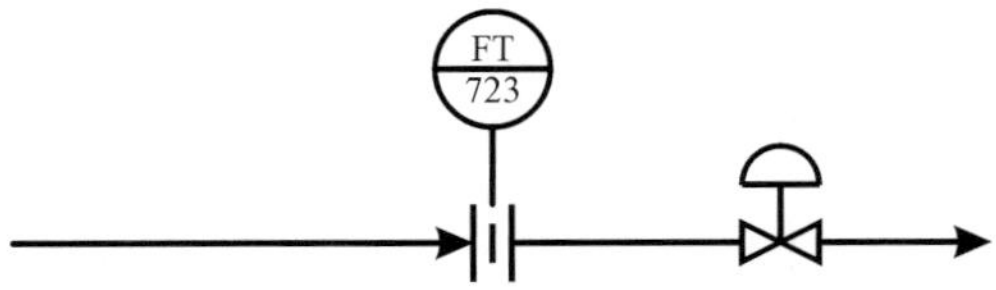

Figure P7.22. Liquid flow measurement.

module having an ADC whose output value (FT723_MEAS) corresponds to the lowest and highest flows as specified in the general instructions prior to P7-1.The FT723_MEAS value should be converted into a flow in lbs/hr (a real number) and placed in the FT723_VAL location.

There are three indicator lights and an alarm on the operator panel that are operated in the following manner, depending on the flow value:

Flow (lb/hr)	FT723_LO	FT723_NM	FT723_HI	FT723_ALM
≤ LL_FLOW	Flashing	Off	Off	See below
> LL_FLOW and ≤ LO_FLOW	On	Off	Off	Off
> LO_FLOW and < HI_FLOW	Off	On	Off	Off
≥ HI_FLOW	Off	Off	On	Off

When a light is flashing, it should flash with an **on** period and **off** period of 0.5 second each.

The LL_FLOW, LO_FLOW, and HI_FLOW values are adjustable by the operator and your ladder does not change them. Assume that LL_FLOW < LO_FLOW < HI_FLOW.

The FT723_ALM is handled in the following manner. FT723_ALM is turned **on** for any of the following conditions:

1. The flow becomes less than or equal to LL_FLOW
2. The alarm has been acknowledged by the operator, but the flow has remained less than or equal to LL_FLOW for ten minutes.

FT723_ALM is turned **off** for the following condition:

The operator acknowledges the alarm by pushing ALM_ACK.

Even if the flow increases above LL_FLOW while FT723_ALM is **on**, the alarm should remain **on** until acknowledged by the operator. If the flow increases above LL_FLOW during the ten minutes after alarm acknowledgment, the timing interval is completely canceled and the next time the flow falls to LL_FLOW or lower, it should be treated as if it is the first time the flow has fallen to LL_FLOW or lower.

The ALM_ACK input is a push button switch. Do not assume it remains **on** for the 10-minute interval that starts when the ALM_ACK switch is pushed. Also, do not assume ALM_ACK is **off** after 10 minutes (it could be taped down). It has to turn **off** in order for it to turn **on** to acknowledge the next alarm.

Assume the following input, output, and internal tags/variables/symbols:

Tag/Var/Symbol Description

ALM_ACK	Alarm acknowledge push button switch; **on** when pushed, **off** when released. Operation described above.
FT723_HI	High flow indication;
FT723_NM	Normal flow indication.
FT723_LO	Low flow indication, low-low when flashing.
FT723_ALM	Flow alarm horn, **on** sounds horn.
FT723_MEAS	Raw FT723 flow measurement, represents 0 to 1000 kg/hr; channel value range listed before P7-1.
FT723_VAL	Flow, in kg/hr (a REAL). Your ladder must calculate this value.
LL_FLOW	Value of flow that triggers low-low alarm. Do not modify this value.
LO_FLOW	Value of flow that triggers low indicator. Do not modify this value.
HI_FLOW	Value of flow that triggers high indicator. Do not modify this value.

The addresses associated with the physical input and output tags/variables/symbols:

Tag/Var./Symbol	ControlLogix	Micro800	MLogix	SLC
ALM_ACK	Local:1:I.Data.0	_IO_EM_DI_00	I:0/0	I:1/0
FT723_HI	Local:2:O.Data.0	_IO_EM_DO_00	O:0/0	O:2/0
FT723_NM	Local:2:O.Data.1	_IO_EM_DO_01	O:0/1	O:2/1
FT723_LO	Local:2:O.Data.2	_IO_EM_DO_02	O:0/2	O:2/2
FT723_ALM	Local:2:O.Data.3	_IO_EM_DO_03	O:0/3	O:2/3
FT723_MEAS	Local:3:I.Ch0Data	_IO_P2_AI_00	I:1.0	I:3.0

The addresses or data types associated with the internal tags/variables/symbols:

Tag/Var./Symbol	CLogix Data Type	Micro800 Data Type	MLogix/SLC Address
FT723_VAL	REAL	REAL	F12:0
LL_FLOW	REAL	REAL	F25:0
LO_FLOW	REAL	REAL	F25:1
HI_FLOW	REAL	REAL	F25:2

P7-23. Flow Indicator and Alarm. Implement a flow indicator and alarm system.

The liquid chemical flow is measured with an orifice plate inserted into the pipe (Figure P7.22). The flow transmitter measures the chemical flow and is calibrated to measure from 0.0 to 1000.0 kg/hr and is connected to an analog input module having an ADC whose output value (FT723_MEAS) corresponds to the lowest and highest flows as specified in the general instructions prior to P7-1.The FT723_MEAS value should be converted into a flow in kg/hr (a real number) and placed in the FT723_VAL location.

There are three indicator lights and an alarm on the operator panel that are operated in the following manner, depending on the flow value:

Flow (kg/hr)	**FT723_LO**	**FT723_NM**	**FT723_HI**	**FT723_ALM**
≤ LO_FLOW	On	Off	Off	Off

> LO_FLOW and < HI_FLOW	Off	On	Off	Off
≥ HI_FLOW and < HH_FLOW	Off	Off	On	Off
≥ HH_FLOW	Off	Off	Flashing	See below

When a light is flashing, it should flash with an **on** period and **off** period of 0.5 second each.

The LO_FLOW, HI_FLOW, and HH_FLOW values are adjustable by the operator and your ladder does not change them. Assume that LO_FLOW < HI_FLOW < HH_FLOW.

The FT723_ALM is handled in the following manner. FT723_ALM is turned **on** for any of the following conditions:

1. The flow becomes greater than or equal to HH_FLOW
2. The alarm has been acknowledged by the operator, but the flow has remained greater than or equal to HH_FLOW for ten minutes.

FT723_ALM is turned **off** for the following condition:

The operator acknowledges the alarm by pushing ALM_ACK.

Even if the flow decreases below HH_FLOW while FT723_ALM is **on**, the alarm should remain **on** until acknowledged by the operator. If the flow decreases below HH_FLOW during the ten minutes after alarm acknowledgment, the timing interval is completely canceled and the next time the flow is at HH_FLOW or higher, it should be treated as if it is the first time the flow has increased to HH_FLOW or higher.

The ALM_ACK input is a push button switch. Do not assume it remains **on** for the 10-minute interval that starts when the ALM_ACK switch is pushed. Also, do not assume ALM_ACK is **off** after 10 minutes (it could be taped down). It has to turn **off** in order for it to turn **on** to acknowledge the next alarm.

Assume the following input, output, and internal tags/variables/symbols:

Tag/Var/Symbol	Description
ALM_ACK	Alarm acknowledge push button switch; **on** when pushed, **off** when released. Operation described above.
FT723_HI	High flow indication;
FT723_NM	Normal flow indication.
FT723_LO	Low flow indication, low-low when flashing.
FT723_ALM	Flow alarm horn, **on** sounds horn.
FT723_MEAS	Raw FT723 flow measurement, represents 0 to 1000 kg/hr; channel value range listed before P7-1.
FT723_VAL	Flow, in kg/hr (a REAL). Your ladder must calculate this value.
LO_FLOW	Value of flow that triggers low indicator. Do not modify this value.
HI_FLOW	Value of flow that triggers high indicator. Do not modify this value.
HH_FLOW	Value of flow that triggers high-high alarm. Do not modify this value.

The addresses associated with the physical input and output tags/variables/symbols are:

Tag/Var./Symbol	ControlLogix	Micro800	MLogix	SLC
ALM_ACK	Local:1:I.Data.0	_IO_EM_DI_00	I:0/0	I:1/0
FT723_HI	Local:2:O.Data.0	_IO_EM_DO_00	O:0/0	O:2/0
FT723_NM	Local:2:O.Data.1	_IO_EM_DO_01	O:0/1	O:2/1
FT723_LO	Local:2:O.Data.2	_IO_EM_DO_02	O:0/2	O:2/2
FT723_ALM	Local:2:O.Data.3	_IO_EM_DO_03	O:0/3	O:2/3
FT723_MEAS	Local:3:I.Ch0Data	_IO_P2_AI_00	I:1.0	I:3.0

The addresses or data types associated with the internal tags/variables/symbols:

Tag/Var./Symbol	CLogix Data Type	Micro800 Data Type	MLogix/SLC Address
FT723_VAL	REAL	REAL	F12:0
LO_FLOW	REAL	REAL	F25:1
HI_FLOW	REAL	REAL	F25:2
HH_FLOW	REAL	REAL	F25:3

P7-24. Temperature Indicator and Alarm. Implement a temperature indicator and alarm system for a reactor vessel.

The reactor vessel temperature is measured with a temperature transducer on the side of the vessel (Figure P7.24). The temperature transmitter is calibrated to measure from 100.0 to 350.0 degrees F and is connected to an analog input module having an ADC whose output value (TT3250_MEAS) corresponds to the lowest and highest values as specified in the general instructions prior to P7-1. The TT3250_MEAS value should be converted into a temperature in degrees Fahrenheit (a real number) and placed in the TT3250_VAL location.

There are three indicator lights and an alarm on the operator panel that are operated in the following manner, depending on the temperature value:

Temp. (°F)	**TT3250_LO**	**TT3250_NM**	**TT3250_HI**	**TT3250_ALM**
< 120	On	Off	Off	Off
≥ 120 and ≤ 275	Off	On	Off	Off
> 275 and ≤ 300	Off	Off	On	Off
> 300	Off	Off	Flashing	See below

When a light is flashing, it should flash with an **on** period and **off** period of 0.5 second each.

The TT3250_ALM is handled in the following manner. TT3250_ALM is turned **on** for any of the following conditions:

1. The temperature becomes greater than 300 degress.
2. The alarm has been acknowledged by the operator, but the temperature has remained above 300 degrees for five minutes.

TT3250_ALM is turned **off** for the following condition:

The operator acknowledges the alarm by pushing ALM_ACK.

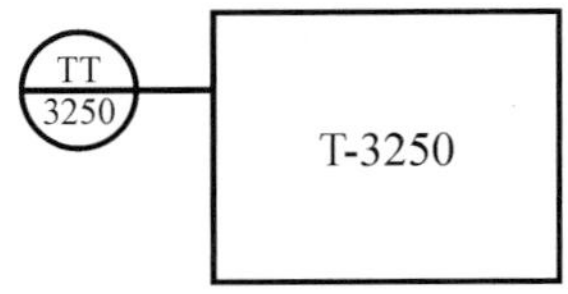

Figure P7.24. Reactor temperature measurement.

Even if the temperature decreases to 300 degrees or below while TT3250_ALM is **on**, the alarm should remain **on** until acknowledged by the operator. If the temperature decreases to 300 degrees or below during the five minutes after alarm acknowledgment, the timing interval is completely canceled and the next time the temperature goes above 300 degrees is treated as if it is the first time the temperature has gone above 300 degrees.

The ALM_ACK input is a push button switch. Do not assume it remains **on** for the 5-minute interval that starts when the ALM_ACK switch is pushed. Also, do not assume ALM_ACK is **off** after 5 minutes (it could be taped down). It has to turn **off** in order for it to turn **on** to acknowledge the next alarm.

Assume the following input, output, and internal tags/variables/symbols:

Tag/Var/Symbol	Description
ALM_ACK	Alarm acknowledge push button switch; **on** when pushed, **off** when released. Operation described above.
TT3250_HI	High temperature indication, high-high when flashing.
TT3250_NM	Normal temperature indication.
TT3250_LO	Low temperature indication.
TT3250_ALM	Temperature alarm horn, **on** sounds horn.
TT3250_MEAS	Raw T-3250 temperature measurement, represents 100 to 350 degrees Fahrenheit; channel value range listed before P7-1.
TT3250_VAL	Tank temperature, in degrees Fahrenheit (a REAL). Your ladder must calculate this value.

The addresses associated with the physical inputs and outputs are:

Tag/Var./Symbol	ControlLogix	Micro800	MLogix	SLC
ALM_ACK	Local:1:I.Data.0	_IO_EM_DI_00	I:0/0	I:1/0
TT3250_HI	Local:2:O.Data.0	_IO_EM_DO_00	O:0/0	O:2/0
TT3250_NM	Local:2:O.Data.1	_IO_EM_DO_01	O:0/1	O:2/1
TT3250_LO	Local:2:O.Data.2	_IO_EM_DO_02	O:0/2	O:2/2
TT3250_ALM	Local:2:O.Data.3	_IO_EM_DO_03	O:0/3	O:2/3
TT3250_MEAS	Local:3:I.Ch0Data	_IO_P2_AI_00	I:1.0	I:3.0

The addresses or data types associated with the internal tag/variable/symbol:

Tag/Var./Symbol	CLogix Data Type	Micro800 Data Type	MLogix/SLC Address
TT3250_VAL	REAL	REAL	F12:0

P7-25. Use ladder logic to implement a voltage indicator and alarm system.

The voltage is measured by a voltage transducer. The voltage transmitter is calibrated to measure from 400.0 to 550.0 AC volts RMS and is connected to an analog input module having an ADC whose output value (VT475_MEAS) corresponds to the lowest and highest values as specified in the general instructions prior to P7-1. The VT475_MEAS value should be converted into a temperature in degrees Fahrenheit (a real number) and placed in the VT475_VAL location.

There are three indicator lights and an alarm on the operator panel that are operated in the following manner, depending on the flow value:

Voltage	**Yellow**	**Green**	**Red**	**VT475_ALM**
≤ LL_VOLT	Flashing	Off	Off	See below
> LL_VOLT and ≤ LO_VOLT	On	Off	Off	Off
> LO_VOLT and < HI_VOLT	Off	On	Off	Off
≥ HI_VOLT and < HH_VOLT	Off	Off	On	Off
≥ HH_VOLT	Off	Off	Flashing	See below

When a light is flashing, it should flash with an **on** period and **off** period of 0.5 second each.

The LL_VOLT, LO_VOLT, HI_VOLT, and HH_VOLT values are adjustable by the operator and your ladder does not change them. Assume that LL_FLOW < LO_VOLT < HI_VOLT < HH_VOLT.

The VT475_ALM is handled in the following manner. VT475_ALM is turned **on** for any of the following conditions:

1. The voltage becomes greater than or equal to HH_VOLT
2. The voltage becomes less than or equal to LL_VOLT
3. The alarm has been acknowledged by the operator, but the voltage has remained greater than or equal to HH_VOLT or less than or equal to LL_VOLT for ten minutes.

VT475_ALM is turned **off** for the following condition:

The operator acknowledges the alarm by pushing ALM_ACK.

Even if the voltage decreases below HH_VOLT (or increases above LL_VOLT) while VT475_ALM is **on**, the alarm should remain **on** until acknowledged by the operator. If the voltage decreases below HH_VOLT (or increases above LL_VOLT) during the ten minutes after alarm acknowledgment, the timing interval is completely canceled and the next time the voltage is at HH_VOLT or higher (or at LL_VOLT or lower), it should be treated as if it is the first time the voltage has increased to HH_VOLT or higher (or decreased to LL_VOLT or lower).

The ALM_ACK input is a push button switch. Do not assume it remains **on** for the 10-minute interval that starts when the ALM_ACK switch is pushed. Also, do not assume ALM_ACK is **off** after 10 minutes (it could be taped down). It has to turn **off** in order for it to turn **on** to acknowledge the next alarm.

Assume the following input, output, and internal tags/variables/symbols:

Tag/Var/Symbol	Description
ALM_ACK	Alarm acknowledge push button switch; **on** when pushed, **off** when released. Operation described above.
VT475_YL	Yellow (low) voltage indicator.
VT475_GN	Green (normal) voltage indicator.
VT475_RD	Red (high) voltage indicator.
VT475_ALM	Voltage alarm horn, **on** sounds horn.
VT475_MEAS	Raw VT475 voltage measurement, represents 400 to 550 volts; channel value range listed before P7-1.
VT475_VAL	Voltage, in volts (a REAL). Your ladder must calculate this value.
LL_VOLT	Value of voltage that triggers low-low alarm. Do not modify this value.
LO_VOLT	Value of voltage that triggers low indicator. Do not modify this value.
HI_VOLT	Value of voltage that triggers high indicator. Do not modify this value.
HH_VOLT	Value of voltage that triggers high-high alarm. Do not modify this value.

The addresses associated with the physical input and output tags/variables/symbols are:

Tag/Var./Symbol	ControlLogix	Micro800	MLogix	SLC
ALM_ACK	Local:1:I.Data.0	_IO_EM_DI_00	I:0/0	I:1/0
VT475_YL	Local:2:O.Data.0	_IO_EM_DO_00	O:0/0	O:2/0
VT475_GN	Local:2:O.Data.1	_IO_EM_DO_01	O:0/1	O:2/1
VT475_RD	Local:2:O.Data.2	_IO_EM_DO_02	O:0/2	O:2/2
VT475_ALM	Local:2:O.Data.3	_IO_EM_DO_03	O:0/3	O:2/3
VT475_MEAS	Local:3:I.Ch0Data	_IO_P2_AI_00	I:1.0	I:3.0

The addresses or data types associated with the internal tags/variables/symbols:

	CLogix	Micro800	MLogix/SLC
Tag/Var./Symbol	Data Type	Data Type	Address
VT475_VAL	REAL	REAL	F12:0
LL_VOLT	REAL	REAL	F25:0
LO_VOLT	REAL	REAL	F25:1
HI_VOLT	REAL	REAL	F25:2
HH_VOLT	REAL	REAL	F25:3

P7-26. Simple Pressure Control. Implement the following calculation and comparison operation for a pressure tank.

The PLC implements a simple pressure control for an air tank that supplies instrument air for the plant section (Figure P7.26). The PT105 pressure sensor measures the pressure. The PLC turns on an air compressor (M-105) and an air inlet valve (XV105) to increase the tank pressure when the pressure is **below** a set

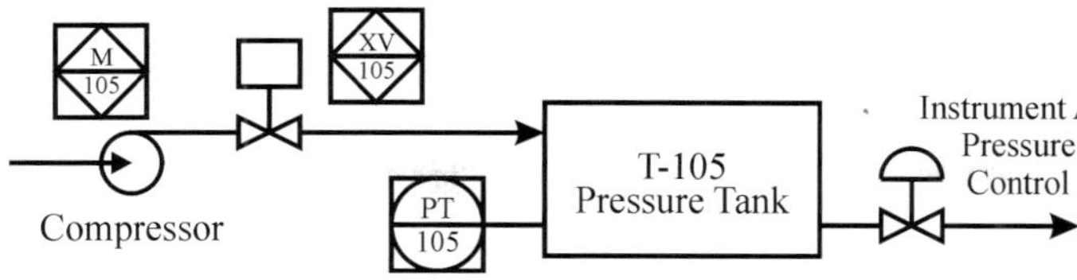

Figure P7.26. Pressure tank system.

minimum and turns off the valve when the pressure is **above** a set maximum. The minimum and maximum pressures are entered by the operator at an HMI. The current pressure, in psi, must be calculated and placed in a certain location for display on the HMI. Additionally, whenever the measured pressure is **greater than** 150% of the set maximum pressure (indicating that the compressor and XV105 valve is stuck on), a high alarm lamp must be turned **on**. Also, whenever the measured pressure is **less than** 75% of the set minimum pressure (indicating that XV105 is stuck on with the compressor off or there is a catastrophic leak), a low alarm lamp must be turned **on**.

Whenever either alarm lamp transitions **on**, an alarm horn must be sounded. However, the operator presses an alarm acknowledge button (ALM_ACK input) to silence the alarm. When silenced with ALM_ACK, the horn remains silent until the next transition of either lamp to **on**.

Also, in order to protect against illegal values entered by the operator, make sure PT105_HI_HMI is no higher than 300 psi and that PT105_LO_HMI is no lower than 25 psi. Also, PT105_HI_HMI should be at least 20 psi higher than PT105_LO_HMI.

The pressure measurement transmitter, PT105, is calibrated to measure from 0 to 300 psi. PT105 is connected to an analog input module having an ADC whose output value (PT105_MEAS) corresponds to the lowest and highest levels as specified in the general instructions prior to P7-1. The PT105_MEAS value should be converted into a pressure in psi (a real number) and placed in PT105_PSI.

Assume the following input, output, and internal tags/variables/symbols:

Tag/Var/Symbol	Description
ALM_ACK	Alarm horn acknowledge, **on** to acknowledge alarm.
M105_RUN	M-105 compressor control, **on** runs compressor.
XV105_SOL	XV105 valve control, **on** opens valve, **off** closes valve.
PT105_LO_LA	Low pressure indicator lamp; **on** when measured pressure < 75% of set minimum desired pressure.
PT105_HI_LA	High pressure indicator lamp; **on** when measured pressure > 150% of set maximum desired pressure.
PT105_HORN	Alarm horn; **on** when either alarm lamp is **on**.
PT105_MEAS	Raw PT105 pressure measurement, represents 0 to300 psi; channel value range listed before P7-1.
PT105_LO_HMI	Minimum pressure in psi, from HMI. Read-only, that is, ladder logic cannot change it.

PT105_HI_HMI	Maximum pressure in psi from HMI. Read-only, that is, ladder logic cannot change it.
PT105_PSI	Current tank pressure in psi (a REAL). To be displayed on HMI.

The addresses associated with the physical input and output tags/variables/symbols:

Tag/Var./Symbol	ControlLogix	Micro800	MLogix	SLC
ALM_ACK	Local:1:I.Data.0	_IO_EM_DI_00	I:0/0	I:1/0
M105_RUN	Local:2:O.Data.0	_IO_EM_DO_00	O:0/0	O:2/0
XV105_SOL	Local:2:O.Data.1	_IO_EM_DO_01	O:0/1	O:2/1
PT105_LO_LA	Local:2:O.Data.2	_IO_EM_DO_02	O:0/2	O:2/2
PT105_HI_LA	Local:2:O.Data.3	_IO_EM_DO_03	O:0/3	O:2/3
PT105_HORN	Local:2:O.Data.4	_IO_EM_DO_04	O:0/4	O:2/4
PT105_MEAS	Local:3:I.Ch0Data	_IO_P2_AI_00	I:1.0	I:3.0

The addresses or data types associated with the internal tags/variables/symbols:

Tag/Var./Symbol	CLogix Data Type	Micro800 Data Type	MLogix/SLC Address
PT105_LO_HMI	REAL	REAL	F12:0
PT105_HI_HMI	REAL	REAL	F12:1
PT105_PSI	REAL	REAL	F12:2

P7-27. Turbine Auxiliary Lube Oil Pump. Using the function chart approach, implement the program for the following auxiliary lube pump for a turbine. This particular problem is part of the controls for a turbine that generates low-pressure high-volume aeration air for wastewater treatment plants.

Figure P7.27 shows the layout of the turbine and the auxiliary lube oil pump. The auxiliary lube oil pump provides oil pressure to that the turbine bearings during turbine start-up and shut-down. When the turbine motor is running, it powers the main lube oil pump. You are implementing ladder logic for the auxiliary lube oil pump only, which is part of the control for the turbine.

In summary, the auxiliary pump is started and runs to force the oil pressure to a desired value, generates a start permissive to the turbine motor drive control, waits for the turbine to reach a minimum speed, and then idles. When the turbine is shut down or trips, the auxiliary pump is run while the turbine slows to a stop and then runs an additional two minutes before stopping. Upon startup, the following steps are executed:

Run auxiliary lube pump until the lube oil pressure reaches the desired value (DES_LPRESS)

Run auxiliary lube pump an additional 2 minutes

Generate permissive to turbine drive motor that turbine can start (LUBE_START_OK). Wait for turbine to reach its desired speed.

After the turbine reaches its desired speed, run auxiliary lube pump an additional 30 seconds

Auxiliary lube pump is now stopped and the process waits for a turbine normal shut down or an emergency trip (abnormal shut down)

Run auxiliary lube pump. Wait for turbine stopped (speed below 5 rpm)

After turbine is stopped, run auxiliary lube pump an additional 30 seconds.

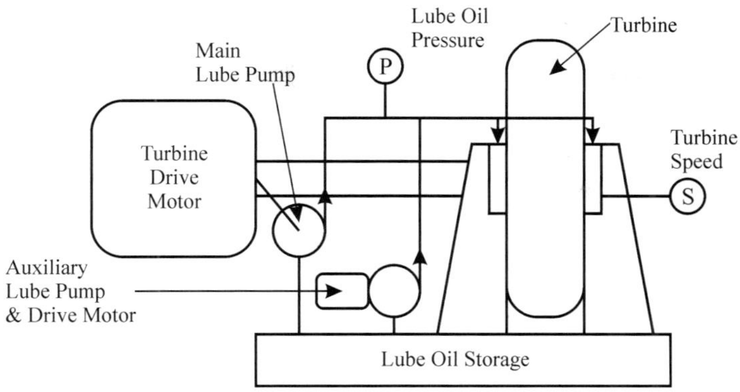

Figure P7.27. Turbine lube system.

Stop auxiliary lube pump (wait for lube oil pressure less than MIN_LPRESS). Also wait for LUBE_RUN to be **off**.

The operation does not repeat. The LUBE_RUN internal coil controls the overall process (more details below).

Auxiliary Lube Pump Motor Control: The auxiliary lube oil pump is driven by a motor, controlled by the AUX_LPUMP_RN output. When the AUX_LPUMP_RN output is energized (**on**), the motor runs and drives the auxiliary pump to increase the lube oil pressure. When the AUX_LPUMP_RN output is de-energized (**off**), the motor coasts to a stop and therefore the auxiliary pump stops. There is a check valve at the pump output so that the main auxiliary pump does not "backfeed" into this pump when the main lube oil pump, driven by the turbine drive motor is running and the auxiliary pump is stopped.

Turbine Drive Motor Permissive: The LUBE_START_OK internal coil (internal Boolean variable/tag) is controlled your part of the program. This permissive is turned **on** 2 minutes after the lube oil pressure has initially reached the desired value. The permissive must continue to be **on** while waiting for the turbine to reach its desired speed. After the turbine reaches its desired speed, LUBE_START_OK should be turned **off** and remain **off**.

Turbine Shutdown and Trip: While the turbine is normally operating, the auxiliary lube oil pump is stopped, waiting for either TURBINE_SHDOWN or TURBINE_TRIP internal variables to be turned **on**. The TURBINE_SHDOWN internal variable is **on** when the turbine is starting a normal shutdown. The TURBINE_TRIP internal variable is **on** when the turbine is initiating an abnormal shutdown.

The DES_LPRESS internal memory location indicates the desired lube oil pressure in pounds per square inch (psi). The MIN_LPRESS internal memory location indicates the minimum lube oil pressure in psi. Assume both values have been placed in the PLC memory from a supervisory PLC or computer. The lube oil pressure is measured by a pressure transducer calibrated to measure from 0 to 200 psi and is connected to an analog input module having an ADC whose output value (PT318_MEAS) corresponds to the lowest and highest pressures as specified in the general instructions prior to P7-1. The PT318_MEAS value should be converted into a pressure, in psi (a real number) and placed in the PT318_VAL variable.

Use the pressure measurement, converted to the proper units to determine when the initial lube oil pressure has been reached and when the pressure is low enough to indicate a stopped pump.

The DES_TSPEED internal memory location indicates the desired turbine speed in revolutions per minute (rpm). Assume this value has been placed in the PLC memory from a supervisory PLC or computer. The turbine speed is measured by a speed transducer calibrated to measure from 0 to 2000 rpm and is connected to an analog input module having an ADC whose output value (ST319_MEAS) corresponds to the lowest and highest rpms as specified in the general instructions prior to P7-1. The ST319_MEAS value should be converted into a speed, in rpm (a real number) and placed in the ST319_VAL variable.

Use the speed measurement, converted to the proper units to determine when the initial turbine speed has been reached and when the speed is low enough to indicate a stopped turbine.

There is no start/stop switch for the control of the auxiliary lube oil pump. Instead, there is a LUBE_RUN internal coil to control the overall operation of the auxiliary pump. When the LUBE_RUN internal coil is turned **on** for the first time only, the control assumes that the operation is starting. When the LUBE_RUN internal coil is turned **off**, the operation should pause at its current step, except when waiting for the turbine to shut down. If LUBE_RUN is turned **off**, while waiting for the turbine to shutdown, the operation must advance to the next step if the turbine is shutdown normally or abnormally. When the operation is paused, all timers must continue to run normally (not stopped/paused) and any physical outputs and internal coils controlled by this part of the program **must not change value**. If LUBE_RUN is turned **on** while the lube oil operation is paused, the operation should resume its suspended step.

There is no reset switch for the auxiliary pump operation.

Do not add any more timed steps to those explicitly stated in the problem. In other words, do not put a timer in a step unless it is stated that the step duration is a specific time. Assume the tolerance on all timer values is ±0.1 seconds.

Assume the following inputs, outputs and internal tags/variables/symbols.

Tag/Var/Symbol	Description
LUBE_RUN	When **on,** auxiliary lube oil pump operation runs. When **off**, pause (explained above).
TURBINE_SDOWN	When **on**, turbine is shutting down normally. Your part of the program does not control this variable.
TURBINE_TRIP	When **on**, turbine is shutting down abnormally (has tripped). Your part of the program does not control this variable.
PT318_MEAS	Raw lube oil pressure measurement, represents 0 to 200 psi.; channel value range listed before P7-1.
ST319_MEAS	Raw turbine speed measurement, represents 0 to 2000 rpm.; channel value range listed before P7-1.
PT318_VAL	Current lube oil pressure, in psi (a REAL). Your ladder must calculate this value.
DES_LPRESS	Desired lube oil pressure, in psi (a REAL). Your ladder must not alter this value.

MIN_LPRESS	Minimum lube oil pressure, in psi (a REAL). Your ladder must not alter this value.
ST319_VAL	Current turbine speed, in rpm (a REAL). Your ladder must calculate this value.
DES_TSPEED	Desired turbine speed, in rpm (a REAL). Your ladder must not alter this value.
AUX_LPUMP_RN	Auxiliary lube oil pump motor, **on** to run pump, **off** to stop pump.
LUBE_START_OK	Lube system has started permissive to main turbine control. **On** indicates that lube oil system is initially running. Your part of the ladder must control this variable.

The addresses associated with the physical input and output tags/variables/symbols are:

Tag/Var./Symbol	ControlLogix	Micro800	MLogix	SLC
PT318_MEAS	Local:3:I.Ch0Data	_IO_P2_AI_00	I:1.0	I:3.0
ST319_MEAS	Local:3:I.Ch1Data	_IO_P2_AI_01	I:1.1	I:3.1
AUX_LPUMP_RN	Local:2:O.Data.0	_IO_EM_DO_00	O:0/0	O:2/0

The addresses and/or data types associated with the internal tags/variables/symbols are:

Tag/Var./Symbol	CLogix Data Type	Micro800 Data Type	MLogix/SLC Address
LUBE_RUN	BOOL	BOOL	B33/20
LUBE_START_OK	BOOL	BOOL	B33/21
TURBINE_SDOWN	BOOL	BOOL	B34/20
TURBINE_TRIP	BOOL	BOOL	B34/21
PT318_VAL	REAL	REAL	F12:0
ST319_VAL	REAL	REAL	F12:1
DES_LPRESS	REAL	REAL	F25:0
MIN_LPRESS	REAL	REAL	F25:1
DES_TSPEED	REAL	REAL	F25:2

P7-28. Using the function chart approach, implement the program for the following station that fills a small amount of gasoline into a lawnmower engine for testing.

Figure P7.28 shows the top, side and end views of a station that fills a small amount of gasoline into a temporary tank on a gasoline engine. At the next station on the assembly line, the engine is started and runs for about a minute to verify proper operation. In summary, the engine is moved into the station, the filling hose is placed into position, the tank is filled, and then the engine is moved out to the next station. Your program is only controlling this station.

Another part of the ladder controls the conveyor. Assume it is always running. This assembly line is asynchronous, that is, each station processes castings at its own speed and does not coordinate its operation with any other station.

Upon initial startup, assume that there are no engines waiting at Engaging Hook 511. The initial states of the main mechanical controls are: engaging hooks in the up position, filling mechanism in the up position, and the filler valve closed. The following major steps are executed:

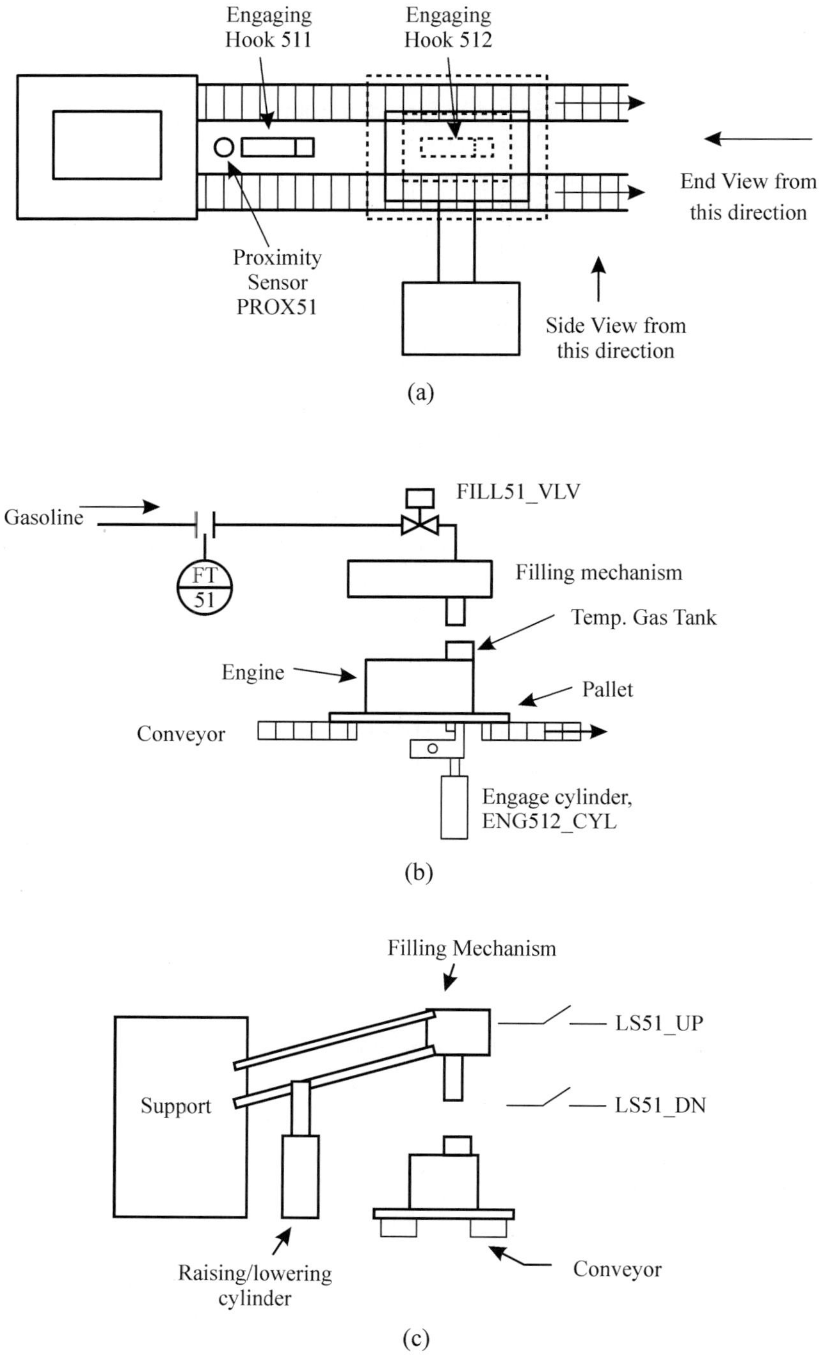

Figure P7.28. Gasoline-filling station: *(a)* top view; *(b)* side view; *(c)* end view.

Wait for an engine to be detected at Engaging Hook 511 (by PROX51),
Activate the ENG511_CYL for 4 seconds to allow only one engine to move into the station (when hook is raised, the next engine is caught by the hook),
Lower the filler assembly to the correct position (sensed by LS51_DN),
Open the filler valve (FILL51_VLV) to allow gasoline to flow into the temporary tank on the engine. The amount of gasoline is calculated by integrating the flow value from a flow sensor (equation given below). When the appropriate amount has been measured, close the filler valve.
Raise the filler assembly, until LS51_UP closes,
Activate the ENG512_CYL for 4 seconds to allow engine to move out.

The operation then repeats. Assume the conveyor is on at all times. The conveyor slides beneath the pallets as they are held by the engaging hooks.

Proximity Sensor: The proximity sensor, PROX51, is inductive and senses the metal pallet the engine is riding on. PROX51 senses the pallet before the pallet reaches the engage position. You must assume that when the pallet is captured by the ENG511_CYL, PROX51 remains **on**.

Engaging Hooks: ENG511_CYL and ENG512_CYL are single action solenoids. Once it is energized, the "hook" moves down and remains in the "down" position as long as power is applied (turned **on**). The "hook" moves up when power is removed (turned **off**). The engaging mechanism works this way to be fail-safe, that is, if power is removed because of a failure, no engines proceed down the conveyor.

Raising/Lowering Cylinder: The mechanism used to lower and raise the filler tube consists of a double-action linear pneumatic cylinder. When the FILL51_DOWN output is energized, the mechanism moves down and continues to move down as long as power is applied (turned **on**). When the FILL51_UP output is energized, the mechanism moves up and continues to move up as long as power is applied (turned **on**). The mechanism stops if neither output is **on**, or if they are energized simultaneously. LS51_UP is **on** when the mechanism is in the fully "up" position. LS51_DN is **on** when the mechanism is in the fully "down" position. There are mechanical stops at the fully up and fully down positions.

Fill Valve: The gasoline filling valve is actuated by a pneumatic single-action cylinder controlled by the FILL51_VLV output. When the FILL51_VLV output is **on**, the valve opens and when FILL51_VLV is **off**, the valve closes.

Gasoline Flow: The gasoline flow through the valve is measured by a flow sensor, FT51. The flow sensor is calibrated to measure from 0 to 2 ml/s and is connected to an analog input module having an ADC whose output integer (FT51_MEAS) corresponds to the lowest and highest values as listed before P7-1. The flow measurement value must be converted into units of ml/s and stored in the FT51_MLS internal memory location (a real number).

Gasoline Volume: In order to determine the volume of gasoline, the flow measurement is *integrated*. In order to do the integrator operation in the PLC, the flow rate is sampled at 0.1-second intervals and the volume is updated accordingly. The volume is updated with each new sample of the flow rate by the formula

$$\mathrm{GAS\_VOL}_i = \mathrm{GAS\_VOL}_{i-1} + \mathrm{T} \times \mathrm{FT\_MLS}_i \qquad (1)$$

In words, the volume is updated by adding to it the product of the current flow rate and the sample period.

What your code needs to do:

(1) Implement equation (1) every 0.1 second (T = 0.1). Just update the current value of GAS_VOL. Do not save the previous value of GAS_VOL.

(2) Sometime **before** opening the FILL51_VLV valve, reset the accumulator, that is, set GAS_VOL to zero.

The DES_GAS_VOL internal memory location indicates the desired amount of gasoline to place in the temporary tank, in milliliters. Assume this value has been placed in the PLC memory from a supervisory PLC or computer.

Alarm: There is an alarm on the step that fills the tank. If the time to do this step is longer than 10 seconds, then FILL51_ALM should be turned **on** and stay **on** for 3 seconds. If the filling step finishes during the 3 seconds FILL51_ALM is **on**, FILL51_ALM should remain **on** until the 3-second interval expires. If the filling step is longer than 13 sec., FILL51_ALM should remain **off** after the 3-second pulse of the alarm is finished. FILL51_ALM is read by the HMI and the HMI generates any indications/horns and handles acknowledgement. The alarm does not alter the operation of the station.

Start/Stop/Pause: The start/stop switches are only for the station. They do not control any other stations, or the conveyor. Upon initial startup, assume there are no engines present in either of the engaging hooks. If the stop switch is pressed at any time, the station operation should **pause**, except when the ENG511_CYL or ENG512_CYL engaging hook controls are activated. The operation **must not** pause when either ENG511_CYL or ENG512_CYL is activated. If the stop switch is pressed when any engaging hook control is activated, the step should complete and the operation should advance to the next step. When the start switch is pressed while the operation is paused, the station should resume the suspended step. When paused, **do not advance** to the next step (exception noted earlier).

When the station is paused, all active outputs (except for the engaging hooks) should be **off**.

Reset: A separate reset switch, RESET_PB, is provided which raises the filling mechanism and resets any internal states. The reset switch should be ignored if the station is running or if paused in one of the steps that is controlling an engaging hook. The start switch should be ignored as long as RESET_PB is pressed or if the filling mechanism is not fully up after the reset operation is initiated. If the station is paused and RESET_PB is **on**, the operator must release the RESET_PB before the start pushbutton switch can be used to restart the station.

Assume the tolerance on all timer values is at most 0.01 second. Do not add any more timed steps than those explicitly stated in the problem.

Assume the following input, output, and internal memory tags/variables/symbols:

Tag/Var/Symbol	Description
START_PB	Start push button, N. O., **on** when starting.
STOP_PB	Stop push button, N. C., **off** when stopping.
RESET_PB	Reset push button, N. O., **on** when restoring station to initial state.
PROX51	Proximity sensor, **on** when pallet at engaging hook 511(position 1).

LS51_DN	Limit switch that closes (**on**) when filler mechanism is in the down position.
LS51_UP	Limit switch that closes (**on**) when filler mechanism is in the up position.
FT51_MEAS	Measurement of gasoline flow rate, represents 0 to 2 ml/sec; range listed above
ENG511_CYL	Engage hook 1 control, **on** to lower hook, **off** raises hook.
ENG512_CYL	Engage hook 2 control, **on** to lower hook, **off** raises hook.
FILL51_DOWN	Filler mechanism lower control, **on** to lower. When this output and FILL51_UP **off**, the cylinder does not move.
FILL51_UP	Filler mechanism raise control, **on** to raise. When this output and FILL51_DOWN **off**, the cylinder does not move.
FILL51_VLV	Opens gasoline filler valve to allow gasoline to flow to engine temporary tank, **on** to open valve, **off** to close..
FT51_MLS	Current gasoline flow rate, in ml/sec. **Your ladder must calculate this value**.
GAS_VOL	Volume of gasoline flowing through valve, in milliliters. Çalculate bu integrating FT51_MLS. **Your ladder must calculate this value**.
DES_GAS_VOL	Desired gasoline volume, in ml (a REAL). **Your ladder does not calculate this value**.
FILL51_ALM	Fill problem alarm, turned **on** when the filling step duration is longer than 10 seconds. This alarm is **on** for 3 ± 0.1 seconds.

The addresses associated with the physical input and output tags/variables/symbols are:

Tag/Var./Symbol	ControlLogix	Micro800	MLogix	SLC
START_PB	Local:1:I.Data.0	_IO_EM_DI_00	I:0/0	I:1/0
STOP_PB	Local:1:I.Data.1	_IO_EM_DI_01	I:0/1	I:1/1
RESET_PB	Local:1:I.Data.2	_IO_EM_DI_02	I:0/2	I:1/2
PROX51	Local:1:I.Data.3	_IO_EM_DI_03	I:0/3	I:1/3
LS51_DN	Local:1:I.Data.4	_IO_EM_DI_04	I:0/4	I:1/4
LS51_UP	Local:1:I.Data.5	_IO_EM_DI_05	I:0/5	I:1/5
FT51_MEAS	Local:3:I.Ch0Data	_IO_P2_AI_00	I:1.0	I:3.0
ENG511_CYL	Local:2:O.Data.0	_IO_EM_DO_00	O:0/0	O:2/0
ENG512_CYL	Local:2:O.Data.1	_IO_EM_DO_01	O:0/1	O:2/1
FILL51_DOWN	Local:2:O.Data.2	_IO_EM_DO_02	O:0/2	O:2/2
FILL51_UP	Local:2:O.Data.3	_IO_EM_DO_03	O:0/3	O:2/3
FILL51_VLV	Local:2:O.Data.4	_IO_EM_DO_04	O:0/4	O:2/4

The addresses and/or data types associated with the internal tags/variables/symbols are:

Tag/Var./Symbol	CLogix Data Type	Micro800 Data Type	MLogix/SLC Address
FILL51_ALM	BOOL	BOOL	B3/134

FT51_MLS	REAL	REAL	F12:0
GAS_VOL	REAL	REAL	F12:1
DES_GAS_VOL	REAL	REAL	F25:0

P7-29. Using the function chart approach, implement the program for the following batch process that mixes a brine solution.

Figure P7.29 shows the layout of a batch process that makes up a brine solution used to regenerate an ion exchanger. In summary, water is pumped into the tank, dry salt is added, more water is added until the conductivity is within a certain band, and the contents of the tank are pumped out to a storage tank. The major steps are:

Operator presses START_PB to start the operation,

Initially charge (fill) the tank with water to a level of 4.0 feet by opening the water inlet valve (controlled by XV4501),

Run airlock motor and bin vibrator to convey the salt into the tank. Run these until the measured conductivity rises to at least 10 mS/cm,

Stop the airlock motor and bin vibrator and then wait an additional 5 seconds for salt to fall down the tube,

Add water until the solution conductivity has decreased to 8.5 mS/cm,

Empty tank to the brine storage by running the discharge pump (controlled by P4514_RUN) and opening the normal outlet valve (controlled by XV4502). Assume the tank is empty when the level drops below 0.08 feet (approximately 1.0 inch).

The operation does not repeat. After one brine solution batch is produced, the operator must press the START_PB in order to make up the next batch of brine solution. After the tank is emptied, no steps should be active.

An airlock removes material from the bottom of the salt bin. An airlock allows material to be removed from the bottom of the bin, as it is added at the top of the bin. (Addition of material to the bin is not part of this problem.) An airlock consists of a vane rotating inside a horizontal cylinder (Figure P7.29*b*). Material enters from the top into a wedge-shaped chamber, is conveyed to the bottom as the vanes rotate, and exits the bottom. The airlock chambers are similar to the wedge-shaped chambers of a Trivial Pursuit game piece that holds the "wedgies". The airlock is driven by a motor that is controlled by the L4513_RUN output. When L4513_RUN is **on**, the airlock rotates, removing material from the bottom of the bin and placing it in the chute. When L4513_RUN is **off**, no material is removed from the bottom of the bin.

A vibrator is attached to the salt bin to keep material from clumping to the sides as it is discharged. The vibrator is driven by a motor that is controlled by the V4512_RUN output. When V4512_RUN is **on**, the vibrator motor rotates. When V4512_RUN is **off**, the vibrator motor is stopped. The vibrator must be run while the airlock motor is running.

There are three valves in this system:

1. Water inlet valve. When XV4501 is **on**, the valve is opened to admit water into the tank. When XV4501 is **off**, the valve is closed.
2. Tank outlet valve to storage. When XV4502 is **on**, the valve is opened to allow brine to be pumped into storage. When XV4502 is **off**, the valve is closed.

3. Tank outlet valve to treatment. When XV4503 is **on**, the valve is opened to allow brine to be pumped to the treatment system. When XV4503 is **off**, the valve is closed.

The P4514_RUN output controls the discharge pump. When P4514_RUN is **on**, the pump runs; when **off**, the pump is stopped. Either XV4502 or XV4503 must also be **on** (open valve) when running the pump.

The tank agitator is controlled by the A4511_RUN output. When A4511_RUN is **on**, the tank agitator is run; when A4511_RUN is **off**, the agitator is stopped. Except for the

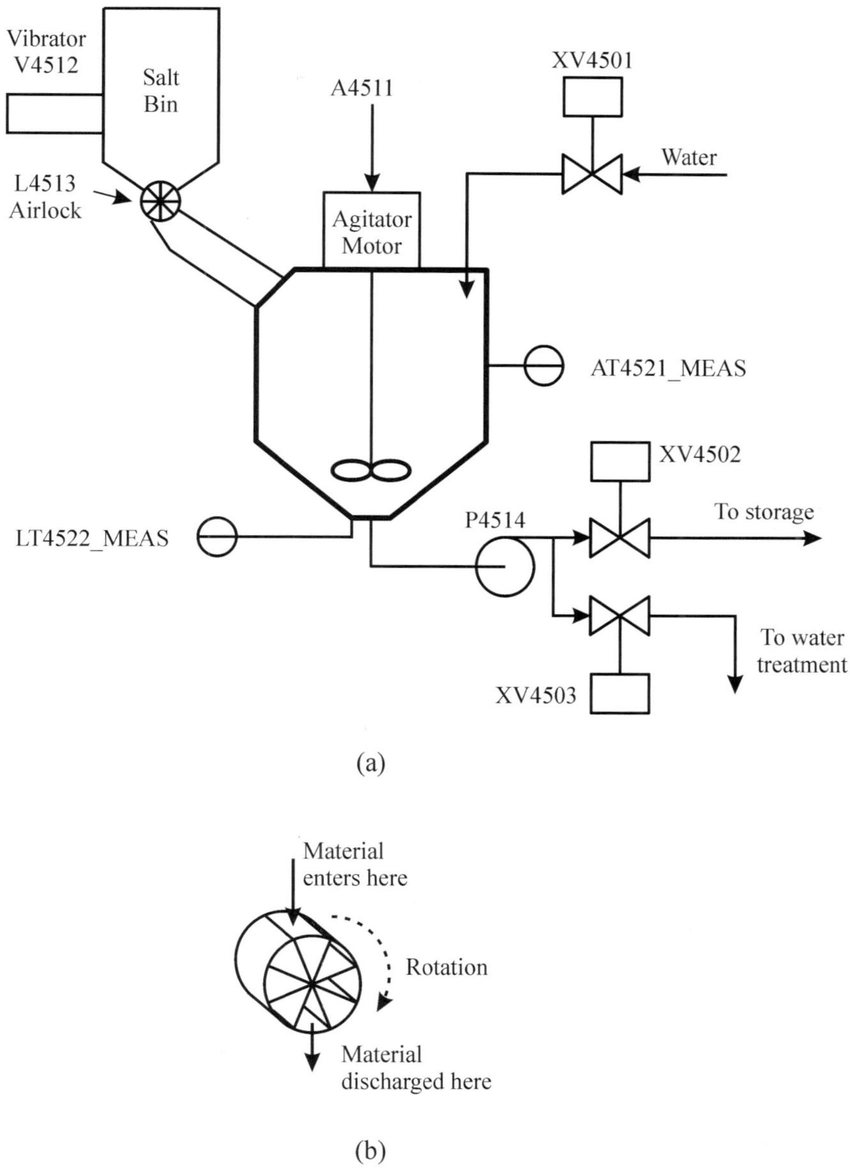

Figure P7.29. Salt brine mixing process: *(a)* equipment; *(b)* airlock valve detail.

initial charge of water, the agitator must be running during all other steps. The agitator must be stopped during the initial charge of water.

The tank level is measured by a level sensor attached to the bottom of the tank. The level sensor is calibrated to measure from 0 to 11 feet and is connected to an analog input module having an ADC whose output integer (LT4522_MEAS) corresponds to the lowest and highest levels as specified before P7-1.

The level measurement value must be converted into units of feet and stored in the LT4522 internal memory location (a real number).

The solution conductivity is measured by a sensor attached to the side of the tank. The sensor is calibrated to measure from 2 to 15 mS/cm (milliSiemens/cm) and is connected to an analog input module having an ADC whose output integer (AT4521_MEAS) corresponds to the lowest and highest conductivity as specified before P7-1. The conductivity measurement value must be converted into units of mS/cm and stored in the AT4521 internal memory location (a real number).

If the stop switch is pressed at any time, the operation should pause. When the start switch is pressed while the operation is paused, the station should resume the suspended step. When the station is paused, all active outputs, **except** for the A4511_RUN output should be **off**. When the operation is paused, timers for a timed step should continue to run and A4511_RUN should not be turned **off** if active. Also, when paused **do not advance** to the next step.

A separate reset switch, RESET_PB, is provided which resets any internal states and empties the tank contents to the water treatment facility. To drain the tank, the P4514_RUN and XV4503 outputs should be **on** until the tank level is below 0.08 feet. Also, the reset switch should turn off the alarm. The reset switch should be ignored if the station is running. The start switch should be ignored when the reset operation is in progress. Also, the start switch should be ignored as long as RESET_PB is pressed.

There is a tank level alarm. If the tank level exceeds 10.0 feet, then LVL_ALARM lamp must be flashed and continue to flash until the RESET_PB is pressed after paused. If the tank level exceeds 10.0 feet (which can happen while adding water to obtain the final conductivity), then there is a problem with the system and it needs to be repaired. The period of the lamp should be 0.60 second **on** and 0.40 second **off**. When this alarm is flashing, then the operator is expected to pause (stop) the operation, reset the station, and fix the problem before restarting the system. Your program is not responsible for stopping the process when LVL_ALARM is flashing. When RESET_PB is pressed with the operation already paused, it should stop flashing LVL_ALARM, even if the tank level is above 10.0 feet.

Assume the tolerance on all timer values is at most 0.01 second.

Assume the following input, output, and internal memory tags/variables/symbols:

Tag/Var/Symbol	Description
START_PB	Start push button, N. O., **on** when starting.
STOP_PB	Stop push button, N. C., **off** when stopping.
RESET_PB	Reset push button, N. O., **on** when restoring station to initial state.
LT4522_MEAS	Measurement of mix tank level, represents 0 to 11.0 feet; integer range listed above
AT4521_MEAS	Measurement of mix tank contents conductivity, represents 2 to 15 mS/cm; integer range listed above

XV4501	Water inlet valve control, **on** to open valve, **off** closes valve.
XV4502	Tank outlet valve to storage control, **on** to open valve, **off** closes valve.
XV4503	Tank outlet valve to drain control, **on** to open valve, **off** closes valve.
A4511_RUN	Agitator motor control; **on** to agitate mix tank contents.
V4512_RUN	Salt bin vibrator motor control; **on** to run vibrator.
L4513_RUN	Salt bin airlock valve control; **on** to run airlock, discharging dry salt to the conveyor.
P4514_RUN	Tank discharge pump control, **on** to run pump.
LVL_ALARM	Level alarm, turned **on** when the tank level is greater than 10.0 feet. Reset switch turns it **off**. **Your ladder must control this**.
AT4521	Current tank contents conductivity. **Your ladder must calculate this value**.
LT4522	Current tank level, in feet. **Your ladder must calculate this value**.

The addresses associated with the physical input and output tags/variables/symbols are:

Tag/Var./Symbol	ControlLogix	Micro800	MLogix	SLC
START_PB	Local:1:I.Data.0	_IO_EM_DI_00	I:0/0	I:1/0
STOP_PB	Local:1:I.Data.1	_IO_EM_DI_01	I:0/1	I:1/1
RESET_PB	Local:1:I.Data.2	_IO_EM_DI_02	I:0/2	I:1/2
AT4521_MEAS	Local:3:I.Ch0Data	_IO_P2_AI_00	I:1.0	I:3.0
LT4522_MEAS	Local:3:I.Ch1Data	_IO_P2_AI_01	I:1.1	I:3.1
XV4501	Local:2:O.Data.0	_IO_EM_DO_00	O:0/0	O:2/0
XV4502	Local:2:O.Data.1	_IO_EM_DO_01	O:0/1	O:2/1
XV4503	Local:2:O.Data.2	_IO_EM_DO_02	O:0/2	O:2/2
A4511_RUN	Local:2:O.Data.3	_IO_EM_DO_03	O:0/3	O:2/3
V4512_RUN	Local:2:O.Data.4	_IO_EM_DO_04	O:0/4	O:2/4
L4513_RUN	Local:2:O.Data.5	_IO_EM_DO_05	O:0/5	O:2/5
P4514_RUN	Local:2:O.Data.6	_IO_EM_DO_06	O:0/6	O:2/6
LVL_ALARM	Local:2:O.Data.7	_IO_EM_DO_07	O:0/7	O:2/7

The addresses and/or data types associated with the internal tags/variables/symbols are:

Tag/Var./Symbol	CLogix Data Type	Micro800 Data Type	MLogix/SLC Address
AT4521	REAL	REAL	F12:0
LT4522	REAL	REAL	F12:1

P7-30. Using the function chart approach, implement the program for the following oven that cures hams.

Figure P7.30 shows the layout of a station that cures hams already hanging on a rack. The station moves the rack into the oven, heats up the oven, holds the temperature, cools it down, and then moves the rack out of the oven. The operator is responsible for rolling the rack into place in front of the oven and then removing the rack once the curing process is complete.

Upon initial startup, assume that the rack is in position in front of the oven and the door is open. The overall operation is first described, followed by the details. When a rack is placed into position by the operator, and the START_PB is pressed, the following steps should be executed:

Move the rack of hams into the oven (RACK_ILS senses).

Close the oven door (DOOR_DLS senses).

The gas burner should be turned **on** to heat up the oven. Wait for the oven temperature to be at least the low control temperature (DES_OVEN_DEG)

Cure (cook) the hams for 9 hours. The burner should be turned **on** and **off** as needed to maintain the desired temperature. Details below.

Cool down the hams by opening the cooling water valve (XV304). Cool the hams until 10 minutes has elapsed **and** the oven temperature is less than 70 degrees.

Open the oven door (DOOR_ULS senses).

Move the rack of hams out of the oven (RACK_OLS senses)

The process is stopped. (It **DOES NOT** repeat.).

Rack Conveyor: The rack of hams is moved into the oven and out of the oven with a chain conveyor driven by a motor. The motor has two controls: RACK_DRV and RACK_IN. When RACK_DRV and RACK_IN are both **on**, the chain conveyor moves the rack into the oven. When RACK_DRV is **on** and RACK_IN is **off**, the rack is moved out of the oven. If RACK_DRV is **off**, the conveyor is stopped. The RACK_ILS limit switch is **on** when the rack is in the proper position within the oven. The RACK_OLS limit switch is **on** when the rack is outside the oven.

Oven Door: The oven door is moved with a rack-and-pinion mechanism driven by a motor. The motor has two controls: DOOR_DRV and DOOR_DN. When DOOR_DRV and DOOR_DN are both **on**, the door is closed. When DOOR_DRV is **on** and DOOR_DN is **off**, the door is opened. If DOOR_DRV is **off**, the door is stopped. The DOOR_DLS limit switch is **on** when the door is closed. The DOOR_ULS limit switch is **on** when the door is open.

Cooling Water Valve: The valve that allows cooling water into a series of spray nozzles above the rack is controlled by the XV304 physical output. When XV304 is energized, the valve opens, directing cooling water into the oven. When XV304 is de-energized, the valve closes. The valve must be open only during the cool-down step. It must not be open for any other part of the operation.

Oven Temperature: The oven temperature is measured by a resistance temperature device (RTD) calibrated to measure from 32 to 300 degrees Fahrenheit and is connected to an analog input module having an ADC whose output integer (TT305_MEAS) corresponds to the lowest and highest temperatures as specified before P7-1. The temperature measurement value must be converted into units of degrees F and stored in the TT305_DEG internal memory location (a real number).

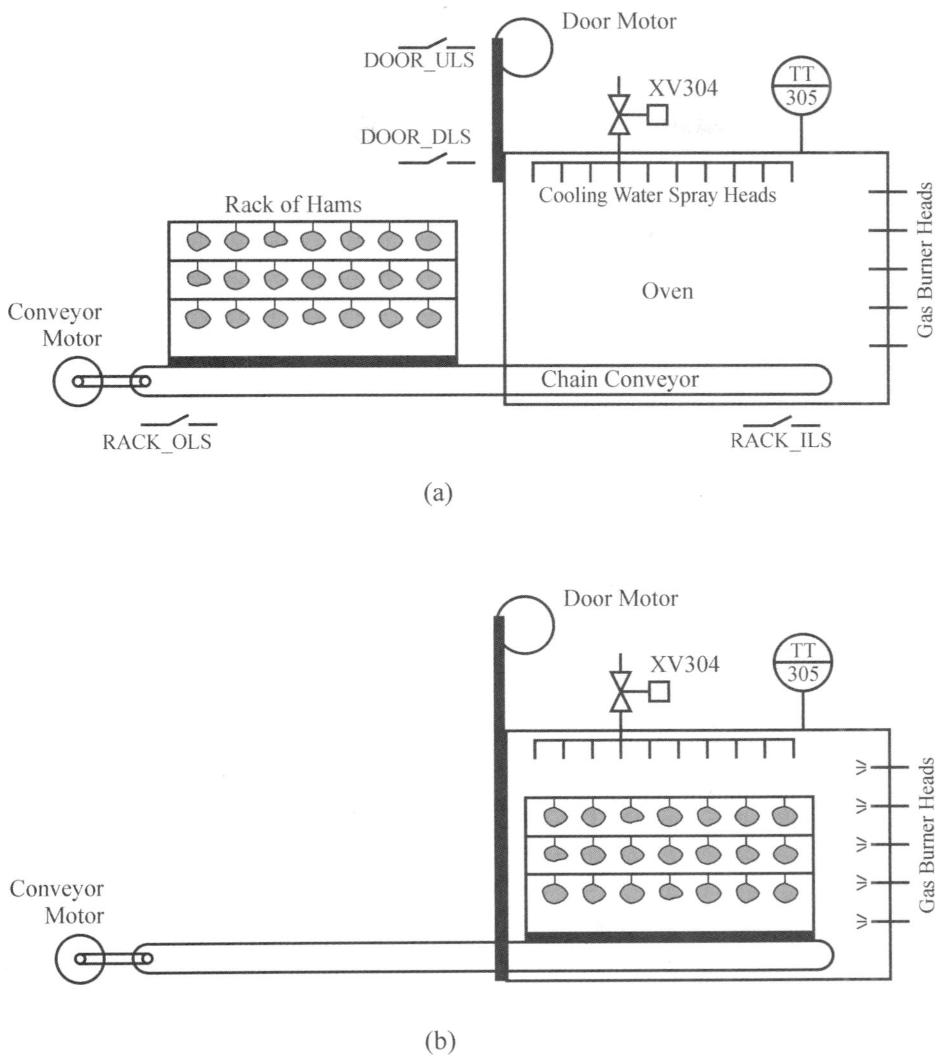

Figure P7.30. Ham curing oven: *(a)* side view; *(b)* rack in oven.

The temperature measurement increases when the gas burner is **on** and decreases when the gas burner is **off** (until it reaches room temperature).

Oven Temperature Control: The temperature control is a simple (non-PID) control. In those parts of the operation where the oven needs to be heated up or maintained at a temperature above room temperature, the oven temperature should be controlled as follows. The BURNER output turns on a gas heater to heat the oven. The BURNER output should be turned **on** when the oven temperature is below DES_OVEN_DEG and should be turned **off** when the oven temperature is above DES_OVEN_DEG + 10. The burner **must be off** during any part of the operation that does not heat up the oven or maintain the oven temperature.

The DES_OVEN_DEG internal memory location indicates the desired oven temperature. Assume this value has been placed in the PLC memory by a supervisory PLC or computer. Your part of the ladder must not modify this value.

Oven Temperature Alarm: There is one alarm indication, OVEN_TEMP_ALM that flashes when the oven temperature is less than DES_OVEN_DEG – 15 or the oven temperature is greater than DES_OVEN_DEG + 15. This alarm is enabled ONLY when curing (cooking) the hams. This indication flashes **on** for 1 second and **off** for 0.5 sec. This alarm must be **off** when in any other step of the operation.

Start/Stop/Pause: When the start switch is pressed (turned **on**) for the first time only, the station assumes there is a rack of hams in position ready to be moved into the oven. When the stop switch is pressed (turned **off**) the operation should pause. **Do not advance** to the next step when paused. When paused, the motor and valve controls must be turned **off**. The oven burner control must not be affected by pause. Also, during a pause, the timer accumulators must continue to increment. Pressing the start switch while the operation of the station is paused causes the station to resume its suspended step.

Reset: A separate reset switch, RESET_PB, is provided which restores the system to its initial state. The reset operation should cause the oven door to open (if closed) and the rack moved out of the oven. Note that the rack must not be moved out of the oven until the door is open. The reset operation is not complete until the oven door is open and the rack is moved outside of the oven. The reset switch should be ignored if the station is running. The start switch should be ignored when the reset operation is in progress. If the station is paused and RESET_PB is **on**, the operator must release the RESET_PB before the start pushbutton switch can be used to restart the station.

Note that the logic for handling the control of the two motors while resetting the station should ensure that the motors do not run beyond a limit switch. For example, once the door is up, the door drive controls must be **off**.

Do not add any more timed steps to those explicitly stated in the problem. In other words, do not put a timer in a step unless it is stated that the step duration is a specific time. Assume the tolerance on all timer values is at most 1 second.

Assume the following physical inputs, outputs, and internal memory tags/variables/symbols.

Tag/Var/Symbol	Description
START_PB	Start push button, N. O., **on** when starting.
STOP_PB	Stop push button, N. C., **off** when stopping.
RESET_PB	Reset push button, N. O., **on** when restoring station to initial state.
RACK_ILS	Limit switch that closes (**on**) when the rack of hams is in the oven.
RACK_OLS	Limit switch that closes (**on**) when the rack is out of the oven.
DOOR_DLS	Limit switch that closes (**on**) when the oven door is down.
DOOR_ULS	Limit switch that closes (**on**) when the oven door is up.
TT305_MEAS	Measurement of oven temperature, represents 32 to 300 degrees F; range listed above.

RACK_DRV	Chain conveyor drive motor run control, **on** to run chain conveyor, **off** stops conveyor. Note the RACK_IN control determines the direction of the conveyor.
RACK_IN	Chain conveyor direction control, **on** to move rack into oven, **off** to move rack out of oven. Note: this output has no effect if RACK_DRV is **off**. However, this control should be **off** when RACK_DRV is **off**.
DOOR_DRV	Oven door drive motor run control, **on** to change door position, **off** stops movement. Note the DOOR_DN control determines the direction of the door movement.
DOOR_DN	Oven door direction control, **on** to move door down, **off** to move door up. Note: this output has no effect if DOOR_DRV is **off**. However, this control should be **off** when DOOR_DRV is **off**.
XV304	Cooling water valve control, **on** to open valve (to direct cooling water into oven), **off** to close valve.
BURNER	Oven gas burner control, **on** to burn gas to heat up the oven, **off** to extinguish the gas flame.
OVEN_TEMP_ALM	Alarm indicator that flashes 1 second **on** and ½ second **off** when the oven temperature is less than DES_OVEN_DEG – 15 or the oven temperature is greater than DES_OVEN_DEG + 15. **Your ladder must control this Boolean**.
TT305_DEG	Current oven temperature, in degrees F. **Your ladder must calculate this value**.
DES_OVEN_DEG	Desired oven temperature. **Your ladder must not change this value.**

The addresses associated with the physical input and output tags/variables/symbols are:

Tag/Var./Symbol	ControlLogix	Micro800	MLogix	SLC
START_PB	Local:1:I.Data.0	_IO_EM_DI_00	I:0/0	I:1/0
STOP_PB	Local:1:I.Data.1	_IO_EM_DI_01	I:0/1	I:1/1
RESET_PB	Local:1:I.Data.2	_IO_EM_DI_02	I:0/2	I:1/2
RACK_ILS	Local:1:I.Data.3	_IO_EM_DI_03	I:0/3	I:1/3
RACK_OLS	Local:1:I.Data.4	_IO_EM_DI_04	I:0/4	I:1/4
DOOR_DLS	Local:1:I.Data.5	_IO_EM_DI_05	I:0/5	I:1/5
DOOR_ULS	Local:1:I.Data.6	_IO_EM_DI_06	I:0/6	I:1/6
TT305_MEAS	Local:3:I.Ch0Data	_IO_P2_AI_00	I:1.0	I:3.0
RACK_DRV	Local:2:O.Data.0	_IO_EM_DO_00	O:0/0	O:2/0
RACK_IN	Local:2:O.Data.1	_IO_EM_DO_01	O:0/1	O:2/1
DOOR_DRV	Local:2:O.Data.2	_IO_EM_DO_02	O:0/2	O:2/2
DOOR_DN	Local:2:O.Data.3	_IO_EM_DO_03	O:0/3	O:2/3
XV304	Local:2:O.Data.4	_IO_EM_DO_04	O:0/4	O:2/4
BURNER	Local:2:O.Data.5	_IO_EM_DO_05	O:0/5	O:2/5
OVEN_TEMP_ALM	Local:2:O.Data.6	_IO_EM_DO_06	O:0/6	O:2/6

The addresses and/or data types associated with the internal tags/variables/symbols are:

Tag/Var./Symbol	CLogix Data Type	Micro800 Data Type	MLogix/SLC Address
TT305_DEG	REAL	REAL	F12:0
DES_OVEN_DEG	REAL	REAL	F25:0

P7-31. Using the function chart approach, implement the program for the following station that tests automobile seats.

Figure P7.31shows the layout of a station that tests foam automobile seats. The station pushes on the seat with a pneumatic ram and measures the force exerted on the seat. If the force is outside a certain band, the seat is marked as defective. This station is one of a series of stations along the conveyor. The roller conveyor is controlled by another part of the ladder, so assume it is always running.

Upon initial startup, assume that the test ram is up, the lift cylinder is down, and there is no seat in the station. The overall operation is first described, followed by the details. The sequence of operation for the station follows these steps:

A seat is sensed by PROX.

Activate LIFT_EXT control for the lift cylinder to raise the seat off of the roller conveyor (LIFT_ULS senses).

Extend the test ram to move the plate down to contact and press into the seat cushion. Since this station must work with various seat sizes, the ram is extended until DIST_MM is at least DES_DIST.

Delay 5 seconds. Four of these seconds allows the force measurement to settle down and the other second is required if the seat is marked as defective. Do not separate the delay into two steps.

Retract the test ram (RAM_ULS senses).

Lower the seat down onto the conveyor (LIFT_DLS senses).

The operation then repeats.

The PROX **proximity sensor** is a retro-reflective photoelectric sensor that senses the presence/absence of a seat in the position where it can be tested. PROX is **on** when a seat is present and **off** when no seat is present. PROX remains **on** as the seat is lifted off the roller conveyor.

Lift Cylinder: The cylinder that lifts the seat off the conveyor is a single-action pneumatic cylinder, controlled by the LIFT_EXT output. When LIFT_EXT is energized, the lift cylinder extends, moving the seat off the conveyor. When LIFT_EXT is de-energized, the cylinder retracts, placing the seat onto the roller conveyor. The LIFT_ULS limit switch is **on** when the lift is in the "up" position (cylinder extended). The LIFT_DLS limit switch is **on** when the lift is in the "down" position.

Test ram cylinder: A double-acting pneumatic cylinder that controls the raising/lowering of the test plate. This cylinder is controlled by two outputs: RAM_EXT and RAM_RET. When the RAM_EXT output is **on**, the test plate is lowered and when the RAM_RET output is **on**, the test plate is raised. The cylinder stops at its current position when both outputs are de-energized (turned **off**). The cylinder will not move if both extend

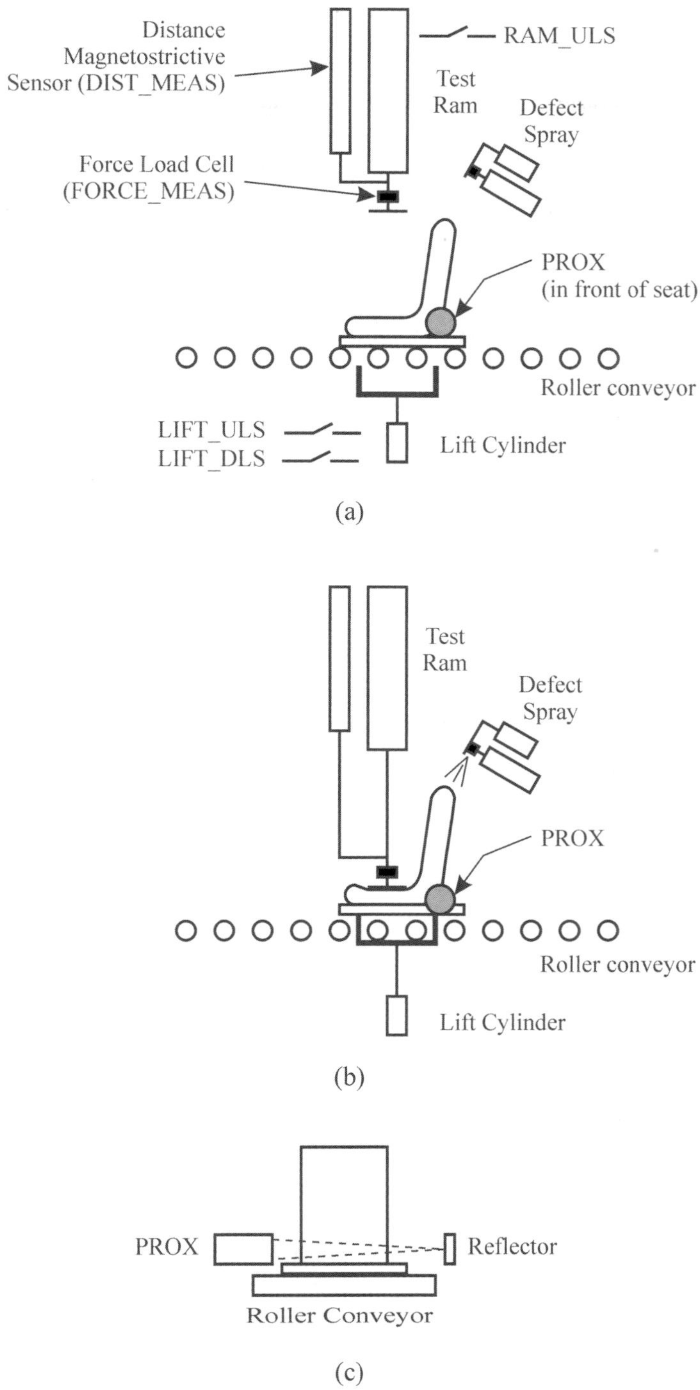

Figure 7.31. Seat testing station: *(a)* side view; *(b)* test ram extended, *(c)* end view, showing PROX location.

and retract controls are energized simultaneously. The RAM_ULS limit switch is **on** when the cylinder is retracted (plate in the "up" position).

Note that the test ram MUST NOT EXTEND past the point at which DIST_MM is at least DES_DIST.

Reject sprayer: The reject sprayer is basically a spray paint can with a cylinder that retracts, pressing on the spray head, sending a burst of paint that marks the seat as defective. When SPRAY_RET is energized, paint is sprayed onto the seat. When SPRAY_RET is de-energized, the paint is not sprayed.

Force test: The force exerted by the seat cushion on the ram is checked 4 seconds after the test ram reaches its position. This delay allows the force measurement to "settle down." If the force exerted by the seat cushion (FORCE_NT) is **NOT** within the boundaries set by FORCE_MIN and FORCE_MAX,

FORCE_MIN <= FORCE_NT <= FORCE_MAX

then SPRAY_RET is activated for one second to mark the seat cushion as defective. The FORCE_MIN and FORCE_MAX internal memory locations have been placed in the PLC memory by a supervisory PLC or computer. Assume that FORCE_MIN < FORCE_MAX.

The test is **only** done at the 4 second mark. If the force happens to fall outside the desired range any time between 4 – 5 seconds of the delay step being active, **do not** activate the sprayer.

Testing Ram Distance: The DES_DIST internal memory location indicates the desired distance the testing ram should be moved down (in mm). Assume this value has been placed in the PLC memory by a supervisory PLC or computer. The testing ram distance is measured by a magnetorestrictive position sensor calibrated to measure from 50 to 850 mm and is connected to an analog input module having an ADC whose output integer (DIST_MEAS) corresponds to the lowest and highest distances as as specified before P7-1. The distance measurement value must be converted into units of mm and stored in the DIST_MM internal memory location (a real number).

The distance measurement increases as the testing ram is extended.

Seat Force: The force exerted by the seat is measured by a tension link load cell on the testing ram shaft calibrated to measure from 0 to 445 newtons and is connected to an analog input module having an ADC whose output integer (FORCE_MEAS) corresponds to the lowest and highest forces as as specified before P7-1. The force measurement value must be converted into units of newtons and stored in the FORCE_NT internal memory location (a real number).

Start/Stop/Pause: The start and stop switches are for this station only. When the start switch is pressed (turned **on**) for the first time only, the station assumes there is are no seats in the station. When the stop switch is pressed (turned **off**) the operation should pause. **Do not advance** to the next step when paused. When paused, the ram cylinder controls must be turned **off**. The lift and spray cylinder controls must not be affected by pause. Also, during a pause, the timer accumulators must continue to increment. Pressing the start switch while the operation of the station is paused causes the station to resume its suspended step.

Note that a pause during the time that a seat is moving into or out of the station effectively disables the test for following seats. The succeeding seats will move through the station without being tested. This is the desired effect of a pause during this part of the operation.

Reset: A separate reset switch, RESET_PB, is provided which restores the system to its initial state. The reset operation should cause the testing cylinder to retract (if down) and the lift cylinder to retract. The reset operation is not complete until the testing cylinder is retracted and the lift cylinder is retracted. **Do not implement a function chart for the reset** operation. Mechanically, there is no interference between these two operations. Assume that an operator is responsible for actually removing the seat from the station after a reset. The reset switch should be ignored if the station is running. The start switch should be ignored when the reset operation is in progress. If the station is paused and RESET_PB is **on**, the operator must release the RESET_PB before the start pushbutton switch can be used to restart the station.

Do not add any more timed steps to those explicitly stated in the problem. In other words, do not put a timer in a step unless it is stated that the step duration is a specific time. Assume the tolerance on all timer values is ±0.01 seconds.

Assume the following physical input and output tags/variables/symbols.

Tag/Var/Symbol	Description
START_PB	Start push button, N. O., **on** when starting.
STOP_PB	Stop push button, N. C., **off** when stopping.
RESET_PB	Reset push button, N. O., **on** when restoring station to initial state.
PROX	Retroreflective proximity switch that is **on** when seat in position to be tested.
RAM_ULS	Limit switch that closes (**on**) when the testing ram is in its fully up position.
LIFT_ULS	Limit switch that closes (**on**) when the lift cylinder is extended and thus the seat is off the conveyor.
LIFT_DLS	Limit switch that closes (**on**) when the lift cylinder is retracted.
DIST_MEAS	Measurement of test ram distance movement in the down direction, represents 50 to 850 mm; range listed above.
FORCE_MEAS	Measurement of compression force on testing ram shaft, represents 0 to 445 newtons; range listed above
LIFT_EXT	Lift cylinder extend control, **on** to extend (to move up), **off** to retract (move down).
RAM_EXT	Testing ram cylinder extend control, **on** to extend (to move down), **off** has no effect on cylinder position.
RAM_RET	Testing ram cylinder retract control, **on** to retract (to move up), **off** has no effect on cylinder position.
SPRAY_RET	Spray paint cylinder control, **on** to spray, **off** to stop spray.
DIST_MM	Current testing ram distance, in mm. **Your ladder must calculate this value.**
FORCE_NT	Current ram force in newtons. **Your ladder must calculate this value.**

DES_DIST	Desired testing ram distance. **Your ladder must not change this value.**
FORCE_MIN	Desired minimum force. **Your ladder must not change this value.**
FORCE_MAX	Desired maximum force. **Your ladder must not change this value.**

The addresses associated with the physical input and output tags/variables/symbols are:

Tag/Var./Symbol	ControlLogix	Micro800	MLogix	SLC
START_PB	Local:1:I.Data.0	_IO_EM_DI_00	I:0/0	I:1/0
STOP_PB	Local:1:I.Data.1	_IO_EM_DI_01	I:0/1	I:1/1
RESET_PB	Local:1:I.Data.2	_IO_EM_DI_02	I:0/2	I:1/2
PROX	Local:1:I.Data.3	_IO_EM_DI_03	I:0/3	I:1/3
RAM_ULS	Local:1:I.Data.4	_IO_EM_DI_04	I:0/4	I:1/4
LIFT_ULS	Local:1:I.Data.5	_IO_EM_DI_05	I:0/5	I:1/5
LIFT_DLS	Local:1:I.Data.6	_IO_EM_DI_06	I:0/6	I:1/6
DIST_MEAS	Local:3:I.Ch0Data	_IO_P2_AI_00	I:1.0	I:3.0
FORCE_MEAS	Local:3:I.Ch1Data	_IO_P2_AI_01	I:1.1	I:3.1
LIFT_EXT	Local:2:O.Data.0	_IO_EM_DO_00	O:0/0	O:2/0
RAM_EXT	Local:2:O.Data.1	_IO_EM_DO_01	O:0/1	O:2/1
RAM_RET	Local:2:O.Data.2	_IO_EM_DO_02	O:0/2	O:2/2
SPRAY _RET	Local:2:O.Data.3	_IO_EM_DO_03	O:0/3	O:2/3

The addresses and/or data types associated with the internal tags/variables/symbols are:

Tag/Var./Symbol	CLogix Data Type	Micro800 Data Type	MLogix/SLC Address
DIST_MM	REAL	REAL	F12:0
FORCE_NT	REAL	REAL	F12:1
DES_DIST	REAL	REAL	F25:0
FORCE_MIN	REAL	REAL	F25:1
FORCE_MAX	REAL	REAL	F25:2

P7-32. Using the function chart approach, implement the program for the following plastic injection molding machine.

Figure P7.32 shows the layout of a station that injection-molds plastic parts. The station clamps the two halves of the mold, injects melted plastic into the mold, allows it to cool and then removes the plastic part from the mold. Note that this problem is a simplified version of an actual machine.

In injection molding, plastic beads fall from a hopper into an auger-type screw. The screw melts (plasticizes) the polymer by virtue of the shearing action on the plastic beads as the screw turns. The heater on the outside of the barrel maintains the temperature, but the screw does the work to melt the plastic. To push the melt into the mold, the entire screw is moved forward.

Upon initial startup, assume that the two halves of the mold are apart, the eject cylinder is retracted, and the extrusion screw is retracted. The overall operation is first described, followed by the details. When E_RUN initially turns **on**, the following steps should be executed:

Clamp the two mold halves together by extending the clamp cylinder (CLAMP_ELS senses).

Feed the melted plastic by moving the extrusion screw forward a desired distance (DES_SCREW_MM).

Pull back the extrusion screw (SCREW_RLS senses)

Hold the mold clamped for 4.5 seconds. This delay allows the plastic to cool and solidify.

Retract the clamp cylinder (CLAMP_RLS senses). The part remains attached to the moveable side of the mold.

Extend the small eject cylinder (EJECT_ELS senses) to push the plastic piece out of the mold and into a chute below.

Retract the eject cylinder (EJECT_RLS senses)

The process repeats.

Assume that the extrusion screw is always running whenever the machine is running.

Mold Clamp Cylinder: The cylinder that controls the extension/retraction of the mold clamp is a double-action hydraulic cylinder, controlled by the CLAMP_EXT and CLAMP_RET outputs. When CLAMP_EXT is energized (turned **on**), the cylinder extends and moves the clamp to push the two mold halves together to form the mold. When CLAMP_RET is energized (turned **on**), the cylinder retracts, pulling the two mold halves apart. When CLAMP_EXT and CLAMP_RET are both **off**, the clamp stops at its present position. When CLAMP_RET and CLAMP_EXT are both **on**, the cylinder operation is not consistent (it may stop or move in either direction slowly). The cylinder has two limit switches, CLAMP_ELS, which is **on** when the cylinder is extended (mold clamped) and CLAMP_RLS, which is **on** when the cylinder is retracted (mold apart).

Note: the CLAMP_EXT output must remain energized while feeding plastic into the mold and while cooling.

Part Eject Cylinder: The cylinder that controls the extension/retraction of the part eject mechanism is a single-action hydraulic cylinder, controlled by the EJECT_EXT output. When EJECT_EXT is energized (turned **on**), the cylinder extends and moves the pins of the eject mechanism to push the molded part out of the moveable part of the mold (and drop into a chute below). When EJECT_EXT is **off**, the cylinder retracts. The cylinder has two limit switches, EJECT_ELS, which is **on** when the cylinder is extended (part pushed out) and EJECT_RLS, which is **on** when the cylinder is retracted.

Extrusion Screw Cylinder: The cylinder that controls the extension/retraction of the extrusion screw is a double-action hydraulic cylinder, controlled by the SCREW_EXT and SCREW_RET outputs. When SCREW_EXT is energized (turned **on**), the cylinder extends and moves the screw head to push melted plastic into the mold. When SCREW_RET is energized (turned **on**), the cylinder retracts. When SCREW_EXT and SCREW_RET are both **off**, the extrusion screw stops at its present position. When SCREW_RET and SCREW_EXT are both **on**, the cylinder operation is not consistent (it may stop or move in either direction slowly). The cylinder has one limit switch, SCREW_RLS, which is **on**

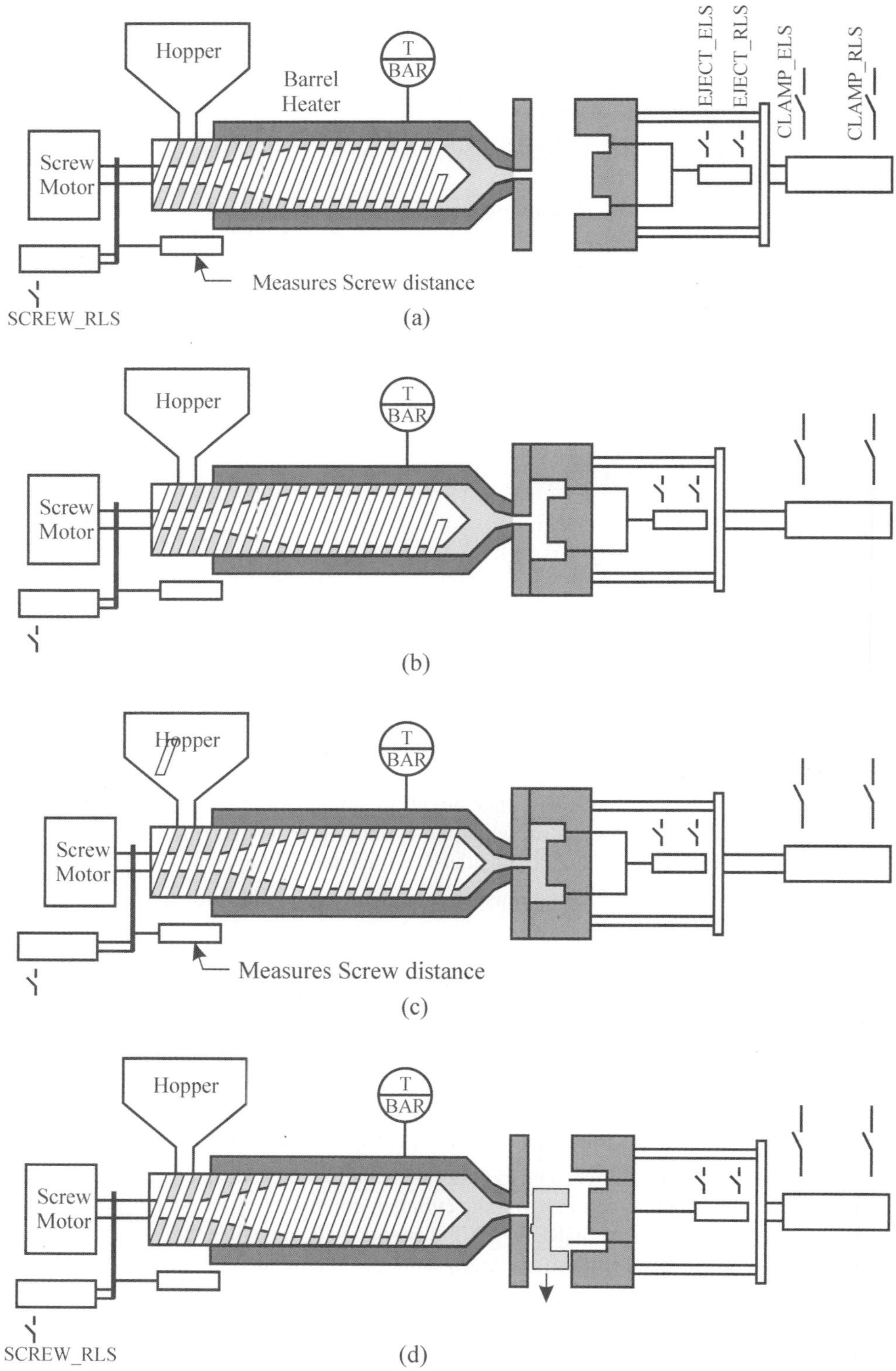

Figure P7.32. Injection molding machine: *(a)* start (mold open); *(b)* mold clamped; *(c)* push plastic into mold; *(d)* eject part.

when the cylinder is retracted. There is no limit switch indicating when the screw is extended.

Extrusion Screw Rotation Motor: This motor is controlled by another part of the ladder logic and beyond the scope of this problem. **You must not have any output coil controlling the screw rotation**.

Extrusion Screw Position: The extrusion screw position is measured by a magnetostrictive position sensor calibrated to measure from 0 to 80 mm and is connected to an analog input module having an ADC whose output integer (SCREWP_MEAS) corresponds to the lowest and highest positions as as specified before P7-1. The position measurement value must be converted into units of millimeters and stored in the SCREWP_MM internal memory location (a real number).

The position measurement increases when SCREW_EXT is **on** and decreases when SCREW_RET is **on** (until it reaches a "stop").

The extrusion screw barrel has electric heater bands surrounding it. The heater bands do not melt the plastic; the process of rotating the extruder screw melts the plastic. The heater bands maintain a uniform temperature profile along the extruder barrel. An actual extruder has at least three temperature control zones. For this problem, only one zone is assumed.

Extruder Barrel Temperature: The extruder barrel temperature is measured by a thermocouple calibrated to measure from 200 to 290 degrees Celsius and is connected to an analog input module having an ADC whose output integer (TBAR_MEAS) corresponds to the lowest and highest temperatures as as specified before P7-1. The temperature measurement value must be converted into units of degrees C and stored in the TBAR_DEG internal memory location (a real number).

Extruder Barrel Temperature Control: The barrel temperature control is done by a separate PID controller. This PID controller is not part of the PLC and has its own temperature sensor.

The DES_TBAR_DEG internal memory location indicates the desired barrel temperature. This value has been placed in the PLC memory by a supervisory PLC.

Barrel Temperature Alarm: There is one alarm indication, TBAR_TEMP_ALM that flashes when the barrel temperature is less than DES_TBAR_DEG –5 or the barrel temperature is greater than DES_TBAR_DEG + 5. This alarm is enabled ONLY when E_RUN is **on**. This indication flashes **on** for 0.5 second and **off** for 1.0 sec. This alarm must be **off** when E_RUN is **off**.

There is no start/stop switch for your part of the ladder. Instead, there is an internal coil, E_RUN that is **on** when the injection molding operation is making parts. When E_RUN is **off**, the operation should pause at its current step except for the following exception:

1. The screw is being advanced (ejecting plastic into mold)
2. The screw is being retracted.

Do not advance to the next step when paused. When paused, all outputs should be **off except** for the following exceptions:

1. The clamp cylinder extend control, CLAMP_EXT must not be turned **off**.
2. The barrel screw cylinder controls must not be turned **off**.

When E_RUN turns back **on** while the operation of the station is paused, the station should resume its suspended step. **DO NOT** retain the timer accumulator when paused.

The start and stop switches are absent from your part of the ladder because they control the overall operation of the extruder machine. In order to reliably start/stop/pause the operation of the machine, the initial fill of the plastic into the screw and barrel must be controlled. This part of the operation is outside the scope of this problem. The other part of the logic controls E_RUN. **You must not have any output coil referring to E_RUN.**

A separate reset internal coil, E_RESET, is provided which restores the extruder part of the machine to its initial state. The reset operation should cause the mold to be unclamped and the eject cylinder to be retracted. The reset operation is not complete until both cylinders are retracted. There is no mechanical interference between these two cylinders. Assume that an operator is responsible for actually removing anything from the station after a reset. E_RESET should be ignored if in **any step where the operation cannot be paused** or if E_RUN is **on**. The E_RUN internal coil used to restart the machine should be ignored when the reset operation is in progress. **You must not have any output coil referring to E_RESET.**

Do not add any more timed steps to those explicitly stated in the problem. In other words, do not put a timer in a step unless it is stated that the step duration is a specific time. Assume the tolerance on all timer values is at most ±0.01 second.

Assume the following physical inputs, outputs, and internal memory tags/variables/symbols.

Tag/Var/Symbol	Description
CLAMP_ELS	Limit switch that closes (**on**) when the mold clamp is extended and holding the two mold halves together.
CLAMP_RLS	Limit switch that closes (**on**) when the mold clamp is retracted and the two mold halves are apart.
EJECT_ELS	Limit switch that closes (**on**) when the part ejector is extended and pushed the part out of the mold.
EJECT_RLS	Limit switch that closes (**on**) when the part ejector is retracted.
SCREW_RLS	Limit switch that closes (**on**) when the extrusion screw is fully retracted.
SCREWP_MEAS	Measurement of extrusion screw position, represents 0 to 80 millimeters; range listed above.
TBAR_MEAS	Measurement of barrel temperature, represents 200 to 290 degrees C; range listed above.
CLAMP_EXT	Mold clamp extend control, **on** to extend clamp, closing mold, **off** stops clamp.
CLAMP_RET	Mold clamp retract control, **on** to retract clamp, opening mold, **off** stops clamp.
EJECT_EXT	Mold eject extend control, **on** to extend pins, forcing part out of mold, **off** retracts pins.
SCREW_EXT	Extrusion screw extend control, **on** to extend screw, forcing plastic into mold, **off** stops movement.

SCREW_RET	Extruction screw retract control, **on** to retract screw, **off** stops movement.

Assume the following internal tags/variables/symbols:

Tag/Var/Symbol	Description
E_RUN	Run/stop/pause control for extrusion part of machine operation; **on** to run extrusion, **off** to stop/pause. Controlled by another part of the ladder logic in the PLC. **You must not have any output coil referring to E_RUN.**
E_RESET	Reset control, **on** to restore extrusion section to initial state. Controlled by another part of the ladder logic in the PLC. **You must not have any output coil referring to E_RESET.**
SCREWP_MM	Screw position, in mm. **Your ladder must calculate this value**.
DES_SCREW_MM	Desired screw position, in mm. **Your ladder must not change this value**.
TBAR_DEG	Current barrel temperature, in degrees C. **Your ladder must calculate this value**.
DES_TBAR_DEG	Desired barrel temperature. **Your ladder must not change this value.**
TBAR_TEMP_ALM	Alarm indicator that flashes ½ second **on** and 1 second **off** when the barrel temperature is less than DES_TBAR_DEG – 5 or the barrel temperature is greater than DES_TBAR_DEG + 5. **Your ladder must control this Boolean**.

The addresses associated with the physical input and output tags/variables/symbols are:

Tag/Var./Symbol	ControlLogix	Micro800	MLogix	SLC
CLAMP_ELS	Local:1:I.Data.3	_IO_EM_DI_03	I:0/3	I:1/3
CLAMP_RLS	Local:1:I.Data.4	_IO_EM_DI_04	I:0/4	I:1/4
EJECT_ELS	Local:1:I.Data.5	_IO_EM_DI_05	I:0/5	I:1/5
EJECT_RLS	Local:1:I.Data.6	_IO_EM_DI_06	I:0/6	I:1/6
SCREW_RLS	Local:1:I.Data.7	_IO_EM_DI_07	I:0/7	I:1/7
SCREWP_MEAS	Local:3:I.Ch0Data	_IO_P2_AI_00	I:1.0	I:3.0
TBAR_MEAS	Local:3:I.Ch1Data	_IO_P2_AI_01	I:1.1	I:3.1
CLAMP_EXT	Local:2:O.Data.0	_IO_EM_DO_00	O:0/0	O:2/0
CLAMP_RET	Local:2:O.Data.1	_IO_EM_DO_01	O:0/1	O:2/1
EJECT_EXT	Local:2:O.Data.2	_IO_EM_DO_02	O:0/2	O:2/2
SCREW_EXT	Local:2:O.Data.3	_IO_EM_DO_03	O:0/3	O:2/3
SCREW_RET	Local:2:O.Data.4	_IO_EM_DO_04	O:0/4	O:2/4

The addresses and/or data types associated with the internal tags/variables/symbols are:

Tag/Var./Symbol	CLogix Data Type	Micro800 Data Type	MLogix/SLC Address
B_RUN	BOOL	BOOL	B3/101
B_RESET	BOOL	BOOL	B3/102
TBAR_TEMP_ALM	BOOL	BOOL	B3/151

SCREWP_MM	REAL	REAL	F12:0
TBAR_DEG	REAL	REAL	F12:1
DES_SCREW_MM	REAL	REAL	F25:0
DES_TBAR_DEG	REAL	REAL	F25:1

P7-33. Using the function chart approach, implement the program for the following railcar wood chip dump station.

Figure P7.33.1 shows the simplified diagram of a station that dumps railcars filled with wood chips into a receiving hopper. The following system is a simplified dump station. Basically, the railcar is clamped, raised, lowered, and unclamped. The rail car is assumed empty when no chips have fallen out of the raised railcar during a certain period of time. The presence of chips falling out of the railcar is sensed by the intensity of a light beam crossing the path of the chips.

A locomotive places a railcar into the dump station. Only one railcar occupies the track in the station and it is uncoupled from any other cars and the locomotive. After the operator verifies that a railcar is in the proper position, he/she presses the STTDR_PB. The PLC program then controls the operation of the dump station. After the operator presses the STTDR_PB, the system must go through the following steps:

Clamp the railcar by activating the clamp cylinder control. CLMP_RLS senses the clamped railcar. Also, before continuing to the next step, the light beam intensity (INTENS_LUX) must be at least 100 lux.

Raise and tip the railcar by activating the DUMP_EXT control to extend the dump cylinder (DUMP_ELS senses).

Let the wood chips fall into the hopper. The appropriate dump cylinder control must be **on** during this step. The light beam intensity is used to sense when all of the chips have fallen out of the railcar. The light beam intensity must be at least 100 lux for 15 seconds before continuing.

Lower the railcar by activating the DUMP_RET control to retract the dump cylinder (DUMP_RLS senses).

Unclamp the railcar. CLMP_ELS senses the unclamped railcar.

The process must now stop itself (and extinguish the running panel lamp).

The operation does not repeat. The operator must press STTDR_PB to initiate the next railcar dump cycle.

Railcar Clamp: The railcar clamp is a hydraulic ram powered by a single action cylinder. When the CLMP_RET output is **on**, the ram retracts until it reaches the fully retracted position, clamping the railcar. When the power is removed (CLMP_RET is **off**), the ram extends until it is in its fully extended (unclamped) position. The CLMP_RET output must remain **on** to hold the railcar while it is being dumped. The CLMP_RLS limit switch is **on** when the ram is fully retracted and the CLMP_ELS limit switch is **on** when the ram is extended.

Dump Cylinder: The dump cylinder is a hydraulic double action cylinder. When the DUMP_EXT output is **on**, the dump cylinder extends. When the DUMP_RET output is **on**, the dump cylinder retracts. When both DUMP_EXT or DUMP_RET outputs are **off** or both outputs are **on**, the dump cylinder ram stops moving and remains at its present position. DUMP_ELS is **on** when the cylinder is in the fully extended (maximum tip) position.

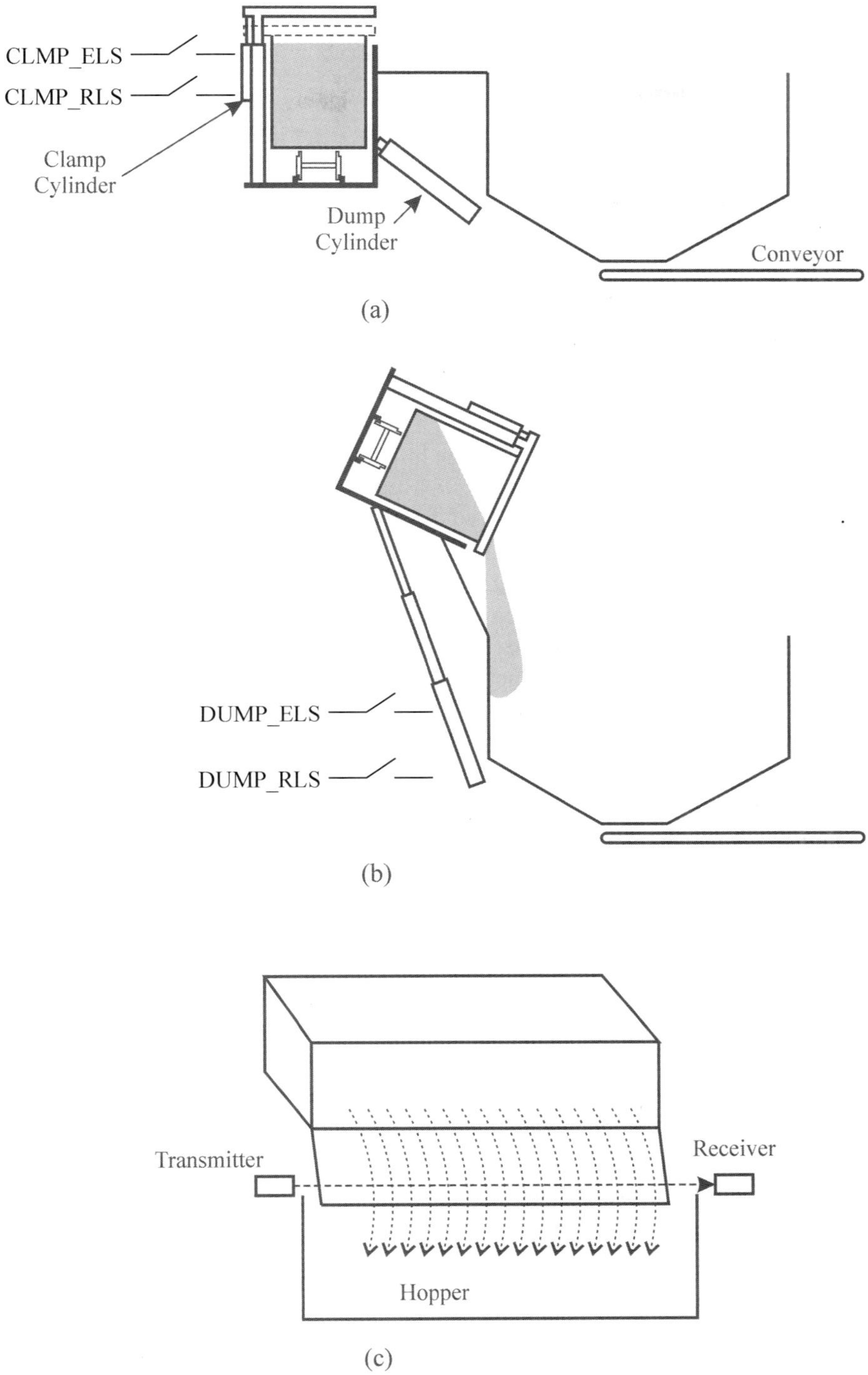

Figure P7.33.1. Railcar dump station: *(a)* side view starting position; *(b)* side view tipped railcar; *(c)* front view showing light intensity sensor.

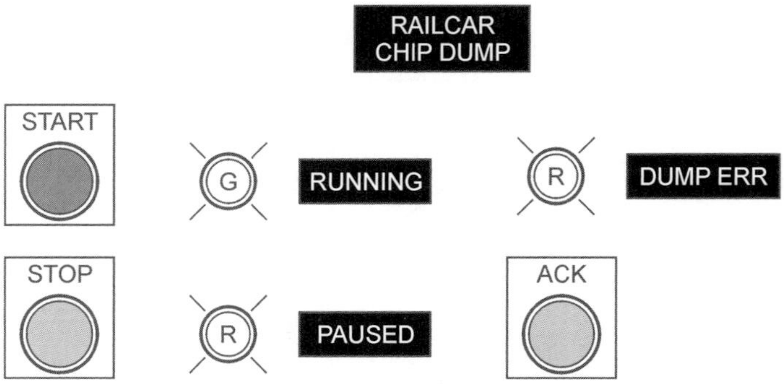

Figure P7.33.2. Railcar dump operator panel.

DUMP_RLS is **on** when the cylinder is in the fully retracted (untipped) position. There are mechanical stops at the fully retracted and fully extended positions. However, the DUMP_EXT output must remain **on** to hold the railcar in position while the wood chips are being dumped.

Chip Conveyor: As the chips are being dumped into the hopper, the conveyor that moves chips out of the hopper must be started. To start the conveyor, the STRT_CONV internal coil must be turned **on** for 1 second. The one-second interval should not start until the railcar is fully raised (and tipped). At the end of the one-second interval, the STRT_CONV internal coil must be **off** and remain **off** until the next railcar is dumped. This internal coil has no effect if the conveyor is already running. It is intended to start the conveyor when the first railcar is dumped. The operator is responsible for stopping the conveyor. Other than STRT_CONV, your ladder has no control over the conveyor.

Light Beam Intensity: The light beam intensity is measured by a light intensity meter. The sensor is calibrated to measure the light received from the light beam transmitter in the range of 10 to 250 lux and is connected to an analog input module having an ADC whose output integer (INTENS_MEAS) corresponds to the lowest and highest readings as specified before P7-1. The light intensity measurement value must be converted into units of lux and stored in the INTENS_LUX internal memory location (a real number).

Alarm: If the dump step does not complete within 2 minutes, the DUMP_ALM lamp must be flashed. This alarm indicates a problem with the light meter (for example, blocked by wood chips). The period of the lamp should be 0.75 second **on** and 0.75 second **off**. The lamp should continue to flash until the ALM_ACK button is pressed (turned **on**). When ALM_ACK is **on**, the DUMP_ALM light must stop flashing and remain off until the next dump step that is longer than 2 minutes. Even if the dump step completes after 2 minutes and before the ALM_ACK button is pressed, the DUMP_ALM lamp **must continue** to flash until acknowledged by the operator.

Start/Stop/Pause: The start/stop switches are only for the dump station. If the stop switch (STPDR_PB) is pressed at any time, the station operation should **pause** at the current step. When the start switch (STTDR_PB) is pressed while the operation is paused, the station should resume the suspended step. When paused, **do not advance** to the next step. If the operation pauses in the dump step, the 2-minute alarm timer should be paused.

When the station is paused, the dump cylinder controls must be **off**. The clamp cylinder control should remain **on** when paused (if active in that step).

Reset: A separate reset switch, RSTDR_PB, is provided which lowers (untips) the railcar and then unclamps the railcar. The clamp must not be released until the railcar has been lowered. The reset switch also resets any internal states. The reset switch should be ignored if the station is running. The start switch should be ignored as long as RSTDR_PB is pressed or if the railcar is still clamped.

Operator Panel: The operator panel for the station is shown in Figure P7.33.2. The "Running" lamp (controlled by the DR_RNLA output) must be **on** when the dump operation is running (and not paused). The "Paused" lamp (controlled by the DR_PALA output) must be **on** when the dump operation is paused. Both lamps cannot be **on** simultaneously. At the end of a dump cycle, both lamps must be **off**. The "Dump Err" indicator is controlled by the DUMP_ALM output, described above.

Tolerance on any timer is at most 0.1 second. Do not add any more timed steps than those explicitly stated in the problem.

Assume the following input, output, and internal memory tags/variables/symbols:

Tag/Var/Symbol	Description
STTDR_PB	Start push button, N. O., **on** when starting.
STPDR_PB	Stop push button, N. C., **off** when stopping.
RSTDR_PB	Reset push button, N. O., **on** when restoring dump station to initial state.
ALM_ACK	Alarm acknowledge button, N. O., **on** to extinguish flashing alarm lamp.
CLMP_ELS	Limit switch that closes (**on**) when clamp cylinder is extended (railcar unclamped).
CLMP_RLS	Limit switch that closes (**on**) when clamp cylinder is retracted (railcar clamped).
DUMP_ELS	Limit switch that closes (**on**) when dump cylinder is extended (railcar raised/tipped).
DUMP_RLS	Limit switch that closes (**on**) when dump cylinder is retracted (railcar lowered).
INTENS_MEAS	Measurement of light meter received light intensity, represents 10 to 250 lux; integer range listed above
DUMP_EXT	Dump cylinder extend control, **on** to move up, **off** stops motion.
DUMP_RET	Dump cylinder retract control, **on** to move down, **off** stops motion.
CLMP_RET	Clamp cylinder retract control; **on** to retract (clamp), **off** to extend (unclamp).
DR_RNLA	Railcar dump "running" lamp. Must be **on** when the station is running (and not paused). Otherwise, it is **off**.
DR_PALA	Railcar dump "paused" lamp. Must be **on** when the station is paused. Otherwise, it is **off**.
DUMP_ALM	Dump step problem alarm indication. Flashes 0.75 sec. **on** and 0.75 sec **off** when dump step requires more than 2 minutes to complete. ALM_ACK turns off indication.

STRT_CONV	Start hopper conveyor internal coil. **Your ladder must control this internal coil.**
INTENS_LUX	Light meter sensor received light intensity, in lux. **Your ladder must calculate this value.**

The addresses associated with the physical input and output tags/variables/symbols are:

Tag/Var./Symbol	ControlLogix	Micro800	MLogix	SLC
STTDR_PB	Local:1:I.Data.0	_IO_EM_DI_00	I:0/0	I:1/0
STPDR_PB	Local:1:I.Data.1	_IO_EM_DI_01	I:0/1	I:1/1
RSTDR_PB	Local:1:I.Data.2	_IO_EM_DI_02	I:0/2	I:1/2
ALM_ACK	Local:1:I.Data.3	_IO_EM_DI_03	I:0/3	I:1/3
CLMP_ELS	Local:1:I.Data.4	_IO_EM_DI_04	I:0/4	I:1/4
CLMP_RLS	Local:1:I.Data.5	_IO_EM_DI_05	I:0/5	I:1/5
DUMP_ELS	Local:1:I.Data.6	_IO_EM_DI_06	I:0/6	I:1/6
DUMP_RLS	Local:1:I.Data.7	_IO_EM_DI_07	I:0/7	I:1/7
INTENS_MEAS	Local:3:I.Ch1Data	_IO_P2_AI_00	I:1.1	I:3.0
DUMP_EXT	Local:2:O.Data.0	_IO_EM_DO_00	O:0/0	O:2/0
DUMP_RET	Local:2:O.Data.1	_IO_EM_DO_01	O:0/1	O:2/1
CLMP_RET	Local:2:O.Data.2	_IO_EM_DO_02	O:0/2	O:2/2
DR_RNLA	Local:2:O.Data.3	_IO_EM_DO_03	O:0/3	O:2/3
DR_PALA	Local:2:O.Data.4	_IO_EM_DO_04	O:0/4	O:2/4
DUMP_ALM	Local:2:O.Data.5	_IO_EM_DO_05	O:0/5	O:2/5

The addresses and/or data types associated with the internal tags/variables/symbols are:

Tag/Var./Symbol	CLogix Data Type	Micro800 Data Type	MLogix/SLC Address
STRT_CONV	BOOL	BOOL	B3/101
INTENS_LUX	REAL	REAL	F12:0

P7-34. Using the function chart approach, implement the program for the following station that heat treats castings.

Figure P7.34 shows the layout of a station that heat-treats one portion of a metal casting in order to harden the outer surface at that portion. In summary, the portion of the casting is heated with an inductive heating coil to a desired treat temperature (typically around 800 °C) and then quenched (quickly cooled) to a desired quench temperature (typically around 50 °C). A thermocouple on a probe cylinder measures the temperature of the heated portion of the casting. The station has a conveyor section that your PLC controls to move the casting into the station and out of the station. The major steps are:

Bring in new casting on carrier in station (PX711 senses),
Stop the conveyor (CV716_RUN),
Simultaneously, move out the heating coil (CY712_ELS senses), extend the temperature probe (CY713_ELS senses) and clamp the carrier to secure the casting (CY714_ELS senses),

Activate the inductive heater to heat the casting to the desired treat temperature (DES_TREAT),

Retract the heating coil (CY712_RLS senses),

Open the water quench valve to spray the casting and cool it down to the desired quench temperature (DES_QUENCH),

Deactivate the quench valve, deactivate the clamp, and deactivate the temperature probe. Use only CY713_RLS to sense that both clamps are retracted, as the temperature probe cylinder is longer and requires more time to retract. Also, this retraction allows the quench spray to dissipate.

Activate the conveyor (CV716_RUN) to move out casting (and bring a new casting into the station).

The operation **repeats**.

Do not use parallel steps. The coil lowering, probe extension, and clamp mechanisms can be activated simultaneously without using parallel steps. All three cylinders have a mechanical stop. You will not damage either mechanism by trying to move it while at the mechanical stop. Do **not** assume that any one finishes first or last.

Also, the retraction of the temperature probe and the clamp can be done simultaneously without parallel steps.

Conveyor: The conveyor section in the station, controlled by the CV716_RUN output, is a metal roller conveyor. You do not have control of the inbound and outbound conveyors, so assume they are always running. Assume the castings are spaced out far enough so that after one casting is processed and moves out, the PX711 sensor will detect the gap between castings.

Proximity Sensor: The casting is on a carrier. Proximity sensor PX711 turns **on** as the casting carrier nears the correct position. If the conveyor is turned **off** when as soon as PX711 transitions **on**, the conveyor will "coast" a little, bringing the casting into the position for heat-treating. Assume PX711 remains **on** during heating and quenching. Assume the carriers are spaced out far enough so that after one casting is processed and moves out, the PX711 sensor will detect the gap between carriers.

The **induction heating** process is a non-contact technique that provides localized heating through custom-designed coils. The heat is transferred to the product via electromagnetic waves. This process allows the right amount of heat to be applied exactly where it is needed, ensuring repeatable performance. The PLC controls the inductive heater through the EX718_ON output, which is **on** to apply heat and **off** to not apply heat.

Inductive Coil Extend/Retract Cylinder: The mechanism used to extend and retract the inductive heating coil consists of a double-action linear hydraulic cylinder. The cylinder is controlled by two outputs. When the CY712_EXT output is energized, the mechanism moves out (extends) and continues to move out as long as power is applied (turned **on**). When the CY712_RET output is energized, the mechanism moves in (retracts) and continues to move in as long as power is applied. The mechanism stops if neither output is **on**, or if they are energized simultaneously. Limit switches CY712_ELS and CY712_RLS indicate the out (extended) and in (retracted) positions, respectively, of the mechanism. The mechanism must be in the extended position in order to properly heat-treat the casting.

Temperature Probe Cylinder: The temperature probe is driven by a single-action pneumatic cylinder. Once the CY713_EXT output is energized, the cylinder extends and pushes the thermocouple probe into a "well" to hold it close to the place that is being heated.

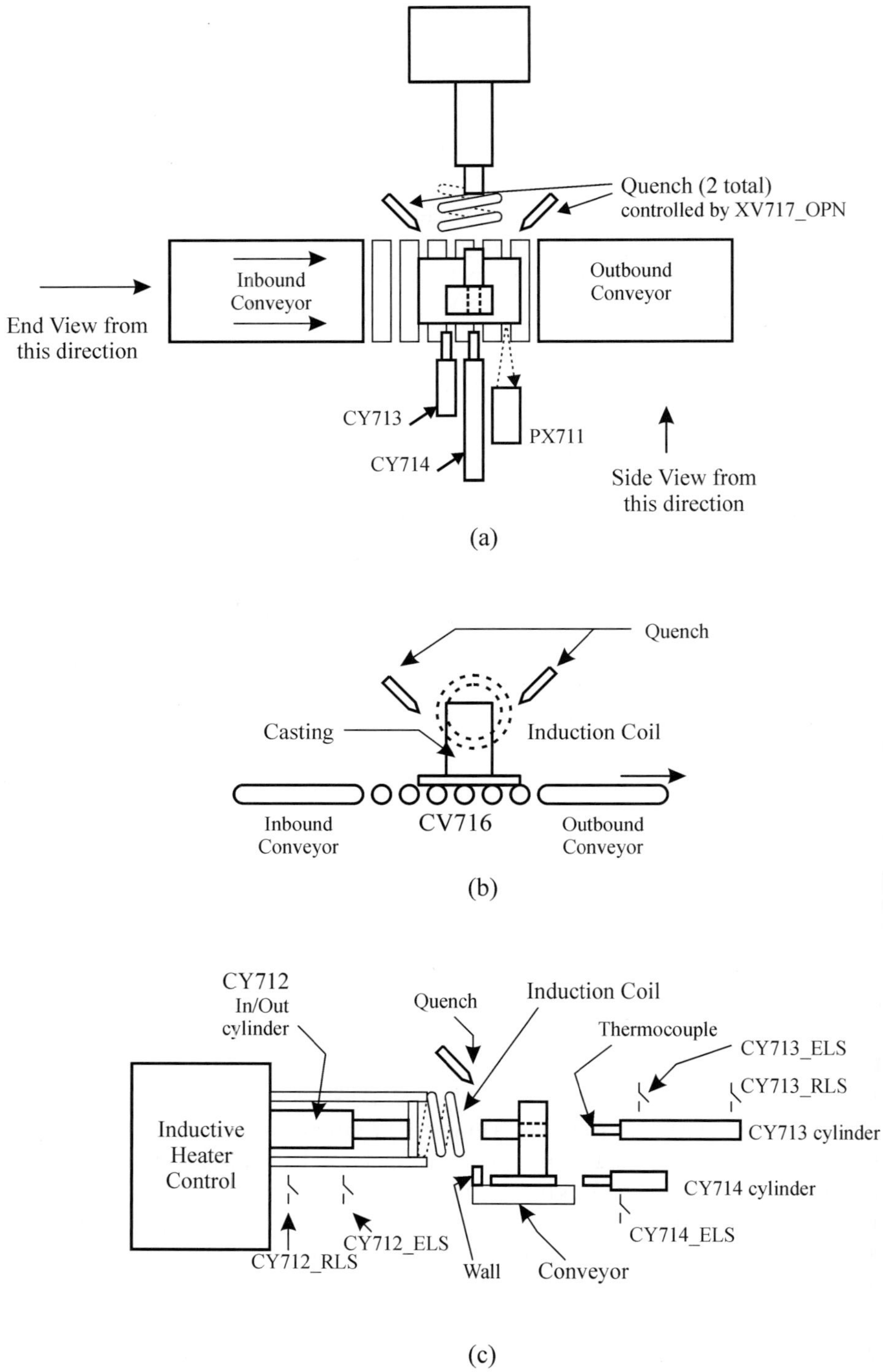

Figure P7-34. Heat treating station: *(a)* top view; *(b)* side view; *(c)* end view.

Two limit switches detect the position of the clamp. CY713_ELS is **on** when the clamp is extended and CY713_RLS is **on** when the clamp is retracted.

Carrier Clamp: The carrier clamp is driven by a single-action pneumatic cylinder. Once the CY714_EXT output is energized, the clamp extends and pushes the carrier about 5 mm into a "wall" to hold the casting during the heat-treating. The carrier is not pushed off the conveyor. One limit switch detects the position of the clamp. CY714_ELS is **on** when the clamp is extended.

Quench Valve: The quench spray is activated by turning **on** the XV717_OPN output which opens a valve to direct two streams of water onto the casting to cool it quickly.

The DES_TREAT and DES_QUENCH internal memory locations indicate the desired treating and quench temperatures, respectively, in °C. Assume these values have been placed in the PLC memory from a supervisory PLC or computer.

The **casting temperature** is measured by a thermocouple on a probe. The thermocouple is connected to an analog input module having an ADC whose output integer (TT719_MEAS) corresponds to the lowest and highest temperatures as specified before P7-1. The temperature measurement value must be converted into units of °C and stored in the TT719_DEGC internal memory location (a real number).

Alarm: There is an alarm for the steps that heat and cool the casting. If it takes longer than 60 seconds to heat the casting **or** longer than 60 seconds to cool the casting, then the HN721 horn must be sounded (beeped) and continue to beep until PB733 is pressed after the station operation is paused. If it takes longer than 60 seconds to heat/cool the casting, then there is a problem with the system and it needs to be repaired. The operation of the station is not altered if this alarm happens. It is only an alert to the operator. If the operation is paused in the step that cools the casting, then the timing interval should be cancelled (not retained), since the quench pump is turned off when paused (to be restarted when the operation is resumed). The period of the horn beep should be 0.5 second **on** and 0.75 second **off**. The horn should commence its first **on** sound as soon as the step exceeds its 60-second interval and not 0.75 seconds later. When this horn is beeping, then the operator is expected to pause (stop) the operation, reset the station, and fix the problem before restarting the system. Your program is not responsible for stopping the process when HN721 is beeping. When PB733 is pressed with the operation already paused, it should stop beeping HN721. If the operation is not paused, pressing PB733 **must not** cancel the beeping horn.

Upon initial startup (PB731 is the start pushbutton), assume there are no castings present at the end of the inbound conveyor. If the stop (PB732) pushbutton switch is pressed at any time, the station operation should pause. However, if the stop switch is pressed when the casting is being heated, the heating step should complete and advance to the next step before the station operation should pause. When the start switch is pressed while the operation is paused, the station should resume the suspended step. When the station is paused, all active outputs **except** for CY713_EXT, CY714_EXT, and EX718_ON are turned **off**. When paused, **do not advance** to the next step (exception noted previously in paragraph.

A separate reset switch, PB733, is provided which resets any internal states so that when the start switch is pressed, no castings are assumed to be in the station. Assume that the casting has been removed before PB731 is pressed. Also, the reset switch should cause the heater coil mechanism to retract (if not already fully retracted) and all other cylinders to retract. The reset switch should be ignored if the casting is being heated or if the station is running. The start switch should be ignored when the reset operation is in progress. Also, the start switch should be ignored as long as PB733 is pressed. The PB733 **must** be ignored if heating the casting.

Assume the tolerance on all timer values is at most ±0.1 seconds.

Assume the following input, output, and internal memory tags/variables/symbols:

Tag/Var/Symbol	Description
PB731	Start push button, N. O., **on** when starting.
PB732	Stop push button, N. C., **off** when stopping.
PB733	Reset push button, N. O., **on** when restoring station to initial state.
PX711	Proximity sensor, **on** (closed) when carrier in station (see above).
CY712_ELS	Limit switch that closes (**on**) when heater coil mechanism is in the fully extended (out) position.
CY712_RLS	Limit switch that closes (**on**) when heater coil mechanism is in the fully retracted (in) position.
CY713_ELS	Limit switch that closes (**on**) when thermocouple probe is extended and thermocouple in position to measure temperature of heated casting.
CY713_RLS	Limit switch that closes (**on**) when thermocouple probe is retracted.
CY714_ELS	Limit switch that closes (**on**) when clamp is extended and carrier is against wall to position it for heat-treating.
TT719_MEAS	Measurement of casting temperature, represents –100 to 1350 °C; range listed above
CY712_EXT	Heat coil mechanism extend control, **on** to move out, **off** has no effect.
CY712_RET	Heat coil mechanism retract control, **on** to move in, **off** has no effect.
CY713_EXT	Thermocouple probe extend control, **on** to move probe out and into position to measure temperature of heated portion of casting, **off** retracts.
CY714_EXT	Casting carrier clamp extend control, **on** to hold carrier against wall to position it for heat-treating, **off** retracts.
CV716_RUN	Station roller conveyor motor control, **on** to move conveyor.
XV717_OPN	Quench valve control, **on** to open valve and therefore direct quench water on the casting, **off** closes valve.
EX718_ON	Inductive heat coil control, **on** to heat casting, **off** does not heat casting.
HN721	Heat/cool alarm horn, beeped **on** when it takes longer than 60 seconds to either heat or cool the casting.
TT719_DEGC	Current coil temperature, in °C. **Your ladder must calculate this value**.

DES_TREAT	Desired heat-treating temperature, °C. Set by the operator. **Your ladder does not calculate this value**.
DES_QUENCH	Desired quench temperature, °C. Set by the operator. **Your ladder does not calculate this value**.

The addresses associated with the physical input and output tags/variables/symbols are:

Tag/Var./Symbol	ControlLogix	Micro800	MLogix	SLC
PB731	Local:1:I.Data.0	_IO_EM_DI_00	I:0/0	I:1/0
PB732	Local:1:I.Data.1	_IO_EM_DI_01	I:0/1	I:1/1
PB733	Local:1:I.Data.2	_IO_EM_DI_02	I:0/2	I:1/2
PX711	Local:1:I.Data.3	_IO_EM_DI_03	I:0/3	I:1/3
CY712_ELS	Local:1:I.Data.4	_IO_EM_DI_04	I:0/4	I:1/4
CY712_RLS	Local:1:I.Data.5	_IO_EM_DI_05	I:0/5	I:1/5
CY713_ELS	Local:1:I.Data.6	_IO_EM_DI_06	I:0/6	I:1/6
CY713_RLS	Local:1:I.Data.7	_IO_EM_DI_07	I:0/7	I:1/7
CY714_ELS	Local:1:I.Data.8	_IO_EM_DI_08	I:0/8	I:1/8
TT719_MEAS	Local:3:I.Ch0Data	_IO_P2_AI_00	I:1.0	I:3.0
CY712_EXT	Local:2:O.Data.0	_IO_EM_DO_00	O:0/0	O:2/0
CY712_RET	Local:2:O.Data.1	_IO_EM_DO_01	O:0/1	O:2/1
CY713_EXT	Local:2:O.Data.2	_IO_EM_DO_02	O:0/2	O:2/2
CY714_EXT	Local:2:O.Data.3	_IO_EM_DO_03	O:0/3	O:2/3
CV716_RUN	Local:2:O.Data.4	_IO_EM_DO_04	O:0/4	O:2/4
XV717_OPN	Local:2:O.Data.5	_IO_EM_DO_05	O:0/5	O:2/5
EX718_ON	Local:2:O.Data.6	_IO_EM_DO_06	O:0/6	O:2/6
HN721	Local:2:O.Data.7	_IO_EM_DO_07	O:0/7	O:2/7

The addresses and/or data types associated with the internal tags/variables/symbols are:

Tag/Var./Symbol	CLogix Data Type	Micro800 Data Type	MLogix/SLC Address
TT719_DEGC	REAL	REAL	F12:0
DES_TREAT	REAL	REAL	F25:0
DES_QUENCH	REAL	REAL	F25:1

P7-35. Using the function chart approach, implement the program for the following station that tests capacitors.

Figure P7.35 shows the layout of a station that tests large capacitors. The station places the capacitor into the test leads, charges the capacitor with a certain voltage, and then measures the voltage after the charging voltage is removed for a certain time. If the voltage is outside a certain band, the capacitor is marked as defective. This station is one of a series of stations along the conveyor. The roller conveyor is controlled by another part of the ladder, so assume it is always running.

Upon initial startup, assume that the lift cylinder is down and there is no capacitor in the station. The overall operation is first described, followed by the details. The sequence of operation for the station follows these steps:

A capacitor is sensed by PROX.
Activate LIFT_EXT control for the lift cylinder to raise the capacitor off of the roller conveyor and to contact the test leads (LIFT_ULS senses).
Connect the charge voltage source to the capacitor by activating the appropriate relay. This step is complete when the capacitor voltage CAP_VOLT is at least START_VOLT.
Delay 21 seconds. Twenty of these seconds are the delay to check for the capacitor holding a voltage and the other second is required if the capacitor is marked as defective. During this step, the charge voltage source is disconnected from the capacitor. **Do not separate the delay into two steps**.
Connect a discharge resistor to the capacitor by activating the appropriate relay. This step is complete when the capacitor voltage CAP_VOLT is less than END_VOLT.
Lower the capacitor down onto the conveyor (LIFT_DLS senses).
The capacitor is moved out.

The operation then repeats.

The PROX **proximity sensor** is a retro-reflective photoelectric sensor that senses the presence/absence of a capacitor in the position where it can be tested. PROX is **on** when a capacitor is present and **off** when no capacitor is present. PROX may or may not remain **on** as the capacitor is lifted off the roller conveyor.

Lift Cylinder: The cylinder that lifts the capacitor off the conveyor is a single-action pneumatic cylinder, controlled by the LIFT_EXT output. When LIFT_EXT is energized, the lift cylinder extends, moving the capacitor off the conveyor. When LIFT_EXT is de-energized, the cylinder retracts, placing the capacitor onto the roller conveyor. The LIFT_ULS limit switch is **on** when the lift is in the "up" position (cylinder extended). The LIFT_DLS limit switch is **on** when the lift is in the "down" position.

Charge voltage contactor: The CHARGE_ON output controls the double-pole contact that connects a voltage source to the capacitor contacts. When the CHARGE_ON output is **on**, contacts close to connect the voltage source to the capacitor contacts and when CHARGE_ON is **off**, the voltage source is not connected to the capacitor.

Discharge resistor contactor: The DISCHARGE_ON output controls the double-pole contact that connects a discharge resistor to the capacitor contacts. When the DISCHARGE_ON output is **on**, contacts close to connect the resistor to the capacitor contacts and when DISCHARGE_ON is **off**, the resistor is not connected to the capacitor.

Note that the controls for the charge voltage and discharge resistor MUST NOT BE ON SIMULTANEOUSLY.

Reject sprayer: The reject sprayer is basically a spray paint can with a cylinder that retracts, pressing on the spray head, sending a burst of paint that marks the capacitor as defective. When SPRAY_RET is energized, paint is sprayed onto the capacitor. When SPRAY_RET is de-energized, the paint is not sprayed.

Capacitor Voltage: The capacitor voltage is measured by a voltmeter that provides a 4 to 20 mA output calibrated to measure from 5 to 240 volts and is connected to an analog input module having an ADC whose output integer (VOLT_MEAS) corresponds to the lowest and highest values as specified before P7-1. The measurement value must be converted into units of volts and stored in the CAP_VOLT internal memory location (a real number).

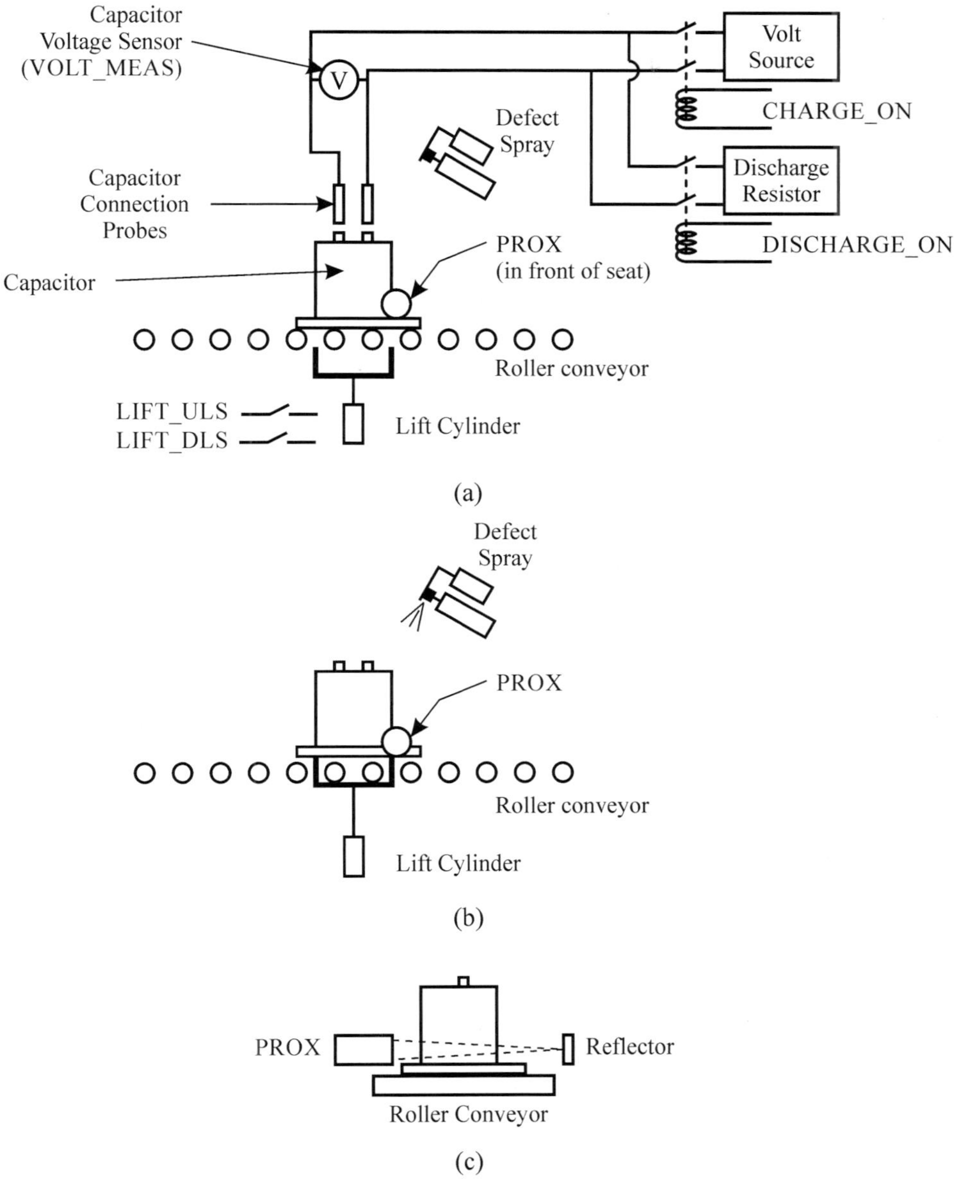

Figure P7.35. Capacitor testing station: *(a)* side view; *(b)* capacitor up, *(c)* end view, showing PROX location.

The voltage measurement increases as the capacitor is charged and decreases as the capacitor is discharged.

Initial Charge Voltage: The START_VOLT internal memory location indicates the desired value of the capacitor charge voltage (in volts). Assume this value has been placed in the PLC memory by a supervisory PLC or computer.

Voltage test: The capacitor voltage is checked 20 seconds after the charging source is disconnected from the capacitor. This delay allows for some discharge and to allow the voltage measurement to "settle down." If the capacitor voltage (CAP_VOLT) is **NOT** within the boundaries set by VOLT_MIN and VOLT_MAX,

VOLT_MIN <= CAP_VOLT <= VOLT_MAX

then SPRAY_RET is activated for one second to mark the capacitor as defective. The VOLT_MIN and VOLT_MAX internal memory locations have been placed in the PLC memory by a supervisory PLC or computer. Assume that VOLT_MIN < VOLT_MAX.

The test is **only** done at the 20 second mark. If the CAP_VOLT happens to fall outside the desired range any time between 20 – 21 seconds of the test step being active, **do not** activate the sprayer.

Alarm: There is a lamp that flashes if a capacitor in the station is defective. If a capacitor is being marked as defective, then the CAP_ALARM must flash and continue to flash until the capacitor is moved out. The operation of the station is not altered if the lamp flashes. It is only an alert to the operator. The period of the lamp flash should be 0.5 second **on** and 0.75 second **off**. The lamp should commence its first **on** flash as soon as the defective capacitor is detected and not 0.75 seconds later. The lamp continues its cycle until the capacitor is lowered to the conveyor and then stops its flashing. The lamp must not stop flashing until the capacitor is on the roller conveyor.

Start/Stop/Pause: The start and stop switches are for this station only. When the start switch is pressed (turned **on**) for the first time only, the station assumes there is no capacitor in the station. When the stop switch is pressed (turned **off**) the operation should pause, except when paused during the testing step. The testing step must advance to the next step if the station is paused and the 21-second time completes. Otherwise, **do not advance** to the next step when paused. When paused, the charging contact and discharge contact controls must be turned **off**. The lift and spray cylinder controls must not be affected by pause. Also, during a pause, the timer accumulators must continue to increment. Pressing the start switch while the operation of the station is paused causes the station to resume its suspended step.

Note that a pause during the time that a capacitor is moving into or out of the station effectively disables the test for following capacitors. The succeeding capacitors will move through the station without being tested. This is the desired effect of a pause during this part of the operation.

Reset: A separate reset switch, RESET_PB, is provided which restores the system to its initial state. The reset operation should cause the charging and discharging contacts to open and the lift cylinder to retract. The reset operation is not complete until the lift cylinder is retracted. **Do not implement a function chart for the reset** operation. Assume that an operator is responsible for actually removing the capacitor from the station after a reset. The reset switch should be ignored if the station is running. The start switch should be ignored when the reset operation is in progress. If the station is paused and RESET_PB is **on**, the operator must release the RESET_PB before the start pushbutton switch can be used to restart the station.

Do not add any more timed steps to those explicitly stated in the problem. In other words, do not put a timer in a step unless it is stated that the step duration is a specific time. Assume the tolerance on all timer values is ±0.01 seconds.

Assume the following input, output and internal memory tags/variables/symbols.

Tag/Var/Symbol	Description
START_PB	Start push button, N. O., **on** when starting.
STOP_PB	Stop push button, N. C., **off** when stopping.
RESET_PB	Reset push button, N. O., **on** when restoring station to initial state.
PROX	Retroreflective proximity switch that is **on** when capacitor in position to be tested.
LIFT_ULS	Limit switch that closes (**on**) when the lift cylinder is extended and thus the capacitor is off the conveyor.
LIFT_DLS	Limit switch that closes (**on**) when the lift cylinder is retracted.
VOLT_MEAS	Measurement of capacitor voltage, represents 5 to 240 volts; range listed above
LIFT_EXT	Lift cylinder extend control, **on** to extend (to move up), **off** to retract (move down).
CHARGE_ON	Charge voltage relay coil control, **on** to close contacts and connect charge voltage source to capacitor (when capacitor in up position), **off** opens contacts.
DISCHARGE_ON	Discharge resistor relay coil control, **on** to close contacts and connect discharge resistor to capacitor (when capacitor in up position), **off** opens contacts.
SPRAY_RET	Spray paint cylinder control, **on** to spray, **off** to stop spray.
CAP_ALARM	Capacitor defective lamp, **flashes** as specified in the problem if the capacitor in the station is defective.
CAP_VOLT	Capacitor voltage measurement (when capacitor in up position) in volts. **Your ladder must calculate this value**.
START_VOLT	Desired capacitor charge voltage. **Your ladder must not change this value.**
END_VOLT	Desired capacitor discharge voltage. **Your ladder must not change this value.**
VOLT_MIN	Desired minimum VOLT. **Your ladder must not change this value.**
VOLT_MAX	Desired maximum VOLT. **Your ladder must not change this value.**

The addresses associated with the physical input and output tags/variables/symbols are:

Tag/Var./Symbol	ControlLogix	Micro800	MLogix	SLC
START_PB	Local:1:I.Data.0	_IO_EM_DI_00	I:0/0	I:1/0
STOP_PB	Local:1:I.Data.1	_IO_EM_DI_01	I:0/1	I:1/1
RESET_PB	Local:1:I.Data.2	_IO_EM_DI_02	I:0/2	I:1/2
PROX	Local:1:I.Data.3	_IO_EM_DI_03	I:0/3	I:1/3
LIFT_ULS	Local:1:I.Data.4	_IO_EM_DI_04	I:0/4	I:1/4
LIFT_DLS	Local:1:I.Data.5	_IO_EM_DI_05	I:0/5	I:1/5

VOLT_MEAS	Local:3:I.Ch0Data	_IO_P2_AI_00	I:1.0	I:3.0
LIFT_EXT	Local:2:O.Data.0	_IO_EM_DO_00	O:0/0	O:2/0
CHARGE_ON	Local:2:O.Data.1	_IO_EM_DO_01	O:0/1	O:2/1
DISCHARGE_ON	Local:2:O.Data.2	_IO_EM_DO_02	O:0/2	O:2/2
SPRAY _RET	Local:2:O.Data.3	_IO_EM_DO_03	O:0/3	O:2/3
CAP_ALARM	Local:2:O.Data.4	_IO_EM_DO_04	O:0/4	O:2/4

The addresses and/or data types associated with the internal tags/variables/symbols are:

Tag/Var./Symbol	CLogix Data Type	Micro800 Data Type	MLogix/SLC Address
CAP_VOLT	REAL	REAL	F12:0
START_VOLT	REAL	REAL	F25:0
END_VOLT	REAL	REAL	F25:1
VOLT_MIN	REAL	REAL	F25:2
VOLT_MAX	REAL	REAL	F25:3

8 Other Ladder Logic Blocks

Chapter Topics:

- Bit shift registers
- First-in/first-out (FIFO) and last-in/first out (LIFO) queues
- Array functions
- Program control
- Shift register and FIFO/LIFO queue applications

OBJECTIVES

- Upon completion of this chapter, you will be able to:
- Describe the function of shift registers and FIFO/LIFO queues
- Describe the function and application of special array function blocks
- Apply shift registers and FIFO/LIFO queues to industrial situations
- Understand the use of program control and its use in applications

Scenario: Displaying the step number of a sequence.

A step number needs to be determined for a sequential system implemented with the technique of Chapter 6. A ControlLogix processor implements the sequential operation, where the step-in-progress bits are LC_Step[0].1 through LC_Step[1].31 (steps 1 - 63). The initial step is indicated by LC_Step[0].0. The step number that corresponds to a particular step-in-progress bit needs to be determined so that it can be displayed on an operator interface.

***Solution*:** The ControlLogix has a function block called FBC (file bit comparison) that can be used for this purpose. Figure 8.1 shows the ladder logic to accomplish this task and an example determination of the step number. The FBC block basically compares two bit arrays and reports the bit number(s) that are dissimilar. The reference bit array of 64 zero bits is set up in Step_Ref0[0] and Step_Ref0[1]. The step-in-progress bits are all in DINTs LC_Step[0] and LC_Step[1], which makes up an array of up to 64 bits. Only one bit is set and so the one mismatch is recorded in LC_Stp_Rslt, a one-element result array. The number in LC_Stp_Rslt[0] is thus the current step number of the sequential operation. The N.C. contact toggles the function block execution so that it executes on alternate program scans.

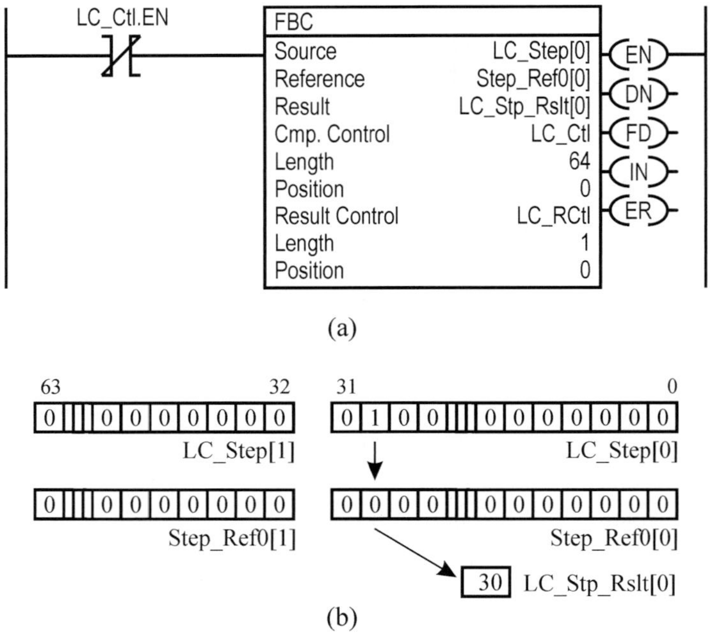

Figure 8.1. Determining current step number: *(a)* ladder logic; *(b)* example.

8.1 INTRODUCTION

The shift register, FIFO/LIFO, and special array ladder logic function blocks described in this chapter are used less frequently but are necessary for certain applications. In addition, the instructions that modify ladder logic execution are described. This chapter describes these types of function blocks for the PLCs covered by this text. As in previous chapters, the function blocks for a particular PLC are described in separate sections. That is, only the section(s) pertaining to the PLC(s) of interest need to be studied. Following these descriptions, simple examples illustrate their application.

The shift register function blocks provide the ability to track the flow of material in a manufacturing system. The shift register can be implemented as an array of Booleans (bits) or as an array of integers. A bit array shift register is often used to track the progress of a generic item (e.g., can, bottle, defective part) as it moves down the line. Depending on how items are removed from the integer array shift register, it is called a first-in/first-out (FIFO) queue or a last-in/first-out (LIFO) queue. A FIFO queue can track parts represented by part numbers or an assembly code as parts move down the line. In a LIFO queue, items are removed in reverse order, and so it is useful when tracking items having a serial number that are being stacked and unstacked.

Many of the PLCs covered by this text provide special functions to handle array calculations or other special array operations. Program control function blocks provide a way to modify the ladder logic scanning.

8.2 OTHER CONTROLLOGIX BLOCKS

The ControlLogix/CompactLogix PLC provides a rich set of ladder function blocks for bit shifting, FIFO/LIFO queues, special array functions, and program control. The basic operation and simple applications of these function blocks are described in this section. Longer bit shifting and FIFO/LIFO applications are presented in section 8.5.

Many of these instructions use the CONTROL data type structure, shown in Figure 3.2*a*. The parts of this structure that are pertinent to each function block are described in conjunction with the function block.

8.2.1 Shift Bit in Register

The bit shifting function blocks and the operation that occurs for each one are:

Block Name	Operation Result
BSL	Bit shift left
BSR	Bit shift right

An example BSL (bit shift left) function block is shown in Figure 8.2*a*, and an example BSR (bit shift right) function block is shown in Figure 8.3*a*. The operands of both of these blocks are described as follows:

Array – Start of bit array. Must be an element of a DINT array data type.
Control – CONTROL structure data type (Figure 3.2*a*).
Source Bit – Bit (Boolean) that will be shifted into the bit array
Length – Number of bits in the bit array.

For the "Control" tag defined in Figure 8.2*a*, the individual parts (fields) of the CONTROL structure are addressed as follows:

Reference	Description
Cnv1_PCtl.UL	Unload (bit that was shifted out)
Cnv1_PCtl.EN	Enable
Cnv1_PCtl.ER	Error (length < 0)
Cnv1_PCtl.LEN	Length of bit array

For each **off**-to-**on** transition of the BSL input, the bit array is shifted left one position. The "Source Bit" value replaces the rightmost bit, and the .UL bit is set to the value of the (shifted out) leftmost bit in the bit array. The .EN bit is identical to the block input condition. The .ER bit is set if the .LEN field is less than zero. The value of the array length (.LEN) can be accessed as a DINT. An example application of the BSL function block is shown in Figure 8.2*a*, and example bits in the bit array before and after a shift are shown in Figure 8.2*b*.

For each **off**-to-**on** transition of the BSR input, the bit array is shifted right one position. The "Source Bit" value replaces the leftmost bit, and the .UL bit is set to the value of the (shifted out) rightmost bit in the bit array. The .EN and .ER bits as well as the .LEN field have the same function as for the BSL block. An example application of the BSR function block is shown in Figure 8.3*a*, and the bits in the bit array before and after a shift are shown in Figure 8.3*b*.

To rotate the bits in a shift register, the "Source Bit" is set to the appropriate bit at the end of the bit array that is copied to the .UL bit when the bits are shifted. For example, for

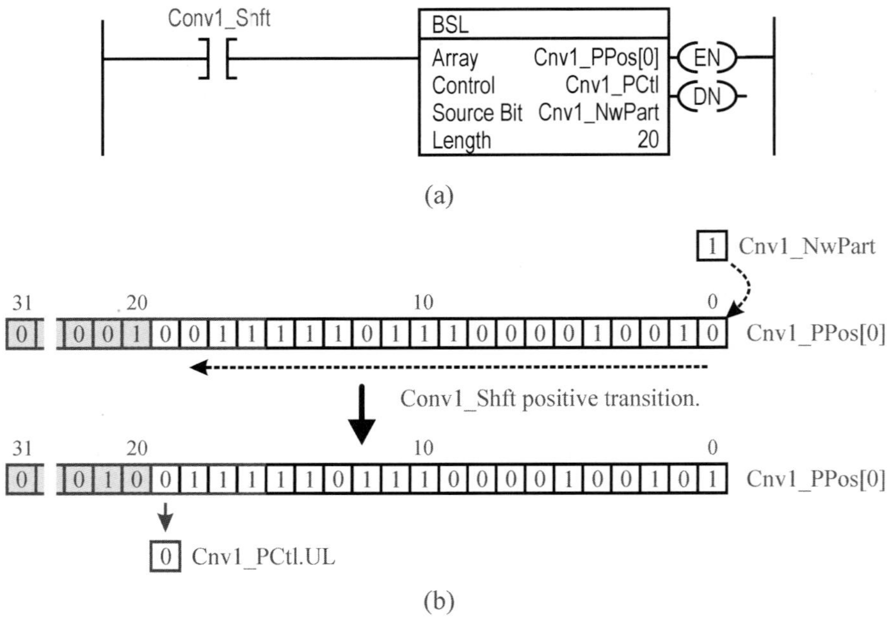

Figure 8.2. ControlLogix bit-shift left: *(a)* example BSL block; *(b)* bit array before and after shift.

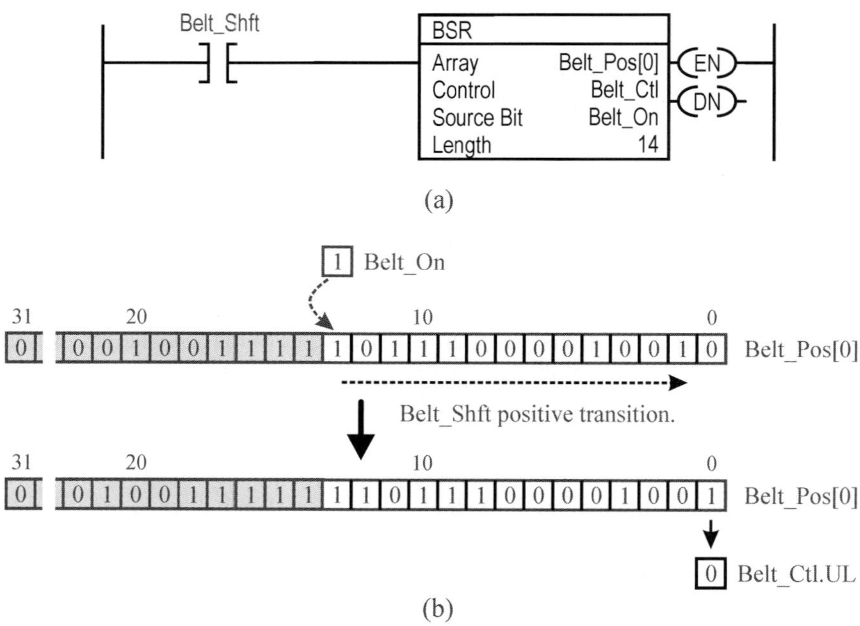

Figure 8.3. ControlLogix bit-shift right: *(a)* example BSR block; *(b)* bit array before and after shift.

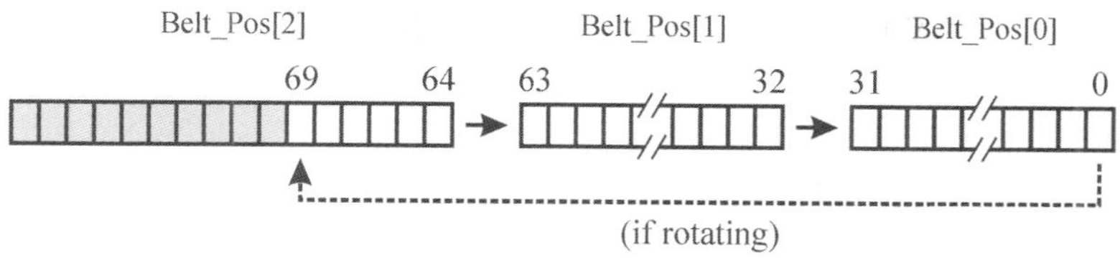

Figure 8.4. Multi-DINT shift register.

the BSL in Figure 8.2*a*, using Cnv1_PPos[0].19 for the "Source Bit" will convert the left shift into a left rotate. In Figure 8.3*a*, using Belt_Pos[0].0 for the "Source Bit" will change the right shift into a right rotate.

Shift registers longer than 32 bits are handled automatically since the "Array" operand is already treated as an array of DINTs. For example, the right shift of Figure 8.3*a* can easily be a 70-bit shift register by defining Belt_Pos as a 3 DINT array and changing the length to 70. In this case, the shift register is constructed as in Figure 8.4. The BSR function block handles the transfer of bits between DINTs automatically. To clear a shift register consisting of multiple DINTs, use the FLL block described in section 8.2.3.

8.2.2 FIFO and LIFO Queues

The array (queue) shifting function blocks and the operation that occurs for each one are:

Block Name	Operation Result
FFL	Load value to the next FIFO position
FFU	Unload value from first FIFO position
LFL	Load value to the next LIFO position
LFU	Unload value from last LIFO position

The basic difference between the operation of a first-in/first-out (FIFO) and a last-in/first-out (LIFO) array is the order in which the array unloads elements. For the FIFO array, elements unload in the same order as they are loaded. For the LIFO array, elements unload in the opposite order as they are loaded.

Normally, the FFL and FFU are used together to construct a FIFO data array. An example application of the FFL and FFU blocks is shown in Figure 8.5. For the FFL and FFU function blocks, the operands are described as follows:

FIFO – First element in a DINT or REAL array data type.
Control – CONTROL structure data type (Figure 3.2*a*).
Length – Maximum length of the FIFO array.
Position – Position where the next data will be loaded.
Source – (FFL only) Tag of the data that will be stored in the array.
Dest – (FFU only) Tag where the data will be placed after it is unloaded from the FIFO.

Typically, the FFL and FFU refer to the same "Control" tag. For the control tag defined in Figure 8.5, the individual fields of the CONTROL structure are addressed as follows:

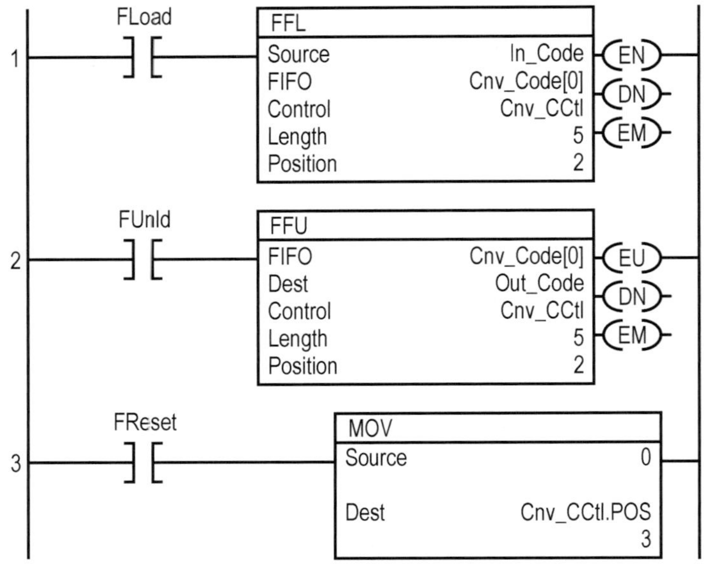

Figure 8.5. ControlLogix FIFO array function blocks.

Reference	Description
Cnv_CCtl.EN	Load enabled
Cnv_CCtl.EU	Unload enabled
Cnv_CCtl.DN	FIFO full
Cnv_CCtl.EM	FIFO empty
Cnv_CCtl.LEN	FIFO array length
Cnv_CCtl.POS	Next entry will load here

For each **off**-to-**on** transition of the FFL input, the "Source" value is placed in the FIFO array at the current position, and the position (.POS) is incremented by one. The .EN bit is identical to the FFL block input condition. The .DN bit is set if the .POS is equal to the array length. If data is added to an already full FIFO, the new data replaces the previously added data entry. The position value of the FIFO (.POS) can be accessed as DINT. In order to reset a FIFO array, set the .POS to zero.

For each **off**-to-**on** transition of the FFU input, the first element in the FIFO array is copied to "Dest", and the remaining data in the array is shifted one position toward the beginning of the array. The position is also decremented by one. The .EU bit is identical to the FFU block input condition. The .EM bit is set if the FIFO is empty (.POS=0). A zero is placed into "Dest" if one attempts to unload an already empty FIFO.

An example operation of the FFL and FFU of Figure 8.5 is shown in Figure 8.6. This example assumes the FIFO starts empty. Five DINTs are loaded onto the FIFO and then unloaded. The contents of the FIFO before and after the numbered scans are also shown. The arrow on the left of the array indicates the .POS value, and the arrow on the right indicates the value unloaded into Out_Code. The FIFO is loaded with 67, 14, 37, 25, and 40. When the FIFO is unloaded, the elements unload in the order 67, 14, 37, 25, and 40. The action at each of the numbered scans is as follows:

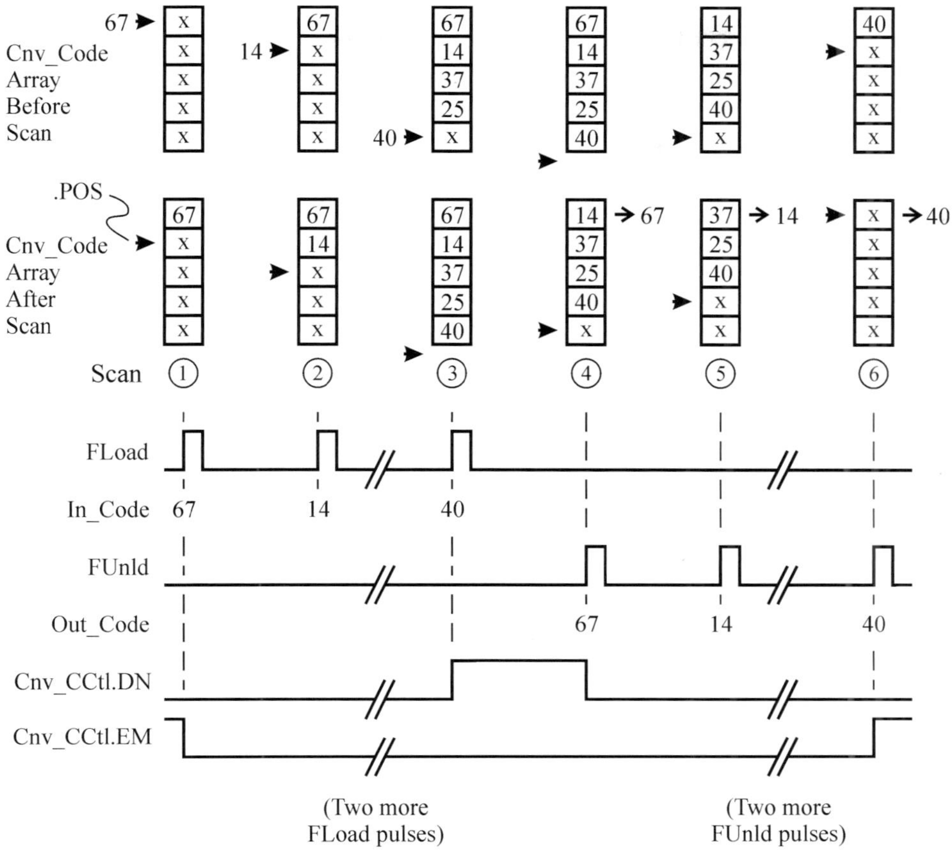

Figure 8.6. ControlLogix FIFO array operation.

1. Load stack with In_Code = 67, .EM turned **off**.
2. Load stack with In_Code = 14
   (37 and 25 loaded between scans 2 and 3)
3. Load stack with In_Code = 40, .DN turned **on**.
4. Unload stack, Out_Code set to 67, .DN turned **off**.
5. Unload stack, Out_Code set to 14
   (37 and 25 unloaded between scans 5 and 6)
6. Unload stack, Out_Code set to 40, .EM turned **on**.

Normally, the LFL and LFU are used together to construct a LIFO data array. An example application of the LFL and LFU blocks is shown in Figure 8.7. The operands for these instructions are identical to the FFL/FFU except for the "LIFO" operand, which is the first element of the LIFO array.

For each **off**-to-**on** transition of the LFL input, the "Source" value is placed in the LIFO array at the current position, and the position (.POS) is incremented by one. The .EN bit is identical to the block input condition. The .DN bit is set if the .POS is equal to the array length. If data is added to an already full LIFO, the new data replaces the previously added

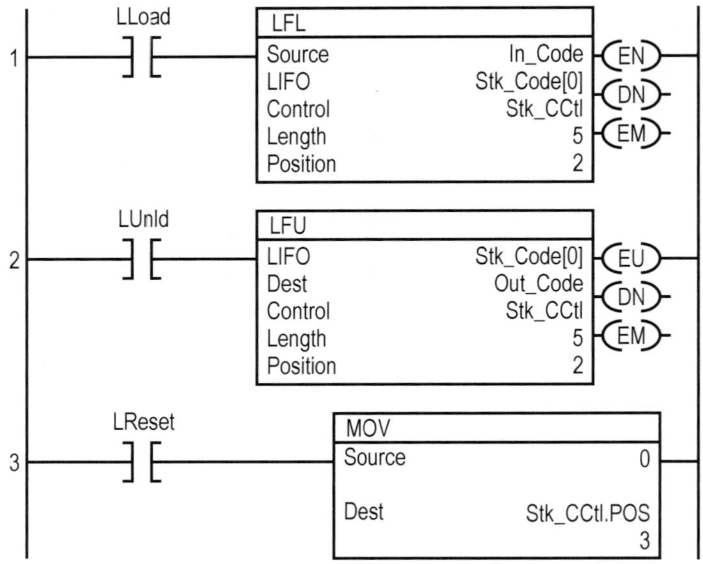

Figure 8.7. ControlLogix LIFO array function blocks.

data entry. The position value of the LIFO (.POS) can be accessed as DINT. In order to reset a LIFO array, set the .POS to zero.

For each **off**-to-**on** transition of the LFU input, the last element in the LIFO array is copied to "Dest", and the position is decremented by one. The .EU bit is identical to the block input condition. The .EM bit is set if the LIFO is empty (.POS=0). A zero is placed into "Dest" if one attempts to unload an already empty LIFO.

Figure 8.8 shows an example operation of the LFL and LFU of Figure 8.7. The LIFO starts empty. Five DINTs are loaded onto the LIFO and then unloaded. The contents of the LIFO before and after the numbered scans are also shown. The LIFO is loaded with 67, 14, 37, 25, and 40. When the LIFO is unloaded, the elements are unloaded in the order 40, 25, 37, 14, and 67. The action at each of the numbered scans is as follows:

1. Load stack with In_Code = 67, .EM turned **off**.
2. Load stack with In_Code = 14
   (37 and 25 loaded between scans 2 and 3)
3. Load stack with In_Code = 40, .DN turned **on**.
4. Unload stack, Out_Code set to 40, .DN turned **off**.
5. Unload stack, Out_Code set to 25
   (37 and 14 unloaded between scans 5 and 6)
6. Unload stack, Out_Code set to 67, .EM turned **on**.

8.2.3 Other Array Function Blocks

Other specialized numerical and bit array function blocks and the operation that occurs for each one are:

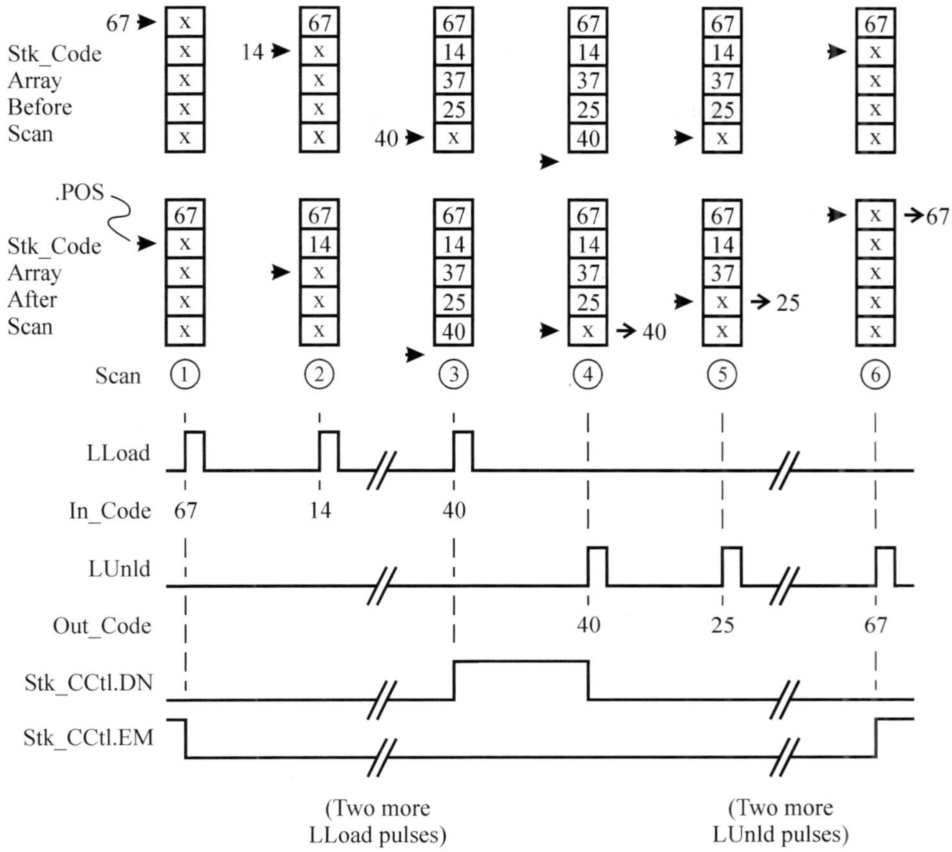

Figure 8.8. ControlLogix LIFO array operation.

Block Name	Operation Result
FAL	Array (file) arithmetic and logic
FSC	Array (file) search and compare
FLL	Fill array with a value
COP	Copy values or structures
FBC	Find mismatches between two bit arrays
DDT	Find and correct mismatched bit arrays

The operation of each of these blocks is explained and illustrated with short examples. Other array blocks are available (e.g., array sorting and averaging) but are not covered in this text.

The FAL (file arithmetic/logic) function block performs the same operations on arrays as the CPT (compute) function block performs on individual elements. In addition, this block can be programmed to perform the operation on the entire array when it executes or incrementally, that is, perform only a few operations each time the block is scanned. The latter mode is useful when processing large arrays or data that are not time-critical. Only

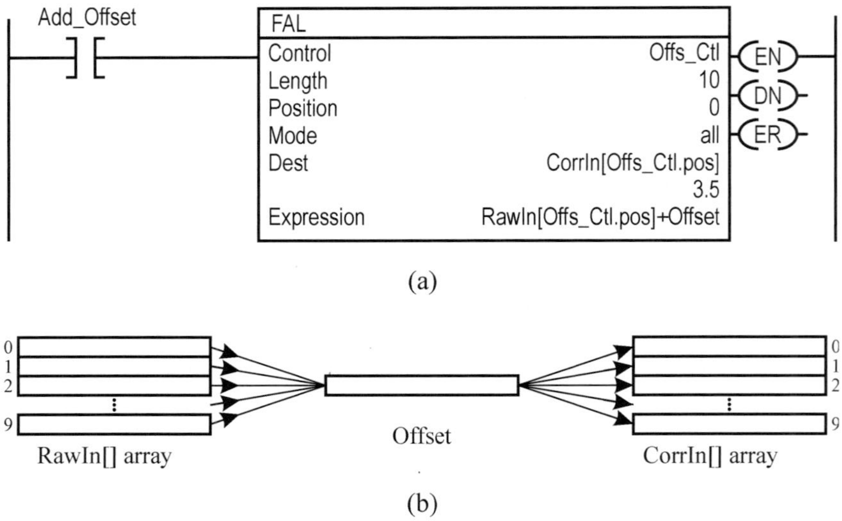

Figure 8.9. ControlLogix FAL array plus constant example: *(a)* ladder logic; *(b)* processing of array elements.

simple applications are described in this section. The interested reader should consult the instruction set reference (Rockwell Automation, 2022a) for more details.

An example FAL function block is shown in Figure 8.9. The operands are as follows:

Control – CONTROL structure data type (Figure 3.2*a*).
Length – Number of elements in the array to be manipulated.
Position – Current position in the array, initially it is typically zero.
Mode – How array elements are processed (all, inc, or a number).
Dest – Tag of the expression result.
Expression – A valid arithmetic, logical, conversion, and copy operation.

For the "Expression" operand, the instruction set reference (Rockwell Automation, 2022a) lists valid expressions. The .POS value is often used as an array index for the "Dest" and/or the "Expression" operands.

For the control tag defined in Figure 8.9*a*, the individual fields of the CONTROL structure are addressed as follows:

Reference	**Description**
Offs_Ctl.EN	FAL enabled
Offs_Ctl.DN	Operation done
Offs_Ctl.ER	Overflow for one array element
Offs_Ctl.LEN	Number of elements in array
Offs_Ctl.POS	Index of current element being processed

The possible values of "Mode" are:

all	When block input condition changes from **off** to **on**, all specified array elements are processed.

inc — One specified array element processed for each **off**-to-**on** transition of the block input condition. The .DN bit is set when all array elements have been processed.

[number] — When block input condition changes from **off** to **on**, [number] of array elements are processed on each scan of this instruction until all specified array elements are processed. The .DN bit is set when all array elements have been processed. The [number] is any valid non-zero positive DINT.

Figure 8.9 shows an example FAL used to add a constant value (Offset) to every element in the RawIn array and store the results in the CorrIn array. The operation is in the "all" mode, meaning all 10 specified additions are processed for each **off**-to-**on** transition of the FAL input. If the "Mode" operand is set to "3", then it will require 4 program scans to process all 10 array elements. If the "Mode" operand is set to "inc", then 10 **off**-to-**on** transitions of Add_Offset are required to process all 10 elements of the arrays.

Figure 8.10 shows an example FAL used to do an element-by-element multiplication of one array by another array to produce a third array. Each element of RawVals is multiplied by the corresponding element in AdjMult to produce an element of the AdjVals array. Since this is a large array, the "Mode" operand is set to "20", and it will require 50 program scans to process all 1000 array elements.

Design Tip

The FAL function block is useful for performing a wide variety of arithmetic operations on data arrays.

The FSC (file search/compare) function block performs a similar operation on arrays as the CMP (compare) function block performs on individual elements, except that it finds the

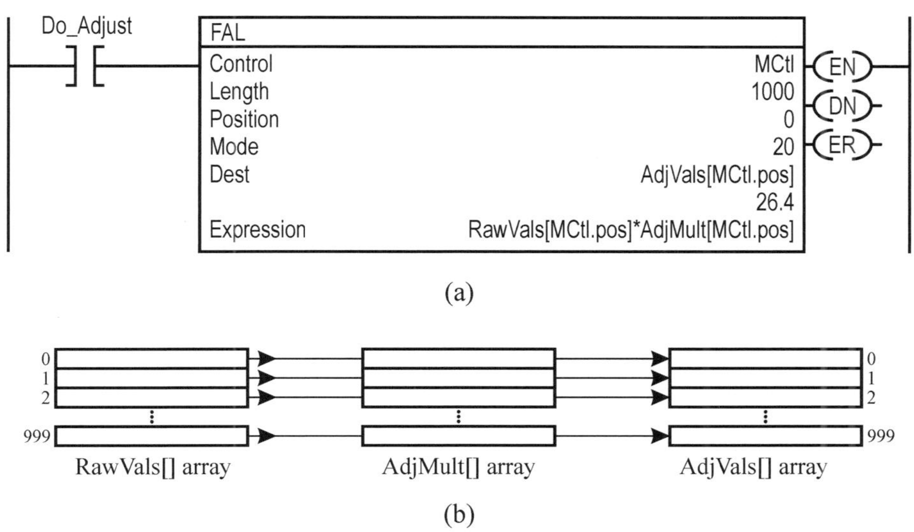

Figure 8.10. ControlLogix FAL array multiplication example: *(a)* ladder logic; *(b)* processing of array elements.

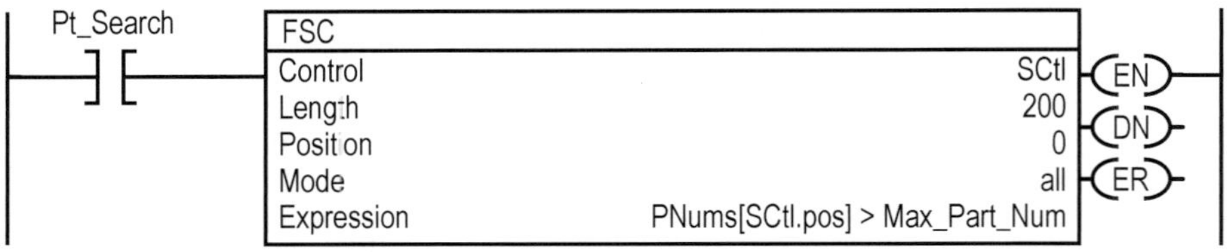

Figure 8.11. ControlLogix FSC example.

first array element that satisfies the comparison. Like the FAL block, the FSC block can be programmed to perform the comparison on the entire array when it executes or incrementally, that is, perform only a few comparisons each time the block is scanned. The latter mode is useful when processing large arrays or data that are not time-critical.

An example FSC function block is shown in Figure 8.11. The operands are as follows:

Control – CONTROL structure data type (Figure 3.2*a*).
Length – Number of elements in the array to be searched.
Position – Current position in the array, initially it is typically zero.
Mode – How array elements are processed (same as FAL "Mode").
Expression – A valid comparison operation.

The .POS value is often used as an array index for the "Expression" operand.

For the control tag defined in Figure 8.11, the individual fields of the CONTROL structure are addressed as follows:

Reference	Description
SCtl.EN	FSC enabled
SCtl.DN	Searched to end of array
SCtl.IN	Inhibit bit, **on** when true comparison detected; must clear to continue search
SCtl.FD	Found bit, **on** when true comparison detected
SCtl.LEN	Number of elements in array
SCtl.POS	Index of current element being processed

Figure 8.11 shows an example FSC used to search an array of part numbers for the first one that is greater than the value in Max_Part_Num. The operation is in the "all" mode, meaning that all 200 array elements may be searched for each **off**-to-**on** transition of the FSC input. The SCtl.FD and SCtl.IN bits will be set if a part number that matches the expression is found. The value of SCtl.POS is the index of the PNums array that has the part number matching the search criterion. If one wanted to search for another match, leave SCtl.POS unmodified, unlatch the SCtl.IN bit, and re-initiate the FSC function block.

The FLL (fill) function block fills the elements of an array with the same value and is useful for initialization. An example FLL block is shown in Figure 8.12. The "Source" operand is either the tag containing the value to copy or a constant value. The "Dest" is the tag of the first element in an integer or real array data type. The "Length" is the number of elements to fill. For best results, the data type of the "Source" should match the data type of each element of the "Dest" array.

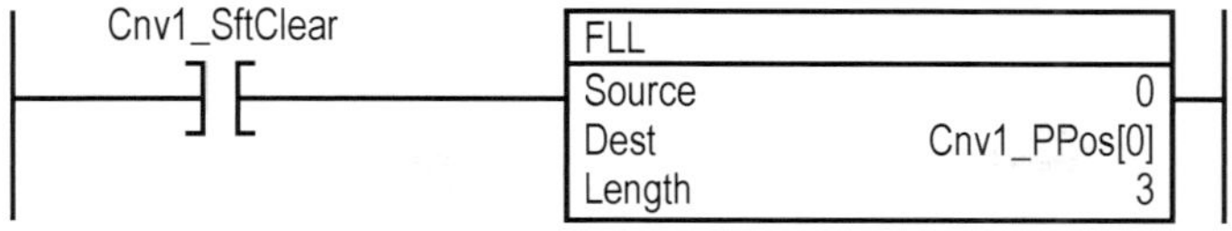

Figure 8.12. ControlLogix FLL example.

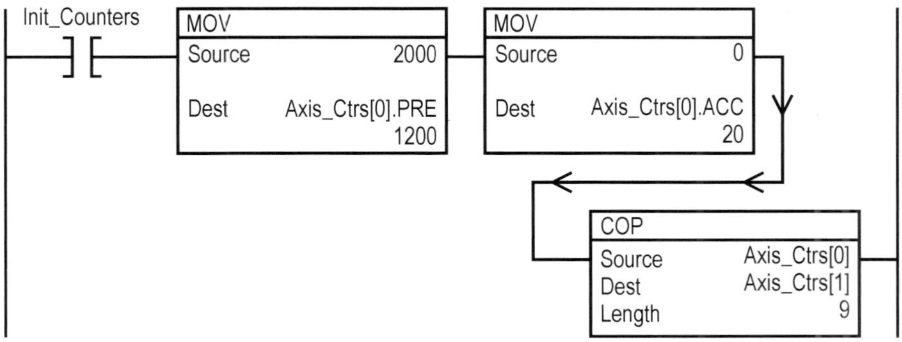

Figure 8.13. ControlLogix COP example.

The COP (copy) function block is similar to the FLL function block, except it will also initialize arrays of structures (e.g., an array of counters or timers). An example application of the COP block is shown in Figure 8.13 that initializes an array of counters. The first counter in the array is initialized and then it is copied to the other 9 counters in the array. The "Source" and "Dest" operands are basically the same as for the FLL block, except that the "Source" can be a structure data type and the "Dest" can be the element of an array of structures.

The FBC (file bit compare) function block compares two bit arrays and records the bit number of each difference in a result array. An example FBC block is shown in Figure 8.14*a*. The operands of this block are described as follows:

Source – Bit array compared to reference. First location of DINT array.
Reference – Reference bit array. First location of DINT array.
Result – Array to store search results. First location of DINT array.
Cmp. Control – CONTROL structure for compare.
Length – Number of bit locations to compare
Position – Current position in source array. Typically initialized to zero.
Result Control – CONTROL structure for the results.
Length – Number of storage locations in result array.
Position – Current position in result array. Typically initialized to zero.

The "Cmp. Control" and "Result Control" operand must be different tags and a CONTROL data type (Figure 3.2*a*). For the "Cmp. Control" tag defined in Figure 8.14*a*, the individual fields of the CONTROL structure are addressed as follows:

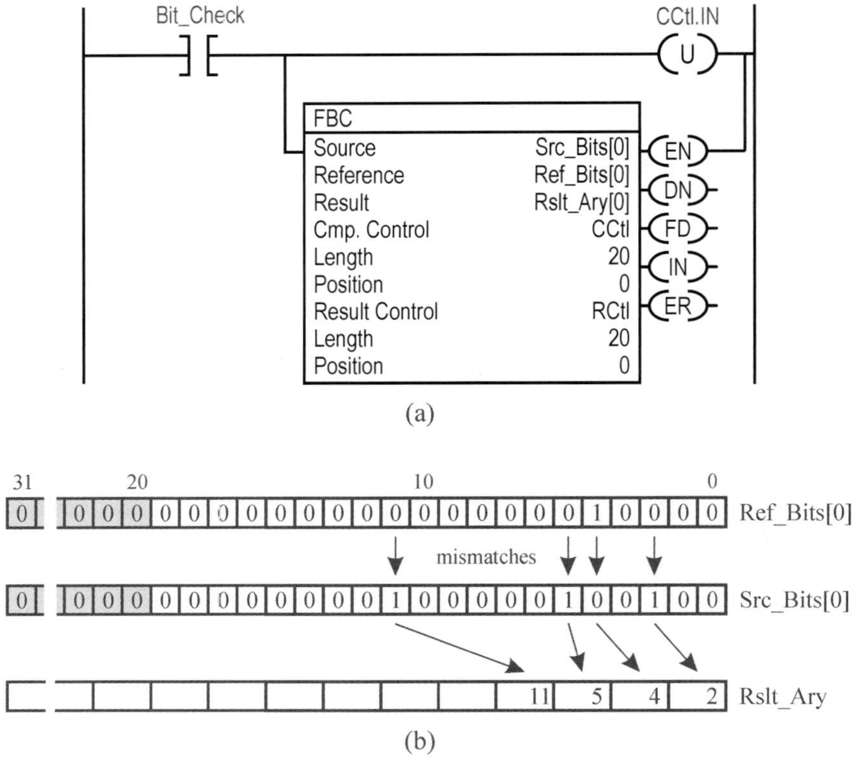

Figure 8.14. ControlLogix FBC example: *(a)* ladder logic; *(b)* FBC execution results.

Reference	Description
CCtl.EN	FBC enabled
CCtl.DN	Set when last bit in array compared.
CCtl.FD	Set when mismatch recorded (one-at-a-time operation) or all mismatches recorded (all-per-scan) operation.
CCtl.IN	Search mode bit, **off** to search for all mismatches, **on** to search for next mismatch only.
CCtl.ER	Error bit, **on** when parameters illegal.
CCtl.LEN	Number of bits to compare.
CCtl.POS	Index of current bit.

For the "Result Control" tag defined in Figure 8.14*a*, the individual fields of the CONTROL structure are addressed as follows:

Reference	Description
RCtl.DN	Set when result array full.
RCtl.LEN	Length of result array.
RCtl.POS	Current position in result array.

For the "Source," "Reference," and "Result" tags, one should not use .POS as the array subscript.

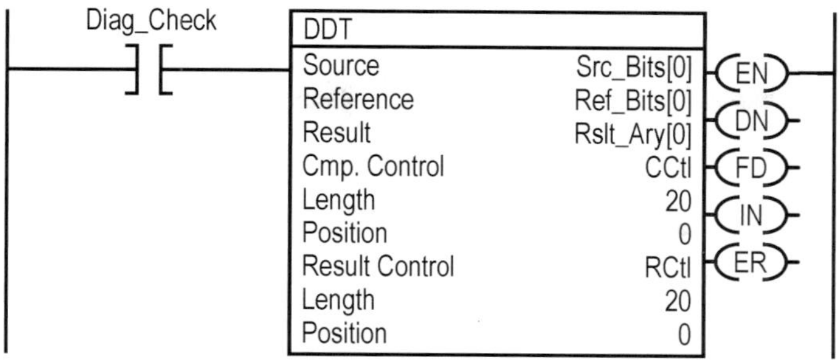

Figure 8.15. ControlLogix DDT function block.

Figure 8.14*a* shows an example FBC used to compare 20 bits of Src_Bits[0] with 20 bits of Ref_Bits[0]. The CCtl.IN bit is **off** to search for all mismatches, and the results are placed in Rslt_Ary. For the particular values of the bit arrays, the result of the function block execution is shown in Figure 8.14*b*.

The DDT (diagnostic detect) function block compares two bit arrays, records the bit number of each difference in a result array, and then changes the bit in the reference array to match the bit in the source array. The DDT function block functions in the same manner as the FBC function block except that the DDT function block changes the reference bits to match the source bits. An example DDT block is shown in Figure 8.15.

8.2.4 Program Execution Control

The ControlLogix PLC offers a number of instructions that can control ladder logic execution. The names and the operation that occurs for each one are:

Block Name	Operation Result
JMP	Jump to a defined label
LBL	Jump destination
JSR	Jump to subroutine, passing data
SBR	Pass data to subroutine
RET	Return results from subroutine
MCR	Master control relay; disable logic section
FOR	Repeatedly execute a routine
BRK	Terminate the repeated execution of a routine.
AFI	Disable rung logic
NOP	No operation
TND	Mark temporary end of ladder

Depending on the particular instruction, it may be a contact, or a coil, or a function block. The operation of each of these instructions is explained and illustrated with short examples.

The JMP (jump) coil and LBL (label) contact are used together to skip portions of ladder logic. When the JMP coil is turned **on**, the ladder scan skips to the referenced LBL contact. One can jump either forward or backward. Jumping backward lets the controller iterate certain sections of logic. However, jumping backward an excessive number of times will cause the controller to fault if the scan time is too long. Jumping forward decreases the

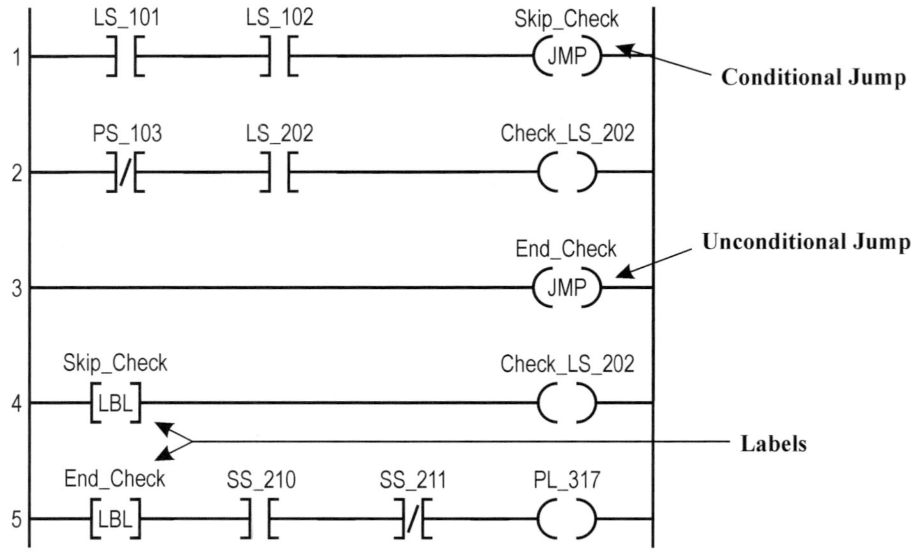

Figure 8.16. ControlLogix JMP and LBL example.

scan time by skipping logic segments until needed. One drawback to jumping forward is that the logic between the jump coil and the label contact is not scanned, so any output coils are not changed. An example ladder segment with two jump and two label instructions is shown in Figure 8.16.

The JSR (jump to subroutine), SBR (subroutine), and RET (return) function blocks are used to execute routines other than the main program routine. Large programs are often broken into subroutines which are called from the main routine. When enabled, the JSR function block directs logic execution to the first rung of the specified routine, and if needed, passes parameters to the routine. If parameters are passed to the subroutine, the first rung of the subroutine must have a SBR function block. Otherwise, if no parameters are passed to the subroutine, the SBR block is not needed. Likewise, the RET function block appears on the last rung of a subroutine if parameters are passed back to the calling routine. Actually, the RET block can have conditions to enable it and occur before the last subroutine rung. In this case, if the RET block is not executed, ladder logic execution continues after the rung that contains the RET block. If the RET block is omitted, the

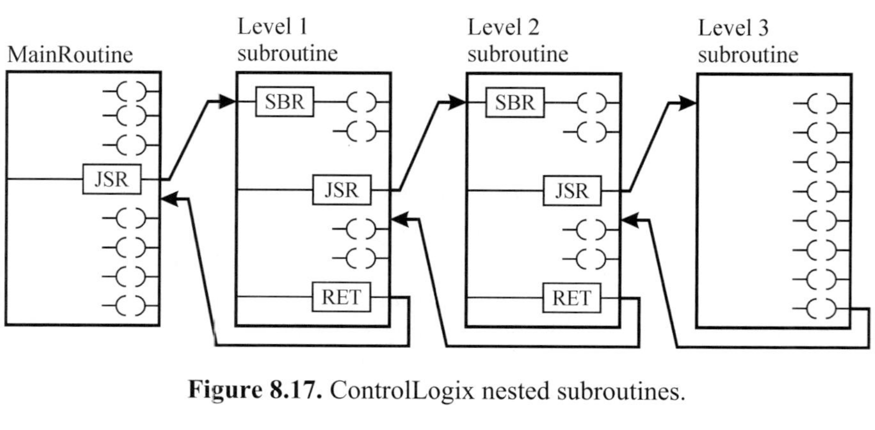

Figure 8.17. ControlLogix nested subroutines.

subroutine automatically returns back to the calling routine after the last subroutine rung is scanned, but no parameters are passed back.

Subroutines can call other subroutines, often called *nested subroutines*, as shown in Figure 8.17. Other than controller memory, there is no limit on the number of nested subroutines.

The parameters are passed into and out of a subroutine by value. The values are copied back and forth between the memory of the calling routine and the subroutine, which increases execution time. Directly accessing program-scoped and controller-scoped tags from within the subroutine will reduce execution time. However, the subroutines may not be easily reused in another PLC program because it will require certain globally-defined tags. Array elements, structures, and entire arrays can all be passed as parameters. In all cases, the data type of the tag passed to a subroutine should match the data type of the subroutine tag that receives the tag.

Figure 8.18 shows an example use of a subroutine. The main routine calls the Add_Two_Num subroutine, passing two values, Real_1 and Real_2. The Add_Two_Num subroutine receives the two values and places them into Num_A and Num_B. The result of

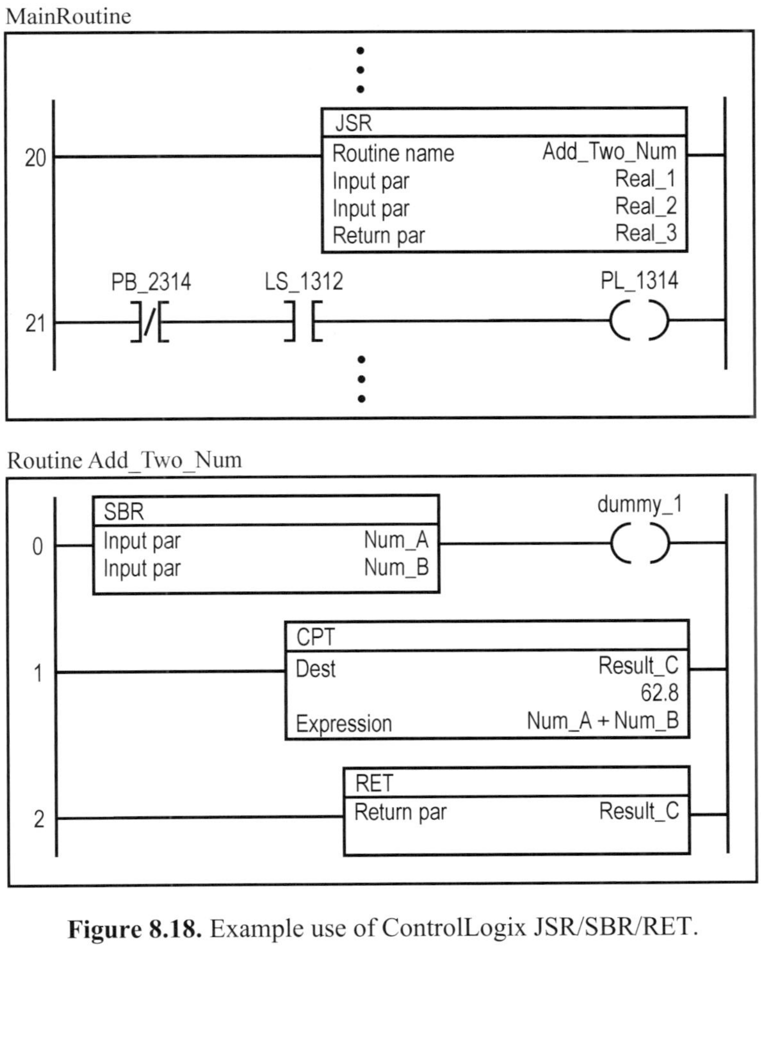

Figure 8.18. Example use of ControlLogix JSR/SBR/RET.

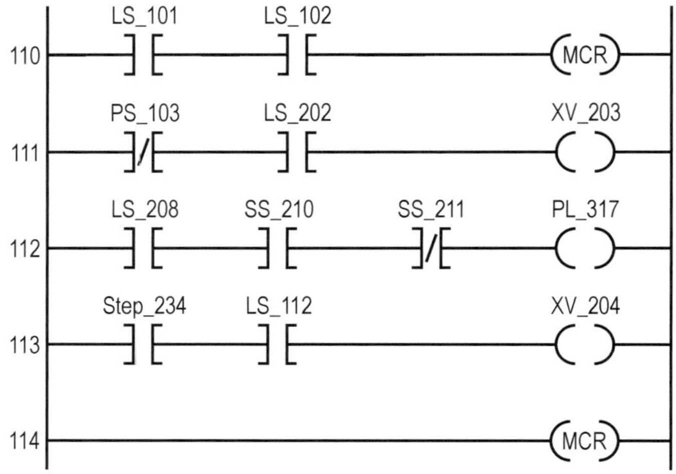

Figure 8.19. Example use of ControlLogix MCR.

the calculation is placed in Result_C. The RET block then passes the value in Result_C back to the main routine, which places the value into Real_3.

The MCR (master control relay) output coil instruction functions like a software master control relay, disabling non-retentive outputs within a range of rungs. This coil is used in pairs to create "zones" of the ladder program where all non-retentive outputs may be disabled. One MCR defines the start of the zone, and one MCR defines the end of the zone. Rung logic on the rung of the first MCR controls execution of the zone. When this rung logic turns **on** the MCR coil (called "enabling the zone"), all rungs in the zone are scanned normally. When the logic of the first MCR rung causes the MCR coil to be **off**, all rungs in the zone are scanned but only to turn **off** all non-retentive outputs in the zone.

There are some cautions when using an MCR zone:

1. The zone must be ended with an unconditional MCR output coil.
2. MCR zones cannot be nested.
3. Do not jump into or out of a MCR zone.

Also, the MCR instruction is not a substitute for a hard-wired master control relay that provides an emergency power shutdown to the I/O.

An example MCR zone is shown in Figure 8.19. When LS_101 and LS_102 are both **on**, the MCR zone is enabled, and rungs 111 to 113 are scanned normally determining the values of the output coils XV_203, PL_317, and XV_204. When either LS_101 or LS_102 is **off**, the MCR zone is disabled. In this case, XV_203, PL_317, and XV_204 are all turned **off**, regardless of the state of the input conditions for rungs 111 to113.

The FOR function block is used to repeatedly call a subroutine. The BRK output coil is used within the subroutine to exit the subroutine and exit the FOR loop. A RET output coil in the subroutine exits the subroutine but does not exit the FOR loop (unless the index is greater than or equal to the terminal value). An example FOR function block is shown in Figure 8.20. When enabled, the FOR function block repeatedly executes the "Routine name" until the "Index" value is greater than or equal to the "Terminal value". After each

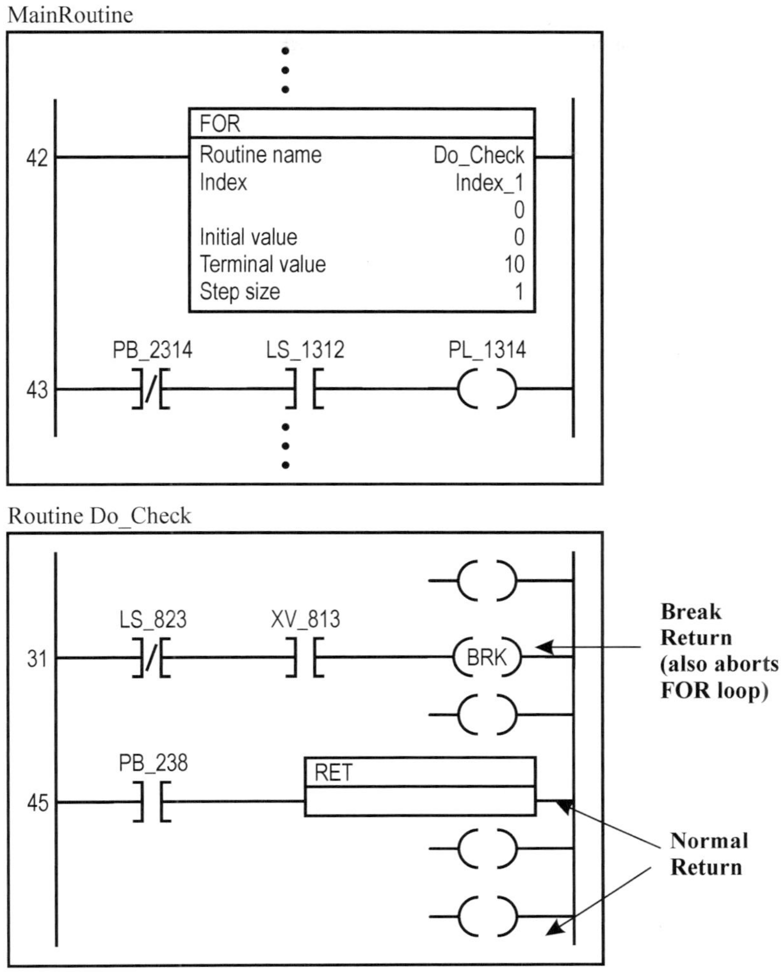

Figure 8.20. Example use of ControlLogix FOR and BRK.

execution of the subroutine, the "Step size" is added to the "Index" and then compared to the "Terminal value".

The FOR block in Figure 8.20 executes the Do_Check subroutine ten times. This example also shows the use of the BRK output coil and the RET block in the subroutine called by the FOR function block. If rung 31 is scanned and LS_823 is **off** and XV_813 is **on**, then the BRK output coil causes the Do_Check subroutine to be exited and the FOR loop to be exited. The next rung scanned is rung 43 in the main routine. If Do_Check rung 45 is scanned and PB_238 is **on**, then the subroutine is exited, but the FOR loop is not exited (unless this is the tenth time Do_Check has been executed).

The AFI and NOP contacts and the TND coil are useful for troubleshooting and debugging. Figure 8.21 shows sample uses of all of these ladder logic instructions. The AFI (always false instruction) contact is used to temporarily disable rung logic. Its result is always false, and so any output coils are turned **off**, and any output function blocks are not

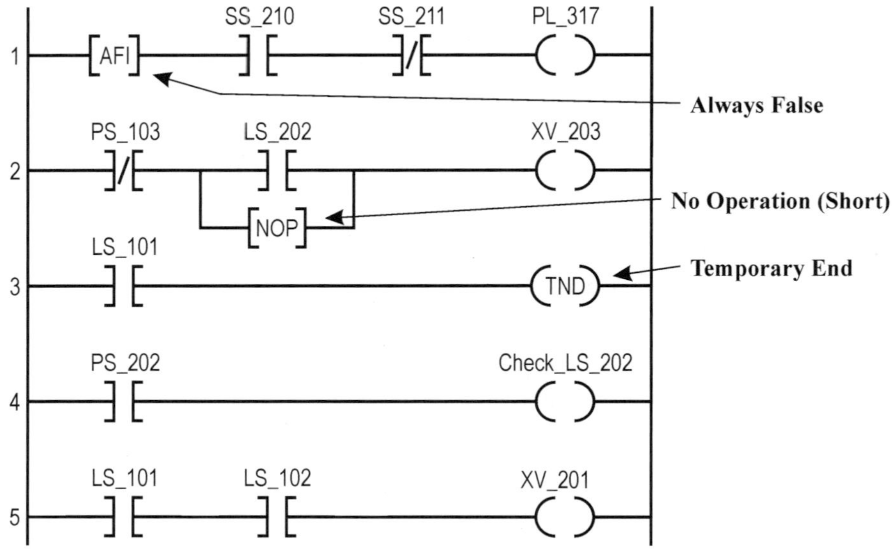

Figure 8.21. Example use of ControlLogix AFI, NOP, and TND.

executed. For rung 1 of Figure 8.21, the AFI contact disables the other rung logic and keeps PL_317 **off**. The NOP (no operation) contact acts like a short when placed in parallel with other contacts. In rung 2, the NOP shorts across LS_202 and essentially bypasses it, forcing the logic to function as if LS_202 is **on**. When enabled, the TND (temporary end) coil acts like the end of a routine. If the TND coil is in a subroutine, control passes to the calling routine. When LS_101 is **on** in Figure 8.21, the TND output coil on rung 3 is enabled and the remainder of the routine rungs are not scanned. When used for troubleshooting, the TND coil is placed at the end of the debugged code. As each new section is debugged, the TND is moved on through the rung

> **Design Tip**
>
> The AFI and NOP contacts and the TND coil are primarily used for troubleshooting.

8.3 OTHER MICRO800 BLOCKS

The Micro800 PLCs provide ladder function blocks for bit shifting, FIFO/LIFO queues, special array functions, and program control. The basic operation and simple application of these function blocks are described in this section. Longer bit shifting and FIFO/LIFO applications are presented in section 8.5.

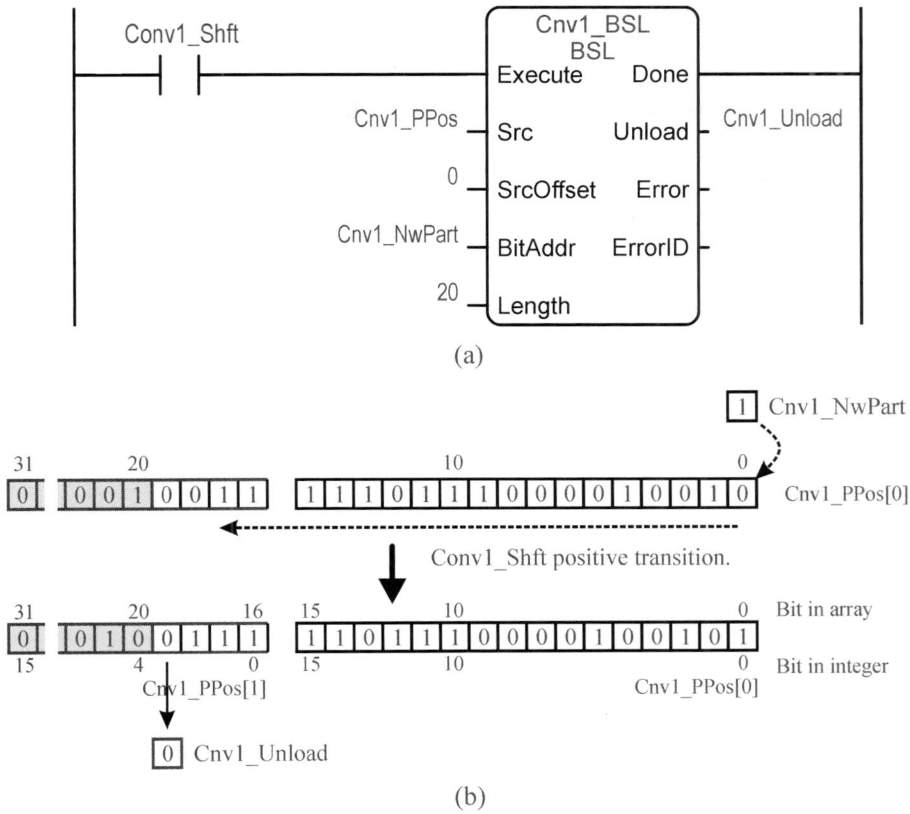

Figure 8.22. Micro800 bit-shift left: *(a)* example BSL block; *(b)* bit array before and after shift.

8.3.1 Shift Bit in Register

The bit shifting function blocks and the operation that occurs for each one are:

Block Name	Operation Result
BSL	Bit shift left
BSR	Bit shift right
ROL	Bit rotate left
ROR	Bit rotate right
SHL	Bit shift left, zeroes filled from right
SHR	Shift right, zeroes filled from left

The BSL and BSR blocks can shift bits within any of the 16- or 32-bit integer or word simple data types, an array of these types, or an array of BOOLs. An example BSL (bit shift left) function block is shown in Figure 8.22*a*, and an example BSR (bit shift right) function block is shown in Figure 8.23*a*. The inputs of both of these blocks are described as follows:

Execute – a positive transition (false-to-true) does the shift.

Src – Start of bit array, a 16- or 32-bit integer or word, an array of these types, or a BOOL array.

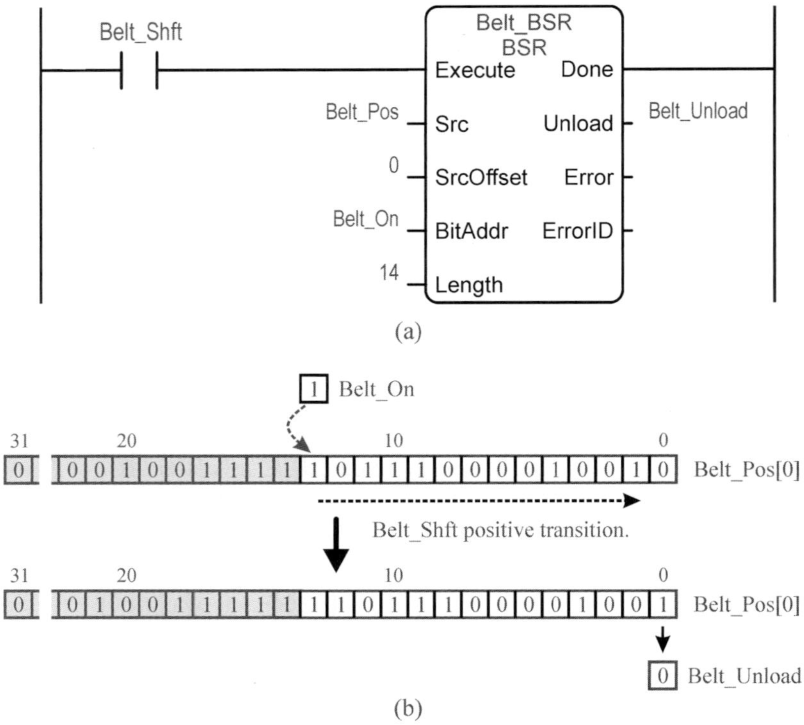

Figure 8.23. Micro800 bit-shift right: *(a)* example BSR block; *(b)* bit array before and after shift.

SrcOffset – Index of Src array element that is the start of the bit array. If Src is an elementary variable, set to zero.

BitAddr – Bit (BOOL) that will be shifted into the bit array

Length – Number of bits in the bit array. Maximum is 2048.

The outputs of both of these blocks are described as follows:

Done – True if operation completed successfully, otherwise false.

Unload – Bit shifted out of the bit array.

Error – If true, error occurred. Remains true as long as the Execute input is true.

ErrorID – Error code if Error is true.

For each **off**-to-**on** transition of the BSL Execute input, the bit array is shifted left one position. The BitAddr input value replaces the rightmost bit, and the Unload output is set to the value of the (shifted out) leftmost bit in the bit array. Possible ErrorID values are listed in Rockwell Automation (2022b). An example application of the BSL function block is shown in Figure 8.22*a*, and example bits in the bit array before and after a shift are shown in Figure 8.22*b*. This particular example shows the operation when the length of the array is longer than an individual array element. The Cnv1PPos is an INT [0..3] data type. Note that for this example, the length is less than 33 and so the shifting is restricted to the first two array elements. There is no shifting in Cnv1PPos[2]. If the SrcOffset input is one, the rightmost position of the bit array is bit zero of Cnv1PPos[1]. The particular example could be handled as one DINT variable for the Src input.

For each **off**-to-**on** transition of the BSR Execute input, the bit array is shifted right one position. The BitAddr input value replaces the leftmost bit, and the Unload output is set to the value of the (shifted out) rightmost bit in the bit array. The other inputs and outputs have the same function as for the BSL block. An example application of the BSR function block is shown in Figure 8.23*a*, and the bits in the bit array before and after a shift are shown in Figure 8.23*b*.

To rotate the bits in a shift register, the BitAddr input is the same variable as the Unload output. For example, for the BSL in Figure 8.22*a*, using Cnv1_Unload for the BitAddr input will convert the left shift into a left rotate. In Figure 8.23*a*, using Belt_Unload for the BitAddr input will change the right shift into a right rotate.

The ROL, ROR, SHL, and SHR blocks only operate on DINT variables. The inputs of the ROL and ROR blocks are described as follows:

IN – Integer to be rotated.
NbR – Number of 1-bit rotations.

The inputs of the SHL and SHR blocks are described as follows:

IN – Integer to be rotated.
NbS – Number of 1-bit shifts.

The output with the name of the block contains the rotated/shifted integer.

An example ROL block is shown in Figure 8.24*a* and an example SHR block is shown in Figure 8.24*c*. Figure 8.24 shows the result of rotating the double integer AD two bits left (Figure 8.24*b*) and shifting the same integer two bits right (Figure 8.24*d*).

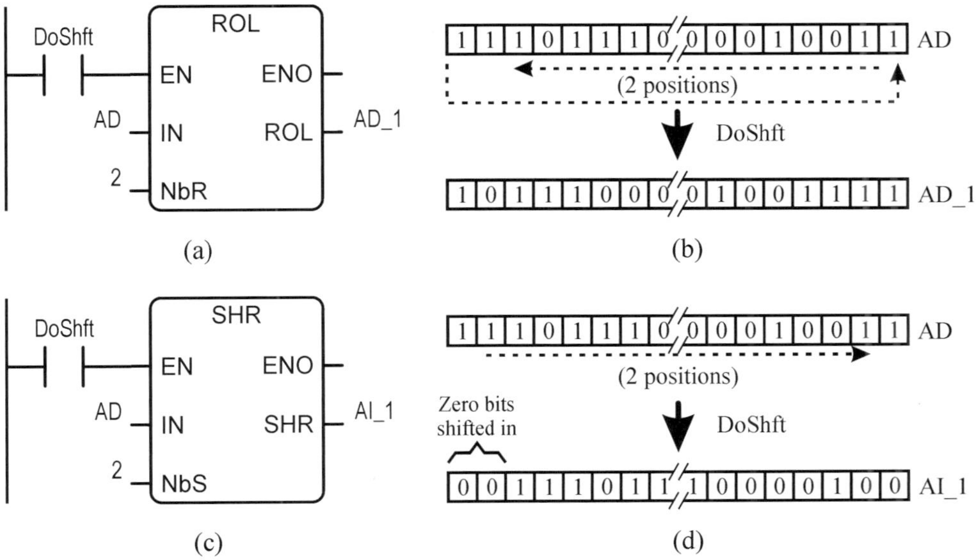

Figure 8.24. Micro800 bit-shift examples: (a) ROL block; (b) ROL input and shift result; (c) SHR block; (d) SHR input and shift result.

8.3.2 FIFO and LIFO Queues

The array (queue) shifting function blocks and the operation that occurs for each one are:

Block Name	Operation Result
FFL	Load value to the next FIFO position
FFU	Unload value from first FIFO position
LFL	Load value to the next LIFO position
LFU	Unload value from last LIFO position

The basic difference between the operation of a first-in/first-out (FIFO) and a last-in/first-out (LIFO) array is the order in which the array unloads elements. For the FIFO array, elements unload in the same order as they are loaded. For the LIFO array, elements unload in the opposite order as they are loaded.

Normally, the FFL and FFU are used together to construct a FIFO data array. An example application of the FFL and FFU blocks is shown in Figure 8.25. For the FFL block, the inputs are described as follows:

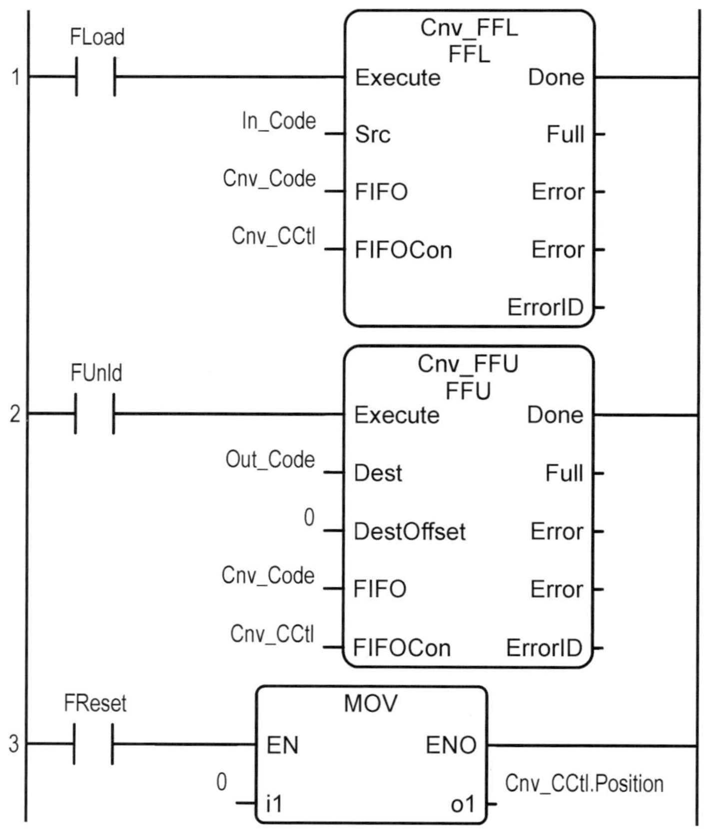

Figure 8.25. Micro800 FIFO array function blocks.

Execute – a positive transition (false-to-true) does the shift.
Src – Variable placed in next FIFO position.
FIFO – FIFO array, an array of any elementary type except for STRING.
FIFOCon – Configuration and control for FIFO, FF_LF_CON data type.

For the FFU block, the inputs are described as follows:

Execute – a positive transition (false-to-true) does the shift.
Dest – Variable unloaded from FIFO placed in this variable.
DestOffset – Index of Dest array element. If Dest is not an array, set to zero.
FIFO – FIFO array, an array of any elementary type except for STRING.
FIFOCon – Configuration and control for FIFO, FF_LF_CON data type.

The outputs of both of these blocks are described as follows:

Done – True if operation completed successfully, otherwise false.
Full – True when FIFO is full, Position is equal to Length.
Empty – True when FIFO is empty, Position is equal to zero.
Error – If true, error occurred. Remains true as long as the Execute input is true.
ErrorID – Error code if Error is true.

Possible ErrorID values are listed in Rockwell Automation (2022b). The FFL and FFU have the same FIFO and FIFOCon variable. For the Cnv_CCtl variable that is the FIFOCon input to both the FFL and FFU blocks in Figure 8.25, the individual fields of the CnvCCtl structure are addressed as follows:

Reference	Description
Cnv_CCtl.Length	FIFO array length
Cnv_CCtl.Position	Next entry will load here

For each **off**-to-**on** transition of the FFL Execute input, the Src value is placed in the FIFO array at the current position, and the position (Cnv_CCtl.Position) is incremented by one. The Cnv_FFL.Full and Cnv_FFU.Full Booleans are set true if the .Position is equal to the array length. If the FIFO is full, an error is reported and nothing is added to the FIFO array. The position value of the FIFO (Cnv_CCtl.Position) can be accessed as UINT. In order to reset a FIFO array, set the .Position to zero.

For each **off**-to-**on** transition of the FFU Execute input, the first element in the FIFO array is copied to the Dest input variable, and the remaining data in the array is shifted one position toward the beginning of the array. The position is also decremented by one. The Cnv_FFL.Empty and Cnv_FFU.Empty Booleans are set true if the FIFO is empty (Cnv_CCtl.Position=0). An error is reported and the Dest variable is not changed if one attempts to unload an already empty FIFO.

An example operation of the FFL and FFU of Figure 8.25 is shown in Figure 8.26. This example assumes the FIFO starts empty and Cnv_CCtl.Length has an initial value of 5. Five DINTs are loaded onto the FIFO and then unloaded. The contents of the FIFO before and after the numbered scans are also shown. The arrow on the left of the array indicates the .POS value, and the arrow on the right indicates the value unloaded into Out_Code. The FIFO is loaded with 67, 14, 37, 25, and 40. When the FIFO is unloaded, the elements unload in the order 67, 14, 37, 25, and 40. The action at each of the numbered scans is as follows:

1. Load stack with In_Code = 67, .Empty turned **off**.
2. Load stack with In_Code = 14

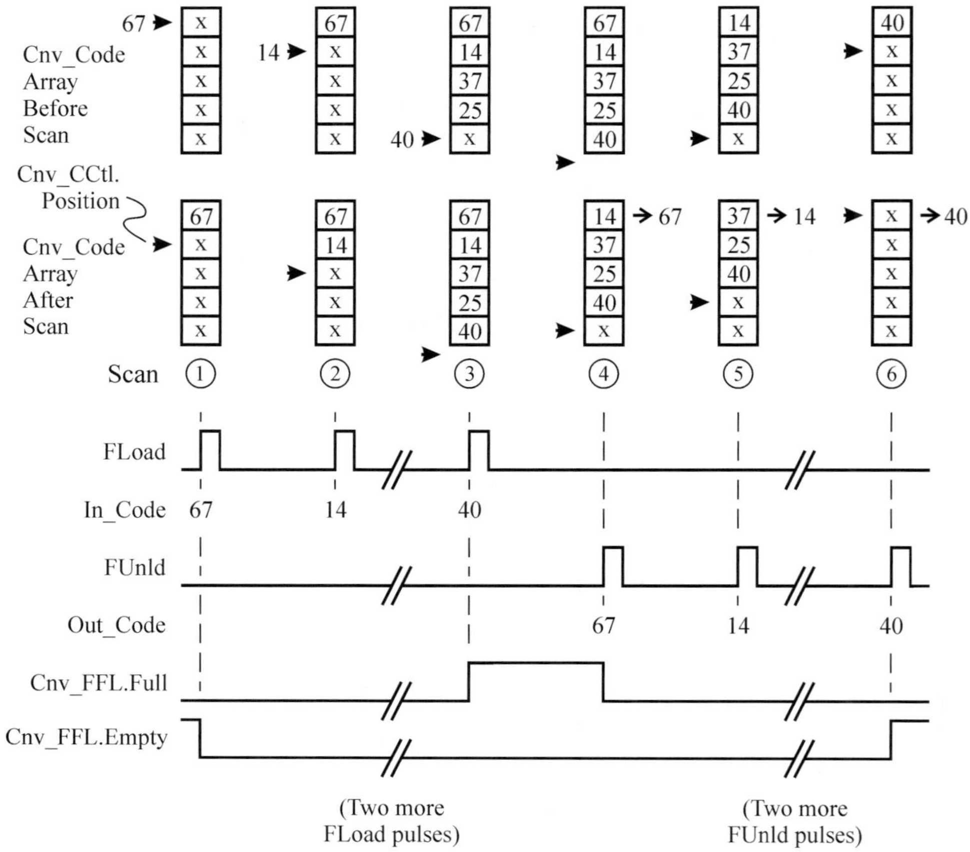

Figure 8.26. Micro800 FIFO array operation.

(37 and 25 loaded between scans 2 and 3)

3. Load stack with In_Code = 40, .Full turned **on**.
4. Unload stack, Out_Code set to 67, .Full turned **off**.
5. Unload stack, Out_Code set to 14
   (37 and 25 unloaded between scans 5 and 6)
6. Unload stack, Out_Code set to 40, .Empty turned **on**.

Normally, the LFL and LFU are used together to construct a LIFO data array. An example application of the LFL and LFU blocks is shown in Figure 8.27. For the LFL block, the inputs are described as follows:

Execute – a positive transition (false-to-true) does the shift.
Src – Variable placed in next LIFO position.
LIFO – LIFO array, an array of any elementary type except for STRING.
LIFOCon – Configuration and control for LIFO, FF_LF_CON data type.

For the LFU block, the inputs are described as follows:

Execute – a positive transition (false-to-true) does the shift.
Dest – Variable unloaded from LIFO placed in this variable.

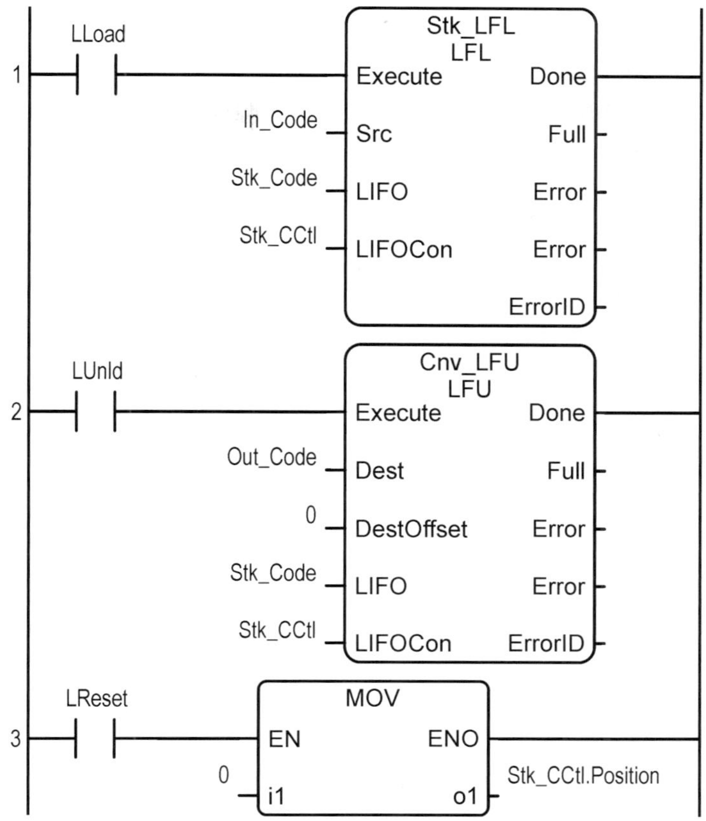

Figure 8.27. Micro800 LIFO array function blocks.

DestOffset – Index of Dest array element. If Dest is not an array, set to zero.
LIFO – LIFO array, an array of any elementary type except for STRING.
LIFOCon – Configuration and control for LIFO, FF_LF_CON data type.

The outputs of both of these blocks are described as follows:

Done – True if operation completed successfully, otherwise false.
Full – True when LIFO is full, Position is equal to Length.
Empty – True when LIFO is empty, Position is equal to zero.
Error – If true, error occurred. Remains true as long as the Execute input is true.
ErrorID – Error code if Error is true.

Possible ErrorID values are listed in Rockwell Automation (2022b). As for the FFL and FFU blocks, the LFL and LFU have the same LIFO and LIFOCon variable.

For each **off**-to-**on** transition of the LFL Execute input, the Src value is placed in the LIFO array at the current position, and the position (.Position) is incremented by one. The .Full output bit is set true if the .Position is equal to the array length. If data is added to an already full LIFO, an error is reported and nothing is added to the LIFO. In order to reset a LIFO array, set the .Position to zero.

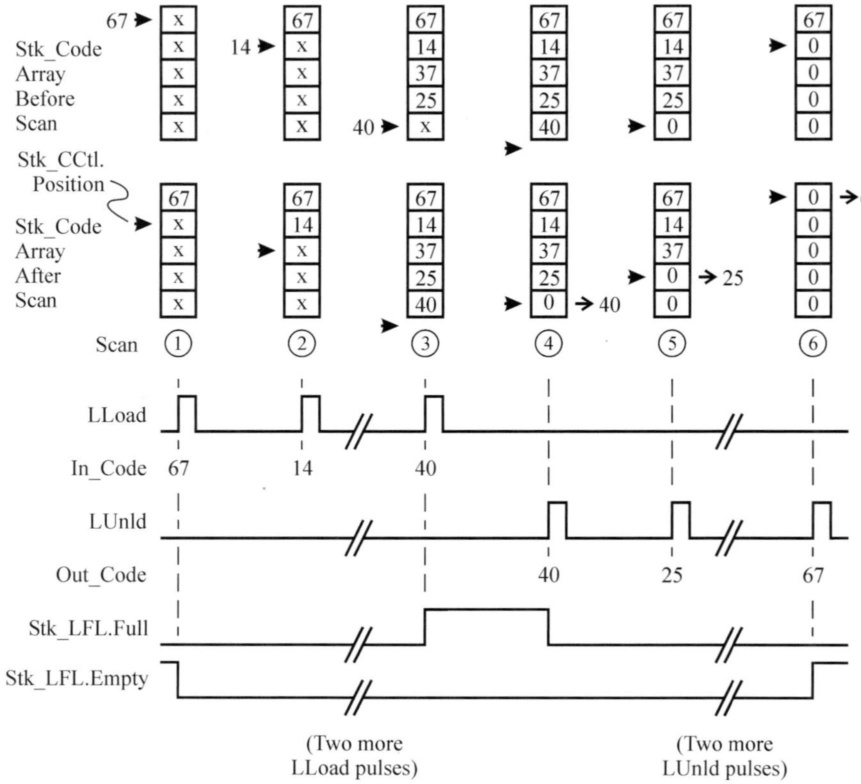

Figure 8.28. Micro800 LIFO array operation.

For each **off**-to-**on** transition of the LFU Execute input, the last element in the LIFO array is copied to the Dest input variable, and the position is decremented by one. The .Empty output is set true if the LIFO is empty (.Position=0). A zero is placed into "Dest" if one attempts to unload an already empty LIFO.

Figure 8.28 shows an example operation of the LFL and LFU of Figure 8.27. The LIFO starts empty and Stk_CCtl.Length has an initial value of 5. Five DINTs are loaded onto the LIFO and then unloaded. The contents of the LIFO before and after the numbered scans are also shown. The LIFO is loaded with 67, 14, 37, 25, and 40. When the LIFO is unloaded, the elements are unloaded in the order 40, 25, 37, 14, and 67. The action at each of the numbered scans is as follows:

1. Load stack with In_Code = 67, .Empty turned **off**.
2. Load stack with In_Code = 14
   (37 and 25 loaded between scans 2 and 3)
3. Load stack with In_Code = 40, .Full turned **on**.
4. Unload stack, Out_Code set to 40, .Full turned **off**.
5. Unload stack, Out_Code set to 25
   (37 and 14 unloaded between scans 5 and 6)

6. Unload stack, Out_Code set to 67, .Empty turned **on**.

8.3.3 Other Data Manipulation Function Blocks

Other specialized numerical and bit array function blocks and the operation that occurs for each one are:

Block Name	Operation Result
AVERAGE	Calculate running average of real values
STACKINT	Manage stack of DINT integers

The operation of each of these blocks is explained and illustrated with short examples.

An example AVERAGE block is shown in Figure 8.29. The particular example calculates a 10-sample running average of FT101. The block inputs are described as follows:

RUN – Execution control.
XIN – REAL variable to be averaged.
N – Number of samples to be averaged. Maximum of 127.

The outputs are described as follows:

ENO – Same value as RUN input.
XOUT – Running average of XIN.

The block maintains a FIFO of N elements. When the RUN input is true, the average of the values in the FIFO is calculated and copied to the XOUT output. When the RUN input is false, the XIN input value is copied to all of the FIFO elements and thus the XOUT output is the same as the XIN value, resetting the calculation. This block executes on every scan and there is no way to "pause" the calculation without destroying all of the previous samples. Hence, this block should be placed in a program organization unit (POU) controlled by a selectable timed interrupt (STI).

The STACKINT maintains a stack of integers and an example is shown in Figure 8.30. The block inputs are described as follows:

PUSH – A positive transition (false-to-true) adds the IN value to the top of the stack.
POP – A positive transition (false-to-true) deletes the value at the top of the stack and the element below becomes the top of the stack.
R1 – When true, resets the stack to its empty state.
IN – On rising edge of PUSH input, this variable is pushed onto the stack.
N – Number of elements in stack. Minimum of 1 and maximum of 128.

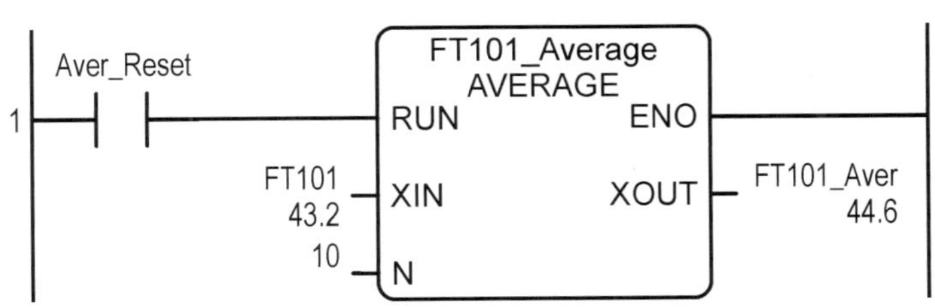

Figure 8.29. Micro800 sample AVERAGE block.

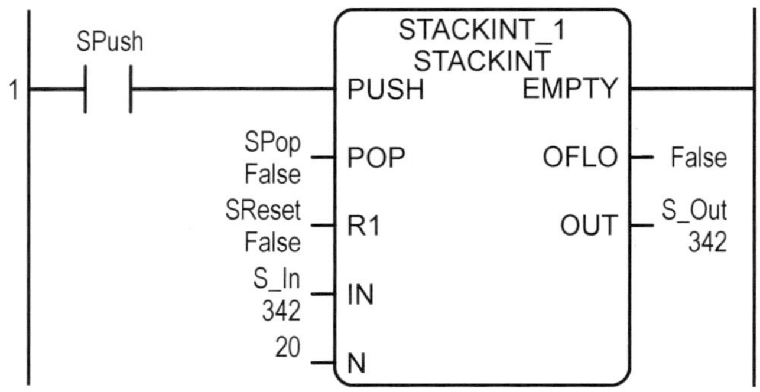

Figure 8.30. Micro800 sample STACKINT block.

The outputs are described as follows:

EMPTY – True when stack is empty.
OFLO – True if stack has overflowed, otherwise false.
OUT – Value at the top of the stack.

Note that the stack must be reset in order to function properly.

8.3.4 Program Execution Control

The Micro800 PLC offers a number of instructions that can control ladder logic execution. The names and the operation that occurs for each one are:

Block Name	Operation Result
>>Label	Jump to a defined label
AFI	Disable rung logic
NOP	No operation
RETURN	Conditional end of a ladder program
SUS	Suspend
TND	Stop current program

Depending on the particular instruction, it may be a contact, or a coil, or a function block. The operation of each of these instructions is explained and illustrated with short examples. For information on the blocks that control the operation of the interrupts, see Chapter 17 of Rockwell Automation (2022b).

The >> (jump) coil and rung label are used together to skip portions of ladder logic. When the >> coil is turned **on**, the ladder scan skips to the referenced rung label. One can jump either forward or backward. Jumping backward lets the controller iterate certain sections of logic. However, jumping backward an excessive number of times will cause the controller to fault if the scan time is too long. Jumping forward decreases the scan time by skipping logic segments until needed. One drawback to jumping forward is that the logic between the jump coil and the label contact is not scanned, so any output coils are not changed. An example ladder segment with two jumps and two labeled rungs is shown in Figure 8.31.

In contrast to the other processors covered by this text, the Micro800 has no subroutines. A project may contain up to 256 programs (called POUs). A POU can call

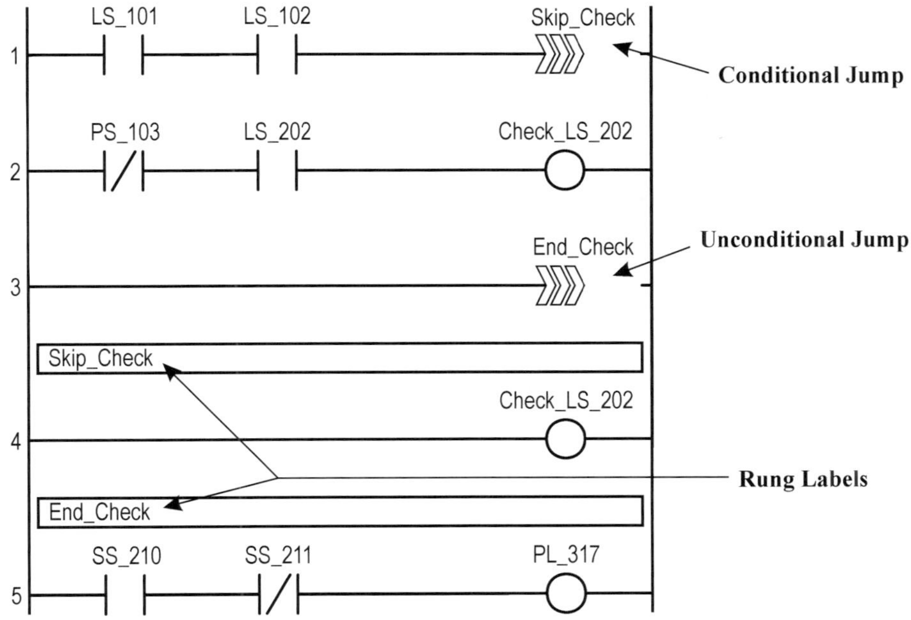

Figure 8.31. Micro800 jump and label example.

either a user defined function (UDF) or a user defined function block (UDFB). A UDF is similar to a traditional subroutine and is limited to have only one output connection. A UDF cannot call other UDFs or other UDFBs. In contrast, UDFBs can call other UDFBs, but the maximum nesting depth is five. The calling of UDFs and UDFBs is illustrated in Figure 8.32.

The parameters are passed into and out of a function block by value or by reference. Function blocks can also reference global variables, and input/output channel variables. However, a function block may not be easily reused in another PLC program if it references globally-defined variables. Array elements, structures, and entire arrays can all be passed as parameters. In all cases, the data type of the variable connected to a UDF or UDFB should match the data type of the block parameter definition.

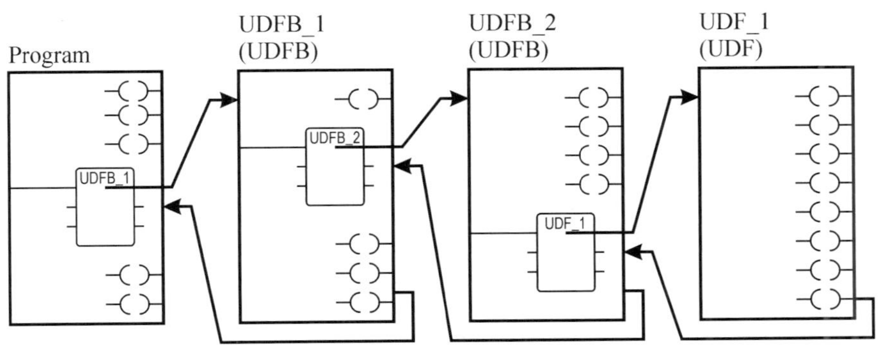

Figure 8.32. Micro800 calling function blocks.

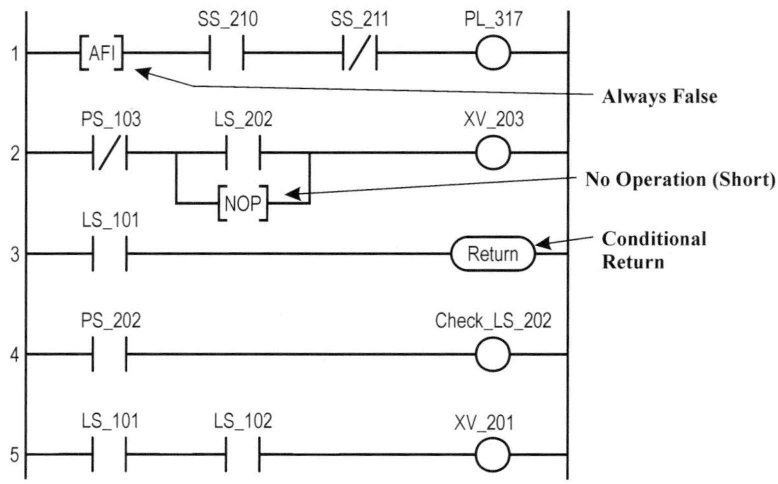

Figure 8.33. Example use of Micro800 AFI, NOP, and RETURN.

The AFI and NOP contacts and the RETURN coil are useful for troubleshooting and debugging. Figure 8.33 shows sample uses of all of these ladder logic instructions. The AFI (always false instruction) contact is used to temporarily disable rung logic. Its result is always false, and so any output coils are turned **off**, and any output function blocks are not executed. For rung 1 of Figure 8.33, the AFI contact disables the other rung logic and keeps PL_317 **off**. The NOP (no operation) contact acts like a short when placed in parallel with other contacts. In rung 2, the NOP shorts across LS_202 and essentially bypasses it, forcing the logic to function as if LS_202 is **on**. When enabled, the RETURN coil acts like the end of a program, UDF, or UDFB. When LS_101 is **on** in Figure 8.33, the RETURN coil on rung 3 is enabled and the remainder of the ladder rungs in the program are not scanned. When used for troubleshooting, the RETURN coil is placed at the end of the debugged code. As each new section is debugged, the RETURN coil is moved on through the rungs.

Design Tip

The AFI and NOP contacts and the RETURN coil are primarily used for troubleshooting.

The SUS (suspend) and TND (temporary stop) blocks are shown in Figure 8.34. When the Enable input of the SUS block is true, the controller execution is suspended indefinitely. The controller remains in the Run mode, but the Run processor status lamp is set to off. The value of the variable at the SusID input is copied to the _SYSVA_SUSPEND_ID system variable. To resume program execution, the Enable input must be false. Then one must change the processor to program mode and then back to run mode. The TND block stops the current scan cycle of the user program. All POU's are affected. After the output scan, input scan, and housekeeping tasks are complete, the program starts execution with the first POU.

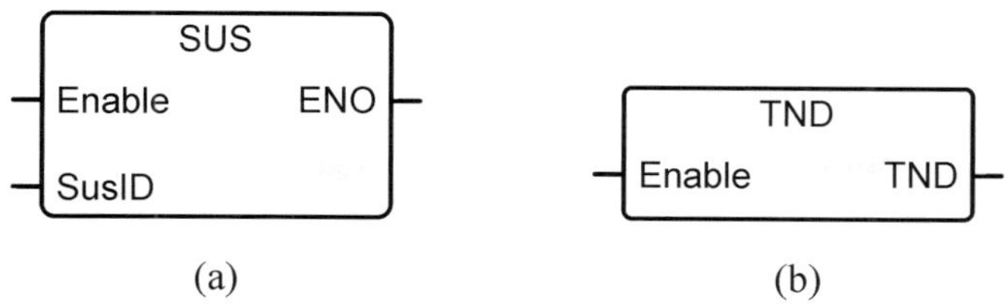

Figure 8.34. Suspend blocks: *(a)* SUS; *(b)* TND.

8.4 OTHER MICROLOGIX/SLC-500 BLOCKS

The MicroLogix/SLC-500 PLCs provide ladder function blocks for bit shifting, FIFO/LIFO queues, special array functions, and program control. The basic operation and simple application of these function blocks are described in this section. Longer bit shifting and FIFO/LIFO applications are presented in section 8.5.

The MicroLogix/SLC-500 function blocks described in this section are very similar to the ControlLogix/CompactLogix function blocks of the same name. However, there are enough differences between the operation of the corresponding ControlLogix and MicroLogix and SLC-500 function blocks that the operation of each block is described instead of just referring to the description in section 8.2. One major difference is that the block parameters are shown as addresses and not as symbols. The symbols can be used when entering the blocks into the ladder logic program, but the programming software displays the addresses in the function block.

Table 8.1 Function Blocks Supported by MicroLogix and SLC-500

Block Name	Processor Model: MicroLogix	Processor Model: SLC-500
BSL, BSR	X	X
FFL, FFU	X	X[1]
LFL, LFU	X	X[1]
FLL, COP	X	X
CPW		X[2]
SWP	X	X[2]
FBC, DDT	X[3]	X[4]
JMP, LBL	X	X
JSR, SBR, RET	X	X[5]
MCR	X	X
TND	X	X

Notes:
1. SLC-5/02 and higher
2. Not MicroLogix 1000
3. Not MicroLogix 1400
4. Series C SLC-5/03 and higher
5. Limits on subroutine nesting

The specific function blocks that are supported vary among the different processors. Table 8.1 shows which function blocks are supported by the MicroLogix and SLC-500 PLCs. None of them support the ControlLogix FAL, FSC, FOR, NXT, BRK blocks or the NOP (no operation) and AFI (always false instruction) contacts. Many of these function blocks use the file control data file element (R6:*n*) structure, shown in Figure 3.19*c*. The parts of this structure that are pertinent to each function block are described in conjunction with the function block.

8.4.1 Shift Bit in Register

The bit shifting function blocks and the operation that occurs for each one are:

Block Name	Operation Result
BSL	Bit shift left
BSR	Bit shift right

An example BSL function block is shown in Figure 8.35*a*, and an example BSR function block is shown in Figure 8.36*a*. The operands of both of these blocks are described as follows:

File – Start of bit array. Must be the address of a 16-bit word.
Control – Address of control structure (Figure 3.19*c*).
Bit Address – Bit (Boolean) that will be shifted into the bit array
Length – Number of bits in the bit array.

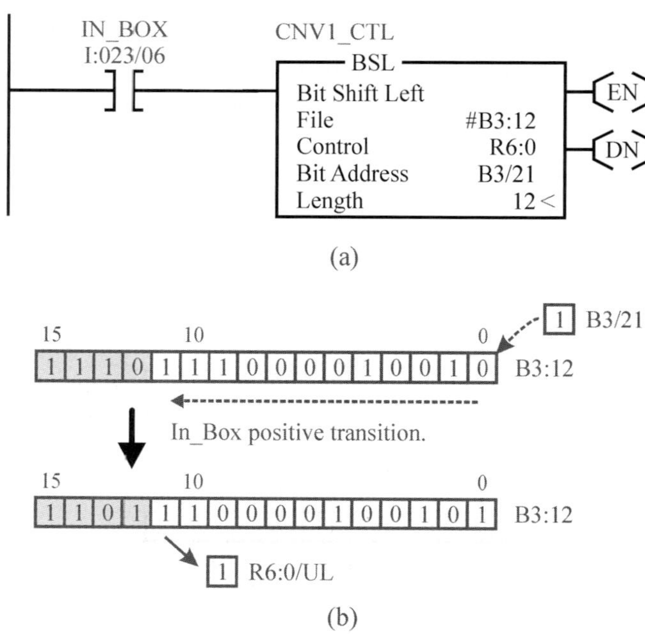

Figure 8.35. MicroLogix/SLC-500 bit-shift left: *(a)* example BSL block; *(b)* bit array before and after shift.

The least-significant bit of the bit array is always bit 0 of the "File" word address. For the "Control" address defined in Figure 8.35*a*, the individual parts (fields) of the structure are addressed as follows:

Symbol Reference	Address Reference	Description
CNV1_CTL/UL	R6:0/UL	Unload (bit shifted out)
CNV1_CTL/EN	R6:0/EN	Enable
CNV1_CTL/ER	R6:0/ER	Error (length < 0)
CNV1_CTL.LEN	R6:0.LEN	Bit array length

When entering the symbol reference into a block parameter, a "/" or "." can be used to separate the symbol or address from the field name, but the above table shows how they are displayed.

For each **off**-to-**on** transition of the BSL input, the bit array is shifted left one position. The "Bit Address" value replaces the rightmost bit, and the /UL bit is set to the value of the (shifted out) leftmost bit in the bit array. The /EN bit is identical to the block input condition. The /ER bit is set if the .LEN field is less than zero. The value of the array length (.LEN) can be accessed as an integer. An application of the BSL function block is shown in Figure 8.35*a*, and the bits in the bit array before and after a shift are shown in Figure 8.35*b*. Note that the symbol for the "Control" address is shown above the BSL and BSR function blocks. For this example, the array is contained in one word. The BSR example shows how multi-word bit arrays are constructed.

For each **off**-to-**on** transition of the BSR input, the bit array is shifted right one position. The "Bit Address" value replaces the leftmost bit, and the /UL bit is set to the value of the

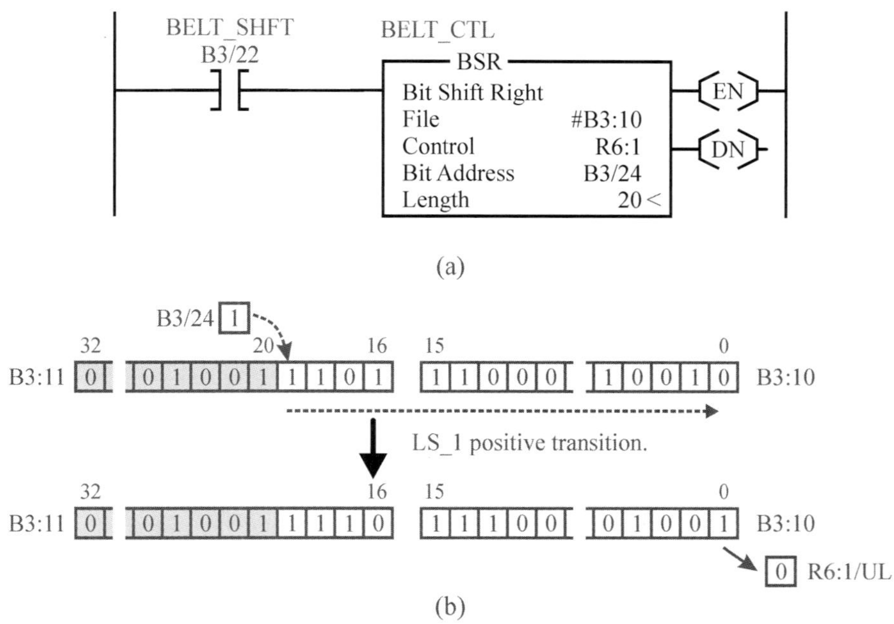

Figure 8.36. MicroLogix/SLC-500 bit-shift right: *(a)* example BSR block; *(b)* bit array before and after shift.

(shifted out) rightmost bit in the bit array. An example application of the BSR function block is shown in Figure 8.36*a*, and the bits in the bit array before and after a shift are shown in Figure 8.36*b*. The bit array for this example is longer than one word, so the next word in the B3 file is used for the upper 4 bits of the array. The BSR function block handles the transfer of bits between the words automatically. To clear a shift register consisting of multiple words, use the FLL block described in section 8.4.3.

To rotate the bits in a shift register, the "Bit Address" is set to the appropriate bit at the end of the bit array that is copied to the /UL bit when the bits are shifted. For example, for the BSL in Figure 8.35*a*, using B3:12/11 for the "Bit Address" will convert the left shift into a left rotate. In Figure 8.36*a*, using B3:10/0 for the "Bit Address" will change the right shift into a right rotate.

8.4.2 FIFO and LIFO Queues

The array (queue) shifting function blocks and the operation that occurs for each one are:

Block Name	Operation Result
FFL	Load value to the next FIFO position
FFU	Unload value from first FIFO position
LFL	Load value to the next LIFO position
LFU	Unload value from last LIFO position

The LFL and LFU blocks are not valid for some SLC-500 processors.

Typically, the FFL and FFU are used together to construct a FIFO data array. An example application of the FFL and FFU blocks is shown in Figure 8.37. For the FFL and FFU function blocks, the parameters are described as follows:

FIFO – Address of first integer in FIFO array.
Control – Address of control structure (Figure 3.19*c*).
Length – Maximum length of the FIFO array.
Position – Position where the next data will be loaded.
Source – (FFL only) Address of the integer that will be stored in the array.
Dest – (FFU only) Address where the integer will be placed after it is unloaded from the FIFO.

Normally, the FFL and FFU refer to the same "Control" address. For the control address defined in Figure 8.37, the individual fields of the CONTROL structure are addressed as follows:

Symbol Reference	Address Reference	Description
CNV_CCTL/EN	R6:2/EN	Load enabled
CNV_CCTL/EU	R6:2/EU	Unload enabled
CNV_CCTL/DN	R6:2/DN	FIFO full
CNV_CCTL/EM	R6:2/EM	FIFO empty
CNV_CCTL.LEN	R6:2.LEN	FIFO array length
CNV_CCTL.POS	R6:2.POS	Next entry will load here

For each **off**-to-**on** transition of the FFL input, the "Source" value is placed in the FIFO array at the current position, and the position (.POS) is incremented by one. The /EN bit is

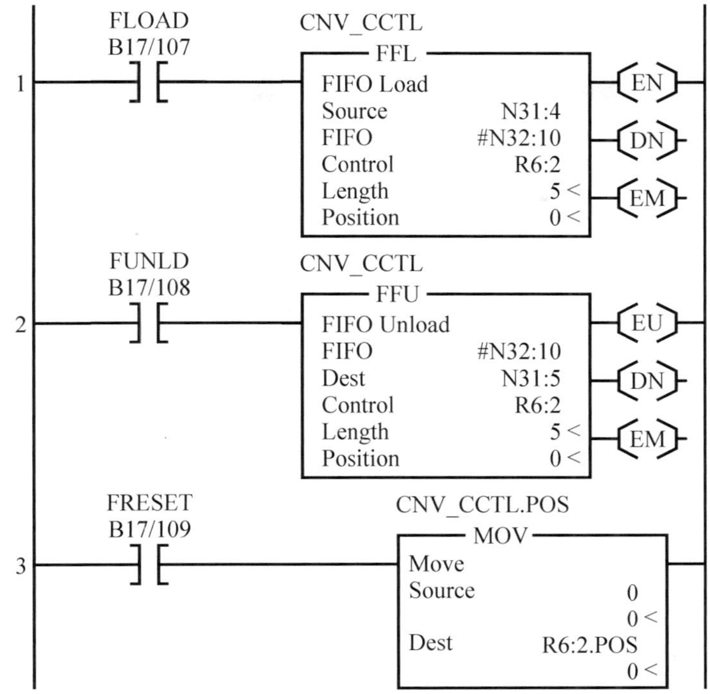

Figure 8.37. MicroLogix/SLC-500 FIFO array function blocks.

identical to the FFL block input condition. The /DN bit is set if the .POS is equal to the array length. If data is added to an already full FIFO, the new data replaces the previously added data entry. The position value of the FIFO (.POS) can be accessed as an integer. In order to reset a FIFO array, set the .POS to zero.

For each **off**-to-**on** transition of the FFU input, the first element in the FIFO array is copied to "Dest", and the remaining data in the array is shifted one position toward the beginning of the array. The position is also decremented by one. The /EU bit is identical to the FFU block input condition. The /EM bit is set if the FIFO is empty (.POS=0). A zero is placed into "Dest" if one attempts to unload an already empty FIFO.

An example operation of the FFL and FFU of Figure 8.37 is shown in Figure 8.38. This example assumes the FIFO starts empty. Five integers are loaded onto the FIFO and then unloaded. The FIFO array starts at N32:10 and ends at N32:14. The contents of the FIFO before and after the numbered scans are also shown. The arrow on the left of the array indicates the .POS value, and the arrow on the right indicates the value unloaded into Out_Code. The FIFO is loaded with 67, 14, 37, 25, and 40. When the FIFO is unloaded, the elements unload in the order 67, 14, 37, 25, and 40. The action at each of the numbered scans is as follows:

1. Load stack with In_Code = 67, /EM turned **off**.
2. Load stack with In_Code = 14
   (37 and 25 loaded between scans 2 and 3)
3. Load stack with In_Code = 40, /DN turned **on**.
4. Unload stack, Out_Code set to 67, /DN turned **off**.

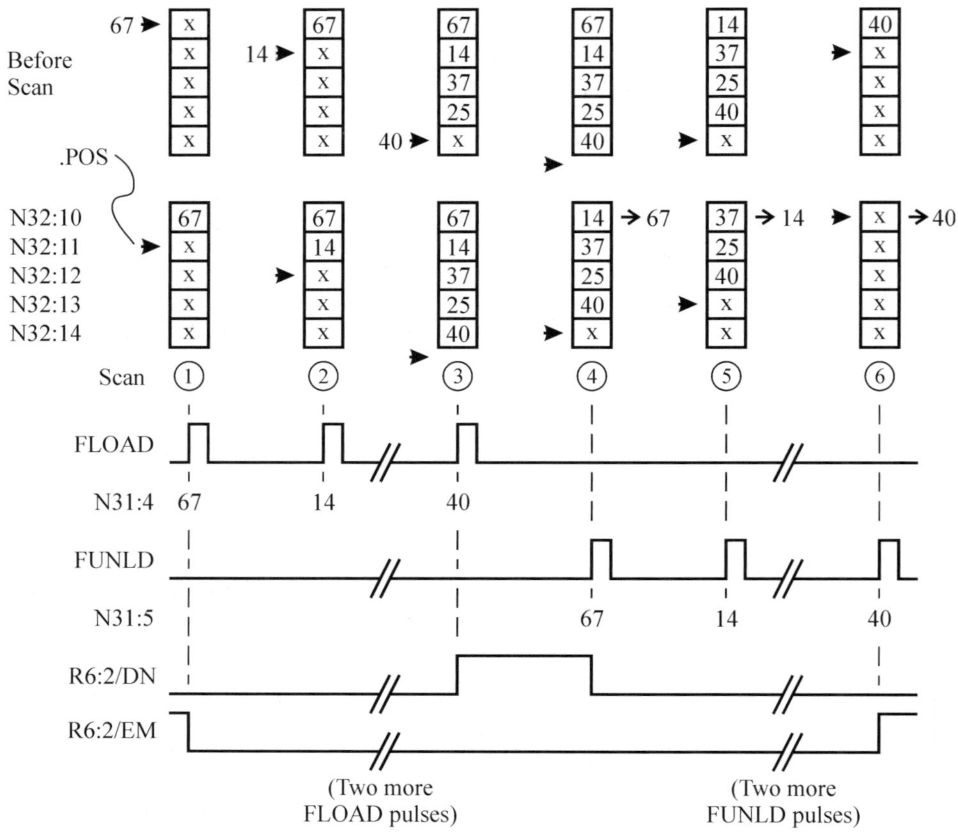

Figure 8.38. MicroLogix/SLC-500 FIFO array operation.

5. Unload stack, Out_Code set to 14
   (37 and 25 unloaded between scans 5 and 6)
6. Unload stack, Out_Code set to 40, /EM turned **on**.

Normally, the LFL and LFU are used together to construct a LIFO data array. An example application of the LFL and LFU blocks is shown in Figure 8.39. The operands for these instructions are identical to the FFL/FFU except for the "LIFO" operand, which is the first element of the LIFO array.

For each **off**-to-**on** transition of the LFL input, the "Source" value is placed in the LIFO array at the current position, and the position (.POS) is incremented by one. The /EN bit is identical to the LFL block input condition. The /DN bit is set if the .POS is equal to the array length. If data is added to an already full LIFO, the new data replaces the previously added data entry. The position value of the LIFO (.POS) can be accessed as DINT. In order to reset a LIFO array, set the .POS to zero.

For each **off**-to-**on** transition of the LFU input, the last element in the LIFO array is copied to "Dest", and the position is decremented by one. The /EU bit is identical to the LFU block input condition. The /EM bit is set if the LIFO is empty (.POS=0). A zero is placed into "Dest" if one attempts to unload an already empty LIFO.

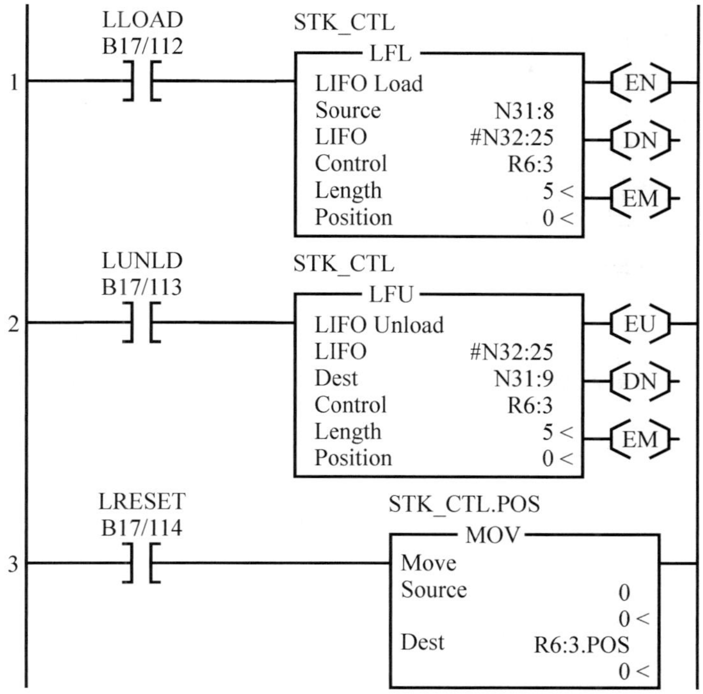

Figure 8.39. MicroLogix/SLC-500 LIFO array function blocks.

Figure 8.40 shows an example operation of the LFL and LFU of Figure 8.39. The LIFO starts empty. Five DINTs are loaded onto the LIFO and then unloaded. The contents of the LIFO before and after the numbered scans are also shown. The LIFO is loaded with 67, 14, 37, 25, and 40. When the LIFO is unloaded, the elements are unloaded in the order 40, 25, 37, 14, and 67. The action at each of the numbered scans is as follows:

1. Load stack with In_Code = 67, /EM turned **off**.
2. Load stack with In_Code = 14
   (37 and 25 loaded between scans 2 and 3)
3. Load stack with In_Code = 40, /DN turned **on**.
4. Unload stack, Out_Code set to 40, /DN turned **off**.
5. Unload stack, Out_Code set to 25
   (37 and 14 unloaded between scans 5 and 6)
6. Unload stack, Out_Code set to 67, /EM turned **on**.

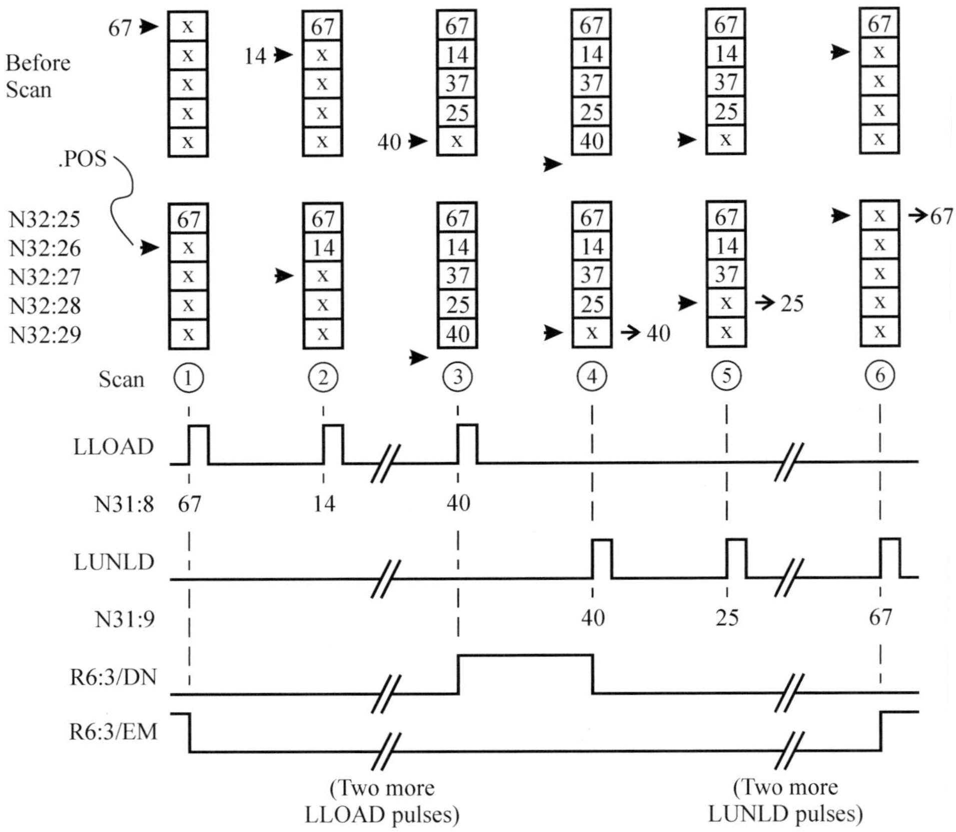

Figure 8.40. MicroLogix/SLC-500 LIFO array operation.

8.4.3 Other Array Function Blocks

Other specialized numerical and bit array function blocks and the operation that occurs for each one are:

Block Name	Operation Result
FLL	Fill array with a value
COP	Copy values, arrays, or structures
CPW	Copy words of data
SWP	Swap high and low bytes of word array
FBC	Find mismatches between two bit arrays
DDT	Find and correct mismatched bit arrays

The operation of each of these blocks is explained and illustrated with short examples.

The FLL (fill) function block fills the elements of an array with the same value and is useful for initialization. An example FLL block is shown in Figure 8.41. The "Source" operand is either the address containing the value to copy or a constant value. The "Dest" is the address of the first element in an integer or floating-point array. The address must be prefixed with the "#" character. The "Length" is the number of elements to fill. The symbol

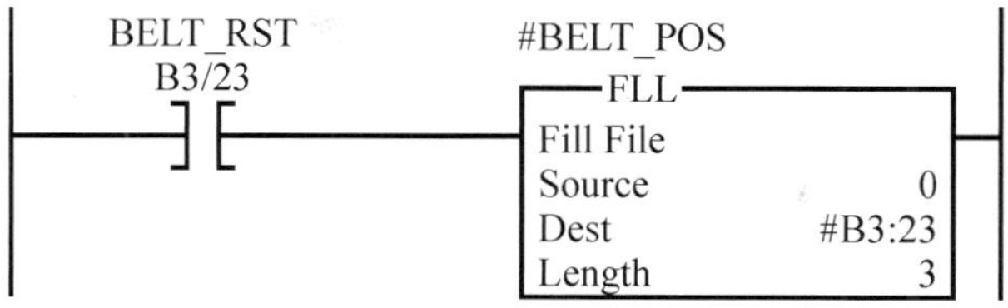

Figure 8.41. MicroLogix/SLC-500 FLL example.

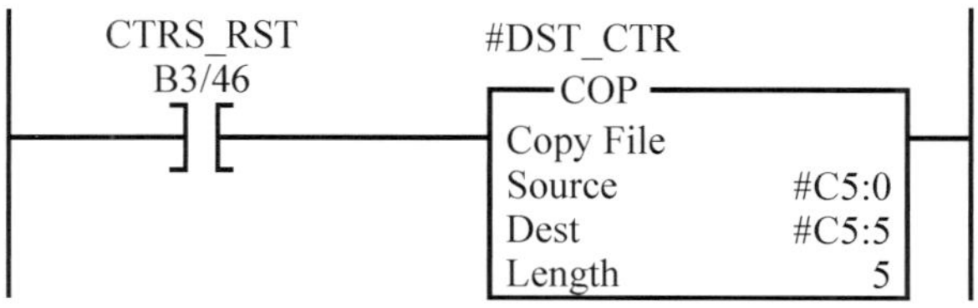

Figure 8.42. MicroLogix/SLC-500 COP example.

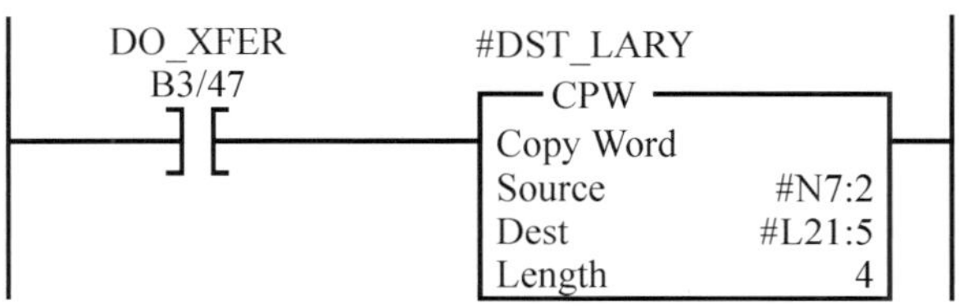

Figure 8.43. MicroLogix/SLC-500 CPW example.

shown above the block is the symbol for the "Dest" address. The "Source" and "Dest" should be the same data type since there is no conversion between data types.

The COP (copy) function block copies arrays. The array can be a simple type (e.g., integer, word, floating point) or an array of structures (e.g., an array of counters or timers). An example application of the COP block is shown in Figure 8.42 that copies an array of 5 counters to another array of 5 counters. The "Source" and "Dest" operands are addresses of the source and destination data arrays, respectively. The "Length" is the number of elements to write in the destination file. The symbol shown above the block is the symbol for the "Dest" address. As for the FLL, the "Source" and "Dest" should be the same data type since there is no conversion between data types, with the exception that words can be copied between B and N file types. The valid data file types are O, I, B, T, C, R, N, F, L, and ST. Also, a maximum of 128 words can be copied.

The CPW (copy words) function block also copies data, but allows different source and destination data types. Also, function files can be used as a source or destination, but not both. Only one (source or destination) can be a function file. An example application of the CPW block is shown in Figure 8.43 that copies an array of 4 integers into an array of 2 long words. The "Source" and "Dest" operands are addresses of the source and destination arrays, respectively. The "Length" is the number of **words** to be transferred to the destination file. The symbol shown above the block is the symbol for the "Dest" address.

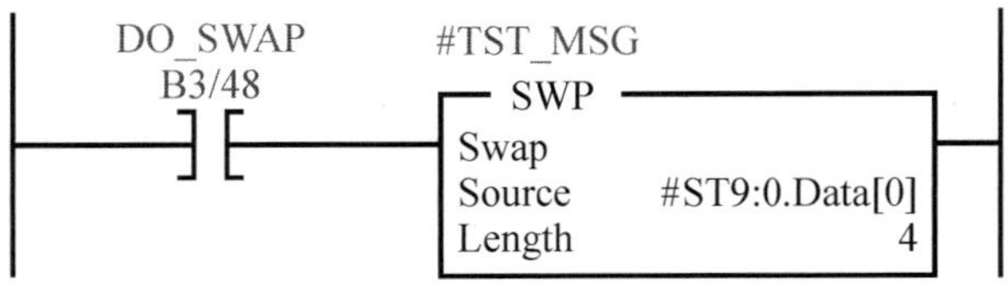

Figure 8.44. MicroLogix/SLC-500 SWP example.

The valid data file types are O, I, B, N, F, L, and PLS. Valid function file types are RTC, HSC, PTO, PWM, STI, and EII. In addition, BHI, MMI, and TPI function files may be used as only as a source, not a destination. A maximum of 128 words can be copied.

Note that the CPW does not convert data types. For example, for the CPW in Figure 8.43, N7:2 is copied into the lower word of L21:5 and N7:3 is copied into the higher word of L21:5. For example, if N7:2 is 255 (E1H) and N7:3 is 366 (150H), then when the CPW block executes, L21:5 becomes (15000E1H).

The SWP (swap words) function block swaps the high and low bytes in a specified word array. An example application of the SWP block is shown in Figure 8.44 that swaps the bytes for the first 4 words of a particular string. The "Source" operand is the address of the word array to be operated on. The "Length" is the number of **words** to be swapped. The symbol shown above the block is the symbol for the "Source" address. The valid data file types are B, N, and ST. A maximum of 128 words can be swapped, though for ST file types a maximum of 41 words can be swapped, as one cannot cross a string boundary.

The FBC (file bit compare) function block compares two bit arrays and records the bit number of each difference in a result array. An example FBC block is shown in Figure 8.45. The parameters of this block are described as follows:

Source – Start address of bit array compared to reference.
Reference – Start address of reference bit array.
Result – Start address of integer array to store search results.
(Cmp.) Control – Control structure for compare.
Length – Number of bit locations to compare
Position – Current position in source array. Typically initialized to zero.
(Result) Control – Control structure for the results.
Length – Number of storage locations in result array.
Position – Current position in result array. Typically initialized to zero.

The two "Control" operands must be different addresses of a control structure (Figure 3.19*c*). The first "Control" address is for the comparison and for the address defined in Figure 8.45, the individual fields of the control structure are addressed as:

Symbol Reference	**Address Reference**	**Description**
C_CTL/EN	R6:8/EN	FBC enabled
C_CTL/DN	R6:8/DN	Set when last bit in array compared.
C_CTL/FD	R6:8/FD	Set when mismatch recorded (one-at-a-time operation) or all mismatches recorded (all-per-scan) operation.

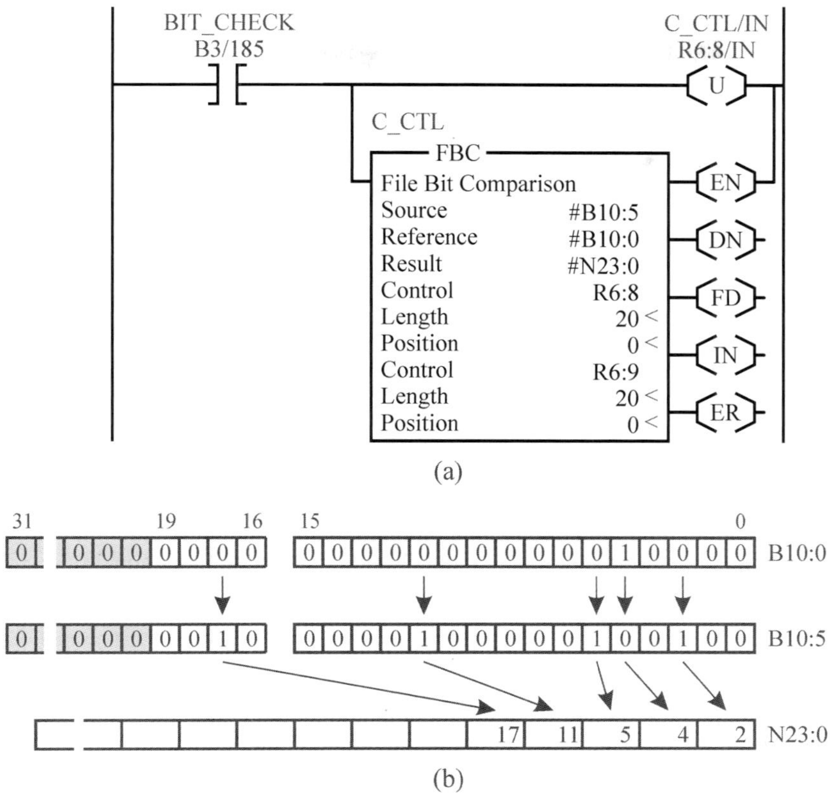

Figure 8.45. MicroLogix/SLC-500 FBC example: *(a)* ladder logic; *(b)* FBC execution results.

C _CTL/IN	R6:8/IN	Search mode bit, **off** to search for all mismatches, **on** to search for next mismatch only.
C_CTL.LEN	R6:8.LEN	Number of bits to compare.
C_CTL.POS	R6:8.POS	Index of current bit.

The second "Control" address is for the result array. If the address of this operand in Figure 8.45 is defined as R_CTL, the individual fields of the control structure are addressed as:

Symbol Reference	**Address Reference**	**Description**
R_CTL/DN	R6:9/DN	Set when result array full.
R_CTL.LEN	R6:9.LEN	Length of result array.
R_CTL.POS	R6:9.POS	Current position in result array.

Figure 8.45*a* shows an example FBC used to compare 20 bits of the array starting at B10:5 with 20 bits of the array starting at B10:0. The R6:8/IN bit is **off** to search for all mismatches, and the results are placed in the integer array starting at N23:0. For particular values of the bit arrays, the result of the function block execution is shown in Figure 8.45*b*.

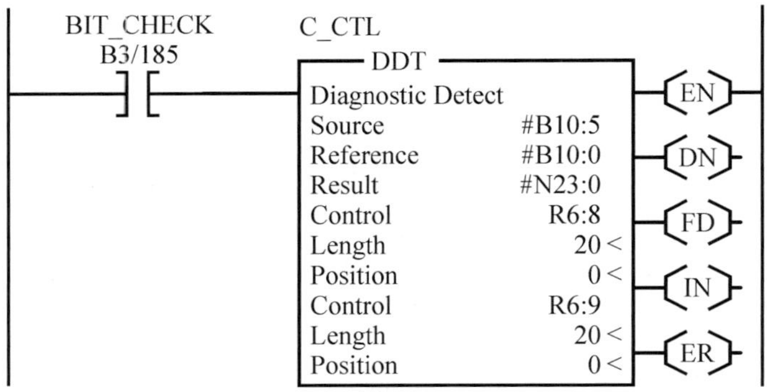

Figure 8.46. MicroLogix/SLC-500 DDT function block.

The DDT diagnostic detect function block (Figure 8.46) compares two bit arrays, records the bit number of each difference in a result array, and then changes the bit in the reference array to match the bit in the source array. The DDT function block functions in the same manner as the FBC function block except that the DDT function block changes the reference bits to match the source bits.

8.4.4 Program Execution Control

The MicroLogix/SLC-500 processors offer a number of instructions that can control ladder logic execution. The names and the operation that occurs for each one are:

Block Name	Operation Result
JMP	Jump to a defined label
LBL	Jump destination
JSR	Jump to subroutine
SBR	First rung of subroutine file
RET	Return from subroutine
SUS	Suspend
MCR	Master control relay; disable logic section
TND	Mark temporary end of ladder

Depending on the particular instruction, it may be a contact, or a coil, or a function block. The operation of each of these instructions is explained and illustrated with short examples.

The JMP (jump) coil and LBL (label) contact are used together to skip portions of ladder logic. When the JMP coil is turned **on**, the ladder scan skips to the referenced LBL contact. The labels are numbers, 0-255. The label is displayed as Q*n*:*x*, where *n* is the ladder logic file number and *x* is the label number. One can jump either forward or backward. Jumping backward lets the controller iterate certain sections of logic. However, jumping backward an excessive number of times will cause the controller to fault if the scan time is too long. Jumping forward decreases the scan time by skipping logic segments until needed. One drawback to jumping forward is that the logic between the jump coil and the label contact is not scanned, so any output coils are not changed. An example ladder segment with two jump and two label instructions is shown in Figure 8.47. For simplicity, the ordinary contact and coil addresses are not shown.

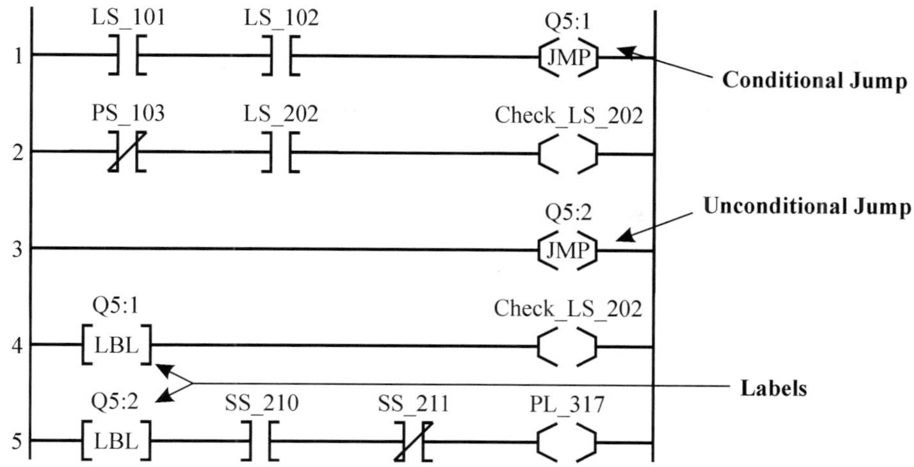

Figure 8.47. MicroLogix/SLC-500 JMP and LBL example.

The JSR (jump to subroutine), SBR (subroutine), and RET (return) function blocks are used to execute program files other than the main ladder program file. Large programs are often broken into subroutine program files, which are called from the main program. When enabled, the JSR function block directs logic execution to the first rung of the specified program file. No parameters are passed to the file. The first rung of the subroutine file must have a SBR function block. Normally, the RET function block appears on the last rung of a subroutine file. Actually, the RET block can occur before the last subroutine rung and have conditions to enable it and . In this case, if the RET block is not executed, ladder logic execution continues after the rung that contains the RET block. If the RET block is omitted, the subroutine automatically returns back to the calling routine after the last subroutine rung is scanned.

Subroutines can call other subroutines, often called *nested subroutines*, as shown in Figure 8.17. A program can have up to eight levels of nested subroutines (four for SLC 5/01).

Figure 8.48 shows an example use of a subroutine. The main program file (file 2) calls subroutine (program) file 10. The subroutine executes its function and then returns control back to the main program file.

The SUS (suspend) block is a debugging aid, providing a way to suspend (halt) the processor and identify the reason for the suspension. Figure 8.49 shows an example ladder logic using the SUS block. When the input to the block is energized, the processor is placed in the suspend mode, the "Suspend ID" integer constant is copied to S:7 and the program file number where the SUS block is located is placed in S:8. In the suspend mode, all outputs are de-energized and the processor stops scanning the I/O and the ladder logic. The user must place the processor in run mode to resume its operation.

The MCR (master control relay) output coil instruction functions like a software master control relay, disabling non-retentive outputs within a range of rungs. This coil is used in pairs to create "zones" of the ladder program where all non-retentive outputs may be disabled. One MCR defines the start of the zone, and one MCR defines the end of the zone. Rung logic on the rung of the first MCR controls execution of the zone. When this rung logic turns **on** the MCR coil (called "enabling the zone"), all rungs in the zone are scanned

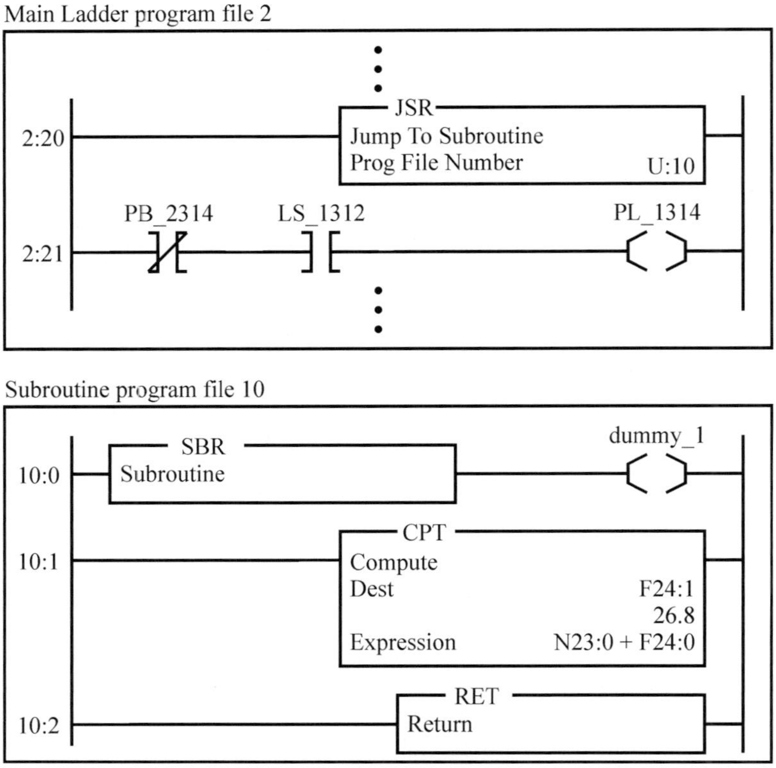

Figure 8.48. Example use of MicroLogix/SLC-500 JSR/SBR/RET.

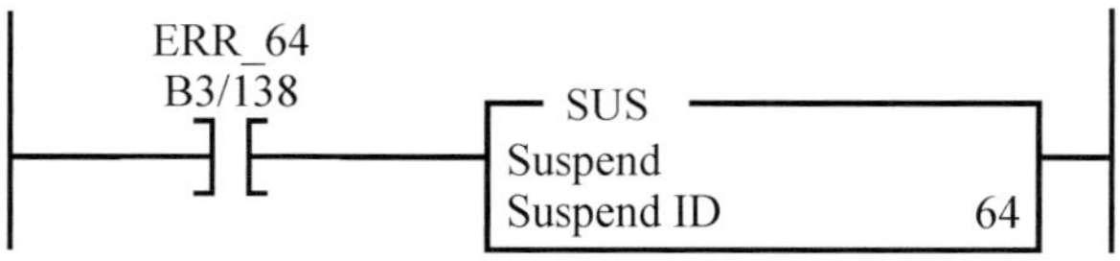

Figure 8.49. Example use of MicroLogix/SLC-500 SUS block.

normally. When the logic of the first MCR rung causes the MCR coil to be **off**, all rungs in the zone are scanned, but only to turn **off** all non-retentive outputs in the zone.

There are some cautions when using an MCR zone:

1. The zone must be ended with an unconditional MCR output coil.
2. MCR zones cannot be nested.
3. Do not jump into or out of a MCR zone.

Also, the MCR instruction is not a substitute for a hard-wired master control relay that provides an emergency power shutdown to the I/O.

An example MCR zone is shown in Figure 8.50. When LS_101 and LS_102 are both **on**, the MCR zone is enabled and rungs 2:111 to 2:113 are scanned normally determining the values of the output coils XV_203, PL_317, and XV_204. When either LS_101 or

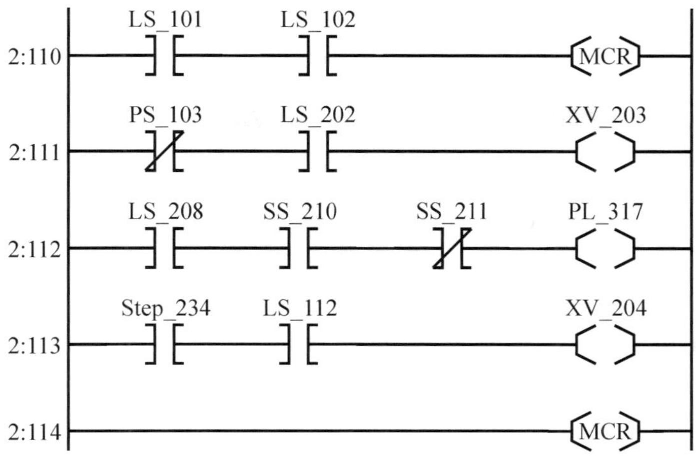

Figure 8.50. Example use of MicroLogix/SLC-500 MCR.

LS_102 is **off**, the MCR zone is disabled. In this case XV_203, PL_317, and XV_204 are all turned **off**, regardless of the state of the input conditions for rungs 2:111 to 2:113.

The TND coil is useful for troubleshooting and debugging. Figure 8.51 shows a sample use of this coil. When enabled, the TND (temporary end) coil acts like the end of the ladder file. If the TND coil is in a subroutine file, control passes to the calling routine. When LS_101 is **on** in Figure 8.51, the TND output coil on rung 23 is enabled and the remainder of the routine rungs are not scanned. When used for troubleshooting, the TND coil is placed at the end of the debugged code. As each new section is debugged, the TND is moved on through the rungs.

Design Tip

The TND coil is primarily used for troubleshooting.

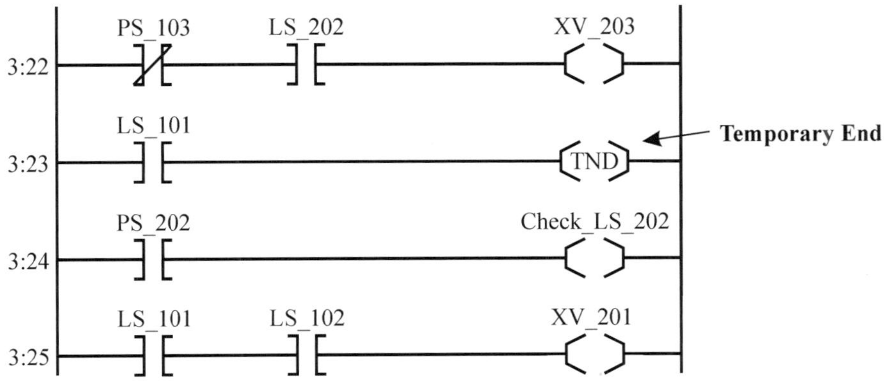

Figure 8.51. Example use of MicroLogix/SLC TND.

8.5 EXAMPLES

The following three examples illustrate the application of shift registers and FIFO/LIFO queues to track material as it moves down a production line. The first example uses a shift register to track part defects. The second example tracks part numbers as they move on a conveyor and as they are stacked and unstacked. Modifying the system to allow parts to be placed on intermediate stations on the transport conveyor produces the third example.

> **Design Tip**
>
> Shift registers and FIFO/LIFO queues are useful for tracking the status or identification of a part as it moves down an assembly line.

Example 8.1. Defect tracking. Implement a defect tracking system for a production system. The system is modeled as 25 stations (Figure 8.52*a*) and a part moves along the system, spending 2 seconds in each station and then 0.5 seconds to move between stations. If a part is determined to be defective at any station, a defect indication is set that follows the part as it moves through the remaining stations. The last station (#25) is the reject station. If the part is marked as defective at the reject station, a reject solenoid is activated for 1.0 seconds to kick the part off the conveying system.

Assume the following physical input, physical output, and internal memory assignments:

Tag/Var./Symbol	Description
Run	Internal coil, **on** when system running (controlled by another part of the ladder).

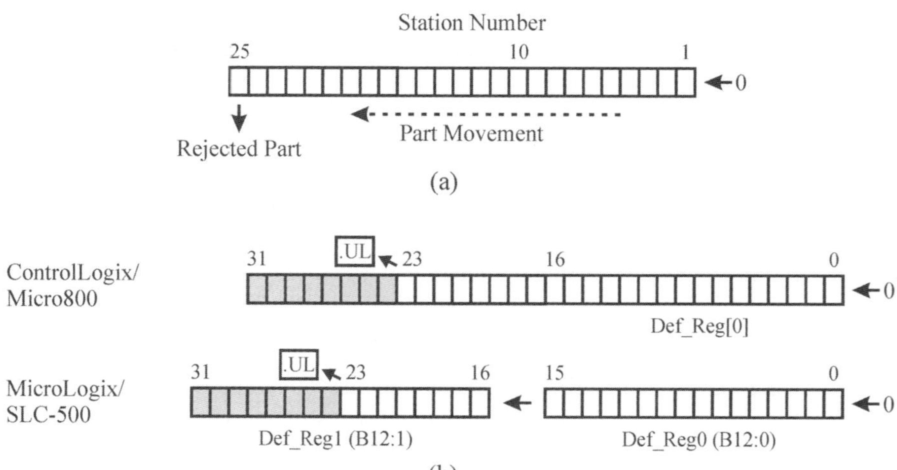

Figure 8.52. Production system defect tracking: *(a)* model of stations; *(b)* shift registers for implementation.

Int_Reset	Internal coil, **on** to clear all defect indications. Assume that Run is already **off** when this turns **on** (controlled by another part of the ladder).
REJECT_SOL	Reject solenoid; **on** for 1.0 second to push defective part off conveying system.

The address associated with the physical output is:

Tag/Var./Symbol	ControlLogix	Micro800	MLogix	SLC
REJECT_SOL	Local:2:O.Data.15	_IO_EM_DO_09	O:0/11	O:1/11

The addresses or data types associated with the internal tags/variables/symbols:

Tag/Var./Symbol	CLogix Data Type	Micro800 Data Type	MLogix/SLC Address
Run	BOOL	BOOL	B3/100
Int_Reset	BOOL	BOOL	B3/101

Solution. A 25-bit shift register will be used to store the defect indication. The appropriate bit is set by the defect-checking logic at individual stations, and the defect indication must shift with the part as it moves down the line. The logic to generate the signals that indicate when the part is moving (Mv_Part) and when it is being processed (Dwell) will also be shown, though that is not part of the problem. As an example, the logic to set the defect indication for station 10 is shown. The logic for any other station that sets a defect bit is similar.

The structure of the shift register for each of the PLCs is shown in Figure 8.52*b*. The solutions are shown in Figures 8.53 and 8.54. For all of the implementations, the shift register is shifted at the start of the dwell time, and if a defective part is present at station 25, the reject solenoid is activated. A station defect indication is set at the start of part movement (end of dwell), allowing the dwell time to test for a defect.

The ControlLogix solution for this problem is shown in Figure 8.53, and the Micro800 solution is shown in Figure 8.54. The MicroLogix solution is very similar to the ControlLogix solution. The ControlLogix and Micro800 solutions are similar, and the function of each rung is:

1. Generate Dwell (part processing) and Mv_Part (part moving).
2. Set bit to be shifted in to always **off** (false).
3. Shift defect indications left one position on rising edge of Dwell. The shift register is 24 bits. The 25th station for defect reject is represented by the Def_RCtl.UL bit.
4. Clear shift register on reset.
5. Eject part if station 25 defect bit turned on.
6. Set station 10 defect if defect detected.

For these solutions, the addresses or data types associated with the internal tags/variables/symbols:

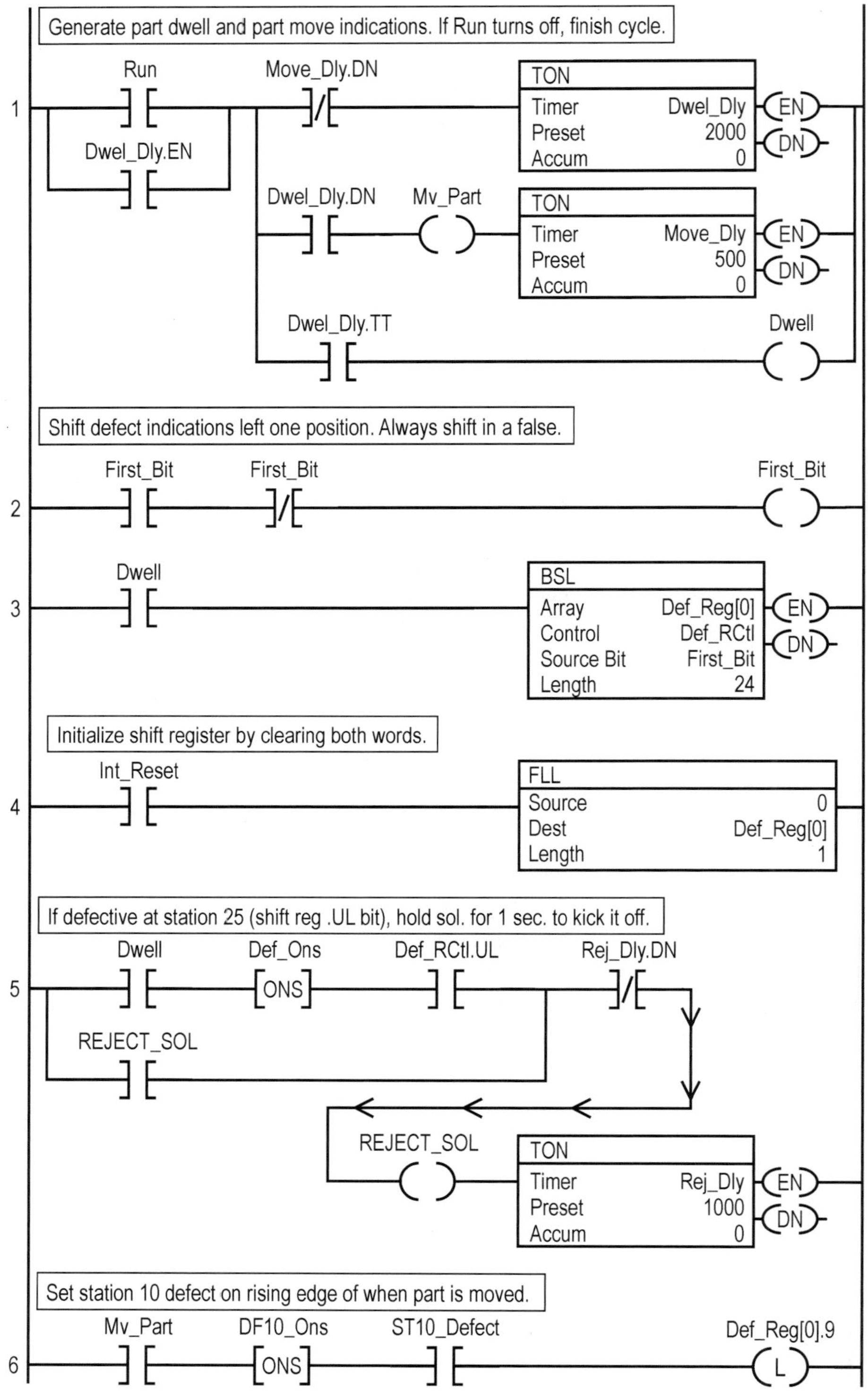

Figure 8.53. ControlLogix solution to Example 8.1.

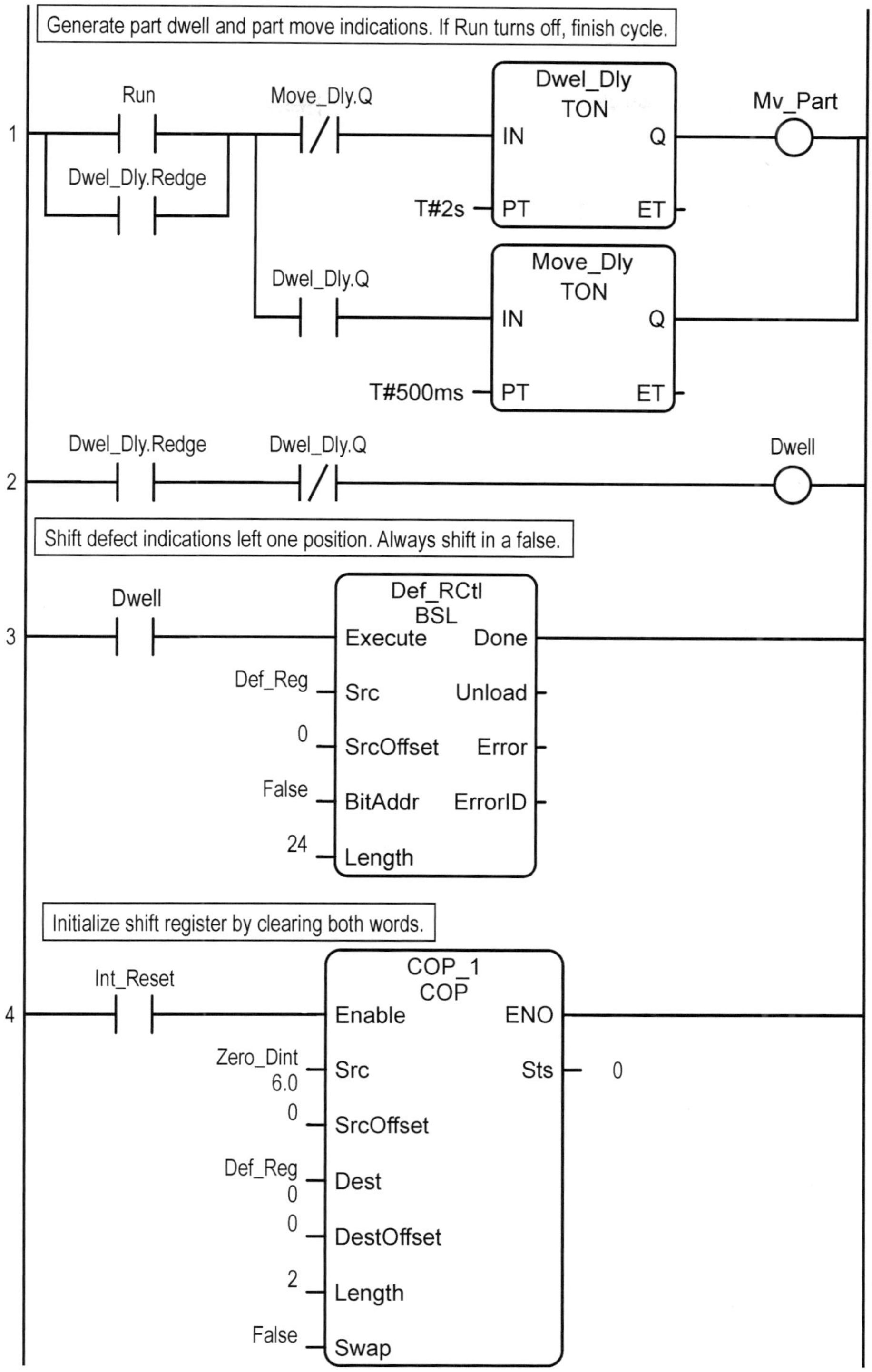

Figure 8.54. Micro800 solution to Example 8.1. *(continued)*

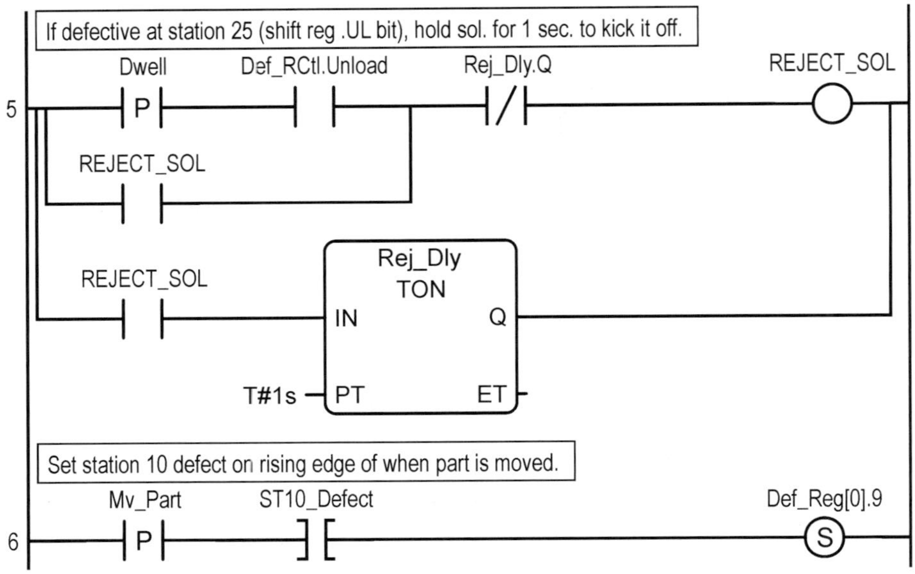

Figure 8.54. *(continued)*

Tag/Var./Symbol	CLogix Data Type	Micro800 Data Type	MLogix/SLC Address
COP_1	n/a	COP	n/a
Def_Ons	BOOL	n/a	B3/94
Def_RCtl	CONTROL	BSL	R6:0
Def_Reg	DINT[1]	DINT[0..1]	B12:0
DF10_Ons	BOOL	n/a	B3/121
Dwel_Dly	TIMER	TON	T4:0
Dwell	BOOL	BOOL	B3/91
First_Bit	BOOL	n/a	B3/95
Move_Dly	TIMER	TON	T4:1
Mv_Part	BOOL	BOOL	B3/92
Rej_Dly	TIMER	TON	T4:2
St10_Defect	BOOL	BOOL	B3/120

Example 8.2. Part tracking and stacking. Implement a system to track the parts as they move along a transport conveyor (Figure 8.55). At the end of the conveyor, the parts are stacked into a group of 10, and then unstacked onto the outbound conveyor. A part can be placed on the transport conveyor no faster than once every two seconds and the conveyor requires 40 seconds for one part to tranverse the transport conveyor. If a part is placed on the transport conveyor, a non-zero number is placed into Conv_Source. A zero in Conv_Source indicates that no part was placed on the transport conveyor. When a part is unstacked, place its number in Unstack_Dest. When unstacking, remove a part number every second. Also, turn **on** the Unstacking internal coil, and keep the transport conveyor stopped when

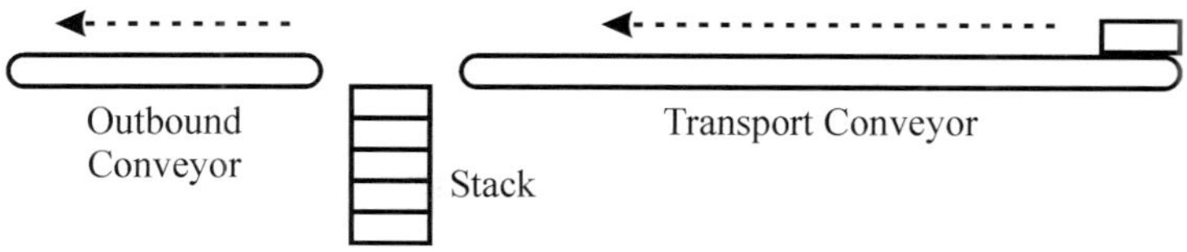

Figure 8.55. System of Example 8.2.

unstacking. Do not be concerned with the logic to place/remove and stack/unstack. Only track the parts on the transport conveyor and on the stack.

Assume the following physical input, physical output, and internal memory assignments:

Tag/Var./Symbol	Description
Conv_Run	Internal coil, **on** when transport conveyor may run (set by another part of the ladder).
Int_Reset	Internal coil, **on** to reset all queues. Assume that Conv_Run is already **off** when this turns **on** (set by another part of the ladder).
Conv_Source	Number of part being placed on transport conveyor.
Unstack_Dest	Number of part being unstacked from stack.
Unstacking	Indicates stack is being unstacked; **on** when unstacking (set by this part of ladder)
M1	Transport conveyor motor; **on** to run conveyor.

The address associated with the physical output is:

Tag/Var./Symbol	ControlLogix	Micro800	MLogix	SLC
M1	Local:2:O.Data.15	_IO_EM_DO_09	O:0/11	O:1/11

The addresses or data types associated with the internal tags/variables/symbols:

Tag/Var./Symbol	CLogix Data Type	Micro800 Data Type	MLogix/SLC Address
Conv_Run	BOOL	BOOL	B3/0
Int_Reset	BOOL	BOOL	B3/1
Unstacking	BOOL	BOOL	B3/3
Conv_Source	DINT	DINT	N14:0
Unstack_Dest	DINT	DINT	N14:50

Solution. The transport conveyor is modeled by a 20-element FIFO queue, and the stack is modeled by a 10-element LIFO queue. The FIFO queue is initialized to empty, and then when M1 (the conveyor motor) is running, the value of Conv_Source (part number) is loaded into the FIFO every 2 seconds. When the FIFO is full, then a part number is unloaded into Conv_Dest. If this number is non-zero, then a part is at the end of the conveyor, and it is loaded into the LIFO queue. This process continues until the LIFO is full. When the LIFO is full, the transport conveyor is stopped and the LIFO is unloaded, once per second until it is empty. Then the transport conveyor is allowed to run, and the process continues.

A timer (Conv_Tmr) generates a two-second tick to coordinate the FIFO load and unload. In order to simplify the logic, the handling of the situation when Conv_Run is turned off during this two-second interval is not considered. The Unld_Tmr generates the one-second tick to unload the LIFO.The ControlLogix solution for this problem is shown in Figure 8.56, and the Micro800 solution is shown in Figure 8.57. The MicroLogix solution is very similar to the ControlLogix solution. The ControlLogix and Micro800se solutions are very similar, and the function of each rung is:

1. Generate FIFO load pulse.
2. Load FIFO. If full, unload it. If unloaded integer is non-zero, load it into the LIFO.
3. Control of Unstacking internal coil.
4. Generate LIFO unload pulse and unload LIFO.
5. Control of M1. Must be **off** when unstacking.
6. Empty both queues when reset.

For these solutions, the addresses or data types associated with the internal tags/variables/symbols:

Tag/Var./Symbol	CLogix Data Type	Micro800 Data Type	MLogix/SLC Address
Conv_Dest	DINT	DINT	N14:1
Conv_FFL	n/a	FFL	n/a
Conv_FFU	n/a	FFU	n/a
Conv_Que	DINT[20]	DINT[0..19]	N14:2
Conv_Tmr	TIMER	TON	T4:1
FIFO_Ctl	CONTROL	FF_LF_CON	R6:1
LIFO_Ctl	CONTROL	FF_LF_CON	R6:2
Stak_LFL	n/a	LFL	n/a
Stak_LFU	n/a	LFU	n/a
Stak_Que	DINT[10]	DINT[0..9]	N14:30
Unld_Tmr	TIMER	TON	T4:2
Unstack_Dest	DINT	DINT	N14:50

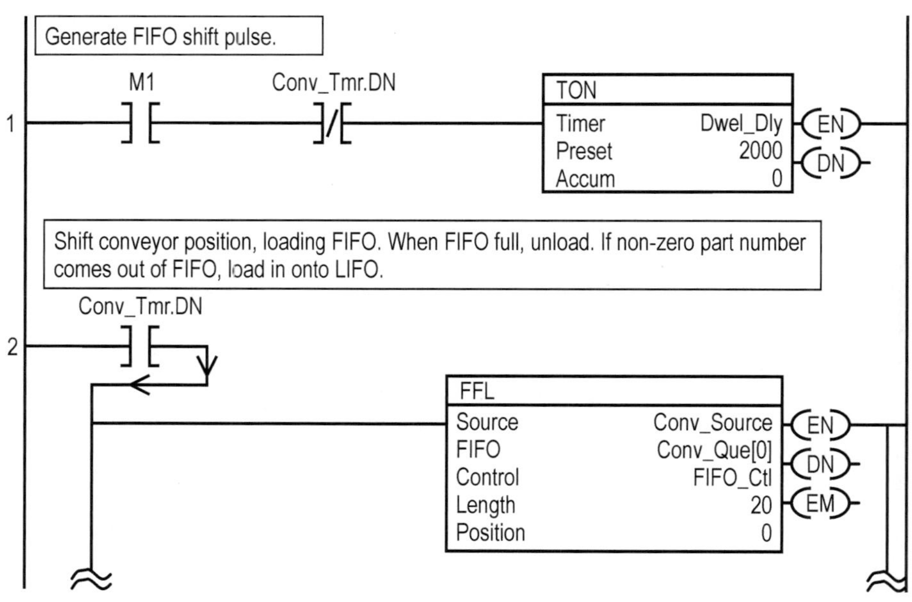

Figure 8.56. ControlLogix solution to Example 8.2. *(continued)*

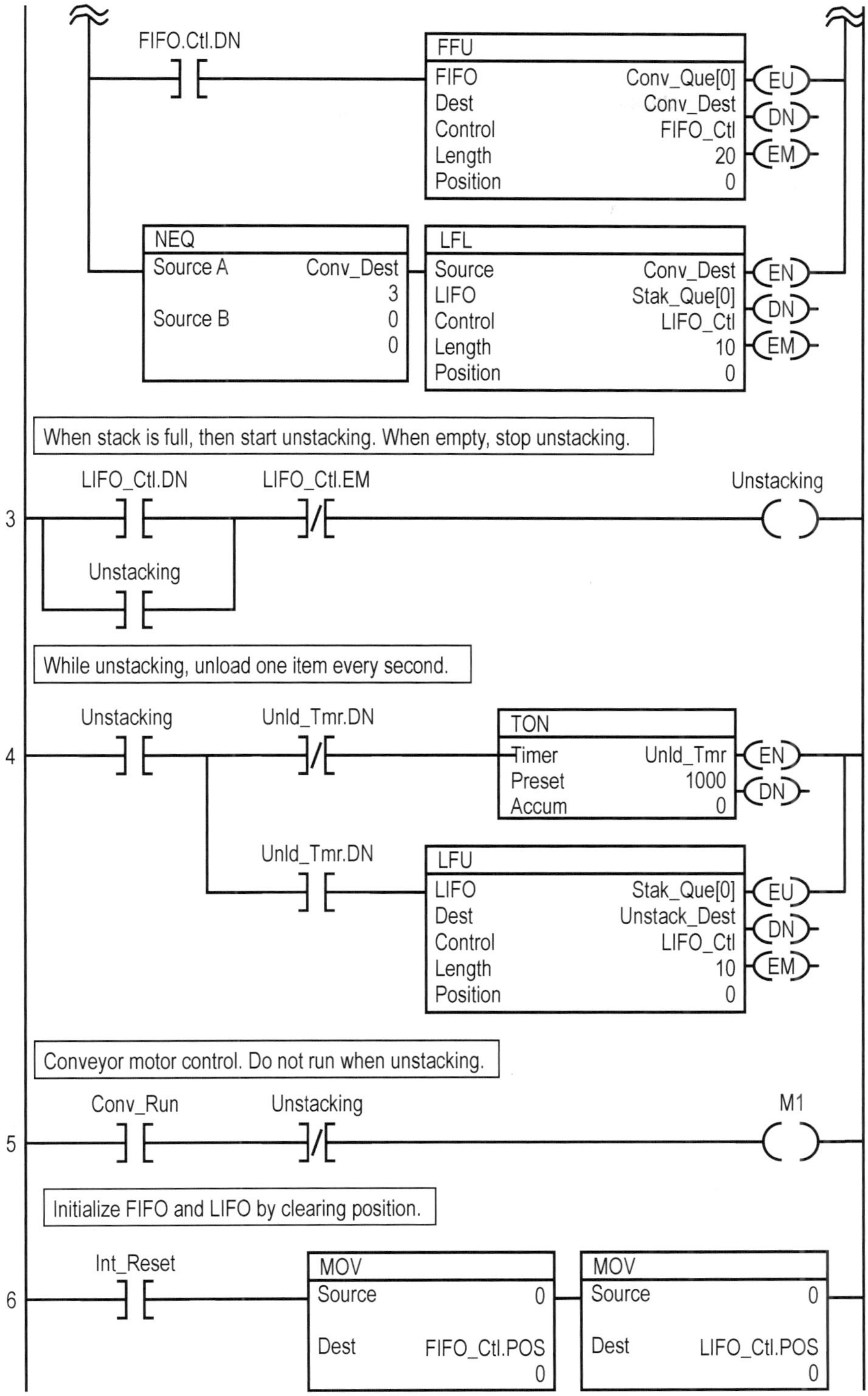

Figure 8.56. *(continued)*

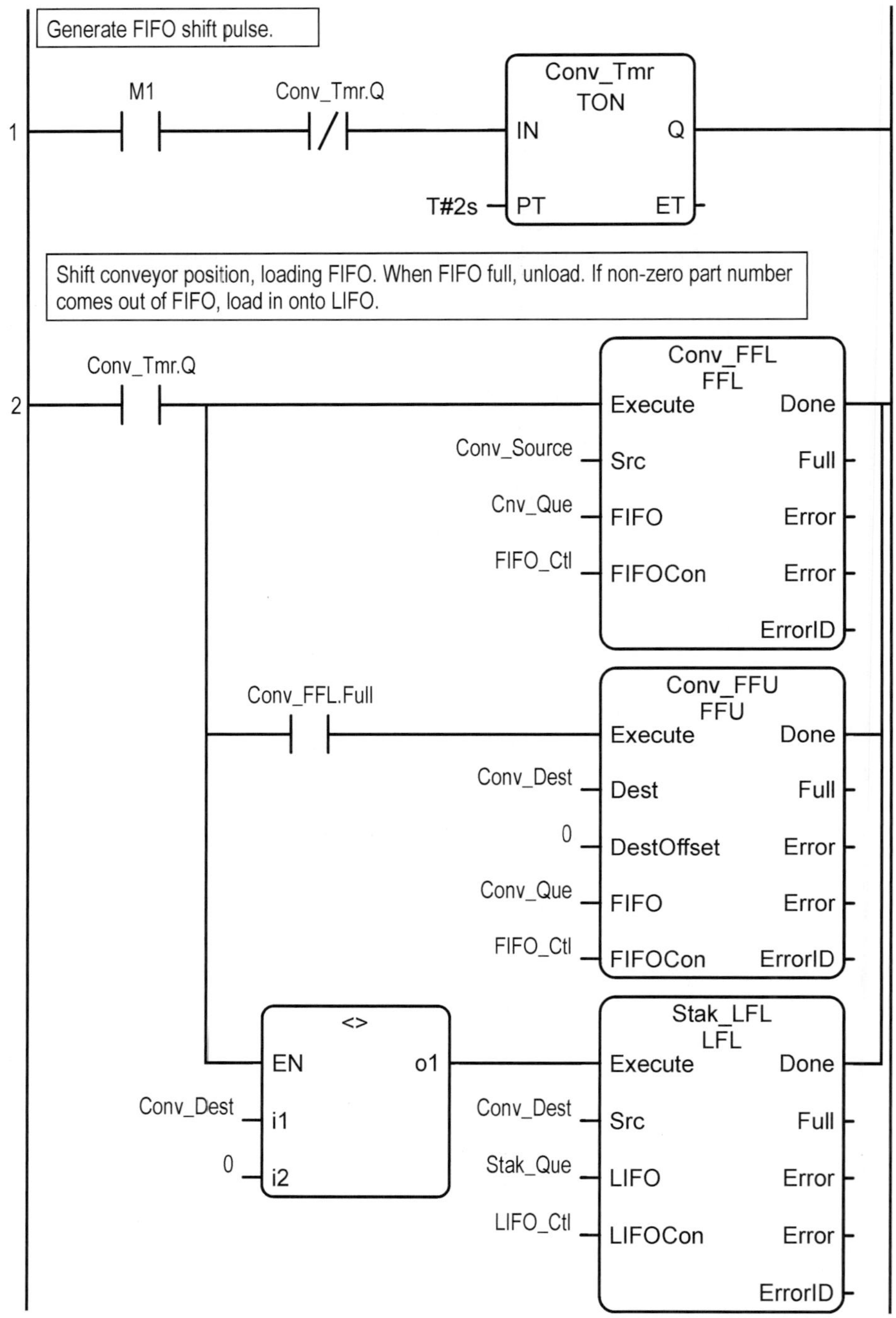

Figure 8.57. Micro800 solution to Example 8.2. *(continued)*

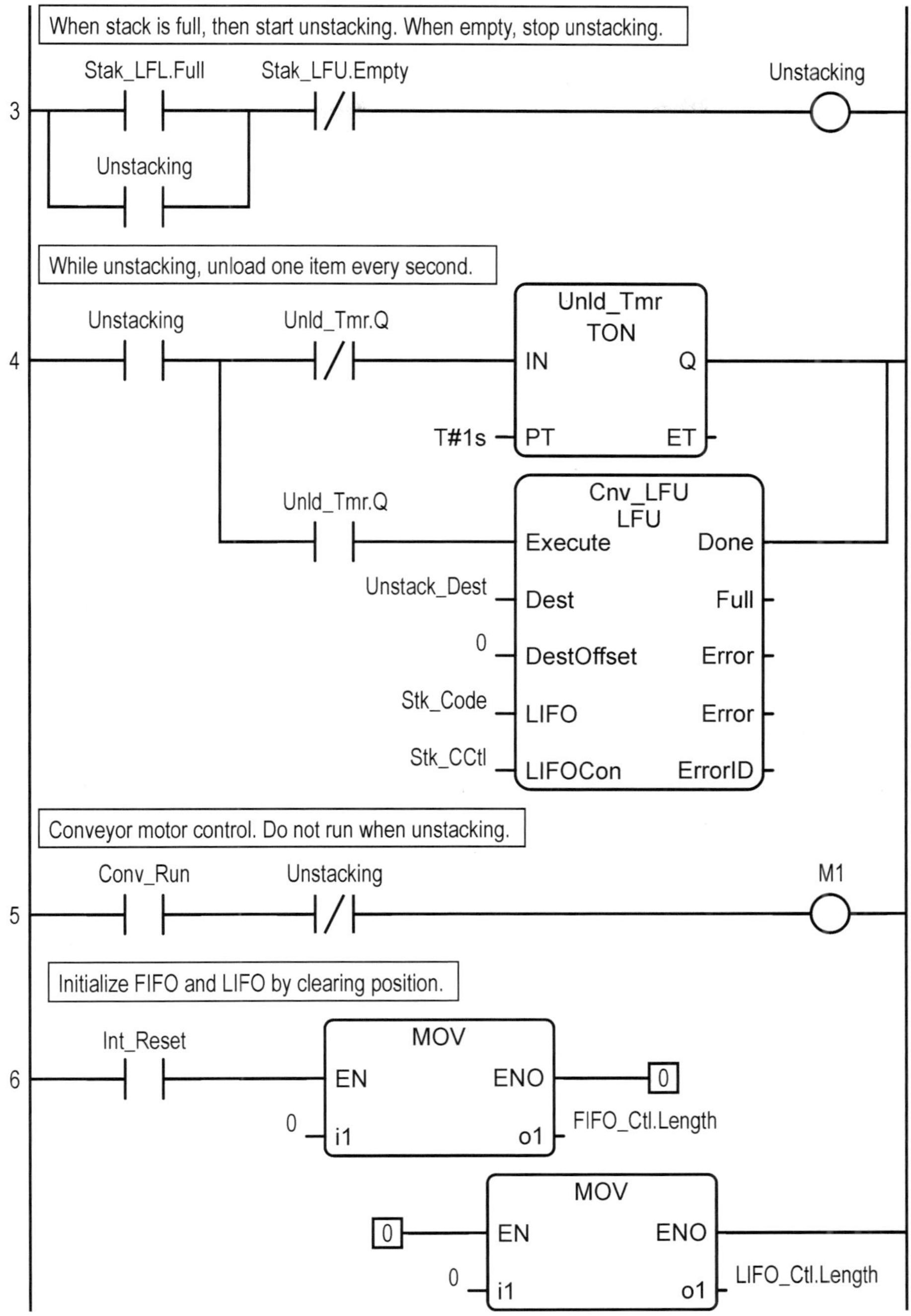

Figure 8.57. *(continued)*

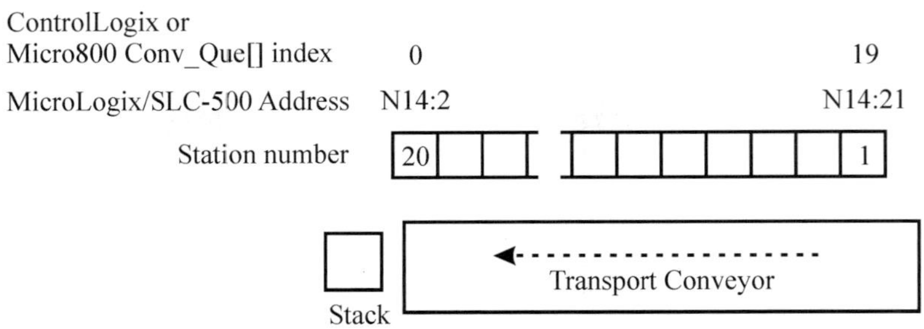

Figure 8.58. Mapping of stations to FIFO queue elements for Example 8.3.

Example 8.3. Modified part tracking and stacking. The transport conveyor of Example 8.2 is modified to be the conveyor that transports manufactured parts from a series of stations to the stacker. When a station places a part on the transport conveyor, the part number is copied to the element in the FIFO corresponding to that station.

Solution. As in Example 8.2, the transport conveyor is modeled by a 20-element FIFO queue, and the stack is modeled by a 10-element LIFO queue. Using the same definitions as for the solution to Example 8.2, the addresses of the FIFO queue elements that correspond to conveyor stations are shown in Figure 8.58. Looking at the FIFO and LIFO queues, the only difference between the solution for Example 8.3 and the solution for Example 8.2 is the FIFO queue initialization. In this example, the FIFO queue is initialized to a full condition, and all entries are zeroed. In addition, Conv_Source is always zero.

For the ControlLogix solution (Figure 8.56) rung 6 changes to the rung shown in Figure 8.59. The rung for the MicroLogix/SLC-500 solution is very similar, except that the blocks are in parallel. The changed rung for the Micro800 solution is shown in Figure 8.60. There is no simple way to initialize an array to all zeros and so an array that is already filled with zeros is copied to the conveyor FIFO to initialize it.

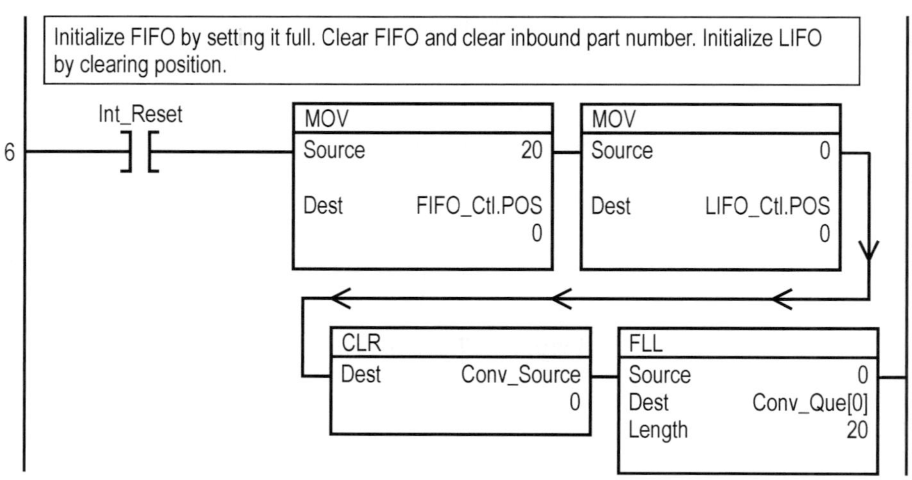

Figure 8.59. Modified ControlLogix rung 6 for Example 8.3 solution.

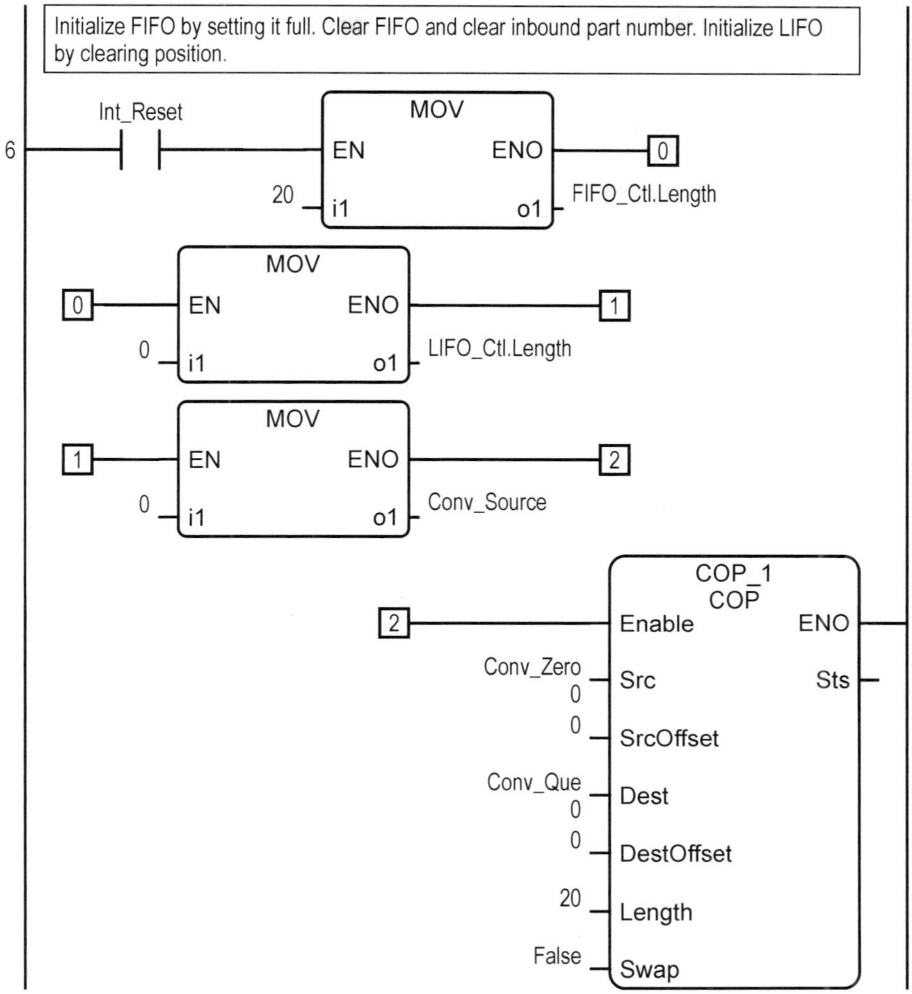

Figure 8.60. Modified Micro800 rung 6 for Example 8.3 solution.

8.6 CHAPTER SUMMARY

For the PLCs covered by this text, this chapter describes shift register, FIFO/LIFO, and special array ladder logic function blocks. Though used less frequently, these function blocks are necessary for certain applications. In addition, the instructions that modify ladder logic execution are described. Following these descriptions, applications of shift registers and FIFO/LIFO queues are illustrated.

REFERENCES

Rockwell Automation, 2008. *SLC Instruction Set Reference Manual*, pub. 1747-RM001G-EN-P, Rockwell Automation Inc., Milwaukee, WI, Nov.

Rockwell Automation, 2011. *MicroLogix 1100 Programmable Controllers Instruction Set Reference Manual*, pub. 1763-RM001D-EN-P, Rockwell Automation Inc., Milwaukee, WI, Sep.

Rockwell Automation, 2014. *MicroLogix™ 1200 and MicroLogix 1500 Programmable Controllers Instruction Set Reference Manual*, pub. 1762-RM001H-EN-P, Rockwell Automation Inc., Milwaukee, WI, Jul.

Rockwell Automation, 2019. *MicroLogix 1400 Programmable Controllers Instruction Set Reference Manual*, pub. 1766-RM001I-EN-P, Rockwell Automation, Inc., Milwaukee, WI, June.

Rockwell Automation, 2022a. *Logix5000 Controllers General Instructions Reference Manual*, pub. 1756-RM003W-EN-P, Rockwell Automation Inc., Milwaukee, WI, Mar.

Rockwell Automation, 2022b. *Micro800 Programmable Controllers General Instructions Reference Manual*, pub. 2080-RM001K-EN-P, Rockwell Automation Inc., Milwaukee, WI, Mar.

PROBLEMS

General instructions for the problems:

Write a ladder logic program for the application and implement it for one or more of the following PLC ladder logic languages:

ControlLogix/CompactLogix, **or**
Micro800, **or**
MicroLogix/SLC-500

If any part of the operation is ambiguous, write down your additional assumptions.

The physical inputs, physical outputs, and internal tags/variables/symbols for each problem are given in the problem. **DO NOT** assign any more physical inputs! Note that for the problems, the ControlLogix addresses assume 1756-, 1734-, 1769-, or 1794-series I/O modules and the CompactLogix addresses assume 5069- or 5094-series I/O modules.

Your solution should include the following:

1. Specify the PLC processor used.
2. Ladder logic diagram (with comments). For consistency among the different PLCs, use only tags/variables/symbols in the ladder logic. Use instructions consistent with the PLC processor.
3. Table listing additional internal memory (tags/variables/symbols) used and a brief description of their use. For the ControlLogix/CompactLogix/Micro800 processors, list the internal tags/variables and the data type. For the MicroLogix/SLC-500 processors, list the internal symbols and the associated memory address.

P8-1. Defect tracking for a diaper-making machine. The production system produces approximately two disposable diapers per second. A diaper is made by combining the various materials onto the base material, called the web. A simplified diagram of the system is shown in Figure P8.1*a*. The web and partially made diapers are checked at various points during processing, but defective diapers are rejected just before they enter the packaging machine. Any detected defect should then be noted and essentially travel with the web to the point at which the diaper containing the defect is removed from line. Model the system as a 200-bit shift register (Figure P8.1*b*) that is shifted upon command (Sync_Pulse transitions from **off** to **on**). The Sync_Pulse internal coil is a pulse train that is assumed on for no longer than one scan and has a nominal 0.2 second period. Sync_Pulse is synchronized to the line speed, and therefore its period may vary.

A finished diaper is represented by 2.5 bits of the reject shift register. There is also some inaccuracy in the defect tracking and so not only is the diaper with the expected defect rejected, but also the one preceding it and the one following it. Therefore, if any of the last 7 bits of the shift register are **on**, the REJECT_ROT is activated for 0.75 seconds to move a rotary arm that removes these diapers. Also, the last 7 bits of the shift register must be cleared when the defective diapers are removed.

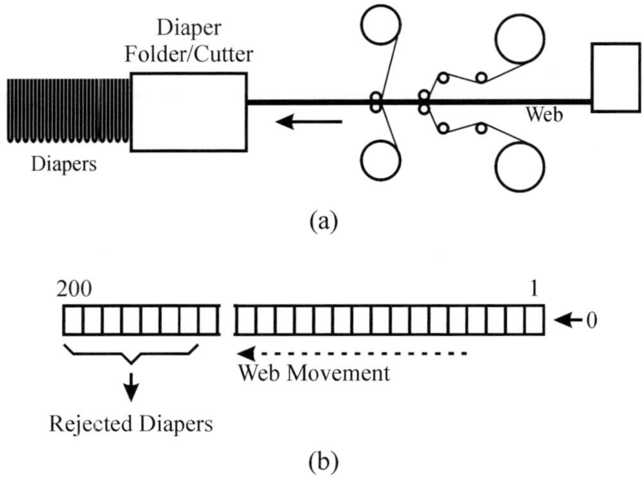

Figure P8.1. Diaper-making machine: *(a)* simplified schematic; *(b)* defect-tracking shift register.

Also, show the logic to set the defect bit at the following positions (position 1 is the start of the line and 200 is the end of the line): 20, 42, 49, 67, 82, 114, and 132. With your solution, specify the tags (ControlLogix/CompactLogix) or addresses (MicroLogix/SLC-500) for the defect bits for these positions.

Assume the following physical input, physical output, and internal memory assignments:

Tag/Symbol	Description
Run	Internal coil, **on** when system running (set by another part of the ladder).
Int_Reset	Internal coil, **on** to clear all defect indications. Assume that Run is already **off** when this turns **on** (set by another part of the ladder).
Sync_Pulse	Line synchronizing pulse. Shift defect-tracking bits when it transitions **on**. Assume it is **on** for only one scan and has a nominal period of 0.2 seconds.
REJECT_ROT	Reject mechanism; **on** for 0.75 second to remove 3 defective diapers from line.

The address associated with the physical output is:

Tag/Symbol	ControlLogix	MicroLogix	SLC-500
REJECT_ROT	Local:2:O.Data.14	O:0/11	O:2/16

The addresses or data types associated with the internal variables:

Tag/Symbol	CLogix Data Type	MLogix Address
Run	BOOL	B3/100
Int_Reset	BOOL	B3/101

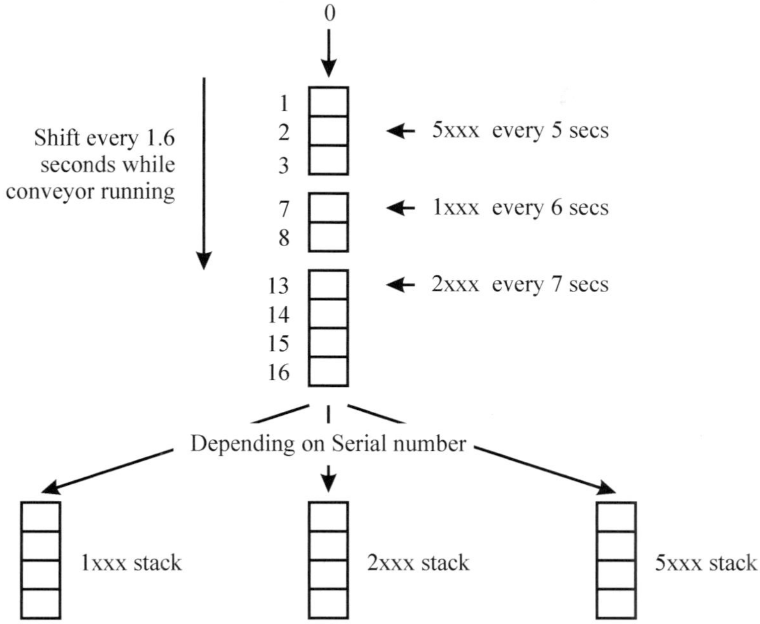

Figure P8.2. Part-tracking system.

P8-2. Part tracking for production line. Figure P8.2 shows a production line that consists of 3 machines. Each machine produces a different part that is loaded onto one of three stacks. When any stack is full, the conveyor is stopped, and the stack is emptied. Your job is to use ladder logic to track the part serial numbers as they proceed down the conveyor and onto the stacks. At the end of the conveyor, each part must be loaded onto the appropriate stack. For this assignment, you are basically tracking the part serial numbers. You are not concerned about controlling the machines that make the part (you will simulate these), physically moving the parts onto the conveyor, or how they are physically stacked and unstacked.

Machine	Part Gen. Interval	Conveyor Position	Serial Number	Stack LIFO	Unload Stack to
A	5 secs	2	5xxx	Stack_5	UL_Stack_5
B	6 secs	7	1xxx	Stack_1	UL_Stack_1
C	7 secs	13	2xxx	Stack_2	UL_Stack_2

The part is not moved onto the conveyor if a part is already occupying the spot. If the conveyor position is already occupied at the end of the part generation interval, the machine waits until the position is unoccupied and then moves a part onto the conveyor. While waiting, the part generation operation pauses. To "move" the part onto the conveyor, move the serial number into the appropriate FIFO position. When the part is moved onto the conveyor, the part generation timer is restarted, and the "xxx" part of the serial number is incremented.

Shift parts on the conveyor FIFO every 1.6 seconds when the conveyor is running. The conveyor is represented by the FIFO from N15:17 (conveyor position 1) to N15:2

(conveyor position 16). The conveyor FIFO is loaded from Conv_Fifo_Ld (a constant zero) and unloaded to Conv_Fifo_Ul.

When the conveyor FIFO is unloaded, the part is placed into one of 3 stacks, depending on the serial #. The stacks are modeled as 10 position LIFO's. When one of the stacks is full, the conveyor should be stopped (and stop shifting the FIFO), and the stack unloaded into the appropriate location. Use a one-second timer to simulate the removal of all of the parts from the stack. When the stack is empty, the conveyor should start, and the tracking of parts on the conveyor resumes.

The start and stop pushbuttons are for the line. When START_PB is pressed, the conveyor moves, and the machines generate parts. When the STOP_PB is pressed, the system operation pauses. When paused, the conveyor should be stopped and any unstacking is also paused. When the operation is paused, the timers used to simulate part generation should be paused and the accumulator retained. When RESET_PB is pressed (and the system is not running), the conveyor and pallet queues are cleared, all machine part interval generation timers are reset, and the "xxx" part of all part serial numbers are set to zero.

The conveyor motor (PLC output) should be driven by your ladder logic.

Assume the following physical input, physical output, and internal memory assignments:

Tag/Var./Symbol	Description
START_PB	Start push button switch, **on** (closed) to start or resume.
STOP_PB	Stop push button switch, **off** (open) to stop or pause.
RESET_PB	Reset push button switch, **on** (closed) resets operation of station (explained above).
CONV_MTR	Conveyor motor starter, **on** to move conveyor.
Conv_Fifo_Ld	Conveyor FIFO is loaded from this location.
Conv_Fifo_Ul	Conveyor FIFO is unloaded to this location.
Conv_Fifo	Conveyor FIFO
Stack_1	Start of LIFO stack for 1xxx parts.
Stack_2	Start of LIFO stack for 2xxx parts.
Stack_5	Start of LIFO stack for 5xxx parts.
UL_Stack_1	LIFO stack for 1xxx parts is unloaded to this location.
UL_Stack_2	LIFO stack for 2xxx parts is unloaded to this location.
UL_Stack_5	LIFO stack for 5xxx parts is unloaded to this location.

The addresses associated with the physical inputs and output are:

Tag/Var./Symbol	ControlLogix	Micro800	MLogix	SLC
START_PB	Local:4:I.Data.0	_IO_EM_DI_01	I:0/0	I:1/0
STOP_PB	Local:4:I.Data.1	_IO_EM_DI_02	I:0/1	I:1/1
RESET_PB	Local:4:I.Data.2	_IO_EM_DI_03	I:0/2	I:1/2
CONV_MTR	Local:5:O.Data.9	_IO_EM_DO_09	O:0/9	O:1/9

The addresses or data types associated with the internal tags/variables/symbols:

Tag/Var./Symbol	CLogix Data Type	Micro800 Data Type	MLogix/SLC Address
Conv_Fifo_Ld	DINT	DINT	N15:0
Conv_Fifo_Ul	DINT	DINT	N15:1
Conv_Fifo	DINT[16]	DINT[0..15]	N15:2
Stack_1	DINT[10]	DINT[0..15]	N15:30
Stack_2	DINT[10]	DINT[0..15]	N15:40
Stack_5	DINT[10]	DINT[0..15]	N15:50
UL_Stack_1	DINT	DINT	N15:71
UL_Stack_2	DINT	DINT	N15:72
UL_Stack_5	DINT	DINT	N15:73

P8-3. Alarm history display. The last ten alarms and the time they occurred need to be displayed on an operator interface screen (Figure P8.3). Program an array of these alarm numbers and their associated time stamp. The time stamp should contain the seconds, minutes, hour, day, and month the alarm occurred. The operator interface will display the number of the alarm, the time and date it occurred, and a text description. Do not generate the text description.

When the Alarm_Trigger transitions from **off**-to-**on**, save the Alarm_Number and the current time in the alarm history. If the history already contains 10 alarms, then the oldest alarm is removed from the list before the new one is added. When Alm_Hist_Rst is **on**, clear the alarm history list, and insert zeros for all alarm numbers and the time stamp.

With your solution, specify the tags (ControlLogix/CompactLogix) or addresses (MicroLogix/SLC-500) for the most recent to the oldest alarm information to be used by the HMI programmer.

Assume the following internal memory assignments:

Tag/Var./Symbol	Description
Alarm_Trigger	Internal coil, **off**-to-**on** transition means the Alarm_Number and the current timestamp should be saved in the history (set by another part of the ladder).
Alarm_Number	Number of alarm to place in alarm history (set by another part of the ladder).
Alm_Hist_Rst	Alarm history reset. When **on**, clear the alarm history list and insert zeros for all alarm numbers and the time stamp (set by another part of the ladder).

Date	Time	No.	Alarm Description
04/07/03	10:19:02	323	T304 at 90% level
04/07/03	10:10:15	125	P124 needs servicing
04/07/03	08:03:23	21	B24 at low level
04/07/03	06:43:32	227	TT235 high temperature

Figure P8.3. Alarm display (partial).

The addresses or data types associated with the internal variables:

Tag/Symbol	CLogix Data Type	MLogix Address
Alarm_Trigger	BOOL	B3/100
Alm_Hist_Rst	BOOL	B3/101
Alarm Number	DINT	N7:24

9 Troubleshooting

Chapter Topics:

- General fault diagnosis techniques
- Diagnosing input/output problems
- PLC processor status indications
- Finding programming errors
- Diagnosing PC-to-PLC communication errors

OBJECTIVES

Upon completion of this chapter, you will be able to:

- Understand the general fault diagnosis method
- Diagnose problems with the PLC hardware and program
- Diagnose PC-to-PLC communication problems

Scenario: Problem with unplanned water pump trips.

Your facility has a jet water pump used to pressurize water to cut coke away from the coker unit. The pump has been intermittently shutting off during the past four months. By the time the plant requested engineering assistance, the unplanned outages have cost a few million dollars. The problem is intermittent, making troubleshooting difficult.

Solution: The jet pump schematics indicate that the jet pump could trip due to numerous causes, one of which is a PLC-initiated trip. Examining the PLC program, it looks to be poorly written and cumbersome. To eliminate or confirm the PLC as the cause of the problem, you locate the output coil for the PLC-initiated trip and insert a coil in parallel with that output, latching an unused internal bit. At the next pump trip, you find this internal bit **on**, indicating the trip was initiated by the PLC. You sigh, because now you will have to decode the PLC logic. The PLC trip could be initiated by a number of conditions, and each of these conditions could be triggered by other conditions (rung 237 of Figure 9.1). To determine the cause of the trip, you go one step backward and insert a latch coil at each of the trip conditions, like rung 451 of Figure 9.1. After the next pump trip, you know which logic path back you need to follow, by which bit (for example, Cond_4_Bit) is latched. So, you insert another set of latch coils on the conditions that drive Condition_4 (Cond_4A_Bit, Cond_4B_Bit, and so on). On the next trip, you determine that the pump tripped due to the signal (or lack of signal) from one of the isolation valve limit switches or one of two selector

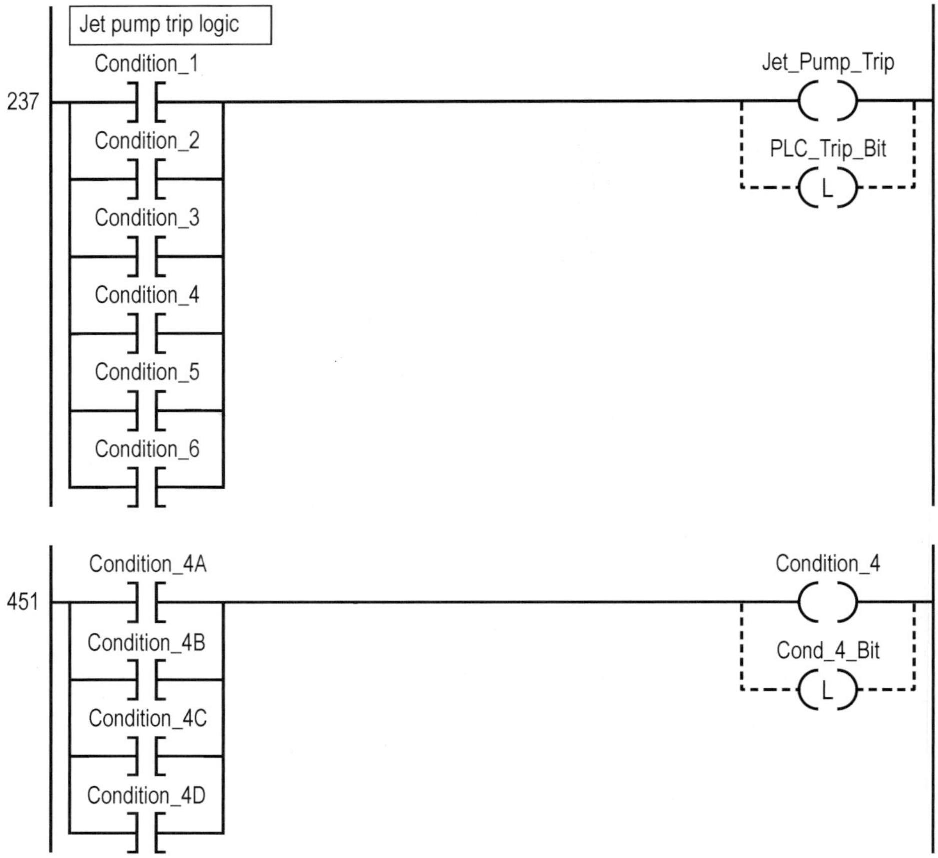

Figure 9.1. Latch coils added to existing logic to aid troubleshooting.

switches. At this point, rather than wait for another failure, you have the electricians investigate all of the isolation valve limit switches and the selector switches. While removing the conduit union to one of the solenoids for the isolation valve, the electricians notice one of the solenoid wires had been smashed during installation a few years earlier (Figure 9.2). Since the defective conductor was still mostly insulated, the short only occurred under certain vibration or temperature conditions. Sometimes the solenoid failure caused a fuse to blow and during other failures, the fuse did not blow. The solenoid is replaced and the problem is resolved, likely saving at least another million dollars.

9.1 INTRODUCTION

Troubleshooting is the process of locating and eliminating the source of trouble in a system. Furthermore, the source of the problem must be found and fixed in a minimum amount of time. This chapter provides an introduction to troubleshooting methods and provides general troubleshooting information for PLC systems. In the area of troubleshooting, one needs (1) good analytical skills, and (2) experience. With good analytical skills, one becomes an expert after years of experience.

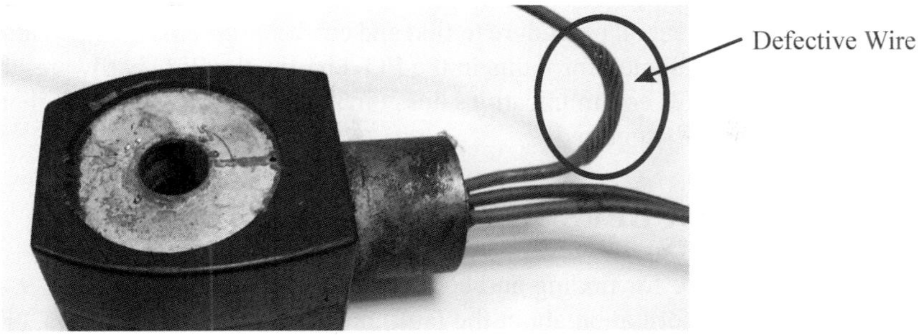

Figure 9.2. Solenoid wire that caused fault.

One knack of finding faults quickly is knowing where to look first. Given a number of possible sources of the failure, one should know the most likely places and eliminate them first. Figure 9.3 shows the failure probability of the different parts of the PLC system hardware (Parr, 2001). This figure shows that the majority of hardware faults in a PLC system are actually a failure in the field devices and not in the PLC. Within the PLC itself, the majority of failures are in the I/O modules. Program errors should be found and fixed before the system is commissioned. After commissioning, the number of failures due to a program error should be small.

Compounding the process of finding the source of a problem is the lack of current documentation. When diagnosing faults, be prepared for:

Out-of-date or missing engineering drawings

Out-of-date or no program listings

Obsolete equipment (no documentation)

Most companies keep a set of documentation that was complete when the equipment was installed. However, as changes are made to the system, the documentation (or at least all copies of it) may not be kept current. Also, manufacturer's manuals may be inadvertently lost or destroyed.

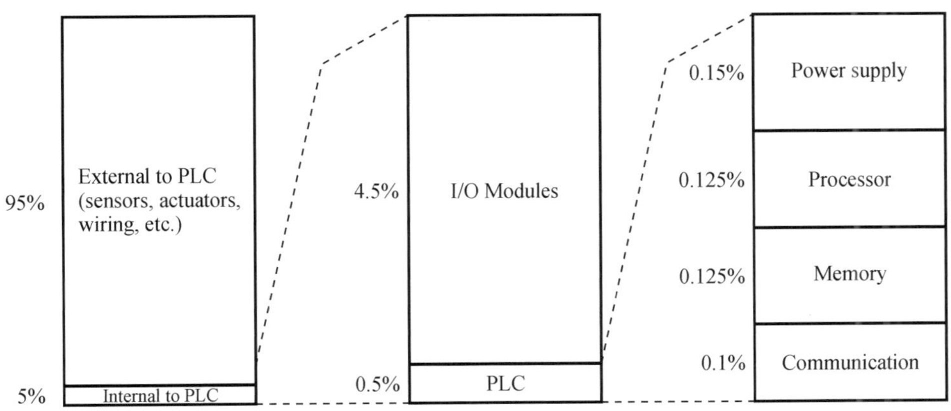

Figure 9.3. Hardware failures in a typical PLC system.

After presenting the general procedure to find and correct problems in the system, more specific information is presented for faults in the I/O devices, the PLC hardware, the PLC program, and the PC-to-PLC communications link. Finally, troubleshooting aids that can be designed into a system are suggested.

9.2 GENERAL TROUBLESHOOTING PROCEDURES

The general procedure for finding and correcting faults is shown in Figure 9.4. With the evidence and other information about the fault, one constructs a hypothesis, or guess, about the source of the problem. In order to verify the source of the problem, one must test the hypothesis. If the test does not confirm the hypothesis, then more information may need to be gathered. If the test corroborates the hypothesis, then the source has been found and then the fault can be corrected or repaired. If the source of the fault is a PLC hardware module, then the module is replaced. Usually, the faulty module is identified and not the actual component within the module. The module is replaced with a spare module and the faulty module is sent to the manufacturer or service depot for repair. Simple programming problems are usually corrected while the processor is still in the run mode and with minimum disruption. More extensive program modifications may require waiting to download the changed program until the next scheduled downtime.

9.2.1 Gathering Information About the Fault

Often, experience is the best source of the type of information that should be gathered. At a minimum, the following should be known:

What happened?

Where is the problem?

What time did it occur? Does another event happen at the same time?

When does the problem occur? Does it repeat?

It is also important to eliminate irrelevant information. However, knowing what is relevant often comes with experience.

9.2.2 Generating a Hypothesis

The hardest part of the procedure in Figure 9.4 is forming one or more hypotheses to test. Composing a hypothesis is based on experience, equipment documentation, or by tracing through the system.

Past ***experience*** is the most common means of generating a hypothesis. One knows the problem and its solution because it has been seen before. If there are a number of possible hypotheses, the guess that is first checked is often the one that is easiest to test or is the most probable one. Experience is primarily learned on the job: the more one has worked, the more experience one has. Good maintenance records can indicate past problems with a particular piece of hardware. However, the maintenance records need to be searchable in order to be useful. A device that has been recently replaced or a part of the program that has been recently modified can also provide hypotheses.

If the type of problem is a new one, then ***troubleshooting charts*** or other information sources (including other people) can be consulted to find new hypotheses. This chapter

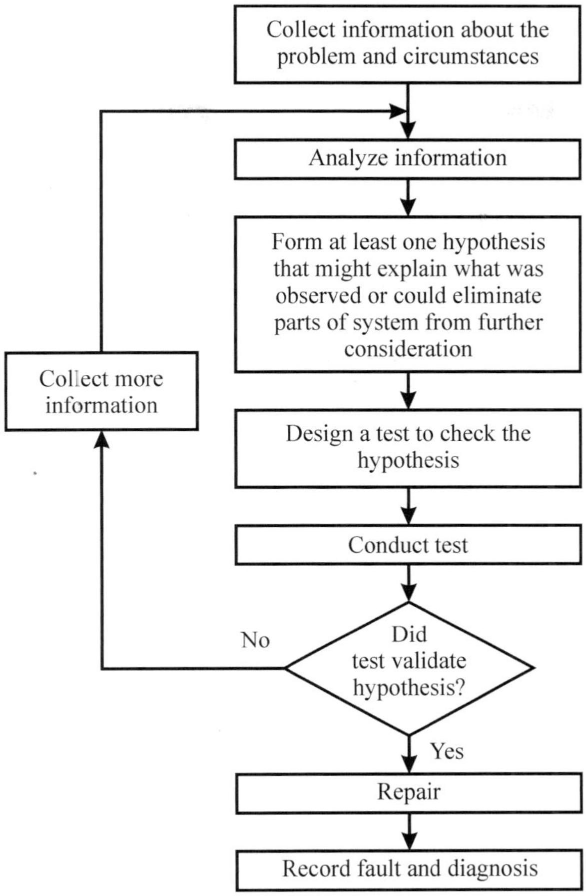

Figure 9.4. General troubleshooting procedure.

contains general troubleshooting information and the manufacturer's documentation is another source.

If experience and other troubleshooting information do not provide a hypothesis, then the system operation will need to be ***traced***. One way to trace the system is to start from the source of the signal and check its status as it is processed by each subsystem. For example, a discrete input can be traced in this order: field device, input module, processor, and program. The scenarios at the beginning of this chapter and Chapters 4 and 7 are all examples of tracing a system to determine the fault.

9.2.3 Testing a Hypothesis

After a hypothesis is constructed, it must be tested. Aside from using a multimeter to test voltages and/or currents, one may replace suspected modules, remove questionable parts of the system, or set a trap to catch the fault.

When diagnosing discrete and analog input/output modules, a digital multimeter is often adequate to find the fault. However, for more complicated modules, such as motion controllers, an oscilloscope or high-speed data aquisition system may be needed. One can ***substitute*** a known good module and verify that the problem has disappeared. One of the disadvantages with this method of testing is that the system operation may be disrupted. For example, in order to replace a suspected defective motion control module when one axis fails, all of the other axis controls on that module must be temporarily disabled when the module is replaced. Normally, this is not an operation that is attempted while the system is operating.

In a loosely coupled system, ***removing*** the suspected module is another way of testing the hypothesis. Several PLCs communicating with each other is a common example of a loosely coupled system. An incorrect message in one of the PLCs in such a system can cause data corruption in the other PLCs. In this case, the suspected PLC is disconnected from the network or placed in program mode to test the hypothesis.

For a spurious or transient problem, setting a trap is often the only effective method of testing its existence and checking for other conditions. Even when the system logs and archives data, it may not be sampled frequently enough to catch the event. In this case, the PLC program is often modified to monitor for the suspected event and record other conditions when it occurs (like the scenario at the beginning of this chapter). In certain systems, the processor can be halted when the event occurs so that all of the PLC data can be examined. The scenario at the beginning of Chapter 7 is an example of an event that caused an automatic halt of the processor and then the PLC data was examined to determine the actual cause of the fault and the subsequent changes in the PLC program to prevent future faults.

9.2.4 Erratic Behavior

If the PLC system exhibits a large number of failures or erratic behavior, the most likely cause is a problem with the power supply. Another cause is the presence of electro-magnetic interference.

First, the integrity of the power and ground connection to the PLC should be checked. One should visually check the power and ground wiring for loose or corroded connections and cracked insulation. The ground integrity can be checked by verifying that the voltage between the PLC ground terminal and a known ground is zero.

Second, the source to the power supply should be checked. The power supply input voltages should be within the recommended range. If the source to the power supply is AC voltage and is lower than required (often called "voltage sag"), then the source of the problem should be identified. If the low voltage is a recurrent problem, then a constant-voltage transformer may need to be inserted between the AC power supply and the PLC power supply connection. If the PLC power supply input voltage is DC, then the power supply that generates this voltage should be checked for problems. In this case, the AC ripple of the DC voltage should also be small.

Thirdly, the outputs of the PLC power supply should be checked that the voltages are within the recommended ranges. Unfortunately, these outputs may be difficult to check due to the inability to connect a meter to the power supply outputs.

Another cause of erratic behavior is the presence of electromagnetic interference (EMI) or radio frequency interference (RFI). Unfortunately, it may be difficult to determine the source since one must correlate the problem with an EMI or RFI event. Possible EMI or RFI events are (1) a large motor starting, (2) an unshielded electric solenoid actuating, (3) arc welding nearby, (4) a lightening strike, or (5) use of a handheld radio transmitter. Solutions to EMI or RFI problems usually involve properly grounding, shielding, or conditioning the PLC power source.

9.3 TROUBLESHOOTING I/O MODULES

The most common troubleshooting problem in a PLC system has to do with the inputs and outputs. When troubleshooting I/O problems, the primary goal is to determine why the internal status of the PLC memory does not agree with the external device. The mapping of the status/value of an I/O channel to the internal PLC memory is covered in Chapter 3 and is not repeated here. A problem with an I/O channel is due to a failure in one of four places:

1. the device,
2. the PLC I/O module, or
3. the power supply for the I/O module or device, or
4. the wiring between the device and the PLC I/O module.

An internal failure of the PLC processor generally affects the values from multiple I/O modules and/or more serious problems and is covered in the next section of this chapter. If the problem appears to occur in multiple I/O modules and these modules are connected to the processor by some communication link, then check the communication link for failures.

When diagnosing I/O failures, a meter that can accurately measure voltage, current, and resistance is often essential. Also, a device (e.g., a notebook computer) that can be used to monitor the PLC memory is necessary.

Discrete I/O modules generally have status indicators on the module, one for each channel, as shown in the circuit diagrams in Chapter 4. Whether the status indicator is on the device side of the module isolation or on the PLC side of the isolation barrier is relevant only to those finding the failed component in the module. If the failure is in the I/O module, the troubleshooting information here will allow one to isolate the failure to a particular module. The failed module is then repaired by the manufacturer or by a service center.

Troubleshooting charts for discrete input and output modules are shown in Table 9.1 and Table 9.2. Since the most likely cause of a failure is the I/O device the charts starts with the state of the I/O device, checking the state of the channel indicator, and then the PLC on-line status of the I/O channel. For example, if the discrete input device is off/open/deactivated, the channel status indicator on the discrete input module is **on**, and the PLC on-line status of the channel is **on**, then the most likely cause is one of three possibilities:

1. Off-state leakage current of the input device is high enough to trigger the threshold detector in the module (AC types only), or
2. Input device is damaged, or
3. Input module circuit is damaged.

The "recommended action" column contains information in order to confirm or eliminate the probable cause.

Troubleshooting analog modules is usually more difficult than for discrete modules because the modules generally have no indication of the analog channel voltage or current. One can connect small panel meters to the analog channels to monitor the voltage/current, but these are generally not used for voltage signals since a voltmeter can easily be connected

Table 9.1. Discrete Input Diagnosis Guide.

Input Device is On/Closed/Activated			
Channel LED is	And	Probable Cause	Recommended Action
On	On-line status shows the input as **off** or ladder program operates as though it is **off**	Input is forced off in program	Remove input force for channel.
		Input circuit is damaged	Verify proper wiring. Try another input channel. Replace module.
	Input device will not turn **off**.	Device is shorted or damaged	Verify device operation. Replace device.
	Program operates as though device is momentarily **off**.	Momentary dip in power supply voltage	Increase DC power supply wattage rating.
Off	On-line status shows the input as **off** or the ladder program operates as though it is **off** and/or the input circuit will not turn **on**.	Low voltage across the input	Check voltage between input channel and common. Check source voltage.
		Incorrect wiring or open circuit	Check wiring and common connections.
		Input signal turn-on time too fast for input circuit	Check module and device timing specifications.
		Input circuit is incompatible	Check specification and sink/source compatibility (if DC input).
		Input circuit is damaged	Verify proper wiring. If possible, try other input channel. Replace module.

Input Device is Off/Open/Deactivated			
Channel LED is	And	Probable Cause	Recommended Action
On	On-line status shows the input as **on** or the ladder program operates as though it is **on** and/or the input circuit will not turn **off**.	Input device off-state leakage current exceeds specification (AC inputs)	Check device and input circuit specifications. Use load resistor to bleed-off circuit.
		Input device is shorted or damaged	Verify device operation. Replace device.
		Input circuit is damaged	Verify proper wiring. If possible, try other input channel. Replace module.
Off	On-line status shows the input as **on** or ladder program operates as though it is **on**.	Input is forced on in program	Remove input force for channel.
		Input circuit is damaged	Verify proper wiring. If possible, try another input channel. Replace module.
	Input device will not turn **on**.	Device is shorted or damaged	Verify device operation. Replace device.

Table 9.2. Discrete Output Diagnosis Guide.

Output Device is On/Energized

Channel LED is	And	Probable Cause	Recommended Action
On	On-line status shows the output as **off** and the output circuit will not turn **off**.	Programming error	Check for duplicate output coils using the search function.
On	On-line status shows the output as **off** and the output circuit will not turn **off**.	Output is forced on in program	Remove output force for channel.
On	On-line status shows the output as **off** and the output circuit will not turn **off**.	Output circuit is damaged	Use the force function to force output off. If the output device does not turn off, the output circuit is damaged. Try another output channel. Replace module.
Off	On-line status shows the output as **off** and the output device will not turn **off**.	Output circuit off-state leakage current may exceed output device specification.	Check device and output circuit specifications. Use load resistor to bleed-off current. Neon lamps as load on 120v output modules need a bleed resistor.
Off	On-line status shows the output as **off** and the output device will not turn **off**.	Incorrect wiring	Check wiring. Disconnect from PLC and verify device operation.
Off	On-line status shows the output as **off** and the output device will not turn **off**.	Output device is shorted or damaged	Verify device operation. Replace device.
Off	On-line status shows the output as **off** and the output device will not turn **off**.	Output device is incompatible	Check specification and sink/source compatibility (if DC input).
Off	On-line status shows the output as **off** and the output device will not turn **off**.	Output circuit is damaged	Verify proper wiring. Try other output channel. Replace module.

Output Device is Off/De-energized

Channel LED is	And	Probable Cause	Recommended Action
On	On-line status shows the output as **on** and the output device will not turn **on**.	Low voltage across the load	Measure the source voltage and check specifications.
On	On-line status shows the output as **on** and the output device will not turn **on**.	Incorrect wiring or open circuit	Check wiring and COMMON connections.
On	On-line status shows the output as **on** and the output device will not turn **on**.	Output device is incompatible	Check specification and sink/source compatibility (if DC input).
On	On-line status shows the output as **on** and the output device will not turn **on**.	Output circuit is damaged	Verify proper wiring. Try other output channel. Replace module.
Off	On-line status shows the output as **on** and the output device will not turn **on**.	Programming error	Check for duplicate output coils using the search function.
Off	On-line status shows the output as **on** and the output device will not turn **on**.	Output is forced off in program	Remove output force for channel.
Off	On-line status shows the output as **on** and the output device will not turn **on**.	Output circuit is damaged	Use the force function to force output off. If the output device does not turn off, the output circuit is damaged. Try another output channel. Replace module.

Table 9.3. Analog Input Diagnosis Guide.

Value of physical quantity is in the normal range			
PLC value is	And	Probable Cause	Recommended Action
Min.	Measurement of voltage or current at module channel is zero	Bad sensor power supply	Check sensor power supply. Replace if necessary.
		Inadequate power supply voltage (current meas.)	Loop power supply voltage should be at least (number of loop devices)x5V + wire resistance voltage drop at 20 mA.
		Broken wire	Check wiring between sensor and module.
	Measurement of voltage or current at module is consistent with sensor value.	Input circuit is damaged	If possible, try another input channel Replace module.
Max.	Measurement of voltage or current at module is at or above maximum value for channel	Incorrect wiring or short circuit to power supply.	Check wiring for proper module connections and shorts.
		Incorrect voltage/ current selection	Check voltage/current selection jumpers or module configuration.
	Measurement of voltage or current at module is consistent with sensor value.	Incorrect voltage/ current selection	Check voltage/current selection jumpers or module configuration.
		Input circuit is damaged	If possible, try another input channel Replace module.
In Range	Value in PLC memory inconsistent with sensor measurement.	Incorrect module configuration	Check channel range, channel min. and max. scaling range values.
		Incorrect resistor value	Check resistor used to convert current to voltage for input channel.
		Module or sensor not calibrated.	Disconnect sensor wiring from module. Connect a known good voltage/current source to module channel and set it to min., max., and mid-range channel values. If PLC values correct, recalibrate sensor. Otherwise, recalibrate module.
	Value in PLC memory varying while sensor is constant (differential voltage input).	Common mode voltage too high.	Check voltage between both differential voltage connections and module common. If higher than allowed, check power supply and sensor grounding.

without disturbing the wiring. However, when current loops are used, these small panel meters will greatly hasten the troubleshooting process because one will not need to break the loop in order to connect and disconnect a meter. But, milliamp clamp meters are available. Breaking an analog output loop causes the device to sense a zero current, which may lead to disastrous consequences. At the very least, a small (100 ohm) resistor should be inserted in the loop so that the loop current can be measured without disturbing the wiring.

Troubleshooting charts for analog modules are shown in Table 9.3 and Table 9.4. Both of these troubleshooting charts start from a valid source of the analog value. For an analog input, the physical quantity must be in the valid range and then the PLC value is checked. If

Table 9.4. Analog Output Diagnosis Guide.

PLC value is in the normal range			
Actuator acts as if value is	And	Probable Cause	Recommended Action
Min.	Measurement of voltage at module channel is zero (voltage output)	Short circuit in wiring or sensor	Disconnect wiring from output channel and recheck voltage. If not zero, check wiring for short.
	Measurement of current at module channel is zero (current output)	Broken wire to sensor	Disconnect wiring from output channel, install 250 ohm resistor across channel. If current not zero, check for broken wire.
		Inadequate power supply voltage	Check loop power supply voltage (may not be possible, depending on module).
	Measurement of voltage or current at module is zero.	Incorrect voltage/ current selection	Check voltage/current selection jumpers or module configuration.
		Output circuit is damaged	If possible, try another output channel Replace module.
	Measurement of voltage/ current at module is consistent with sensor value.	Actuator circuit is damaged	Test actuator response to different voltages/currents. Replace actuator.
Max.	Measurement of voltage or current at module is at or above maximum value for channel	Incorrect wiring or short circuit to power supply.	Check wiring for proper module connections and shorts.
		Output circuit is damaged	If possible, try another output channel Replace module.
	Measurement of voltage/ current at module is consistent with sensor value.	Actuator circuit is damaged	Test actuator response to different voltages/currents. Replace actuator.
In Range	Actuator response inconsistent with value in PLC memory.	Incorrect module configuration	Check channel range, channel min. and max. scaling range values.
		Module or actuator not calibrated.	Disconnect actuator wiring from module. Measure voltage/current at channel for min., max., and mid-range PLC values. If values correct, recalibrate actuator. Otherwise, recalibrate module.
		Actuator not calibrated.	Test actuator response to different voltages/currents. Replace actuator.

the PLC value is "pegged" at the minimum or maximum value, then the channel voltage/current must be checked to find the source of the problem. If the value in the PLC is in the valid range, but is inconsistent with a panel meter indication (e.g., the physical quantity is about 30 gpm and the PLC value is 20 gpm), then the most likely causes are incorrect configuration and/or calibration of the I/O module or sensor. For an analog output, the PLC value must be in the valid range and then the actuator response is checked. If the actuator is "pegged" at the minimum or maximum, then the channel voltage/current must be checked to find the source of the problem. If the actuator response is in the valid range, but is inconsistent with the PLC value (e.g., the PLC value for a value is 60% and the valve is

40% open), then the most likely causes are configuration and calibration of the I/O module or sensor. Further testing is necessary to isolate calibration problems.

The ability to override the status of a physical discrete input or to override the logic driving a physical output coil and force the output to a desired status is useful for testing. Although there are few exceptions, this override function only applies to physical discrete inputs and physical discrete outputs. This function is called input/output forcing.

An output force is useful for troubleshooting discrete outputs. Forcing an output to the **on** and/or **off** state can be used to test that particular output. Otherwise, output forces should not be used. Using an output force to override the PLC logic should be used only temporarily until the logic can be corrected. An input force is often useful for testing a PLC program. If the PLC is not connected to physical I/O, the forces allow one to simulate the inputs. Other ways of simulating the system without having any physical I/O are described in Chapter 15. An input force can also be used to temporarily bypass a failed discrete input device so that operation may continue while the device is being repaired. However, overriding safety devices in this manner is not recommended.

Safety Caution

Note: Enabling or disabling forces can result in sudden machine movement, possibly injuring personnel. While forces are enabled, a new force value is applied immediately, with no prompt for confirmation. **USE EXTREME CAUTION WHEN USING FORCES!**

Allen-Bradley discrete input/output forcing modifies the program scan as shown in Figure 9.5. The force table functions as a filter. The force table entry for a particular physical discrete input or output can be one of three characters:

. Not forced

1 Forced **on**

0 Forced **off**

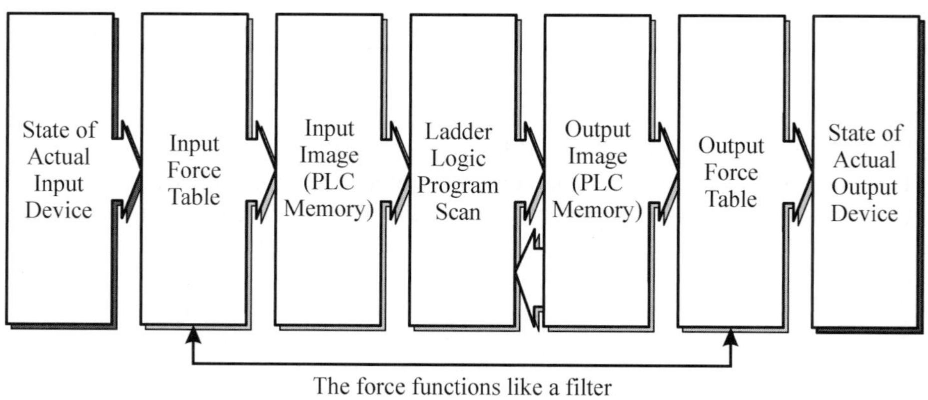

Figure 9.5. Relationship of Allen-Bradley forcing with program scan.

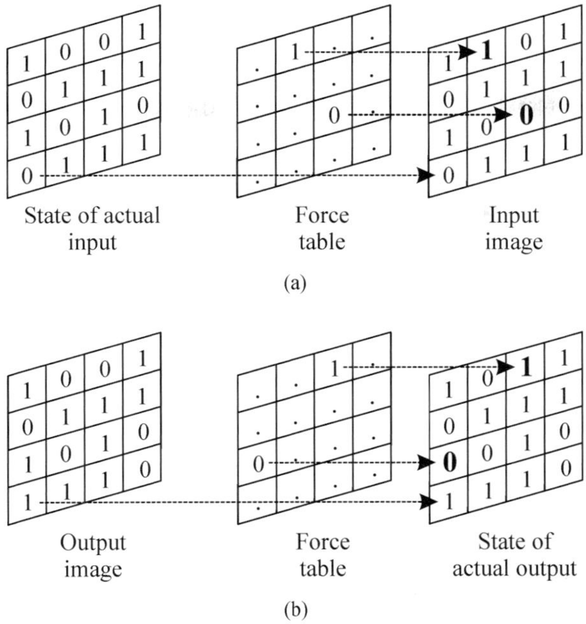

Figure 9.6. Allen-Bradley force function: *(a)* discrete inputs; *(b)* discrete outputs.

When a physical discrete input is forced (Figure 9.6*a*), the value in the force table overrides the actual input device status with the value in the force table. When a particular physical discrete output is forced (Figure 9.6*b*), the value in the force table overrides the transfer of the value in the output image to the physical output channel. Though Figure 9.6 shows forcing for discrete input/output channels, the same process applies to analog channels for the ControlLogix, CompactLogix, and Micro800 processors.

For the ControlLogix, CompactLogix, MicroLogix, and SLC processors, it requires two steps to actually implement a force. First, the particular inputs/outputs are forced to the desired values and secondly all of the forces are enabled. The forces are not actually applied until they are enabled. The Micro800 processor handles forces (what the programming software calls "Lock") differently than the other processors. There is no global "force enable" as in the other processors. Inputs and outputs are locked individually.

For MicroLogix and SLC-500 processors the size of the force tables corresponds to the size of the discrete input and discrete output image files. For these processors, only discrete inputs and discrete outputs may be forced.

For ControlLogix processors, the force table for each discrete I/O module resides in the processor. Therefore, if one controller is forcing a particular input module, other controllers monitoring the input module are not affected. In contrast, output modules are "owned" by a particular processor and the owner processor is the only processor allowed to write values to the module. If the owner forces an output module channel, other processors monitoring the output module will also monitor the forced value.

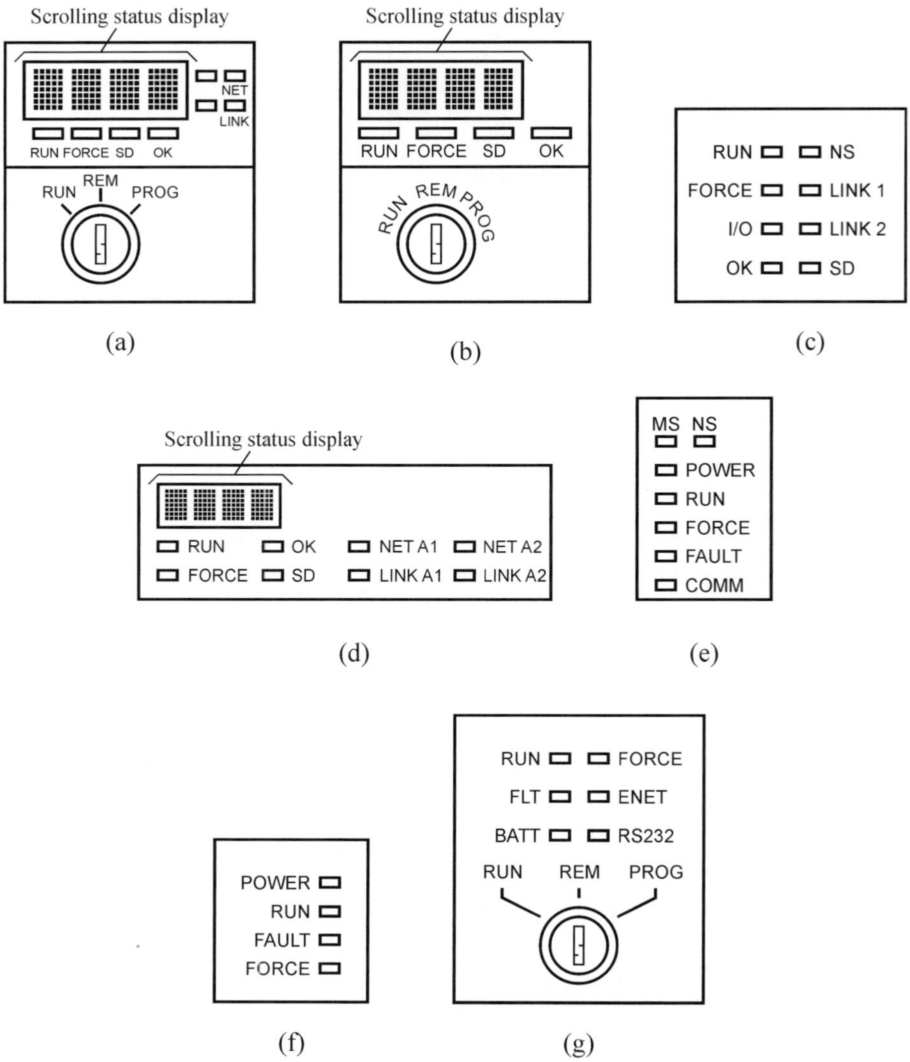

Figure 9.7. Processor diagnostic indicators: *(a)* ControlLogix L8*x*E; *(b)* ControlLogix L7*x*; *(c)* CompactLogix 5370; *(d)* CompactLogix 5380; *(e)* Micro850/870; *(f)* MicroLogix 1100/1400; *(g)* SLC-5/05.

9.4 PROCESSOR STATUS INDICATORS

Most PLC processors have a series of status lights on the front panel of the module. Detailed information about the status lights and the information they convey is found in the manufacturer's manual. A sampling of the status indicators for the PLCs covered by this text are described in this section.

The ControlLogix L8*x* processor indicators are shown in Figure 9.7*a* and are described as:

Indicator	**Description**
RUN	Green when processor in RUN mode
FORCE	Amber when I/O forces enabled and/or active
SD	Green when accessing SD card
	Flashing or solid red when SD card fault
OK	Green when controller OK
	Red when faulted or powering up
NET	Green when controller has IP address and link functioning
	Red when duplicate IP address or invalid configuration
LINK	Flashing green when port activity

Status and fault messages are also displayed on the 4-digit scrolling display.

The ControlLogix L7*x* processor indicators are shown in Figure 9.7*b* and are described as:

Indicator	**Description**
RUN	Green when processor in RUN mode
FORCE	Amber when I/O forces enabled and/or active
SD	Green when accessing SD card
	Flashing or solid red when SD card fault
OK	Green when controller OK
	Red when faulted or powering up

Status and fault messages are also displayed on the 4-digit scrolling display.

The CompactLogix 5370 processor indicators are shown in Figure 9.7*c* and are described as:

Indicator	**Description**
RUN	Green when processor in RUN mode
FORCE	Amber when I/O forces enabled and/or active
I/O	Green when communicating with I/O
	Flashing green when comm. error with module
	Flashing red when faulted
OK	Green when controller OK
	Red when faulted or powering up
NS	Green when controller has IP address and link functioning
	Red when duplicate IP address or invalid configuration
LINK 1/2	Flashing green when port activity
	Solid green when link exists, but no activity
SD	Green when accessing SD card
	Flashing or solid red when SD card fault

The CompactLogix 5380 processor indicators are shown in Figure 9.7*d* and are described as:

Indicator	**Description**
RUN	Green when processor in RUN mode
FORCE	Amber when I/O forces enabled and/or active
OK	Green when controller OK
	Red when faulted or powering up

SD	Green when accessing SD card
	Flashing or solid red when SD card fault
NET A1/2	Green when controller has IP address and link functioning
	Red when duplicate IP address or invalid configuration
LINK A1/2	Flashing green when port activity
	Solid green when link exists, but no activity

The indicators on a Micro850 and Micro870 processors are shown in Figure 9.*7e* and are described as:

Indicator	Description
MS	Green when processor operational
	Red when processor faulted
NS	Green when controller has IP address and link functioning
	Red when duplicate IP address or invalid configuration
POWER	Green when power applied to processor
RUN	Green when processor in RUN mode
FORCE	Amber when I/O forces enabled and/or active
FAULT	Flashing red when controller has fault
	Steady red when fatal error (no communications)
COMM	Flashing green when communicating on serial port

The indicators on a MicroLogix 1100 and 1400 processor are shown in Figure 9.7*f* and are described as:

Indicator	Description
POWER	Green when power applied to processor
RUN	Green when processor in RUN mode
FAULT	Flashing red when controller has fault
	Steady red when fatal error (no communications)
FORCE	Amber when I/O forces enabled and/or active

The 1400 processor has an additional LCD display showing other status information.The MicroLogix 1200 processor has additional COMM0 and DCOMM indicators. The MicroLogix 1500 has additional BAT LO, COMM 0, COMM 1, and DCOMM status indications.

The indicators on a SLC-5/05 processor are shown in Figure 9.9*g* and are described as:

Indicator	Description
RUN	Green when processor in RUN mode
FLT	Flashing red when controller has fault
	Steady red when fatal error (no communications)
BATT	Red when battery has low voltage or is missing
FORCE	Amber when I/O forces enabled and/or active
ENET	Green when communicating on Ethernet
	Flashing red when faulted
RS232	Green when communicating with PC on serial link

For the SLC-5/03, the ENET indicator is replaced by a DH485 indicator for the DH-485 network and for the SLC-5/04, the ENET indicator is replaced by DH+ for the DH+ network. The other SLC processors have similar indicators.

To diagnose processor faults or to access more detailed information about the processor; one must access the processor status through the programming software. In many PLC processors, the status indicators and fault information is accessible to the PLC programs and the PLC program can be programmed to take certain actions. For example, the ControlLogix, CompactLogix, Micro800, MicroLogix and SLC-500 processors can be programmed to execute a fault routine when a fault is detected. Normally, the processor will halt when a major fault is detected. For these processors, a fault routine can record the fault, clear the fault and restart the processor so that operation can be continued without interruption.

9.5 PROGRAM PROBLEMS

In order to aid in troubleshooting program problems, PLC programming software packages provide two useful tools: (1) data monitor, and (2) cross reference. Data monitoring functions allow one to monitor and/or modify specified program variables. A program cross reference provides a list of the program variables and for each variable indicates where it is used and how it is used (contact, coil, block, etc.).

The data monitoring functions vary among the vendors from a primitive display of the data tables to a graphical display similar to a human-machine interface (HMI). Most vendors provide a data monitor that allows the user to specify the variables/tags/symbols to be monitored and/or modified. Table 9.5 summarizes the various data monitoring functions.

When troubleshooting, the cross reference information is very useful when it is searched while online with the processor. References to a particular variable can easily be determined and then the user can jump to other parts of the program that use that variable. The user can trace the operation backwards by finding the places where the variable is written (coil, move, etc.). Alternatively, a user can trace the operation forward by finding the places where a particular output coil or calculation result is used in the program. An example cross reference is shown in Figure 9.8. In this cross-reference, the Int_Reset tag is used in the program in five places: once as an ordinary output coil (OTE), twice as a normally open contact (XIC) and twice as a normally closed contact (XIO).

Table 9.5. Data Monitoring Functions.

Processor	Name of Data Monitoring Function	Display Format
ControlLogix/ CompactLogix	Data Monitor	All program or routine tags displayed in table.
	Watch Window	Routine tags or selected tags displayed in table.
Micro800	Variable Monitor	Variables displayed in table - tabs for global, local, system, and I/O.
MicroLogix/ SLC-500	Data file monitoring	Display of data file contents in table.
	Custom Data Monitor	User specifies symbols in table.
	Custom Graphical Monitor[1]	HMI-like graphical display.

[1]Only available in professional versions of programming software.

ENG2_RET	0	BOOL	Invert
Engage Hook 2 retract control			
AliasFor:	Local:2:O.Data.1		
Base Tag:	Local:2:O.Data.1		
*ENG2_RET - MainProgram/MainRoutine - *22(OTE)*			
Eng2_Tmr		TIMER	Invert
*Eng2_Tmr - MainProgram/MainRoutine - *14(TON)*			
Engr_Tmr.DN - MainProgram/MainRoutine - 14(XIC)			
GRIP_CLOS	0	BOOL	Invert
Gripper close control			
AliasFor:	Local:2:O.Data.6		
Base Tag:	Local:2:O.Data.6		
*GRIP_CLOS - MainProgram/MainRoutine - *27(OTE)*			
Int_Reset	0	BOOL	Invert
Reset in progress			
*Int_Reset - MainProgram/MainRoutine - *15(OTE), 0(XIO), 15(XIC), 16(XIC), 20(XIO)*			

Figure 9.8. Sample cross reference (ControlLogix).

A troubleshooting chart for common programming problems is shown in Table 9.6. Before checking for programming errors, first verify that it is not a problem with a physical input/output. Use Tables 9.1 to 9.4 to eliminate that possibility. For beginning programmers, a repeated ordinary coil is a common programming error.

It may also be possible that the controller does not behave properly in a given situation because the situation was unforeseen in the original program. In this case, the program needs to be modified to detect the new conditions and control the process accordingly.

For a spurious or transient problem, setting a trap is often the only effective method of testing its existence and checking for other conditions. In this case, the PLC program is modified to monitor for the suspected event. When the suspected event occurs, an internal coil can be latched and other conditions can be stored. That is the technique illustrated by the scenario at the start of this chapter. In extreme conditions, the processor can be halted, though this will obviously disrupt the operation of other parts of the process.

Depending on the particular PLC, there may be ladder logic instructions useful for troubleshooting program problems. All of the processors covered by this text have jump instructions that can be used to skip sections of code. The ControlLogix, CompactLogix, and Micro800 processors have an AFI ladder logic contact that disables the logic on a particular rung. These instructions are covered in Chapter 8.

9.6 COMMUNICATION PROBLEMS

This section focuses on diagnosing communication problems between a personal computer and a PLC. Troubleshooting communication problems between two PLCs or to remote I/O is vendor-dependent and is not covered in this text. A personal computer is connected to a PLC in one of two ways: (1) directly through the computer serial or USB port (Figure 9.9), or (2) through a communication network (Figure 9.10).

A personal computer (PC) is often directly connected to a PLC with a serial (usually RS-232) or USB cable. However, depending on the vendor, the software connection between the programming software to the PC serial or USB port may be direct (Figure 9.9*a*)

Table 9.6. Program Diagnosis Guide.

Symptom	Probable Cause	Recommended Action
Discrete input seems to intermittently change.	Problem with physical input	Follow Table 15.1 diagnosis guide.
	Discrete input used as coil	Use cross reference to verify that output coil does not refer to input.
	HMI writing to discrete input.	Check for reference to input in HMI. Verify that HMI only reads variable.
Discrete output or internal coil does not turn **on** when it should.	Repeated coil output (not set/reset)	Use cross reference to check that coil or negated coil does not refer to same variable..
	Rung not being executed	Verify that jump-to-subroutine is being executed. Check for jumps.
	Problem in logic that determines state of contacts	Use cross reference to find coils referring to contacts in the rung being examined. Check logic driving these coils.
Timing interval too short or non-existant	Timer preset is zero	Check timer preset.
	Retentive timer not being reset	Check timer reset logic (may be missing).
	Timer preset equal to 1 (PLC-5/SLC)	Timer preset equal to 1 may actually time-out immediately.
Items not being counted	Incorrect preset	Check counter preset. Some counters do not count beyond the preset value. Check any logic that loads the preset value.
	Counter not allowed to count	Check counter reset and/or enable logic.
	Counter input not transitioning	Check counter input logic. Most counters count transitions at the input.
Counter seems to count more items than expected	Incorrect counter input logic	Counter is counting extra transitions. For A-B counters, also check for logic turning off CU bit in counter structure.

or through a communication server (Figure 9.9*b*). The use of a separate software package (FactoryTalk Linx or RSLinx) to manage the connection between the programming software and the PLC is common to Rockwell Automation products. In either case, most problems involve either the USB/serial cable or the configuration of the PC serial or USB port. Assuming the serial cable has the correct connections, the cable should be checked for:

Bad solder or crimp connections
Bent or pushed-in connector pins
Faulty grounding connection (for long cables)

The serial port should be checked for the correct:

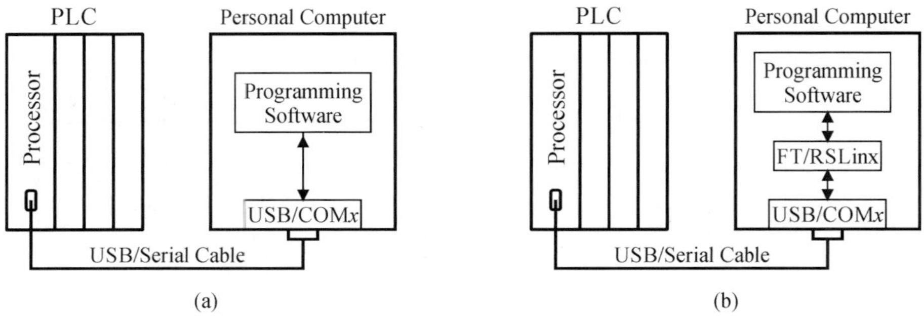

Figure 9.9. Direct PC-to-PLC communications: *(a)* General connection; *(b)* Rockwell Automation software connection.

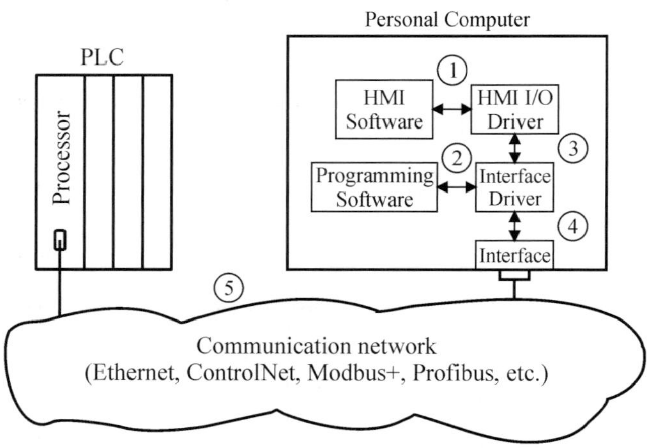

Figure 9.10. PC-to-PLC connection through communication network.

Baud rate
Parity (even, odd, or none)
Number of stop bits
Error checking method
Duplex (full-duplex or half-duplex)
Flow control (Xon/Xoff, etc.)

With the RSLinx software one can auto-configure the serial connection and thus automatically find the proper serial port configuration (assuming the PLC serial port exists and is properly configured).

A common problem when downloading a program to the PLC for the first time is that the programmer forgets to set the PLC serial port baud rate in the PLC program configuration to match the baud rate of the PC serial port. When that happens, the program will generally download without error, but communications between the PC and the PLC will terminate at the end of the download. This problem is common when programming PLC-5 processors because the default PLC serial port baud rate of 2400 bits/sec. is much slower than most users want to communicate with the PLC. In addition, the programming software does not detect that the new program has a different PLC serial port configuration. For a SLC-500, the problem of downloading a program with different serial port parameters is less prevalent since the programming software detects the changed communication parameters and displays a warning message.

A less common problem is that the PLC or PC serial port may be defective or the PLC serial port may have been accidentally disabled. To test the PC serial port, check that it can communicate with another PLC. To check for a disabled port, try communication to the PLC through another communication port (e.g., Ethernet). As a last resort, setting the PLC to a default state will determine if the PLC serial port is defective. Unfortunately, this method will interrupt the operation of whatever the PLC is controlling and the PLC program will need to be restored. To set the PLC to its default state, power must be disconnected to the PLC and the processor battery must be removed. Depending on the vendor, one may need to also use a temporary jumper to discharge a hold-up capacitor that maintains the voltage to the PLC volatile memory. In the default state, the PLC serial port parameters are set to a default configuration as documented by the manufacturer.

When the connection between the PLC and the PC is not via a serial or USB cable, but through a communication network (Figure 9.10), diagnosing a problem becomes more complex. In this case, the PC usually also serves as an operator interface and the same communication network is used for the PLC programming and the operator interface. The problem can occur in any of the numbered connections in Figure 9.10. Usually, one must check each "link" in the communication chain by examining each of the software drivers in the communication path.

9.7 DESIGNING FOR FAULT DIAGNOSIS

Since most equipment will fail at some time, the designer should add indicators and fault information to the operator interface to allow common faults to be diagnosed quickly. For example, a simple conveyor motor starter requires two discrete inputs (start and stop push buttons) and one output (motor contactor). However, when a fault occurs, maintenance personnel may spend excessive time tracing the source of the fault. With a few additional inputs and some indicator lamps, the time to repair will be reduced. For this conveyor motor control, the total number of inputs and outputs could be:

Inputs	**Outputs**
Start push button	Motor running
Stop push button	Motor stopped
Contactor auxiliary contact	Conveyor running
Motor overload contact	Conveyor fault
480 volt supply on	
Conveyor drive speed switch	
Conveyor follower speed switch	
Motor current	

The speed switches on the drive and follower conveyor shafts allow quick diagnosis of the major conveyor mechanical problems. The motor current measurement can be used to indicate a potential problem with the conveyor before the motor overheats or a shaft breaks. The fault lamp only indicates that a fault has been detected. The specific fault can be identified by the PLC and indicated on an operator interface screen. For example, the above inputs could be used to indicate the following specific faults:

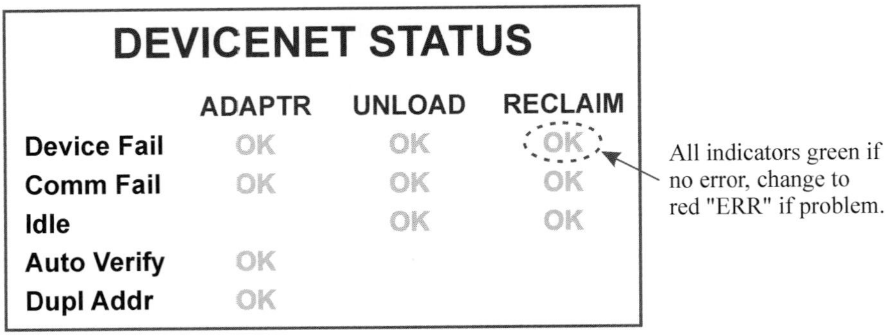

Figure 9.11. Example communication adapter diagnostic operator interface display.

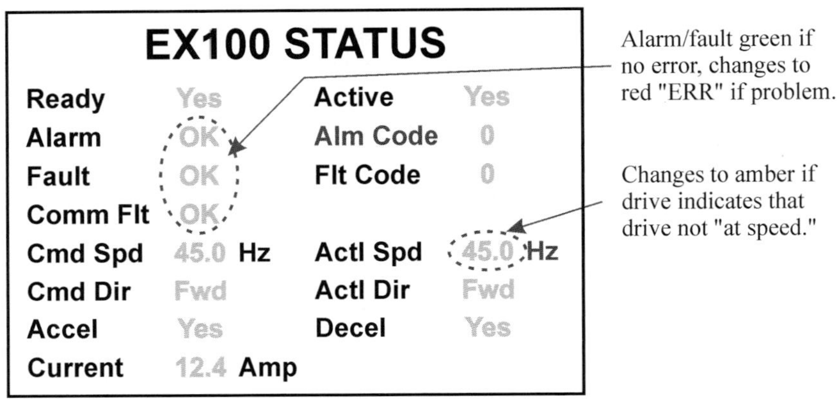

Figure 9.12. Example variable-speed drive diagnostic operator interface display.

Auxiliary contact not closed in 10 seconds after start
Motor overload
480 volt supply fault
Motor current excessive (precursor to eventual overload)
Conveyor drive failure (auxiliary contact closed, drive speed switch off)
Conveyor belt failure (drive speed switch on, follower speed switch off)

Other types of diagnostic screens on the operator interface include displays that show status information as in Figures 9.11 – 9.13. The status information for a DeviceNet communication network adapter is shown in Figure 9.11. In this case the status information for the adapter module and the two nodes on the DeviceNet network are indicated. Figure 9.12 shows a sample diagnostic display for a variable-speed drive connected to the processor through a network. Drive status information as well as the status of the communication network are displayed. Status information for a 5069-series discrete input module is shown in Figure 9.13. This type of diagnostic display for discrete and analog input/output modules duplicates the status indicators on the modules and is especially useful in diagnosing a problem when the I/O modules are remotely located.

LOCAL SLOT 4

		STATUS	SYMBOL	DESCRIP
Chan 0	4:I.Pt00		G2C2PS	Gate 2 Panel Switch - To Conv 2
Chan 1	4:I.Pt01		G2B1PS	Gate 2 Panel Switch - To Bunk 1
Chan 2	4:I.Pt02		G2B1S	Gate 2 Jog Switch - To Bunk 1
Chan 3	4:I.Pt03		G2C2S	Gate 2 Jog Switch - To Conv 2
Chan 4	4:I.Pt04		G2JSR	Gate 2 Remote Switch
Chan 5	4:I.Pt05		G2SS	Gate 2 Safety Switch
Chan 6	4:I.Pt06		G2B1	Gate 2 Bunk 1 Position Switch
Chan 7	4:I.Pt07		G2C2	Gate 2 Conv 2 Position Switch
Chan 8	4:I.Pt08		G2B1P	Gate 2 Bunk 1 Plug
Chan 9	4:I.Pt09		G2P	Gate 2 Plugged
Chan 10	4:I.Pt10		B1HLP	Bunker 1 High
Chan 11	4:I.Pt11		B1FLP	Bunker 1 Full
Chan 12	4:I.Pt12		C2JSJ	Conveyor 2 Jog Switch
Chan 13	4:I.Pt13		C2JSR	Conveyor 2 Remote Switch
Chan 14	4:I.Pt14		C2STPS	Conv 2 Stop Panel Switch
Chan 15	4:I.Pt15		C2STS	Conv 2 Start Panel Switch

Figure 9.13. Example discrete input module diagnostic operator interface display.

9.8 CHAPTER SUMMARY

This chapter provides an introduction to troubleshooting methods and provides general troubleshooting information for PLC systems. Troubleshooting information is presented for the I/O devices, the PLC hardware, the PLC program, and the PC-to-PLC communications link.

REFERENCES

Allen-Bradley, 1995. *SLC 500 TM Modular Hardware Style – Installation and Operation Manual*, pub. 1746-6.2, Allen-Bradley Company, Inc, Milwaukee, WI.

Allen-Bradley, 1996. *SLC 500 TM Analog I/O Modules: User Manual*, pub. 1746-6.4, Allen-Bradley Company, Inc, Milwaukee, WI.

Allen-Bradley, 1998. *Analog Output Module: User Manual*, pub. 1771-6.5.30, Allen-Bradley Company, Inc, Milwaukee, WI.

Allen-Bradley, 1999. *Analog Input Module: User Manual*, pub. 1771-6.5.115, Allen-Bradley Company, Inc, Milwaukee, WI.

Rockwell Automation, 1998. *Industrial Automation Wiring and Grounding Guidelines*, pub. 1770-4.1, Rockwell Automation, Inc., Milwaukee, WI.

Rockwell Automation, 2008a. *ControlLogix™ Controller and Memory Board: Installation Instructions*, pub. 1756-IN101L-EN-P, Rockwell Automation, Inc., Milwaukee, WI, Apr.

Rockwell Automation, 2008b. *SLC 500™ Modular Hardware Style: User Manual*, pub. 1747-UM011G-EN-P, Rockwell Automation, Inc., Milwaukee, WI, June.

Rockwell Automation, 2015. *ControlLogix Analog I/O Modules: User Manual*, pub. 1756-UM009D-EN-P, Rockwell Automation, Inc., Milwaukee, WI, Mar.

Rockwell Automation, 2017. *ControlLogix Controllers User Manual*, pub. 1756-UM001P-EN-P, Rockwell Automation, Inc., Milwaukee, WI, May.

Rockwell Automation, 2018. *CompactLogix 5370 Controllers: User Manual*, pub. 1769-UM021I-EN-P, Rockwell Automation, Inc., Milwaukee, WI, June.

Rockwell Automation, 2022a. *Compact 5000 I/O Analog Modules User Manual*, pub. 5069-UM005C-EN-P, Rockwell Automation, Inc., Milwaukee, WI, Mar.

Rockwell Automation, 2022b. *Compact 5000 I/O Digital Modules User Manual*, pub. 5069-UM004E-EN-P, Rockwell Automation, Inc., Milwaukee, WI, Jan.

Rockwell Automation, 2022c. *CompactLogix 5380 and Compact GuardLogix 5380 Controllers: User Manual*, pub. 5069-UM001I-EN-P, Rockwell Automation, Inc., Milwaukee, WI, Apr.

Rockwell Automation, 2022d. *ControlLogix 5580 and GuardLogix 5580 Controllers: User Manual*, pub. 5069-UM543M-EN-P, Rockwell Automation, Inc., Milwaukee, WI, May.

Rockwell Automation, 2022e. *Logix5000 Controllers I/O and Tag Data Programming Manual*, pub. 1756-PM004J-EN-P, Rockwell Automation, Inc., Milwaukee, WI, Apr.

Rockwell Automation, 2022f. *Micro830,Micro850, and Micro870 Programmable Controllers User Manual*, pub. 2080-UM002M-EN-P, Rockwell Automation, Inc., Milwaukee, WI, Apr.

10 PID Control

Chapter Topics:

- General features of feedback control
- PID controller theory and tuning
- Operational aspects of PID control
- PLC PID control implementation

OBJECTIVES

Upon completion of this chapter, you will be able to:

- Understand the basic operation of a PID controller
- Understand how to tune a PID controller
- Implement PID control in a PLC

Scenario: Peak shaving load control. A university has contracted with you to install an automatic load control for Generator #3, their diesel-powered generator. The power plant mainly supplies steam to heat and cool campus buildings. However, it has two electric generators. Generator #1 is driven by a non-condensing steam turbine and supplies a maximum of 250 kW. Generator #3 is a 1000 kW diesel generator normally used to supply power to the power plant equipment when the utility power source fails. Generator #2 was removed a few years ago.

In order to save money, Generator #3 is being used to "shave" the peak electricity demand in the summer. The cost of the electricity is a certain rate as long as the campus demand stays below a certain peak level (determined weekly). If the campus demand remains above this peak for at least 15 minutes, the utility adds a hefty "demand charge" to the monthly bill. Even though the cost of electricity from the diesel generator is more than the normal charge from the utility, there is substantial cost savings to running generator #3 to keep the campus demand below the peak.

The generator load is controlled by a potentiometer on the power plant panel. When the campus demand approaches the peak demand, the operator starts the diesel, synchronizes it to the bus, connects it to the bus, and then adjusts the generator load to keep the campus demand just below the peak. An operator has other duties, so s/he does not always watch the campus demand meter. As a consequence, the campus has incurred demand charges because the operator was busy with other duties when the demand initially exceeded the peak, or when the load control needed to be increased enough to keep the demand below the peak during the day.

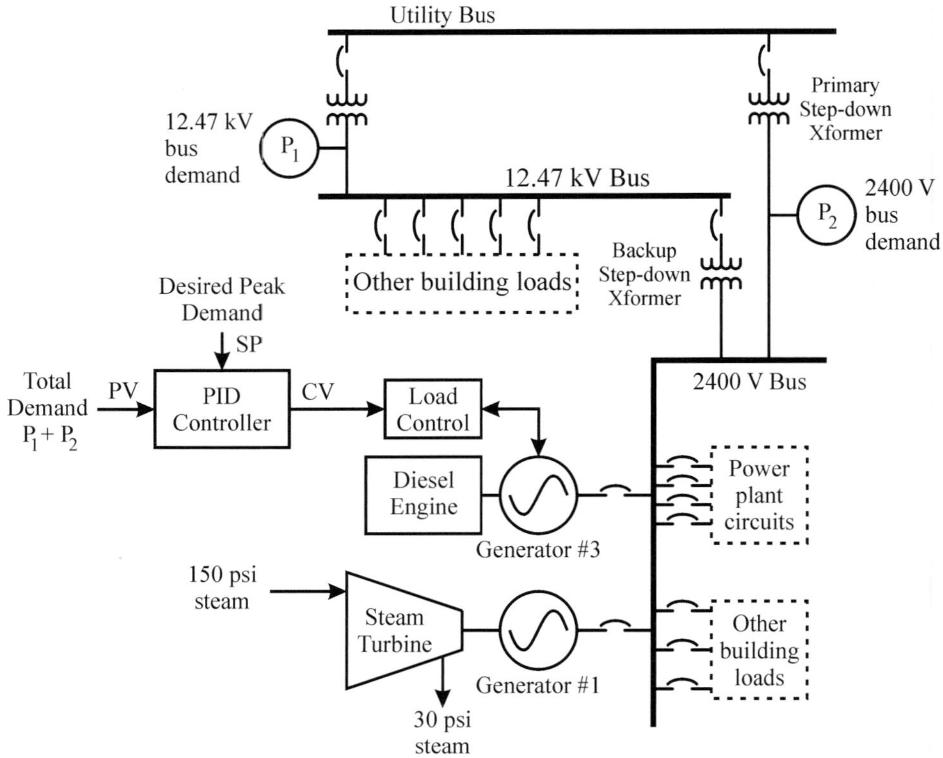

Figure 10.1. Simplified diagram of peak shaving system.

Therefore, you added functionality to the existing ControlLogix to automatically control the Generator #3 load to keep the campus demand below the peak. This scenario focuses on the tuning of the PID controller that controls the generator load to keep the campus demand below the peak. Modification of the control scheme to prevent negative power flow through the primary step-down transformer is covered in Example 10.7 of Erickson (2023).

Figure 10.1 shows how the PID control fits into the system. The setpoint (SP) is the desired peak demand, the process variable (PV) measurement is the total demand, and the control variable (CV) is the desired generator load, 0 – 1000 kW. You did some initial tests to determine how fast the generated power changed when the load control was changed by a small amount (50 kW) and calculated the initial PID controller parameters as $K_P = 1.0$, $K_I = 6$, and $K_D = 0.0$. The ladder logic and PID parameters are shown in Figure 10.2.

When you install the system and observe its operation during commissioning, the control seems to hold the demand reasonably close to the desired peak. The campus demand has some fluctuation, and so the generated power fluctuates as well. However, the next day you are called because the generated power seems to be fluctuating more than it should be. You cannot come immediately, and so they revert to manual control of the generator load to finish the day.

The next day, you investigate the PID controller a little more carefully when it is controlling the generator load. The meter displaying the generated power has some

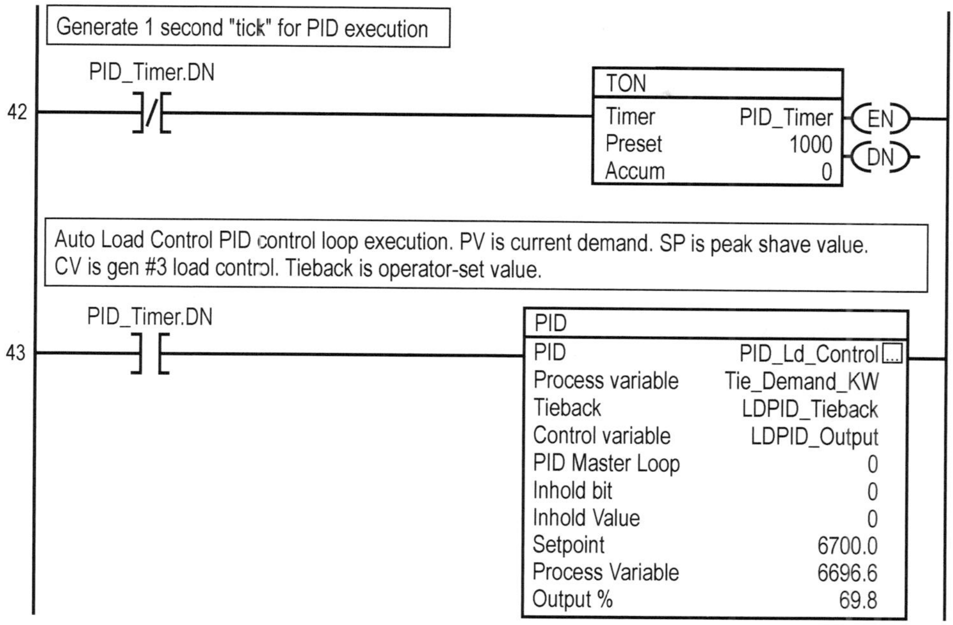

PID - PID_Ld_Control Configuration

Scaling		**Configuration**		**Tuning**	
PV Unscaled Max:	100.0	PID Equation:	Independent	Setpoint (SP):	6700.0
PV Unscaled Min:	0.0	Control Action:	SP - PV	Set Output:	0.0 %
Engineering Unit Max:	9600.0	Derivative Of:	PV	Output Bias:	0.0 %
Engineering Unit Min:	0.0	Loop Update:	1.0 secs.	Proportional Gain (Kc):	1.0
CV Max (at 100%):	100.0	CV High Limit:	100 %	Integral Gain (Ki)	
CV Min (at 0%):	0.0	CV Low Limit:	0 %	[/secs] :	6.0
Tieback Max (at 100%):	100.0	Deadband Value:	0	Derivative Time (Td)	
Tieback Min (at 0%):	0.0	PV Tracking:	No	[secs] :	0.00
		Cascade Loop:	No	Manual:	No
		Cascade Type:	N/A	Software Manual:	No

Figure 10.2. ControlLogix ladder logic for PID controller.

reasonable fluctuation. However, when you examine the PID controller CV, it is definitely oscillating, from 10% to 60%, with a period of about 25 seconds. What should you do?

***Solution*:** As a rule of thumb, any time one observes oscillation, the gain of the controller, K_P, should be reduced by at least half. You reduce both the proportional and integral gain to $K_P = 0.5$, $K_I = 3$. The oscillation disappears, and the generator load levels out. The control of the campus demand is not as "tight," that is, it seems to fluctuate a little more around the desired peak, but that is acceptable since the desired peak value is below the demand charge value.

Design Tip

As a rule of thumb, any time the PID control exhibits oscillation, reduce the controller gain to no more than half of its value. Depending on the form of the algorithm, other parameters may be adjusted. When the oscillation has disappeared, reevaluate the controller tuning.

10.1 INTRODUCTION

The PLC was originally developed for the discrete-control aspects of discrete and batch control applications. However, as mentioned in Chapter 1, most programmable logic control systems are a combination of continuous and discrete control elements. The proportional-integral-derivative (PID) controller is the basic tool for regulatory control of continuous processes. This chapter provides a foundation in the use of the PID controller function in PLCs.

For background, general approaches to control are described and then followed by the features of feedback control. The PID control algorithm is described from an intuitive approach, and then the most popular tuning methods are described. Enhancements to PID control, cascade and feedforward control are covered in Erickson (2023) and Erickson and Hedrick (1999). Operational aspects of PID control are described and then followed by a description of the PID function blocks for the PLCs covered by this text.

In continuous control, the *process* is the system to be controlled. A manufacturing plant may be considered as consisting of a series of processes, or the plant itself may be viewed as one process. Since PID controllers control only one variable, the definition of a process is restricted to those processes having only one controlled variable. In general, a process will have a number of inputs and outputs as shown in Figure 10.3. The signals are defined as follows:

Process Variable (PV) - Process output which is to be maintained at a desired value by adjustment of a process input.

Control Variable (CV) - The process input that is adjusted to maintain the process variable at the setpoint. This input is also commonly called the manipulated variable (MV) or valve out (VO).

Disturbance – A process input other than the control variable that affects the process variable.

A simple flow process is shown in Figure 10.4*a*, and its corresponding block diagram representation is shown in Figure 10.4*b*. The process variable is the flow, and the control variable is valve position. The upstream and downstream pressures are both disturbances,

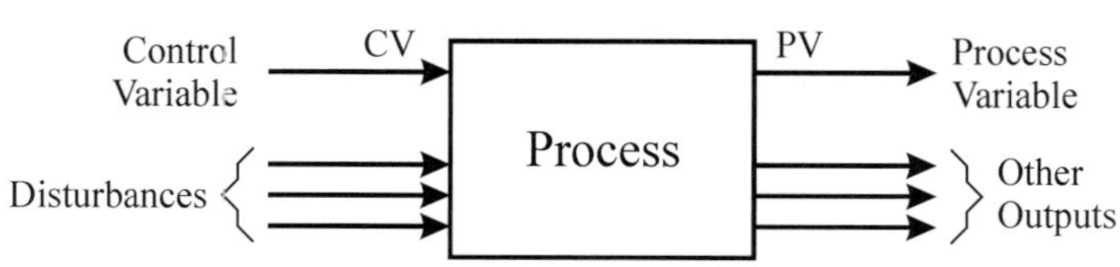

Figure 10.3. Block diagram of a general process.

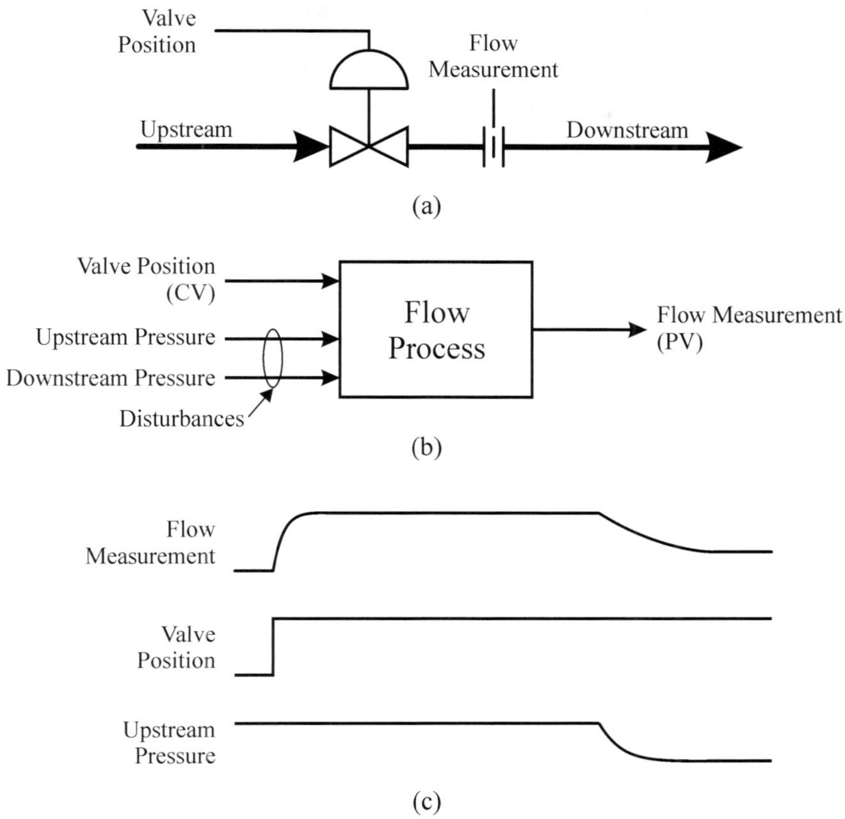

Figure 10.4. Example flow process: *(a)* process; *(b)* as block diagram; *(c)* effect of valve position and upstream pressure on flow.

since either one can affect the flow. As illustrated in Figure 10.4*c*, both the valve position (control variable) and the upstream pressure (disturbance) determine the actual flow (process variable).

In order to place the material in this chapter in its proper context, the most common strategies for regulatory control are described and related to each other.

Open-Loop Control. If a process has no disturbances, or their effect on the process variable is negligible, then open-loop control is sufficient. Based on an indicated value of the process variable, the control variable is adjusted by a plant operator. In this control strategy, the control variables are simply set to their design values and held there. Many people consider this strategy as no control, but by setting the control variables, the control system is influencing the process. For example, consider the control of a heat exchanger, depicted in Figure 10.5, which uses steam to heat cold water and produce hot water. The goal is to control the hot water temperature to a desired value, ±2°C. An open-loop control strategy for this hot water heater is shown in Figure 10.6. Given a desired temperature, the reference selector produces a steam valve position either by calculation or table look-up. Obviously, any variations in the cold-water temperature or steam pressure, among other external

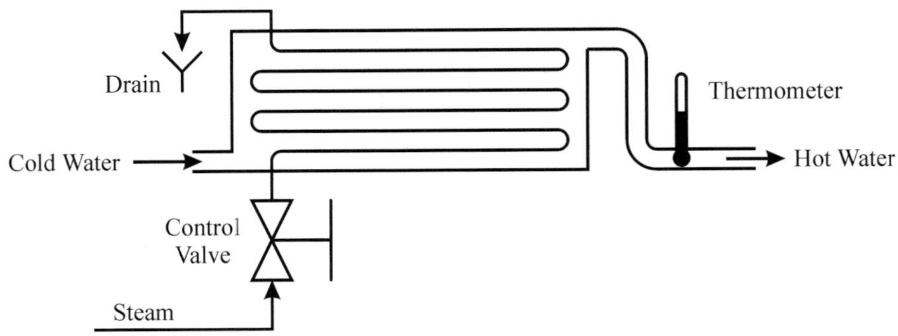

Figure 10.5. Example heat exchanger.

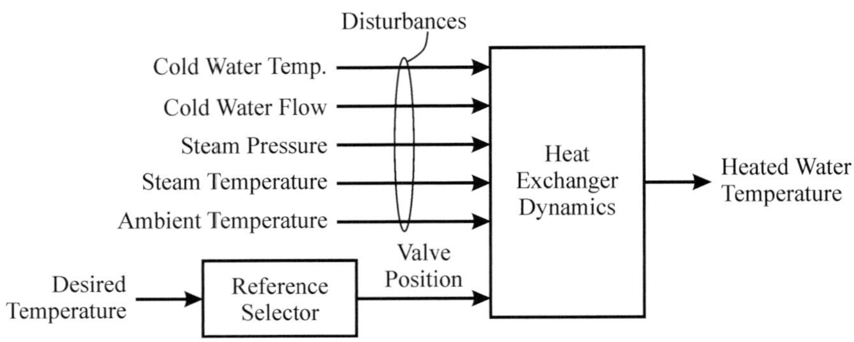

Figure 10.6. Open-loop control strategy for heat exchanger.

influences, will affect the hot water temperature and the reference selector will not account for them.

This strategy is useful for those systems that have no or small disturbance effects. Obviously, if the system is subject to significant disturbances, then some more sophisticated control is needed.

Feedforward Control. If one knows the types of disturbances that can influence a process, then the reference selector can be modified to measure these disturbances and modify the calculation of the control variable. For example, the heat exchanger of Figure 10.5 is subject to the following disturbances:

Cold water temperature	Steam temperature
Cold water flow	Ambient temperature
Steam pressure	

All of these influences are called *disturbances* because they affect the output but are not manipulated. To construct a feedforward controller, the important disturbances are measured and used by the reference selector, along with the desired value of the output to calculate the control variable value. For example, Figure 10.7 shows a possible feedforward controller for the heat exchanger. The important disturbances, cold water temperature, cold water flow, and steam pressure are considered to be either the disturbances that have the

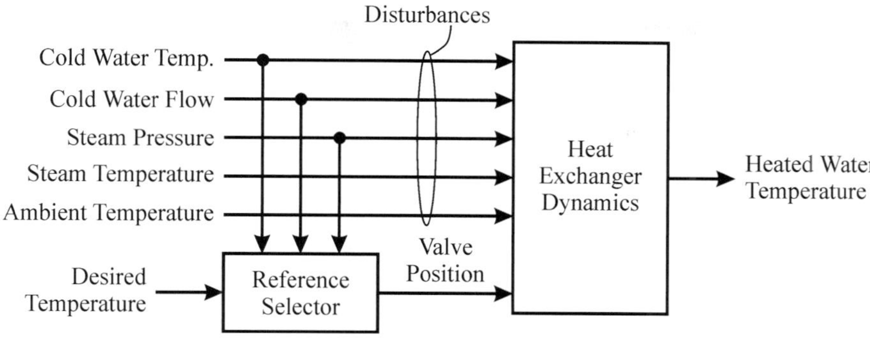

Figure 10.7. Feedforward control strategy for heat exchanger.

biggest effect or the ones most likely to change. The other disturbances are unmeasured and still influence the process to some extent.

This strategy is useful for those cases where the process operating point does not change and the process is subject to significant disturbances whose effect on the process variable is known. The disturbances not used in this strategy must have small effects on the process variable.

Manual Feedback Control. The strategy of using knowledge of the output to take corrective action is called *feedback*. In manual feedback control, operating personnel monitor the process variable of interest and take corrective action in order to maintain the desired value. A manual feedback control scheme for the heat exchanger is shown in Figure 10.8. An operator periodically monitors the hot-water temperature and adjusts the steam valve position in order to maintain the hot-water temperature.

Manual control is useful in a large number of applications whose process variable is not critical, whose disturbances are few, or where the cost of automation is too high to justify automatic (open-loop or feedback) control.

Automatic Feedback Control. When the operator in a manual feedback strategy is replaced by a controller device that continuously measures the process variable, then one has automatic feedback control. Automatic control helps to eliminate disturbance effects. To maintain the process variable at the setpoint, the control variable is adjusted each time there

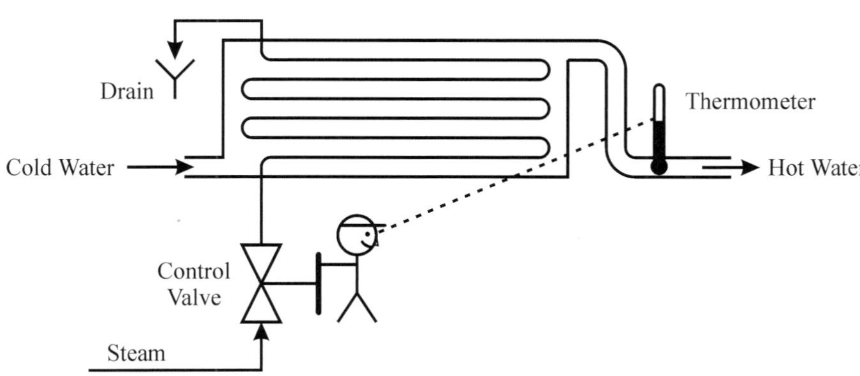

Figure 10.8. Manual feedback control for heat exchanger.

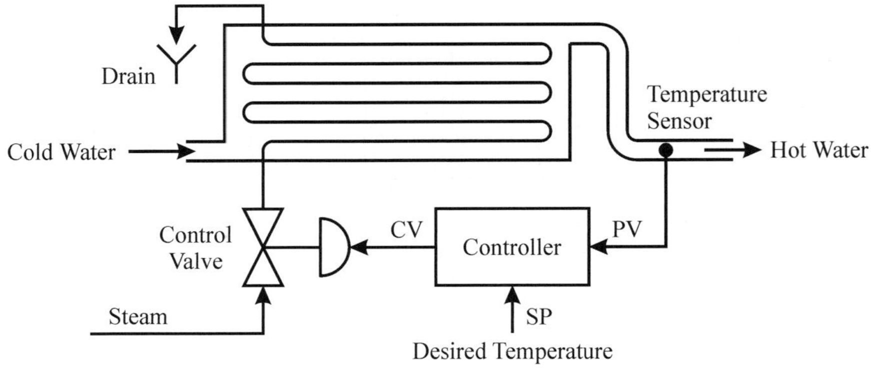

Figure 10.9. Automatic feedback control for heat exchanger.

is a change in the process variable. For the heat exchanger example, the operator is replaced by an automatic feedback controller (Figure 10.9). The *setpoint* (SP) is the desired value of the process variable, the hot-water temperature. The combination of feedback and feedforward control is considered later in this chapter. The only drawback to automatic feedback control is its cost. Beyond the hardware cost, an installation, tuning, and maintenance cost is incurred.

Feedback control also applies to discrete processes. For example, a solenoid valve has only two valid states, open or closed. Often limit switches are attached to the valve to determine if the valve reaches the commanded state. If the controller commands the valve to open and the open limit switch is still **off** after 10 seconds, then the valve (or the limit switch) has failed, and the system must take corrective action, such as shutting down the system. At the very least, an alarm is generated to inform the operator of the problem. In this example, the limit switches provide the feedback to the discrete controller.

10.2 FEEDBACK CONTROL PERFORMANCE

The block diagram of a typical feedback control system for a single-input, single-output (SISO) control system is shown in Figure 10.10. A controller compares the process variable measurement with the setpoint and calculates a control variable to maintain the setpoint. For simplicity, the final control element, process, and measuring device are often lumped together and called the "process," as in Figure 10.11.

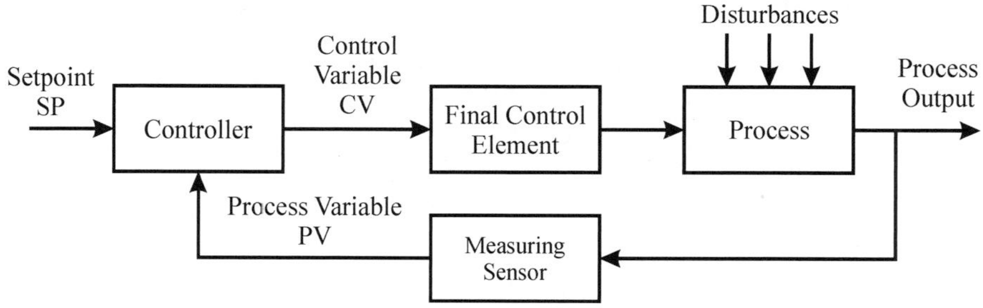

Figure 10.10. General single-input, single-output (SISO) control system.

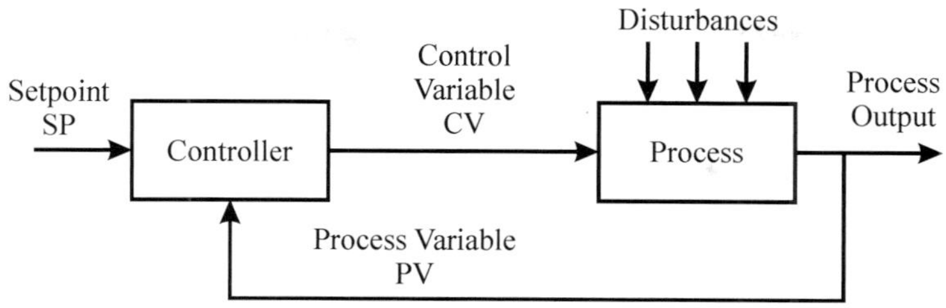

Figure 10.11. Simplified SISO control system.

Regardless of the particular controller algorithm, there are four general advantages of a feedback control system compared with an open-loop system (Dorf and Bishop, 2017; Erickson and Hedrick, 1999; Phillips and Harbor, 1999):

1. Increased speed of response
2. Reduced error
3. Disturbance rejection
4. Reduced sensitivity to modeling errors

Ultimately, the controller forces the process variable to follow a setpoint value by adjusting the control variable. Sometimes, the setpoint is not a constant but a trajectory. For most systems, it is impossible to make the process variable perfectly follow the setpoint. Hence, quantitative control performance measures are needed to judge the quality of the controller. These performance measures are based on the response to the two external inputs to the system: setpoint changes and disturbance changes.

10.2.1 Setpoint Input Changes

The most common control loop performance measures are based on changes to the setpoint. In many processes, the setpoint is held constant for long periods of time, as in the pressure of a steam header. In others, the setpoint is changed at frequent intervals, as in the temperature of a batch reactor. In still other cases, the setpoint assumes a profile. For example, in metal annealing, the furnace temperature is ramped from an initial temperature to a final temperature, held constant for a period of time, and then ramped back to the initial temperature. However, the common control performance measures are based on a step change in the setpoint, which corresponds to the situation where an operator or supervisory controller makes an abrupt change in the setpoint and waits for the process variable to reach a steady-state value. Typical control variable and process variable responses are shown in Figure 10.12. The responses in Figure 10.12 are indicative of an *underdamped* process. The following aspects of these curves are used to evaluate the control system performance.

Offset. The difference between the setpoint and the process variable when the system has reached steady state is called the offset. For most systems, the desired offset is zero.

Rise Time. The time required for the process variable to go from 10% to 90% of the steady-state change is the rise time T_r. Generally, a short rise time is desired, although a short rise time often requires a rapid control variable change, leading to increased wear of the final control element. The short rise time is only relative and usually refers to a rise time shorter than the rise time of the system with no control. The rise time cannot be arbitrarily short, since given the constraints on the magnitudes of the control variable, there is a shortest achievable rise time.

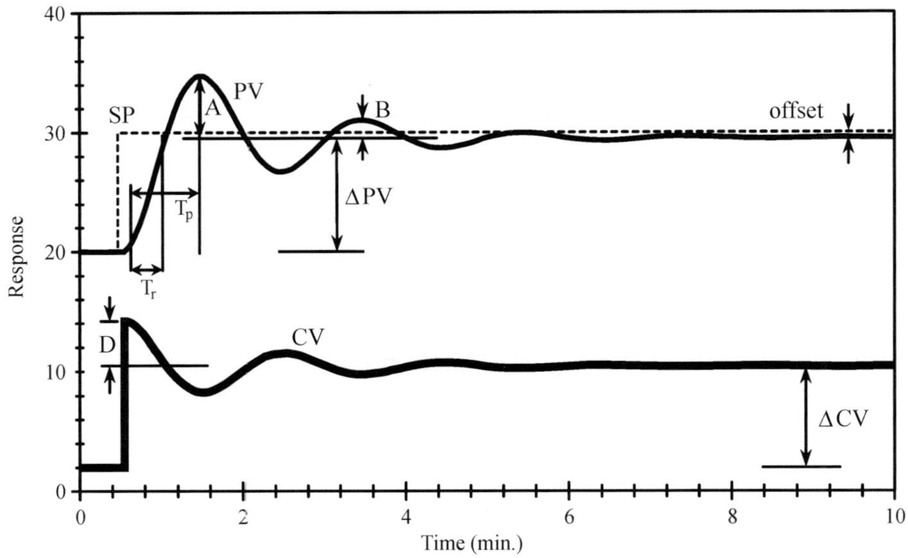

Figure 10.12. Typical process and control variable response to step setpoint change.

Percent Overshoot. The amount that the process variable proceeds beyond its final value is called the overshoot (*A* in Figure 10.12). Obviously, if the process variable response does not contain a peak whose value exceeds the final value (commonly called an overdamped response), the overshoot is zero. The *percent overshoot* (PO), defined as

$$\text{PO} = \frac{\text{A}}{\Delta\text{PV}} \times 100\%$$

is often used as the performance measure in place of the overshoot, since it is a relative measure that does not depend on the setpoint change magnitude. Generally, little or no overshoot is desirable in a process control system since overshoot has a detrimental effect on downstream processes.

Peak Time. The peak time T_p is the time from the setpoint step change to the time of the first peak of an underdamped process variable response. The peak time and the rise time are closely related, and so usually only one of them is used.

Damping Ratio. If the process variable response is underdamped, the damping ratio is the ratio of the second peak to the first peak in the response (*B*/*A* in Figure 10.12). Usually, a small ratio is desired.

Settling Time. The settling time T_s is the time from the setpoint change to the time that the process variable response has settled within a certain percentage band of the final value, usually 2 or 5%. Generally, a short rise time also means a short settling time. As for the rise time, a short settling time is desired.

Integral Error. The cumulative deviation of the process variable from the setpoint can be measured a number of ways. The two most frequently used measurements are:

Integral of the absolute error (IAE):

$$\text{IAE} = \int_0^\infty |\,\text{SP}(t) - \text{PV}(t)\,|\,dt \tag{10.1}$$

Integral of the product of time and the absolute error (ITAE):

$$\text{ITAE} = \int_0^\infty t \left| \text{SP}(t) - \text{PV}(t) \right| dt \tag{10.2}$$

The IAE is the most common performance measurement in this group. When the process variable is a quality measurement, like concentration, the IAE basically measures the total amount of off-specification material produced by the system. The ITAE penalizes deviations that endure for a long time.

Control Variable Percent Overshoot. The control (manipulated) variable overshoot is calculated in the same manner as the percent overshoot of the process variable. It is an indication of the amount of control effort needed to produce the closed-loop process variable response. A higher control variable percent overshoot (CVPO) usually means more wear on the final control element, a valve in many process control systems:

$$\text{CVPO} = \frac{\text{D}}{\Delta\text{CV}} \times 100\%$$

10.2.2 Disturbance Input Changes

Disturbances are uncontrolled inputs to the process that affect the process variable. These inputs generally cause significant deviations from the setpoint if corrective action is not taken. The major types of disturbances encountered in process control are step disturbances, stochastic disturbances, and ramp disturbances.

Step Disturbance. A disturbance characterized by a large, abrupt change in its value is modeled as a step signal. Typical manipulated variable and process variable responses to a step disturbance are shown in Figure 10.13. The performance measures for the process variable response are largely the same, except that the rise time and percent overshoot are replaced by the *maximum deviation*, which has meaning only in the context of a disturbance response. The maximum deviation measures the maximum effect of the disturbance on the process variable response, and obviously, a small value is desired. In the heat exchanger (Figure 10.5), an abrupt change in the heated water flow is a step disturbance.

Stochastic Disturbance. Stochastic disturbances are those seemingly random upsets that disturb the system from the steady state. Their average value is zero, so they do not have the

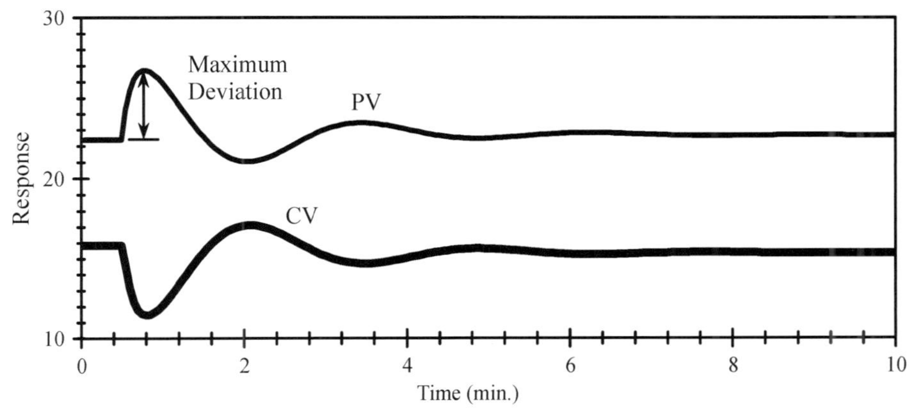

Figure 10.13. Typical process and control variable response to step disturbance change.

major effects on the process variable that step disturbances have. Nevertheless, their effect can be dramatic on the expected performance as will be demonstrated by the controller tuning examples. The only stochastic disturbance considered in this text is measurement noise, since a small amount of it is present in any real system and the quantization process used to convert the measurement into a value to be used by the digital control algorithm introduces its own noise.

Ramp Disturbance. Ramp-like disturbances generally correspond to environmental factors such as ambient outside temperature. They will not be directly used in the design procedure.

10.3 PID CONTROLLER

The most common industrial controller is based on the PID algorithm. After explaining each part of the basic algorithm, other parameters that must be set in a commercial PID controller are explained. The operational modes of a PID controller are described and then guidance concerning which terms should be selected is presented before the tuning procedures are presented.

Even though all of the PLC vendors implement some form of the digital PID controller, both the analog and digital PID terms are described. Historically, the analog PID controller has existed much longer. Also, its equations are simpler and therefore easier to understand.

10.3.1 PID Controller Terms

The three parts (often called *terms*) of the PID controller are (1) proportional, (2) integral, and (3) derivative. These terms are first explained from an intuitive viewpoint and then the specific equations are described. Each part of the algorithm is explained as follows:

1. *Proportional*. The control variable is calculated as a constant (gain) multiplied by the error (setpoint minus process variable). The proportional gain is analogous to the operator in Figure 10.8 adjusting the valve position in an amount proportional to the difference between the thermometer reading and the hot-water temperature. For example, if the hot-water temperature needs to be increased 5 degrees, the valve is opened one percent. A proportional-only controller for the heat exchanger is shown in Figure 10.14. The equation for the proportional term is commonly expressed as

$$\mathrm{CV}(t) = K_P(\mathrm{SP}(t) - \mathrm{PV}(t)) = K_P e(t)$$

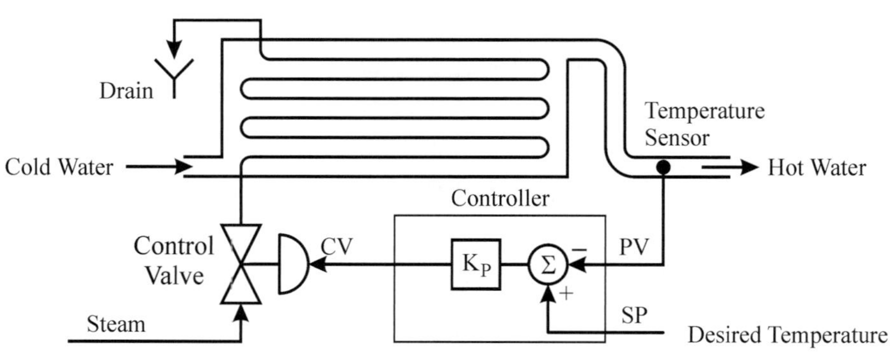

Figure 10.14. Proportional control for heat exchanger.

where K_P is the proportional gain and e is the error. The proportional action reduces the error between the setpoint and the process variable but does not eliminate it. Increasing the proportional gain also speeds up the response of the closed-loop system. However, there is usually a practical upper limit because increasing the proportional gain often causes increased oscillation. If the proportional gain is too high, the system is usually unstable.

2. *Integral*. The control variable calculation is based on the integrated error. The integral term is analogous to the operator in Figure 10.8 returning periodically to check the hot-water temperature. Each time the hot-water temperature is checked, the valve position is adjusted by an amount proportional to the difference between the thermometer reading and the desired hot-water temperature. For example, every hour the hot-water temperature is checked and the valve is opened one percent for every 3 degrees low and closed one percent for every 3 degrees high. As long as the hot-water temperature is not equal to the desired temperature, the steam valve position is changed. Integral action is added to the heat exchanger controller in Figure 10.15. Integral action drives the error to zero. Integral action is also called reset. The equation for integral action is commonly

$$\mathrm{CV}(t) = \frac{1}{T_I}\int_0^t e(t)dt \qquad \text{or} \qquad \mathrm{CV}(t) = K_I\int_0^t e(t)dt$$

where T_I is called the integral time constant in units of minutes or seconds. On some controllers, the integral gain is specified, which is related to the integral time constant as

$$K_I = \frac{1}{T_I}$$

and whose units are repeats per minute or repeats per second. In digital controllers, the integration is approximated by a summation,

$$\mathrm{CV}(n) = \frac{T}{T_I}\sum_{i=0}^{n} e(i) = \mathrm{CV}(n-1) + \frac{T}{T_I}\left[e(n)\right]$$

where T is the sample period, n is the current sample number, and $e(n)$ is the error at the nth sampling instant.

The integral term is mainly used to force the offset to zero. However, it generally tends to slow the system down, in contrast to the proportional term. As the integral time constant

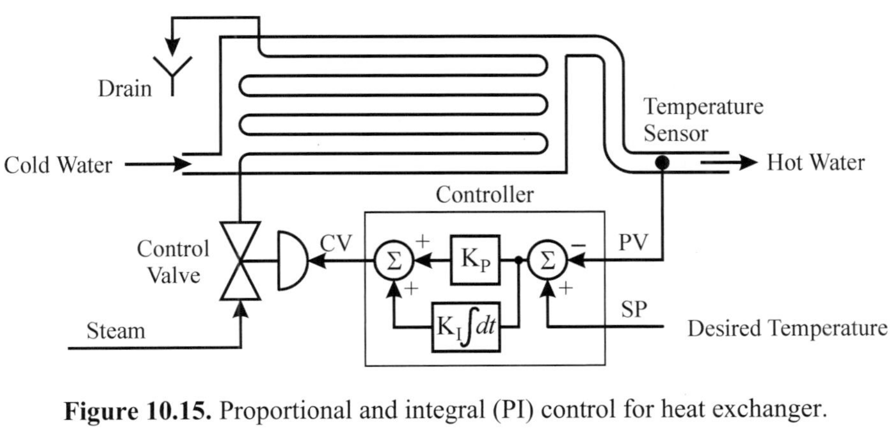

Figure 10.15. Proportional and integral (PI) control for heat exchanger.

is reduced (integral gain increased), the system response becomes more oscillatory. A small enough value of the integral time constant will cause the system to become unstable.

When a controller has integral action, a persistent error will cause the integral term to increase (or decrease) to a value of large magnitude. The integral term does not become smaller until the sign of the error changes. This situation is called integral windup. Many commercial controllers limit the value of the integral term and/or modify the integral action when this situation is encountered to temporarily decrease the integral gain as the magnitude of the error decreases but before the error changes sign.

3. *Derivative*. Also called *rate action*, the control variable calculation is based on the rate of change of the process variable. The derivative term is analogous to the operator in Figure 10.8 "spiking" the steam valve position in an amount proportional to the difference between the thermometer reading and the new hot-water temperature. For example, if the hot-water temperature needs to be increased quickly 5 degrees, the valve is opened 5 percent for one minute and then closed 4 percent. Derivative action is added to the heat exchanger controller in Figure 10.16. The equation for derivative action is

$$\mathrm{CV}(t) = T_D \frac{d}{dt}[e(t)] \qquad \text{or} \qquad \mathrm{CV}(t) = K_D \frac{d}{dt}[e(t)]$$

where T_D is called the derivative time constant in minutes or seconds. On some controllers, the derivative gain, K_D, is specified, which is equal to the derivative time constant. In reality, a pure derivative amplifies any measurement noise so controllers implement a filtered derivative,

$$T_D s \approx \frac{T_D s}{\alpha s + 1} \qquad \text{or} \qquad T_D s \approx \frac{T_D s + 1}{\alpha T_D s + 1}, \tag{10.3}$$

where α has a value of about 10. In digital controllers, the derivative may be approximated by a difference,

$$\mathrm{CV}(n) = \frac{T_D}{T}[e(n) - e(n-1)] = \frac{T_D}{T}\Delta e(n)$$

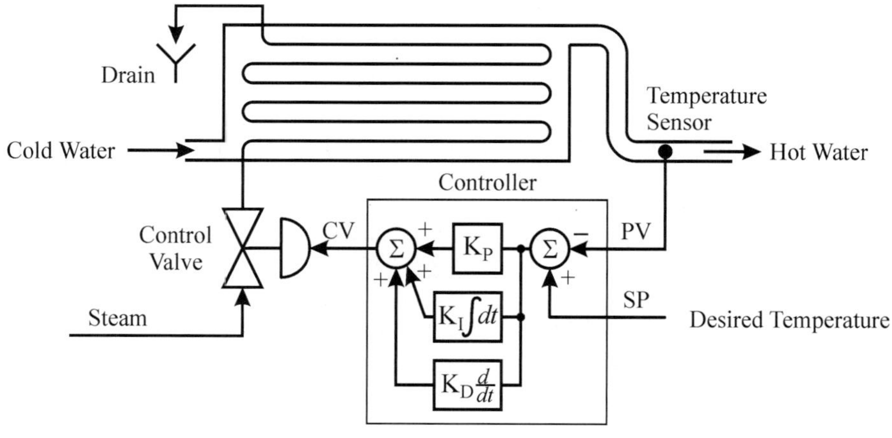

Figure 10.16. Proportional-integral-derivative (PID) control for heat exchanger.

where T is the sample period and n is the current sample number. Most PLC vendors use a discrete equivalent of (10.3) as the derivative term for the digital PID controller.

Major changes in disturbances are anticipated by derivative (rate) action. The derivative term also tends to cancel the effect of the integral action in slowing down the system response. However, the use of derivative action should be restricted to those processes where the process variable, the controlled variable measurement, is noise free. Derivative action tends to amplify any noise in the measurement.

Of course, there are many possible combinations of these three actions. For example, a P controller has only proportional action, a PI controller has proportional and integral action, and so on. The K_P, T_I, and T_D (or K_P, K_I, and K_D) are called the PID tuning parameters.

10.3.2 Forms of the Equation

There are essentially eight forms of the basic PID equation. Depending on the vendor, one may or may not have a chance to select the particular PID algorithm form. The various choices for algorithm type are:

Dependent gains (ISA standard) versus independent gains (parallel)
Derivative-on-error versus derivative-on-measurement
Positional versus velocity

In the independent gains form of the PID equation, all three terms are calculated independently and summed to obtain the manipulated variable. In the dependent gains form, the K_P modifies the integral and derivative terms. In most industrial situations, derivative action on the measurement is generally preferred to derivative action on the error, because then setpoint changes do not cause a large change in the control variable. In the positional form of the PID algorithm, the value of the control variable is calculated. In order to initialize the algorithm correctly, an initial bias term must be specified. In the velocity form of the PID, the change in the control variable is calculated. The velocity form may also be called the incremental form. Block diagrams of all combinations of gains and derivative action are shown in Figures 10.17 and 10.18 for the positional form of the algorithm. For the continuous (analog) PID, the four possible equations for the positional forms of the algorithm are:

Dependent gains, derivative on error (Figure 10.17*a*):

$$\text{CV}(t) = K_P\left(e(t) + \frac{1}{T_I}\int e(t) + T_D \frac{d}{dt} e(t) \right) + \text{Bias}$$

Dependent gains, derivative on measurement (Figure 10.17*b*):

$$\text{CV}(t) = K_P\left(e(t) + \frac{1}{T_I}\int e(t) - T_D \frac{d}{dt} \text{PV}(t) \right) + \text{Bias}$$

Independent gains, derivative on error (Figure 10.18*a*):

$$\text{CV}(t) = K_P e(t) + \frac{1}{T_I}\int e(t) + T_D \frac{d}{dt} e(t) + \text{Bias}$$

Independent gains, derivative on measurement (Figure 10.18*b*):

$$\text{CV}(t) = K_P e(t) + \frac{1}{T_I}\int e(t) - T_D \frac{d}{dt} \text{PV}(t) + \text{Bias}$$

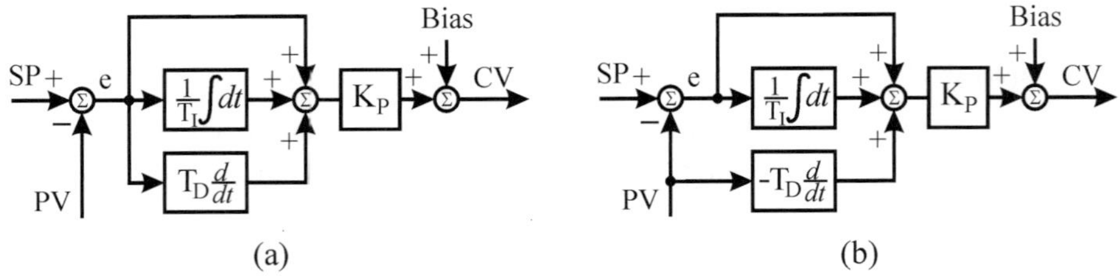

Figure 10.17. Forms of the dependent gains positional PID algorithm: *(a)* derivative on error; *(b)* derivative on measurement.

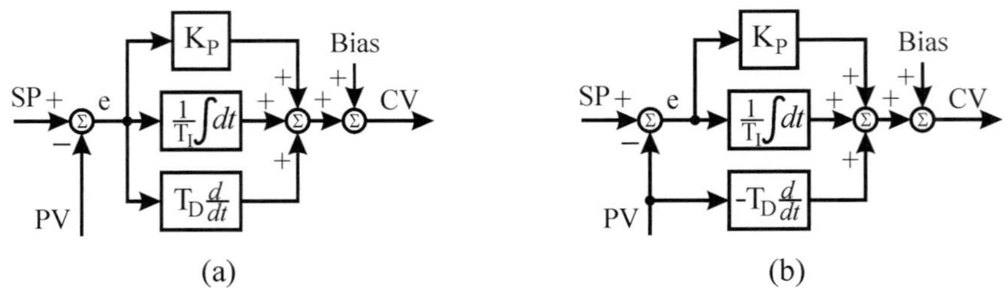

Figure 10.18. Forms of the independent gains positional PID algorithm: *(a)* derivative on error; *(b)* derivative on measurement.

When implemented in a PLC, the four possibilities for the positional form of the digital PID equation are:

Dependent gains, derivative on error (Figure 10.17*a*):

$$\mathrm{CV}(n) = K_P\left(e(n) + \frac{T}{T_I}\sum_i e(i) + \frac{T_D}{T}\Delta e(n)\right) + \mathrm{Bias}$$

Dependent gains, derivative on measurement (Figure 10.17*b*):

$$\mathrm{CV}(n) = K_P\left(e(n) + \frac{T}{T_I}\sum_i e(i) - \frac{T_D}{T}\Delta \mathrm{PV}(n)\right) + \mathrm{Bias}$$

Independent gains, derivative on error (Figure 10.18*a*):

$$\mathrm{CV}(n) = K_P e(n) + \frac{T}{T_I}\sum_i e(i) + \frac{T_D}{T}\Delta e(n) + \mathrm{Bias}$$

Independent gains, derivative on measurement (Figure 10.18*b*):

$$\mathrm{CV}(n) = K_P e(n) + \frac{T}{T_I}\sum_i e(i) - \frac{T_D}{T}\Delta \mathrm{PV}(n) + \mathrm{Bias}$$

For the continuous PID controller, the velocity forms of the equations are not possible. So, the velocity form only applies to the digital PID algorithm. The four types of PID equations are obtained by subtracting CV*(n-1)* from CV*(n)*. The block diagrams of these equations look similar to the diagrams for the positional PID (Figures 10.17 and 10.18) except that the *Bias* signal is replaced by CV*(n-1)*. For brevity, only the equation for the dependent gains, derivative on error (Figure 10.17*a*) is shown:

$$\mathrm{CV}(n) = \mathrm{CV}(n-1) + K_P\left(\Delta e(n) + \frac{T}{T_I}e(n) + \frac{T_D}{T}\left[e(n) - 2e(n-1) + e(n-2)\right]\right)$$

Note that actual implementations of the digital PID controller are more complex than the equations presented here. As indicated earlier, the derivative term is generally filtered. Also, the integral gain is often modified to handle integral windup.

Concerning these PID algorithm choices, only the derivative-on-error versus derivative-on-measurement may be critical to the control engineer. If the user needs the derivative action, then derivative-on-measurement is preferred because a setpoint change does not induce a large change in the controller output, which happens if the derivative is on the error. The vendor has often already made the choice of positional versus velocity forms of the equations. The particular choice will have an impact on such issues as initialization and bumpless transfer. The choice between dependent versus independent gains affects the tuning rule equations. For example, if the tuning rules calculate the parameters for an dependent gain PID, divide the T_I by K_P and multiply the T_D by K_P to obtain the independent gain parameters.

10.3.3 Other Controller Parameter Selections

Besides the main PID parameters K_P, T_I, and T_D and the choices for the PID equation form, there are other parameter selections that must be made before the controller can be operational.

Direct and Reverse Action. The sign relationship of the process variable to the control variable is specified through the direct/reverse selection. Direct and reverse action refers to the sign of the gain of the controller. The control action should be set as follows:

Specify the control action as *direct acting* if the CV must increase to correct for an increasing PV.

Specify the control action as *reverse acting* if the CV must decrease to correct for an increasing PV.

For a reverse-acting controller (Figures 10.17 and 10.18), the error is

Error = SP - PV

For a direct-acting controller, the error is

Error = PV - SP

Note that a reverse-acting controller is connected to a direct-acting process, that is, an increase in CV causes an increase in the PV measurement. A direct-acting controller is connected to a reverse-acting process, that is, an increase in CV causes a decrease in the PV measurement.

The difference between reverse- and direct-acting controllers is illustrated in the two possible level control schemes of Figure 10.19. Assume the valves are increase-open, that is, an increase in the CV output from the controller causes the valve to open further. In this case, the PID controller in Figure 10.19*a* is direct acting. When the level measurement is above the setpoint, the outlet valve needs to be opened to decrease the level to the setpoint. If the control valve is moved to control the inlet flow (Figure 10.19*b*), the controller is reverse acting. In this case, when the level measurement is above the setpoint, the inlet valve needs to be closed to decrease the level to the setpoint.

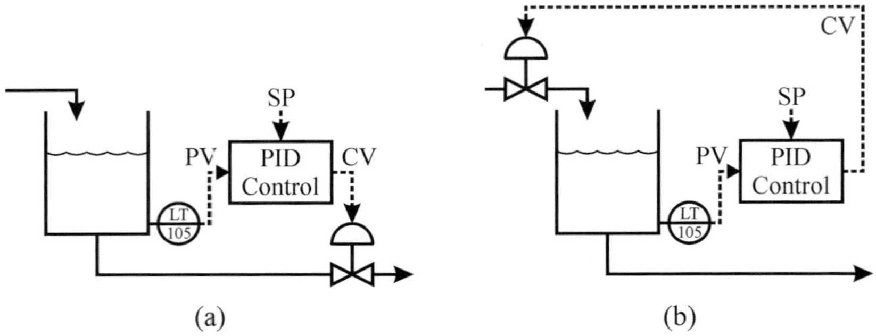

Figure 10.19. Direct- versus reverse-acting control (increase-open valves): *(a)* direct-acting control; *(b)* reverse-acting control.

The type of valve is selected based on whether the valve should fail open or fail closed if the control system fails and/or system power is removed. Note that if the valves in Figure 10.19 are increase-close, then the type of controller also changes. For an increase-close valve, the controller in Figure 10.19*a* is reverse acting, and the controller in Figure 10.19*b* is direct acting.

If one happens to set this parameter to the wrong value, the manipulated variable will often move to its maximum or minimum value and remain there, often causing the process variable to go to one extreme, generally an undesirable response.

Design Tip

Make sure you properly set the reverse/direct parameter.

Bumpless Transfer. This important parameter is often overlooked when configuring a controller. Without bumpless transfer, when the controller mode is switched from manual to automatic, the error between the process variable and setpoint generally causes an abrupt change in the control variable. This abrupt change can damage equipment (e.g., "bang valves") and/or upset downstream processes with the process variable oscillations that often result. With bumpless transfer, when the controller mode is switched from manual to automatic, the initial CV output is not changed, and the CV is ramped to the calculated value. The equipment and the process better tolerate this gradual change in the control variable. Depending on the controller vendor, bumpless transfer may be part of the algorithm and cannot be disabled. For others, specifying a parameter enables/disables bumpless transfer.

Loop Sample Period. The PID algorithm should be executed at a fixed, periodic interval called the loop sample period or loop sample time. The loop sample period is related to the dynamics of the controlled process. For example, a slow process, such as a temperature process will need a loop sample period of one second or longer. A flow loop generally requires a loop sample period of about 0.25 seconds. Motion control may require a loop sample period of milliseconds, which is why motion control is often not done in the PLC but in dedicated motion control modules.

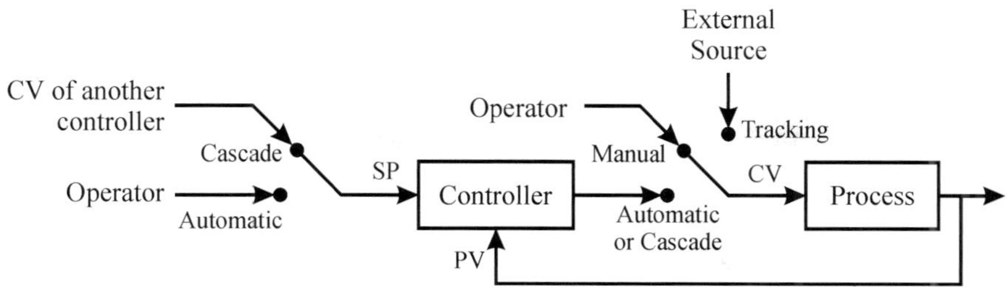

Figure 10.20. Controller setpoint and control variable source determined by mode switch.

10.3.4 Basic Controller Modes

The term *mode* indicates the state of the controller. The PID algorithm really only has two states: manual and automatic. Nevertheless, control vendors usually assign modes to the algorithm that indicate the source of the setpoint and/or the source of the controller output, the control variable. The most common modes are manual, automatic, tracking, and cascade. These modes indicate who or what has control of the setpoint and/or control variable and generally provide some protection against inadvertent manipulation of the setpoint or control variable by another device or operator. When the controller is in the automatic mode, the controller calculates the manipulated variable according to the PID equation, and the operator changes the setpoint. When the controller is in the manual mode, the operator sets the control variable, and the PID equation is not executed. When the controller is in the tracking mode, the control variable tracks another variable, usually the output of another controller. When in the manual or tracking modes, the internal variables of the PID algorithm are modified so that the calculated CV matches the tracked variable so there is a bumpless transfer when the mode is changed to automatic. When the controller is in the cascade mode, the setpoint is supplied from another controller or device (not the operator) and the PID algorithm calculates the control variable. The cascade mode is prevalent among the DCS vendors and not common in PLCs. The mode can be considered as a switch at the controller output and/or setpoint, as in Figure 10.20.

Ordinarily, when the controller starts up, it is in the manual mode. When the controller mode is switched from manual to automatic, most controllers implement some form of bumpless transfer, so that the control variable does not make an abrupt change due to any error between the process variable and the setpoint. When the controller mode is switched from automatic to manual, the control variable is retained at its last calculated value or some default value (if the controller can be configured in this manner).

10.3.5 PID Term Selection

Depending on the controller manufacturer, different combinations of proportional, integral, and derivative terms may be selected. Also, by setting the derivative gain to zero, the controller reduces to a PI controller. By setting the proportional gain and the derivative gain of a independent gain PID controller to zero, the PID controller reduces to an I-only controller. The selection of controller terms should be based on the type of process, that is, how the process variable reacts to a change in the control variable.

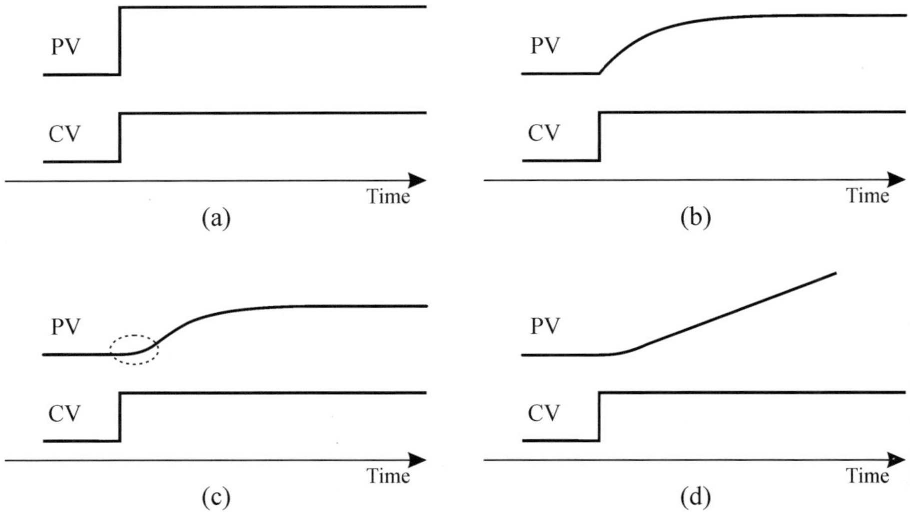

Figure 10.21. Process responses: *(a)* gain-only; *(b)* first-order lag; *(c)* second-order lag; *(d)* integrating.

Gain-Only Process: I Controller. When the response of the process variable to a change in the manipulated parameter is instantaneous (Figure 10.21*a*), the process is a gain only, and only the integral term is required. Example processes that are essentially only a gain are:

Speed control of centrifugal compressors

Liquid flow control when a fast actuator is used

First-Order Lag Process: PI Controller. When the process can be adequately represented as a first-order lag (Figure 10.21*b*), both proportional and integral terms should be used. The majority of industrial processes fall into this category. Example first-order processes are:

Gas pressure control

Liquid flow control with slow actuators

Concentration control

Second-Order Low-Noise Process: PID Controller. When the process is best represented as a second-order lag (Figure 10.21*c*) ***and*** the process variable contains little noise, then the derivative term should be used with the proportional and integral terms. It is important to remember that the derivative term will amplify any noise present on the process variable. Therefore, the derivative term is generally used sparingly. An example low-noise second-order process is:

Temperature control

Note that the second-order response is distinguished from the first-order response by the more gradual initial change in the process variable.

Integrating Process: P Controller. If the process is best represented as an integrator (Figure 10.21*d*), the process is not self-regulating, and only the proportional term is required. However, some integral action may be needed if the integration time is long (slope of process variable response curve is small). An example integrating process is:

Tank level control by regulation of the outflow or inflow.

10.4 PID CONTROLLER TUNING

There are many techniques of controller tuning. The ones presented here are the most popular techniques and will allow one to determine the parameters for reasonably good control. In many cases, the performance of the system may be adequate when tuned with these methods. In practice, one will use one or more of the following techniques to get the parameters "in the ballpark" and then adjust the parameters until the desired performance is obtained, or until one runs out of time.

Most processes may be approximated as being linear over a small operating range. However, over a wide range of operation, most processes will exhibit some nonlinearity in the form of a process gain change. As a general rule, tuning should be done at the operating point where the maximum gain occurs. Then when the process moves to an operating point where the gain is lower, the control becomes more sluggish but does not become unstable.

The specification of the desired system response should include both the desired process variable and control (manipulated) variable responses. Some general rules follow:

- There is a trade-off between speed of response and the amount of overshoot in the process response. Usually, a fast process variable response also causes the process variable to overshoot its desired value before settling to the desired value.
- The required control variable changes to achieve the desired process variable response should be considered. A process variable response that closely follows a setpoint generally requires vigorous control variable action and excessive wear on the final control element. If the apparent process gain increases because the process is at a new operating point, the system may become unstable.
- For a given set of tuning parameters, the response to a setpoint change will not be the same as the response to a disturbance change.

There are two major categories of tuning methods: (1) closed-loop and (2) open-loop tuning. Closed-loop tuning refers to tuning the controller while it is operating in the automatic mode, and thus is operating in the closed loop. Open-loop tuning methods base the initial tuning of the PID parameters on the empirical first-order-plus-deadtime process model, obtained while the controller is in the manual mode. Both of these methods have their advantages and disadvantages.

Closed-loop tuning, whether done by the operator or autotuned by the controller, is probably the most popular technique of controller tuning. Autotuning for the Modicon PLC is provided by a standard function block and is built-in for the ControLogix PIDE function block (Chapter 11). For the other PLCs discussed in this text, autotuning is provided by third-party vendors and is not discussed. All tuning techniques, including autotuning, require that the process be "bumped" in order to evaluate the effect of the tuning parameters and/or calculate a new set of tuning parameters. Closed-loop tuning techniques are applicable to reasonably fast processes that can tolerate process variable oscillations. On the other hand, open-loop tuning techniques work well for slow processes but require a process model. In the author's opinion, the Fertik open-loop tuning method yields the best response for process control since the overshoot is minimal.

> ## Design Tip
>
> Be mindful of the unit used to measure time. Whatever unit (seconds or minutes) the vendor uses should be used throughout the calculations for any of the tuning methods. For example, if T_I is expressed in seconds, then any time that is measured as part of the tuning procedure should be measured in seconds.

10.4.1 Settings for Plant Startup

Initial controller parameter values that may be used during a plant startup are given in Table 10.1. The values are conservative and will normally result in a sluggish process variable response. These parameter values are general guidelines and serve to get a plant operating, after which the tuning techniques outlined in the succeeding sections should be used to change the parameters to provide the desired response at normal operating conditions.

Table 10.1 Initial PID Controller Parameter Values

	K_P	T_I (min.)	T_D (min.)
Flow	0.3	1.0	0
Temperature	1.3	0.5	0.05
Liquid level	2.0	50.0	0
Gas pressure	5.0	50.0	0

10.4.2 Effects of Controller Terms

The following tuning techniques are based on the following observations about the process (closed-loop) response:

- As the controller proportional gain is increased, the response to setpoint changes becomes more oscillatory, commonly called underdamped (Figure 10.22*a*).
- At some greater gain, the response of the control loop will become a steady-state oscillation (Figure 10.22*b*). The manipulated variable and process variable will be one-half of a cycle (180°) out of phase at this gain. The system is called "marginally stable."
- If the gain is increased past the point where steady-state oscillation is observed, the control loop will become unstable, and the oscillations will increase in amplitude (Figure 10.22*c*). The frequency of the oscillation is determined by the process dynamics.

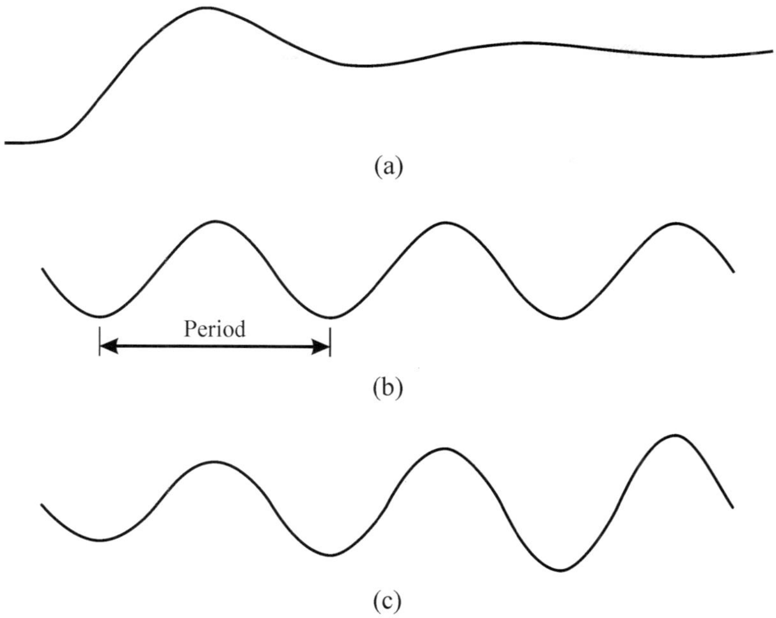

Figure 10.22. Types of oscillatory response: *(a)* underdamped; *(b)* sustained oscillation; *(c)* unstable.

The effect of the proportional, integral, and derivative gains on controller gains at different frequencies may be illustrated by examining the contribution of each controller term to a slow change in the error signal (low frequency) and a fast change in the error signal (high frequency), as shown in Figure 10.23:

The effect of the proportional gain is the same for all frequencies.

Integral gain is more responsive to low frequencies.

Derivative gain is more responsive to high frequencies.

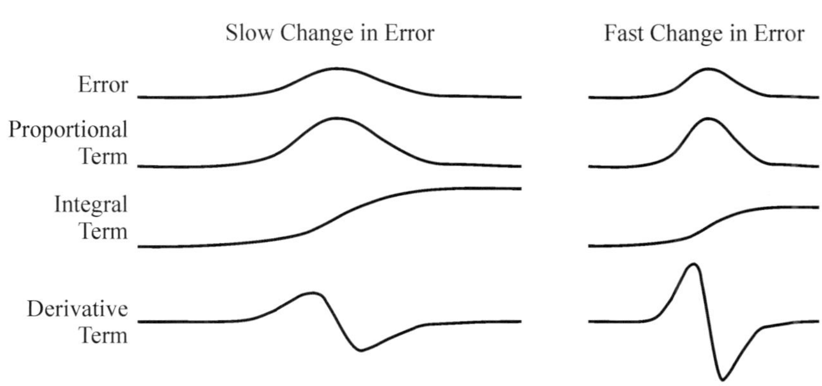

Figure 10.23. Contribution of each controller term to a change in error.

10.4.3 Closed-Loop Tuning Methods

Tuning may be performed while the controller is in the automatic mode, that is, operating in the closed loop. The problem with closed-loop tuning is its effect on downstream processes, since the tuning method produces oscillations. Depending on the particular process, these oscillations may or may not be acceptable. The one responsible for tuning will need to make that determination.

If the process responds fairly rapidly to changes in the control variable, then tuning may be accomplished by observing the response to changes in the controller parameters. When a process responds slowly to changes in the control variable, then the open-loop tuning methods are more appropriate.

10.4.3.1 Ziegler-Nichols Closed-Loop Tuning. This method was originally proposed by Ziegler and Nichols (1942). The tuning process is characterized by finding the gain at which the system is marginally stable and the frequency of oscillation at this point. From these two parameters, the controller parameters are calculated. The parameter calculations are intended to produce a closed-loop damping ratio of 1/4 to a setpoint change (Figure 10.24). The method is as follows:

1. Start any trending (plotting) of the process variable.
2. If the process is listed in Table 10.1, set the K_P to the initial gain in Table 10.1. Otherwise, set K_P to a small value. Set T_D to zero and the T_I time constant to its maximum value (or K_I to zero). Now place the controller in the automatic mode.
3. If one is not sure how the process will react, increase the proportional gain in small steps. If small changes in K_P are not causing oscillation, make larger changes, but no larger than doubling the K_P. After each adjustment, observe the process variable or control variable response to a setpoint change. When sustained oscillations (**Figure 10.22*b***) are observed, note the value of the proportional gain and the period of the oscillations:

   G_u = proportional gain for sustained oscillations

   P_u = period of oscillations (in appropriate time units)
4. Calculate the controller settings as shown in Table 10.2.
5. Make any final adjustments in K_P, T_I, or T_D to obtain the desired process variable response.

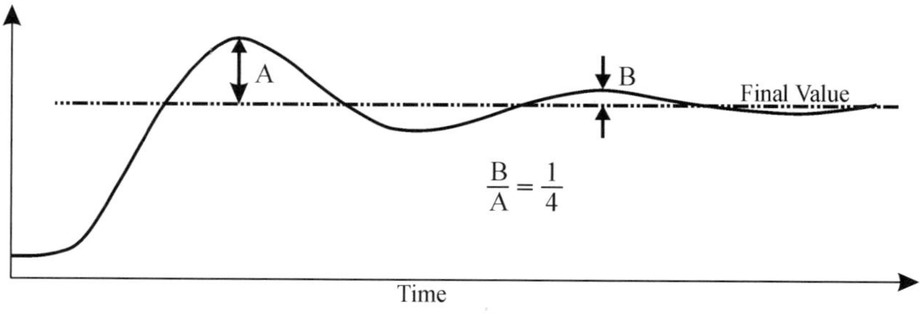

Figure 10.24. One-fourth damping ratio.

Table 10.2 Ziegler-Nichols Closed-Loop Tuning PID Parameters

	Dependent Gains Eqn.	Independent Gains Eqn.
P controller		
K_P	$0.5G_u$	$0.5G_u$
PI controller		
K_P	$0.45G_u$	$0.45G_u$
T_I	$P_u/1.2$	$P_u/(1.2K_P)$
PID controller		
K_P	$0.6G_u$	$0.6G_u$
T_I	$P_u/2.0$	$P_u/(2.0K_P)$
T_D	$P_u/8.0$	$(P_uK_P)/8.0$

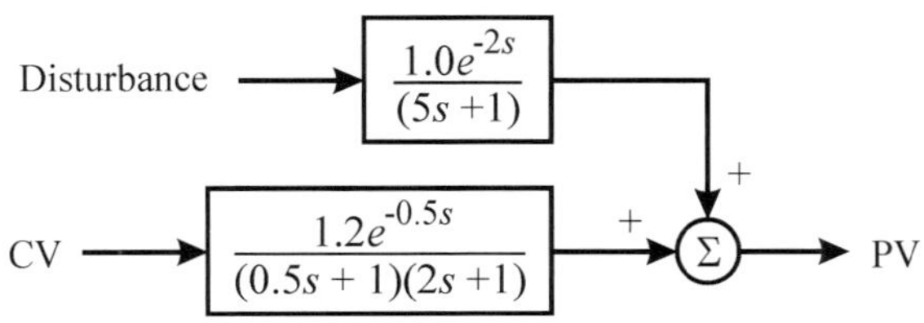

Figure 10.25. Process used for self-regulating tuning examples.

Example 10.1. For the process shown in block diagram in Figure 10.25, where the time constants are measured in minutes, tune a PI controller to control this process using the Ziegler-Nichols closed-loop tuning method.

Solution. The starting values of the controller parameters are

$$K_P = 0.1,\ T_I = 1000 \text{ min.},\ T_D = 0 \text{ min.}$$

Using proportional-only control, the gain K_P is increased until the response exhibits a sustained oscillation when the setpoint is changed by a small amount. The value of K_P for sustained oscillation is 5.0, and the sustained oscillation appears as in Figure 10.26. Actually, the oscillation is growing slightly for this gain. However, this situation corresponds to tuning under actual process conditions, since one does not usually have the time to observe very many oscillations. Thus the values of the ultimate gain and ultimate period are

$$G_u = 5.0$$
$$P_u = 3.1 \text{ min.}$$

If a PI controller is desired, the tuning parameters are:

$$K_P = 0.45\ G_u = 0.45\times5.0 = 2.25$$
$$T_I = P_u\,/1.2 = 3.1/1.2 = 2.58 \text{ min.}$$

The setpoint change response of the system with this PI controller is shown in Figure 10.27. The damping ratio is actually smaller than 1/4.

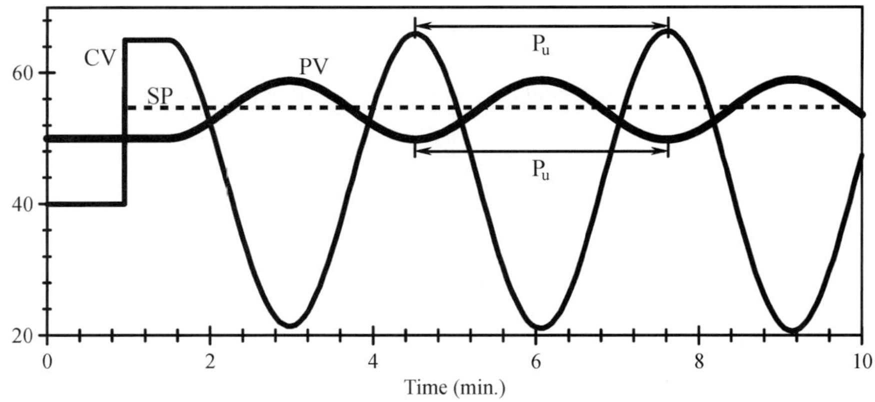

Figure 10.26. Sustained oscillations used in Zeigler-Nichols closed-loop tuning.

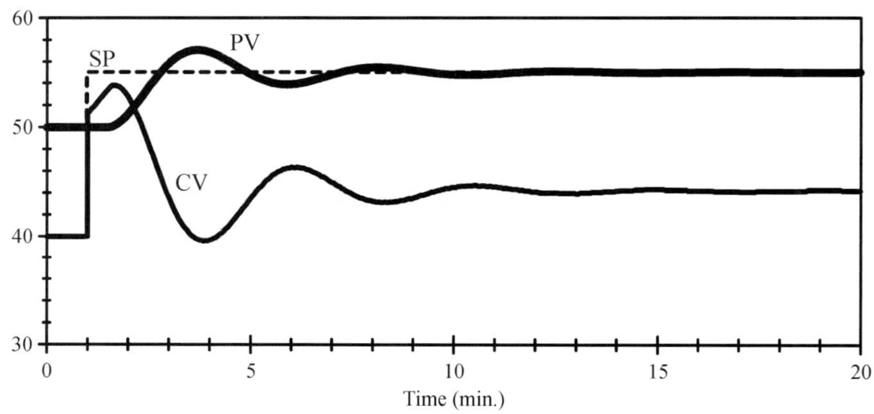

Figure 10.27. Response of system to a setpoint change when PI controller is tuned with Ziegler-Nichols closed-loop method.

10.4.3.2 Damped Oscillation Tuning Method. When it is undesirable to allow sustained oscillations, the following method is used. The process is characterized by finding the proportional gain at which the system has a one-fourth damping ratio (Figure 10.24) and the frequency of oscillation at this point. Similar to the Ziegler-Nichols method, the controller parameters are calculated from the gain and oscillation frequency. The method is as follows:

1. Start any trending (plotting) of the process variable.
2. If the process is listed in Table 10.1, set the K_P to the initial gain in Table 10.1. Otherwise, set K_P to a small value. Set T_D to zero and the T_I time constant to its maximum value (or K_I to zero). Now place the controller in the automatic mode.
3. Increase the proportional gain in small steps. After each adjustment, observe the process variable response to a setpoint change. When a one-fourth damping ratio is observed, note the value of the proportional gain and the period of the oscillations.

Table 10.3 Damped Oscillation Closed-Loop Tuning PID Parameters

	Dependent Gains Eqn.	Independent Gains Eqn.
P controller		
K_P	$1.1G_d$	$1.1G_d$
PI controller		
K_P	$1.1G_d$	$1.1G_d$
T_I	$P_d/2.6$	$P_d/(2.6K_P)$
PID controller		
K_P	$1.1G_d$	$1.1G_d$
T_I	$P_d/3.6$	$P_d/(3.6K_P)$
T_D	$P_d/9.0$	$(P_dK_P)/9.0$

G_d = proportional gain for damping ratio of 1/4
P_d = period of oscillations (in appropriate time units)

4. Calculate the controller settings as shown in Table 10.3.
5. Make any final adjustments in K_P, T_I, or T_D to obtain the desired process variable response.

Example 10.2. For the same process as in Example 10.1, tune a PI controller to control this process using the damped oscillation tuning method.

Solution. The starting values of the controller parameters are:

$$K_P = 0.1,\ T_I = 1000 \text{ min.},\ T_D = 0 \text{ min.}$$

Using proportional-only control, the gain K_P is increased until the response exhibits a response with quarter-wave damping when the setpoint is changed by a small amount. The value of K_P for one-fourth damped oscillation is 2.75 and the period of the oscillation is 3.7 min. Thus,

$$G_d = 2.75$$
$$P_d = 3.7 \text{ min.}$$

If a PI controller is desired, the tuning parameters are:

$$K_P = 1.1G_d = 1.1 \times 2.75 = 3.0$$
$$T_I = P_d/2.6 = 3.7/2.6 = 1.42 \text{ min.}$$

The setpoint change response of the system with this PI controller is shown in Figure 10.28. Obviously, in this case the method does not produce acceptable results. Comparing the PI parameters to the previous example, the gain is too high and the integral time constant is too low. However, as indicated in the introductory remarks of this chapter, all of the tuning methods in this chapter are only designed to provide the engineer with a good starting point from which the parameters are further refined.

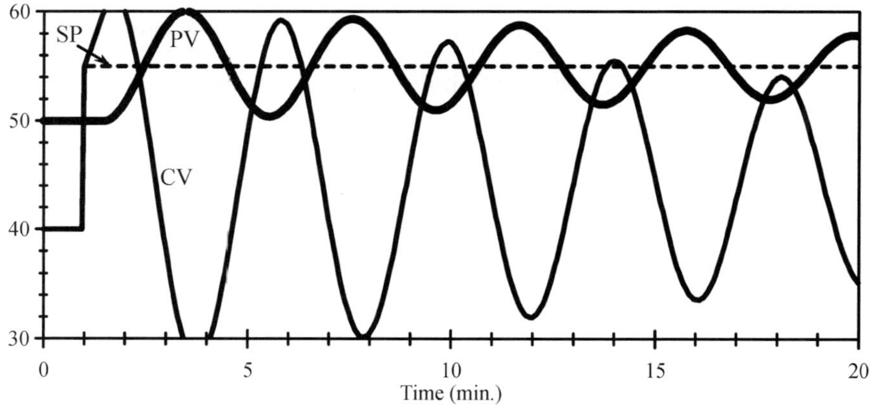

Figure 10.28. Response of system to a setpoint change when PI controller is tuned with damped oscillation method.

10.4.4 Open-Loop Tuning Methods

Tuning may be determined based on the process response to a step change in the control variable while the controller is in manual mode. However, the process is not being controlled during such tests. This technique can be used to determine the tuning of slow processes. All of the open-loop tuning methods for self-regulating processes (reaches a steady-state value after a step change in the manipulated variable) use a first-order-plus-deadtime model of the process. An integrator-plus-deadtime model is used to tune non-self-regulating processes. Open-loop tuning methods are most amenable to slow processes or for those processes for which a model has been determined.

10.4.4.1 First-Order Model of Self-Regulating Process. The transfer function of a general first-order-plus-deadtime (FODT) model is

$$G_{FODT}(s) = \frac{Ke^{-s\theta_D}}{\tau s + 1} \tag{10.4}$$

where

K = process gain,
θ_D = apparent process deadtime,
τ = apparent process time constant.

Many processes are higher than first order (contain more than one lag term) and so any first-order model will only be an approximation. However, the approximation is sufficient for tuning PID controllers.

The graphical procedure for determining the model is based on Ziegler and Nichols (1942). This method is based on features of the response of (10.4) to a step input of magnitude A,

$$y(t) = KA\left(1 - e^{-(t-\theta_D)/\tau}\right)$$

This response is evaluated when the exponent is one ($t = \theta_D + \tau$). When the exponent is one, the value of the response is

$$y(t) = KA\left(1 - e^{-1}\right) = 0.632KA$$

In other words, the response reaches 63.2% of its change when $t = \theta_D + \tau$. Given a response plot as shown in Figure 10.29, the procedure to find the gain, apparent deadtime, and apparent time constant is as follows:

1. Start any trending (plotting) of the PV and CV. The units of the PV and CV must be in percent of span since that is the unit used in the controller calculation.
2. Place the controller in manual mode and allow the process to reach steady state (PV is a "flat line").
3. Impose a step change in the controller output (CV).
4. Allow the process to reach steady state.
5. Stop any trending of the PV and CV and get a hard copy of the response curves.
6. Draw a tangent to the process variable response curve at its point of maximum slope, the point of inflection (Figure 10.29). In practice, this step may be tricky, especially if the response contains noise.
7. The term T_1 is the time at which the step change in the manipulated variable was imposed. T_2 is the time at which the tangent drawn in the previous step intersects the time axis, *drawn horizontally from the beginning of the process variable response curve*. The apparent process deadtime $\theta_D = T_2 - T_1$.
8. The term T_3 is the time at which the process variable response has made 63% of its total change. The apparent process time constant $\tau = T_3 - T_2$.
9. The process gain is the change in the process variable divided by the change in the control variable,

$$K = \frac{\Delta\text{PV}}{\Delta\text{CV}} = \frac{\text{PV}_2 - \text{PV}_1}{\text{CV}_2 - \text{CV}_1}$$

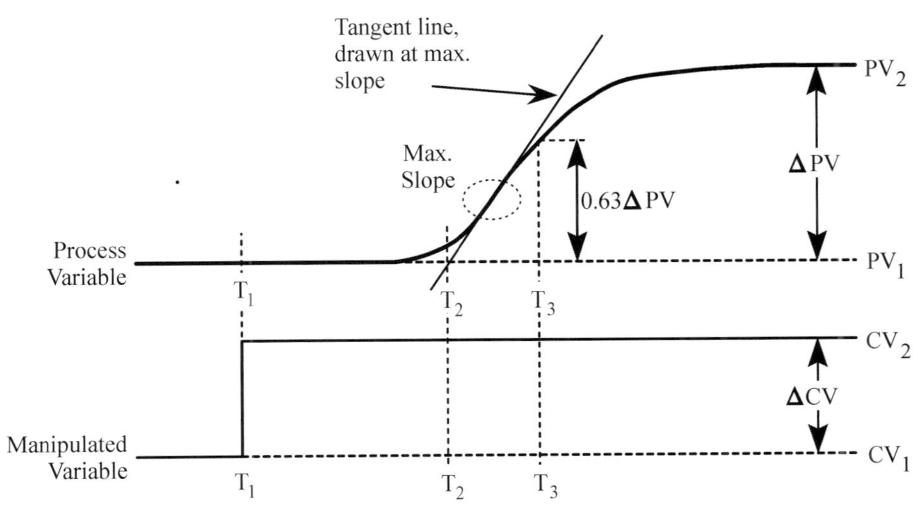

Figure 10.29. Determination of first-order approximate model.

10.4.4.2 Open-Loop Tuning of Self-Regulating Process. Given a first-order-plus-deadtime model that approximates the self–regulating process, there are a few methods one can use to calculate the tuning parameters. The Ziegler-Nichols open-loop tuning method (Ziegler and Nichols, 1942) and the two Cohen-Coon tuning methods (Cohen and Coon, 1953) result in similar tuning parameters. These three methods usually yield a set of parameters that generates an oscillatory response to a setpoint change. A fourth method, proposed by Fertik (1975), yields a different set of tuning parameters. The Fertik method generally yields a nonoscillatory response to a setpoint change, which is desirable for most process control applications. The rationales behind each of these methods are:

Ziegler-Nichols open-loop	Quarter-wave damping
Cohen-Coon	Quarter-wave damping
Cohen-Coon with constraints	Minimum IAE (Eqn. 10.1)
Fertik	Minimum ITAE (Eqn 10.2), <8% overshoot

Each method assumes a first order plus deadtime process model of the form of (10.4). The Ziegler-Nichols and Cohen-Coon open-loop tuning methods assume a process controllability, α,

$$\alpha = \frac{\theta_D}{\tau}$$

Table 10.4 Open-loop Tuning PID Parameters for Self-regulating Process

	Ziegler-Nichols open-loop	Cohen-Coon	Cohen-Coon with constraints
P controller			
K_P	$\frac{1}{K\alpha}$	$\frac{1}{K}\left(\frac{1}{\alpha}+0.333\right)$	$\frac{1.208}{K}(\alpha)^{-0.956}$
PI controller			
K_P	$\frac{0.9}{K\alpha}$	$\frac{1}{K}\left(\frac{0.9}{\alpha}+0.082\right)$	$\frac{0.928}{K}(\alpha)^{-0.946}$
T_I	$3.33\theta_D$	$\theta_D\left[\frac{3.33+0.333\alpha}{1+2.2\alpha}\right]$	$0.928\tau(\alpha)^{+0.583}$
PID controller			
K_P	$\frac{1.2}{K\alpha}$	$\frac{1}{K}\left(\frac{1.35}{\alpha}+0.27\right)$	$\frac{1.37}{K}(\alpha)^{-0.95}$
T_I	$2\theta_D$	$\theta_D\left[\frac{2.5+0.5\alpha}{1+0.6\alpha}\right]$	$0.740\tau(\alpha)^{+0.738}$
T_D	$0.5\theta_D$	$\theta_D\left[\frac{0.37}{1+0.2\alpha}\right]$	$0.365\tau(\alpha)^{+0.95}$

The K_P, T_I, or T_D are calculated from the first-order-plus-deadtime model parameters. The calculations also depend on the type of controller (P only, PI, or PID). Table 10.4 summarizes these calculations for the dependent gains formof the PID equation.

Example 10.3. For the same process as in Example 10.1, tune a PI and a PID controller to control this process using the Ziegler-Nichols open-loop tuning method. Show the response of the system to setpoint and disturbance changes.

Solution. First, a first-order-plus-deadtime approximation to this process must be determined. The controller is placed in manual mode, and the PV response to a step change in CV is recorded in Figure 10.30. The approximate first-order-plus-deadtime model is obtained using the method outlined in section 10.4.4.1 and the pertinent information is marked on the response in Figure 10.30.

$$\Delta \text{CV} = 8$$

$$\Delta \text{PV} = 44.6 - 35 = 9.6$$

$$T_1 = 12 \text{ min.}$$

$$T_2 \approx 12.7 \text{ min.}$$

$$T_3 \approx 15.1 \text{ min.} \qquad (0.63\Delta\text{PV} + \text{PV}_1 = 0.63(9.6) + 35 = 41.05)$$

The model parameters are calculated as:

$$\theta_D = T_2 - T_1 = 12.7 - 12 = 0.7 \text{ min.}$$

$$\tau = T_3 - T_2 = 15.1 - 12.7 = 2.4 \text{ min.}$$

$$\text{Gain} = K = \frac{\Delta\text{PV}}{\Delta\text{CV}} = \frac{9.6}{8} = 1.2$$

The approximate model is thus

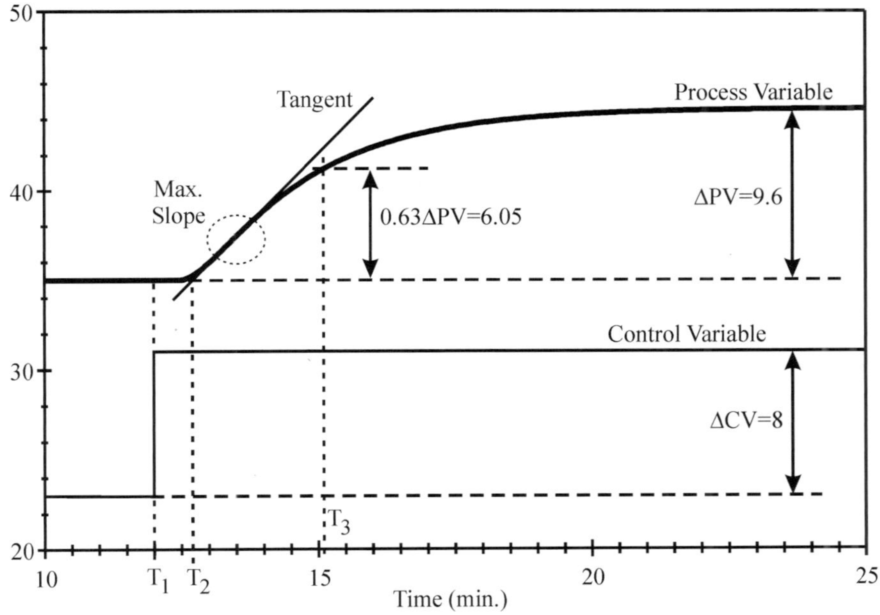

Figure 10.30. First-order process model from graphical data for Example 10.3.

$$G_{FODT}(s) = \frac{1.2e^{-0.7s}}{2.4s + 1}$$

Using the equations in Table 10.4, the dependent gains PI controller parameters are determined to be

$K_P = 2.57$, $T_I = 2.33$ min., $T_D = 0$ min.

and the dependent gains PID controller parameters are

$K_P = 3.43$, $T_I = 1.4$ min., $T_D = 0.35$ min.

The setpoint change response of the system with this PI controller is shown in Figure 10.31 and has more overshoot than the PI controller in Example 10.1. The PID controller response is shown in Figure 10.32 and illustrates the magnification of the measurement noise by the derivative term. The CV makes small changes even when there are no apparent PV changes. The addition of the derivative term gives a better process variable response at

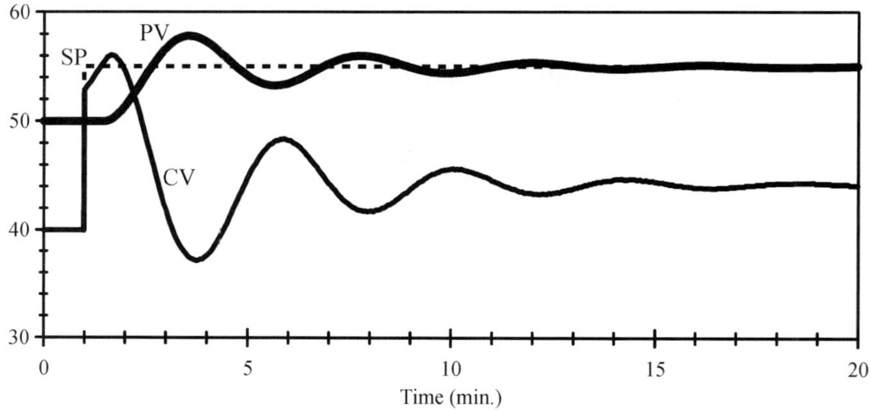

Figure 10.31. Response of system to a setpoint change when PI controller is tuned with Ziegler-Nichols open-loop method.

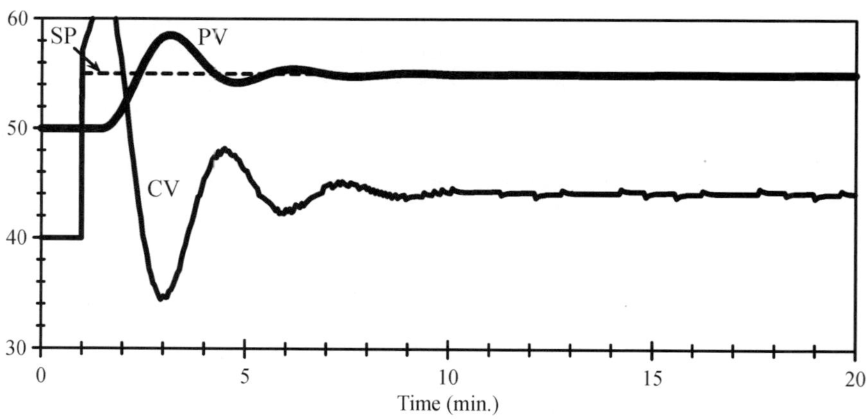

Figure 10.32. Response of system to a setpoint change when PID controller is tuned with Ziegler-Nichols open-loop method.

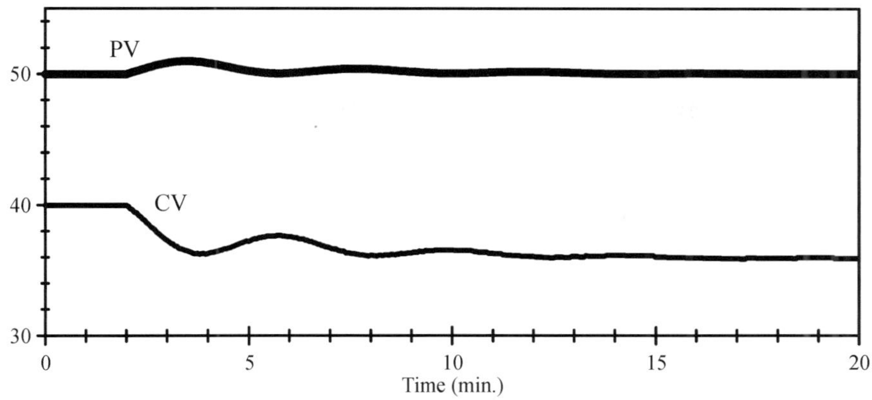

Figure 10.33. Response of system to a step disturbance change of +5 when PI controller is tuned with Ziegler-Nichols open-loop method.

the expense of more vigorous control variable action. The response to a disturbance of +5 is shown in Figure 10.33 for the PI controller.

Tuning using the Fertik method entails determining the type of controller (PI or PID), calculating the Fertik controllability, α_F, as

$$\alpha_F = \frac{\theta_D}{\theta_D + \tau} = \frac{T_d}{T_{ps}},$$

$$T_d = \theta_D$$

$$T_{ps} = \theta_D + \tau$$

and then reading the normalized parameters from the set of graphs in Figures 10.34 – 10.38. The parameters may be optimized for setpoint response or for disturbance response. Note that the PID controller is not recommended for those processes whose Fertik controllability is greater than 0.5. These processes are dominated by deadtime.

Example 10.4. For the same process as in Example 10.1, tune a PI and a PID controller to control this process using the Fertik open-loop tuning method. Tune the controller for an "optimal" setpoint response and show its response to a setpoint change. Also tune the controller for an "optimal" disturbance response and demonstrate its response to a step disturbance.

Solution. The first-order-plus-deadtime approximation to this process is determined in the same manner as for Example 10.3

$$G_{FODT}(s) = \frac{Ke^{-s\theta_D}}{\tau s + 1} = \frac{1.2e^{-0.7s}}{2.4s + 1}$$

The Fertik controllability is determined as $\alpha_F = 0.23$. The value of $T_{ps} = 3.1$. Using the graphs in Figures 10.34 and 10.35, the PI controller parameters for a setpoint change are

$$KK_P \approx 1.4 \Rightarrow K_P = 1.4/1.2 = 1.17$$

$$T_I / T_{ps} \approx 0.86 \Rightarrow T_I = 0.86 \times 3.1 = 2.67 \text{ min.}$$

and the PID controller parameters for a setpoint change are determined using the graphs in Figures 10.36 - 10.38:

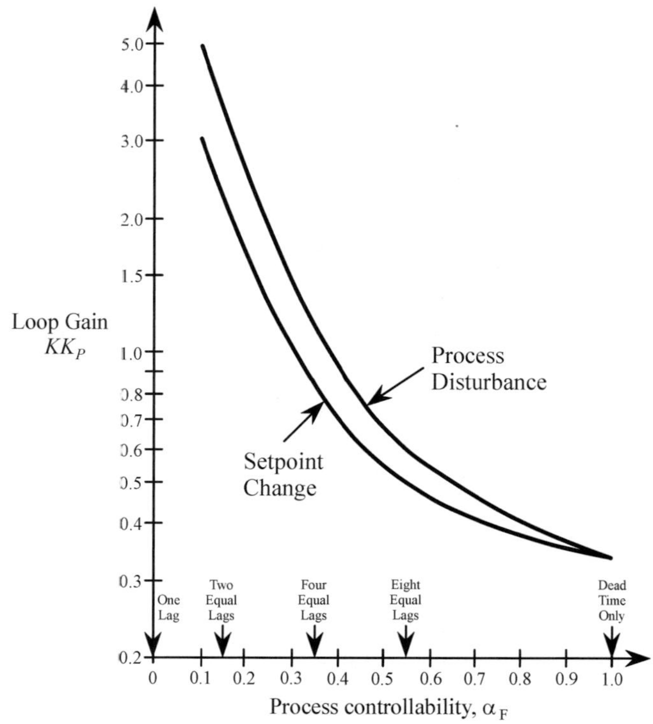

Figure 10.34. Fertik controller gain for PI control. (Copyright 1975, ISA - The International Society of Automation. All rights reserved. Used with permission of ISA.)

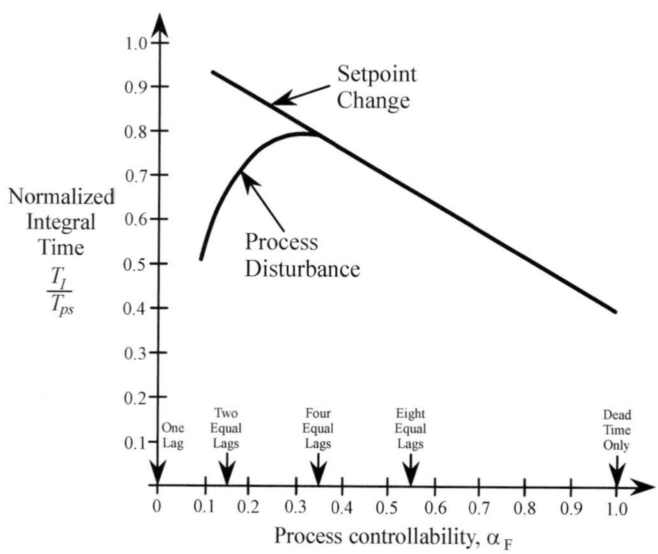

Figure 10.35. Fertik controller integral time for PI control. (Copyright 1975, ISA - The International Society of Automation. All rights reserved. Used with permission of ISA.)

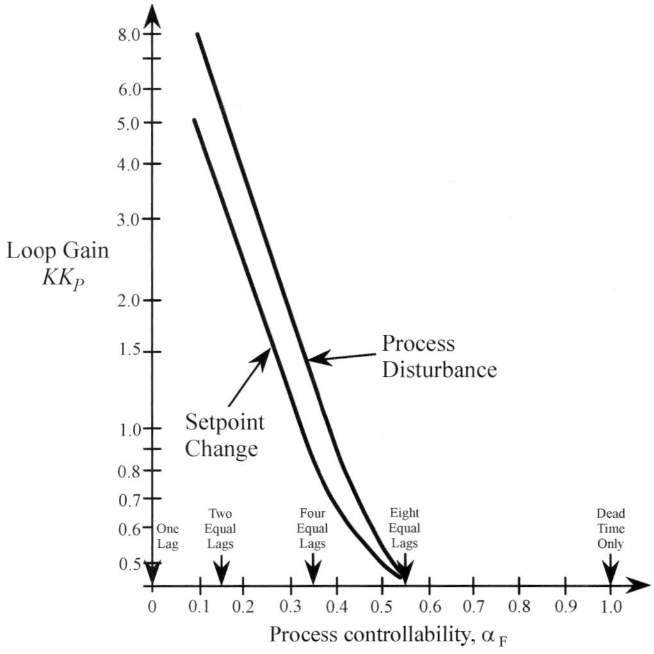

Figure 10.36. Fertik controller gain for PID control.

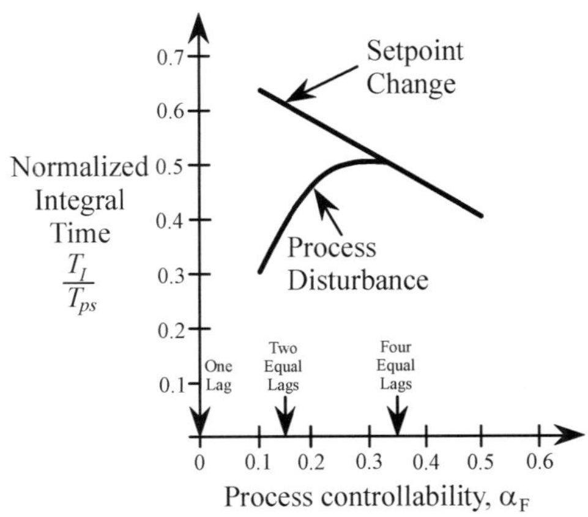

Figure 10.37. Fertik controller integral time for PID control.

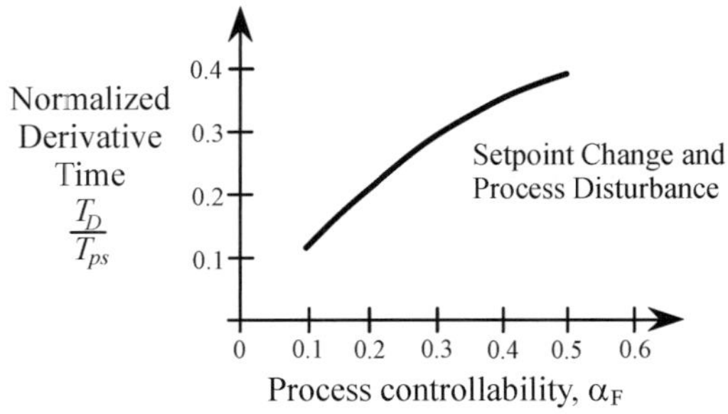

Figure 10.38. Fertik controller derivative time for PID control. (Copyright 1975, ISA - The International Society of Automation. All rights reserved. Used with permission of ISA.)

$KK_P \approx 1.9 \Rightarrow K_P = 1.9/1.2 = 1.58$

$T_I / T_{ps} \approx 0.57 \Rightarrow T_I = 0.57 \times 3.1 = 1.77$ min.

$T_D / T_{ps} \approx 0.24 \Rightarrow T_D = 0.24 \times 3.1 = 0.74$ min.

The setpoint change response of the system with the PI controller is shown in Figure 10.39 and for the PID controller is shown in Figure 10.40. In contrast to the other methods presented in this chapter, the process variable exhibits a small overshoot to a setpoint change, but a slower rise time. In a similar manner, the PI parameters for good disturbance response are:

$K_P = 1.67$, $T_I = 2.39$ min., $T_D = 0.0$ min.

and the PID parameters for good disturbance response are:

$K_P = 2.5$, $T_I = 1.52$ min., $T_D = 0.74$ min.

The response of the system to a step disturbance of +5 with the PI controller is shown in Figure 10.41. Compared with the parameters determined with Ziegler-Nichols tuning, the parameters determined by the Fertik tuning method give a system response with smaller deviation but a slightly longer rise time.

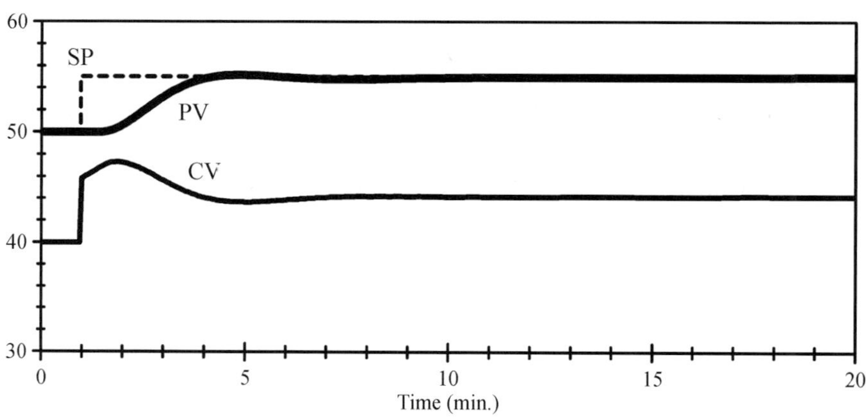

Figure 10.39. Response of system to a setpoint change when PI controller is tuned according to the Fertik open-loop method.

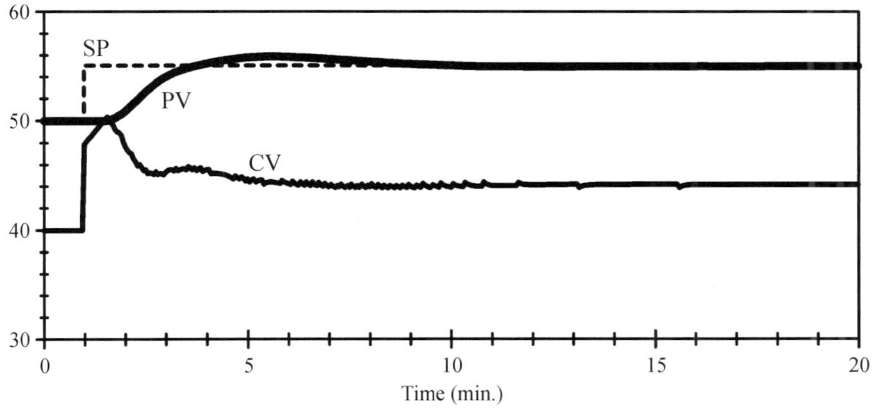

Figure 10.40. Response of system to a setpoint change when PID controller is tuned according to the Fertik open-loop method.

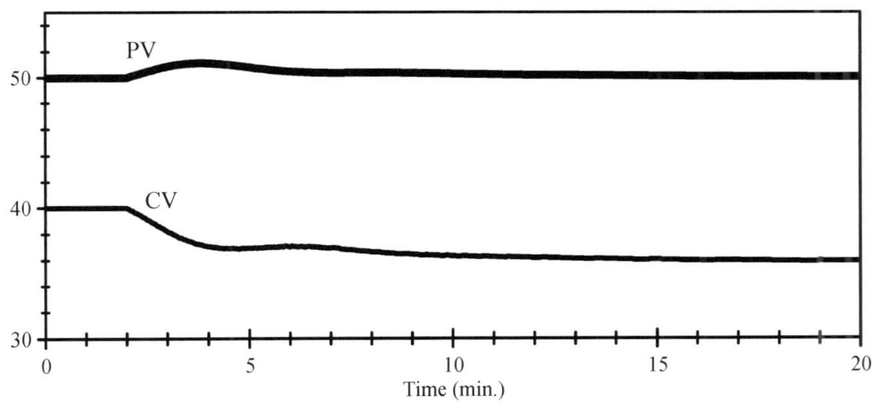

Figure 10.41. Response of system to a step disturbance change of +5 when PI controller is tuned with the Fertik open-loop method.

10.4.4.3 Integrating Processes. This method was originally proposed by Ziegler and Nichols (1942) and may be used to determine the tuning parameters for a non-self-regulating process.

1. Start any trending (plotting) of the PV and CV.
2. Place the controller in manual mode and allow the process to reach steady state.
3. Impose a step change in the controller output (CV).
4. When there is enough trend information, put the controller back in automatic mode (so the process does not "run away").
5. Using the recorded process response curve, draw a straight line tangent to the process response curve at its point of maximum rate of ascent as shown in Figure 10.42.
6. The term T_1 is the time at which the step change in the manipulated variable was imposed. T_2 is the time at which the tangent drawn in the previous step intersects

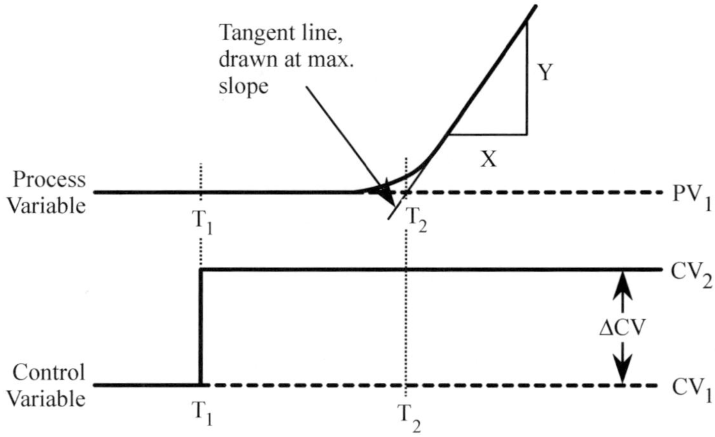

Figure 10.42. Determination of parameters for non-self-regulating process PID tuning.

the time axis, *drawn horizontally from the beginning of the process variable response curve*. The apparent process deadtime $\theta_D = T_2 - T_1$.

7. The integrator gain is the slope of the process variable response divided by the change in the control variable,

$$K_S = \frac{\text{slope of tangent}}{\Delta\text{CV}} = \frac{\text{Y}/\text{X}}{\text{CV}_2 - \text{CV}_1}$$

The controller settings are shown in Table 10.5. As usual, make any final adjustments in the K_P, T_I, or T_D to obtain the desired process variable response.

Table 10.5 Open-loop Tuning PID Parameters for Integrating Process

	Dependent Gains Eqn.	Independent Gains Eqn.
P controller		
K_P	$1/(K_S\theta_D)$	$1/(K_S\theta_D)$
PI controller		
K_P	$0.9/(K_S\theta_D)$	$0.9/(K_S\theta_D)$
T_I	$3.3\theta_D$	$3.3\theta_D/K_P$
PID controller		
K_P	$1.2/(K_S\theta_D)$	$1.2/(K_S\theta_D)$
T_I	$3.3\theta_D$	$3.3\theta_D/K_P$
T_D	$0.5\theta_D$	$0.5\theta_D K_P$

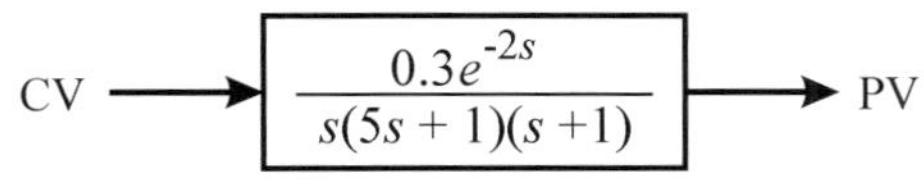

Figure 10.43. Process used for non-self-regulating tuning example.

Example 10.5. Assume the process is represented by the block diagram in Figure 10.43, where the time constants are measured in minutes. Tune a PI and a PID controller to control this process using the Ziegler-Nichols open-loop tuning method for integrating processes.

Solution. First, an approximate model must be determined. In order to keep the process variable from moving too far away from the operating point, as soon as the response seems to assume a constant slope, the control variable is changed to a step in the negative direction and for the same duration as the original change. This change brings the process back to its original operating point.

The appropriate information needed to determine the approximate process is marked in Figure 10.44. The model parameters are thus

$$\theta_D = T_2 - T_1 \approx 8 - 2 = 6 \text{ min.}$$

$$K_S = \frac{\text{slope of tangent}}{\Delta CV} \approx \frac{4/(12-8.8)}{5} = 0.25$$

The PI parameters are calculated to be

$$K_P = \frac{0.9}{K_S \theta_D} = \frac{0.9}{0.25 \times 6} = 0.6$$

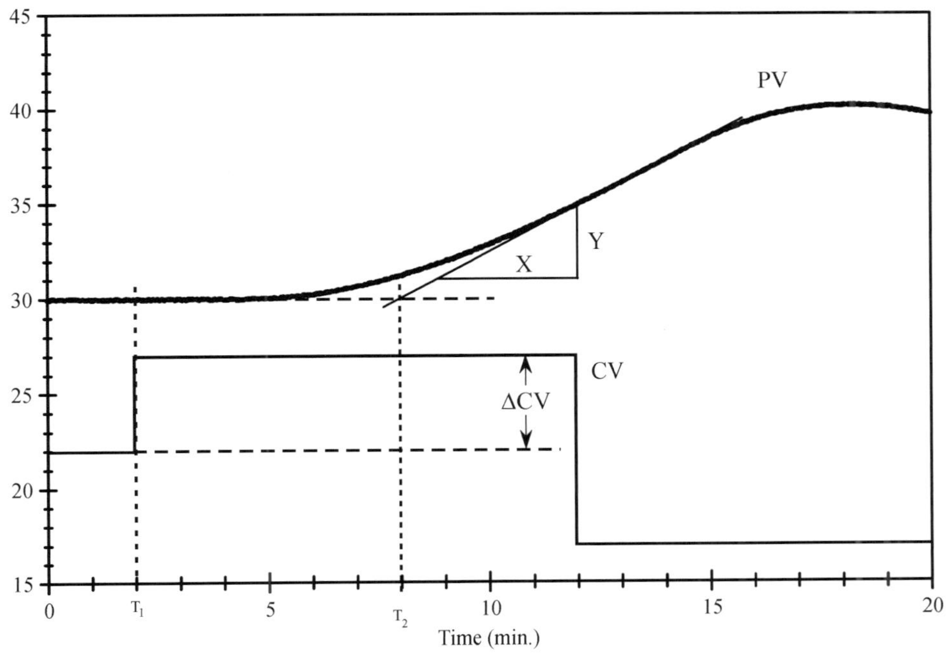

Figure 10.44. Non-self-regulating process model parameters for Example 10.5.

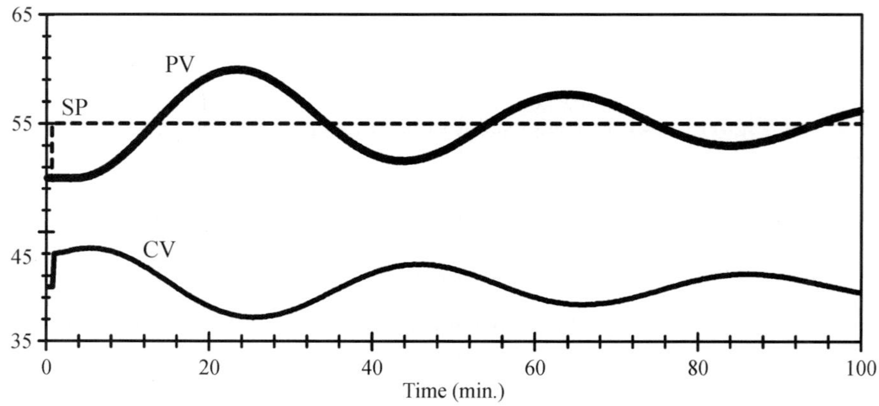

Figure 10.45. Response of non-self-regulating system to a setpoint change when the PI controller is tuned according to the Ziegler-Nichols open-loop method.

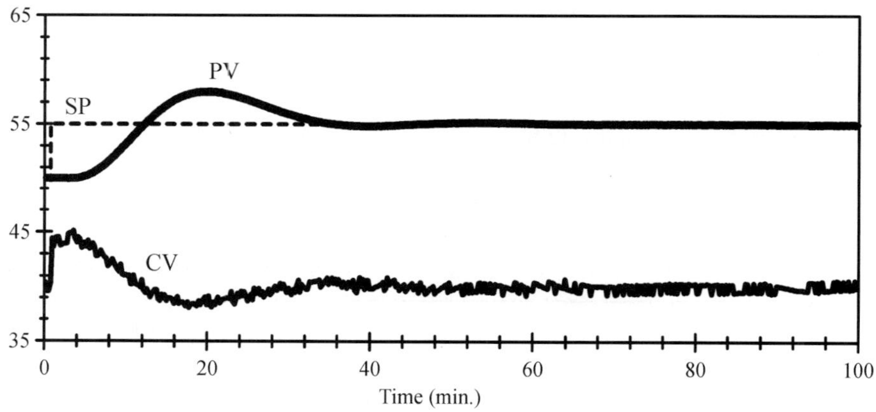

Figure 10.46. Response of non-self-regulating system to a setpoint change when the PID controller is tuned according to the Ziegler-Nichols open-loop method.

$$T_I = 3.3\theta_D = 3.3 \times 6 = 19.8 \text{ min.}$$

The PID parameters are calculated to be

$$K_P = \frac{1.2}{K_S \theta_D} = \frac{1.2}{0.25 \times 6} = 0.8$$

$$T_I = 3.3\theta_D = 3.3 \times 6 = 19.8 \text{ min.}$$

$$T_D = 0.5\theta_D = 0.5 \times 6 = 3 \text{ min.}$$

The responses of this system to a setpoint change are shown for the PI and PID controllers in Figures 10.45 and 10.46, respectively. As seen in other examples, the derivative term speeds up the response but amplifies the small amount of measurement noise.

10.5 OPERATIONAL ASPECTS

The previous section introduced some of the functions external to the PID controller algorithm that must be employed in certain applications. In general, there are a number of external functions that must be used in conjunction with a PID controller. These functions are grouped into measurement conditioning, setpoint management, and output processing. In addition, alternative controllers are also described.

Measurement conditioning. Depending on the type of measurement, it may need to be conditioned before it can be used for control. This type of conditioning also includes the blocks needed to implement feedforward control.

Scaling – Changes the scale of a raw analog input value into appropriate units or changes the CV into an analog output value. Depending on the processor, this function may also be handled in the analog input or output module.

Filtering – Filters noise from a measurement, making it "smoother." Sometimes called a lag function block. Heavy filtering of the PV will affect the performance of the PID controller and may cause it to be unstable.

Square root – Converts the differential pressure measurement from orifice-plate flow meters into a volumetric flow rate. This conversion is a weighted square root calculation.

Mass flow calculation – Converts a volumetric gas flow rate into a mass flow rate. It requires a temperature and pressure measurement.

Totalizer – Integrates a flow rate to calculate the volume of material that has passed through the flow meter. Basically, an integrator with more features.

Differentiator – Takes the derivative of a variable, for example, to calculate acceleration from a velocity measurement. It may be combined into with some filtering and called a lead function block.

Integrator – Integrates a variable, for example, to calculate the position from a velocity measurement.

Setpoint management. Depending on the application, the setpoint may be more complicated than a constant value.

Bumpless transfer – Smooths transfer between an operator-generated setpoint and a program-generated setpoint.

Ramp/soak generation – Generates the temperature setpoint for applications (e.g., metal processing) that require a ramp/soak profile (Figure 10.47).

Ratio – Calculates a setpoint by multiplying one variable by a ratio.

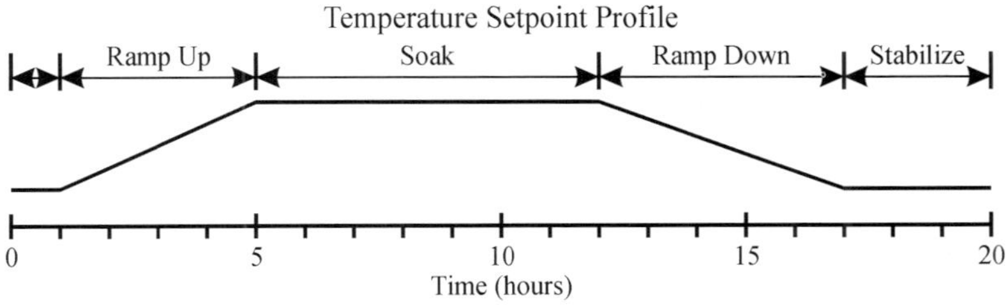

Figure 10.47. Typical temperature ramp/soak profile.

Output processing. Depending on the final control element, the controller output may need to be modified rather than used directly.

Velocity limiter – limits the rate of change of the control variable. This function may be integrated into the PID control algorithm.

High/low selector – Selects the maximum/minimum among multiple signals.

Split range – Implements split-ranging where two (or more) actuators cover the full range of the control variable.

Servo control – If the valve is controlled by an electric servomotor, the CV must be converted to two logic outputs to the motor, typically open/close or raise/lower.

Pulse-width modulation – Convert the CV analog output into a binary output signal where the pulse width is proportional to the CV value.

Manual station – Allows an operator to switch between automatic and manual modes. Coordinates the source of the CV in automatic and manual modes. This function may be integrated into the PID function block.

Alternative controllers. If the process variable measurement is an analog signal and the final control element is a discrete device, then there are simpler alternatives to the full PID algorithm. For example, a home heating system is a simple on-off control. The temperature is the measurement and the only actuating signal turns the heater (and fan) **on** when the temperature is below a setpoint and turns the heater **off** when the temperature is some value above the setpoint. This approach can be extended to a system that has two discrete actuators, for example, one actuator to control heating and another to control cooling.

10.6 PLC PID FUNCTION BLOCKS

As done in previous chapters, the PID control and related function blocks for a particular PLC are briefly described in separate sections. The reader is referred to the appropriate documentation for more details. Each section illustrates the ladder logic to implement a single PID loop.

10.6.1 ControlLogix/CompactLogix PID function block

The ControlLogix and CompactLogix PID function block fully implements the PID control algorithm and includes anti-reset windup and accommodations for feedforward compensation. An example PID loop is shown in Figure 10.48.

This example uses a timer to trigger execution of the PID block. This trigger method is recommended only if the sample period (called the update time) is at least a second and is several times longer than the continuous task scan time. For processes that require shorter sample periods, the PID block should be placed in a periodic task whose execution rate is the desired sample period. In either case, the analog input module should be producing the process variable tag five to ten times faster than the loop sample period. The real time sampling (RTS) feature of the analog input modules is the most accurate way of executing a PID block. When the data timestamp for the analog input module changes, then execute the PID block. In this case, the real time sampling rate becomes the loop update rate.

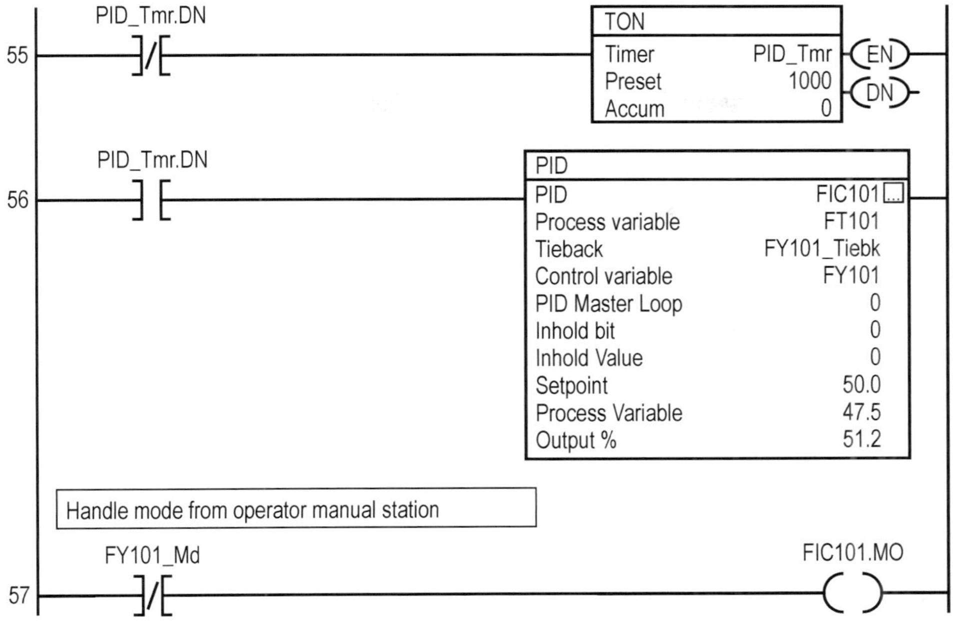

PID - FIC101 Configuration

Scaling		**Configuration**		**Tuning**	
PV Unscaled Max:	100.0	PID Equation:	Dependent	Setpoint (SP):	1.2
PV Unscaled Min:	0.0	Control Action:	SP - PV	Set Output:	0.0 %
Engineering Unit Max:	2.0	Derivative Of:	PV	Output Bias:	0.0 %
Engineering Unit Min:	0.0	Loop Update:	1.0 secs.	Proportional Gain (Kc):	1.17
CV Max (at 100%):	100.0	CV High Limit:	100 %	Reset Time (Ti)	
CV Min (at 0%):	0.0	CV Low Limit:	0 %	[mins/repeat] :	2.67
Tieback Max (at 100%):	100.0	Deadband Value:	0	Derivative Time (Td)	
Tieback Min (at 0%):	0.0	PV Tracking:	No	[mins] :	0.00
		Cascade Loop:	No	Manual:	No
		Cascade Type:	N/A	Software Manual:	No

Figure 10.48. Example ControlLogix PID loop.

The operands of the PID block are described as follows:

PID – PID structure tag.

Process variable – Tag of process variable.

Tieback – Tag of value copied to CV when in manual mode.

Control variable – Tag of control variable.

PID Master Loop – Tag of PID structure for master (outer) loop in cascade control if this loop is the slave (inner) loop.

Inhold bit – Inhold bit from analog output module that receives the CV. Supports bumpless restart when controller changes to run mode.

Inhold Value – Data readback from analog output module to support bumpless restart.

Setpoint – Displays current value of setpoint.

Process Variable – Displays current value of scaled process variable.

Output % – Displays current value of control variable, in % of range.

The "Process variable" operand should be the tag corresponding to the unscaled analog input channel reading. Also, the "Control variable" operand should be the tag corresponding to the unscaled analog output channel value. When cascading two PID loops, the slave PID block automatically converts the output from the master loop into the correct units for the setpoint of the slave loop.

For the PID tag defined in Figure 10.48, the individual fields of the PID structure that are useful for programmatic control of the PID loop are addressed as follows:

Reference	Description
FIC101.SP	Setpoint
FIC101.MO	Manual mode: **on** for manual, **off** for automatic
FIC101.SWM	Software manual mode: **on** for manual, **off** for automatic
FIC101.SO	CV (in %) when SWM bit is **on**.
FIC101.BIAS	Bias term added to output. Feedforward added here.
FIC101.OUT	CV, in percent (0-100)

The PID block offers two manual modes. When the .MO bit is set, the value in the Tieback tag is scaled and copied to the controller output. When the .SWM bit is set, the value in the .SO field is copied to the controller output. The .MO bit overrides the .SWM bit if both are set. The .MO bit is intended to be set from an operator manual station and the tieback value is the analog input readback of the operator-set control variable, as shown in Figure 10.49. FY101_Md is **on** when the controller should be in automatic mode. The transfer of the mode from the operator station to the PID block is on the third rung of Figure 10.48. If the particular loop does not have an operator manual station, then the .MO bit alone can be used to programmatically control the controller mode.

To implement a feedforward correction, copy the feedforward value to the BIAS field. However, keep in mind that if the controller does not use integral action when the mode is changed from manual to automatic, the BIAS term is modified to make the calculated CV output match the tieback (or SO value) when the mode is changed. In this case, any feedforward correction should be disabled when in the manual or software manual mode and not re-enabled until some time after changing to the automatic mode. Setting the

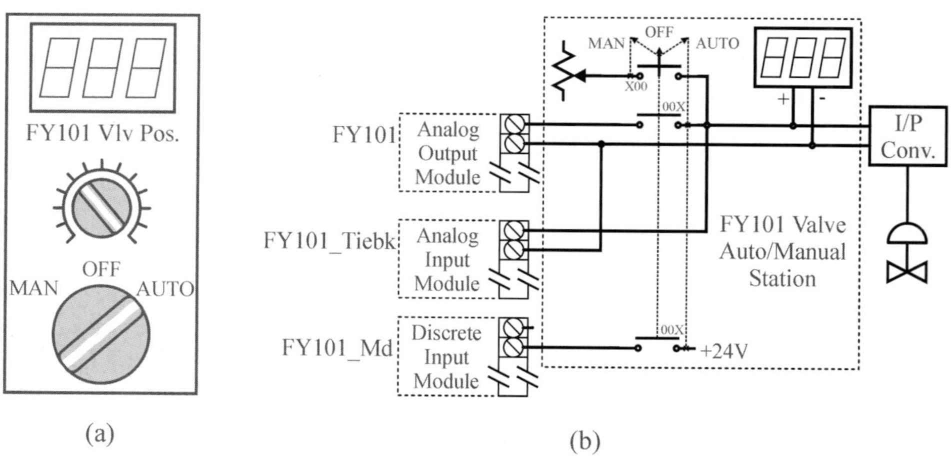

Figure 10.49. Operator manual station: *(a)* panel; *(b)* PLC connections.

.NOBC bit will disable the bias term back-calculation, but then there is no bumpless transfer when the controller does not have integral action.

An example PID configuration is shown in Figure 10.48. The tuning parameters are those for the PI controller of Example 10.4. The configuration in the PID structure includes:

K_P, T_I, and T_D or K_P, K_I, and K_D
Dependent or independent gains form of the algorithm
Reverse/direct action
Derivative on PV/error
Loop update time
Upper and lower limits of the CV calculation (may be smaller range than bounds)
Deviation deadband – if the error magnitude is within the deadband, the controller considers the error to be zero.
Alarms: PV high, PV low, and deviation
Upper and lower bounds of the PV
Upper and lower bounds of the CV
Upper and lower bounds of tieback value

The actual algorithms used to calculate the CV are very similar to the PID algorithm equations in section 10.3.2. The positional form of the algorithm is used. The user may select either the dependent gains (ISA standard) or independent gains form of the algorithm. For the dependent gains, the tuning parameters are K_P, T_I, and T_D, and time is measured in minutes. For the independent gains form of the algorithm, the tuning parameters are K_C, K_I, and K_D, and time is measured in seconds.

Regardless of the method used to trigger the PID block (timer, periodic task, or real-time sampling), the loop update time is specified as a separate configuration parameter. The loop update time must match the actual sample period or the algorithm calculation will be incorrect.

The ControlLogix does not have ladder logic function blocks to handle signal conditioning, setpoint management, or output processing. These functions are provided by the FBD function blocks, described in Chapter 11 of Erickson (2023).

10.6.2 Micro800 PID function blocks

The Micro800 has two PID function blocks that implement the PID control algorithm. The IPIDCONTROLLER block includes anti-reset windup. The PID block is more basic and is similar to the MicroLogix PID block.

IPIDCONTROLLER block

An example IPIDCONTROLLER loop is shown in Figure 10.50. The inputs of this block are described as follows:

EN – When true, PID calculation is executed. When false, the block is not executed.
Process – Process variable.
SetPoint – Setpoint.
FeedBack – Value of Control Variable applied to process.
Auto – Controller auto/manual mode. If true, controller in auto mode. If false, in manual mode.

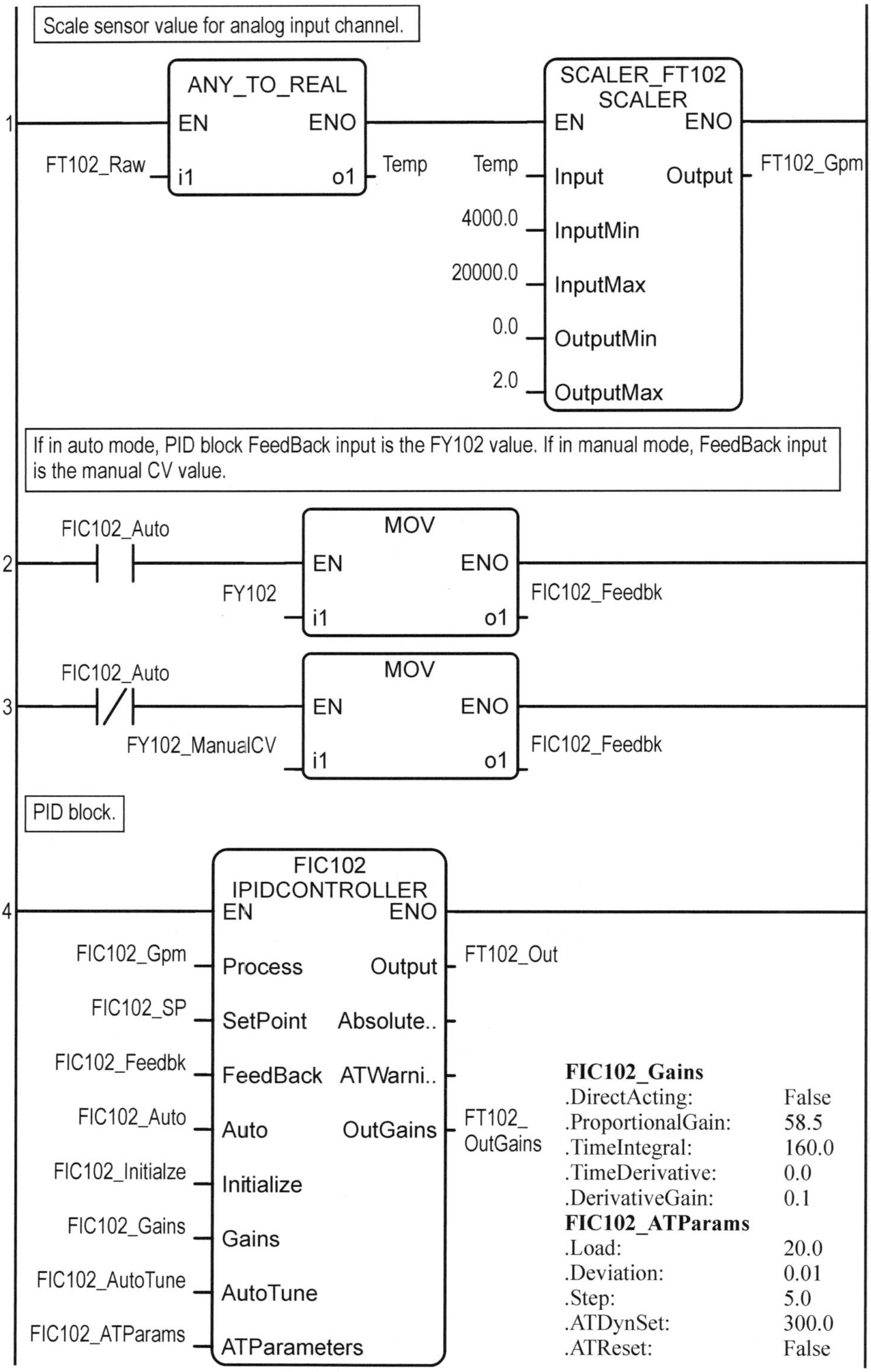

Figure 10.50. Example Micro800 PID loop using the IPIDCONTROLLER block *(continued)*.

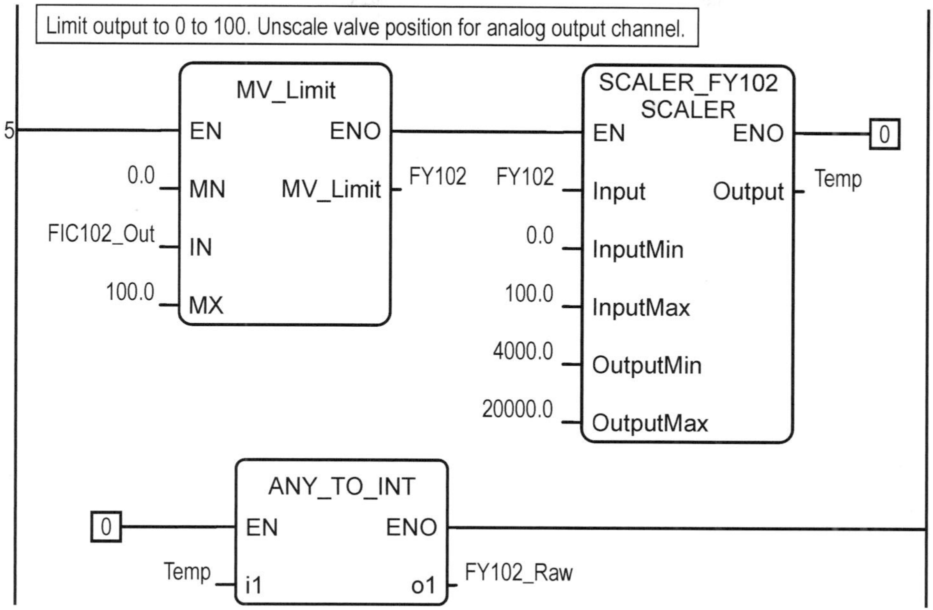

Figure 10.50. *(continued)*

Initialize – A false-to-true or true-to-false causes the controller to clear any integral action and initialize an AutoTune sequence.

Gains – Variable of type GAIN_PID that holds the K_P, T_I, and T_D

AutoTune – When true, initiate an AutoTune sequence if both Auto and Initialize are false.

ATParameters – Variable of type AT_Param that contains the auto-tune parameters.

The outputs of this block are described as follows:

Output – Control Variable, result of PID control calculation.

AbsoluteError – Absolute value of the difference between the Process and SetPoint inputs.

ATWarning – Status of AutoTune sequence.

OutGains – Gains calculated as a result of the AutoTune sequence. GAIN_PID data type.

The elements of the GAIN_PID data type are:

DirectActing – Sign of controller gain. If true, the controller is direct-acting. If false, the controller is reverse-acting.

ProportionalGain – K_P , = 0.0001. If < 0.0001, it is set to 0.0001.

TimeIntegral – T_I in seconds, = 0.0001. If < 0.0001, it is set to 0.0001.

TimeDerivative – T_D in seconds, = 0.0. If < 0.0, it is set to 0.0. When 0.0, derivative action is disabled.

Derivative Gain – Modifies the derivative action dynamics, = 0.0. If < 0.0, it is set to 0.1. A higher value makes the system more responsive to changes in the PV, but increases the possibility of instability

The elements of the AT_Param data type are:

Load – Starting value of Output (CV) when starting AutoTune.

Deviation – Standard deviation of the measurement. A reasonable approximation is one-half of the variation in the PV when the CV is held constant.

Step – Change in CV for the AutoTune sequence. The AutoTune sequence basically determines the dynamic response of the system to this step change.

ATDynamSet – Assigned time for AutoTune, in seconds. Basically longer than the expected time for the PV to settle to a constant value after a step change in the CV. If the PV does not settle to a constant value within this period, the AutoTune sequence is stopped and an error is reported.

ATReset – If true, the CV is set to 0.0 at the conclusion of the AutoTune. If true, the CV remains at the Load value when AutoTune completes.

The actual algorithms used to calculate the CV are very similar to the PID algorithm equations in section 10.3.2. The velocity form of the algorithm with dependent gains is used and time is measured in seconds. Derivative action is on the measurement and there is no option for derivative on error. The derivative implementation is a filtered derivative whose transfer function is

$$\frac{T_D}{\left(\frac{T_D}{DG}s+1\right)}$$

where DG is the DerivativeGain.

The IPIDCONTROLLER block does not limit the Output (CV), so any limiting must be handled with extra logic. This feature allows one to implement anti-reset windup using the FeedBack block input. To implement anti-reset windup, the output of the limiting function is “fed back”, that is, connected to the FeedBack input of the IPIDCONTROLLER block. The block algorithm detects that the actual CV is not the same as the calculated CV and so assumes the CV is saturated, at a limit. In this case, the integral term in the controller is recalculated so that the calculated CV equals the actual CV. The sample loop shown in Figure 10.50 implements anti-reset windup.

When in manual mode, the value of the FeedBack input to the block should be set to the desired CV (called the “Tieback” in the ControlLogix PID block), as shown in Figure 10.50.

This block can use a timer to trigger execution. This trigger method is recommended only if the sample period is at least a second and is several times longer than the continuous task scan time. For processes that require shorter sample periods, the IPIDCONTROLLER block should be placed in a POU controlled by an STI (selectable timed interrupt) whose period is the desired sample period. Placing the block in a POU controlled by an STI is recommended for autotuning. Regardless of the method used to trigger the IPIDCONTROLLER block, it internally determines the sample rate. The loop update time does not need to be specified as for the ControlLogix.

An example PID loop using the IPIDCONTROLLER block is shown in Figure 10.50. This logic is placed in a POU controlled by an STI (selectable timed interrupt) set for 100 milliseconds. The analog input channel and analog output channels are in expansion modules, the signals are 4 to 20 mA, and the data format is “Engineering Units,” which means a range of 4000 to 20000 for the INT from the analog input channel and the INT to the analog output channel. The CV is limited to the range of 0.0 to 100.0 and this limited value is connected to the FeedBack input of the controller block.

The tuning parameters are those for the PI controller of Example 10.4. The values in the GAIN_PID variable are shown in the figure. Note that the gain, ProportionalGain is the value in Figure 10.48 times 50. In the ControlLogix PID block the PV and SP are both scaled to a percentage of the PV engineering units range before calculating the CV, which is also in percentage of the CV engineering units range and then it is scaled to the CV in engineering units. This scaling is not done for the Micro800 PID blocks. Because the range of the CV (0.0 to 100.0) is 50 times the range of the PV (0.0 to 2.0), the proportional gain must be appropriately adjusted. The units of the TimeIntegral parameter is seconds whereas in Figure 10.48 the Reset Time (T_i) has units of minutes. Hence, the integral time of the PID block in Figure 10.50 is 60 times the T_i in Figure 10.48.

The IPIDCONTROLLER function block has autotuning capabilities. When initiated, the CV makes a step change and the response to this step is analyzed to determine the PID tuning parameters. Rockwell Automation (2022b) describes the autotuning function.

PID Block

An example PID loop using the PID block is shown in Figure 10.51. The inputs of this block are described as follows:

Enable – When true, PID calculation is executed. When false, the block is not executed and the CV is set to zero.

PV – Process variable.

SP – Setpoint.

AutoManual – Controller auto/manual mode. If true, controller in auto mode. If false, in manual mode.

CVManual – CV when in manual mode.

CVMax – Maximum limit for CV

CVMin – Minimum limit for CV

Gains – Variable of type PID_GAINS that holds the K_P, T_I, and T_D

Control – Sign of controller gain. If true, the controller is direct-acting. If false, the controller is reverse-acting.

The outputs of this block are described as follows:

Active – Block status: true if active; otherwise false.

CV – Control Variable, result of PID control calculation.

AbsoluteError – Absolute value of the difference between the Process and SetPoint inputs.

Error – If true, error occurred.

ErrorID – Error code if Error is true.

Possible error codes are listed in Rockwell Automation (2022b).

The elements of the PID_GAINS data type are:

Kc – Proportional gain, K_P , = 0.0001. If $<$ 0.0001, an error is reported.

Ti – Integral time constant, T_I in seconds, = 0.0001. If $<$ 0.0001, an error is reported.

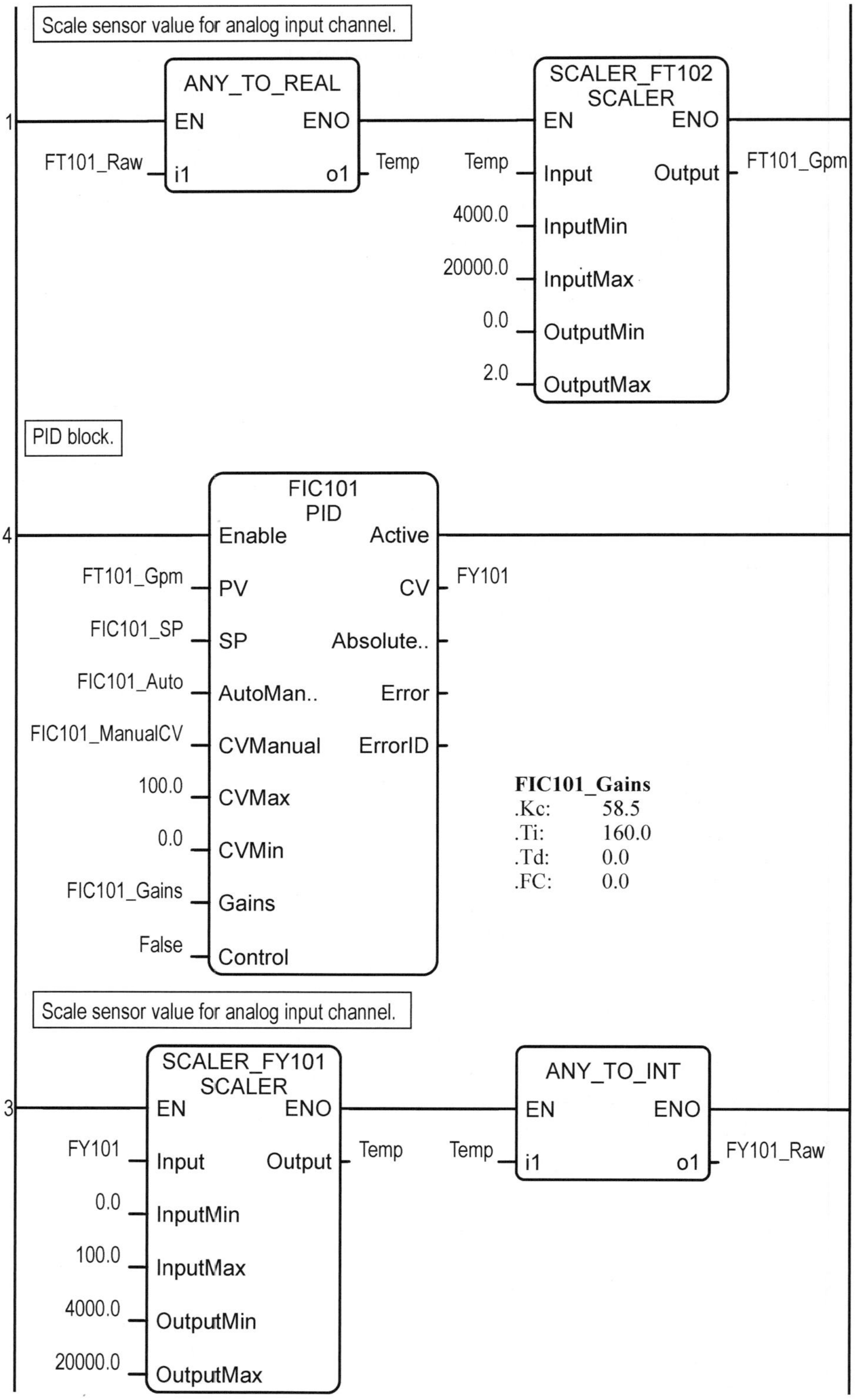

Figure 10.51. Example Micro800 PID loop using the PID block.

TdimeDerivative – T_D in seconds, = 0.0. If < 0.0, it is set to 0.0. When 0.0, derivative action is disabled.

FC – Derivative filter constant to smooth derivative action, = 0.0. If < 0.0, an error is reported. A higher value smooths the derivative action. Recommended range is zero to 20.

The PID block should be placed in a POU controlled by an STI (selectable timed interrupt) whose period is the desired sample period. Using a timer block to trigger execution will lead to unacceptable results, as when the block is scanned when the trigger is false (the majority of the scans), the CV output is set to zero. The loop update time does not need to be specified as for the ControlLogix.

Other Function Blocks

The Micro800 has some function blocks to handle signal conditioning and output processing:

Block Name	Description
HYSTER	Performs hysteresis on a difference of two REALs
LIM_ALRM	Alarming with hysteresis for a REAL variable
PWM	Controls a pulse-width modulated discrete output channel
SCL	Scale REAL variable with limits and alarms

The SCL block is similar to the SCALER block but with limiting and alarm features and shown in Figure 10.52. The inputs of this block are described as follows:

Enable – When true, scaling is executed.
In – Analog input variable.
InRawMax – Maximum value of In input
InRawMin – Minimum value of In input
InEUMax – Value of Out (scaled value) that corresponds to InRawMax
InEUMin – Value of Out (scaled value) that corresponds to InRawMin
Limiting – When true, Out is limited to be between InEUMin and InEUMax.

The outputs of this block are described as follows:

Done – True if operation completed successfully. False if operation encountered an error, or the Enable input is false.
EnableOut – Equal to the Enable input, though when Out overflows, it is set to false.
Out – Scaled value of the In input.
MaxAlarm – True when In > InRawMax
MinAlarm – True when In < InRawMin
Status – Status (DINT) of conversion. If not zero, an error occurred.
Error – If true, error occurred.
ErrorID – Error code if Error is true.

Possible values of the Status and ErrorID outputs are listed in Rockwell Automation (2022b). If the Limiting input is false, the Out value is correct, even if the In input is outside the range of InRawMin to InRawMax.

An example SCL function block is shown in Figure 10.52 that performs the conversion in Example 7.3 for an expansion module analog input. Since the analog input channel value

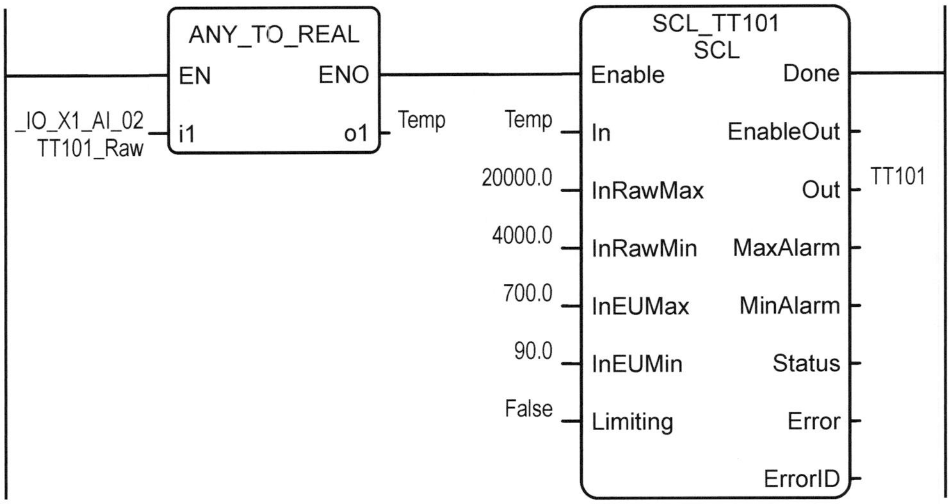

Figure 10.52. Example Micro800 SCL block for expansion analog input channel.

is an INT, an ANY_TO_REAL block converts the channel value into a REAL variable that the SCL block can use. The SCL block implements the scaling equation (7.1). In terms of the block inputs and outputs, the equation is

$$\text{Output} = \frac{\text{In} - \text{InRawMin}}{\text{InRawMax} - \text{InRawMin}} \times (\text{InEUMax} - \text{InEUMin}) + \text{InEUMin}$$

The LIM_ALRM block (Figure 10.53*a*) provides alarming functions with hysteresis. The inputs of this block are described as follows:

EN – When true, block executes its function.
H – High alarm value
X – Variable to be alarmed
L – Low alarm value
EPS – Value of hysteresis

The outputs of this block are described as follows:

QH – High alarm. true if X above H input.
Q – Alarm, true if either QH or QL is true.
QL – Low alarm, true if X below L input.

The operation of this block is illustrated in Figure 10.53*b*. If X becomes higher than H, the QH and Q outputs are set to true and are not set to false until X becomes lower than H-EPS. If X becomes lower than L, the QL and Q outputs are set to true and are not set to false until X becomes higher than L+EPS.

The HYSTER block (Figure 10.54*a*) provides hysteresis status. The inputs of this block are described as follows:

EN – When true, block executes its function.
XIN1 – Variable to be tested
XIN2 – High limit for test

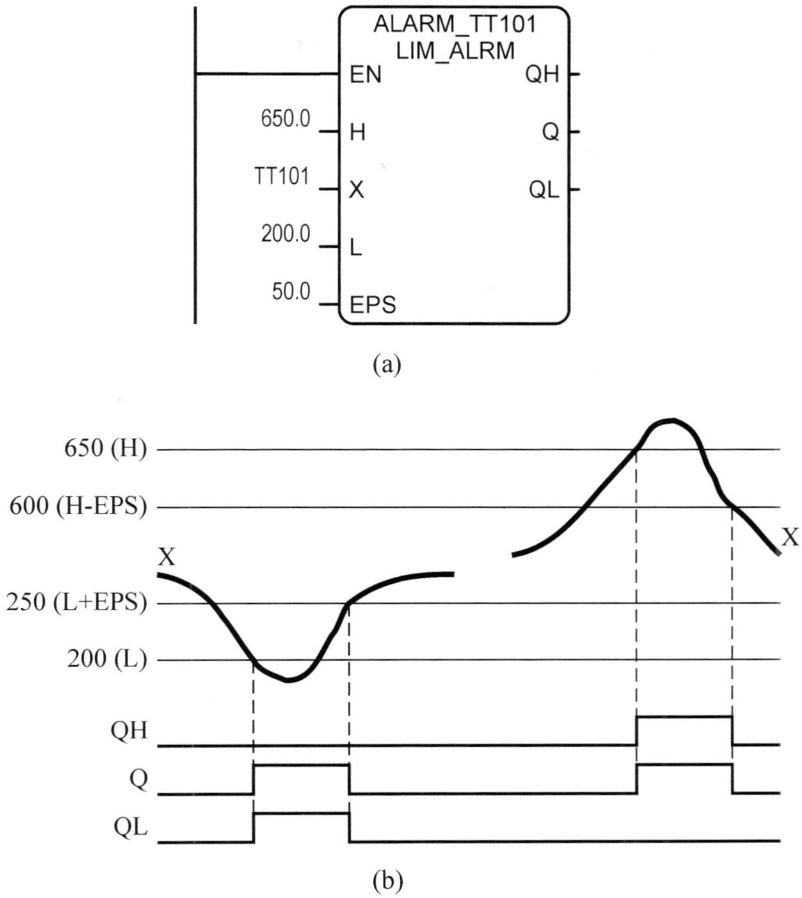

Figure 10.53. Example Micro800 LIM_ALRM block: *(a)* ladder logic example; *(b)* operation example.

EPS – Value of hysteresis

The Q output of this block is true when the XIN1 input has exceeded XIN2+EPS but has not gone lower than XIN2-EPS. The operation of this block is illustrated in Figure 10.54*b*.

The PWM block converts the CV analog output into a discrete signal where the pulse width is proportional to the CV value. Unfortunately, this block may currently be used in one particular catalog number, 2080-LC20-20QBB, of Micro800 controllers. An example block is shown in Figure 10.55. The inputs of this block are described as follows:

Enable – When true, block executes its function.

On – Controls the operation of the pulse-width modulated (PWM) discrete output. When true, the PWM is active. When false, the discrete output is inactive.

Freq – Pulse frequency. UDINT data type. The period is the inverse of the frequency.

DutyCycle – Pulse duty cycle, 0 to 1000, which represents 0 to 100% of the period.

ChType – Type of discrete output. Currently, this input may only be zero.

ChSlot – Channel slot. Currently, this input may only be zero.

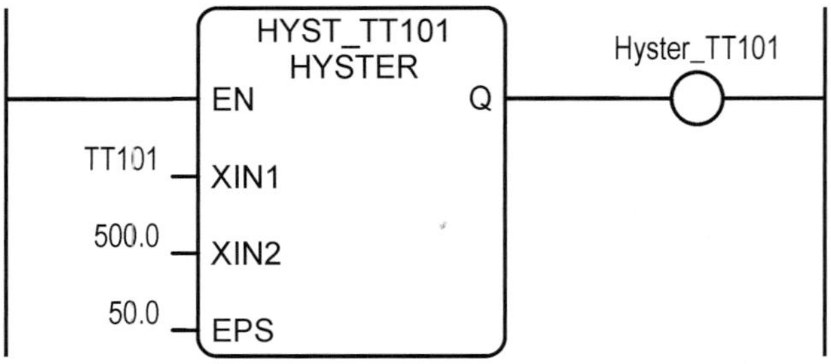

(a)

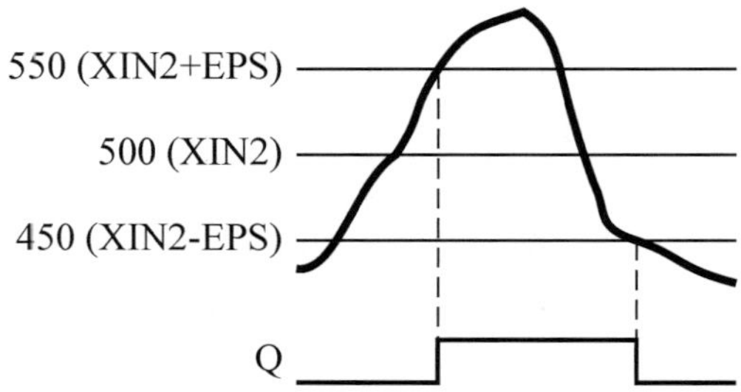

(b)

Figure 10.54. Example Micro800 HYSTER block: *(a)* ladder logic example; *(b)* operation example.

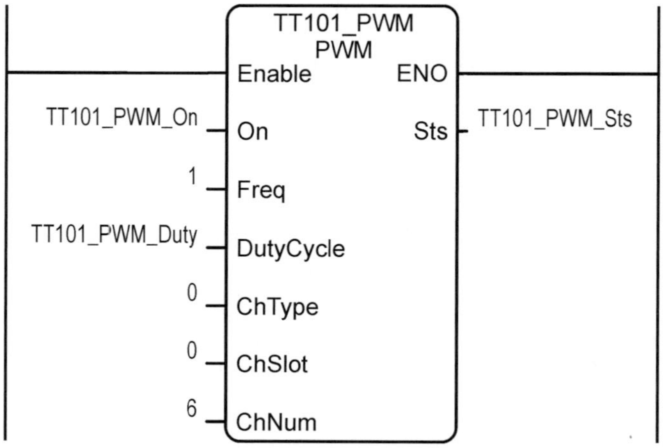

Figure 10.55. Micro800 PWM block.

ChNum – Channel number. Currently, this input may only be 6.

The Sts output of this block is a UINT that indicates the state of the PWM function. See Rockwell Automation (2022b) for a list of the possible values.

10.6.3 MicroLogix/SLC-500 PID function block

The MicroLogix PID function block fully implements the PID control algorithm and includes anti-reset windup and accommodations for feedforward compensation. An example PID loop is shown in Figure 10.56.

Note the lack of a timer to initiate execution. If placed in a normal program file, the "Time Mode" parameter is set to "Timed," which causes the PID block to execute at rate set in the "Loop Update" value (0.01 to 10.23 seconds). If placed in a STI program file, the "Time Mode" parameter is set to "STI," and the "Loop Update" value is set to the STI time interval. The SLC-500 PID block only executes the dependent gains form of the PID equation (time units in minutes) and has only one manual mode.

The operands of the PID block are described as follows:

PID File – Address of PID structure.
Process variable – Address of process variable (must be a word).
Control variable – Address of control variable (must be a word).

The "Process Variable" operand should be the address of an analog input channel configured for PID scaling in the range of 0 to 16383. Also, the "Control Variable" calculated by the PID block is in the range of 0 to 16383 which corresponds to the range of an analog output channel configured for PID scaling.

The PID structure must be an element of a PD data file. For the "PID File" address defined in Figure 10.56, the individual fields of the PD structure that are useful for programmatic control of the PID loop are addressed as follows:

Reference	Address Reference	Description
FIC101.SP	PD19:0.SP	Setpoint
FIC101/AM	PD19:0/AM	Manual mode: **on** for manual, **off** for automatic
FIC101.CV	PD19:0.CV	CV, in range of 0 to 16383..
FIC101.FF	PD19:0.FF	Feedforward bias term added to output (-16383 to 16383).
FIC101.CVP	PD19:0.CVP	CV, in percent (0 to 100)

The PID block offers one manual mode. When the /AM bit is set, other logic in the processor can move a value into the .CV field. The /AM bit is intended to be set from an operator manual station and the .CV value is set to the analog input readback of the operator-set control variable, as shown in Figure 10.49. In Figure 10.56, FY101_MD is **on** when the controller should be in automatic mode. The transfer of the mode from the operator station to the PID block is on the second rung of Figure 10.56 and the transfer of the analog readback is also on this rung. If the particular loop does not have an operator manual station, then the /AM bit alone can be used to programmatically control the controller mode. To implement a feedforward correction, copy the feedforward value to the .FF field.

Example PID configuration is shown in Figure 10.56. The tuning parameters are those for the PI controller of Example 10.4. The configuration in the PD structure includes:

Derivative on PV/error
Reverse/direct action
Loop update time

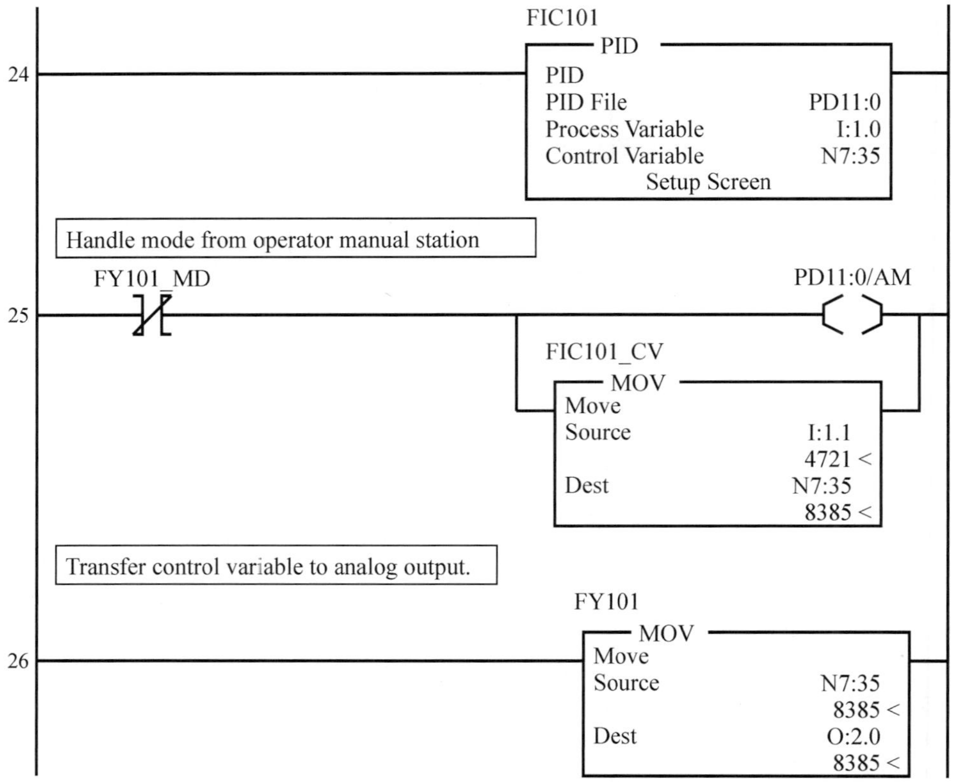

PID - Rung #2:24 - PD11:0

Controller Gain, Kc: 1.2
Reset Term, Ti: 2.7
Rate Term, Td: 0.00
Loop Update Time: 1.00
Control Mode: E = SP - PV
PID Control: Manual
Time Mode: Timed
Output Limiting: Yes
Reset and Gain (RG): 0(No)

Setpoint: 120
Maximum Setpoint Limit: 200
Minimum Setpoint Limit: 0
Control Variable Percent (CV%): 60
Output Max CV(%): 100
Output Min CV(%): 0
Derivative Action (DA): 0 (No)
Deadband: 0

Figure 10.56. Example MicroLogix PID loop.

Upper and lower bounds of the PV
Upper and lower limits of the CV calculation
K_P, T_I, and T_D

The actual algorithms used to calculate the CV are very similar to the PID algorithm equations in section 10.3.2. The positional form of the algorithm with dependent gains is used. The tuning parameters are K_P, T_I, and T_D and time is measured in minutes.

Regardless of the method used to trigger the PID block (timed or STI program), the loop update time is specified as a separate configuration parameter. The loop update time must match the actual sample period or the algorithm calculation will be incorrect.

The MicroLogix does not have ladder logic function blocks to handle signal conditioning, setpoint management, or output processing. The available arithmetic function blocks must be used to provide these functions.

The PID function block for the SLC-500 is slightly different from the MicroLogix. An example SLC-500 PID block is shown in Figure 10.57. The "PID File" field in the MicroLogix PLC is named "Control Block" for the SLC-500 and contains the starting address of a 23-element integer block (for example, N10:0). In addition, there is another field called the "Control Block Length," fixed at 23 words. To access the various parts of the PID control block, one must use the integer and/or bit address; there are not pre-defined fields as for the MicroLogix PD structure.

The individual fields of the control block that are useful for programmatic control of the PID loop are addressed as follows:

MicroLogix Reference	SLC-500 Address	Description
FIC101.SP	N10:2	Setpoint
FIC101/AM	N10:0/1	Manual mode: on for manual, off for automatic
FIC101.CV	not avail.	CV, in range of 0 to 16383.
FIC101.FF	N10:6	Feedforward bias term added to output (-16383 to 16383).
FIC101.CVP	N10:16	CV, in percent (0 to 100)

Note that one cannot directly transfer the analog input readback of the operator-set control variable (Figure 10.49) to a value in the PID control block. However, if the CV from the PID block is placed in an integer location, one can overwrite this integer location with the operator-set CV when the controller is in manual mode. The PID block monitors this CV change and will implement bumpless transfer when the mode is changed from manual mode to automatic mode. As shown in Figure 10.57, the PID block CV is stored in N7:35 and this value is overwritten with the operator-set CV when in manual mode. In rung 26, the CV is transferred to the analog output channel.

10.7 CHAPTER SUMMARY

This chapter describes PID control in the context of a PLC implementation. After introducing general features of feedback control, the PID control algorithm is described from an intuitive approach. Operational aspects of PID control are also delineated. The most popular methods of calculating the PID tuning parameters are described and illustrated. Operational aspects of PID control are described and then followed by a description of the PID function blocks for the PLCs covered by this text.

REFERENCES

Cohen, G. H. and G. A. Coon, 1953. "Theoretical Considerations of Retarded Control," *Trans. of the ASME*, **75** (7), 827-834.

Dorf, R.C. and R. H. Bishop, 2017. *Modern Control Systems*, 13th ed., Pearson, New York.

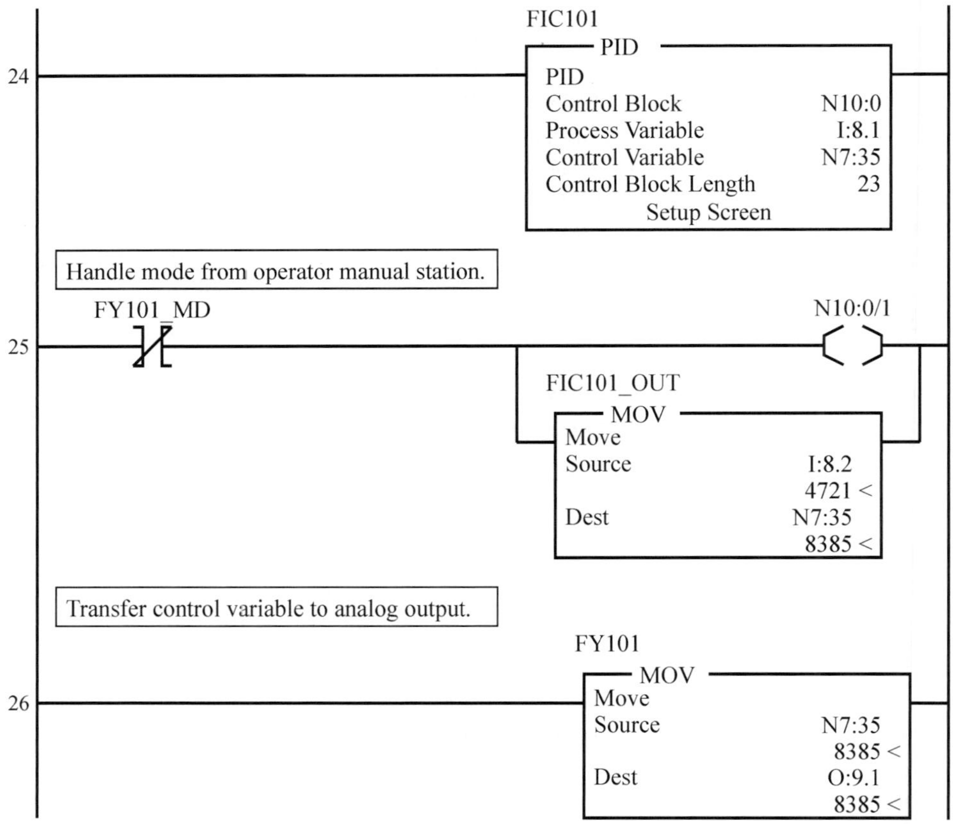

PID - Rung #2:24 - N10:0

Controller Gain, Kc: 1.2
Reset Term, Ti: 2.7
Rate Term, Td: 0.00
Loop Update Time: 1.00
Control Mode: E = SP - PV
PID Control: Manual
Time Mode: Timed
Output Limiting: Yes
Reset and Gain (RG): 0 (No)

Setpoint: 120
Maximum Setpoint Limit: 200
Minimum Setpoint Limit: 0
Control Variable Percent (CV%): 60
Output Max CV(%): 100
Output Min CV(%): 0
Derivative Action (DA): 0 (No)
Deadband: 0

Figure 10.57. Example SLC-500 PID loop.

Erickson, Kelvin T., 2023. *Programmable Logic Controllers: An Emphasis on Design and Application,* 4th Edition, Dogwood Valley Press, Rolla, MO.

Erickson, K. T. and J. L. Hedrick, 1999. *Plantwide Process Control*, John Wiley & Sons, New York.

Fertik, H. A., 1975. "Tuning Controllers for Noisy Processes," *ISA Trans.*, **14** (4), 292-304.

Marlin, T. E, 2000. *Process Control: Designing Processes and Control Systems for Dynamic Performance*, 2nd Ed., McGraw-Hill, New York.

Phillips, C.L. and R. D. Harbor, 1999. *Feedback Control Systems*, 4th ed., Prentice Hall, New York.

Riggs, J. B, 1999. *Chemical Process Control*, Ferret Pub., Lubbock, TX.

Rockwell Automation, 2008. *SLC Instruction Set Reference Manual*, pub. 1747-RM001G-EN-P, Rockwell Automation Inc., Milwaukee, WI, Nov.

Rockwell Automation, 2011. *MicroLogix 1100 Programmable Controllers Instruction Set Reference Manual*, pub. 1763-RM001D-EN-P, Rockwell Automation Inc., Milwaukee, WI, Sep.

Rockwell Automation, 2014. *MicroLogix™ 1200 and MicroLogix 1500 Programmable Controllers Instruction Set Reference Manual*, pub. 1762-RM001H-EN-P, Rockwell Automation Inc., Milwaukee, WI, Jul.

Rockwell Automation, 2019. *MicroLogix 1400 Programmable Controllers Instruction Set Reference Manual*, pub. 1766-RM001I-EN-P, Rockwell Automation, Inc., Milwaukee, WI, June.

Rockwell Automation,2020. *Logix5000 Controllers Process Control and Drives and Phase and Sequence Instruction Reference Manual*, pub. 1756-RM006L-EN-P, Rockwell Automation Inc., Milwaukee, WI, Sep.

Rockwell Automation, 2022a. *Logix5000 Controllers General Instructions Reference Manual*, pub. 1756-RM003W-EN-P, Rockwell Automation Inc., Milwaukee, WI, Mar.

Rockwell Automation, 2022b. *Micro800 Programmable Controllers General Instructions Reference Manual*, pub. 2080-RM001K-EN-P, Rockwell Automation Inc., Milwaukee, WI, Mar.

Ziegler, J. G. and N. B. Nichols, 1942. "Optimum Settings for Automatic Controllers," *ASME Trans.*, **64**, 759-768.

PROBLEMS

> **General instructions for the tuning problems**:
>
> The controller tuning problems assume that you have access to a PID tuning demonstrator program, like the "PIDTune" program provided by the author. You can also use SIMULINK to construct a process simulation and generate graphs of the responses.

P10-1. Tune a PID controller for the following first-order process:

Gain = 1.1
Deadtime = 1.5 minutes
First time constant = 3.5 minutes

(a) Find the PI and PID controller parameters using the Ziegler-Nichols closed-loop method. Attach the plot of the sustained oscillations used to determine the controller parameters. Also, attach the plot of the closed-loop setpoint response of the system with a PI controller whose parameters have been determined with the Ziegler-Nichols closed-loop method. Use a setpoint change of plus or minus 5.

(b) Find the PI and PID controller parameters using the four open-loop methods (Ziegler-Nichols, Cohen-Coon, Cohen-Coon with constraints, and Fertik). For the Fertik method, tune for a **setpoint** response. Attach the plot of the closed-loop setpoint response of the system with a PI controller tuned with Fertik's method.

P10-2. Tune a PID controller for the following first-order process:

Gain = 1.05
Deadtime = 1.3 minutes
First time constant = 4.2 minutes

(a) Find the PI and PID controller parameters using the Ziegler-Nichols closed-loop method. Attach the plot of the sustained oscillations used to determine the controller parameters. Also, attach the plot of the closed-loop setpoint response of the system with a PI controller whose parameters have been determined with the Ziegler-Nichols closed-loop method. Use a setpoint change of plus or minus 5.

(b) Find the PI and PID controller parameters using the four open-loop methods (Ziegler-Nichols, Cohen-Coon, Cohen-Coon with constraints, and Fertik). For the Fertik method, tune for a **setpoint** response. Attach the plot of the closed-loop setpoint response of the system with a PI controller tuned with Fertik's method.

P10-3. Tune a PID controller for the following first-order process:

Gain = 1.0
Deadtime = 3.0 minutes
First time constant = 2.0 minutes

(a) Find the PI and PID controller parameters using the Ziegler-Nichols closed-loop method. Attach the plot of the sustained oscillations used to determine the controller parameters. Also, attach the plot of the closed-loop setpoint response of the system with a PI controller whose parameters have been determined with the Ziegler-Nichols closed-loop method. Use a setpoint change of plus or minus 5.

(b) Find the PI and PID controller parameters using the four open-loop methods (Ziegler-Nichols, Cohen-Coon, Cohen-Coon with constraints, and Fertik). For the Fertik method, tune for a **setpoint** response. Attach the plot of the closed-loop setpoint response of the system with a PI controller tuned with Fertik's method.

P10-4. Tune a PID controller for the following first-order process:

Gain = 1.4
Deadtime = 1.2 minutes
First time constant = 2.5 minutes

(a) Find the PI and PID controller parameters using the Ziegler-Nichols closed-loop method. Attach the plot of the sustained oscillations used to determine the controller parameters. Also, attach the plot of the closed-loop setpoint response of the system with a PI controller whose parameters have been determined with the Ziegler-Nichols closed-loop method. Use a setpoint change of plus or minus 5.

(b) Find the PI and PID controller parameters using the four open-loop methods (Ziegler-Nichols, Cohen-Coon, Cohen-Coon with constraints, and Fertik). For the Fertik method, tune for a **setpoint** response. Attach the plot of the closed-loop setpoint response of the system with a PI controller tuned with Fertik's method.

P10-5. Tune a PID controller for the following first-order process:

Gain = 0.9
Deadtime = 4.5 minutes
First time constant = 5.5 minutes

(a) Find the PI and PID controller parameters using the Ziegler-Nichols closed-loop method. Attach the plot of the sustained oscillations used to determine the controller parameters. Also, attach the plot of the closed-loop setpoint response of the system with a PI controller whose parameters have been determined with the Ziegler-Nichols closed-loop method. Use a setpoint change of plus or minus 5.

(b) Find the PI and PID controller parameters using the four open-loop methods (Ziegler-Nichols, Cohen-Coon, Cohen-Coon with constraints, and Fertik). For the Fertik method, tune for a **setpoint** response. Attach the plot of the closed-loop setpoint response of the system with a PI controller tuned with Fertik's method.

P10-6. Tune a PID controller for the following first-order process:

Gain = 1.3
Deadtime = 2.0 minutes
First time constant = 3.0 minutes

(a) Find the PI and PID controller parameters using the Ziegler-Nichols closed-loop method. Attach the plot of the sustained oscillations used to determine the controller parameters. Also, attach the plot of the closed-loop setpoint response of the system with a PI controller whose parameters have been determined with the Ziegler-Nichols closed-loop method. Use a setpoint change of plus or minus 5.

(b) Find the PI and PID controller parameters using the four open-loop methods (Ziegler-Nichols, Cohen-Coon, Cohen-Coon with constraints, and Fertik). For the Fertik method, tune for a **setpoint** response. Attach the plot of the closed-loop setpoint response of the system with a PI controller tuned with Fertik's method.

P10-7. Tune a PID controller for the following first-order process:

Gain = 1.1
Deadtime = 4.0 minutes
First time constant = 5.0 minutes

(a) Find the PI and PID controller parameters using the Ziegler-Nichols closed-loop method. Attach the plot of the sustained oscillations used to determine the controller parameters. Also, attach the plot of the closed-loop setpoint response of the system with a PI controller whose parameters have been determined with the Ziegler-Nichols closed-loop method. Use a setpoint change of plus or minus 5.

(b) Find the PI and PID controller parameters using the four open-loop methods (Ziegler-Nichols, Cohen-Coon, Cohen-Coon with constraints, and Fertik). For the Fertik method, tune for a **setpoint** response. Attach the plot of the closed-loop setpoint response of the system with a PI controller tuned with Fertik's method.

P10-8. Tune a PID controller for the following first-order process:

Gain = 1.1
Deadtime = 2.0 minutes
First time constant = 4.0 minutes

(a) Find the PI and PID controller parameters using the Ziegler-Nichols closed-loop method. Attach the plot of the sustained oscillations used to determine the controller parameters. Also, attach the plot of the closed-loop setpoint response of the system with a PI controller whose parameters have been determined with the Ziegler-Nichols closed-loop method. Use a setpoint change of plus or minus 5.

(b) Find the PI and PID controller parameters using the four open-loop methods (Ziegler-Nichols, Cohen-Coon, Cohen-Coon with constraints, and Fertik). For the Fertik method, tune for a **setpoint** response. Attach the plot of the closed-loop setpoint response of the system with a PI controller tuned with Fertik's method.

P10-9. Tune a PID controller for the following second-order process:

Gain = 1.2
Deadtime = 2.0 minutes
First time constant = 4.0 minutes
Second time constant = 4.5 minutes

(a) Find the PI controller parameters using the Ziegler-Nichols closed-loop method. Attach the plot of the sustained oscillations used to determine the controller parameters. Also, attach the plot of the closed-loop response of the system with a PI controller to a disturbance change of plus or minus 10.

(b) Find the approximate first-order model of the process. Attach the open-loop response of the system used to obtain the approximate model.

(c) Using your first-order approximate model, find the PI controller parameters using the four open-loop methods (Ziegler-Nichols, Cohen-Coon, Cohen-Coon with constraints, and Fertik). For the Fertik method, tune for a **disturbance** response. Attach the plot of the closed-loop disturbance response (a disturbance change of plus or minus 10) of the system with a PI controller tuned with Fertik's method. **DO NOT set the process to the approximate first-order model.** The simulated process should still be the second-order process given above.

P10-10. Tune a PID controller for the following second-order process:

Gain = 1.3
Deadtime = 1.5 minutes
First time constant = 3.0 minutes
Second time constant = 5.0 minutes

(a) Find the PI controller parameters using the Ziegler-Nichols closed-loop method. Attach the plot of the sustained oscillations used to determine the controller parameters. Also, attach the plot of the closed-loop response of the system with a PI controller to a disturbance change of plus or minus 10.

(b) Find the approximate first-order model of the process. Attach the open-loop response of the system used to obtain the approximate model.

(c) Using your first-order approximate model, find the PI controller parameters using the four open-loop methods (Ziegler-Nichols, Cohen-Coon, Cohen-Coon with constraints, and Fertik). For the Fertik method, tune for a **disturbance** response. Attach the plot of the closed-loop disturbance response (a disturbance change of plus or minus 10) of the system with a PI controller tuned with Fertik's method. **DO NOT set the process to the approximate first-order model.** The simulated process should still be the second-order process given above.

P10-11. Tune a PID controller for the following second-order process:

Gain = 1.07
Deadtime = 2.0 minutes
First time constant = 3.5 minutes
Second time constant = 4.5 minutes

(a) Find the PI controller parameters using the Ziegler-Nichols closed-loop method. Attach the plot of the sustained oscillations used to determine the controller parameters. Also, attach the plot of the closed-loop response of the system with a PI controller to a disturbance change of plus or minus 10.

(b) Find the approximate first-order model of the process. Attach the open-loop response of the system used to obtain the approximate model.

(c) Using your first-order approximate model, find the PI controller parameters using the four open-loop methods (Ziegler-Nichols, Cohen-Coon, Cohen-Coon with constraints, and Fertik). For the Fertik method, tune for a **disturbance** response. Attach the plot of the closed-loop disturbance response (a disturbance change of plus or minus 10) of the system with a PI controller tuned with Fertik's method. **DO NOT set the process to the approximate first-order model.** The simulated process should still be the second-order process given above.

P10-12. Tune a PID controller for the following second-order process:

Gain = 1.4
Deadtime = 2.0 minutes
First time constant = 3.0 minutes
Second time constant = 4.0 minutes

(a) Find the PI controller parameters using the Ziegler-Nichols closed-loop method. Attach the plot of the sustained oscillations used to determine the controller parameters. Also, attach the plot of the closed-loop response of the system with a PI controller to a disturbance change of plus or minus 10.

(b) Find the approximate first-order model of the process. Attach the open-loop response of the system used to obtain the approximate model.

(c) Using your first-order approximate model, find the PI controller parameters using the four open-loop methods (Ziegler-Nichols, Cohen-Coon, Cohen-Coon with constraints, and Fertik). For the Fertik method, tune for a **disturbance** response. Attach the plot of the closed-loop disturbance response (a disturbance change of plus or minus 10) of the system with a PI controller tuned with Fertik's method. **DO NOT set the process to the approximate first-order model.** The simulated process should still be the second-order process given above.

P10-13. Tune a PID controller for the following second-order process:

Gain = 1.5
Deadtime = 1.0 minutes
First time constant = 2.0 minutes
Second time constant = 2.5 minutes

(a) Find the PI controller parameters using the Ziegler-Nichols closed-loop method. Attach the plot of the sustained oscillations used to determine the controller parameters. Also, attach the plot of the closed-loop response of the system with a PI controller to a disturbance change of plus or minus 10.

(b) Find the approximate first-order model of the process. Attach the open-loop response of the system used to obtain the approximate model.

(c) Using your first-order approximate model, find the PI controller parameters using the four open-loop methods (Ziegler-Nichols, Cohen-Coon, Cohen-Coon with constraints, and Fertik). For the Fertik method, tune for a **disturbance** response. Attach the plot of the closed-loop disturbance response (a disturbance change of plus or minus 10) of the system with a PI controller tuned with Fertik's method. **DO NOT set the process to the approximate first-order model.** The simulated process should still be the second-order process given above.

P10-14 Tune a PID controller for the following second-order process:

Gain = 0.9
Deadtime = 2.0 minutes
First time constant = 3.5 minutes
Second time constant = 4.0 minutes

(a) Find the PI controller parameters using the Ziegler-Nichols closed-loop method. Attach the plot of the sustained oscillations used to determine the controller parameters. Also, attach the plot of the closed-loop response of the system with a PI controller to a disturbance change of plus or minus 10.

(b) Find the approximate first-order model of the process. Attach the open-loop response of the system used to obtain the approximate model.

(c) Using your first-order approximate model, find the PI controller parameters using the four open-loop methods (Ziegler-Nichols, Cohen-Coon, Cohen-Coon with constraints, and Fertik). For the Fertik method, tune for a **disturbance** response. Attach the plot of the closed-loop disturbance response (a disturbance change of plus or minus 10) of the system with a PI controller tuned with Fertik's method. **DO NOT set the process to the approximate first-order model.** The simulated process should still be the second-order process given above.

11 Sensors and Actuators

Chapter Topics:

- Discrete and analog sensors
- Discrete and analog actuators

OBJECTIVES

Upon completion of this chapter, you will be able to understand:

- The types of sensors and their applications
- The types of actuators

Scenario 1: Using photoelectric proximity sensors to detect silicon wafer cassettes. A particular application processes "cassettes" of 150 mm silicon wafers in a series of chemical baths. Each cassette holds up to 20 wafers and two cassettes are processed simultaneously. A gantry is used to move the cassettes from the loading station to the first of a series of chemical baths. Figure 11.1 shows the major steps to extract the cassettes from a bath. In this figure, the second cassette is behind the cassette shown. The cassettes remain in a chemical bath for a certain amount of time and then they are moved to the next bath. Eventually, they are moved from the last chemical bath to an unloading station, where they are removed by an operator.

As part of the failure handling, the system needs to verify that cassettes are properly being removed from a bath. There is a reliability problem with this type of end effecter. Occasionally, it will not pick up one or both of the cassettes. A similar system using this type of end effecter has two sensors in each bath. A sensor beneath each cassette position is used to verify that cassettes are being picked up. However, this method cannot be used in this system because of the chemicals in the baths. So, diffuse reflective photoelectric proximity sensors are mounted on one of the end effecter paddles (Figure 11.1*d*). However, this application has significant background reflections from the liquid in the baths and from the white plastic beneath the cassettes in the loading and unloading stations. Before mounting the photoelectric sensors, you test a sensor and find that it can be adjusted to be insensitive to the liquid background, as long as the sensor is only checked when the gantry is fully up. It is still sensitive to the white plastic bench top at the loading and unloading stations. However, placing a gray pad at the loading and unloading stations solved that problem. Encouraged, both sensors are mounted and adjusted. However, when operating the system, you note that it now appears to be sensitive to the liquid background when the gantry is above a bath. The sensor indicates a cassette in the end effecter when there is no cassette.

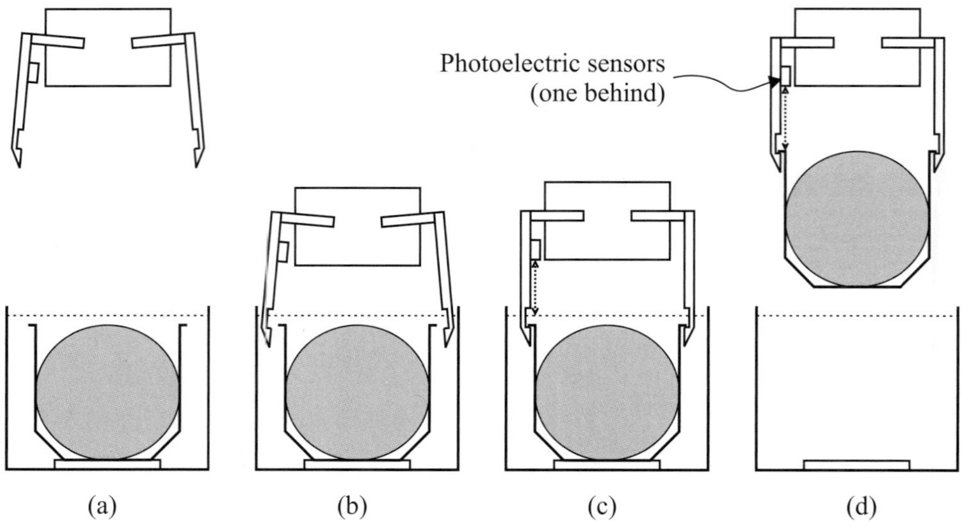

Figure 11.1. Gantry picking up cassettes: (a) up with clamp open; (b) down, ready to clamp cassettes; (c) cassettes clamped; (d) up with cassettes.

Solution: Obviously, a different sensor must be used. After researching diffuse photoelectric sensors, you discover that there are certain diffuse reflective photoelectric sensors that suppress background reflections with triangulation. You purchase and install these sensors. These sensors work reliably. The presence/absence of a cassette must still be checked when the gantry is in the up position since the cassette must be out of the bath for the sensor to detect the cassette.

Scenario 2: Thermocouple measurements in an aluminum heat-treating furnace. Heating an aluminum plate to a temperature just below the melting temperature in order to alter the alloy grain structure is a common operation in aluminum production. When examining the aluminum plates in a certain batch, the metal grain structure is not correct, indicating a problem with the furnace temperature.

Solution: Obviously, something is wrong with the furnace temperature control. Examining the temperature logs for that batch of aluminum plates processed by the furnace reveals no anomalies. The furnace zone temperatures are monitored by thermocouples mounted in the furnace ducts that supply the heated air. To verify the operation of the furnace temperature control, a furnace survey is performed. Thermocouples mounted on a steel plate are placed inside the furnace and these thermocouples monitor the temperatures in the various parts of the furnace during a normal heat-treating operation. The actual temperatures, as determined by the thermocouples mounted on the steel plate, are about 10 °F higher than the temperatures reported by the control system. Investigating the thermocouple measurement system, you find that there is no cold-junction compensation, but all of the thermocouple junctions with the thermocouple input modules are at an isothermal block. The temperature of the isothermal block is about 40 °F, and it should be 32 °F. It is then that someone notices that the temperature of the equipment room where the thermocouple input modules are located is about 10 °F higher than normal. It is normally around 60 °F and someone had set the thermostat to 72 °F, which is more amenable to working in the room. The control signal for the isothermal block cooling control is at its maximum, indicating that it is not able to maintain the isothermal block at its proper temperature of 32 °F. The equipment room thermostat is changed to 60 °F. After about an hour, the isothermal block temperature is back at 32 °F. A subsequent survey of the furnace reveals no temperature measurement problems. A locking cover is placed over the room thermostat to prevent this problem in the future.

11.1 INTRODUCTION

In order for the PLC to be useful, it must be connected to the process. Sensors are inputs to the PLC and actuators are the outputs that manipulate the process. A typical PLC system has connections to both discrete and analog devices. Discrete and analog sensors are described first since there are a large number of possible sensors. Lastly, discrete and analog actuators are explained.

11.2 DISCRETE SENSORS

The output of a discrete sensor has two states: off/on, open/closed, or 0/1. The most common application of a discrete sensor is to sense the presence or absence of an object. Limit switches and proximity sensors fulfill this role. Many motion control applications use encoders to sense the position and/or velocity of an object. Other discrete sensors detect the value of a physical quantity such as a high temperature or a low pressure. Although all of these types of sensors are described, proximity sensors are emphasized due to their wide variety of implementation technologies.

11.2.1 Limit switches

Limit switches are generally contact sensors; in other words, the object to be sensed comes in contact with the sensor and moves a lever arm. Two example limit switches are shown in Figure 11.2. The roller lever actuator shown in Figure 11.2*a* is often used to detect large objects that are widely spaced, for example, boxes moving down a conveyor belt. This

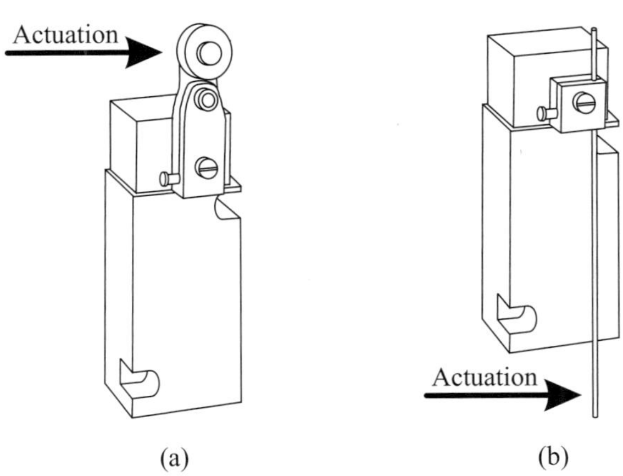

Figure 11.2. Example limit switches: *(a)* roller lever-actuated; *(b)* gravity return.

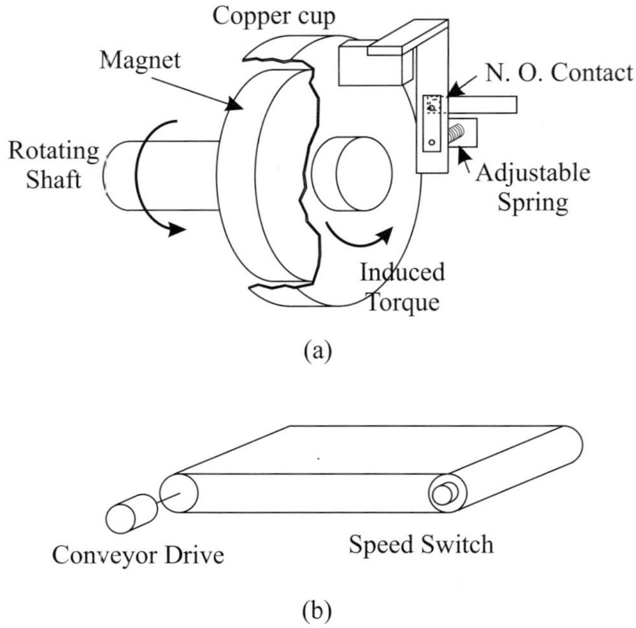

Figure 11.3. Speed switch: *(a)* mechanism; *(b)* typical placement.

type of limit switch is mounted so that the object pushes the lever right (or left) to change the contact. The gravity return actuation in Figure 11.2*b* is often used to detect light objects moving down a conveyor. This type of limit switch is mounted above the conveyor.

One special form of a limit switch is a speed switch. A speed switch detects the speed or direction of a rotating shaft. When mounted on the non-driven shaft of a conveyor belt, the speed switch detects both a broken belt or a broken drive shaft. The magnetic induction type of operating mechanism is illustrated in Figure 6.3*a*. When the shaft rotates, the rotating magnet induces eddy currents in the copper cup. These eddy currents produce a torque proportional to the shaft speed. The cup rotates a lever that operates a contact. As the shaft speed increases, the torque overcomes an opposing spring force and the contact closes. The contact remains closed as the shaft speed increases to its normal value. When the shaft speed decreases, a speed is reached where the torque produced by the eddy current no longer overcomes the spring force and the contact opens. Speed switches can alternatively use a mechanical or hydraulic coupling to drive the switch contact.

11.2.2 Proximity sensors

Proximity switches are non-contact sensors. The object to be sensed does not need to come in contact with the sensor, though it may need to be close (depending on the type of proximity switch). Proximity sensors are based on electromagnetic field, electric field, ultrasonic, or photoelectric technologies.

The major types of proximity sensors are summarized in Table 11.1. For each one, the sensed materials, the typical distance, and problem environments are listed. Note that some of the technologies can be used for distance measurement.

Table 11.1 Comparison of Discrete Proximity Sensors

Sensor Type	Sensed Materials	Maximum Distance	Measure Distance?	Problem Environments
Inductive	Metal	15 mm	yes, nonlinear	
Capacitive	Metal, Nonmetal	30 mm	yes, nonlinear	dirty, wet
Hall effect	Magnetic	3 mm	no	dirty, wet
Reed switch	Magnetic	60 mm	no	
Ultrasonic	Any	10 m	yes	
Photoelectric	Any	0.3m-100 m[1]	yes	dusty

Notes:

1. Depends on sensing mode (through-beam, retro-reflective, diffuse)

Inductive

Inductive proximity switches detect ferrous and nonferrous metallic objects. They operate on the principle that metals interact with an electromagnetic field. Inductive proximity switches are either unshielded (Figure 11.4*a*) or shielded (Figure 11.4*b*). Shielded sensors are used when the sensor is mounted flush in surrounding metal without causing the mount to trigger the sensor. The electromagnetic field is generated by a coil of wire on a ferrite core. The coil is part of an oscillator circuit.

The operation of an inductive proximity switch is shown in Figure 11.5. As a metal object enters the electromagnetic field, eddy currents are induced in the metal object. These eddy currents absorb some of the electromagnetic energy, causing the oscillator amplitude

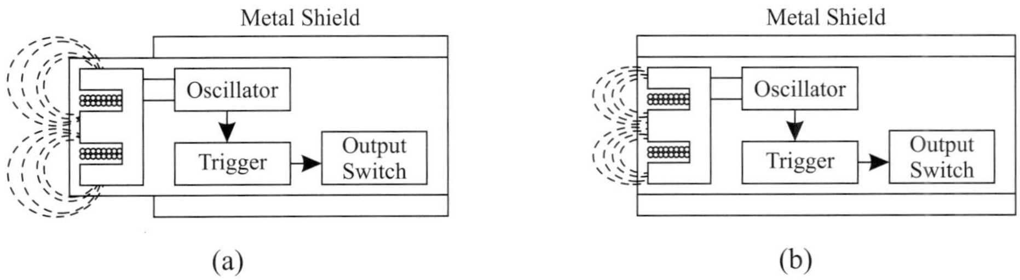

Figure 11.4. Inductive proximity switch block diagram: *(a)* unshielded; *(b)* shielded.

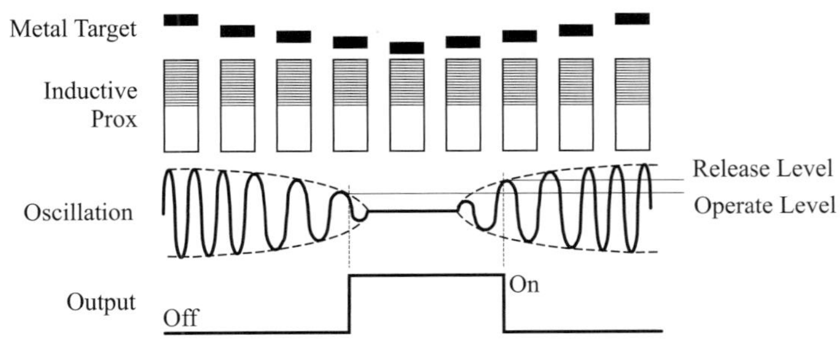

Figure 11.5. Inductive proximity switch operation.

Table 11.2 Inductive Proximity Correction Factors

Target Material	Correction Factor
Mild Steel	1.0
Stainless Steel	0.85
Brass	0.50
Aluminum	0.45
Copper	0.40

to decrease. A trigger circuit senses the decreased oscillation amplitude and changes the state of the output to **on**. When the oscillation amplitude then increases beyond the release level, the state of the output is changed to **off**. Note that the operate and release levels are different, called the hysteresis. Hysteresis helps prevent "chattering" (output rapidly turning **on** and **off**) when the sensor is vibrated or when the object is stationary at the nominal sensing distance.

The sensing distance is a function of the object size and material, the area of the electromagnetic field, which is determined by the face area of the sensor, and whether the sensor is shielded or unshielded. For example, an unshielded cylindrical inductive proximity switch with an 18 mm diameter from a certain vendor has a nominal sensing distance of 8 mm for a mild steel object that is 18 mm square. If this sensor is shielded, the nominal sensing distance drops to 5 mm. If the unshielded sensor has a diameter of 12 mm, the nominal sensing distance decreases to 4 mm and if the unshielded sensor has a diameter of 30 mm, the nominal sensing distance increases to 15 mm. To determine the sensing distance for materials other than mild steel, a correction factor must be used to determine the actual sensing range:

$$\text{Sensing Range} = (\text{Nominal Sensing Range})\times(\text{Correction Factor})$$

A sample table of correction factors is shown in Table 11.2.

There are special versions of inductive proximity switches for specialized applications. Ferrous-selective proximity switches only detect metals containing iron. Non-ferrous selective proximity switches only detect metals without iron (for example, copper, aluminum). Other inductive proximity switches are designed for welding environments and other applications where large magnetic fields are present.

The size and shape of the target object also affects the sensing distance. For example, a rounded target usually reduces the sensing distance. An object smaller than the sensor face area reduces the sensing distance and an object larger than the face area increases the sensing distance.

Inductive proximity switches are useful for detecting the piston position for pneumatic and hydraulic cylinders. Figure 11.6 shows an example application. The piston is magnetized and the cylinder walls are non-magnetic.

Capacitive

Capacitive proximity switches detect metallic and nonmetallic objects. They operate on the principle that materials interact with an electrostatic field. The electrostatic field is generated by a plate that acts as one electrode of a capacitor (Figure 11.7). The detected object acts as the other capacitor electrode. The capacitor plate is part of an oscillator circuit.

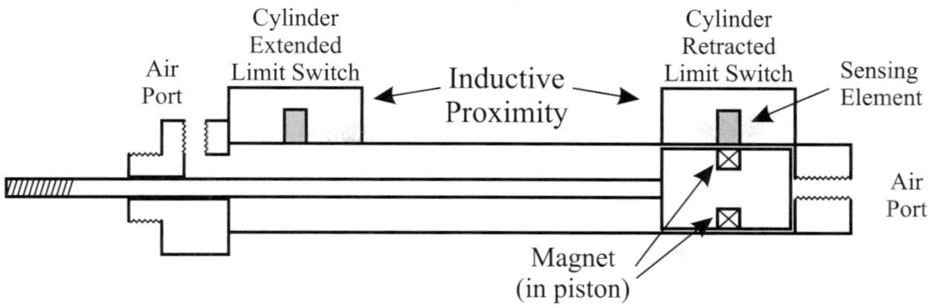

Figure 11.6. Inductive proximity sensor as cylinder piston limit switches.

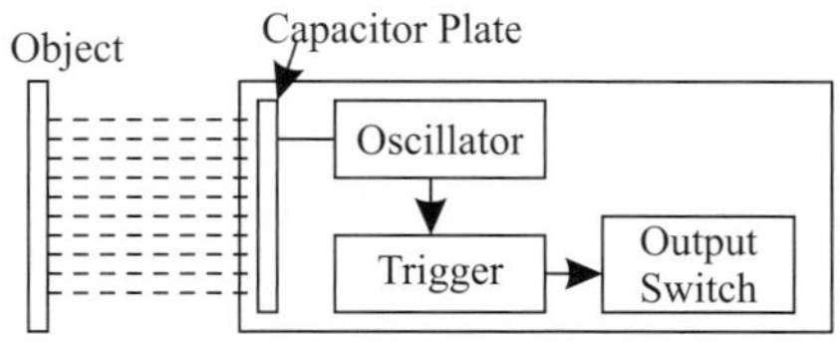

Figure 11.7. Capacitive proximity switch block diagram.

The operation of a capacitive proximity switch is shown in Figure 11.8. As an object enters the electrostatic field, the capacitance increases, causing the oscillator to start oscillating and the amplitude to increase. A trigger circuit senses the increased oscillation amplitude and changes the state of the output to **on** when the amplitude is at the operate level. When the oscillation amplitude then decreases below the release level, the state of the output is changed to **off**. Note that the operate and release levels are different, called the hysteresis. Hysteresis helps prevent "chattering" (output rapidly turning **on** and **off**) when the sensor is vibrated or when the object is stationary at the nominal sensing distance.

The sensing distance is a function of the object size and material; and the area of the electrostatic field, which is determined by the face area of the sensor. For example, a cylindrical capacitive proximity switch with an 18 mm diameter from a certain vendor has a nominal sensing distance of 1 mm for a glass object that is 18 mm square. If the object is polypropylene, the nominal sensing distance is 0.5 mm. If the sensor has a diameter of 12 mm, the nominal sensing distance decreases to 0.5 mm and if the sensor has a diameter of 30 mm, the nominal sensing distance increases to 2 mm. To determine the sensing distance for

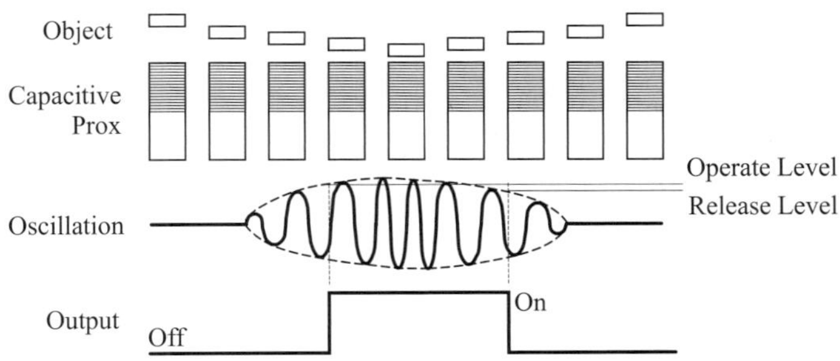

Figure 11.8. Capacitive proximity switch operation.

Table 11.3 Capacitive Proximity Correction Factors

Target Material	Correction Factor
Alcohol	0.85
Glass	0.20 - 0.55
Marble	0.50
Polypropylene	0.10
Wood, dry	0.10 - 0.40
Wood, wet	0.60 - 0.85

various materials, the nominal sensing distance is multiplied by a correction factor to determine the actual sensing range. Sample correction factors are shown in Table 11.3.

The size and shape of the target object also affects the sensing distance. For example, a rounded target usually reduces the sensing distance. An object smaller than the sensor face area reduces the sensing distance and an object larger than the face area increases the sensing distance.

Although capacitive proximity sensors are not sensitive to the target material magnetic properties, they are sensitive to the medium in the gap between the sensor and the target. Anything that affects the dielectric constant of the gap will change the proximity detection. Thus, capacitive proximity sensors are not recommended for applications that have excessive dust, water, or machining fluids.

Hall Effect

Hall effect sensors are used to detect magnetic materials. If a magnetic field is applied perpendicular to a current-carrying conductor, it generates a voltage perpendicular to the current flow. The sensor is manufactured as a plate of a semiconductor material because the Hall effect is more prominent in semiconductors. The operation of a Hall sensor is illustrated in Figure 11.9. When a DC current flows through the semiconductor plate, no transverse voltage is generated (Figure 11.9*a*). However, when a magnetic field is applied perpendicular to the plate current flow, the current flow curves toward one edge of the plate, generating a small voltage (Figure 11.9*b*). This small voltage is amplified, and then when above a threshold, the magnetic object is detected. As for the inductive and capacitive

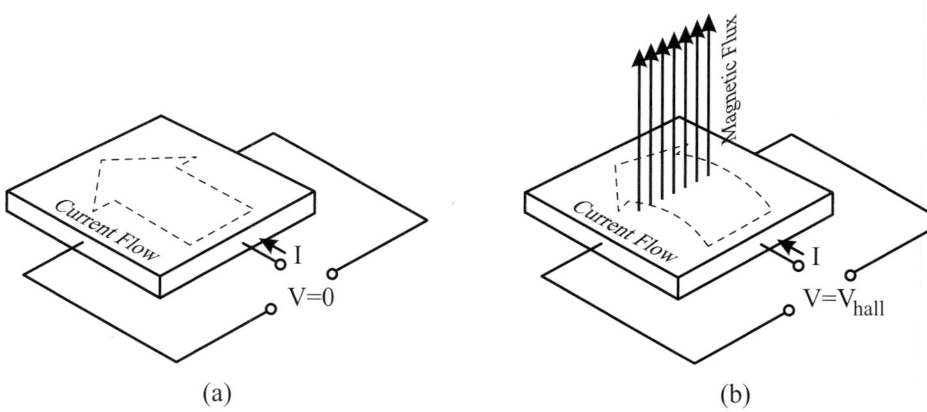

Figure 11.9. Hall effect principle: *(a)* no magnetic field; *(b)* magnetic field present.

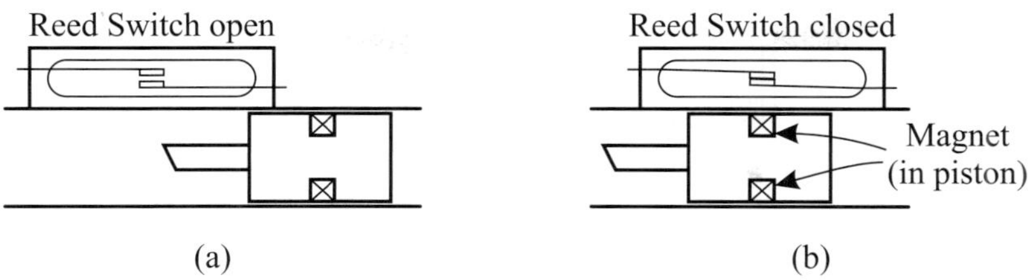

Figure 11.10. Magnetic reed switch operation: *(a)* no magnetic field; *(b)* magnetic field present.

proximity switches, Hall effect proximity switches have hysteresis. The magnetic field strength needed to turn on the switch is higher than the field strength required to sustain the on switch indication. Hall sensors are frequently used to detect the position of a cylinder piston, as for inductive proximity sensors (Figure 11.7).

Magnetic Reed Switch

As an alternative the Hall effect sensor, a magnetic reed switch can also be used to detect the position of a magnetic cylinder piston. Operation of a magnetic reed switch is illustrated in Figure 11.10. A reed switch contains hermetically-sealed contacts. When the piston magnet moves close to the switch, the contacts are magnetized and pull together (Figure 11.10*b*). When the magnet moves away from the contacts, they open. The advantage of the reed switch is that it can handle AC signals.

Ultrasonic

Ultrasonic transducers use sound waves to detect an object. They can be used to measure distance or proximity. The basic principle is to emit a sound wave "pulse" and then measure the time for the pulse to be detected by the receiver. There are two basic configurations as shown in Figure 11.11. The transmitter and receiver can be separate piezo-electric elements focused on the same point, Figure 11.11*a*. In this case, the operation is shown in Figure 11.12. The wave pulse is generated by the transmitter, reflected by the object and detected by the receiver. The alternative configuration of an ultrasonic transducer (Figure 11.11*b*) combines the transmitter and receiver. In this case, the transmitter and receiver do not need to be focused on the same point. The operation of the combined transmitter and receiver ultrasonic transducer is shown in Figure 11.13. The wave pulse is generated by the piezo-electric element, and then the element is switched to its

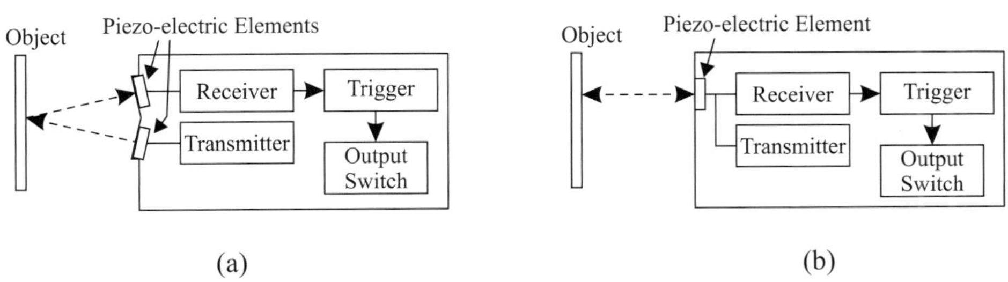

Figure 11.11. Ultrasonic proximity switch block diagram: *(a)* separate transmitter and receiver element; *(b)* combined transmitter/receiver element.

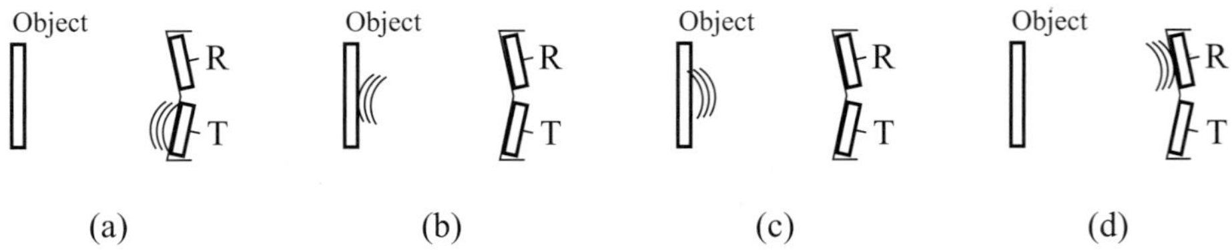

Figure 11.12. Operation of separate transmitter and receiver: *(a)* wave launched; *(b)* wave impinges object; *(c)* wave reflected; *(d)* wave impinges receiver.

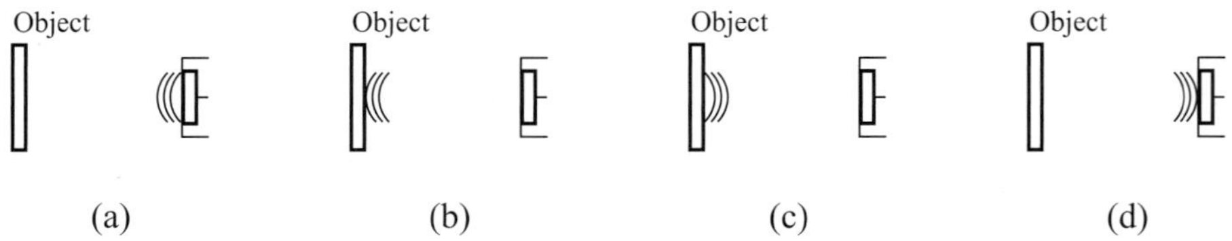

Figure 11.13. Operation of combined transmitter/receiver: *(a)* wave launched; *(b)* wave impinges object; *(c)* wave reflected; *(d)* wave impinges receiver.

receiving mode. The wave is reflected by the object and then detected by the element. In this type of ultrasonic transducer, there is a minimum sensing distance. The trailing edge of the pulse must leave the transducer before the leading edge of the reflected pulse can be detected. If the target is too close, the leading edge of the reflected wave pulse will be impinging on the transducer before it has completely transmitted the wave pulse.

Because ultrasonic sensors depend on receiving a reflected sound wave for proper operation, the target must be a good reflector of sound waves. Preferably, the target should be a smooth, flat surface. Rounded or uneven objects may be detected, but at a reduced maximum sensing distance. Soft materials, such as foam rubber and fabric are difficult to detect since they are not strong sound reflectors.

Ultrasonic proximity switches have a longer sensing range than inductive or capacitive proximity switches. Typical sensing ranges are from 30 cm to 100 cm.

Photoelectric Sensors

Most photoelectric sensors operate in the same basic manner, shown in Figure 11.14. An oscillator drives a light-emitting diode (LED) to generate a modulated light beam. The modulation frequency is typically several kilohertz. A phototransistor responds to this light

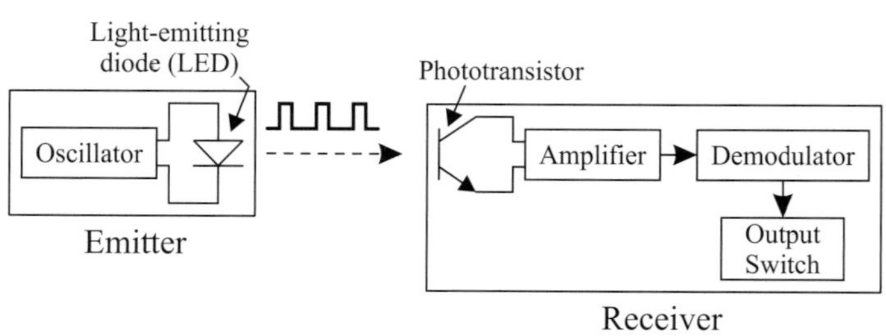

Figure 11.14. Photoelectric sensor block diagram.

beam, and its response is amplified. The demodulator essentially responds only to the modulated light beam, effectively blocking any ambient light. Phototransistors are more sensitive in the infrared or visible red range of light frequencies and so the emitter LED is typically in this range. Photoelectric sensors that differentiate between colors require a visible light source. For certain applications, a visible LED may be used and for others that sense a wide range of colors, an incandescent light source is required.

There are some applications where only ambient light is sensed. Hot metal or glass objects emit infrared light. When the infrared light is more intense than the surrounding light level, a receiver with a filter that only admits infrared light can be used. No emitter is required for these applications.

Photoelectric sensors operate in one of two ways: light-operate or dark-operate. A light-operate photoelectric sensor energizes its output when the received light intensity increases above a threshold. A dark-operate sensor energizes its output when the received light intensity decreases below a threshold.

The major difference between types of photoelectric sensors is the way the emitter and receiver are packaged. The major types are: through-beam, retroreflective, and diffuse reflective. Polarizing and convergent-beam photoelectric sensors are used for special applications.

Through-beam. In the through-beam configuration, the emitter and receiver are positioned opposite to each other so that the beam from the emitter is aimed directly at the receiver (Figure 11.15). An object is detected when it interrupts the beam. As long as the sensed object is opaque, the through-beam configuration is the most reliable method of using photoelectric sensors. The distance between the emitter and receiver can be more than 100 meters. Through-beam sensors are best for sensing non-metallic objects through heavy dirt, dust, moisture, and condensation.

There are few applications where the through-beam configuration is not the best. It is generally not recommended when detecting clear films or containers. If the object does not completely interrupt the beam, then apertures, lenses, or fiber optics may be used to shape the beam to match the part profile. If the part is small and passes at a predictable distance from the sensor, then a convergent-beam reflective sensor works better. If the emitter and receiver are close, an infrared beam tends to pass through thin opaque materials like paper, cloth, and plastics. In this case, the light beam is attenuated with apertures or the emitter and receiver are intentionally misaligned. Alternatively, the material may be opaque to a visible LED emitter.

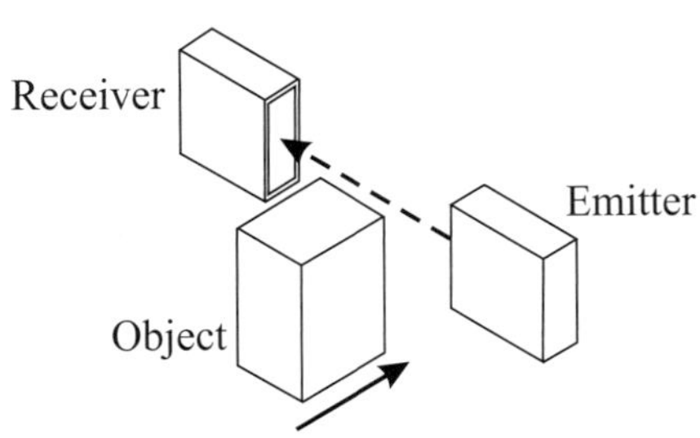

Figure 11.15. Through-beam photoelectric sensor operation.

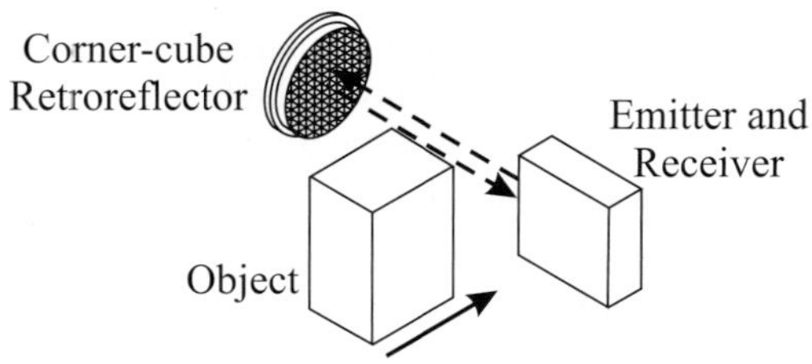

Figure 11.16. Retroreflective photoelectric sensor operation.

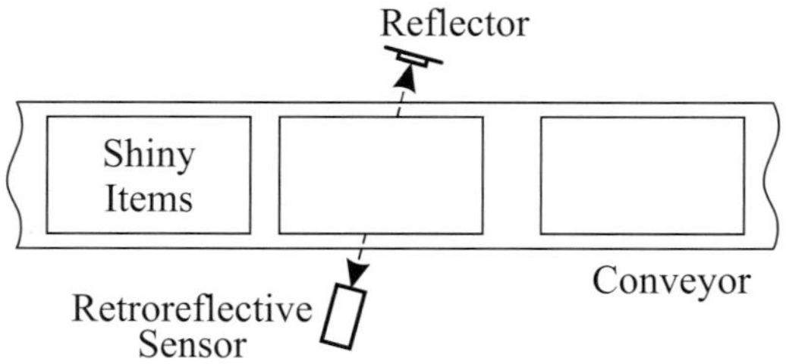

Figure 11.17. Skewing retroreflective photoelectric sensor.

Retro-reflective. In the retro-reflective configuration, the emitter and receiver are in the same package and a reflector is used to return light to the receiver (Figure 11.16). An object is detected when it interrupts the beam. The retro-reflective configuration is used instead of the through-beam configuration when sensing is possible from only one side. The reflector is generally constructed of corner-cube prisms (like a bicycle reflector) that reflect the light parallel to the incident beam. This configuration is most popular in conveyor application where the objects are large (for example, cartons) and where the environment is relatively clean. The sensing distance is typically less than 5 meters, significantly less than the distance for the through-beam configuration. The retro-reflective configuration does not work for those applications that are a challenge for the through-beam configuration.

If the sensed objects are shiny and flat (for example, a shrink-wrapped box), the sensor is skewed so that light reflected from the object does not reach the receiver (Figure 11.17). For shiny rounded objects, (for example, an aluminum can), a special type of retro-reflective sensor is used to ignore reflections from the object. A corner-cube prism reflector has the property that the retroreflected light waves are rotated 90° to the incident wave. Polarizing filters are placed in front of the emitter and receiver lenses. The two filters are oriented so that the polarization planes are right angles to each other (Figure 11.18). Light reflected by the shiny object is blocked by the receiver polarizing filter because it has not been rotated 90° by the retroreflector.

Diffuse reflective. The diffuse reflective configuration is similar to the retro-reflective configuration except that the sensed object reflects the light beam. An object is sensed when it reflects light to the receiver (Figure 11.19). Though this method of sensing is attractive since it has the fewest number of components, it may be challenging to make it sense reliably and consistently. To use the diffuse reflective photoelectric sensor, the reflective properties of the sensed object must be drastically different than the background. It will be a challenge to sense a shiny object in front of a reflective background. As a general rule, the

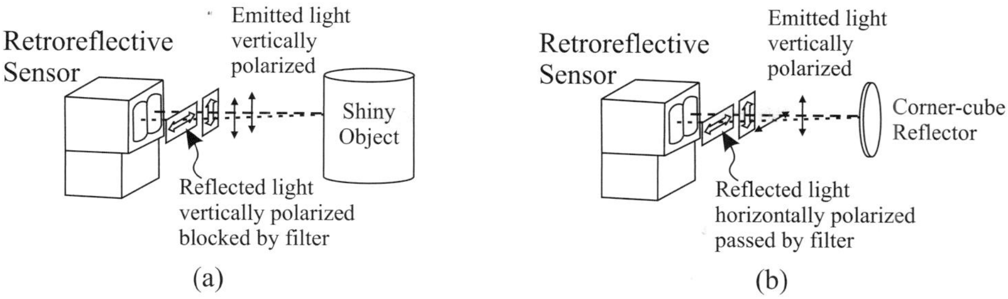

Figure 11.18. Polarized retroreflective photoelectric sensor: *(a)* shiny object in path; *(b)* no object in path.

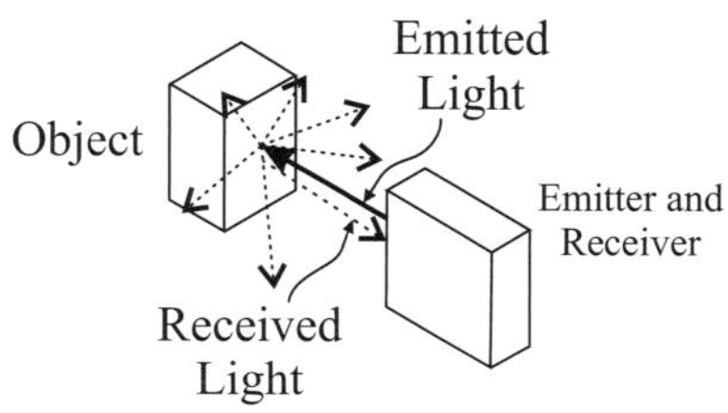

Figure 11.19. Diffuse reflective photoelectric sensor operation.

distance from the diffuse reflective sensor to the background should be at least four times the distance to the sensed object (Banner Engineering, 1994). This rule assumes that the background is less reflective than the sensed object. Any attempt to adjust the amplifier gain in order to suppress the background will not improve the situation. In this case, an alternative sensor configuration is the only viable alternative.

Diffuse reflective photoelectric sensors can suppress background reflections with triangulation. In this arrangement, there are two receivers, each aimed at a different target distance (Figure 11.20). Receiver R1 is aimed at the sensed target and R2 is aimed at the background. The focusing lens is adjusted for R1. When the reflected light intensity sensed by R1 is greater than the intensity sensed by R2, the target object is recognized. Triangulation ignores objects beyond the target.

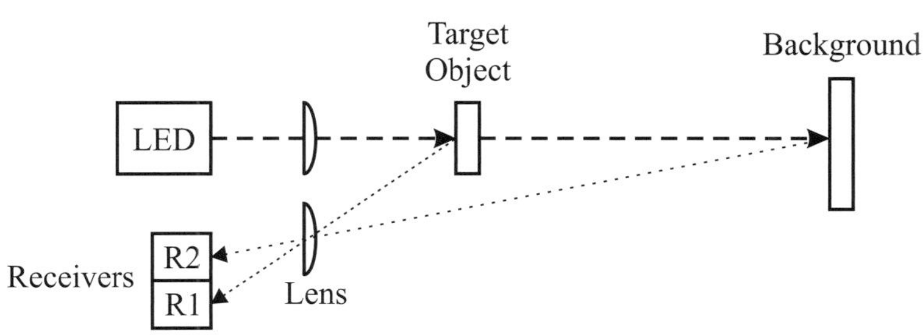

Figure 11.20. Diffuse reflective photoelectric sensor operation using triangulation to suppress background.

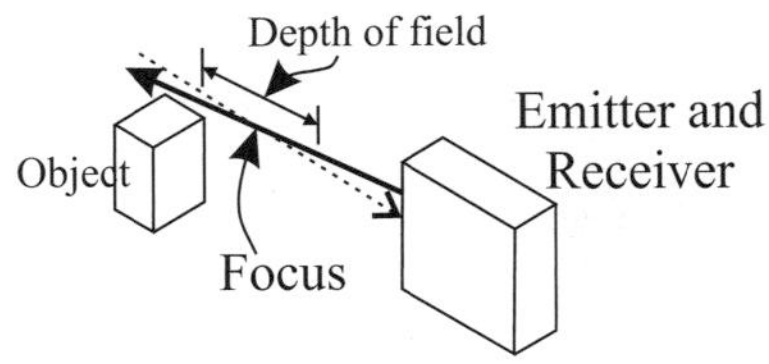

Figure 11.21. Convergent beam reflective photoelectric sensor operation.

Convergent-beam. A convergent-beam configuration uses lenses to focus both the emitter and receiver to the same point (Figure 11.21). This configuration produces a small, intense area at a fixed distance from the sensor lens. This configuration is useful for sensing small objects and for sensing objects with low reflectivity.

Registration marks. Special diffuse reflective photoelectric sensors detect registration marks on packages. The registration mark aids in aligning the material for printing or packaging. These sensors have a fixed focus and detect the registration color mark by sensing the gray-scale contrast between the mark and the background. Depending on the background and mark colors, the user typically selects either a red or green LED as the emitter.

Clear objects. Sensing clear materials, such as clear plastic films or clear glass bottles is a challenging application for photoelectric sensors. Photoelectric diffuse reflective sensors without lenses can detect clear materials at short distances (up to 50 mm). Without the lenses, the receiver is less dependent on receiving light reflected parallel to the emitted beam. One vendor uses coaxial optics and circular polarization to maximize the contrast between the object and the background.

Fiber optics. In certain situations, space may be too restrictive or the environment too hostile for the photoelectric sensor components. In these situations, fiber optic cables are used to couple the photoelectric sensor to the sensing situation (Figure 11.22). For the through-beam configuration, two separate fiber optic cables are used – one to transmit the

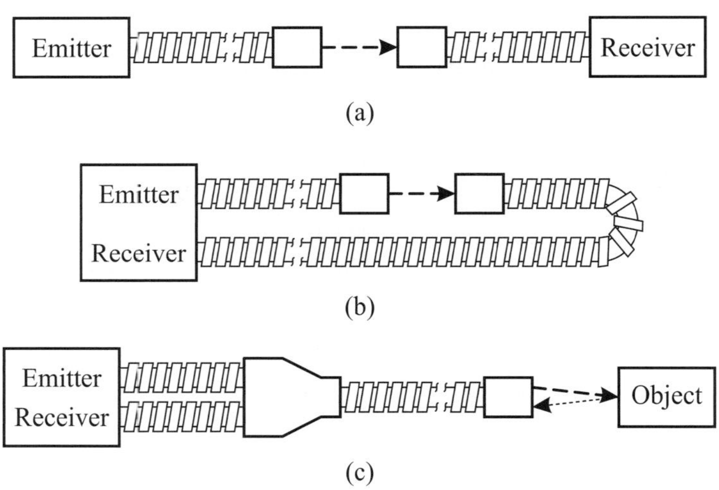

Figure 11.22. Photoelectric sensors with fiber optics: *(a)* individual emitter and receiver; *(b)* combined emitter and receiver; *(c)* bifurcated fiber for reflective operation.

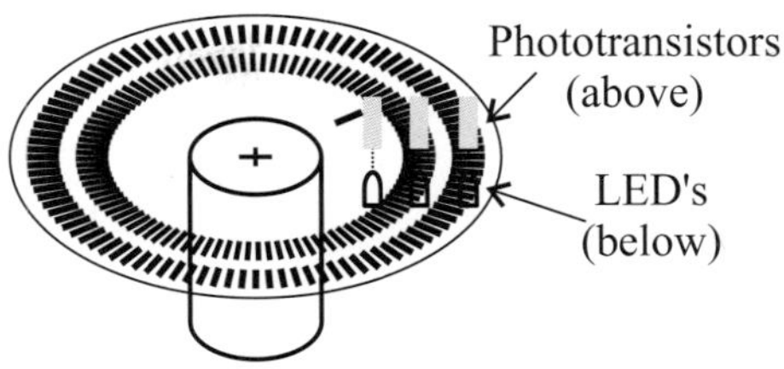

Figure 11.23. Simplified encoder.

emitted beam and the other to conduct the received beam (Figure 11.22*a, b*). For diffuse or retroreflective configurations, a single fiber optic cable is split into two cables, called a bifurcated cable (Figure 11.22*c*). Half of the fiber optic cable strands conduct the emitted light and the other half of the strands conduct the reflected light.

11.2.3 Encoders

Motion applications use encoders to determine the position and/or velocity of a rotary shaft. An encoder consists of a disk with lines in one or more bands or tracks (Figure 11.23). For each band there is a miniature through-beam photoelectric sensor that senses the lines passing as the disk rotates. The number of lines on the encoder disk determines the encoder resolution. Encoders with more than 1000 lines per track are common.

There are two basic types of encoders: incremental and absolute. Incremental encoders are more common since they require fewer disk tracks and fewer interface signals to the input module.

The two basic types of incremental encoder disks are shown in Figure 11.24. A single-track encoder has only one output signal that is a square wave (Figure 11.24*a*). The direction of travel cannot be determined with this type of encoder. The travel distance is determined by counting the pulses and the velocity is determined by measuring the frequency of the square wave signal. A two-phase quadrature encoder has three tracks (Figure 11.24*b*). The relative phasing of the A and B signals indicate the direction of encoder rotation. If the positive edge of A leads the positive edge of B, the encoder is rotating clockwise (Figure 11.25*a*). For counterclockwise rotation, the positive edge of A

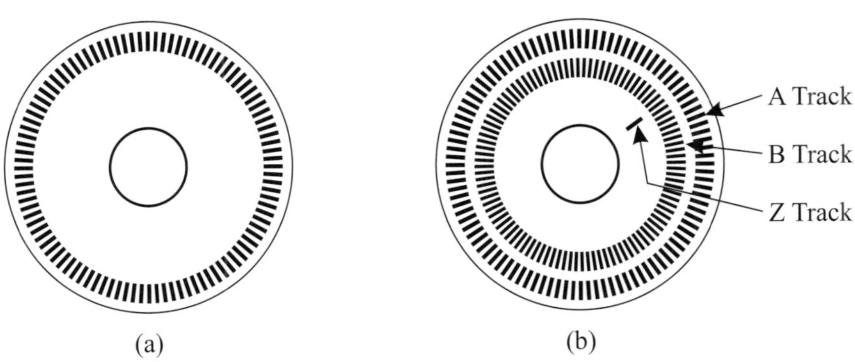

Figure 11.24. Incremental encoder disk: *(a)* single track; *(b)* two-phase quadrature encoder tracks.

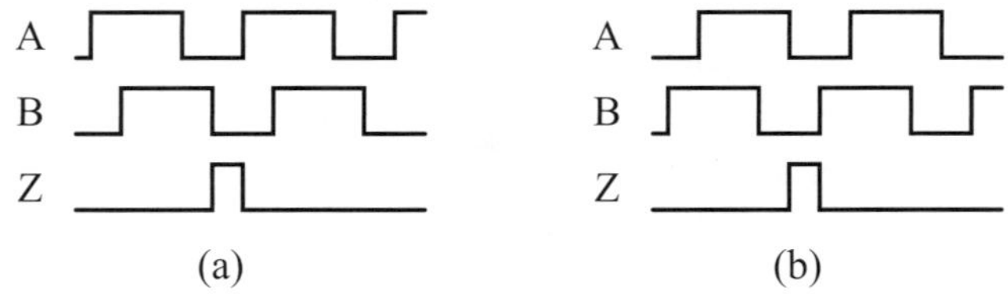

Figure 11.25. Quadrature encoder signals: *(a)* clockwise rotation; *(b)* counterclockwise rotation.

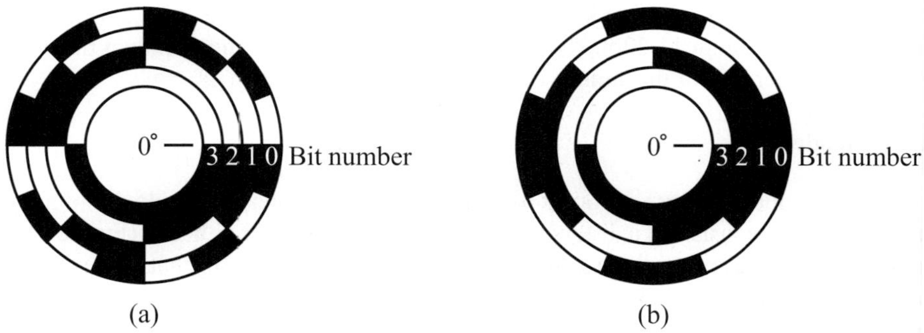

Figure 11.26. Absolute encoder disks: *(a)* binary code; *(b)* Gray code.

lags the positive edge of B (Figure 11.25*b*). The Z signal is the index pulse, which pulses once for each encoder revolution.

An absolute encoder provides multiple outputs, usually eight or sixteen, that provide the position of the disk relative to a zero point. Example absolute encoders that provide a 4-bit output are shown in Figure 11.26. These encoders provide the position of the disk relative to the zero position shown in the figure. Binary and Gray code representations of the position are the most popular coding schemes. The binary and Gray code representations of the disk positions are shown in Table 11.4. Gray code has the advantage that only one track changes during any transition. If there is any ambiguity because the phototransistor is reading the "edge" of one of the bands, the indicated position is in error by only one. For example, suppose the disk is at a clockwise rotation of 180 degrees, between sectors 7 and 8. The Gray coding would indicate either sector 7 or 8 because only one bit is uncertain. For binary coding, all four signals change. Therefore, any sector between 0 and 15 could be indicated.

11.2.4 Other discrete sensors

Other discrete sensors detect a particular value of a physical quantity such as high level, low temperature, and low pressure. These types of sensors are used when knowing the value is not necessary, only that it has reached a certain value. For example, a temperature switch can have a bimetallic contact that opens at a certain temperature (Figure 11.27*a*). A pressure switch is usually attached to a diaphragm with an opposing spring (Figure 11.27*b*). The switch opens or closes at a certain pressure difference between the two pressure ports.

Table 11.4 Absolute Encoder Outputs

Clockwise rotation angle (degrees)	Sector number	Binary code	Gray code
0.0 - 22.5	0	0000	0000
22.5 - 45.0	1	0001	0001
45.0 - 67.5	2	0010	0011
67.5 - 90.0	3	0011	0010
90.0 - 112.5	4	0100	0110
112.5 - 135.0	5	0101	0111
135.0 - 157.5	6	0110	0101
157.5 - 180.0	7	0111	0100
180.0 - 202.5	8	1000	1100
202.5 - 225.0	9	1001	1101
225.0 - 247.5	10	1010	1111
247.5 - 270.0	11	1011	1110
270.0 - 292.0	12	1100	1010
292.5 - 315.0	13	1101	1011
315.0 - 337.5	14	1110	1001
337.5 - 360.0	15	1111	1000

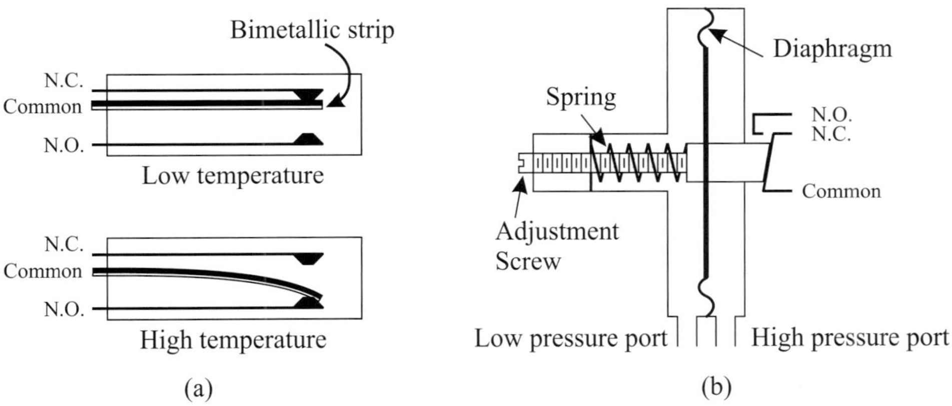

Figure 11.27. Other discrete sensors: *(a)* high temperature switch; *(b)* high pressure switch.

11.3 ANALOG SENSORS

An analog sensor measures the value of a physical quantity than can have an infinite number of values. Table 11.5 lists the most common physical quantities and the prevalent type of sensor technology used to measure the quantity. A sensor converts the physical phenomenon to an appropriate signal that is connected to an analog input channel. Many sensors contain both a transducer and a transmitter (Figure 11.28). In general terms, a transducer is a device that converts energy from one form to another. In the context of PLCs, a transducer converts a physical phenomenon (for example, temperature, flow, pressure, level. etc.) to a low-level electrical signal. The transmitter amplifies the millivolt or microampere signal from the transducer to the volt or milliampere signal required by the

Table 11.5 Common Transducer Types

Physical Quantity	Transducer
Temperature	Thermocouple Resistance Temperature Detector (RTD) Thermistor
Displacement	Potentiometer Linear Variable Differential Transformer (LVDT)
Force, Weight	Strain Gauge
Pressure	Force on diaphragm Capacitance
Flow	Pressure drop across orifice Coriolis

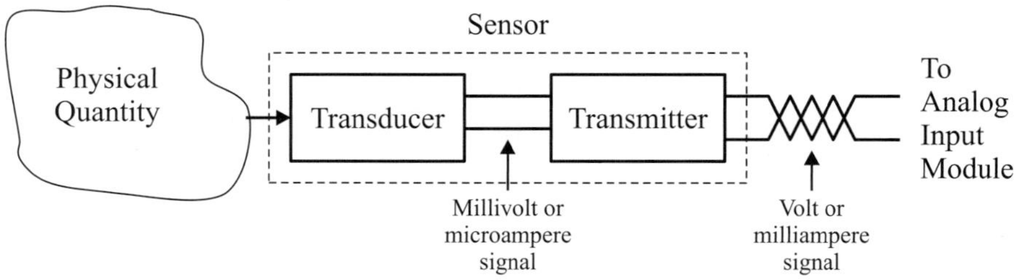

Figure 11.28. Typical sensor block diagram.

analog input module. Not all transducers require a transmitter. For example, analog input modules that directly measure the millivolt thermocouple voltage are common.

When characterizing continuous sensors, the terms accuracy and repeatability are used to describe their performance. The *accuracy* of a sensor is its ability to produce the correct output. The *repeatability* of a sensor is its ability to produce the same output when the measured value is the same. The difference between accuracy and repeatability is illustrated in Figure 11.29. Sensor 1 has better repeatability than sensor 2. However, the accuracy of sensor 2 is better than sensor 1 since its range of values is closer to the true value.

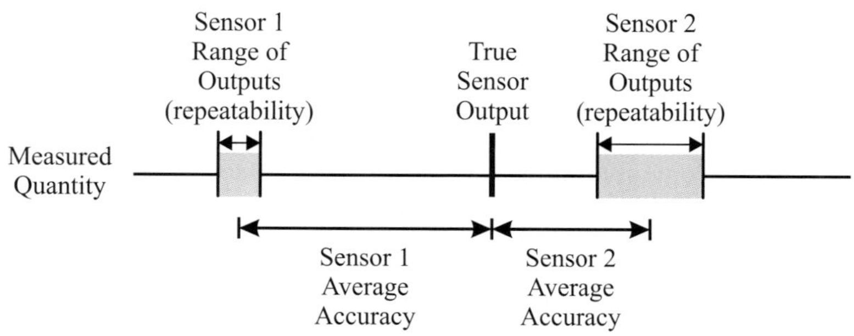

Figure 11.29. Difference between accuracy and repeatability.

After introducing the bridge circuit which is common in transducer circuitry, the various types of transducers used to measure temperature, displacement, force, pressure, flow, and level are described.

11.3.1 Bridge circuit

Many of the transducers listed in Table 11.5 rely on a bridge circuit to convert resistance changes to voltage changes. One may well ask why one cannot measure the resistance directly with an ohmmeter. There are two main reasons resistance is typically not directly measured:

1. Temperature compensation
2. Typical resistance changes are small

Temperature compensation is an important consideration. Often the resistance changes due to temperature can be as significant as the resistance changes due to the measured phenomenon.

A voltage sensitive bridge is shown in Figure 11.30. The resistances are typically in the 1000's of ohms. A current sensitive bridge is very similar, except that resistances are typically around 10 ohms. If R_S approaches zero (a perfect source) and R_D approaches infinity (a perfect detector) then,

$$V_a = V_S \frac{R_3}{R_1 + R_3} \text{ and } V_b = V_S \frac{R_4}{R_2 + R_4}$$

$$\begin{aligned} V = V_a - V_b &= V_S \left[\frac{R_3}{R_1 + R_3} - \frac{R_4}{R_2 + R_4} \right] \\ &= V_S \left[\frac{R_3 (R_2 + R_4) - R_4 (R_1 + R_3)}{(R_1 + R_3)(R_2 + R_4)} \right] \\ &= V_S \left[\frac{R_2 R_3 - R_1 R_4}{(R_1 + R_3)(R_2 + R_4)} \right] \end{aligned} \tag{11.1}$$

If the bridge is balanced, $V = 0$, and

$$R_2 R_3 - R_1 R_4 = 0 \text{ and } \frac{R_1}{R_2} = \frac{R_3}{R_4}$$

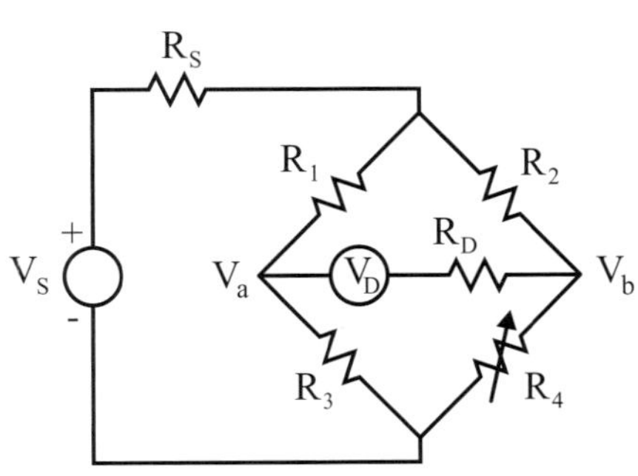

Figure 11.30. Voltage sensitive bridge.

Typically, $R_1 = R_2$, and so $R_3 = R_4$ when the bridge is balanced. When the bridge is balanced, what happens when R_4 (the transducer) changes to unbalance the bridge?

Let $R_1 = R_2$, and $R_4 = R_3 + \Delta R$

$$V = V_S \left[\frac{R_1 R_3 - R_1 (R_3 + \Delta R)}{(R_1 + R_3)(R_1 + R_3 + \Delta R)} \right] = V_S \left[\frac{-R_1 (\Delta R)}{(R_1 + R_3)(R_1 + R_3 + \Delta R)} \right] \quad (11.2)$$

If $R_1 + R_3 >> \Delta R$

$$V \approx V_S \left[\frac{-R_1 (\Delta R)}{(R_1 + R_3)^2} \right] \quad (11.3)$$

If one is interested in measuring changes in the resistance of R_4 (ΔR),

$$\Delta R \approx -V \left[\frac{(R_1 + R_3)^2}{V_S R_1} \right] \quad (11.4)$$

11.3.2 Temperature sensors

The five most common temperature sensors are the thermocouple, resistance temperature device (RTD), thermistor, integrated circuit, and optical pyrometer. The temperature ranges of these five sensors and their relative accuracy are compared in Table 11.6. For the thermocouple, RTD, and thermistor, a typical device spans part of the entire range specified in Table 11.6. The relative advantages and disadvantages of these sensors is shown in Table 11.7. The basic responses of each sensor type are compared in Figure 11.31.

Regardless of the temperature sensor, many applications require the sensor be mounted inside a thermowell (Figure 11.32) that protects the sensor from the process chemicals. The thermowell also allows the temperature sensor to be replaced without interrupting the process stream. When measuring the temperature of a chemical stream or the chemical inside a process vessel, thermowells are required. Thermowells are not utilized when the space is limited (for example, motor bearings and compressor internals), when a fast response is needed, when measuring the air temperature, or when measuring surface temperature. The thermowell does introduce a measurement lag into the process. For example, a thermowell may add a 5 second lag into the process which translates into an effective time delay of 20 to 25 seconds.

Table 11.6 Temperature Sensor Comparison

Sensor	Measuring Range	Accuracy
Thermocouple	-270 to 1700 °C[1]	±0.75%[1]
RTD	-250 to 750 °C	±0.5%
Thermistor	-80 to 150 °C	±0.2 °C
Integrated circuit	-30 to 85 °C	±1°C
Noncontact optical	-20 to 4000 °C	±1 to 2%[2]

Notes:
1. Depends on thermocouple type
2. Percentage of full scale

Table 11.7 Temperature Sensor Advantages and Disadvantages

Sensor	Advantages	Disadvantages
Thermocouple	Self-powered Simple Rugged Inexpensive Wide temperature range	Nonlinear Low voltage Needs reference Least stable Least sensitive
RTD	Most stable Most accurate More linear than thermocouple	Expensive Requires power source Small ΔR Self-heating
Thermistor	High output Fast	Nonlinear Limited temperature range Fragile Requires power source
Integrated circuit	Most linear Highest output Inexpensive	Low temperature range Requires power source Slow Self-heating
Optical pyrometer	Non-contact Highest maximum temperature	Expensive Many mechanical parts

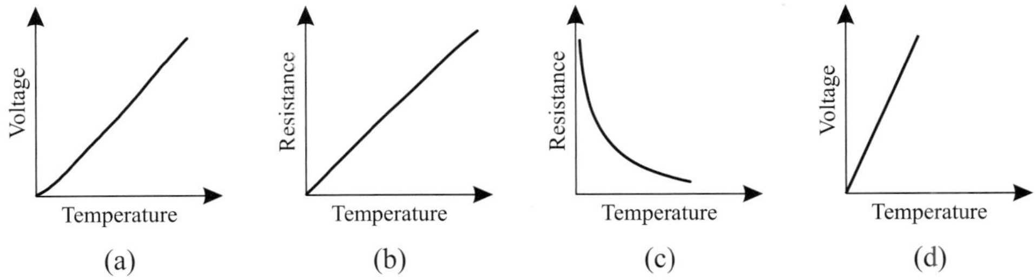

Figure 11.31. Typical response curves: *(a)* thermocouple; *(b)* RTD; *(c)* thermistor; *(d)* integrated circuit.

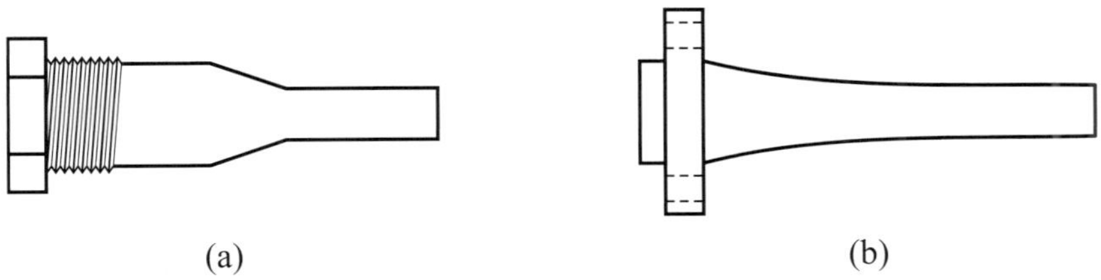

Figure 11.32. Thermowell: (a) threaded well; (b) flanged well.

Thermocouple

Thermocouples are the most versatile temperature-measuring element. They can be used over a wide temperature range. They are much more rugged than the other types of sensors and are often welded to a metal part or clamped under a screw. They can be manufactured on-site by soldering or welding. The only major drawback is the low voltage measurement and so one must be careful when the thermocouple is a long distance from the analog input module.

The operation of a thermocouple is based on the Seebeck effect which is a property of junctions of dissimilar metals. When two wires that are different metals are joined at both ends and one end is heated, an electromotive force (emf) is generated. For example, in Figure 11.33*a*, as $T_1 - T_2$ increases, the emf also increases. If the circuit is broken at the cold end (Figure 11.33*b*), the open circuit voltage, v_{ab}, is a function of the temperature difference between T_1 and T_2.

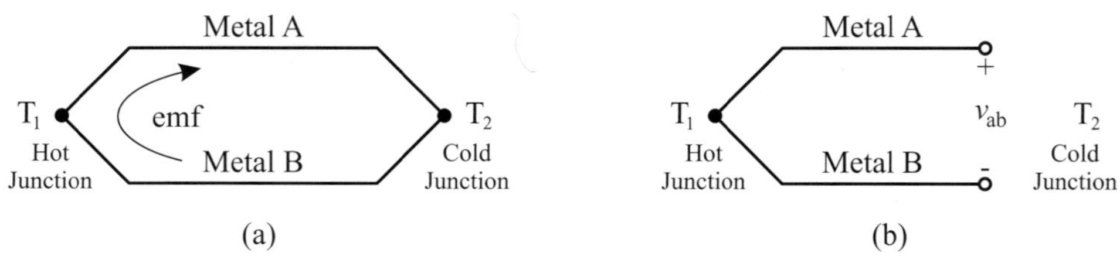

Figure 11.33. Basic thermocouple operation: *(a)* Seebeck effect; *(b)* Seebeck voltage.

Table 11.8 Thermocouple Comparison

Type	+ Wire	- Wire	Temp. Range (°C)	Advantages	Disadvantages
B	Pt70-Rh30	Pt94-Rh6	0 to 1820	High temp.	
E	Chromel (Ni-Cr)	Constantan (Cu-Ni)	-270 to 1000	Most sensitive	Larger drift
J	Iron	Constantan[1] (Cu-Ni)	-210 to 1200	Cheapest	
K	Chromel (Ni-Cr)	Alumel (Ni-Al)	-270 to 1372	Stability	
N	Nicrosil (Ni-Cr-Si)	Nisil (Ni-Si-Mn)	-270 to 1300	More stable than K	
R	Pt87-Rh13	Platinum	-50 to 1768	Linear at high temperature	More expensive than K
S	Pt90-Rh10	Platinum	-50 to 1768	Linear at high temperature	More expensive than K
T	Copper	Constantan (Cu-Ni)	-270 to 400	Good resistance to corrosion from moisture	Limited temperature range

Note:
1. Not the same copper-nickel alloy as for the E and T thermocouples.

Table 11.8 lists the most common thermocouple types and the materials used to make the junction. The letter designations follow the recommendations of the International Society of Automation (ISA) organization and the American Society for Testing and Materials (ASTM). The letter type, for example, type J, identifies a specific temperature-voltage relationship and not a particular composition of the two wires. The wires for thermocouples of a given type are allowed to have variations in composition as long as the temperature-voltage relationship remains within specified tolerances (Burns, et. al., 1993). The voltage generated by these thermocouples as a function of temperature is shown in Figure 11.34*a* (Burns, et. al, 1993). The Seebeck coefficient, shown in Figure 11.34*b* is the slope of the temperature-voltage curves in Figure 11.34*a*. The Seebeck coefficient clearly shows the nonlinear nature of thermocouples. The K, R, and S thermocouples are reasonably linear for temperatures greater than 0 °C.

In practice, the junction voltage cannot be measured directly because the voltmeter leads introduce additional thermocouple junctions (Figure 11.35). In the case of the K thermocouple in Figure 11.35, there exist chromel-copper and alumel-copper junctions in addition to the chromel-alumel junction. The two new junctions introduced by the measurement process produce their own voltage, dependent on the temperature of the junctions. The standard thermocouple voltage-to-temperature conversions specify that the reference junctions are at 0 °C. In the past, an ice bath was used to keep the reference

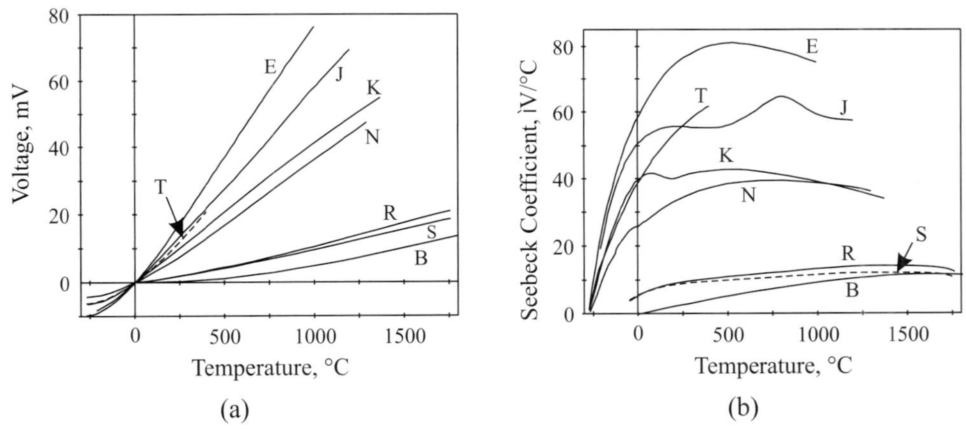

Figure 11.34. Thermocouple characteristics as a function of temperature: *(a)* voltage; *(b)* Seebeck coefficient.

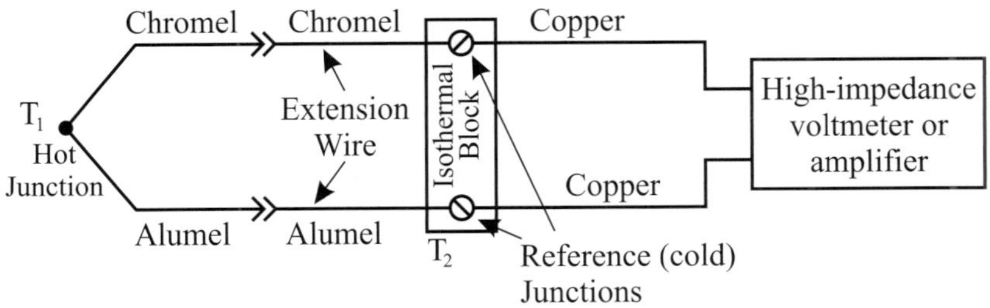

Figure 11.35. K thermocouple voltage measurement.

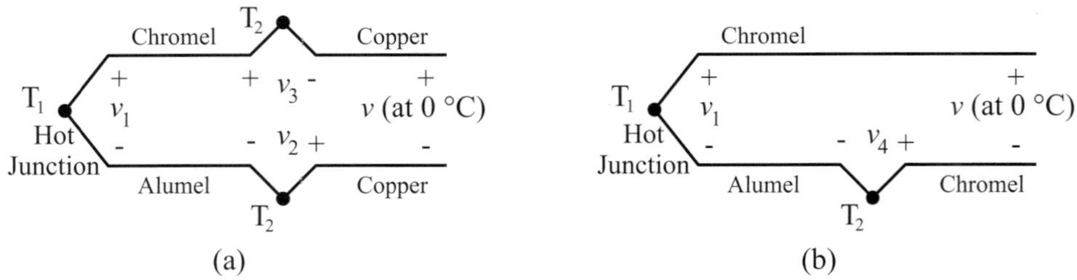

Figure 11.36. Cold junction temperature effects: *(a)* cold junction thermocouples; *(b)* equivalent circuit.

junctions at 0 °C. However, with *cold-junction compensation* the temperature of the cold junction is measured and then used to adjust the measured voltage. With the advent of microprocessor-based analog input modules, the ice bath is not necessary because the microprocessor can implement the cold-junction compensation.

If the reference (cold) junction is not at 0 °C, the effect of thermocouple junctions at the isothermal block of Figure 11.35 is shown in Figure 11.36*a*. Both of these junctions produce a voltage which alters the measured voltage. Since the copper-copper connections inside the voltmeter do not produce any thermocouple junctions, the equivalent thermocouple circuit is shown in Figure 11.36*b*. The effect of a cold junction is essentially another thermocouple (of the same type) wired in opposition to the thermocouple being used for measurement. In order to compensate for the cold junction temperature, one needs to measure the temperature of the analog input module terminal block (the cold junction). The terminal block temperature is often measured with a thermistor or integrated circuit temperature sensor (Figure 11.37). The equivalent thermocouple voltage for the cold junction is added to the measured voltage to obtain an equivalent voltage that is then converted to the temperature, using a table or polynomial equation (Burns, et. al, 1993; Omega, 2010). In summary, the procedure for cold junction compensation is:

1. Measure analog input voltage (thermocouple junction), V_{TC} .
2. Measure reference junction temperature.
3. Find thermocouple junction voltage for this reference temperature, V_{CJ} .
4. Add reference junction voltage to measured analog voltage to obtain cold-junction compensated input, $V = V_{TC} + V_{CJ}$.
5. Convert V to temperature using a table or polynomial equation.

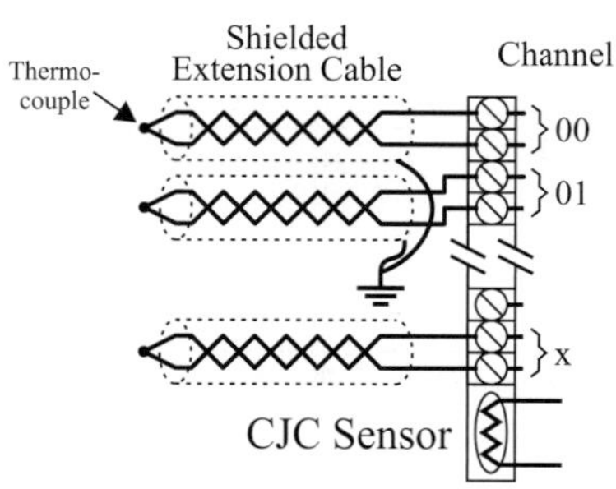

Figure 11.37. Typical thermocouple input module with cold-junction compensation.

This procedure will be illustrated by two examples, but first the conversion of the junction voltage to temperature will be explained. The voltage generated by these thermocouples as a function of temperature is presented in tabular form in Burns et. al. (1993). Given a measured voltage, one can look up the temperature, using interpolation to calculate the temperature when the voltage falls between two temperature entries. The linear interpolation formula is developed as follows. If V_1 is the voltage for temperature T_1 and V_2 is the voltage for temperature T_2, and V falls between V_1 and V_2 then T, the temperature that corresponds to V, is calculated as

$$T = \frac{V - V_1}{V_2 - V_1}(T_2 - T_1) + T_1 \tag{11.5}$$

The standard thermocouple tables are quite large since they report the voltage for every degree in the temperature range. For this reason, Burns et al (1993) also provide approximate polynomial functions that can calculate the temperature, given the voltage. The general equation for the polynomial is

$$T = c_0 + c_1 V + c_2 V^2 + c_3 V^3 + c_4 V^4 + \ldots + c_n V^n \tag{11.6}$$

where V is the voltage in mV and n is the polynomial order. There are also polynomials that can calculate the thermocouple voltage for a given temperature of the form

$$V = c_0 + c_1 T + c_2 T^2 + c_3 T^3 + c_4 T^4 + \ldots + c_n T^n \tag{11.7}$$

For example, for a J thermocouple producing a voltage between 0.0 and 42.919 mV, the coefficients of the 7th order polynomial to convert the voltage to the temperature, in °C, are:

$c_0 = 0.0$ $\quad$ $c_4 = -2.549687\times10^{-4}$

$c_1 = 1.978425\times10^{1}$ $\quad$ $c_5 = 3.585153\times10^{-6}$

$c_2 = -2.001204\times10^{-1}$ $\quad$ $c_6 = -5.344285\times10^{-8}$

$c_3 = 1.036969\times10^{-2}$ $\quad$ $c_7 = 5.09890\times10^{-10}$

The error in the calculated temperature value is between –0.04 to 0.04 °C. The amount of computation is a major drawback to using polynomial functions for the conversion. If one does not need a small error range, a lookup table can be constructed from selected points from the standard. The temperature is obtained by interpolating between the entries. For example, one control equipment vendor used a table of 40 entries to obtain a maximum error of 1 °C.

Example 11.1. The wiring diagram of a T thermocouple is shown in Figure 11.38. The voltage across the thermocouple wires, measured at the terminals of the analog input channel is 14.420 mV. The temperature of the cold junction is 24 °C. What is the temperature (in °C) of the hot junction? Find your answers to the nearest one-hundredth of a degree. Do the voltage-to-temperature conversion using:

(a) the T thermocouple table

(b) the National Institute of Standards and Technology (NIST) polynomial coefficients:

Solution. (a) A partial T thermocouple reference table is shown in Figure 11.39. Note that for positive temperatures, the temperature increases by one degree as one proceeds across the row.

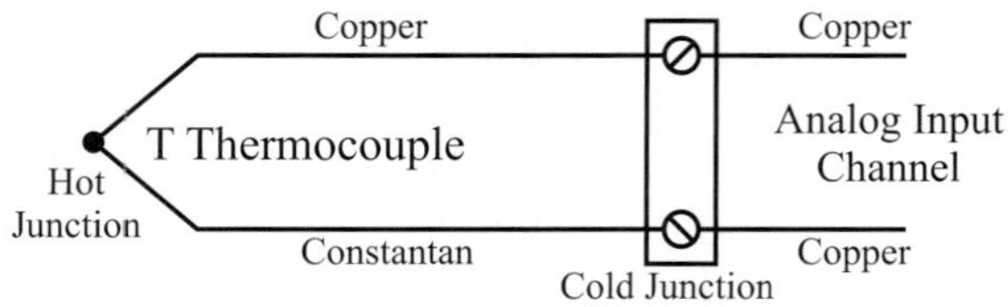

Figure 11.38. T thermocouple measurement example.

ITS-90 Table for type T thermocouple

°C	0	-1	-2	-3	-4	-5	-6	-7	-8	-9	-10
					Thermoelectric Voltage in mV						
-270	-6.258										
-260	-6.232	-6.236	-6.239	-6.242	-6.245	-6.248	-6.251	-6.253	-6.255	-6.256	-6.258
...											
-140	-4.419	-4.443	-4.466	-4.489	-4.512	-4.535	-4.558	-4.581	-4.604	-4.626	-4.648
-130	-4.177	-4.202	-4.226	-4.251	-4.275	-4.300	-4.324	-4.348	-4.372	-4.395	-4.419
-120	-3.923	-3.949	-3.975	-4.000	-4.026	-4.052	-4.077	-4.102	-4.127	-4.152	-4.177
-110	-3.657	-3.684	-3.711	-3.738	-3.765	-3.791	-3.818	-3.844	-3.871	-3.897	-3.923
...											
-10	-0.383	-0.421	-0.459	-0.496	-0.534	-0.571	-0.608	-0.646	-0.683	-0.720	-0.757
0	0.000	-0.039	-0.077	-0.116	-0.154	-0.193	-0.231	-0.269	-0.307	-0.345	-0.383
°C	0	-1	-2	-3	-4	-5	-6	-7	-8	-9	-10

ITS-90 Table for type T thermocouple

°C	0	1	2	3	4	5	6	7	8	9	10
					Thermoelectric Voltage in mV						
0	0.000	0.039	0.078	0.117	0.156	0.195	0.234	0.273	0.312	0.352	0.391
10	0.391	0.431	0.470	0.510	0.549	0.589	0.629	0.669	0.709	0.749	0.790
20	0.790	0.830	0.870	0.911	0.951	0.992	1.033	1.074	1.114	1.155	1.196
30	1.196	1.238	1.279	1.320	1.362	1.403	1.445	1.486	1.528	1.570	1.612
40	1.612	1.654	1.696	1.738	1.780	1.823	1.865	1.908	1.950	1.993	2.036
...											
290	14.283	14.341	14.399	14.456	14.514	14.572	14.630	14.688	14.746	14.804	14.862
300	14.862	14.920	14.978	15.036	15.095	15.153	15.211	15.270	15.328	15.386	15.445
310	15.445	15.503	15.562	15.621	15.679	15.738	15.797	15.856	15.914	15.973	16.032
320	16.032	16.091	16.150	16.209	16.268	16.327	16.387	16.446	16.505	16.564	16.624
...											
400	20.872										
°C	0	1	2	3	4	5	6	7	8	9	10

Figure 11.39. T thermocouple temperature-voltage table (partial).

From the problem, $V_{TC} = 14.42$ mV. The cold junction temperature is 24 °C, and the corresponding thermocouple voltage is read from the table as 0.951 mV, meaning $V_{CJ} = 0.951$ mV. Therefore, $V = V_{TC} + V_{CJ} = 15.371$ mV.

Reading the table, the temperature that corresponds to this junction voltage is between 308 and 309 C. The temperatures and corresponding junction voltages are

Temperature	Voltage
308 °C	15.326 mV
309 °C	15.386 mV

To use the formula in equation (11.5), $T_1 = 308$, $T_2 = 309$, $V_1 = 15.326$, and $V_2 = 15.386$.

$$T(°\mathrm{C}) = \frac{15.371 - 15.326}{15.386 - 15.326}(309 - 308) + 308 = \frac{0.045}{0.06}(1) + 308 = 308.75\ °\mathrm{C}$$

(b) For the T thermocouple, the NIST polynomial coefficients for converting temperature in the range of 0 to 400 °C to junction voltage (in mV) are

$c_0 = 0.0$ $c_5 = 1.0996880928\times10^{-11}$
$c_1 = 3.8748106364\times10^{-2}$ $c_6 = -3.0815758772\times10^{-14}$
$c_2 = 3.3292227880\times10^{-5}$ $c_7 = 4.5479135290\times10^{-17}$
$c_3 = 2.0618243404\times10^{-7}$ $c_8 = -2.7512901673\times10^{-20}$
$c_4 = -2.1882256846\times10^{-9}$

Using these coefficients in equation (11.7), the cold junction voltage for 24 °C is 0.951 mV, meaning $V_{CJ} = 0.9513$ mV. Therefore, $V = V_{TC} + V_{CJ} = 15.371$ mV.

The NIST polynomial coefficients for converting junction voltage in the range of 0 to 20.872 mV to the temperature (in °C) are

$c_0 = 0.0$ $c_4 = -2.165394\times10^{-3}$
$c_1 = 2.592800\times10^{1}$ $c_5 = 6.048144\times10^{-5}$
$c_2 = -7.602961\times10^{-1}$ $c_6 = -7.293422\times10^{-7}$
$c_3 = 4.637791\times10^{-2}$

The conversion error is ±0.03 °C. Using equation (11.6), the conversion of V into the temperature is

$$\begin{aligned} T(°C) &= c_0 + c_1V + c_2V^2 + c_3V^3 + c_4V^4 + c_5V^5 + c_6V^6 \\ &= 0.0 + 398.5393 - 179.6334 + 168.4293 - 120.8775 + 51.8959 - 9.6193 \\ &= 308.7343\ °C \end{aligned}$$

Rounded to hundredths, $T = 308.73$ °C, which is within the error tolerance, as compared to the value obtained from the table in part (a).

Example 11.2. The wiring diagram of a T thermocouple is shown in Figure 11.38. The voltage across the thermocouple wires, measured at the terminals of the analog input channel is –4.790 mV. The temperature of the cold junction is 21 °C. What is the temperature (in °C) of the "hot" junction? Find your answers to the nearest one-hundredth of a degree. Do the voltage-to-temperature conversion using:

(a) the T thermocouple table

(b) the National Institute of Standards and Technology (NIST) polynomial coefficients:

Solution. (a) A partial T thermocouple reference table is shown in Figure 11.39. Note that for negative temperatures, the temperature *decreases* by one degree as one proceeds across the row.

From the problem, $V_{TC} = -4.790$ mV. The cold junction temperature is 21 °C, and the corresponding thermocouple voltage is read from the table as 0.830 mV, meaning $V_{CJ} =$ 0.830 mV. Therefore, $V = V_{TC} + V_{CJ} = -3.960$ mV.

Reading the table, the temperature that corresponds to this junction voltage is between –121 and –122 °C. The temperatures and corresponding junction voltages are

Temperature	Voltage
–122 °C	–3.975 mV
–121 °C	–3.949 mV

To use the formula in equation (11.5), $T_1 = -122$, $T_2 = -121$, $V_1 = -3.975$, and $V_2 = -3.949$.

$$T(°\text{C}) = \frac{-3.960 - (-3.975)}{-3.949 - (-3.975)}(-121 - (-122)) + (-122) = \frac{0.015}{0.026}(1) - 122 = -121.42\ °\text{C}$$

(b) For the T thermocouple, the NIST polynomial coefficients for converting temperature in the range of 0 to 400 °C to junction voltage (in mV) are given in Example 11.1. Using these coefficients in equation (11.7), the cold junction voltage for 21 °C is 0.8299 mV, meaning V_{CJ} = 0.8299 mV. Therefore, $V = V_{TC} + V_{CJ}$ = –3.9601 mV.

The NIST polynomial coefficients for converting junction voltage in the range of –5.603 to 0 mV to the temperature (in °C) are

$c_0 = 0.0$ $\qquad c_4 = 4.2527777\times10^{-1}$
$c_1 = 2.5949192\times10^{1}$ $\qquad c_5 = 1.3304473\times10^{-1}$
$c_2 = -2.1316967\times10^{-1}$ $\qquad c_6 = 2.0241446\times10^{-2}$
$c_3 = 7.9018692\times10^{-1}$ $\qquad c_7 = 1.2668171\times10^{-3}$

The conversion error is –0.02 to +0.04 °C. Using equation (11.6), the conversion of V into the temperature is

$$\begin{aligned} T(°\text{C}) &= c_0 + c_1V + c_2V^2 + c_3V^3 + c_4V^4 + c_5V^5 + c_6V^6 + c_7V^7 \\ &= 0.0 - 102.7588 - 3.3428 - 49.0699 + 104.5812 - 129.5608 + 78.0571 - 19.3455 \\ &= -121.4496\ °\text{C} \end{aligned}$$

Rounded to hundredths, T = –121.44 °C, which is within the error tolerance, as compared to the value obtained from the table in part (a).

Other thermocouple configurations are shown in Figure 11.40. When multiple thermocouples are connected in parallel (Figure 11.40*a*), the net voltage measured at the cold junction is indicative of the *average* of the measured temperatures. When two thermocouples are connected in series, but in opposite polarities, the net voltage measured at the cold junction indicates the *difference* of the two hot junction temperatures. For example, in Figure 11.40*b*, the voltage at the cold junction measures $T_1 - T_2$. A *thermopile* (Figure 11.40*c*) is a series connection of thermocouples. All hot junctions are at the same temperature and all cold junctions are at the same temperature. This arrangement effectively amplifies the thermocouple voltage. For the arrangement in Figure 11.40*c*, the measurement is three times the voltage of one thermocouple. However, cold-junction

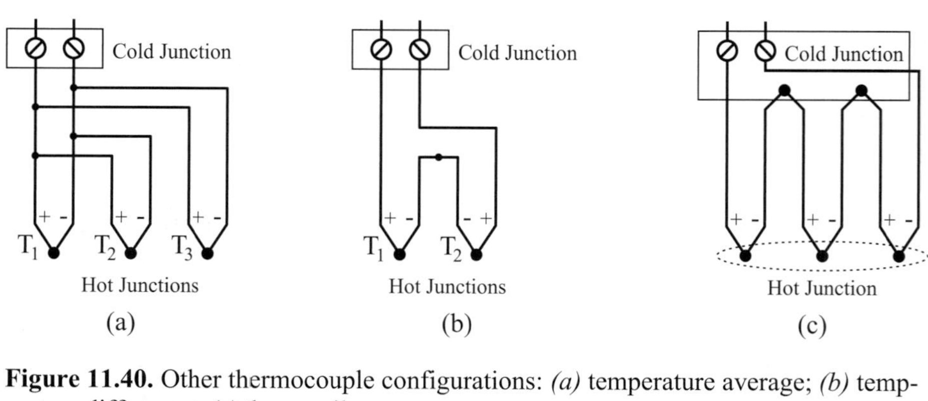

Figure 11.40. Other thermocouple configurations: *(a)* temperature average; *(b)* temperature difference; *(c)* thermopile.

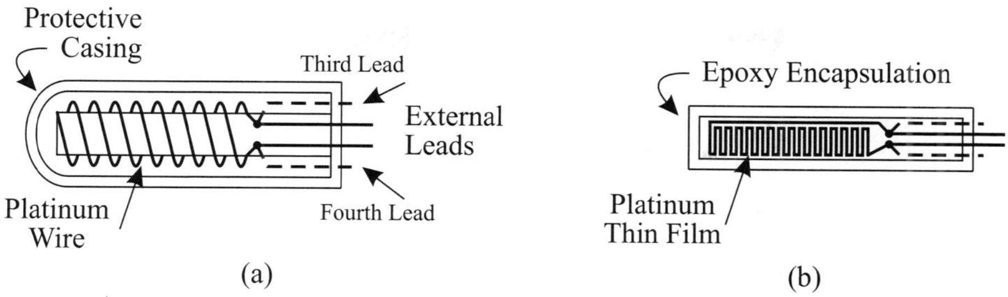

Figure 11.41. Typical RTD construction: *(a)* wire wound; *(b)* thin film.

compensation is slightly more complicated since for each thermocouple in series, one must also add the cold-junction compensation voltage to obtain the correct voltage for the voltage-to-temperature conversion. Thermopiles are used as the detector in some infrared radiation pyrometers.

Resistance Temperature Device (RTD)

A resistance temperature device (RTD) is somewhat less versatile than a thermocouple due to its smaller temperature range. However, it is more linear and more stable than a thermocouple.

The operation of a RTD is based on the temperature dependence of wire resistance. The resistance of a metal wire changes with temperature. Figure 11.41 shows two typical methods of RTD construction. In Figure 11.41*a*, the platimum wire is wrapped around a support and then encapsulated inside a ceramic or glass casing. Since platinum is expensive, most RTDs are now constructed as a thin film deposited on a small ceramic substrate and then encapsulated (Figure 11.41*b*). Platinum is the metal of choice for RTDs due to its stability and relatively large change in resistance with temperature changes. Copper and nickel are other common RTD metals. Balco (a nickel-iron alloy) and tungsten are also used to construct RTD, though rarely used. A typical RTD resistance versus temperature is shown in Figure 11.42.

RTDs are commonly classified by their nominal resistance at 0 °C. Platinum wire RTDs are typically 100 ohms at 0 °C. The nominal resistance of platinum thin film RTDs

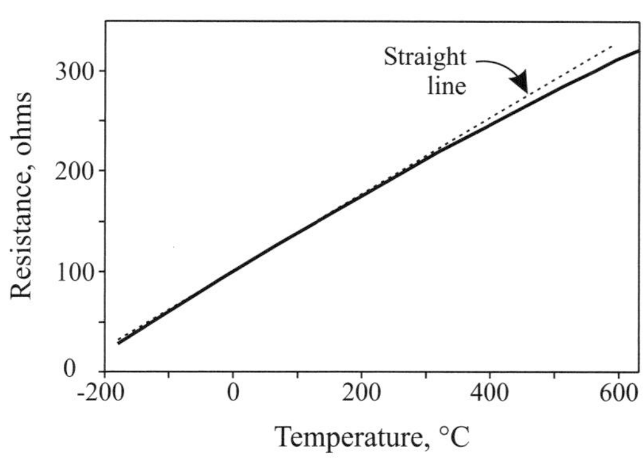

Figure 11.42. Typical RTD resistance characteristic as a function of temperature.

Table 11.9 Callendar-Van Dusen Coefficients for Common RTDs

Standard	Temperature Coeff. (α)	A	B	C (T < 0)
DIN 43760	0.003850	3.9080×10^{-3}	-5.8019×10^{-7}	-4.2735×10^{-12}
IEC 60751	0.0038505	3.9083×10^{-3}	-5.7750×10^{-7}	-4.183×10^{-12}
American	0.003911	3.9692×10^{-3}	-5.8495×10^{-7}	-4.2325×10^{-12}

range from 200 to 1000 ohms at 0 °C. One common method of expressing the relationship between temperature (T) and resistance (R_{RTD}) is the Callendar-Van Dusen equation:

$$R_{RTD} = R_0 + R_0\alpha\left[T - \delta(0.01T - 1)(0.01T) - \beta(0.01T - 1)(0.01T)^3\right]$$

where R_0 is the resistance at 0 °C, T is the temperature in °C, and α, β, and δ are constants. The constant α is the temperature coefficient in units of ohms/ohms/°C. This relationship is typically expressed as the cubic equation

$$R_{RTD} = R_0\left[1 + AT + BT^2 + C(T - 100)^3\right] \tag{11.8}$$

This equation is valid over the range of -183 °C to 630 °C. The coefficients for common RTDs are shown in Table 11.9. Note that the C coefficient is zero for temperatures above 0 °C. The different standards specify different wire purity and composition. The DIN 43760 and IEC 60751 standard thermocouples are more prevalent than the older American standard. Since the C coefficient in equation (11.8) is zero for temperatures above 0 °C, the resulting quadratic equation can be solved for the temperature given a measurement of the RTD resistance. In this situation, the temperature can be calculated as

$$T = \frac{1}{2}\left[-\frac{A}{B} - \sqrt{\left(\frac{A}{B}\right)^2 - 4\left(\frac{R_0 - R_{RTD}}{BR_0}\right)}\right] \tag{11.9}$$

When used with a bridge circuit, RTD resistance can be measured as in Figure 11.43. However, if the RTD is far from the bridge, then the lead resistance can be significant, leading to errors in the resistance measurement. To compensate for lead resistance, a 3-wire RTD can be used where a third wire is connected to one lead of the RTD element. A bridge with a 3-wire RTD is shown in Figure 11.44. The value of R_{L3} is irrelevant since the current through it is very small.

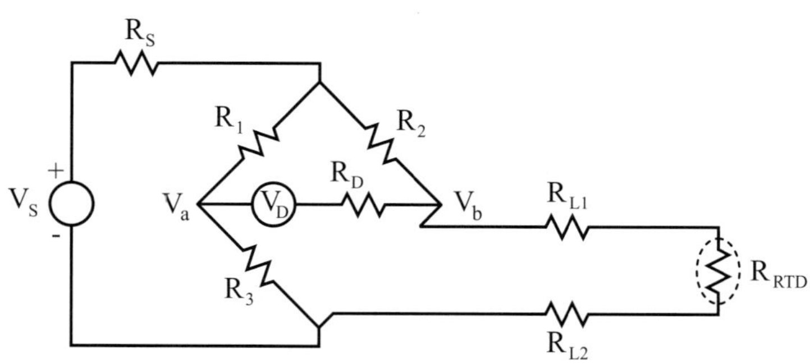

Figure 11.43. Wheatstone bridge with two-wire RTD.

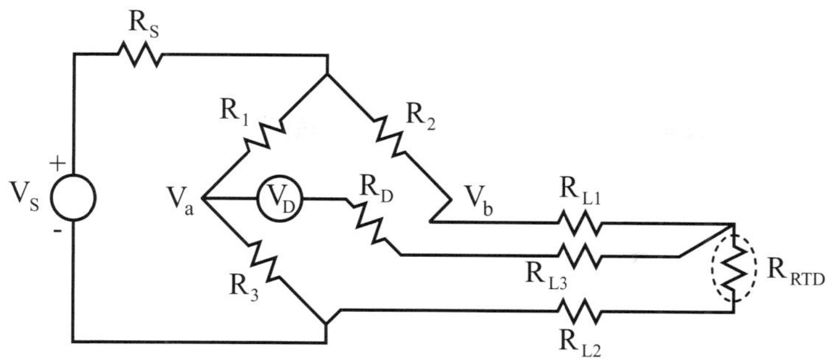

Figure 11.44. Wheatstone bridge with three-wire RTD.

The main problem with measuring RTD resistances is that the bridge circuit only works well with small changes in temperature. An RTD whose resistance is 100 ohms at 0 °C will have a resistance of 175.86 ohms at 200 °C. This resistance change violates the assumption of relatively small resistance changes in order to use equation (11.4) to calculate the resistance. One must use equation (11.2) which means there is a nonlinear relationship between resistance changes (ΔR) and the measured voltage, V_D.

The most accurate method of measuring RTD resistance uses a current source with either a 3-wire or 4-wire RTD. Figure 11.45 illustrates the method for a four-wire RTD. A current source establishes a constant current through the RTD. Since the amplifier has high input impedance, the current through the sense wires is very small and hence the effect of the sense wire resistances, R_{L3} and R_{L4}, is negligible. Note that the lead wire resistances, R_{L1} and R_{L2}, have no effect on the measurement. The RTD resistance is then

$$R_{RTD} = \frac{V}{I_{RTD}}$$

A current source can be used with a three-wire RTD to produce an accurate measurement of RTD resistance as shown in Figure 11.46 (Rockwell Automation, 2022). The sense lead is used to subtract the effect of both lead resistances. The amplifier outputs are

$$V_1 = I_{RTD}\left(R_{RTD} + 2R_L\right)$$
$$V_2 = 2I_{RTD}R_L$$
$$V = V_1 - V_2 = I_{RTD}R_{RTD}$$

and the RTD resistance is calculated with the same equation as for the four-wire RTD.

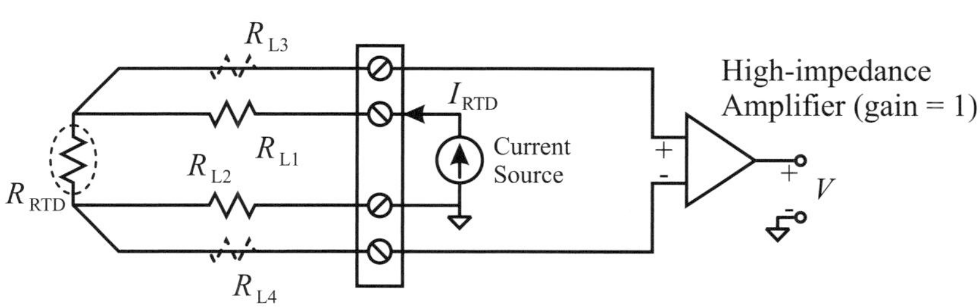

Figure 11.45. Accurate temperature measurement with four-wire RTD.

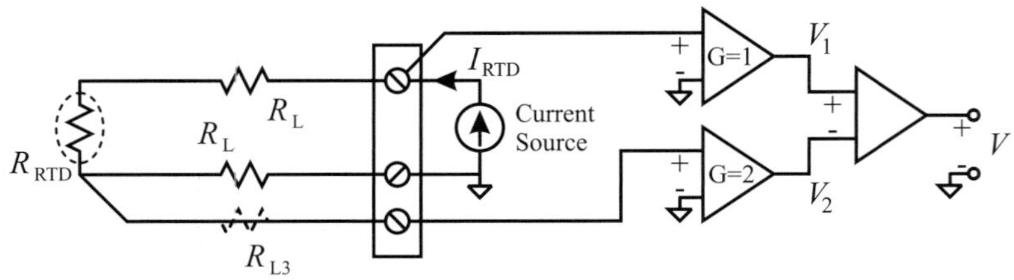

Figure 11.46. Accurate temperature measurement with three-wire RTD.

The accuracy of a RTD is affected by the magnitude of the current through it. Because the RTD is a resistance, current through it produces a small amount of heat, called the *self-heating effect.* The current through the RTD needs to be small (less than 1 mA) to minimize self-heating.

Example 11.3. A three-wire RTD measures temperature and the analog input uses the circuit of Figure 11.46 to measure the resistance. The RTD is a platinum RTD whose nominal resistance (R_0) is 100 ohms at 0 °C. The RTD has been manufactured according to the IEC 60751 standard. The measured RTD resistance is 180 ohms. What is the temperature of the RTD? Find your answers to the nearest one-hundredth of a degree. Do the voltage-to-temperature conversion using the Callendar-Van Dusen equation.

Solution. Since the measured resistance is greater than 100 ohms, the RTD temperature is greater than 0 C. R_0 = 100 and R_{RTD} = 180. The RTD temperature is calculated using equation (11.9) with the appropriate coefficients in Table 11.9 that correspond to the IEC 60751 standard

$$T = \frac{1}{2}\left[-\frac{3.9083\times10^{-3}}{-5.775\times10^{-7}} - \sqrt{\left(\frac{3.9083\times10^{-3}}{-5.775\times10^{-7}}\right)^2 - 4\left(\frac{100-180}{\left(-5.775\times10^{-7}\right)100}\right)}\right]$$

and $T = 211.29$ °C.

Thermistor

A thermistor is also a temperature-sensitive resistor. However, the thermistor differs from the RTD in one important feature – it is more sensitive to temperature changes. Over a narrow temperature range, the thermistor resistance can change as much as a few percent. Most thermistors are made from a semiconductor material that exhibits a negative temperature coefficient, that is, its resistance decreases with temperature.

Most thermistors are constructed as a ceramic bead with two lead wires (Figure 11.47). To protect it from environmental factors, a thermistor is encapsulated in epoxy or glass. A

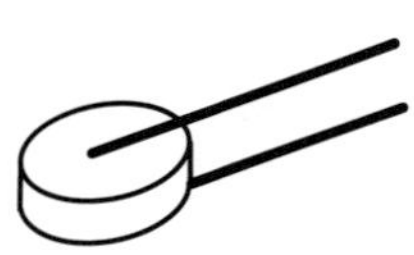

Figure 11.47. Typical thermistor construction.

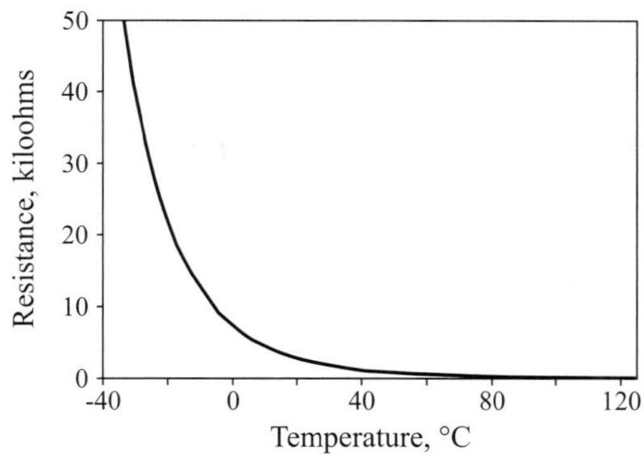

Figure 11.48. Typical thermistor resistance characteristic as a function of temperature.

typical temperature-resistance relationship for a thermistor is shown in Figure 11.48. Note that for lower temperatures, the resistance change is greater for a given change in temperature. The Steinhart-Hart equation is a common method of expressing the relationship between the resistance and temperature:

$$\frac{1}{T} = a + b(\ln R) + c(\ln R)^3$$

where a, b, and c are constants, R is the resistance in ohms, and T is the temperature in Kelvins (K = °C+273.15). The manufacturer specifies the Steinhart-Hart constants since thermistor curves have not been standardized to the same extent as for thermocouples and RTDs.

The circuits used to measure thermistor resistance are basically the same as for an RTD. However, compensation for lead resistance is generally not necessary since the lead resistance is very small compared to the thermistor resistance changes.

Integrated Circuit Temperature Sensor

Two temperature sensors that are integrated circuits are described here. The National Semiconductor LM335 device produces a voltage proportional to temperature and the Analog Devices AD590 produces a current proportional to temperature. Both devices are based on the property that the base-emitter voltage of a bipolar transistor is proportional to temperature. In contrast to the thermocouple and thermistor, the temperature range of the integrated circuit temperature sensors is limited to the range of -55 °C to 150 °C.

The LM335 temperature sensor (National Semiconductor, 2008) operates as a two-terminal zener diode with a breakdown voltage directly proportional to its absolute temperature. Figure 11.49*a* shows the symbol and connections to the device. The ADJ terminal is used to calibrate the sensor and increase the accuracy. The LM335 is used as a basic temperature sensor as shown in Figure 11.49*b*. If smaller error is desired, then the ADJ terminal is used as in Figure 11.49*c*. Depending on the device within the LM335 series, the uncalibrated device error can be as high as 9 °C. Calibration reduces this error to 1 °C. The R_1 resistor determines the bias current which must be within the range 0.4 to 5 mA. The bias current should be low to prevent self-heating errors.

The AD590 temperature sensor (Analog Devices, 2013) is a two-terminal device that produces an output current proportional to the absolute temperature. With a supply voltage

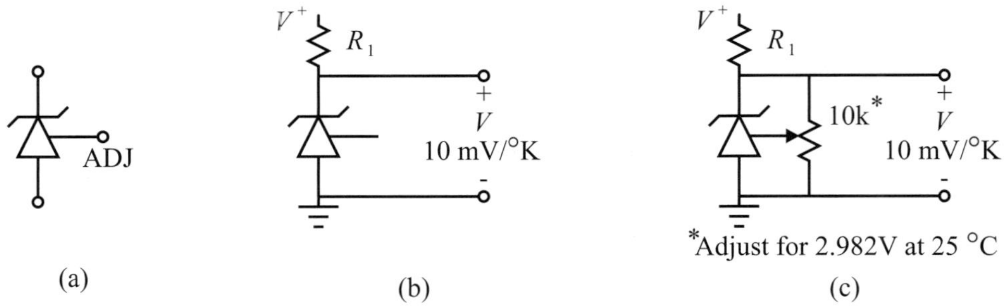

Figure 11.49. LM335 temperature sensor: *(a)* symbol; *(b)* basic circuit; *(c)* calibrated output.

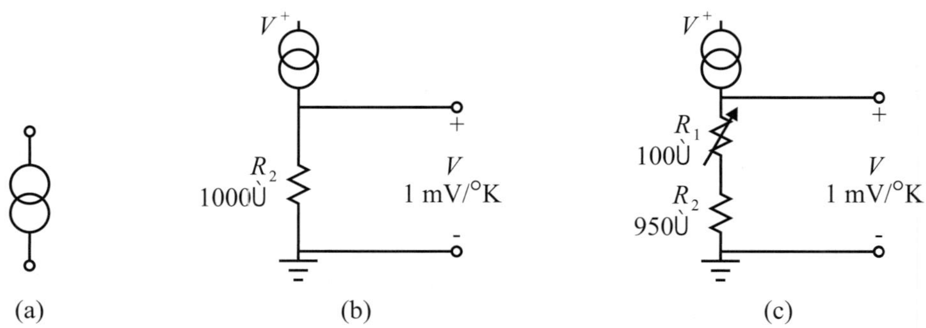

Figure 11.50. AD590 temperature sensor: *(a)* symbol; *(b)* basic circuit; *(c)* calibrated output.

between 4 to 30 volts, the device acts like a high-impedance constant current regulator producing 1 μA/°K. Figure 11.50*a* shows the symbol and connections to the device. The AD590 is used as a basic temperature sensor as shown in Figure 11.50*b*. If smaller error is desired, then a variable resistor is added as in Figure 11.50*c*. The R_1 resistor is adjusted to produce the proper voltage at the output. Depending on the particular AD590 device, the uncalibrated device error ranges from 1.7 °C to 5.5 °C. Calibration reduces this error to 1.0 °C to 2.0 °C.

Noncontact Temperature Measurement

A pyrometer is a noncontact method of temperature measurement. It is commonly used when measuring the temperature of a target that is inaccessible or in a hostile environment. Common applications are measuring flame temperature or the temperature of a moving strip of hot metal. There are two basic types of noncontact temperature measurement devices. The first is infrared thermometers that measure the infrared radiation (wavelength 0.7 to 80 μm) emitted from the hot object. The second type, optical pyrometers, measure temperature in the visible light spectrum (wavelength 0.4 to 0.7 μm).

Both of these types of noncontact temperature measurement utilize the property that all materials having a temperature above absolute zero (-273 °C) emit radiant energy. Noncontact temperature measurement devices measure this radiant energy by comparing it to an ideal target, called a *blackbody*. A blackbody emits all of its energy. Emissivity is defined as the ratio between the actual energy emitted from a target and the energy emitted from a blackbody.

$$\text{Emissivity} = \frac{\text{Energy emitted from target}}{\text{Energy emitted from blackbody}}$$

Emissivity is dependent on both wavelength and temperature. So, in order to measure the temperature of an object, its emissivity must be known. However, for many materials, the emissivity changes rapidly as the target temperature changes. This phenomenon complicates the temperature measurement and is an inherent weakness in noncontact temperature measurement.

For a given application, physical factors other than the temperature of the target influence the measured radiation. These factors include: target material, target size, target distance, interfering gases, interfering film (for example, glass or plastic), and the ambient temperature. All noncontact thermometers measure all of the energy in their field of view within the specified spectral band. Additional reflected energy sources in the field of view are measured along with the target energy source.

A typical infrared thermometer is shown in Figure 11.51. Using infrared filters, an optical lens system focuses energy onto an infrared detector, which converts the energy to an electrical voltage. Fiber optics are typically used to separate the electronics from the application. Single-color infrared thermometers are the most common noncontact measurement device. These devices measure temperature within a narrow or wide segment of the infrared spectrum. Multicolor infrared thermometers measure the temperature at two or more segments of the infrared spectrum. If the emissivity values at the two segments are the same, the temperature can be accurately measured.

One of the major drawbacks to earlier infrared thermometers is that the operator must estimate the target emissivity in order to determine the target temperature. Recent instruments add a pulsed laser to a narrow bandwidth single-color infrared thermometer to determine the emissivity. The pulsed laser wavelength is within the same radiation bandwidth used to measure the target temperature. At the detector, the pulsed laser radiation (square wave) is superimposed on the relatively constant target radiation. With signal processing, the two components are separated and the pulsed laser radiation provides a measurement of the emissivity.

In a typical optical pyrometer (Figure 11.52), the unknown intensity of the light radiated from the hot target at 0.655 μm is compared with the brightness of the lamp filament. The temperature is determined by manually (or automatically) adjusting the lamp current until the brightness of the filament matches the brightness of the target (Figure 11.53). Optical pyrometers are better suited for manual operation or for applications that do not require fast sample rates.

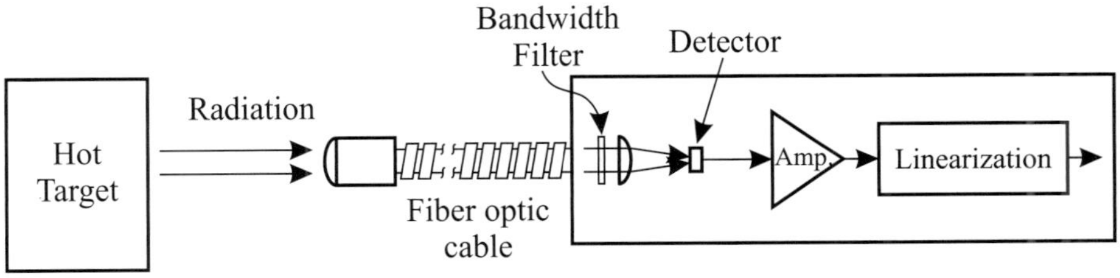

Figure 11.51. Infrared radiation detector.

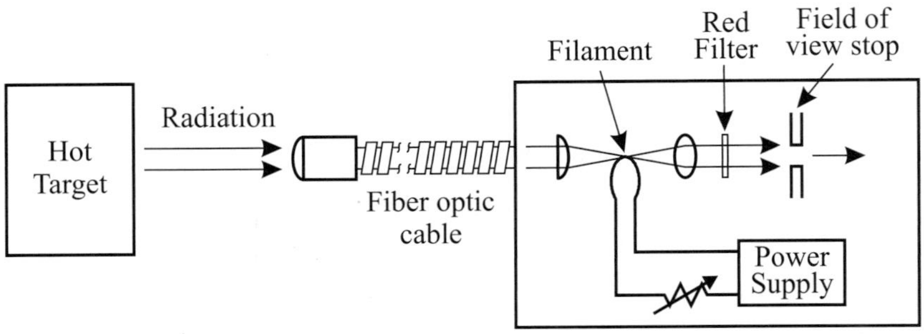

Figure 11.52. Optical pyrometer.

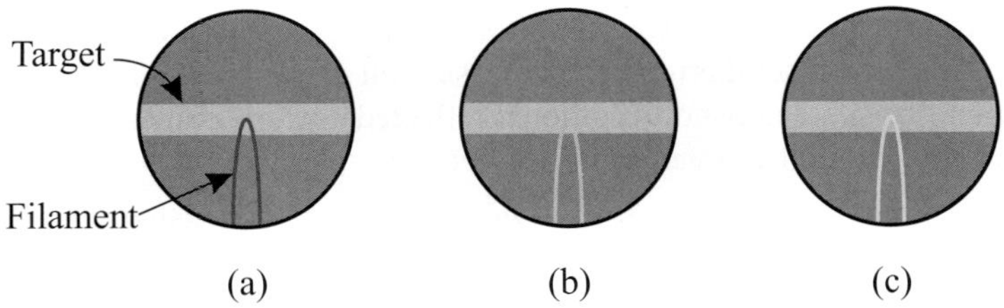

Figure 11.53. Disappearing-filament pyrometer: *(a)* filament cooler; *(b)* filament same temperature; *(c)* filament hotter.

11.3.3 Displacement Sensors

One of the easiest ways to measure linear displacement is with a potentiometer, as shown in Figure 11.54. It is relatively inexpensive, though less accurate than other ways to measure displacement. A rotary potentiometer can be used to measure rotary position.

A linear variable differential transformer (LVDT) is the most common method of measuring accurate linear position. As shown in Figure 11.55*a*, a LVDT is a transformer with two secondary windings and a moveable core. The two secondary windings are separated along the core and are connected in such a way that one secondary (secondary A) is in phase with the primary and the other secondary winding (secondary B) is out of phase with the primary. When the two secondary voltages are added and rectified, the output voltage, V, is proportional to the position of the moveable core. This relationship is shown

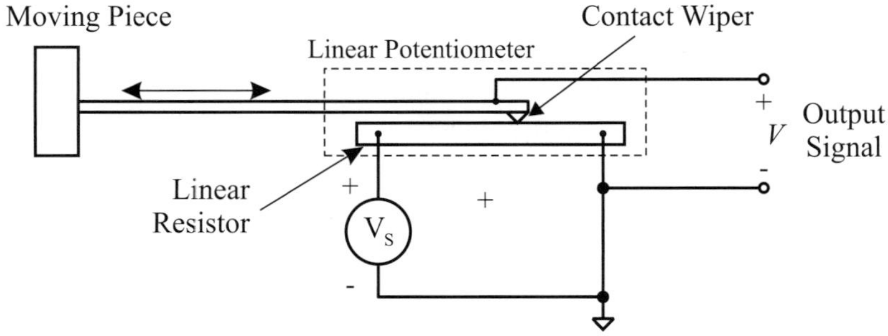

Figure 11.54. Linear potentiometer position sensor.

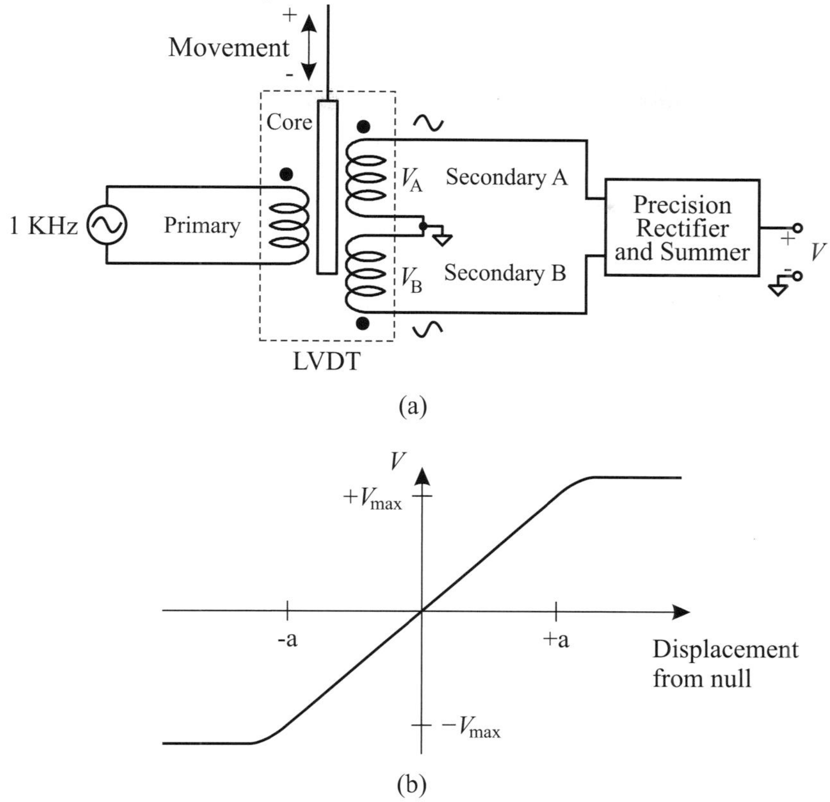

Figure 11.55. Linear variable differential transformer (LVDT): *(a)* general circuit; *(b)* output as function of position.

in Figure 11.55*b*. When the core is in the middle position, the two secondary voltages exactly balance each other and the output voltage is zero. This position is also called the null position. When the core is moved up from the null (Figure 11.55*a*), the coupling between the primary and secondary A is stronger and thus the output voltage is positive. When the core is moved down from the null, the coupling between the primary and secondary B is stronger and thus the output voltage is negative. In reality, the linear relationship between the position and the output voltage is only valid over a certain range (Figure 11.55*b*).

The typical accuracy of a LVDT is 0.025%. Though generally used for small displacements (< 5 cm), stroke lengths of up to 60 cm are possible. For larger displacements, a LVDT is prohibitively expensive.

A magnetostrictive position sensor uses the interaction between a magnet and a waveguide to determine the position of the magnet. It is a noncontact method of measuring position, that is, the object being sensed does not contact the sensor. Figure 11.56 illustrates the operation of a magnetostrictive position sensor.

Magnetostriction is the property of ferromagnetic materials (iron, nickel, cobalt) to change shape in the presence of a magnetic field. This effect is optimized by selecting the proper alloy and subjecting it to certain metallurgical operations. A wire made from a

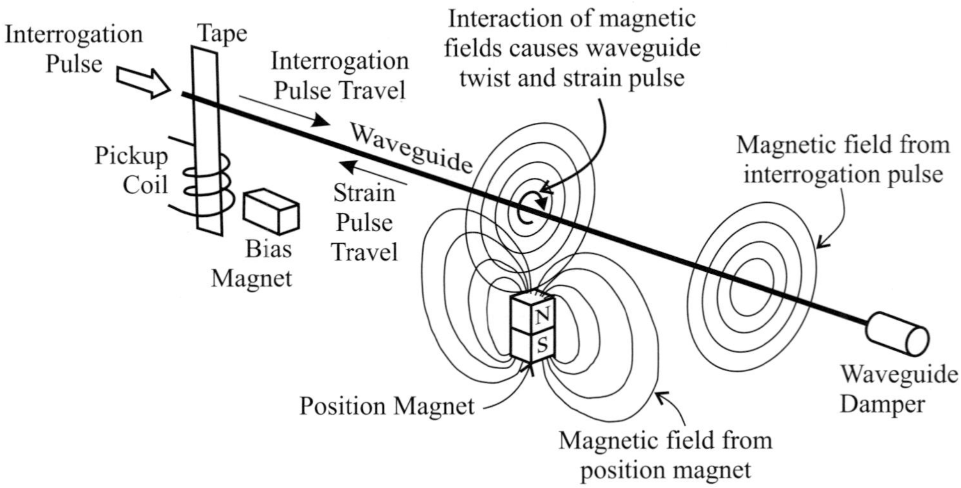

Figure 11.56. Magnetorestrictive position sensor operation.

magnetostrictive material has the Wiedermann effect. When a magnetic field is aligned with the wire and a current is passed through the wire, a twisting occurs at the location of the magnetic field.

A magnetostrictive sensor uses the Wiedermann effect. An interrogation current pulse is launched down the waveguide wire. When the magnetic field generated by the pulse travels down the waveguide and encounters the magnetic field of the position magnet, it generates a mechanical strain pulse. The strain pulse travels along the waveguide at sonic speed. The strain pulse is sensed through the Villari effect: applying stress to a magnetostrictive material changes its magnetic permeability. A small piece of magnetostrictive material, called the tape, is welded to the waveguide at the same end that the interrogation pulse is generated. A pickup coil is wrapped around the tape and a biasing magnet is placed next to this coil. The sonic strain pulse from the waveguide passes down the tape, changing its magnetic permeability and inducing a pulse in the pickup coil. The elapsed time between the launch of the interrogation pulse and the arrival of the pulse at the pickup coil indicates the magnetic position. Since the strain pulse travels in both directions along the waveguide, it must be damped at the waveguide end opposite the pickup.

Ultrasonic transducers (Figure 11.11) are generally used when sensing long distances. The distance is the time difference between the transmitted and received "ping" divided by the velocity of sound in the medium (usually air).

Rotary encoders can also be used to measure linear position. In this case, there must be a mechanical mechanism to convert the linear motion into rotary motion.

11.3.4 Force Sensors

Force transducers are frequently used to measure weight. Strain gauges are the most common way to measure force along an axis. Strain gauges are based on the principle that the resistance of a conductor changes as it is stretched. There are two types of strain gauges: unbonded and bonded.

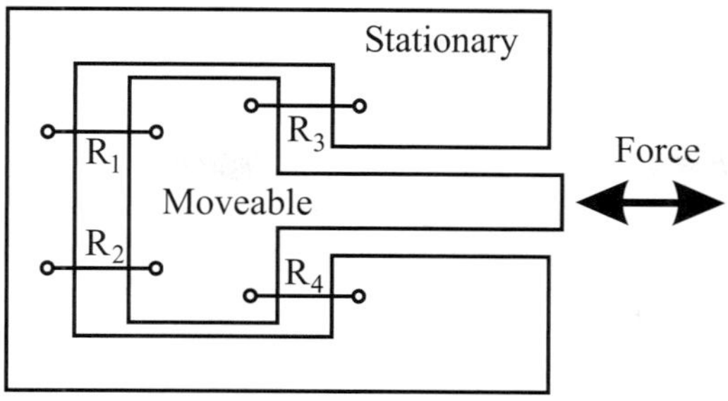

Figure 11.57. Unbonded strain gauge.

An unbonded strain gauge is shown in Figure 11.57. The moveable part is attached to a stationary part with four wires, labeled $R_1 - R_4$. Actually, each "wire" consists of a length of wire wrapped around the pair of nonconductive posts multiple times in order to increase the resistance change. As the moveable part moves right or left, one pair of wires stretches (increasing the resistance) and the other pair contracts (decreasing the resistance). The resistance changes are converted to voltage change with a voltage sensitive bridge (Figure 11.30).

More commonly, force is measured with a bonded strain gauge (Figure 11.58*a*). The strain gauge is composed of wire or foil mounted on a flexible backing which is then bonded to the material whose strain is being measured. The tensile or compressive force along the material causes it to stretch or shrink. The strain gauge is subject to the same movement and stretches or shrinks along the axis of the wire path. This movement changes the resistance of

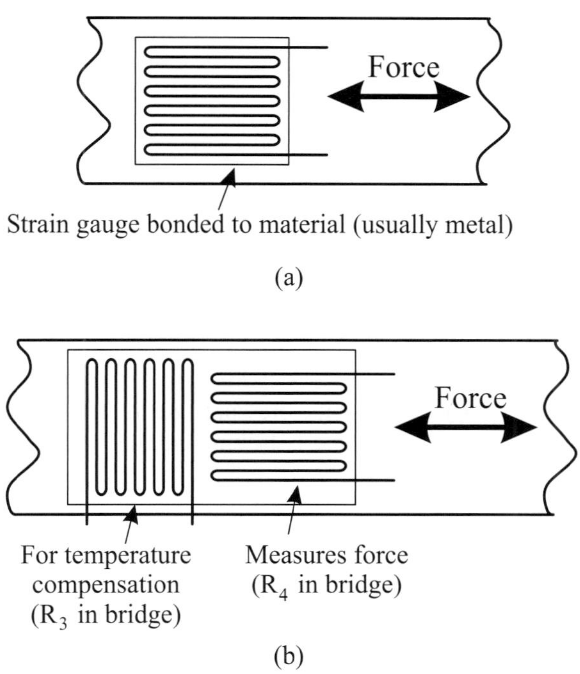

Figure 11.58. Bonded strain gauge: *(a)* element; *(b)* with temperature compensation.

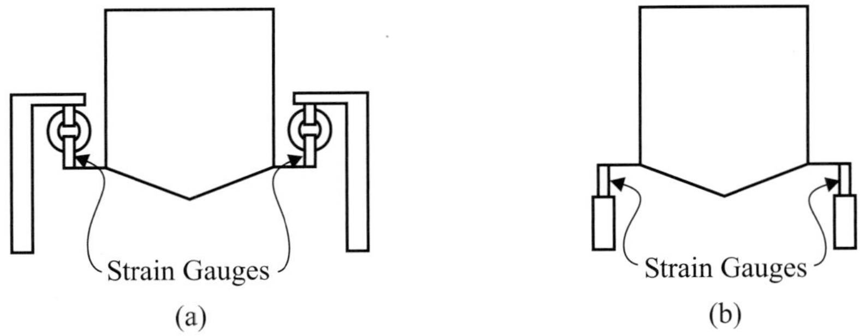

Figure 11.59. Vessel weight measurement with strain gauge: *(a)* tension; *(b)* compression.

the strain gauge. This resistance change is small, and so a Wheatstone bridge (Figure 11.30) converts the resistance change into a voltage change.

The strain gauge resistance is also dependent on the temperature, so to compensate for temperature changes, another strain gauge is mounted with its axis perpendicular to the force being measured. Since the foil of this strain gauge has very little material oriented in the direction of the measured force, its resistance is dependent on the temperature. This second strain gauge is connected as R_3 in Figure 11.30.

Strain gauges are commonly used to measure the weight of vessels, as shown in Figure 11.59. Though only two strain gauges are shown, the vessel will have three or four gauges measuring the weight. The sum of the weights is the weight of the vessel.

Strain gauges can also measure the force in multiple directions. In this case, each direction has its own strain gauge and bridge circuit, like Figure 11.58*b*.

11.3.5 Pressure Sensors

Pressure is always measured with respect to a reference pressure. There are basically three types of pressure-measuring situations (Figure 11.60):

1. Absolute pressure, where the reference is a complete vacuum.
2. Gage pressure, where the reference is atmospheric pressure.
3. Differential pressure, which is the difference between two pressures.

Note that gage pressure is a differential pressure.

In process control, the three situations are often differentiated by suffixing the pressure units with "a" (absolute pressure), "g" (gage pressure), or "d" (differential pressure). Also note that the standard atmospheric pressure is 14.7 pounds per square inch (psi).

The most common way of measuring pressure is with a capacitive transducer (Figure 11.61*a*). In a capacitive transducer, the difference in pressure at the two isolating diaphragms is coupled to a sensing diaphragm. The sensing diaphragm is fixed between two capacitor plates and as the sensing diaphragm moves, the capacitance changes. The capacitance is directly related to the pressure difference, and an electronic circuit converts the measured capacitance to the pressure. The entire pressure meter that contains the measurement cell is shown in Figure 11.61*b*.

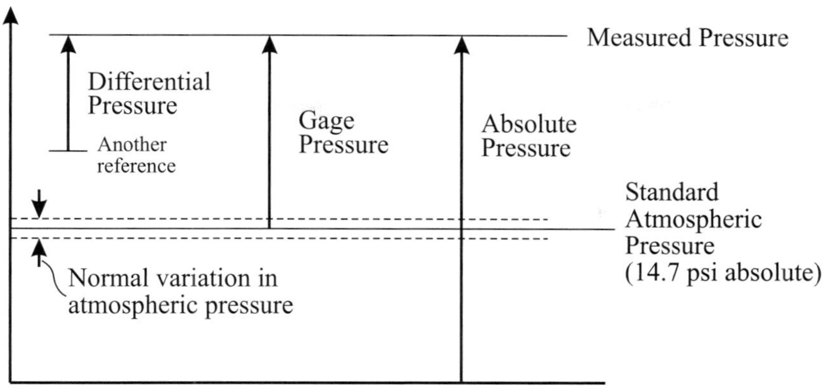

Figure 11.60. Relationship of absolute, gage, and differential pressures.

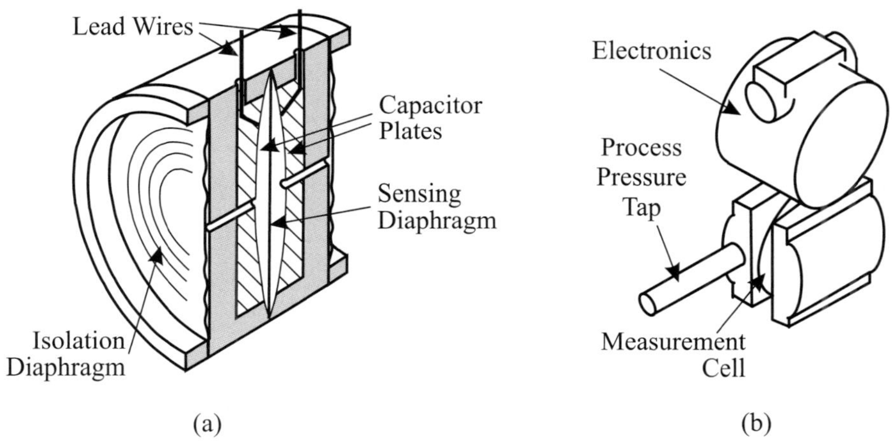

Figure 11.61. Capacitive pressure measurement: *(a)* measurement cell; *(b)* differential pressure meter.

Alternatively, a strain gauge can be bonded directly to the diaphragm (Figure 11.62) to measure pressure. The spiral foil in the center measures the tangential strain and the radial foil around the outside measures the radial strain. The measured force divided by the diaphragm area is the pressure.

Another method of measuring pressure involves a piezoresistive bridge mounted on a thin silicon diaphragm (Figure 11.63). Each of the bridge resistors is a resistor constructed from silicon. The bridge resistors are arranged so that when pressure is applied to the diaphragm, R_{P1} and R_{P2} increase in value and R_{V1} and R_{V2} decrease in value. These changes cause a millivolt output across the S+ and S- terminals.

Regardless of the particular method of measuring the pressure, one side of the diaphragm is coupled to the process fluid. How the other side of the diaphragm is handled determines the type of measurement. If a chamber on the other side of the diaphragm is completely evacuated, the sensor measures absolute pressure. If the other side of the

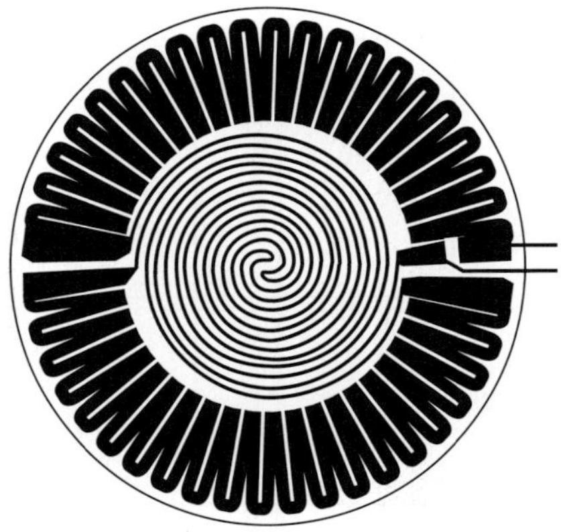

Figure 11.62. Pressure measurement with strain gauge on diaphragm.

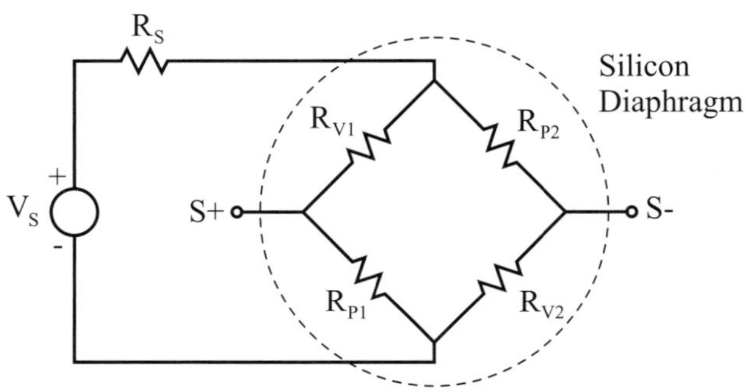

Figure 11.63. Semiconductor pressure sensor.

diaphragm is exposed to the atmosphere, the sensor measures gage pressure. A differential pressure sensor connects the other side of the diaphragm to the reference pressure.

11.3.6 Flow Sensors

There are many methods of measuring the flow of gases, liquids, and solids. Only the most common methods are described here. Gas and fluid flow is emphasized because of the larger variety of measuring methods.Table 11.10 compares common flow sensors.

Fluid and gas flow sensors can be categorized by the manner in which they measure flow:

1. *Flow sensors with wetted moving parts* (such as turbine meters). These sensors have precision-machined moving parts that are in the process stream. For this reason, they are generally only applicable to clean fluids.
2. *Flow sensors with no wetted moving parts* (such as orifice and vortex). The lack of moving parts is an advantage, though they are subject to excessive wear if measuring the flow of a very dirty or abrasive stream. These sensors require an obstruction that is inserted in the flow stream.

Table 11.10 Flow Sensor Comparison

Sensor Type	Accuracy[1]	Gas[3]	Liquids	Slurries[3]	Comments
Orifice Plate	3%	Y	Y	N	Clean liquids
Venturi Tube	3%	Y	Y	S	Clean liquids
Pitot Tube	4%	Y	Y	N	Air and water
Turbine	0.5%[2]	S	Y	N	Clean liquids
Magnetic	0.5%[2]	N	Y	Y	
Vortex	1%[2]	Y	Y	N	Clean liquids
Transit Ultrasonic	4%	S	Y	N	Clean liquids
Doppler Ultrasonic	4%	N	Y	S	Dirty liquids
Coriolis	0.4%[2]	S	Y	S	

Notes:
1. Percentage of full scale.
2. Percentage of flow rate.
3. Y=yes; N=no; S=sometimes, suitable under certain conditions.

3. *Flow sensors that do not obstruct the stream* (such as magnetic and Coriolis). These sensors are suitable for dirty and abrasive fluids. They are also suitable for very viscous fluids.
4. *Flow sensors mounted externally to the stream* (such as a clamp-on ultrasonic). These sensors do not come in contact with the stream, but because of their limitations, they are not used in many applications.

Fluid and gas flow sensors may also be categorized as to what they measure directly:

1. *Velocity*, such as turbine and orifice meters. The volumetric flow is determined by multiplying the velocity by the cross-section area.
2. *Volumetric*, such a positive-displacement meter. These types of meters are not discussed here.
3. *Mass*, such as Coriolis mass flow meters.

Differential Pressure Flowmeters

A very common method of measuring fluid or gas flow is to measure the pressure drop across an orifice plate or venturi tube inserted into the stream (Figure 11.64). The orifice plate or venturi tube is called the primary element and the differential pressure (DP) meter is called the secondary element.

The orifice plate (Figure 11.64*a*) is a flat piece of metal with a hole of a certain size bored in it. The hole is sharp-edged in order to obtain the sharpest drop in pressure. The orifice diameter is typically around one-half of the pipe diameter. The most common location for the two DP meter pressure taps are 25 mm to either side of the orifice plate. Another common location for the pressure taps, called vena contracta taps, are located one pipe diameter upstream and at the vena contracta (point of minimum pressure) 0.3 to 0.8 pipe diameters downstream of the orifice. Orifice plates are easy to install, have no moving parts, and application data is generally more extensive than for other differential pressure flow meters.

The venturi tube (Figure 11.64*b*) consists of a section of pipe with a conical entrance, a short straight throat, and a conical outlet. The DP meter taps are upstream of the inlet and at

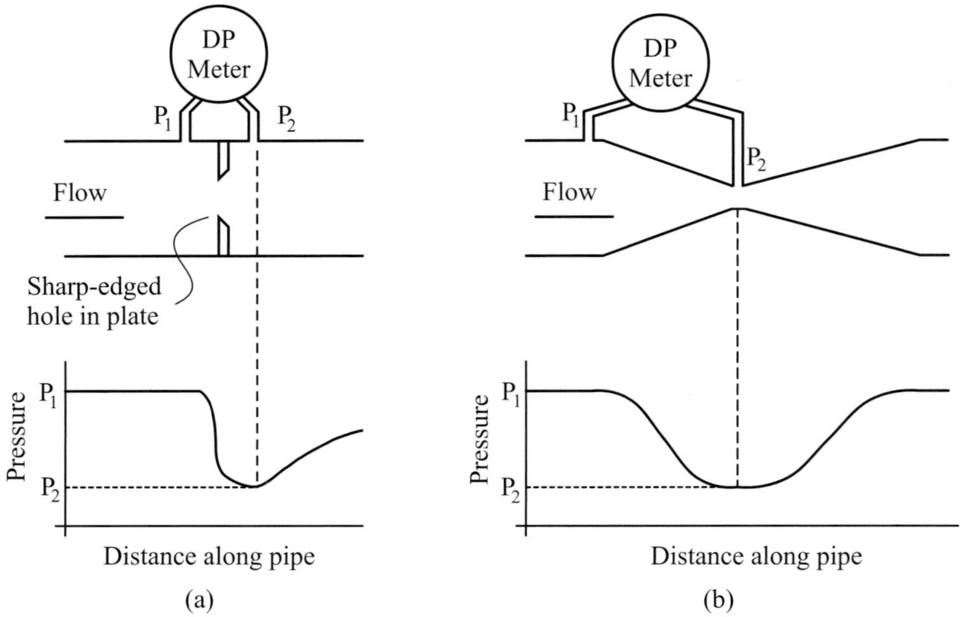

Figure 11.64. Flow measurement with differential pressure sensors: *(a)* orifice plate; *(b)* venturi tube.

the throat. A venturi tube will measure more flow than a comparable orifice plate, but is larger and more expensive than an orifice plate.

When measuring the flow through large air or gas ducts, a pitot tube (Figure 11.65) is used. The probe consists of two parts. The impact pressure is sensed with a tube bent toward the flow, and the static pressure is sensed by another closed tube with a hole in the side away from the flow. Since the velocity is not evenly distributed across the duct, a more accurate measurement of velocity is obtained with an averaging pitot tube. An averaging pitot tube has four or more impact pressure taps located at mathematically defined locations (Figure 11.65*b*).

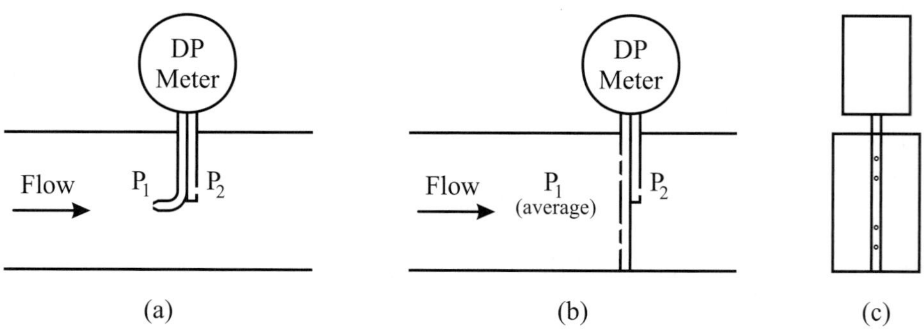

Figure 11.65. Flow measurement with pitot tube: *(a)* non averaging tube; *(b)* averaging tube; *(c)* end view of averaging tube.

One of the disadvantages of all differential pressure flow meters is their low accuracy (typically 3% of full scale) and low turndown ratio. The turndown ratio is the maximum flow rate divided by the minimum flow rate. The turndown ratio for differential flow meters is 4:1 at best.

For all of the differential pressure flow measurement techniques,

$$\text{Velocity of fluid (or gas)} = k\sqrt{\Delta P} = k\sqrt{P_2 - P_1}$$

The volumetric flow rate, Q, is thus

$$Q = \text{velocity} \times \text{area} = Ak\sqrt{\Delta P} = K\sqrt{\Delta P}$$

For this reason, square-root conversions are common in process control applications.

When measuring gas flow rates, temperature and pressure compensation is required if the temperature and pressures change from their original design values (Battikha, 2007). Liquids are incompressible, but will require temperature compensation for accurate measurements.

All differential pressure flow meters measure volumetric flow rate, that is, volume/time. For gases, the volumetric flow rate can be converted to mass flow rate using the ideal gas law:

$$PV = nRT$$

Converting to flow rates,

$$P\frac{dV}{dt} = \frac{dn}{dt}RT \tag{11.10}$$

where dV/dt is the volumetric flow rate, dn/dt is the mass flow rate, R is the gas constant, P is pressure and T is temperature. Using Q for the volumetric flow rate, equation (11.10) can be rearranged to calculate the mass flow rate, given the volumetric flow rate

$$\frac{dn}{dt} = \frac{P}{RT}Q \tag{11.11}$$

For liquids, one needs to know the liquid density, ρ, to calculate the mass flow rate from the volumetric flow rate

$$\frac{dn}{dt} = \rho Q \tag{11.12}$$

Turbine Flowmeters

A turbine flowmeter (Figure 11.66) has a rotor immersed in the liquid stream that turns as the fluid moves through it. A magnetic pickup mounted above the turbine senses the passage of each rotor blade, generating a pulse waveform. Since each pulse represents the passage of a certain amount of fluid, the volumetric flow rate is proportional to the frequency of the pulses. Pelton wheel and paddlewheel flow meters are similar to a turbine flow meter except that the rotor is only partially submerged in the flow stream.

Turbine flowmeters are good for low flow rates and are not sensitive to changes in fluid density. In addition, by counting the pulses, the amount of fluid passing through the turbine can be determined. However, they are not suitable for dirty or viscous fluids. In addition, they are affected by any entrained gas in the fluid. Because they have moving parts, turbine flowmeters are more susceptible to wear.

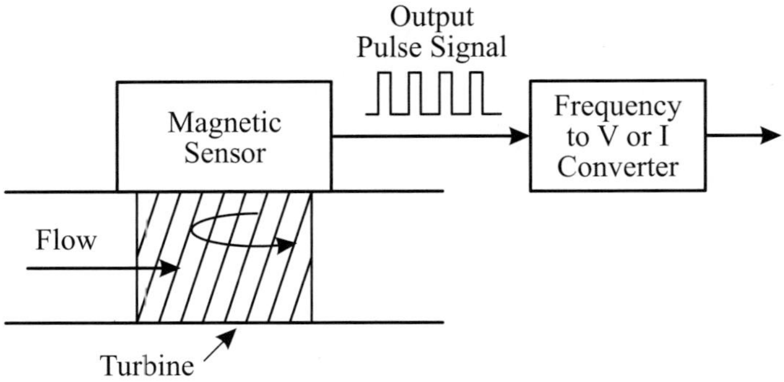

Figure 11.66. Turbine flowmeter.

Magnetic Flowmeters

A magnetic flowmeter is based on Faraday's law. When a conductor is moved perpendicular through a magnetic field, a voltage is induced in the conductor that is proportional to the velocity of the conductor. In a magnetic flowmeter (Figure 11.67), the flow is perpendicular to a magnetic field. The voltage induced by the moving fluid is sensed by two electrodes that are aligned perpendicular to both the flow and the magnetic field. The voltage is proportional to the average velocity of the fluid. The fluid must be electrically conductive. However, the magnetic meter has no moving parts and is unaffected by changes in the fluid viscosity and pressure.

Vortex Flowmeters

In a vortex flowmeter (Figure 11.68) an obstruction, called a bluff body, is placed in the flow stream. This obstruction is a particular shape and sets up vortices in the flow stream. The frequency of the vortices is proportional to the volumetric flow rate. Vortex vibrations are often sensed by strain gauges, capacitive sensors, or magnetic pickups. The vibration frequency is then converted to a volumetric flow rate. The vortex flow meter is suitable for fluids that are not too viscous and that do not contain a significant amount of solids.

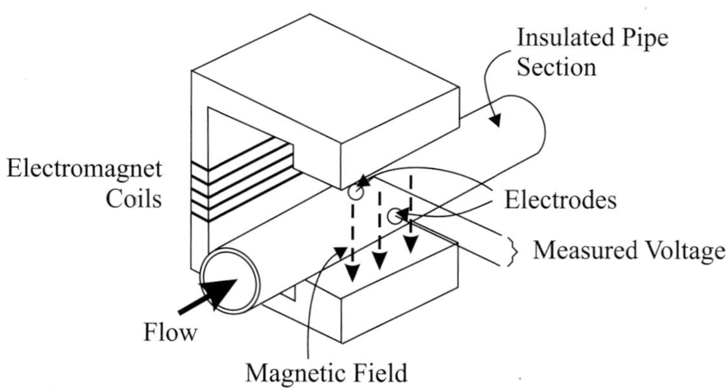

Figure 11.67. Magnetic flowmeter.

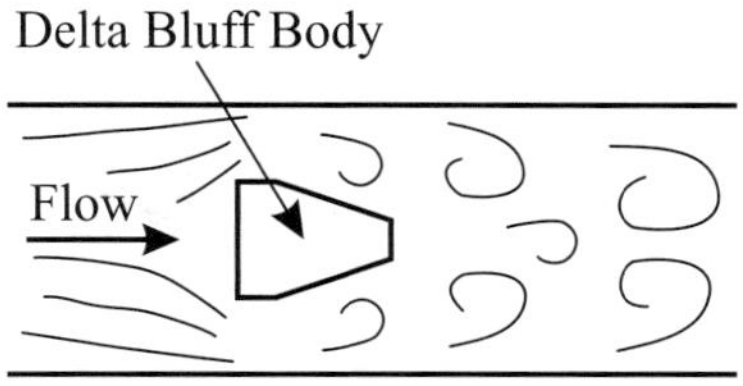

Figure 11.68. Vortex flowmeter.

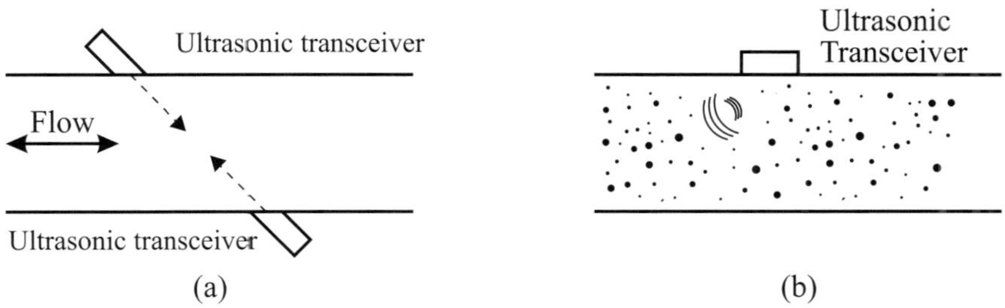

Figure 11.69. Ultrasonic flowmeters: *(a)* transit-time; *(b)* Doppler.

Ultrasonic Flowmeters

There are two types of ultrasonic flowmeters. Both the transit-time and Doppler ultrasonic flowmeters do not obstruct the flow and can measure flow in either direction. Both of them can be clamped to the outside of a pipe which is an advantage when adding a flowmeter to an existing application.

The transit-time flowmeter (Figure 11.69*a*) uses two ultrasonic transceivers placed at a 45 degree angle to the flow stream. Both send out ultrasonic pulses that are detected by the other. The difference in transit time between the two beams is proportional to the flow rate. The ultrasonic pulses that are traveling in the direction of the flow travel faster than the pulses that are traveling against the flow. This type of flowmeter cannot be used with porous pipes (such as cast iron or cement pipes).

The ultrasonic Doppler flowmeter (Figure 11.69*b*) transmits ultrasonic pulses into the flow stream. Bubbles and other foreign matter in the flow stream reflect the pulses which are received by the transceiver. The velocity of the particles change the frequency of the reflected pulses. This frequency shift is proportional to the flow rate. The reflected pulses are averaged to obtain an average flow rate. In order for this type of flowmeter to operate, the flow stream must have entrained gas or other particles.

Coriolis Flowmeter

The previous flowmeters measured either the stream velocity or volumetric flow rate. By also measuring the temperature (and pressure for gas streams), one can calculate the mass flow rate. The Coriolis flow meter measures the mass flow rate directly.

In its most common configuration, a Coriolis flowmeter consists of two U-shaped tubes through which the fluid flows (Figure 11.70). Note that the full flow of the fluid travels through the tubes. The "pipe" at the top of the flowmeter is for support. Figure 11.71 illustrates the operation of this flowmeter. Each tube is forced to oscillate at its natural frequency (on the order of 80 times a second). The motion in Figure 11.71*a* is exaggerated.

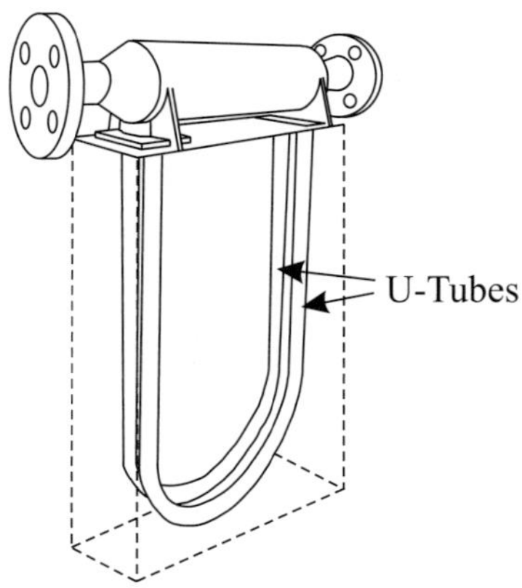

Figure 11.70. Coriolis flowmeter.

The actual motion is a few millimeters. As the tube moves right, incoming fluid resists by pushing left. Outgoing fluid pushes right because it resists being moved left and out. These forces set up a twist in the tube. When the tube moves left in the second half of the vibration cycle, the twist is in the opposite direction. This combination of forces that causes the tube twist is called the Coriolis effect. The mass flow is proportional to the maximum twist angle.

The Coriolis flow meter measures most streams, including viscous and abrasive streams. However, gas pockets in the stream affect the measurement. Since it measures mass flow rate directly and fairly accurately, the Coriolis flowmeter is becoming popular in blending applications. It is relatively costly, the only real reason its use is not more widespread.

Solid Flow

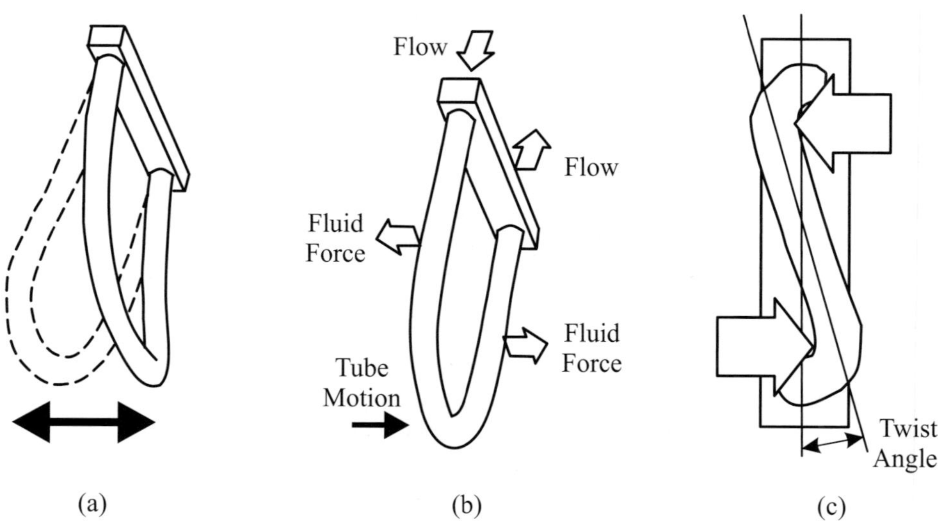

Figure 11.71. Coriolis flowmeter operation: (a) tube vibration; (b) forces on tube; (c) resultant tube twist.

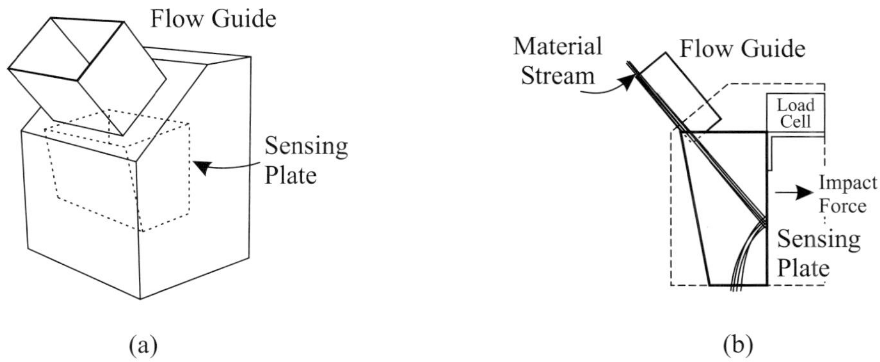

Figure 11.72. Dry solids flowmeter: *(a)* overall view; *(b)* principle of operation.

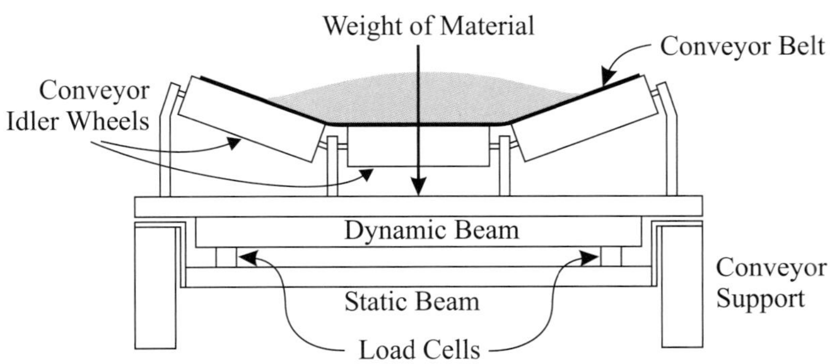

Figure 11.73. Conveyor belt scale for solid flow measurement.

There are two common ways of measuring the flow of solid material. Figure 11.72 shows a flow sensor that measures material being gravity-fed. The material passes down a flow guide at a certain angle and then impinges on a sensing plate. The force on the sensing plate, measured by the load cell, is proportional to the flow rate (Siemens, 2015b).

When the material is being conveyed on a belt, an easy method of measuring the material flow is to place a weigh scale below one of the idler assemblies (Figure 11.73). In this case, the weight on the idler wheels is transferred to a dynamic beam (Siemens, 2015a). Strain-gauge load cells between the dynamic beam and the static beam measure the weight. The weigh scale is calibrated to "factor out" the belt weight, so the weight is zero when the belt contains no material. Multiplying the weight by the belt speed (sensed by an encoder mounted on an idler wheel) calculates the solid flow rate.

11.3.7 Level Sensors

Level measurements are key indicators of material inventory within a system. Technically, level sensors are used to measure the position of the boundary between two materials. Commonly, the interface is between air and a solid or between air and a liquid. However, there are some applications where the interface between two liquids (for example, water and oil) is of interest.

Table 11.11 Level Sensor Comparison

Sensor Type	Accuracy[1]	Liquid	Solid	Limitations
Differential Pressure	0.5 - 2%	Y	N	Plugging of lines is a problem
Bubbler	1 - 2%	Y	N	
Capacitance	1 - 2%	Y	Y	Material buildup on probe
Ultrasonic	0.25 - 0.5%	Y	Y	Foam and powders
Radar	0.5 - 5 mm[2]	Y	Y	

Notes:
1. Percentage of full scale.
2. Absolute accuracy.

The three main categories of level measurement are:

1. Pressure head
2. Position of the boundary
3. Weight of the material

The first two categories are covered in this section. Figure 11.59 illustrates vessel weight measurement. The most common methods of measuring level are listed and compared in Table 11.11.

Differential Pressure

Using a pressure sensor is a common way of measuring the level of a liquid tank. If the tank is open at the top, a gage pressure sensor at the bottom of the tank (Figure 11.74*a*) measures the *hydrostatic* head, the pressure exerted by the fluid. If the tank is closed, then the differential pressure relative to the pressure of the atmosphere on top of the liquid must be used (Figure 11.74*b*). The pressure divided by the density is the height, though the pressure transmitter can be calibrated to produce an output in height units. Since it is usually difficult to perfectly align the sensor element with the tank bottom, the zero point of the pressure transmitter is often adjusted in order to measure the height of the fluid relative to the tank bottom.

This type of level measurement is easy to install and has a wide range of measurement. The pressure sensor diaphragm material must be compatible with the fluid in the tank. Also,

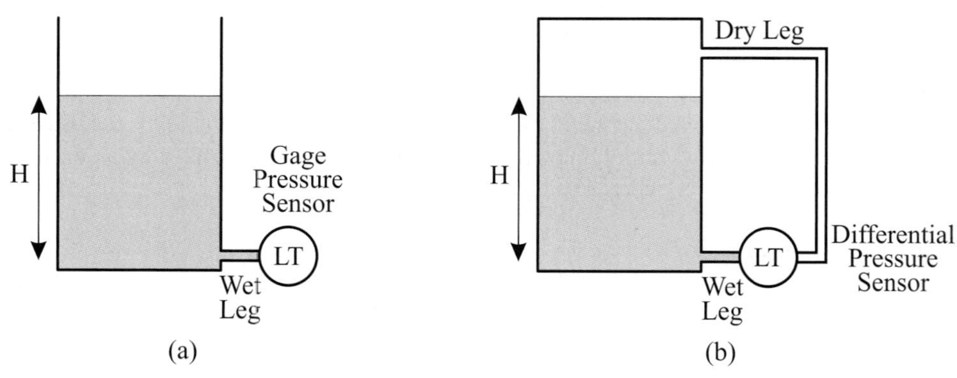

Figure 11.74. Liquid level measurement with pressure sensor: *(a)* open tank; *(b)* closed tank.

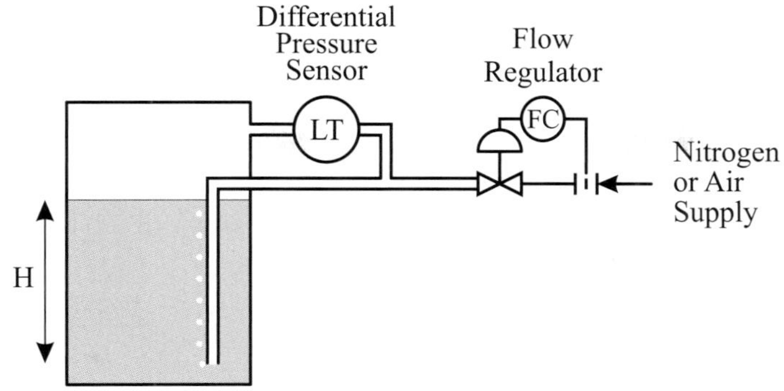

Figure 11.75. Liquid level measurement with bubbler tube.

small diameter piping between the pressure meter and the tank is susceptible to plugging. This type of level measurement assumes the fluid density remains constant.

Bubbler Tube

When the fluid in the tank is highly corrosive, an alternate method of measuring liquid height involves a bubbler tube. In this method (Figure 11.75), a small amount of air or inert gas is forced through a tube that extends nearly to the bottom of the tank. The end of the tube is notched to produce uniform small bubbles. The pressure required to force air bubbles from the bottom of the tube is the hydrostatic pressure exerted by the fluid level. A flow controller provides a constant air flow (5 to 50 m^3/hr.) to the bubbler tube. The pressure of the air supply should be at least 10 psi greater than the maximum hydrostatic pressure when the tank is full.

Overall, this scheme costs more than a simple differential pressure meter. If an inert gas is required, then it must be provided. The additional flow controller adds to the cost. In some cases, a pressure regulator must be added to the air supply to the flow controller to provide smoother level measurements.

Capacitance

A capacitance level probe measures the electrical capacitance between the probe and the vessel wall (Figure 11.76). This capacitance varies as the tank level varies. Capacitance level probes are often called RF probes due to their use of radio-frequency signals (a few MHz) to measure the capacitance. Capacitance probes can be used for both liquids and solids. There are three basic configurations. If the tank is metal and the tank contents are non-conductive, the probe consists of a metal rod (Figure 11.76*a*). In this case, the rod and the tank wall are the capacitor plates and the material in the tank is the dielectric. If the tank contents are conductive, then the probe is insulated (Figure 11.76*b*). In this case, the probe and the tank contents are the capacitor plates and the insulation is the dielectric. If the tank is an irregular shape (not a vertical cylinder) or constructed of plastic, the probe consists of a rod surrounded by a cylinder (Figure 11.76*c*). The outer cylinder serves as the other capacitor plate. This configuration has the advantage that the probe electronics can be calibrated before the probe is placed inside the tank.

Capacitance probes are relatively easy to install though calibration may be time consuming. The buildup of conductive material on the outside surface of the insulated

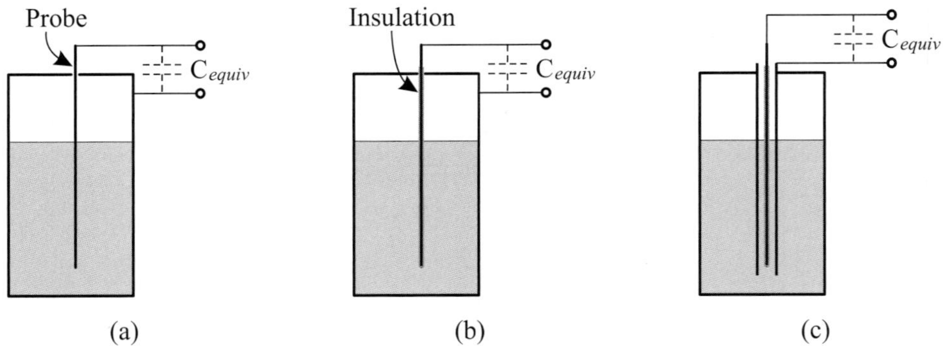

Figure 11.76. Capacitance level measurement: *(a)* non-conductive material; *(b)* conductive material; *(c)* nonmetallic tank.

probe has the most detrimental effect on the accuracy of a capacitance level sensor. Materials that produce conductive coatings include latex, carbon black, and fine metal powders.

Ultrasonic

Ultrasonic level sensors use sound waves to detect an object. The basic principle is to emit a "pulse" of a frequency slightly above the normal range of audible sound and then measure the time for the pulse to be detected by the receiver. The sensor usually combines the transmitter and receiver in one piezo-electric element. The operation of ultrasonic level sensor is shown in Figure 11.77. The wave pulse is generated by the piezo-electric element and then the element is switched to its receiving mode. The wave is reflected by the object and then detected by the element. The drawback to this approach is that there is a minimum sensing distance. The trailing edge of the pulse must leave the transducer before the leading edge of the reflected pulse can be detected. Note that with the sensor configuration in Figure 11.77, the distance measured is the air space over the material. The distance must be subtracted from the tank height to obtain the material height. Many sensors are calibrated to do this conversion automatically. The sensor can be mounted at the bottom of the tank, but this configuration exposes the sensor to the material.

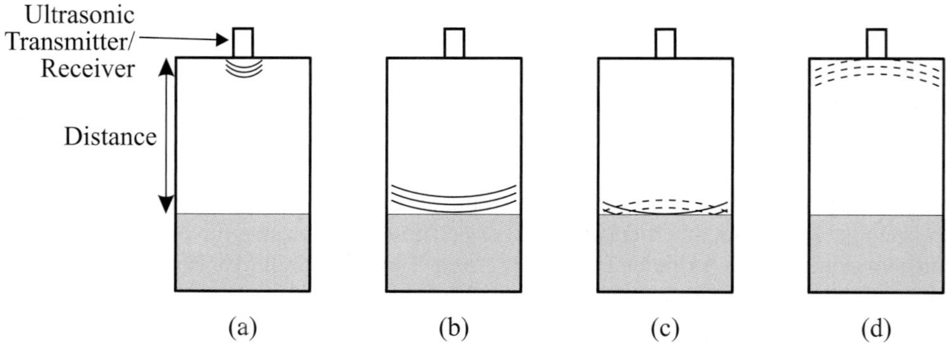

Figure 11.77. Ultrasonic level measurement: *(a)* wave launched; *(b)* wave impinges surface; *(c)* wave reflected; *(d)* wave impinges receiver.

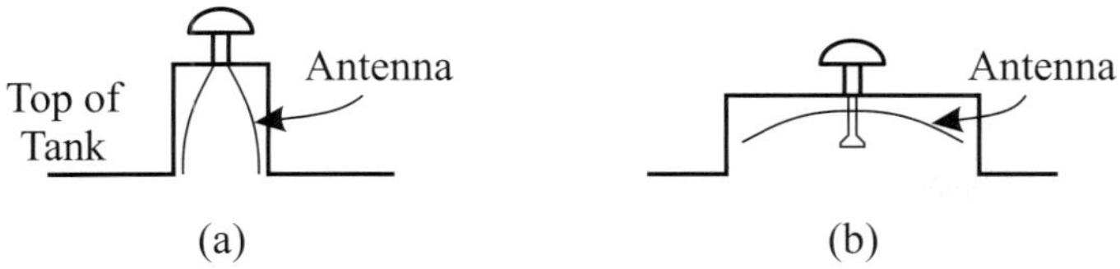

Figure 11.78. Radar level sensors: *(a)* horn antenna; *(b)* parabolic antenna.

Because ultrasonic sensors depend on receiving a reflected sound wave for proper operation, the material top surface should be a good reflector of sound waves. Preferably, the material should have a smooth and flat surface, that is, liquid. However, most solid materials are fine, but at a reduced maximum sensing distance. This type of sensor has problems when the liquid has foam on top or the solid is powdery. Also, since the speed of sound is dependent on temperature, an accurate measurement requires temperature compensation.

Radar

Radar level sensors are similar in operation to ultrasonic level sensors except that the frequency is typically around 10 GHz. Radar level sensors either "pulse" the wave or frequency-modulate a continuous wave. The operation of the "pulse" type radar sensor is similar to the ultrasonic level sensors described previously. The continuous wave radar sensor sends out a frequency modulated signal, usually in successive linear ramps. The frequency difference between the transmitted and received signals indicates the distance. Radar level sensors are mounted on top of the tank and measure the distance of the air space over the material. Sample radar transmitter/receiver antennas are shown in Figure 11.78.

As for the ultrasonic sensors, the reflective properties of the material surface affect the strength of the received radar signal. Liquids are good reflectors and solids are not good reflectors. In contrast to ultrasonic level sensors, radar can detect the liquid level beneath airy foam or dust. However, if the foam or dust is thick, its level will be detected instead of the underlying liquid. Compared with ultrasonic level sensors, the composition of the air space does not affect the performance of radar sensors. However, ultrasonic sensors work better in dirty applications or with solids whose grain size is greater than 20 mm.

11.3.8 Chemical Sensors

Chemical sensors measure some chemical property aspect of a process control material. The three most common chemical measurements are pH, oxidation-reduction potential (ORP), and conductivity. All three of these types of measurements are briefly described in this section.

pH

pH is a measure of the acidity or alkalinity of a water solution. The relative number of hydrogen (H^+) and hydroxide (OH^-) ions determines the acidity or alkalinity of a water solution. If the number of hydrogen ions is higher, the solution is acidic. If the number of hydroxide ions is higher, the solution is alkaline (also called basic). If the number of hydrogen and hydroxide ions is the same, the solution is neutral. In any water-based solution a very small number of the water molecules will disassociate to form hydrogen and hydroxide ions:

$$H_2O = H^+ + OH^-$$

The concentration of the hydrogen ions multiplied by the concentration of hydroxide ions is a constant. Expressed as an equation,

$$K_W = [H^+]\,[OH^-]$$

where the brackets signify molar concentration and K_W is the dissociation constant for water. The value of K_W is dependent on temperature and is 1.00×10^{-14} at 25 °C. In its strict definition, pH is the negative logarithm of the hydrogen ion activity. In practice, the pH is another way of expressing the hydrogen ion concentration, defined as

$$pH = -\log[H^+] \tag{11.13}$$

For example, if the hydrogen ion concentration is 1.00×10^{-5} moles/liter, the pH is 5.00. A neutral solution has a pH of 7.0, if the temperature is 25 °C. An acidic solution has a pH greater than 0.0 and less than 7.0. A smaller pH value indicates a stronger acid. A basic solution has a pH greater than 7.0 and less than 14.0. A larger pH value indicates a stronger base.

Using equation (11.13) as a basis for measuring pH means that one measures the hydrogen ion concentration over 15 decades, which is not practical. Instead, the measurement of pH is based on the use of a pH-sensitive electrode, a reference electrode, and a temperature sensor.

A simplified pH measurement cell is shown in Figure 11.79. The potential difference is measured between the pH-sensitive electrode (called the measuring electrode) and the reference electrode. The temperature sensor, usually an RTD or thermistor, is used for temperature compensation. The measuring electrode, the reference electrode, and the temperature sensor are often packaged together, as shown in Figure 11.80. The measuring electrode and the reference electrode are both electrochemical half-cells. The measuring electrode consists of a silver wire coated with silver chloride immersed in a chloride solution of fixed concentration and thus fixed pH. The chloride solution is contained in a bulb of pH-sensitive glass that allows hydrogen ions to diffuse into the chloride solution. Electric potentials are developed at the interfaces between the wire and the filling solution, the filling solution and the glass, and the glass and the liquid whose pH is being measured.

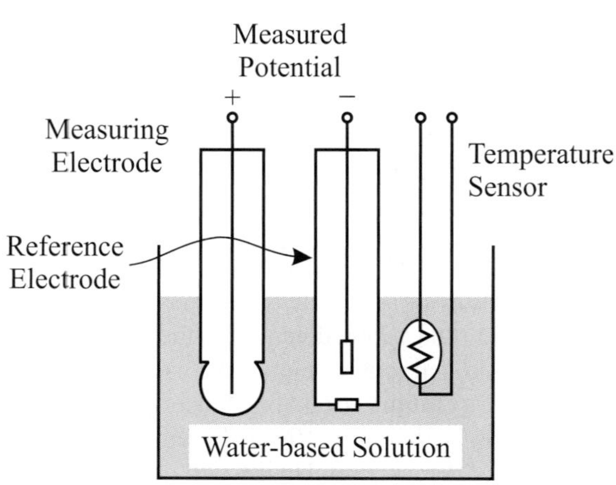

Figure 11.79. Simplified pH measurement cell.

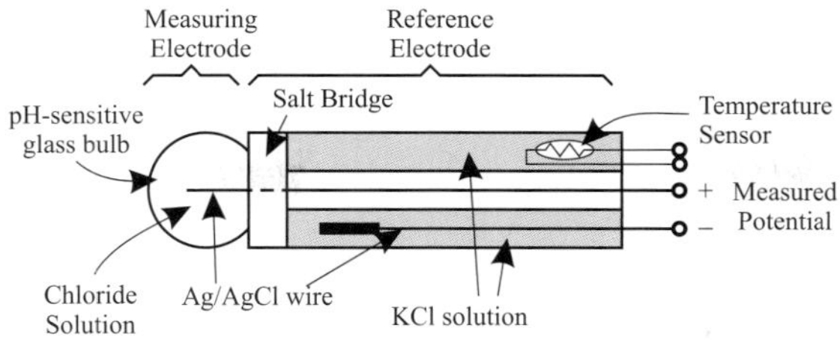

Figure 11.80. Combination pH probe construction.

The potential developed on the outside of the bulb depends on the pH of the test liquid. Since the other two potentials are fixed, the overall potential developed by the measuring electrode half-cell is dependent on the test liquid pH. The reference electrode consists of a silver wire coated with silver chloride immersed in a potassium chloride solution within a glass tube. For most industrial pH probes, the reference solution is a gel. The salt bridge consists of a porous material and it provides the electrical contact between the liquid whose pH is being measured and the reference electrode. Electric potentials are developed at the interfaces between the wire and the potassium chloride solution and across the salt bridge. The electric potentials of both of these interfaces are constant and hence the potential developed by the reference half-cell is constant.

Although the value of pH can range from 0 to 14, solutions whose pH is near either end of the range are better measured with a conductivity sensor (Rosemount Analytical, 2001). Solutions with high pH (>13) quickly destroy glass probes. Since the salt bridge allows ion exchange between the reference cell and the measured liquid, certain chemicals, such as bisulfites and ammonia, can poison the reference cell. Multiple junction reference electrodes can counteract this effect. Undissolved solids in the measured liquid can plug the salt bridge or coat the sensor. In this case, on-line or manual cleaning may be needed.

Oxidation-Reduction Potential (ORP)

In certain applications, the oxidizing or reducing nature of a chemical is of importance. In drinking water and wastewater treatment, the oxidizing or reducing nature of water measures the ability of the water to support (or not support) life. In a metal plating application, the oxidation-reduction potential measures the metal depletion in the plating bath.

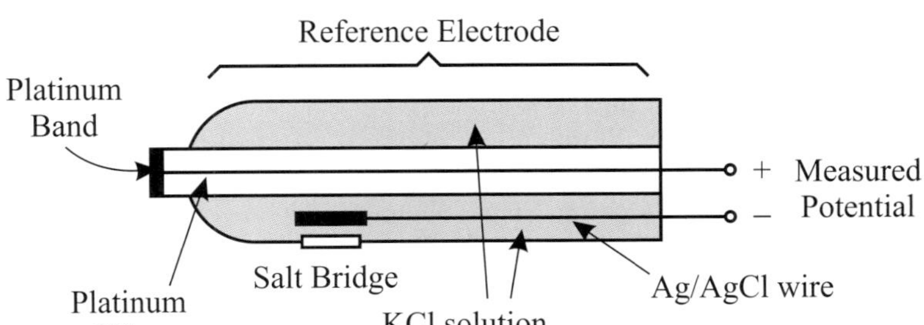

Figure 11.81. Combination ORP probe construction.

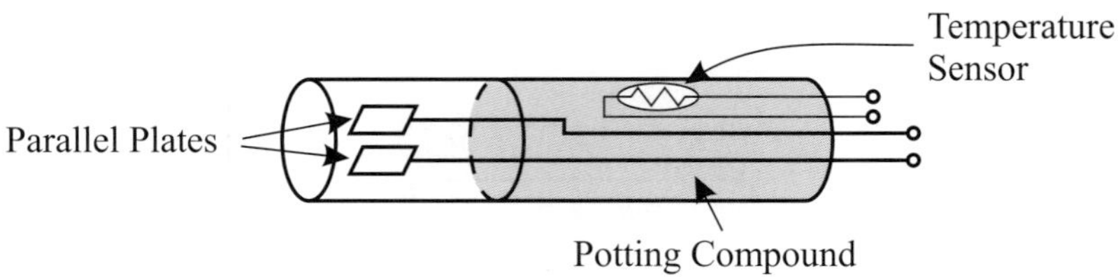

Figure 11.82. Conductivity probe construction.

An oxidation-reduction potential (ORP) sensor measures the ratio of concentrations of oxidized and reduced substances. An ORP measurement sensor is an electrochemical cell similar to a pH sensor; the basic difference is the measurement electrode (Figure 11.81). The measurement electrode consists of a band or disk of a noble metal (usually platinum or gold) attached to the base of a sealed glass tube. A wire of the same metal is welded to the band to provide a connection to the electrode lead. The glass tube is not filled. The cell voltage is the ORP of the measured solution.

Conductivity

Conductivity is the ability of a solution to conduct an electric current between two electrodes. The current between the two probes is conducted by ions in the solution. When the solution has a larger number of ions, the conductivity is higher. The electric current is an alternating current to prevent complete ion migration to the two electrodes. Conductivity is the reciprocal of resistivity and is usually measured in a unit of measurement called the "Siemen/cm", abbreviated as "S." For many solutions, the conductivity is in the range of μS or mS.

An example conductivity probe is shown in Figure 11.82. The electrodes are insulated by the potting compound and the surface and spacing of the electrodes determines the cell constant. A cell constant of 1.0 implies two electrodes spaced one centimeter apart with each electrode one square centimeter in area. Since the conductivity of common solutions varies as much as 1-3% per °C, a temperature sensor must be included to provide accurate conductivity measurements.

11.4 DISCRETE ACTUATORS

Discrete actuators convert the on/off PLC output signals into physical phenomena. Cylinders, motors, and valves are described in this section. In mechanical systems, the physical phenomenon is most often motion. Cylinders provide linear or rotary motion. Electric motors also control rotary or linear motion. Only rotating motors are described here. On/off valves control the flow/no-flow of gases and liquids. Electric motors also drive pumps to provide gas or liquid flow.

11.4.1 Pneumatic and Hydraulic Actuators

Air or hydraulic fluid commonly provides actuation force to move an object. Though electric solenoids can be used to move small objects, they cannot compare to the large actuation force and high efficiency of pneumatic and hydraulic cylinders (cylinders are also called solenoids). Pneumatic and hydraulic actuators are often combined and called "fluid power." This section concentrates on pneumatic systems, but the concepts apply equally to

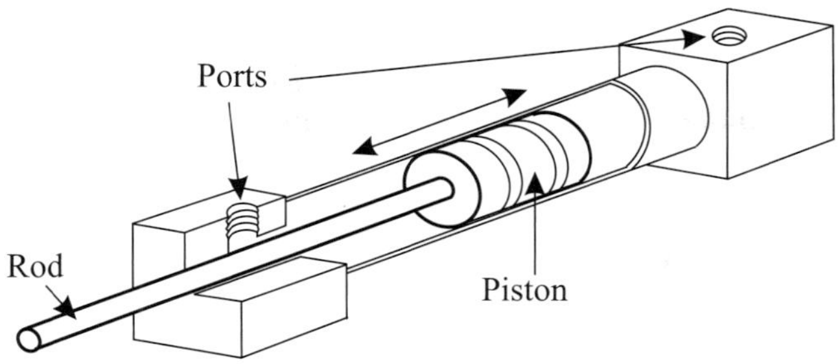

Figure 11.83. Typical linear cylinder construction.

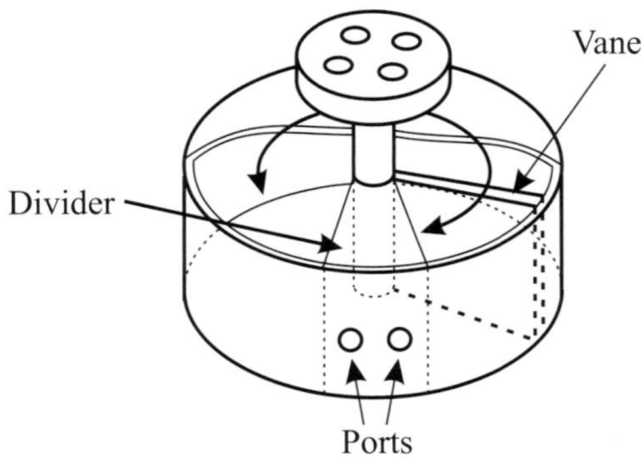

Figure 11.84. Typical rotary cylinder construction.

hydraulic systems. In general, pneumatic actuators are for low force applications and hydraulic actuators are for high force applications. Since the actuation force is pressure times the cylinder area, larger cylinders are needed for larger actuation force.

There are two major components in a fluid power actuator: a cylinder and a control valve. The cylinder provides the actuation movement and the control valve directs the flow of air or hydraulic fluid to the cylinder. The major categories of cylinders are discussed first, followed by the types of control valves.

Cylinders are classified in two ways: (1) linear vs. rotary, and (2) the number of control ports. A linear cylinder provides straight-line motion and a rotary cylinder provides rotational motion. A typical linear cylinder is shown in Figure 11.83. A piston slides within a metal cylinder and is attached to a rod. Pressurized air (or hydraulic fluid) is admitted to a port to move the piston and hence the rod. Unfortunately, the piston is free to twist. So-called nonrotating cylinders have a square or hexagonal rod. A rotary cylinder (Figure 11.84) consists of a vane attached to a shaft that is free to rotate within a wide, short cylinder. Pressurized air enters one of the two ports to move the vane in either the clockwise or counterclockwise direction. Because of the placement of the two ports and the divider, the maximum rotation is about 270°.

Cylinders are also classified as to whether they have one control port (single-acting) or two control ports (double-acting). When pressure is applied to the port of a single-acting spring-return cylinder (Figure 11.85*a*), the piston moves right, extending the rod and

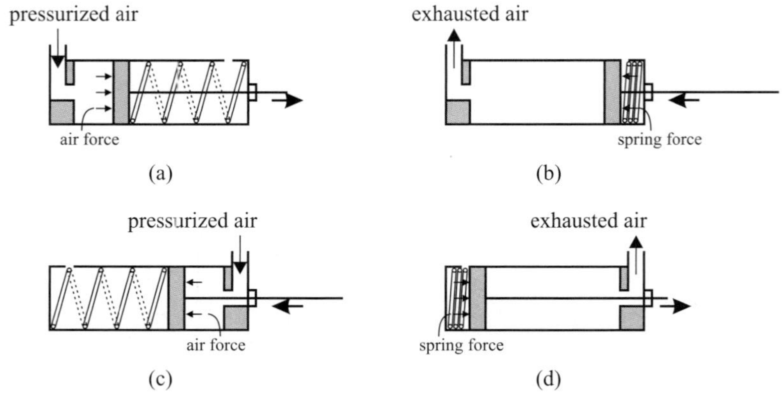

Figure 11.85. Single acting cylinder operation: *(a)* spring return, port pressurized; *(b)* spring return, port exhausted; *(c)* spring extend, port pressurized; *(d)* spring extend, port exhausted.

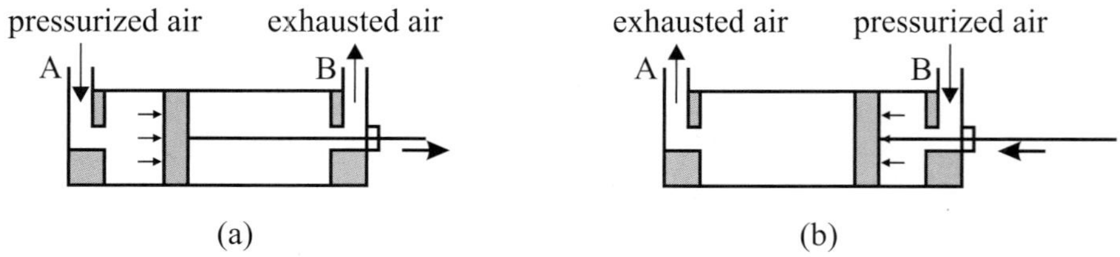

Figure 11.86. Double acting cylinder operation: *(a)* port A pressurized and port B exhausted; *(b)* port A exhausted and port B pressurized.

compressing the spring. When the port is exhausted to the atmosphere (Figure 11.85*b*), the spring forces the piston to the left, retracting the rod. When pressure is applied to the port of a single-acting spring extend cylinder (Figure 11.85*c*), the piston moves left, retracting the rod and compressing the spring. When the port is exhausted (Figure 11.85*d*), the spring causes the piston to move right and extend the rod. Note that the spring defines the default position of the cylinder. A double-acting cylinder (Figure 11.86) has two control ports, labeled A and B. When port A is pressurized and port B is exhausted, the piston moves right. If port A is exhausted and port B is pressurized, the piston moves left. Rotary cylinders are almost always double-acting due to the complications in mounting a spring. In this case, if one port is pressurized and the other is exhausted, the shaft moves clockwise. If the port connections are reversed, the shaft moves counterclockwise.

A double-acting cylinder opens up other control possibilities. Ports A and B can both be exhausted (freeing the piston), both pressurized, or both closed. The latter two possibilities allow the piston to be held at an intermediate position, allowing position control. In addition, the piston extension and retraction velocities can be controlled with separate speed control valves.

A single-acting cylinder with spring return/retract will not impart as much actuating force since a portion of the force required to move the piston must overcome the opposing

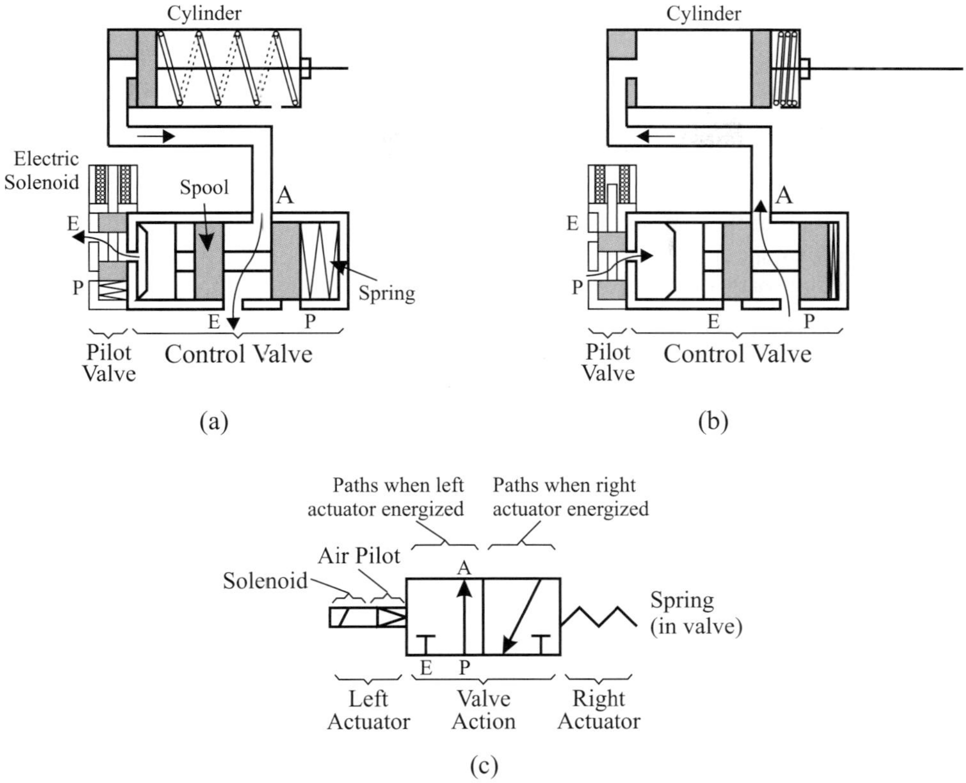

Figure 11.87. Three port directional valve: *(a)* port A exhausted; *(b)* port A pressurized; *(c)* valve symbol.

spring force. The control valves for single-acting cylinders are simpler. However, control valves for double-acting cylinders can also be used for single action cylinders.

Directional control valves are classified as to the number of ports and the number of positions. The number of ports refers to the number of air (or hydraulic) connections to the valve and the number of positions is the number of distinct states of the valve. Each position defines one set of flow paths.

The control of a single-acting cylinder normally requires a three-port two-position control valve, illustrated in Figure 11.87. The three ports are the port for the cylinder ("A"), the pressure port ("P"), and the exhaust port ("E"). The pressure port is connected to a source of pressurized air and the exhaust port is connected to the atmosphere. The spool is the main component of the control valve and its position determines the flow paths. In Figure 11.87*a*, the right actuator (spring) holds the spool in the left position and port A is connected to the exhaust port. Therefore, the cylinder spring causes the cylinder piston to move left and retract the rod. In Figure 11.87*b*, the solenoid is energized, moving the spool to the right position and port A is connected to the pressure port. In this case, the cylinder piston moves right, extending the rod. The control valve has a pilot valve (controlled by an electric solenoid) which directs pressurized air to provide the force to move the spool to the right position. When the solenoid is de-energized, the spool piston is exhausted and a spring provides the force to move the spool to the left position. Using a pilot valve to control the spool position is a common technique. The force required to move the spool can be surprisingly large and requires a larger electric solenoid for direct control of the spool. A pilot valve only needs a small electric solenoid to control its position and it directs pressurized air to move a piston that controls the spool. The pilot valve E and P ports are normally internally connected to the E and P valve ports.

The symbol for the three-port two-position valve in Figure 11.87*a* is shown in Figure 11.87*c*. The two spool positions are shown in the two boxes in the middle part of the symbol. The flow path(s) between the ports for each spool position is shown in its flow box. The symbols for the valve spool actuators are shown next to the end of the flow boxes. The actuator that controls a flow path is next to the flow box. In Figure 11.87*c*, the combination of the solenoid and air pilot valve is on the left side and sets the flow paths to the paths indicated in the left flow box. The spring sets the flow paths to what is shown in the right flow box since it is the actuator on the right side.

The control of a double-acting cylinder normally needs a four-port or five-port control valve. A five-port two-position control valve is shown in Figure 11.88. Two ports are for the cylinder ("A" and "B"), two ports are for exhaust ("EA" and "EB"), and one port is for pressure ("P"). In a four-port control valve, the two exhaust ports are connected internally so there is only one external exhaust connection, but the spool is identical to the spool in Figure 11.88. In Figure 11.88*a*, the solenoid is not actuated and so the flow paths are set by the spring, the right actuator. Port A is connected to an exhaust port and port B is connected to the pressure port, forcing the cylinder piston to move left and retract the rod. In Figure 11.88*b*, the solenoid (left actuator) is energized, moving the spool to the right. Port A is connected to the pressure port and port B is connected to an exhaust port and the cylinder piston moves right, extending the rod. As for the three-port valve, a pilot valve directs

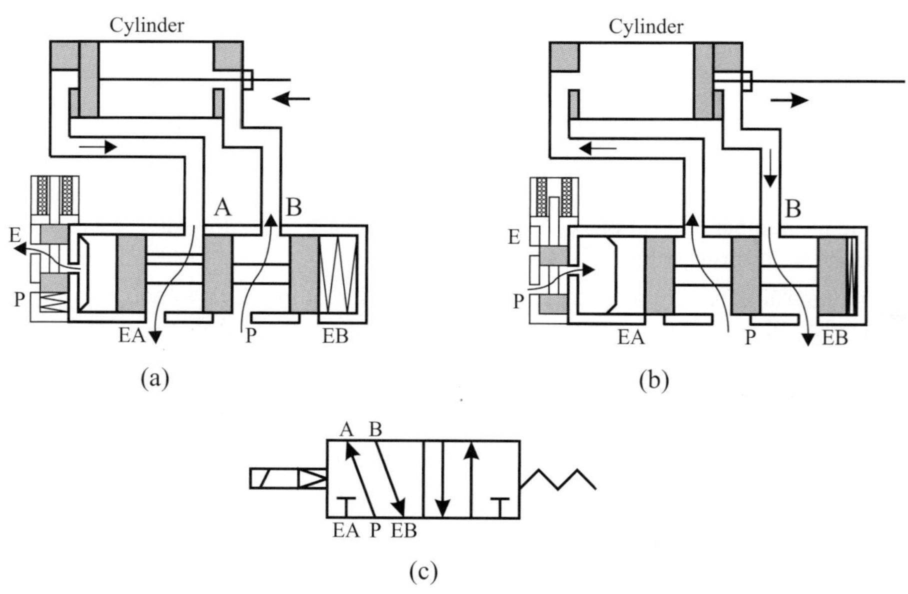

Figure 11.88. Five port two-position directional valve: *(a)* port A exhausted and port B pressurized; *(b)* port A pressurized and port B exhausted; *(c)* valve symbol.

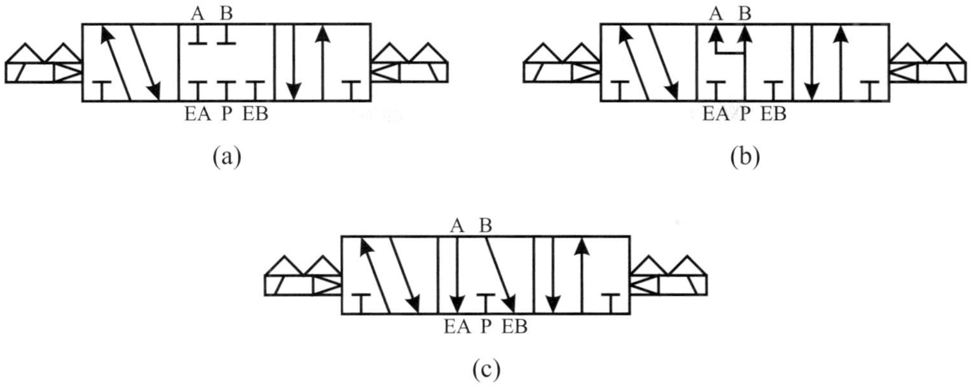

Figure 11.89. Five port three-position directional valve: *(a)* closed center; *(b)* pressure center; *(c)* exhaust center.

pressurized air to provide the force to move the spool to the right position. The symbolic representation for this five-port two-position valve appears in Figure 11.88*c*.

Five-port three-position valves provide other control options for double-acting cylinders. The three different types are shown symbolically in Figure 11.89. All of these valves have two control solenoids. The center position shows the flow paths when neither control solenoid is energized. When the right solenoid is energized, the rightmost position shows the flow paths. When the right solenoid is then deenergized, a spring returns the spool to the center position. When the left solenoid is energized, the leftmost position shows the flow paths. If both solenoids are energized, the spool position is undefined. Typically, the spool assumes the center position in this circumstance. For all three valves shown in

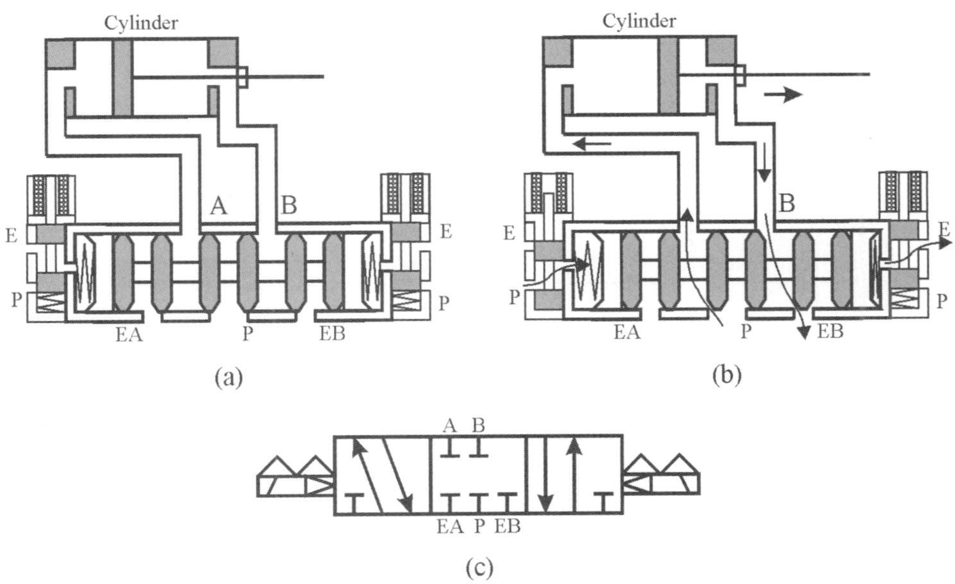

Figure 11.90. Five port three-position closed center valve: *(a)* neither solenoid energized; *(b)* left solenoid energized; port A pressurized and port B exhausted; *(c)* valve symbol.

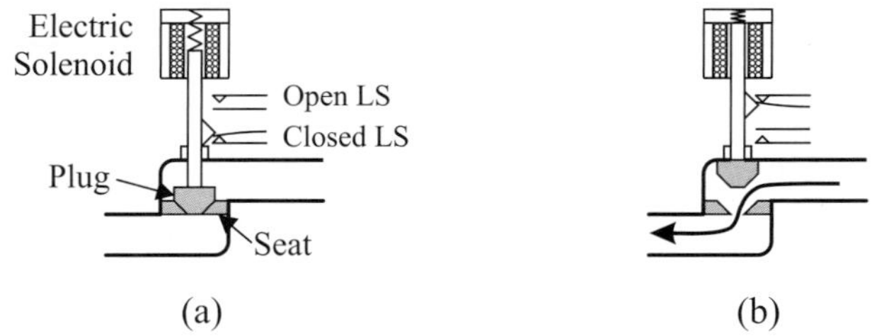

Figure 11.91. Solenoid-actuated discrete valve: *(a)* closed; *(b)* open.

Figure 11.89, an electric solenoid controls a pilot valve. The differences between the valves shown in Figure 11.89 are the paths when the valve is in the center position. Both ports A and B can be closed (Figure 11.89*a*), pressurized (Figure 11.89*b*), or exhausted (Figure 11.89*c*). When both ports A and B are exhausted, the cylinder piston is free to move. When both ports are pressurized, the cylinder piston will stop immediately. When both ports are closed, the cylinder piston also ceases motion, though it may continue to move a small distance if the load is light. Figure 11.90 shows the operation of a closed-center valve when neither solenoid is energized and when the left solenoid is energized.

PLC discrete outputs are used to operate the directional control valves. A two-position valve requires only a single discrete output channel. A three-position valve needs two discrete output channels. The logic to control a three-position valve should include logic that prevents both outputs from being energized simultaneously.

11.4.2 Discrete Valves

Discrete valves provide on-off control of liquid and gas flows. A typical electric-solenoid valve is shown in Figure 11.91. When the solenoid is energized, it moves a valve stem and a plug off a seat, allowing the process fluid or gas to flow. Depending on the situation, the valve may contain limit switches indicating that the valve is open and/or closed. If the valve is large, a pilot valve may be used to provide a larger actuating force. Alternatively, a reversing motor with gearing can provide more actuating force.

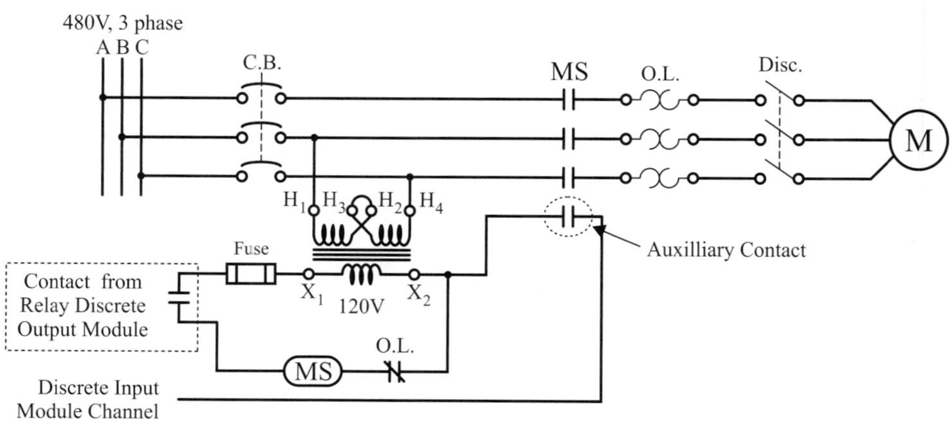

Figure 11.92. Motor contactor.

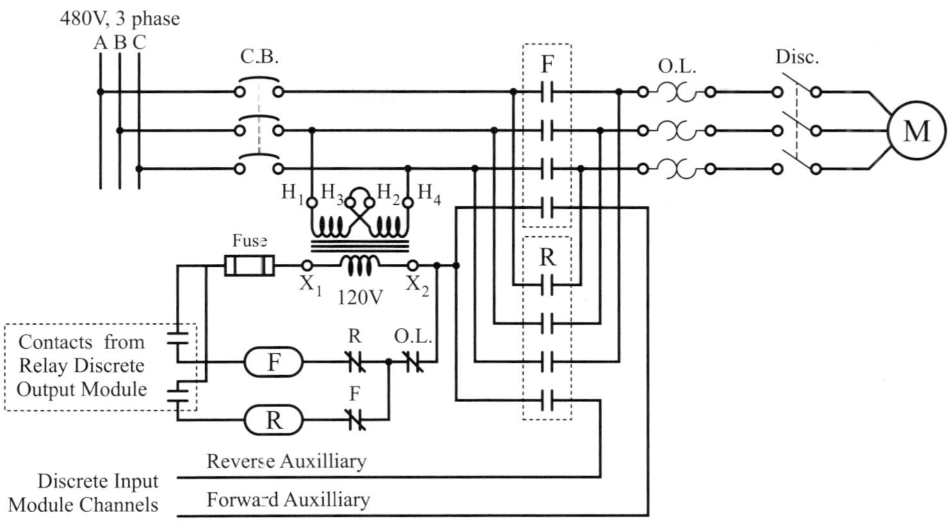

Figure 11.93. Reversing motor contactor.

11.4.3 Motor Contactors

Single-speed motors are often controlled with a motor contactor as shown in Figure 11.92. The example shown is for a three-phase AC motor. A PLC relay output channel is used to control the contactor coil since the coil power source is derived from the 440-volt bus and is isolated from the PLC. The contactor often has a fourth contact, called the auxiliary contact. This contact provides a discrete input feedback to the PLC that indicates the motor contactor is closed. Note that the auxiliary contact also opens when there is an overload. Contactors can also provide a normally-closed overload contact to the PLC to detect overload problems.

A contactor for a reversing motor is shown in Figure 11.93. In this case, there are two discrete outputs from the PLC, one for each direction. To reverse the motor, the A and C phases are switched. The contactor also has protection to prevent both forward and reverse directions being energized simultaneously since this action would result in a short between the A and C phases. Each direction has its own auxiliary contact.

To provide an operator override for a motor, a hand-off-auto (HOA) switch is often used in conjunction with PLC control. Hand-off-auto switches and their use with motor contactors are discussed in section 4.2.2 and an example is shown in Figure 4.23.

11.5 ANALOG ACTUATORS

Instead of just on/off control, there are many situations where a physical quantity must be regulated. In this case, the PLC has an analog output channel connected to an actuator. The most common actuators are an electric motor or a control valve. These are described first, followed by a survey of devices to control the feed rate of solid materials.

11.5.1 Electric Motors

Electric motors often provide the power to move mechanical systems, to pump process fluids, to compress gases, and to pneumatically convey powders. Before modern electronic motor controllers, most applications that required a variable-speed motor used a DC motor. However, AC motors are typically more reliable (no brushes to wear out) and are less expensive. Modern solid-state drive electronics have allowed the AC motor to basically supplant the DC motor in many variable-speed applications. Most modern motor controllers (called *motor drives*, or just *drives*) use electronics to provide efficient control of the motor speed and/or torque.

Variable-speed DC drives are described first, followed by variable-speed AC drives. The drive for a stepper motor, a special kind of brushless motor, is described next. This section on motor drives concludes with the control of servomotors, a specially-designed DC or AC motor.

Variable-Speed DC Drives

Traditionally, DC motors have been used in variable-speed applications. The speed of a DC motor is easily regulated by changing the armature current while holding the field current constant. The torque is controlled by changing the field current with a constant armature current. Before electronics, the speed of a DC motor was regulated by placing a variable resistor in series with the armature power supply. This method is inefficient because of the power loss in the resistor, especially under heavy mechanical loads.

A simplified DC motor drive is shown in Figure 11.94. The three-phase power is rectified into a DC voltage that drives the armature. One phase is rectified to produce the field voltage. The rectifiers are silicon-controlled rectifiers (SCRs). An SCR has the property that when a positive pulse is applied to the gate terminal (labeled G_X in Figure 11.94) and there is a positive anode-to-cathode voltage, the rectifier conducts as long as the anode-to-cathode voltage remains positive. When used in a bridge configuration to rectify an AC voltage, the SCR conducts during one half-cycle of the AC waveform. The application of the gate pulse is often called "firing" the gate. If the gate pulse is delayed so that the SCR conducts during only part of the AC half-cycle, then only that portion is

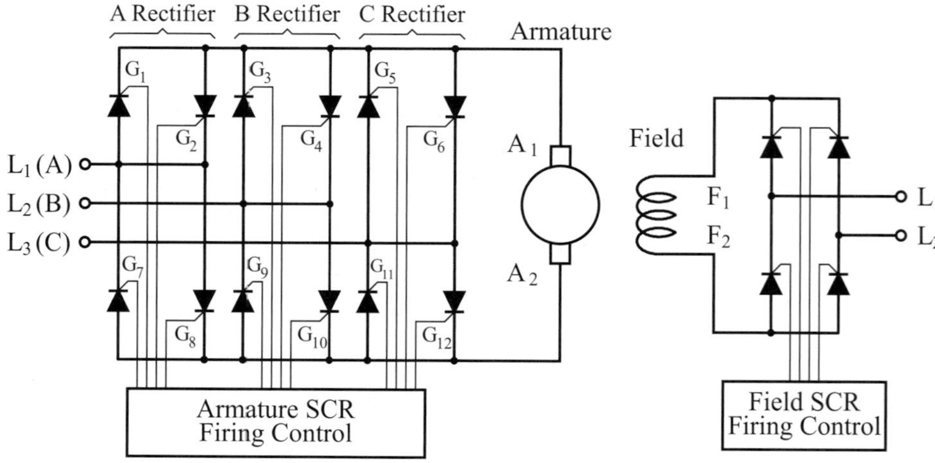

Figure 11.94. Simplified DC motor drive.

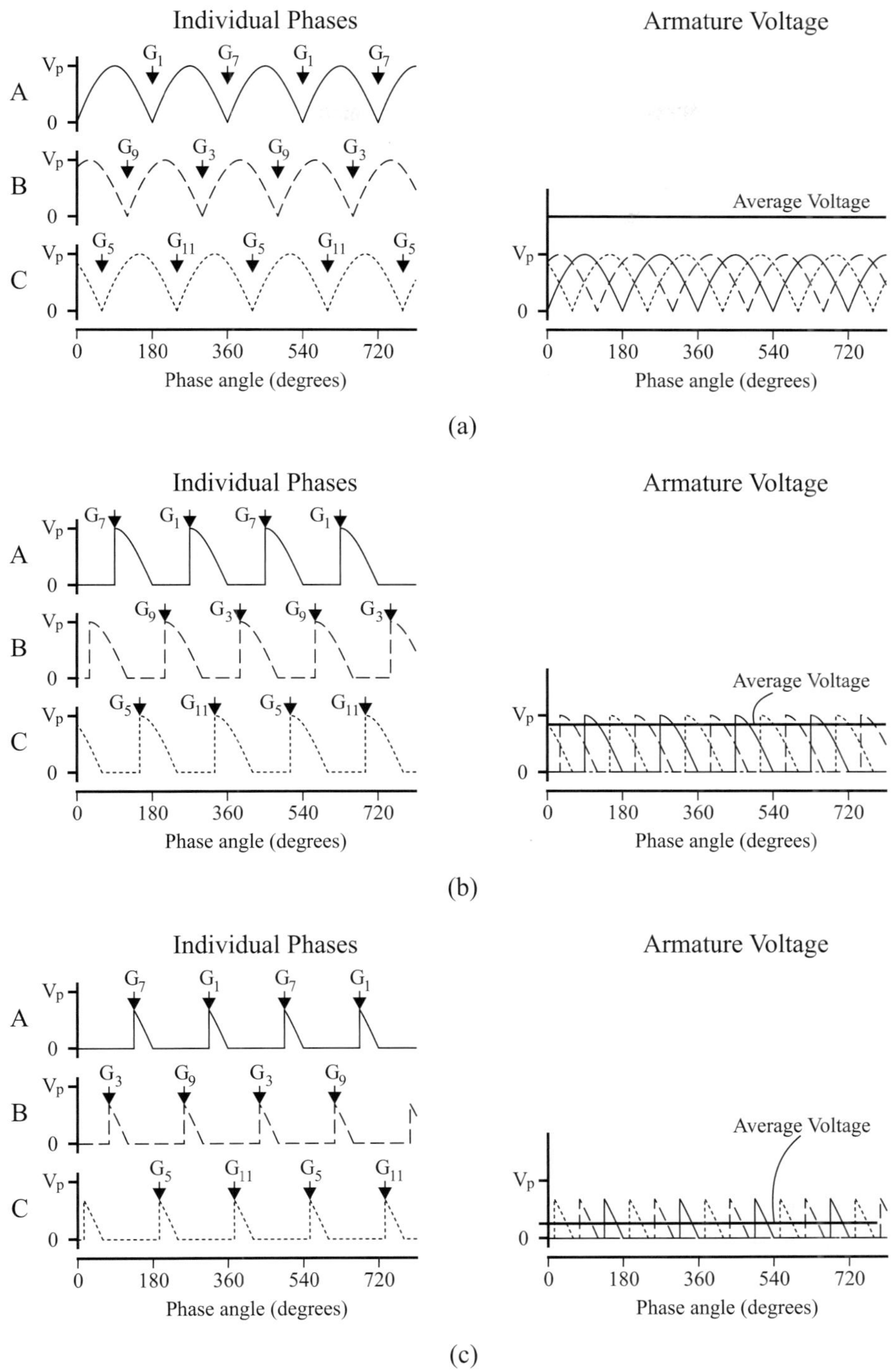

Figure 11.95. DC armature voltage as a function of SCR firing angle: *(a)* zero degrees; *(b)* 90 degrees; *(c)* 135 degrees.

conducted to the DC load. The result of delaying the gate pulse is a reduction in the average DC voltage. In this manner, the armature voltage can be varied from zero to the maximum. SCRs can conduct high currents with little loss and so are much more efficient than variable resistors when controlling the armature voltage.

The operation of the three-phase rectifier in Figure 11.94 is illustrated in Figure 11.95. The odd-numbered SCRs are used to provide a positive voltage to the armature (where A_1 is considered the positive terminal) and the even-numbered SCRs are used to provide a negative voltage to the armature. For the sake of describing the operation, only the odd-numbered SCRs are considered. The SCR firing delay (called the *firing delay*) is expressed in degrees starting from the beginning of the AC waveform. One AC cycle is 360 degrees. When the appropriate SCRs are fired at the beginning of each AC half-cycle (Figure 11.95*a*), the resultant armature voltage is at its maximum, $1.65 \times V_p$, where V_p is the peak phase voltage (Fitzgerald, Kingsley, and Umas, 2003). If all three phases are summed, the waveform is "rippled." However, the large inductance of the armature smoothes out these variations, producing a nearly constant (over a few AC cycles) average voltage. Note that within each phase bridge, the firing of the lower SCR is delayed 180 degrees from the upper SCR to steer the negative portion of the AC waveform. When the SCR firing angles are delayed 90 degrees (Figure 11.95*b*), only half of the AC waveform is conducted to the armature, halving the average voltage and halving the speed. When the SCR firing angles are delayed 135 degrees (Figure 11.95*c*), the average voltage is less than 15% of the maximum.

The field rectifier shown in Figure 11.94 allows the field voltage to be varied, providing torque control. If a variable field voltage is not needed, then ordinary rectifiers are used in place of the SCRs.

Variable-Speed AC Drives

The speed of an AC induction motor is determined by its construction and the frequency of the voltage applied to it. Therefore, to control the speed, the frequency of the voltage must be varied. However, if the frequency is decreased while holding the peak voltage constant, the motor stator will magnetically saturate, causing an increase in the line current (Hubert, 2002). To prevent this problem, the line voltage must be reduced to maintain a fixed ratio of the frequency to the line voltage.

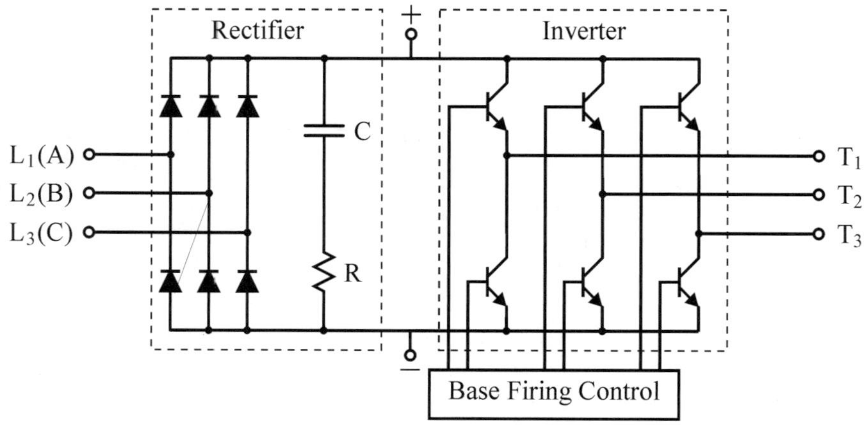

Figure 11.96. Simplified AC motor drive.

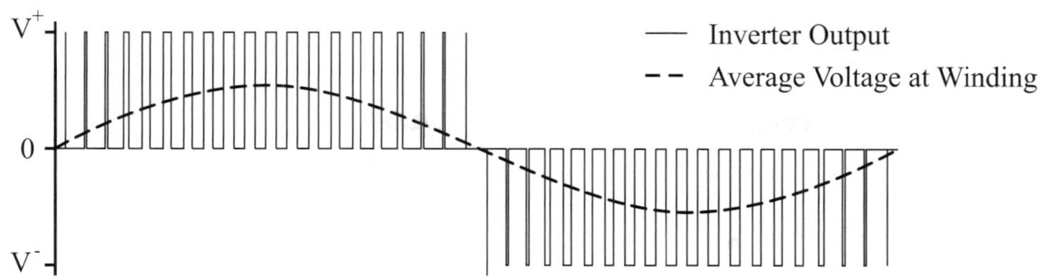

Figure 11.97. Pulse-width modulation to produce AC voltage and frequency for one phase.

A simplified AC motor drive is shown in Figure 11.96. As for the DC drive, the three-phase power is rectified into a DC voltage. However, there is no voltage control. A capacitor provides some filtering of the rectified voltage. The inverter uses pulse-width modulation to produce the three-phase waveforms at the appropriate frequency and voltage. The inverter operation for one phase is shown in Figure 11.97. In this case, the frequency is 30 Hz and so the motor voltage must be one-half of the line voltage. The appropriate transistor bases are sequenced to produce positive pulse-width modulated voltages during the first half of the cycle and negative voltages during the last half of the cycle. During each half cycle, the duty cycle of the pulses start at zero, increase to 50%, and then decrease to zero. The duty cycle is defined to be the percent of time the pulse is nonzero during its period. Since the maximum duty cycle is 50%, the peak voltage is 50% of the line voltage.

Stepper Motor Drives

A stepper motor is a brushless permanent-magnet type motor that translates incoming pulses into mechanical motion. Because the stepper motor makes a fixed angular motion in response to the pulses, stepper motors are often used in an open-loop configuration. In this case, the motion is controlled by generating the proper pulses to the motor, and encoders are not used. Figure 11.98 shows a simplified stepper driver and stepper motor. The stepper driver is similar to the inverter section of an AC motor drive. In this case, the stepper has four phases. Typical pulses for this stepper motor when driven in the half-step mode are shown in Figure 11.99. The overlapping phase pulses provide the ability to position the

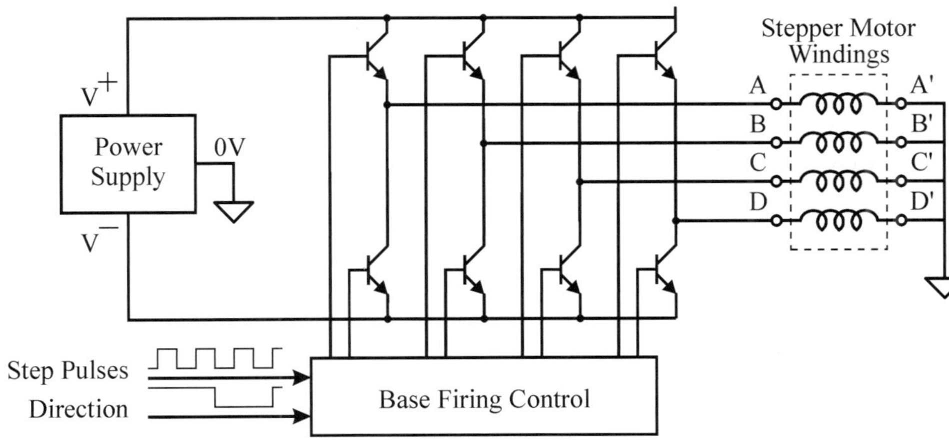

Figure 11.98. Simplified stepper motor drive.

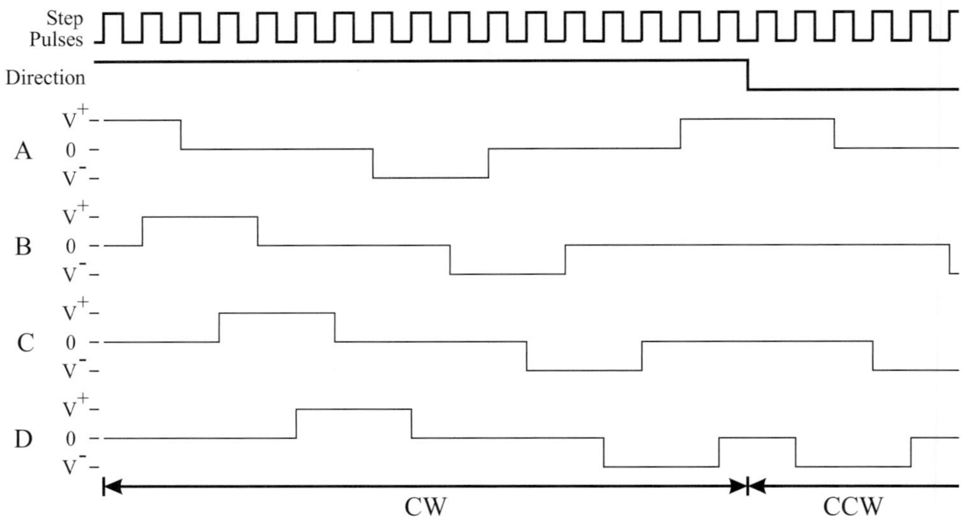

Figure 11.99. Phase signals for stepper motor operation.

motor between poles. In the example shown, the motor turns clockwise for 18 half-steps and then reverses direction.

Stepper motors have two disadvantages. First, if the stepper motor encounters a torque load higher than its rating, the motor "cogs" and does not move. For this reason, position feedback is provided by encoders in high-performance motion applications. Second, one must be careful when using encoders in a stepper motor application. Because the motor is driven with pulses, the actual shaft rotational motion is also "pulsed" and not smooth as for a servomotor. If the stepper motor is driving a flexible load (for example, a belt), the pulsed shaft motion is magnified and causes false encoder pulses. For this reason, servomotors are often a better choice.

Servomotor Drives

A servomotor is a specially-designed brushed DC motor or brushless AC motor. The traditional servomotor is a brushed DC motor, where the coil is wound on the armature and the stator is a permanent magnet. Though the motor controller of Figure 11.94 can be used, servomotor applications are typically lower power. In this case, the variable armature voltage is often provided by a variable-output DC power supply whose voltage is automatically adjusted to maintain a desired speed (measured by a tachometer). A brushless AC servomotor is a smaller version of larger AC motors. Hence, the drive electronics are basically the same as for the variable-speed AC drive of Figure 11.96.

11.5.2 Control Valves

In chemical process control systems, control valves regulate the flow of fluids or gases. The typical parts of a globe valve are shown in Figure 11.100. The two basic components of a control valve are the valve body and the actuator. The valve may have other accessory components such as a current-to-pneumatic converter, a positioner, a position sensor, and limit switches.

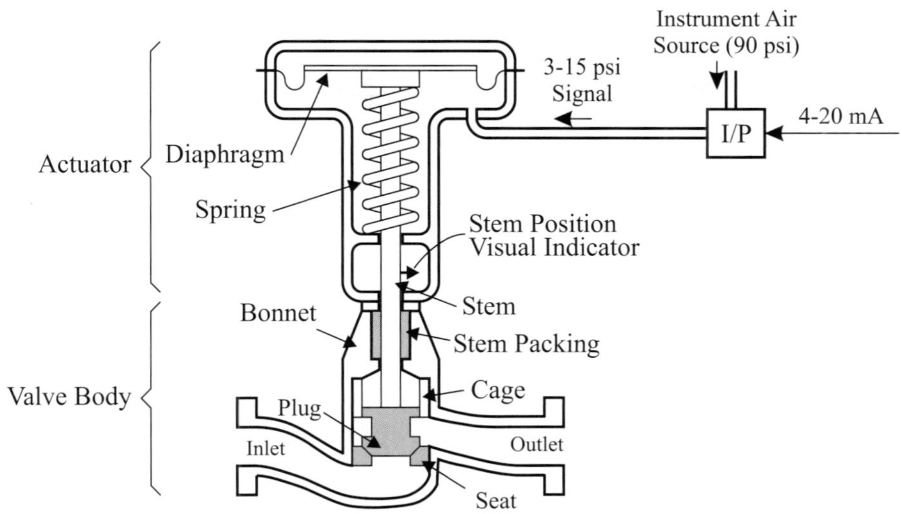

Figure 11.100. Typical pneumatic-actuated control valve (air-to-open).

The valve shown in Figure 11.100 has linear control motion, that is, the stem moves up and down to change the position of the plug and thus vary the flow opening. When the plug is down against the seat, the valve is closed. A cage guides the plug movement. The bonnet contains the stem packing which prevents process fluid leakage from around the stem, and yet allows the stem to move. The actuator in Figure 11.100 is a pneumatic diaphragm. Pressurized air admitted to the lower chamber moves the diaphragm up and thus raises the valve stem. A current-to-pneumatic (I/P) converter changes the 4-20 mA signal from the PLC to a 3-15 psi "signal" to the valve actuator.

The actuator for the control valve in Figure 11.100 is commonly called an *air-to-open* or *fail-close* actuator. If air pressure is lost, the valve automatically closes. Whether a valve

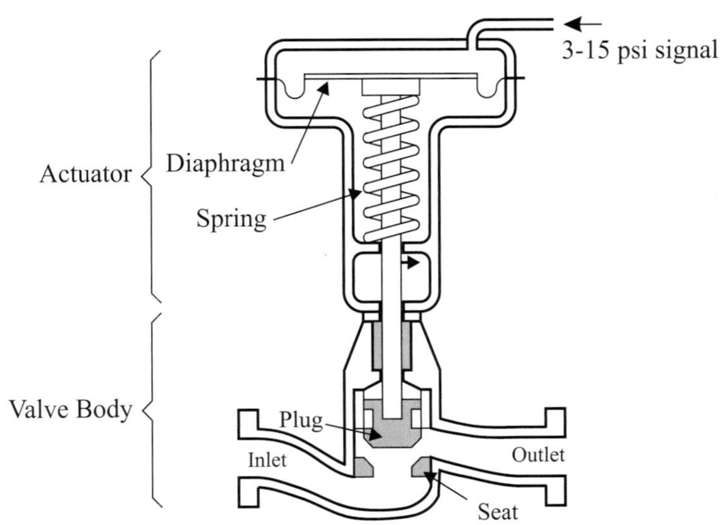

Figure 11.101. Typical pneumatic-actuated air-to-close control valve.

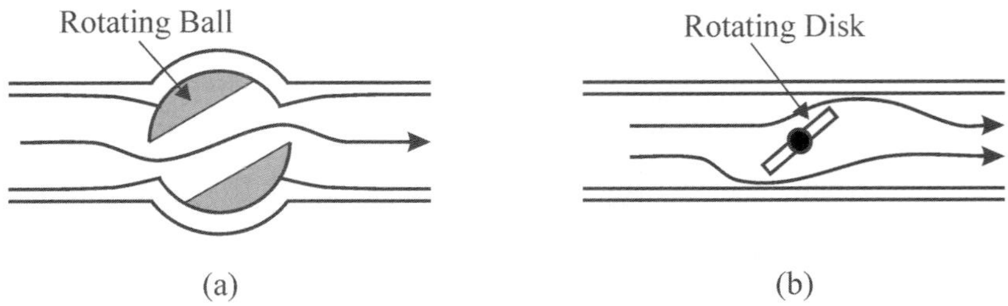

Figure 11.102. Rotary-actuated valves: *(a)* ball valve; *(b)* butterfly valve.

should fail in the open or closed position is usually dictated by safety considerations. An *air-to-close* or *fail-open* actuator is shown in Figure 11.101. Note that its basic difference with the air-to-open actuator is that the pressurized air is admitted to the upper chamber and lowers the valve stem in response to increasing pressure.

A fail-close type of actuator is used when the fluid is a fuel or a material input to a exothermic (heat-producing) reaction. Most valves that handle chemicals are the fail-close type. A fail-open type of actuator is appropriate when the manipulated stream is cooling fluid to a reactor or other type of system. Generally, one must ask the question, "If the air pressure to this valve is lost, what is the safe position of this valve?"

Though the globe valve is common, other types of valves may be appropriate. Two common rotary-action valves, ball and butterfly valves are shown in Figure 11.102. Both of these valves have less turbulence at maximum flow. In addition, the actuator must supply a rotary action for these valves. Rotary valves are generally less costly than globe valves in normal applications (Baumann, 2009).

The valve is sized appropriately for the given application. The valve sizing coefficient, C_V, is defined to be

$$C_v = q\sqrt{G_f/\Delta P}$$

where q is the flow rate, G_f is the specific gravity of the process fluid, and ΔP is the pressure drop across the valve (Baumann, 2009). Calculating the required C_V for the maximum flow often requires guesswork since the inlet pressure may unknown. The usual result is an oversized valve and its maximum opening is less than 50% at the maximum flow rate (Baumann, 2009).

Besides calculating the valve C_V, choosing the valve flow characteristic is another important design decision. A valve's flow characteristic is the flow rate through the valve body as a function of the position as it changes from 0% to 100%. The *inherent* flow characteristics are theoretical and are determined for a constant pressure drop across the valve. The *installed* flow characteristics are the actual flow characteristics when the valve is in the actual system. In general, the installed flow characteristic is not the same as the inherent flow characteristic. There are three flow characteristics: (1) linear, (2) equal percentage, and (3) quick opening. The flow rate through the valve as a function of the stem travel for each of these characteristics is shown in Figure 11.103. For a *linear* flow characteristic, the flow rate is directly proportional to the valve stem travel. An *equal percentage* flow characteristic has the property that equal increments of stem travel result in equal percentage changes in the flow. For example, if the flow doubles as the stem moves

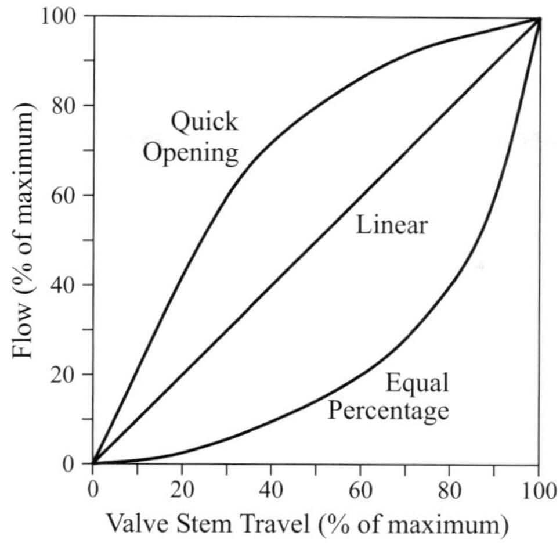

Figure 11.103. Valve flow characteristics.

from 30% to 40%, the flow will also double as the stem moves from 50% to 60%. A *quick opening* characteristic has large flow changes with minimal changes in the stem position when the valve is nearly closed. As the valve opens, the effect of increases in the stem position lessens. This type of flow characteristic is not suitable for regulatory control.

Baumann (2009) uses the following rule of thumb when deciding whether to choose a valve with an inherent linear or equal percentage flow characteristic. Find the ratio between the pressure drop across the valve at its minimum flow and the pressure drop at its maximum flow. If this ratio is less than 2:1, choose a linear inherent valve characteristic. Otherwise, use an equal percentage flow characteristic. Baumann cites two other reasons why an equal percentage valve is often a better choice:

1. Better utilization of the available valve control range. Since most valves are oversized, the typically flow is around 40% of the maximum valve flow. At this flow, the stem position for a linear valve is about 40% and the stem position for an equal percentage valve is about 75%.
2. Better response at low flows. The minimum flow is usually around 5% of the maximum. A small process upset at this flow for a linear valve characteristic may cause the plug to bump against the seat. However, for an equal percentage characteristic, the stem position is around 15%, and a small process upset will not cause a problem.

Ideally, one wants a linear installed flow characteristic. If the stem position changes by 10% of its travel, the flow should also change by 10% of the maximum. However, this situation is unusual over the full range of the flow. Since most control loops operate within a narrower range, it is only necessary that the relationship between the stem position and flow is "reasonably linear" over the range of interest. Most PID tuning rules will tolerate a change in the "valve gain" of ±50%.

The control valve is often the critical component in a process control loop and has a major influence on loop stability. The major problem is dead time, that is, the time required for the actuator/valve to respond to a change in the PLC analog output signal. This characteristic is often called the "dead band" or "stiction" of the actuator/valve combination. The biggest problem is the friction in the valve stem movement due to the packing. The packing must provide a good seal. But if too tight, the stem will not move until

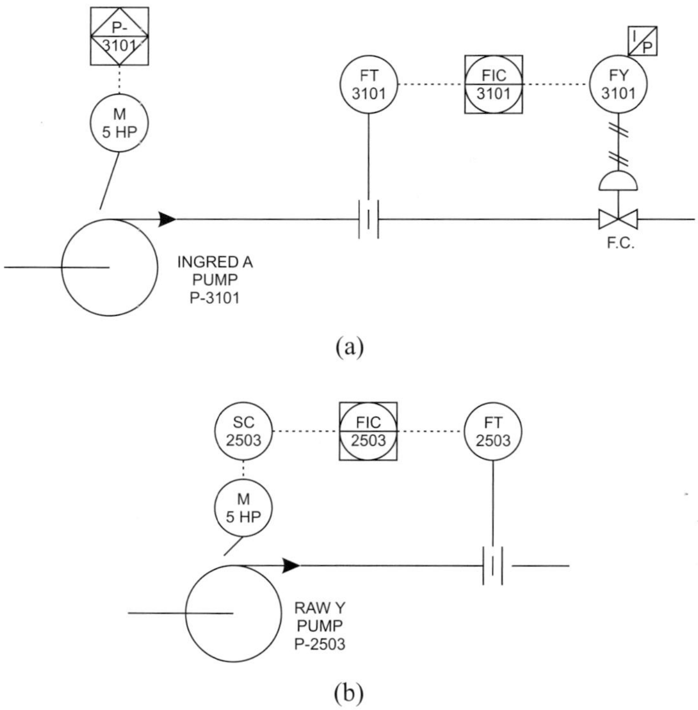

Figure 11.104. Flow control: *(a)* fixed-speed pump; *(b)* variable-speed pump.

the pressure to the actuator builds up enough to overcome the friction. The ideal valve has an operating dead band of less than 1% of the maximum travel (Baumann, 2009).

A motor is an alternate valve actuator. There are two major types: (1) AC motor gear drive, and (2) stepper motor. In the AC motor drive actuator, the motor speed is reduced by a set of gears to a few revolutions per minute. A stepper motor provides more accurate positioning, but its torque is limited. Both types of motor actuators provide more precise positioning, but both often have slower dynamics when compared with pneumatic actuators. In addition, motor actuators do not have fail-close or fail-open capability.

When controlling the flow of a liquid, there are two possible basic control configurations. Conventionally, a fixed-speed motor driving a pump provides the pressure head and a control valve provides the throttling to control the flow. This scheme is shown in Figure 11.104*a*. With the introduction of variable-speed AC motor drives, using a variable-speed motor driving a pump to control liquid flow became possible (Figure 11.104*b*). The throttling action of a valve is thermodynamically irreversible and thus the pressure drop across the valve is wasted energy. The perceived energy savings is a rationale for using a variable-speed drive instead of a valve to control the flow. However, the energy savings can only be justified when the pump is deliberately oversized (Baumann, 1981). There are some applications where a speed-controlled pump is clearly the best choice. When pumping a highly corrosive fluid, a speed-controlled pump avoids the expense of a special alloy valve. If valve dead band is a problem, a speed-controlled pump can improve the control.

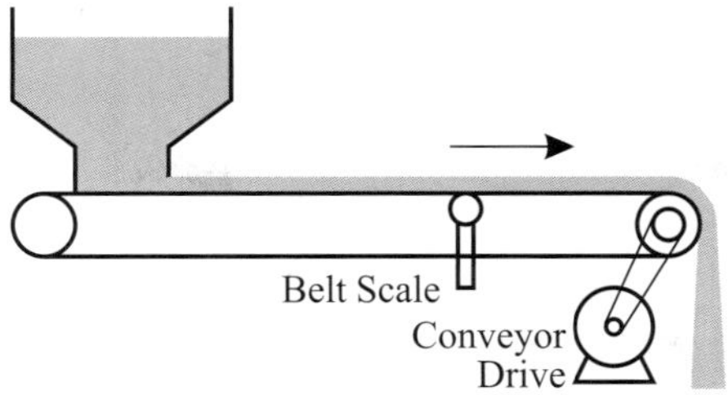

Figure 11.105. Weigh belt feeder.

11.5.3 Solid Material Feeders

Controlling the feed rate of solid material is a little more difficult than controlling the flow of a liquid or gas. Common methods of solid material flow include weigh belt feeders, vibratory feeders, rotary valves, and loss-in-weight feeders.

A conveyor belt scale (Figure 11.73) forms the basis for a weigh belt feeder, illustrated in Figure 11.105. A bed of material is formed on the conveyor belt as it is discharged from a hopper and the belt speed is controlled in order to maintain the desired flow rate. Weigh belt feeders are typically used for granular materials that do not generate much dust.

A vibratory feeder replaces the conveyor belt of Figure 11.105 with a vibrating tray. A vibrating tray is more efficient when moving mid-sized materials (0.3 to 50 mm in diameter). On a vibratory feeder, particles are "thrown" up and forward so that the particles drop further down the surface. Figure 11.106 illustrates how a vibrating tray moves one particle (Eriez, 2007). During one cycle of the vibrating tray, the tray surface travels horizontally and vertically from the bottom of the stroke to the top of the stroke and then returns to its starting position. An individual particle starts at position "A" on the tray. At position "B," the particle leaves the tray with a trajectory that goes through "C" and then drops to "D." When the tray is at the top part of the stroke, the particle is in position "C" and the tray is back at the bottom part of the stroke when the particle drops to "D."

The vibration amplitude, frequency and angle of deflection are chosen to move a certain material at a specific rate. The actual vibrations are often generated by an electric motor driving an eccentric (out-of-balance) shaft. The vibration created by the eccentric is amplified through a system of springs. An alternative, more efficient design uses a vibrating electromagnet to generate the vibrations (Eriez, 2007).

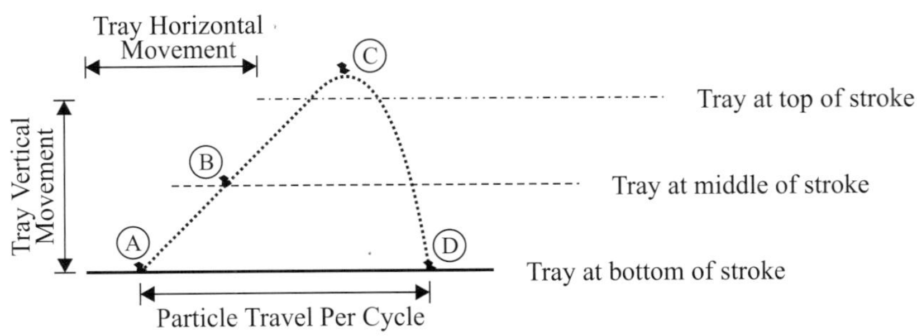

Figure 11.106. Vibratory feeder operation.

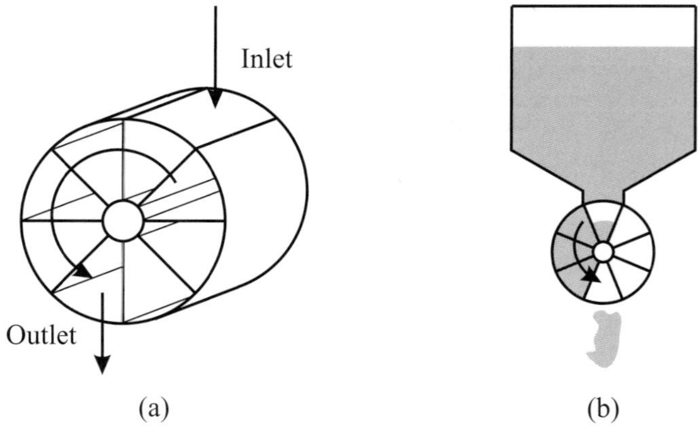

Figure 11.107. Rotary valve: *(a)* vanes of valve; *(b)* feeding material from hopper.

Rotary valves are a common way to control the flow of powders from a bulk tank. A rotary valve (Figure 11.107) consists of a series of vanes contained within a cylinder. Material enters a wedge-shaped compartment from the top, is rotated 180 degrees and then drops from the inverted compartment. An attached motor whose speed is controlled by a variable-frequency drive controls the valve rotation speed. The material flow rate is a function of the size of each compartment and the rotation speed. Often, the hopper is pressurized to drive the material into the compartments and to prevent "bridging" of the material at the bottom of the hopper. The rotary valve also provides an airlock in systems where the powder is pneumatically conveyed to the hopper and then fed from the hopper into the process.

In a loss-in-weight feeder, the weight of the entire hopper containing the material is continuously measured. The rate of hopper weight loss is the flow rate. A separate device (for example, screw feeder, rotary valve, or vibrating pan) is manipulated to control the feed rate. The disadvantage of this type of feeder is that the hopper must be periodically refilled. When the hopper is being refilled, the calculation for the weight loss rate must be suspended. If an accurate measurement of material leaving the hopped is desired, then material feeding must be stopped while the material is being refilled.

11.6 CHAPTER SUMMARY

This chapter describes common sensors and actuators that appear in PLC-controlled systems. Both discrete and analog devices are covered.

REFERENCES

Allen-Bradley, 1993. *Bulletin 1333 Adjustable Frequency AC Motor Drives (Series B/C): User Manual*, pub. 1395-5.1, Allen-Bradley Company, Inc., Milwaukee, WI, April.

Analog Devices, 2013. *Two Terminal IC Temperature Transducer, AD590*, Rev. E, Analog Devices, Inc., Norwood, MA, http://www.analog.com.

Anderson, M., and J. Herdan, 1999. "Oxidation-reduction Potential," International Scientific Communications, Inc., www.iscpubs.com/articles/aln/n9912and.pdf, Dec.

Banner Engineering, 1994. *Handbook of Photoelectric Sensing*, 2nd ed., part number 03190, Banner Engineering Corp., Minneapolis, MN.

Battikha, N. E., 2007. *The Condensed Handbook of Measurement and Control*, 3rd ed., ISA, Research Triangle Park, NC.

Baumann, H. D., 1981. "Control Valve vs. Variable-Speed Pump," *Chemical Engineering* (June 29): 81-84.

Baumann, H. D., 2009. *Control Valve Primer: A Users' Guide*, 4th ed., ISA, Research Triangle Park, NC.

Burns, G. W., M. G. Scroger, G. F. Strouse, M. C. Croarkin, and W. F. Guthrie, 1993. *Temperature-Electromotive Force Reference Functions and Tables for the Letter-Designated Thermocouple Types Based on the ITS-90*, NIST Monograph 175, National Institute of Standards and Technology, Washington, D.C.

Carr, Joseph J., 1993. *Sensors and Circuits: Sensors, Transducers, and Supporting Circuits for Electronic Instrumentation, Measurement and Control*, Prentice-Hall, Englewood Cliffs, NJ.

Cejer, Mark. "Resistive Temperature Detectors: An Alternative to Thermocouples for Precise, Repeatable Temperature Measurements," Keithley Instruments, Inc., Cleveland, Ohio.

Eriez, 2007. *How to Choose and Use Vibratory Feeders and Conveyors*, Eriez Magnetics.

Fitzgerald, A. E., Charles Kingsley, Jr., and Stephen D. Umans. 2003. *Electric Machines*, 6th ed., McGraw-Hill, New York, NY.

GLI International. n. d. "Electrolytic (Contacting) Conductivity Measurement," Technical Bulletin TB-C1, Rev. 2-201, GLI International, Inc., Milwaukee, WI.

Hubert, Charles I., 2002. *Electric Machines: Theory, Operation, Applications, Adjustment, and Control*, 2nd ed., Prentice-Hall, Upper Saddle River, NJ.

Maron, Jeffrey, 1999. "Selecting Noncontact Pyrometers and Thermometers," *Process Heating*, Business News Publishing Co., September.

National Semiconductor, 2008. *LM135/LM235/LM335, LM135A/LM235A/LM335A Precision Temperature Sensors*, National Semiconductor Corporation, http://www.national.com.

Omega Engineering, 2010. *OMEGA Complete Temperature Measurement Handbook and Encyclopedia®*, Volume MMXIV™ 7th Edition, Omega Engineering, Inc., Stamford, CT, http:/www.omega.com.

Prisciandaro, J., Butchko, D., 2004. *Introduction to Pneumatics and Pneumatic Circuit Problems for FPEF Trainer*, Fluid Power Education Foundation, Milwaukee, WI.

Rockwell Automation, 2007. *Bulletin 1395 Digital DC Drive: User Manual*, pub. 1395-5.40-MAY07, Rockwell Automation, Milwaukee, WI, May.

Rockwell Automation, 2022. *ControlLogix Analog I/O Modules: User Manual*, pub. 1756-UM009E-EN-P, Rockwell Automation, Inc., Milwaukee, WI, Sep.

Rosemount Analytical, 2010a. *The Theory of pH Measurement*, pub. ADS 43-002/rev. C, Rosemount Analytical, Inc., Irvine, CA, November.

Rosemount Analytical, 2010b. *Theory and Practice of pH Measurement*, pub. PN 44-6033/rev. D, Rosemount Analytical, Inc., Irvine, CA, December.

Schultz, Warren, 1997. "Interfacing Semiconductor Pressure Sensors to Microcomputers," *Motorola Application Note AN1318*, Motorola, Inc.

Siemens, 2015a. *Belt Scales Product Application Guide*, Siemens Level and Weighing, Peterborough, ON.

Siemens, 2015b. *Flowmeters Product Application Guide*, Siemens Level and Weighing, Peterborough, ON.

12 Communication Networks

Chapter Topics:

- General communication network concepts
- Standard and proprietary factory networks
- PLC-to-PLC communication function blocks

OBJECTIVES

Upon completion of this chapter, you will be able to understand:

- Typical communication networks
- Major features of standard and proprietary factory networks
- Implementation of PLC-to-PLC communications with the PLCs covered by this text

Scenario: Detecting a failure in a remote I/O module

A particular application processes "cassettes" of 150 mm silicon wafers in a series of chemical baths (Figure 12.1). The current system uses a ControlLogix PLC to control a gantry that moves the cassettes between the baths. The operator places a pair of cassettes into the loading station and the gantry moves the cassettes to the first of a series of chemical baths. The cassettes remain in a chemical bath for a certain amount of time and then they are moved to the next bath. Eventually, they are moved from the last chemical bath to an unloading station, where they are removed by an operator. The operator places the cassettes into a "spin" dryer and spins the cassettes for 3 minutes to remove any water remaining on

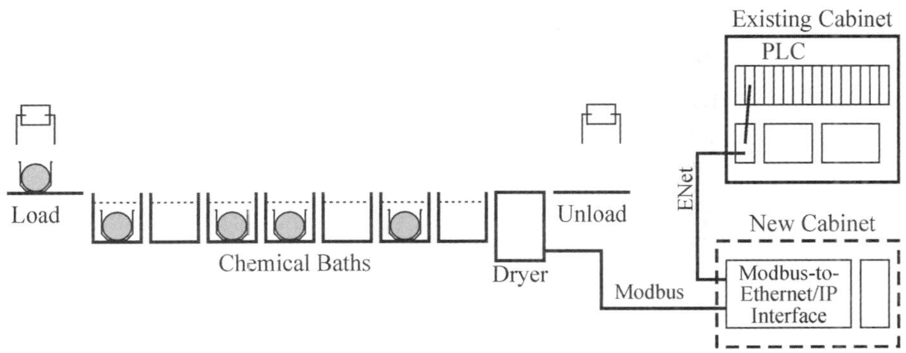

Figure 12.1. Silicon wafer processing system.

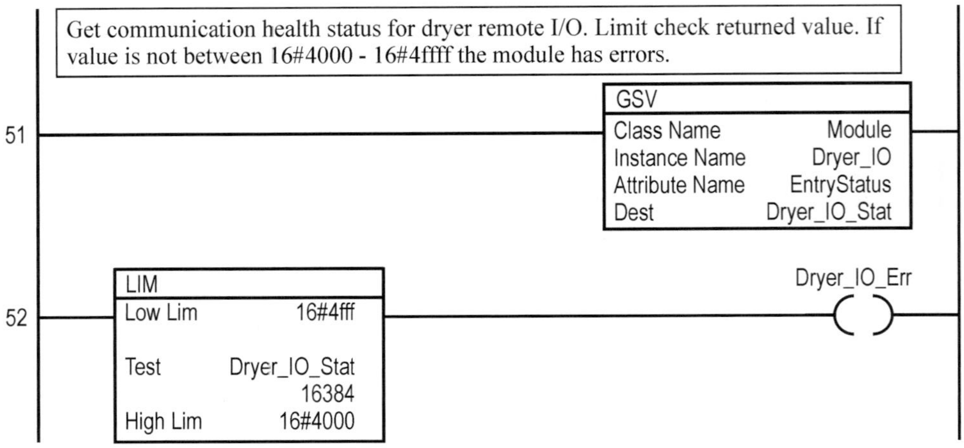

Figure 12.2. Ladder logic to check for presence of remote module.

the surface of the wafers. When the drying cycle is finished, the operator removes the cassettes from the dryer.

As an improvement to the process, the PLC is upgraded to load and unload the cassettes from the dryer and to control the dryer operation. The dryer has a Modbus interface and a Modbus-to-Ethernet/IP interface module is purchased since your ControlLogix PLC already has an Ethernet/IP communication module. A new cabinet houses the dryer interface module and a power supply (Figure 12.1). The Modbus-to-Ethernet/IP interface module is configured as a remote I/O module that has 16 discrete inputs and 16 discrete outputs, and they are therefore PLC processor tags. No special PLC message blocks are required to communicate with the Modbus-to-Ethernet module.

The system is installed as designed. The PLC dryer control works as it should. However, during your testing of changes in the motion control, you need to hit the E-stop button. When power is restored, the PLC cannot seem to control the dryer. After some troubleshooting, you discover a blown fuse for the Modbus-to-Ethernet module . However, the PLC has not detected a lack of communication with the module.

Solution: You add ladder logic (Figure 12.2) to read the communication status of the Ethernet I/P node. If the communication status is outside a certain range of values, then a communication error indication is set for the module.

12.1 INTRODUCTION

This chapter is an overview of communication networks available for use in Allen-Bradley PLC systems. A *local area network* (LAN) is a high-speed medium-distance communication system linking multiple devices. Each device on a LAN is often called a *node*. In most factory automation systems there are multiple communication networks, often connected in a hierarchical manner (Figure 12.3). The backbone network is generally Ethernet and connects the business and office computers. A gateway device connects an automation network to the backbone. Most systems have multiple automation networks, though only one is shown in Figure 12.3. An automation network is optimized for the needs of automation: short messages and fast response. The automation network links PLCs and

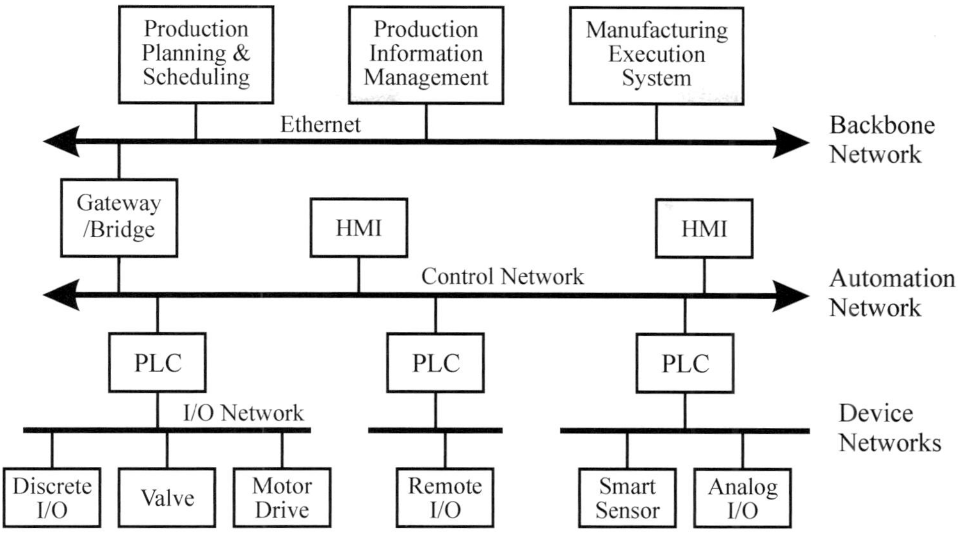

Figure 12.3. Factory communications hierarchy.

operator terminals (HMI). The gateway provides the business-related function access to the PLCs and their information. Below the PLCs, the third level of the hierarchy contains the input/output (I/O) networks. These networks range from remote I/O networks to distributed I/O networks. A remote I/O network is basically a proprietary extension of the normal chassis-based I/O where a chassis of I/O modules is located remotely from the processor. A distributed I/O network is one in which the individual discrete I/O or analog I/O devices are connected to a standard fieldbus network.

The various networks described in this chapter generally fit into the hierarchy as listed in Table 12.1. However, note that Ethernet is often used as the networking technology for all levels of the hierarchy, even down to the I/O network. Standard network protocols are covered first, generally starting from the top of the hierarchy and proceeding to the bottom. Then the proprietary Allen-Bradley protocols are described. Next, the function blocks that implement PLC-to-PLC communication are explained. Lastly, heartbeat logic to verify the communication channel and operation of the partner processors is described.

Table 12.1 Categorization of Networks

Backbone Network	Automation Network	Device Network
Ethernet	Ethernet/IP	Ethernet/IP
Ethernet/IP	ControlNet	DeviceNet
Modbus/TCP	PROFINET	AS-i
	Modbus+	PROFIBUS-DP/DP
	DH+	INTERBUS
	FF HSE	PROFINET
		FF H1

12.2 NETWORK PROTOCOLS

With the advent of computers and the need to interconnect them, communication networks were developed. In like manner, not long after the first PLCs were developed, the need arose to communicate with remote I/O systems and then with other PLCs. However, it was not until personal computers became widespread that the need for standard local area networks emerged. In response to this need, the Institute of Electrical and Electronic Engineers (IEEE) Computer Society established the Standards Project 802 in 1980. The group of standards developed by this organization, commonly called the 802 standards, allowed equipment from different manufacturers to communicate through a local area network.

The development of standards for factory communications took a more tortuous route (Felser, 2002). After many years of work, the attempt to standardize a universal fieldbus at the international level in 1996 resulted in a mixture of Foundation Fieldbus and WorldFIP. Due to national interests, this standard was not ratified and the standardization process degenerated into a political and economic battle. In 1999, the major parties in the debate compromised to create the IEC 61158 standard which encompassed all fieldbus systems. The eight fieldbus specifications in IEC 61158 are called "types." This standard was ratified in 2000 and more types have been added since (IEC, 2019a). It became evident that the collection of fieldbus specifications in the IEC 61158 standard proved impractical for implementation. Therefore a guideline to implementing a functioning system was compiled as a separate standard, IEC 61784 (IEC, 2019b). In IEC 61784, a functioning system is called a "profile."

Network nodes follow a set of rules, called a *protocol*, to communicate with each other. A protocol includes everything from the voltage levels on the connection wires (if it is a wired network) to the interpretation of the data packets by an application program. A protocol often includes:

Network medium
Access to the medium by multiple nodes
Handling communication errors
Information encoding
Message interpretation

To provide a guideline to protocol development, the International Standards Organization (ISO) published the Open Systems Interconnection (OSI) reference model in 1979. The OSI reference model divides the protocol functions into seven hierarchical layers (called a protocol "stack"), as shown in Figure 12.4. The data link layer is divided into two sublayers, medium access control (MAC) and logical link control (LLC). Each of these layers performs some function of the protocol. For efficiency, most factory automation networks use only layers 1, 2, and 7. The other layers add unnecessary complexity and delay.

In operation, the use of the OSI model appears as in Figure 12.5. A program sends a message to the application layer of the protocol in its node. The message proceeds downward through the protocol stack, is placed on the network medium by the physical layer, is received by the physical layer of the destination node, and then proceeds up the protocol stack of the destination node. The operation of a protocol is analogous to the process of sending a letter. The letter is typed (application, presentation), sealed into an addressed envelope (session, transport), picked up by the mail carrier and placed in a mail

Layer	Name		
7	Application		Interface to user
6	Presentation		Data format conversion
5	Session		System-to-system connection; log-on and log-off
4	Transport		Reliable data transfer between end devices
3	Network		Outgoing messages divided into packets. Incoming packets assembled into messages. Routing done here
2	Data Link	2B Logical Link Control 2A Medium Access Control	Reliable delivery of packet to local node. Error detection/correction.
1	Physical		Encodes and physically transfers packet to another local node.

Figure 12.4. Open Systems Interconnection (ISO) reference model.

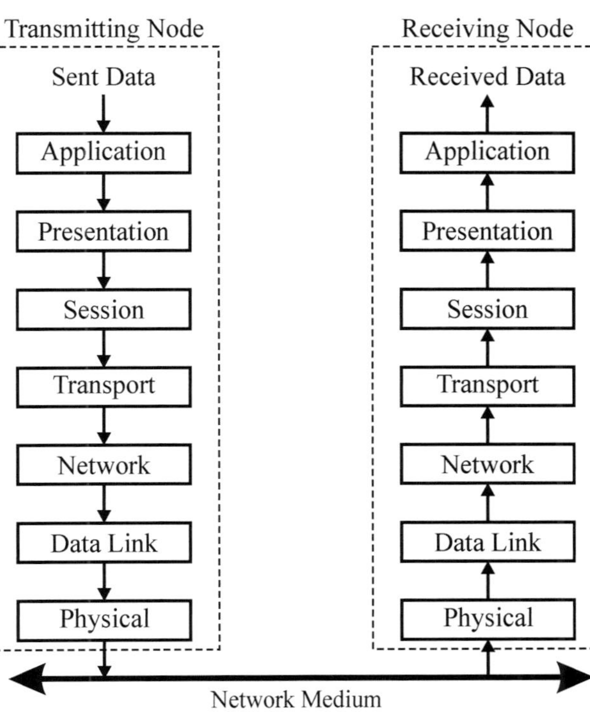

Figure 12.5. Protocol operation.

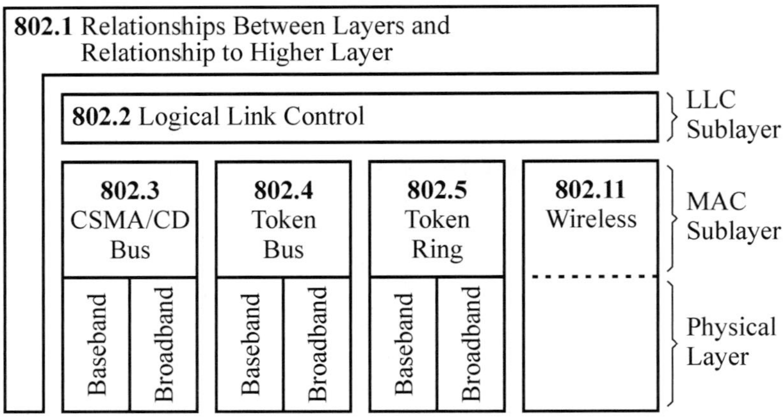

Figure 12.6. IEEE 802 standards.

sack going to the appropriate destination (network, data link), and transported to destination post office (physical). At the destination city, the process is reversed.

The IEEE 802 protocols cover the lower two layers of the OSI model. Some of the parts of the 802 standard are shown in Figure 12.6. The 802.2 standard specifies a common logical link control (LLC) sublayer for the other parts of the standard. The most common standards for office networks are 802.3, which defines the Ethernet protocol, and 802.11,

Table 12.2. IEC 61784 Profiles and IEC 61158 Protocols

IEC 61784 Profile	IEC 61158 Protocols			Protocol Name
	Physical	Data Link	Application	
CPF-1/1	Type 1	Type 1	Type 9	Foundation Fieldbus (H1)
CPF-1/2	Ethernet	TCP/UDP/IP	Type 5	Foundation Fieldbus (HSE)
CPF-1/3	Type 1	Type 1	Type 9	Foundation Fieldbus (H2)
CPF-2/1	Type 2	Type 2	Type 2	ControlNet
CPF-2/2	Ethernet	TCP/UDP/IP	Type 2	Ethernet/IP
CPF-3/1	Type 3	Type 3	Type 3	PROFIBUS-DP
CPF-3/2	Type 1	Type 3	Type 3	PROFIBUS-PA
CPF-3/3	Ethernet	TCP/UDP/IP	Type 10	PROFInet
CPF-4/1	Type 4	Type 4	Type 4	P-Net RS-485
CPF-5/1	Type 1	Type 7	Type 7	WorldFIP (MPS, MCS)
CPF-5/2	Type 1	Type 7	Type 7	WorldFIP (MPS, MCS, SubMMS)
CPF-5/3	Type 1	Type 7	Type 7	WorldFIP (MPS)
CPF-6/1	Type 8	Type 8	Type 8	INTERBUS
CPF-6/2	Type 8	Type 8	Type 8	INTERBUS TCP/IP
CPF-6/3	Type 8	Type 8	Type 8	INTERBUS Subset
CPF-7/1	Type 6	Type 6	-	Swiftnet transport
CPF-7/2	Type 6	Type 6	Type 6	Swiftnet full stack

which defines wireless networks. The token bus (802.4) protocol was developed for factory communications. Though this standard was later withdrawn, many of its concepts were used in the Foundation Fieldbus standard. The token ring (802.5) protocol was adopted by IBM, but not currently used since Ethernet was extended to use fiber optic media.

The protocols in the IEC 61158 standard cover the physical, data link, and application layers of the OSI model. The IEC standard is organized by these layers. The individual types are thus described for each layer. In order to guide practical implementations of the protocols, guidelines were compiled into IEC 61784 as *profiles*. The relationship between the IEC 61784 profiles, the IEC 61158 types, and the protocol names is shown in Table 12.2. There are basically seven different main profiles in the original standard. Before describing the individual networks, general topologies, media access methodologies, and network media need to be discussed.

12.2.1 Network Topologies

When constructing a local area network, there are three main network topologies, that is, how devices are physically connected. The bus topology (Figure 12.7*a*) is by far the most common topology in factory automation communications. In a bus network, the devices are all connected to the same media and theoretically every PLC can communicate with any other PLC. The network connection is commonly called a "tap." Depending on the particular network, the tap may be a simple BNC "T" connector, a terminal strip, or a multi-pin connector. In some networks, the cable may be "daisy-chained" between PLCs. The star topology, shown in Figure 12.7*b* is used in some older, proprietary networks. In this case, each PLC is connected to a central network controller and any PLC-to-PLC messages pass through the network controller. The ring topology (Figure 12.7*c*) was developed for early fiber-optic networks. In a ring topology, each PLC is connected to and communicates with its two neighbors. An individual PLC-to-PLC message may have to make multiple "hops" in order to reach its destination.

The star topology is still used in Ethernet networks. In this case, the "network controller" is a hub or switch. Even though the Ethernet physical topology is most like a multi-level star, it logically acts like a bus connection, that is, each network device communicates with another network device and is not cognizant of any intervening hardware. The ring topology has basically been supplanted by the fiber optic version of Ethernet. In addition, some of the bus networks allow fiber optics to replace cable segments.

12.2.2 Media Access

With a bus topology, one must have some method of controlling a particular device's access to the bus. There are three main methods of bus arbitration:

1. Master/slave
2. Carrier sense multiple access with collision detection
3. Token passing

Though the bus arbitration for a particular network is often not one of these three, it has elements of at least one of them.

With master/slave arbitration, one of the network nodes is the designated master. The master polls each network node to check its presence (or node status, "health") and to allow the node to transmit messages to other nodes. Master/slave is the simplest arbitration

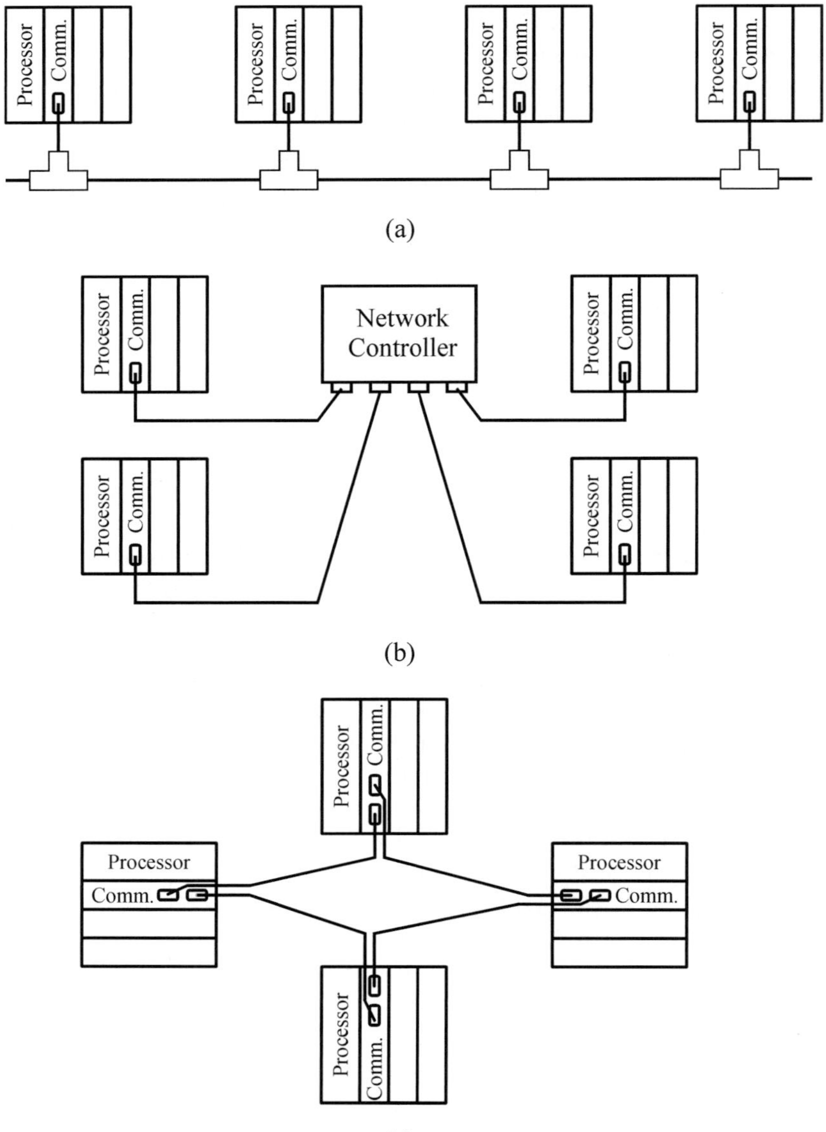

Figure 12.7. Network topologies: *(a)* bus; *(b)* star; *(c)* ring.

method. However, if the master node fails, then the network fails. A backup master will solve this problem. This method of bus arbitration was common in the earlier distributed process control systems. Though not used for PLC-to-PLC communications, master/slave bus arbitration is still utilized in I/O networks.

Carrier sense multiple access with collision detection (CSMA/CD) is a masterless arbitration scheme. When a node wants to send a message, it waits for no network traffic and then attempts to send a message. If two nodes transmit simultaneously, then a *collision* occurs and each of the offending nodes waits a random amount of time before checking for

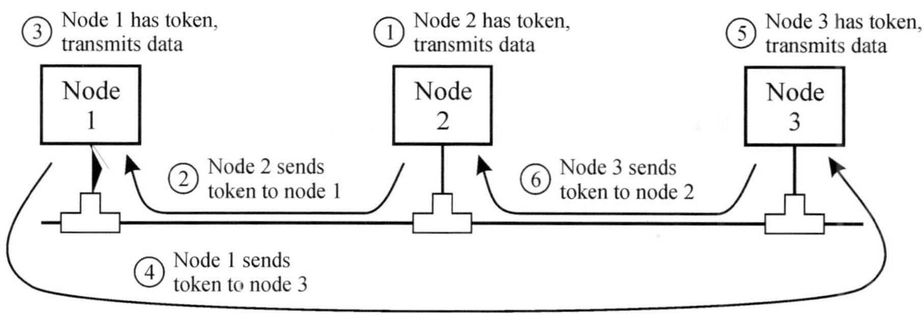

Figure 12.8. Token passing operation.

no network traffic in an attempt to resend the message. Ethernet uses CSMA/CD as its basic bus access method. While CSMA/CD is a simple masterless protocol, performance rapidly deteriorates if the network traffic is heavy. With the original Ethernet specification, when the attempted network transmissions exceed about 60% of the available capacity, the transmission delay increases dramatically (Ciminiera et al, 1988). Higher baud rate and the use of Ethernet switches in place of hubs have overcome this disadvantage (Lian et al, 2001).

In token-passing arbitration, a special message, called the *token*, is circulated among the network nodes. Possessing the token grants the node access to the network and the node can transmit messages. When finished with the transmissions, the station passes the token to the next station in the list. The token is analogous to the baton that is passed from one runner to the next in a relay race. The runner possessing the baton is the one racing. Token passing is illustrated in Figure 12.8. The token rotation is 2-1-3. The steps to generate the initial token and rotation list are not shown. The token starts in node 2. When node 2 finishes transmitting messages, it sends a special message to node 1, granting it the token. Node 1 transmits its messages and then passes the token to node 3. When node 3 finishes its message transmission, it passes the token to node 2 and the cycle recommences. Token passing guarantees delivery of some messages under high network traffic, even beyond the available capacity. This arbitration is the most complicated, having to deal with such issues as generating the initial token and allowing stations to leave and enter the rotation.

When the Ethernet (CSMA/CD) and token bus protocols were being developed in the 1980's as the IEEE 802.3 and 802.4 standards, there was a fierce debate as to which one was more suitable for factory communications. Ethernet was generally viewed as not acceptable since message delivery could not be guaranteed under high network loads. When things "go wrong" in a factory, the network traffic is substantial due to the alarm messages being transmitted from the PLCs to the operator terminals. The debate has largely subsided because of three reasons:

1. The cost of an Ethernet connection declined faster than the cost of a token bus connection.
2. The baud rate of the Ethernet network increased 1000-fold.
3. The development of Ethernet switches

However, the token bus arbitration method still survives in the Foundation Fieldbus, ControlNet and Profibus networks.

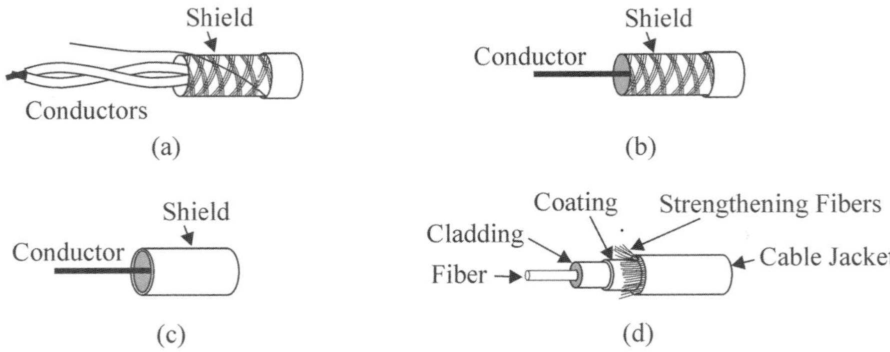

Figure 12.9. Network media: *(a)* twisted pair; *(b)* baseband coaxial; *(c)* broadband coaxial; *(d)* fiber optics.

12.2.3 Network Media

There are five main categories of network media:

Twisted pair
Baseband coaxial cable
Broadband coaxial cable
Fiber optic
Radio frequency

As its name implies, a twisted pair cable consists of pairs of wires twisted around each other. The twisting provides some noise immunity. Shielding the cable further increases the noise immunity. Figure 12.9*a* shows the construction of an example shielded twisted pair cable. Twisted pair cables are generally used for lower speed networks. A baseband coaxial cable is a flexible cable having a shield around a single conductor (Figure 12.9*b*). Baseband coaxial cable can handle higher data rates and longer distances than twisted pair cable. Broadband RG-59 coaxial cable is the medium commonly used for cable TV. It is a rigid coaxial cable and the shield forms the outside of the cable (Figure 12.9*c*). For each channel, there is a transmitting frequency and a receiving frequency. One end of the cable is called the head end. All nodes transmit to the head end, which retransmits the signal at the receiving frequency. This medium supports multiple channels on the same physical medium and thus each channel is its own LAN. With repeaters, a network with this medium can be extended up to 30 miles. Though it can support very large networks, the connection cost for each node is relatively high due to the high-frequency modulators and demodulators. Fiber optic cable (Figure 12.9*d*) is immune to electrical interference and so is used in very noisy industrial environments. Though initially this medium was costly, its cost is now comparable to wired media. However, it is still more complex to install since it cannot be easily spliced. In manufacturing, it is typically used to extend wired busses long distances and through electrically noisy areas. Wireless communication networks rely on radio frequency transmission. This type of medium is becoming popular since it lowers the installation cost. However, it is only useful for short distances and objects in the transmission path can disrupt communications. In addition, there are security concerns since the signal can be monitored remotely.

12.2.4 Redundancy

Having a redundant system means that one can tolerate failures. Redundant network media is fairly common, but it only protects against a cut cable or fiber. Redundant network cables with redundant communication modules protects against more failures, but at the expense of more complexity. Those automation networks that support redundancy are indicated in the following sections.

12.3 ETHERNET

Ethernet is by far the most prevalent backbone network. In the early 1980's, General Motors spearheaded the development of GM MAP (General Motors Manufacturing Automation Protocol), which at the time was expected to be the backbone network for factory communications. The medium was broadband coaxial cable. A few systems were installed, but the connection cost for each node was high. The cost of Ethernet network connections continued to decline, essentially making GM MAP prohibitively expensive.

Physical Layer

The original 802.3 specification called for two types of physical media, thin coaxial cable (RG-8) and thick coaxial cable (RG-58). The most popular media today is CAT-5 or CAT-6 (category 5 or category 6) unshielded twisted pair cable with 4 twisted pairs or duplex fiber. Wireless modems are also available. The usual structure of an Ethernet network is that of a tree/star (Figure 12.10*a*). At the switch (or hub), the network resembles a star topology. However, further up the hierarchy, the network looks more like a tree with routers connecting to multiple switches. Rockwell Automation recently introduced the Ethernet ring topology for a device networks, called device-level ring (DLR) for Ethernet/IP (Rockwell Automation, 2019a), shown in Figure 12.10*b*.

Data Link Layer

Logically, the Ethernet network more resembles a bus. As described earlier, CSMA/CD (carrier-sense multiple access with collision detection) is the bus access method.

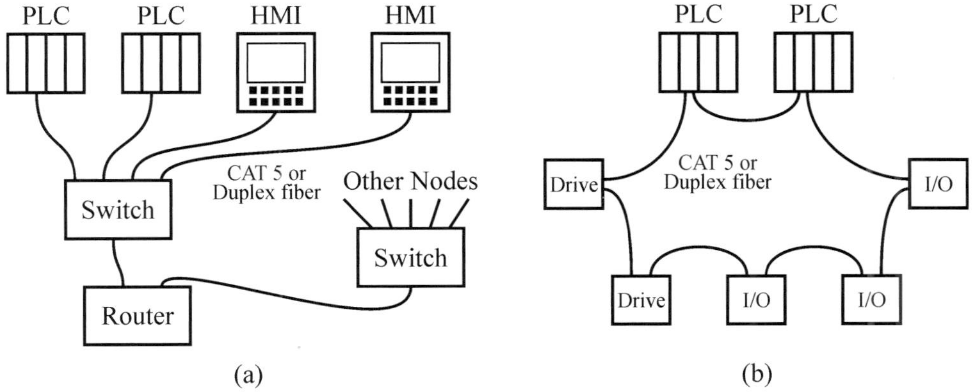

Figure 12.10. Ethernet topologies: *(a)* tree/star; *(b)* device-level ring.

Application Layer

Multiple application protocols have been built on top of the Ethernet data link and physical layers. Ethernet/IP is described in subsection 12.5.3. Foundation Fieldbus HSE, Modbus/TCP, and PROFINET are described in Erickson (2023).

12.4 CIP-RELATED PROTOCOLS

The DeviceNet, ControlNet, and Ethernet Industrial Protocol (Ethernet/IP) protocols are all related because they share the Common Industrial Protocol (CIP) for the upper layers (Figure 12.11). The differences between these protocols are at layers 1, 2, and 3 of the OSI model.

ControlNet was originally developed by Allen-Bradley and Honeywell in the early 1990's to transmit time-critical I/O and peer-to-peer data. In parallel with this effort, Allen-Bradley also developed the DeviceNet protocol for device communications. Both of these protocol specifications were published in the mid-1990's. In 1997, the administration of the ControlNet protocol was transferred to ControlNet International, a multi-vendor association. Adminstration of the DeviceNet protocol was turned over to the Open DeviceNet Vendors Association in 1995. Ethernet/IP was developed recently to provide an acceptable application layer and industrially-hardened connectivity built on top of standard Ethernet)ODVA, 2015b).

The DeviceNet and ControlNet physical and data link layers are described first, followed by a brief introduction to the Ethernet/IP and CIP layers.

12.4.1 DeviceNet

The DeviceNet protocol was developed for device networks. It is relatively low speed but efficient at handling the short messages to/from I/O modules.

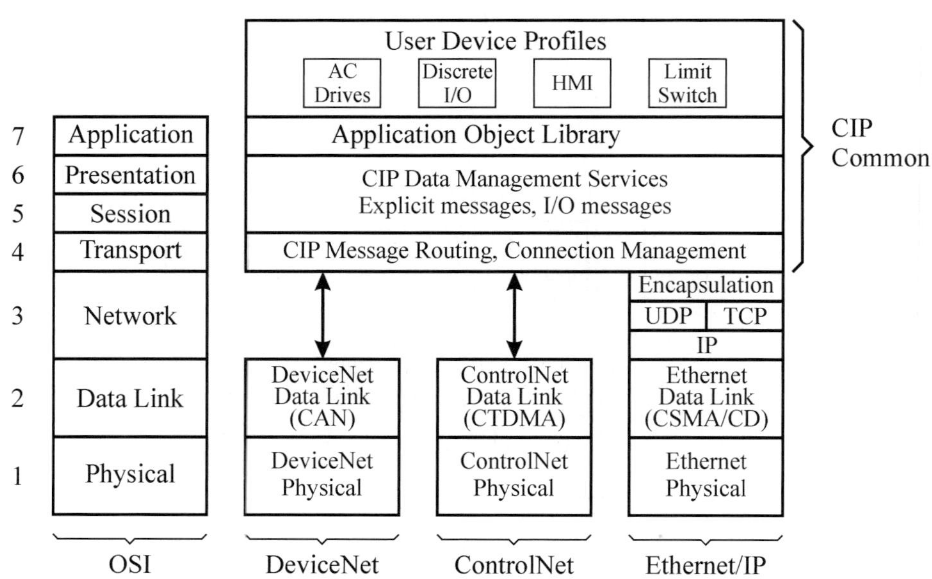

Figure 12.11. Communication model for CIP-related protocols.

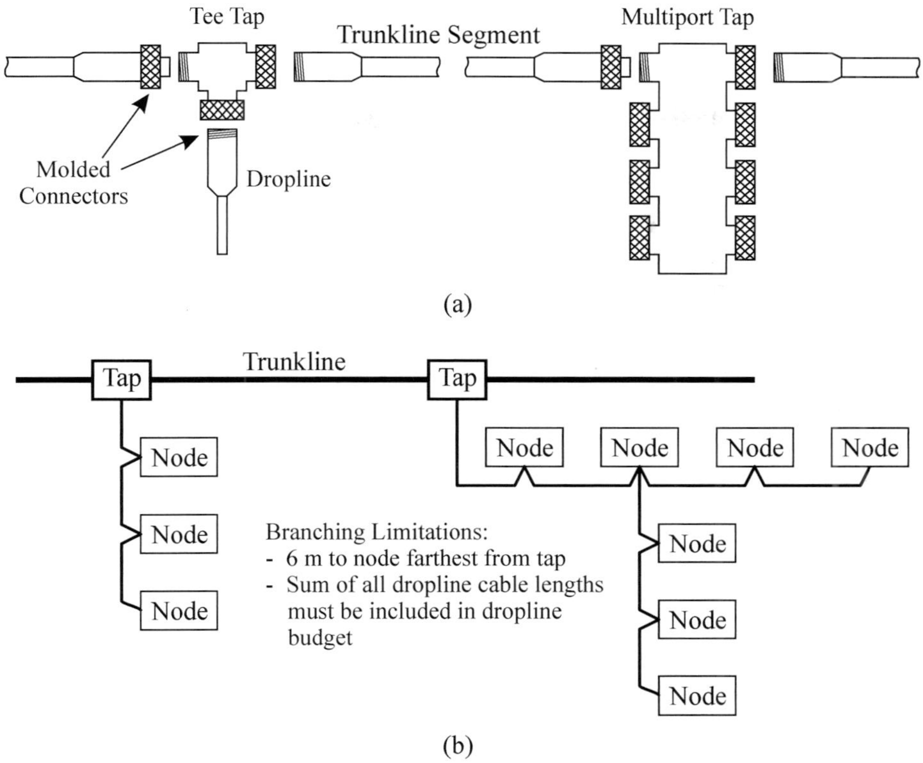

Figure 12.12. DeviceNet network topology: *(a)* trunkline taps; *(b)* daisy-chained drop lines.

Physical Layer

The physical layer of DeviceNet is the same as the Controller Area Network (CAN) developed by Bosch in the early 1980's as an automotive network standard. The DeviceNet medium is a 5 conductor shielded cable, two twisted pairs plus a shield drain. One twisted pair carries the signal and the other twisted pair carries the power. The network topology is trunkline/dropline (Figure 12.12). A trunkline tap can be for one or multiple drop lines (Figure 12.12*a*). In addition, a dropline can be daisy-chained between devices, as shown in Figure 12.12*b*. A DeviceNet network can have up to 64 nodes. There are three data transmission rates: 125, 250, and 500 Kbit/s. The maximum trunkline length depends on the data rate. For the thick trunkline cable the maximum distance is 500 m for the 125 Kbit/s data rate down to 100 m for the 500 Kbit/s data rate. If the thin (drop line) cable is used for the trunkline, the maximum length of the trunkline is 100 m. The maximum drop length is 6 m and the sum of all drop cable lengths must not exceed a maximum.

Several different connector types can be used on DeviceNet. Both sealed and unsealed connectors are available. For devices installed on the factory floor, mini-style and micro-style pluggable, sealed connectors are used. For nodes installed inside cabinets, open-style connectors can be used.

Devices are ordinarily powered from the DeviceNet cable. Since the trunkline current rating is 8 amps, the power supply conductor is broken into segments, each segment powered by a separate power supply.

Data Link Layer

The method of bus arbitration is similar to the CSMA/CD used on Ethernet, but without the collisions. If two or more nodes start transmitting simultaneously, the bus conflict is resolved by a bit-wise arbitration. A logical "0" is dominant on the wire and overrides a logical "1." When a node transmits a "1," but hears a "0," it immediately stops transmitting. The "winning" node continues to transmit its message to completion. With this method of arbitration, neither information nor time is lost. Since the first data in a DeviceNet message is an 11-bit identifier, the winning node is the one with the lowest-numbered identifier. The message contains at most 8 bytes of data, though there is a fragmentation protocol that permits larger amounts of data to be broken into multiple messages.

Network and Transport

In accordance with CIP, a connection must be established between two devices before data can be exchanged. The 11-bit CAN identifier is used to define the connection identifier. The uniqueness of connection identifiers is required to take full advantage of producer/consumer capabilities.

DeviceNet devices may be clients or servers or both. For a typical client device, its connection produces requests and consumes responses. For a typical server device, its connections consume requests and produce responses. However, a client or server connection may only consume messages. These connections would be the destination for Cyclic or Change-of-State messages. Similarly, a client or server connection may only produce messages. These connections would be the source for Cyclic or Change-of-State messages.

DeviceNet provides for a simplified communication scheme based on a master/slave relationship. This connection scheme called the Predefined Master/Slave Connection Set, simplifies the movement of the I/O messages most often used in device-level networks. Most sensors and actuators perform some predetermined function in which the type and amount of data produced and/or consumed by the device is known. The Predefined Master/Slave Connection Set provides connections that are almost entirely configured when the device is added to the network or after the network is powered up. In this case, a "master" device must claim ownership of this predefined connection set within its "slave" to start the data flow. Multiple masters cannot share slave devices. The slave devices produce data in one of the following manners:

Polled - Slave receives "output" data from the master device in an order defined by the scan list in the master. For a given system configuration, the polled method produces deterministic behavior.

Cyclic - Slave device produces its data at a precisely defined interval. Depending on the application this method of data production can reduce the amount of traffic on the wire and more efficiently use the available bandwidth.

Change-of-state - Slave produces data whenever it changes, or at a base heartbeat rate. With the heartbeat, the consuming device knows that the producer is still alive and active.

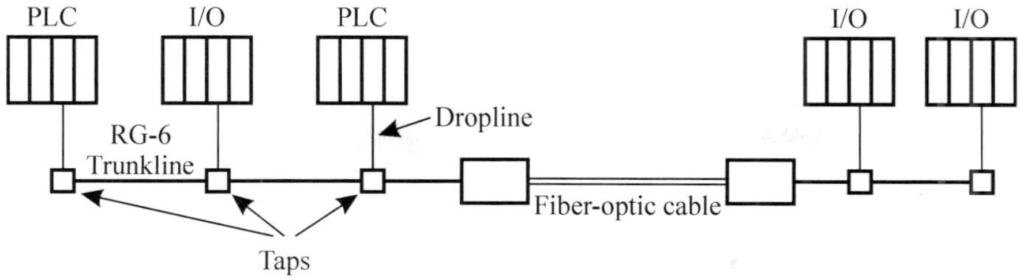

Figure 12.13. ControlNet trunkline/dropline topology.

12.4.2 ControlNet

The ControlNet protocol was developed as an automation network (Figure 12.3). It is designed to efficiently handle time-critical data transfers between PLCs.

Physical Layer

The ControlNet medium can either be RG-6 coaxial cable or fiber optic cable. The usual network topology is trunkline/dropline (Figure 12.13) using RG-6 coaxial cable. Fiber optic cables are used as point-to-point links to extend the network. ControlNet also supports tree and star topologies with fiber optic repeaters. Both coaxial cable and fiber optic systems can be designed to meet the requirements of intrinsic safety environments. A ControlNet network can have up to 99 nodes. Without signal repeaters, it can extend to 1000 m with two nodes and 250 m with 48 nodes. With repeaters, the trunk can extend up to 25 km. ControlNet supports redundant media. If redundant media is used, nodes transmit on both cables and listen continuously to both cables. When listening, each node independently decides which cable has the best signal. Data is transmitted at the rate of 5 Mbit/s and is Manchester encoded.

Data Link Layer

ControlNet uses Concurrent Time Domain Multiple Access (CTDMA) to control node access to the bus. CDTMA is designed so that time-critical data transfer, such as I/O or peer-to-peer inter-locking data, is not hindered by other data transfers such as PLC programming. The operation of CDTMA is shown in Figure 12.14. Time is divided into network update times (NUTs). The duration of a NUT is configured to be between 2 ms and 100 ms. Within the first part of each NUT, an implicit token rotation algorithm grants media access. The token rotation starts from the lowest configured node address and proceeds to the highest configured node address. There is no centralized bus scheduler. All nodes are continuously synchronized so that they exactly know when they may access the media to broadcast their data. Transmission time for real-time data is also allocated in the NUT. The remaining transmission time within a NUT supports non-time-critical data.

Though CDTMA does not have a centralized scheduler, the network must be configured by some network management device. The biggest drawback to this method of bus arbitration is that the entire network must be reconfigured when any node adds a new scheduled transmission. During network configuration all running PLC programs must be stopped.

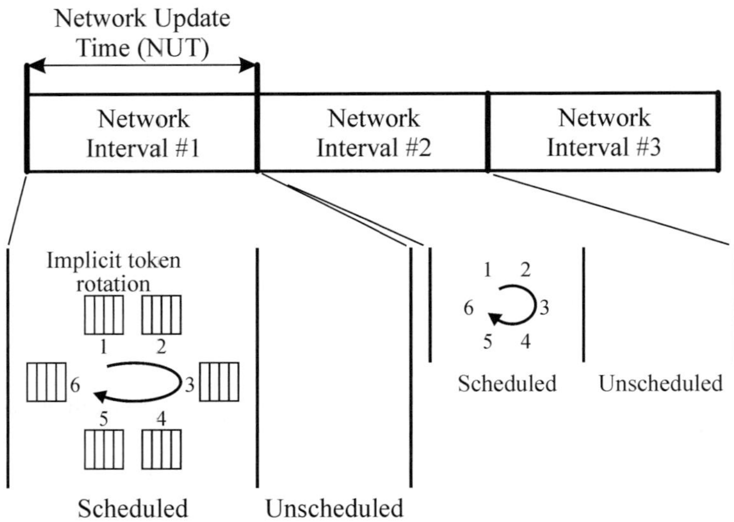

Figure 12.14. ControlNet bus arbitration.

Network and Transport Layers

ControlNet requires a formal connection to be established between application entities before information can be transferred. A connection defines a path or virtual circuit between application end points for data transfer. Connections can be either peer-to-peer or multicast (multiple destination nodes read the same data packet). Data may be transmitted based on "cyclic", "change-of-state" or "application" triggers.

12.4.3 Ethernet/IP

EtherNet/IP uses standard Ethernet and TCP/IP (transport control protocol/Internet protocol) protocols to transport CIP messages. As for both DeviceNet and ControlNet, EtherNet/IP provides a producer/consumer model for the exchange of time-critical control data. EtherNet/IP uses IP multicast so that many EtherNet/IP devices can receive the same produced information from a single producing device. EtherNet/IP uses standard IEEE 802.3 technology. To improve determinism, EtherNet/IP recommends the use of commercial Ethernet switches with 100 Mbps transmission rate and full-duplex operation.

As shown in Figure 12.11, CIP messages are encapsulated into either a Transport Control Protocol (TCP) or a User Datagram Protocol (UDP) connection. The encapsulation protocol defines a reserved TCP port number and a reserved UDP port number that is supported by all EtherNet/IP devices. Depending on the particular message, it may be sent by either UDP or TCP. Some encapsulated messages can only be sent by TCP. Other messages may be sent by either UDP or TCP (ControlNet Int., 2001b).

On Ethernet, it is possible for a data packet to be lost, for example due to excessive collisions. Ethernet does not guarantee the delivery of a packet. Producers continue to send data at the specified rate and the consumers receive the next packet after the lost packet. The degree to which lost packets can be tolerated is application-specific. Ethernet is unsuitable for those applications that cannot tolerate lost packets.

12.4.4 CIP Application Layer

The Common Industrial Protocol (CIP) provides common communication protocol layers on top of DeviceNet, ControlNet, and Ethernet/IP (Figure 12.11). CIP has two primary purposes:

Transport of control-oriented data

Transport of other information, such as configuration, PLC program, and diagnostics.

The CIP is object-oriented. Both communication and application entities are referenced as objects. The general CIP object model is shown in Figure 12.15 (ODVA, 2015). The objects in this model are:

Unconnected Message Manager - Processes CIP unconnected explicit messages.

Connection Class - Allocates and manages internal resources associated with both Explicit and Input/Output messaging connections.

Connection Object - Manages the communication-specific aspects associated with a particular application-to-application network relationship.

Network-Specific Link Object - Provides the configuration and status of a physical CIP network connection (DeviceNet, ControlNet, and Ethernet/IP objects).

Message Router - Distributes Explicit Request Messages to the appropriate handler.

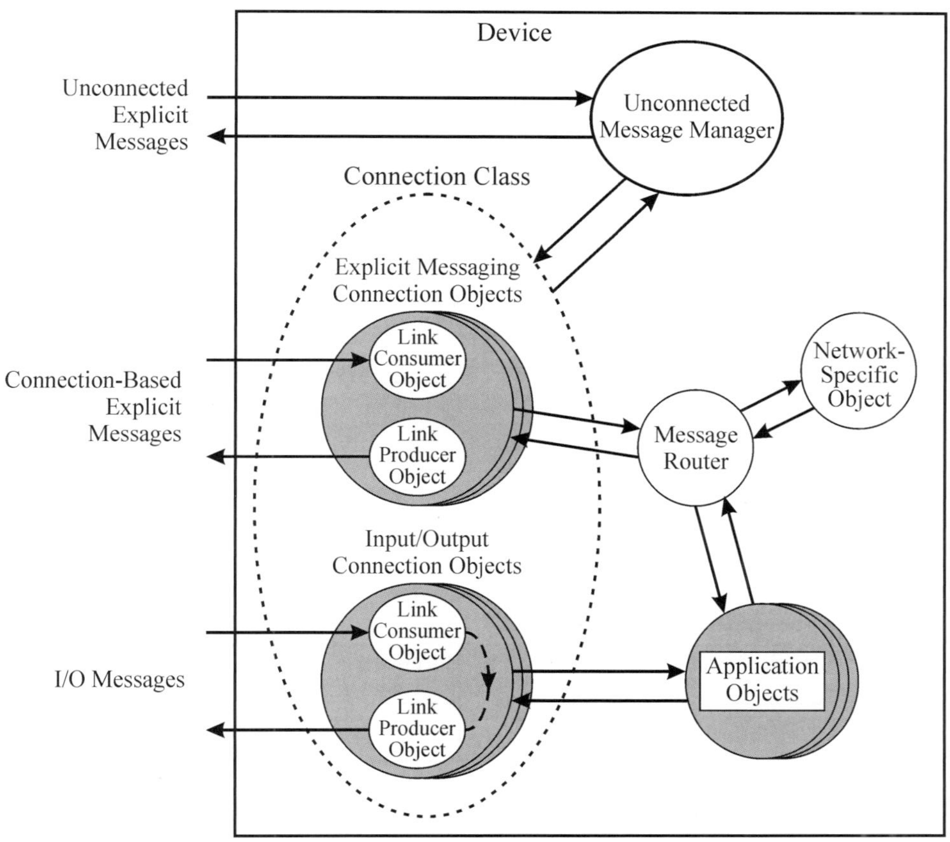

Figure 12.15. CIP object model.

Application Objects - Implement the intended purpose of the product.

The Unconnected Message Manager (UCMM) is responsible for processing the messages that establish both explicit messaging and I/O connections. Explicit messaging connections are point-to-point, that is exist between only two devices. I/O connections can be either point-to-point or *multicast*. Multicast connections allow a single transmission to be heard by many nodes.

12.5 PROFIBUS (DP, PA)

The PROFIBUS (PROcess FIeld BUS) protocol is the result of an effort funded by the German Federal Ministry for Research and Technology in 1987-1990 to build a process automation digital field bus. PROFIBUS has three parts:

1. PROFIBUS-DP (decentral periphery)
2. PROFIBUS-PA (process automation)
3. PROFIBUS-FMS (fieldbus message specification)

PROFIBUS-FMS and PROFIBUS-PA were published as the DIN 19245 standard, parts 1 and 2. In 1993, PROFIBUS-DP was added to this standard as part 3. PROFIBUS became a part of the EN 50170 standard in 1996. Though PROFIBUS-FMS was the first PROFIBUS communication protocol, it was omitted from the IEC 61158 standard and so it is not discussed here. PROFINET is a more recent development by the PROFIBUS trade organization, PNO (PROFIBUS Nutzerorganisation) to answer the trend toward increasing utilization of Ethernet in factory automation.

The relationship between PROFIBUS-DP, PROFIBUS-PA, and PROFINET are shown in Figure 12.16. PROFIBUS-PA is essentially a subset of the data link layer of PROFIBUS-DP with a different physical layer. PROFINET is not the PROFIBUS application layer on top of Ethenet. It basically uses standard information technology functions.

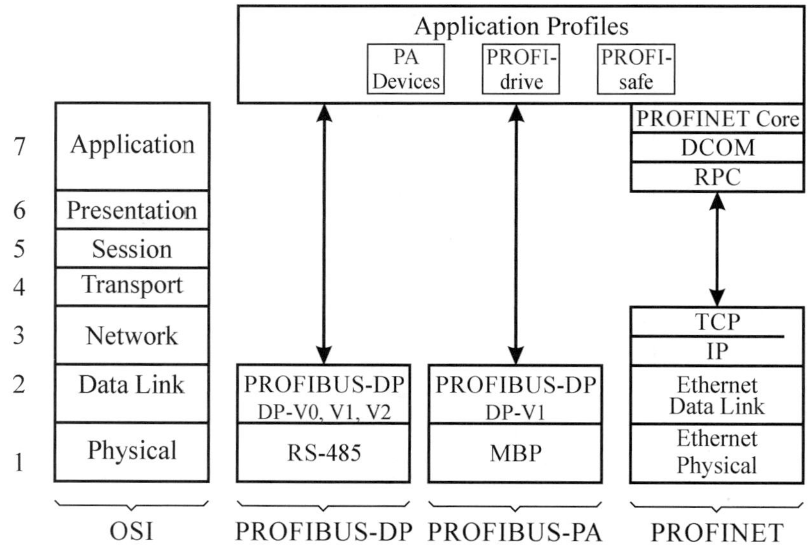

Figure 12.16. Communication model for PROFIBUS protocols.

The PROFIBUS-DP network is described first, followed by PROFIBUS-PA. PROFINET is not discussed in this chapter, but is covered in Erickson (2023).

12.5.1 PROFIBUS-DP

The PROFIBUS-DP communication protocol is widely used in PLC automation in Europe. It was developed for a mix of discrete and analog devices. It appears to be primarily used for distributed I/O, though Siemens also uses a subset of it (called Multipoint Interface) for PLC-to-PLC communication.

Physical Layer

PROFIBUS-DP uses RS-485 as the physical layer. The medium is shielded twisted pair cable usually with 9-pin D-sub connectors. Though 9-pin D-sub connectors are common, sealed connectors are also available. The network topology is a bus (Figure 12.7*a*), terminated on each end with an active terminator. A PROFIBUS-DP segment can have up to 32 nodes. With repeaters connecting network segments, there can be a maximum of 126 devices on the network. The allowable transmission rates range from 9.6 Kbit/s to 12 Mbit/s. Obvously, all nodes must use the same transmission rate. The maximum bus length depends on the data rate and the type of cable. With a type-A cable and a transmission rate of 9.6 Kbit/s the maximum distance is 1200 m. If the transmission rate is 12 Mbit/s, the maximum segment distance drops to 100 m. The signals on the bus are non-return-to-zero (NRZ) encoded.

For intrinsic safety applications, there is a version of the physical layer, called RS-485-IS. In RS-485-IS, the current on the bus must not exceed a certain limit, which depends on the voltage level. For RS-485-IS, the medium is a shielded twisted 4-wire cable and its maximum transmission rate is 1.5 Mbit/s. Fiber optic cabling options are also available. The fiber can replace a network point-to-point cable, can allow a star and ring configurations (Siemens, 2009).

Data Link Layer

PROFIBUS-DP defines three versions of its data link layer:

DP-V0: Basic DP functions
DP-V1: Enhancements for process automation
DP-V2: Enhancements for motion control

DP-V0 is the original PROFIBUS data link layer and the enhancements are recent additions. DP-V0 is described first and then the enhancements are discussed.

DP-V0 supports both master/slave and peer-to-peer communications. There are three classes of network devices:

DP Master Class 1 (DPM1)
DP Master Class 2 (DPM2)
DP Slave

A DPM1 device cyclically exchanges information with slaves at a specified message cycle. PLCs are typically DPM1 devices. A DPM2 device is also a bus master, but is usually an engineering or configuration device. A DPM2 does not exchange information with a slave device. A slave device is typically a remote I/O device (for example, discrete I/O or variable-speed drive), but can be HMI devices.

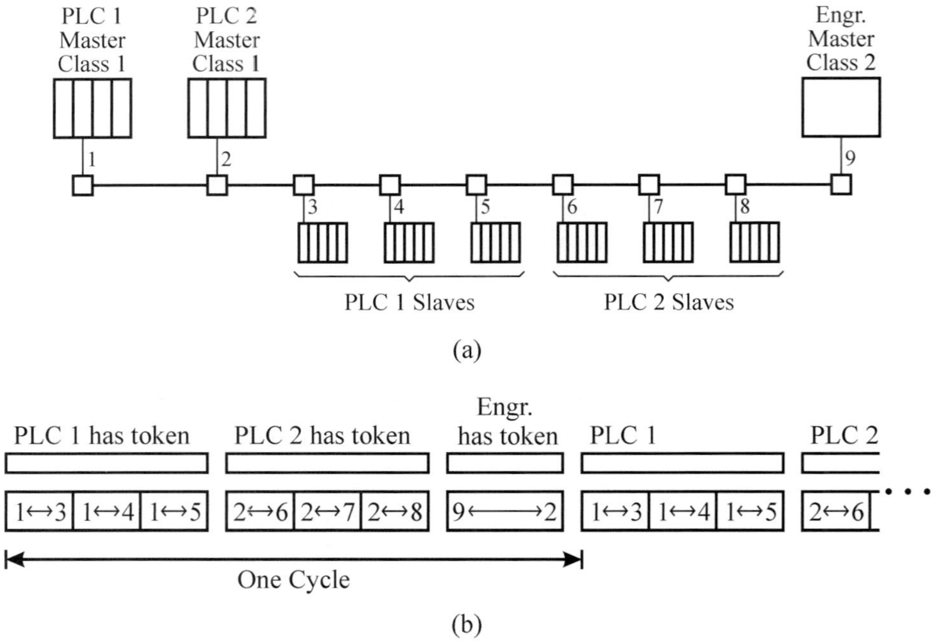

Figure 12.17. Example DP-V0 operation: *(a)* system; *(b)* message timing.

In a mono-master system, there is only one master and it controls the bus. In a multi-master system (Figure 12.17*a*), multiple masters share the bus. Since only one master can control the bus, a token is circulated among the masters. The master that possesses the token controls the bus and communicates with its slaves, or with one of the other masters. Slaves respond only to direct queries from a master device. Though all masters can read the input and output images of any slave, only one DPM1 device can write to a slave's output image. This master is assigned during configuration. In Figure 12.17*a*, PLC 1 is the master of slave devices 3 - 5 and PLC 2 is the master of slaves 6 - 8. When a particular PLC has the token, it communicates with its slaves, checking their configuration and then transferring the data (Figure 12.17*b*). When the engineering device (DPM2) has the token, it communicates peer-to-peer with one of the PLCs.

DP-V1 adds acyclic capability to DP-V0. Transmission of any acyclic data is done between the last cyclic exchange with a slave and the end of a cycle. When a DPM2 device has the token during this period, it can transmit data to any other master or slave. At the end

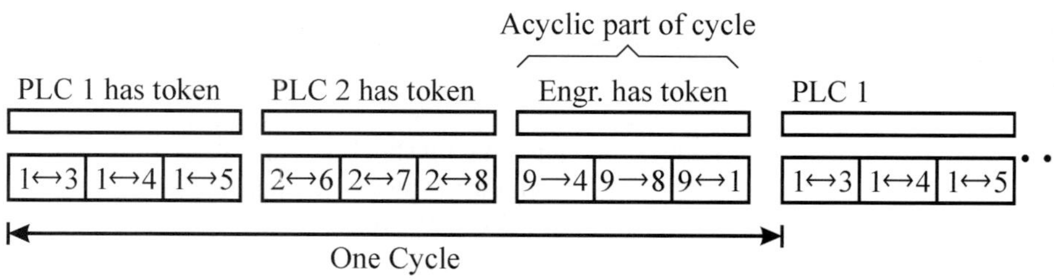

Figure 12.18. Example DP-V1 operation timing.

of the cycle, the token is always returned to the first DPM1 device. If there is a large amount of acyclic data, its transmission may span multiple cycles. The total cycle time in DP-V1 is longer than in DP-V0. In DP-V1, a DPM2 master can communicate to any slave. For example, a DP-V1 timing of the system in Figure 12.17*a* is shown in Figure 12.18. In this case, during the acyclic part of the period, station 9 (DPM2) sends information to two slaves and has a short exchange with PLC 1. DP-V1 also defines alarms and portable function blocks.

DP-V2 adds functions needed for high-speed motion control to DP-V1. In DP-V2, a slave can communicate with another slave directly without having a master as an intermediary. In this case, a slave broadcasts (publishes) its data on the network and other slaves can receive (subscribe) this message without needing to receive it from a master. With DP-V2, one can configure the network for isochronous mode where the clock deviation between devices is less than a microsecond. All participating device cycles are synchronized to the bus master cycle.

12.5.2 PROFIBUS-PA

PROFIBUS-PA was designed as a substitute for 4-20 mA and HART signal transmission in process control applications. PROFIBUS-PA uses the same physical layer as for Foundation Fieldbus H1. In PROFIBUS, this transmission method is called Manchester-encoded, bus powered (MBP). Lower power (called MBP-LP) and intrinsically safe (MBP-IS) versions of this physical layer are also available.

At the data link layer, PROFIBUS-PA uses the DP-V1 protocol, which allows acyclic data communication for operator interfaces and for device configuration. The DP-V1 protocol also defines alarms. At the application layer, it uses function blocks designed for the needs of process control applications.

12.6 AS-I

The actuator sensor interface (AS-i) network was developed in the early 1990's by a consortium of European companies. It was developed as a low-level bus to interface discrete sensors and actuators. It was later extended to include analog sensors and actuators. It was ratified as EN 50295-2 in 1998 and IEC 62026-2 in 2000 (IEC, 2008). The AS-i protocol covers layers 1, 2, and 7 of the OSI model, though only the physical and data link layers are described here.

Physical Layer

AS-i uses a two-wire unshielded cable as the medium. The usual cable is yellow and has a trapezoidal cross-section (Figure 12.19), though a conventional round cable is also available. The network topology may be a linear bus, tree, star, or some combination of the three (Figure 12.20). An AS-i network can have a maximum of 96 slave devices. The total cable length is 100 m, though with repeaters the cable length can be extended to 300 m. The transmission rate is a fixed 167 Kbit/s. The two-conductor cable supplies both power and signal, though if a device draws more than 200 mA a separate two-conductor black cable with the same trapezoidal cross-section supplies additional 24-volt DC power. A red version of the cable supplies 240-volt AC power.

There are four versions of AS-i. The original AS-i specification (V1) allowed 31 slave devices, each with a maximum of 4 discrete inputs and 4 discrete outputs and so the network

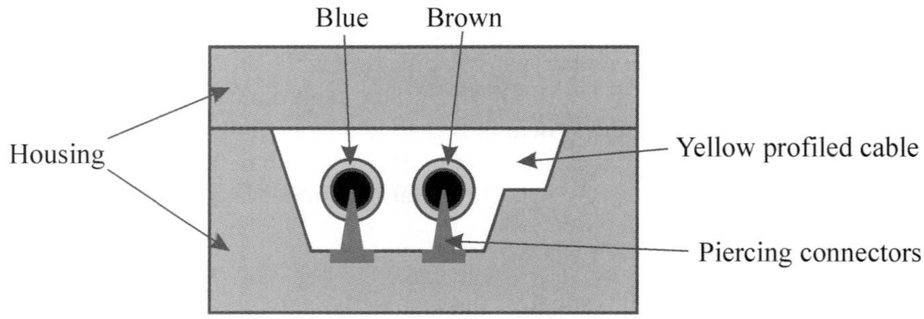

Figure 12.19. AS-i network cable.

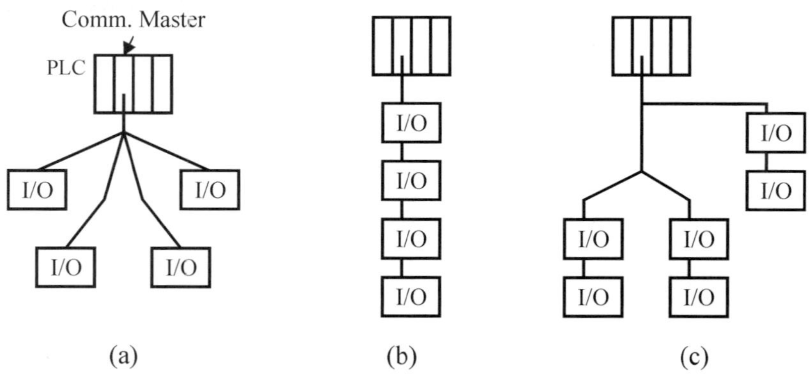

Figure 12.20. AS-i topologies: *(a)* star; *(b)* linear bus; *(c)* tree.

can support 124 discrete inputs and 124 discrete outputs. Version 2.1 of the AS-i specification introduced extended addressing and the ability to handle analog inputs and outputs. In V2.1, there can be a maximum of 62 slaves, addressed as 31 A slaves and 31 B slaves. An AS-i V2.1 network can support a maximum of 248 discrete inputs and 186 discrete outputs. ASI-3 increased the maximum number of discrete inputs and outputs to 496 each and added Combi modules with enhanced capabilities. ASI-5 increased the number of nodes to 96 with up to 16 bits of data per node. The system can have up to 1536 discrete inputs and 1536 discrete outputs.

Data Link Layer

The AS-i protocol is a master/slave protocol. The master polls the slaves and receives a response. In V1, the master sends a request message to a slave with the four output bits and the slave responds with a message containing the four discrete input bits. A fully loaded (31 slaves) V1 network scans all slaves within 5 ms.

For an AS-i V2.1 network, the master scans each slave in the A mode and then in the B mode. A V2.1 network with 62 slaves has a 10 ms scan time. Analog signals are handled by an AS-i V2.1 master by using multiple scans to transmit parts of the analog data. Five cycles are required to transmit 12-bit data and seven cycles are needed for 16-bit data.

In ASI-5, 96 nodes can be scanned with a 5 ms cycle time. This version also added acyclic communication.

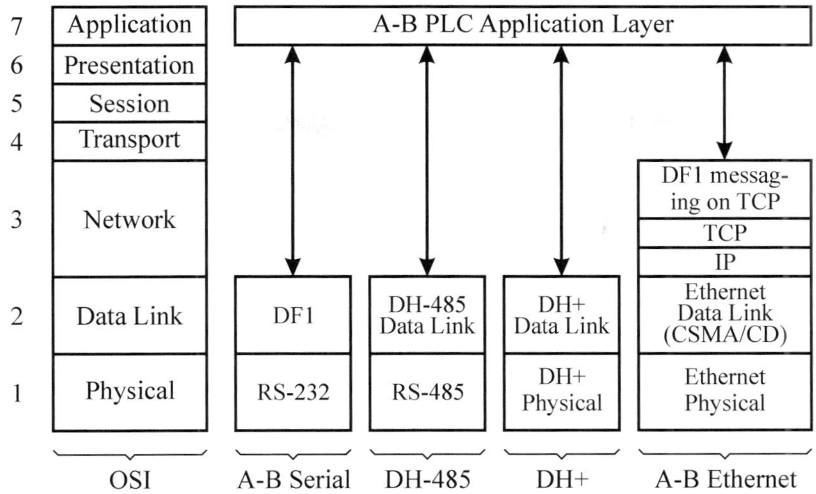

Figure 12.21. Communication model for Allen-Bradley protocols.

12.7 ALLEN-BRADLEY PROPRIETARY NETWORKS

Allen-Bradley introduced its first peer-to-peer network, Data Highway, in 1979, for a multi-drop network of PLCs. Later, it was enhanced and named Data Highway+ (DH+). This network became the standard network between PLC-5 processors. At about the same time, the DH485 network was developed for the SLC-500 processors. When PLC-5 Ethernet processors became available, the same messaging was implemented on top of Ethernet. These four implementations are related as shown in Figure 12.21. All of these protocols are considered proprietary, though the DF1 protocol and message structure are openly available (Allen-Bradley, 1996). For all four of these protocols, the structure of the messages is the same.

12.7.1 A-B Serial

Communications over a serial link with any Allen-Bradley ControlLogix, MicroLogix, or SLC-500 processor uses the DF1 protocol. The DF1 protocol is a data-link layer protocol that combines features of subcategories D1 (data transparency) and F1 (two-way simultaneous transmission with embedded responses) of the ANSI x3.28 specification (ANSI, 1976). The DF1 protocol has two categories:

half-duplex for master/slave communication, and
full-duplex for peer-to-peer communication.

12.7.2 A-B Data Highway Plus

The Data Highway Plus (DH+) medium is shielded twisted pair cable. The network topology is a daisy-chained or trunkline/dropline linear bus terminated at both ends (Allen-Bradley, 1994). Each trunkline can have up to 64 nodes. The transmission rate may be 57.6, 115.2, or 230.4 kbit/s. Earlier processors only supported the lowest transmission rate. The maximum trunkline length is 3050 m. The DH+ cables can be redundant, but there is no built-in function to switch between cables when one of them fails.

Networks can be joined by bridges to form larger networks. The routing of the message is not automatic. The application layer (ladder logic program) must specify the routing path in the message.

The data link layer implements a token passing protocol. When a node has sent all of its messages or used all of its token-hold time, the token is passed to the node with the next highest address. When there is no node with a higher address, then it is passed to the node with the lowest address and the cycle starts again.

12.7.3 A-B Data Highway-485

The Data Highway-485 (DH-485) network was developed to connect SLC-500 processors. It uses a three-conductor shielded cable as its medium (Allen-Bradley, 1994). The network topology is a daisy-chained linear bus with special taps. The network can have up to 32 nodes. The transmission rate can range from 1.2 to 19.2 kbit/s. The maximum segment length is 1220 m, though with two 1747-AIC+ modules back-to-back, two segments can be connected to make a total network length of 2440 m (Rockwell Automation, 2006).

The data link layer implements a token passing protocol. Certain low-level devices (for example, a bar-code reader) are respond-only nodes and do not participate in the token passing.

12.8 COMMUNICATION BLOCKS

Ethernet, ControlNet, DeviceNet, and Data Highway Plus interface modules are available for most of the Allen-Bradley PLCs, though CompactLogix processors only support Ethernet, DeviceNet, and DH-485. PROFIBUS-DP master modules are available for ControlLogix and SLC-500 processors. AS-i master modules are available for ControlLogix and SLC-500 processors. Most of the modules that interface to the sensor networks (DeviceNet, PROFIBUS, and AS-i) map directly into the processor input/output image. The table is a fixed length and a portion of that image is dedicated to each slave device on the network.

ControlLogix and CompactLogix Ethernet modules that support Ethernet/IP can function as scanners for I/O modules connected directly to Ethernet. In this case, the Ethernet module and I/O modules are added to the I/O configuration of the controller and the module channels are automatically mapped to processor tags. To the processor, the Ethernet I/O modules appear similar to chassis-based I/O modules.

The MSG block provides the peer-to-peer communications function for Allen-Bradley PLCs. This block is described and then the procedure for setting up the producer/consumer interaction for ControlLogix and CompactLogix processors is illustrated. Lastly, the communication blocks for Micro800 processors are described.

MSG Block

The various forms of the MSG block are shown in Figure 12.22. The MSG operation is triggered on the false-to-true transition of its input. In general, the MSG block allows data to be exchanged between processors on any of the peer-to-peer networks (Ethernet,

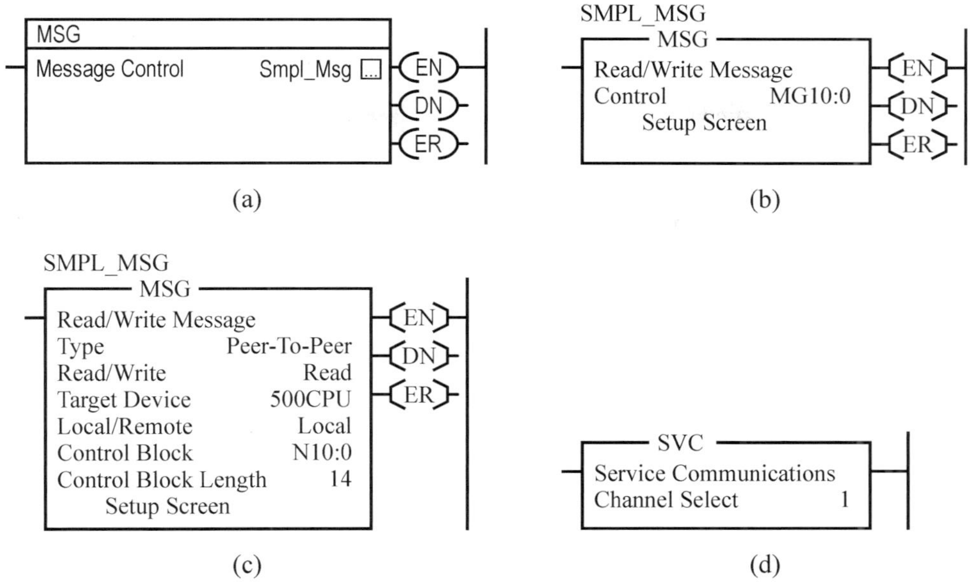

Figure 12.22. MSG blocks: *(a)* ControLogix; *(b)* MicroLogix; *(c)* SLC-500; *(d)* MicroLogix/SLC-500 SVC.

ControlNet, DH+, DH-485) for which the processor has an interface. The SVC block (Figure 12.22*d*) is used by the MicroLogix and SLC-500 processors to force an immediate processing of any outstanding communication requests, rather than wait for the request to be processed after the ladder logic scan is complete. See the appropriate documentation for details on the use of this block (Rockwell Automation, 2008, 2011, 2014b, and 2019b).

The network used to illustrate the configuration of the MSG block is shown in Figure 12.23. All examples use the subnet for processor-to-processor communications. The Ethernet network at the top of the figure is the automation network and is used by the operator terminals and engineering workstations to access the processors. No I/O modules are shown so that the examples can concentrate on the PLC-to-PLC communication. Examples with ControlNet and DH+ are in Erickson (2016) and Erickson (2013).

For the ControlLogix PLC, the MSG block allows one to read/write tags in another ControlLogix, CompactLogix, or Micro800 processor or to read/write data addresses in the MicroLogix, SLC-500, PLC-5, PLC-3, and PLC-2 processors. Assuming the processor being programmed is named Reactor1 (slot 0 of the ControlLogix-1 backplane), the I/O configuration is set up as shown in Figure 12.24. The I/O modules are not shown. Note that the communication interface modules are shown in the tree structure. A module resides in the backplane slot indicated by the bracketed number. The module name is after the module catalog number. Any ControlLogix or CompactLogix processors for which communications is desired should appear in the I/O configuration tree. Communication with the other processors in Figure 12.23 is possible, even though they cannot be added to the I/O Configuration of a ControlLogix or CompactLogix processor.

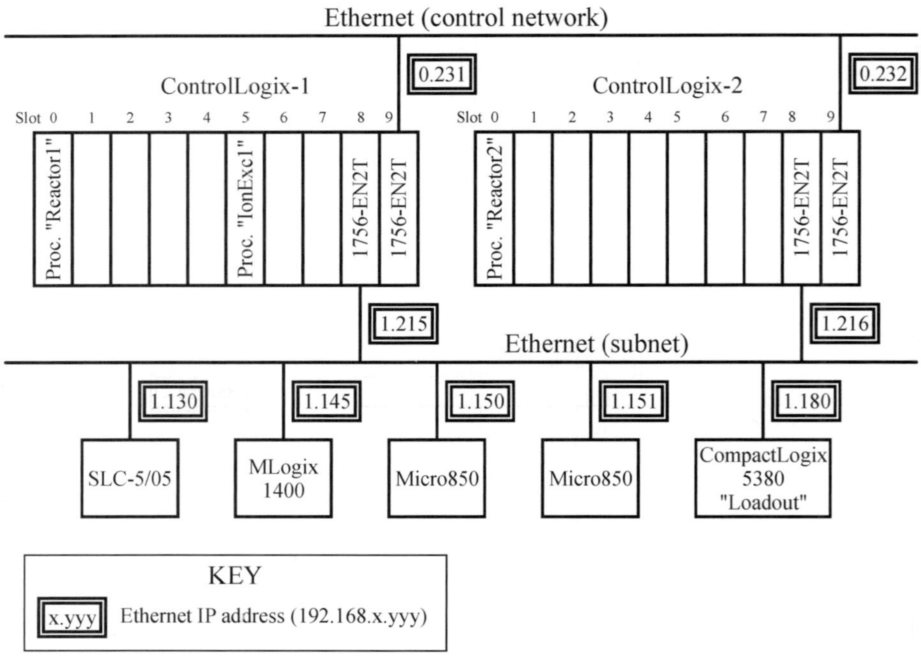

Figure 12.23. Networks for MSG block examples.

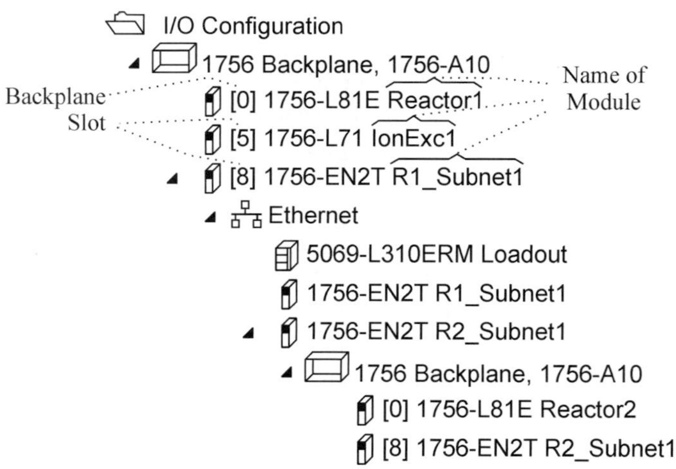

Figure 12.24. ControlLogix I/O configuration tree for Reactor1 processor.

The Message Control tag specifies the messaging details. The fields in the MSG configuration (Figure 12.25) are as follows:

Message Type - The three main categories of messages are data table read/write, CIP generic and block transfer read/write. The data table messages transfer information between processors and the other processor is specified as a CIP (ControlLogix), PLC-5, SLC-500, PLC-3 or PLC-2. The block transfer messages

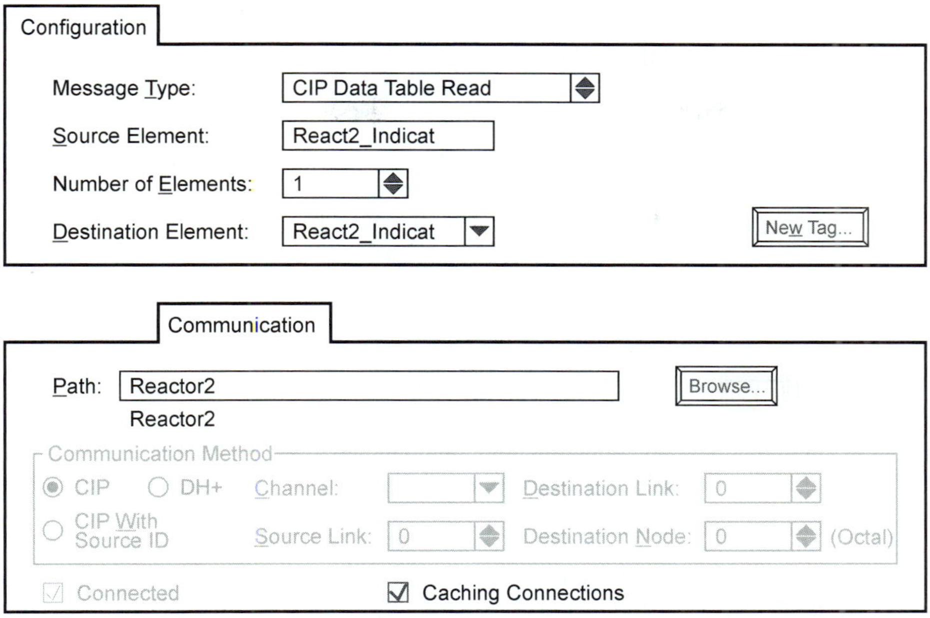

Figure 12.25. ControlLogix MSG configuration for reading from Reactor2 processor.

are used to communicate with a block-transfer I/O module (for example, a PLC-5 analog input module) and the CIP generic messages are for configuring and testing I/O modules and for other special uses.

Source Element - The source of the data. If the message is a read, this tag (ControlLogix) or address (other processors) is in the other processor. If the message is a write, this tag is in the processor.

Number of Elements - The number of elements to transfer, normally one. If the source and destination are arrays, this is the number of elements to read/write.

Destination Element - The destination of the data. If the message is a read, this tag is in the processor. If the message is a write, this tag (ControlLogix) or address (other processors) is in the other processor.

Path - If the other processor is a ControlLogix processor, this is the name of the processor as specified in the I/O configuration tree. If the other processor is not a ControlLogix processor, the path is specified as

comm_module, *port, address, port, address,* etc.

The "comm_module" is the name of the module in the I/O configuration that connects to the network of the other processor. Each port-address pair specifies the port by which the message exits the module and the address is the address of the next module in the path. Valid port-address pairs are described in Rockwell Automation (2022b).

Cache Connections - If checked, the connection is not closed after the transfer is completed.

Communication Method - This set of fields is only active when the message type is not one of the CIP types (CIP Data Table Read, CIP Data Table Write, or CIP Generic). The CIP method is the communication method between ControlLogix,

CompactLogix, and Micro800 processors. The DH+ method is used if the message is routed through a DH+ network. In this case the other fields are used as follows:

Channel - Channel (A or B) on DHRIO module connected to the DH+ network.

Source Link - Link ID for the local DH+ link (0-199). Used only if a routing table has been configured in the DHRIO module.

Destination Link - Link ID for the remote DH+ link where the target device resides (0 - 199). Used only if a routing table has been configured in the DHRIO module.

Destination Node - DH+ address of remote processor (0-77 octal).

The CIP with Source ID method is used for an unsolicited message that is routed through an Ethernet network or a ControlNet network to an operator interface that requires the source link to identify the data. The use of the other fields is described in Rockwell Automation (2022b).

When reading a tag named React2_Indicat in the processor named Reactor2 into the React2_Indicat tag of the processor, the MSG configuration appears as shown in Figure 12.25. The name of the other processor (Reactor2) appears in the Path field. If the same information was read from the CompactLogix 5380 processor named "Loadout," the only change would be that the path is set to "Loadout." An example writing data to a SLC-5/05 processor is shown in Figure 12.26. In this case, an INT (R1_Cmd) is written to N7:10 in the SLC-5/05. The path is "R1_Subnet1,2,192.168.1.130" and the Communication Method is set to "CIP."

MicroLogix 1100 and 1400 processors have a built-in Ethernet port. These processors can communicate with other Ethernet processors, such as PLC-5/xxE, SLC-5/05,

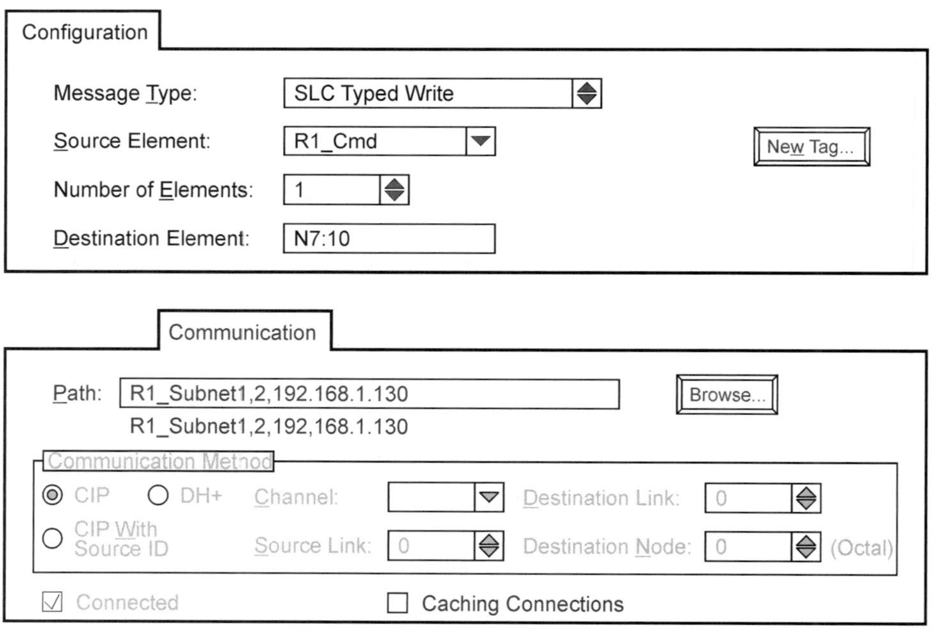

Figure 12.26. ControlLogix MSG configuration for writing to SLC-500 on the Ethernet subnet.

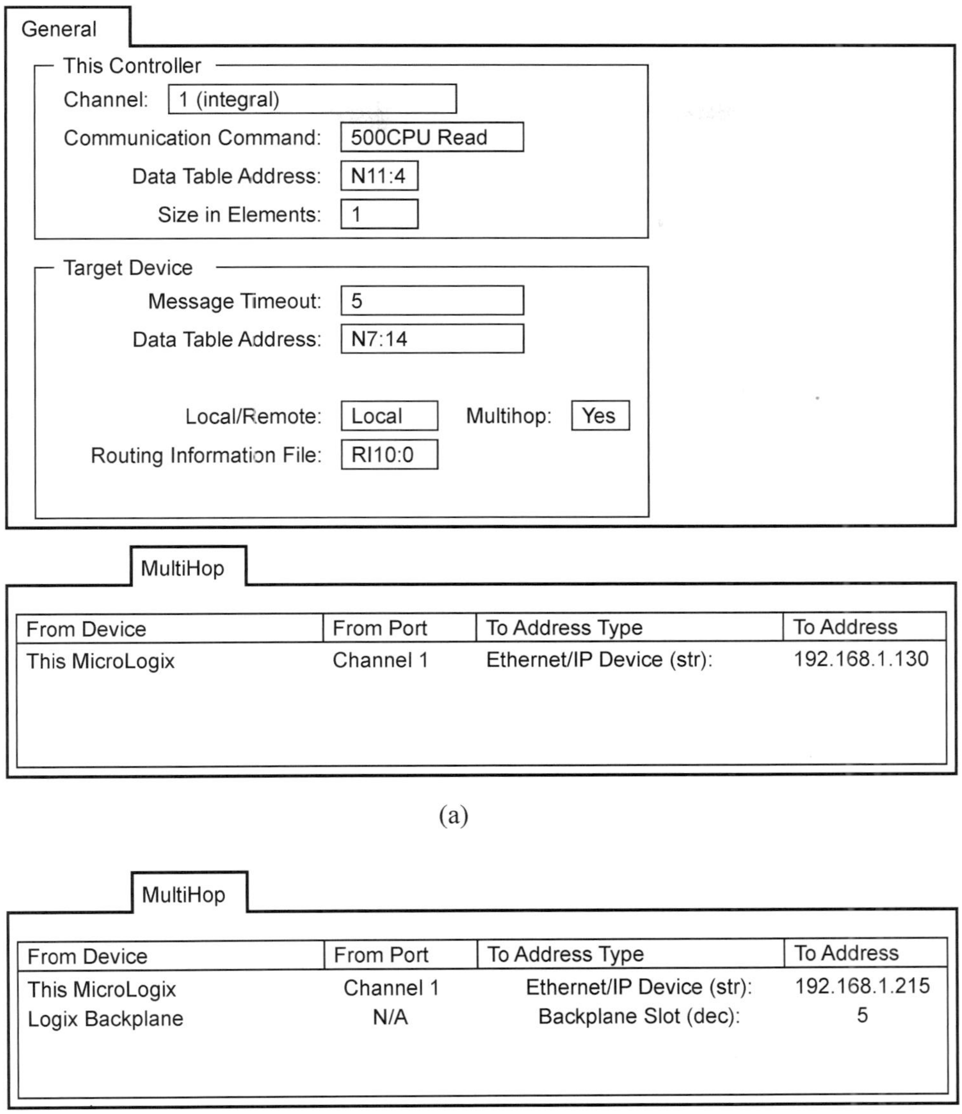

Figure 12.27. MicroLogix 1400 MSG configuration: *(a)* for reading from a SLC-5/05 on Ethernet; *(b)* for reading from a ControlLogix processor.

ControlLogix, and CompactLogix. For example, consider communication between the MicroLogix 1400 controller and the SLC-5/05 controller in Figure 12.23. The MSG configuration for the MicroLogix 1400 reading an integer from SLC-5/05 is shown in Figure 12.27*a*. The fields for this (local) PLC are:

Channel – The processor communication port. The possible channel numbers are:
0 – Serial
1 – Ethernet (MicroLogix 1100, 1400) or Serial (MicroLogix 1200, 1500)

Communication Command - Specifies the type of transfer (read/write) and the type of the other processor (MicroLogix/SLC-500, PLC-5 or PLC-2).

Data Table Address – Address of the data in this processor. If the message is a read, this address is the destination of the data. If the message is a write, this address is the source of the data.

Size in Elements – The number of elements to transfer, starting with the given data table address.

The available ports depend on the particular processor. The fields for the other processor (target device) are:

Message Timeout – Length of time, in seconds, the message must complete its operation.

Data Table Address - Address of the data in the other processor. If the message is a read, this address is the source of the data. If the message is a write, this address is the destination of the data.

Local/Remote - Set to "Local" if the target processor is on the same network segment as this processor. Set to "Remote" if the target device is on another network segment.

If the target device is on a remote network, additional fields must be specified (Rockwell Automation, 2008):

Multihop – Automatically "Yes" if the Channel is 1 (Ethernet)

Routing Information File (RI) – RI file type holding the Ethernet address information.

The address information is specified on the "MultiHop" tab.

A MicroLogix 1100 and 1400 can also communicate with a ControlLogix or CompactLogix controller. For example, the information in the "MultiHop" tab in Figure 12.27*b* reads an integer from the IonExc1 processor in slot 5 of the ControlLogix-1 chassis. Note that the MicroLogix controller is accessing address "N7:14" and one must configure the IonExc1 processor to map the N7 data file address to a ControlLogix tag.

The configuration of the MSG block for a SLC-500 processor (Figure 12.22*c*) is similar to the configuration of the MicroLogix MSG block. There are two basic differences: (1) a SLC-500 can only communicate with another SLC-500 or PLC-5 processor, and (2) the control block must be in an integer data file.

Producer/Consumer

When processors are connected with Ethernet communication modules that support Ethernet/IP or with ControlNet, the exchange of data between processors can be set up as a producer/consumer. It is easiest to configure between ControlLogix/CompactLogix processors, but it is also possible between ControlLogix and PLC-5 processors and between PLC-5 processors. The configuration for ControlLogix processors is described here.

The producer/consumer connection is configured with the tags. No MSG block is needed to control the communication. If ControlNet is part of the connection, the ControlNet network must be configured (with RSNetworx), so each produced tag is

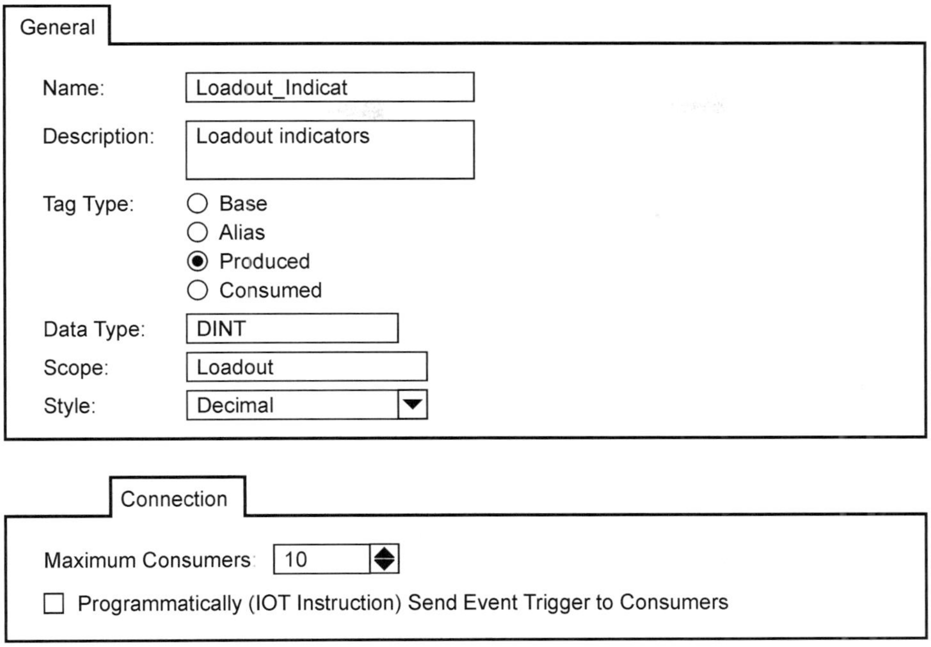

Figure 12.28. ControlLogix tag configuration for producer (Loadout).

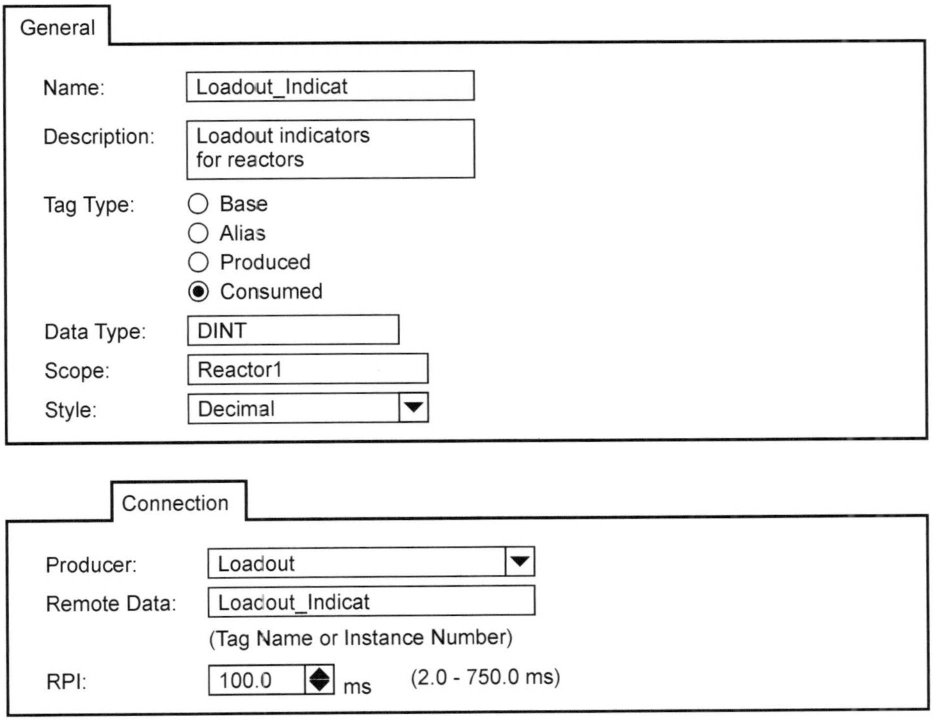

Figure 12.29. ControlLogix tag configuration for consumer (Reactor1).

assigned a time slot to be sent and a connection number that is configured in the nodes that consume that tag.

As an example, suppose a tag named Loadout_Indicat in the processor named Loadout needs to be read by the Reactor1 and Reactor2 processors. This tag is set up to be *produced* by the Loadout processor and *consumed* by the Reactor1 and Reactor2 processors. The tag properties of Loadout_Indicat in the Loadout processor are shown in Figure 12.28. The properties of the consumed tag, also called Loadout_Indicat in the Reactor1 processor are shown in Figure 12.29. The Reactor1 I/O configuration has the tree of Figure 12.24 in order to identify the Producer. The properties of the consumed tag in the Reactor2 processor will appear identical to Figure 12.29, except that the scope is "Reactor2". The I/O configuration in the Reactor2 project will also need the I/O configuration of Figure 12.24 in order to identify the location of the Loadout processor.

Micro800 Communication Blocks

The Micro800 blocks for communication and the operation that occurs for each one are:

Block Name	Operation
MSG_CIPGENERIC	Send a CIP generic explicit message
MSG_CIPSYMBOLIC	Send a CIP symbolic explicit message
MSG_MODBUS	Send a Modbus message
MSG_MODBUS2	Send a Modbus/TCP message on Ethernet

Only the MSG_CIPSYMBOLIC block with Ethernet is described, as it performs the same function as the MSG block for the other processors. The other blocks are described in Rockwell Automation (2022c). Also, this information is mainly applicable to those Micro800 processors with an Ethernet port. Figure 12.30 shows an example MSG_CIPSYMBOLIC, as well as logic to initiate it and transfer the received data to a variable. The block inputs are described as follows:

IN – Execution control. Rising edge to start execution.

CtrlCfg –Block configuration, of type CIPCONTROLCFG.

SymbolicCfg – Tag name or variable name of data in target (other) processor, of type CIPSYMBOLICCFG.

TargetCfg – Defines target (other) processor, of type CIPTARGETCFG.

Data – Message data. If reading from target, received data stored here. If writing to target, data to be sent is stored here. Data type is USINT[490].

The outputs are described as follows:

Q – Finish status. True if block completed successfully

Status – Status of block execution, of type CIPSTATUS

DataLength – Number of data bytes read if reading from target. Otherwise, it is zero.

Possible Status information is described in Rockwell Automation (2022c). The elements of the CIPCONTROLCFG data type are:

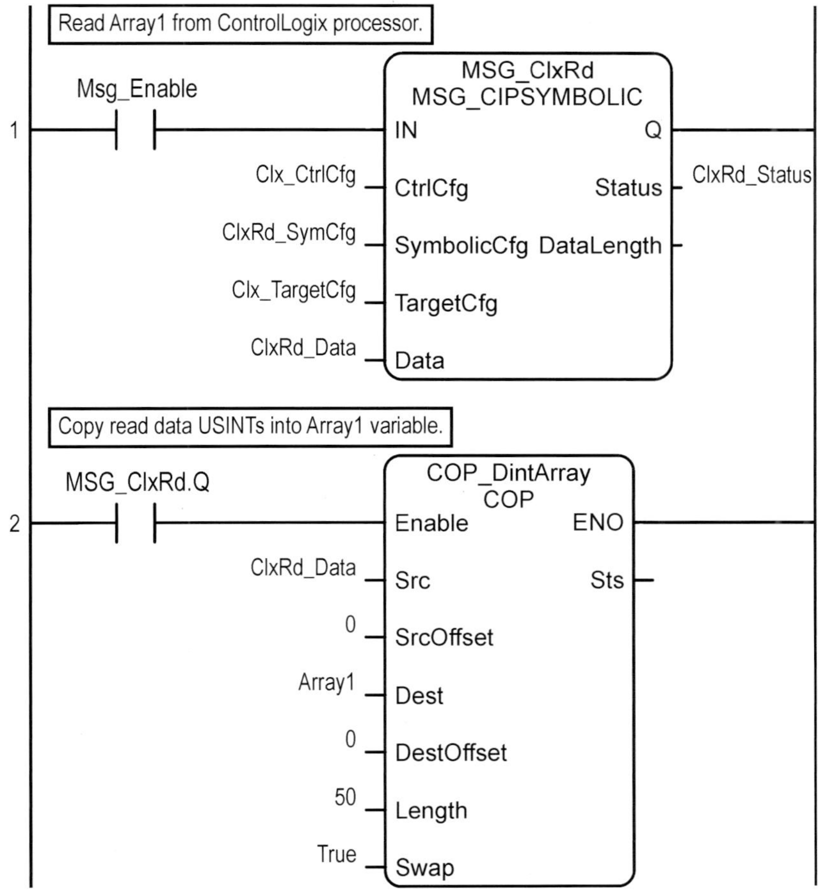

Figure 12.30. Micro800 MSG_CIPSYMBOLIC reading Array1 from ControlLogix.

Cancel – True to cancel operation.

TriggerType – If zero, one message is sent when IN changes from false to true. If a number, the cyclic value in milliseconds and the message is sent as long as IN is true

StrMode – Reserved for future use.

The elements of the CIPTARGETCFG data type are:

Path – Path to target. A maximum of two hops. Specified as

4, *dest_IP_address, port, address,* etc.

The “4" identifies the Ethernet port. The subsequent port, address are the same format as for the ControlLogix MSG path (Rockwell Automation, 2022b).

CipConnMode – CIP connection type: 0 = unconnected (default); 1 = Class3 connection.

UcmmTimeout – Unconnected message timeout in milliseconds. If zero, the timeout is 3 seconds. If nonzero, but less than 250, set to 250. If greater than 10,000, set to 10,000.

ConnMsgTimeout – Class3 connection timeout in milliseconds. If zero, the timeout is 10 seconds. If nonzero, but less than 800, set to 800. If greater than 10,000, set to 10,000.

ConnClose – Connection closing behavior. If true, connection is closed when the message completes. If false, the connection remains open. Default is false.

The elements of the CIPSYMBOLICCFG data type are:

Service – 0 for read, 1 for write.

Symbol – Name of the tag/variable in the target; maximum of 80 characters.

Count – Number of elements to read/write.

Type – Data type of variable being read or written. Possible values:

193 – BOOL	197 – LINT	201 – ULINT
194 – SINT	198 – USINT	202 – REAL
195 – INT	199 – UINT	203 – LREAL
196 – DINT	200 – UDINT	204 – STRING

Offset – For future use. Set to zero.

The maximum number of data bytes in the message is limited to 490 bytes, which means one is limited to 245 16-bit variables, 122 32-bit variables, and 61 64-bit variables.

For the example in Figure 12.30, the Micro800 at address 192.168.1.150 reads Array1, a 50-element DINT array, from the processor in slot 0 of the ControlLogix-1 chassis. The configuration tags for the MSG_CIPSYMBOLIC block are set to the following values:

Clx_CtrlCfg:

.Cancel	False
.TriggerType	1000

Clx_TargetCfg:

.Path	'4,192.168.1.215,1,0'
.CipConnMode	0
.UcmmTimeout	2000
.ConnMsgTimeout	2000
.ConnClose	False

ClxRd_SymCfg:

.Service	0
.Symbol	'Array1'
.Count	50
.DataType	196

Note that the COP block Swap input must be set to True, as the received data is not in the correct byte order. The first two parts of the path in the ClxRd_TargetCfg variable identify the "hop" to the Ethernet adapter in the ControlLogix-1 chassis. The last two parts identify the path from the adapter to the processor in slot 0.

Figure 12.31 shows the ladder logic for the Micro 850 at address 192.168.1.150 writing a REAL variable, FT101, to the processor in slot 0 of the ControlLogix-1 chassis. The CtrlCfg and TargetCfg tags for the MSG_CIPSYMBOLIC block are the same as for Figure 12.30. The configuration tag for the SymbolicCfg is as follows:

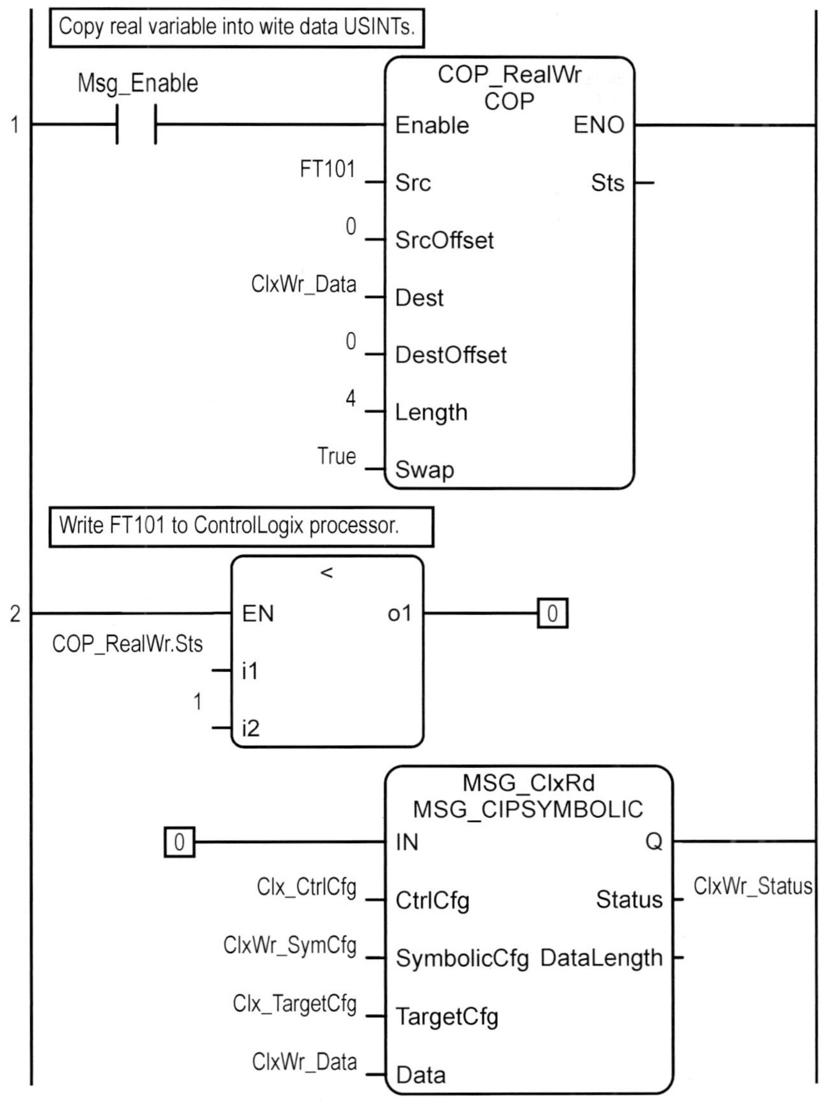

Figure 12.31. Micro800 MSG_CIPSYMBOLIC writing FT101 to ControlLogix.

ClxWr_SymCfg:

.Service	1
.Symbol	'FT101'
.Count	1
.DataType	202

In order to communicate with another Micro800, the only part of the configuration that needs to change is the path. For example, in order to change the example in Figure 8.30 so that the Micro850 at 192.168.1.150 reads Array1 from the Micro850 at 192.168.1.151, the only change is that Clx_TargetCfg.Path is modified to '4,192.168.1.151'

12.9 HEARTBEAT LOGIC

Communication between the PLCs and between the PLC and OI (operator interface) are often vital. Though most communication modules have status indicators that indicate when the communication channel has errors, they do not indicate whether the partner PLC or OI is still running. Most communication interfaces operate somewhat independently of the processor scanning its program. A processor can read and write the memory of another processor even when the other processor is in "program" mode or is faulted. Hence, programming, which is often called heartbeat logic, is used to verify that partner processors are "live." The heartbeat programming is located in the PLCs and the OI. If either the PLC or PC fails to set their given bit within a certain period (a few seconds), an alarm will occur and be indicated at the PLC and the OI panels. When communications between partner processors or the operator interface cease, certain actions may be initiated.

The general structure of PLC-to-PLC heartbeat logic with message-type communication is shown in Figure 12.32*a*. Both PLCs are assumed to be periodically reading an integer block of data from the other PLC. In each integer block, a bit is identified as the "heartbeat" bit. Heartbeat1 is a bit in the integer array in PLC 1 and read by PLC 2 and Heartbeat2 is a bit in the integer array in PLC 2 and read by PLC 1. PLC 2 copies Heartbeat1 to Heartbeat2 and PLC 1 copies the invert of Heartbeat2 to Heartbeat1. As a result of these operations, both Heartbeat1 and Heartbeat2 oscillate between **on** and **off**. The on and off period is somewhere between one to two times the message read period. For example, if each message in Figure 12.32*a* is executed every 1 second, the on and off period of Heartbeat1 and Heartbeat2 is between 1 and 2 seconds, depending on the synchronization between the timers in the two PLCs. If Heartbeat1 (or Heartbeat2) is **on** for more than 3 seconds or **off** for more than 3 seconds, communication has ceased and a "heartbeat" alarm should be activated. The heartbeat alarm will activate if either PLC is disconnected from the network or if the other PLC is switched to program mode.

The general structure of PLC-to-PLC heartbeat logic with producer/consumer (publisher/subscriber) communication is shown in Figure 12.32*b* and is similar to a PLC-to-PLC heartbeat. Each PLC has a produced integer that is consumed by the other PLC. Heartbeat1 is one bit of an integer produced by PLC 1 and consumed by PLC 2. Heartbeat2 is one bit of an integer produced by PLC 2 and consumed by PLC 1.

The general structure of PLC-to-OI heartbeat logic is shown in Figure 12.32*c* and is similar to a PLC-to-PLC heartbeat. Since it generally not possible for a PLC to read/write to the OI, the OI controls the communication. A script in the OI periodically runs to copy the integer read from the PLC to another integer that is written to the PLC. Depending on the OI, it may not be possible for the OI to do this operation.

12.10 CHAPTER SUMMARY

This chapter reviews standard and proprietary factory communication networks available for Allen-Bradley PLC systems. Standard network protocols are covered first and then the proprietary Allen-Bradley protocols are described. Next, the function blocks that implement PLC-to-PLC communication are explained. Lastly, the heartbeat method of checking the operation of the partner processors is described

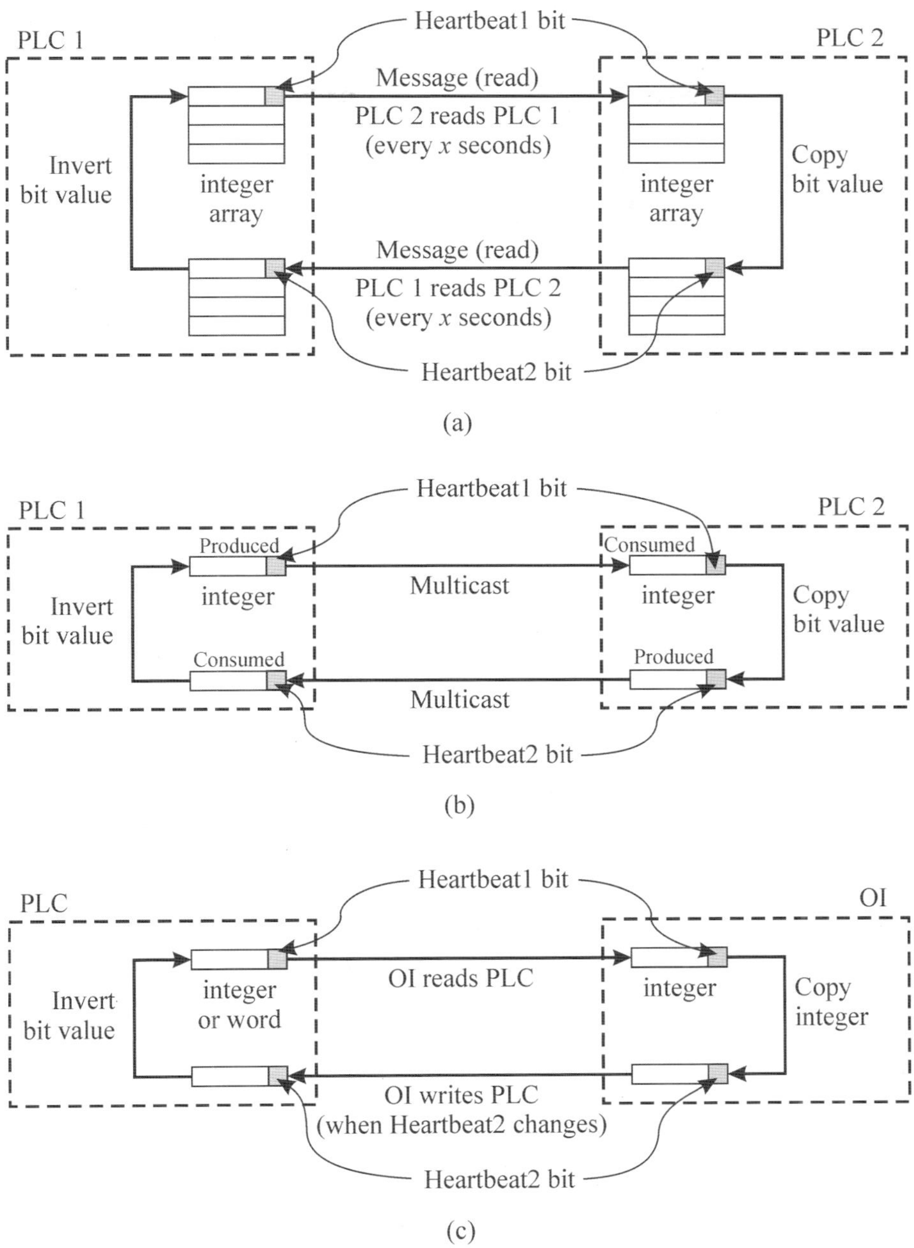

Figure 12.32. Heartbeat concept: *(a)* PLC-to-PLC with MSG; *(b)* PLC-to-PLC with producer/consumer (publisher/subscriber); *(c)* PLC-to-OI.

REFERENCES

Allen-Bradley, 1994. *Data Highway/Data Highway Plus/Data Highway II/Data Highway-485 Cable: Installation Manual*, pub. 1770-6.2.2, Allen-Bradley Company, Inc., Milwaukee, WI, April.

Allen-Bradley, 1996. *DF1 Protocol and Command Set: Reference Manual*, pub. 1770-6.5.16, Allen-Bradley Company, Inc., Milwaukee, WI, October.

ANSI, 1976. *Procedures for the Use of the Communication Control Characters of American National Standard Code for Information Interchange in Specified Data Communication Links*, ANSI.X3-28.1976, American National Standards Institute, Washington, D.C.

Ciminiera, Luigi, Claudio DeMartini, and Adriano Valenzano, 1988. "Industrial IEEE 802.3 Networks with Short Delivery Time for Urgent Messages," *IEEE Transactions on Industrial Electronics*, vol. 35, no.1 (February): 18-25.

Erickson, Kelvin T., 2013. *Allen-Bradley PLCs: An Emphasis on Design and Application,* Dogwood Valley Press, Rolla, MO.

Erickson, Kelvin T., 2016. *Programmable Logic Controllers: An Emphasis on Design and Application,* 3rd Ed., Dogwood Valley Press, Rolla, MO.

Felser, M., 2002. "The Fieldbus Standard: History and Structure," Technology Leadership Day 2002, organized by MICROSWISS Network, HTA Luzern, Oct. 10.

IEC, 2003a. *Digital Data Communications for Measurement and Control - Fieldbus for Use in Industrial Control Systems - Part 3: Data Link Service Definition*, IEC 61158-3, Ed. 3.0, International Electrotechnical Commission, Geneva, Switzerland.

IEC, 2003b. *Digital Data Communications for Measurement and Control - Fieldbus for Use in Industrial Control Systems - Part 4: Data Link Protocol Specification*, IEC 61158-4, Ed. 3.0, International Electrotechnical Commission, Geneva, Switzerland.

IEC, 2003c. *Digital Data Communications for Measurement and Control - Fieldbus for Use in Industrial Control Systems - Part 5: Application Layer Service Definition*, IEC 61158-5, Ed. 3.0, International Electrotechnical Commission, Geneva, Switzerland.

IEC, 2003d. *Digital Data Communications for Measurement and Control - Fieldbus for Use in Industrial Control Systems - Part 6: Application Layer Protocol Specification*, IEC 61158-6, Ed. 3.0, International Electrotechnical Commission, Geneva, Switzerland.

IEC, 2008. *Low-Voltage Switchgear and Controlgear - Controller-Device Interfaces (CDIs) - Part 2: Actuator Sensor Interface (AS-I)*, IEC 62026-2, Ed. 2.0, International Electrotechnical Commission, Geneva, Switzerland.

IEC, 2010. *Industrial communication networks – Profiles Part 2: Additional Fieldbus Profiles for Real-Time Networks Based on ISO/IEC 8802-3*, IEC 61784-2, Ed. 2.0, International Electrotechnical Commission, Geneva, Switzerland.

IEC, 2019a. *Industrial communication networks – Fieldbus specifications* - Part 1: Overview and guidance for the IEC 61158 and IEC 61784 series, IEC 61158-1, Ed. 2.0, International Electrotechnical Commission, Geneva, Switzerland, April.

IEC, 2019b. *Industrial communication networks – Profiles Part 1: Fieldbus profiles*, IEC 61784-1, Ed. 5.0, International Electrotechnical Commission, Geneva, Switzerland, April.

IEEE, 2022. *IEEE Standard for Information technology—Telecommunications and information exchange between systems—Local and metropolitan area networks—Common specifications—Part3: Carrier-Sense Multiple Access with Collision Detection (CSMA/CD) Access Method and Physical Layer Specifications*, IEEE 802.3™-2022, Institute of Electrical and Electronics Engineers, New York, NY.

Lian, Feng-Li, James R. Moyne, and Dawn M. Tilbury, 2001. "Performance Evaulation of Control Networks: Ethernet, ControlNet, and DeviceNet," *IEEE Control Systems Magazine*, (February): 66-83.

ODVA, 2013a. *CIP Networks Library - Vol. 3: DeviceNet Adaptation of CIP*, Edition 1.14, Open DeviceNet Vendor Assoc. and ControlNet Int., Ltd., Ann Arbor, MI, Nov.

ODVA, 2013b. *CIP Networks Library - Vol. 4: ControlNet Adaptation of CIP*, Edition 1.8, Open DeviceNet Vendor Assoc. and ControlNet Int., Ltd., Ann Arbor, MI, Apr.

ODVA, 2015a. *CIP Networks Library - Vol. 1: Common Industrial Protocol*, Edition 3.18, Open DeviceNet Vendor Assoc. and ControlNet Int., Ltd., Ann Arbor, MI.

ODVA, 2015b. *CIP Networks Library - Vol. 2: EtherNet/IP Adaptation of CIP*, Edition 1.19, Open DeviceNet Vendor Assoc. and ControlNet Int., Ltd., Ann Arbor, MI, June 5.

Profibus, 2016. *PROFIBUS System Description Technology and Application*, PROFIBUS Nutzerorganisation, e.V., Karlsruhe, Germany, April.

Rockwell Automation, 2006. *AIC+ Advanced Interface Converter: User Manual*, pub. 1761-UM004B-EN-P, Rockwell Automation, Milwaukee, WI, June.

Rockwell Automation, 2008. *SLC Instruction Set Reference Manual*, pub. 1747-RM001G-EN-P, Rockwell Automation Inc., Milwaukee, WI, Nov.

Rockwell Automation, 2011. *MicroLogix 1100 Programmable Controllers Instruction Set Reference Manual*, pub. 1763-RM001D-EN-P, Rockwell Automation Inc., Milwaukee, WI, Sep.

Rockwell Automation, 2014a. *ControlLogix Data Highway Plus-Remote I/O Communication Interface Module User Manual*, pub. 1756-UM514C-EN-P, Rockwell Automation Inc., Milwaukee, WI, June.

Rockwell Automation, 2014b. *MicroLogix™ 1200 and MicroLogix 1500 Programmable Controllers Instruction Set Reference Manual*, pub. 1762-RM001H-EN-P, Rockwell Automation Inc., Milwaukee, WI, Jul.

Rockwell Automation, 2019a. *EtherNet/IP Network Configuration User Manual*, pub. ENET-UM001Q-EN-P, Rockwell Automation Inc., Milwaukee, WI, Mar.

Rockwell Automation, 2019b. *MicroLogix 1400 Programmable Controllers Instruction Set Reference Manual*, pub. 1766-RM001I-EN-P, Rockwell Automation, Inc., Milwaukee, WI, June.

Rockwell Automation, 2020. *Logix5000 Controllers Produced and Consumed Tags Programming Manual*, pub. 1756-PM011K-EN-P, Rockwell Automation Inc., Milwaukee, WI, Sep.

Rockwell Automation, 2021. *ControlLogix5000 Ethernet/IP Network Devices User Manual*, pub. 1756-UM004C-EN-P, Rockwell Automation Inc., Milwaukee, WI, June.

Rockwell Automation, 2022a. *1756 ControlLogix Communication Modules Specifications Technical Data*, pub. 1756-TD003Q-EN-P, Rockwell Automation Inc., Milwaukee, WI, Mar.

Rockwell Automation, 2022b. *Logix5000 Controllers General Instructions Reference Manual*, pub. 1756-RM003W-EN-P, Rockwell Automation Inc., Milwaukee, WI, Mar.

Rockwell Automation, 2022c. *Micro800 Programmable Controllers General Instructions Reference Manual*, pub. 2080-RM001K-EN-P, Rockwell Automation Inc., Milwaukee, WI, Mar.

Siemens, 2006. *SIMATIC NET AS-Interface – Introduction and Basics Manual*, Release 04/2006, pub. C79000-G8976-C089-04, Siemens AG, Nuernberg, Germany.

Siemens, 2009. *SIMATIC NET PROFIBUS Networks Manual*, Edition 04/2009, pub. C79000-G8976-C124-03, Siemens AG, Nuernberg, Germany.

13 Human-Machine Interface

Chapter Topics:

- General operator interface concepts
- Operator panel and screen design

OBJECTIVES

Upon completion of this chapter, you will be able to understand

- Operator interface types
- Good operator panel design
- Good operator computer screen design

Scenario 1: Operator-initiated starts with permissives.

A certain batch control process is started from a touch-screen HMI (human-machine interface). However, the process does not start because not all of the permissives are met. When a detailed status screen is displayed, the operator discovers that two cooling valves are open. They should have been closed in the last step of the previous batch. Maintenance personnel determine that the valves are being commanded closed, but the valves are still in the open position. One valve is quickly repaired, but the other one needs to be replaced. The process cannot be operated without this valve, so they wait a day for a new valve to be received from the vendor and replaced. However, as soon as power is restored to the valve limit switches, the batch process immediately starts filling the initial ingredient. The operator has not initiated the batch. Nothing is hurt by this action. However, this action is unexpected, and you are called in to find the reason for the unexpected start.

Solution: At first, you are puzzled by the cause. After interviewing the maintenance personnel who handled the valve repair and the operator who reported the valve problem, your suspect a problem with the batch start. When questioned, the operator does not remember stopping the batch after reporting the problem with the two cooling valves. An initial check of the PLC program uncovers no problem. Next, you examine the touch-screen programming. The start button is set to be a latching button, that is, when the button icon is depressed, a bit is set in the PLC. This bit remains set when the button icon is released. The stop button is configured to be an unlatching button. The batch started unexpectedly because the operator start request was still active when the power was restored to the valve limit switch. The PLC detected a closed valve (all permissives met), and the first step of the batch process was initiated. The operator start and stop requests are changed to act like momentary push buttons. The operator start is changed to a momentary switch that sets a bit in the PLC as long as the button icon is pressed. Otherwise, the bit is cleared. The operator stop is changed to a momentary switch that clears a bit in the PLC as long as the button icon is pressed. Otherwise, the bit is set. Also, the batch start/stop logic is changed to be more like Figure 2.32. To prevent this problem in the future, the programming standard is changed to require all HMI operator starts to be momentary push buttons. In addition, a small project is initiated to correct this problem in all processes operated with a touch-screen HMI.

Scenario 2: Hydrofluoric acid spill.

A certain process contains a hydrofluoric (HF) acid bath (Figure 13.1). The tank contains a high level switch and a high-high level switch. The PLC control maintains the HF tank level between the high level switch and the high-high level switch. When the tank level drops below the high level switch (switch turns **on**), the HF fill valve is turned **on** to fill the tank until the high-high level switch turns **off**. The operator can override the automatic HF fill valve control and add acid to the tank, but not beyond the high-high level switch. If the high-high level switch is **off**, the HF fill valve is closed.

The system that places and removes items in the HF bath is upgraded. During the testing of this upgrade, this bath is empty. The high-high level switch is not working, so it is removed from the tank, but still connected to the PLC discrete input module. One night, the HF acid monitor on the plant sewer discharge sounds, indicating excess HF acid in the sewer discharge. Only a few units in the plant use HF acid and so the source is quickly located to this process. When the personnel arrive, HF acid has overflowed the tank, covering part of the floor and flowing into the floor drain. The HF lockout valve is quickly closed to prevent further loss and the area is cleaned up.

Solution: There are three faults here. The obvious one is the absence of the high-high level switch. In addition, the HF lockout valve should have been closed since HF acid was not necessary for the testing and the high-high level switch was not operational. The third fault was the operator-initiated HF tank fill. No operator opened the HF fill valve since it was still being tested and not operational. The most likely cause was an HF tank fill initiated by a person cleaning the touch screen. In order to help prevent this problem in the future, the HMI is changed to add a confirmation pop-up window whenever the operator attempts to open the HF fill valve by touching its icon on the screen. This change will at least prevent

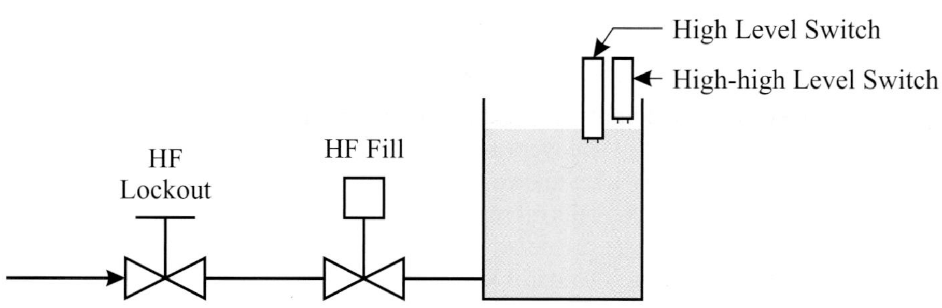

Figure 13.1. Hydrofluoric acid tank.

inadvertent addition of HF acid, and could have prevented this accident in spite of the other failures.

13.1 INTRODUCTION

In the not-too-distant past, the typical user interface to a control system consisted of a panel with industrial-sized switches, push buttons, panel lamps, gauges, PID controllers, and strip chart recorders. The size ranged from a small panel with a few switches and lamps to a panel along one or more walls of a room. The larger panels typically had a graphical representation of the process with the switches and lamps in appropriate locations. All of the panel devices were hard wired to the field devices and to PLC modules. With the advent of low-cost personal computers, the larger hard-wired panels were replaced by a computer screen with process graphics and operator commands entered via a keyboard. Today, a typical user interface is a color graphic touch screen. Panel displays have not disappeared entirely, though. They are sometimes still used as a hot backup if the computer system fails. In addition, some equipment has local operator control panels. For example, a conveyor motor may have a hand-off-auto (HOA) switch located close to the motor that allows maintenance personnel to override the action of the control system.

The human-machine interface (HMI) is a window to the process and is called by various other names:

- operator interface (OI)
- man-machine interface (MMI)
- touch panel

In this chapter, the general structure and functions of a user interface are reviewed. Good design practice for operator panels and operator screens is also discussed. The discussion is general though some examples are presented.

13.2 HMI TYPES

Depending on the application, an operator interface may be:

Mechanical buttons/switches and panel lamps
Stack lights
Small text displays (one to four lines)
Large scrolling text displays
Proprietary graphic displays
Proprietary HMI software on a personal computer
Open software-based systems (for example, Visual Basic) on a personal computer
Web servers embedded in a PLC

Mechanical switches and lights are used for simple applications. For example, a wood chip truck dump station and its local control panel are shown in Figure 13.2. The trailer is raised with a hydraulic cylinder, and the wood chips are conveyed to the storage bin. The panel has start/stop switches for the hydraulic pump and the conveying system and a joy stick to control the up/down of the lift. There are "running" indicators for the hydraulic pump and the conveying system. In addition, there is an individual "running" indicator for each of the conveyors in the conveying system.

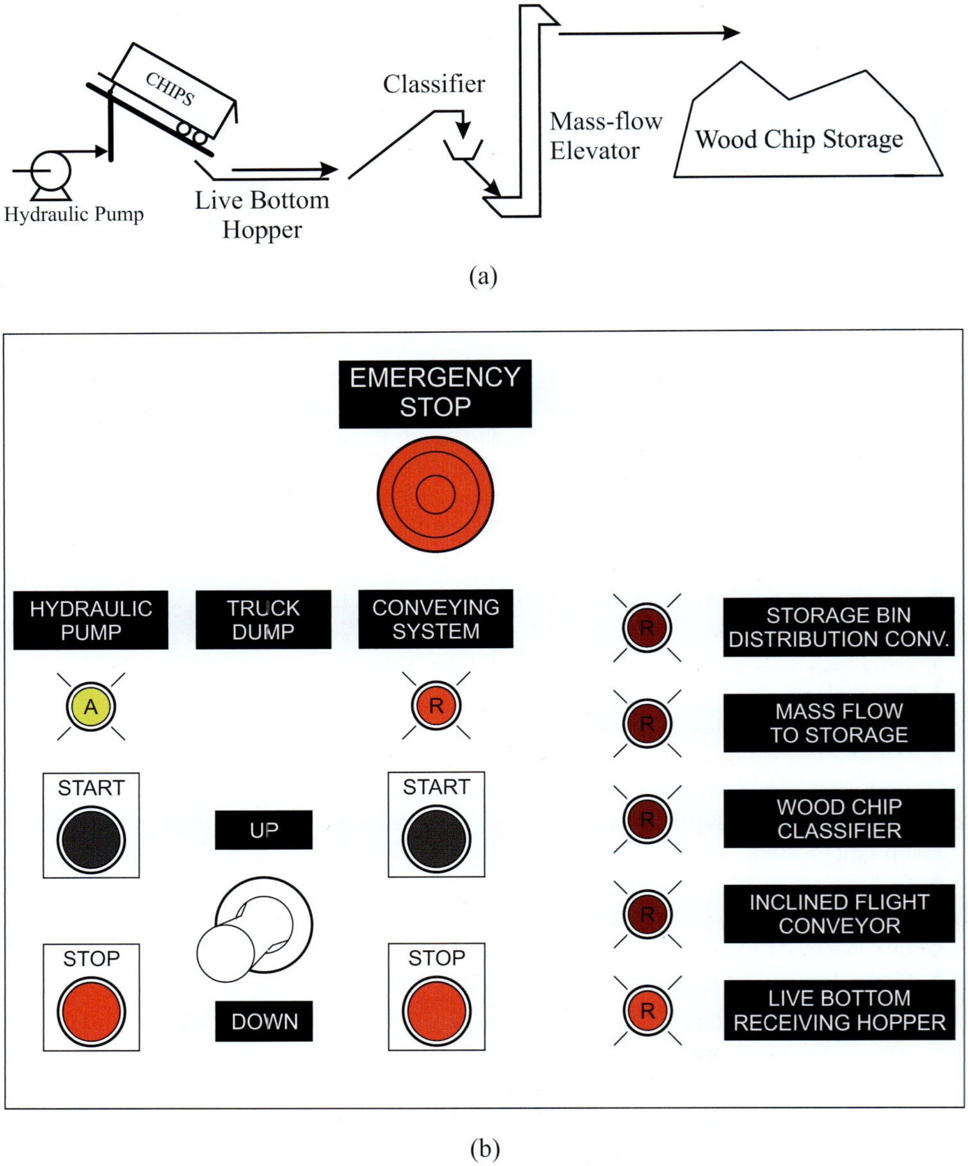

Figure 13.2. Wood chip truck dump: *(a)* system; *(b)* local control panel.

Stack lights consist of cylindrical lamps 50-70 mm diameter and 60-70 mm high stacked on top of one another (Figure 13.3). The lamps can be steady, flashing, or strobed. In addition, horn or siren modules can be in the stack. Stack lights are typically used on machine applications to allow an operator to quickly assess the overall status of a machine or process. For example, green could indicate a normally running machine. Yellow could indicate that a part feeder is low and needs refilling, and red flashing could indicate a stopped machine. Typically, a touch screen or text display is used in conjunction with a stack light for more detailed status information.

Figure 13.3. Stack light.

Text displays range from simple one- or two-line displays to large scrolling-text displays. The small displays are typically used to display status and error information for a machine or portion of the process. The small text displays are less expensive than comparable-sized touch screens. The large scrolling-text displays are often used on assembly lines to display messages for the line personnel, for example, current production rate or problems with certain stations. Text displays operate in one of two basic methods. The PLC can send the text message or a message number to the display device. In the latter method, the display device is programmed with a table that contains the text message for each corresponding number.

Proprietary graphic displays are prepackaged systems that have a monochrome or color LCD (liquid-crystal display) screen. The screen may or may not be touch-sensitive. If not touch-sensitive, the screen is often surrounded by function keys or other buttons allowing operator input. These systems often communicate directly with the PLC to read and write memory locations. One possible drawback of this type of display is that they only work with the PLCs of the vendor.

Proprietary HMI software running on a personal computer is generally more complicated than the proprietary graphic displays. However, they have more capabilities and can tie into other applications, for example, batch control and the storage of production data in a SQL database. Most of these HMI packages can be used with the PLCs of many vendors. The advantage of these systems is that the platform (a personal computer) is relatively inexpensive. Many of these systems are placed in control rooms. However, if the system must be deployed in the industrial environment where computer hard-disk drives will often fail, there are alternatives. The proprietary HMI system can be implemented with distributed clients (a diskless personal computer) communicating back to a centralized server using terminal services, sockets, or a web browser.

The general structure of proprietary HMI software is shown in Figure 13.4. There are three major components of the HMI package: a database, one or more graphics screens, and one or more servers. The database is the heart of the HMI software. The graphic screen objects are programmed to interact with the data base variables, and the database variables interact with the servers. In general, a different server is required for each type of PLC and communication network. For example, for a particular PLC, there is one server for communicating through its serial port and another server when communicating through its

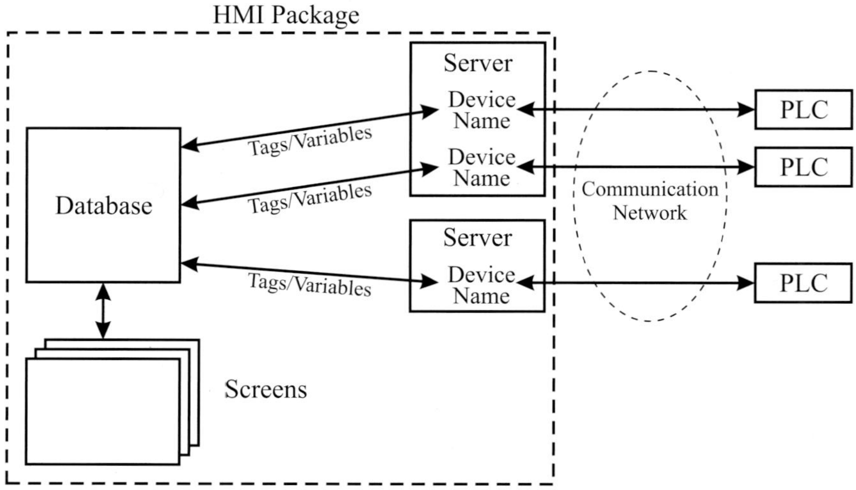

Figure 13.4. General structure of HMI software.

Ethernet port. Within the server, a "device name" is basically the communication channel defined for a particular PLC. Often, one can define multiple device names for one PLC, especially if some of the PLC data needs to be acquired frequently and most of the data can be acquired at a much slower rate. Communication between the database and the server is generally modeled on the dynamic data exchange (DDE) or object linking and embedding (OLE) method of data exchange.The current value of the database tags/variables are acquired from the server which communicates with the PLC. If the operator changes a tag/variable in the database, its value is sent to the server, which writes it to the PLC.

Open software-based systems (for example, Visual Basic) on a personal computer can be used in place of the proprietary HMI software. One still needs to construct a database and either write or purchase PLC communication drivers (I/O servers). These types of systems can have an overall lower cost since one does not need to pay for an annual license fee for every computer that is running the HMI. However, the initial development costs tend to be higher. Like the proprietary HMI, open software-based systems can be implemented with distributed clients.

An alternate operator interface is the use of an embedded web server in the PLC (Figure 13.5). The web server is part of the Ethernet interface module, and its pages are stored either within the Ethernet interface module or within the PLC data memory. The web server provides an alternate path to view and/or modify the PLC data memory. The types of web pages that can be constructed range from text displays to graphical displays.

13.3 HMI PANEL DESIGN

When the operator interface is a panel consisting of simple switches and indicators, they should be laid out in an orderly and consistent manner. The example in Figure 13.2*b* shows a simple panel with 6 controls: emergency stop, two start push buttons, two stop push buttons, and a two-position joy stick. This particular panel also shows the "running" status of the hydraulic pump and the conveying system. Additionally, the "running" status of each part of the conveying system is displayed on the right side. Another simple panel is shown

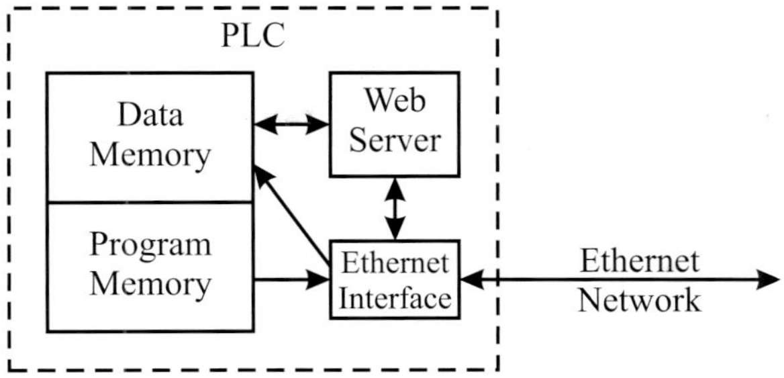

Figure 13.5. Embedded web server.

in Figure 13.6. This is a control for a cooling water sump system. There is an overall on/off rotary switch, individual rotary switches for the three system pumps, and three keyed switches for the three cooling tower pumps. There are also various alarm and other indicator lamps, as well as an alarm acknowledgement push button.

The intuitive placement and operation of various controls (often called *operators*) is shown in Figure 13.7. Pushbuttons should be placed in a position corresponding to the desired action (Figure 13.7*a*). According to the NFPA 79 standard (NFPA, 2021), start pushbuttons should be mounted above or to the left of their associated stop pushbutton. Up is also the intuitive direction for forward and increasing speed. For sideways directions, the push buttons should mimic the expected motion. For rotating switches, the intuitive direction for start is in the clockwise direction (Figure 13.7*b*). For a rotating knob control (for example, to adjust a desired speed), the clockwise direction is the intuitive one to increase the quantity (Figure 13.7*c*). On a toggle operator control, the up direction should be used for raising or increasing (Figure 13.7*d*).

According to the NFPA 79 standard (NFPA, 2021), panel operators and indicators should be color coded in accordance with Table 13.1. In general, red and amber indicators signify that a potential hazardous condition exists. Emergency stop switches should have a red-colored palm or mushroom head. In addition, the background around the switch should be yellow. For any panel indicator, some method of testing the indicator should be provided. Some indicators are "push-to-test," that is, whenever they are pushed, the lamp is lit. For indicators that are not "push-to-test" and for illuminated pushbuttons, the panel should have a lamp test pushbutton.

The distance between panel operators should be large enough so that a gloved hand or finger used to actuate one button/switch does not accidentally trigger another button/switch. The minimum distance between buttons is typically 40 to 50 mm.

In most manufacturing control panels, a red pilot light indicates an abnormal condition. However, in power plants or power systems, the opposite is generally true. For a power system, closed circuit breakers are indicated by red pilot lights and open circuit breakers are indicated by green pilot lights.

Operators expect some kind of feedback to their actions. If the result cannot be heard or seen by some movement, then a panel indication should be turned on.

Panels can also have analog indications, for example, flows, temperatures, and pressures. These can be displayed with digital or analog meters and there are advantages for each one. Digital meters can display a value with a smaller resolution. For example, a

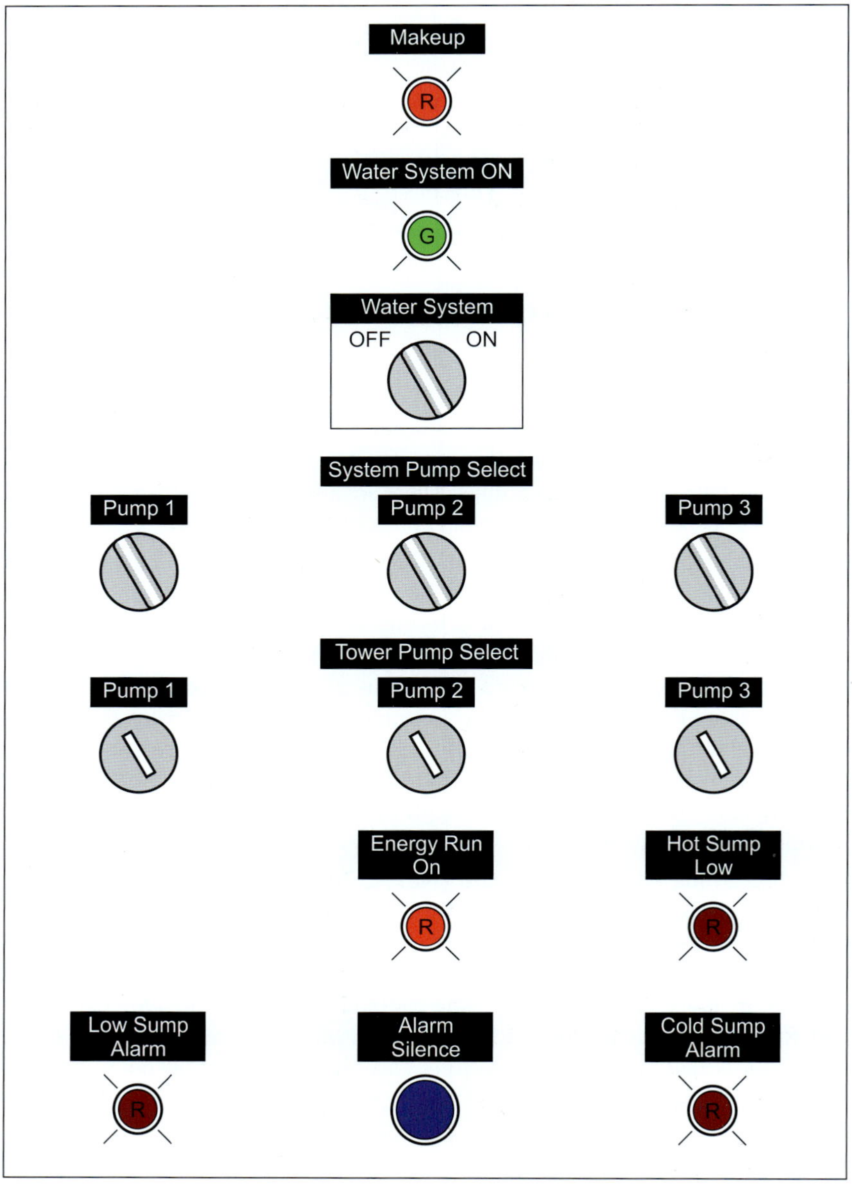

Figure 13.6. Operator panel for cooling water system.

4-digit meter has a resolution of 0.1% full-scale. An analog meter has a resolution of about 1%, regardless of the signal accuracy. However, an operator can more easily detect abnormal patterns from an analog meter. For example, one can quickly scan a row of pressure dial indicators and pick out the one or two that are approaching an abnormal condition before an alarm sounds. If a panel requires many analog indicators, it is usually more cost-effective to show them on a graphic display.

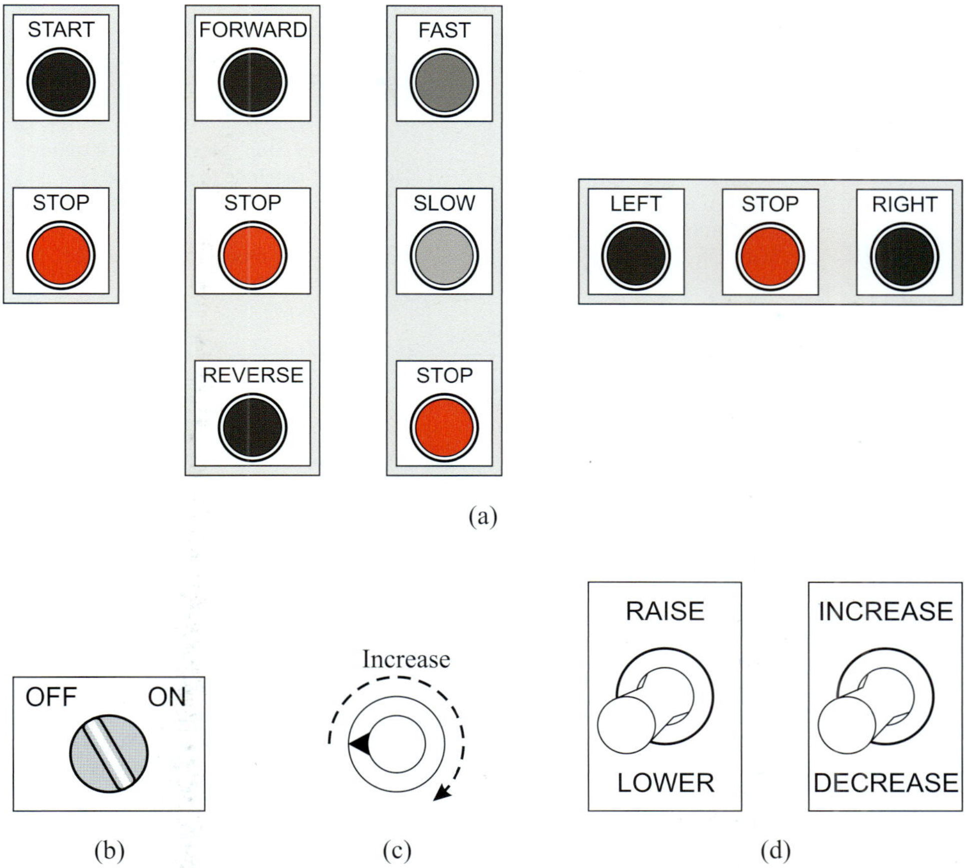

Figure 13.7. Intuitive operators: *(a)* push buttons; *(b)* selector switch; *(c)* rotating knob; *(d)* toggle.

13.4 GRAPHICAL HMI DESIGN

Many PLC-based control systems use graphical displays as the primary operator interface. Guidelines for their design are presented after describing the general types of graphical displays. These guidelines are based on the information in Hollifield et al. (2008), Hexatec (2002), and ISA (2015).

13.4.1 Screen Design General Considerations

Display clarity is very important in the control room where there are multiple graphical displays, each one showing some aspect of the plant. The operator must be able to easily scan the displays from a distance and notice if any problems are present. The ability to note problems from a distance also applies to the operator screens present on the plant floor. This section presents some guidelines for screen design (Hexatec, 2002 and Hollifield et al.,

Table 13.1 Color Coding for Pushbuttons and Pilot Lights

Color	Typical Function	Examples
Pushbutton		
RED	Stop, Emergency Stop, Off	Emergency Stop, Master stop of motors
AMBER	Return, Emergency return, Intervention - suppress abnormal condition	Return of machine to safe state. Continue after fault
GREEN	Start, On	General start, motor start, sequence start, energize control
BLACK	Other functions	Jog, test, reset
BLUE/GRAY	Any function not covered by above	
Pilot Light		
RED	Danger or alarm, abnormal condition requiring immediate attention	
AMBER	Attention, caution/marginal condition Change or impending change	Automatic cycle or motors running. Temperature is approaching limit.
GREEN	Machine ready, safety	Safe condition or authorization to proceed. Cycle complete, ready for restart.
WHITE	Normal condition confirmation	Gate in position after command to move.
BLUE/GRAY	Any function not covered by above	
Illuminated Pushbutton[1]		
RED	Stop, Emergency Stop, Off	Machine stalled because of overload. E-Stop switch actuated.
AMBER	Caution/start operation intended to avoid dangerous conditions	Pressure is approaching limit; pressing. button starts pressure-relief operation..
GREEN	Machine ready/Start or On	Start or on after authorization by light.
WHITE	Confirmation of energized circuit or function started/Start next step or preselected function	Energizing of the machine working cycle; pressing button starts feed
BLUE/GRAY	Any function not covered by above	

Notes:
1. Function(s) of the light is separated from the function(s) of the button by slash (/).

2008). These design guidelines apply to graphical screens in the control room as well as local operator panels out in the process or mounted on a machine.

Screen Layout

Screens should be arranged to facilitate easy scanning. Generally, operators scan a screen like they would scan a newspaper or magazine. In the U.S. and Europe, one starts in the upper left corner, proceeds to the upper right corner, and then on down the page. Since a graphical screen does not have lines, most operators will only do a few incomplete scans of the screen (Figure 13.8). With this consideration, important items should be positioned along the top and right sides of the display.

When the screen contains a graphical representation of the process, it should be a simplified version of the P&ID diagrams. Material flow should mimic the P&ID diagram and is usually left-to-right and/or top-to-bottom.

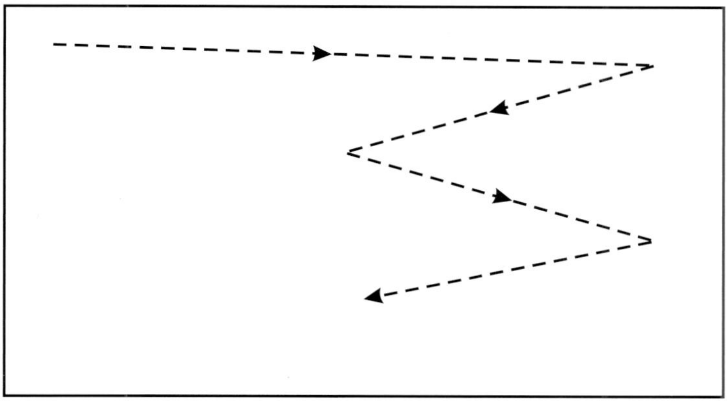

Figure 13.8. Typical scan of operator interface screen.

Background Color

Although black and white provide good color contrast for text, they are not good screen backgrounds since they are not suitable for a range of colors. The background should be light gray. Gray provides a good contrast for the alarm and mitigates the problems of glare, color interference and eye fatigue. Bright colors are very tiring on the eyes. Since control rooms are brightly lighted, the displays should be more like reading paper documents rather than consist of brightly glowing text, lines, and objects on a dark background. The use of subtle shading to create the illusion of raised or lowered panels on the screen is an effective way to differentiate areas of the display.

One can experiment with different values of gray backgrounds and foregrounds to determine the best one for the particular control room and display hardware. Figure 13.9 illustrates a typical 10-level gray background and foreground. A 20% to 30% black level is often a good background choice. A screen element similar to that shown in Figure 13.9 can be used to test the background and foreground choices. Since the appearance of printed color graphics is not the same as screen color graphics, the background and foreground choices should be tested in the actual control room and with the computer displays and personnel that will use the displays.

Though photographs are sometimes used as backgrounds, they generally produce cluttered backgrounds that do not have enough contrast with the foreground colors. Photographs are appropriate for maintenance and diagnostic displays, providing useful information for the personnel without having to find the manual for a piece of equipment.

Foreground Colors

Color is a powerful tool for enhancing the visibility of key data, but excessive use of color can be confusing or overwhelming. Therefore, the number of colors should be kept to a minimum. Thirty years ago, most HMI systems had only four colors, and the operators were happy with that!

Equipment outlines and process flow lines should be dark gray or black. Color should not be used to indicate the type of material (for example, white for steam, brown for natural gas) as consistency for an entire process is difficult and operators will probably not memorize them. Bright colors should draw the operator's attention to abnormal conditions

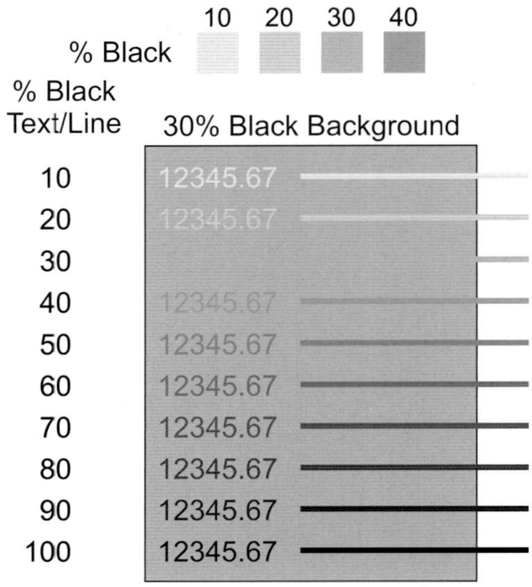

Figure 13.9. Gray background and foreground contrast.

and alarms. If the process is running correctly, there should be little color on the display. A color convention for status and alarms is covered in a later subsection.

Large areas of primary colors (for example, red) will cause a color-vision effect called *successive contrast*. If one looks at a bright-red image for about 30 seconds and then looks at a white surface, an *afterimage* consisting of the same image but in a green color is perceived. In addition, about 8% of men have some degree of color blindness. For these people, there is some confusion in the perception of reds and greens (also yellows, oranges, and browns). Therefore, one should not solely rely on color to indicate plant condition. Because the center of the human retina cannot easily sense blue (Murch, 1984), blue should be avoided for small objects and thin lines.

Static Graphics

Many operator interface screens contain a graphical representation of the process plant. The graphical representation helps the operator to visualize the process and the measurement locations. However, there is a tendency to place too much detail on the display and therefore make it difficult to discern important dynamic information. Therefore, keep the displays simple. Instead of putting the entire process on one screen, important overall information belongs on the top level screen, and then the operator selects the appropriate unit-level screen display for more detail.

According to Hollifield et al. (2008), process lines should be depicted as

Dark gray or black lines.

Main process lines should be 3 pixels wide and secondary lines should be 1 pixel wide.

No more than two or three line types (solid, dotted, dashed).

Muted tones are best for process equipment representations. Large areas of strong colors are distracting. Outlining an object in black will make it noticeable. Many HMI

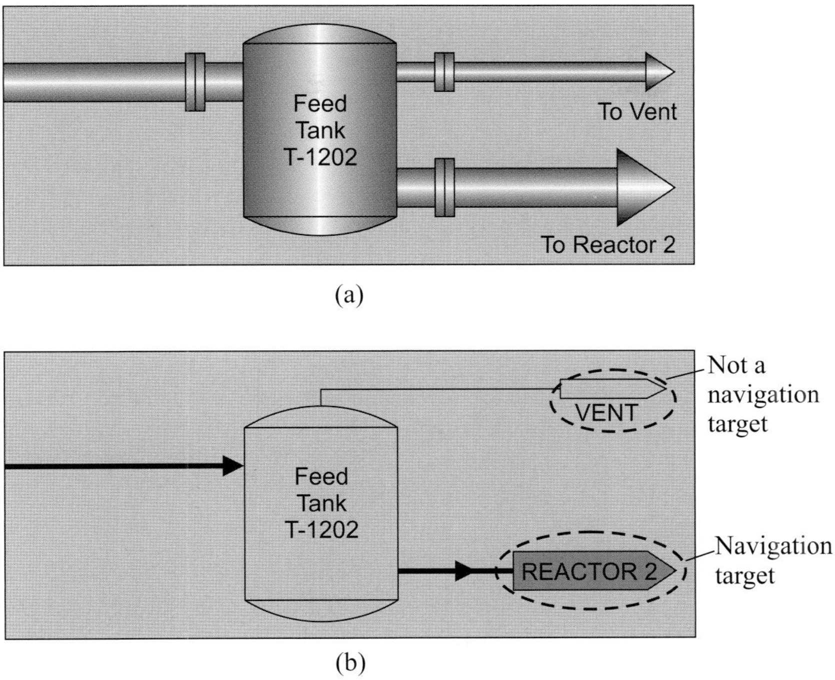

Figure 13.10. Lines and vessels: *(a)* three-dimensional (not recommended); *(b)* two-dimensional.

packages have graphical object libraries that contain three-dimensional objects such as tanks conveyors, towers, and so on. While nice in appearance, three-dimensional objects are distracting and so two-dimensional objects should be used exclusively (Figure 13.10). Other attributes:

The vessel interior should be the same as the background color.

The vessel shape should be representative of the actual object, but without much detail.

The vessel size should be relative to its importance

There should be no animation of any internal equipment, such as an agitator or flame.

Process flow should generally be from left to right. Vapors generally flow up and liquids down, although pumps affect this depiction. Process line entry and exit should be consistently portrayed. Entry and exit points that are navigation links to other screens should be consistently presented and be different from non-navigation links.

Text

The text on a display is an important way of communicating information to the operator. Unfortunately, a common complaint about displays is that the text is difficult to read. There are many text fonts available, but the following guidelines will avoid problems:

1. Select a common font that will exist on all computers, such as Arial. This guideline is especially important if the HMI is being developed for the Windows operating

system. If the screens are developed with an unusual font, the screen font may be mapped to a different font when it is transferred to the operator station PC.

2. Use a sans-serif type of font, such as Arial. Fonts are classified as serif or sans-serif. A serif refers to the extra decorative lines at the ends and corners of each letter. Books, magazine, and newspapers are normally set in a serif font because a serif font is easier to read. However, because computer screens do not clearly render the detail of a serif font, it is better to use a sans-serif font.
3. Select a font size large enough for most operators. Remember that many operators are older than 40 and need reading glasses for small text. A 16-point font is a good starting point. Up to two larger point sizes (for example, 20-point and 24-point) should be selected for labels and headings.
4. Avoid the temptation to use more than one font and more than three point sizes. If more detail is needed, use a "pop-up" window selected by the operator to view the additional text.
5. The amount of text should be minimized. The screen is not a complete description of the process.
6. Hollifield et al. (2008) recommend dark gray for text color, not black.
7. Reserve upper-case text for headings. TOO MUCH UPPER-CASE TEXT CAUSES EYE STRAIN AND CAN BE DIFFICULT TO READ, ESPECIALLY IF IT IS UNDERLINED.
8. Most text should be in lower case, with the first letter of each word capitalized, like a book title. Some people favor capitalizing only the first letter of the leading word. Whether "Boiler Stack Temperature" is preferred over "Boiler stack temperature" is the user's choice. However, whichever one is chosen, be consistent. Using only lower case text is distracting to the operator.

Data Values

Live data values should be shown differently than static text. Hollifield et al. (2008) recommends a bold, dark blue for live data values. Data values should be logically grouped on the screen. An operator will find it difficult to scan a random distribution of data values around a process graphic. If the operator needs to compare values, they should be next to each other. Units should be displayed sparingly. For example, if flows are displayed in pounds/hour, placing a "lbs/hr" next to each flow value clutters the display. It is also important to use the same set of units throughout. For example, one should decide whether temperatures should be displayed in °C, °K, °F, or °R and then maintain those units throughout. The displayed data resolution should be appropriate for the measurement. For example, displaying decimal digits for a temperature (e.g., 104.23 °C) is often unnecessary since the measurement accuracy is often greater than +/- 1 °C.

For overview displays, analog values should not be displayed as numbers, as numbers are hard to assess at a glance. Instead, analog indicators should be as shown in Figure 13.11, an example display of a compressor unit (Hollifield et al., 2008). A pictorial view of the same compressor is shown in Figure 13.12. Notice how easily an operator can determine what parts of the compressor are causing problems with the information on screen in Figure 13.11 as compared to the data shown in Figure 13.12. The parts of the basic analog indicator suggested by Hollifield at al (2008) is shown in Figure 13.13*a*. Alarm indicators and

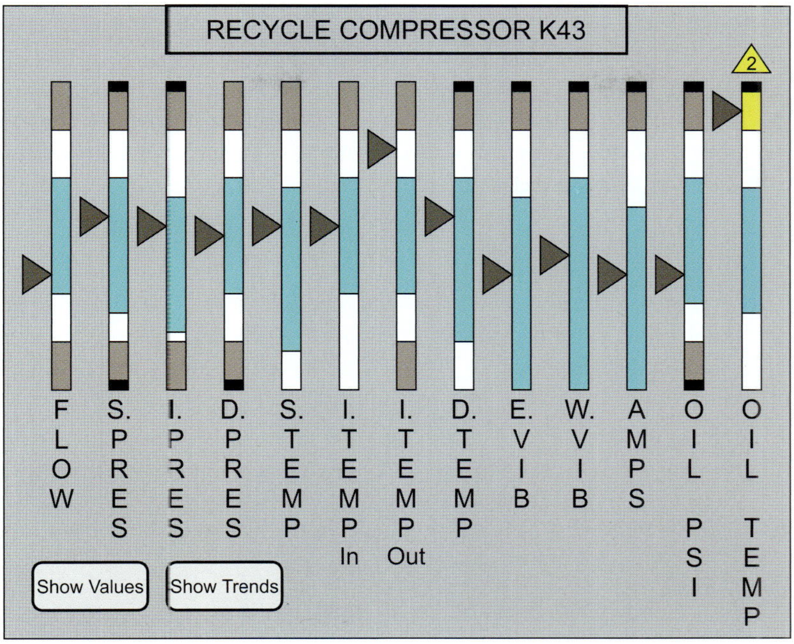

Figure 13.11. Analog indicators on compressor overview screen.

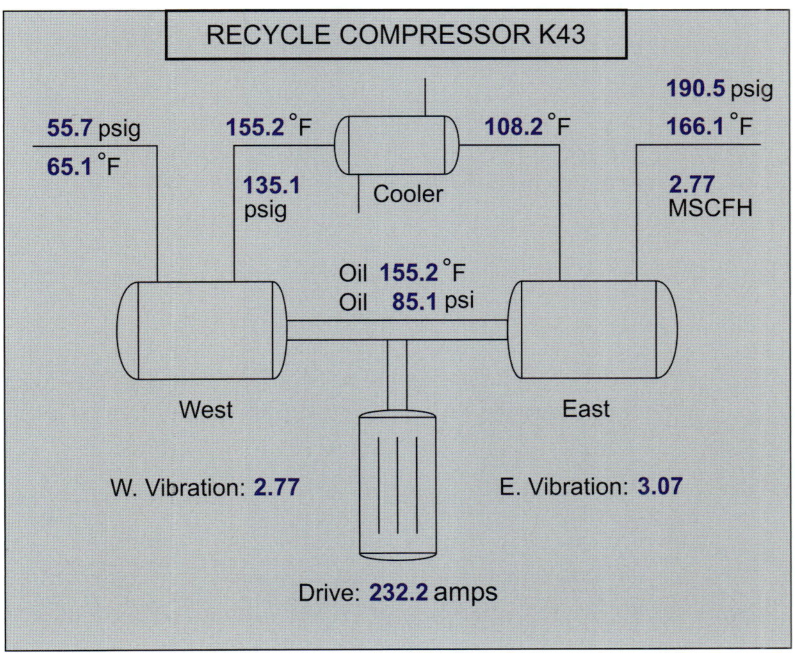

Figure 13.12. Conventional compressor overview screen.

interlocks are described later in this section. Possible enhancements to the basic analog indicator are shown in Figure 13.13*b*.

The analog indication of vessel levels is often poorly portrayed. The indication should display the level without excessive importance or distraction. Figure 13.14*a* shows two examples of poor vessel level indication. Two examples of better level indication are shown in Figure 13.14*b*. For some vessels, a trend line may be best.

Trends

The basic analog indicator only shows the current value. For chemical processes, displaying a graph of the value backwards from now to a previous time is often very useful. For example, the four trends shown in Figure 13.15 have the same current value, but very

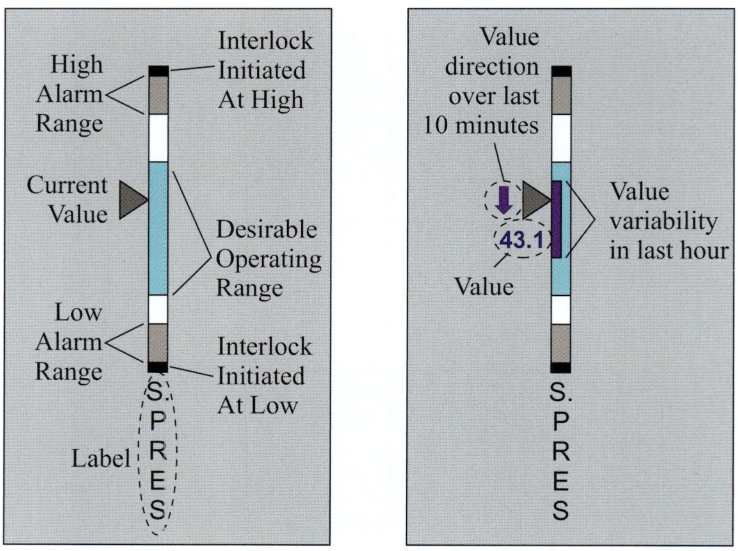

(a) (b)

Figure 13.13. Analog indicator: *(a)* basic information; *(b)* enhancements.

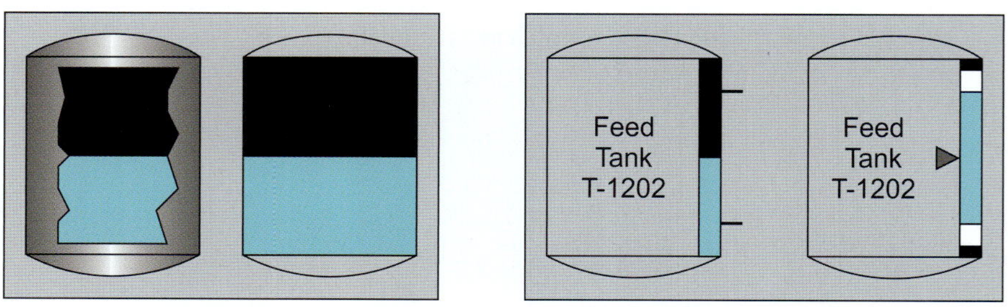

(a) (b)

Figure 13.14. Vessel level indicator: *(a)* poor examples; *(b)* better examples.

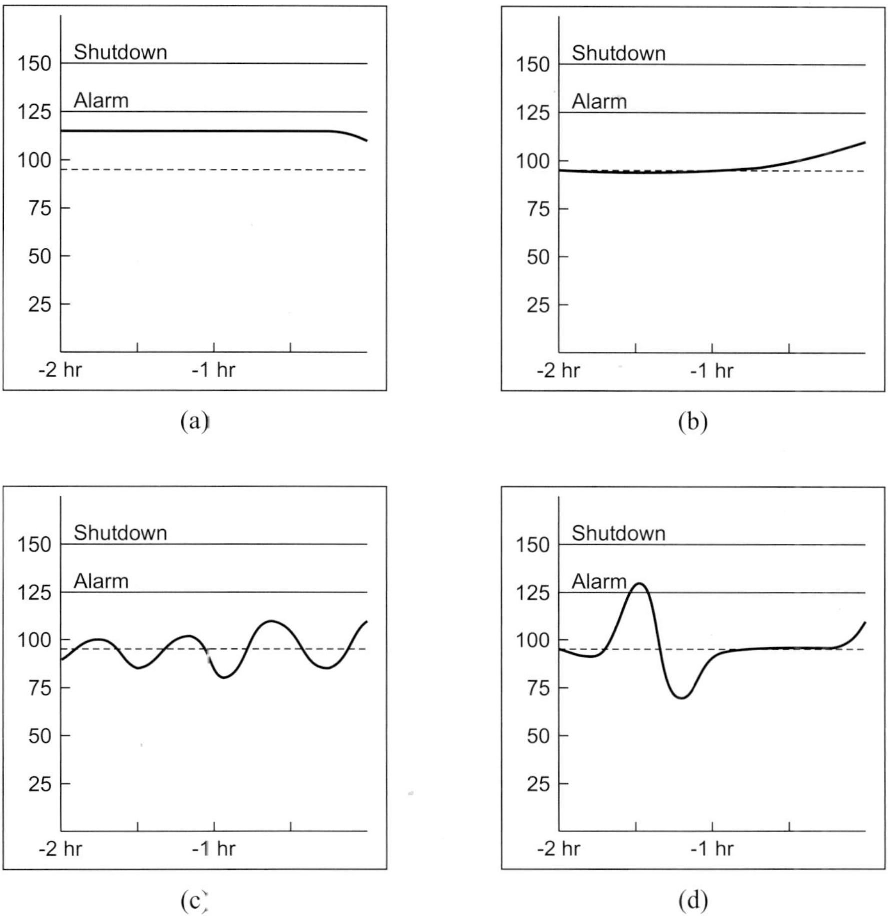

Figure 13.15. Example trends: *(a)* decreasing toward desired value; *(b)* increasing toward alarm; *(c)* oscillating; *(d)* previous upset, heading toward another alarm.

different trajectories leading to the current value. The trend in Figure 13.15*a* indicates there is no problem, but the operator may need to take some action for the other three trends.

Important trends should be part of every Level 1 and Level 2 graphic screen. Dedicated Level 2 displays showing only trends and PID controller faceplates are often useful for a process control unit. If a needed trend is not displayed on the screen, then an operator should be able to access it with a minimum of keystrokes or mouse clicks.

Alarms

In addition to the normal three-priority system of alarms, Hollifield et al. (2008) recommend a fourth level for diagnostic-type alarms. The diagnostic alarm indicates an equipment malfunction that cannot be resolved by the operator and therefore requiring maintenance. The suggested colors and associated shapes for the priorities are:

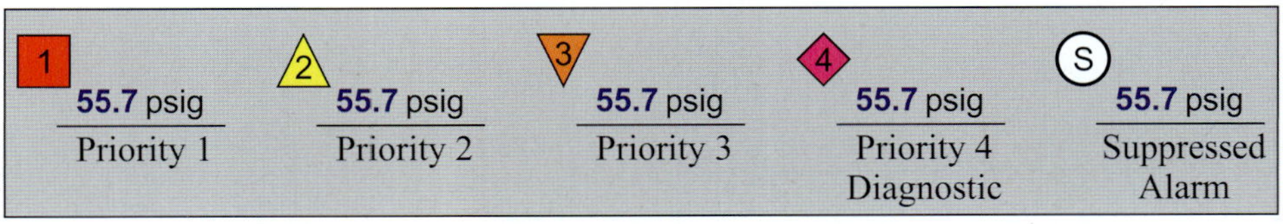

Figure 13.16. Alarm indicators.

Priority 1 – Red square
Priority 2 – Yellow triangle
Priority 3 – Orange inverted triangle
Priority 4 – Magenta diamond

The alarm priority is also redundantly indicated within the shape. These indications are shown in Figure 13.16. A suppressed alarm is shown as a white circle enclosing an "S".

The indicator can be placed near the value (if displayed as text) or the analog indicator. These alarm indicators have the disadvantage that the specific type of alarm is not indicated, especially when the indicator is beside the text value. When placed with the analog indicator, some alarm types (for example, high/low) may be obvious, but others (such as deviation) may not be as obvious. Generally, the operator will need to navigate to another screen to determine the specific alarm.

The alarm type could also be indicated within the shape of the alarm indication (for example, follow the priority number with a "H" for a high value). This type of alarm is not recommended (Hollifield et al., 2008) because it leads to overcrowding and confusing displays.

Every value or indicator with an alarm should have an associated Level 2 or 3 display that aids the operator in the diagnosis and handling of event causing the alarm. The associated display should be accessed with a single mouse-click of touch-screen press. The appropriate Level 2 or 3 display should also highlight any items associated with the alarm that the operator will need to use to react to the alarm.

Alarm indicators also indicate whether an alarm is acknowledged. The three possible states of an alarm and how they are usually indicated are

State of Alarm	Indicator
Active, not acknowledged	Blinking
Active, acknowledged	Visible, not blinking
No longer active, but not acknowledged	Blinking slowly

If the alarm is critical, an audible alert should sound. Audible alerts impart importance since they are heard by multiple personnel. A system that can create multiple horn pitches and tones can transmit the importance of the alarm. A high-pitch, fast-pulsing sound conveys urgency, whereas a low-pitch, slow-pulsing sound is less urgent (Hexatec, 2002).

Standard Objects

Depicting standard objects such as valves, pumps, and PID controllers deserve consideration.

The running/stopped state of motors, pumps, or other motor-driven equipment is shown as being filled with a muted color (Figure 13.17*a*) and not with a bright color such as

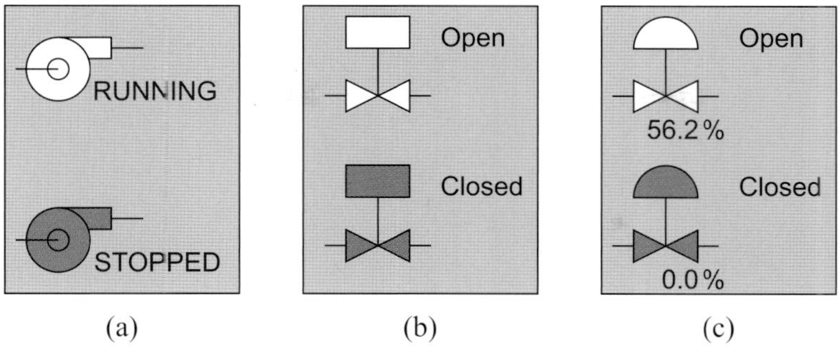

Figure 13.17. Equipment state indications: *(a)* pump or motor; *(b)* discrete valve; *(c)* regulatory valve.

red or green. Color should bring attention to an abnormal condition. A simple text string next to the motor or pump can also indicate the state, but is not as easily noticed as color. Clicking on the object should bring up a faceplate with commands and more detailed status, such as Figure 13.18*a*.

The open/closed state of discrete (on/off) and regulatory values is also shown as being filled with a muted color (Figure 13.17*b* and 13.17*c*). The percent open of a regulatory control valve can be optionally shown with the valve. As for motor devices, clicking on a discrete valve object should bring up a faceplate with commands and more detailed status, such as Figure 13.18*b*. Regulatory control valves are associated with a controller and so clicking on valve depiction in the display generally brings up the controller faceplate.

The depiction of process controllers, typically PID controllers, is often a challenge. The controller generally has a large number of parameters, however an effective display

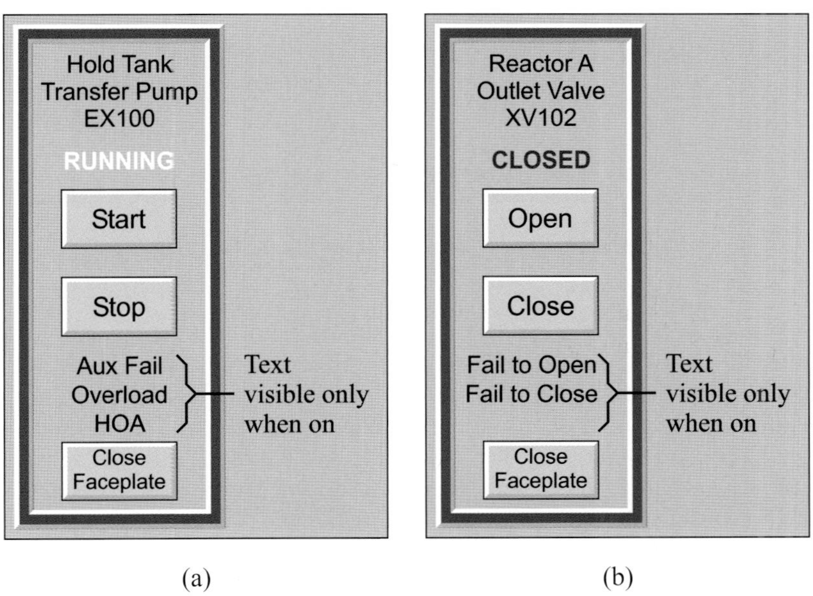

Figure 13.18. Typical discrete device control faceplates: *(a)* motor; *(b)* valve.

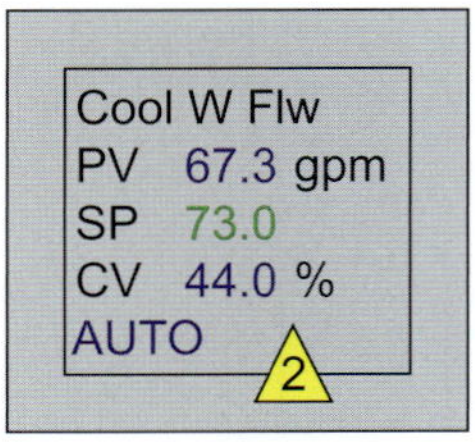

Figure 13.19. Basic controller element.

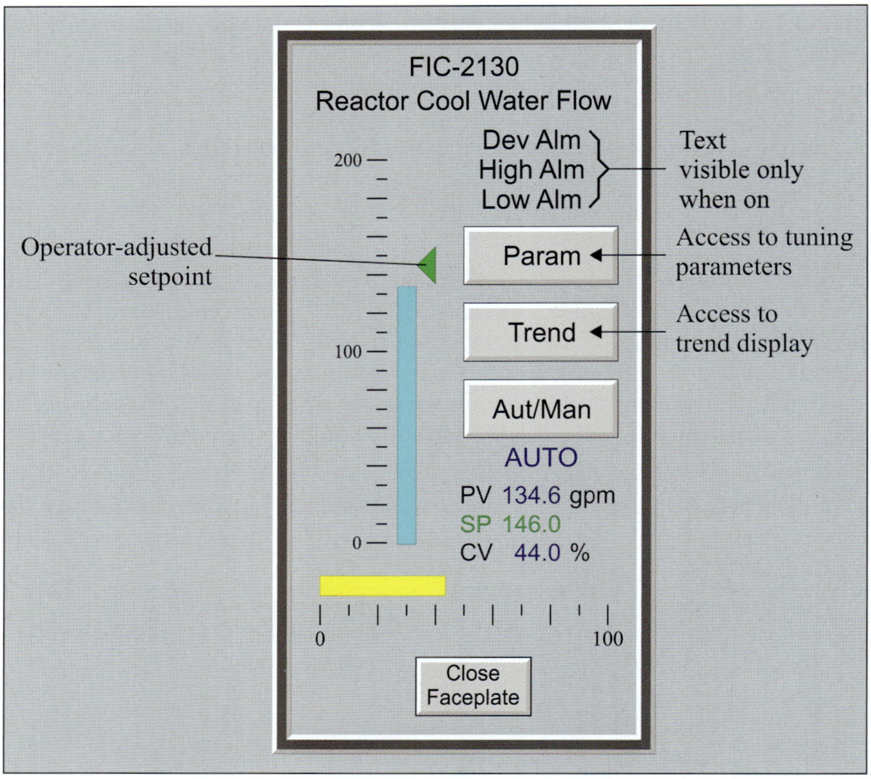

Figure 13.20. Typical PID control loop faceplate.

only needs to show a few of them. Details are shown only when the operator or engineer calls up a detail display screen. The representation presented here follows what Hollifield et al. (2008) suggest. The basic controller element shows the process variable (PV), setpoint (SP), controller output (CV or VO), the controller mode, and any alarm indication. An example is shown in Figure 13.19. The "live" values of PV and CV are shown in dark blue and the SP is shown as a dark green value. A click of the basic controller element should bring up a faceplate as shown in Figure 13.20. Clicking the "Param" button brings up other detail displays of controller parameters. Clicking on the "Trend" button displays a trend of the PV, SP, and CV.

13.4.2 Design of Displays

Graphical HMI displays should be constructed as a hierarchy to provide different extents of operating detail to aid personnel in performing their tasks. The general hierarchy (Hollifield et al., 2008 and ISA, 2015) is:

Level 1 – Area overview
Level 2 – Unit control
Level 3 – Unit subsystem detail
Level 4 – Unit support and diagnostic displays

and is shown in Figure 13.21. The type of screen is often very different for the various levels of the control hierarchy. At the lowest level the display is very simple touch screen with graphical representations of the switches and indicators. In a control room, a graphical representation of the process presents overall status information about the unit. Detailed information about an object is accessed by touching the object. At the higher levels, managers are more interested in overall operation of the plant, and the HMI typically consists of numbers and graphs of production information, such as production rate, percent rejects, utilization rates, material flow between plant units, and so on.

Proceeding from the top (Level 1) to the bottom (Level 4), the four levels represent increasing levels of detail. The typical features of the displays at each level is described and illustrated with examples.

Level 1 – Area Overview

The overview display should show the most extensive view of the process under the control of the operator. It should provide a clear indication of the performance of the unit and track the key performance indicators (KPIs). The overview display shows key features such as:

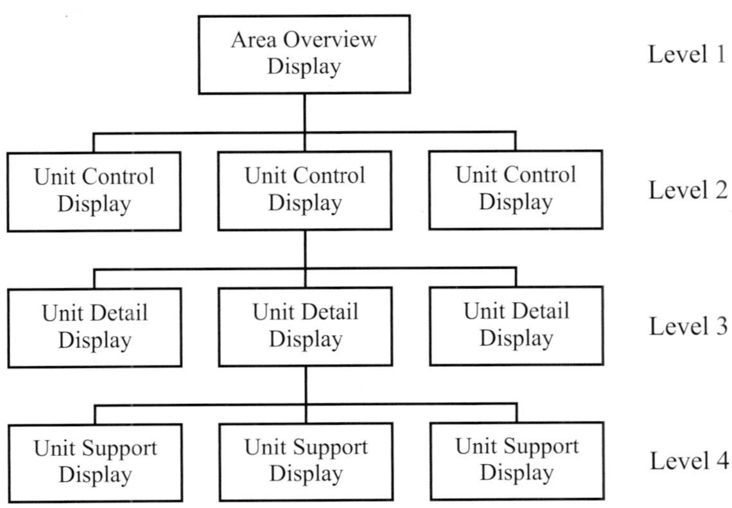

Figure 13.21. Display Hierarchy.

Values and trends of high level KPIs such as production rate, product quality, efficiency, and safety

Trends of important process variables

Alarms of the two highest priorities, including acknowledgement status

Major equipment status

Advanced control performance and status

Indications of abnormal situations

Figure 13.22 shows an example Level 1 overview display for the chemical process described in section 22.3.

There may be more than one overview display designed for different operating states. For example, in addition to the overview display for normal operation, there may be overview displays for startup and shutdown.

There may also be an overview display available for management, typically called a "dashboard." This display shows the overall state of the process, current value and trend on KPIs and possibly key alarms, but this display is generally not used by operators.

Level 2 – Unit Control

A Level 2 graphic screen should contain all of the information and controls associated with a particular unit. A unit is defined as the set of equipment modules and control modules usually centered on a major piece of equipment, for example, a reactor (ISA, 2010). The Level 2 screen associated with a particular unit should be the operator's primary display for both routine changes and monitoring. During the process of designing the HMI screens, the Level 2 displays should be designed first. The Level 2 displays should capture the operator's mental model of the unit. While a piping and instrumentation diagram (P&ID) is a tool used to define the scope of the unit, the display should not look like a P&ID. All alarms of the top two priorities relevant to the unit should be displayed.

Figure 13.23 shows an example Level 2 display for one of the reactor units for the chemical process described in section 22.3 of Erickson (2023). Important variables are trended. The agitator current trend is shown as a "sparkline" (Hollifield et al., 2008) as precision is not essential. Clicking on the controller for the ingredient S flow control brings up a faceplate similar to one shown in Figure 13.20. Access to important safety controls (Hold, Shutdown, and Emergency Shutdown) are given prominence. One alarm is in effect, a diagnostic alarm with one of the valves for the Clean-In-Place system. Navigation buttons are along the bottom of the display. These buttons provide one-click access to the most likely displays to operator may need to invoke.

Some units may have other Level 2 displays to handle specific situations, such as startup and shutdown. These displays contain information for the specific task. An example display for the startup of a reactor unit is shown in Figure 13.24. The display depicts the expected progress of a startup procedure. The trends show the expected progression of the variable along with the past values. The display also contains a checklist showing the progression of the startup steps.

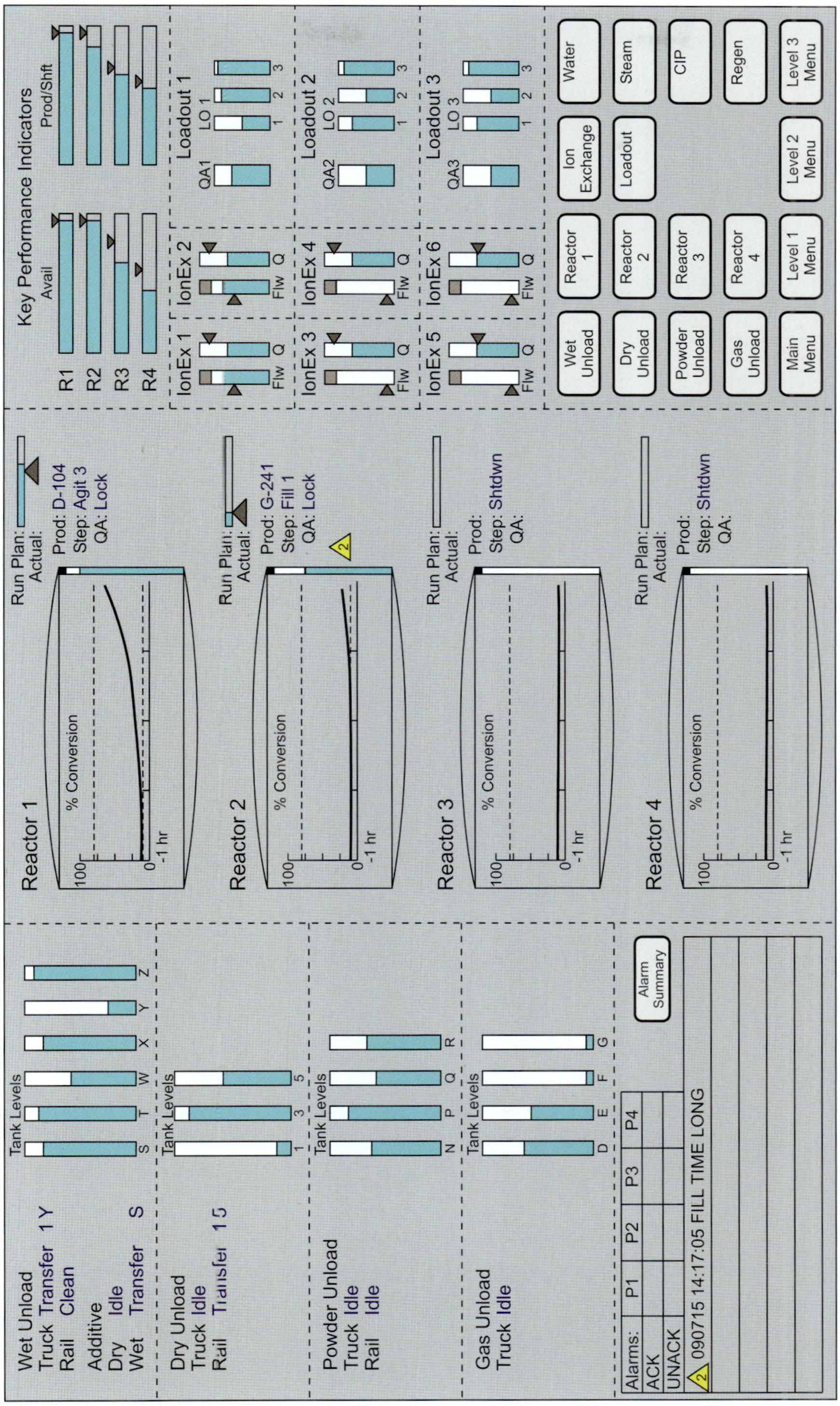

Figure 13.22. Example Level 1 process overview display.

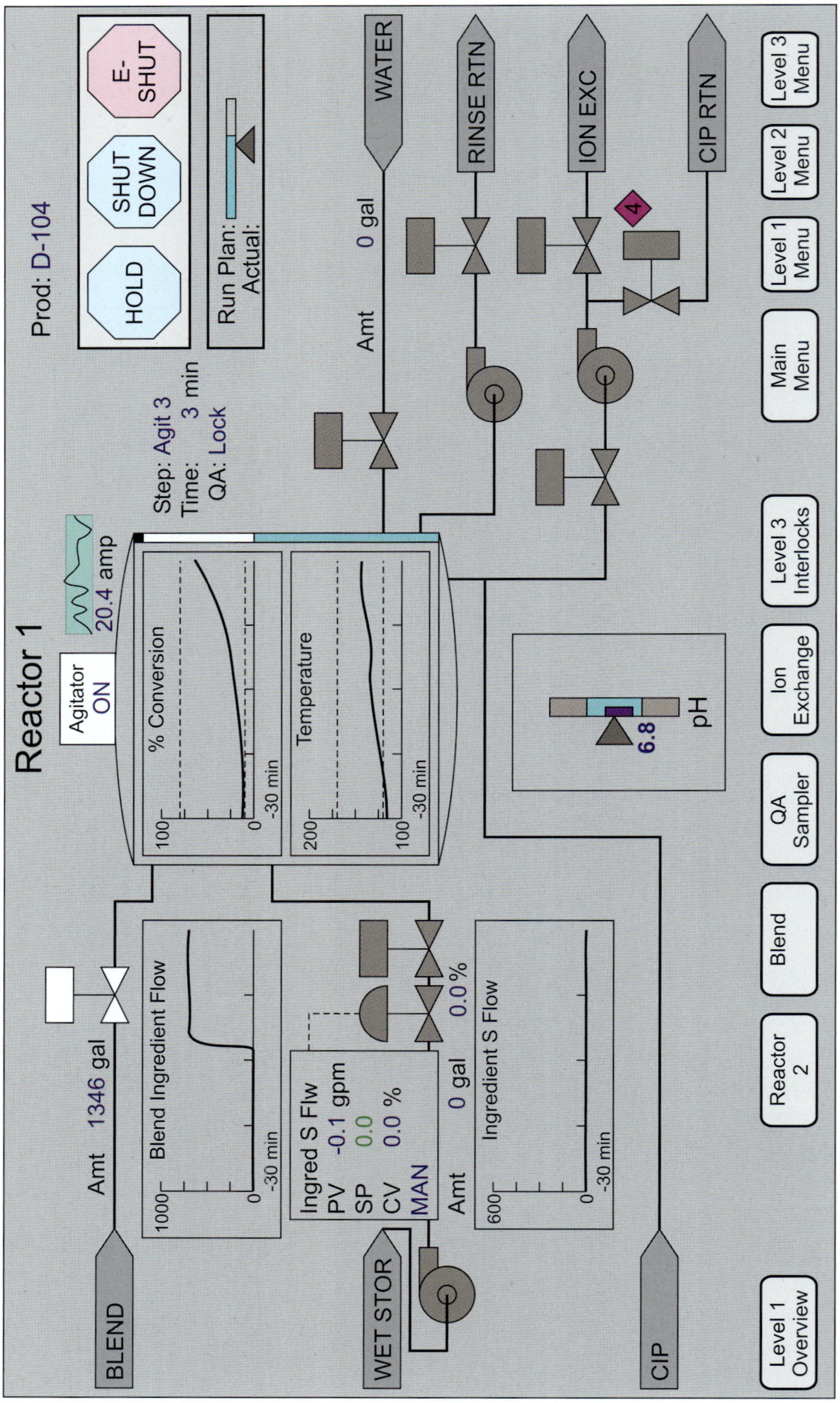

Figure 13.23. Example Level 2 unit display.

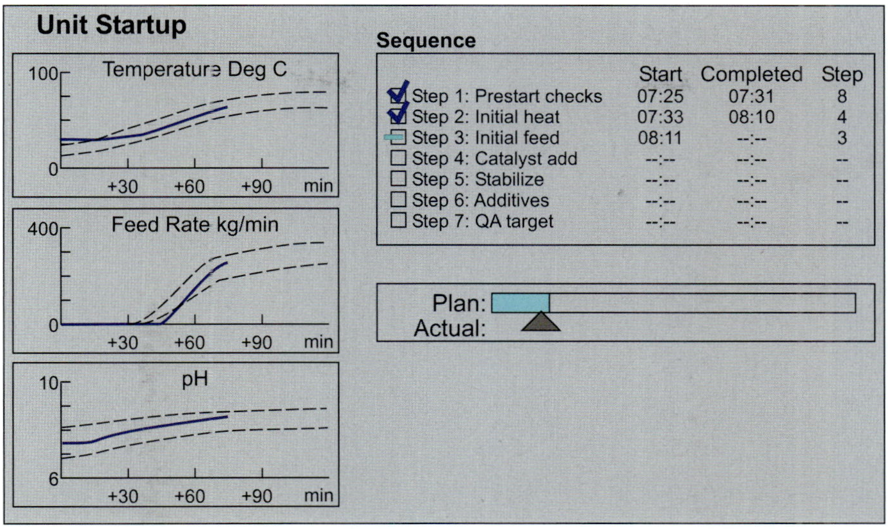

Figure 13.24. Example Level 2 startup display.

Level 3 – Unit Subsystem Detail

A Level 3 screen should contain all of the information and controls associated with a particular subsystem that is part of a unit. A Level 3 screen is intended for troubleshooting or for manipulating items not shown on a Level 2 screen. Figure 13.25 is a sample Level 3 screen for an ion exchanger. All indicators, controllers, alarms, and interlocks are shown on the display. A click on the controller for regeneration or water flow brings up a faceplate similar to one shown in Figure 13.20. A click on a discrete valve brings us a faceplate similar to one shown in Figure 13.18*b*. Trends of all variables and interlock information that might be needed for troubleshooting a problem are shown.

Level 4 – Unit Support and Diagnostics

Level 4 displays are accessed from the other display types. Some of the more common types of Level 4 displays include:

1. *Configuration/Setup*. Detailed process command parameters and parameters used to set up and/or configure control modules are on this type of display. For example, PID loop tuning parameters are displayed and modified on this type of display. This type of display tends to be more textual. Also, security measures are often attached to this screen so that only lead operators or engineers can change the parameters.

2. *Alarm Summary*. The alarm summary presents a complete list of time-stamped active alarms. It includes alarms that have been acknowledged by an operator as well as those that have not been acknowledged by an operator. An alarm remains in the alarm summary until the condition causing the alarm has cleared and an operator has acknowledged it. The three possible states of an alarm in this summary are thus (1) active, not acknowledged; (2) active, acknowledged; and (3) no longer active, but not acknowledged. Generally, a different text/background color or blinking/not blinking is used to distinguish the three categories of alarms. Alarms may also be prioritized. The alarm summary list may be arranged by priority

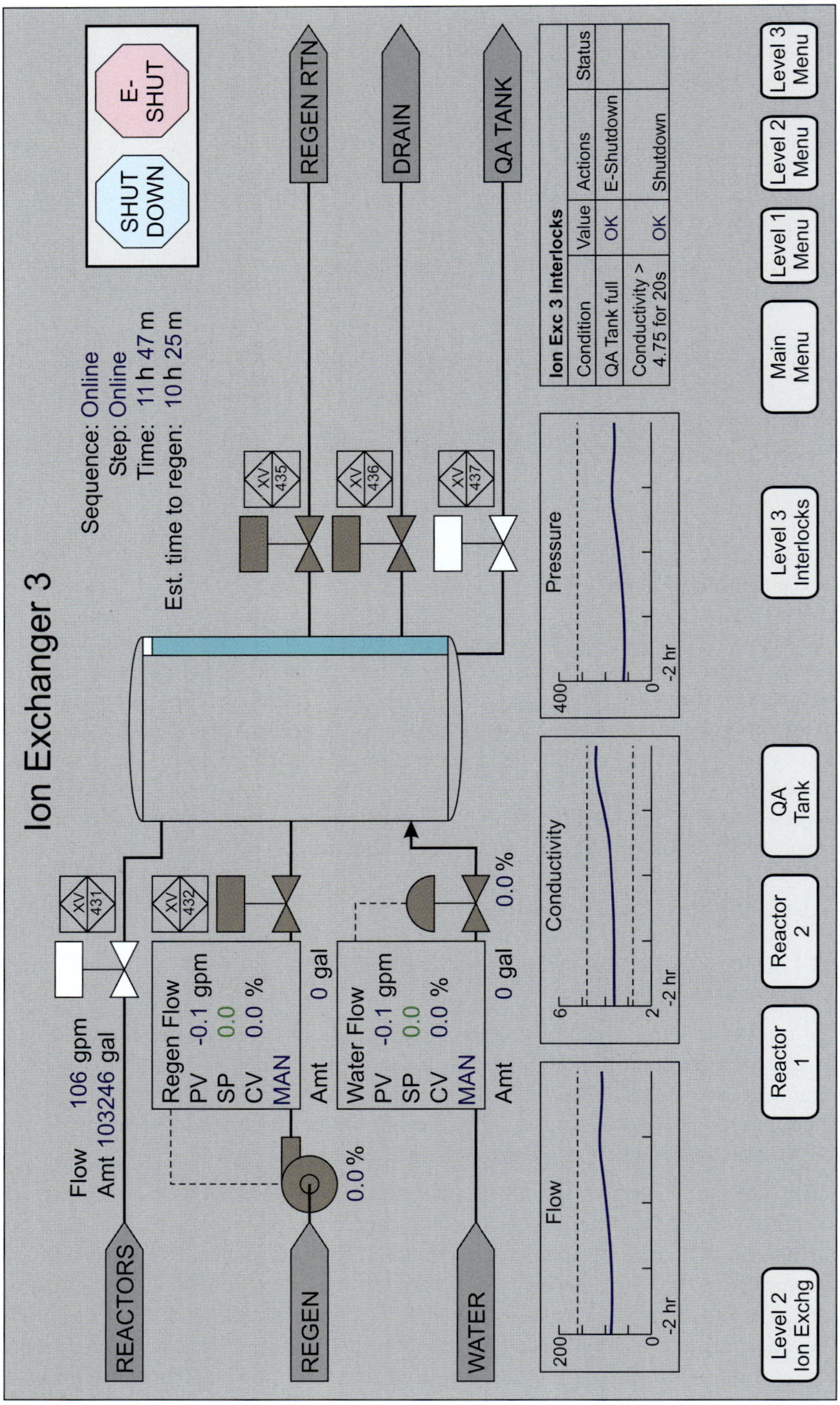

Figure 13.25. Example Level 3 display.

(highest at top), by the time stamp (most recent at top), by the acknowledgment status (unacknowledged alarms at top), or in some combination of these groupings.

3. *Event History*. An event history presents a time-stamped list of all significant events that have occurred in the process. Besides alarms, events such as motors starting/stopping, operators logging in/out, and changes in configuration parameters are recorded. The type of information that may be logged depends on the user interface vendor. Generally, the event history is maintained in more permanent storage and eventually copied to an archival device.

4. *Diagnostic*. These screens are used by maintenance personnel to diagnose equipment failures and usually contain information more detailed than alarms. For example, most variable-speed motor drives (controllers) have a wealth of diagnostic information. A diagnostic display could be set up to display this information when requested. For example, with this display an operator could diagnose a drive fault as an "overcurrent fault on acceleration" without needing to walk to the physical drive and access the information from the drive front panel.

5. *Operating Procedures*. This screen is especially useful to document procedures the operator is expected to perform infrequently.

6. *Abnormal Situation Response Procedure*. This type of display is called from an alarm indication on another display or from the alarm summary. Having this information will decrease the time required to resolve the alarm.

Interlocks

An interlock is designed to detect an abnormal process condition and then override normal control actions to prevent an undesirable or hazardous event. The display of interlocks is given special consideration, as they generally activate infrequently and so the operator is very likely to forget how to handle them. An activated interlock should be clearly shown on the appropriate Level 2 and Level 3 displays.

An example interlock for an analog process value on a Level 2 display is shown in Figure 13.26. The particular interlock protects against a reactor vessel being overfilled. A high vessel level will first activate a Priority 2 alarm when the level is at 4.5 meters and activate an interlock at 4.8 meters. The interlock closes both feed valves. The depiction for normal operation (no alarm and no interlock) is shown in Figure 13.26a. The level is shown as an analog depiction with black sections at the top and bottom, indicating that an interlock will activate at high and low level. Figure 13.26b shows the high level alarm activated and since the alarm is a precursor to the interlock, an interlock diamond appears. The diamond is not connected to anything and shows the operator that if the level is not reduced, an interlock action will occur. Clicking on the diamond will bring up more information about the interlock, including the effect of the interlock. If the high level is not resolved, the interlock activates and appears as shown in Figure 13.26c. The display shows an additional priority 2 indication which shows the tripped interlock. The actions taken by the interlock, commanding both feed controllers to manual mode and closing the valves, are enclosed in yellow rectangles.

An example interlock for a discrete process is shown in Figure 13.27. The particular interlock is a belt misalignment on the common conveyor, C-1505, detected by AS1506. In this case, there is no precursor condition and as soon as the misalignment is detected, the interlock immediately shuts off the conveyor C-1505. However, there is an interlock (I-3)

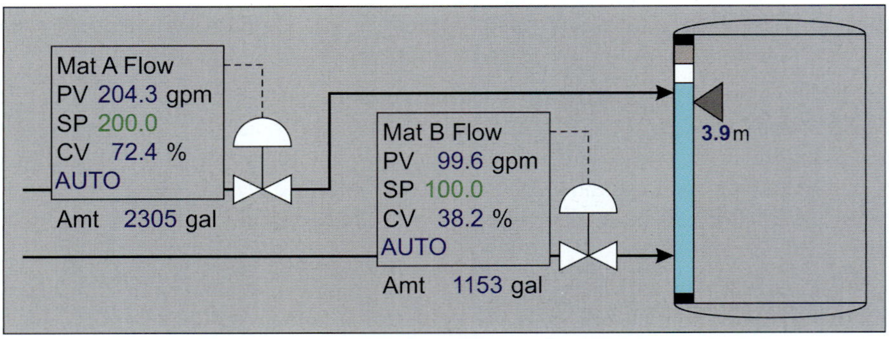

(a)

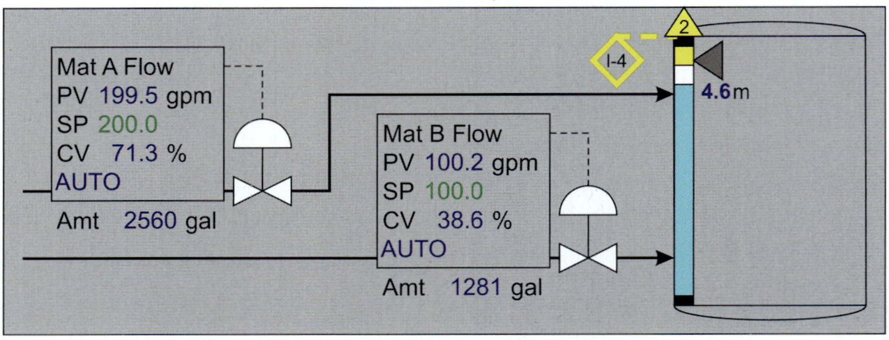

(b)

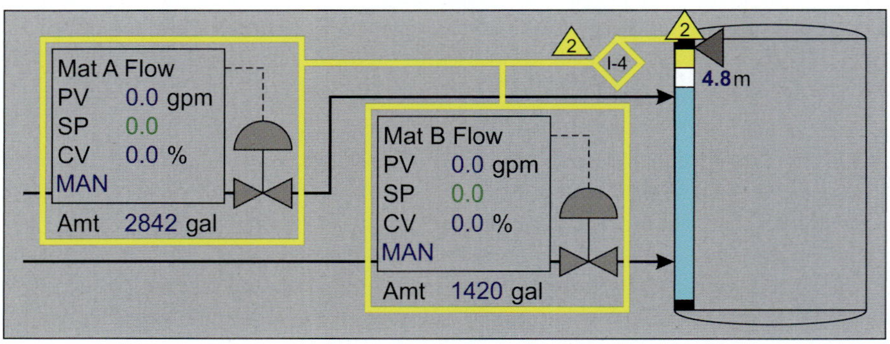

(c)

Figure 13.26. Example analog interlock: *(a)* normal operation; *(b)* alarm precursor to interlock; *(c)* interlock activated.

detecting the stop on C-1505 that immediately shuts off the two feeding conveyors, C-1503 and C-1504. In turn, each of the feeding conveyors has an interlock that immediately shuts its dump slide gate. Rather than show additional interlock tripped indications, only the initiating condition (belt misalignment) is shown so the operator can immediately discern the real problem.

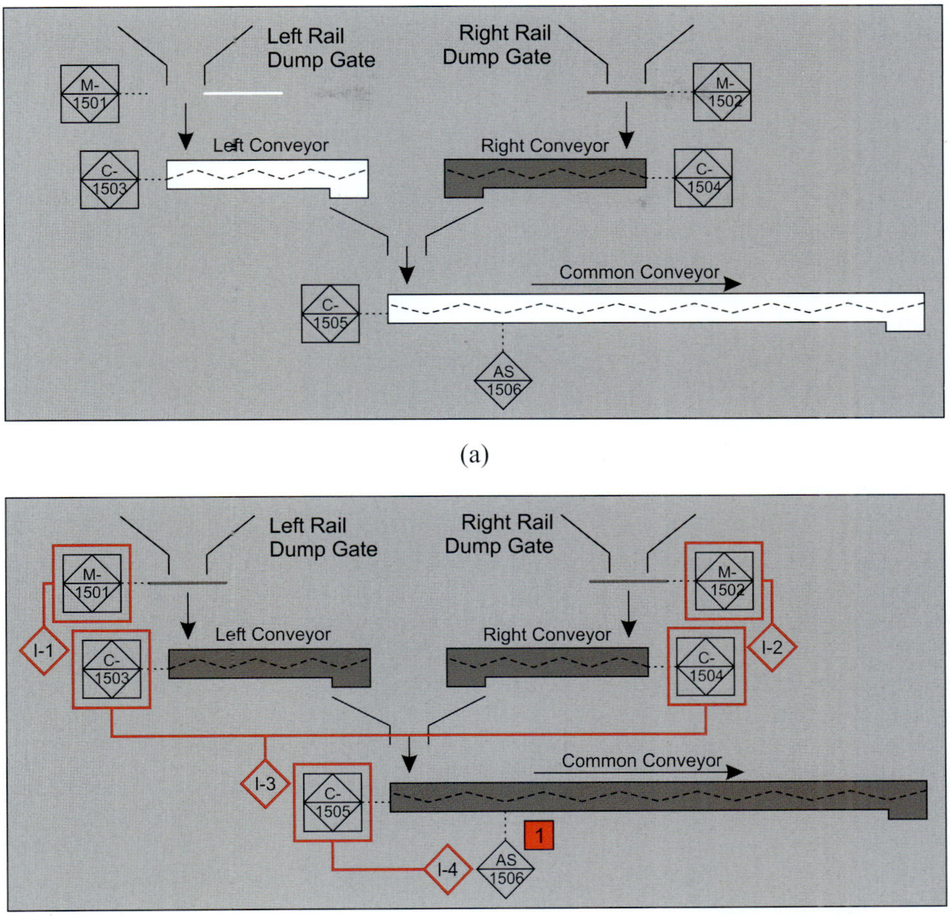

Figure 13.27. Example discrete interlock: *(a)* normal operation; *(b)* interlock activated.

Navigation

To use a system that has multiple screens, the operator must be able to change between screens quickly and easily. In the manufacturing environment, touch screens are the preferred method since they eliminate the less-reliable keyboard. In a control room, the pointing device is often a mouse.

Regardless of the pointing device, the screen "hotspot" (touch area) must be obvious and large enough to be hit easily. There are two main ways of denoting a hotspot. A button on the display is an obvious hotspot. The button should be clearly labeled to indicate the effect of "pressing" the button. Alternatively, the hotspot can be embedded in the graphical picture. For example, in a process overview display showing a schematic layout of the process, "pressing" on an area will change the screen to a more detailed display of the area.

The displays are normally organized in a hierarchical, or tree, structure. The top-level display is the overview of the process. Moving between the displays should be simple and obvious. The lower right hand part of the screen is a good place for the "Next lower display" button. It should also be possible to backtrack and return to the previous screen. All displays should have a button that displays the overview screen and a button that accesses the detailed alarm screen.

If possible, the navigation buttons should be grouped together so the operator can move between screens with little movement of the pointing device. If the operator is expected to routinely switch between two or more screens, the button for the next screen in the sequence should be located at same point on each screen, so that the operator can quickly change displays without moving the pointing device.

When buttons are used for device control, then a pop-up window should be used to confirm the action selected, for example, "Open XV103 Valve — OK or Cancel." As shown in scenario 2 at the start of this chapter, the confirmation avoids accidentally activating the device.

When using a touch screen, one must be careful that hotspots are not too close to prevent accidentally selecting an adjacent hotspot. When the HMI is in the field and the operators have gloves, hotspots may need to be spaced 50 mm or more center-to-center.

13.5 GRAPHICAL HMI DEVELOPMENT

A lifecycle model of HMI development is presented in the ANSI/ISA-101 standard (ISA, 2015) and shown in Figure 13.28. This model is more applicable to the HMI development for larger systems. The foundation of the lifecycle is the development of system standards. The main stages in the lifecycle are design and review, implement, and operate. Across the entire lifecycle, there are continuous work processes.

Within the lifecycle model, the activity to build displays is further explained and the basic types of display animation are described. This section concludes with some helpful hints about developing HMI applications.

13.5.1 HMI Lifecycle

The lifecycle is entered at two points. A new system or major change in an existing system which includes migration to a new HMI platform or vendor starts with the development of system standards. A design for new displays or changes in existing displays enters the lifecycle at the design and review stage.

System standards start with an overall philosophy and lead to a style guide and toolkits of the graphical symbols and other items needed to implement the style guide. The HMI philosophy provides the guiding principles and conceptual fundamentals for the HMI design. The style guide applies the philosophy and provides guidance and implementation examples. The toolkit contains the graphical symbols and other components as needed to implement the style guide.

The design of the HMI system starts with the design of the console, especially if the HMI will contain multiple displays. Then user, task, and functional requirements along with any applicable standards are used for the system design and then leads to the display design.

The implementation stage takes the design information from the previous stage and builds the actual display screens and the system hardware. After a test of the system and user training (often using a simulated process), the HMI is commissioned by a final test of the HMI with the real system. Some industries will require specific verification of the system before it is placed in operation.

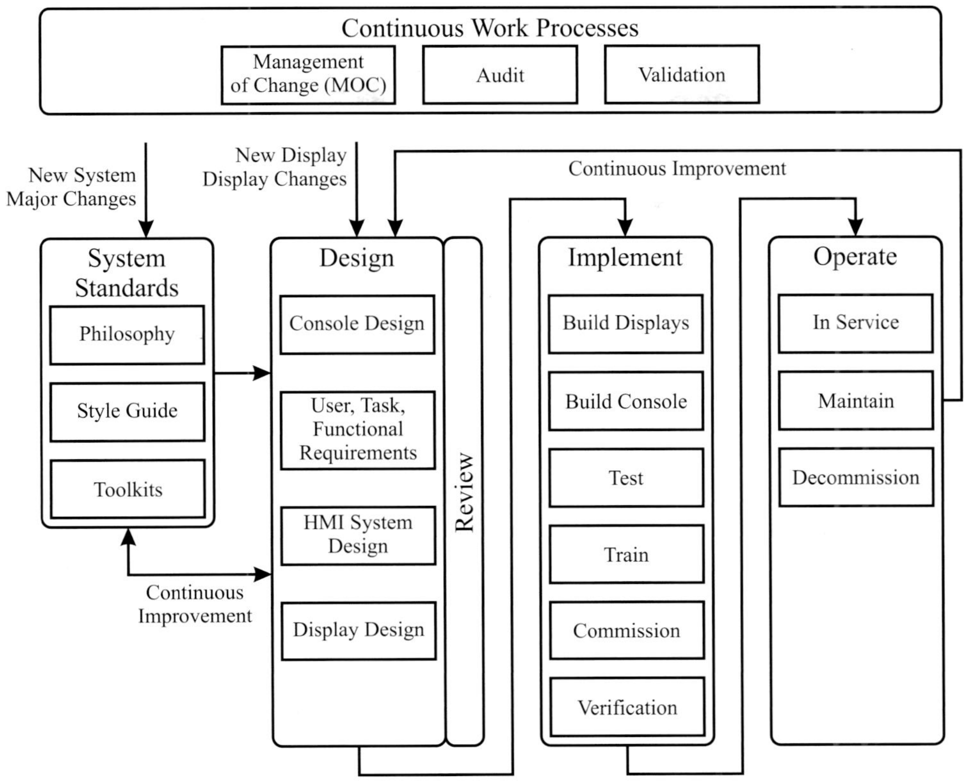

Figure 13.28. HMI Lifecycle.

Once the HMI has been commissioned, the lifecycle moves into the operate stage and the HMI is considered to be "in service." However, the system still requires maintenance which consists of operating system, security, and HMI platform software updates, corrections to HMI errors, and changes in the process. Decommissioning is the process of removing part or all of the HMI from service.

The lifecycle model has two continuous improvement loops. The first loop is a cycle between the design and system standards stage. This loop acknowledges that as improvements are made to the design, system standards are updated. The second loop is a cycle between the operate and design stages. This loop shows the improvements that are made based on the operation of the HMI.

After the HMI is in service the management of change activity is used to handle changes to the HMI. Periodic audits assure the HMI standards are being followed and some industries may require specific validation plans.

Security issues are often overlooked when designing an HMI. The computer-based user interface allows access to a large amount of process information and parameters. Thus, it is necessary to restrict access to certain functions. Typically, a user logs on to the system with a *user name* and *password,* which grants him or her an *access level.* A higher access level permits greater access to the restricted functions. For example, an operator generally is not permitted to change recipes or PID tuning parameters. However, an engineer logging onto the same system can access these functions provided his or her access level is high enough.

13.5.2 Build Display Activity

The build display activity in the implement stage of the HMI lifecycle model (Figure 13.28) involves the following four major tasks:

1. Set up the communication with the PLCs.
2. Create the tag database.
3. Draw the graphical objects on the windows.
4. Animate the objects.

Setting up the communication with the PLCs often involves configuring the servers and access names shown in Figure 13.4. However, for a simple graphical HMI package designed to interface with a limited set of PLCs, the communication setup is usually uncomplicated and may consist of only specifying the type of interface to the PLC.

The tag database is the heart of the HMI software. To ease the task of building the database, most HMI packages provide a way to export and import the database. One can configure the database entries with a spreadsheet program and then import the database into the HMI application. Most HMI vendors also provide a way to import tags from the PLC programming software.

The graphical drawing tools available in most HMI packages are generally simple. More complicated objects are generally imported from a drawing package, or imported from a library of objects.

Object animation is basically the process of attaching properties to a screen object that allow it to react to an operator touch or to change its color, shape, and so on, dependent on a tag value. The types of animation are described below.

Most people develop an application by doing the tasks in the order above. However, one can draw each graphical object first, and then animate it, letting the system prompt the user to define any undefined tag. This process is known as creating a tag "on the fly."

13.5.3 Animation Types

There are two basic types of animation: (1) user input, and (2) display. A user input type of animation allows an operator to change tag values. A display animation allows a value to be displayed as well as allow an object to change shape, position, color, and so on.

User Input Animation

Touch-sensitive. This type of object animation allows an operator to input data into the system. For example, the operator may open or close a valve, enter a new alarm setpoint, or type a password for access to maintenance functions. These objects are identified with a frame or highlight around the object when the mouse cursor passes over it. There are generally three types of touch-sensitive inputs:

1. Discrete - Controls the value of a Boolean tag. When touched, a dialog box appears, prompting the operator to select one of two commands, for example, "Open Valve" or "Close Valve."

2. Analog - Used to input the value of an analog (integer or real) tag. When activated, an input box appears, prompting the operator to enter a value. The value may be entered from the standard keyboard or from an optional on-screen keypad.
3. String - Used to input the value of a text string type of tag. When touched, an input box appears, prompting the operator to enter the string. The characters may be entered from the standard keyboard or from an optional on-screen keyboard.

Pushbutton. Perform an operation when the object is touched (touch-screen display) or when clicked with the mouse. The most common operations are:

1. Discrete value change - The object acts like a pushbutton that controls the state of a Boolean tag. Available pushbutton actions include set (latch), reset (unlatch), toggle (change state), momentary on, and momentary off.
2. Execute script - Allows a small program to be executed when the object is touched. Depending in the HMI package, there may be up to three different scripts: "When Pressed," "When Released," and "While Down." A "When Pressed" script executes once when the object is pressed and a "When Released" script executes once when the object is released. A "While Down" script executes continuously on a set time interval as long as the object continues to be pressed. This type of script waits one time interval before executing, so if the script must start execution immediately, create a duplicate "When Pressed" script. Scripts are generally written in Visual Basic, or another similar language. Scripts can be used to set tags to specific values, show and/or hide screens, start and control other applications, print the screen, and so on.

Slider. This type of object is a symbol that can be moved around the window with a finger on a touch screen or a mouse cursor. As the object "handle" is moved, it alters the value of the tag attached to it. The slider provides an alternative way to input analog values into the system albeit with less precision than a touch-sensitive user input. The slider may be horizontal or vertical. In Figure 13.29*a*, the slider object is the "handle." Some HMI packages allow one to define both a horizontal and vertical slider user input as shown in Figure 13.29*b*. In this case, the slider is inside a rectangle and its position determines the value of two variables, for example, a x-position and a y-position.

Display Animation

Value Display. Displays the value of a tag or expression on the screen. If the tag is a Boolean, then one user-defined text message is displayed if the tag is true and another user-defined text message is displayed if the tag is false.

Line, Fill and Text Color. Changes the line color, fill color, or text color attributes of an object. There are four types of color animation:

1. Discrete - The color is linked to a Boolean tag or expression. In this case, the color can be one of two values.
2. Analog - The color is linked to a real or integer tag or expression. The HMI may allow a continuous color gradation, or may define a fixed number of value ranges defined by breakpoints. In the latter case, each range has a different color.
3. Alarm - The color of an object can be linked to the alarm state of a tag or group of alarms. If the alarm state is a Boolean, the color can be one of two values; one for the normal state and one for the alarm state. If the tag has multiple alarm conditions

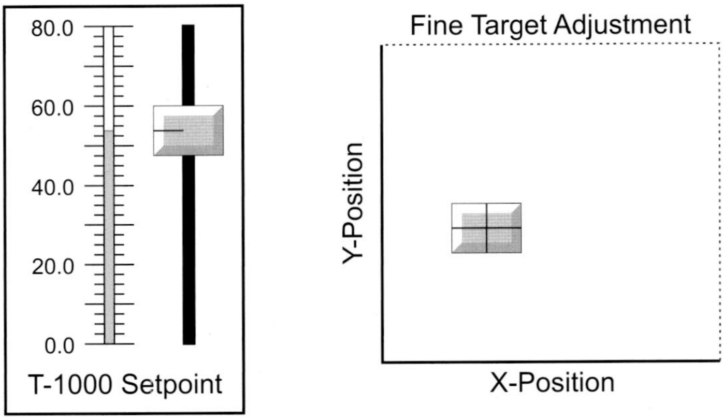

(a) (b)

Figure 13.29. Slider user input: *(a)* one dimension; *(b)* two dimensions.

(for example, low-low, low, high, high-high), then the normal state and each alarm condition can have a different color.

Object Percent Fill. Varies the fill level of a filled shape according to the value of an integer or real expression. The fill can be vertical or horizontal. This type of animation is often used to indicate an analog value (Figure 13.13), though it can also be used to build a horizontal bar graph.

Object Size. Varies the height and/or width of an object, depending on the value of an integer or real expression.

Object Location. Changes the location of an object, depending on the value of an integer or real expression. This type of animation is useful for displaying the movement of a gantry crane.

Visibility. Controls the visibility (whether it is shown or not) of an object according the value of a Boolean tag or expression.

Blink. Makes an object blink based on the value of a Boolean tag or expression.

Rotation. Makes an object rotate based on the value of an integer or real tag or expression.

Disable. Disables the touch functionality of an object based on the value of a Boolean tag or expression. The disable animation is very useful for security features. For example, a touch-sensitive or pushbutton input can be disabled based upon the logged-on operator's access level or name.

13.5.4 Helpful Hints

When initially programming an HMI application, it is helpful to verify the communication configuration. Configure a few tags (for example, one Boolean, one integer, and one real), draw and animate a few objects with these tags, and then test for proper operation. When using an HMI package for the first time, this step may save one many hours of correcting minor mistakes in the display elements.

When defining a tag, one should make sure to select the correct data type. After one has saved a tag definition or attached the tag to a screen object, some HMI packages will not allow one to easily change the basic type. For example, if one incorrectly defines an integer as a discrete data type, it may not be easy to correct the data type. In some HMI packages, the process to delete a tag involves a few steps. If one needs to correct the data type for a tag already attached to one or more screen objects, then these screen objects will need to be modified to remove all references to the tag. Then the tag can be deleted and redefined.

Any tag that refers to a physical input may need to be set to "Read Only" (depending on the HMI package and PLC).

Complex objects are assembled from many simple objects (lines, filled shapes, and text) that are either combined or grouped together. When simple objects are *combined*, the assemblage is treated as a single object. Any changes in object attributes (for example, fill color or line color) are applied to all of the components in the combination. When simple objects are *grouped*, each object retains its attributes and animation. Object grouping is useful for creating PID or discrete controller faceplates. Once a set of objects is grouped, the HMI software may not allow the attributes and animation of the individual objects to be changed. In this case, to change the appearance or operation of a grouping, it must be "ungrouped" and the individual object(s) changed before regrouping. When a group is duplicated, one generally just needs to change the tags associated with the animation.

Multiple status indicators for the same device can be "stacked" on top of each other, each one having a different visibility animation. For example, the five alarm indications of Figure 13.16 can be placed in the same position, with priority 1 on top and a suppressed alarm on the bottom. The visibility animation for each alarm object is programmed so that at most one of these indicators is displayed. The drawback to stacking objects is when an object's animation needs to be changed. The objects on top will need to be "unstacked" before the bottom objects can be altered.

Frequently used objects (including their animation) should be placed in libraries. When these objects are imported, only the tag references need to be changed.

13.6 CHAPTER SUMMARY

This chapter reviews the general structure and functions of a user interface. Design guidelines for operator panels and operator screens are also presented. The guidelines are illustrated with examples. The chapter concludes with a description of the lifecycle model of HMI development and additional information about display animation.

REFERENCES

Erickson, Kelvin T., 2023. *Programmable Logic Controllers: An Emphasis on Design and Application,* 4th Edition, Dogwood Valley Press, Rolla, MO.

Hexatec, 2002. *Operator Screen (HMI) Guidelines*, Hexatec, Inc, Northumberland, UK.

Hollifield, Bill R, Dana Oliver, Ian Nimmo, and Eddie Habibi, 2008. *The High Performance HMI Handbook: A Comprehensive Guide to Designing, Implementing and Maintaining Effective HMIs for Industrial Plant Operations*, PAS, Houston, TX.

Invensys Systems, Inc., 2014. *InTouch® HMI Concepts and Capabilities Guide*, Invensys Systems, Inc., Lake Forest, CA.

ISA, 2010. *ANSI/ISA-S88.00.01-2010, Batch Control, Part 1: Models and Terminology*, International Society of Automation, Research Triangle Part, NC.

ISA, 2015. *ANSI/ISA-101.01-2015, Human Machine Interfaces for Process Automation Systems*, International Society of Automation, , Research Triangle Part, NC, July.

Murch, Gerald, 1984. " Physiological Principles for the Effective Use of Color," *IEEE Computer Graphics and Applications*, vol. 4, no.11 (November): 49-54.

NFPA, 2021. *NFPA 79: Electrical Standard for Industrial Machinery*, 2015 Edition, National Fire Protection Association, Quincy, MA.

Rockwell Automation, 2022a. *FactoryTalk View Machine Edition User's Guide,* Doc. ID VIEWME-UM004Q-EN-E, Rockwell Automation, Inc., Milwaukee, WI, Mar.

Rockwell Automation, 2022b. *FactoryTalk View Site Edition User's Guide,* Doc. ID VIEWSE-UM006P-EN-E, Rockwell Automation, Inc., Milwaukee, WI, August.

Tufte, Edward, 2001. *The Visual Display of Quantitative Information*, 2nd Ed., Graphics Press.

14 Selecting a PLC

Chapter Topics:

- General PLC selection criteria
- PLC families

OBJECTIVES

Upon completion of this chapter, you will be able to understand:

- Considerations when selecting a PLC for a given application
- The range of PLC families offered by Rockwell Automation, Siemens, Schneider Electric and GE

Scenario: Conversion of power plant coal- and woodchip-handling PLC.

In a previous project, you supervised the conversion of a power plant coal-handling system from a relay-controlled system to a PLC-controlled system with a SLC-5/05 processor. The installation is successful and they are pleased with the increased flexibility..

Later, the wood-chip handling relay-controlled system is converted to a PLC-controlled system and its functions are added to the coal handler PLC. The current panel layout (Figure 14.1*a*) did not have enough space to add the needed additional I/O channels for the woodchip-handling functions. Furthermore, the wood-chip handling system is split into two parts. Half of the terminal blocks for the wood-chip I/O channels are in an inside panel 12 meters away and the other half are in an outside panel with a 200 meter cable distance. Since the SLC-5/05 is now discontinued, it is time to upgrade to a ControlLogix 5580 or a CompactLogix 5380 system. However, the 200 meter distance to the remote panel is too long for a standard copper Ethernet cable. A fiber optic network extender is a possibility, but rejected because maintenance personnel will need to learn how to repair fiber optic cable. Thus, the existing DeviceNet cables will need to be retained. There is a ControlLogix DeviceNet scanner module (1756-DNB), but the cost of a ControlLogix system is significantly more than an equivalent CompactLogix system. There is no existing DeviceNet scanner for the CompactLogix 5380. Is there a solution that will permit the use of the less expensive CompactLogix 5380?

Solution: Even though there is no DeviceNet scanner module for the CompactLogix 5380, there is an Ethernet-to-DeviceNet linking module (1788-EN2DNR) that will serve the needed function (Figure 14.1*b*). The module is added to the CompactLogix I/O configuration and programmed much like the ControlLogix DeviceNet scanner to map the input and output bytes of the DeviceNet nodes into DINTs of the DeviceNet linking module. Even though the cost of the 1788-EN2DNR is more than the cost of the 1756-DNB,

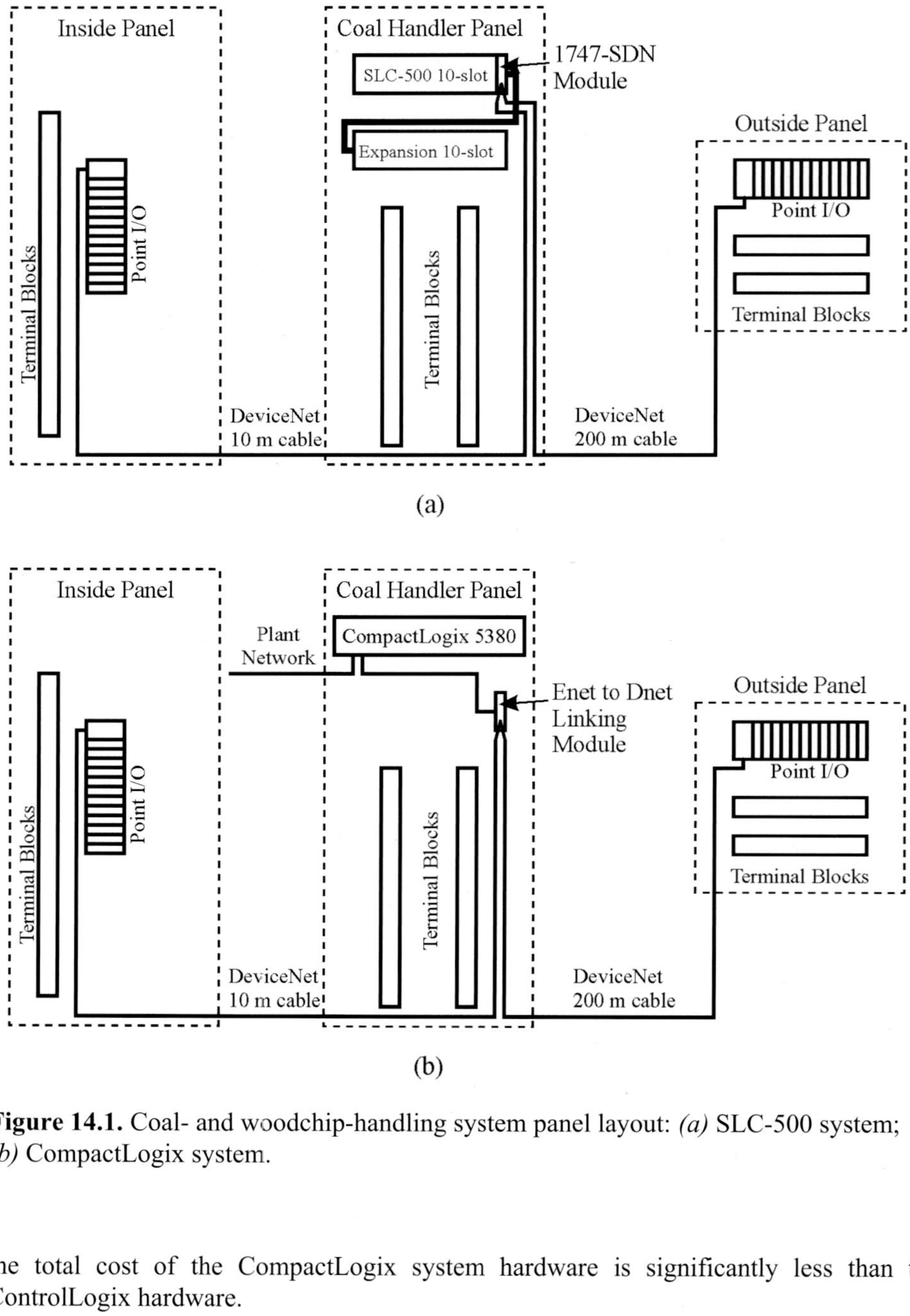

Figure 14.1. Coal- and woodchip-handling system panel layout: *(a)* SLC-500 system; *(b)* CompactLogix system.

the total cost of the CompactLogix system hardware is significantly less than the ControlLogix hardware.

14.1 INTRODUCTION

Most of this text has been concerned with programming PLCs. Before one can program a PLC, it must first be selected. Actually, the first choice is whether to choose a PLC or a

PC-based control system. That comparison is discussed in Chapter 1 and is not addressed here. General selection factors are described first and then the PLC families of the major manufacturers available in North America are characterized.

14.2 SELECTION FACTORS

When selecting a PLC, the biggest consideration is the application, which ranges from a small stand-alone relay replacement to a large system containing hundreds of thousands of I/O points, requiring multiple PLCs and coordination among them. PLCs can be classified according to the range of I/O channels needed by the application as shown in Figure 14.2. As indicated in Figure 14.2, as the number of I/O channels that a particular PLC can handle increases, the amount of program memory and available functions also increases. The medium-sized PLCs (and larger) generally have specialized I/O modules and more communication options. A few years ago, PID control was not supported by the smaller PLCs. However, that function is now available even in the smallest PLCs. The smaller PLCs generally only support ladder logic programming whereas the larger PLCs support multiple languages.

The smallest (nano) PLCs are often a compact unit with built-in I/O channels, though some have the capability to add a few expansion I/O channels. As one progresses to larger PLCs, they progress to DIN-rail mounted modules that are connected together and then up to chassis-based systems where the modules are plugged into a backplane. The

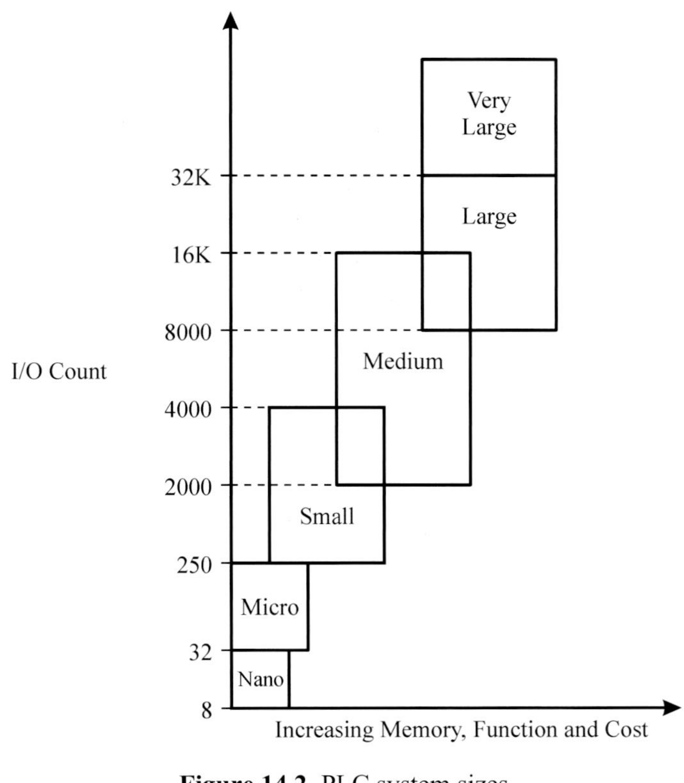

Figure 14.2. PLC system sizes.

medium-sized (and larger) PLCs tend to support some kind of remote I/O, whether it is with a proprietary network and/or with a standard network.

Though the number of I/O channels is a big selection factor, other factors that determine the type of PLC chosen include:

- Specialized I/O - Does the application demand high-speed counters, motion control, TTL digital I/O, or other specialized interface?
- Remote I/O - Is some of the I/O located some distance from the processor?
- Expandability - Can more I/O be added to the system in the future?
- Networking - Does data need to be exchanged between processors or between processors and computers?
- Redundancy - Does the system need to function even when the processor fails?
- Safety - Does the system need to meet certain safety standards?
- Programming language - Does it support languages other than ladder logic?
- Manufacturer reputation - If the system fails, are replacement parts readily available?
- Standardization - Has the company standardized on PLCs from a certain manufacturer?
- Price

The cost of a PLC system is much more than just the cost of the PLC hardware. The overall cost includes programming, installation, maintenance, and training. The relatively high cost of training means that companies generally limit the allowable PLCs to those of one vendor. The $500 savings on a single PLC from a different vendor is overshadowed by the costs required to train even one technician. The maintenance cost also includes the cost of spare modules kept in stock for speedy replacement. This consideration often limits one to keep the types of I/O modules to a minimum. For example, instead of using 8-, 16-, and 32-channel 24-V discrete input modules, it is often a wiser choice to use only 16-channel modules.

When a PLC is used for automatic safety protection, it must meet certain safety standards (IEC, 2010a; ISA, 2004). The safety integrity level (SIL) defines the safety performance of a critical control system in terms of its risk, or probability to fail. The four safety levels and their availabilities are described in Table 14.1. The availability is the percent of time that the system operates correctly. For example, if a high-level trip for a tank has an availability of 99%, it is predicted to fail once for every 100 times the tank reaches the high level. Depending on the chemicals involved, this may or may not be an acceptable

Table 14.1 Safety Integrity Levels

Safety Integrity Level	Availibility Required	Probability to Fail on Demand	Generalized View
1	90.00% - 99.00%	10^{-2} - 10^{-1}	Minor property and production protection.
2	99.00% - 99.90%	10^{-3} - 10^{-2}	Major property and production protection. Possible employee injury
3	99.90% - 99.99%	10^{-4} - 10^{-3}	Employee and community impact.
4	99.99% - 99.999%	10^{-5} - 10^{-4}	Catastrophic community impact.

risk. A higher SIL level indicates greater safety performance and lower risk. The higher SIL levels are realized with redundancy, frequent testing, and self-diagnostics.

When a process hazards analysis concludes that automatic safety protection (commonly called a safety instrumented system), is required, the SIL is assigned by a variety of techniques (Summers, 1998).

14.3 ALLEN-BRADLEY PLC FAMILIES

Though there are a large number of vendors of PLC equipment, this text covers only the Allen-Bradley (Rockwell Automation) PLCs. In addition, this section is further limited to the PLC families available at the time of this writing. Only the major features of these families are described. More information can be obtained from Rockwell Automation (2022).

The general families of Allen-Bradley PLCs are summarized in Table 14.2. These systems span the entire range of applications. The ControlLogix and CompactLogix are the most prominent platforms and replace the discontinued PLC-5 and SLC-500 systems. However, due to the large installed base of PLC-5 and SLC-500 systems, they will continue to be supported for many years.

The Micro800 controllers are designed for low-cost standalone applications. The Micro810 controller is basically a self-contained replacement for industrial relays. All but the Micro810 controller can support additional I/O through plug-in modules. The Micro830 controller can accommodate up to 4 expansion modules, further increasing the number of

Table 14.2 Allen-Bradley PLC Families

Processor Family	Maximum Discrete		Maximum Analog		Notes
	Inputs	Outputs	Inputs	Outputs	
Micro800	28+268[1]Exp	20+268[1]Exp	4[2]+70[3]Exp	0+70[3]Exp	Max. 8 exp. modules
MicroLogix 1400	20+Exp	12+Exp	4+Exp	0+Exp	Max. 7 exp. modules
CompactLogix					
5370 L1	16+Exp	16+Exp	0+Exp	0+Exp	Max. 8 exp. modules
5370 L2	16+Exp	16+Exp	4+Exp	2+Exp	Max. 4 exp. modules
5370 L3	960[1]	960[1]	240[3]	240[3]	Max. 30 local modules
5380	496[1]	496[1]	248[3]	248[3]	Max. 31 local modules
5480	496[1]	496[1]	248[3]	248[3]	Max. 31 local modules
ControlLogix	128K[4]	128K[4]	4000[5]	4000[5]	Specialized I/O
SmartGuard 600	16+Exp	8+Exp	0	0	Max. 32 I/O modules
Compact					
GuardLogix 5370	960[1]	960[1]	240[3]	240[3]	Max. 30 local modules
GuardLogix 5380	496[1]	496[1]	248[3]	248[3]	Max. 31 local modules
GuardLogix	128K[4]	128K[4]	4000[5]	4000[5]	Specialized I/O

Notes:
1. Maximum number total discrete inputs plus outputs, assuming no analog I/O.
2. Certain Micro810/820 only. Each analog input uses one discrete input channel.
3. Maximum number total analog inputs plus outputs, assuming no discrete I/O.
4. Maximum number total discrete inputs plus outputs.
5. Maximum number total analog inputs plus outputs.

possible I/O channels. The Micro850 and Micro870 can accommodate up to 4 and 8 expansion modules, respectively.

Until recently, the MicroLogix processors were the default controllers for small applications. However, as of this writing, the MicroLogix 1400 is currently the only one still available and it will probably be discontinued shortly. This controller is programmed in ladder logic and mounts on a DIN rail. The MicroLogix 1400 PLCs can have expansion I/O modules and have a built-in Ethernet port.

The CompactLogix processors are intended for small or medium-sized systems. The processor and I/O modules mount on a DIN rail. There are currently five versions of CompactLogix processors: 5370 L1, 5370 L2, 5370 L3, 5380, and 5480. A CompactLogix 5370 L1 processor has some built-in input/output channels and uses 1734-series modules for any local expansion I/O. A CompactLogix 5370 L2 processor has some built-in input/output channels and uses 1769-series modules for any local expansion I/O. A CompactLogix 5370 L3 processor has no built-in input/output channels and uses 1769-series modules for any local expansion I/O. A CompactLogix 5380 processor has no built-in input/output channels and uses 5069-series modules for any local expansion I/O. A CompactLogix 5480 controller is like a CompactLogix 5380 with a Windows 10 Enterprise commercial operating system running in parallel with the logic functions. All of the CompactLogix processors have built-in Ethernet ports for remote I/O modules and those ports are used for remote I/O or remote modules (for example, variable-frequency drives).

The ControlLogix processors support the needs of very large systems. This architecture supports multiple processors in a chassis. The ControlLogix 5580 and ControlLogix 5570 processors are the current versions, though the 5580 processors are recommended for new applications. The older ControlLogix 5560 processors are discontinued.The ControlLogix hardware is a modular chassis-based system. The system can be expanded beyond the local I/O limits with Ethernet/IP, ControlNet, and DeviceNet scanners. Additionally, the ControlLogix system has Remote I/O scanners, providing a connection to older systems.

A redundant ControlLogix system can be configured by populating two ControlLogix chassis identically with one or two processors, up to seven enhanced Ethernet or ControlNet communication modules, and one System Redundancy module. The two System Redundancy modules are connected with a fiber-optic cable. All I/O modules must be in chassis remote from the two redundant controller chassis. For added redundancy, the Ethernet and ControlNet networks can also be redundant. If ControlNet is used, the network must have two nodes additional to the redundant controller chassis.

As an alternative, there is also hot backup capability for ControlLogix and certain CompactLogix processors (Rockwell Automation, 2017). Ladder logic in the primary and secondary controller switches control of the outputs which are all located in remote chassis. There are no hardware modules to perform this function. Both controllers are active and additional logic in both controllers inhibits the output connections in the peer (secondary) controller so that only one controller owns the output modules. This solution provides backup capabilities for CompactLogix systems which does not have the redundancy options available for a ControlLogix system. However, the transfer is not bumpless, as the outputs may temporarily revert to the default value during switchover and motion control is not supported.

The SmartGuard 600 processor is designed for application in systems up to SIL 3 according to IEC-61508 without restrictions. This processor is for micro-sized safety systems and can communicate with up to 32 distributed I/O safety modules over DeviceNet. One of the versions of the SmartGuard 600 offers Ethernet/IP connectivity. The SmartGuard 600 processors have redundant microprocessors in one controller and self-test of the I/O channels during operation. Errors in an I/O channel are automatically detected and if the error affects safe operation of the channel, the respective channel is forced to the **off** state. The processor is programmed according to IEC 61131-3 function block language.

The GuardLogix processors are special ControlLogix and CompactLogix processors that add safety functions within a regular ControlLogix or CompactLogix system. For the ControlLogix system, the primary controller part number has an "S" suffix and is used with a safety partner module to achieve SIL 3. Without the safety partner, the processor achieves SIL 2. The Compact GuardLogix 5370 and 5380 processors have SIL 2 and SIL 3 versions. The SIL 3 version has a redundant built-in processor. For the GuardLogix controllers, the safety-related I/O is connected via Ethernet/IP, DeviceNet, or ControlNet. For the CompactLogix GuardLogix controller, the safety-related I/O may only be connected via Ethernet/IP.

14.4 CHAPTER SUMMARY

This chapter reviews the various factors one should consider when selecting a PLC for a particular application. After summarizing general selection criteria, the Allen-Bradley PLC families are described.

REFERENCES

IEC, 2010a. *Functional safety of electrical/electronic/programmable electronic safety- related systems - Part 1: General requirements*, IEC 61508-1:2010, International Electrotechnical Commission, Geneva, Switzerland.

IEC, 2010b. *Functional safety of electrical/electronic/programmable electronic safety- related systems - Part 2: Requirements for electrical/electronic/programmable electronic safety-related systems*, IEC 61508-2:2010, International Electrotechnical Commission, Geneva, Switzerland.

IEC, 2010c. *Functional safety of electrical/electronic/programmable electronic safety- related systems - Part 3: Software requirements*, IEC 61508-3:2010, International Electrotechnical Commission, Geneva, Switzerland.

IEC, 2010d. *Functional safety of electrical/electronic/programmable electronic safety- related systems - Part 4: Definitions and abbreviations*, IEC 61508-4:2010, International Electrotechnical Commission, Geneva, Switzerland.

IEC, 2010e. *Functional safety of electrical/electronic/programmable electronic safety- related systems - Part 5: Examples of methods for the determination of safety integrity levels*, IEC 61508-5:2010, International Electrotechnical Commission, Geneva, Switzerland.

IEC, 2010f. *Functional safety of electrical/electronic/programmable electronic safety-related systems - Part 6: Guidelines on the application of IEC 61508-2 and IEC 61508-3*, IEC 61508-6:2010, International Electrotechnical Commission, Geneva, Switzerland.

IEC, 2010g. *Functional safety of electrical/electronic/programmable electronic safety- related systems - Part 7: Overview of techniques and measures*, IEC 61508-7:2010, International Electrotechnical Commission, Geneva, Switzerland.

ISA, 2004. *ANSI/ISA-84.00.01-2004 Part 1 (IEC 61511-1 Mod), Functional Safety: Safety Instrumented Systems for the Process Industry Sector - Part 1: Framework, Definitions, System, Hardware and Software Requirements*, International Society of Automation, Research Triangle Park, NC.

Rockwell Automation, 2017. *Hot Backup Version 4.0 – Application Technique*, pub. HIGHAV-AT001C-EN-P, Rockwell Automation, Inc., Milwaukee, WI, Aug.

Rockwell Automation, 2022. *Rockwell Automation Product Catalog*, Rockwell Automation, Milwaukee, WI, https://configurator.rockwellautomation.com/#/browse.

Summers, Angela E., 1998. "Techniques for Assigning a Target Safety Integrity Level," *ISA Transactions*, 37: 95-104.

15 Control Projects

Chapter Topics:

- Phases of a typical control project
- Example control requirements
- Testing

OBJECTIVES

Upon completion of this chapter, you will be able to understand:

- What happens in a typical factory automation project
- The importance of a control requirements definition
- How to test a program with a process simulation

Scenario: A successful project begins with a thorough understanding of control improvements and with the ability to use the right frame of reference.

On one project, you are confronted with a typical late design situation. The process is a waste filtration system. The control design specifies the need for 400 highly interactive interlocks. By experience, you know that this design will be very expensive to implement and commission. You need to reduce the costs of the design.

Solution: Initially, you have a hard time understanding why there are so many interlocks. After consulting with the customer you discover the problem. The original designer's frame of reference was in terms of PID loops and interlocks. For all but a handful of interlocks, the control objective is to stop the inflow if the effluent flow is stopped. By taking a different view, most of the interlocks were replaced with a control scheme that stops the appropriate inflows when the outflows stop. You are able to significantly reduce costs by focusing on the requirements rather than control details developed from the wrong frame of reference.

15.1 INTRODUCTION

A control system is designed and implemented as part of a capital project. During the lifecycle of a plant there are numerous control projects. To properly understand the role of control engineering on a capital project, a plant lifecycle view should be considered (Erickson and Hedrick, 1999). Though capturing some aspects of a lifecycle view, this chapter focuses on the control project.

Though all phases of a typical control design project are discussed in this chapter, certain aspects of a successful project are elaborated. In order to shorten the installation and commissioning phases of a project, the program must be tested as early as possible. The use of process simulation is essential to this goal since it allows one to test the program without the process and I/O hardware. The control requirements developed during the preliminary engineering phase are important and so an example control requirements definition is presented as the last major section.

15.2 TYPICAL CONTROL DESIGN PROJECT

Project teams execute control projects to build and improve manufacturing facilities. Control projects range from small equipment modification projects to large, greenfield projects; from process de-bottlenecks to new product lines.

One very important aspect of a control project is the *extent of automation*, depicted in Figure 15.1. As depicted in this figure, a plant is composed of three major areas as it relates to a capital project:

- Control and information technology
- Plant personnel
- Process technology

The "extent of automation" line is the boundary between the plant personnel and the other two major areas. All control projects must consider personnel as part of the system. The extent of automation is defined in the control requirements definition, developed in the preliminary design.

The process technology area contains the equipment that processes the materials to make product. In this area, automation provides the physical leverage to plant personnel. The extent of automation, therefore, includes mechanical and pneumatic moving equipment, such as pumps, valves, and conveyors.

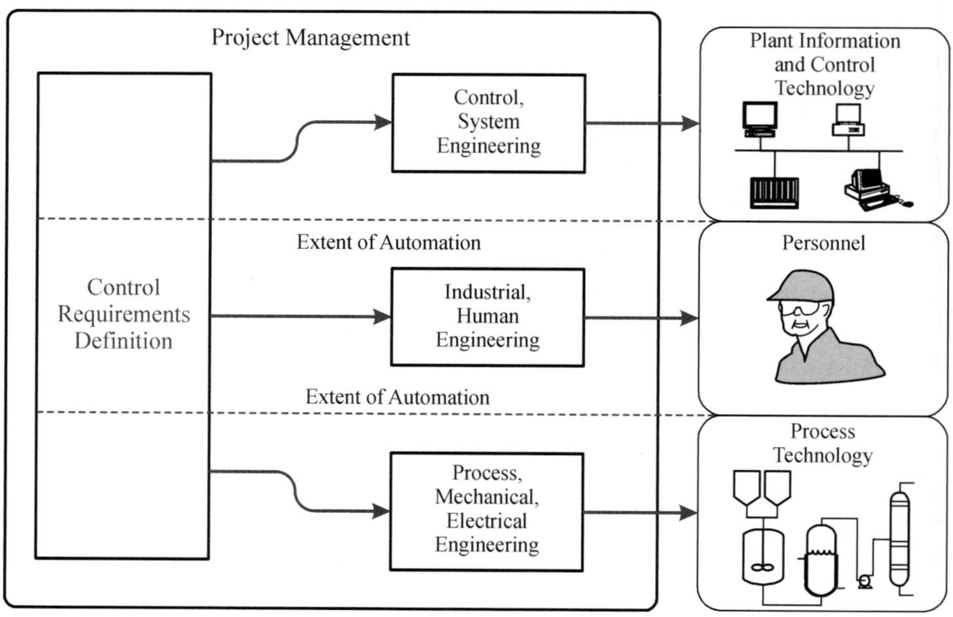

Figure 15.1. Extent of automation on control projects.

The control and information technology area contains the PLCs and computers that implement the control strategy. The extent of automation here is the amount of operating activities that are performed by the PLCs and computers. The automation may be fully-automatic, replacing manual activities, or it may be semi-automatic, augmenting the manual activities.

The process technology and the control and information technology support the plant operating personnel. The technology also provides the mental and physical leverage to operate the plant effectively. The user interface lies at the boundary between the personnel and the other two major areas. Thus, the user interface is a critical consideration in capital project execution.

In the context of this text, control projects contain three major aspects:

1. PLCs – control strategy implementation,
2. HMI – interface between the operator and the automation,
3. Electrical and Instrumentation (E & I) – direct interface to the process equipment.

Regardless of the size, the eight major phases of every control project are (Figure 15.2):

Planning
Preliminary Design
Detailed Design
Implementation
Installation
Commissioning
Startup
Training

Though all phases are described in this section, the most emphasis is placed on the preliminary design phase.

15.2.1 Planning

Regardless of the size, every capital project begins with some planning. There are two major products of the planning phase:

- First draft of project requirements
- Estimate of project resources

The overall flow on a capital project begins with the definition of the project requirements. Omission of this step often leads to project failure. This step also requires a multidisciplinary team approach. The team should include plant operation experts and engineering experts familiar with the application and with the plant's current standards (equipment, documentation, etc.).

The project requirements form the basis of the first estimate of the needed project resources. This first estimate is often very rough. A more accurate estimate is often produced by the next step, preliminary engineering.

Once the requirements are clearly stated, they can be distributed to the appropriate engineering disciplines for the design and implementation phases. Control engineers are involved in meeting the requirements for the control and information technology. Industrial and human engineering experts handle the plant organization requirements. Of course, the

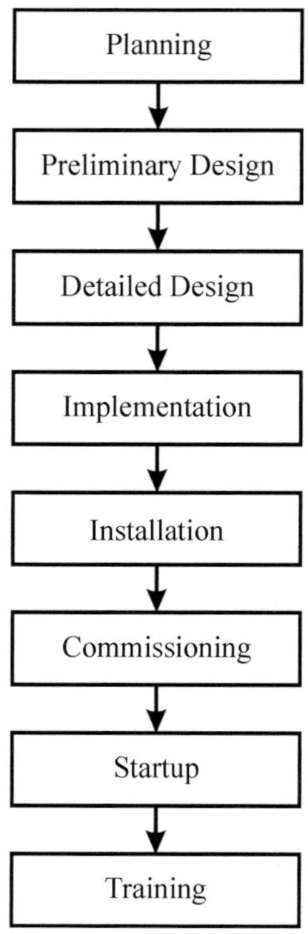

Figure 15.2. Control project steps.

process technology is the responsibility of the process engineering experts. All of these disciplines are orchestrated by the project management.

15.2.2 Preliminary Design

During preliminary engineering, the primary objective is to first understand how to operate the process and then define the control requirements to support the operations.

There are two groups involved in the preliminary design. The process engineers are concerned with the process, that is, what is to be controlled. This group is responsible for the process technology. The control engineers are involved with the control requirements and thus are responsible for the control technology.

As depicted in Figure 15.3, the process engineer's preliminary design activities are summarized as follows:

- Prepare the conceptual process design.
- Prepare the initial process technology budget.

The conceptual process design produces the design specifications for the physical plant equipment. For a discrete-parts manufacturing process, a conceptual process design includes production rates and material flow patterns. For a chemical process, it includes data about any chemical unit operations and material/energy flow properties. The latter are specified at various operating points in the process for the various steady-state operating conditions.

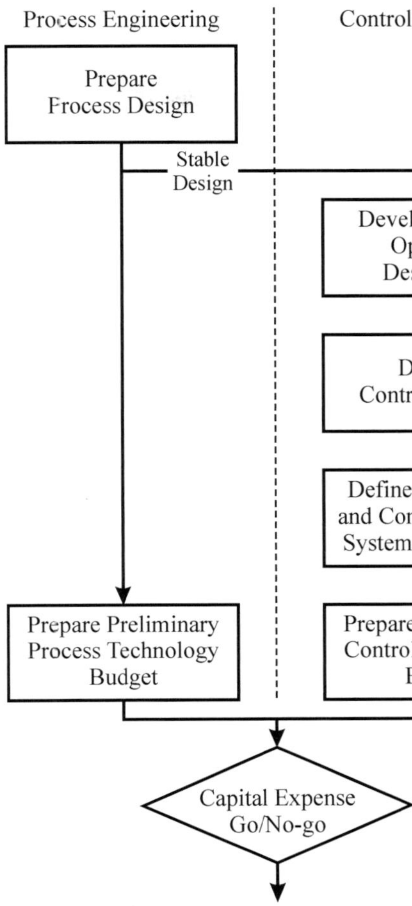

Figure 15.3. Preliminary engineering tasks.

The process design is usually represented by a process flow diagram (PFD) with an associated equipment item list for the major plant systems. The process technology capital budget is determined from this information.

For the control aspects of a process design, the preliminary engineering activities for the project (Figure 15.3), in order, are as follows:

- Develop the process operation description (POD).
- Develop the control concept.
- Define the preliminary automation and control strategy.
- Define the preliminary system architecture.
- Prepare the preliminary control technology budget.

Collectively, the POD, the control concept, and the control strategy are called the control requirements definition (CRD). An example CRD for a coal-handling system is presented in section 15.4.

Preliminary engineering design is an iterative process. Several iterations are required at various review levels before the requirements for the control system emerge. This process ensures that the control design is consistent with the operating philosophy of the plant. The control requirements must be well understood before the control strategy and system architecture is attempted.

With the information from the preliminary engineering tasks, the preliminary project budget can be refined. With this information, the last task in the preliminary design stage is usually a go/no-go decision about the project. Up to this point, relatively little money has been spent on the capital project. The bulk of the project budget is spent on the detailed design and installation phases.

The control technology for any plantwide control strategies should be part of the conceptual process design package (Erickson and Hedrick, 1999). Traditionally, plantwide control is considered late in the detailed design phase. This late entry usually means the budget for plantwide control is too low to realize any benefits. The design is already "cast in CAD" and it is often too late to raise the level of control to accommodate plantwide schemes. Taking plantwide control into consideration in the early design phase forces the team to consider "design for operability," which can significantly reduce the lifecycle costs of a process plant.

Within the preliminary design phase, the control engineer should first develop a process operation description (POD). The POD defines the operating strategy for the proposed plant design and therefore operating personnel are key participants in its development. The POD includes a description of the operating modules and defines the boundaries of the process areas, cells, and units to be operated (Erickson and Hedrick, 1999). It also defines the operating states and the activities to be performed for each. If the activities contain steps, the steps are listed, and their sequence is indicated. The control objectives are identified as part of the operating state description. The POD should describe all operating activities whether or not they will be automated.

In preparing a POD, the typical questions to be addressed include (Erickson and Hedrick, 1999):

- What are the basic steps or divisions within the process operation? How is each operated?
- What are the significant process conditions and/or events that impact quality assurance, production, or protection of equipment, people, and the environment?
- What are the routine, special, and emergency response operations in each operating state?
- What types of process information must be collected and saved for performance monitoring (reports, data archiving, etc.)?
- What data must be supplied to other systems, including business data systems?

After developing the POD, the control engineers need to determine the amount of control to be implemented, captured as the control concept. The control concept contains:

- Requirements to be met by the control and information technology.
- Extent of automation required to operate the plant and achieve enterprise objectives.

The control concept forms the basis for the control system design. It defines the extent of automation in terms of control requirements for the areas, cells, and units. The control concept is also used to verify the delivered system during system testing and commissioning. It should be prepared without a particular implied implementation.

In preparing the control concept the typical questions to be considered include (Erickson and Hedrick, 1999):

- What operating module activities and control objectives will be manual and what should be automated (i.e., the extent of automation)?
- How flexible must the automation be (e.g., multiple products, frequent equipment changes)?
- How do control activities interact (e.g., concurrent operation, shared resources)?
- How do control activities and operating personnel interact (e.g., data presentation, equipment selections, control commands, data entry)?

When developing the control concept, one should not spend much time considering details such as the control logic of individual valves, blowers, pumps, and motors. While important details are often unearthed while developing the control concept, the details should be kept to a minimum. It is proper to note them and save them for later use. The overall design process is less efficient if too much detail is covered too soon (particularly if the process design is still unstable). Control details should be left for consideration during detailed design.

After the control requirements are understood and documented, the project team develops the preliminary control strategy and system architecture. The control strategy defines the control types and interrelations to fulfill the control requirements. The system architecture defines the hardware and software platform that will be required to execute the control strategies.

The control concept and control strategy are developed using the reference model described in Chapter 3 of Erickson and Hedrick (1999). Operating modules as defined by the POD help segment these documents in order to relate them to the process design.

A preliminary CRD package, containing a first pass at the POD and the control concept, should be produced as soon as practical. The level of detail is obviously limited to the information available at the time. Certain items might simply be noted as TBD (to be determined).

Likewise, the initial control strategy and system architecture based on the control concept is helpful in reviewing various control concept candidates early in the project. For both the control strategy and system architecture, the control algorithms and hardware/software requirements are very high-level or conceptual. This philosophy provides flexibility in the budgeting activity.

The CRD sections are valuable inputs in developing the preliminary P&ID in the detail design effort. Later, as the process design becomes stable and the detail design is finalized, the CRD package is updated to produce a final version. All TBD references, for instance, should be resolved. This final version becomes the final project control documentation, an important project deliverable.

The CRD documentation is only the beginning of the control design package. It is used as a design basis for preparing the other follow-on project deliverables. For example, in the detailed design phase, the control strategy must be further described and detailed prior to beginning implementation. Similarly, the control system schematics may also be influenced by items mentioned in the documents.

A CRD has many benefits, both direct and indirect. The primary benefit is the early involvement of those persons most familiar with the plant's operation in developing the control system requirements. This method forces the participants to contemplate and describe the exact process area operation. Many issues relating to basic operations will surface *prior* to the selection of the control system equipment.

There are also several secondary benefits from the process. These include:

- Identification of the extent of automation early in a project, which provides a better baseline for budget control
- Initial definition of alarm management requirements to avoid unnecessary alarms in inactive operating states
- Initial definition of operator interface span of vision and control

15.2.3 Detailed Design

The detailed design begins after project approval, either near or at the end of the preliminary engineering activity. The detailed design phase requires many times the effort spent in preliminary design. Some detailed design is usually considered when developing the control concept in the preliminary design phase. A new or unusual requirement may require a check on whether or not the requirement can be implemented.

Detailed design adds sufficient detail to the initial control strategy and system architecture to provide a firm basis for implementation. It is important that the detailed specifications fulfill the control requirements as defined by the control concept. Detailed design also includes defining the system test, education, and documentation requirements for the system.

The detailed design phase develops the following:

- Measurement and final control functional specifications
- Safety instrumented system functional specifications
- Discrete and regulatory control functional specifications
- Procedural control functional specifications
- Process information data models
- User interface functional specifications

These specifications include the control technology specifications for the system hardware, embedded software, and application software programming. These specifications should address the size, performance, reliability, and environmental conditions to be met by the technology in this application.

Many projects must also address the process and control technology considerations of package process units. These units are pre-engineered packages that supply both the process and control technology for a plant system. These systems should be required to meet the same level of detail design specifications as the project-engineered systems. How these systems interface with the rest of the control technology is important.

A detailed design is represented by several design information packages. These individual packages tend to be highly interrelated. The types of design documentation include:

- P&ID
- Loop diagrams

- Electrical schematics
- Panel layouts
- Sequence control description
- Human Machine Interface (HMI) descriptions
- Implementation standards

For chemical process control, the P&ID traditionally contained the functional specification of the plant controls. However, modern control strategies are too complex to adequately represent on the P&ID. The P&ID still represents the measurement and final control systems, but many projects use other representations for the higher level controls.

The measurement and final control systems are still largely distinct components that must be wired to the controllers. Loop diagrams and electrical schematics provide the interconnect details for implementation and installation.

If procedural control automation is required, a sequence description is prepared. This sequence description provides the instructions for programming the controllers to provide the required level of automation.

The user interface is defined by panel layout drawings if field operation is required. In current computer-based systems, a description of the HMI is prepared. The HMI description is a critical but often overlooked design statement. The user interface is the most important aspect of a control system design. The best control strategy will fail if the user interface is not utilized.

Overall, one should minimize "reinventing" the wheel. If there is a design, or part of a design, that has proven itself, do not reengineer it. More importantly, there should be a system of capturing reusable strategies that can be shared in an enterprise.

15.2.4 Implementation

The implementation phase involves the procurement of the measurement and final control hardware and the control system hardware with its embedded software. It also includes the configuration and programming of the programmable logic controllers and the graphical user interfaces according to the detail design. This activity is complete after the system is verified in system testing. As part of this system testing, the system may be "staged," or set up in a temporary facility, to more completely test the system before installation.

The implementation of the system architecture requires hardware considerations such as power, grounding, wiring, and protection. The degree of concern is a function of the actual site environment of a process plant.

Standard code to implement sequential and device control plays an important role in the implementation phase. Encapsulating standard control functions in a function block is one means to this goal. At the very least, standard ladder logic code segments should be saved in a library for reuse.

The final step in the implementation activity is to verify the correct operation of the system. It begins with module testing followed by preliminary system testing by the implementation team. More detail about this testing is presented in section 15.3.

15.2.5 Installation

Installation is the activity that situates and powers the system at the plant site. This activity includes the interconnection wiring and system interconnection check-out. Each measurement and final control device is checked. The PLC system and operator interfaces are made operational to help in this effort.

After measurement and final control system wring-out, the control system performance is checked to make sure communication over the network is acceptable. Once system operation is checked and the installation is verified, the system is ready for commissioning.

15.2.6 Commissioning

For this phase, each plant system is brought on-line and the system is checked to make sure it works as required to support the operation of the process technology. The initial control settings are established and the system is readied for the first production run. At this point, the system is ready to make product. For a chemical process, this phase is usually done in conjunction with verifying the process technology. The combination is commonly called water testing or first chemical testing.

15.2.7 First Production Startup and Turnover

With the previous step, the control engineering team should have demonstrated that the control requirements have been met and the system is ready to be used. As the first production begins, at least some of the engineering team should remain with the system. The operating personnel should be running the process while the engineers monitor the performance and operation. The transition from the engineering team to the operating team is critical to the successful completion of the project. After the operating personnel take ownership, the project can be declared finished. The only activity left is to make sure the design information package is updated to an as-built state and turned over to the production and maintenance personnel.

15.2.8 Training

During the latter stages of the implementation activity, training of the plant personnel that must use the system should start. Training activities proceed from this point to the end of the project and may extend beyond the end of the project as new personnel join the organization. Engineers, operators, supervisors, management, and maintenance personnel all need to be trained to use the new system. Of course, the type of training differs according to the position and duties of the personnel. If a staged system was developed for testing, it often plays an important role in training operators since an emergency situation can be simulated on the staged system without fear of actual equipment damage or personnel injury if the operator makes a mistake.

15.3 TESTING

Before the system is installed, the PLC program should be tested for proper operation and to verify the operator interfaces. This testing is performed without any connection to the actual process. In order to appropriately test the system operation, the actual process should be simulated. For most processes, the simulation does not need to be an accurate model of

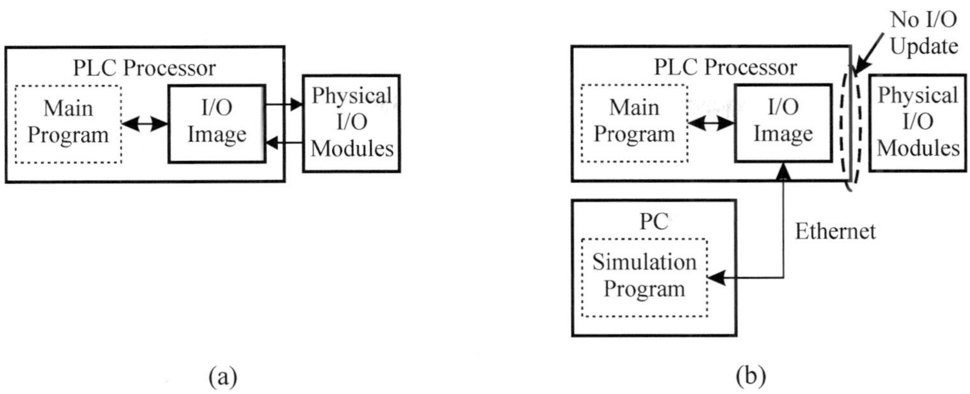

Figure 15.4. External process simulation: *(a)* normal I/O update; *(b)* PC-based simulation..

the process, just enough so that the PLC operation can be verified. If the unit is a complicated chemical process, then the expense of designing a reasonably faithful dynamic model may be justified. The process simulation can also be used for operator training; both before and after the actual system is commissioned.

There are two primary ways to simulate the process:

1. External process simulator, or
2. Simulation program in the PLC processor.

The separate process simulator has the advantage that it does not require any of the PLC program memory, but has the disadvantage that it requires extra hardware (a PC). Formerly, an additional interface module was required but with the adoption of Ethernet as a standard interface to the controller as well as for remote I/O communication, the need for this interface module has disappeared. For both of these simulation methods, the actual PLC processors are used in the testing.

For most of the PLCs covered by this text, one can purchase an *emulator* that runs the PLC program on a PC, thus not requiring a processor to verify the program operation. An emulator has the disadvantages:

- Program executes slower than the PLC processor, meaning that any critical timing issues cannot be accurately tested.
- Some processor communication interfaces may not be modeled, possibly restricting the operator interface and PLC-to-PLC communication tests.

Emulators are often useful for the initial program testing, but may not be acceptable for full-scale system testing.

The operation of an external process simulator is shown in Figure 15.4. The process simulator interacts with the I/O image directly and the actual I/O is not used. The simulator must be configured (or programmed) to set the emulated physical inputs to their proper values in response to the physical outputs commanded by the PLC processor. An external process simulator is most effective when needing to simulate a chemical process with moderate fidelity.

The use of the PLC to simulate the process is a powerful tool in the testing of a PLC program. It does not require separate hardware, though it uses some of the processor memory. Using the PLC to simulate a chemical process is less effective, as it is generally easier to build a simulation with an external process simulator. Depending on the particular processor, the process of constructing a simulation can be very easy, to moderately difficult. In all cases, the normal I/O update is disabled or circumvented, and a simulation program in the PLC "reads" the usual physical outputs and "writes" the usual physical inputs. This simulation logic can be called "tie-back logic." After explaining the general philosophy behind simulation logic and illustrating it with some examples, the specific methods of implementing simulation logic for each of the PLCs covered by this text are described.

15.3.1 General Simulation Concepts

The general concept behind the process simulation is that the PLC outputs are the "process" inputs and the PLC inputs are the "process" outputs. In the simulation logic the "inputs" are thus the output coils or analog output channel values. Only the physical PLC outputs should be used to drive the simulated physical PLC inputs. One should not use any internal tags/variables of the program being simulated. The simulation ***should not*** depend on a particular implementation. With this rule, the tested program could be completely rewritten, and the simulation could still be used to test it. To formulate the simulation logic for a particular physical input one should ask the question, "In the actual process, what physical output(s) causes this input to change?" The following examples illustrate the formulation of simulation logic.

Example 15.1. *Simulation of Metal Shear of Example 6.2.* The simulation logic does not include the start and stop switches. These are still assumed under operator control and can easily be forced even when the physical input modules are absent. The simulation logic needs to control the following inputs:

Tag/Var./Symbol	Description
PROX	Proximity sensor, **on** when strip in shearing position
DOWN_LS	Limit switch, N. O., **on** (closed) when blade fully down
UP_LS	Limit switch, N. O., **on** (closed) when blade fully up

The following outputs are available to control the inputs:

Tag/Var./Symbol	Description
CONV1_MTR	Conveyor 1 control, **on** to move material on conveyor 1
CONV2_MTR	Conveyor 2 control, **on** to move material on conveyor 2
SHEAR_CYL_RET	Shear cylinder control, **on** to retract cylinder and move blade down

Solution. The logic for the two limit switches is straightforward. They are both controlled by SHEAR_CYL_RET (or at least respond in accordance with this output) in the actual process. When SHEAR_CYL_RET is turned **on**, the UP_LS should turn **off** after a very short delay and DOWN_LS should turn **on** after a few seconds. When SHEAR_CYL_RET is turned **off**, the DOWN_LS should turn **off** after a very short delay and UP_LS should turn **on** after a few seconds. The ControlLogix logic for these two limit switches is shown in rungs 1 and 2 of Figure 15.5. For the purposes of simulation, the "very short delay" when turning **off** the limit switches is zero, through the use of ordinary output coils.

The logic for PROX requires a little more thought. Ultimately, PROX responds when one or both of the conveyors are run. PROX should turn **on** a few seconds after both

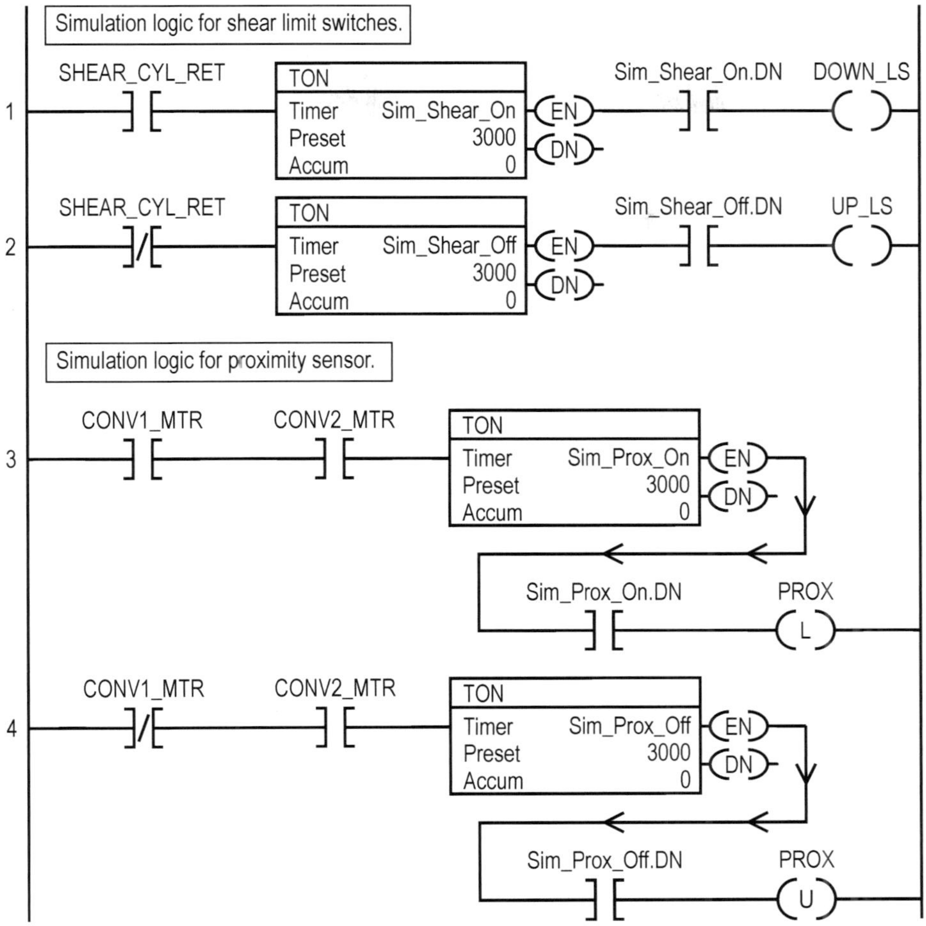

Figure 15.5. ControlLogix simulation logic for metal shear of Example 6.2.

conveyors start running. PROX should turn **off** a few seconds after only conveyor 2 is running. The ladder logic for PROX is shown in rungs 3 and 4 of Figure 15.5. Note the use of latch/unlatch for PROX.

Example 15.2. *Simulation of Tub Loader Control of Example 6.3.* The simulation logic needs to control the following inputs:

Tag/Var./Symbol	Description
PE272	Photoelectric sensor, **off** when part passes.
TUB_PROX	Proximity sensor, **on** (closed) when tub is present, though not necessarily in position to receive parts.
GATE1_OPLS	Limit switch, **on** (closed) when Gate 1 is open.
GATE1_CLLS	Limit switch, **on** (closed) when Gate 1 is closed.
GATE2_OPLS	Limit switch, **on** (closed) when Gate 2 is open.
GATE2_CLLS	Limit switch, **on** (closed) when Gate 2 is closed.

The following outputs are available to control the inputs:

Tag/Var./Symbol	Description
BELT_RUN	Belt conveyor control, **on** to run conveyor to move parts from milling machine to the parts tub.
TROLL_RUN	Powered roller conveyor control, **on** to run conveyor to move parts tub.
GATE1_RET	Gate 1 cylinder control, **on** to retract cylinder and open gate; **off** closes gate.
GATE2_RET	Gate 2 cylinder control, **on** to retract cylinder and open gate; **off** closes gate.

Solution. The logic for the gate limit switches is straightforward and are treated in a similar fashion as done in Example 15.1. Both limit switches for a gate 1 are controlled by GATE1_RET (or at least respond in accordance with this output) in the actual process. When GATE1_RET is turned **on**, the GATE1_CLLS should turn **off** immediately and GATE1_OPLS should turn **on** after a few seconds. When GATE1_RET is turned **off**, the GATE_OPLS should turn **off** immediately and GATE1_CLLS should turn **on** after a few seconds. The gate 2 limit switches are handled in a similar fashion. The ControlLogix logic for these four limit switches is shown in rungs 1 – 4 of Figure 15.6.

The logic for TUB_PROX requires a little more thought. Ultimately, TUB_PROX responds when the roller conveyor is running. TUB_PROX should be latched **on** a few seconds after the conveyor is running and Gate 2 is open, simulating a new tub coming in. TUB_PROX should be unlatched **off** a few seconds after the conveyor is running with Gate 1 open, simulating the tub moving out. The ladder logic for TUB_PROX is shown in rungs 5 and 6 of Figure 15.6.

PE272 is simulated by pulsing it off every 0.15 seconds when BELT_RUN is **on**. The ladder logic for PE272 is shown in rung 7 of Figure 15.6.

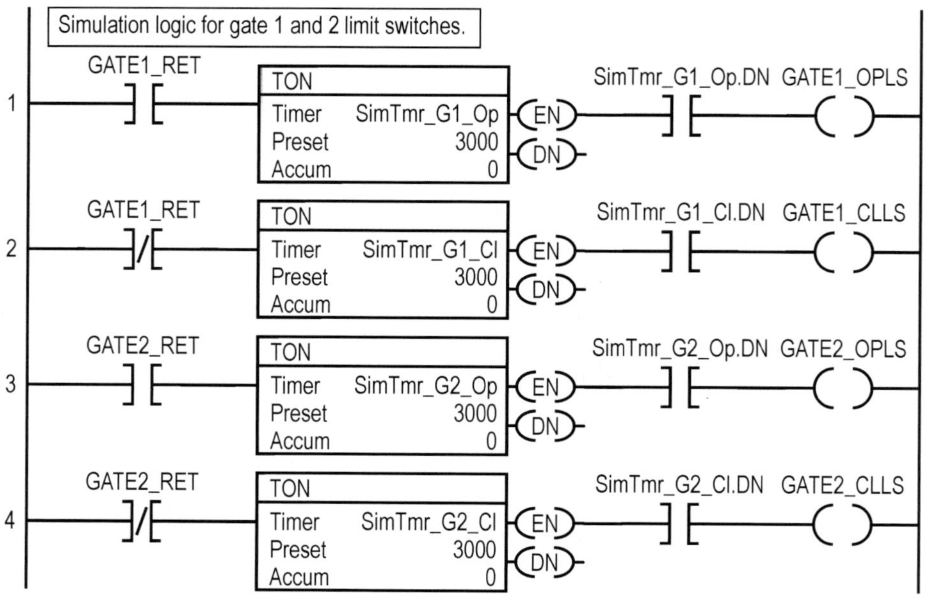

Figure 15.6. ControlLogix simulation logic for tub loader control of Example 6.3. *(continued)*

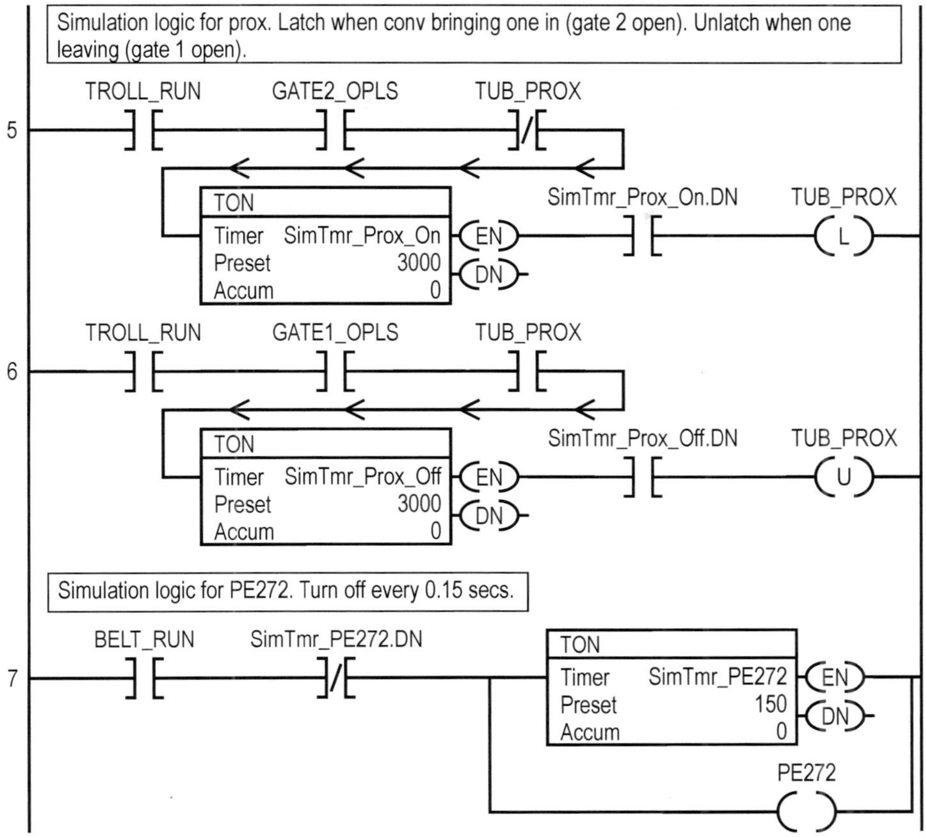

Figure 15.6. *(continued)*

15.3.2 ControlLogix Simulation Logic

To implement simulation logic in a ControlLogix PLC, there are two possible methods:

1. Inhibit the input modules and/or remote communication scanners, or
2. Remove the "alias for" for those tags that are an alias for a physical input.

For either method, a simulation program is added to the main task (Figure 15.7). The logic in the simulation program reads the physical output tags and writes to the physical input tags. The simulation program can remain as part of the controller, even when not simulating. Note that the output modules do not need to be inhibited, as the simulation does not alter the outputs, just the inputs.

Inhibiting the input modules is the easiest method. However, for CompactLogix 5370 processors, if the project has any expansion I/O modules, all of the expansion I/O modules and the embedded I/O must be inhibited, or the processor will fault. For a ControlLogix processor, the I/O modules are not needed. Only a chassis with a power supply and the processor in the correct slot is required. Therefore, the chassis must have enough slots to place the processor in the correct slot. If testing the HMI, then any required communication modules (for example, Ethernet or ControlNet) must be added. For the CompactLogix 5380 and 5480 processors, only the processor is required.

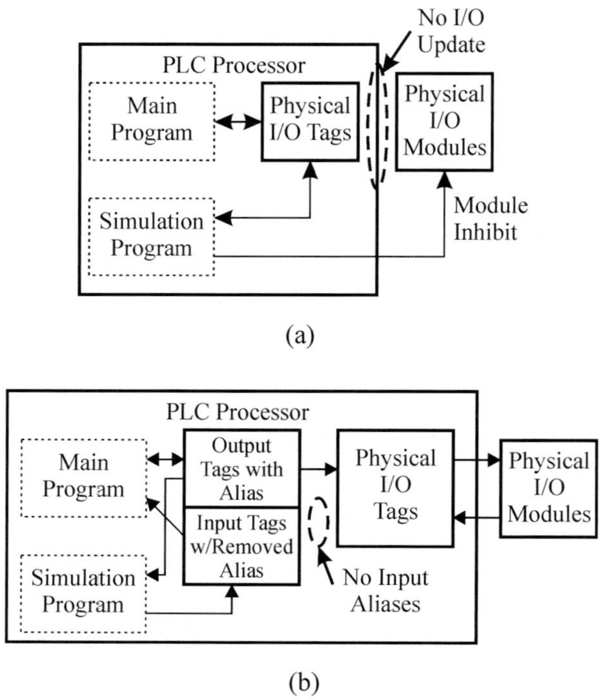

Figure 15.7. ControlLogix process simulation methods: *(a)* module inhibits; *(b)* removed input aliases.

The modules can be inhibited programmatically (Rockwell Automation, 2020c) as shown in Figure 15.8. This figure shows the logic for the main routine of the simulation program. If the Simulate_System controller tag is true, then bit 2 of the Mode attribute of the two discrete input modules (named Disc_In_1 and Disc_In_2) are set, inhibiting the modules. If Simulate_System is false, then bit 2 of the Mode attribute of the two modules are reset, uninhibiting the modules. The simulation logic is contained in the routine named "Simulation" and is executed only if Simulate_System is true.

If the modules are accessed through a scanner module (Ethernet, ControlNet, DeviceNet, etc.) in a ControlLogix system, then inhibiting the scanner module will inhibit all of the remote modules. If the modules are accessed through the embedded Ethernet port on a 5580 processor, the Ethernet port can be programmatically inhibited for simulation purposes (Rockwell Automation, 2022a). It is also possible to inhibit the Ethernet ports on CompactLogix 5370 and 5380 processors (Rockwell Automation, 2020b), but will likely disable any HMI communication, as well.

When redefining the aliases, the general idea is to export the tag database as a comma-separated-value (CSV) file, redefine the aliases for the tags that refer to physical inputs, and then import these changes into the project. To setup the simulation logic, the procedure is as follows:

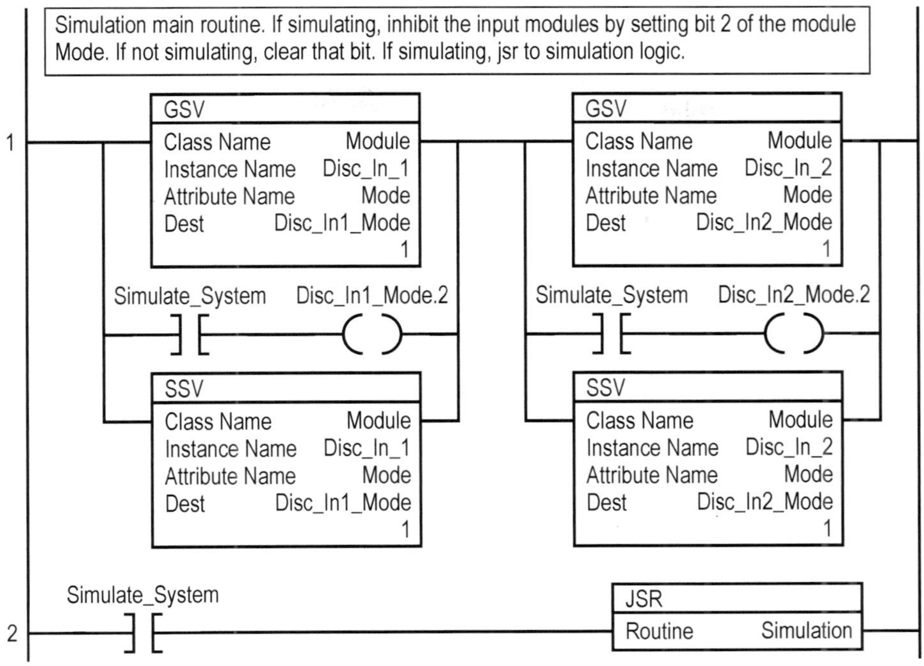

Figure 15.8. ControlLogix simulation main routine with programmatic input module inhibit.

1. Export the tags as an "RSLogix 5000 Import/Export File (*.CSV)" file. Only controller tags need to be exported.
2. With a text editor or spreadsheet program, delete all of the lines (rows) of the tags that are not the physical inputs. Keep the first seven lines of header information and save this file. This file contains the aliased physical input definitions.
3. For the file created in the previous step, define the aliases as regular tags. In the first column (spreadsheet column A), replace "ALIAS" with "TAG." Put the appropriate data type in the fifth column (column E). Delete the I/O address in the sixth column (column F). Save this new file with a different name. This file contains the simulated input definitions.
4. Import the simulated input tags as an "RSLogix 5000 Import/Export File (*.CSV)" file. You will receive a warning since the import writes over existing variable definitions.
5. Program the simulation logic in a separate program in the main (continuous) taks and make sure it is scheduled.

An example exported physical input tag alias definition and the corresponding simulated input tag definition for a CompactLogix 5370 L3 processor is shown in Figure 15.9. This example is for the engine inverter station of Example 6.4. The spreadsheet format of the files is shown. The relevant information appears in columns A through F and so column G is not shown.

When finished with the simulation logic testing, the original physical input tag aliases are restored and the simulation program is disabled:

	A	B	C	D	E	F
1	remark	CSV-Import-Export				
2	remark	Date =				
3	remark	Version = RSLogix 5000 v33.00				
4	remark	Owner = K. Erickson				
5	remark	Company =				
6	0.3					
7	TYPE	SCOPE	NAME	DESCRIPTION	DATATYPE	SPECIFIER
8	ALIAS		PALL_UPLS	Limit switch$NInd. pall off conyr		Local:1:I.Data.4
9	ALIAS		PROX1	Prox Sensor$NPall at Eng hook1		Local:1:I.Data.3
10	ALIAS		RESET_PB	Reset Pushbutton		Local:1:I.Data.2
11	ALIAS		ROTR_CCWLS	Rotator Cclockwise$Nlimit switch		Local:1:I.Data.8
12	ALIAS		ROTR_CWLS	Rotator Clockwise$Nlimit switch		Local:1:I.Data.7
13	ALIAS		ROTR_DNLS	Rotator Down$Nlimit switch		Local:1:I.Data.6
14	ALIAS		ROTR_UPLS	Rotator Up$Nlimit switch		Local:1:I.Data.5
15	ALIAS		START_PB	Start Pushbutton		Local:1:I.Data.0
16	ALIAS		STOP_PB	Stop Pushbutton		Local:1:I.Data.1

(a)

	A	B	C	D	E	F
1	remark	CSV-Import-Export				
2	remark	Date =				
3	remark	Version = RSLogix 5000 v33.00				
4	remark	Owner = K. Erickson				
5	remark	Company =				
6	0.3					
7	TYPE	SCOPE	NAME	DESCRIPTION	DATATYPE	SPECIFIER
8	TAG		PALL_UPLS	Limit switch$NInd. pall off conyr	BOOL	
9	TAG		PROX1	Prox Sensor$NPall at Eng hook1	BOOL	
10	TAG		RESET_PB	Reset Pushbutton	BOOL	
11	TAG		ROTR_CCWLS	Rotator Cclockwise$Nlimit switch	BOOL	
12	TAG		ROTR_CWLS	Rotator Clockwise$Nlimit switch	BOOL	
13	TAG		ROTR_DNLS	Rotator Down$Nlimit switch	BOOL	
14	TAG		ROTR_UPLS	Rotator Up$Nlimit switch	BOOL	
15	TAG		START_PB	Start Pushbutton	BOOL	
16	TAG		STOP_PB	Stop Pushbutton	BOOL	

(b)

Figure 15.9. CompactLogix 5370 input tag definitions for Example 6.4: *(a)* physical input aliased tags; *(b)* physical inputs with no alias.

1. Import the aliased physical input tags saved in step (2) of the simulation logic setup as an "RSLogix 5000 Import/Export File (*.CSV)" file. You will receive a warning since the import writes over existing tag definitions.
2. In the main (continuous) task, move the simulation program to the Unscheduled category.

In order to re-enable the simulation routine, import the simulated input tags and schedule the simulation program.

For RSLogix 5000 version 13.0 and later, the simulation program can be saved as a library file (L5X file) so that it can be removed and then imported later for more testing. For RSLogix versions prior to version 13.0, there is no easy way to save a simulation routine so that it can be removed and then imported later for more testing, though Erickson (2013) and Erickson (2005) detail a procedure.

15.3.3 Micro800 Simulation Logic

The Micro800 processor is one of the more difficult PLCs to implement simulation logic since the I/O scan cannot be disabled and the variables associated with the physical inputs cannot be redefined as an internal address (like ControlLogix/CompactLogix). To circumvent these difficulties, the input image is transferred to a duplicate input image that is used by the program in place of the actual input image (Figure 15.10). To test the program with the simulation logic, the transfer to the duplicate input image is disabled and the simulation function block writes to the duplicate input image.

Any plug-in and expansion I/O modules in the project must be present. If testing the HMI, then any required communication modules (for example, Ethernet) must be added.

In the Connected Components Workbench programming software, the variables aliased to a physical I/O channel are basically for program documentation. The program code uses the physical I/O channel addresses and not the variable alias. When a variable is changed to be the alias for a different I/O channel, any references to that variable are deleted from the program and replaced by the formerly aliased I/O address. For example, suppose the variable named "Prox_237" is aliased to the discrete input _IO_EM_DI_06. Any contact or coil uses of Prox_237 appear as "Prox_237(_IO_EM_DI_06)" in the ladder logic. If Prox_237 is changed to be aliased to _IO_EM_DI_07, all references to Prox_237 are replaced by "_IO_EM_DI_06".

The variables that should be aliased to the physical inputs are defined as global variables and these variables, essentially a duplicate of the physical inputs, are used in the

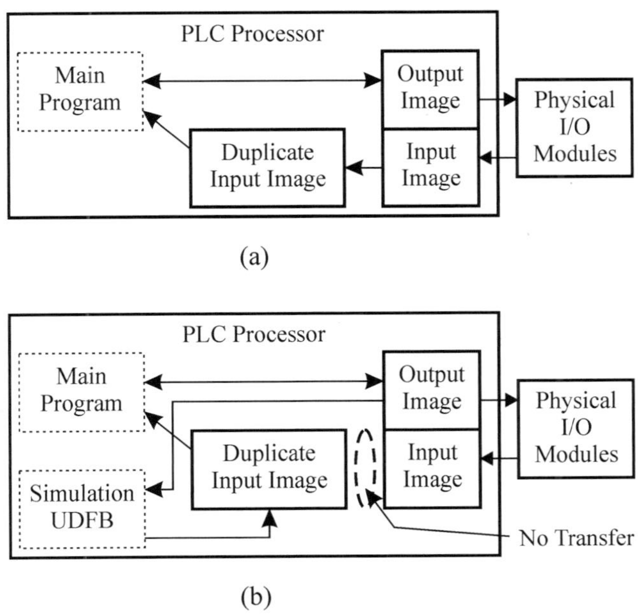

Figure 15.10. MicroLogix routine to duplicate input image for Example 6.4.

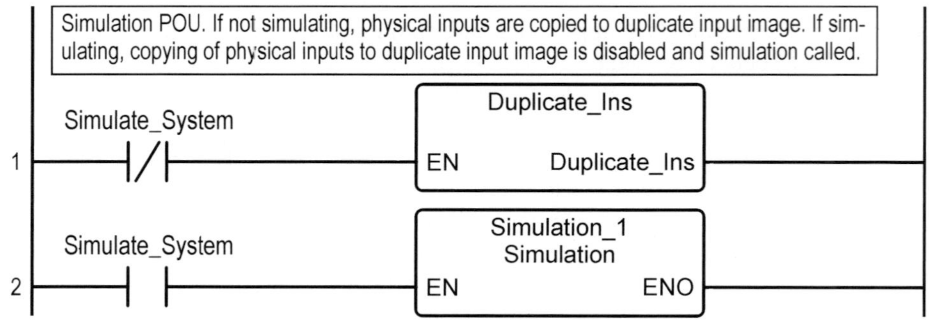

(a)

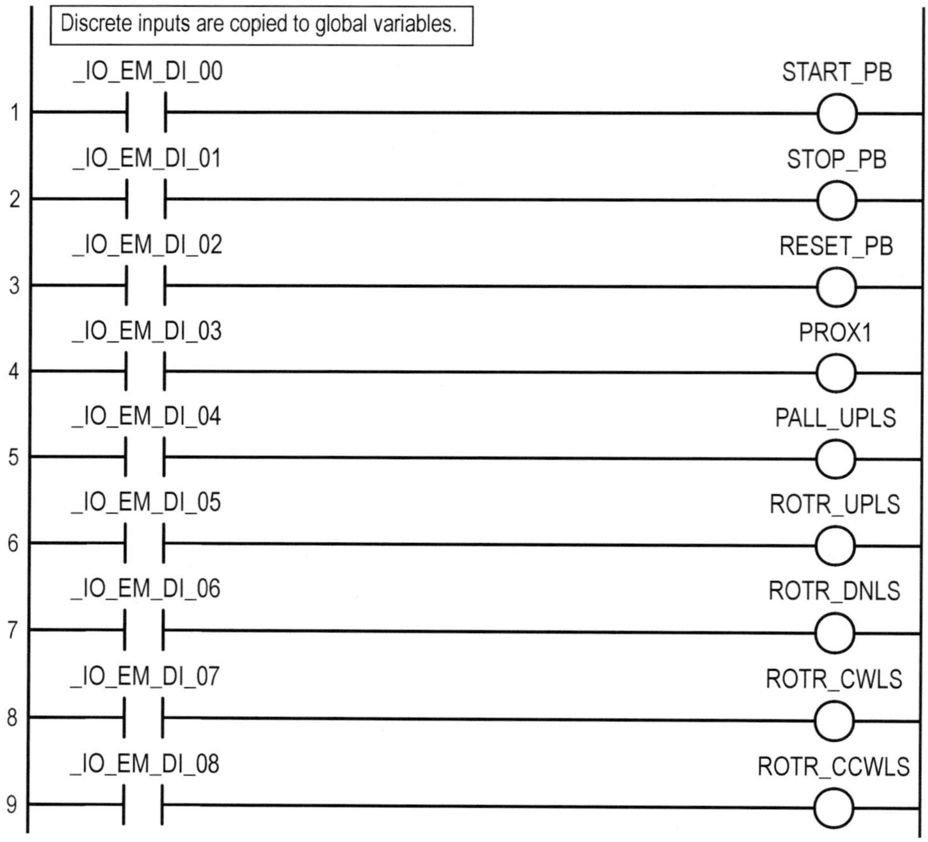

(b)

Figure 15.11. Micro800 simulation POU: *(a)* control of duplicate image and simulation; *(b)* Duplicate_Ins routine for Example 6.4.

ladder logic. A function block is programmed to copy the value of each physical input to the appropriate global variable. The simulation logic is programmed in a separate function block. The simulation function block and the function that copies the physical inputs to the duplicate physical inputs must be controlled so that only one of them writes to the duplicate physical inputs. In the example shown in Figure 15.11*a*, the Simulate_System internal variable controls this process. If Simulate_System is true, the physical inputs are not copied to the duplicate physical inputs and the simulation function block controls the duplicate physical inputs. If Simulate_System is false, the physical inputs are copied to the duplicate physical inputs (Duplicate_Ins function block) and the simulation function block is not called. The ladder logic for the Duplicate_Ins function block for Example 6.4 is shown in Figure 15.11*b*.

15.3.4 MicroLogix Simulation Logic

The MicroLogix processor is also one of the more difficult PLCs to implement simulation logic since the I/O scan cannot be disabled and the symbols associated with the physical inputs cannot be redefined as an internal address (like ControlLogix/CompactLogix). To circumvent these difficulties, the input image is transferred to a duplicate input image that is used by the program in place of the actual input image (Figure 15.12). To test the program with the simulation logic, the transfer to the duplicate input image is disabled and the simulation routine writes to the duplicate input image. Any expansion I/O modules are not needed. Only a power supply and the processor are required. If testing the HMI, then any required communication modules (for example, Ethernet) must be added.

In the RSLogix500 programming software, the symbols are for program documentation. The program code uses the addresses and not the symbols. When the address associated with a symbol is changed, the address in the program does not change. Before programming the processor, the physical input symbols must be associated with

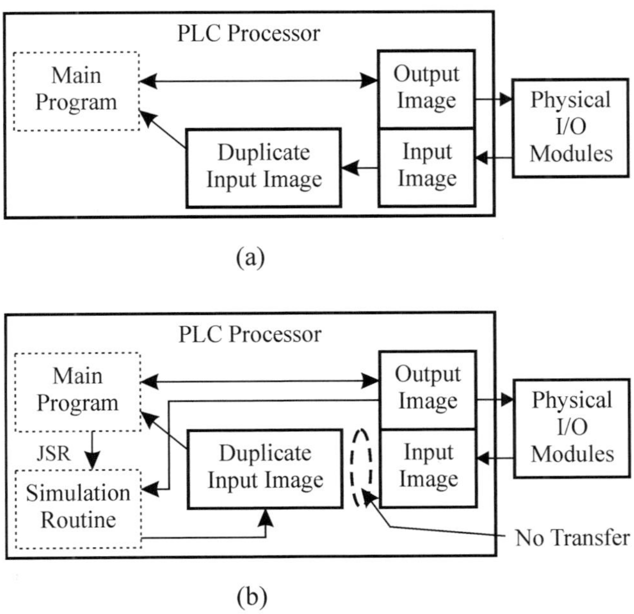

Figure 15.12. MicroLogix process simulation: *(a)* normal transfer to duplicate input image; *(b)* simulation logic.

internal addresses, the duplicate input image. As an example, the addresses associated with the physical inputs for the engine inverter station of Example 6.4 are redefined as

Symbol	Duplicate Addr.	Physical Addr.
START_PB	B99:0/0	I:0/0
STOP_PB	B99:0/1	I:0/1
RESET_PB	B99:0/2	I:0/2
PROX1	B99:0/3	I:0/3
PALL_UPLS	B99:0/4	I:0/4
ROTR_UPLS	B99:0/5	I:0/5
ROTR_DNLS	B99:0/6	I:0/6
ROTR_CWLS	B99:0/7	I:0/7
ROTR_CCWLS	B99:0/8	I:0/8

The function that copies the physical input image to the duplicate input image is shown in Figure 15.13*b*.

The simulation logic is programmed in a separate routine (for example, U:100, in Figure 15.13*a*). The simulation function and the function that copies the physical input image to the duplicate input image must be controlled so that only one of them writes to the duplicate input image. In the example shown in Figure 15.13, an internal symbol, SIMULATE_SYSTEM, controls the duplicate input image. If SIMULATE_SYSTEM is **on**, the physical input image is not copied to the duplicate input image and the simulation function controls the duplicate input image. If SIMULATE_SYSTEM is **off**, the physical input image is copied to the duplicate input image and the simulation function is not called.

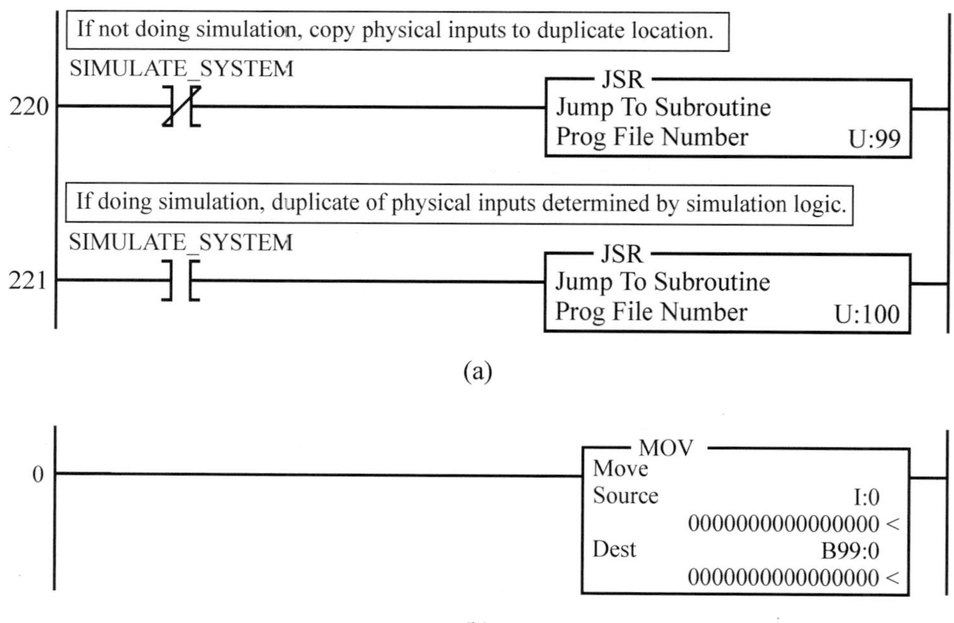

Figure 15.13. MicroLogix simulation: *(a)* control of duplicate image and simulation; *(b)* Duplicate_Ins subroutine for Example 6.4.

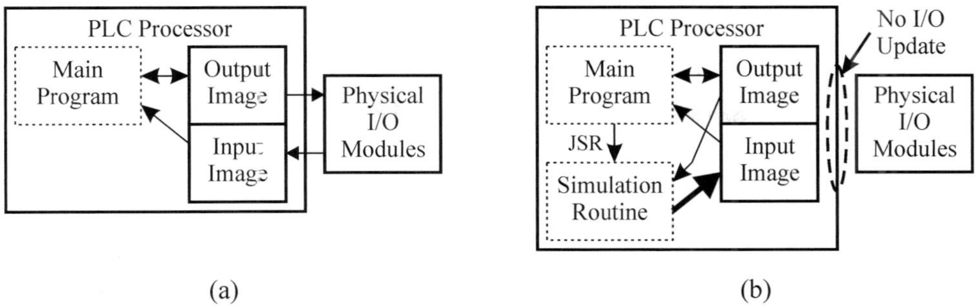

Figure 15.14. SLC-500 process simulation: *(a)* normal I/O update; *(b)* simulation logic.

The RSLogix500 software does not support library files, but the simulation ladder logic rungs can be "copied" into the paste buffer and then "pasted" into an empty project in order to save it for later use. After saving the simulation program file, it can be deleted from the original project.

15.3.5 SLC-500 Simulation Logic

Simulation logic is also easy to implement on the SLC-500. To implement simulation logic in a SLC-500, the I/O scan must be inhibited and a simulation logic routine added to the program (Figure 15.14). The logic in the simulation routine reads the discrete output image and writes to the discrete input image. The simulaton logic routine does not need to be removed from the processor.

For a SLC-500, the I/O modules are not needed, though the system must contain the processor, all chassis, and all power supplies for the application, but the I/O modules are not necessary. When testing the HMI, then any required communication modules (for example, ControlNet or Remote I/O) must be added.

The simulation logic is set up as follows:

1. Open the processor status and select the "IO" tab. Set all I/O slot enables (except for slot 0) to "0." An example is shown in Figure 15.15.
2. Program the simulation logic in a separate program file.
3. In the main program file, place a JSR block that calls the simulation program file.

When finished with the simulation logic testing, the I/O scan is re-enabled and the simulation program file JSR is disabled:

1. Place a contact that is always false before the JSR block that calls the simulation program file.
2. Open the processor status and select the "IO" tab. Set all I/O slot enables to "1."

The RSLogix500 software does not support library files, but the simulation ladder logic rungs can be "copied" into the paste buffer and then "pasted" into an empty project in order to save it for later use. After saving the simulation program file, it can be deleted from the original project.

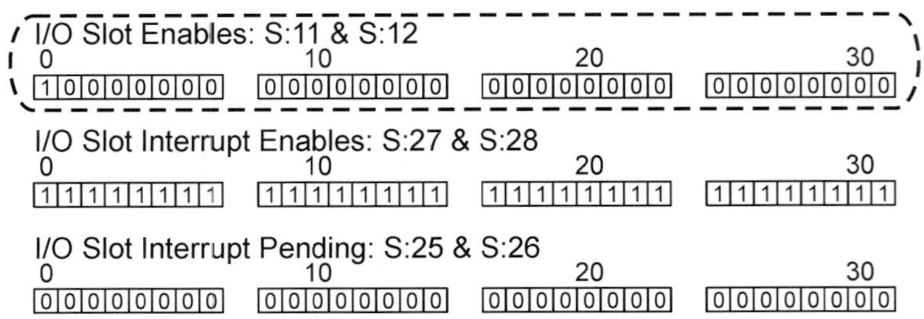

Figure 15.15. SLC-500 I/O slot enables.

15.4 COAL HANDLING SYSTEM EXAMPLE

The coal handling system manages the transfer of coal from either a truck or the reclaim coal pile to one of three bunkers. An overall schematic of the process is shown in Figure 15.16. This section presents the control requirements definition (CRD) that contains the process operation description, the control concept, and the control strategy for the coal handling system. The accompanying CD contains the detailed design documentation.

15.4.1 Process Operation Description

Purchased coal is unloaded into a truck dump hopper. The truck trailer tilts up, allowing the coal to fall out into a hopper, where it is conveyed to the reclaim coal pile or to one of the coal bunkers inside the power plant. When the truck dump hopper is empty, the coal must come from the reclaim hopper, which is loaded from the reclaim coal pile. The primary control objective of the coal handling unit is to convey coal from one of two sources (truck hopper or reclaim hopper) to one of four destinations (reclaim, bunker 1, bunker 2, or bunker 3). The operating states of this unit are shown in Figure 15.15. Normal operation is intermittent, running only when a truck needs to unload or a coal bunker is nearly empty. The states are described as follows:

Operating State Name: Idle

Routine Activities

Clean equipment as needed.

Perform inspection.

Exception Handling

None.

Primary Control Objectives

None.

Performance Information

None.

State End Conditions

Startup requested.

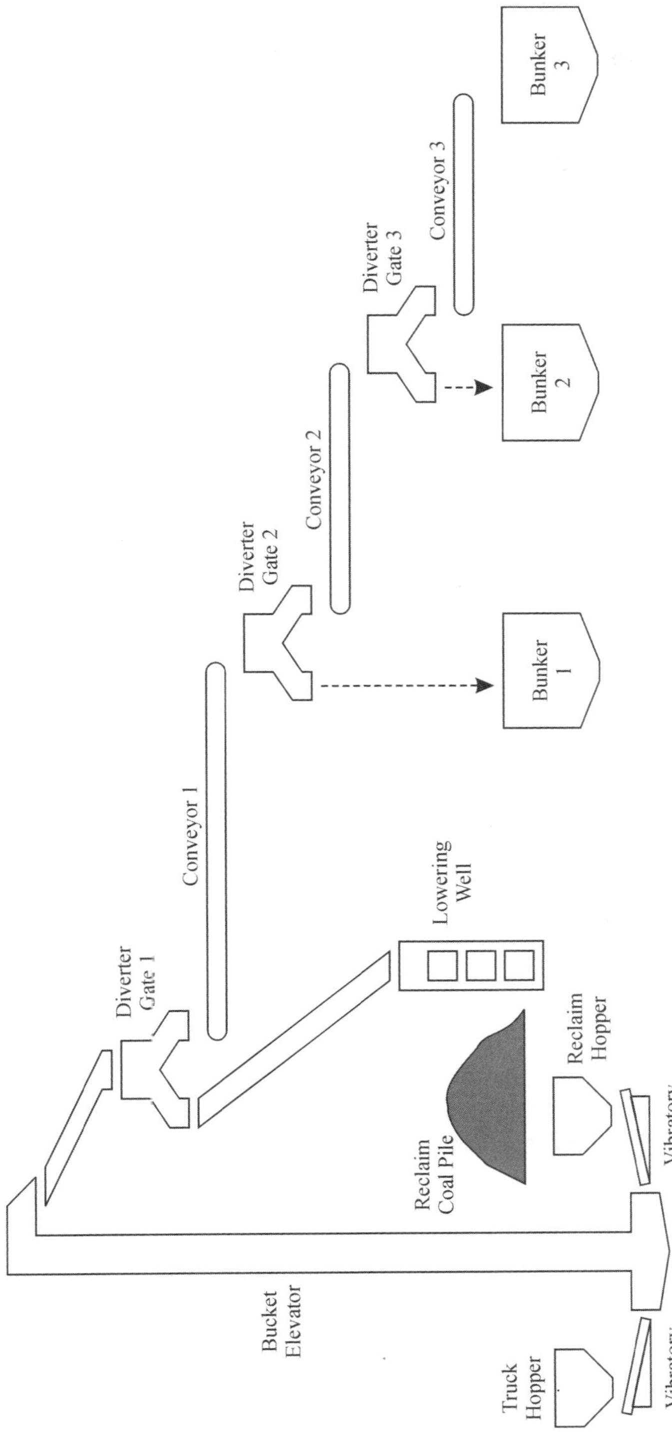

Figure 15.16. Overall process.

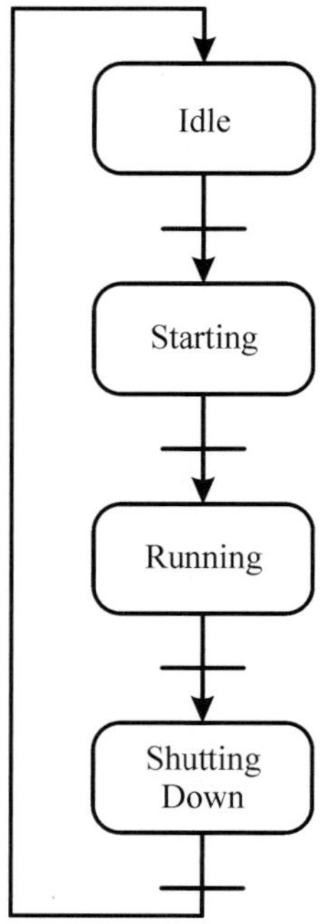

Figure 15.17. Coal handler states.

Operating State Name: Starting

Routine Activities

Set diverter gates for the proper source/destination.

Start conveyors for a source/destination.

Start bucket elevator

Start appropriate feeder

Exception Handling

Shutdown if:

Conveyor motor failure.
Conveyor belt mis-alignment.
Conveyor pull cord e-stop
Coal plug detected.

Primary Control Objectives

None.

Performance Information

None.

State End Conditions

Conveyors for a source/destination path started.

Operating State Name: Running

Routine Activities

Periodic checks of level in Bunker #1, #2, and Bunker #3 (if being filled).

Exception Handling

Shutdown if:

Conveyor motor failure.
Conveyor belt mis-alignment.
Conveyor pull cord e-stop
Coal plug detected.
High level on a destination bunker.
High reclaim coal pile level.

Primary Control Objectives

Level of coal in bunker.

Performance Information

None.

State End Conditions

Shutdown requested either by operator, or when bunker 3 is full.

Operating State Name: Shutting Down

Routine Activities

Shutdown all path conveyors.

Stop vibratory feeders

Exception Handling

None.

Primary Control Objectives

None.

Performance Information

None.

State End Conditions

All conveyors stopped.

15.4.2 Control Concept

Unit Supervision

Extent of Automation

Provide control of feeders, bucket elevator, conveyors, and diverter gates.

Provide automatic startup and shutdown for the transfer of coal to bunker 3.

Flexibility of Automation

None.

Control Activity Coordination

None.

Interaction with Operating Personnel

Operators can control feeders, bucket elevator, conveyors, and diverter gates.

Path selection done by operator. Interlocks prevent illegal selections.

Display alarm status information from devices.

15.4.3 Control Strategy

Two conveyor paths can be automatically controlled:

1. Unloading feeder to bunker 3
2. Reclaim feeder to bunker 3

All other paths are controlled by the operator.

When a path is started, the equipment is started in backward order. For example, the last conveyor in a path is started first. When a path is shutdown, the equipment is stopped so that all material on a conveyor is removed before that conveyor is stopped.

The operator uses switches on the control panel to manually control the conveyor and gate devices. Local control switches located near each gate and conveyor provide override control.

The operator can select "automatic mode" from the control panel. In automatic mode, the path to bunker 3 must be manually set. When the bunker 3 level is low, the conveyors, the bucket elevator, and the selected source feeder (reclaim or unload) are appropriately started. When the bunker 3 level is full, the conveyors are appropriately stopped.

Devices

Unloading feeder:

Start/stop by auto path control to bunker 3 or operator.

Reclaim feeder:

Start/stop by auto path control to bunker 3 or operator.

Bucket elevator:

Start/stop by auto path control to bunker 3 or operator.

Diverter gate 1:

Change position by operator.

Conveyor 1:

Start/stop by auto path control to bunker 3 or operator.

Diverter gate 2:

Change position by operator.

Conveyor 2:

Start/stop by auto path control to bunker 3 or operator.

Diverter gate 3:

Change position by operator.

Conveyor 3:

Start/stop by auto path control to bunker 3 or operator.

Interlocks

Do not allow start of conveyor if destination bunker level is high.

Do not allow start of bucket elevator if diverter gate set to "reclaim" and coal pile level is high.

Backward startup and forward shutdown is enforced even when manually controlled by the operator.

If a downstream conveyor is stopped because of a failure, the upstream conveyors are emergency-stopped.

15.4.4 Detailed Design

Documentation for the detailed design is in the accompanying CD. The following items are included:

Detailed control narrative which details the process operation and device control
Operation description of the OI panel that displays the alarms
Test plan
Electrical wiring diagrams
Panel layout
Operator graphical control panel design
PLC program
InTouch project screens used to test program before installation
OI panel project files

15.5 CHAPTER SUMMARY

This chapter described the typical phases in a control design project. The control requirements and testing aspects of a project are elaborated.

REFERENCES

Erickson, Kelvin T., 2005. *Programmable Logic Controllers: An Emphasis on Design and Application,* Dogwood Valley Press, Rolla, MO.

Erickson, Kelvin T., 2013. *Allen-Bradley Programmable Logic Controllers: An Emphasis on Design and Application,* Dogwood Valley Press, Rolla, MO.

Erickson, K. T. and Hedrick, J. L., 1999. *Plantwide Process Control*, John Wiley.

ISA, 2010. *ANSI/ISA-S88.01-2010, Batch Control, Part 1: Models and Terminology*, International Society of Automation, Research Triangle Park, NC.

Rockwell Automation, 2008. *SLC 500 Modular Hardware Style User Manual*, pub. 1747-UM011G-EN-P, Rockwell Automation, Inc., Milwaukee, WI, June.

Rockwell Automation, 2017. *ControlLogix Controllers User Manual*, pub. 1756-UM001P-EN-P, Rockwell Automation, Inc., Milwaukee, WI, May.

Rockwell Automation, 2020a. "Disabling the Embedded Ethernet Port on the 5580 Controllers", Knowledgebase Answer ID QA36352, Rockwell Automation, Inc., Milwaukee, WI, Mar.

Rockwell Automation, 2020b. "Disabling the Embedded Ethernet Port on the 5370 or 5380 Controllers", Knowledgebase Answer ID QA36360, Rockwell Automation, Inc., Milwaukee, WI, Mar.

Rockwell Automation, 2020c. "Programmatically Inhibiting a Module with RSLogix/Studio 5000", Knowledgebase Answer ID QA7686, Rockwell Automation, Inc., Milwaukee, WI, Feb.

Rockwell Automation, 2022. *Logix5000 Controllers I/O and Tag Data Programming Manual*, pub. 1756-PM004J-EN-P, Rockwell Automation, Inc., Milwaukee, WI, Apr.

PROBLEMS

General instructions for the problems:

Implement the solution and include the simulation logic for one of the following PLC processors:

ControlLogix/CompactLogix, **or**
Micro800, **or**
MicroLogix, **or**
SLC-500

Simulation logic specifications:

1. Do not simulate START_PB, STOP_PB, and RESET_PB in your simulation logic. Define these physical inputs as internal tags/variables/symbols so they can be manipulated without having to force them.
2. The simulation logic cannot use any internal logic (step internal coils, etc.) of the program being tested. Only use the defined physical I/O.
3. If the application contains a RESET_PB, or some other tag/variable/symbol used to reset the process, it may be used to set any initial states of the simulation logic (for example, set LS3 **on**).

When your solution is downloaded to a processor and the processor is placed in Run mode, the operator will modify the appropriate tags/variables/symbols to start the process. For most of the problems, the test is started with the following procedure:

1. The START_PB, STOP_PB, and RESET_PB are initially **off**.
2. The RESET_PB is turned **on** and then **off**.
3. The STOP_PB is turned **on**.
4. The START_PB is turned **on** and then **off**.

The application should start. If the process repeats its operation, it should continue until stopped by the operator (for example, by turning **off** the STOP_PB).

Your solution should include the following:

1. Function chart
2. Listing of program
3. Project files

Additional Specifications for a ControlLogix/CompactLogix Implementation:

1. Place simulation logic in programnamed “Simulation.”
2. Tag names for any timers used in the simulation logic need to start with “SimTmr_” and must be program tags (scope is “Simulation” program), not controller tags

Additional Specifications for a MicroLogix Implementation:

1. Place logic for transferring or not transferring to the duplicate input image in a POU named "Simulation_Prog."
2. Simulation logic must be in a User-Defined Function Block named "Simulation".
3. Names for any timers used in the simulation logic need to start with "SimTmr_" and must be local variables (scope is "Simulation" UDFB), not global variables.

Additional Specifications for a MicroLogix Implementation:

1. The routine to copy the physical input addresses to the duplicate input image is in program file #99, named "DUPLICATE_INS"
2. The duplicate input image is in data file B99.
3. Place simulation logic in program file #100, named "SIMULATE."
4. Any simulation timers should go in data file T100.
5. Any internal bits for the simulation should be in data file B101.

Additional Specifications for a SLC-500 Implementation:

1. Place simulation logic in program file #100, named "SIMULATE."
2. Any simulation timers should be in data file T100.
3. Any internal bits for the simulation should be in data file B101.

P15-1. Marker Ink Injector Control. Implement the highlighter ink injector control of problem P6-1 and include the simulation logic.

P15-2. Part Remover Machine Control. Implement the control for a machine that unloads a part from a weld die of problem P6-2 and include the simulation logic.

P15-3. Barcode Label Applier Control. Implement the control for the station that places a barcode label on a screwdriver of problem P6-3 and include the simulation logic.

P15-4. Six-pack Carton Erector Control. Implement the control for the machine that unfolds a six-pack carton of problem P6-4 and include the simulation logic.

P15-5. Stacked Items Pressing Machine Control. Implement the control for the machine that presses multiple stacked items into an assembly of problem P6-5 and include the simulation logic.

P15-6. Screw Turner Machine Control. Implement the screw turning machine control of problem P6-6 and include the simulation logic.

P15-7. Disc Polishing Machine Control. Implement the disc polishing machine control of problem P6-7 and include the simulation logic.

P15-8. Plastic Injection Molding Machine Control. Implement the injection molding machine control of problem P6-8 and include the simulation logic.

P15-9. BBQ Grill Handle Welder Control. Implement the grill handle welder control of problem P6-9 and include the simulation logic.

P15-10. Disk Install Machine Control. Implement the disk installation machine control of problem P6-10 and include the simulation logic.

P15-11. Soap Packaging Machine Control. Implement the soap packaging machine control of problem P6-11 and include the simulation logic.

P15-12. Cap Striping Control. Implement the control for a machine that puts a stripe on a cap of problem P6-12 and include the simulation logic.

P15-13. Thermoforming Machine Control. Implement the thermoforming machine control of problem P6-13 and include the simulation logic.

P15-14. Metal Part Former Machine Control. Implement the metal part forming machine control of problem P6-14 and include the simulation logic.

P15-15. Turbine Auxiliary Lube Oil Pump Control. Implement the turbine auxiliary lube oil pump control of problem P7-27 and include the simulation logic.

P15-16. Curing Oven Control. Implement the control for the ham-curing oven of problem P7-30 and include the simulation logic.

P15-17. Automobile Seat Tester. Implement the seat testing station control of problem P7-31 and include the simulation logic.

P15-18. Plastic Injection Molding Machine Control. Implement the injection molding machine control of problem P7-32 and include the simulation logic.

P15-19. Casting Heat Treating Station Control. Implement the control for the heat treating station of problem P7-34 and include the simulation logic.

P15-20. Capacitor Test Station Control. Implement the capacitor testing station control of problem P7-35 and include the simulation logic.

A Number Systems and Conversions

Since PLCs are basically specialized computers, they ultimately use binary numbers for their calculations. This appendix reviews the representation of numbers in binary, octal, hexadecimal, and binary-coded decimal (BCD) formats. In addition, this appendix covers the conversion of decimal numbers into binary, octal, hexadecimal, and BCD numbers and the conversion of binary, octal, and hexadecimal numbers into decimal numbers. Only whole numbers, that is, numbers without a fractional part are considered.

Binary, octal, decimal, and hexadecimal number systems are all positional number systems. Each digit position represents a weight that is a power of the number system *base*. The value of a digit is determined by multiplying the digit by the weight for that position. In a decimal number, the base is 10. The leftmost digit has a weight of 10^0; the next digit left has a weight of 10^1, the next digit left has a weight of 10^2, and so on. For example, the decimal number 2596 means $2\times10^3 + 5\times10^2 + 9\times10^1 + 6\times10^0$.

Binary, octal, and hexadecimal numbers have a base of 2, 8, and 16, respectively. BCD numbers have a base of 10, but the individual digits are represented by the 4-bit binary pattern for the decimal digit. A subscript shows the base of a number when it is not obvious. For example, 42_{10} means the number is decimal, while 42_8 is octal and 42_{16} is hexadecimal. The "H" or "h" suffix is an alternate way of showing a hexadecimal number, such as 42H or 42h. Since 8 and 16 are powers of 2, conversions between binary and octal or between binary and hexadecimal numbers are simple.

A.1 BINARY NUMBERS

In a binary number, the only digits allowed are 0 and 1. The base is 2 and the powers of 2 increase from 0 as shown in Figure A.1. One digit in a binary number is called a *bit*, 8 bits are called a *byte*, and 16 bits are called a *word* (Figure A.2). The *least significant* bit has the smallest weight and the *most significant* bit has the smallest weight.

For signed integers, negative numbers are represented in binary most commonly in the *two's complement* representation. To negate a binary number, each digit is complemented ($1 \rightarrow 0$ and $0 \rightarrow 1$) and then one is added to the complement. In the two's complement

	15		7	6	5	4	3	2	1	0	Digit position
• •	2^{15}	• •	2^7	2^6	2^5	2^4	2^3	2^2	2^1	2^0	Weight as power of 2
• •	32,768	• •	128	64	32	16	8	4	2	1	Weight as decimal number

Figure A.1. Weight of binary digits.

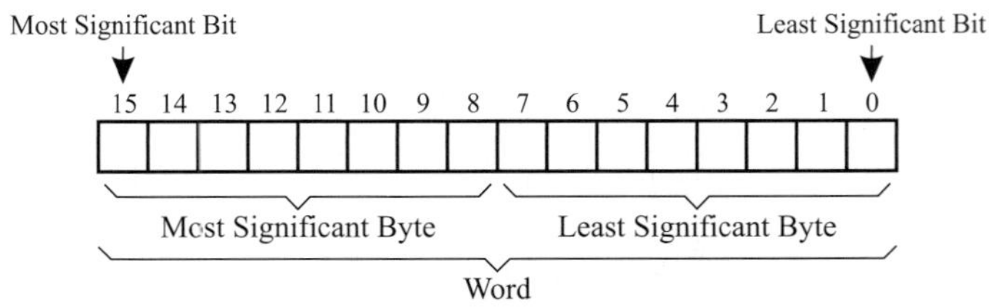

Figure A.2. Sixteen-bit binary number.

representation, a negative binary number is recognized by a 1 in the most significant (leftmost) position. For example, the 16-bit integer number 0000000001011001_2 is negated in the following manner:

0000000001011001_2	original number
↓	
1111111110100110_2	each digit complemented
↓	
1111111110100111_2	add one to the complement

Binary-to-Decimal Conversion Converting a positive binary number to a decimal number is easy. Each digit is multiplied by its weight and then all of the values are summed to obtain the result. This method is equivalent to adding the powers of two for which the digit is a one. For example, to convert the 8-bit number 01011001_2 to decimal,

$$\begin{aligned} 01011001_2 &= 0\times2^7 + 1\times2^6 + 0\times2^5 + 1\times2^4 + 1\times2^3 + 0\times2^2 + 0\times2^1 + 1\times2^0 \\ &= 0 + 64 + 0 + 16 + 8 + 0 + 0 + 1 \\ &= 89_{10} \end{aligned}$$

If the binary number is negative, one should negate it (converting it to a positive number) and then convert the positive number to its decimal equivalent. The negative sign is then prefixed to the result. For example, to convert the 8-bit signed integer number 11001010_2 to decimal,

	Since the leftmost digit is 1, this is a negative binary number. Therefore, it must first be negated
11001010_2	original (negative) number
↓	
00110101_2	each digit complemented
↓	
00110110_2	add one to the complement

$$\begin{aligned} 00110110_2 &= 0\times2^7 + 0\times2^6 + 1\times2^5 + 1\times2^4 + 0\times2^3 + 1\times2^2 + 1\times2^1 + 0\times2^0 \\ &= 0 + 0 + 32 + 16 + 0 + 4 + 2 + 0 \\ &= 54_{10} \end{aligned}$$

So, $11001010_2 = -54_{10}$

Decimal-to-Binary Conversion Converting a decimal number to a binary number is a little more complicated. For a positive decimal number, the procedure is as follows:

1. Repeatedly divide the decimal number by 2 and note the remainders. Continue until the result is zero.
2. The binary number is formed by writing the remainders in reverse order, starting with the last remainder as the most-significant digit.
3. Prefix the result from step 2 with enough 0 digits to produce the number of digits needed in the result (usually 8, 16, or 32 digits).

For example, to convert 121_{10} to a 16-bit binary integer,

$121 \div 2 = 60$ + remainder of 1
$60 \div 2 = 30$ + remainder of 0
$30 \div 2 = 15$ + remainder of 0
$15 \div 2 = 7$ + remainder of 1
$7 \div 2 = 3$ + remainder of 1
$3 \div 2 = 1$ + remainder of 1
$1 \div 2 = 0$ + remainder of 1

Write remainders in this order

So, $121_{10} = 1111001_2$. Prefixing it with zeros to produce a 16-bit integer, the result is

$$121_{10} = 0000000001111001_2$$

To convert a negative decimal number to a binary number,

1. Ignoring the sign of the decimal number, convert it to a binary number.
2. Prefix the result from step 1 with enough 0 digits to produce the number of digits needed in the result (usually 8, 16, or 32 digits).
3. Take the twos complement of the binary number determined in step 2.

For example, to convert -121_{10} to a 16-bit binary integer,

From the previous example, $121_{10} = 0000000001111001_2$

0000000001111001_2	121_{10}
↓	
1111111110000110_2	each digit complemented
↓	
1111111110000111_2	add one to the complement

So, $-121_{10} = 1111111110000111_2$

A.2 OCTAL NUMBERS

In an octal number, only the digits 0 - 7 are allowed. The base is 8 and the powers of 8 for the digit weights increase from 0 as shown in Figure A.3. The relationship between octal numbers and binary numbers is shown in Table A.1. An octal digit represents 3 binary

	5	4	3	2	1	0	Digit position
• •	8^5	8^4	8^3	8^2	8^1	8^0	Weight as power of 8
• •	32,768	4096	512	64	8	1	Weight as decimal number

Figure A.3. Weight of octal digits.

Table A.1. Octal-Binary Number Equivalents

Octal	Binary
0	000
1	001
2	010
3	011
4	100
5	101
6	110
7	111

digits. Negative octal numbers are represented in twos complement, but with the binary digits grouped by three digits to form each octal digit.

Octal-to-Decimal Conversion Converting a positive octal number to a decimal number is easy. Each digit is multiplied by its weight and then all of the values are summed to obtain the result. For example, to convert the 6-digit number 042736_8 to decimal,

$$\begin{aligned}042736_8 &= 0\times8^5 + 4\times8^4 + 2\times8^3 + 7\times8^2 + 3\times8^1 + 6\times8^0 \\ &= 0 + 16384 + 1024 + 448 + 24 + 6 \\ &= 17886_{10}\end{aligned}$$

Decimal-to-Octal Conversion Converting a decimal number to an octal number is similar to the method used to convert a decimal number to binary. For a positive decimal number, the procedure is as follows:

1. Repeatedly divide the decimal number by 8 and note the remainders. Continue until the result is zero.

2. The octal number is formed by writing the remainders in reverse order, starting with the last remainder as the most-significant digit.

For example, to convert 12107_{10} to octal,

$$\begin{aligned}12107 \div 8 &= 1513 + \text{remainder of } 3 \\ 1513 \div 8 &= 189 + \text{remainder of } 1 \\ 189 \div 8 &= 23 + \text{remainder of } 5 \\ 23 \div 8 &= 2 + \text{remainder of } 7 \\ 2 \div 8 &= 0 + \text{remainder of } 2\end{aligned}$$

Write remainders in this order (↑)

So, $12107_{10} = 27513_8$. Alternatively, one could convert the decimal number into a binary number and then starting from the right, write the octal equivalent of each group of 3 binary digits, for example,

$$
\begin{aligned}
12107_{10} &= 10111101001011_2 \\
&= 10\ 111\ 101\ 001\ 011_2 \\
&= 2\quad 7\quad 5\quad 1\quad 3_8
\end{aligned}
$$

A.3 HEXADECIMAL NUMBERS

In a hexadecimal number, the digits 0 - 9, A - F are allowed. The base is 16 and the digit weights are powers of 16 as shown in Figure A.4. The relationship between hexadecimal numbers and binary numbers is shown in Table A.2. A hexadecimal digit represents four binary digits. Negative hexadecimal numbers are represented in twos complement, but with the binary digits grouped by four digits to form each hexadecimal digit.

Hexadecimal-to-Decimal Conversion Converting a positive hexadecimal number to a decimal number is easy. Each digit is multiplied by its weight and then all of the values are summed to obtain the result. The digits A - F need to be converted to their decimal equivalent (Table A.2) before executing the multiplication. For example, to convert the 4-digit number $4A7E_{16}$ to decimal,

$$
\begin{aligned}
4A7E_{16} &= 4\times16^3 + A\times16^2 + 7\times16^1 + E\times16^0 \\
&= 4\times16^3 + 10\times16^2 + 7\times16^1 + 14\times16^0 \\
&= 16384 + 2560 + 112 + 14 \\
&= 19070_{10}
\end{aligned}
$$

Decimal-to-Hexadecimal Conversion Converting a decimal number to a hexadecimal number is similar to the method used to convert a decimal number to binary or octal. For a positive decimal number, the procedure is as follows:

	5	4	3	2	1	0	Digit position
• •	16^5	16^4	16^3	16^2	16^1	16^0	Weight as power of 16
• •	1,048,576	65536	4096	256	16	1	Weight as decimal number

Figure A.4. Weight of hexadecimal digits.

Table A.2. Hexadecimal-Binary Number Equivalents

Hexadecimal	Binary	Decimal	Hexadecimal	Binary	Decimal
0	0000	0	8	1000	8
1	0001	1	9	1001	9
2	0010	2	A	1010	10
3	0011	3	B	1011	11
4	0100	4	C	1100	12
5	0101	5	D	1101	13
6	0110	6	E	1110	14
7	0111	7	F	1111	15

1. Repeatedly divide the decimal number by 16 and note the remainders. Convert the remainders that are greater than 9 to their hexadecimal equivalent. Continue until the result is zero.
2. The hexadecimal number is formed by writing the remainders in reverse order, starting with the last remainder as the most-significant digit.

For example, to convert 12107_{10} to hexadecimal,

$$
\begin{aligned}
12107 \div 16 &= 756 + \text{remainder of } 11 \rightarrow \text{B}\\
756 \div 16 &= 47 + \text{remainder of } 4 \rightarrow 4\\
47 \div 16 &= 2 + \text{remainder of } 15 \rightarrow \text{F}\\
2 \div 16 &= 0 + \text{remainder of } 2 \rightarrow 2
\end{aligned}
$$

Write remainders in this order (↑)

So, 12107_{10} = 2F4BH. Alternatively, one could convert the decimal number into a binary number and then starting from the right, write the hexadecimal equivalent of each group of 4 binary digits, for example,

$$
\begin{aligned}
12107_{10} &= 10111101001011_2\\
&= 10\ 1111\ 0100\ 1011_2\\
&= 2\quad \text{F}\quad 4\quad \text{B}_{16}
\end{aligned}
$$

A.3 BINARY-CODED DECIMAL NUMBERS

Many earlier PLCs stored integer numbers internally in the BCD format. This format was used because early operator display panels used thumbwheel switches to enter numeric values and the pattern of discrete inputs from a thumbwheel switch is in the BCD format. Recent PLCs internally store integer values in the binary format. Nevertheless, thumbwheel switches are still used as an inexpensive method of operator-input of numeric values and these BCD values must be converted to a normal binary number before they can be used in calculations.

BCD numbers have a base of 10, but the individual digits are represented by the 4-bit binary pattern for the decimal digit. For example, the decimal number 7395 is represented as a BCD number:

7	3	9	5
↓	↓	↓	↓
0111	0011	1001	0101

The bit pattern is the same as 0111001110010101_2 which is equivalent to $29{,}589_{10}$. Note that the display of a BCD number is the same as a hexadecimal number (except that there are no A-F digits).

To convert a BCD number into binary, the BCD number is first converted to a decimal number and then the decimal number is converted to its equivalent binary representation. Most PLCs have a function block that handles this conversion.

B Electrical Diagram Symbols

Table B.1. Selector Switches

Example Two-position Selector Switches	
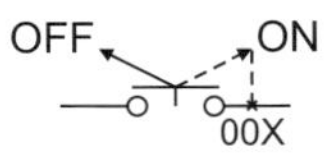	Left maintain - spring return from right Contact closed in right position, open in left
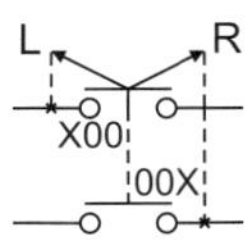	Left and right maintain Top contact closed in left position, open in right Bottom contact closed in right position, open in left
Example Three-position Selector Switches	
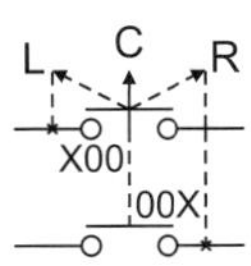	Center maintain - spring return from left and right Top contact closed in left position, open otherwise Bottom contact closed in right position, open otherwise
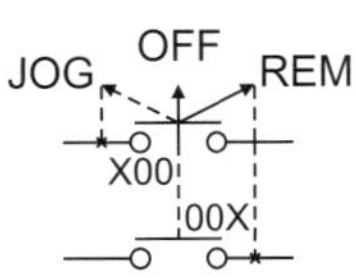	Center and right maintain - spring return from left Top contact closed in left position, open otherwise Bottom contact closed in right position, open otherwise
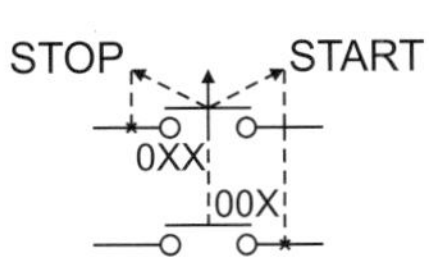	Center maintain - spring return from left and right Top contact open in left position, closed otherwise Bottom contact closed in right position, open otherwise

Table B.2. Discrete Input Devices

Input Type	Symbol	Description
Push Button	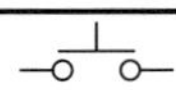	NO
		NC
		Mushroom head
	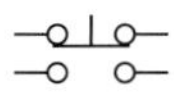	NC and NO
Limit Switch		NO
		NO - held closed
		NC
		NC - held open
Time Delay Contact	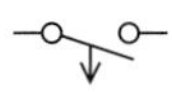	NO - delay opening
		NO - delay closing
	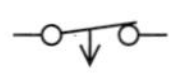	NC - delay closing
		NC - delay opening
Proximity Switch		Closes in position
	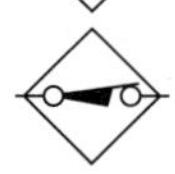	Opens in position
Speed Switch	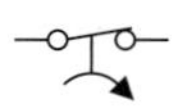	Opens on increase
	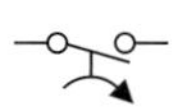	Closes on increase
Temperature Switch		Opens on rise
	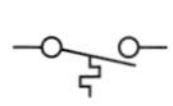	Closes on rise
Liquid Level Switch	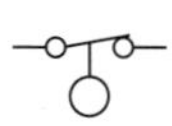	Opens on increase
	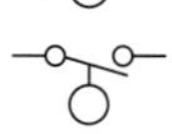	Closes on increase
Pressure Switch	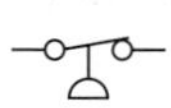	Opens on increase
	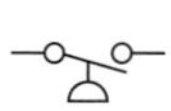	Closes on increase
Flow Switch	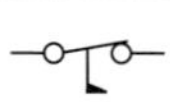	Opens on increase
	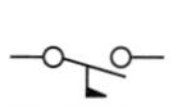	Closes on increase

Notes:
NC - Normally closed contact
NO - Normally open contact

Table B.3. Protective Devices

Symbol	Description
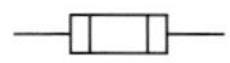	Fuse
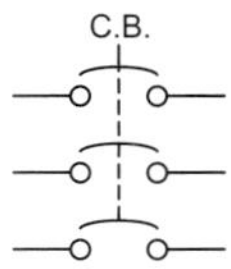	Circuit breaker
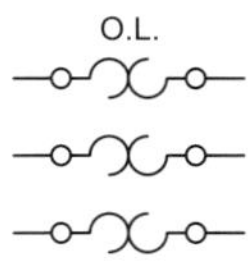	Motor overload
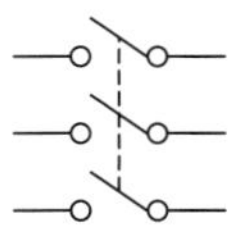	Disconnect

Table B.4. Discrete Output Devices

Symbol	Description	
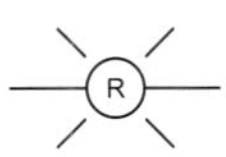	Pilot Light	Letter is lens color A - amber R - red B - blue W - white G - green
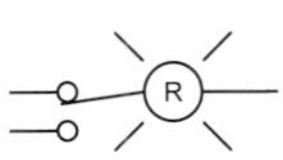	Pilot Light (Push-to-test)	
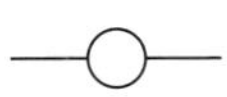	Relay Coil	
	Solenoid	
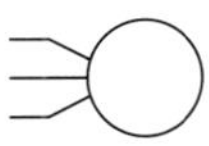	Three-phase motor	
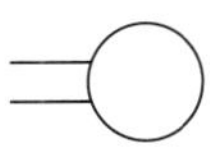	Single-phase motor	

C Piping and Instrumentation Diagram (P&ID) Symbols

The symbols used in the P&ID's in this text are explained in this appendix. For the most part, they adhere to the ISA standard, S5.1 (ISA, 2022).

Each measuring device and control equipment module on a P&ID is identified with a tag that is two or three letters followed by a number. The letter abbreviations are used to identify the various types of measuring devices and control equipment. The first letter of the tag indicates the variable type and the succeeding letter(s) provide more information about the function being performed. The abbreviations used in this text are summarized in Table C.1. The only letter that does not conform to the S5.1 standard is the letter "X", which is unclassified in S5.1. Examples of common abbreviations are:

FT	Flow transmitter
LAH	Level alarm, high level
FIC	Flow controller with the measurement indicated
XV	Discrete (on/off) valve
EX	Electric motor

Each tag is located inside a circle (regulatory entity) or a diamond (discrete entity). A square is placed around the circle or diamond if the entity performs control functions. These symbols are summarized in Figure C.1. A diamond inside a square represents a device, a discrete control object. A diamond alone represents a status, a discrete value. A circle inside a square represents a loop, a regulatory control object. A circle alone represents an indicator, an analog value, which is typically an analog transmitter.

In addition, interlock causes and actions are attached to the circles and/or diamonds. An interlock cause is indicated with by the letter "I" followed by a number within a smaller diamond. The associated interlock action uses the same identification as the cause, except within a small square. The specific interlock cause and action information is usually documented as a note the diagram or in another document.

The signal connections between control entities are indicated as different line styles, as shown in Figure C.2. Symbols for common process elements are shown in Figure C.3.

REFERENCE

ISA, 2022 *ANSI/ISA-S5.1-2022, Instrumentation Symbols and Identification*, International Society of Automation, Research Triangle Park, NC.

Table C.1 Identification Letters

	First Letter	Suceeding Letters
A	Analyzer	Alarm
C	Concentration	Control
D	Density	
E	Voltage	
F	Flow	
H		High
I		Indicating
L	Level	Low
P	Pressure	
Q		Totalize
S	Speed, slide gate	Switch
T	Temperature	Transmitter
V		Valve
W	Weight, force	
X	Discrete (on-off)	Motor
Y		Compute, convert

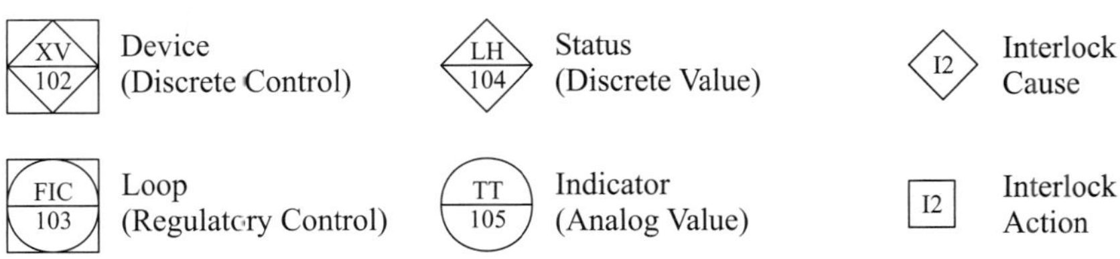

Figure C.1. Symbols for control entities.

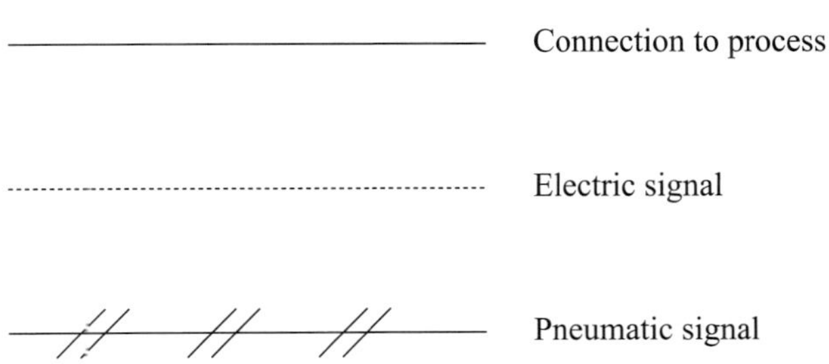

Figure C.2. Control signals.

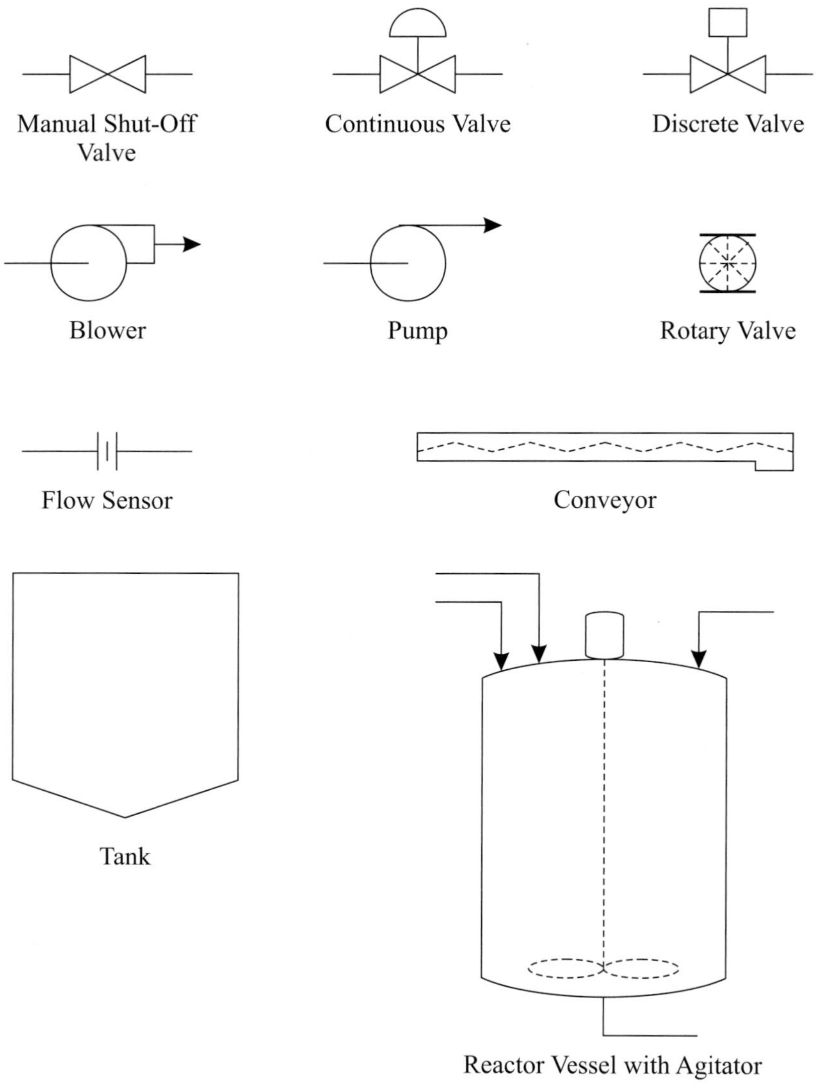

Figure C.3. Process equipment symbols.

Glossary

Note: Definitions quoted from other sources are so indicated with the source. References with similar definitions are also noted. Italicized words in a definition are defined in this glossary.

absolute pressure. The pressure of a gas or liquid measured relative to a complete vacuum.

accuracy. The ability of an *analog sensor* to produce the correct output.

action. Defines the outputs or other behaviors that are active in a *function chart* or *sequential function chart* step.

actuator. A device that manipulates a physical input to the process (for example, an air valve or a motor). Its input is connected to a PLC *output module*.

ADC. Abbreviation for *analog-to-digital converter*.

address. The alphanumeric string that defines the location of an input *channel*, output channel, or internal processor memory.

air-to-close/air-to-open. Describes control valve fail-safe action. When the air supply fails, an air-to-close valve will open so flow through the valve is unobstructed, and an air-to-open valve will close, obstructing flow though the valve.

American National Standards Institute. A private, non-profit organization that administers and coordinates voluntary standardization in the U. S.

analog actuator. An *actuator* used to regulate a physical quantity. The most common analog actuators are an electric motor or a control valve.

analog input module. Receives a measured quantity (voltage or current) from an *analog sensor* and other devices that produce *analog signals*.

analog output module. Generates a voltage or current signal that drives an *analog actuator* to manipulate a physical quantity.

analog sensor. A *sensor* that measures the value of a physical quantity than can have an infinite number of values.

analog signal. Represents a physical quantity that can have an infinite number of values.

analog-to-digital converter. A device that converts the analog voltage into an integer number, which has a finite resolution.

AND. A logical operation that requires all input signals to be true (1, on) to produce a true output.

ANSI. Abbreviation for *American National Standards Institute*.

Array. An indexed collection of data elements, all of the same data type. Each element is designated by its index, or position in the array. For example, if an array is named "Int_Ary", one of its elements is designated by "Int_Ary[4]," where "4" is the index, also called the subscript. Arrays can have more than one dimension and each dimension has its own index. A two-dimensional array of numbers is arranged as a table and has two indices. For example "Two_Ary[3,2]" refers to the element in the third row and second column.

ASCII. Abbreviation for American Standard Code for Information Interchange. A seven-bit code used to represent alphanumeric characters, punctuation marks, and control codes.

AS-i. Abbreviation for Acuator Sensor Interface. Originally developed as a low-level bus to interface to discrete sensors and actuators.

ASTM. Abbreviation for American Society for Testing and Materials.

automatic feedback control. *Feedback control* in which a controller device monitors the controlled variable of interest and commands a *manipulated variable* in order to maintain a desired value of the *controlled variable*.

automatic mode. The state of a PID controller where it calculates the *manipulated variable* according to the PID equation, and the operator changes the *setpoint*.

baseband coaxial cable. A *coaxial cable* that sends signals at one frequency.

batch. "1.) The material that is being produced or that has been produced by a single execution of a batch process. 2.) An entity that represents the production of a material at any point in the process." (ISA, 2015)

batch control. Control functions for a *batch process*.

batch process. Finite quantities (batches) of material are produced by subjecting quantities of input materials to a defined order of processing actions using one or more pieces of equipment. Batch processes are discontinuous processes from a material flow standpoint. Batch processes are neither discrete nor continuous, though they have characteristics of both.

baud rate. The number of binary bits transmitted per second over a serial communication link.

BCD. Abbreviation for *binary-coded decimal*.

binary-coded decimal. A way of representing decimal numbers in a binary format. The numbers have a base of 10, but the individual digits are represented by the 4-bit binary pattern for the decimal digit.

binary number. A number system whose base is 2 and uses only the digits 0 and 1. Proceeding from the right-most (least significant) position, the digit weights are 1, 2, 4, 8, 16, 32, and so on.

bit. The smallest unit of binary information, it can have a value of 0 or 1.

block diagram. A representation of the process where major groups of equipment are shown as rectangular blocks with the material and/or information flow between them.

Boolean. The *data type* of a *bit*.

branch. A parallel path within a ladder logic *rung*.

bridge. A device that connects two *local area networks* that have the same or similar *protocols*.

bridge circuit. Converts *sensor* resistance changes to voltage or current changes, providing temperature compensation.

broadband coaxial cable. A *coaxial cable* that sends signals at multiple frequencies.

bubbler tube. Used to measure the height of corrosive liquids. A small amount of air or inert gas is forced through this tube. The pressure required to force air bubbles from the tube is proportional to the fluid level.

bumpless transfer. When the PID controller is switched from *manual mode* to *automatic mode*, the calculated *manipulated variable* change is either filtered, or ramped to the calculated value. This gradual change in the *manipulated variable* is better tolerated by the equipment and the process.

bus. A conductor or group of conductors serving as a common power or communication connection.

bus topology. A communication network topology where all nodes are connected to the same medium and thus share access. A node on the bus can transmit or receive data from any of the other nodes.

byte. A group of eight binary digits.

CAN. Abbreviation for *Controller Area Network*.

capacitive pressure sensor. A pressure sensor where the sensing diaphragm is fixed between two capacitor plates and as the sensing diaphragm moves with changes in pressure, the capacitance changes.

capacitive proximity sensor. A *proximity sensor* that detects metallic and nonmetallic objects by the principle that materials interact with an electrostatic field.

capital project. A project in which manufacturing facilities are constructed and/or improved.

carrier-sense multiple access with collision detection. A masterless method of controlling access to a *local area network* with bus topology.

cascade control. Control scheme where the *manipulated variable* output of one controller becomes the *setpoint* of another controller. (ISA, 1993)

cascade mode. The state of a PID controller where the *setpoint* is supplied from another controller or device (not the operator) and the PID algorithm calculates the *control variable*.

cell. A logical grouping of equipment required to process one stream or manufacture one product or group of products. A process cell is a set of cooperating *units*. (ISA, 2015)

central processing unit. The part of the PLC that interprets and executes the *firmware*.

channel. The connection of one *signal* to an I/O module. The physical connection involves one or more electrical terminals.

closed-loop control. Same as *feedback control*.

coaxial cable. A flexible or rigid cable having a shield around a single conductor.

coil. A *ladder diagram* symbol that represents a discrete output or bit in internal memory.

cold junction compensation. A method of measuring the temperature of the *thermocouple* cold junction and using it to adjust the measured thermocouple hot junction voltage.

conductivity. The ability of a solution to conduct an electric current between two electrodes.

contact. A *ladder diagram* symbol that represents a discrete input or bit in internal memory.

continuous control. A control system whose signal inputs and outputs are *continuous-time signals*. The control devices are either analog electronic or pneumatic controllers.

continuous process. A process in which material passes in a continuous stream through the processing equipment. (ISA, 2015)

continuous-time signal. A signal whose values are defined for all time. See also *discrete-time signal*.

Control Concept. A section of the *Control Requirement Definition* that defines the extent of automation for the plant system.

control loop. See *loop*.

Control Requirement Definition. Control information package containing the *Process Operation Description*, *Control Concept*, *control strategy* and detailed control design.

control strategy. Defines the control types and interrelations to fulfill the control requirements.

control variable. A quantity varied by the controller in order to affect the *controlled variable*. Also called *manipulated variable*.

controlled variable. "In a control loop, the variable the value of which is sensed to originate a feedback signal." (ISA, 1993) Measured by the *process variable*.

Controller Area Network. The physical layer of *DeviceNet*. Originally developed as an automotive network.

ControlNet. A network originally developed for time-critical I/O and peer-to-peer data transmission.

convergent-beam photoelectric sensor. A *photoelectric sensor* that uses lenses to focus both the emitter and receiver at the same point.

Coriolis flowmeter. Measures the mass flow rate directly with the Coriolis effect. Fluid moving through a vibrating U-shaped tube sets up a twist in the tube. The mass flow is proportional to the maximum twist angle.

CPU. Abbreviation for *central processing unit*.

critically damped. The time response of the system to a step signal input has the fastest time response possible without *overshoot*. (ISA, 1993)

CRD. Abbreviation for *Control Requirement Definition*.

cross reference. A list of program variables and for each variable, indicates where it is used and how it is used.

CSMA/CD. Abbreviation for *carrier-sense multiple access with collision detection*.

current sinking. See *sinking*.

current sourcing. See *sourcing*.

CV. Abbreviation for *control variable*.

DAC. Abbreviation for *digital-to-analog converter*.

damping ratio. The ratio of the second peak to the first peak in the *process variable* response. Only defined if the process variable response is *underdamped*. (ISA, 1993)

dark-operate sensor. A *photoelectric sensor* that energizes its output when the received light intensity decreases below a threshold.

Data Highway. One of the first PLC-to-PLC communication protocols; developed by Allen-Bradley.

Data Highway +. An enhanced version of *Data Highway*.

data type. A classification that identifies data as one of various types, such as Boolean, integer, or real and states the valid range of values and its storage format.

DCS. Abbreviation for *distributed control system*.

deadtime. "The interval of time between initiation of an input change or stimulus and the start of the resulting observable response." (ISA, 1993)

decimal number. A number system whose base is 10 and uses the digits 0 - 9. Proceeding from the right-most (least significant) position, the digit weights are 1, 10, 100, 1000, and so on.

derivative action. The controller *manipulated variable* calculation is based on the rate of change of the *process variable*. Also called rate action. (ISA, 1993)

derivative gain. For a PID controller, the derivative gain is the proportionality constant relating the change in controller output to the *derivative action*.

derivative time constant. Same as the *derivative gain*.

DeviceNet. A relatively low-speed network for distributed I/O modules and other devices.

DH+. Abbreviation for *Data Highway+*.

DH-485. Abbreviation for Data Highway-485, a version of *Data Highway* for the SLC-500 processors.

differential pressure. The difference between two pressure measurements.

differential pressure flowmeter. A sensor that measures fluid or gas flow by measuring the pressure drop across an *orifice plate* or *venturi tube*.

diffuse reflective photoelectric sensor. A *photoelectric sensor* where the emitter and receiver are in the same package and the sensed object returns the light beam to the receiver.

digital-to-analog converter. Converts the integer number from the PLC into a voltage signal.

digital control. A control system whose control devices are digital controllers or digital computers. Technically, PID controllers in a PLC are digital controllers, though they behave like an analog electronic controller. Also called discrete-time control, the signal inputs and outputs are *discrete-time signals*.

direct-acting. A process or controller is called direct-acting if an increase in the *input signal* causes an increase in the *output signal*.

direct action. See *direct-acting*.

directly represented variable. Address of a memory or I/O location in the PLC.

discrete actuator. Convert the on/off PLC output signal into a physical phenomena.

discrete control. Control consisting of *discrete sensors* and *discrete actuators*.

discrete input module. An *input module* that senses the status of a device that has only two states, on/off, open/closed, running/stopped, and so on.

discrete output module. An *output module* that switches a device that has only two states, on/off, open/closed, and so on.

discrete parts manufacturing process. A type of process where a specified quantity of material moves as a unit (part or group of parts) between work stations, and each unit maintains its unique identity (ISA, 2015).

discrete sensor. A *sensor* whose output has two states: off/on, open/closed, or 0/1.

discrete signal. Represents a physical quantity that has only two states, on/off, open/closed, and so on.

discrete-time control. See *digital control.*

discrete-time signal. A signal whose values are defined only at the sampling instants. Any process measurements must be converted to discrete-time signals before being used by a digital controller. See also *continuous-time signal.*

distributed control system. Developed to provide *regulatory control* functions for factory automation. See also *programmable electronic system.*

disturbance. Process influence that affects the *process variable*, but is not manipulated by the controller. (ISA, 1993)

Doppler ultrasonic flowmeter. Transmits ultrasonic pulses into the flow stream and monitors the reflected pulses. The fluid flow rate is proportional to the frequency shift of the reflected pulses.

double-acting cylinder. A pneumatic or hydraulic cylinder with two controls.

dwell time. The time a process remains halted in order for another process to occur. For example, a carousel is commonly used to move a part between machining stations. The part is loaded onto the carousel, the carousel moves the part to the individual stations, pausing while each machining operation is performed, and eventually the part is removed from the carousel. The *dwell time* is the time the carousel pauses to allow the stations to process the part.

electromagnetic interference. An electromagnetic disturbance that disrupts or degrades the performance of electronic equipment.

EMI. Abbreviation for *electromagnetic interference*.

empirical model. Mathematical model of a process based on experimental data.

encoder. A discrete sensor used in position control applications to determine the position and/or velocity of a rotary shaft.

Ethernet. A popular, inexpensive *local area network* originally developed for the office environment, but is becoming a popular network for factory automation.

Ethernet/IP. A network that provides a control application layer on top of standard *Ethernet*.

expansion chassis. A chassis containing only *I/O modules* and a communication interface, and not the PLC processor.

fail-safe. A failure in the system causes the process to assume a safe condition, normally **off**. For example, a fail-safe output channel will assume the **off** condition when there is a failure in the module, PLC processor, or power source.

fault. An internal processor error that causes the processor to halt its operation.

FBD. Abbreviation for *function block diagram*.

feedback control. Control scheme that uses knowledge of the output to take corrective action. See *automatic feedback control* and *manual feedback control*. (ISA, 1993)

feedforward control. Control scheme that eliminates or reduces the *disturbance* effect on the *process variable* by using a measurement of the *disturbance* to modify the calculation of the *manipulated variable*. (ISA, 1993)

firewall. A communication network device that filters all network packets to determine if they are legitimate requests to be passed to the destination. A firewall protects a local area network from unauthorized users.

firmware. In a PLC processor, the software that is embedded in the processor that executes the user program and handles the communication tasks. The user program cannot modify the firmware. Other modules, for example, communication interfaces or motion control modules also contain firmware. In earlier processors, the firmware was contained in read-only memory (ROM) integrated circuit chips. Most recent processors store the firmware in electrically-erasable read-only memory (EEROM) integrated circuit chips that can allow firmware updates without disassembling the module.

FODT. Abbreviation for first-order plus deadtime, a type of *empirical model.*

force. To override the status of a physical discrete input or to override the logic driving a physical output coil and make the output be a desired value.

Foundation Fieldbus H1. A *local area network* for device-level communications originally developed as a substitute for 4-20 mA signal transmission in process control applications.

Foundation Fieldbus HSE. A *local area network* that provides the Foundation Fieldbus application layer on top of standard *Ethernet.*

function block. (1) In a ladder logic program, any entity that is not a contact or coil, for example, timer and counter, (2) The entities that are interconnected in a function block diagram.

function block diagram. One of the standard PLC languages. A set of interconnected *function blocks* and signal connections between the blocks.

function chart. A diagram of interconnected steps, actions, and transitions used for sequential control. A simplified version of the *sequential function chart.*

fuse. A device consisting of easily melted metal that interrupts electrical current when the current is too high.

gage pressure. The pressure of a gas or liquid measured relative to atmospheric pressure.

gain. The ratio of the change in the *steady-state* output to a step change in the input, provided the output does not saturate (ISA, 1993).

gateway. A device that connects two *local area networks* that have significantly different *protocols.*

global data table. Data that is shared between PLC processors.

global variable. A variable accessible to all elements contained in a configuration, resource, or program. For example, a global program variable is accessible to all function blocks and routines in the program.

ground. A connection made from the equipment to earth, or zero potential, for safety purposes.

GUI. Abbreviation for graphical user interface. See *human-machine interface.*

Hall effect proximity sensor. A *proximity sensor* that detects magnetic materials based on the Hall effect. When a magnetic field is applied perpendicular to a DC current flowing through a semiconductor plate, a small voltage difference perpendicular to the current flow is generated. When this voltage is above a threshold, the magnetic object is detected.

hexadecimal number. A number system whose base is 16 and uses the digits 0 - 9 and letters A - F. Proceeding from the right-most (least significant) position, the digit weights are 1, 16, 256, 4096, and so on.

high-signal selector. A device that automatically compares two or more input signals and allows the signal with the higher value to be the output.

HMI. Abbreviation for *human-machine interface.*

HOA switch. Type of switch having three positions: (1) Hand, (2) Off, and (3) Auto. It is often used between the PLC and a discrete device, such as a motor. In the "Hand" position, the operator manually starts the device. In the "Off" position, the device is manually stopped. In the "Auto" position, the PLC starts and stops the device.

human-machine interface. The interface between the human operator and the control system.

hydraulic. Refers to a system or component that is moved with fluid pressure.

hydraulic actuator. A cylinder with a piston that provides the actuation force with hydraulic fluid pressure.

hysteresis. A dead band purposely introduced in the operation of discrete sensors to eliminate "chattering" when the sensed phenomenon is at the on/off boundary. For a proximity sensor, the object is sensed at a certain distance. The object must move further away to change the sensor output.

IAE. Abbreviation for the integral of the absolute error, a control loop performance measure.

IEC. Abbreviation for International Electrotechnical Commission, an international standards-making organization.

IEC 61131. Abbreviation for *IEC 1131: Programmable Logic Controllers*, International Electrotechnical Commission, 1993.

IEC 61131-3. Part 3 of the IEC 61131 (formerly 1131) standard. Defines the standard PLC languages: *ladder logic*, *sequential function chart*, *function block diagram*, *structured text*, and *instruction list*.

IEEE. Abbreviation for The Institute of Electrical and Electronics Engineers.

IEEE 802.3. The IEEE standard that defines the *Ethernet* data communication protocol, which is a *carrier sense multiple access with collision detection* type of protocol.

IEEE 802.4. The IEEE standard that defines a *token bus* communication protocol.

IL. Abbreviation for *instruction list*.

image table. The PLC processor memory that contains a copy of the input and output *channel* values.

increase close/increase open. Controller parameter provided to compensate for valve actuator action (air-to-close or air-to-open). Use increase close for an air-to-close valve and increase open for an air-to-open valve.

indexed addressing. A data addressing scheme where the value of an index register is added to the address to obtain a new address.

indirect addressing. A data addressing scheme where the contents of one memory location specify the location of another data item. For example, the integer "Ary_Idx" could specify which element in the *array* "Int_Ary" needs to be displayed for the operator.

inductive proximity sensor. A *proximity sensor* that detects ferrous and nonferrous metallic objects by the principle that metals interact with an electromagnetic field.

infrared thermometer. A noncontact temperature sensor that measures temperature by focusing the infrared radiation from the hot target onto an infrared detector, which converts the energy to an electrical voltage.

input module. Provides the physical interface between the PLC and an *input signal*.

input signal. "A *signal* applied to a device, element, or system." (ISA, 1993)

instruction. (1) In a ladder diagram, it generically refers to the contact and coil symbols as well as to the names of the function blocks, (2) a statement in an instruction list program.

instruction list. One of the standard PLC languages. A low-level textual language similar to microprocessor assembly language.

integral action. The controller *manipulated variable* calculation is based on the integrated error, the *setpoint* minus the *process variable*. (ISA, 1993)

integral gain. For PI or PID controllers, the integral gain is the proportionality constant relating the change in controller output to the *integral action*. (ISA, 1993)

integral time constant. The reciprocal of the *integral gain*. (ISA, 1993)

integral windup. When a controller has *integral action*, a persistent error will cause the integral term to increase (or decrease) to a value of large magnitude.

integrated circuit temperature sensor. A temperature sensor that is based on the property that the base-emitter voltage of a bipolar transistor is proportional to temperature.

INTERBUS. Originally developed by a German consortium, it is a *local area network* with a ring topology.

interlock. Designed to detect an abnormal process condition and take action to prevent an undesirable or hazardous event.

I/O module. A plug-in component that provides the physical interface between the PLC processor and the field devices, such as switches, lamps, and valves.

IP rating. The IEC enclosure designation consisting of the letters "IP" followed by two digits.

ISA. Abbreviation for the International Society of Automation.

isolation. Provides the interface between the field and PLC voltage levels, preventing ground loops and noisy electrical signals from interfering with the PLC operation. Often provided by *optical isolation*.

ITAE. Abbreviation for the integral of the product of time and the absolute error, a control loop performance measure.

jog. A momentary **on** state often used by the operator to override the automatic motion control.

jump. Program execution control that allows one to skip ladder rungs, instruction list statements, or SFC steps.

ladder diagram. One of the standard PLC languages. A graphical language resembling an electrical wiring diagram with contacts, coils and blocks.

ladder logic. A symbolic programming language traditionally used in *programmable logic controllers*. It was developed from the relay ladder logic wiring diagram.

LAN. Abbreviation for *local area network*.

latch. (1) A ladder logic output coil that is energized when any rung path passes power and remains energized when no rung path passes power, (2) an electronic or electromechanical device that energizes its output and maintains the output after the input signal is turned off.

LCD. Abbreviation for *liquid crystal display*.

LD. Abbreviation for *ladder diagram*.

LED. Abbreviation for *light-emitting diode*.

light-emitting diode. A semiconductor diode that emits light when current passes through the junction in the forward direction.

light-operate sensor. A *photoelectric sensor* that energizes its output when the received light intensity increases above a threshold.

limit switch. A *discrete sensor* that indicates the presence/absence of an object. Often, the sensed object moves a lever arm to close a switch contact.

liquid crystal display. A low-power display that uses a liquid crystal to display the graphics and text for an operator interface.

local area network. A high-speed medium-distance communication system linking multiple devices.

local variable. A variable defined at the software element (for example, function block or subroutine) that can only be accessed by the software element.

lockout. (1) Completely disabling all energy sources to a system and applying a locking device to each energy source to prevent accidental startup, (2) in a program, a condition that stops a device and does not allow it to restart until the condition is cleared.

loop. A single-input, single-output controller that monitors a transmitter and manipulates a physical quantity (usually a control valve position) in order to force a process variable to the desired setpoint. The controller is typically a *PID controller*. (ISA, 1993)

loss-in-weight feeder. A solid material feeder that infers the flow rate from the rate of hopper weight loss. A separate device (for example, screw feeder) is manipulated to control the feed rate.

low-signal selector. A device that automatically compares two or more input signals and allows the signal with the lower value to be the output.

magnetic flowmeter. Measures fluid flow based on Faraday's law. The moving fluid induces a voltage across two electrodes that are aligned perpendicular to both the flow and the magnetic field. The voltage is proportional to the average velocity of the fluid.

magnetic reed switch. A *proximity sensor* consisting of sealed contacts. When a magnet moves close to the switch, the contacts are magnetized and pull together.

man-machine interface. Same as *human-machine interface.*

manipulated variable. A quantity varied by the controller in order to affect the *controlled variable.* (ISA, 1993) Also called the *control variable.*

manual feedback control. *Feedback control* in which operating personnel monitor the *controlled variable* of interest and take corrective action in order to maintain the desired value.

manual mode. The state of a PID controller where the operator sets the *manipulated variable*, and the PID equation is not executed.

mask. A logical function that only allows only certain bits in the word to be changed or compared. For example, in a masked word move, a "1" in the mask means the particular bit is passed from the source to the destination and a "0" in the mask means the particular bit is blocked.

master. In a *master/slave* type of controlling access to a *local area network*, this node controls how the other nodes (*slave* nodes) access the network. The master polls the slave nodes, permitting them to transmit messages.

master control relay. (1) A hard-wired relay that de-energizes physical discrete control devices when an emergency-stop switch is pressed; (2) a ladder logic instruction that enables/disables non-retentive outputs within a range of rungs.

master/slave. A method of controlling access to a *local area network* with bus topology where one *node* is designated the master that controls how other nodes access the network.

MCR. Abbreviation for *master control relay.*

memory. The part of the PLC processor that stores data and program either permanently or temporarily.

memory map. The allocation of the PLC processor memory into segments for certain purposes. For example, certain segments may be allocated for data and others for program. A particular data segment may only contain one type of data (for example, bits, timers, etc.).

Modbus. One of the first PLC-to-PLC communication protocols; developed by Modicon.

Modbus Plus. The Modbus communication protocol implemented on top of a *token bus* network.

Modbus/TCP. The Modbus communication protocol implemented on top of *Ethernet.*

mode. (1) For a PLC processor, it defines how the processor scans the program (see *program mode*, *run mode*, and *test mode*); (2) for a PID controller, it indicates the state of the control algorithm (see *automatic mode*, *cascade mode*, *manual mode*, and *tracking mode*).

motor contactor. An output actuator used to control a single-speed motor that is basically a three- or four-pole relay that handles the high motor current.

motor drive. A modern motor controller that uses electronics to provide efficient control of the motor speed and/or torque.

MV. Abbreviation for *manipulated variable.*

NAND. A logical operation that requires all input signals to be true (1, on) to produce a false output.

NEMA. Abbreviation for National Electrical Manufacturers Association.

NFPA. Abbreviation for National Fire Protection Association.

node. Device on a *local area network.*

nonretentive timer. A timer that resets its accumulator value when the input signal is false (0, off).

NOR. A logical operation that requires all input signals to be false (0, off) to produce a true output.

normally-closed contact. A contact that passes power if its control is unenergized. The contact blocks power if its control is energized.

normally-open contact. A contact that passes power if its control is energized. The contact blocks power if its control is unenergized.

NOT. A logical operation that yields a false (0, off) output if the input is true (1, on) and a true output if the input is false.

octal number. A number system whose base is 8 and uses the digits 0 - 7. Proceeding from the right-most (least significant) position, the digit weights are 1, 8, 64, 512, and so on.

off-delay timer. A timer that delays the turn-off of a signal and does not delay the turn-on.

off-line programming. Programming the PLC while not connected to the PLC processor through a communication port. Later, the program is downloaded to the processor.

offset. The difference between the *setpoint* and the *process variable* when the system has reached *steady-state*. (ISA, 1993)

on-delay timer. A timer that delays the turn-on of a signal and does not delay the turn-off.

on-line programming. Programming the PLC while connected to the PLC processor through a communication port. Most processors will allow one to make changes while the processor is in the *run mode*.

one-shot contact. See *transitional contact*.

open-loop control. A control strategy where the *manipulated variables* are set to their design values and held constant. (ISA, 1993)

open-loop response. The *process variable* response to a step change in the *manipulated variable* while the controller is in the *manual mode*.

operating point. The normal operating value for the system variables.

operating state. The current condition of the equipment entity, e.g., *process cell* or *unit*. The operating state also defines how the equipment entity will operate and how it will respond to commands.

operator interface. Same as *human-machine interface*.

optical isolation. Isolation provided by optical couplers that consist of a photodiode transferring digital information to a phototransistor. There is no electrical connection between the photodiode and phototransistor.

optical pyrometer. A noncontact temperature sensor that measures temperature by comparing the intensity of the light radiated from the hot target with the brightness of an incandescent lamp filament.

OR. A logical operation that requires at least one input signal to be true (1, on) to produce a true output.

orifice plate. A flat piece of metal with a sharp-edged hole of a certain size bored in it used to measure gas or liquid flow rate.

ORP. Abbreviation for *oxidation-reduction potential*.

output module. Provides the physical interface between the PLC and an *output signal*.

output signal. "A *signal* delivered by a device, element or system." (ISA, 1993)

overdamped. The time response of the system to a step signal input has no *overshoot* and is slower than *critically damped*. (ISA, 1993)

override control. Scheme used to protect process equipment or personnel. See *high-signal selector* and *low-signal selector*.

overshoot. "The maximum excursion beyond the final *steady-state* value of output as the result of an input change." (ISA, 1993)

oxidation-reduction potential. The oxidizing or reducing nature of a chemical.

P&ID. Abbreviation for *Piping and Instrumentation Diagram*.

P-NET. A process automation *local area network* developed by Proces-Data A/S.

parallel circuit. A connection of *contacts*, *coils*, or *function blocks* where all right ends are connected together and all left ends are connected together.

password. A string of characters chosen by a computer or control system user and used to authenticate the user in order to prevent unauthorized access to the computer or PLC.

peak time. The time from the *setpoint step* change to the time of the first peak of an underdamped *process variable* response.

peer-to-peer. Communication that occurs between similar devices on *a local area network*. In factory automation, generally refers to PLC-to-PLC communication.

percent overshoot. The amount of the *overshoot*, expressed as a percentage of the process variable final value.

permissive. A condition that prevents a device from being started.

PFD. Abbreviation for *Process Flow Diagram.*

pH. A measure of the acidity or alkalinity of a water solution.

photoelectric sensor. A sensor that uses a phototransistor receiver to respond to a modulated light beam to determine proximity or distance. The sensor contains an emitter to generate the light beam or may sense ambient light.

PI control. Abbreviation for Proportional-plus-Integral control. A controller whose output is the sum of *proportional action* and *integral action.*

PID control. Abbreviation for Proportional-plus-Integral-plus-Derivative control. A controller whose output is the sum of *proportional action, integral action*, and *derivative action.*

Piping and Instrumentation Diagram. A symbolic drawing of a process, or portion of a process, showing the piping and instrumentation.

PLC. Abbreviation for *programmable logic controller.*

pneumatic. Refers to a system or component that is moved with air pressure.

pneumatic actuator. A cylinder with a piston that provides the actuation force with air pressure.

POD. Abbreviation for *Process Operation Description.*

process. "Physical or chemical change of matter or conversion of energy; e.g., change in pressure, temperature, speed, electrical potential, etc." (ISA, 1993) For this text, the process is the system to be controlled.

process cell. See *cell.*

Process Flow Diagram. An abstract drawing of a process or plant showing equipment groupings (areas, cells, units, etc.) and the material and/or information flows between them.

process input. When considering the process, an input is generally material moving into the process. From the standpoint of control, the process input is a *manipulated variable.*

Process Operation Description. Design information package that defines the operating strategy and control objectives of the plant systems.

process output. When considering the process, an output is generally material moving out of the process. From the standpoint of control, the process output is a *controlled variable* or a *process variable.*

process unit. See *unit.*

process variable. The measured value of the *controlled variable* which, along with the setpoint, is used by the controller to calculate a value of the *manipulated variable.*

PROFIBUS. Acronym for PROcess FIeld BUS, a process automation *local area network.* There are three parts: *PROFIBUS-DP*, *PROFIBUS-PA*, and *PROFIBUS-FMS.*

PROFIBUS-DP. Acronym for PROcess FIeld BUS - Decentral Periphery, a process automation *local area network* developed for a mix of discrete and analog devices.

PROFIBUS-FMS. Acronym for PROcess FIeld BUS - Fieldbus Messaging Specification, an application layer messaging protocol.

PROFIBUS-PA. Acronym for PROcess FIeld BUS - Process Automation, a process automation *local area network* designed as a substitute for 4-20 mA signal transmission in process control applications.

PROFInet. A protocol developed for distributed automation applications whose lower layers are standard *Ethernet.*

program. In IEC 61131-3, it generally consists of an interconnection of *function blocks*, each of which may be written in any of the IEC languages. In general, a program consists of all of the executable user-written software in the PLC.

program mode. The PLC processor does not scan the program. Depending on the particular PLC processor the physical inputs may be copied into the input image, but the physical outputs are disabled.

programmable electronic system. Any system that implements the control functions. A programmable electronic system encompasses the functions of both a traditional *distributed control system* (DCS) and a traditional *programmable logic controller* (PLC).

programmable logic controller. Developed to provide *discrete control* functions for factory automation. See also *programmable electronic system*.

proportional action. The controller *manipulated variable* is calculated as a constant (called the *proportional gain*) multiplied by the error, the *setpoint* minus the *process variable*. (ISA, 1993)

proportional band. The change in error required to produce a full range, 100%, change in the manipulated variable, due to *proportional action*. It is equal to 100 divided by the *proportional gain*. (ISA, 1993)

proportional gain. The ratio of the change in the *manipulated variable* due to *proportional action* to the change in the error. See also *proportional band*. (ISA, 1993)

protocol. A set of rules that govern how nodes communicate with each other over a *local area network*.

proximity sensor. A noncontact type of *discrete sensor* that senses an object with electromagnetic field, electric field, ultrasonic, or photoelectric technologies.

proximity switch. See *proximity sensor*.

PV. Abbreviation for *process variable*.

radar sensor. A sensor that uses radio waves to measure the distance to an object.

RAM. Abbreviation for *random-access memory*.

random-access memory. Alterable storage for the PLC program and data.

rate action. See *derivative action*.

rated voltage. The voltage range for which a system is designed to operate.

ratio control. Control scheme used in blending or other applications where two streams need to be held in constant ratio to each other.

recipe. "The necessary set of information that uniquely defines the production requirements for a specific product." (ISA, 2015)

regulatory control. The control type used to maintain *process variables* at desired values. It is often referred to as *continuous control* or *discrete-time control*.

relay. An electromechanical device consisting of a coil of wire magnetically coupled to a moveable piece of iron (armature), held in its resting position with a spring. The armature is mechanically connected to electrical contacts. When sufficient current passes through the coil, the armature moves, changing the contact position.

reliability. The ability of a system to perform its function without error, commonly expressed as mean time between failures (MTBF) or the failure probability.

repeatability. The ability of a sensor to produce the same output when the measured value is the same.

reset. A ladder logic output coil that is de-energized when any rung path passes power and remains de-energized when no rung path passes power.

reset action. See *integral action*.

reset windup. See *integral windup*.

resistance temperature device. A temperature sensor that consists of a length of metal (usually platinum) wire. The resistance of the metal wire increases with increasing temperature.

resolution. The smallest change that can be sensed by an analog input channel.

resource. In a *configuration*, it provides the support functions for the execution of *programs*.

retentive timer. A timer that retains its accumulator value when the input signal is false (0, off).

retro-reflective photoelectric sensor. A *photoelectric sensor* where the emitter and receiver are in the same package and a reflector is used to return the light beam to the receiver.

reverse-acting. A process or controller is called reverse-acting if an increase in *the input signal* causes a decrease in the *output signal.*

reverse action. See *reverse-acting.*

rise time. The time required for the process variable to go from 10% to 90% of the *steady state* change. (ISA, 1993)

rotary valve. A solid material feeder that uses wedge-shaped compartments within a cylinder to control material flow from a tank. The flow rate is controlled by the valve rotation speed.

RS-232. A common serial communication standard, it specifies the signal voltages, signal timing, signal function, a protocol, and mechanical connectors.

RTD. Abbreviation for *resistance temperature device.*

run mode. The PLC processor transfers the physical inputs to the input image, scans the program, and transfers the output image to the physical outputs.

rung. The arrangement of contacts, coils, and function blocks to control one of more outputs in a *ladder diagram.*

Safety Instrumented System. The control technology for automatic safety protection.

scan. The process by which a PLC processor reads the physical inputs, executes the user program, and writes the physical outputs.

scan time. The time required to complete the three tasks of a *scan.*

SCR. Abbreviation for *silicon-controlled rectifier.*

seal circuit. A ladder logic circuit that controls a device that must run continuously, but is started and stopped with momentary switches.

sensitivity. See *resolution.*

sensor. A device that monitors a physical phenomenon of the process. Its output is connected to a PLC *input module.*

sequence. The order in which process steps are executed.

sequence control. Enforces the correct operation of process equipment in a step-wise fashion.

sequencer. A type of function block that implements a *sequence.*

sequential function chart. One of the standard PLC languages. A diagram of interconnected *steps*, *actions*, and *transitions* used to for sequential control.

serial communication. A method of communication between two devices whereby the message is sent one bit at a time.

Seriplex. Abbreviation for SERIal multiPLEXed control bus. Originally developed as a low-level bus to interface to discrete components such as panel switches.

servomotor. A specially-designed brushed DC motor or brushless AC motor for precision closed-loop motion control.

set. A ladder logic output coil that is energized when any rung path passes power and remains energized when no rung path passes power.

setpoint. "An input variable which sets the desired value of the *controlled variable*." (ISA, 1993)

settling time. The time from the *setpoint* change to the time that the process variable response has settled within a certain percentage band of the final value, usually 2% or 5%. (ISA, 1993)

SFC. Abbreviation for *sequential function chart.*

shift register. A set of binary bits arranged so that data can be shifted from one bit to another. A shift register may be the bits in a word, or may be the bits in consecutive words.

signal. "In process instrumentation, physical variable, one or more parameters of which carry information about another variable (which the signal represents)." (ISA, 1993)

silicon-controlled rectifier. A semiconductor device that has the property that when a positive pulse is applied to the gate terminal and there is a positive anode-to-cathode voltage, the rectifier conducts as long as the anode-to-cathode voltage remains positive.

single-acting cylinder. A pneumatic or hydraulic cylinder with one control.

sinking. The designation of a device connection in which the current (power) flows into the device when it is active. An output is called sinking if current flows **in** when the output is active. An input is called sinking if current flows **in** when the input is active.

SIS. Abbreviation for *Safety Instrumented System.*

SISO. Abbreviation for single-input, single-output.

slave. In a *master/slave* type of controlling access to a *local area network*, this node transmits only when polled by the master.

solenoid. (1) An electromechanical device consisting of a coil of wire wrapped around a moveable iron plunger, held in its resting position with a spring. When sufficient current passes through the coil, the plunger moves, effecting some motion. (2) A pneumatic or hydraulic cylinder.

sourcing. The designation of a device connection in which the current (power) flows out of the device when it is active. An output is called sourcing if current flows **out** when the output is active. An input is called sourcing if current flows **out** when the input is active.

SP. Abbreviation for *setpoint.*

speed switch. A special form of a *limit switch* that switch detects the speed or direction of a rotating shaft.

split range control. Control scheme when process has one *process variable* and more than one *manipulated variable*.

ST. Abbreviation for *structured text.*

stable system. A system is considered to be stable if a bounded *input signal* always results in an *output signal* that is also bounded.

state. The current condition of the physical or control entity. The state also defines how the entity will operate and how it will respond to commands. (ISA, 2015)

steady-state. The long-term output response of a system after it has been disturbed. Generally refers to the long-term output due to a *manipulated variable* or *disturbance change*. (ISA, 1993)

step. In a *function chart* or *sequential function chart*, it is an operation spanning a length of time, the state of a process.

stepper motor. A brushless permanent-magnet type motor that makes a fixed angular motion in response to pulsed input signals.

strain gauge. A *sensor* that measures force along an axis based on the principle that the resistance of a conductor changes as it is stretched.

structured text. One of the standard PLC languages. A textual language useful for calculation-intensive functions.

SwiftNet. A *local area network* developed as a very high speed flight data bus.

tagout. The placement of warning tags on energy sources to help ensure the equipment is not restarted until maintenance or repair functions are finished. See also *lockout*.

task. Within a *resource*, it is set up to control one or more programs and/or function blocks to execute periodically or to execute upon the occurrence of a specified trigger.

TBD. Abbreviation for "To be determined".

test mode. The PLC processor transfers the physical inputs to the input image and scans the program (updating the output image), but the physical outputs remain disabled.

thermistor. A temperature sensor most often made from a semiconductor material that has a negative temperature coefficient, that is, its resistance decreases with temperature.

thermocouple. A temperature sensor that consists of a junction of two dissimilar metals. The voltage across the junction changes with the temperature.

thermopile. A series connection of *thermocouples* used to amplify the measured thermocouple voltage.

thermowell. A temperature sensor is mounted inside it to protect the sensor from the process chemicals.

through-beam photoelectric sensor. A *photoelectric sensor* where the emitter and receiver are positioned opposite to each other so that the light beam from the emitter is aimed directly at the receiver.

thumbwheel switch. A switch consisting of a rotary wheel with positions numbered 0 through 9. Each switch has four connections to a *discrete input module*. Thumbwheel switches can be placed adjacent to each other to allow inexpensive operator input of numerical values.

tie-back. For PID control, the operator-supplied *control variable* when in manual mode.

time constant. In an expression for linear system time response, the time constant is the value, τ, in the response term, $Ae^{-s\tau}$. In a *transfer function*, the time constant is the value, τ, in the denominator term, $1+s\tau$. For the output of a first-order system whose input is a step signal, the time constant is the time required to complete 63.2% of the total output change. (ISA, 1993)

toggle switch. A switch consisting of a projecting lever that can assume at least two positions. Each position either opens or closes an electrical circuit.

token passing. A method of controlling access to a *local area network* with bus topology that involves circulating a special message, called the token, among the *nodes*.

topology. The physical structure of a communication network.

tracking mode. The state of a PID controller where its *manipulated variable* tracks another variable, usually the output of another controller. When in this mode, the internal variables of the PID algorithm are modified so that the calculated control variable matches the tracked variable so there is a bumpless transfer when the mode is changed to automatic.

transducer. Same as *sensor*.

transfer function. For a continuous-time system, the transfer function is the ratio of the Laplace transform of the output variable to the Laplace transform of the input variable, with all initial conditions assumed to be zero. (ISA, 1993)

transit-time ultrasonic flowmeter. Uses two ultrasonic (sound) transceivers placed at a 45 degree angle to the flow stream. Both send out ultrasonic pulses that are detected by the other. The difference in transit time between the two beams is proportional to the flow rate.

transition. See *transition condition*.

transition condition. The condition that causes control to pass from one step to the next in a *function chart* or *SFC*.

transitional contact. A contact that passes power for one scan of the rung when its control changes state. The contact may be activated for a positive transition (**off**-to-**on**) or a negative transition (**on**-to-**off**) of the control.

turbine flowmeter. A rotor immersed in the liquid stream that turns as the fluid moves through it. The flow rate is inferred from the rotor speed.

ultrasonic flowmeter. See *transit-time ultrasonic flowmeter* and *Doppler ultrasonic flowmeter*.

ultrasonic sensor. A sensor that uses sound waves to detect the proximity of an object or to measure the distance to an object.

underdamped. The time response of the system to a step signal input has *overshoot*. (ISA, 1993)

unit. A set of equipment modules and control modules usually centered on a major piece of equipment, e.g., a reactor. A unit combines all necessary physical processing and control equipment required to perform one or more major processing activities, e.g., react or separate. (ISA, 2015)

unlatch. A ladder logic output coil that is de-energized when any rung path passes power and remains de-energized when no rung path passes power.

variable-speed AC drive. Converts a constant AC source voltage into a variable frequency three-phase AC voltage. Varying the frequency varies the AC motor speed.

variable-speed DC drive. Converts a constant AC source voltage into a variable DC armature voltage with silicon-controller rectifiers. Varying the armature voltage varies the DC motor speed.

venturi tube. A section of pipe with a conical entrance, a short straight throat, and a conical outlet used to measure gas or liquid flow rate.

vibratory feeder. A solid material feeder that uses a tray with a certain vibrating motion to move the solid particles.

vortex flowmeter. An obstruction, placed in the fluid flow stream, sets up vortices in the flow stream. The frequency of the vortices is proportional to the volumetric flow rate.

weigh belt feeder. A conveyor belt with a weigh scale to control the flow of solid material. The belt speed is controlled in order to maintain the desired flow rate.

word. A group of sixteen binary digits.

WorldFIP. A process automation *local area network* developed by the French.

References

ISA, 1993 *ISA-S51.1-1979 (R1993), Process Instrumentation Terminology*, International Society of Automation, Research Triangle Park, NC.

ISA. 2015. *ANSI/ISA-S88.01-2010, Batch Control, Part 1: Models and Terminology*, International Society of Automation, Research Triangle Park.

Index

R